SEISMIC RAY THEORY

Seismic Ray Theory presents the most comprehensive treatment of the seismic ray method available. This method plays an important role in seismology, seismic exploration, and the interpretation of seismic measurements.

The book presents a consistent treatment of the seismic ray method, based on the asymptotic high-frequency solution of the elastodynamic equation. At present, this is the most general and powerful approach to developing the seismic ray method. High-frequency seismic body waves, propagating in complex three-dimensional, laterally varying, isotropic or anisotropic, layered and block structures are considered. Equations controlling the rays, travel times, amplitudes, Green functions, synthetic seismograms, and particle ground motions are derived, and the relevant numerical algorithms are proposed and discussed. Many new concepts, which extend the possibilities and increase the efficiency of the seismic ray method, are included. The book has a tutorial character: derivations begin with a relatively simple problem in which the main ideas are easier to explain and then advance to more complex problems. Most of the derived equations in the book are expressed in algorithmic form and may be used directly for computer programming. The equations and proposed numerical procedures find broad applications in numerical modeling of seismic wavefields in complex 3-D structures and in many important inversion methods (tomography and migration among others).

Seismic Ray Theory will prove to be an invaluable advanced textbook and reference volume in all academic institutions in which seismology is taught or researched. It will also be an invaluable resource in the research and exploration departments of the petroleum industry and in geological surveys.

Vlastislav Červený is a Professor of Geophysics at Charles University, Praha, Czech Republic. For 40 years he has researched and taught seismic ray theory worldwide. He is a member of the editorial boards of several journals, including the *Journal of Seismic Exploration* and the *Journal of Seismology*. In 1997 he received the Beno Gutenberg medal from the European Geophysical Society in recognition of his outstanding theoretical contribution in the field of seismology and seismic prospecting. In 1999 he received the Conrad Schlumberger Award from the European Association of Geoscientists and Engineers in recognition of his many contributions to asymptotic wave theory and its applications to seismic modeling. He is an Honorary Member of the Society of Exploration Geophysicists. He has published two previous books: *Theory of Seismic Head Waves* (with R. Ravindra, 1971) and *Ray Method in Seismology* (with I. A. Molotkov and I. Pšenčík, 1977). He has also published more than 200 research papers.

SEISMIC RAY THEORY

V. ČERVENÝ

Charles University

CAMBRIDGE UNIVERSITY PRESS
Cambridge, New York, Melbourne, Madrid, Cape Town, Singapore, São Paulo

Cambridge University Press
The Edinburgh Building, Cambridge CB2 2RU, UK

Published in the United States of America by Cambridge University Press, New York

www.cambridge.org
Information on this title: www.cambridge.org/9780521366717

First published 2001
This digitally printed first paperback version 2005

A catalogue record for this publication is available from the British Library

Library of Congress Cataloguing in Publication data
Červený, Vlastislav.
Seismic ray theory / V. Červený
p.   cm.
Includes bibliographical references and index.
ISBN 0-521-36671-2
1. Seismic waves – Mathematical models.   I. Title.
QE538.5 .C465   2001
551.22′01′5118 – dc21   00-040355

ISBN-13 978-0-521-36671-7 hardback
ISBN-10 0-521-36671-2 hardback

ISBN-13 978-0-521-01822-7 paperback
ISBN-10 0-521-01822-6 paperback

Contents

Preface   *page* vii

1   INTRODUCTION   1

2   THE ELASTODYNAMIC EQUATION AND ITS SIMPLE SOLUTIONS   7

2.1. Linear Elastodynamics   8
2.2. Elastic Plane Waves   19
2.3. Elastic Plane Waves Across a Plane Interface   37
2.4. High-Frequency Elastic Waves in Smoothly Inhomogeneous Media   53
2.5. Point-Source Solutions. Green Functions   73
2.6. Application of Green Functions to the Construction
of More General Solutions   89

3   SEISMIC RAYS AND TRAVEL TIMES   99

3.1. Ray Tracing Systems in Inhomogeneous Isotropic Media   102
3.2. Rays in Laterally Varying Layered Structures   117
3.3. Ray Tracing   124
3.4. Analytical Ray Tracing   129
3.5. Ray Tracing in Curvilinear Coordinates   137
3.6. Ray Tracing in Inhomogeneous Anisotropic Media   148
3.7. Ray Tracing and Travel-Time Computations in 1-D Models   160
3.8. Direct Computation of Travel Times and/or Wavefronts   178
3.9. Perturbation Methods for Travel Times   189
3.10. Ray Fields   199
3.11. Boundary-Value Ray Tracing   217
3.12. Surface-Wave Ray Tracing   228

4   DYNAMIC RAY TRACING. PARAXIAL RAY METHODS   234

4.1. Dynamic Ray Tracing in Ray-Centered Coordinates   237
4.2. Hamiltonian Approach to Dynamic Ray Tracing   259
4.3. Propagator Matrices of Dynamic Ray Tracing Systems   278
4.4. Dynamic Ray Tracing in Isotropic Layered Media   289
4.5. Initial Conditions for Dynamic Ray Tracing   310
4.6. Paraxial Travel-Time Field and Its Derivatives   322
4.7. Dynamic Ray Tracing in Cartesian Coordinates   331
4.8. Special Cases. Analytical Dynamic Ray Tracing   341

4.9. Boundary-Value Ray Tracing for Paraxial Rays   348
4.10. Geometrical Spreading in a Layered Medium   356
4.11. Fresnel Volumes   372
4.12. Phase Shift Due to Caustics. KMAH Index   380
4.13. Dynamic Ray Tracing Along a Planar Ray. 2-D Models   384
4.14. Dynamic Ray Tracing in Inhomogeneous Anisotropic Media   400

5   **RAY AMPLITUDES**   417

5.1. Acoustic Case   419
5.2. Elastic Isotropic Structures   449
5.3. Reflection/Transmission Coefficients for Elastic Isotropic Media 477
5.4. Elastic Anisotropic Structures   504
5.5. Weakly Dissipative Media   542
5.6. Ray Series Method. Acoustic Case   549
5.7. Ray-Series Method. Elastic Case   567
5.8. Paraxial Displacement Vector. Paraxial Gaussian Beams   582
5.9. Validity Conditions and Extensions of the Ray Method   607

6   **RAY SYNTHETIC SEISMOGRAMS**   621

6.1. Elementary Ray Synthetic Seismograms   622
6.2. Ray Synthetic Seismograms   630
6.3. Ray Synthetic Seismograms in Weakly Dissipative Media   639
6.4. Ray Synthetic Particle Ground Motions   643

APPENDIX A **FOURIER TRANSFORM, HILBERT TRANSFORM,
            AND ANALYTICAL SIGNALS**   661

A.1. Fourier Transform   661
A.2. Hilbert Transform   663
A.3. Analytical Signals   664

References   667
Index   697

Preface

The book presents a consistent treatment of the seismic ray method, applicable to high-frequency seismic body waves propagating in complex 3-D laterally varying isotropic or anisotropic layered and block structures. The seismic ray method is based on the asymptotic high-frequency solution of the elastodynamic equation. For finite frequencies, the ray method is not exact, but it is only approximate. Its accuracy, however, is sufficient to solve many 3-D wave propagation problems of practical interest in seismology and seismic exploration, which can hardly be treated by any other means. Moreover, the computed rays may be used as a framework for the application of various more sophisticated methods.

In the seismic ray method, the high-frequency wavefield in a complex structure can be expanded into contributions, which propagate along rays and are called elementary waves. Individual elementary waves correspond, for example, to direct P and S waves, reflected waves, various multiply reflected/transmitted waves, and converted waves. A big advantage of the ray method is that the elementary waves may be handled independently. In the book, equations controlling the rays, travel times, amplitudes, Green functions, seismograms, and particle ground motions of the elementary waves are derived, and the relevant numerical algorithms are developed and discussed.

In general, the theoretical treatment in the book starts with a relatively simple problem in which the main ideas of the solution are easier to explain. Only then are the more complex problems dealt with. That is one of the reasons why pressure waves in fluid models are also discussed. All the derivations for pressure waves in fluid media are simple, clear, and comprehensible. These derivations help the reader to understand analogous derivations for elastic waves in isotropic and anisotropic solid structures, which are often more advanced. There is, however, yet another reason for discussing the pressure waves in fluid media: they have often been used in seismic exploration as a useful approximation for P waves in solid media.

Throughout the book, considerable attention is devoted to the seismic ray theory in inhomogeneous anisotropic media. Most equations derived for isotropic media are also derived for anisotropic media. In addition, weakly anisotropic media are discussed in some detail. Special attention is devoted to the qS wave coupling.

A detailed derivation and discussion of paraxial ray methods, dynamic ray tracing, and ray propagator matrices are presented. These concepts extend the possibilities of the ray method and the efficiency of calculations. They can be used to compute the travel times and slowness vectors not only along the ray but also in its vicinity. They also offer numerous other important applications in seismology and seismic exploration such as solution of boundary-value ray tracing problems, computation of geometrical spreading, Fresnel volumes, Gaussian beams, and Maslov-Chapman integrals.

Most of the final equations are expressed in algorithmic form and may be directly used for programming and applied in various interpretation programs, in numerical modeling of seismic wave fields, tomography, and migration.

I am greatly indebted to my friends, colleagues, and students from Charles University and from many other universities and institutions for helpful suggestions and valuable discussions. Many of them have read critically and commented on certain parts of the manuscript. I owe a special debt of thanks to Peter Hubral, Bob Nowack, and Colin Thomson for advice and constructive criticism. I am further particularly grateful to Ivan Pšenčík and Luděk Klimeš for everyday discussions and to Eva Drahotová for careful typing of the whole manuscript. I also wish to express my sincere thanks to the sponsors of the Consortium Project "Seismic Waves in Complex 3-D Structures" for support and to Schlumberger Cambridge Research for a Stichting Award, which was partially used in the preparation of the manuscript. Finally, I offer sincere thanks to my family for patience, support, and encouragement.

CHAPTER ONE

Introduction

The propagation of seismic body waves in complex, laterally varying 3-D layered structures is a complicated process. Analytical solutions of the elastodynamic equations for such types of media are not known. The most common approaches to the investigation of seismic wavefields in such complex structures are (a) methods based on direct numerical solutions of the elastodynamic equation, such as the finite-difference and finite-element methods, and (b) approximate high-frequency asymptotic methods. Both methods are very useful for solving certain types of seismic problems, have their own advantages and disadvantages, and supplement each other suitably.

We will concentrate here mainly on high-frequency asymptotic methods, such as the ray method. The high-frequency asymptotic methods are based on an asymptotic solution of the elastodynamic equation. They can be applied to compute not only rays and travel times but also the ray-theory amplitudes, synthetic seismograms, and particle ground motions. These methods are well suited to the study of seismic wavefields in smoothly inhomogeneous 3-D media composed of thick layers separated by smoothly curved interfaces. The high-frequency asymptotic methods are very general; they are applicable both to isotropic and anisotropic structures, to arbitrary 3-D variations of elastic parameters and density, to curved interfaces arbitrarily situated in space, to an arbitrary source-receiver configuration, and to very general types of waves. High-frequency asymptotic methods are also appropriate to explain typical "wave" phenomena of seismic waves propagating in complex 3-D isotropic and anisotropic structures. The amplitudes of seismic waves calculated by asymptotic methods are only approximate, but their accuracy is sufficient to solve many 3-D problems of practical interest.

Asymptotic high-frequency solutions of the elastodynamic equation can be sought in several alternative forms. In the ray method, they are usually sought in the form of the so-called *ray series* (see Babich 1956; Karal and Keller 1959). For this reason, the ray method is also often called the *ray-series method*, or the *asymptotic ray theory* (ART).

The seismic ray method can be divided into two parts: *kinematic* and *dynamic*. The kinematic part consists of the computation of seismic rays, wavefronts, and travel times. The dynamic part consists of the evaluation of the vectorial complex-valued amplitudes of the displacement vector and the computation of synthetic seismograms and particle ground motion diagrams.

The most strict approach to the investigation of both kinematic and dynamic parts of the ray method consists of applying asymptotic high-frequency methods to the elastodynamic equations. The kinematic part of the ray method, however, may also be attacked by some simpler approaches, for example, by variational principles (Fermat principle). It is even possible to develop the whole kinematic part of the seismic ray method using the well-known Snell's law. Such approaches have been used for a long time in seismology and have given a number of valuable results. There may be, however, certain methodological objections to their application. In the application of *Snell's law*, we must start from a model consisting of homogeneous layers with curved interfaces and pass from this model to a smoothly varying model by increasing the number of interfaces. Such a limiting process offers very useful seismological insights into the ray tracing equations and travel-time computations in inhomogeneous media, but it is more or less intuitive. The *Fermat principle* has been used in seismology as a rule independently for P and S waves propagating in inhomogeneous media. The elastic wavefield, however, can be separated into P and S waves only in homogeneous media (and perhaps in some other simple structures). In laterally varying media with curved interfaces, the wavefield is not generally separable into P and S waves; the seismic wave process is more complicated. Thus, we do not have any exact justification for applying the principle independently to P and S waves. In media with larger velocity gradients, the ray method fails due to the strong coupling of P and S waves. Only the approach based on the asymptotic solution of the elastodynamic equation gives the correct answer: the separation of the seismic wavefield in inhomogeneous media into two independent wave processes (P and S) is indeed possible, but it is only *approximate, in that it is valid for high frequencies and sufficiently smooth media only.*

Similarly, certain properties of vectorial complex-valued amplitudes of seismic body waves can be derived using energy concepts, particularly using the expressions for the energy flux. Such an approach is again very useful for intuitive physical understanding of the amplitude behavior, but it does not give the complete answer. The amplitudes of seismic body waves have a vectorial complex-valued character. The waves may be elliptically polarized (S waves) and may include phase shifts. These phase shifts influence the waveforms. The energy principles do not yield a complete answer in such situations. Consequently, they cannot be applied to the computation of synthetic seismograms and particle ground motion diagrams.

Recently, several new concepts and methods have been proposed to increase the possibilities and efficiency of the standard ray method; they include *dynamic ray tracing, the ray propagator matrix*, and *paraxial ray approximations*. In the standard ray method, the travel time and the displacement vector of seismic body waves are usually evaluated along rays. Thus, if we wish to evaluate the seismic wavefield at any point, we must find the ray that passes through this point (boundary value ray tracing). The search for such rays sometimes makes the application of the standard ray method algorithmically very involved, particularly in 3-D layered structures. The paraxial ray methods, however, allow one to compute the travel time and displacement vector not only along the ray but also in its paraxial vicinity. It is not necessary to evaluate the ray that passes exactly through the point. The knowledge of the ray propagator matrix makes it possible to solve analytically many complex wave propagation problems that must be solved numerically by iterations in the standard ray method. This capability greatly increases the efficiency of the ray method, particularly in 3-D complex structures.

The final ray solution of the elastodynamic equation is composed of elementary waves corresponding to various rays connecting the source and receiver. Each of these elementary waves (reflected, refracted, multiply reflected, converted, and the like) is described by its own ray series. In practical seismological applications, the higher terms of the ray series have not yet been broadly used. In most cases, the numerical modeling of seismic wavefields and the interpretation of seismic data by the ray method have been based on the zeroth-order leading term of the ray series. In this book, mainly the zeroth-order terms of the ray series are considered. These zeroth-order terms, however, are treated here in a great detail. Concise expressions for the zeroth-order ray-theory Green function for a point source and receiver situated at any place in a general 3-D, layered and blocked, structure are derived. For a brief treatment of the higher-order terms of the ray series for the scalar (acoustic) and vectorial (elastic) waves see Sections 5.6 and 5.7.

As is well known, the ray method is only approximate, and its applications to certain seismological problems have some restrictions. Recently, several new extensions of the ray method have been proposed; these extensions overcome, partially or fully, certain of these restrictions. They include the method of summation of Gaussian beams, the method of summation of Gaussian wave packets, and the Maslov-Chapman method. These methods have been found very useful in solving various seismological problems, even though certain aspects of these methods are still open for future research.

The whole book may be roughly divided into five parts.

In the first part, the main principles of the asymptotic high-frequency method as it is used to solve the elastodynamic equation in a 3-D laterally varying medium are briefly explained and discussed. A particularly simple approach is used to derive and discuss the most important equations and related wave phenomena from the seismological point of view. It is shown how the elastic wavefield is approximately separated into individual elementary waves. These individual waves propagate independently in a smoothly varying structure, their travel times are controlled by the eikonal equation, and their amplitudes are controlled by the transport equation. Various important phenomena of seismic wavefields connected with 3-D lateral variations and with curved interfaces are derived and explained, both for isotropic and anisotropic media. Great attention is devoted to the differences between elastic waves propagating in isotropic and anisotropic structures. Exact and approximate expressions for acoustic and elastodynamic Green functions in homogeneous media are also derived. See Chapter 2.

The second part is devoted to ray tracing and travel-time computation in 3-D structures. The ray tracing and travel-time computation play an important role in many seismological applications, particularly in seismic inversion algorithms, even without a study of ray amplitudes, polarization, and wavelet shape. In addition to individual rays, the ray fields are also introduced in this part. The singular regions of the ray fields and related wave phenomena are explained. Special attention is devoted to the definition, computation, and physical meaning of the geometrical spreading. See Chapter 3.

The third part is devoted to dynamic ray tracing and paraxial ray methods. The paraxial ray methods can be used to compute the travel time and other important quantities not only along the ray but also in its vicinity. Concepts of dynamic ray tracing and of the ray propagator matrix are explained. The dynamic ray tracing is introduced both in ray-centered and Cartesian coordinates, for isotropic and anisotropic structures. Various important applications of the paraxial ray method are explained. See Chapter 4.

The fourth part of the book discusses the computation of ray amplitudes. Very general expressions for ray amplitudes of an arbitrary multiply reflected/transmitted (possibly converted) seismic body wave propagating in acoustic, elastic isotropic, elastic anisotropic, laterally varying, layered, and block structures are derived. The medium may also be weakly dissipative. Both the source and the receiver may be situated either in a smooth medium or at a structural interface or at the Earth's surface. Final equations for the amplitudes of the ray-theory elastodynamic Green function is laterally varying layered structures are derived. Great attention is also devoted to the ray-series solutions, both in the frequency and the time domain. The seismological applications of higher-order terms of the ray series are discussed. See Chapter 5.

The fifth part explains the computation of ray synthetic seismograms and ray synthetic particle ground motions. Several possibilities for the computation of ray synthetic seismograms are proposed: in the frequency domain, in the time domain, and by the summation of elementary seismograms. Advantages and disadvantages of individual approaches are discussed. Certain of these approaches may be used even for dissipative media. The basic properties of linear, elliptic, and quasi-elliptic polarization are described. The causes of quasi-elliptic polarization of S waves are briefly summarized. See Chapter 6.

This book, although very extensive, is still not able to cover all aspects of the seismic ray method. This would increase its length inadmissibly. To avoid this, the author has not discussed many important subjects regarding the seismic ray method (or related closely to it) or has discussed them only briefly. Nevertheless, the reader should remember that the main aim of this book is to present a detailed and complete description of the seismic ray method with a real-valued eikonal for 3-D, laterally varying, isotropic or anisotropic, layered, and block structures. The author, however, does not and had no intention of including all the extensions and applications of the seismic ray method and all the problems related closely to it. We shall now briefly summarize several important topics that are related closely to the seismic ray method but that will not be treated in this book or that will be treated more briefly than they would deserve.

1. Although the seismic ray method developed in this book plays a fundamental role in various inverse problems of seismology and of seismic exploration and in many interpretational procedures, the actual inversion and interpretational procedures are not explicitly discussed here. These procedures include seismic tomography, seismic migration, and the location of earthquake sources, among others.

2. The seismic ray method has found important applications in forward and inverse scattering problems. With the exception of a brief introduction in Section 2.6.2; the scattering problems themselves, however, are not discussed here.

3. The seismic ray method may be applied only to structural models that satisfy certain smoothness criteria. The construction of 2-D and 3-D models that would satisfy such criteria is a necessary prerequisite for the application of the seismic ray method, but is not discussed here at all. Mostly, it is assumed that the model is specified in Cartesian rectangular coordinates. Less attention is devoted to models specified in curvilinear coordinate systems (including spherical); see Section 3.5.

4. The seismic ray method developed here may be applied to high-frequency seismic body waves propagating in deterministic, perfectly elastic, isotropic or anisotropic media. Other types of waves (such as surface waves) are only briefly mentioned. Moreover, viscoelastic, poroelastic, and viscoporoelastic models are not considered. The exception is a weakly dissipative (and dispersive) model that does not require complex-valued ray tracing; see Sections 5.5 and 6.3.5. In Sections 2.6.4 and 5.6.8, the space-time ray method and the ray method with a complex eikonal are briefly discussed, even though they deserve considerably more attention. The computation of complex-valued rays in particular (for example, in dissipative media, in the

caustic shadow, and in some other singular regions) may be very important in applications. Actually, the seismic ray method, without considering complex rays, is very incomplete.

5. Various extensions of the seismic ray method have been proposed in the literature. These extensions include the asymptotic diffraction theory, the method of edge waves, the method of the parabolic wave equation, the Maslov-Chapman method, and the method of summation of Gaussian beams or Gaussian wave packets, among others. Here we shall treat, in some detail, only the extensions based on the summation of paraxial ray approximations and on the summation of paraxial Gaussian beams; see Section 5.8. The method based on the summation of paraxial ray approximations yields integrals close or equal to those of the Maslov-Chapman method. The other extensions of the seismic ray method are discussed only very briefly in Section 5.9, but the most important references are given there.

6. No graphical examples of the computation of seismic rays, travel times, ray amplitudes, synthetic seismograms, and particle ground motions in 3-D complex models are presented for two reasons. First, most figures would have to be in color, as 3-D models are considered. Second, the large variety of topics discussed in this book would require a large number of demonstration figures. This would increase the length and price of the book considerably. The interested readers are referred to the references given in the text, and to the www pages of the Consortium Project "Seismic Waves in 3-D Complex Structures"; see *http://seis.karlov.mff.cuni.cz/ consort/main.htm* for some examples.

The whole book has a tutorial character. The equations presented are (in most cases) derived and discussed in detail. For this reason, the book is rather long. Owing to the extensive use of various matrix notations and to the applications of several coordinate systems and transformation matrices, the resulting equations are very concise and simply understandable from a seismological point of view. Although the equations are given in a concise and compact form, the whole book is written in an algorithmic way: most of the expressions are specified to the last detail and may be directly used for programming.

To write the complicated equations of this book in the most concise form, we use mostly matrix notation. To distinguish between 2×2 and 3×3 matrices, we shall use the circumflex (ˆ) above the letter for 3×3 matrices. If the same letter is used for both 2×2 and 3×3 matrices; for example, $\mathbf{M}$ and $\hat{\mathbf{M}}$, matrix $\mathbf{M}$ denotes the 2×2 left upper submatrix of $\hat{\mathbf{M}}$:

$$\hat{\mathbf{M}} = \begin{pmatrix} M_{11} & M_{12} & M_{13} \\ M_{21} & M_{22} & M_{23} \\ M_{31} & M_{32} & M_{33} \end{pmatrix}, \qquad \mathbf{M} = \begin{pmatrix} M_{11} & M_{12} \\ M_{21} & M_{22} \end{pmatrix}.$$

Similarly, we denote by $\hat{\mathbf{q}} = (q_1, q_2, q_3)^T$ the 3×1 column matrix and by $\mathbf{q} = (q_1, q_2)^T$ the 2×1 column matrix. The symbol T as a superscript denotes the matrix transpose. Similarly, the symbol $-1T$ as a superscript denotes the transpose of the inverse, $\mathbf{A}^{-1T} = (\mathbf{A}^{-1})^T$, $\hat{\mathbf{A}}^{-1T} = (\hat{\mathbf{A}}^{-1})^T$.

In several places, we also use 4×4 and 6×6 matrices. We denote them by boldface letters in the same way as the 2×2 matrices; this notation cannot cause any misunderstanding.

In parallel with matrix notation, we also use component notation where suitable. The indices always have the form of right-hand suffixes. The uppercase suffixes take the values

1 and 2, lowercase indices 1, 2 and 3, and greek lowercase indices 1, 2, 3 and 4. In this way, M_{IJ} denote elements of **M** and M_{ij} elements of $\hat{\mathbf{M}}$. We also denote $f(x_i) = f(x_1, x_2, x_3)$, $f(x_I) = f(x_1, x_2)$, $[f(x_i)]_{x_k=0} = f(0, 0, 0)$, $[f(x_i)]_{x_K=0} = f(0, 0, x_3)$, $[f(x_i)]_{x_1=0} = f(0, x_2, x_3)$. The Einstein summation convention is used throughout the book. Thus, $M_{IJ}q_J = M_{I1}q_1 + M_{I2}q_2$ ($I = 1$ or 2), $M_{ij}q_j = M_{i1}q_1 + M_{i2}q_2 + M_{i3}q_3$ ($i = 1, 2$ or 3). Similarly, M_{iJ} denotes the elements of the 3×2 submatrix of matrix $\hat{\mathbf{M}}$.

We also use the commonly accepted notation for partial derivatives with respect to Cartesian coordinates x_i (for example, $\lambda_{,i} = \partial\lambda/\partial x_i$, $u_{i,ji} = \partial^2 u_i/\partial x_j \partial x_i$, $\sigma_{ij,j} = \partial\sigma_{ij}/\partial x_j$). In the case of velocities, we shall use a similar notation to denote the partial derivatives with respect to the ray-centered coordinates. For a more detailed explanation, see the individual chapters.

In some equations, the classical vector notation is very useful. We use arrows above letters to denote the 3-D vectors. In this way, any 3-D vector may be denoted equivalently as a 3×1 column matrix or as a vectorial form.

In complex-valued quantities, $z = x + iy$, the asterisk is used as a superscript to denote a complex-conjugate quantity, $z^* = x - iy$. The asterisk between two time-dependent functions, $f_1(t) * f_2(t)$, denotes the time convolution of these two functions, $f_1(t) * f_2(t) = \int_{-\infty}^{\infty} f_1(\tau) f_2(t - \tau) d\tau$.

The book does not give a systematic bibliography on the seismic ray method. For many other references, see the books and review papers on the seismic ray method and on some related subjects (Červený and Ravindra 1971; Červený, Molotkov, and Pšenčík 1977; Hubral and Krey 1980; Hanyga, Lenartowicz, and Pajchel 1984; Bullen and Bolt 1985; Červený 1985a, 1985b, 1987a, 1989a; Chapman 1985, in press; Virieux 1996; Dahlen and Tromp 1998). The ray method has been also widely used in other branches of physics, mainly in electromagnetic theory (see, for example, Synge 1954; Kline and Kay 1965; Babich and Buldyrev 1972; Felsen and Marcuvitz 1973; Kravtsov and Orlov 1980).

CHAPTER TWO

The Elastodynamic Equation and Its Simple Solutions

The seismic ray method is based on asymptotic high-frequency solutions of the elastodynamic equation. We assume that the reader is acquainted with linear elastodynamics and with the simple solutions of the elastodynamic equation in a homogeneous medium. For the reader's convenience, we shall briefly discuss all these topics in this chapter, particularly the plane-wave and point-source solutions of the elastodynamic equation. We shall introduce the terminology, notations, and all equations we shall need in the following chapters. In certain cases, we shall only summarize the equations without deriving them, mainly if such equations are known from generally available textbooks. This applies, for example, to the basic concepts of linear elastodynamics. In other cases, we shall present the main ideas of the solution, or even the complete derivation. This applies, for example, to the Green functions for acoustic, elastic isotropic and elastic anisotropic homogeneous media.

In addition to elastic waves in solid isotropic and anisotropic models, we shall also study pressure waves in fluid models. In this case, we shall speak of the *acoustic case*. There are two main reasons for studying the acoustic case. The first reason is tutorial. All the derivations for the acoustic case are very simple, clear, and comprehensible. In elastic media, the derivations are also simple in principle, but they are usually more cumbersome. Consequently, we shall mostly start the derivations with the acoustic case, and only then shall we discuss the elastic case. The second reason is more practical. Pressure waves in fluid models are often used as a simple approximation of P elastic waves in solid models. For example, this approximation is very common in seismic exploration for oil.

The knowledge of plane-wave solutions of the elastodynamic equation in homogeneous media is very useful in deriving approximate high-frequency solutions of elastodynamic equation in smoothly inhomogeneous media. Such approximate high-frequency solutions in smoothly inhomogeneous media are derived in Section 2.4. In the terminology of the ray-series method, such solutions represent the zeroth-order approximation of the ray method. The approach we shall use in Section 2.4 is very simple and is quite sufficient to derive all the basic equations of the zeroth-order approximation of the ray method for acoustic, elastic isotropic, and elastic anisotropic structures. In the acoustic case, the approach yields the eikonal equation for travel times and the transport equation for scalar amplitudes. In the elastic case, it yields an approximate high-frequency decomposition of the wave field into the separate waves (P and S waves in isotropic; qP, qS1, and qS2 in anisotropic media). Thereafter, it yields the eikonal equations for travel times, the transport equations for amplitudes, and the rules for the polarization of separate waves.

Note that Section 2.4 deals only with the zeroth-order approximation of the ray-series method. The higher-order terms of the ray series are not discussed here, but will be considered in Chapter 5 (Sections 5.6 for the acoustic case and Section 5.7 for the elastic case). From the systematic and theoretical points of view, it would be more convenient to start the whole treatment directly with the ray-series method, and only then discuss the zeroth-order approximation, as a leading term of the ray-series method. The reason why we have moved the ray-series treatment to Sections 5.6 and 5.7 is again tutorial. The complete treatment of rays, ray-theory travel times, and paraxial methods in Chapters 3 and 4 is based on eikonal equations only. Similarly, all the treatments of ray amplitudes in Sections 5.1 through 5.5 are based on transport equations only. Thus, we do not need to know the higher-order terms of the ray series in Chapters 3 and 4 and in Sections 5.1 through 5.5; the results of Section 2.4 are sufficient there. Consequently, the whole ray-series treatment, which is more cumbersome than the derivation of Section 2.4, can be moved to Sections 5.6 and 5.7. Most of the recent applications of the seismic ray method are based on the zeroth-order approximation of the ray series. Consequently, most readers will be interested in the relevant practical applications of the seismic ray method, such as ray tracing, travel time, and ray amplitude computations. These readers need not bother with the details of the ray-series method; the zeroth-order approximation, derived in Section 2.4, is sufficient for them. The readers who wish to know more about the ray-series method and higher order terms of the ray series can read Sections 5.6 and 5.7 immediately after reading Section 2.4. Otherwise, no results of Sections 5.6 and 5.7 are needed in the previous sections.

Section 2.5 discusses the point-source solutions and appropriate Green functions for homogeneous fluid, elastic isotropic, and elastic anisotropic media. In all three cases, exact expressions for the Green function are derived uniformly. For elastic anisotropic media, exact expressions are obtained only in an integral form. Suitable asymptotic high-frequency expressions are, however, given in all three cases. These expressions are used in Chapter 5 to derive the asymptotic high-frequency expressions for the ray-theory Green function corresponding to an arbitrary elementary wave propagating in a 3-D laterally varying layered and blocked structure (fluid, elastic isotropic, elastic anisotropic).

The Green function corresponds to a point source, but it may be used in the representation theorem to construct considerably more complex solutions of the elastodynamic equation. If we are interested in high-frequency solutions, the ray-theory Green function may be used in the representation theorems. For this reason, representation theorems and the ray-theory Green functions play an important role even in the seismic ray method. The representation theorems are derived and briefly discussed in Section 2.6. The same section also discusses the scattering integrals and the first-order Born approximation. These integrals contain the Green function. If we use the ray-theory Green function in these integrals, the resulting scattering integrals can be used broadly in the seismic ray method and in relevant applications. Such approaches have recently found widespread applications in seismology and seismic exploration.

2.1 Linear Elastodynamics

The basic concepts and equations of linear elastodynamics have been explained in many textbooks and papers, including some seismological literature. We refer the reader to Bullen (1965), Auld (1973), Pilant (1979), Aki and Richards (1980), Hudson (1980a), Ben-Menahem and Singh (1981), Mura (1982), Bullen and Bolt (1985), and Davis (1988), where many other references can be found. For a more detailed treatment, see Love (1944), Landau and Lifschitz (1965), Fung (1965), and Achenbach (1975). Here we shall introduce only the most useful terminology and certain important equations that we shall need later. We shall mostly follow and use the notations of Aki and Richards (1980).

To write the equations of linear elastodynamics, some knowledge of tensor calculus is required. Because we wish to make the treatment as simple as possible, we shall use Cartesian coordinates x_i and Cartesian tensors only.

We shall use the Lagrangian description of motion in an elastic continuum. In the Lagrangian description, we study the motion of a particle specified by its original position at some reference time. Assume that the particle is located at the position described by Cartesian coordinates x_i at the reference time. The vector distance of a particle at time t from position $\vec{x}$ at the reference time is called the *displacement vector* and is denoted by $\vec{u}$. Obviously, $\vec{u} = \vec{u}(\vec{x}, t)$.

We denote the Cartesian components of the *stress tensor* by $\tau_{ij}(\vec{x}, t)$ and the Cartesian components of the *strain tensor* by $e_{ij}(\vec{x}, t)$. Both tensors are considered to be symmetric,

$$\tau_{ij} = \tau_{ji}, \qquad e_{ij} = e_{ji}. \tag{2.1.1}$$

The strain tensor can be expressed in terms of the displacement vector as follows:

$$e_{ij} = \tfrac{1}{2}(u_{i,j} + u_{j,i}). \tag{2.1.2}$$

The stress tensor $\tau_{ij}(\vec{x}, t)$ fully describes the stress conditions at any point $\vec{x}$. It can be used to compute *traction* $\vec{T}$ acting across a surface element of arbitrary orientation at $\vec{x}$,

$$T_i = \tau_{ij} n_j, \tag{2.1.3}$$

where $\vec{n}$ is the unit normal to the surface element under consideration.

The *elastodynamic equation* relates the spatial variations of the stress tensor with the time variations of the displacement vector,

$$\tau_{ij,j} + f_i = \rho \ddot{u}_i, \qquad i = 1, 2, 3. \tag{2.1.4}$$

Here f_i denote the Cartesian components of body forces (force per volume), and ρ is the density. The term with f_i in elastodynamic equation (2.1.4) will also be referred to as *the source term*. Quantities $\ddot{u}_i = \partial^2 u_i / \partial t^2$, $i = 1, 2, 3$, represent the second partial derivatives of u_i with respect to time (that is, the Cartesian components of *particle acceleration $\ddot{\vec{u}}$*). In a similar way, we shall also denote the Cartesian components of *particle velocity* $\partial u_i / \partial t$ by v_i or $\dot{u}_i$.

The introduced quantities are measured in the following units: stress τ_{ij} and traction T_i in pascals (Pa; that is, in kg m^{-1} s^{-2}), the components of body forces f_i in newtons per cubic meter (N/m^3; that is, in kg m^{-2} s^{-2}), density ρ in kilograms per cubic meter (kg m^{-3}), and displacement components u_i in meters (m). Finally, strain components e_{ij} are dimensionless.

2.1.1 Stress-Strain Relations

In a linear, anisotropic, perfectly elastic solid, the constitutive stress-strain relation is given by the *generalized Hooke's law*,

$$\tau_{ij} = c_{ijkl} e_{kl}. \tag{2.1.5}$$

Here c_{ijkl} are components of the elastic tensor. The elastic tensor has, in general, $3 \times 3 \times 3 \times 3 = 81$ components. These components, however, satisfy the following *symmetry relations*:

$$c_{ijkl} = c_{jikl} = c_{ijlk} = c_{klij}, \qquad\qquad (2.1.6)$$

which reduce the number of independent components of the elastic tensor from 81 to 21.

The components c_{ijkl} of the elastic tensor are also often called *elastic constants, elastic moduli, elastic parameters, or stiffnesses*. In this book, we shall mostly call them elastic moduli. They are measured in the same units as the stress components (that is, in Pa $=$ $\text{kg m}^{-1} \text{s}^{-2}$).

If we express e_{kl} in terms of the displacement vector components, see Equation (2.1.2), and take into account symmetry relations (2.1.6), we can also express Equation (2.1.5) in the following form:

$$\tau_{ij} = c_{ijkl} u_{k,l}. \qquad\qquad (2.1.7)$$

The components of elastic tensor c_{ijkl} are also often expressed in an abbreviated Voigt form, with two indices instead of four. We shall denote these components by capital letters C_{mn}. C_{mn} is formed from c_{ijkl} in the following way: m corresponds to the first pair of indices, i, j and n to the second pair, k, l. The correspondence $m \to i, j$ and $n \to k, l$ is as follows: $1 \to 1, 1; 2 \to 2, 2; 3 \to 3, 3; 4 \to 2, 3; 5 \to 1, 3; 6 \to 1, 2$.

Due to symmetry relations (2.1.6), the 6×6 matrix C_{mn} fully describes the elastic moduli of an arbitrary anisotropic elastic medium. It is also symmetric, $C_{mn} = C_{nm}$ and is commonly expressed in the form of a table containing 21 independent elastic moduli:

$$\begin{pmatrix} C_{11} & C_{12} & C_{13} & C_{14} & C_{15} & C_{16} \\ & C_{22} & C_{23} & C_{24} & C_{25} & C_{26} \\ & & C_{33} & C_{34} & C_{35} & C_{36} \\ & & & C_{44} & C_{45} & C_{46} \\ & & & & C_{55} & C_{56} \\ & & & & & C_{66} \end{pmatrix}. \qquad (2.1.8)$$

The elastic moduli C_{mn} below the diagonal $(m > n)$ are not shown because the table is symmetrical, $C_{mn} = C_{nm}$. The diagonal elements in the table are always positive for a solid medium, but the off-diagonal elements may be arbitrary (positive, zero, negative). Note that C_{mn} is not a tensor.

A whole hierarchy of various anisotropic symmetry systems exist. They are described and discussed in many books and papers; see, for example, Fedorov (1968), Musgrave (1970), Auld (1973), Crampin and Kirkwood (1981), Crampin (1989), and Helbig (1994). The most general is the *triclinic symmetry*, which may have up to 21 independent elastic moduli. In simpler (higher symmetry) anisotropic systems, the elastic moduli are invariant to rotation about a specific axis by angle $2\pi/n$ (n-fold axis of symmetry). We shall briefly discuss only two such simpler systems which play an important role in recent seismology and seismic exploration: orthorhombic and hexagonal.

In the *orthorhombic symmetry system*, three mutually perpendicular twofold axes of symmetry exist. The number of significant elastic moduli in the orthorhombic system is reduced to nine. If the Cartesian coordinate system being considered is such that its axes coincide with the axes of symmetry, the table (2.1.8) of C_{ij} reads:

$$
\begin{pmatrix}
C_{11} & C_{12} & C_{13} & 0 & 0 & 0 \\
 & C_{22} & C_{23} & 0 & 0 & 0 \\
 & & C_{33} & 0 & 0 & 0 \\
 & & & C_{44} & 0 & 0 \\
 & & & & C_{55} & 0 \\
 & & & & & C_{66}
\end{pmatrix}. \tag{2.1.9}
$$

In the *hexagonal symmetry system*, one sixfold axis of symmetry exists. The number of significant elastic moduli in the hexagonal system is reduced to five. If the Cartesian coordinate system being considered is such that the x_3-axis coincides with the sixfold axis of symmetry, the table of nonvanishing elastic moduli is again given by (2.1.9), but C_{ij} satisfy the following four relations: $C_{22} = C_{11}$, $C_{55} = C_{44}$, $C_{12} = C_{11} - 2C_{66}$, $C_{23} = C_{12}$. If the sixfold axis of symmetry coincides with the x_1-axis, the four relations are as follows: $C_{22} = C_{33}$, $C_{55} = C_{66}$, $C_{23} = C_{22} - 2C_{44}$, $C_{13} = C_{12}$.

It is possible to show that the invariance of elastic moduli to rotation by $\pi/3$ in the hexagonal system (sixfold axis of symmetry) implies general invariance to rotation *by any angle*. From this point of view, the hexagonal symmetry system is equivalent to a *transversely isotropic medium* in which the elastic moduli do not change if the medium is rotated about the axis of symmetry by any angle. Traditionally, the vertical axis of symmetry of the transversely isotropic medium has been considered. At present, however, the transversely isotropic medium is considered more generally, with an arbitrarily oriented axis of symmetry (inclined, horizontal). If it is vertical, we speak of *azimuthal isotropy* (Crampin 1989).

In seismology, the most commonly used anisotropy systems correspond to the hexagonal symmetry. Systems more complicated than orthorhombic have been used only exceptionally. Moreover, the anisotropy is usually weak in the Earth's interior. A suitable notation for elastic moduli in weakly anisotropic media was proposed by Thomsen (1986).

As we have seen, the abbreviated Voigt notation C_{mn} for elastic moduli is useful in discussing various anisotropy symmetries, particularly if simpler symmetries are involved. If, however, we are using general equations and expressions, such as the constitutive relations or the elastodynamic equation and its solutions, it is more suitable to use the elastic moduli in the nonabbreviated form of c_{ijkl}. Due to the Einstein summation convention, all general expressions are obtained in a considerably more concise form. For this reason, we shall consistently use the notation c_{ijkl} for elastic moduli rather than C_{mn}. As an exercise, the reader may write out the general expressions for some simpler anisotropy symmetries, using the notation C_{mn} for the elastic parameters.

Elastic anisotropy is a very common phenomenon in the Earth's interior (see Babuška and Cara 1991). It is caused by different mechanisms. Let us briefly describe three of these mechanisms. Note, of course, that the individual mechanisms may be combined.

1. **Preferred orientation of crystals.** Single crystals of rock-forming minerals are intrinsically anisotropic. If no preferred orientation of mineral grains exists, polycrystalline aggregates of anisotropic material behave macroscopically isotropically. In case of preferred orientation of these grains, however, the material behaves macroscopically anisotropically. Preferred orientation is probably one of the most important factors in producing anisotropy of dense aggregates under high pressure and temperature conditions.

2. **Anisotropy due to aligned inclusions.** The presence of aligned inclusions (such as cracks, pores, or impurities) can cause effective anisotropy of rocks, if observed at long wavelengths. The most important two-phase systems with distinct anisotropic behavior are a cracked solid and a poroelastic solid.

3. **Anisotropy due to regular sequences of thin layers.** Regular sequences of isotropic thin layers of different properties are very common in the Earth's interior, at least in the upper crust (foliation, bedding, and so on). If the prevailing wavelength of the wave under consideration is larger than the thickness of the individual thin layers, the regular sequences of thin layers behave anisotropically.

Let us give several typical examples of actual anisotropy symmetries important in seismology:

a. Hexagonal symmetry. (i) Vertical axis of symmetry: periodic horizontal thin layering. This symmetry is also known as VTI (vertical transverse isotropy) symmetry. (ii) Horizontal axis of symmetry: parallel vertical cracks. This symmetry is also known as HTI (horizontal transverse isotropy) symmetry.

b. Orthorhombic symmetry. (i) Olivine (preferred orientation of crystals) and (ii) Combination of periodic thin layering with cracks perpendicular to the layering.

In the *isotropic solid*, the components of elastic tensor c_{ijkl} can be expressed in terms of two independent elastic moduli λ and μ as follows:

$$c_{ijkl} = \lambda\delta_{ij}\delta_{kl} + \mu(\delta_{ik}\delta_{jl} + \delta_{il}\delta_{jk}); \tag{2.1.10}$$

see Jeffreys and Jeffreys (1966) and Aki and Richards (1980). Here δ_{ij} is the Kronecker symbol,

$$\delta_{ij} = 1 \quad \text{for } i = j, \qquad \delta_{ij} = 0 \quad \text{for } i \neq j. \tag{2.1.11}$$

Elastic moduli λ, μ are also known as Lamé's elastic moduli; μ is called the *rigidity* (or shear modulus).

For an isotropic medium stress-strain relation, Equation (2.1.5) can be expressed in the following form:

$$\tau_{ij} = \lambda\delta_{ij}\theta + 2\mu e_{ij}, \qquad \theta = e_{kk} = u_{k,k} = \nabla \cdot \vec{u}. \tag{2.1.12}$$

Equation (2.1.12) represents the famous Hooke's law. Quantity θ is called the *cubical dilatation*. If we replace θ and e_{ij} by the components of the displacement vector, we obtain

$$\tau_{ij} = \lambda\delta_{ij}u_{k,k} + \mu(u_{i,j} + u_{j,i}). \tag{2.1.13}$$

Instead of Lamé's elastic moduli λ and μ, some other elastic parameters are also often used in isotropic solids: the bulk modulus (or incompressibility) k, the Young modulus E, the Poisson ratio σ, and compressibility κ. They are related to λ and μ as follows:

$$\begin{aligned} k &= \lambda + \tfrac{2}{3}\mu, & E &= 3\mu(\lambda + \tfrac{2}{3}\mu)/(\lambda + \mu), \\ \kappa &= k^{-1} = \left(\lambda + \tfrac{2}{3}\mu\right)^{-1}, & \sigma &= \tfrac{1}{2}\lambda/(\lambda + \mu). \end{aligned} \tag{2.1.14}$$

The physical meaning of these parameters is explained in many textbooks (Bullen and Bolt 1985; Auld 1973; Pilant 1979).

If $\mu = 0$, we speak of a *fluid medium* (Bullen and Bolt 1985). In this case, $\sigma = \tfrac{1}{2}, k = \lambda$, and Hooke's law (2.1.12) reads $\tau_{ij} = \lambda\delta_{ij}\theta$. Commonly, we then use pressure p instead of τ_{ij},

$$\tau_{ij} = -p\delta_{ij}, \qquad p = -\tfrac{1}{3}\tau_{ii}. \qquad\qquad (2.1.15)$$

Constitutive relation (2.1.12) then reads

$$p = -\lambda\theta = -k\theta. \qquad\qquad (2.1.16)$$

The fluid medium is commonly used in seismic prospecting for oil as an approximation of the solid medium (the so-called acoustic case).

Pressure p and elastic moduli λ, μ, k, and E are measured in $\mathrm{Pa} = \mathrm{kg\,m^{-1}\,s^{-2}}$, the compressibility is in $\mathrm{kg^{-1}m\,s^2}$, and σ is dimensionless.

Constitutive relations (2.1.5), (2.1.12), and (2.1.16) are related to small deviations from a *"natural" reference state*, in which both stress and strain are zero. Such a natural reference state, however, does not exist within the Earth's interior due to the large lithological pressure caused by self-gravitation. It is then necessary to use some other reference state, for example, the *state of static equilibrium*. By definition the strain is zero in this state, but the stress is nonzero. It is obvious that constitutive relations (2.1.5), (2.1.12), and (2.1.16), are not valid in this case. If we study the deviations from the state of static equilibrium, we can, however, work with small *incremental stresses* instead of actual stresses. The incremental stress is defined as the difference between the actual stress and the stress corresponding to the static equilibrium state. This incremental stress is zero in the state of static equilibrium and satisfies constitutive relations (2.1.5), (2.1.12), and (2.1.16) if the deviations from the state of static equilibrium are small. In the following text, we shall not emphasize the fact that τ_{ij} actually represents the incremental stress, and simply call τ_{ij} the stress. See the detailed discussion in Aki and Richards (1980).

All the elastic moduli are, in general, functions of position. If they are constant (independent of position) in some region, we call the medium *homogeneous* in that region.

Even more complex linear constitutive relations than those given here can be considered. In linear viscoelastic solids, the elastic moduli depend on time. In the time domain, the constitutive relations are then expressed in convolutional forms. Alternatively, in the frequency domain, the elastic moduli are complex-valued and may depend on frequency. Such constitutive relations are commonly used to study wave propagation in dissipative media; see Kennett (1983). Unless stated otherwise, we shall consider only frequency- and time-independent elastic moduli here. For weakly dissipative media, see Sections 5.5 and 6.3.

2.1.2 Elastodynamic Equation for Inhomogeneous Anisotropic Media

Inserting relation (2.1.7) into elastodynamic equation (2.1.4), we obtain the elastodynamic equation for an unbounded anisotropic, inhomogeneous, perfectly elastic medium:

$$(c_{ijkl}u_{k,l})_{,j} + f_i = \rho\ddot{u}_i, \qquad i = 1, 2, 3. \qquad\qquad (2.1.17)$$

The elastodynamic equation (2.1.17) represents a system of three coupled partial differential equations of the second order for three Cartesian components $u_i(x_j, t)$ of the displacement vector $\vec{u}$.

Alternatively, the elastodynamic equation may be expressed in terms of 12 partial differential equations of the first order in three Cartesian components $v_i(x_j, t)$ of the particle velocity vector $\vec{v} = \dot{\vec{u}}$, and 9 components τ_{ij} of the stress tensor. We use Equation (2.1.4) for $\dot{v}_i = \ddot{u}_i$ and the time derivative of (2.1.7) for $\dot{\tau}_{ij}$. We obtain

$$\rho\dot{v}_i = \tau_{ij,j} + f_i, \qquad \dot{\tau}_{ij} = c_{ijkl}v_{k,l} - \dot{m}_{ij}. \qquad (2.1.18)$$

Function m_{ij} is nonvanishing in regions where the generalized Hooke's law (2.1.7) is not valid, for example, in the source region. It is closely related to the so-called stress-glut and to the moment tensor density. See Kennett (1983, p. 76). System (2.1.18) is alternative to (2.1.17), if we replace f_i in (2.1.17) by $f_i - m_{ij,j}$ (equivalent force system).

In this book, we shall use the elastodynamic equation in the form of (2.1.17) consistently. If $m_{ij,j} \neq 0$, it is understood that f_i represents $f_i - m_{ij,j}$. Equations (2.1.18) will be used in Section 5.4.7 only. It is, however, also possible to build the seismic ray method completely on the elastodynamic equation in the form of (2.1.18). See, for example, Chapman and Coates (1994) and Chapman (in press).

2.1.3 Elastodynamic Equation for Inhomogeneous Isotropic Media

As in the previous case, we insert (2.1.13) into (2.1.4) and obtain the elastodynamic equation for the unbounded isotropic, inhomogeneous, perfectly elastic medium,

$$(\lambda u_{j,j})_{,i} + [\mu(u_{i,j} + u_{j,i})]_{,j} + f_i = \rho\ddot{u}_i, \qquad i = 1, 2, 3. \qquad (2.1.19)$$

If we perform the derivatives, we obtain,

$$(\lambda + \mu)u_{j,ij} + \mu u_{i,jj} + \lambda_{,i}u_{j,j} + \mu_{,j}(u_{i,j} + u_{j,i}) + f_i = \rho\ddot{u}_i,$$
$$i = 1, 2, 3. \qquad (2.1.20)$$

This equation is often written in vectorial form:

$$(\lambda + \mu)\nabla\nabla \cdot \vec{u} + \mu\nabla^2\vec{u} + \nabla\lambda\nabla \cdot \vec{u}$$
$$+ \nabla\mu \times \nabla \times \vec{u} + 2(\nabla\mu \cdot \nabla)\vec{u} + \vec{f} = \rho\ddot{\vec{u}}. \qquad (2.1.21)$$

As we can see, the elastodynamic equation (2.1.17) for the anisotropic inhomogeneous medium is formally simpler than elastodynamic equations (2.1.19) or (2.1.20) for isotropic inhomogeneous media. This is, of course, due to the summation convention. Consequently, we shall often prefer to work with (2.1.17), and only then shall we specify the results for isotropic media.

2.1.4 Acoustic Wave Equation

The elastodynamic equations remain valid even in a fluid medium where $\mu = 0$. However, rather than working with displacement vector u_i, it is then more usual to work with pressure $p = -\frac{1}{3}\tau_{ii}$ and with particle velocity $v_i = \dot{u}_i$. The acoustic wave equations for nonmoving fluids are then usually expressed as

$$p_{,i} + \rho\dot{v}_i = f_i, \qquad v_{i,i} + \kappa\dot{p} = q. \qquad (2.1.22)$$

Functions $f_i = f_i(\vec{x}, t)$ and $q = q(\vec{x}, t)$ represent source terms; f_i has the same meaning as in (2.1.4) and q is the *volume injection rate density*; see Fokkema and van den Berg (1993). Brekhovskikh and Godin (1989) call the same term q the *volume velocity source*. For a detailed derivation of acoustic wave equations, even for moving media, see Brekhovskikh and Godin (1989). Equations (2.1.22) correspond to the elastodynamic equations (2.1.18) for $\mu = 0$, τ_{ij} given by (2.1.15) and $q = \frac{1}{3}\kappa\dot{m}_{ii}$.

If we eliminate the particle velocity v_i from (2.1.22), we obtain the *scalar acoustic wave equation for pressure* $p(\vec{x}, t)$,

$$(\rho^{-1}p_{,i})_{,i} + f^p = \kappa \ddot{p}, \qquad \text{where } f^p = \dot{q} - (\rho^{-1}f_i)_{,i}. \qquad (2.1.23)$$

Alternatively, (2.1.23) reads

$$\nabla \cdot (\rho^{-1}\nabla p) + f^p = \kappa \ddot{p}. \qquad (2.1.24)$$

If density ρ is constant, Equation (2.1.24) yields

$$\nabla^2 p + \rho f^p = c^{-2}\ddot{p}, \qquad c = \sqrt{1/\rho\kappa} = \sqrt{k/\rho}. \qquad (2.1.25)$$

Equation (2.1.25) represents the standard mathematical form of the scalar wave equation, as known from mathematical textbooks; see Morse and Feshbach (1953), Jeffreys and Jeffreys (1966), Bleistein (1984), and Berkhaut (1987), among others. Quantity $c = c(\vec{x})$ is called the *acoustic velocity*. The acoustic velocity $c(\vec{x})$ is a model parameter, $c = (\rho\kappa)^{-1/2}$. It does not depend at all on the properties of the wavefield under consideration, but only on ρ and κ. Here we shall systematically use the acoustic wave equation with a variable density (2.1.23) or (2.1.24). We will usually consider $f_i = 0$, so that $f^p = \dot{q}$.

We can also eliminate the pressure from (2.1.22). We then obtain the *vectorial acoustic wave equation for the particle velocity* $v_i(\vec{x}, t)$

$$(kv_{j,j})_{,i} + f_i^v = \rho \ddot{v}_i, \qquad \text{where } f_i^v = \dot{f}_i - (kq)_{,i}.$$

Here k is the bulk modulus. This wave equation is a special case of the time derivative of the elastodynamic equation (2.1.19) for $\mu = 0$. For this reason, we shall not study it separately. By the *acoustic wave equation*, we shall understand the scalar acoustic wave equation for pressure (2.1.23).

2.1.5 Time-Harmonic Equations

In this book, we shall mostly work directly in the time domain, with transient signals. Nevertheless, it is sometimes useful to discuss certain problems in the frequency domain. Let us consider the general elastodynamic equation (2.1.17) with a time-harmonic source term $\vec{f}(\vec{x}, t) = \vec{f}(\vec{x}, \omega) \exp[-i\omega t]$. Displacement vector $\vec{u}(\vec{x}, t)$ is then also time-harmonic,

$$\vec{u}(\vec{x}, t) = \vec{u}(\vec{x}, \omega) \exp[-i\omega t]. \qquad (2.1.26)$$

Here $\omega = 2\pi f = 2\pi/T$ is the circular frequency, where f is frequency and T is period. To keep the notation as simple as possible, we do not use new symbols for $\vec{u}(\vec{x}, \omega)$ and $\vec{f}(\vec{x}, \omega)$; we merely distinguish them from $\vec{u}(\vec{x}, t)$ and $\vec{f}(\vec{x}, t)$ by using argument ω instead of t. Moreover, we shall only use arguments ω and t where confusion might arise. Instead of circular frequency ω, we shall also often use the frequency $f = \omega/2\pi$.

The elastodynamic equation (2.1.17) in the frequency domain reads

$$(c_{ijkl}u_{k,l})_{,j} + \rho\omega^2 u_i = -f_i, \qquad i = 1, 2, 3, \qquad (2.1.27)$$

where $u_i = u_i(\vec{x}, \omega)$, $f_i = f_i(\vec{x}, \omega)$. Elastodynamic equations (2.1.19) through (2.1.21) can be expressed in the frequency domain in the same way.

For acoustic media, we introduce the time-harmonic quantities $p(\vec{x}, \omega), q(\vec{x}, \omega), \vec{v}(\vec{x}, \omega)$, and $\vec{f}(\vec{x}, \omega)$ as in (2.1.26). The general acoustic equations (2.1.22) then read

$$p_{,i} - i\omega\rho v_i = f_i, \qquad v_{i,i} - i\omega\kappa p = q. \qquad (2.1.28)$$

The acoustic wave equation for pressure (2.1.24) in the frequency domain reads as follows:

$$\nabla \cdot (\rho^{-1}\nabla p) + \kappa \omega^2 p = -f^p, \qquad f^p = -i\omega q - (\rho^{-1}f_i)_{,i}. \qquad (2.1.29)$$

If density ρ is constant, wave equation (2.1.25) yields

$$\nabla^2 p + k^2 p = -\rho f^p, \qquad k = \omega/c, \qquad (2.1.30)$$

where k is the *wave number*. Equation (2.1.30) is known as the *Helmholtz equation*.

Instead of the time factor $\exp[-i\omega t]$ in (2.1.26), it would also be possible to use factor $\exp[i\omega t]$. We will, however, use $\exp[-i\omega t]$ consistently, even in the Fourier transform.

The Fourier transform pair for a transient signal $x(t)$ will be used here in the following form:

$$x(f) = \int_{-\infty}^{\infty} x(t)\exp[i2\pi ft]dt, \qquad x(t) = \int_{-\infty}^{\infty} x(f)\exp[-i2\pi ft]df. \qquad (2.1.31)$$

Function $x(f)$ is the Fourier spectrum of $x(t)$. We again use the same symbol for the transient signal $x(t)$ (in the time domain) and for its Fourier spectrum $x(f)$ (in the frequency domain); we distinguish between the two by arguments t and f, if necessary. The *sign convention* used in Fourier transform (2.1.31) is common in books and papers related to the seismic ray method. For more details on the Fourier transform, see Appendix A.

Note that the Fourier spectrum of the transient displacement vector satisfies elasto-dynamic equation (2.1.27), where $\omega = 2\pi f$ and $\vec{f}$ is the Fourier spectrum of the body force. Similarly, the Fourier spectrum of pressure satisfies the acoustic wave equation in the frequency domain (2.1.29). The elastodynamic equation (2.1.27) in the frequency domain remains valid even for viscoelastic media, where c_{ijkl} are complex-valued and frequency-dependent. See Kennett (1983) and Section 5.5. Similarly, (2.1.29) remains valid for complex-valued frequency-dependent κ.

For a real-valued transient signal $x(t)$, the Fourier spectrum $x(f)$ of $x(t)$ satisfies the relation $x(-f) = x^*(f)$. For this reason, we shall mostly present the spectra $x(f)$ only for $f \geq 0$. Unless otherwise stated, we shall use the following convention: the presented spectrum $x(f)$ corresponds to $f \geq 0$ and can be determined from the relation $x(-f) = x^*(f)$ for negative frequencies.

2.1.6 Energy Considerations

The strictest and most straightforward approach to deriving the basic equations of the seismic ray theory is based on the asymptotic solution of the elastodynamic equation. Energy need not be considered in this procedure. Nevertheless, it does provide a very useful physical insight into all derived equations. We shall, therefore, also briefly discuss some basic energy concepts. Some of them may also find direct applications in seismology, particularly in the investigation of the mechanism of the seismic source.

In studying seismic wave propagation, the deformation processes are mostly considered to be adiabatic. The density of *strain energy* W is then given by the relation

$$W = \tfrac{1}{2}\tau_{ij}e_{ij}; \qquad (2.1.32)$$

see Aki and Richards (1980). Note that W is also called the *strain energy function*. In view of linear constitutive relation (2.1.5),

$$W = \tfrac{1}{2}c_{ijkl}e_{ij}e_{kl}. \qquad (2.1.33)$$

In solids, the strain energy function satisfies the *condition of strong stability* for any non-vanishing strain tensor e_{ij} (see Backus 1962),

$$W = \tfrac{1}{2}c_{ijkl}e_{ij}e_{kl} > 0. \tag{2.1.34}$$

In a fluid medium ($\mu = 0$), the *condition of weak stability*, $W \geq 0$, is satisfied.

The density of *kinetic energy* K is given by the relation,

$$K = \tfrac{1}{2}\rho \dot{u}_i \dot{u}_i. \tag{2.1.35}$$

The sum of W and K is called the density of *elastic energy* and is denoted by E,

$$E = W + K = \tfrac{1}{2}c_{ijkl}e_{ij}e_{kl} + \tfrac{1}{2}\rho \dot{u}_i \dot{u}_i. \tag{2.1.36}$$

All these equations can be found in any textbook on linear elastodynamics; see, for example, Auld (1973).

For completeness, we shall also give the expression for the density of *elastic energy flux* $\vec{S}$, with Cartesian components S_i,

$$S_i = -\tau_{ij}\dot{u}_j = -c_{ijkl}e_{kl}\dot{u}_j = -c_{ijkl}u_{k,l}\dot{u}_j. \tag{2.1.37}$$

See, for example, Kogan (1975), Burridge (1976), Petrashen (1980), and a very detailed discussion in Auld (1973). Elastic energy flux vector $\vec{S}$ is an analogue of the well-known Poyinting vector in the theory of electromagnetic waves. It is not difficult to prove, using elastodynamic equation (2.1.4), that the introduced energy quantities satisfy the *energy equation*,

$$\partial E/\partial t + \nabla \cdot \vec{S} = \vec{f} \cdot \dot{\vec{u}}. \tag{2.1.38}$$

We merely multiply (2.1.4) by $\dot{u}_i$ and rearrange the terms.

The energy quantities $W, K, E, \vec{S}$ introduced here depend on the spatial position and on time t. In high-frequency signals and/or high-frequency harmonic waves, all these quantities vary rapidly with time. It is, therefore, very useful to study certain time-averaged values of these quantities, not only instantaneous values. The time averaging can be performed in several ways. In a time-harmonic wavefield, we usually consider quantities time-averaged over one period. We denote the energy quantities averaged over one period by a bar above the letter, $\overline{W}, \overline{K}, \overline{E}$ and $\overline{S}_i$. For example,

$$\overline{W}(x_i) = \frac{1}{T} \int_0^T W(x_i, t)\mathrm{d}t, \tag{2.1.39}$$

where T denotes the period of the time-harmonic wave under consideration. A more general definition of the time-averaged energy values, applicable even to quasi-harmonic waves, was proposed by Born and Wolf (1959). The time-averaged value of $W(x_i, t)$ is, in this case, given by the relation

$$\overline{\overline{W}}(x_i, t) = \frac{1}{t_2 - t_1} \int_{t_1}^{t_2} W(x_i, t)\mathrm{d}t, \tag{2.1.40}$$

where time interval $t_2 - t_1$ is large (that is, considerably larger than the prevailing period of the wavefield). Averaged quantities $\overline{\overline{K}}(x_i)$, $\overline{\overline{E}}(x_i)$, and $\overline{\overline{S}}_i(x_i)$ can be defined in a similar way. If the wavefield is strictly harmonic, definition (2.1.40) yields (2.1.39) for $t_2 - t_1 \gg T$.

In seismology, however, harmonic and quasi-harmonic wavefields do not play the important role that they play in other fields of physics; we usually consider signals of a finite duration, including short signals. Quantity $\overline{\overline{W}}(x_i)$ would then depend on the choice of t_1 and t_2. Assume that the signal under consideration is effectively concentrated in time interval $t_{min} < t < t_{max}$, where t_{min} represents the effective onset and t_{max} stands for the effective end of the signal. Outside time interval (t_{min}, t_{max}), the signal effectively vanishes. It would then be natural to take $t_1 = t_{min}$ and $t_2 = t_{max}$ in (2.1.40). If, however, we take $t_1 < t_{min}$ and/or $t_2 > t_{max}$, relation (2.1.40) would lead to a distorted result because the time averaging would also be performed over time intervals in which the signal is zero.

In certain seismological applications, however, it may be useful to consider a slightly modified definition (2.1.40) even in the case of shorter high-frequency signals. We modify it by removing constant factor $(t_2 - t_1)^{-1}$ and take infinite limits in the integral. As the signal effectively vanishes outside time interval $(t_1 = t_{min}, t_2 = t_{max})$, we make no mistake if we take $(-\infty, +\infty)$ instead of (t_1, t_2). However, we do not obtain time-averaged quantities because we have removed the averaging factor $(t_2 - t_1)^{-1}$. We shall call these modified quantities *time-integrated energy quantities* and denote them by a circumflex over the letter:

$$\hat{W}(\vec{x}) = \int_{-\infty}^{\infty} W(\vec{x}, t)\mathrm{d}t = \tfrac{1}{2}c_{ijkl}\int_{-\infty}^{\infty} e_{ij}(\vec{x}, t)e_{kl}(\vec{x}, t)\mathrm{d}t,$$

$$\hat{K}(\vec{x}) = \int_{-\infty}^{\infty} K(\vec{x}, t)\mathrm{d}t = \tfrac{1}{2}\rho\int_{-\infty}^{\infty} \dot{u}_i(\vec{x}, t)\dot{u}_i(\vec{x}, t)\mathrm{d}t,$$

$$\hat{E}(\vec{x}) = \int_{-\infty}^{\infty} E(\vec{x}, t)\mathrm{d}t = \hat{W}(\vec{x}) + \hat{K}(\vec{x}),\tag{2.1.41}$$

$$\hat{S}_i(\vec{x}) = \int_{-\infty}^{\infty} S_i(\vec{x}, t)\mathrm{d}t = -c_{ijkl}\int_{-\infty}^{\infty} e_{kl}(\vec{x}, t)\dot{u}_j(\vec{x}, t)\mathrm{d}t.$$

These time-integrated energy quantities will be very useful in our treatment. Time-averaged quantities $\overline{\overline{W}}, \overline{\overline{K}}, \overline{\overline{E}}, \overline{\overline{S}}_i$ can be obtained from time-integrated quantities $\hat{W}, \hat{K}, \hat{E}, \hat{S}_i$ in a very simple way: we merely divide them by the effective length of the signal.

For an acoustic medium ($\mu = 0$), the individual energy quantities W, K, E and S_i can easily be obtained from those given before:

$$W = \tfrac{1}{2}\kappa p^2, \qquad K = \tfrac{1}{2}\rho v_i v_i, \qquad E = W + K, \qquad S_i = p v_i. \tag{2.1.42}$$

Here p is the pressure, v_i are components of the particle velocity, and κ is the compressibility. See Section 2.1.4. The time-averaged and time-integrated energy quantities in the acoustic case can be constructed from (2.1.42) in the same way.

Energy quantities $W, K, E, \overline{W}, \overline{K}, \overline{E}$ and $\overline{\overline{W}}, \overline{\overline{K}}, \overline{\overline{E}}$ represent the energy density and are measured in J m^{-3} (joule per cubic meter) $=$ Pa $=$ kg m^{-1} s^{-2}. Time-integrated quantities $\hat{W}, \hat{K}$, and $\hat{E}$ are then measured in Pa s $=$ kg m^{-1}s^{-1}. Finally, the densities of energy flux $S_i, \overline{S}_i$ and $\overline{\overline{S}}_i$ are measured in J m^{-2} s^{-1} $=$ W m^{-2} (watt per square meter) $=$ kg s^{-3}, and $\hat{S}_i$ in J m^{-2} $=$ Pa m $=$ kg s^{-2}.

In physics, the quantities with dimension of energy $\times$ time are called the *action*; consequently, $\hat{W}, \hat{K}$, and $\hat{E}$ represent the density of action. We shall, however, not use this terminology but shall refer to $\hat{W}, \hat{K}$, and $\hat{E}$ as the time-integrated energy quantities.

A physical quantity that plays an important role in many wave propagation applications is the *velocity vector of the time-averaged energy flux*. We shall denote it $\vec{\mathcal{U}}^E$, and its component $\mathcal{U}_i^E$. The components are $\mathcal{U}_i^E$ are defined by the relation

$$\mathcal{U}_i^E = \overline{S}_i / \overline{E}. \tag{2.1.43}$$

It is easy to see that an alternative definition of $\mathcal{U}_i^E$ is

$$\mathcal{U}_i^E = \hat{S}_i / \hat{E}. \tag{2.1.44}$$

2.2 Elastic Plane Waves

In this section, we shall discuss the properties of plane waves propagating in homogeneous, perfectly elastic, isotropic, or anisotropic media. Plane waves are the simplest solutions of the elastodynamic equation. The procedure used here to derive the properties of plane waves will be used in Section 2.4 to study wave propagation in smoothly inhomogeneous elastic media. Before dealing with plane waves propagating in elastic media, we shall also briefly discuss acoustic plane waves.

2.2.1 Time-Harmonic Acoustic Plane Waves

We shall seek the plane-wave solutions of the time-harmonic acoustic wave equation (2.1.30), with c constant and real-valued and with vanishing source term ρf^p. We shall describe the acoustic pressure plane wave by the equation

$$p(\vec{x}, t) = P \exp[-i\omega(t - T(\vec{x}))], \tag{2.2.1}$$

where p is pressure, ω is the circular frequency ($\omega = 2\pi f$), P is some scalar constant which may be complex-valued, and $T(\vec{x})$ is a linear homogeneous function of Cartesian coordinates x_i,

$$T(\vec{x}) = p_i x_i. \tag{2.2.2}$$

Here p_i are real-valued or complex-valued constants. If p_i are real-valued, plane wave (2.2.1) is called *homogeneous* (even for complex-valued P). For complex-valued p_i, plane wave (2.2.1) is called *inhomogeneous*. Here we shall consider only real-valued p_i; inhomogeneous waves will be discussed in Section 2.2.10. Coefficients p_i are not arbitrary but must satisfy the relations following from the acoustic wave equation. Inserting (2.2.1) into (2.1.30) with $\rho f^p = 0$, we obtain the condition

$$p_1^2 + p_2^2 + p_3^2 = 1/c^2, \tag{2.2.3}$$

where $c = (\rho\kappa)^{-1/2}$. Equation (2.2.3) represents a necessary condition for the existence of nontrivial plane waves (2.2.1) propagating in an acoustic homogeneous medium. This condition is known by many names; here we shall call (2.2.3) the *existence condition*. Pressure $p(\vec{x}, t)$, given by (2.2.1), is a solution of the acoustic wave equation and represents a plane wave only if constants p_1, p_2, and p_3 satisfy existence condition (2.2.3). The scalar complex-valued constant P in (2.2.1) may be arbitrary.

The pressure $p(\vec{x}, t)$, given by (2.2.1) through (2.2.3), is constant along planes

$$T(\vec{x}) = p_i x_i = \text{const.} \tag{2.2.4}$$

For t varying, equation $t = T(\vec{x})$ represents a moving plane, here called the *wavefront*. This is the reason why wave (2.2.1) is called a plane wave. We shall denote the unit normal to the wavefront $\vec{N}$, $N_i = p_i/\sqrt{p_k p_k}$. Hence,

$$\vec{p} = \vec{N}/\mathcal{C}, \tag{2.2.5}$$

where $\mathcal{C}$ represents the velocity of the propagation of the wavefront in the direction perpendicular to it. This velocity is usually called the *phase velocity*. By inserting (2.2.5) into existence condition (2.2.3), we obtain $\mathcal{C} = c$. Thus, the phase velocity of the pressure plane wave propagating in a homogeneous fluid medium equals $c = (\rho\kappa)^{-1/2}$. It does not depend on the frequency or on the direction of propagation $\vec{N}$. Vector $\vec{p}$ with Cartesian components

p_i, given by (2.2.5), is called the *slowness vector*; it is perpendicular to the wavefront, and its length equals slowness $1/c$.

In addition, it is also usual to introduce *phase velocity vector* $\vec{C}$, which has the same direction as slowness vector $\vec{p}$ but length C:

$$\vec{C} = C\vec{N} = \vec{p}/(\vec{p} \cdot \vec{p}). \tag{2.2.6}$$

The plane-wave solution (2.2.1) of the acoustic wave equation is complex-valued. Both the real and imaginary parts of (2.2.1), and any linear combination thereof, are again solutions of (2.1.30) with $f^p = 0$.

Plane-wave solution (2.2.1) can be expressed in many alternative forms; for example,

$$p(\vec{x}, t) = P \exp[ik_n x_n] \exp[-i\omega t],$$

where $k_n = (\omega/c)N_n$ are Cartesian components of *wave vector* $\vec{k} = \omega \vec{p}$.

The plane-wave solution of acoustic equation (2.1.28) with $f_i = q = 0$ can be written as

$$p(\vec{x}, t) = P \exp[-i\omega(t - T(\vec{x}))], \qquad \vec{v}(\vec{x}, t) = \frac{\vec{N}P}{\rho c} \exp[-i\omega(t - T(\vec{x}))]. \tag{2.2.7}$$

Here P is again an arbitrary constant; all other quantities have the same meaning as defined earlier. Note that the product of velocity and density ρc is also known as the *wave impedance*.

2.2.2   Transient Acoustic Plane Waves

Transient pressure plane waves can be obtained from the time-harmonic plane waves using Fourier transform (2.1.31). We can, however, also work directly with transient signals. For this purpose, we shall use analytical signals $F(\zeta)$. We shall consider a general, complex-valued expression for the transient acoustic plane wave

$$p(\vec{x}, t) = PF(t - T(\vec{x})), \tag{2.2.8}$$

where P is again an arbitrary complex-valued constant and $T(\vec{x})$ is a linear function of the coordinates given by (2.2.2). For a given real-valued signal $x(t)$, the analytical signal $F(\zeta)$ is defined by the relations

$$F(\zeta) = x(\zeta) + ig(\zeta), \qquad g(\zeta) = \frac{1}{\pi}\mathcal{P}.\mathcal{V}. \int_{-\infty}^{\infty} \frac{x(\sigma)}{\sigma - \zeta}d\sigma, \tag{2.2.9}$$

where $\mathcal{P}.\mathcal{V}.$ stands for the principal value. The two functions $x(\zeta)$ and $g(\zeta)$ form a Hilbert transform pair. More details on analytical signals can be found in Appendix A. A well-known example of the Hilbert transform pair are functions

$$x(\zeta) = \cos \omega \zeta, \qquad g(\zeta) = -\sin \omega \zeta, \tag{2.2.10}$$

so that $F(\zeta) = \exp(-i\omega\zeta)$. Thus, time-harmonic plane wave (2.2.1) is a special case of transient plane wave (2.2.8). Another important example of the Hilbert transform pair is

$$x(\zeta) = \delta(\zeta), \qquad g(\zeta) = -1/\pi\zeta, \tag{2.2.11}$$

where $\delta(\zeta)$ is the Dirac delta function. The corresponding analytical signal, called the analytic delta function, can be expressed as

$$F(\zeta) = \delta^{(A)}(\zeta) = \delta(\zeta) - i/\pi\zeta. \tag{2.2.12}$$

As in the case of time-harmonic plane wave (2.2.1), transient plane wave (2.2.8) is a solution of the acoustic wave equation only if constants p_i in (2.2.2) satisfy existence condition (2.2.3). Constant P and analytical signal $F(\zeta)$, however, are arbitrary. For this reason, we prefer to work with the analytical signal solutions and not with the time-harmonic solutions. In the same way as in the case of the plane time-harmonic wave, we can again introduce the wavefront, the slowness vector, and the like.

In realistic applications, of course, we can, generally, consider either the real or imaginary part of solution (2.2.8). For example,

$$p(\vec{x}, t) = \text{Re}\{PF(t - T(\vec{x}))\}, \qquad (2.2.13)$$

or

$$p(\vec{x}, t) = \tfrac{1}{2}\{PF(t - T(\vec{x})) + P^*F^*(t - T(\vec{x}))\}. \qquad (2.2.14)$$

Here the asterisk denotes a complex-conjugate function. Moreover, any linear combination of the real and imaginary parts of (2.2.8) is again a solution of the acoustic wave equation.

2.2.3 Vectorial Transient Elastic Plane Waves

In a homogeneous elastic, isotropic, or anisotropic medium, the procedure for determining plane waves and their characteristic parameters is more involved than in the acoustic case. The main complication is that we can obtain several types of plane waves, propagating in the same direction with different velocities. The velocities can be determined by solving eigenvalue problems for certain 3×3 matrices.

We shall describe the vectorial transient plane elastic wave by the equation

$$\vec{u}(\vec{x}, t) = \vec{U}F(t - T(x_i)), \qquad (2.2.15)$$

where $\vec{u}$ is the displacement vector, $\vec{U}$ a complex-valued vectorial constant, and F, t, T have the same meaning as before. Once again, we shall use $F(\zeta)$ for the analytical signal as in the expression for the transient pressure plane wave in the acoustic case; see (2.2.8). We can make this assignment because we shall treat the acoustic case independent of the elastic case throughout the book. Actually, the analytical signals corresponding to displacement, particle velocity, and pressure are mutually related. If the analytical signal for displacement is $F(\zeta)$ (see (2.2.15)), then the analytical signals for the particle velocity and pressure are $\dot{F}(\zeta)$.

We wish to determine the number of plane waves with different velocities that can propagate along a specified direction in the medium, as well as their velocities, slowness vectors, and polarizations. All this can be determined by inserting (2.2.15) into the elastodynamic equation. We shall consider elastodynamic equation (2.1.17) with elastic moduli c_{ijkl} constant, density ρ constant, and $f_i = 0$. The elastodynamic equation then reads

$$a_{ijkl}u_{k,lj} = \ddot{u}_i, \qquad i = 1, 2, 3, \qquad (2.2.16)$$

where we have used the notation

$$a_{ijkl} = c_{ijkl}/\rho. \qquad (2.2.17)$$

Thus, a_{ijkl} are *density-normalized elastic moduli* of dimension $\text{m}^2\,\text{s}^{-2}$ (velocity squared).

It seems surprising to start our treatment with anisotropic and not with isotropic media. Due to various symmetries and the Einstein summation convention, however, certain steps of the mathematical treatment for the anisotropic medium are formally simpler than for the isotropic medium, which is, in some ways, a degenerate case of the anisotropic medium.

Inserting ansatz plane-wave solution (2.2.15) into (2.2.16) and assuming $F'' \neq 0$, we obtain the following equations:

$$a_{ijkl} p_j p_l U_k - U_i = 0, \qquad i = 1, 2, 3. \tag{2.2.18}$$

This is a system of three linear equations for U_1, U_2, and U_3. If U_i and p_i satisfy (2.2.18), expression (2.2.15) is then a solution of elastodynamic equation (2.2.16) and represents a transient elastic plane wave.

2.2.4 Christoffel Matrix and Its Properties

System of equations (2.2.18) can be simplified if we introduce a 3×3 matrix $\hat{\Gamma}$ with real-valued components Γ_{ik} given by the relation

$$\Gamma_{ik} = a_{ijkl} p_j p_l. \tag{2.2.19}$$

Then (2.2.18) can be expressed in the following simple form:

$$\Gamma_{ik} U_k - U_i = 0, \qquad i = 1, 2, 3. \tag{2.2.20}$$

Matrix $\hat{\Gamma}$, given by (2.2.19), will be referred to as the *Christoffel matrix*, and Equation (2.2.20) will be called the *Christoffel equation*. Traditionally, the term Christoffel matrix has usually been connected with matrix $c_{ijkl} N_j N_l$; see Helbig (1994). The notation (2.2.19), which contains the components of slowness vector $\vec{p}$, has been broadly used in the ray method of seismic waves propagating in inhomogeneous anisotropic media, as it is very suitable for the application of the Hamiltonian formalism. We hope no confusion will be caused if we use the term Christoffel matrix for $\hat{\Gamma}$ given by (2.2.19).

We shall now list four important properties of Christoffel matrix $\hat{\Gamma}$.

1. Matrix $\hat{\Gamma}$ is symmetric,

   $$\Gamma_{ik} = \Gamma_{ki}.$$

2. The elements of matrix $\hat{\Gamma}$, Γ_{ik}, are homogeneous functions of the second degree in p_i. By homogeneous function $f(x_i)$ of the kth degree in x_i, we understand a function $f(x_i)$, which satisfies the relation

   $$f(ax_i) = a^k f(x_i), \tag{2.2.21}$$

   for any nonvanishing constant a. It is obvious from (2.2.19) that Γ_{ik} satisfies the relation

   $$\Gamma_{ik}(ap_j) = a^2 \Gamma_{ik}(p_j) \tag{2.2.22}$$

   so that Γ_{ik} is a homogeneous function of the second degree in p_i.
3. Matrix $\hat{\Gamma}$ satisfies the relation

   $$p_j \partial \Gamma_{ik} / \partial p_j = 2\Gamma_{ik}. \tag{2.2.23}$$

   This relation follows from Euler's theorem for homogeneous functions $f(x_i)$ of the kth degree, which reads

   $$x_j \partial f(x_i) / \partial x_j = k f(x_i). \tag{2.2.24}$$

   Euler's theorem immediately yields (2.2.23). It is also not difficult to derive (2.2.23) directly from (2.2.19).

4. Matrix $\hat{\Gamma}$ is positive definite. This means that Γ_{ik} satisfies the inequality

$$\Gamma_{ik}a_i a_k > 0, \tag{2.2.25}$$

where a_j are the components of any nonvanishing real-valued vector $\vec{a}$. Property (2.2.25) follows from the condition of strong stability (2.1.34). As e_{ij} and e_{kl} may be quite arbitrary, we can take $e_{ij} = a_i p_j$, $e_{kl} = a_k p_l$. Then (2.1.34) yields (2.2.25).

We shall now discuss the *eigenvalues and eigenvectors* of the matrix $\hat{\Gamma}$. We denote the eigenvalues by the letter G. Eigenvalues G are defined as the roots of the *characteristic equation*

$$\det(\Gamma_{ik} - G\delta_{ik}) = 0, \tag{2.2.26}$$

that is

$$\det \begin{pmatrix} \Gamma_{11} - G & \Gamma_{12} & \Gamma_{13} \\ \Gamma_{12} & \Gamma_{22} - G & \Gamma_{23} \\ \Gamma_{13} & \Gamma_{23} & \Gamma_{33} - G \end{pmatrix} = 0. \tag{2.2.27}$$

This equation represents the cubic algebraic equation

$$G^3 - PG^2 + QG - R = 0, \tag{2.2.28}$$

where P, Q, and R are invariants of matrix $\hat{\Gamma}$:

$$P = \operatorname{tr}\hat{\Gamma}, \qquad R = \det\hat{\Gamma},$$
$$Q = \det \begin{pmatrix} \Gamma_{11} & \Gamma_{12} \\ \Gamma_{12} & \Gamma_{22} \end{pmatrix} + \det \begin{pmatrix} \Gamma_{22} & \Gamma_{23} \\ \Gamma_{23} & \Gamma_{33} \end{pmatrix} + \det \begin{pmatrix} \Gamma_{11} & \Gamma_{13} \\ \Gamma_{13} & \Gamma_{33} \end{pmatrix}. \tag{2.2.29}$$

It is obvious from (2.2.28) that matrix $\hat{\Gamma}$ has three eigenvalues. We denote them G_m, $m = 1, 2, 3$. Because matrix $\hat{\Gamma}$ is symmetric and positive definite, all the three eigenvalues G_1, G_2, and G_3 are real-valued and positive.

It is not difficult to conclude from (2.2.26) that eigenvalues G_1, G_2, G_3 are homogeneous functions of the second degree in p_j, similar to Γ_{ik}. Thus,

$$G_m(ap_j) = a^2 G_m(p_j), \tag{2.2.30}$$

and

$$p_j \partial G_m / \partial p_j = 2G_m. \tag{2.2.31}$$

The second relation follows from Euler's theorem (2.2.24).

Now we shall discuss the eigenvectors of $\hat{\Gamma}$. We shall denote the eigenvector corresponding to eigenvalue G_m by $\vec{g}^{(m)}$, $m = 1, 2, 3$. It is defined by the relations

$$(\Gamma_{ik} - G_m\delta_{ik})g_k^{(m)} = 0, \qquad i = 1, 2, 3, \tag{2.2.32}$$

with the normalizing condition,

$$g_k^{(m)} g_k^{(m)} = 1 \tag{2.2.33}$$

(no summation over m). As we can see from (2.2.33), we take the eigenvectors as unit vectors. Thus, matrix $\hat{\Gamma}$ has three eigenvalues G_m ($m = 1, 2, 3$), and three corresponding eigenvectors $\vec{g}^{(m)}$ ($m = 1, 2, 3$). Unit vectors $\vec{g}^{(m)}$ are also mutually perpendicular.

Using (2.2.32) and (2.2.33), we can find an expression for G_m in terms of Γ_{ik} and $g_i^{(m)}$, which will be useful in the following. Multiplying (2.2.32) by $g_i^{(m)}$ and taking into account (2.2.33), we obtain

$$G_m = \Gamma_{ik} g_i^{(m)} g_k^{(m)} = a_{ijkl} p_j p_l g_i^{(m)} g_k^{(m)}. \qquad (2.2.34)$$

In the *degenerate case* of two identical eigenvalues, the direction of the two corresponding eigenvectors cannot be determined from (2.2.32) and (2.2.33); only the plane in which they are situated can be determined. This plane is perpendicular to the remaining eigenvector.

2.2.5 Elastic Plane Waves in an Anisotropic Medium

Conditions (2.2.20) following from the elastodynamic equation can be expressed as follows:

$$(\Gamma_{ik} - \delta_{ik})U_k = 0, \qquad i = 1, 2, 3. \qquad (2.2.35)$$

Equation (2.2.35) represents a system of three homogeneous linear algebraic equations for U_1, U_2, and U_3. The system has a nontrivial solution only if the determinant of the system vanishes,

$$\det(\Gamma_{ik} - \delta_{ik}) = \det \begin{pmatrix} \Gamma_{11} - 1 & \Gamma_{12} & \Gamma_{13} \\ \Gamma_{21} & \Gamma_{22} - 1 & \Gamma_{23} \\ \Gamma_{31} & \Gamma_{32} & \Gamma_{33} - 1 \end{pmatrix} = 0. \qquad (2.2.36)$$

Here again, $\Gamma_{ik} = a_{ijkl} p_j p_l$; see (2.2.19). This equation plays a very important role in investigating the propagation of plane waves in a homogeneous elastic medium. Plane wave (2.2.15) is a solution of the elastodynamic equation if, and only if, its slowness vector $\vec{p}$ (with Cartesian components p_i) satisfies Equation (2.2.36). As in the acoustic case, we shall refer to (2.2.36) as the *existence condition* for plane waves in homogeneous anisotropic media. It represents a polynomial equation of the sixth degree in the components of the slowness vector p_i. In the phase space with coordinates p_1, p_2, and p_3, Equation (2.2.36) represents a surface, known as the *slowness surface*. For more details on the slowness surface, refer to Section 2.2.8. The slowness surface has three sheets, corresponding to three eigenvalues, which will be discussed later. Existence condition (2.2.36) also represents a cubic equation for the squares of phase velocities $C^{(m)}$, $m = 1, 2, 3$, of the plane waves, which can propagate in the medium described by density-normalized elastic moduli a_{ijkl}, in any selected direction specified by unit vector $\vec{N}$, normal to the wavefront. Since $p_i = N_i/C^{(m)}$, $\Gamma_{ik} = a_{ijkl} N_j N_l/(C^{(m)})^2$.

Here, we shall discuss Equation (2.2.35) not only from the point of view of the existence condition, but also from a more general point of view. We also wish to determine the direction of displacement vector amplitude $\vec{U}$. In this discussion, we shall apply the eigenvalue formalism. This approach, based on the solution of the eigenvalue problem for matrix Γ_{ij}, will be applied in Section 2.4 to a more complicated problem of wave propagation in smoothly inhomogeneous media.

Equation (2.2.35) constitutes a typical eigenvalue problem; see (2.2.32). Comparing (2.2.35) with (2.2.32), we can conclude that (2.2.35) is satisfied if, and only if, one of the three eigenvalues of matrix $\hat{\Gamma}$ equals unity,

$$G_m(p_i) = 1, \qquad m = 1 \text{ or } 2 \text{ or } 3. \qquad (2.2.37)$$

In the phase space with coordinates p_1, p_2, and p_3, these equations represent three branches of the slowness surface (2.2.36). The corresponding eigenvector $\vec{g}^{(m)}$ then determines the direction of $\vec{U}$ so that

$$\vec{U} = A\vec{g}^{(m)}, \tag{2.2.38}$$

where A is a complex-valued constant scalar amplitude quantity.

As matrix $\hat{\Gamma}$ has three eigenvalues G_m, Equations (2.2.35) are satisfied in three cases. These three cases specify the three plane waves, which can generally propagate in a homogeneous anisotropic medium in the direction of $\vec{N}$.

Eigenvectors $\vec{g}^{(m)}$ determine the polarization of the individual waves so that they can be referred to as *polarization vectors*. If $G_1 \neq G_2 \neq G_3$, they can be strictly determined from (2.2.32) and (2.2.33). If two eigenvalues are equal, the relevant polarization vectors cannot be determined uniquely from (2.2.32) and (2.2.33); some other conditions must be taken into account. We shall speak of the *degenerate case* of two equal eigenvalues. The most important degenerate case is the case of *isotropic media* (see Section 2.2.6); the other degenerate case is the case of the so-called *shear wave singularities* in anisotropic media.

Let us now determine the phase velocities $\mathcal{C}^{(1)}$, $\mathcal{C}^{(2)}$, and $\mathcal{C}^{(3)}$ of the three plane waves propagating in an anisotropic homogeneous medium in the direction specified by N_i. Putting

$$p_i = N_i/\mathcal{C}^{(m)} \tag{2.2.39}$$

for the mth plane wave ($m = 1$ or 2 or 3) and inserting it into (2.2.37) yields

$$\mathcal{C}^{(m)^2} = G_m(N_i), \tag{2.2.40}$$

due to (2.2.30). Here $G_m(N_i)$ is obtained from standard eigenvalue $G_m(p_i)$, if p_i is replaced by N_i. Thus, $G_m(N_i)$ is the eigenvalue of matrix $\bar{\Gamma}_{ik} = a_{ijkl}N_jN_l$. Matrix $\bar{\Gamma}_{ik}$ has the same properties as Γ_{ik}; only p_i is replaced by N_i in all relations. It is obvious that $G_m(N_i)$ is always positive. Both matrices Γ_{ik} and $\bar{\Gamma}_{ik}$ have the same eigenvectors $\vec{g}^{(m)}$, $m = 1, 2, 3$.

Equation (2.2.40) yields the square of the phase velocity, not the phase velocity itself. We shall only consider the positive values of $\mathcal{C}^{(m)}$:

$$\mathcal{C}^{(m)} = \sqrt{G_m(N_i)}. \tag{2.2.41}$$

Thus, the phase velocities of plane waves in a homogeneous anisotropic medium depend on the direction of propagation of wavefront N_i.

From a practical point of view, the three phase velocities $\mathcal{C}^{(1)}$, $\mathcal{C}^{(2)}$, and $\mathcal{C}^{(3)}$ for a fixed direction N_i are determined by the solution of cubic equation (2.2.28), where P, Q, and R are given by (2.2.29), with p_i replaced by N_i (that is, Γ_{ik} is replaced by $\bar{\Gamma}_{ik} = a_{ijkl}N_jN_l$). The square roots of the three solutions yield the phase velocities of the three waves.

The plane wave with the highest phase velocity is usually called the quasi-compressional wave (or quasi-P or qP wave). The remaining two plane waves are called quasi-shear waves (quasi-S1 and quasi-S2 waves, or qS1 and qS2 waves).

If we combine Equations (2.2.15), (2.2.2), (2.2.38), and (2.2.39), we obtain the following final expression for any of the three plane waves propagating in a homogeneous anisotropic medium in a fixed direction N_i:

$$\vec{u}(x_i, t) = A\vec{g}^{(m)}F\left(t - N_ix_i/\mathcal{C}^{(m)}\right), \tag{2.2.42}$$

where A is an arbitrary complex-valued constant and $\mathcal{C}^{(m)}$ is given by (2.2.41). The plane wave is linearly polarized.

2.2.6 Elastic Plane Waves in an Isotropic Medium

In isotropic media, the phase velocities can be determined analytically. The elements of matrix $\hat{\Gamma}$ are simply obtained by inserting (2.1.10) into (2.2.19):

$$\Gamma_{ik} = \frac{\lambda + \mu}{\rho} p_i p_k + \frac{\mu}{\rho} \delta_{ik} p_l p_l. \tag{2.2.43}$$

Thus, eigenvalues G can be determined from relation (2.2.27), which takes the following form:

$$\det \begin{pmatrix} \frac{\lambda+\mu}{\rho} p_1^2 + \frac{\mu}{\rho} p_i p_i - G & \frac{\lambda+\mu}{\rho} p_1 p_2 & \frac{\lambda+\mu}{\rho} p_1 p_3 \\ \frac{\lambda+\mu}{\rho} p_1 p_2 & \frac{\lambda+\mu}{\rho} p_2^2 + \frac{\mu}{\rho} p_i p_i - G & \frac{\lambda+\mu}{\rho} p_2 p_3 \\ \frac{\lambda+\mu}{\rho} p_1 p_3 & \frac{\lambda+\mu}{\rho} p_2 p_3 & \frac{\lambda+\mu}{\rho} p_3^2 + \frac{\mu}{\rho} p_i p_i - G \end{pmatrix} = 0.$$

This determinant can be expanded in powers of $(\rho^{-1}\mu p_i p_i - G)$. It is easy to see that the terms with $(\rho^{-1}\mu p_i p_i - G)^n$ vanish for $n = 1$ and 0 so that

$$\det(\ldots) = \left(\frac{\mu}{\rho} p_i p_i - G\right)^3 + \left(\frac{\mu}{\rho} p_i p_i - G\right)^2 \frac{\lambda+\mu}{\rho} p_k p_k,$$

which leads to

$$\det(\ldots) = \left(\frac{\mu}{\rho} p_i p_i - G\right)^2 \left(\frac{\lambda + 2\mu}{\rho} p_i p_i - G\right).$$

Putting

$$\alpha = \left(\frac{\lambda + 2\mu}{\rho}\right)^{1/2} = \left(\frac{k + \frac{4}{3}\mu}{\rho}\right)^{1/2}, \qquad \beta = \left(\frac{\mu}{\rho}\right)^{1/2} \tag{2.2.44}$$

finally yields

$$\det(\ldots) = (\alpha^2 p_i p_i - G)(\beta^2 p_i p_i - G)^2.$$

Thus, in an isotropic homogeneous medium, matrix $\hat{\Gamma}$ has the following three eigenvalues:

$$G_1(p_i) = G_2(p_i) = \beta^2 p_i p_i, \qquad G_3(p_i) = \alpha^2 p_i p_i. \tag{2.2.45}$$

Two of these eigenvalues, G_1 and G_2, are identical. If we replace p_i by N_i and take into account $N_i N_i = 1$, we obtain

$$G_1(N_i) = G_2(N_i) = \beta^2, \qquad G_3(N_i) = \alpha^2. \tag{2.2.46}$$

Equations (2.2.41) then yield the following three phase velocities $C^{(m)}$:

$$C^{(1)} = C^{(2)} = \beta, \qquad C^{(3)} = \alpha. \tag{2.2.47}$$

Thus, the isotropic medium represents a degenerate case of an anisotropic medium, with two identical phase velocities. Only two plane waves can propagate in an isotropic medium in a fixed direction N_i. The faster, propagating with phase velocity α, is called the *compressional wave* (P wave). The slowness vector components p_i corresponding to this wave must satisfy the *existence condition of* P *waves* everywhere:

$$p_1^2 + p_2^2 + p_3^2 = 1/\alpha^2; \tag{2.2.48}$$

see (2.2.45) for $G_3 = 1$. The slower plane wave propagates with velocity β and is called the *shear wave* (S *wave*). The slowness vector components p_i corresponding to this wave must satisfy the *existence condition of* S *waves* everywhere:

$$p_1^2 + p_2^2 + p_3^2 = 1/\beta^2. \tag{2.2.49}$$

Velocities α and β do not depend on the position or the direction of propagation.

As we can see from (2.2.48) and (2.2.49), the slowness surfaces in the phase space p_1, p_2, and p_3 of both P and S waves are spheres. The radius of the sphere is $1/\alpha$ for P waves (the slowness of P waves) and $1/\beta$ for S waves (the slowness of S waves).

We will now determine the eigenvectors of Γ_{ik} in isotropic media. As we know, we can determine strictly only the eigenvector $\vec{g}^{(3)}$, but not eigenvectors $\vec{g}^{(1)}$ and $\vec{g}^{(2)}$, since $G_1 = G_2$. Eigenvector $\vec{g}^{(3)}$ satisfies relation (2.2.32), with Γ_{ik} given by (2.2.43) and with $G_3 = 1$,

$$\left[\frac{\lambda + \mu}{\rho} p_i p_k + \frac{\mu}{\rho} \delta_{ik} p_l p_l - \delta_{ik} \right] g_k^{(3)} = 0.$$

If we insert $p_i = N_i/\alpha$ and multiply the equation by $g_i^{(3)}$, we obtain $(N_i g_i^{(3)})^2 = 1$. This result yields

$$\vec{g}^{(3)} = \pm \vec{N}. \tag{2.2.50}$$

Thus, the displacement vector of the plane P wave is perpendicular to the wavefront.

Whereas the eigenvector $\vec{g}^{(3)}$ corresponding to the P wave is uniquely determined, eigenvectors $\vec{g}^{(1)}$ and $\vec{g}^{(2)}$ corresponding to the plane shear wave cannot be strictly determined. We can only say that they are perpendicular to $\vec{g}^{(3)}$ and that they are also mutually perpendicular. Thus, they are situated in the wavefront.

The plane P wave can then finally be expressed as

$$\vec{u}(\vec{x}, t) = A\vec{N}F(t - N_i x_i/\alpha), \tag{2.2.51}$$

where A is an arbitrary complex-valued constant. Similarly, for the plane S wave,

$$\vec{u}(\vec{x}, t) = (B\vec{e}_1 + C\vec{e}_2)F(t - N_i x_i/\beta), \tag{2.2.52}$$

where B and C are some complex-valued constants and $\vec{e}_1$ and $\vec{e}_2$ are two mutually perpendicular unit vectors, both perpendicular to $\vec{N}$.

Thus, the plane P wave in an isotropic medium is always linearly polarized in the direction of $\vec{N}$, perpendicular to the wavefront. The polarization of the S wave (2.2.52) is, however, more complicated. Because B and C are generally complex-valued, we are unable to find a real-valued unit vector $\vec{e}_s$ for which it would be possible to put $B\vec{e}_1 + C\vec{e}_2 = D\vec{e}_s$. This means that the S plane waves are not, in general, linearly polarized. If the plane S wave is time-harmonic, $F(\zeta) = \exp(-i2\pi f\zeta)$, it may be elliptically polarized in the plane of the wavefront. For general transient signals, we speak of quasi-elliptical polarization, again in the plane of the wavefront. The elliptical or quasi-elliptical polarization is reduced to linear polarization only if B or C vanishes, or if

$$\arg B - \arg C = k\pi, \qquad k = 0, \pm 1, \dots. \tag{2.2.53}$$

As a special case of (2.2.53), the S wave is linearly polarized if both B and C are real-valued. For a more detailed treatment, see Section 6.4.

Note that the compressional (P) plane wave in a homogeneous medium is also known as the longitudinal, irrotational, or dilatational wave, and the shear (S) plane wave in a homogeneous medium is known as the transverse, equivoluminal, or rotational wave. These terms describe the main features of the two kinds of plane waves propagating in homogeneous media (i.e., the irrotational character of P waves and equivoluminal character of S waves).

2.2.7 Energy Considerations for Plane Waves

In this section, we shall specify energy equations presented in Section 2.1.6 for plane waves. We shall start with the acoustic case; see (2.1.42). Because the energy expressions are nonlinear in pressure p and particle velocity v_i, we cannot use complex-valued expressions (2.2.8) but must use expressions (2.2.14). To abbreviate the equations, we shall omit the argument $\zeta = t - T(\vec{x})$ of the analytical signal. Then, in view of (2.2.14) and (2.2.7),

$$p = \tfrac{1}{2}(PF + P^*F^*), \qquad v_i = \tfrac{1}{2}(V_iF + V_i^*F^*), \tag{2.2.54}$$

where $V_i = p_i P/\rho$. Here p denotes pressure, and p_i stands for Cartesian components of the slowness vector $\vec{p}$. The asterisk denotes complex-conjugate quantities. Equations (2.1.42) then yield

$$
\begin{aligned}
W &= \tfrac{1}{8}\kappa(PF + P^*F^*)^2, \\
K &= \tfrac{1}{8}\rho(V_iF + V_i^*F^*)(V_iF + V_i^*F^*), \\
E &= W + K, \\
S_i &= \tfrac{1}{4}(PF + P^*F^*)(V_iF + V_i^*F^*).
\end{aligned}
\tag{2.2.55}
$$

To derive the expressions for the time-integrated quantities $\hat{W}$, $\hat{K}$, $\hat{E}$, and $\hat{S}_i$ for plane waves, we shall need to know several integral properties of the analytical signal. For details refer to Appendix A, particularly (A.2.9), (A.2.10), (A.3.12), and (A.3.13). Integrating (2.2.55) over time from $-\infty$ to ∞ yields

$$
\begin{aligned}
\hat{W} &= \tfrac{1}{2}\kappa PP^*f_p, \qquad \hat{K} = \tfrac{1}{2}\kappa PP^*f_p, \\
\hat{E} &= \kappa PP^*f_p, \qquad \hat{S}_i = \rho^{-1}p_iPP^*f_p.
\end{aligned}
\tag{2.2.56}
$$

Here f_p takes the form

$$f_p = \int_{-\infty}^{\infty} x^2(t)\,dt. \tag{2.2.57}$$

For time-harmonic waves, $\overline{W}$, $\overline{K}$, $\overline{E}$, and $\overline{S}_i$ are again given by (2.2.56), only f_p is replaced by f_H,

$$f_H = \frac{1}{T}\int_0^T x^2(t)\,dt, \tag{2.2.58}$$

where T is the period. For $x(t) = \cos\omega t$ or $x(t) = \sin\omega t$, (2.2.58) yields $f_H = \tfrac{1}{2}$.

In the *elastic case*, the treatment is very similar to the acoustic case. We use the relation

$$u_i = \tfrac{1}{2}(U_iF + U_i^*F^*) \tag{2.2.59}$$

and obtain

$$
\begin{aligned}
u_{i,j} &= -\tfrac{1}{2}(U_ip_j\dot{F} + U_i^*p_j\dot{F}^*), \\
e_{ij} &= -\tfrac{1}{4}(U_ip_j + U_jp_i)\dot{F} - \tfrac{1}{4}(U_i^*p_j + U_j^*p_i)\dot{F}^*, \\
\tau_{ij} &= -\tfrac{1}{2}c_{ijkl}p_l(U_k\dot{F} + U_k^*\dot{F}^*).
\end{aligned}
$$

Considering (A.3.12) and (A.3.13) for $\dot{F}$ instead of F, we obtain

$$\hat{W} = \tfrac{1}{2}\rho\Gamma_{ik}U_iU_k^*f_c, \qquad \hat{K} = \tfrac{1}{2}\rho U_iU_i^*f_c,$$

$$\hat{E} = \hat{W} + \hat{K}, \qquad \hat{S}_i = \tfrac{1}{2}\rho a_{ijkl}p_l(U_jU_k^* + U_j^*U_k)f_c. \tag{2.2.60}$$

Γ_{ik} are elements of matrix $\hat{\boldsymbol{\Gamma}}$; see (2.2.19). f_c takes the form

$$f_c = \int_{-\infty}^{\infty} \dot{x}^2(t)dt. \tag{2.2.61}$$

For time-harmonic waves, $\overline{W}, \overline{K}, \overline{E}$, and $\overline{S}_i$ are again given by (2.2.60); only f_c is replaced by f_A,

$$f_A = \frac{1}{T}\int_0^T \dot{x}^2(t)dt, \tag{2.2.62}$$

where T is the period. For $x(t) = \cos\omega t$ or $x(t) = \sin\omega t$, (2.2.62) yields

$$f_A = \tfrac{1}{2}\omega^2 = 2\pi^2 f^2. \tag{2.2.63}$$

Equations (2.2.60) can be simplified if we take into account (2.2.38) and write $U_i = Ag_i^{(m)}$ and $U_i^* = A^*g_i^{(m)}$. Then, for a general anisotropic medium, $\Gamma_{ik}U_iU_k^* = \Gamma_{ik}g_i^{(m)}g_k^{(m)}AA^* = G_mAA^* = AA^*$, as $G_m = 1$ for the selected mth plane wave; see (2.2.34) and (2.2.37). Similarly, $U_iU_i^* = AA^*$ and $a_{ijkl}p_l(U_jU_k^* + U_j^*U_k) = 2a_{ijkl}p_lg_j^{(m)}g_k^{(m)}AA^*$. Equations (2.2.60) then read

$$\hat{W} = \hat{K} = \tfrac{1}{2}\rho AA^*f_c,$$

$$\hat{E} = 2\hat{K} = \rho AA^*f_c, \tag{2.2.64}$$

$$\hat{S}_i = \rho a_{ijkl}p_lg_j^{(m)}g_k^{(m)}AA^*f_c.$$

The result (2.2.64) is very interesting. For any plane wave propagating in a homogeneous anisotropic medium, the time-integrated quantities of the strain and kinetic energies are equal, $\hat{W} = \hat{K}$. Let us emphasize that this result is valid for plane waves propagating in a homogeneous medium, but it is not generally the case. Later on, we shall prove that the result remains *approximately* valid even for high-frequency waves propagating in slightly inhomogeneous media. Note that the result $\hat{W} = \hat{K}$ can also be obtained directly from (2.2.60), if we use $\Gamma_{ik}U_iU_k^* = U_iU_i^*$, following from (2.2.35).

For time-harmonic plane waves propagating in a homogeneous medium, the result (2.2.64) remains valid, only f_c is replaced by f_A; see (2.2.62).

We will now determine the *velocity vector of energy flux* $\vec{\mathcal{U}}^E$ of the plane wave in a homogeneous anisotropic medium; see (2.1.44). For plane waves, the velocity vector of the energy flux is called the *group velocity vector* and is denoted $\vec{\mathcal{U}}$. It is given by the relation

$$\mathcal{U}_i = \hat{S}_i/\hat{E} = a_{ijkl}p_lg_j^{(m)}g_k^{(m)} = \tfrac{1}{2}\partial G_m/\partial p_i; \tag{2.2.65}$$

see (2.1.44). The relation $a_{ijkl}p_lg_j^{(m)}g_k^{(m)} = \tfrac{1}{2}\partial G_m/\partial p_i$ follows immediately from (2.2.34).

As we can see from (2.2.65) and (2.2.6), the group velocity vector $\vec{\mathcal{U}}$ is different from the phase velocity vector $\vec{C}$ in anisotropic media, both in direction and in magnitude. In other words, in anisotropic media, the energy of plane waves does not propagate perpendicular to the wavefront. Moreover, the group velocity $\mathcal{U}$ is different from the phase velocity C, where $\mathcal{U}$ and C are given by relations

$$\mathcal{U} = (\mathcal{U}_i\mathcal{U}_i)^{1/2}, \qquad C = (C_iC_i)^{1/2} = (p_ip_i)^{-1/2}. \tag{2.2.66}$$

We can, however, derive simple, but very important relations between the group and phase velocity vectors. Equations (2.2.34), (2.2.37), and (2.2.65) yield

$$\mathcal{U}_i p_i = 1. \tag{2.2.67}$$

This equation plays a fundamental role in the investigation of elastic waves propagating in anisotropic media. Alternatively, using (2.2.5) and (2.2.6), we obtain

$$\mathcal{U}_i N_i = \mathcal{C}, \qquad \mathcal{U}_i C_i = \mathcal{C}^2, \tag{2.2.68}$$

where $\mathcal{C}$ is the phase velocity.

Similar relations for $\hat{W}$, $\hat{K}$, $\hat{E}$, and $\hat{S}_i$ can be simply obtained even for plane waves in homogeneous isotropic media. We present them here without deriving them:

$$\hat{W} = \tfrac{1}{2}[(\lambda + \mu)(U_k p_k)(U_i^* p_i) + \mu(U_k U_k^*)p_i p_i]f_c,$$
$$\hat{K} = \tfrac{1}{2}\rho(U_i U_i^*)f_c, \qquad \hat{E} = \hat{W} + \hat{K},$$
$$\hat{S}_i = \tfrac{1}{2}[(\lambda + \mu)p_k(U_i U_k^* + U_i^* U_k) + 2\mu(U_k U_k^*)p_i]f_c.$$

For P waves, $\vec{U} = A\vec{N}$. This yields

$$\hat{W} = \hat{K} = \tfrac{1}{2}\rho A A^* f_c, \qquad \hat{E} = \rho A A^* f_c, \qquad \hat{S}_i = \rho \alpha N_i A A^* f_c. \tag{2.2.69}$$

For S waves, $\vec{U} = B\vec{e}_1 + C\vec{e}_2$, and we obtain

$$\hat{W} = \hat{K} = \tfrac{1}{2}\rho(BB^* + CC^*)f_c, \qquad \hat{E} = \rho(BB^* + CC^*)f_c,$$
$$\hat{S}_i = \rho \beta N_i (BB^* + CC^*)f_c. \tag{2.2.70}$$

To determine the time-averaged quantities for time-harmonic waves, we can again use (2.2.69) and (2.2.70), but we must substitute f_A for f_c; see (2.2.61) and (2.2.62).

Finally, the components of group velocity vector $\mathcal{U}_i$ and of phase velocity vector C_i are given by the following relations:

$$\mathcal{U}_i = C_i = \alpha N_i \text{ for P waves,} \qquad \mathcal{U}_i = C_i = \beta N_i \text{ for S waves;} \tag{2.2.71}$$

see (2.1.44) and (2.2.6). Thus, the group velocity vector and the phase velocity vector coincide for plane waves propagating in a homogeneous isotropic medium. Equations (2.2.67) and (2.2.68) are, of course, satisfied even in this case.

2.2.8 Phase and Group Velocity Surfaces. Slowness Surface

The phase velocity $\mathcal{C}$, group velocity $\mathcal{U}$, phase slowness $1/\mathcal{C}$, and group slowness $1/\mathcal{U}$ of plane waves propagating in a homogeneous anisotropic medium depend on the direction of propagation of the wave. Several important surfaces have been introduced to demonstrate clearly these directional variations. These surfaces have found many important applications in the solution of wave propagation problems in anisotropic media and will be needed in the following sections.

As before, we shall specify the direction of propagation of the wavefront by unit normal $\vec{N}$ with components N_i. Unit vector $\vec{N}$ also specifies the direction of slowness vector $\vec{p} = \vec{N}/\mathcal{C}$ and of phase velocity vector $\vec{C} = \mathcal{C}\vec{N}$. The relevant group velocity vector $\vec{\mathcal{U}}$ has the direction of unit vector $\vec{t}$, specified by components $t_i = \mathcal{U}_i/\mathcal{U}$; see (2.2.65) for $\mathcal{U}_i$. Similarly, the vector of the group slowness has the direction of $\vec{t}$ and is given by relation $\mathcal{U}^{-1}\vec{t}$. It follows from (2.2.67) or (2.2.68) that $\vec{N}$ and $\vec{t}$ satisfy the relation $\vec{t} \cdot \vec{N} = \cos \gamma = C/\mathcal{U}$, where γ specifies the angle between $\vec{p}$ and $\vec{\mathcal{U}}$.

Let us consider an arbitrary point O in a homogeneous anisotropic medium and introduce arbitrarily a local Cartesian coordinate system at O. We shall introduce the following three surfaces at O:

1. **Phase velocity surface.** The phase velocity surface is defined as the locus of the end points of phase velocity vector $\vec{C}$, constructed at O in all possible directions. Consequently, it represents a polar graph of phase velocity C as a function of the two take-off angles that specify the direction of $\vec{N}$. The phase velocity surface is also called the *normal surface*, or simply the *velocity surface*.
2. **Slowness surface.** The slowness surface is defined as the locus of the end points of all slowness vectors $\vec{p}$, constructed at O in all directions. Consequently, it represents a polar graph of phase slowness $1/C$ as a function of the two take-off angles that specify the direction of $\vec{N}$. The slowness surface is also sometimes called the *phase slowness surface* to distinguish it from the analogous group slowness surface. Here we shall not use the group slowness surfaces at all so that we can use the abbreviated term slowness surface for the phase slowness surface.
3. **Group velocity surface.** The group velocity surface is defined as the locus of end points of all group velocity vectors $\vec{\mathcal{U}}$, constructed at O in all possible directions. Consequently, it represents a polar graph of group velocity $\mathcal{U}$ as a function of the two take-off angles that specify the direction of $\vec{t}$. The group velocity surface is also known as the *wave surface*.

It may be useful to introduce also some other surfaces, see Helbig (1994). In the following sections, we shall need only the three surfaces introduced above.

The individual surfaces are mutually related by simple geometric relationships, and each of them can be constructed geometrically from any other. For example, it is possible to show that the phase velocity surface can be obtained from the slowness surface (and vice versa) as the *geometric inverse* using reciprocal radii. Similarly, the slowness surface and the wave surface are mutually related by the so-called *polar reciprocity*. For a detailed description of these relationships, see Helbig (1994).

In an isotropic homogeneous medium, all these surfaces are spherical. This is, however, not the case for anisotropic homogeneous media, where the surfaces are more complex. In Figure 2.1, several typical forms of slowness surfaces and the relevant group velocity surfaces are shown in a planar section passing through O. Figure 2.1(a) corresponds to an isotropic medium, and Figures 2.1(b) through 2.1(d) correspond to a transversely isotropic medium with an axis of symmetry coinciding with p_3.

To explain certain pecularities of the slowness surfaces and group velocity surfaces, we shall discuss them in greater theoretical detail. We shall start with the *slowness surface* and consider a local Cartesian system p_1, p_2, p_3 with its origin at O. It can be shown that the equation

$$\det(a_{ijkl} p_j p_l - \delta_{ik}) = 0 \qquad (2.2.72)$$

represents the *slowness surface*. As explained in Section 2.2.5, $\det(a_{ijkl} p_j p_l - \delta_{ik}) = (G_1(p_i) - 1)(G_2(p_i) - 1)(G_3(p_i) - 1)$, where $G_1(p_i)$, $G_2(p_i)$, and $G_3(p_i)$ are the three eigenvalues of Christoffel matrix $\Gamma_{ik} = a_{ijkl} p_j p_l$. Consequently, the slowness surface

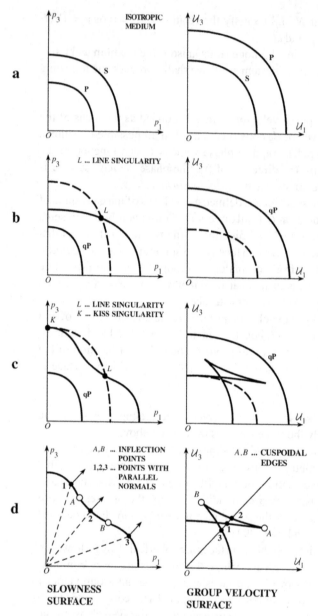

Figure 2.1. Examples of slowness surfaces (left) and group velocity surfaces (right). (a) Isotropic medium; (b)–(d) transversely isotropic medium whose axis of symmetry coincides with the p_3-axis. Only the sections passing through the axis of symmetry are presented. K and L are the kiss and line singularities of the slowness surface. The line singularity is a circle about the axis of symmetry passing through L. In (d), the slowness surface is concave between inflection points A and B. The group velocity surface is then multivalued and forms a loop with cuspoidal points A and B. The normals to the slowness surface specify the directions of the group velocity vectors. The directions of the normals to the slowness surface at points 1, 2, and 3 are the same; therefore; the relevant points 1, 2, and 3 in the group velocity surface lie on a straight line passing through O.

SLOWNESS SURFACE   **GROUP VELOCITY SURFACE**

(2.2.72) has three sheets, specified by implicit equations

$$G_m(p_1, p_2, p_3) = 1, \qquad m = 1, 2, 3. \tag{2.2.73}$$

These three sheets correspond to the three plane waves propagating in an anisotropic homogeneous medium – qP, qS1, and qS2.

The equations of the slowness surfaces (2.2.72) or (2.2.73) may be expressed in several alternative forms. We can, for example, solve the equation $G_m(p_1, p_2, p_3) = 1$ for p_3 and express the equation for the mth sheet of the slowness vector in *explicit form*: $p_3 = \theta(p_1, p_2)$. In this explicit form, we shall discuss the slowness surfaces in Section 2.5.5. The equations of the slowness surfaces can also be expressed in a simple *parameteric form.* If we insert $p_i = |\vec{p}|N_i$ into (2.2.73) and take into account that $G_m(p_1, p_2, p_3)$ is a homogeneous function of the second order in p_i, we obtain

$$G_m(p_1, p_2, p_3) = |\vec{p}|^2 G_m(N_1, N_2, N_3)$$
$$= \left(|\vec{p}|C^{(m)}(N_1, N_2, N_3)\right)^2 = 1; \qquad (2.2.74)$$

see (2.2.40). Thus, (2.2.73) yields $|\vec{p}| = 1/C^{(m)}$ along the slowness surface, and, consequently,

$$p_i(N_1, N_2, N_3) = N_i/C^{(m)}(N_1, N_2, N_3), \qquad i = 1, 2, 3. \qquad (2.2.75)$$

This equation represents the parameteric equation of the slowness surface and corresponds fully to the definition of the slowness surface introduced in this section. As a consequence, (2.2.72) and (2.2.73) represent the implicit analytic equations for the complete slowness surface and for one of its sheets, respectively.

We shall denote by $\vec{N}^S$ the outer unit normal vector to any sheet of slowness surface $G_m(p_1, p_2, p_3) = 1$. Normal vector $\vec{N}^S$ has the direction of $\nabla G_m(p_1, p_2, p_3)$ so that $N_i^S \sim \partial G_m/\partial p_i$. Using (2.2.65), we obtain a very important relation,

$$\mathcal{U}_i = a \, N_i^S, \qquad (2.2.76)$$

where a is a constant of proportionality. Thus, the group velocity vector $\vec{\mathcal{U}}$ corresponding to slowness vector $\vec{p}(N_i)$ has the direction of the normal $\vec{N}^S$ to the slowness surface, constructed at the end point of slowness vector $\vec{p}(N_i)$. A similar property can also be proved conversely: the normal to the group velocity surface determines the direction of the relevant slowness vector. These two properties, together with the relation $\vec{p} \cdot \vec{\mathcal{U}} = 1$, express the polar reciprocity of the slowness surface and of the group velocity surface and may be used to construct one surface from the other.

The slowness surface in an anisotropic medium has three sheets. Each sheet is a closed continuous surface, symmetric with respect to the origin, with continuous normals along the surface. The innermost (fastest) sheet, corresponding to qP waves, is always convex. In most practically important anisotropy symmetries, it is completely separated from the two outer sheets, corresponding to two qS waves. These two outer sheets are often concave in certain regions of N_i, particularly in strongly anisotropic media. See Figures 2.1(c) and 2.1(d). In the planar section of the slowness surface passing through O, the convex and concave parts are separated by points of inflection at which the radii of curvature of the slowness surface change sign. Due to the existence of the concave parts of the slowness surface, the normals $\vec{N}^S$ to one sheet of the slowness surface may have the same direction at several different points of this sheet. See Figure 2.1(d).

As we can see in Figures 2.1(b) and 2.1(c), the individual sheets of the slowness surface may have common points. These points are known as *singular points*, and the relevant directions $\vec{N}$ are called *singular directions*. Singular directions are very common for qS sheets. In fact, the two qS sheets of the slowness surface are *always* connected by singular points. As the two phase velocities coincide at singular points, the two eigenvalues of the Christoffel matrix also coincide at these points. Consequently, the relevant eigenvectors of the two qS waves cannot be uniquely determined. Thus, the polarization of shear waves at singular points is singular and may be anomalous in the vicinity of these singular points. For details, see Crampin (1981), Crampin and Yedlin (1981), Grechka and Obolentseva (1993), Helbig (1994), and Rümpker and Thomson (1994).

There are three types of such singularities.

a. Point singularity. A point singularity is represented by a point intersection of two sheets of the slowness surface. The two slowness sheets have a shallow conical shape in the vicinity of the intersection point, with vertices at that point. In optics, these singularities are usually called *conical points*. For a detailed treatment of the point singularity, see Rümpker and Thomson (1994).

b. Kiss singularity. A kiss singularity is a common point of two sheets of the slowness surface, at which the two sheets touch tangentially. See point K in Figure 2.1(c).

c. Line singularity. This singularity corresponds to a line along which two sheets of the slowness surface intersect. Line singularities are common in transversely isotropic media where the slowness surfaces are axially symmetric. Consequently, even the line singularities are axially symmetric in this case (circles about the axis). See Crampin and Yedlin (1981). Note that points L in Figures 2.1(b) and 2.1(c) correspond to the intersection of the singular line with the presented sections. Because the medium in Figure 2.1 is presumably transversely isotropic with an axis of symmetry coinciding with p_3, the line singularity corresponds to a circle about the axis of symmetry p_3, intersecting the presented sections at L.

We shall now briefly discuss the intersections of the slowness surface with an arbitrarily oriented straight line. These intersections play a very important role in the solution of the problem of reflection/transmission of plane waves at a plane interface between two anisotropic media. We shall consider an oriented straight line, its orientation specified by unit vector $\vec{n}^L$. Any straight line intersects the complete slowness surface *at six points at the most*. The number of intersections is always even. At any point of intersection, we can construct the outer normal to the slowness surface, represented by the relevant group velocity vector $\vec{\mathcal{U}}$. The scalar products $\vec{\mathcal{U}} \cdot \vec{n}^L$ are then always positive at one half of the intersection points and negative at the other half. For example, in the very common case of six intersections, three relevant group velocities are oriented "along $\vec{n}^L$" (that is, $\vec{\mathcal{U}} \cdot \vec{n}^L > 0$), and three group velocities "against $\vec{n}^L$" (that is, $\vec{\mathcal{U}} \cdot \vec{n}^L < 0$). In general, certain roots corresponding to the intersection of the slowness surface with the straight line may be complex-valued. In this case, pairs of complex-conjugate roots are obtained.

We shall now discuss the *group velocity surfaces*. In this case, we introduce a local Cartesian coordinate system $\mathcal{U}_1$, $\mathcal{U}_2$, $\mathcal{U}_3$ with its origin at O. The group velocity surfaces often have a considerably more complicated shape than the slowness surfaces, particularly if the slowness surface is concave for certain N_i. The group velocity surface is then multivalued in the relevant regions and forms loops with cuspidal edges. The multivaluedness of the group velocity surface is caused by the normals to the slowness surface, which have the same direction at several points of the surface if the surface is locally concave. See Figure 2.1(d). The cusps of the loops in the group velocity surface correspond to the inflection points on the slowness surface. Note that the group velocity surface of the qP wave is always single-valued. This is the consequence of the convexity of the qP slowness sheet. The slowness surface and the relevant group velocity surface are polarly reciprocal; therefore, the normals to the group velocity surface specify the direction of the relevant slowness vectors. See points 1, 2, and 3 in Figure 2.1(d).

In certain applications, it is useful to compute the Gaussian curvatures of the slowness surface K^S and of the relevant group velocity surface K^G. For more details on K^S, K^G, and other related quantities in anisotropic media, see Section 4.14.4.

We shall finish this section with an important remark. The slowness surface and group velocity surface, as introduced here, are fully *plane-wave concepts*. However, they play an important role also in the investigation of the elastic wavefields generated by a point source in homogeneous anisotropic media; see Section 2.5.5. For high-frequency waves, the group velocity surface represents a wavefront, corresponding to a point source situated at O in a homogeneous anisotropic medium for travel time $T = 1$ s. This can be understood intuitively as the seismic energy propagates from the point source at O along straight lines in a homogeneous medium and travels the distance corresponding to the group velocity $\mathcal{U}$ in travel time $T = 1$ s.

2.2.9 Elastic Plane Waves in Isotropic and Anisotropic Media: Differences

In this section, we shall summarize the most important differences between the properties of plane elastic waves propagating in homogeneous isotropic and anisotropic media. The fluid medium (acoustic case) is considered a special case of the elastic isotropic medium.

1. **Shear wave splitting.** In an anisotropic homogeneous medium, *three* types of plane waves can propagate in a direction specified by N_i: one quasi-compressional (qP) wave and two quasi-shear (qS1 and qS2) waves. These three waves have, in general, different properties, particularly different velocities of propagation. In an isotropic medium, only two types of plane waves can propagate: one compressional (P) and one shear (S) wave. Because only one shear wave propagates in isotropic media, and two relevant quasi-shear waves in anisotropic media, we usually speak of shear wave splitting in anisotropic media. In a fluid medium (acoustic case), only one type of plane wave can propagate. This is the compressional (P) wave. Shear waves do not propagate in a fluid medium.

2. **Relation between the phase and group velocity vectors.** The phase and group velocity vectors of any individual plane wave propagating in a homogeneous anisotropic medium have different directions and different magnitudes. Because the phase velocity vector is perpendicular to the wavefront and the group velocity vector is parallel to the energy flux, we can conclude that the energy flux is not perpendicular to the wavefront. Later on, we shall identify the trajectories along which the high-frequency part of seismic energy flows as seismic rays. Thus, we can also say that the rays are not perpendicular to the wavefronts in anisotropic media. In isotropic media, the phase and group velocity vectors of the plane wave under consideration coincide both in direction and in magnitude. Both vectors are perpendicular to the wavefront. Thus, in isotropic homogeneous media, the rays of plane waves are perpendicular to the wavefronts. This, of course, also applies to the acoustic case.

3. **Dependence of the phase and group velocities on the direction of propagation.** Both the phase and group velocities of plane waves in a homogeneous anisotropic medium depend on the direction of propagation of the wavefront, $\vec{N}$. In isotropic media, the phase and group velocities of propagation of a plane wave not only coincide but also do not depend on the direction of propagation of the plane wave.

4. **Polarization.** The displacement vector of any of the three plane waves propagating in a homogeneous anisotropic medium is strictly determined as the relevant

eigenvector of matrix $\hat{\boldsymbol{\Gamma}}$. The direction of the eigenvectors is, in general, different from the direction of propagation (unit vector $\vec{N}$), from the direction of group velocity vector $\vec{\mathcal{U}}$, and from any direction tangent to the wavefront. The polarization of all three plane waves propagating in an anisotropic solid is linear (assuming that the qS1 and qS2 waves are separated). In isotropic media, the situation is different. For P waves, the direction of the phase velocity vector, group velocity vector, and eigenvector coincide. For S waves, two vectors coincide (phase and group velocity vectors), and the eigenvectors are perpendicular to them. The polarization of the compressional plane wave is always linear, but the polarization of the S wave is, in general, elliptical or quasi-elliptical in a plane perpendicular to $\vec{N}$. The particle velocity motion in a fluid medium (acoustic case) is always perpendicular to the wavefront, as is the displacement vector of compressional waves.

5. **Shear wave singularities.** For certain specific directions of propagation, the phase velocity vectors of both qS1 and qS2 waves coincide. The behavior of qS waves in the vicinity of such directions is particularly complex; we speak of *shear wave singularities*. These do not exist in isotropic media. See Section 2.2.8 and Figure 2.1.

2.2.10 Inhomogeneous Plane Waves

In the previous sections, we assumed that $T(x_i) = p_i x_i$ and that p_k are real-valued. It is not difficult to generalize the results for complex-valued p_k and T. Instead of standard homogeneous plane waves, we then obtain inhomogeneous plane waves.

The inhomogeneous plane waves play an important role in various wave propagation problems, particularly in problems of reflection and transmission of plane waves at plane structural interfaces. See Caviglia and Morro (1992).

Let us assume that quantities $T(x_i)$ and p_k are complex-valued,

$$T(\vec{x}) = T^R(\vec{x}) + iT^I(\vec{x}), \qquad p_k = p_k^R + ip_k^I. \tag{2.2.77}$$

We can then put

$$T^R(\vec{x}) = p_k^R x_k, \qquad T^I(\vec{x}) = p_k^I x_k. \tag{2.2.78}$$

Let us first consider *time-harmonic acoustic wave* (2.2.1). Then

$$\begin{aligned}
F(t - T(\vec{x})) &= \exp[-i\omega(t - T(\vec{x}))] \\
&= \exp(-\omega T^I(\vec{x})) \exp[-i\omega(t - T^R(\vec{x}))], \tag{2.2.79}
\end{aligned}$$

and pressure $p(\vec{x}, t)$ is given by the relation

$$p(\vec{x}, t) = P \exp(-\omega T^I(\vec{x})) \exp[-i\omega(t - T^R(\vec{x}))]. \tag{2.2.80}$$

Thus, the amplitude factor of the inhomogeneous plane wave is not constant but decays exponentially with increasing $\omega T^I(\vec{x})$. We will call planes $T^R(\vec{x}) = p_k^R x_k = $ const. *planes of constant phases* (or, alternatively, wavefronts) and planes $T^I(\vec{x}) = p_k^I x_k = $ const. *planes of constant amplitudes*. The real-valued vectors $\vec{p}^R$ and $\vec{p}^I$, perpendicular to these planes, are usually called the propagation vector and the attenuation vector, respectively.

Inhomogeneous plane wave (2.2.80) is a solution of the acoustic wave equation if complex-valued constants p_i satisfy existence condition (2.2.3). If we use (2.2.77) for p_k, (2.2.3) becomes

$$p_i^R p_i^R - p_i^I p_i^I = 1/c^2, \qquad 2p_i^R p_i^I = 0. \tag{2.2.81}$$

From the second equation of (2.2.81), we immediately see that the planes of constant phases $p_k^R x_k =$ const. and the planes of constant amplitudes $p_k^I x_k =$ const. are mutually perpendicular. Thus, the gradient of amplitudes of inhomogeneous waves is parallel to the phase fronts.

The actual velocity of propagation of phase fronts c_R of inhomogeneous plane waves in the direction of $\vec{p}^R$ is given by the relation

$$1/c_R^2 = p_i^R p_i^R = 1/c^2 + p_i^I p_i^I > 1/c^2. \tag{2.2.82}$$

This equation shows that the velocity of propagation of the phase front of the inhomogeneous wave c_R is lower than velocity $c = \sqrt{k/\rho}$ of the homogeneous plane wave.

We have considered only real-valued bulk modulus k. If we are interested in viscoelastic media, we can use complex-valued k. Then $1/c^2$ is complex-valued, and the second equation of (2.2.81) reads $2p_i^R p_i^I = \operatorname{Im}(1/c^2)$. This relationship shows that the planes of constant amplitudes and the planes of constant phases are not mutually perpendicular in viscoelastic media. A special terminology related to plane waves is used in viscoelastic media. The plane wave is called *homogeneous*, if $\vec{p}^R$ and $\vec{p}^I$ are parallel, and *inhomogeneous*, if $\vec{p}^R$ and $\vec{p}^I$ are not parallel.

We shall now discuss the general expressions for *inhomogeneous acoustic plane waves in the time domain*. The analytical signal $F(t - T)$ must be determined for a complex-valued $T = T^R + iT^I$ in this case; see Appendix A.3:

$$
\begin{aligned}
F(t - T) = {} & \frac{1}{\pi} \int_{-\infty}^{\infty} \frac{T^I x(u)\,du}{(T^I)^2 + (t - T^R - u)^2} \\
& - \frac{i}{\pi} \int_{-\infty}^{\infty} \frac{(t - T^R - u)x(u)\,du}{(T^I)^2 + (t - T^R - u)^2}.
\end{aligned}
\tag{2.2.83}
$$

These equations can also be expressed in a more condensed form:

$$F(t - T) = -\frac{i}{\pi} \int_{-\infty}^{\infty} \frac{x(u)}{t - T - u}\,du = -\frac{i}{\pi(t - T)} * x(t). \tag{2.2.84}$$

The inhomogeneous vectorial plane waves propagating in an *elastic isotropic non-dissipative medium* have properties similar to those of the inhomogeneous scalar acoustic plane waves discussed earlier. Both for P and S plane waves, the planes of constant phases are perpendicular to the planes of constant amplitudes. Equations (2.2.77) through (2.2.84) remain valid even in the elastic case. Only acoustic velocity c should be replaced by α (for P waves) or by β (for S waves), and the amplitude factor in (2.2.80) should be replaced by the appropriate vectorial amplitude factor; see (2.2.51) and (2.2.52).

The situation is, however, more complex in an *elastic anisotropic nondissipative medium*. In an anisotropic medium, the planes of constant amplitudes of inhomogeneous plane waves are not perpendicular to the planes of constant phases; however, they are perpendicular to the planes perpendicular to the group velocity vector $\vec{\mathcal{U}}$. In other words, the group velocity vector $\vec{\mathcal{U}}$ and vector $\vec{p}^I$ are mutually perpendicular, $\vec{\mathcal{U}} \cdot \vec{p}^I = 0$. Even in an anisotropic medium, the phase velocity of the inhomogeneous plane wave is less than the phase velocity of the homogeneous plane wave.

2.3 Elastic Plane Waves Across a Plane Interface

Detailed knowledge of the process of reflection and transmission (R/T process) of plane elastic waves at a plane structural interface between two homogeneous, isotropic or anisotropic media is very useful in the seismic ray theory. As we shall see later, the plane wave R/T laws may be used locally even in considerably more general structures. We shall, therefore, briefly explain the main principles of the solution of the problem.

Before we treat the elastic case, we shall demonstrate the R/T process in a simple acoustic case. The reason is not only tutorial but also practical, because the acoustic case plays an important role in methods of seismic prospecting for oil. Only after this shall we consider both isotropic and anisotropic elastic media.

We will adopt a nontraditional approach to the treatment of the R/T problem, which will be convenient when we study high-frequency seismic wave fields in 3-D layered structures later. More specifically, our approach does not require the introduction of a local Cartesian coordinate system connected with the interface and the plane of incidence. The approach is particularly useful in anisotropic media, because it does not require the elastic moduli to be transformed to local Cartesian coordinates. However, it also has convenient applications in isotropic media because it does not require the rotation of the S wave components into the plane of incidence and into a direction perpendicular to the plane of incidence.

We will consider a plane interface, $\Sigma(x_i) = 0$, with unit normal $\vec{n}$ oriented to either side of the interface. We denote one of the two homogeneous halfspaces as halfspace 1 and the other as halfspace 2. We assume that the halfspaces are in *welded contact* across interface Σ. This means that displacement components u_i and traction components T_i, acting across surface Σ, are the same on both sides of the interface. We remind the reader that traction components T_i can be expressed in terms of the displacement components as

$$T_i = c_{ijkl} n_j u_{k,l}. \tag{2.3.1}$$

In an isotropic medium, the traction components are given by the relation

$$T_i = \lambda n_i u_{k,k} + \mu n_j (u_{i,j} + u_{j,i}). \tag{2.3.2}$$

For more details, see, for example, Auld (1973), Pilant (1979), and Aki and Richards (1980).

In the acoustic case, the six conditions of the welded contact reduce to two boundary conditions: continuity of pressure and of the normal component of the particle velocity across Σ; see Brekhovskikh and Godin (1989). If we wish to work with pressure only, and not with the particle velocity, we can express the particle velocity in terms of the pressure gradient. Instead of v_i, we use $\dot{v}_i = -\rho^{-1} p_{,i}$; see (2.1.22) with $f_i = 0$. Thus, we require the continuity of pressure and of the normal component of the density-normalized pressure gradient.

In Sections 2.3.1 through 2.3.3, only time-harmonic plane waves (with circular frequency ω) will be considered. It is easy to generalize the results to transient plane waves using the Fourier transform; see Section 2.3.4.

2.3.1 Acoustic Case

We denote the velocity and density in halfspace 1 by c_1 and ρ_1 and in halfspace 2 by c_2, ρ_2. We assume that a homogeneous time-harmonic pressure plane wave is incident at interface Σ from halfspace 1. We shall express the incident plane wave as

$$p(\vec{x}, t) = P \exp[-i\omega(t - p_i x_i)], \qquad p_i = N_i / c_1. \tag{2.3.3}$$

Here N_i are Cartesian components of the normal to the wavefront of the incident wave, p_i are Cartesian components of the slowness vector of the incident wave, and P is a scalar amplitude.

We remind the reader that normal $\vec{n}$ to interface Σ may be oriented to either side of interface Σ (that is, into halfspace 1 or into halfspace 2). We distinguish these two cases by orientation index ϵ,

$$\epsilon = \text{sgn}(\vec{p} \cdot \vec{n}), \tag{2.3.4}$$

where $\vec{p}$ is the slowness vector of the incident wave. Thus, $\epsilon = -1$ denotes the situation in which normal $\vec{n}$ is oriented into the first medium ("against the incident wave"), whereas $\epsilon = 1$ denotes the situation in which the normal is oriented into the second medium ("along the incident wave").

The incident wave itself cannot satisfy the two boundary conditions [continuity of pressure p and the density-normalized normal component of the pressure gradient $(\nabla p \cdot \vec{n})/\rho$]. We introduce two new time-harmonic plane waves generated at the interface: one reflected wave $p^r(\vec{x}, t)$ and one transmitted wave $p^t(\vec{x}, t)$,

$$p^r(\vec{x}, t) = P^r \exp\left[-i\omega(t - p_i^r x_i)\right], \qquad p_i^r = N_i^r / c_1, \tag{2.3.5}$$

$$p^t(\vec{x}, t) = P^t \exp\left[-i\omega(t - p_i^t x_i)\right], \qquad p_i^t = N_i^t / c_2. \tag{2.3.6}$$

Here P^r, P^t, $\vec{N}^r$, $\vec{N}^t$, $\vec{p}^r$, and $\vec{p}^t$ have obvious meanings. These quantities are not yet known. We will prove that the three plane waves (incident, reflected, and transmitted) satisfy the two boundary conditions and determine the unknown functions and quantities of (2.3.5) and (2.3.6) from these boundary conditions.

The normal component of the density-normalized pressure gradient for the incident wave is then

$$\rho_1^{-1} \nabla p \cdot \vec{n} = i\omega \rho_1^{-1}(\vec{p} \cdot \vec{n}) P \exp[-i\omega(t - p_i x_i)]. \tag{2.3.7}$$

Similar expressions are obtained also for the reflected and transmitted waves. Thus, the two boundary conditions along Σ can now be expressed as

$$P \exp[i\omega p_i x_i] + P^r \exp\left[i\omega p_i^r x_i\right] = P^t \exp\left[i\omega p_i^t x_i\right],$$

$$\rho_1^{-1} P(\vec{p} \cdot \vec{n}) \exp[i\omega p_i x_i] + \rho_1^{-1} P^r (\vec{p}^r \cdot \vec{n}) \exp\left[i\omega p_i^r x_i\right] \tag{2.3.8}$$

$$= \rho_2^{-1} P^t (\vec{p}^t \cdot \vec{n}) \exp\left[i\omega p_i^t x_i\right].$$

The factors $\exp[-i\omega t]$ and $i\omega$ have been omitted in (2.3.8). It is obvious that the amplitudes of the R/T plane waves, P^r and P^t, should be constants, independent of $\vec{x}$; otherwise, (2.3.5) and (2.3.6) would not represent plane waves. If (2.3.8) is to yield P^r and P^t independent of $\vec{x}$, the following relations must be satisfied along the interface:

$$\exp[i\omega p_i x_i] = \exp\left[i\omega p_i^r x_i\right] = \exp\left[i\omega p_i^t x_i\right]. \tag{2.3.9}$$

Boundary conditions (2.3.8) then greatly simplify:

$$P + P^r = P^t, \qquad P(\vec{p} \cdot \vec{n})/\rho_1 + P^r(\vec{p}^r \cdot \vec{n})/\rho_1 = P^t(\vec{p}^t \cdot \vec{n})/\rho_2. \tag{2.3.10}$$

Now we shall discuss (2.3.9) in more detail and propose procedures to determine $\vec{p}^r$ and $\vec{p}^t$. Equations (2.3.9) yield

$$\vec{p} \cdot \vec{x} = \vec{p}^r \cdot \vec{x} = \vec{p}^t \cdot \vec{x} \tag{2.3.11}$$

along Σ. These relations imply that the travel times of reflected waves, $T^r(\vec{x}) = \vec{p}^r \cdot \vec{x}$, and the travel times of transmitted waves, $T^t(\vec{x}) = \vec{p}^t \cdot \vec{x}$, equal the travel times of the incident wave, $T(\vec{x}) = \vec{p} \cdot \vec{x}$, along interface Σ. Consequently, the tangential components of slowness vectors $\vec{p}$, $\vec{p}^r$ and $\vec{p}^t$ along Σ are the same.

We will decompose the slowness vectors into tangential and normal components:

$$\vec{p} = \vec{a} + \sigma\vec{n}, \qquad \vec{p}^r = \vec{a}^r + \sigma^r\vec{n}, \qquad \vec{p}^t = \vec{a}^t + \sigma^t\vec{n}. \tag{2.3.12}$$

Here σ, σ^r, and σ^t are constants, and $\vec{a}$, $\vec{a}^r$, and $\vec{a}^t$ are the vectorial tangential components of the appropriate slowness vectors. σ, σ^r, and σ^t can be expressed as

$$\sigma = \vec{p} \cdot \vec{n}, \qquad \sigma^r = \vec{p}^r \cdot \vec{n}, \qquad \sigma^t = \vec{p}^t \cdot \vec{n}. \tag{2.3.13}$$

Using (2.3.12) and (2.3.13), we obtain the relations for the tangential components $\vec{a}$, $\vec{a}^r$, and $\vec{a}^t$ of the slowness vector:

$$\vec{a} = \vec{p} - \vec{n}(\vec{p} \cdot \vec{n}), \qquad \vec{a}^r = \vec{p}^r - \vec{n}(\vec{p}^r \cdot \vec{n}), \qquad \vec{a}^t = \vec{p}^t - \vec{n}(\vec{p}^t \cdot \vec{n}). \tag{2.3.14}$$

Equations (2.3.14) allow us to express the relation $\vec{a} = \vec{a}^r = \vec{a}^t$ in the following way:

$$\vec{p}^r - \vec{n}(\vec{p}^r \cdot \vec{n}) = \vec{p}^t - \vec{n}(\vec{p}^t \cdot \vec{n}) = \vec{p} - \vec{n}(\vec{p} \cdot \vec{n}). \tag{2.3.15}$$

Equations (2.3.15) are not just valid for homogeneous acoustic waves but are valid very generally. They are also valid for elastic waves in isotropic or anisotropic media, given an arbitrarily oriented interface Σ and an arbitrarily oriented plane incident wave. They remain valid even for inhomogeneous plane waves.

Equations (2.3.15) are not sufficient to determine the slowness vectors of reflected and transmitted waves $\vec{p}^r$ and $\vec{p}^t$. In addition, we must consider existence condition (2.2.3). Identical existence conditions are satisfied also for plane waves in isotropic solids so that the final equations will also be valid for P and S waves in an isotropic medium; see (2.2.48) and (2.2.49). For plane waves in an anisotropic medium, however, the existence condition has a different form; see (2.2.36).

Here we shall only determine $\vec{p}^t$; the expression for $\vec{p}^r$ will then be obtained as a special case. Because $\vec{a}^t = \vec{a}$, (2.3.12) yields $\vec{p}^t = \vec{a} + \sigma^t\vec{n}$. The only unknown quantity in this relation is σ^t. It can be determined from (2.2.3), which reads $\vec{p}^t \cdot \vec{p}^t = 1/c_2^2$. Inserting $\vec{p}^t = \vec{a} + \sigma^t\vec{n}$ yields

$$\vec{a} \cdot \vec{a} + (\sigma^t)^2 = 1/c_2^2.$$

Thus

$$\sigma^t = \pm\left(1/c_2^2 - \vec{a} \cdot \vec{a}\right)^{1/2} = \pm\left[1/c_2^2 - 1/c_1^2 + (\vec{p} \cdot \vec{n})^2\right]^{1/2}. \tag{2.3.16}$$

As we can see in (2.3.16), σ^t is either real-valued (for $1/c_2^2 > \vec{a} \cdot \vec{a}$) or purely imaginary (for $1/c_2^2 < \vec{a} \cdot \vec{a}$). We shall discuss these two cases independently.

a. For $1/c_2^2 > \vec{a} \cdot \vec{a}$, σ^t is real-valued. The sign in (2.3.16) is selected in the following way. As the wave is transmitted,

$$\operatorname{sgn}\sigma^t = \operatorname{sgn}(\vec{p}^t \cdot \vec{n}) = \operatorname{sgn}\sigma = \operatorname{sgn}(\vec{p} \cdot \vec{n}) = \epsilon, \tag{2.3.17}$$

where ϵ is given by (2.3.4). Thus,

$$\sigma^t = \epsilon\left[1/c_2^2 - 1/c_1^2 + (\vec{p} \cdot \vec{n})^2\right]^{1/2}.$$

The final equation for the slowness vector of the transmitted wave is then

$$\vec{p}^t = \vec{p} - \left\{(\vec{p} \cdot \vec{n}) - \epsilon\left[(1/c_2)^2 - (1/c_1)^2 + (\vec{p} \cdot \vec{n})^2\right]^{1/2}\right\}\vec{n}. \tag{2.3.18}$$

b. For $1/c_2^2 < \vec{a} \cdot \vec{a}$, σ^t is purely imaginary, and the transmitted wave is inhomogeneous. The sign of σ^t must be chosen in such a way to obtain the wave that *decays exponentially from the interface*. We denote by x_i^0 the Cartesian coordinates of an arbitrarily selected point on Σ and express the argument of the exponential function of the transmitted wave (2.3.6) as follows:

$$-\mathrm{i}\omega(t - p_i^t x_i) = -\mathrm{i}\omega[t - p_i^t x_i^0 - a_i(x_i - x_i^0)] - \omega n_i(x_i - x_i^0)\operatorname{Im}\sigma^t.$$

Thus, we require that the following condition be satisfied:

$$n_i(x_i - x_{i0})\operatorname{Im}\sigma^t > 0. \tag{2.3.19}$$

As the wave is transmitted and propagates in the halfspace 2, $\operatorname{sgn}(n_i(x_i - x_{i0})) = \epsilon$. This result yields the final form of the condition (2.3.19):

$$\epsilon\operatorname{Im}\sigma^t > 0. \tag{2.3.20}$$

Condition can be modified, introducing directly the square root instead of $\operatorname{Im}\sigma^t$. We obtain

$$\left(1/c_2^2 - \vec{a} \cdot \vec{a}\right)^{1/2} = +\mathrm{i}\left(\vec{a} \cdot \vec{a} - 1/c_2^2\right)^{1/2} \quad \text{for } \vec{a} \cdot \vec{a} > 1/c_2^2. \tag{2.3.21}$$

Thus, for an inhomogeneous transmitted wave, (2.3.18) reads

$$\vec{p}^t = \vec{p} - \left\{(\vec{p} \cdot \vec{n}) - \mathrm{i}\epsilon\left[(1/c_1)^2 - (1/c_2)^2 - (\vec{p} \cdot \vec{n})^2\right]^{1/2}\right\}\vec{n}. \tag{2.3.22}$$

Here $1/c_2^2 < \vec{a} \cdot \vec{a}$, with $\vec{a} \cdot \vec{a} = 1/c_1^2 - (\vec{p} \cdot \vec{n})^2$.

The reflected acoustic wave is always homogeneous, with σ^r real-valued. For reflected waves, the choice of sign should be as follows:

$$\operatorname{sgn}\sigma^r = \operatorname{sgn}(\vec{p}^r \cdot \vec{n}) = -\operatorname{sgn}\sigma = -\operatorname{sgn}(\vec{p} \cdot \vec{n}) = -\epsilon.$$

For reflected acoustic waves, we substitute c_2 for c_1 in (2.3.18) and obtain

$$\vec{p}^r = \vec{p} - 2(\vec{p} \cdot \vec{n})\vec{n}. \tag{2.3.23}$$

These are the final equations for $\vec{p}^t$ and $\vec{p}^r$, the slowness vectors of R/T waves.

We shall now solve (2.3.10) for P^r and P^t. In view of (2.3.23),

$$(\vec{p}^r \cdot \vec{n}) = -(\vec{p} \cdot \vec{n}).$$

The solution of (2.3.10) is then

$$P^r = \frac{\rho_2(\vec{p} \cdot \vec{n}) - \rho_1(\vec{p}^t \cdot \vec{n})}{\rho_2(\vec{p} \cdot \vec{n}) + \rho_1(\vec{p}^t \cdot \vec{n})}P, \qquad P^t = \frac{2\rho_2(\vec{p} \cdot \vec{n})}{\rho_2(\vec{p} \cdot \vec{n}) + \rho_1(\vec{p}^t \cdot \vec{n})}P. \tag{2.3.24}$$

Quantities P^r/P and P^t/P are usually called *acoustic reflection and transmission coefficients for pressure*; $R^r = P^r/P$ is the reflection coefficient, and $R^t = P^t/P$ the transmission coefficient.

Let us summarize. We have derived the general equations for the slowness vectors of both generated waves, (2.3.18) and (2.3.23). We have also derived the expressions for the amplitudes of both generated waves; see (2.3.24). Thus, the problem has been solved completely. For more details on acoustic reflection/transmission coefficients, see Section 5.1.4.

In the whole treatment of the R/T process, we have not used certain traditional concepts such as the plane of incidence, angles of incidence, and angles of reflection and transmission. We have also not introduced a local Cartesian coordinate system at the interface. Moreover, we have not had to compute certain trigonometric functions, but we have replaced them by scalar products. In the acoustic case, these differences have, more or less, a formal meaning. They will, however, play a considerably more important role in the elastic case.

For completeness, we shall also briefly explain certain of these traditional concepts used in the investigation of the R/T process. *The plane of incidence* is introduced as a plane specified by unit normal vector $\vec{n}$ and by the slowness vector of the incident wave, $\vec{p}$. Equations (2.3.15) then imply that the slowness vectors of R/T waves, $\vec{p}^{\,r}$ and $\vec{p}^{\,t}$, are also situated in the plane of incidence. *Angle of incidence* i^{inc} is defined as the acute angle between normal $\vec{n}$ and the slowness vector of the incident wave. The angles of reflection i^r and transmission i^t are introduced similarly. Taking the scalar product of (2.3.18) with the unit vector tangent to the interface, we obtain

$$\frac{\sin i^t}{c_2} = \frac{\sin i^{inc}}{c_1}. \tag{2.3.25}$$

This is the famous *Snell's law*. We also find, as a special case of (2.3.25), that the angle of reflection equals the angle of incidence. Using angles i^{inc} and i^t, R/T coefficients (2.3.24) can be expressed in a more familiar way,

$$
\begin{aligned}
R^r &= \frac{\rho_2 c_2 \cos i^{inc} - \rho_1 c_1 \cos i^t}{\rho_2 c_2 \cos i^{inc} + \rho_1 c_1 \cos i^t}, \\
R^t &= \frac{2\rho_2 c_2 \cos i^{inc}}{\rho_2 c_2 \cos i^{inc} + \rho_1 c_1 \cos i^t};
\end{aligned}
\tag{2.3.26}
$$

see (2.3.24).

These are the final expressions for the acoustic R/T coefficients for pressure. We can, however, also introduce the *acoustic R/T coefficients for particle velocity*, R^r_{pv} and R^t_{pv}. For the incident pressure plane wave given by (2.3.3), particle velocity $\vec{v}(\vec{x}, t)$ can be expressed in the following form:

$$\vec{v}(\vec{x}, t) = V\vec{N}F(t - p_i x_i), \qquad V = P/(\rho_1 c_1);$$

see (2.2.7). Similar expressions can also be given for reflected and transmitted waves, with $V^r = P^r/(\rho_1 c_1)$ and $V^t = P^t/(\rho_2 c_2)$. It is easy to see that $R^r_{pv} = V^r/V = P^r/P = R^r$ but that $R^t_{pv} = V^t/V = (\rho_1 c_1/\rho_2 c_2)P^t/P = (\rho_1 c_1/\rho_2 c_2)R^t$. Thus, the acoustic R/T coefficients for particle velocity are given by the relations

$$
\begin{aligned}
R^r_{pv} &= R^r = \frac{\rho_2 c_2 \cos i^{inc} - \rho_1 c_1 \cos i^t}{\rho_2 c_2 \cos i^{inc} + \rho_1 c_1 \cos i^t}, \\
R^t_{pv} &= \frac{\rho_1 c_1}{\rho_2 c_2} R^t = \frac{2\rho_1 c_1 \cos i^{inc}}{\rho_2 c_2 \cos i^{inc} + \rho_1 c_1 \cos i^t}.
\end{aligned}
$$

2.3.2 Isotropic Elastic Medium

We will now discuss the reflection and transmission of plane elastic waves at a plane interface between two homogeneous isotropic elastic halfspaces. We denote the velocities of the P and S waves and the densities in halfspace 1 by α_1, β_1, and ρ_1 and in halfspace 2 by α_2, β_2, and ρ_2. Similarly, Lamé's elastic moduli in halfspace 1 are λ_1 and μ_1 and in halfspace 2 are λ_2 and μ_2. We assume that a homogeneous plane wave is incident at the interface from halfspace 1. The displacement vector of the incident P wave will be expressed as

$$\vec{u}(\vec{x}, t) = A\vec{N} \exp[-i\omega(t - p_i x_i)], \tag{2.3.27}$$

where $\vec{N}$ is the unit normal to the wavefront, $p_i = N_i/\alpha_1$. For the incident S wave,

$$\vec{u}(\vec{x}, t) = (B\vec{e}_1 + C\vec{e}_2) \exp[-i\omega(t - p_i x_i)], \tag{2.3.28}$$

where $p_i = N_i/\beta_1$. We assume that $\vec{e}_1$, $\vec{e}_2$, and $\vec{N}$ form a right-handed, mutually orthogonal triplet of unit vectors for each of the considered waves. B and C are two components of the displacement vector of the incident S wave in the direction of unit vector $\vec{e}_1$ (S1 component) and of unit vector $\vec{e}_2$ (S2 component). Thus, components B and C depend on the definition of $\vec{e}_1$ and $\vec{e}_2$. If $\vec{e}_1$ and $\vec{e}_2$ are rotated about $\vec{N}$ to different $\vec{e}_1'$ and $\vec{e}_2'$, we, of course, obtain a different B and C.

The incident plane wave itself cannot satisfy the six conditions of the welded contact at interface Σ, and reflected P and S and transmitted P and S plane waves must be introduced. Four of them are of the same type as the incident wave (P → P, S → S) and are called *unconverted* reflected/transmitted waves. Four others are of a type different to the incident wave (P → S, S → P) and are called *converted* reflected/transmitted waves. Expressions for all the generated plane R/T waves are similar to those for the incident wave; see (2.3.27) and (2.3.28). We shall only denote their amplitude functions A^r, B^r, and C^r for the reflected waves and A^t, B^t, and C^t for the transmitted waves. The slowness vectors $\vec{p}$ of the four generated waves are, of course, different; consequently, $\vec{e}_1$ and $\vec{e}_2$ should also be different. In the same way as for the acoustic waves in Section 2.3.1, we shall determine their slowness vectors $\vec{p}$ from the interface conditions and from the existence conditions. If we insert all expressions for the incident, reflected, and transmitted waves into the boundary condition of the welded contact, we find that argument $p_i x_i$ should be equal for all the waves along the interface. This finding and the existence conditions allow us to compute the slowness vectors for all generated waves from the slowness vector of the incident wave. Let us use V to denote the velocity of the incident wave (that is, α_1 for the incident P wave and β_1 for the incident S wave) and $\tilde{V}$ to denote the velocity of any selected generated wave (that is, α_1, β_1, α_2, or β_2 according to the type of selected wave). We also denote by $\tilde{\vec{p}}$ the slowness vector of the selected R/T wave. Then, in view of (2.3.18) and (2.3.23),

$$\tilde{\vec{p}} = \vec{p} - \left\{ (\vec{p} \cdot \vec{n}) \mp \epsilon \left[(1/\tilde{V})^2 - (1/V)^2 + (\vec{p} \cdot \vec{n})^2 \right]^{1/2} \right\} \vec{n}. \tag{2.3.29}$$

Here $\vec{p}$ is the slowness vector of the incident wave, $\vec{n}$ is the unit normal to interface Σ. The upper sign (−) corresponds to the transmitted wave, the lower sign (+) refers to the reflected wave, and ϵ is the orientation index; see (2.3.4). Equation (2.3.29) is very general, it is valid for any type of reflected or transmitted P or S wave and for any arbitrarily oriented interface Σ. For the unconverted reflected wave ($V = \tilde{V}$), Equation (2.3.29) simplifies to

$$\tilde{\vec{p}} = \vec{p} - 2(\vec{p} \cdot \vec{n})\vec{n}. \tag{2.3.30}$$

For the converted reflected wave, however, general expression (2.3.29) must be used.

The generated R/T wave may be inhomogeneous. We again denote the slowness vector of the selected R/T wave by $\tilde{p}_i = a_i + \sigma n_i$. If the selected wave is inhomogeneous, σ is purely imaginary. The inhomogeneous wave must decay exponentially from the interface. The sign condition can be determined in the same way as in Section 2.3.1. It reads:

$$\pm \epsilon \, \text{Im} \, \sigma > 0. \tag{2.3.31}$$

The upper sign corresponds to a transmitted wave, and the lower sign refers to a reflected wave. Alternatively, we can use the condition

$$(1/\tilde{V}^2 - \vec{a} \cdot \vec{a})^{1/2} = +\mathrm{i}(\vec{a} \cdot \vec{a} - 1/\tilde{V}^2)^{1/2} \qquad \text{for } \vec{a} \cdot \vec{a} > 1/\tilde{V}^2. \qquad (2.3.32)$$

Condition (2.3.32) is the same for reflected and transmitted waves and for any orientation index ϵ. Thus, the slowness vector $\tilde{\vec{p}}$ of any R/T inhomogeneous wave is given by the relation

$$\tilde{\vec{p}} = \vec{p} - \{(\vec{p} \cdot \vec{n}) \mp \mathrm{i}\epsilon[(1/V)^2 - (1/\tilde{V})^2 - (\vec{p} \cdot \vec{n})^2]^{1/2}\}\vec{n}; \qquad (2.3.33)$$

see (2.3.29). Thus, reflected waves also may be inhomogeneous for $\tilde{V} > V$.

As we can immediately see from (2.3.29), the slowness vectors of all four generated R/T plane waves are situated in the plane of incidence, determined by normal $\vec{n}$ to interface Σ and by slowness vector $\vec{p}$ of the incident wave. Taking the scalar product of (2.3.29) with the unit vector tangent to the interface, we obtain the standard Snell's law:

$$\frac{\sin i^{rp}}{\alpha_1} = \frac{\sin i^{rs}}{\beta_1} = \frac{\sin i^{tp}}{\alpha_2} = \frac{\sin i^{ts}}{\beta_2} = \frac{\sin i^{inc}}{V}. \qquad (2.3.34)$$

Here $V = \alpha_1$ for the incident P wave, and $V = \beta_1$ for the incident S wave. All angles i are defined as acute angles between normal $\vec{n}$ and the slowness vectors of the relevant waves. The type of wave is indicated by the superscript (r and t stand for reflected and transmitted; p and s refer to P and S waves).

The boundary conditions require the continuity of the displacement and of the traction vectors across interface Σ. We can express these boundary conditions in general Cartesian coordinates; local Cartesian coordinates will not be introduced at this step. The displacement components can be expressed in a straightforward way. The traction components in an isotropic solid can be expressed in terms of $\vec{u}$ using (2.3.2). For the incident P wave, the traction components are given by the following expressions:

$$T_i = \mathrm{i}\omega A[\lambda_1 n_i(N_k p_k) + 2\mu_1 N_i(n_k p_k)] \exp[-\mathrm{i}\omega(t - p_i x_i)], \qquad (2.3.35)$$

where $\vec{p}$ is the relevant slowness vector of the incident P wave, and $N_i = \alpha_1 p_i$. For the incident S wave, the analogous expression is

$$T_i = \mathrm{i}\omega[B\mu_1 n_j(e_{1i} p_j + e_{1j} p_i) + C\mu_1 n_j(e_{2i} p_j + e_{2j} p_i)] \exp[-\mathrm{i}\omega(t - p_i x_i)]. \qquad (2.3.36)$$

Here $\vec{p}$ is the slowness vector of the incident S wave.

The expressions for the traction components of the generated R/T waves are practically the same as (2.3.35) and (2.3.36); only the individual quantities in these expressions must be properly specified for each of the waves. The six boundary equations then read

$$\begin{aligned} A^t N_i^t + B^t e_{1i}^t + C^t e_{2i}^t - A^r N_i^r - B^r e_{1i}^r - C^r e_{2i}^r &= D_i, \\ A^t X_i^t + B^t Y_i^t + C^t Z_i^t - A^r X_i^r - B^r Y_i^r - C^r Z_i^r &= E_i. \end{aligned} \qquad (2.3.37)$$

where $i = 1, 2, 3$. Here quantities X_i, Y_i, and Z_i have the following meaning:

$$\begin{aligned} X_i &= \lambda n_i(N_k p_k) + 2\mu N_i(n_k p_k), \\ Y_i &= \mu n_j(e_{1i} p_j + e_{1j} p_i), \\ Z_i &= \mu n_j(e_{2i} p_j + e_{2j} p_i); \end{aligned} \qquad (2.3.38)$$

see (2.3.35) and (2.3.36). Notations (2.3.38) are formally the same for the reflected waves (X_i^r, Y_i^r, Z_i^r) and for the transmitted waves (X_i^t, Y_i^t, Z_i^t), but all the quantities must be properly specified for these waves. In system (2.3.37), unit vectors $\vec{e}_1$, $\vec{e}_2$, and $\vec{N}$ are denoted by superscripts r and t, which indicate whether they refer to reflected or transmitted waves. The slowness vector for each wave should be determined using (2.3.29).

Note that $\vec{N}^r, \vec{e}_1^r$, and $\vec{e}_2^r$ in (2.3.37) do not form a mutually orthogonal triplet of unit vectors because they correspond to different reflected waves with different rays: $\vec{N}^r$ corresponds to the reflected P wave, and $\vec{e}_1^r, \vec{e}_2^r$ corresponds to the reflected S wave. The same also applies to the transmitted wave.

The right-hand-side (RHS) expressions in (2.3.37), D_i and E_i, correspond to the incident wave. For the incident P wave,

$$D_i = AN_i, \qquad E_i = AX_i. \tag{2.3.39}$$

For the incident S wave, we write

$$D_i = Be_{1i} + Ce_{2i}, \qquad E_i = BY_i + CZ_i. \tag{2.3.40}$$

Here X_i, Y_i, and Z_i are again given by (2.3.38), with appropriately specified quantities.

The unknown amplitude factors of the reflected and transmitted plane waves, A^r, B^r, C^r, A^t, B^t, and C^t, can be determined using the system of six equations (2.3.37).

System (2.3.37) is very general. As we know, unit vector $\vec{e}_3 \equiv \vec{N}$ is perpendicular to the relevant wavefront for each wave, and the three unit vectors $\vec{e}_1, \vec{e}_2, \vec{e}_3 \equiv \vec{N}$, corresponding to the wave under consideration, must form a right-handed, mutually orthogonal triplet of unit vectors. Otherwise, however, unit vectors $\vec{e}_1$ and $\vec{e}_2$ may be chosen arbitrarily. They need not be perpendicular or parallel to the plane of incidence. In general, they can be arbitrarily rotated about $\vec{N}$.

For example, let us consider the incident S wave and denote the relevant unit vectors $\vec{e}_1^s$ and $\vec{e}_2^s$ and $\vec{e}_3^s \equiv \vec{N}^s$. Unit vector $\vec{N}^s$ can be strictly determined, but $\vec{e}_1^s$ and $\vec{e}_2^s$ may be chosen in different ways in the plane perpendicular to $\vec{N}^s$. System (2.3.37) with (2.3.40) remains valid for any of these choices of $\vec{e}_1^s$ and $\vec{e}_2^s$. The incident wave may even be elliptically polarized. System (2.3.37) is also valid for different choices of the unit vectors $\vec{e}_1$ and $\vec{e}_2$ for reflected and transmitted S waves.

We can also introduce reflection/transmission coefficients, if we formally consider the S1 and S2 components of the displacement of S waves as independent waves. We introduce the *reflection coefficient R_{mn} $(m, n = 1, 2, 3)$ for displacement* so that m specifies the type of incident wave, and n indicates the type of generated reflected wave. Indices m and n are determined as follows:

$m = 1$, S1 component of the incident S wave (unit vector $\vec{e}_1$),

$m = 2$, S2 component of the incident S wave (unit vector $\vec{e}_2$),

$m = 3$, P incident wave (unit vector $\vec{N}$).

The definition of n is similar but is related to the reflected waves. For example,

$$R_{12} = C^r/B, \qquad R_{11} = B^r/B, \qquad R_{33} = A^r/A.$$

The whole set of nine reflection coefficients R_{mn} forms a 3×3 *matrix of reflection coefficients* $\hat{\mathbf{R}}^r$ *for displacement*. In very much the same way, we also obtain the 3×3 *matrix of transmission coefficients for displacement*. The total number of reflection and transmission coefficients is 18.

The analytical solution of system (2.3.37) for the individual coefficients is cumbersome; the system should be solved numerically. The system, however, can be simplified if we consider particular unit vectors $\vec{e}_1$ and $\vec{e}_2$. We will choose $\vec{e}_2$ perpendicular to the plane of incidence for all waves. Unit vector $\vec{e}_1$ for all waves can then be calculated using the relation $\vec{e}_1 = \vec{e}_2 \times \vec{N}$. In this particular case, it is convenient (but not necessary) to introduce a local Cartesian coordinate system wherein the x_2-axis coincides with $\vec{e}_2$ and the x_3-axis coincides with the normal to the interface. Traditionally, the x_3-axis is vertical so that the interface is horizontal; the S2 component of the S wave is also horizontal. Consequently, it is usual in seismology to call the S2 component of the S wave the SH component (S horizontal), or even the SH wave. Similarly, the S1 component is called the SV component (S vertical), or the SV wave. This terminology is a little confusing for several reasons. First, SH and SV are components of the S wave, not waves. Second, the plane interface may be inclined so that SH is not necessarily horizontal. Third, even if the interface is horizontal, SV is not vertical. Nevertheless, the terminology is very common, so we shall occasionally use it here. We will, however, prefer to speak of S1 and S2 components of S waves.

Using the foregoing specification for $\vec{e}_2$ and of the relevant local Cartesian coordinate system, we obtain

$$N_2 = N_2^r = N_2^t = 0, \qquad e_{12} = e_{12}^r = e_{12}^t = 0,$$
$$n_1 = n_2 = 0, \qquad n_3 = 1, \qquad\qquad\qquad\qquad (2.3.41)$$
$$e_{22} = e_{22}^r = e_{22}^t = 1, \qquad e_{21} = e_{21}^r = e_{21}^t = e_{23} = e_{23}^r = e_{23}^t = 0.$$

System (2.3.37) can then be decomposed into two subsystems. The first contains only four linear equations for A^t, B^t, A^r, and B^r,

$$A^t N_1^t + B^t e_{11}^t - A^r N_1^r - B^r e_{11}^r = D_1,$$
$$A^t N_3^t + B^t e_{13}^t - A^r N_3^r - B^r e_{13}^r = D_3,$$
$$A^t X_1^t + B^t Y_1^t - A^r X_1^r - B^r Y_1^r = E_1, \qquad\qquad (2.3.42)$$
$$A^t X_3^t + B^t Y_3^t - A^r X_3^r - B^r Y_3^r = E_3.$$

Note that D_1, D_3, E_1, and E_3 in (2.3.42) are given by (2.3.39) for the incident P wave and by (2.3.40) for the incident S wave. In (2.3.40), the terms with C vanish due to (2.3.41). The second system consists of only two equations, for C^r and C^t,

$$C^t - C^r = C,$$
$$C^t \rho_2 \beta_2 N_3^t - C^r \rho_1 \beta_1 N_3^r = C \rho_1 \beta_1 N_3. \qquad\qquad (2.3.43)$$

Systems (2.3.42) and (2.3.43) have certain important consequences. Let us consider three possible incident waves:

a. Incident P wave ($A \neq 0$, $B = C = 0$). Then $D_2 = E_2 = 0$, and only P reflected, SV reflected, P transmitted and SV transmitted waves are generated.
b. Incident SV wave ($B \neq 0$, $A = C = 0$). Again $D_2 = E_2 = 0$, and only P reflected, SV reflected, P transmitted and SV transmitted waves are generated.
c. Incident SH wave ($C \neq 0$, $A = B = 0$). Then $D_1 = D_3 = E_1 = E_3 = 0$, and only SH reflected, and SH transmitted waves are generated.

Thus, the SH waves are fully separated from P and SV waves in the process of reflection and transmission of plane waves at a plane interface. For this reason, system (2.3.42) is also called the *P-SV system*, and system (2.3.43) is called the *SH system*.

In this case, the matrices of reflection and transmission coefficients $\hat{\mathbf{R}}^r$ and $\hat{\mathbf{R}}^t$ for displacement are also simpler as $R_{12} = R_{21} = R_{23} = R_{32} = 0$. For $\hat{\mathbf{R}}^r$, we then obtain

$$\hat{\mathbf{R}}^r = \begin{pmatrix} R_{11} & 0 & R_{13} \\ 0 & R_{22} & 0 \\ R_{31} & 0 & R_{33} \end{pmatrix}. \qquad (2.3.44)$$

Here the reflection coefficients have the traditional meaning of displacement reflection coefficients: R_{11}, SV → SV; R_{22}, SH → SH; R_{33}, P → P; R_{13}, SV → P; and R_{31}, P → SV. The literature related to these reflection coefficients (and to alternative transmission coefficients) is quite extensive. The analytical expressions for these R/T coefficients are well-known. They will also be given in Section 5.3, where the most important references can be found. Individual R/T coefficients are also discussed there from a seismological point of view. Note that the total number of R/T coefficients is, in this case, 10 (5 reflection coefficients and 5 transmission coefficients).

Thus, to compute the amplitudes of plane waves, both reflected and transmitted, at a plane interface, we can use one of two procedures:

a. Solving the system of six equations (2.3.37) numerically.
b. Solving decomposed systems (2.3.42) and (2.3.43). For an incident P wave, this possibility is always simpler. For an incident S wave, the situation is not as straightforward. It may also prove very convenient if $\vec{e}_2$ is situated perpendicularly to the plane of incidence. If $\vec{e}_2$ is oriented generally, the application of (2.3.42) and (2.3.43) requires preliminary rotation of the S components about $\vec{N}$.

In Section 5.3, the suitability of system (2.3.37) and of decomposed systems (2.3.42) and (2.3.43) will be discussed from the point of view of the ray method and its applications to complex structures. It will be demonstrated that decomposed systems (2.3.42) and (2.3.43) are more convenient in 1-D and mostly in 2-D media. In general 3-D structures, however, the application of general system (2.3.37) may be more efficient. Section 5.3 also gives analytical expressions for the individual R/T coefficients and some numerical examples.

One question is still open: how to choose $\vec{e}_1$ and $\vec{e}_2$ for reflected and transmitted waves. In general, they may be chosen arbitrarily, but they must be perpendicular to the normal $\vec{N}$ to the wavefront of the wave under consideration. (Vectors $\vec{N}$ of all generated waves are known from Snell's law.) Some choices, however, may be more convenient. We shall describe one such choice briefly, and call it the *standard choice*. We denote by $\vec{e}_1$, $\vec{e}_2$ and $\vec{N}$ the right-handed orthogonal triplet corresponding to the incident wave, by $\vec{e}_1^r$, $\vec{e}_2^r$ and $\vec{N}^r$ the analogous triplet corresponding to the arbitrarily selected reflected/transmitted wave, and by $\vec{i}_2$ the unit vector perpendicular to the plane of incidence. In the standard choice, it is required that the angle between $\vec{e}_2^r$ and $\vec{i}_2$ is the same as the angle between $\vec{e}_2$ and $\vec{i}_2$. The standard choice has certain important consequences. For example, it implies that $\vec{e}_1^t$ and $\vec{e}_2^t$ approach $\vec{e}_1$ and $\vec{e}_2$ if $\alpha_1 \to \alpha_2$ and $\beta_1 \to \beta_2$. Thus, in the limiting case of a vanishing interface, $\vec{e}_1$ and $\vec{e}_2$ are continuous across the interface. The standard choice also implies the reciprocity of determination of $\vec{e}_1^r$ and $\vec{e}_2^r$.

We shall now describe how to determine $\vec{e}_2^r$ and $\vec{e}_2^r$ for an arbitrarily selected reflected wave, corresponding to the standard choice. The results will also be valid for the transmitted waves. We assume that vector $\vec{e}_1$ and $\vec{e}_2$ of the incident wave are known and that $\vec{N}$ and $\vec{N}^r$ are known. We denote $\vec{i}_2$ the unit vector perpendicular to the plane of incidence,

$\vec{i}_2 = (\vec{n} \times \vec{N})/|\vec{n} \times \vec{N}|$. Now we determine auxiliary unit vectors $\vec{e}_1'$, $\vec{e}_2'$ and $\vec{e}_1^{r\prime}$, $\vec{e}_2^{r\prime}$, which specify the planes perpendicular to $\vec{N}$ and $\vec{N}^r$,

$$\vec{e}_2' = \vec{e}_2^{r\prime} = \vec{i}_2, \qquad \vec{e}_1' = \vec{i}_2 \times \vec{N}, \qquad \vec{e}_1^{r\prime} = \vec{i}_2 \times \vec{N}^r.$$

We can put

$$\vec{e}_2 = A\vec{e}_1' + B\vec{e}_2'$$

so that $A = \vec{e}_2 \cdot \vec{e}_1'$ and $B = \vec{e}_2 \cdot \vec{e}_2'$. In the standard choice, $\vec{e}_2^r$ must satisfy the equivalent relation,

$$\vec{e}_2^r = A\vec{e}_1^{r\prime} + B\vec{e}_2^{r\prime}.$$

Thus, the final solution is

$$\vec{e}_2^r = (\vec{e}_2 \cdot \vec{e}_1')\vec{e}_1^{r\prime} + (\vec{e}_2 \cdot \vec{e}_2')\vec{e}_2^{r\prime}, \qquad \vec{e}_1^r = \vec{e}_2^r \times \vec{N}^r. \tag{2.3.45}$$

If unit vector $\vec{e}_2$ of the incident wave equals $\vec{i}_2$, standard choice (2.3.45) simplifies to

$$\vec{e}_2^r = \vec{e}_2, \qquad \vec{e}_1^r = \vec{e}_2^r \times \vec{N}^r. \tag{2.3.46}$$

This is indeed the option we have used to decompose general system (2.3.37) into systems (2.3.42) and (2.3.43). This choice can be interpreted simply: in the plane of incidence, the SV components of all waves (incident, reflected, transmitted) must point to the same side of the relevant slowness vectors. For more details, see Section 5.3.

2.3.3　Anisotropic Elastic Medium

In this section, we shall solve the problem of reflection and transmission of plane elastic waves at a plane interface Σ between two homogeneous *anisotropic* halfspaces. We shall denote the elastic moduli and the density in halfspace 1 by $c_{ijkl}^{(1)}$ and $\rho^{(1)}$ and in the halfspace 2 by $c_{ijkl}^{(2)}$ and $\rho^{(2)}$. We shall also use density-normalized elastic moduli given by relations $a_{ijkl}^{(1)} = c_{ijkl}^{(1)}/\rho^{(1)}$ and $a_{ijkl}^{(2)} = c_{ijkl}^{(2)}/\rho^{(2)}$. We assume that a homogeneous plane wave is incident at interface Σ from halfspace 1 and that it is one of the three possible types: a quasi-compressional (qP) or one of the two quasi-shear waves (qS1 or qS2). The slowness vector $\vec{p}$ of the selected incident wave must satisfy existence condition (2.2.36). The phase velocity $C^{(m)}$ of the incident wave is then given by the relation $C^{(m)} = (G_m(N_i))^{1/2}$, where $G_m(N_i)$ is the eigenvalue of matrix $\bar{\Gamma}_{ik} = a_{ijkl}^{(1)} N_j N_l$. Unit vector $\vec{N}$ again denotes the unit normal to the wavefront, $p_i = N_i/C^{(m)}$. The displacement vector of the incident wave points in the same direction as the relevant eigenvector $\hat{g}^{(m)}$ of matrix $\hat{\Gamma}$, and the group velocity vector is given by relation (2.2.65).

In the following discussion, we shall omit the superscript (m) specifying the selected type of incident wave ($m = 1, 2$, or 3) and tacitly understand that the incident wave may be of any type. Thus, in view of (2.2.42), the displacement vector of the incident wave can be expressed as

$$\vec{u}(\vec{x}, t) = A\vec{g} \exp[-i\omega(t - p_i x_i)]. \tag{2.3.47}$$

Here all the symbols have their standard meanings.

As we know, three different types of plane waves, with different velocities, may propagate in a homogeneous anisotropic medium. Thus, six plane waves will, in general, be generated at interface Σ. Any one of them can be described by a relation similar to (2.3.47),

$$\vec{u}^k(\vec{x}, t) = A^k \vec{g}^k \exp\left[-i\omega\left(t - p_i^k x_i\right)\right], \tag{2.3.48}$$

with $k = 1, 2, \ldots, 6$ (no summation over k). The first three correspond to the first medium (reflected waves; $k = 1, 2, 3$), and the last three correspond to the second medium (transmitted waves; $k = 4, 5, 6$). The unknown quantities in (2.3.48) will be determined from known parameters of the incident wave (2.3.47) and from the boundary conditions of the welded contact.

The boundary conditions also indicate that the arguments of the exponential functions of all waves must be the same along interface Σ. In other words, the tangential components of the slowness vector must be equal for all waves along Σ. This yields

$$\vec{p}^k - \vec{n}(\vec{p}^k \cdot \vec{n}) = \vec{p} - \vec{n}(\vec{p} \cdot \vec{n}); \tag{2.3.49}$$

see (2.3.15). These equations are not yet sufficient to determine $\vec{p}^k$; they only determine the tangential components of $\vec{p}^k$. To determine the normal component of $\vec{p}^k$, we must also use existence condition (2.2.36). A specific procedure of determining the slowness vectors of all six generated waves will be discussed later; now we shall return to the boundary conditions.

The boundary conditions require the displacement and traction to be continuous across Σ. The displacement is straightforward; see (2.3.47) and (2.3.48). For the traction, we shall use (2.3.1). We then obtain the following six equations, representing the boundary conditions at Σ:

$$\begin{aligned} A^1 g_i^1 + A^2 g_i^2 + A^3 g_i^3 - A^4 g_i^4 - A^5 g_i^5 - A^6 g_i^6 &= -A g_i, \\ A^1 X_i^1 + A^2 X_i^2 + A^3 X_i^3 - A^4 X_i^4 - A^5 X_i^5 - A^6 X_i^6 &= -A X_i, \end{aligned} \tag{2.3.50}$$

$i = 1, 2, 3$. Here

$$\begin{aligned} X_i^k &= c_{ijnl}^{(1)} n_j g_n^k p_l^k, \qquad k = 1, 2, 3, \\ X_i^k &= c_{ijnl}^{(2)} n_j g_n^k p_l^k, \qquad k = 4, 5, 6 \end{aligned} \tag{2.3.51}$$

(no summation over k). The right-hand sides of (2.3.50) correspond to the incident wave, and X_i is given by the same expression as X_i^k, $k = 1, 2, 3$. In this case, values g_n^k and p_l^k correspond to the incident wave and, of course, must be properly specified.

System (2.3.50) represents the final system of six linear algebraic equations in six unknown amplitude factors of the generated reflected/transmitted, qP, qS1, and qS2 waves.

System (2.3.50) can be used to determine the reflection/transmission coefficients for displacement A^k/A. There are nine reflection coefficients and nine transmission coefficients. The 3×3 matrices of reflection and transmission coefficients $\hat{\mathbf{R}}^r$ and $\hat{\mathbf{R}}^t$ for displacement now have a clear physical meaning; as qS1 and qS2 represent independent waves, not just two components S1 and S2 of one wave. We remind the reader that S1 and S2 are two components of the S wave in isotropic media.

System (2.3.50) can be decomposed into two independent systems only in exceptional cases of some planes of symmetry. Analytical solutions are also very exceptional and cumbersome. In most cases, the direct numerical solution of (2.3.50) is the simplest and most straightforward.

Before solving (2.3.50), however, we must know the slowness vectors and the eigenvectors of all generated waves. It is considerably more complicated to determine the slowness vectors and eigenvectors of R/T waves in anisotropic media than in isotropic media, and it is also more complicated than the numerical solution of (2.3.50).

Let us now return to the determination of slowness vectors $\vec{p}^r$ and $\vec{p}^t$, corresponding to reflected and transmitted waves. They must satisfy (2.3.49), as well as existence conditions (2.2.36). Geometrically, the existence conditions represent slowness surfaces (2.2.72). They are different for reflected and transmitted waves. For reflected waves, the density-normalized elastic parameters a_{ijkl} in (2.2.72) correspond to $a_{ijkl}^{(1)}$; for transmitted waves, they correspond to $a_{ijkl}^{(2)}$. To simplify the treatment, we assume that the normal $\vec{n}$ to interface Σ is oriented *into halfspace 1*. We shall first consider only reflected waves, with $a_{ijkl} = a_{ijkl}^{(1)}$ in (2.2.72).

a. REFLECTED WAVES

The procedure for determining the slowness vectors of the three reflected waves follows. We express the slowness vector of reflected wave $\vec{p}^r$ in the form $\vec{p}^r = \vec{a} + \sigma\vec{n}$, where $\vec{a}$ is the component of the slowness vector tangent to the interface. This is the same for all three reflected waves, including the incident wave, and can be considered known. Because the normal vector $\vec{n}$ to the interface is also known, the only unknown quantity in the relation $\vec{p}^r = \vec{a} + \sigma\vec{n}$ is σ. We can determine it by inserting $\vec{p} = \vec{a} + \sigma\vec{n}$ into the slowness surface equation for reflected waves (2.2.72), where $a_{ijkl} = a_{ijkl}^{(1)}$. We obtain an algebraic equation of the sixth order in σ. The equation has six roots $\sigma_i, i = 1, 2, \ldots, 6$. Geometrically, the described procedure is equivalent to the computation of intersections of the slowness surface (2.2.72) of reflected waves with a straight line $\vec{p} = \vec{a} + \sigma\vec{n}$. The problem was discussed in Section 2.2.8. Some of the six roots $\sigma_1, \sigma_2, \ldots, \sigma_6$ may be double and are then considered as two coinciding roots. Complex-valued roots may also occur, always appearing as pairs of complex-conjugate roots. Consequently, the number of real-valued roots is always even or zero.

The determination of all six roots $\sigma_1, \sigma_2, \ldots, \sigma_6$ is a standard numerical problem. The final, but more important step consists of selecting the three physical solutions corresponding to the actual reflected waves from the six available roots $\sigma_1, \sigma_2, \ldots, \sigma_6$.

We shall first discuss the *real-valued roots*. In isotropic media, the selection of roots is simple; it is based on the sign of σ_i. The waves with positive σ_i propagate into the halfspace to which the unit normal is pointing, and the waves with negative σ_i propagate to the other halfspace. This division makes it simple to select the correct solutions. In anisotropic media, the selection is more complex. The direction of the slowness vector is different from the direction of the energy flux represented by group velocity vector $\vec{\mathcal{U}}$. The slowness vector of the plane R/T wave being considered may point to one side of the interface, and the relevant group velocity vector may point to the other side of the interface. The rule for selecting proper roots σ_i is based on *the direction of the group velocity vector*, not on the direction of the phase velocity vector. Thus, for reflected waves, we must select such σ_i for which group velocity vector $\vec{\mathcal{U}}$ points into the first medium. The relevant condition is

$$\vec{\mathcal{U}} \cdot \vec{n} > 0. \tag{2.3.52}$$

This condition is often called *the radiation condition of reflected waves for real-valued roots*. There is *always* just one half of the real-valued roots σ_i for which the radiation condition is satisfied; see Section 2.2.8. The other roots σ_i, which do not satisfy (2.3.52), are of no importance in the problem of reflection of plane waves and should be excluded from the list.

If all six roots σ_i are real-valued, radiation condition (2.3.52) solves the problem completely. An even number of roots, however, may be complex-valued (pairs of complex-conjugate roots). Only one physical root of any pair of complex-conjugate roots should be selected. The proper *radiation condition of reflected waves for complex-conjugate roots* can be expressed in two alternative forms:

$$\text{Im}(\vec{p}) \cdot \vec{n} > 0, \qquad \text{Im}(\sigma) > 0. \tag{2.3.53}$$

The plane wave selected by (2.3.53) is inhomogeneous, with amplitudes decreasing exponentially with increasing distance from the interface to the first halfspace. Let us emphasize again that the normal $\vec{n}$ in radiation conditions (2.3.52) and (2.3.53) is oriented into halfspace 1.

The radiation conditions of reflected waves, (2.3.52) and (2.3.53), select three physical roots σ^k, $k = 1, 2, 3$, of the six roots σ_i, $i = 1, 2, \ldots, 6$. Using σ^k, the relevant slowness vectors $\vec{p}^k = \vec{a} + \sigma^k \vec{n}$, $k = 1, 2, 3$, can be determined and used in (2.3.50). Thus, we *always have three reflected waves*, some of them homogeneous and some inhomogeneous. For singular directions, two of these roots may coincide, and the number of reflected waves may reduce to two.

b. TRANSMITTED WAVES

For transmitted waves, the procedure is analogous to that for reflected waves. Slowness surface (2.2.72) with $a_{ijkl} = a_{ijkl}^{(2)}$ should be considered in this case. Six relevant roots $\sigma_1, \sigma_2, \ldots, \sigma_6$ are determined in the same way as for the reflected waves. They are, of course, different from the analogous quantities for reflected waves, because the slowness surfaces of reflected and transmitted waves are different. The selection of the three physical roots $\sigma^4, \sigma^5, \sigma^6$ of roots $\sigma_1, \sigma_2, \ldots, \sigma_6$ is practically the same as for the reflected waves. If the roots are real-valued, the group velocity vectors must point into the second medium. Similarly, if the roots are complex-valued, the amplitudes of the inhomogeneous plane wave must decrease exponentially with increasing distance from the interface into the second halfspace. This property yields the following *radiation conditions of transmitted waves*:

$$\vec{\mathcal{U}} \cdot \vec{n} < 0, \qquad \text{Im}(\vec{p}) \cdot \vec{n} < 0 \quad \text{or} \quad \text{Im}(\sigma) < 0. \tag{2.3.54}$$

The first radiation condition corresponds to the real-valued roots σ; the other two alternative conditions correspond to the complex-valued roots σ. Finally, three transmitted waves are obtained with slowness vectors $\vec{p}^4$, $\vec{p}^5$, and $\vec{p}^6$, given by the relation $\vec{p}^k = \vec{a} + \sigma^k \vec{n}$.

Let us again emphasize that the group velocity vectors of the selected reflected waves always point into the first medium, and the group velocity vectors of the selected transmitted waves point into the second medium (if the roots are real-valued). The slowness vectors of reflected waves, however, may point into the second medium, and the slowness vectors of the transmitted waves may point into the first medium. We remind the reader that the normal $\vec{n}$ in radiation conditions (2.3.54) is oriented into halfspace 1.

For many other details regarding the reflection and transmission of plane elastic waves at a plane interface between two homogeneous anisotropic halfspaces and for numerical examples of R/T coefficients see, for example, Fedorov (1968), Henneke (1972), Rokhlin, Bolland, and Adler (1976), Keith and Crampin (1977), Daley and Hron (1977, 1979), Petrashen (1980), Šílený (1981), Payton (1983), Gajewski and Pšenčík (1987a), Wright (1987), Graebner (1992), Schoenberg and Protázio (1992), Kim, Wrolstad, and Aminzadeh (1993), Blangy (1994), Chapman (1994, in press), Thomson (1996a, 1996b), and Rueger (1997). See also Section 5.4.7.

2.3.4 Transient Plane Waves

In previous sections, the problem of reflection and transmission of plane elastic waves at a plane interface between two homogeneous halfspaces was solved for time-harmonic waves. The generalization of the results for transient plane waves is straightforward.

Let us first discuss the acoustic case. Assume that the incident pressure plane wave is given by the relation

$$p(\vec{x}, t) = P F(t - p_i x_i).   \tag{2.3.55}$$

Here $F(\zeta)$ is the analytical signal, $F(\zeta) = x(\zeta) + \mathrm{i} g(\zeta)$, where $x(\zeta)$ and $g(\zeta)$ form a Hilbert transform pair. Quantities P and p_i have the same meaning as in (2.3.3). Because the problem of reflection/transmission is linear, we can apply the Fourier transform and obtain the following expressions for reflected and transmitted plane waves:

$$p^r(\vec{x}, t) = R^r P F\bigl(t - p_i^r x_i\bigr), \qquad p^t(\vec{x}, t) = R^t P F\bigl(t - p_i^t x_i\bigr).   \tag{2.3.56}$$

Here R^r and R^t are the reflection and transmission coefficients given by (2.3.24) or (2.3.26). Slowness vectors $\vec{p}^r$ and $\vec{p}^t$ can be obtained using (2.3.23) and (2.3.18).

Thus, the analytical signals corresponding to reflected and transmitted acoustic plane waves equal the analytical signal of the incident wave. This very important conclusion follows from the fact that R/T coefficients R^r and R^t are frequency-independent.

For elastic media (both isotropic and anisotropic), the conclusion is the same. The analytical signals corresponding to all reflected and transmitted elastic plane waves equal the analytic signal of the incident wave.

These conclusions, however, do not imply that the shapes of the real-valued signals of R/T waves cannot be different from the shape of the real-valued signal of the incident wave. To study the actual shapes of the real-valued signals, we must pass from the complex-valued representations (2.3.55) and (2.3.56) to real-valued representations; see (2.2.13). We then obtain

$$
\begin{aligned}
p(\vec{x}, t) &= \mathrm{Re}(P)\,\mathrm{Re}(F(t - p_i x_i)) - \mathrm{Im}(P)\,\mathrm{Im}(F(t - p_i x_i)), \\
p^r(\vec{x}, t) &= \mathrm{Re}(R^r P)\,\mathrm{Re}\bigl(F\bigl(t - p_i^r x_i\bigr)\bigr) - \mathrm{Im}(R^r P)\,\mathrm{Im}\bigl(F\bigl(t - p_i^r x_i\bigr)\bigr), \\
p^t(\vec{x}, t) &= \mathrm{Re}(R^t P)\,\mathrm{Re}\bigl(F\bigl(t - p_i^t x_i\bigr)\bigr) - \mathrm{Im}(R^t P)\,\mathrm{Im}\bigl(F\bigl(t - p_i^t x_i\bigr)\bigr).
\end{aligned}
$$

$$\tag{2.3.57}$$

Assume now that the incident plane wave is homogeneous (with real-valued p_i), with a real-valued amplitude P. Then $\mathrm{Im}(P) = 0$, and (2.3.57) yields $p(\vec{x}, t) = P x(t - p_i x_i)$. Thus, the shape of the signal of the incident wave is represented by function $x(t - p_i x_i)$. The shapes of the signals of the reflected and transmitted plane waves will be different from $x(\zeta)$ in the following two cases.

a. If the reflection/transmission coefficient of the generated wave is complex-valued, the shape of the generated wave is then a linear combination of $x(\zeta)$ and its Hilbert transform $g(\zeta)$. This also applies to homogeneous plane waves.

b. If the slowness vector components of the generated wave are complex-valued, the generated wave is then inhomogeneous, and (2.2.83) must be used to determine $\mathrm{Re}(F(\zeta))$ and $\mathrm{Im}(F(\zeta))$.

Equations (2.3.57) are sufficiently general to calculate the shape of the signal of R/T waves even in the case of an inhomogeneous wave with complex-valued R/T coefficients. They also allow incident inhomogeneous plane waves to be considered.

It is simple to show that the acoustic R/T coefficients become complex-valued for the so-called postcritical angle of incidence $i^{inc} > i^*$, where $i^* = \arcsin(c_1/c_2)$; see (2.3.25). For $i^{inc} > i^*$, $\cos i^t$ is imaginary, which implies that the shape of the signals of the R/T waves is different from the shape of the signal of the incident wave for postcritical angles of incidence.

If $i^* < i^{inc} < 90°$, the reflected acoustic plane wave is homogeneous, but the transmitted plane wave is inhomogeneous. In this case, there is a basic difference between the signals of the reflected and transmitted plane waves. The signal of the reflected homogeneous plane wave remains fixed as the wave propagates from the interface, but the signal of the transmitted inhomogeneous plane wave varies; see (2.2.83) with T^I varying.

We shall now consider the reflection and transmission of elastic waves. Several critical angles of incidence correspond to the individual generated waves. All the R/T coefficients become complex-valued when the angle of incidence is larger than the minimum critical angle. Moreover, certain transmitted and reflected waves may be inhomogeneous for postcritical angles of incidence (for example, the converted reflected S → P wave). For more details, refer to Section 5.3.

2.4 High-Frequency Elastic Waves in Smoothly Inhomogeneous Media

In Section 2.2, we studied the propagation of elastic plane waves in homogeneous, isotropic, and anisotropic unbounded media. In all these cases, we determined the types of plane waves that can propagate in such media and their most important characteristics. In the acoustic case, only one type of plane wave of a scalar character may propagate: the pressure wave. In an elastic homogeneous isotropic medium (with $\mu \neq 0$), two types of plane waves of a vectorial character may exist: the compressional (P) and the shear (S) wave. Finally, in an elastic homogeneous anisotropic medium, three types of plane waves of a vectorial character are admissible: one quasi-compressional (qP) and two quasi-shear (qS) waves. We have already shown how to determine the velocities of propagation for these waves. In addition, we have determined the polarization of the plane waves of vectorial character.

In inhomogeneous media, the solution of the elastodynamic equation is considerably more complex. The main problem in solving the elastodynamic equation in inhomogeneous media resides in the fact that the wavefield cannot, in general, be resolved into several independent waves. For example, the elastic wavefield in an *inhomogeneous isotropic* medium cannot be strictly separated into compressional and shear waves.

The Earth's interior, however, is inhomogeneous, and P and S waves have been successfully observed on seismological records. This seems to contradict the previous statement. The explanation follows. Actually, in smoothly inhomogeneous media, the high-frequency elastic waves separate into P and S waves *approximately*. The separated waves, however, do not satisfy the elastodynamic equation exactly, but only approximately. The properties of these high-frequency P and S waves propagating in smoothly inhomogeneous elastic media are locally very similar to the properties of the P and S plane waves propagating in homogeneous media.

The most popular method of studying these high-frequency waves propagating in smoothly inhomogeneous media is the *ray-series method* (also simply called the *ray method*). It is based on the *asymptotic solution of the elastodynamic equation* in the form of a ray series. In the frequency domain, the ray series corresponds to a series in inverse powers of frequency ω. The ray-series method will be discussed in more detail in

Sections 5.6 and 5.7. See also Babich (1956), Babich and Alekseyev (1958), Karal and Keller (1959), Alekseyev and Gel'chinskiy (1959), Alekseyev, Babich, and Gel'chinskiy (1961), Červený and Ravindra (1971), Červený, Molotkov, and Pšenčík (1977), Červený and Hron (1980), and Achenbach, Gautesen, and McMaken (1982), among others. In practical applications in seismology and in seismic prospecting, the first (leading) term of the ray series is primarily the only one used. This approximation is also often called the *zeroth-order approximation of the ray method*. The higher order terms of the ray series have been used only occasionally.

Thus, we are mostly interested in the zeroth-order ray approximation, and not in the higher order terms of the ray series. In this section, we attempt to explain the basic properties and derive the basic equations of the ray method in a very simple way, considering only the zeroth-order terms of the ray series. The derivation is straightforward and easy to understand. The main purpose is to explain the approximate separation of the wavefield into individual waves and to derive the equations for their travel times (eikonal equation), for the amplitudes in the zeroth-order approximation (transport equation), and for the polarization of these waves. All complications connected with the higher order terms of the ray series, the principal and additional components of these terms, and the transport equation of higher order are discussed in Sections 5.6 and 5.7.

The approach used here to derive the approximate high-frequency expressions for seismic body waves propagating in smoothly varying media is based on a simple generalization of the plane-wave approach used in Section 2.2. The solution is again assumed to have the form of (2.2.15), in which vectorial complex-valued amplitude $\vec{U}$ and travel time T are arbitrary slowly varying functions of coordinates, $\vec{U} = \vec{U}(x_i)$ and $T = T(x_i)$. We remind the reader that $\vec{U}$ is a constant vector and $T(x_i)$ is a linear function of coordinates in the case of plane waves. Let us first consider the time-harmonic solution. Inserting it into the elastodynamic equation yields a polynomial in ω, consisting of three terms. Because we are looking for high-frequency solutions, we can only consider the terms with the highest powers of ω. The first term immediately yields the *eikonal equation*, which controls the travel-time function $T(x_i)$, and the second term yields the *transport equation*, which controls the amplitude function. The third term (with the lowest power of ω) is, in fact, neglected.

In the ray-series method, see Sections 5.6 and 5.7, the third term is compensated by the higher order terms of the ray series. The zeroth-order term, however, is not affected at all by neglecting this term. Thus, all the equations we shall derive in this section for travel time $T(x_i)$ and for the zeroth-order term of the ray series $\vec{U}(x_i)$ are also valid in the ray-series method.

Instead of the time-harmonic high-frequency waves, we can also use high-frequency signals. We call the analytical signal $F(t)$ given by (2.2.9) the high-frequency analytical signal, if its Fourier spectrum $F(\omega)$ effectively vanishes for low frequencies,

$$|F(\omega)| = 0, \qquad \text{for } 0 \le \omega \le \omega_0, \tag{2.4.1}$$

where ω_0 is high. Then

$$|\ddot{F}(t)| \gg |\dot{F}(t)| \gg |F(t)|. \tag{2.4.2}$$

Thus, we can neglect the term that contains $|F(t)|$ with respect to the terms containing $|\dot{F}(t)|$ and $|\ddot{F}(t)|$. This is physically simple to understand.

The condition that the frequency should be high has only a very qualitative meaning. Without going into details, we can roughly explain this condition in terms of a (prevailing) wavelength. Wavelength λ can be expressed by the equation $\lambda = 2\pi c/\omega$, where ω is the (prevailing) frequency and c is the average propagation velocity at a given point. The approximate high-frequency solution of the elastodynamic equation in a smooth medium requires that the *appropriate material parameters of the medium* (ρ, c, λ, μ, c_{ijkl}) *not vary greatly over distances of the order of wavelength* λ. We also require that the slowness vector and the amplitude vector of the wave under consideration not vary greatly over the same distance of λ. The validity conditions of the ray method will be discussed in Section 5.9.

2.4.1 Acoustic Wave Equation

It is simple to derive the approximate HF solutions of the acoustic wave equation. Let us first consider an acoustic wave equation (2.1.25) for pressure p, with a variable velocity $c = c(x_i)$, a constant density, and no source term,

$$\nabla^2 p = \frac{1}{c^2(x_i)}\ddot{p}. \tag{2.4.3}$$

We shall try to find an approximate time-harmonic high-frequency solution of this equation in the following form:

$$p(x_i, t) = P(x_i)\exp[-i\omega(t - T(x_i))]. \tag{2.4.4}$$

We assume that frequency ω is high, $\omega \gg 0$. Both $P(x_i)$ and $T(x_i)$ are presumably smooth scalar functions of coordinates. For $P(x_i) = $ const. and $T(x_i) = p_i x_i$, Equation (2.4.4) represents a plane-wave solution. Equation $t = T(x_i)$ represents the moving wavefront of the wave under consideration, which is, in general, curved.

If we take into account the vectorial identity $\nabla \cdot a\vec{b} = \vec{b} \cdot \nabla a + a\nabla \cdot \vec{b}$, we obtain

$$\begin{aligned}\nabla^2 p &= \nabla \cdot \nabla p \\ &= \{i\omega(\nabla P + i\omega P\nabla T) \cdot \nabla T \\ &\quad + (\nabla^2 P + i\omega\nabla T \cdot \nabla P + i\omega P\nabla^2 T)\}\exp[-i\omega(t - T(x_i)].\end{aligned}$$

We can then express acoustic equation (2.4.3) in the following form:

$$-\omega^2 P[(\nabla T)^2 - 1/c^2] + i\omega[2\nabla P \cdot \nabla T + P\nabla^2 T] + \nabla^2 P = 0. \tag{2.4.5}$$

Because Equation (2.4.5) should be satisfied for any frequency ω, the expressions with ω^2, ω^1, and ω^0 must vanish. We have, however, three expressions and only two unknown functions, $T(x_i)$ and $P(x_i)$, which presumably do not depend on ω. Thus, (2.4.5) cannot be satisfied exactly. For high frequencies ω, however, the most important terms will be the first (with ω^2) and the second (with ω). Because we are interested in the HF solutions of (2.4.5), we shall require that the two first terms in (2.4.5) vanish. We then arrive at the *eikonal equation*,

$$(\nabla T)^2 = 1/c^2, \tag{2.4.6}$$

and the *transport equation*,

$$2\nabla P \cdot \nabla T + P\nabla^2 T = 0. \tag{2.4.7}$$

Both equations play a fundamental role in the ray method. The eikonal equation represents a nonlinear partial differential equation of the first order for travel time $T(x_i)$. It has usually been solved by *ray tracing* and by subsequent computation of T along the rays;

see Chapter 3. The transport equation represents a linear partial differential equation of the first order in $P(x_i)$. It simplifies very much if it is solved along the rays. It then reduces to an ordinary differential equation of the first order for $P(x_i)$ and can be solved analytically in terms of the ray Jacobian. See Section 3.10.

But what about term $\nabla^2 P$ in (2.4.5), which has not been considered in our treatment? If we assume the solution in the form of (2.4.4), we can do nothing. Solution (2.4.4) can, however, be generalized. The amplitude may be considered as a series in inverse powers of frequency,

$$P(x_i, \omega) = P_0(x_i) + \frac{1}{i\omega} P_1(x_i) + \frac{1}{(i\omega)^2} P_2(x_i) + \cdots \qquad (2.4.8)$$

Inserting (2.4.8) into (2.4.4), we obtain the *ray-series solutions*. These solutions will be discussed in more detail in Section 5.6, where a system of transport equations of higher order for $P_1(x_i), P_2(x_i), \ldots$, will be derived. The system can be used to compute successively $P_0(x_i), P_1(x_i), P_2(x_i), \ldots$. The leading term $P_0(x_i)$, of course, satisfies transport equation (2.4.7).

The ray-series solution (2.4.8) eliminates the last term in (2.4.5) quite strictly. Several other attempts to eliminate the last term of (2.4.5) have also been reported in the literature. In all these cases, $\nabla^2 P$ is considered to be a correction, a small term. We shall briefly mention three such attempts.

1. It is possible to consider the last term of (2.4.5) as a source term in the acoustic wave equation and take it into account in some sort of *generalized Born approximation*, similar to perturbation methods. The regions of high $\nabla^2 P$ then formally represent nonphysical sources of the scattered wavefield. See Section 2.6.2 for more details.
2. It is possible to combine the last term of (2.4.5) with its second term and construct a *frequency-dependent transport equation*. This transport equation can be solved by iterations, keeping $\nabla^2 P$ on the RHS.
3. Finally, it is possible to combine the last term with the first (eikonal) term. This yields a *frequency-dependent eikonal* (sometimes called the hypereikonal) and *frequency-dependent rays*. See Biondi (1992) and Zhu and Chun (1994b).

The approximate HF equations are simple even if the acoustic wave equation contains a variable density, ρ,

$$\nabla \cdot \frac{1}{\rho} \nabla p = \frac{1}{\rho c^2} \ddot{p}. \qquad (2.4.9)$$

Using the ansatz solution (2.4.4), we obtain an alternative to (2.4.5):

$$-\omega^2 P \left[(\nabla T)^2 - \frac{1}{c^2} \right] + i\omega \left[2\nabla P \cdot \nabla T + P\nabla^2 T - \left(\frac{P}{\rho} \right) \nabla T \cdot \nabla \rho \right]$$
$$+ \rho \nabla \cdot \frac{1}{\rho} \nabla P = 0.$$

Thus, the eikonal equation is again obtained in the form of (2.4.6). The transport equation is, however, slightly different:

$$2\nabla P \cdot \nabla T + P\nabla^2 T - (P/\rho)\nabla T \cdot \nabla \rho = 0. \qquad (2.4.10)$$

We can give it the same form as (2.4.7), if we use $P/\sqrt{\rho}$ instead of P:

$$2\nabla T \cdot \nabla(P/\sqrt{\rho}) + (P/\sqrt{\rho})\nabla^2 T = 0. \qquad (2.4.11)$$

Thus, the difference is only formal. The variable density does not complicate the approximate solutions at all. Consequently, in the following text we shall primarily consider the acoustic wave equation with the variable density.

We have discussed the time-harmonic solution, (2.4.4). It is not difficult to work directly with the analytical signals. Instead of (2.4.4), we shall express the solution in the following form:

$$p(x_i, t) = P(x_i)F(t - T(x_i)),$$

where $F(\zeta)$ is a high-frequency analytic signal. Equation (2.4.5) now reads

$$P[(\nabla T)^2 - 1/c^2]\ddot{F}(\zeta) - [2\nabla P \cdot \nabla T + P\nabla^2 T]\dot{F}(\zeta) + \nabla^2 P F(\zeta) = 0,$$

with $\zeta = t - T(x_i)$. Because the equation must be satisfied identically for any ζ, it again yields eikonal equation (2.4.6) and transport equation (2.4.7).

Transport equations (2.4.7) and (2.4.11) are sometimes expressed in different alternative forms; instead of P, we use PP^*, where P^* denotes the complex conjugate of P. For real-valued travel times T, transport equation (2.4.7) is also valid for P^*:

$$2\nabla P^* \cdot \nabla T + P^*\nabla^2 T = 0.$$

Multiplying (2.4.7) by P^*, multiplying the foregoing equation by P, and adding the two products yields

$$\nabla \cdot (PP^*\nabla T) = 0. \tag{2.4.12}$$

Similarly, for a variable density, (2.4.11) can be expressed as

$$\nabla \cdot \left(\frac{PP^*}{\rho}\nabla T\right) = 0. \tag{2.4.13}$$

Both equations (2.4.12) and (2.4.13), of course, also remain valid if P is real-valued; in this case, $PP^* = P^2$.

2.4.2 Elastodynamic Equation for Isotropic Inhomogeneous Media

In principle, the derivation of the basic equations for high-frequency seismic body waves propagating in isotropic smoothly inhomogeneous media remains the same as in the acoustic case. There is only one important difference: the elastodynamic equation is vectorial, whereas the acoustic wave equation for pressure is scalar. We have two options in treating the elastodynamic equation: in vectorial form, see (2.1.21), or in component form, see (2.1.20). Here we shall use the component form to be consistent with the anisotropic case, where the component form is more convenient. As in the acoustic case, we shall not consider the source term.

Our ansatz solution for the displacement vector $\vec{u}(x_j, t)$ will read

$$u_i(x_j, t) = U_i(x_j)F(t - T(x_j)), \tag{2.4.14}$$

where $F(\zeta)$ represents a high-frequency analytical signal, $\zeta = t - T(x_j)$. Solution (2.4.14) again represents a generalization of the plane-wave solution, with U_i and T varying arbitrarily (but slowly) with the coordinates. Before we insert (2.4.14) into elastodynamic equation (2.1.20), we shall compute several derivatives:

$$u_{i,j} = U_{i,j}F - U_i T_{,j}\dot{F},$$
$$u_{i,jm} = U_{i,jm}F - U_{i,j}T_{,m}\dot{F} - U_{i,m}T_{,j}\dot{F} - U_i T_{,jm}\dot{F} + U_i T_{,j}T_{,m}\ddot{F},$$
$$\ddot{u}_i = U_i \ddot{F}.$$

Inserting these expressions into elastodynamic equation (2.1.20) yields

$$N_i(\vec{U})\ddot{F} - M_i(\vec{U})\dot{F} + L_i(\vec{U})F = 0, \tag{2.4.15}$$

where

$$N_i(\vec{U}) = -\rho U_i + (\lambda + \mu)U_j T_{,i}T_{,j} + \mu U_i T_{,j}T_{,j},$$
$$M_i(\vec{U}) = (\lambda + \mu)[U_{j,i}T_{,j} + U_{j,j}T_{,i} + U_j T_{,ij}] + \mu[2U_{i,j}T_{,j} + U_i T_{,jj}]$$
$$\qquad + \lambda_{,i}U_j T_{,j} + \mu_{,j}U_i T_{,j} + \mu_{,j}U_j T_{,i},$$
$$L_i(\vec{U}) = (\lambda + \mu)U_{j,ij} + \mu U_{i,jj} + \lambda_{,i}U_{j,j} + \mu_{,j}(U_{i,j} + U_{j,i}). \tag{2.4.16}$$

Equations (2.4.16) look rather cumbersome, but their derivation is easy and straightforward. They will also simplify considerably later on.

We shall again try to satisfy (2.4.15) approximately, for high-frequency wavefields. As in the acoustic case, we put

$$N_i(\vec{U}) = 0 \tag{2.4.17}$$

and

$$M_i(\vec{U}) = 0. \tag{2.4.18}$$

Equation (2.4.17) will yield the approximate separation of the high-frequency wavefield into two wavefields corresponding to P and S waves. It will also yield the eikonal equations corresponding to both waves. Equation (2.4.18) may then be used to determine the transport equations for the amplitudes of the P and S waves.

Now we shall discuss Equation (2.4.17) in more detail. It can be altered to read

$$(\Gamma_{ij} - \delta_{ij})U_j = 0, \qquad i = 1, 2, 3, \tag{2.4.19}$$

where

$$\Gamma_{ij} = \frac{\lambda + \mu}{\rho}T_{,i}T_{,j} + \frac{\mu}{\rho}\delta_{ij}T_{,k}T_{,k}. \tag{2.4.20}$$

As we can see, (2.4.19) represents a system of three linear algebraic equations in U_1, U_2, and U_3. It coincides formally with plane-wave equations (2.2.35) and (2.2.43), if we put

$$\vec{p} = \nabla T. \tag{2.4.21}$$

We shall again call $\vec{p}$ the *slowness vector*. The slowness vector, however, is not constant now (as it was in the case of plane waves) but is rather a function of position. Similarly, ρ, λ, and μ are also functions of coordinates. Nevertheless, Equation (2.4.19) can be discussed in exactly the same way as for the plane waves, realizing that (2.4.19) represents a typical eigenvalue problem. We remind the reader that matrix Γ_{ij} has three eigenvalues in isotropic media:

$$G_1(x_i, p_i) = G_2(x_i, p_i) = \beta^2(x_i)p_i p_i, \qquad G_3(x_i, p_i) = \alpha^2(x_i)p_i p_i. \tag{2.4.22}$$

Thus, two eigenvalues coincide. Here $\alpha(x_i)$ and $\beta(x_i)$ are given by the relations

$$\alpha(x_i) = \left[\frac{\lambda(x_i) + 2\mu(x_i)}{\rho(x_i)}\right]^{1/2}, \qquad \beta(x_i) = \left[\frac{\mu(x_i)}{\rho(x_i)}\right]^{1/2}. \qquad (2.4.23)$$

The definition of $\alpha(x_i)$ and $\beta(x_i)$ is formally the same as in the case of plane waves; however, α and β are not constant but rather *functions of coordinates*.

The eigenvectors of Γ_{ij} ($\vec{g}^{(1)}$, $\vec{g}^{(2)}$, and $\vec{g}^{(3)}$) have the following directions: eigenvector $\vec{g}^{(3)}$ corresponding to eigenvalue $G_3 = \alpha^2 p_i p_i$ has the same direction as the slowness vector $\vec{p} = \nabla T$. Eigenvectors $\vec{g}^{(1)}$ and $\vec{g}^{(2)}$, corresponding to the two coinciding eigenvalues $G_1 = G_2 = \beta^2 p_i p_i$, are mutually perpendicular unit vectors, also perpendicular to $\vec{p} = \nabla T$. Their direction in the plane perpendicular to $\vec{p}$, however, cannot be determined uniquely from (2.4.19).

We can draw the following conclusions. Two different types of high-frequency seismic body waves can propagate in a smoothly inhomogeneous isotropic elastic medium.

1. P waves. The travel-time field of the P waves satisfies equation $G_3(x_i, p_i) = 1$ so that

$$\nabla T \cdot \nabla T = 1/\alpha^2(x_i). \qquad (2.4.24)$$

This is the eikonal equation for P waves; $\alpha(x_i)$ is given by (2.4.23). The P wave is linearly polarized and the particle motion has the same direction as $\vec{p} = \nabla T$. The slowness vector can also be expressed as

$$\vec{p} = \nabla T = \vec{N}/\alpha, \qquad (2.4.25)$$

where $\vec{N}$ is the unit vector perpendicular to wavefront $T(x_i) = $ const. This means that the displacement vector is polarized along $\vec{N}$,

$$\vec{U}(x_i) = A(x_i)\vec{N}, \qquad (2.4.26)$$

where $A(x_i)$ is a scalar, complex-valued amplitude function of the P waves.

2. S waves. The travel-time field of the S wave satisfies equation $G_1(x_i, p_i) = G_2(x_i, p_i) = 1$, so that

$$\nabla T \cdot \nabla T = 1/\beta^2(x_i). \qquad (2.4.27)$$

This is the eikonal equation for shear waves; $\beta(x_i)$ is given by (2.4.23). The displacement vector is polarized in the plane perpendicular to $\vec{N}$ (that is, in the plane tangent to the wavefront). As in Section 2.2.6, we introduce two mutually perpendicular unit vectors $\vec{e}_1$ and $\vec{e}_2$, which are also perpendicular to $\vec{N}$. Then,

$$\vec{U}(x_i) = B(x_i)\vec{e}_1 + C(x_i)\vec{e}_2, \qquad (2.4.28)$$

where $B(x_i)$ and $C(x_i)$ are scalar, complex-valued amplitude functions of the S waves. The slowness vector of the S wave is given by the relation

$$\vec{p} = \nabla T = \vec{N}/\beta, \qquad (2.4.29)$$

where $\vec{N}$ is the unit vector perpendicular to wavefront $T(x_i) = $ const.

A note concerning the terminology. In inhomogeneous media, the P waves are not purely compressional, longitudinal, or irrotational. Similarly, the S waves are not purely shear, transverse, or equivoluminal. To avoid possible misunderstanding, we shall not use these terms but shall call them P and S waves systematically throughout the text. This terminology corresponds to the classical seismological meaning: P (primae), S (secundae).

In expressions (2.4.26) and (2.4.28), the amplitude functions $A(x_i)$, $B(x_i)$, and $C(x_i)$ are still undetermined. To determine them, we can use (2.4.18). We expect to obtain the *transport equations* for A, B, and C.

Let us again start with P waves. We multiply $M_i(\vec{U})$ by $p_i = T_{,i}$ and use the relations

$$U_j = A\alpha T_{,j}, \qquad U_{j,i} = A_{,i}\alpha T_{,j} + A(\alpha_{,i} T_{,j} + \alpha T_{,ji}).$$

Then

$$M_i(A\vec{N})T_{,i} = (\lambda + 2\mu)[2\alpha^{-1} A_{,i} T_{,i} + A(2\alpha^{-2}\alpha_{,i} T_{,i}$$
$$+ 2\alpha T_{,ji} T_{,j} T_{,i} + \alpha^{-1} T_{,ii})] + \alpha^{-1} A(\lambda + 2\mu)_{,i} T_{,i}.$$

We shall now calculate expressions $T_{,ji} T_{,j} T_{,i}$,

$$T_{,ji} T_{,j} T_{,i} = \tfrac{1}{2}(T_{,j} T_{,j})_{,i} T_{,i} = -\alpha^{-3}\alpha_{,i} T_{,i}$$

and obtain

$$M_i(A\vec{N})T_{,i} = 2\alpha\rho A_{,i} T_{,i} + \alpha\rho A T_{,ii} + \alpha^{-1}(\alpha^2\rho)_{,i} T_{,i} A.$$

Equation (2.4.18) implies that this expression must be zero. This yields the *first form of the transport equation* for P waves, in component form:

$$2\rho\alpha A_{,i} T_{,i} + \alpha\rho A T_{,ii} + \alpha^{-1}(\alpha^2\rho)_{,i} T_{,i} A = 0. \tag{2.4.30}$$

In vector form, (2.4.30) reads

$$2\rho\alpha\nabla A \cdot \nabla T + A(\alpha\rho\nabla^2 T + \alpha^{-1}\nabla T \cdot \nabla(\alpha^2\rho)) = 0. \tag{2.4.31}$$

The transport equation can be simplified if we consider $\sqrt{\rho\alpha^2} A$ instead of A:

$$2\nabla T \cdot \nabla(\sqrt{\rho\alpha^2} A) + \sqrt{\rho\alpha^2} A\nabla^2 T = 0. \tag{2.4.32}$$

This is the *final form of the transport equation* for P waves in inhomogeneous elastic media. It has exactly the same form as the transport equation for acoustic waves if we use $\sqrt{\rho\alpha^2} A(x_i) = \sqrt{\lambda + 2\mu} A(x_i)$ instead of $P(x_i)$; see (2.4.7). Alternatively, it can be expressed in the following form:

$$\nabla \cdot (\rho\alpha^2 A A^* \nabla T) = 0. \tag{2.4.33}$$

It may seem surprising that the transport equation for pressure waves in the acoustic case (2.4.11) and the transport equation for the P waves in elastic isotropic media are formally different; the first of them is for $P/\sqrt{\rho}$, and the second is for $\sqrt{\rho\alpha^2} A$. The explanation is simple. The approximate high-frequency relation between the amplitude of pressure wave P and the amplitude of the displacement vector of P wave A is as follows: $P = \rho\alpha A$ (with $\alpha = c$). Thus, $P/\sqrt{\rho} = \sqrt{\rho\alpha^2} A$, and both the transport equations (2.4.32) and (2.4.11) coincide.

The transport equations for S waves can be derived in a way similar to that for P waves. We insert expression (2.4.28) into Equation (2.4.18) and take the scalar products of (2.4.18) with $\vec{e}_1$ and $\vec{e}_2$:

$$M_i e_{1i} = [(\lambda + \mu)e_{1i} T_{,j} + 2\mu T_{,i} e_{1j}]U_{j,i}$$
$$+ [(\lambda + \mu)T_{,ij} e_{1i} + \mu e_{1j} T_{,kk} + \lambda_{,i} T_{,j} e_{1i} + \mu_{,k} T_{,k} e_{1j}]U_j = 0,$$

$$M_i e_{2i} = [(\lambda + \mu)e_{2i}T_{,j} + 2\mu T_{,i}e_{2j}]U_{j,i}$$
$$+ [(\lambda + \mu)T_{,ij}e_{2i} + \mu e_{2j}T_{,kk} + \lambda_{,i}T_{,j}e_{2i} + \mu_{,k}T_{,k}e_{2j}]U_j = 0.$$

Taking into account that

$$U_j = Be_{1j} + Ce_{2j}, \qquad U_{j,i} = B_{,i}e_{1j} + C_{,i}e_{2j} + Be_{1j,i} + Ce_{2j,i},$$

we obtain

$$2\mu T_{,i}B_{,i} + B(\mu_{,k}T_{,k} + \mu T_{,kk}) + 2C\mu T_{,i}e_{1j}e_{2j,i} = 0,$$
$$2\mu T_{,i}C_{,i} + C(\mu_{,k}T_{,k} + \mu T_{,kk}) + 2B\mu T_{,i}e_{2j}e_{1j,i} = 0. \qquad (2.4.34)$$

These equations represent two transport equations for S waves, particularly for the amplitude functions $B(x_i)$ and $C(x_i)$, expressed in component form. They are valid for arbitrarily chosen unit vectors $\vec{e}_1$ and $\vec{e}_2$, perpendicular to $\vec{N}$. Of course, $\vec{e}_1$ and $\vec{e}_2$ must also be mutually perpendicular. Transport equations (2.4.34) are not as simple as the transport equations of P waves because they are mutually coupled. The coupling is caused by the terms containing $T_{,i}e_{1j}e_{2j,i}$ and $T_{,i}e_{2j}e_{1j,i}$.

In the seismic ray method, unit vectors $\vec{e}_1$ and $\vec{e}_2$ are traditionally assumed to coincide with unit normal $\vec{n}$ and unit binormal $\vec{b}$ to the ray. The transport equations (2.4.34) for this choice of $\vec{e}_1$ and $\vec{e}_2$ can be found in Červený and Ravindra (1971) and Červený, Molotkov, and Pšenčík (1977). This option, however, does not remove the coupling of transport equations for $B(x_i)$, $C(x_i)$, and, consequently, the coupling of both S wave components. Transport equation (2.4.34) can be decoupled and considerably simplified if we choose $\vec{e}_1$ and $\vec{e}_2$ so that

$$T_{,i}e_{1j}e_{2j,i} = 0, \qquad T_{,i}e_{2j}e_{1j,i} = 0. \qquad (2.4.35)$$

It is simple to see that Equations (2.4.35) are satisfied if vectors $T_{,i}e_{1j,i}$ and $T_{,i}e_{2j,i}$ have only one nonvanishing component, oriented in the direction of the slowness vector,

$$T_{,i}e_{1j,i} = aT_{,j}, \qquad T_{,i}e_{2j,i} = bT_{,j}. \qquad (2.4.36)$$

Here a and b are some quantities that should be chosen to guarantee that $\vec{e}_1$ and $\vec{e}_2$ are unit vectors.

It is not difficult to compute $\vec{e}_1$ and $\vec{e}_2$ such that they satisfy (2.4.35) along the ray. We shall discuss the relevant algorithms in Section 4.1.

If $\vec{e}_1$ and $\vec{e}_2$ are selected to satisfy (2.4.35), transport equations (2.4.34) simplify considerably. In component form,

$$2\mu T_{,i}B_{,i} + B(\mu_{,k}T_{,k} + \mu T_{,kk}) = 0,$$
$$2\mu T_{,i}C_{,i} + C(\mu_{,k}T_{,k} + \mu T_{,kk}) = 0. \qquad (2.4.37)$$

In vector form,

$$2\nabla T \cdot \nabla(\sqrt{\rho\beta^2}B) + (\sqrt{\rho\beta^2}B)\nabla^2 T = 0,$$
$$2\nabla T \cdot \nabla(\sqrt{\rho\beta^2}C) + (\sqrt{\rho\beta^2}C)\nabla^2 T = 0. \qquad (2.4.38)$$

As we can see, transport equations (2.4.37) and (2.4.38) for B and C are fully decoupled. The first transport equation for B does not contain C, and vice versa. For this reason, we usually call unit vectors $\vec{e}_1$ and $\vec{e}_2$, satisfying decoupling conditions (2.4.35), the *polarization vectors of* S *waves*.

Transport equation (2.4.38) for S waves can be expressed in an alternative form in a manner similar to that for acoustic waves and elastic P waves:

$$\nabla \cdot (\rho\beta^2 B B^* \nabla T) = 0, \qquad \nabla \cdot (\rho\beta^2 C C^* \nabla T) = 0. \tag{2.4.39}$$

We can also express the transport equation for $B B^* + C C^*$ as

$$\nabla \cdot (\rho\beta^2 (B B^* + C C^*) \nabla T) = 0. \tag{2.4.40}$$

The transport equations (2.4.38) for both components of the high-frequency S waves propagating in a smooth inhomogeneous isotropic medium again have exactly the same form as transport equation (2.4.11) for the acoustic case and as transport equation (2.4.32) for P waves. They are not expressed directly for quantities B and C but for $\sqrt{\rho\beta^2} B = \sqrt{\mu} B$ and for $\sqrt{\rho\beta^2} C = \sqrt{\mu} C$.

2.4.3   Elastodynamic Equation for Anisotropic Inhomogeneous Media

We shall seek approximate high-frequency solutions of the elastodynamic equation (2.1.17) with $f_i = 0$ in an anisotropic, smoothly inhomogeneous medium. We again assume the vectorial solution in the form (2.4.14). Inserting (2.4.14) into elastodynamic equation (2.1.17) again yields (2.4.15), where N_i, M_i, and L_i are given by the relations

$$\begin{aligned}
N_i(\vec{U}) &= c_{ijkl} T_{,l} T_{,j} U_k - \rho U_i, \\
M_i(\vec{U}) &= c_{ijkl} T_{,j} U_{k,l} + (c_{ijkl} T_{,l} U_k)_{,j}, \\
L_i(\vec{U}) &= (c_{ijkl} U_{k,l})_{,j}.
\end{aligned} \tag{2.4.41}$$

For high-frequency wavefields propagating in a smoothly varying medium, we obtain an approximate solution of (2.4.15) if we put $N_i(\vec{U}) = 0$ and $M_i(\vec{U}) = 0$. We shall now discuss these equations.

The equations $N_i(\vec{U}) = 0$ can be expressed in the following form:

$$(\Gamma_{ik} - \delta_{ik}) U_k = 0, \qquad i = 1, 2, 3, \tag{2.4.42}$$

where

$$\Gamma_{ik} = \frac{c_{ijkl}}{\rho} T_{,j} T_{,l}. \tag{2.4.43}$$

In this case all quantities c_{ijkl}, ρ, and $T_{,i}$ are functions of coordinates. We can again put $\vec{p} = \nabla T$. Equations (2.4.42) then take exactly the same form as in the case of plane waves in homogeneous anisotropic media (see (2.2.35)) and can be solved in the same way: using the eigenvalues and eigenvectors of the Christoffel matrix Γ_{ik}.

The Christoffel matrix Γ_{ik} has, in general, three eigenvalues $G_1(x_i, p_i)$, $G_2(x_i, p_i)$, $G_3(x_i, p_i)$ and three relevant eigenvectors $\vec{g}^{(1)}(x_i, p_i)$, $\vec{g}^{(2)}(x_i, p_i)$, and $\vec{g}^{(3)}(x_i, p_i)$. In the same way as for plane waves, we can formulate the following conclusions.

Three seismic body waves (qS1, qS2, and qP) may propagate in a smoothly inhomogeneous anisotropic medium in a specified direction. They correspond to the three eigenvalues $G_m(x_i, p_i)$, $m = 1, 2, 3$, and to the relevant three eigenvectors $\vec{g}^{(1)}$, $\vec{g}^{(2)}$, and $\vec{g}^{(3)}$. Let us consider one of them. The travel-time field of the selected wave satisfies the nonlinear partial differential equation of the first order

$$G_m(x_i, p_i) = 1, \qquad p_i = T_{,i}. \tag{2.4.44}$$

We again call (2.4.44) the *eikonal equation*. The relevant wave with travel time $T(x_i)$ satisfying (2.4.44) is then linearly polarized:

$$\vec{U}(x_j) = A(x_j)\vec{g}^{(m)}(x_j). \tag{2.4.45}$$

Here $\vec{g}^{(m)}$ is the eigenvector corresponding to eigenvalue G_m, and $A(x_j)$ is a scalar, complex-valued, amplitude function. For simplicity, we shall omit superscript m.

We shall now derive the transport equation for $A(x_j)$. We insert (2.4.45) into M_i given by (2.4.41) and multiply it by g_i:

$$
\begin{aligned}
M_i(A\vec{g})g_i &= c_{ijkl}p_j g_i (Ag_k)_{,l} + (c_{ijkl}p_l g_k A)_{,j}g_i \\
&= c_{ijkl}p_j g_i g_k A_{,l} + c_{ijkl}p_j g_i Ag_{k,l} \\
&\quad + (c_{ijkl}Ap_l g_k g_i)_{,j} - c_{ijkl}p_l Ag_k g_{i,j} \\
&= c_{ijkl}p_j g_i g_k A_{,l} + (c_{ijkl}p_l Ag_i g_k)_{,j} \\
&= c_{ijkl}p_j g_i g_k A_{,l} + c_{ijkl}p_l g_k g_i A_{,j} + A(c_{ijkl}p_l g_k g_i)_{,j} \\
&= 2c_{ijkl}p_j g_i g_k A_{,l} + A(c_{ijkl}p_l g_k g_i)_{,j}.
\end{aligned}
$$

Let us now introduce vector $\vec{\mathcal{U}}$ with components

$$\mathcal{U}_i = \rho^{-1}c_{ijkl}p_l g_k g_j. \tag{2.4.46}$$

As we have shown in Sections 2.2 and 2.3, this vector has the meaning of a group velocity vector in the case of plane waves propagating in homogeneous anisotropic media. In the next section, we shall show that it also has the same meaning in the case of high-frequency seismic body waves propagating in smoothly varying anisotropic media. Using this notation, we can express the equation $M_i(A\vec{g})g_i = 0$ in the following form:

$$2\rho\mathcal{U}_i A_{,i} + A(\rho\mathcal{U}_i)_{,i} = 0. \tag{2.4.47}$$

This is one of the possible forms of the *transport equation for an inhomogeneous anisotropic medium*. It can be expressed in many alternative forms. Its vector form is simple:

$$2\rho\vec{\mathcal{U}} \cdot \nabla A + A\nabla \cdot (\rho\vec{\mathcal{U}}) = 0. \tag{2.4.48}$$

Some simple algebra involving ρ yields

$$2\vec{\mathcal{U}} \cdot \nabla(\sqrt{\rho}A) + (\sqrt{\rho}A)\nabla \cdot \vec{\mathcal{U}} = 0. \tag{2.4.49}$$

This is the final form of the transport equation for an anisotropic smoothly inhomogeneous medium that we shall use in the following text. The sought function in (2.4.49) is $\sqrt{\rho}A$, not A. An alternative form of transport equation (2.4.49) is

$$\nabla \cdot (\rho AA^*\vec{\mathcal{U}}) = 0. \tag{2.4.50}$$

Because the anisotropic medium includes, as a special case, the isotropic medium, transport equation (2.4.49) must also include the transport equation for waves propagating in isotropic inhomogeneous media; see (2.4.32) and (2.4.38). We shall prove that this is true for P waves. For P waves in isotropic models, $\vec{\mathcal{U}} = \alpha^2\nabla T$. Inserting this relation into (2.4.49) yields

$$2\alpha^2\nabla T \cdot \nabla(\sqrt{\rho}A) + (\sqrt{\rho}A)\nabla \cdot (\alpha^2\nabla T) = 0.$$

It is not difficult to prove that this equation is fully equivalent to (2.4.32). The proof for S waves is analogous. It is, of course, required that unit vectors $\vec{e}_1$ and $\vec{e}_2$ satisfy relations (2.4.35).

In Section 3.10.6, we shall show that all types of transport equations can be solved simply along rays. The transport equation then takes the form of an ordinary differential equation of the first order.

If we multiply (2.4.46) by p_i and use (2.2.34) and (2.4.44), we obtain an important relation:

$$p_i \mathcal{U}_i = 1. \tag{2.4.51}$$

The same relation was derived earlier for plane waves propagating in a homogeneous anisotropic medium; see (2.2.67). Thus, the relation is valid more generally, even for high-frequency seismic body waves propagating in anisotropic, smoothly inhomogeneous media. The quantities p_i and $\mathcal{U}_i$ in (2.4.51) are now functions of position.

For more details on the ray-series method in inhomogeneous anisotropic media, see Section 5.7. See also Babich (1961a), Červený (1972), Červený, Molotkov, and Pšenčík (1977), and Gajewski and Pšenčík (1987a).

2.4.4 Energy Considerations for High-Frequency Waves Propagating in Smoothly Inhomogeneous Media

The expressions derived in the previous sections for HF seismic body waves propagating in inhomogeneous media differ from the expressions for plane waves propagating in homogeneous media in three ways.

a. Travel-time field $T(x_i)$ is not a linear function of the coordinates (as in the case of plane waves) but is a general function of coordinates. This property also implies that slowness vector $\vec{p} = \nabla T(x_i)$ is not a constant vector but rather depends on the coordinates. Travel time $T(x_i)$ satisfies the eikonal equation.

b. The amplitudes of waves under consideration are not constant (as in the case of plane waves) but vary with the coordinates. They satisfy the transport equation.

c. Finally, the polarization vectors of elastic waves ($\vec{N}$, $\vec{e}_1$, and $\vec{e}_2$ in isotropic inhomogeneous media and $\vec{g}^{(1)}$, $\vec{g}^{(2)}$, and $\vec{g}^{(3)}$ in anisotropic inhomogeneous media) are not constant vectors but vary with the coordinates.

The expressions for the strain energy function and for the energy flux require the differentiation of the displacement vector. Thus, they will also include derivatives of the amplitudes, slowness vectors, and polarization vectors of the waves under consideration. Consequently, they will be more complex than the expressions for the plane waves derived in Section 2.2.7.

Let us demonstrate these expressions on the simple case of a time-harmonic acoustic wavefield with variable density. We remind the reader that the general energy quantities W, K, and S_i are given by relations

$$W = \tfrac{1}{2}\kappa p^2, \qquad K = \tfrac{1}{2}\rho v_i v_i, \qquad S_i = p v_i; \tag{2.4.52}$$

see (2.1.42). Here p is the pressure, and v_i are components of the particle velocity. We shall consider high-frequency acoustic waves in the following form:

$$p(x_i, t) = P(x_i) \exp[-i\omega(t - T(x_i))],$$
$$v_i(x_i, t) = V_i(x_i) \exp[-i\omega(t - T(x_i))].$$

If we take into account Equations (2.1.28) with $f_i = 0$ connecting pressure p and particle velocity v_i, we can express V_i in terms of pressure amplitude P as follows:

$$V_i = \frac{p_i}{\rho} P + \frac{1}{i\omega\rho} \frac{\partial P}{\partial x_i}.$$

Again considering $p(x_i, t)$ and $v_i(x_i, t)$ in real-valued form, we obtain

$$p = \tfrac{1}{2}(P \exp[-i\omega\zeta] + P^* \exp[i\omega\zeta]),$$

$$v_i = \tfrac{1}{2}(V_i \exp[-i\omega\zeta] + V_i^* \exp[i\omega\zeta]).$$

Inserting these expressions into (2.4.52) and time-averaging over one period, we finally arrive at

$$\overline{W} = \tfrac{1}{4}(\rho c^2)^{-1} P P^*,$$

$$\overline{K} = \tfrac{1}{4}(\rho c^2)^{-1}[P P^* + 2p_i c^2 \omega^{-1} \operatorname{Im}(P^* \partial P/\partial x_i)$$
$$+ c^2 \omega^{-2}(\partial P/\partial x_i)(\partial P^*/\partial x_i)],$$

$$\overline{S_i} = \tfrac{1}{2}\rho^{-1}[p_i P P^* + \omega^{-1} \operatorname{Im}(P^* \partial P/\partial x_i)].$$

$$(2.4.53)$$

As we can see, all relations in (2.4.53) for the averaged energy quantities are expressed in terms of pressure, not in terms of particle velocities. Only the expression for $\overline{W}$ corresponds fully to the plane wave expression; see (2.2.56) where f_p being replaced with $f_H = \tfrac{1}{2}$. The expressions for $\overline{K}$ and $\overline{S_i}$ are different; they have more terms.

We can, however, observe a very interesting and important fact. The expressions for $\overline{K}$ and $\overline{S_i}$ are expressed as a series in descending powers of ω. For high-frequency waves, it is sufficient to consider only the leading terms:

$$\overline{W} = \overline{K} = \tfrac{1}{4}(\rho c^2)^{-1} P P^*, \qquad \overline{S_i} = \tfrac{1}{2}\rho^{-1} p_i P P^*. \qquad (2.4.54)$$

Equations (2.4.54) are exactly the same as those obtained for the plane waves in a homogeneous medium. The only difference is that quantities ρ, c, p_i, and P depend on the coordinates in (2.4.54).

In Section 2.2.7, the group velocity vector $\vec{\mathcal{U}}$ of plane waves propagating in a homogeneous medium was introduced as the velocity vector of the energy flux $\vec{\mathcal{U}}^E$. Consequently, the Cartesian components of the group velocity vector $\mathcal{U}_i$ are given by relations $\mathcal{U}_i = \hat{S}_i/\hat{E} = \overline{S_i}/\overline{E}$, where $\hat{E} = \hat{W} + \hat{K}$ and $\overline{E} = \overline{W} + \overline{K}$. In the same way, we can introduce the group velocity vector $\vec{\mathcal{U}}$ even for *high-frequency seismic body waves* propagating in smoothly inhomogeneous media. In this case, the group velocity vector $\vec{\mathcal{U}}$ depends on position.

To calculate the group velocity vector of high-frequency acoustic waves propagating in an inhomogeneous medium, (2.4.54) yields the expected result:

$$\mathcal{U}_i = \frac{\overline{S_i}}{\overline{W} + \overline{K}} = c^2 p_i = c N_i. \qquad (2.4.55)$$

Thus, the group velocity vector of HF acoustic waves is perpendicular to the wavefront and its magnitude is c.

A result similar to that for time-harmonic waves is also obtained for HF transient waves. After neglecting higher order terms, the time-integrated energy quantities become

$$\hat{W} = \hat{K} = \tfrac{1}{2}(\rho c^2)^{-1} P P^* f_p, \qquad \hat{S}_i = \rho^{-1} p_i P P^* f_p, \qquad (2.4.56)$$

where f_p is given by (2.2.57).

The derivation of the energy expressions corresponding to high-frequency *elastic* waves propagating in inhomogeneous media is in principle the same as in the acoustic case. If we retain only the leading terms corresponding to the highest powers of frequency, we obtain the same result as for the plane waves in homogeneous media. For the readers' convenience, we shall summarize the expressions for the time-integrated energy quantities $\hat{W}$ (strain energy), $\hat{K}$ (kinetic energy), $\hat{E}$ (elastic energy), $\hat{S}_i$ (components of the energy flux), and $\mathcal{U}_i$ (components of the group velocity vector) for various situations.

a. HF P waves in inhomogeneous isotropic media:

$$\vec{u}(x_i, t) = A(x_i)\vec{N}F(t - T(x_i)),$$
$$\hat{W} = \hat{K} = \tfrac{1}{2}\rho AA^* f_c, \qquad \hat{E} = 2\hat{W} = \rho AA^* f_c, \tag{2.4.57}$$
$$\hat{S}_i = \rho\alpha N_i AA^* f_c, \qquad \mathcal{U}_i = \alpha N_i.$$

b. HF S waves in inhomogeneous isotropic media:

$$\vec{u}(x_i, t) = (B(x_i)\vec{e}_1 + C(x_i)\vec{e}_2)F(t - T(x_i)),$$
$$\hat{W} = \hat{K} = \tfrac{1}{2}\rho(BB^* + CC^*)f_c,$$
$$\hat{E} = 2\hat{W} = \rho(BB^* + CC^*)f_c, \tag{2.4.58}$$
$$\hat{S}_i = \rho\beta N_i(BB^* + CC^*)f_c, \qquad \mathcal{U}_i = \beta N_i.$$

c. HF waves in inhomogeneous anisotropic media:

$$\vec{u}(x_i, t) = A(x_i)\vec{g}^{(m)}F(t - T(x_i)),$$
$$\hat{W} = \hat{K} = \tfrac{1}{2}\rho AA^* f_c, \qquad \hat{E} = \hat{W} + \hat{K} = \rho AA^* f_c, \tag{2.4.59}$$
$$\hat{S}_i = \rho a_{ijkl}p_l g_j^{(m)} g_k^{(m)} AA^* f_c, \qquad \mathcal{U}_i = a_{ijkl}p_l g_j^{(m)} g_k^{(m)}.$$

In all these expressions, f_c is given by (2.2.61).

2.4.5 High-Frequency Seismic Waves Across a Smooth Interface

The explanation of propagation of seismic body waves in inhomogeneous media would be incomplete without considering the interaction of these waves with structural interfaces. We will consider two inhomogeneous media that are in contact along a curved interface Σ. The exact investigation of the process of reflection and transmission at this interface is rather complicated; analytical solutions are known only for very special cases of some symmetries. Thus, the process can be investigated numerically (finite differences) or approximately. For smooth interfaces, the approximate high-frequency methods provide very useful results. The solutions, of course, have only a *local character*; the solution at one point on the interface may be quite different from the solution at another point on the interface.

Because we shall be discussing the high-frequency solutions, we shall require interface Σ to be only slightly curved. Let us consider a point Q at which the wave under consideration is incident at interface Σ and denote the two main radii of curvature of Σ at Q by $R_1(Q)$ and $R_2(Q)$. It is then required that $R_1(Q)$ and $R_2(Q)$ be considerably larger than the prevailing wavelength λ of the wave incident at Q,

$$R_1(Q) \gg \lambda(Q), \qquad R_2(Q) \gg \lambda(Q). \tag{2.4.60}$$

As in the previous section, we shall consider only the zeroth-order term of the ray series. The complete ray-series solution, including the higher order terms, will be discussed in

Sections 5.6 and 5.7. For the leading term of the ray series, of course, we shall obtain exactly the same results here as in Chapter 5.

We shall again first consider only the *acoustic case*. We shall use the notation of Section 2.3.1 in full. We must, however, remember that all material parameters may vary with the coordinates. Thus, we have velocity $c_1(x_i)$ and density $\rho_1(x_i)$ in the first medium and the corresponding quantities $c_2(x_i)$ and $\rho_2(x_i)$ in the second medium. The incident wave, propagating in the first medium, is given by the relation

$$p(x_i, t) = P(x_i) \exp[-i\omega(t - T(x_i))]. \tag{2.4.61}$$

For reflected and transmitted waves, we shall assume the solutions in the relevant forms:

$$p^r(x_i, t) = P^r(x_i) \exp[-i\omega(t - T^r(x_i))],$$
$$p^t(x_i, t) = P^t(x_i) \exp[-i\omega(t - T^t(x_i))]. \tag{2.4.62}$$

The travel time T, slowness vector $\vec{p} = \nabla T$, and amplitude P of the incident wave are known at point Q. We need to determine travel times T^r and T^t, slowness vectors $\vec{p}^r = \nabla T^r$ and $\vec{p}^t = \nabla T^t$, and amplitudes P^r and P^t at Q. The knowledge of $T^r(Q)$ and $\vec{p}^r(Q)$ will provide sufficient initial conditions to perform ray tracing of the reflected ray from initial point Q and to calculate the travel time along it. Similarly, the knowledge of $P^r(Q)$ will provide the initial condition to determine the amplitude function of the reflected wave along the ray. For the transmitted waves, the situation is similar.

Across the interface, the pressure and the normal component of the particle velocity are continuous. Similarly, as in Section 2.3.1, we shall use the time derivative of the normal component of particle velocity $\dot{v}_n$ instead of v_n. For the incident wave, $\dot{v}_n$ is given by the following relation at Q:

$$\dot{v}_n(x_i, t) = -\rho^{-1}(i\omega P n_i p_i + P_{,i} n_i) \exp[-i\omega(t - T(x_i))];$$

see (2.1.22) with $f_i = 0$ and (2.4.61). Similar expressions are also obtained for the reflected and transmitted waves.

The boundary conditions along interface Σ contain factors $\exp(i\omega T(x_i))$, $\exp(i\omega T^r(x_i))$, and $\exp(i\omega T^t(x_i))$. For $T^r(x_i)$ and $T^t(x_i)$ different from $T(x_i)$ along Σ, the boundary conditions would yield amplitudes P^r and P^t of reflected and transmitted waves depending exponentially on frequency ω and on the coordinates. This conclusion implies that $T^r(x_i)$, $T^t(x_i)$, and $T(x_i)$ must be equal along Σ. The boundary conditions along Σ can then be expressed as

$$P + P^r = P^t,$$
$$\rho_1^{-1}(p_i n_i)P + \rho_1^{-1}(p_i^r n_i)P^r = \rho_2^{-1}(p_i^t n_i)P^t - \Delta. \tag{2.4.63}$$

Here the symbol Δ has the following meaning:

$$\Delta = (i\omega)^{-1}\left[\rho_1^{-1}(P_{,i}n_i) + \rho_1^{-1}(P_{,i}^r n_i) - \rho_2^{-1}(P_{,i}^t n_i)\right]. \tag{2.4.64}$$

It is not simple to solve the system of equations (2.4.63), due to term Δ. We are, however, seeking high-frequency approximate solutions. We can then neglect Δ, because it is of the order of $1/i\omega$.

Thus, for high-frequency waves, we can put $\Delta = 0$ in (2.4.63). Formally, we obtain exactly the same system of equations as for the acoustic plane waves on the plane interface between two homogeneous media (2.3.10). We must, however, remember two basic differences.

a. All quantities in (2.4.63) may vary along the interface, and the interface itself may be curved. System (2.4.63) must be solved at the point of incidence Q, considering all quantities at that point.

b. Equations (2.4.63) are only approximate and are valid for high-frequency waves.

We shall now discuss the relations

$$T(x_i) = T^r(x_i) = T^t(x_i) \tag{2.4.65}$$

along Σ. Equations (2.4.65) express a physically simply understandable fact that the travel times of the reflected and transmitted waves equal the travel time of the incident wave along interface Σ. It is usual to call (2.4.65) the *phase-matching relations* and to call the procedure the *phase matching*. Here we shall use the phase-matching relations to determine $T^r(Q)$, $T^t(Q)$, $\vec{p}^r(Q)$, and $\vec{p}^t(Q)$.

The phase-matching relation immediately implies

$$T(Q) = T^r(Q) = T^t(Q). \tag{2.4.66}$$

If we expand $T(x_i)$, $T^r(x_i)$, and $T^t(x_i)$ into a Taylor series in terms of Cartesian coordinates x_i at Q, phase-matching relation (2.4.65) also implies that the first, second, and higher tangential derivatives (in the plane tangential to interface Σ at Q) of T, T^r, and T^t at Q are equal. The equality of the first tangential derivatives of the travel-time fields along interface Σ has been discussed in more detail for plane waves in Section 2.3.1. Here we shall exploit these results very briefly. As in the case of plane waves, the slowness vectors of reflected wave $\vec{p}^r(Q)$ and of transmitted wave $\vec{p}^t(Q)$ are given by relations (2.3.23) and (2.3.18), where all quantities are taken at point Q.

If we put $\Delta = 0$, the system of equations (2.4.63) simplifies considerably:

$$\begin{aligned} P + P^r &= P^t, \\ \rho_1^{-1}(p_i n_i)P + \rho_1^{-1}(p_i^r n_i)P^r &= \rho_2^{-1}(p_i^t n_i)P^t. \end{aligned} \tag{2.4.67}$$

All quantities in (2.4.67) are taken at Q. The system fully coincides with the system for amplitudes of the plane waves reflected and transmitted at the plane interface between two homogeneous media. This means that the solution can be expressed in terms of the reflection and transmission coefficients of plane waves at a plane interface,

$$P^r(Q) = R^r P(Q), \qquad P^t(Q) = R^t P(Q). \tag{2.4.68}$$

Here reflection coefficient R^r and transmission coefficient R^t are given by (2.3.24) or (2.3.26), where all the quantities are taken at point Q. Thus, in the high-frequency approximation, the ratio of the amplitude of the reflected (transmitted) wave to the amplitude of the incident wave at point Q does not depend on the curvature of the interface, on the curvature of the wavefront of the incident wave, and on the gradients of velocities and densities in the vicinity of Q. It depends only on the local values of the velocities and densities directly at point Q on both sides of Σ ($c_1(Q)$, $c_2(Q)$, $\rho_1(Q)$, and $\rho_2(Q)$) and on the angle of incidence i_1. This ratio is given by a standard reflection (transmission) coefficient of the plane wave at the plane interface between two homogeneous media.

We emphasize one important point: we are speaking of amplitudes of reflected (transmitted) waves only on interface Σ at point Q. The amplitudes of the reflected (transmitted) wave *away from the interface* depend, of course, on all the previously mentioned factors. The transport equation must be used to evaluate the amplitudes of the reflected (transmitted) waves away from the interface. Term $\nabla^2 T^r$ ($\nabla^2 T^t$) in the transport equation for the reflected (transmitted) wave depends considerably on all these factors, particularly on the curvature of the interface.

All the preceding conclusions are valid only for the zeroth-term of the ray series (zeroth-order approximation). In the zeroth-order approximation, term Δ in the system of equations (2.4.63) really does not influence the results. It would, however, also be possible to use the whole ray series; see Section 5.6. Then, the higher order terms of the ray series calculated in this way compensate term Δ in (2.4.63). Thus, the higher order terms depend on $\partial P/\partial x_i$, $\partial P^r/\partial x_i$, and $\partial P^t/\partial x_i$ at Q; see (2.4.64). As discussed earlier, $P^r(x_i)$ and $P^t(x_i)$ in the vicinity of point Q also depend on the curvature of the wavefront of the incident wave, the curvature of the interface, the gradients of velocities ∇c_1 and ∇c_2, and densities $\nabla \rho_1$ and $\nabla \rho_2$ in the vicinity of Q. Consequently, ∇P^r and ∇P^t, computed at Q, also depend on the factors mentioned, and the higher order terms of the ray series of reflected (transmitted) waves will depend on these factors *directly at point Q*. This is the primary difference between the zeroth-order term, discussed here, and the higher order terms.

Let us now consider interface Σ between two *inhomogeneous isotropic elastic media*. The P and S velocities and the density in the first medium are denoted by $\alpha_1(x_i)$, $\beta_1(x_i)$, and $\rho_1(x_i)$, and in the second medium by $\alpha_2(x_i)$, $\beta_2(x_i)$, and $\rho_2(x_i)$. The incident wave may be P or S. Four waves are then generated at the interface: two reflected (P and S), and two transmitted (P and S). We can investigate the process of reflection and transmission in exactly the same way as in the acoustic case. We are again interested only in the high-frequency approximate solution or, more specifically, in the zeroth-order approximation of the ray method. We will not present all the equations here because the procedure is the same as that for acoustic waves. We shall only briefly summarize the main conclusions. We shall not distinguish between velocities α and β but simply use V. The equations will then be valid both for P and S waves. We merely insert $V = \alpha$, if the wave under consideration is P, and $V = \beta$, if it is S. We shall now select one of the four reflected/transmitted waves and denote all quantities corresponding to this wave with a tilde over the relevant symbol. The same quantities for the incident wave are without the tilde. Snell's law may then be expressed in the following way:

$$\frac{\sin i(Q)}{V(Q)} = \frac{\sin \tilde{i}(Q)}{\tilde{V}(Q)}, \tag{2.4.69}$$

where $i(Q)$ is the angle of incidence at Q and $\tilde{i}$ is the angle of reflection/transmission. The equation for the initial value of slowness vector $\tilde{p}(Q)$ of the reflected/transmitted wave reads

$$\vec{\tilde{p}}(Q) = \vec{p}(Q) - \left\{ (\vec{p}(Q) \cdot \vec{n}(Q)) \right.$$
$$\left. \mp \epsilon \left[\tilde{V}^{-2}(Q) - V^{-2}(Q) + (\vec{p}(Q) \cdot \vec{n}(Q))^2 \right]^{1/2} \right\} \vec{n}(Q). \tag{2.4.70}$$

Here ϵ is again the orientation index, given by the relation

$$\epsilon = \mathrm{sgn}(\vec{p}(Q) \cdot \vec{n}(Q)). \tag{2.4.71}$$

The upper sign $(-)$ in (2.4.70) corresponds to the transmitted waves; the lower sign $(+)$ refers to the reflected waves. Formula (2.4.70) is valid both for unconverted R/T waves (P → P, S → S) and for converted R/T waves (P → S, S → P). The validity of Equations (2.4.69) and (2.4.70) is, of course, limited only to high-frequency waves. Only for plane waves, reflected or transmitted at the plane interface between two homogeneous half-spaces, do the equations have an exact meaning.

In the same way as for acoustic waves, the initial values for the amplitudes of the generated waves at Q, $\tilde{A}(Q)$ for P waves and/or $\tilde{B}(Q)$ and $\tilde{C}(Q)$ for S waves, can be calculated using the standard plane wave equations. For example, we can use explicit expressions for plane-wave reflection/transmission coefficients or directly use the system of equations (2.3.37). All the equations have, of course, only a local character, and the quantities included in these equations must be taken at point Q. Amplitudes $\tilde{A}(Q)$, $\tilde{B}(Q)$, and $\tilde{C}(Q)$, corresponding to the individual reflected/transmitted waves, do not depend on the curvature of the interface, on the curvature of the incident wave, or on the gradients of the velocities and densities at Q. They depend only on $\alpha_1(Q)$, $\alpha_2(Q)$, $\beta_1(Q)$, $\beta_2(Q)$ and $\rho_1(Q)$, $\rho_2(Q)$, on the angle of incidence, and on the amplitude of the wave incident at Q.

Finally, the reflection/transmission of nonplanar high-frequency elastic waves at a curved interface between two *inhomogeneous anisotropic media* can again be solved in the same way. In the zeroth-order ray approximation, the general rules derived for plane waves at a plane interface between two homogeneous anisotropic solids in Section 2.3.3 remain valid even now, but must be considered locally, at point Q.

2.4.6 Space-Time Ray Method

Several alternative methods have been used to investigate the propagation of elastic waves in smoothly inhomogeneous layered media. We shall briefly discuss one of them: the space-time ray method. Instead of (2.4.14), we shall use a more general ansatz solution:

$$u_i(x_j, t) = U_i(x_j, t) \exp[i\theta(x_j, t)]. \tag{2.4.72}$$

The solution in this form is usually called the space-time ray solution, or the WKB solution. It is assumed that the complex-valued vectorial amplitude function $U_i(x_j, t)$ is a slowly varying function of x_i and t and that the scalar real-valued phase function $\theta(x_j, t)$ is a rapidly varying function of x_i and t. Solution (2.4.72) represents a *quasi-monochromatic wave packet*. The local instantaneous frequency $\omega(x_i, t)$ and the local instantaneous wave vector $\vec{k}(x_i, t)$ are defined by relations $\omega = -\partial\theta/\partial t$, $k_i = \partial\theta/\partial x_i$. The ansatz solution (2.4.72) is more general than the standard space ray method ansatz solution $u_i(x_j, t) = U_i(x_j) \exp[-i\omega(t - T(x_j))]$, where U_i does not depend on t, and $\theta(x_j, t)$ is a linear function of t, $\theta(x_j, t) = -\omega(t - T(x_j))$.

Various versions of (2.4.72) have been used. Often, a large parameter p is formally introduced into the exponential function of (2.4.72) so that the exponential factor reads $\exp[ip\theta(x_j, t)]$. Function $\theta(x_j, t)$ may then vary slowly. It is also possible to construct the *space-time ray series*, in inverse powers of (ip). In this section, we shall not discuss the space-time ray series solution, but only the zeroth-order approximation (2.4.72).

The space-time ray method can be applied to different wave fields, including elastic, acoustic, and electromagnetic. It has found useful applications in the investigation of waves propagating in viscoelastic media and in moving media, also including nonlinear problems. Typical wave phenomena studied by the space-time ray method are dispersion and absorption of waves. For the most comprehensive treatment of the space-time ray-series approach and for its early historical development see Babich, Buldyrev, and Molotkov (1985), where many other references can be found. For the application of the space-time ray method to viscoelastic media see Thomson (1997a), Section 6.

This book is devoted mainly to the propagation of high-frequency seismic body waves in perfectly elastic models, where the standard space ray method can be applied. For this reason, we shall not consider the space-time ray method in detail. To give the reader a taste of the space-time ray method, we shall apply it to perfectly elastic, inhomogeneous, anisotropic media, controlled by the elastodynamic equation (2.1.17). The treatment of pressure waves in fluid media and elastic waves in isotropic media is quite analogous. For more complex media (viscoelastic and the like), it would, of course, be necessary to consider the appropriate elastodynamic equations.

Inserting (2.4.72) into the elastodynamic equation (2.1.17) for a perfectly elastic, inhomogeneous, anisotropic medium yields

$$-N_i(\vec{U}) + \mathrm{i}M_i(\vec{U}) + L_i(\vec{U}) = 0, \qquad i = 1, 2, 3, \tag{2.4.73}$$

where N_i, M_i, and L_i are given by relations

$$
\begin{aligned}
N_i(\vec{U}) &= c_{ijkl}\theta_{,j}\theta_{,l}U_k - \rho\theta_{,t}^2 U_i, \\
M_i(\vec{U}) &= (c_{ijkl}\theta_{,l}U_k)_{,j} + c_{ijkl}\theta_{,j}U_{k,l} - \rho U_i\theta_{,tt} - 2\rho\theta_{,t}U_{i,t}, \\
L_i(\vec{U}) &= (c_{ijkl}U_{k,l})_{,j} - \rho U_{i,tt}.
\end{aligned}
\tag{2.4.74}
$$

Here $\theta_{,i} = \partial\theta/\partial x_i$, $\theta_{,t} = \partial\theta/\partial t$, and so on. $N_i(\vec{U})$ is the most rapidly varying function; it is quadratic in $\theta_{,i}$ and $\theta_{,t}$. $M_i(\vec{U})$ is linear in $\theta_{,i}$, $\theta_{,t}$, $\theta_{,ij}$, and $\theta_{,tt}$, and $L_i(\vec{U})$ does not depend on θ at all. The elastodynamic equation (2.1.17) with ansatz (2.4.72) is approximately satisfied if $N_i(\vec{U}) = 0$ and $M_i(\vec{U}) = 0$. Equation $N_i(\vec{U}) = 0$ can be used to separate the three waves that can propagate in smoothly inhomogeneous anisotropic media (qS1, qS2, qP) and to find the eikonal equations for phase function θ and the polarization vectors of these three waves. Equation $M_i(\vec{U}) = 0$ can then be used to find the transport equation.

Equation $N_i(\vec{U}) = 0$ can be expressed in the following form:

$$\Gamma_{ik}U_k - \theta_{,t}^2 U_i = 0, \qquad \Gamma_{ik} = a_{ijkl}\theta_{,j}\theta_{,l}. \tag{2.4.75}$$

The 3×3 matrix $\hat{\boldsymbol{\Gamma}}$, given by (2.4.75), is a *space-time Christoffel matrix*. We use the same notation for it as in the space ray method and hope that there will be no misunderstanding. Equations (2.4.75) constitute a 3×3 eigenvalue problem, in the same way as Equations (2.2.35) in the space ray method. We can determine the three eigenvalues G_m ($m = 1, 2, 3$), and the three relevant eigenvectors $\vec{g}^{(m)}$ ($m = 1, 2, 3$) of the space-time Christoffel matrix. These three eigenvalues and eigenvectors correspond to the three waves that can propagate in smoothly varying anisotropic media: qS1, qS2, and qP waves. The *eikonal equation* of the space-time ray method, corresponding to the mth wave ($m = 1, 2, 3$), reads

$$G_m(x_i, \theta_{,j}) = \theta_{,t}^2, \qquad G_m = a_{ijkl}\theta_{,j}\theta_{,l}g_i^{(m)}g_k^{(m)}. \tag{2.4.76}$$

It differs from the eikonal equation (2.4.44) of the space ray method by factor $\theta_{,t}^2$. Eigenvector $\vec{g}^{(m)}$ determines the polarization of the mth wave, $\vec{U} = A\vec{g}^{(m)}$.

As $\omega = -\theta_{,t}$ and $k_i = \theta_{,i}$, the eikonal equation (2.4.76) for a given x_i also represents the relation between ω and k_i. Such relations are called *local dispersion relations*. We shall present two alternative forms of the local dispersion relation for waves propagating in inhomogeneous anisotropic media:

$$\omega^2 = G_m(x_i, k_j), \qquad \omega = \Omega(x_i, k_j), \tag{2.4.77}$$

where $\Omega(x_i, k_j) = [G_m(x_i, k_j)]^{1/2}$.

Before we discuss equation $M_i(\vec{U}) = 0$, it will be useful to derive equations for certain *average energy quantities*, in much the same way as in Section 2.4.4. We are particularly interested in the average strain (potential) energy $\overline{W}$, average kinetic energy $\overline{K}$, average elastic energy $\overline{E} = \overline{W} + \overline{K}$, and average elastic energy flux $\overline{S}_i$. The averaging is performed with respect to θ, over a period of 2π. Moreover, only the leading terms in $\theta_{,i}$ and $\theta_{,t}$ are considered. All derivations are the same as in the space ray method, and we shall not repeat them. Assuming ansatz solution (2.4.72) in real-valued form,

$$u_i(x_j, t) = \tfrac{1}{2}(U_i \exp(i\theta) + U_i^* \exp(-i\theta)), \tag{2.4.78}$$

with real-valued θ and g_i, but complex-valued U_i, we obtain

$$\overline{W} = \tfrac{1}{4}c_{ijkl}\theta_{,j}\theta_{,l}U_iU_k^*, \qquad \overline{K} = \tfrac{1}{4}\rho\theta_{,t}^2U_iU_i^*,$$

$$\overline{E} = \overline{W} + \overline{K}, \qquad \overline{S_i} = -\tfrac{1}{4}c_{ijkl}\theta_{,l}\theta_{,t}(U_jU_k^* + U_j^*U_k). \tag{2.4.79}$$

All the expressions $\overline{W}, \overline{K}, \overline{E},$ and $\overline{S_i}$ depend both on position x_i and time t. Using $N_i(\vec{U}) = 0$, it is also possible to show that $\overline{W} = \overline{K}$. We can then simply determine the average elastic energy $\overline{E}$:

$$\overline{E} = \overline{W} + \overline{K} = 2\overline{K} = \tfrac{1}{2}\rho\theta_{,t}^2U_iU_i^*. \tag{2.4.80}$$

Now we shall return to equation $M_i(\vec{U}) = 0$. Inserting $U_k = Ag_k^{(m)}$ and multiplying the equation by $g_i^{(m)}$ yields

$$M_i\!\left(A\vec{g}^{(m)}\right)g_i^{(m)} = 2W_jA_{,j} + AW_{j,j} - 2\rho\theta_{,t}A_{,t} - \rho\theta_{,tt}A = 0, \tag{2.4.81}$$

where we have used the notation

$$W_j = c_{ijkl}g_i^{(m)}g_k^{(m)}\theta_{,l}. \tag{2.4.82}$$

(2.4.81) represents one form of the transport equation for A, analogous to (2.4.47). It can be expressed in several other forms. By multiplying (2.4.81) with A^* and adding to it the complex-conjugate form, we obtain a simpler transport equation for AA^*:

$$(W_jAA^*)_{,j} = (\rho\theta_{,t}AA^*)_{,t}. \tag{2.4.83}$$

This is a generalization of (2.4.50). Finally, by multiplying (2.4.83) by $\theta_{,t}$ and taking into account $c_{ijkl}\theta_{,j}\theta_{,lt}g_k^{(m)}g_i^{(m)} = 2\rho\theta_{,t}\theta_{,tt}$, we obtain an important and well-known relation:

$$\partial\overline{E}/\partial t + \overline{S}_{i,i} = 0. \tag{2.4.84}$$

Thus, the *average* energy quantities $\overline{E} = \overline{W} + \overline{K}$ and $\overline{S_i}$ satisfy the energy equations (2.1.38) in the space-time ray method.

As usual, we can introduce the group velocity vector $\vec{\mathcal{U}}$ by the relation

$$\mathcal{U}_i = \overline{S_i}/\overline{E} = -a_{ijkl}\theta_{,l}g_i^{(m)}g_k^{(m)}/\theta_{,t}. \tag{2.4.85}$$

Note that $\theta_{,t} = -\omega$, so that the sign in (2.4.85) is $+$ if we use ω instead of $\theta_{,t}$. Using (2.4.85), energy relation (2.4.84) yields the next familiar form:

$$\partial\overline{E}/\partial t + (\overline{E}\mathcal{U}_i)_{,i} = 0. \tag{2.4.86}$$

Because the eikonal equation (2.4.76) of the space-time ray method is known, it is easy to construct the space-time ray tracing systems, representing the characteristics of the eikonal equation. Standard Hamiltonian procedures can be used in this case. As an example, see the surface-wave ray tracing systems in Section 3.12. Similarly, the derived transport equations can be used to determine the amplitude variations along the space-time rays.

For more details on the space-time ray method and on its applications refer to Babich (1979), Kirpichnikova and Popov (1983), and particularly to the monograph by Babich,

Buldyrev, and Molotkov (1985). Considerable attention has also been devoted to the space-time ray method in some other books on the ray method; see Felsen and Marcuvitz (1973) and Kravtsov and Orlov (1980). The space-time ray method may also be applied to solving some problems involving time-dependent boundary conditions, moving sources, and faulting sources among others. This includes the investigation of the reflection of waves from moving bodies and the related Doppler effect. Finally, the complex-valued variant of the space-time ray method can be used to derive and study the Gaussian wave packets propagating along space rays. See Babich and Ulin (1981a, 1981b), Babich, Buldyrev, and Molotkov (1985), and other references given in Section 5.8.2. Gaussian wave packets have found applications in the construction of solutions of the elastodynamic equations, based on the integral superposition of Gaussian wave packets. Such solutions remove certain singularities of the ray method. See the references cited in Section 5.8.5.

An alternative method, close to the space-time ray method, is based on variational principles. In 1965, G. B. Whitham postulated a new variational principle in the theory of wavefields, which has also found applications in seismic wavefields (Whitham 1965, 1974). The principle is known as the *principle of the average Lagrangian*, or as *Whitham's variational principle*. For a detailed comparison of Whitham's variational principle with the space-time ray-series method, see Babich, Buldyrev, and Molotkov (1985, Section 2, §4). For applications of Whitham's variational principle in seismology, see Dahlen and Tromp (1998) and Ben-Hador and Buchen (1999), where other references can also be found. Dahlen and Tromp (1998) use Whitham's variational principle to develop in full the zeroth-order approximation of the space ray method for seismic body waves propagating in isotropic, smoothly inhomogeneous, global Earth models. Moreover, they also formulate the principle for models containing structural interfaces and the boundary of the model (Earth's surface).

2.5 Point-Source Solutions. Green Functions

The simplest solutions of the elastodynamic and acoustic wave equations in homogeneous media are the plane waves studied in Sections 2.2 and 2.3. The next simplest solutions refer to a point source. Whereas the plane-wave solutions cannot be defined for an inhomogeneous medium, the point-source solutions have a well-founded physical meaning even there. Certain strictly defined point-source solutions of the acoustic and elastodynamic wave equations are called acoustic and elastodynamic *Green functions*. They play a very important role in many seismological applications, both in the numerical modeling of seismic wavefields and in the inversion of seismic data. See Section 2.6.

In this section, we shall derive the general expressions for the point-source solutions of the acoustic and elastodynamic wave equations *for homogeneous media*. We shall then determine the Green functions for both cases. For the acoustic case, we shall use two approaches to determine the Green function: the exact approach and the approximate approach. The approximate high-frequency approach is based on the results of Section 2.4, particularly on the eikonal and transport equations. The exact and approximate high-frequency approaches yield the same results in the acoustic case, if one free constant in the approximate solution is properly specified. Both the exact and approximate approaches will also be applied in the elastodynamic case for the isotropic media. See Sections 2.5.3 and 2.5.4. Finally, in Section 2.5.5, we shall derive the asymptotic high-frequency expressions for the elastodynamic Green function in anisotropic homogeneous media.

In all three cases (acoustic, elastic isotropic, and elastic anisotropic), we shall use the same approach to derive the integral representation of the Green functions, based on contour integrals in the complex plane. For the acoustic and elastic isotropic case, the final integrals can be computed exactly using the Weyl integral; see Section 2.5.1. The Weyl integral is, however, not applicable to the anisotropic medium, and the resulting integrals can be calculated only approximately, by the method of stationary phase.

The results of this section will be used in Chapter 5 to derive asymptotic high-frequency ($\omega \to \infty$) expressions for the ray-theory Green functions in general 3-D laterally varying layered structures (acoustic, elastic isotropic, and elastic anisotropic).

2.5.1 Point-Source Solutions of the Acoustic Wave Equation

Let us first consider the acoustic wave equation for a homogeneous medium without the source term

$$\nabla^2 p = c^{-2} \ddot{p}, \qquad p = p(x_i, t). \tag{2.5.1}$$

We shall solve this equation in spherical coordinates (r, θ, ϕ), with $r = 0$ at the origin of Cartesian coordinates. Equation (2.5.1) can be solved easily in spherical coordinates by the Fourier method of separation of variables. For details and results, see, for example, Bleistein (1984). Here, however, we are interested only in radially symmetric solutions, independent of θ and ϕ. Laplacian $\nabla^2 p$ then takes the following form:

$$\nabla^2 p = \frac{1}{r^2} \frac{\partial}{\partial r} \left(r^2 \frac{\partial p}{\partial r} \right) = \frac{1}{r} \frac{\partial^2 (rp)}{\partial r^2}.$$

Inserting this into (2.5.1) yields

$$\frac{\partial^2 (rp)}{\partial r^2} = \frac{1}{c^2} \frac{\partial^2 (rp)}{\partial t^2}. \tag{2.5.2}$$

Quantity rp depends only on r and t so that (2.5.2) represents a one-dimensional wave equation in rp. Its solution is simple:

$$rp(r, t) = a F(t - T(r)) + b F(t + T(r)), \qquad T(r) = r/c. \tag{2.5.3}$$

Here $F(t \mp T)$ is an analytical signal, and a and b are constants that may be complex-valued. Equation (2.5.3) finally yields the radially symmetric solution of the acoustic wave equation (2.5.1),

$$p(r, t) = \frac{a}{r} F\left(t - \frac{r}{c} \right) + \frac{b}{r} F\left(t + \frac{r}{c} \right). \tag{2.5.4}$$

The solution (2.5.4) of acoustic wave equation (2.5.1) represents two spherical waves. Both solutions are constant along spherical surfaces $T(r) =$ const. (that is, $r =$ const.). The solution with the minus sign ($-$) corresponds to the spherical wave with an expanding wavefront (for t increasing), propagating *away* from the origin of coordinates to infinity. Similarly, the plus sign ($+$) corresponds to the spherical wave with the wavefront propagating *toward* the origin of coordinates. The spherical waves with the *expanding* wavefronts are usually called *outgoing* (or *exploding*) spherical waves, and the spherical waves with the *shrinking* wavefronts are called *ingoing* (or *imploding*) spherical waves.

The time-harmonic spherical waves are given by the general relation

$$p(r, t) = \frac{a}{r} \exp\left[-i\omega \left(t - \frac{r}{c} \right) \right] + \frac{b}{r} \exp\left[-i\omega \left(t + \frac{r}{c} \right) \right]. \tag{2.5.5}$$

Equation (2.5.5) can also be expressed in the following form:

$$p(r, t) = \left(\frac{a}{r} \exp[ikr] + \frac{b}{r} \exp[-ikr] \right) \exp[-i\omega t], \tag{2.5.6}$$

where k is the wave number, $k = \omega/c$. The first terms in (2.5.5) and (2.5.6) correspond to the outgoing spherical waves, and the second terms correspond to the ingoing spherical waves.

The outgoing spherical waves correspond to *waves generated by a point source* situated at the origin of coordinates. In this section, we shall consider only these point-source solutions. Thus, instead of (2.5.4), the solution now reads

$$p(r, t) = \frac{a}{r} F\left(t - \frac{r}{c} \right). \tag{2.5.7}$$

Note, however, that the imploding spherical waves also play an important role in various seismological applications and in the backward continuation of wavefields.

For a time-harmonic point source, the solutions are given by the relation

$$p(r, t) = \frac{a}{r} \exp\left[-i\omega\left(t - \frac{r}{c} \right) \right] = \frac{a}{r} \exp[ikr] \exp[-i\omega t]. \tag{2.5.8}$$

Formally, constant a in (2.5.8) represents the amplitude of the spherical time-harmonic wave at unit distance from the source ($r = 1$). It depends on the strength of the source. For the time-domain solution (2.5.7), a does not represent the amplitude at $r = 1$ because $|F|$ may be different from unity.

For certain strictly defined point sources, and, consequently, for a strictly defined constant a and analytical signal F, the solution $p(r, t)$ given by (2.5.7) is called the acoustic Green function. See Section 2.5.2.

Equations (2.5.4) through (2.5.8) yield exact solutions of acoustic wave equation (2.5.1). At large distances from the point source, the wavefront of the spherical wave is locally close to a plane wavefront, and amplitude a/r varies only slowly with respect to Cartesian coordinates x_i. Thus, we can also try to seek the high-frequency solutions of acoustic wave equation (2.5.1) using the approximate method of Section 2.4. The method is based on the ansatz solution $p(x_i, t) = P(x_i)F(t - T(x_i))$, on the determination of phase function $T(x_i)$ from the eikonal equation $(\nabla T)^2 = 1/c^2$, and on the determination of amplitude function $P(x_i)$ from the transport equation $2\nabla T \cdot \nabla P + P\nabla^2 T = 0$; see (2.4.6) and (2.4.7). For a symmetric (omnidirectional) point source situated at the origin of the coordinates, T and P depend on r only, and we have $\nabla T \cdot \nabla T = (dT/dr)^2$. The eikonal equation then yields $T(r) = \pm r/c$. The Laplacian reads

$$\nabla^2 T = \frac{1}{r^2} \frac{d}{dr} \left(r^2 \frac{dT}{dr} \right) = \pm \frac{2}{rc}.$$

Because $2\nabla T \cdot \nabla P = \pm(2/c)dP/dr$, the transport equation reduces to

$$dP/dr = -P/r.$$

This yields the solution $P = a/r$, where a is a constant independent of r. Thus,

$$p(r, t) = \frac{a}{r} F\left(t - \frac{r}{c} \right). \tag{2.5.9}$$

This solution corresponds to the outgoing spherical wave and is the same as (2.5.7).

Thus, in this case, the approximate method based on the eikonal and transport equations yields an exact solution.

In (2.5.9), constant a does not depend on coordinates r, θ, and ϕ. In fact, in the approximate method, a should be independent of r, but it may depend on θ and ϕ. Function $P(r, \theta, \phi) = (1/r)a(\theta, \phi)$ is still a solution of transport equation (2.4.7). Thus, the approximate high-frequency solution of acoustic wave equation (2.5.1) can be expressed as

$$p(r, \theta, \phi, t) = \frac{a(\theta, \phi)}{r} F\left(t - \frac{r}{c}\right). \tag{2.5.10}$$

Function $a(\theta, \phi)$ represents the amplitude of the time-harmonic pressure wave generated by a point source at unit distance ($r = 1$) from the source. Alternatively, we can say that it represents the distribution of amplitudes along a unit spherical surface with its center at the source. Function $a(\theta, \phi)$ may be arbitrary, but it should be smooth. In the text, we shall use the following terminology: a point source with $a(\theta, \phi) = $ const. will be called *omnidirectional*.

We shall now consider an omnidirectional time-harmonic point source. An important role in the theory of wave propagation is played by the *integral expansion of a spherical wave into plane waves*, given by the classical *Weyl integral* (Weyl 1919):

$$\frac{\exp[-i\omega(t - r/c)]}{r} = \frac{i\omega}{2\pi} \iint_{-\infty}^{\infty} \frac{dp_1 dp_2}{p_3} \exp[-i\omega(t - p_i x_i)]. \tag{2.5.11}$$

Here $r = (x_1^2 + x_2^2 + x_3^2)^{1/2}$ and p_i satisfy the relation $p_i p_i = 1/c^2$ so that

$$p_3 = \left(1/c^2 - p_1^2 - p_2^2\right)^{1/2}. \tag{2.5.12}$$

If $p_1^2 + p_2^2 > 1/c^2$, p_3 is imaginary, with the sign taken as follows: Im $p_3 > 0$.

We shall not derive the Weyl integral (2.5.11) because this has been done in many other textbooks on wave propagation. For a detailed derivation, discussion, and seismological applications, see Tygel and Hubral (1987) and DeSanto (1992). The Weyl integral (2.5.11) has been used to solve theoretically various wave propagation problems of great seismological interest. For example, it has been broadly used in computing the wavefield generated by a point source in a 1-D layered structure.

Thus, when a solution of any wave propagation problem in a specified structure is known *for plane waves*, a formal integral solution of the same problem *for a point source* can be obtained using the Weyl integral and will consist of an integral superposition of plane-wave solutions. It should be emphasized that the Weyl integral also includes *inhomogeneous plane waves* for $p_1^2 + p_2^2 > 1/c^2$.

The integral representation (2.5.11) can be expressed in several alternative forms. For axially symmetric point-source solutions, the cylindrical coordinates can be introduced, and one integral in (2.5.11) can be replaced by a Bessel function. This yields the *Sommerfeld integral*, representing an expansion of a spherical wave into cylindrical waves. Many important methods of wave propagation in 1-D layered models are based on the Sommerfeld integral; these methods include the *reflectivity method* (Fuchs 1968a, 1968b; Fuchs and Müller 1971; Müller 1985). For details on the Sommerfeld integral, again refer to Tygel and Hubral (1987) and DeSanto (1992).

2.5.2 Acoustic Green Function

We shall now derive the exact expression for the acoustic Green function in a homogeneous unbounded medium ($\rho = $ const., $c = $ const.) for time harmonic waves. It would be possible to use several alternative methods to derive these expressions. We shall use an approach based on *contour integrals in a complex plane*. The advantage of this method is that we can use it not only for acoustic but also for elastic, isotropic, and anisotropic media. For acoustic models, certain other approaches may yield the acoustic Green function in a simpler and more straightforward way. We wish, however, to use the same method in all cases (acoustic, elastic isotropic, and elastic anisotropic).

We shall first seek the integral solution of the acoustic wave equation (2.1.29) for pressure $p(x_j)$ in a homogeneous medium in the frequency domain, assuming an arbitrary source term $f^p(x_j)$. Only later shall we specify $f^p(x_j)$ to obtain the acoustic Green function. For a homogeneous medium, (2.1.29) reads

$$\frac{1}{\rho}\nabla^2 p(x_j) + \frac{\omega^2}{\rho c^2} p(x_j) = -f^p(x_j). \tag{2.5.13}$$

To solve (2.5.13) for $p(x_j)$, we use the 3-D Fourier transform:

$$p(x_j) = \iiint_{-\infty}^{\infty} \bar{p}(k_j) \exp[ik_n x_n] dk_1 dk_2 dk_3,$$

$$f^p(x_j) = \iiint_{-\infty}^{\infty} \bar{f}^p(k_j) \exp[ik_n x_n] dk_1 dk_2 dk_3. \tag{2.5.14}$$

The inverse Fourier transform reads

$$\bar{p}(k_j) = (8\pi^3)^{-1} \iiint_{-\infty}^{\infty} p(x_j) \exp[-ik_n x_n] dx_1 dx_2 dx_3,$$

$$\bar{f}^p(k_j) = (8\pi^3)^{-1} \iiint_{-\infty}^{\infty} f^p(x_j) \exp[-ik_n x_n] dx_1 dx_2 dx_3. \tag{2.5.15}$$

Multiplying (2.5.13) by $\exp[-ik_n x_n]/8\pi^3$ and taking the integral $\iiint_{-\infty}^{\infty} \ldots dx_1 dx_2 dx_3$ of the resulting equation yields the relation between $\bar{p}(k_j)$ and $\bar{f}^p(k_j)$:

$$\bar{p}(k_j)(\omega^2/c^2 - k_i k_i) = -\rho \bar{f}^p(k_j). \tag{2.5.16}$$

Note that the quantities k_i in (2.5.14) through (2.5.16) have the physical meaning of the components of wave vector $\vec{k}$. Inserting (2.5.16) into (2.5.14) yields the final equation for $p(x_i)$,

$$p(x_i) = \iiint_{-\infty}^{\infty} \frac{\rho \bar{f}^p(k_j)}{k_i k_i - \omega^2/c^2} \exp[ik_n x_n] dk_1 dk_2 dk_3. \tag{2.5.17}$$

This is the formal integral solution of (2.5.13), assuming that source function $f^p(x_j)$ is known. Function $\bar{f}^p(k_j)$ is obtained from $f^p(x_j)$ using (2.5.15).

It will be useful to express the solution (2.5.17) for $p(x_j)$ in a slightly modified form:

$$p(x_j) = \rho \iint_{-\infty}^{\infty} I(k_J, x_3) \exp[ik_N x_N] dk_1 dk_2, \tag{2.5.18}$$

$$I(k_J, x_3) = \int_{-\infty}^{-\infty} \frac{\bar{f}^p(k_j)}{k_3^2 - (\omega^2/c^2 - k_N k_N)} \exp[ik_3 x_3] dk_3. \tag{2.5.19}$$

The summations in $k_N k_N$ and $k_N x_N$ are over $N = 1, 2$.

Integral $I(k_J, x_3)$ can be evaluated by contour integration in complex plane $k_3 = \mathrm{Re}(k_3) + \mathrm{i}\,\mathrm{Im}(k_3)$, using the Jordan lemma. We choose contour C in complex plane $\mathrm{Re}(k_3)$, $\mathrm{Im}(k_3)$ along real axis $\mathrm{Im}(k_3) = 0$ and along a half-circle with its center at point $k_3 = 0$ and with radius R_c in the upper half-plane $\mathrm{Im}(k_3) > 0$. We denote the region inside contour C by C^+. The contribution along the semicircle with radius R_c in the upper half-plane vanishes for $R_c \to \infty$ due to factor $\exp(\mathrm{i}k_3x_3)$ in the integrand. Thus, integral (2.5.19) can be expressed in terms of the residues corresponding to poles situated in the upper half-plane and along the real axis.

If $k_N k_N < \omega^2/c^2$, there are two poles along the real axis: $k_3 = \pm(\omega^2/c^2 - k_N k_N)^{1/2}$, corresponding to outgoing and ingoing waves. If $k_N k_N > \omega^2/c^2$, the pole is on the positive part of the imaginary axis, corresponding to an inhomogeneous wave, exponentially decreasing with increasing x_3. We are interested here only in the outgoing waves, not in the ingoing waves. To eliminate the ingoing waves, we use the classical approach: we assume that ω is complex-valued, with a slight positive imaginary-valued part. Pole $k_3 = +(\omega^2/c^2 - k_N k_N)^{1/2}$ is then shifted from the real-valued axis into the upper half-plane, and pole $k_3 = -(\omega^2/c^2 - k_N k_N)^{1/2}$ moves into the lower half-plane. Integral (2.5.19) reduces to the residue calculated at point $k_3 = +(\omega^2/c^2 - k_N k_N)^{1/2}$. A simple calculation of the residue yields

$$I(k_J, x_3) = \mathrm{i}\pi k_3^{-1} \bar{f}^p(k_j) \exp[\mathrm{i}k_3 x_3]. \tag{2.5.20}$$

To interpret (2.5.18) and (2.5.20) in terms of familiar quantities, we substitute $k_i = \omega p_i$. Thus, instead of variables k_i, which have the meaning of components of wave vector $\vec{k}$, we shall use variables p_i, which have the meaning of the slowness vector components. Inserting $k_N = \omega p_N$ and (2.5.20) into (2.5.18) yields

$$p(x_j) = \mathrm{i}\pi\rho\omega \iint_{-\infty}^{\infty} \frac{\bar{f}^p(k_j)}{p_3} \exp[\mathrm{i}\omega p_n x_n]\,\mathrm{d}p_1\mathrm{d}p_2, \tag{2.5.21}$$

where $p_3 = (1/c^2 - p_N p_N)^{1/2}$, and $\mathrm{Im}\,p_3 > 0$ for $p_1^2 + p_2^2 > 1/c^2$.

Equation (2.5.21) represents the final form of the exact solution of wave equation (2.5.13). It can be used for very general forms of source term $f^p(x_j)$. Frequency ω may again be taken real-valued, eliminating its small imaginary part we considered in calculating (2.5.19).

In the following, we shall specify $\bar{f}^p(k_j)$ in (2.5.21) for a point source corresponding to the acoustic Green function. We shall first introduce the *acoustic Green function in the time domain*. By acoustic Green function in the time domain, we shall understand the solution of the acoustic wave equation for pressure (2.1.23) with $f_i = 0$, in which source term $f^p(x_j, t)$ is specified as follows:

$$f^p(x_j, t) = \partial q(x_j, t)/\partial t = \delta(t - t_0)\delta(\vec{x} - \vec{x}_0). \tag{2.5.22}$$

Here $\delta(t - t_0)$ is a 1-D and $\delta(\vec{x} - \vec{x}_0)$ a 3-D delta function. We shall denote the acoustic Green function by

$$G(\vec{x}, t; \vec{x}_0, t_0). \tag{2.5.23}$$

Thus, the acoustic Green function is a solution of the equation

$$\nabla \cdot (\rho^{-1}\nabla G) - \kappa \ddot{G} = -\delta(t - t_0)\delta(\vec{x} - \vec{x}_0). \tag{2.5.24}$$

From a physical point of view, the acoustic Green function $G(\vec{x}, t; \vec{x}_0, t_0)$ represents the pressure at point $\vec{x}$ and time t caused by a point source situated at point $\vec{x}_0$, representing the first time derivative $\partial q/\partial t$ of injection volume rate density $q(t)$, with a time dependence of $\partial q/\partial t$ corresponding to an impulse delta function applied at time t_0.

Because $\delta(t - t_0)$ has the dimension of s^{-1} and $\delta(\vec{x} - \vec{x}_0)$ has the dimension of m^{-3}, the dimension of the Green function $G(\vec{x}, t; \vec{x}_0, t_0)$ is $kg\,m^{-4}\,s^{-1}$; see (2.5.24). To assign $G(\vec{x}, t; \vec{x}_0, t_0)$ the physical units of pressure, we must multiply $f^P(x_j, t)$ by a constant of unit magnitude and dimension $m^3\,s^{-1}$.

We can also introduce the time-harmonic acoustic Green function $G(\vec{x}, \vec{x}_0, \omega)$, which is the solution of the time-harmonic acoustic wave equation

$$\nabla \cdot (\rho^{-1} \nabla G) + \omega^2 \kappa G = -\delta(\vec{x} - \vec{x}_0). \tag{2.5.25}$$

In this equation, we consider $t_0 = 0$. The dimension of the time-harmonic acoustic Green function $G(\vec{x}, \vec{x}_0, \omega)$ is $kg\,m^{-1}$. To assign $G(\vec{x}, \vec{x}_0, \omega)$ the physical units of pressure, we must multiply the RHS of (2.5.25) by a constant of unit magnitude and dimension s^{-2}.

We shall now derive the expression for the acoustic Green function in the frequency domain, for a homogeneous medium. In this case, (2.5.25) takes the form

$$\frac{1}{\rho} \nabla^2 G + \frac{\omega^2}{\rho c^2} G = -\delta(\vec{x} - \vec{x}_0). \tag{2.5.26}$$

Comparing (2.5.26) with (2.5.13) we obtain $f^P(x_j) = \delta(\vec{x} - \vec{x}_0)$. Equation (2.5.15) then yields $\bar{f}^P(k_j) = (1/8\pi^3) \exp[-i\omega p_n x_{0n}]$. Inserting this into (2.5.21) yields

$$G(\vec{x}, \vec{x}_0, \omega) = \frac{i\rho\omega}{8\pi^2} \iint_{-\infty}^{\infty} \frac{1}{p_3} \exp[i\omega p_n(x_n - x_{0n})] dp_1 dp_2. \tag{2.5.27}$$

Using the Weyl integral (2.5.11) finally yields

$$G(\vec{x}, \vec{x}_0, \omega) = \frac{\rho}{4\pi r} \exp[i\omega r/c], \tag{2.5.28}$$

where $r = |\vec{x} - \vec{x}_0|$. It is also simple to express the acoustic Green function $G(\vec{x}, t; \vec{x}_0, t_0)$ in the time domain:

$$G(\vec{x}, t; \vec{x}_0, t_0) = \frac{\rho}{4\pi r} \delta(t - t_0 - r/c). \tag{2.5.29}$$

These are the final expressions for the acoustic Green functions in the frequency and time domains in a homogeneous medium. As we can see, the amplitudes of the Green function decrease with increasing r as r^{-1} in a 3-D homogeneous medium.

2.5.3 Point-Source Solutions of the Elastodynamic Equation

We will again consider the elastodynamic equation for a homogeneous isotropic medium (with $\lambda = $ const., $\mu = $ const., and $\rho = $ const.), with source term $\vec{f}$ vanishing,

$$(\lambda + \mu)u_{j,ij} + \mu u_{i,jj} = \rho \ddot{u}_i, \qquad i = 1, 2, 3. \tag{2.5.30}$$

We will look for spherical wave solutions using the approximate high-frequency method described in Section 2.4.2. Using Equations (2.4.14), (2.4.26), and (2.4.28), we obtain the approximate solution

$$\vec{u}(x_i, t) = A(x_i)\vec{N}F(t - T^P(x_i)) + (B(x_i)\vec{e}_1 + C(x_i)\vec{e}_2)F(t - T^S(x_i)). \tag{2.5.31}$$

Here $\vec{N}$ is the unit vector normal to the wavefront, $\vec{e}_1$ and $\vec{e}_2$ are two mutually perpendicular unit vectors tangent to the wavefront (perpendicular to $\vec{N}$). Travel-time functions $T^P(x_i)$ and $T^S(x_i)$ satisfy eikonal equations (2.4.24) and (2.4.27). Similarly, amplitude functions $A(x_i)$, $B(x_i)$, and $C(x_i)$ are the solutions of transport equations (2.4.32) and (2.4.38). In the same way as in Section 2.5.1, we can find the solutions of the eikonal and transport equations:

$$
\begin{aligned}
T^P(r) &= r/\alpha, & T^S(r) &= r/\beta, \\
A(r, \theta, \phi) &= a(\theta, \phi)/r, & B(r, \theta, \phi) &= b(\theta, \phi)/r, \\
C(r, \theta, \phi) &= c(\theta, \phi)/r.
\end{aligned}
\tag{2.5.32}
$$

Here a, b, and c are arbitrary smooth functions of two radiation (take-off) angles θ and ϕ. The final approximate high-frequency expression for a wavefield generated by a point source reads

$$
\begin{aligned}
u_i(x_j, t) = & \frac{1}{r}a(\theta, \phi)N_i F\left(t - \frac{r}{\alpha}\right) + \frac{1}{r}b(\theta, \phi)e_{1i} F\left(t - \frac{r}{\beta}\right) \\
& + \frac{1}{r}c(\theta, \phi)e_{2i} F\left(t - \frac{r}{\beta}\right).
\end{aligned}
\tag{2.5.33}
$$

Expression (2.5.33) is, of course, only approximate. It is valid only at larger distances from the source, where the slowness vectors and vectorial amplitudes of the waves generated by a point source do not vary greatly over the distance of the prevailing wavelength. We also speak of the *far-field zone*. In the exact solution of the point-source problem in an isotropic homogeneous medium, some additional terms occur; these terms play an important role in the near-field zone but can be neglected in the far-field zone.

The analytical signals of the P and S waves generated by a point source in (2.5.33) are the same in our treatment. Only then can the complete wavefield be properly separated into the wavefields of P and S waves. It would, however, also be possible to consider three independent solutions, one for the P and the other two for the S1 and S2 waves, both with different analytical signals.

Similarly, as in the acoustic case, we shall speak of *omnidirectional point sources of P, S1 or S2 waves* if $a(\theta, \phi) = $ const., $b(\theta, \phi) = $ const., or $c(\theta, \phi) = $ const., respectively.

2.5.4 Elastodynamic Green Function for Isotropic Homogeneous Media

We shall consider a unit single-force point source situated at point $\vec{x}_0$, oriented along axis x_n. We assume that its time dependence is represented by a spike (delta function) at time t_0. The Cartesian components of force $\vec{f}$ in elastodynamic equation (2.1.17) can then be described by the relation,

$$
f_i(x_j, t) = \delta_{in}\delta(t - t_0)\delta(\vec{x} - \vec{x}_0),
\tag{2.5.34}
$$

and the elastodynamic equation reads

$$
(c_{ijkl}u_{k,l})_{,j} - \rho\ddot{u}_i = -\delta_{in}\delta(t - t_0)\delta(\vec{x} - \vec{x}_0).
\tag{2.5.35}
$$

We shall call the solution of (2.5.35) the elastodynamic Green function and denote it

$$
u_i(x_j, t) = G_{in}(\vec{x}, t; \vec{x}_0, t_0).
\tag{2.5.36}
$$

Thus, *elastodynamic Green function* $G_{in}(\vec{x}, t; \vec{x}_0, t_0)$ is a solution of the equation

$$(c_{ijkl}G_{kn,l})_{,j} - \rho \ddot{G}_{in} = -\delta_{in}\delta(t - t_0)\delta(\vec{x} - \vec{x}_0). \tag{2.5.37}$$

From a physical point of view, $G_{in}(\vec{x}, t; \vec{x}_0, t_0)$ represents the ith Cartesian component of the displacement vector at location $\vec{x}$ and time t due to a point source situated at $\vec{x}_0$, representing a single unit force oriented along the nth Cartesian axis, with the time dependence corresponding to an impulse delta function applied at time t_0.

Time-harmonic elastodynamic Green function $G_{in}(\vec{x}, \vec{x}_0, \omega)$ is a solution of the equation

$$(c_{ijkl}G_{kn,l})_{,j} + \rho \omega^2 G_{in} = -\delta_{in}\delta(\vec{x} - \vec{x}_0). \tag{2.5.38}$$

The foregoing definitions of the elastodynamic Green functions are also valid for inhomogeneous anisotropic media. For isotropic homogeneous unbounded media, however, the elastodynamic equations for the Green functions can be solved analytically. In most other cases, the differential equations should be solved numerically, or approximately.

We shall now consider a homogeneous isotropic unbounded medium. Differential equation (2.5.37) then has the form

$$(\lambda + \mu)G_{jn,ij} + \mu G_{in,jj} - \rho \ddot{G}_{in} = -\delta_{in}\delta(t - t_0)\delta(\vec{x} - \vec{x}_0). \tag{2.5.39}$$

The exact solution of this equation can be found in several alternative ways. For example, Aki and Richards (1980) used an approach based on Lamé's potentials. Here, we shall derive the expressions for the elastodynamic Green function directly for the displacement vector components, not invoking Lamé's potentials at all. We shall use an approach similar to that in Section 2.5.2. The advantage of this approach is that it is quite universal; it can be used for both isotropic and anisotropic media. For anisotropic media, Lamé's potentials cannot be used.

We shall again first seek the solution of elastodynamic equation (2.1.27) in the frequency domain for a general source term $f_i(x_j)$. Only later shall we specify f_i using (2.5.34) to obtain the elastodynamic Green function. Using the 3-D Fourier transform, we can write

$$u_k(x_j) = \iiint_{-\infty}^{\infty} \bar{u}_k(k_j) \exp[ik_n x_n]dk_1 dk_2 dk_3,$$
$$f_k(x_j) = \iiint_{-\infty}^{\infty} \bar{f}_k(k_j) \exp[ik_n x_n]dk_1 dk_2 dk_3. \tag{2.5.40}$$

The inverse 3-D Fourier transform is as follows:

$$\bar{u}_k(k_j) = (8\pi^3)^{-1} \iiint_{-\infty}^{\infty} u_k(x_j) \exp[-ik_n x_n]dx_1 dx_2 dx_3,$$
$$\bar{f}_k(k_j) = (8\pi^3)^{-1} \iiint_{-\infty}^{\infty} f_k(x_j) \exp[-ik_n x_n]dx_1 dx_2 dx_3. \tag{2.5.41}$$

Multiplying (2.1.27) by $\exp(-ik_n x_n)/8\pi^3$ and taking the integral $\iiint_{-\infty}^{\infty} \ldots dx_1 dx_2 dx_3$ of the resulting equations yields the system of three linear equations in $\bar{u}_i(k_j)$ for a homogeneous medium,

$$D_{ik}\bar{u}_k = \rho^{-1}\bar{f}_i, \qquad D_{ik} = a_{ijkl}k_j k_l - \omega^2 \delta_{ik}. \tag{2.5.42}$$

The solution of this system is

$$\bar{u}_k(k_j) = \frac{B_{ki}\bar{f}_i(k_j)}{\rho \det \hat{\mathbf{D}}}, \tag{2.5.43}$$

where B_{ki} are cofactors of D_{ki}. Inserting (2.5.43) into (2.5.40), we obtain the general solution of the elastodynamic equation (2.1.27) for a homogeneous medium in the frequency domain

$$u_k(x_j) = \frac{1}{\rho} \iint_{-\infty}^{\infty} I_k(k_J, x_3) \exp[ik_N x_N] dk_1 dk_2, \tag{2.5.44}$$

where the integral I_k is given by the relation

$$I_k(k_J, x_3) = \int_{-\infty}^{\infty} \frac{B_{ki}\bar{f}_i(k_j)}{\det \hat{\mathbf{D}}} \exp[ik_3 x_3] dk_3. \tag{2.5.45}$$

Equations (2.5.44) and (2.5.45) are valid for both isotropic and anisotropic media, for arbitrary source terms $f_i(x_j)$. Function $\bar{f}_i(k_j)$ must be determined from $f_i(x_j)$ using (2.5.41).

Now we shall discuss the *isotropic homogeneous elastic medium*. The expression for $B_{ki}/\det \hat{\mathbf{D}}$ can be calculated analytically from (2.5.42):

$$\frac{B_{ki}}{\det \hat{\mathbf{D}}} = \frac{(\beta^2 - \alpha^2)k_k k_i + (\alpha^2 k_s k_s - \omega^2)\delta_{ki}}{(\alpha^2 k_s k_s - \omega^2)(\beta^2 k_l k_l - \omega^2)}. \tag{2.5.46}$$

Inserting this into (2.5.45) yields the final expression for $I_k(k_j, x_3)$ for an isotropic homogeneous medium,

$$I_k(k_J, x_3) = \int_{-\infty}^{\infty} \frac{(\beta^2 - \alpha^2)k_k k_i + (\alpha^2 k_s k_s - \omega^2)\delta_{ki}}{(\alpha^2 k_s k_s - \omega^2)(\beta^2 k_l k_l - \omega^2)} \bar{f}_i(k_j) \exp[ik_3 x_3] dk_3. \tag{2.5.47}$$

Integral (2.5.47) can again be computed by contour integration in a complex plane $k_3 = \mathrm{Re}(k_3) + i\,\mathrm{Im}(k_3)$, using the same approach as in Section 2.5.2. Since ω is complex-valued with a small positive imaginary part, integral (2.5.47) has two poles within C^+, k_3^α, and k_3^β:

$$k_3^\alpha = (\omega^2/\alpha^2 - k_N k_N)^{1/2}, \qquad k_3^\beta = (\omega^2/\beta^2 - k_N k_N)^{1/2}. \tag{2.5.48}$$

It is not difficult to compute the residues at poles $k_3 = k_3^\alpha$ and $k_3 = k_3^\beta$. Equation (2.5.47) then yields

$$I_k(k_J, x_3) = \frac{i\pi}{\omega^2} \left\{ \frac{k_k k_i}{k_3^\alpha} \bar{f}_i^\alpha(k_j) \exp\left[ik_3^\alpha x_3\right] \right.$$
$$\left. - \frac{k_k k_i - \omega^2 \delta_{ik}/\beta^2}{k_3^\beta} \bar{f}_i^\beta(k_j) \exp\left[ik_3^\beta x_3\right] \right\}, \tag{2.5.49}$$

where

$$\bar{f}_i^\alpha(k_j) = [\bar{f}_i(k_j)]_{k_3=k_3^\alpha} \qquad \bar{f}_i^\beta(k_j) = [\bar{f}_i(k_j)]_{k_3=k_3^\beta}.$$

Expression (2.5.49) for $I_k(k_J, x_3)$ is also valid if $k_N k_N > \omega^2/\alpha^2$ and/or $k_N k_N > \omega^2/\beta^2$. We then only have to consider $\mathrm{Im}(k_3^\alpha) > 0$ and/or $\mathrm{Im}(k_3^\beta) > 0$.

Inserting (2.5.49) into (2.5.44) yields two integrals. In the first, we introduce new variables p_I^α by substituting $k_I = \omega p_I^\alpha$. Similarly, in the second one, we introduce the variables p_I^β by substituting $k_I = \omega p_I^\beta$. This yields

$$
u_k(x_j) = \frac{\pi i \omega}{\rho} \Bigg\{ \iint_{-\infty}^{\infty} \frac{p_i^\alpha p_k^\alpha}{p_3^\alpha} \bar{f}_i^\alpha \exp[i\omega p_s^\alpha x_s] \mathrm{d}p_1^\alpha \mathrm{d}p_2^\alpha
$$
$$
+ \iint_{-\infty}^{\infty} \frac{\delta_{ik}/\beta^2 - p_i^\beta p_k^\beta}{p_3^\beta} \bar{f}_i^\beta \exp[i\omega p_s^\beta x_s] \mathrm{d}p_1^\beta \mathrm{d}p_2^\beta \Bigg\}. \tag{2.5.50}
$$

Here

$$
p_3^\alpha = (1/\alpha^2 - p_N^\alpha p_N^\alpha)^{1/2}, \qquad \mathrm{Im}(p_3^\alpha) > 0 \quad \text{for } p_N^\alpha p_N^\alpha > 1/\alpha^2,
$$
$$
p_3^\beta = (1/\beta^2 - p_N^\beta p_N^\beta)^{1/2}, \qquad \mathrm{Im}(p_3^\beta) > 0 \quad \text{for } p_N^\beta p_N^\beta > 1/\beta^2. \tag{2.5.51}
$$

The source terms $\bar{f}_i^\alpha$ and $\bar{f}_i^\beta$ in (2.5.50) are obtained from $\bar{f}_i^\alpha(k_j)$ and $\bar{f}_i^\beta(k_j)$ by the appropriate substitution, $k_I = \omega p_I^\alpha$ or $k_I = \omega p_I^\beta$.

Equation (2.5.50) is still very general and is valid for an arbitrary source function $f_i(x_j)$. If $f_i(x_j)$ is known, the functions $\bar{f}_i^\alpha$ and $\bar{f}_i^\beta$ in (2.5.50) can be computed using (2.5.41).

We now wish to compute the *elastodynamic Green function* in the frequency domain. We must consider $f_i(x_j)$ given by the relation $f_i(x_j) = -\delta_{in}\delta(\vec{x} - \vec{x}_0)$; see (2.5.38). Using (2.5.41), we can compute $\bar{f}_i(k_j)$ and obtain $\bar{f}_i^\alpha = (\delta_{in}/8\pi^3)\exp[-i\omega p_s^\alpha x_{0s}]$, $\bar{f}_i^\beta = (\delta_{in}/8\pi^3)\exp[-i\omega p_s^\beta x_{0s}]$. Inserting this into (2.5.50) yields

$$
G_{kn}(\vec{x}, \vec{x}_0, \omega) = \frac{i\omega}{8\pi^2 \rho} \Bigg\{ \iint_{-\infty}^{\infty} \frac{p_k^\alpha p_n^\alpha}{p_3^\alpha} \exp[i\omega p_s^\alpha (x_s - x_{0s})] \mathrm{d}p_1^\alpha \mathrm{d}p_2^\alpha
$$
$$
+ \iint_{-\infty}^{\infty} \frac{\delta_{kn}/\beta^2 - p_k^\beta p_n^\beta}{p_3^\beta} \exp[i\omega p_s^\beta (x_s - x_{0s})] \mathrm{d}p_1^\beta \mathrm{d}p_2^\beta \Bigg\}. \tag{2.5.52}
$$

The integrals in (2.5.52) can be computed using the Weyl integral (2.5.11). Taking the second partial derivatives of (2.5.11) with respect to x_k and x_n yields

$$
\frac{\partial^2}{\partial x_k \partial x_n} \frac{\exp[i\omega r/c]}{r} = -\frac{i\omega^3}{2\pi} \iint_{-\infty}^{\infty} \frac{p_k p_n}{p_3} \exp[i\omega p_s (x_s - x_{0s})] \mathrm{d}p_1 \mathrm{d}p_2, \tag{2.5.53}
$$

where $r = |\vec{x} - \vec{x}_0|$. Using (2.5.53), we obtain this simple result from (2.5.52),

$$
G_{kn}(\vec{x}, \vec{x}_0, \omega) = \frac{\delta_{kn}}{4\pi\rho\beta^2 r} \exp[i\omega r/\beta]
$$
$$
- \frac{1}{4\pi\rho\omega^2} \frac{\partial^2}{\partial x_k \partial x_n} \frac{\exp[i\omega r/\alpha] - \exp[i\omega r/\beta]}{r}, \tag{2.5.54}
$$

with $r = |\vec{x} - \vec{x}_0|$. This is the final exact expression for the elastodynamic Green function in a homogeneous isotropic medium, in the frequency domain. It may be expressed in many alternative forms. For example, we can write it in the form of a finite series in $(-i\omega)^{-1}$:

$$
G_{kn}(\vec{x}, \vec{x}_0, \omega) = G_{kn}^P(\vec{x}, \vec{x}_0, \omega) + G_{kn}^S(\vec{x}, \vec{x}_0, \omega), \tag{2.5.55}
$$

where G_{kn}^P and G_{kn}^S are given by relations

$$
G_{kn}^P(\vec{x}, \vec{x}_0, \omega)
$$
$$
= \left[\frac{1}{4\pi\rho\alpha^2} N_k N_n - \frac{A_{kn}}{(-i\omega)4\pi\rho\alpha r} - \frac{A_{kn}}{(-i\omega)^2 4\pi\rho r^2} \right] \frac{\exp[i\omega r/\alpha]}{r},
$$

$$G_{kn}^S(\vec{x}, \vec{x}_0, \omega)$$

$$= \left[\frac{\delta_{nk} - N_n N_k}{4\pi\rho\beta^2} + \frac{A_{kn}}{(-i\omega)4\pi\rho\beta r} + \frac{A_{kn}}{(-i\omega)^2 4\pi\rho r^2} \right] \frac{\exp[i\omega r/\beta]}{r}.$$

$$(2.5.56)$$

Here $A_{kn} = \delta_{kn} - 3N_k N_n$. Using the Fourier transform, we arrive at the expressions for the elastodynamic Green function in a homogeneous isotropic medium in the time domain:

$$G_{kn}(\vec{x}, t; \vec{x}_0, t_0) = \frac{N_k N_n}{4\pi\rho\alpha^2 r} \delta\left(t - \frac{r}{\alpha}\right) + \frac{\delta_{kn} - N_k N_n}{4\pi\rho\beta^2 r} \delta\left(t - \frac{r}{\beta}\right)$$

$$+ \frac{3N_k N_n - \delta_{kn}}{4\pi\rho r^3} \int_{r/\alpha}^{r/\beta} \tau \delta(t - \tau)\, d\tau. \qquad (2.5.57)$$

Expressions (2.5.54) through (2.5.57) are exact. At large distances r from the source, sufficient accuracy may be achieved if the second and third terms in (2.5.56) are neglected. We then speak of the *far-field approximation* of the elastodynamic Green function

$$G_{kn}(\vec{x}, \vec{x}_0, \omega) \doteq \frac{N_k N_n}{4\pi\rho\alpha^2 r} \exp[i\omega r/\alpha] + \frac{\delta_{kn} - N_k N_n}{4\pi\rho\beta^2 r} \exp[i\omega r/\beta].$$

$$(2.5.58)$$

The far-field approximation is simultaneously a *high-frequency approximation* in view of the factors $1/(-i\omega)$ and $1/(-i\omega)^2$ with the second and third terms.

As we can see, the expression (2.5.33) obtained in Section 2.5.3 corresponds to the high-frequency approximation of the elastodynamic Green function, if constants a, b, and c are chosen as follows:

$$a = \frac{N_n}{4\pi\rho\alpha^2}, \qquad b = \frac{e_{1n}}{4\pi\rho\beta^2}, \qquad c = \frac{e_{2n}}{4\pi\rho\beta^2}. \qquad (2.5.59)$$

Comparing (2.5.58) with (2.5.33), we have taken into account the identity,

$$e_{1n}e_{1i} + e_{2n}e_{2i} = \delta_{in} - N_i N_n. \qquad (2.5.60)$$

The physical interpretation of (2.5.59) is simple. The quantities e_{1n}, e_{2n}, and N_n represent the decomposition of the single unit force oriented along the x_n-axis into unit vectors $\vec{e}_1$, $\vec{e}_2$, and $\vec{e}_3 \equiv \vec{N}$.

2.5.5 Elastodynamic Green Function for Anisotropic Homogeneous Media

To derive the elastodynamic Green function for an anisotropic homogeneous medium, we can start with Equations (2.5.44) and (2.5.45). These equations are exact. Unfortunately, in anisotropic media, we are not able to solve integrals (2.5.44) and (2.5.45) exactly, in terms of simple analytical functions, as in the case of isotropic media. We shall only derive approximate, high-frequency asymptotic expressions ($\omega \to \infty$) for the elastodynamic Green function, using Buchwald's (1959) approach. For alternative approaches, see Lighthill (1960), Duff (1960), Burridge (1967), Yeatts (1984), Hanyga (1984), Kazi-Aoual, Bonnet, and Jouanna (1988), Tverdokhlebov and Rose (1988), Ben-Menahem (1990), Ben-Menahem and Sena (1990), Tsvankin and Chesnokov (1990), Zhu (1992), Kendall, Guest, and Thomson (1992), Wang and Achenbach (1993, 1994, 1995), Every and Kim (1994), and Vavryčuk and Yomogida (1996).

We assume that the source is situated at S, the receiver is situated at R, and distance $\overline{SR}$ is large with respect to the wavelength. Following Buchwald (1959), we first rotate the axes x_1, x_2, and x_3 in (2.5.44) and (2.5.45) so that the positive x_3-axis is in the direction from S to R. Integrals (2.5.44) and (2.5.45) then read

$$u_k(R) = \frac{1}{\rho} \iint_{-\infty}^{\infty} I_k \mathrm{d}k_1 \mathrm{d}k_2, \qquad I_k = \int_{-\infty}^{\infty} \frac{B_{ki}\bar{f}_i(k_j)}{\det \hat{\mathbf{D}}} \exp[\mathrm{i}k_3 x_3]\mathrm{d}k_3.$$

$$(2.5.61)$$

As in Section 2.5.4, we shall calculate integral I_k of (2.5.61) by contour integration in complex plane $k_3 = \mathrm{Re}(k_3) + \mathrm{i}\,\mathrm{Im}(k_3)$. Contour C is again composed of two parts; the first part is along the real axis and the second is along a semicircle in the upper half-plane, $\mathrm{Im}(k_3) > 0$. The residues correspond to the roots of the equation $\det \hat{\mathbf{D}} = 0$. Because $\det \hat{\mathbf{D}}$ is a sixth-order polynomial in k_3, the equation $\det \hat{\mathbf{D}} = 0$ has six roots for k_3. Some of them may be real-valued, and others may be complex-valued.

Let us first discuss the complex-valued roots. If they are situated in the lower half-plane, they are outside region C^+ and do not contribute to the integral. The complex-valued roots with the positive imaginary parts, situated in the upper half-plane, yield inhomogeneous waves, which become exponentially small at large distances $\overline{SR}$ as $\omega \to \infty$. Because we are interested only in regular asymptotic high-frequency contributions (for $\omega \to \infty$), we can omit the complex-valued roots in our treatment. Thus, we shall consider only the real-valued roots, situated along the real k_3-axis.

As in Sections 2.5.2 and 2.5.4, the poles situated along the real k_3-axis represent outgoing and ingoing waves. We wish to consider only the outgoing waves, so that we shall shift the poles corresponding to $k_3 > 0$ from the real k_3-axis into region C^+ (upper half-plane). We can follow Buchwald (1959) and consider ω with a small positive imaginary part, $\omega + \mathrm{i}\epsilon$. This shifts the poles situated along the positive part of real axis k_3 into region C^+, but the poles situated along the negative part of the real k_3-axis are outside region C^+.

Finally, we shall consider only the situations in which the individual poles are well separated. The situations of coinciding poles and of poles situated close to other poles require special treatment.

We shall now select one of the roots of $\det \hat{\mathbf{D}} = 0$ for k_3 in region C^+ and denote it $k_3 = \theta(k_1, k_2)$. Because we are not considering inhomogeneous waves in our treatment, function $\theta(k_1, k_2)$ is real-valued and positive. Calculating the residue of integral I_k in (2.5.61) for root $k_3 = \theta(k_1, k_2)$ yields

$$I_k = 2\pi\mathrm{i} \left\{ \frac{B_{ki}\bar{f}_i(k_j)}{\partial \det \hat{\mathbf{D}}/\partial k_3} \exp[\mathrm{i}k_3 x_3] \right\}_{k_3 = \theta(k_1, k_2)}. \qquad (2.5.62)$$

Now we shall return to the real-valued ω, neglecting ϵ. The final expressions for $u_k(R)$ is obtained from (2.5.61),

$$u_k(R) = \frac{2\pi\mathrm{i}}{\rho} \iint_{-\infty}^{\infty} \left\{ \frac{B_{ki}\bar{f}_i(k_j)}{\partial \det \hat{\mathbf{D}}/\partial k_3} \exp[\mathrm{i}k_3 x_3] \right\}_{k_3 = \theta(k_1, k_2)} \mathrm{d}k_1 \mathrm{d}k_2. \qquad (2.5.63)$$

We now substitute $k_I = \omega p_I$. Then $\det \hat{\mathbf{D}} = \omega^6 \det(\Gamma_{in} - \delta_{in})$, where Γ_{in} are components of the well-known Christoffel matrix $\Gamma_{in} = a_{ijnl} p_j p_l$. Similarly, $B_{ki} = \omega^4 S_{ki}$, where S_{ki}

are cofactors of $\Gamma_{ki} - \delta_{ki}$. Hence, (2.5.63) yields

$$u_k(R) = \frac{2\pi i\omega}{\rho} \iint_{-\infty}^{\infty} A_{ki}(p_1, p_2) \bar{f}_i^{\,\theta}(p_1, p_2) \exp[i\omega\theta(p_1, p_2)x_3]dp_1dp_2,$$

(2.5.64)

$$A_{ki}(p_1, p_2) = \left\{ \frac{S_{ki}}{\partial \det(\Gamma_{in} - \delta_{in})/\partial p_3} \right\}_{p_3=\theta(p_1,p_2)},$$

(2.5.65)

$$\bar{f}_i^{\,\theta}(p_1, p_2) = \{\bar{f}_i(\omega p_j)\}_{p_3=\theta(p_1,p_2)}.$$

Quantity $p_3 = \theta(p_1, p_2)$ is a positive real-valued root of the equation $\det(\Gamma_{in} - \delta_{in}) = 0$. The equation $\det(\Gamma_{in} - \delta_{in}) = 0$ represents the complete slowness surface, including the slowness surface branches corresponding to qP, qS1, and qS2 waves. Thus, we can say that integral (2.5.64) represents the integral over slowness surface $p_3 = \theta(p_1, p_2)$.

Expression (2.5.65) for $A_{ki}(p_1, p_2)$ can be expressed in an alternative useful form. Substituting $k_i = \omega p_i$ into (2.5.42), we obtain the equation for $\bar{u}_k$ in the following form: $(\Gamma_{ik} - \delta_{ik})\bar{u}_k = \omega^{-2}\rho^{-1}\bar{f}_i$. We shall seek solution $\bar{u}_k$ using the relation $\bar{u}_k = \omega^{-2}Ag_k$, where $\vec{g}$ is the eigenvector of Γ_{ik}, corresponding to eigenvalue G. (We do not denote them $\vec{g}^{(m)}$ and G_m, because this relation can be used for any sheet of the slowness surface.) Multiplying the equation by g_i yields

$$A(\Gamma_{ik}g_ig_k - 1) = A(G - 1) = \rho^{-1}g_i\bar{f}_i;$$

see (2.2.34). Consequently, one of the solutions of equation $(\Gamma_{ik} - \delta_{ik})\bar{u}_k = \omega^{-2}\rho^{-1}\bar{f}_i$ reads $\bar{u}_k = \omega^{-2}\rho^{-1}g_ig_k\bar{f}_i/(G - 1)$. Using this new form of solution in integrals (2.5.60) through (2.5.63) again yields (2.5.64), where $A_{ki}(p_1, p_2)$ is given by the relation,

$$A_{ki}(p_1, p_2) = g_kg_i/(\partial G/\partial p_3) = \tfrac{1}{2} g_kg_i/\mathcal{U}_3.$$

(2.5.66)

Expression (2.5.64) with (2.5.65) or (2.5.66) is still valid for any source term $f_i(x_j)$. It will represent the elastodynamic Green function $G_{kn}(R, S, \omega)$ if we put $f_i(x_j) = \delta_{in}\delta(\vec{x} - \vec{x}(S))$; see (2.5.38). Then $\bar{f}_i^{\,\theta}(p_1, p_2)$ is given by the relation,

$$\bar{f}_i^{\,\theta}(p_1, p_2) = (8\pi^3)^{-1}\delta_{in} \exp[-i\omega x_{03}\theta(p_1, p_2)],$$

where $x_{03} = x_3(S)$; see (2.5.41) and (2.5.65). Inserting this expression for $\bar{f}_i^{\,\theta}(p_1, p_2)$ into (2.5.64) yields

$$G_{kn}(R, S, \omega) = \frac{i\omega}{4\pi^2\rho} \iint_{-\infty}^{\infty} A_{kn}(p_1, p_2) \exp[i\omega(x_3 - x_{03})\theta(p_1, p_2)]dp_1dp_2.$$

(2.5.67)

The expression (2.5.67) does not consider the inhomogeneous waves generated by the point source, but it is highly accurate. Using (2.5.67), the Green function $G_{kn}(R, S, \omega)$ may be calculated in three ways. The first is based on analytical computation. This is, however, possible only occasionally for very simple types of anisotropy. The second consists in the numerical treatment of (2.5.67). Finally, the third option is to compute (2.5.67) asymptotically, for $|\omega(x_3 - x_{03})| \to \infty$.

Here we shall treat (2.5.67) asymptotically for $|\omega(x_3 - x_{03})| \to \infty$, using the method of stationary phase. We denote the stationary point by $p_1 = p_1^r$ and $p_2 = p_2^r$. It satisfies the relations

$$[\partial \theta(p_1, p_2)/\partial p_1]_{p_1^r, p_2^r} = 0, \qquad [\partial \theta(p_1, p_2)/\partial p_2]_{p_1^r, p_2^r} = 0.$$

Geometrically, this means that the stationary points are situated at points (p_1^r, p_2^r) of the slowness surface such that the plane tangent to the slowness surface is perpendicular to the p_3-axis (in the p_i-space). In other words, stationary points (p_1^r, p_2^r) are points of the slowness surface, at which the normals to the slowness surface are parallel to the source-receiver direction. This can also be interpreted in terms of the group velocity vector $\vec{U}$, of which we know that it is perpendicular to the slowness surface. At stationary points, the group velocity vector is parallel to the source-receiver direction.

To calculate the stationary contributions in a very simple and objective way, we shall follow Buchwald (1959) and rotate axes p_1 and p_2 into the direction of the principal curvatures of the slowness surface at stationary point $p_{1,2} = p_{1,2}^r$. In the vicinity of the stationary point,

$$\theta(p_1, p_2) = p_3^r + \tfrac{1}{2}k_1(p_1 - p_1^r)^2 + \tfrac{1}{2}k_2(p_2 - p_2^r)^2 + \ldots, \qquad (2.5.68)$$

where k_1 and k_2 are the principal curvatures of the slowness surface along the p_1- and p_2-axes, at $p_{1,2} = p_{1,2}^r$. The linear terms are missing in expansion (2.5.68) because point (p_1^r, p_2^r) is stationary. The slowness surface is convex at $p_{1,2} = p_{1,2}^r$ for direction SR if $k_1 < 0$ and $k_2 < 0$. Similarly, it is concave if $k_1 > 0$ and $k_2 > 0$. Point $p_{1,2} = p_{1,2}^r$ is called elliptical if $k_1 > 0$ and $k_2 > 0$ or if $k_1 < 0$ and $k_2 < 0$. It is called hyperbolic if $k_1 > 0$ and $k_2 < 0$ or $k_1 < 0$ and $k_2 > 0$. Finally, it is called parabolic if $k_1 = 0$ and $k_2 \neq 0$ or $k_2 = 0$ and $k_1 \neq 0$. The parabolic points are excluded from our asymptotic high-frequency treatment; they are singular in the ray method.

Inserting (2.5.68) into (2.5.67) yields approximately

$$G_{kn}(R, S, \omega) = (i\omega/4\pi^2 \rho) A_{kn}(p_1^r, p_2^r) \exp[i\omega p_3^r(x_3 - x_{03})]$$

$$\times \left(\int_{-\infty}^{\infty} \exp\left[\tfrac{1}{2}i\omega k_1(p_1 - p_1^r)^2(x_3 - x_{03})\right] dp_1 \right)$$

$$\times \left(\int_{-\infty}^{\infty} \exp\left[\tfrac{1}{2}i\omega k_2(p_2 - p_2^r)^2(x_3 - x_{03})\right] dp_2 \right). \qquad (2.5.69)$$

We shall now use the well-known Poisson integral:

$$\int_{-\infty}^{\infty} \exp[iku^2] du = \sqrt{\pi/|k|} \exp\left[\tfrac{1}{4}i\pi \, \mathrm{sgn}\, k\right]. \qquad (2.5.70)$$

Equation (2.5.69) then yields

$$G_{kn}(R, S, \omega)$$
$$= \frac{1}{2\pi\rho(x_3 - x_{03})\sqrt{|K^S|}} A_{kn}(p_1^r, p_2^r) \exp\left[i\sigma_0 \tfrac{\pi}{2} + i\omega p_3^r(x_3 - x_{03})\right].$$
$$(2.5.71)$$

Here $K^S = k_1 k_2$ is the Gaussian curvature of the slowness surface at $p_{1,2} = p_{1,2}^r$, and

$$\sigma_0 = 1 + \tfrac{1}{2}\,\mathrm{sgn}\, k_1 + \tfrac{1}{2}\,\mathrm{sgn}\, k_2. \qquad (2.5.72)$$

Thus, $\sigma_0 = 0$ if both $\mathrm{sgn}\, k_1$ and $\mathrm{sgn}\, k_2$ are negative (convex elliptic point), $\sigma_0 = 2$ if both of them are positive (concave elliptic point), and $\sigma_0 = 1$ if k_1 and k_2 have opposite signs.

Inserting (2.5.66) into (2.5.71) yields a new expression for G_{kn}:

$$G_{kn}(R, S, \omega) = \frac{g_k g_n}{4\pi\rho(x_3 - x_{03})\mathcal{U}_3\sqrt{|K^S|}} \exp\left[i\tfrac{\pi}{2}\sigma_0 + i\omega p_3^r(x_3 - x_{03})\right].$$

$$(2.5.73)$$

This is the final equation for the ray-theory elastodynamic Green function $G_{kn}(R, S, \omega)$. In its derivation, we have chosen the x_3-axis along the direction SR and axes x_1 and x_2 along the main directions of the slowness surface. It is not difficult to modify (2.5.73) for general orientation of axes x_1, x_2, and x_3. We only need to replace $x_3 - x_{03}$ by $r = \overline{SR}$ and $\mathcal{U}_3$ by $|\mathcal{U}|$. The Gaussian curvature K^S is invariant so that we can retain it. We can also express σ_0 in the alternative form:

$$\sigma_0 = 1 - \tfrac{1}{2}\operatorname{Sgn}\mathbf{D}^S.$$

$$(2.5.74)$$

Here $\operatorname{Sgn}\mathbf{D}^S$ is the signature of the 2×2 *curvature matrix* $\mathbf{D}^S$ *of the slowness surface*. The reader is reminded that the signature of $\mathbf{D}^S$ equals the number of positive eigenvalues of $\mathbf{D}^S$ minus the number of negative eigenvalues of $\mathbf{D}^S$. Thus, it equals 2, 0 or -2. If we again use the positive Cartesian x_3-axis in the direction from S to R and define the slowness surface by the equation $p_3 = \theta(p_1, p_2)$, then $D_{11}^S = -\partial^2\theta/\partial p_1^2$, $D_{22}^S = -\partial^2\theta/\partial p_2^2$, and $D_{12}^S = D_{21}^S = -\partial^2\theta/\partial p_1\partial p_2$. Suitable expressions for $\mathbf{D}^S$ can also be found directly from the equations for the slowness surface (2.2.72) or (2.2.73) or from the known expression for the Hamiltonian; see (4.14.28). In isotropic media, $D_{11}^S = D_{22}^S = V$ and $D_{12}^S = D_{21}^S = 0$. If the slowness surface is locally convex (outward from S), then $\operatorname{Sgn}\mathbf{D}^S = 2$ and $\sigma_0 = 0$, as in isotropic media. In the concave parts of the slowness surface, $\operatorname{Sgn}\mathbf{D}^S = -2$, $\sigma_0 = 2$, and $\exp[i\tfrac{\pi}{2}\sigma_0] = -1$. Finally, when one eigenvalue of $\mathbf{D}^S$ is positive and the other is negative, $\operatorname{Sgn}\mathbf{D}^S = 0$ and $\sigma_0 = 1$. The last two "anomalous" possibilities cannot accur in isotropic media. Note that K^S can be expressed in terms of $\mathbf{D}^S$ as $K^S = \det\mathbf{D}^S$. The final expression for $G_{kn}(R, S, \omega)$ then reads

$$G_{kn}(R, S, \omega) = \frac{g_k g_n}{4\pi\rho r\mathcal{U}\sqrt{|K^S|}} \exp\left[i\tfrac{\pi}{2}\sigma_0 + i\omega T(R, S)\right].$$

$$(2.5.75)$$

Here all the quantities σ_0, K^S, p_i, and $\mathcal{U}$ are taken at the relevant stationary points $p_{1,2} = p_{1,2}^r$ of the slowness surface at which group velocity vector $\vec{\mathcal{U}}$ is parallel to direction $\overrightarrow{SR}$, and $T(R, S) = p_i^r(x_i - x_{0i})$ is the travel time from the source S to the receiver R. To obtain the complete Green function, contributions (2.5.75) must be added for all stationary points on the slowness surface.

In the time domain, the ray-theory elastodynamic Green function for a homogeneous anisotropic medium reads

$$G_{kn}(R, t; S, t_0) = \frac{g_k g_n}{4\pi\rho r\mathcal{U}\sqrt{|K^S|}} \exp\left[i\tfrac{\pi}{2}\sigma_0\right]\delta^{(A)}(t - t_0 - T(R, S)).$$

$$(2.5.76)$$

The expressions (2.5.75) and (2.5.76) for the ray-theory elastodynamic Green function in a homogeneous anisotropic medium can be modified in several alternative ways. Instead of the curvature matrix $\mathbf{D}^S$ of the slowness surface and its Gaussian curvature K^S, some alternative quantities (curvature of the wavefront, curvature of the group velocity surface, and the like) can be used. Many useful relations for $\mathbf{D}^S$ and K^S can be found in Section 4.14.4.

2.6 Application of Green Functions to the Construction of More General Solutions

In previous sections, we have discussed the plane-wave and point-source solutions of acoustic and elastodynamic equations. Both solutions have played a very important role in the construction of more general solutions of acoustic and elastodynamic wave equations. There is, however, a basic difference between the two. *Plane-wave solutions* apply only to homogeneous media, which may contain planar nonintersecting interfaces. Any inhomogeneity of the medium and/or any complexity of the interface (edges or curvature) destroys planar wavefronts and generates a more complex wavefield. *Point-source solutions* and relevant Green functions, however, are well defined even in laterally varying layered structures with nonplanar interfaces. The problem, of course, consists in their computation. The computation of Green functions is simple for homogeneous media (see Section 2.5) but may be more involved for complex structures.

In this book, we exploit broadly both the plane-wave and point-source solutions. The application of the plane-wave solutions is, more or less, methodological. They are extremely useful in the development of approximate high-frequency solutions of acoustic and elastodynamic equations in smoothly inhomogeneous media (see Section 2.4) and in the derivation of basic equations of the ray method. In fact, the zeroth-order approximation of seismic ray method consists in the local application of plane-wave solutions. This is well demonstrated on the zeroth-order ray theory treatment of the reflection and transmission of seismic body waves on a curved interface; see Section 2.4.5. Thus, the plane waves themselves cannot be used in laterally varying media containing curved interfaces, but the ray method, which represents a local extension of plane waves, can be used there. In other words, many plane-wave rules can be applied locally even in laterally varying layered structures.

The application of point-source solutions and relevant Green functions is quite different. Using Green functions, solutions of acoustic and elastodynamic equations for a very general distribution of sources and complex boundary conditions can be constructed using the *representation theorems*; see Section 2.6.1. The Green functions represent "building blocks" of such solutions. If the Green functions in the representation theorems are treated exactly, the representation theorems yield *exact solutions*. In complex structures, exact expressions for Green functions are not usually feasible. If we are, however, interested in *asymptotic, high-frequency solutions*, we can apply the ray-theory Green functions in representation theorems. This reasoning explains why the derivation of ray-theory Green functions for complex, laterally varying, isotropic or anisotropic layered structures is one of the main aims of this book. See Chapter 5 for such derivations.

The Green functions have also found important applications in the investigation of the scattering of acoustic and elastic waves. Just as in representation theorems, the Green functions represent building blocks of scattering integrals. Among various forms of scattering integrals, the so-called Born integrals, representing the single-scattering approximation, are very popular. The Born approximation plays a very important role both in direct and inverse problems of seismology, particularly in seismic exploration for oil. See Section 2.6.2 for a brief treatment of the Born approximation. The Born approximation has also been generalized and used to increase the accuracy of the zeroth-order approximation of the ray method. See also Section 2.6.2.

As an example of the representation theorems derived in Section 2.6.1, we shall derive expressions for the wavefield generated by a line source in a homogeneous medium in

Section 2.6.3. Such expressions will be used in Chapter 5 to derive the Green functions for 2-D models, corresponding to a line source parallel to the axis of symmetry of the model. The 2-D Green functions may find applications in 2-D computations similar to the 3-D Green functions in 3-D models.

The applications of Green functions are too numerous to be treated here exhaustively. The Green functions have found extensive applications in the inversion of seismic data, but here we are mainly interested in forward modeling. Our aim is modest: just to derive suitable expressions for the ray-theory Green functions that could be used in such applications (see Chapter 5) and to outline briefly how more general solutions of acoustic and elastodynamic equations can be constructed using the Green functions.

2.6.1 Representation Theorems

We shall work in the frequency domain and follow the derivation given by Kennett (1983). For a time-domain treatment, see Aki and Richards (1980) or Hudson (1980a).

We multiply (2.1.27) by $G_{in}(\vec{x}, \vec{x}_0, \omega)$ and subtract the product of (2.5.38) with $u_i(\vec{x}, \omega)$. Then we integrate the result over a volume V, which includes point $\vec{x}_0$. We obtain

$$u_n(\vec{x}_0, \omega) = \int_V f_i G_{in} dV(\vec{x}) + \int_V [(c_{ijkl}u_{k,l})_{,j} G_{in} - (c_{ijkl}G_{kn,l})_{,j} u_i] dV(\vec{x}).$$
(2.6.1)

Here $f_i = f_i(\vec{x}, \omega)$, $u_k = u_k(\vec{x}, \omega)$, $G_{in} = G_{in}(\vec{x}, \vec{x}_0, \omega)$, $G_{kn,l} = \partial G_{kn}/\partial x_l$, $dV(\vec{x}) = dx_1 dx_2 dx_3$. If we consider symmetry relations (2.1.6), particularly $c_{ijkl} = c_{klij}$, we obtain a new version of (2.6.1):

$$u_n(\vec{x}_0, \omega) = \int_V f_i G_{in} dV(\vec{x}) + \int_V (\tau_{ij} G_{in} - u_i H_{ijn})_{,j} dV(\vec{x}).$$
(2.6.2)

Here τ_{ij} is the stress tensor at $\vec{x}$ and H_{ijn} is the "Green function" stress tensor at $\vec{x}$, due to a single-force point source situated at $\vec{x}_0$ and oriented along the x_n-axis:

$$\tau_{ij} = \tau_{ij}(\vec{x}, \omega) = c_{ijkl}(\vec{x})\partial u_k(\vec{x}, \omega)/\partial x_l,$$

$$H_{ijn} = H_{ijn}(\vec{x}, \vec{x}_0, \omega) = c_{ijkl}(\vec{x})\partial G_{kn}(\vec{x}, \vec{x}_0, \omega)/\partial x_l.$$

The second integral in (2.6.2) can be transformed into the surface integral over surface S of V using the Gauss divergence theorem:

$$u_n(\vec{x}_0, \omega) = \int_V f_i G_{in} dV(\vec{x}) + \int_S (\tau_{ij} G_{in} - u_i H_{ijn}) n_j dS(\vec{x})$$

$$= \int_V f_i G_{in} dV(\vec{x}) + \int_S (T_i G_{in} - u_i h_{in}) dS(\vec{x}).$$
(2.6.3)

Here $\vec{n}$ is the outward normal to S, $T_i = T_i(\vec{x}, \omega) = \tau_{ij} n_j$ is the ith component of traction $\vec{T}(\vec{x}, \omega)$ corresponding to the normal $\vec{n}$ to S at $\vec{x}$, and $h_{in} = h_{in}(\vec{x}, \vec{x}_0, \omega) = H_{ijn}(\vec{x}, \vec{x}_0, \omega) n_j$ is the "Green function" traction component at $\vec{x}$.

At this point, it is usual to interchange the role of $\vec{x}$ and $\vec{x}_0$. In addition, we shall use $\vec{x}'$ instead of $\vec{x}_0$. Equation (2.6.3) then yields the final form of the representation theorem:

$$u_n(\vec{x}, \omega) = \int_V f_i(\vec{x}', \omega) G_{in}(\vec{x}', \vec{x}, \omega) dV(\vec{x}')$$

$$+ \int_S [T_i(\vec{x}', \omega) G_{in}(\vec{x}', \vec{x}, \omega) - u_i(\vec{x}', \omega) h_{in}(\vec{x}', \vec{x}, \omega)] dS(\vec{x}').$$

$$(2.6.4)$$

The first integral in (2.6.4) represents the displacement at $\vec{x}$ due to the single-force point sources $\vec{f}(\vec{x}', \omega)$ distributed within V. The second, surface integral represents the displacement at $\vec{x}$ due to the boundary conditions along S: (a) due to displacement $\vec{u}(\vec{x}', \omega)$ along S and (b) due to traction $\vec{T}(\vec{x}', \omega)$ along S.

A certain disadvantage of the representation theorem (2.6.4) is that the Green functions $G_{in}(\vec{x}', \vec{x}, \omega)$ and $h_{in}(\vec{x}', \vec{x}, \omega)$ in (2.6.4) correspond to the "source" at $\vec{x}$ and "receiver" at $\vec{x}'$. It would be physically more understable to have the arguments $\vec{x}'$ and $\vec{x}$ in the opposite order. It is possible to prove (see Aki and Richards 1980) that $G_{in}(\vec{x}', \vec{x}, \omega)$ satisfies the following reciprocity relation:

$$G_{in}(\vec{x}', \vec{x}, \omega) = G_{ni}(\vec{x}, \vec{x}', \omega), \qquad (2.6.5)$$

if G_{in} satisfies homogeneous boundary conditions on S. For details on homogeneous boundary conditions, see Aki and Richards (1980, p. 25). Then (2.6.4) yields

$$u_n(\vec{x}, \omega) = \int_V f_i(\vec{x}', \omega) G_{ni}(\vec{x}, \vec{x}', \omega) dV(\vec{x}')$$

$$+ \int_S [T_i(\vec{x}', \omega) G_{ni}(\vec{x}, \vec{x}', \omega) - u_i(\vec{x}', \omega) h_{ni}(\vec{x}, \vec{x}', \omega)] dS(\vec{x}').$$

$$(2.6.6)$$

If the Green functions in the representation theorems are treated exactly, the representation theorems yield an exact solution of the elastodynamic equation, corresponding to the arbitrary distribution of single-force point sources $\vec{f}(\vec{x}, \omega)$ in V and to the boundary conditions along S. If we use the ray-theory Green functions, the representation theorems will only yield an approximate result. In this case, however, the representation theorem (2.6.6) can always be used because the *ray-theory Green function always satisfies the reciprocity relations (2.6.5)*. This will be proved in Chapter 5.

Surface S may even represent a structural interface, the Earth's surface, or the like. The advantage of the representation theorems in the form of (2.6.4) and (2.6.6) is that all quantities under the integrals (T_i, u_i, G_{in}, and h_{in}) are continuous across the structural interface.

Similar representation theorems can also be derived in the acoustic case for pressure p in fluid media. Multiplying (2.5.25) by $p(\vec{x}, \omega)$ and (2.1.29) by $G(\vec{x}, \vec{x}_0, \omega)$, we obtain

$$p(\vec{x}, \omega) = \int_V f^p(\vec{x}', \omega) G(\vec{x}', \vec{x}, \omega) dV(\vec{x}')$$

$$+ \int_S n_i \rho^{-1}(\vec{x}') [p_{,i}(\vec{x}', \omega) G(\vec{x}', \vec{x}, \omega)$$

$$- p(\vec{x}', \omega) G_{,i}(\vec{x}', \vec{x}, \omega)] dS(\vec{x}'), \qquad (2.6.7)$$

as in the elastodynamic case. Here $p_{,i}$ and $G_{,i}$ denote derivatives with respect to x_i'. This is the representation theorem for pressure in fluid media. In the case of homogeneous boundary conditions, we can use the reciprocity relation

$$G(\vec{x}, \vec{x}', \omega) = G(\vec{x}', \vec{x}, \omega). \tag{2.6.8}$$

The representation theorem then reads

$$p(\vec{x}, \omega) = \int_V f^p(\vec{x}', \omega)G(\vec{x}, \vec{x}', \omega)\mathrm{d}V(\vec{x}')$$

$$+ \int_S n_i \rho^{-1}(\vec{x}')[p_{,i}(\vec{x}', \omega)G(\vec{x}, \vec{x}', \omega)$$

$$- p(\vec{x}', \omega)G_{,i}(\vec{x}, \vec{x}', \omega)]\mathrm{d}S(\vec{x}'). \tag{2.6.9}$$

Representation theorems (2.6.7) and (2.6.9) are applicable even if S is an interface. Quantities $p_{,i}$ and $G_{,i}$, however, are not continuous across the interface. It is possible to modify (2.6.7) and (2.6.9) to contain only quantities continuous across the interface. We use the notation

$$\dot{v}^{(n)} = -\rho^{-1}n_i p_{,i}, \qquad h = -\rho^{-1}n_i G_{,i}. \tag{2.6.10}$$

Here $\dot{v}^{(n)}$ is the time derivative of the normal component of particle velocity, and h is the relevant Green function. Both $\dot{v}^{(n)}$ and h are continuous across the interface. Inserting (2.6.10) into (2.6.7) yields

$$p(\vec{x}, \omega) = \int_V f^p(\vec{x}', \omega)G(\vec{x}', \vec{x}, \omega)\mathrm{d}V(\vec{x}')$$

$$+ \int_S [p(\vec{x}', \omega)h(\vec{x}', \vec{x}, \omega) - \dot{v}^{(n)}(\vec{x}', \omega)G(\vec{x}', \vec{x}, \omega)]\mathrm{d}S(\vec{x}').$$

$$\tag{2.6.11}$$

All the quantities in the surface integral of (2.6.11) ($\dot{v}^{(n)}$, p, G, and h) are now continuous across a structural interface. If reciprocity relation (2.6.8) is valid, we can replace $G(\vec{x}', \vec{x}, \omega)$ and $h(\vec{x}', \vec{x}, \omega)$ in (2.6.11) by $G(\vec{x}, \vec{x}', \omega)$ and $h(\vec{x}, \vec{x}', \omega)$.

As in the elastodynamic case, the representation theorems (2.6.7), (2.6.9), and (2.6.11) are exact if the Green function G is treated exactly. They yield only an approximate solution if the ray-theory Green functions are used. It will be proved in Section 5.1 that the *ray-theory Green function always satisfies the reciprocity relation (2.6.8)*. Consequently, the representation theorem can be used in the form of (2.6.9).

For more details on the representation theorems see Burridge and Knopoff (1964), Achenbach (1975), Pilant (1979), Aki and Richards (1980), Hudson (1980a), Kennett (1983), and Bleistein (1984). The surface integrals in the representation theorems (2.6.7), (2.6.9), and (2.6.11) for pressure are known as *Kirchhoff integrals*, or acoustic Kirchhoff integrals; see Section 5.1.11. Analogously, the surface integral in the representation theorem (2.6.6) for the displacement vector is called the *Kirchhoff integral for elastic waves* (or also simply the *elastic Kirchhoff integral*); see Section 5.4.8. Kirchhoff integrals have also been successfully used in the inversion of seismic data, particularly in seismic exploration for oil (Kirchhoff migration); see Bleistein (1984), Carter and Frazer (1984), Kuo and Dai (1984), Wiggins (1984), Kampfmann (1988), Keho and Beydoun (1988), Docherty (1991), Gray and May (1994), and Tygel, Schleicher, and Hubral (1994) among others. See also Haddon and Buchen (1981), Sinton and Frazer (1982), Scott and Helmberger (1983), Frazer and Sen (1985), Zhu (1988), Ursin and Tygel (1997), and Chapman (in press).

In earthquake seismology, the representation theorems have found important applications in the earthquake source theory. For a very detailed treatment, see Aki and Richards (1980) and Kennett (1983). Note that representation theorems (2.6.4) and (2.6.6) remain valid even for viscoelastic models, with complex-valued, frequency-dependent elastic parameters and for prestressed media. For prestressed media, the expressions for T_i and h_{ni} must, of course, be modified. See details in Kennett (1983).

2.6.2 Scattering Integrals. First-Order Born Approximation

Let us consider a structural model $\mathcal{M}$ described by elastic parameters $c_{ijkl}(\vec{x})$ and density $\rho(\vec{x})$. The source distribution in model $\mathcal{M}$ is specified by $f_i(\vec{x}, \omega)$. We wish to determine the displacement vector $\vec{u}(\vec{x}, \omega)$, which satisfies elastodynamic equation (2.1.27). Often, it may be considerably simpler to find the solution of elastodynamic equation (2.1.27) for the same source distribution but for some other reference model $\mathcal{M}^0$ close to $\mathcal{M}$. Denote the elastic parameters and the density in the reference model $\mathcal{M}^0$ $c_{ijkl}^0(\vec{x})$ and $\rho^0(\vec{x})$ and introduce quantities $\Delta c_{ijkl}(\vec{x})$ and $\Delta\rho(\vec{x})$ as follows:

$$c_{ijkl}(\vec{x}) = c_{ijkl}^0(\vec{x}) + \Delta c_{ijkl}(\vec{x}), \qquad \rho(\vec{x}) = \rho^0(\vec{x}) + \Delta\rho(\vec{x}). \qquad (2.6.12)$$

We call $\Delta c_{ijkl}(\vec{x})$ and $\Delta\rho(\vec{x})$ the *perturbations* of elastic parameters and density and assume that they are small. In other words, we assume that reference model $\mathcal{M}^0$ is close to model $\mathcal{M}$. Reference model $\mathcal{M}^0$ is also called the *nonperturbed* or *background model*, and model $\mathcal{M}$ is called the *perturbed model*.

We shall assume that the solution $\vec{u}^0(\vec{x}, \omega)$ of the elastodynamic equation (2.1.27) for reference model $\mathcal{M}^0$ and for the source distribution $\vec{f}(\vec{x}, \omega)$ is known. We express the unknown solution $\vec{u}(\vec{x}, \omega)$ of elastodynamic equation (2.1.27) for the perturbed model $\mathcal{M}$ and for the same source distribution $f_i(\vec{x}, \omega)$ in terms of $\vec{u}^0(\vec{x}, \omega)$ as follows:

$$\vec{u}(\vec{x}, \omega) = \vec{u}^0(\vec{x}, \omega) + \Delta\vec{u}(\vec{x}, \omega). \qquad (2.6.13)$$

We again assume that the perturbations of the displacement vector $\Delta\vec{u}(\vec{x}, \omega)$ are small. We now wish to find $\Delta\vec{u}(\vec{x}, \omega)$. Having done so, we can also easily determine $\vec{u}(\vec{x}, \omega)$ from (2.6.13).

Inserting (2.6.12) and (2.6.13) into (2.1.27) and subtracting (2.1.27) for the reference model, we obtain

$$\left(c_{ijkl}^0 \Delta u_{k,l}\right)_{,j} + \rho^0 \omega^2 \Delta u_i = -\left(u_{k,l}^0 \Delta c_{ijkl}\right)_{,j} - \omega^2 u_i^0 \Delta\rho$$
$$- (\Delta c_{ijkl} \Delta u_{k,l})_{,j} - \omega^2 \Delta\rho \Delta u_i. \qquad (2.6.14)$$

Here $\Delta u_{k,l}$ denotes $(\Delta u_k)_{,l}$, the partial derivative of the perturbation Δu_k with respect to x_l.

The last two terms in (2.6.14) contain products of two, presumably small, perturbations. We shall neglect them and consider only the *first-order perturbation theory*. Hence,

$$\left(c_{ijkl}^0 \Delta u_{k,l}\right)_{,j} + \rho^0 \omega^2 \Delta u_i = -f_i^0(\vec{x}, \omega), \qquad (2.6.15)$$

where $f_i^0(\vec{x}, \omega)$ is given by the relation

$$f_i^0(\vec{x}, \omega) = \left[u_{k,l}^0(\vec{x}, \omega) \Delta c_{ijkl}(\vec{x})\right]_{,j} + \omega^2 u_i^0(\vec{x}, \omega) \Delta\rho(\vec{x}). \qquad (2.6.16)$$

But (2.6.15) represents the elastodynamic equation (2.1.27) for $\Delta\vec{u}(\vec{x}, \omega)$ in the reference, background model $\mathcal{M}^0$, the source term $\vec{f}^0(\vec{x}, \omega)$ being given by (2.6.16). Because we assume that $\vec{u}^0(\vec{x}, \omega)$ is known, source term $\vec{f}^0(\vec{x}, \omega)$ is also known. It depends on perturbations Δc_{ijkl} and $\Delta\rho$. For $\Delta c_{ijkl} = 0$ and $\Delta\rho = 0$, $\vec{f}^0(\vec{x}, \omega) = 0$.

The solution of elastodynamic equation (2.6.15) can be found using representation theorem (2.6.4). Considering only the volume integral and reciprocity relations (2.6.5), we arrive at

$$\Delta u_n(\vec{x}, \omega) = \int_V \left[(u_{k,l}^0(\vec{x}', \omega)\Delta c_{ijkl}(\vec{x}'))_{,j} + \omega^2 u_i^0(\vec{x}', \omega)\Delta\rho(\vec{x}') \right]$$
$$\times G_{ni}^0(\vec{x}, \vec{x}', \omega)\mathrm{d}V(\vec{x}'). \tag{2.6.17}$$

The integration is taken over volume V in which $\Delta c_{ijkl}(\vec{x}')$ and $\Delta\rho(\vec{x}')$ are nonvanishing, and $\mathrm{d}V(\vec{x}') = \mathrm{d}x_1'\mathrm{d}x_2'\mathrm{d}x_3'$. The Green function $G_{ni}^0(\vec{x}, \vec{x}', \omega)$ corresponds to reference medium $\mathcal{M}^0$. Integral (2.6.17) contains the derivatives of perturbations $\Delta c_{ijkl}(\vec{x}')$. It can, however, be expressed in a more suitable form, which does not contain the derivatives of $\Delta c_{ijkl}(\vec{x}')$. Using

$$\left(u_{k,l}^0\Delta c_{ijkl}\right)_{,j} G_{ni}^0 = \left(u_{k,l}^0 G_{ni}^0 \Delta c_{ijkl}\right)_{,j} - u_{k,l}^0 G_{ni,j}^0 \Delta c_{ijkl},$$

inserting this into (2.6.17) and transforming the volume integral over the first term into a surface integral, we obtain, for an unbounded medium,

$$\Delta u_n(\vec{x}, \omega) = \int_V \left[\omega^2 u_i^0(\vec{x}', \omega)\Delta\rho(\vec{x}')G_{ni}^0(\vec{x}, \vec{x}', \omega) \right.$$
$$\left. - u_{k,l}^0(\vec{x}', \omega)\Delta c_{ijkl}(\vec{x}')G_{ni,j}^0(\vec{x}, \vec{x}', \omega) \right] \mathrm{d}V(\vec{x}'). \tag{2.6.18}$$

The wavefield $\Delta u_n(\vec{x}, \omega)$ given by (2.6.17) or (2.6.18) is called the *scattered wavefield*, and integrals (2.6.17) and (2.6.18) are called the *scattering integrals*. Because we have neglected two terms in (2.6.14), the scattering integrals yield only an approximate result (the so-called single-scattering approximation). We speak of the *first-order Born approximation* for the scattered wavefield.

As we can see from (2.6.16), perturbations Δc_{ijkl} and $\Delta\rho$ form the source term $\vec{f}^0(\vec{x}, \omega)$, which generates the scattered wavefield. These sources, however, have a character different from the actual physical sources $\vec{f}(\vec{x}, \omega)$. They generate a scattered wavefield only if they are excited by an incident wave. In other words, they wait passively for an incident wave and generate the scattered wavefield only after the incidence of the wave. For this reason, we also speak of *passive sources*, compared to the *active sources* described by $\vec{f}(\vec{x}, \omega)$.

In the same way as in elastic media, we can also derive scattering integrals for pressure wavefields in fluids. We consider the reference model $\mathcal{M}^0$, described by compressibility $\kappa^0(\vec{x})$ and density $\rho^0(\vec{x})$, and the perturbed model $\mathcal{M}$, described by compressibility $\kappa(\vec{x})$ and density $\rho(\vec{x})$. We consider arbitrary "physical" sources $f^p(\vec{x}, \omega)$. We denote the (known) solution of the acoustic wave equation (2.1.29) with source term $f^p(\vec{x}, \omega)$ in reference model $\mathcal{M}^0$ by $p^0(\vec{x}, \omega)$ and in perturbed model $\mathcal{M}$ by $p(\vec{x}, \omega)$. We introduce perturbations $\Delta\kappa$, $\Delta\rho^{-1}$, and Δp as

$$\kappa(\vec{x}) = \kappa^0(\vec{x}) + \Delta\kappa(\vec{x}), \qquad \rho^{-1}(\vec{x}) = \rho^{0-1}(\vec{x}) + \Delta\rho^{-1}(\vec{x}),$$
$$p(\vec{x}, \omega) = p^0(\vec{x}, \omega) + \Delta p(\vec{x}, \omega). \tag{2.6.19}$$

We assume that $\Delta\kappa$, $\Delta\rho^{-1}$, and Δp are small and that $p^0(\vec{x}, \omega)$ is known. The first-order Born approximation for the scattered pressure wavefield in an unbounded medium then reads

$$\Delta p(\vec{x}, \omega) = \int_V \left[\left(p_{,i}^0(\vec{x}', \omega) \Delta \rho^{-1}(\vec{x}') \right)_{,i} + \omega^2 p^0(\vec{x}', \omega) \Delta \kappa^0(\vec{x}') \right]$$
$$\times G^0(\vec{x}, \vec{x}', \omega) dV(\vec{x}'). \tag{2.6.20}$$

In the derivation of (2.6.20), we have neglected the nonlinear terms with $\Delta p \Delta \kappa$ and $\Delta p \Delta \rho^{-1}$ and have used the reciprocity relation (2.6.8) for the acoustic Green function. As for the elastic case, we can eliminate the derivatives of perturbations $\Delta \rho^{-1}$. Hence,

$$\Delta p(\vec{x}, \omega) = \int_V \left[\omega^2 p^0(\vec{x}', \omega) \Delta \kappa^0(\vec{x}') G^0(\vec{x}, \vec{x}', \omega) \right.$$
$$\left. - p_{,i}^0(\vec{x}', \omega) \Delta \rho^{-1}(\vec{x}') G_{,i}^0(\vec{x}, \vec{x}', \omega) \right] dV(\vec{x}'). \tag{2.6.21}$$

The Born approximation yields a linear relation between Δu_n (or Δp) on the one hand and Δc_{ijkl}, $\Delta \rho$ (or $\Delta \kappa$, $\Delta \rho^{-1}$) on the other. See (2.6.18) and (2.6.21). This property is very attractive in the inversion of seismic data.

The Born approximation has been known for a long time in theoretical physics, where it has been used to solve problems of potential scattering. For a more detailed treatment and many other references see Hudson and Heritage (1981), Bleistein (1984), Chapman and Orcutt (1985), Cohen, Hagin, and Bleistein (1986), Bleistein, Cohen, and Hagin (1987), Beydoun and Tarantola (1988), Beydoun and Mendes (1989), Wu (1989a, 1989b), Beylkin and Burridge (1990), Coates and Chapman (1990a), Ben-Menahem and Gibson (1990), Gibson and Ben-Menahem (1991), Červený and Coppoli (1992), and Ursin and Tygel (1997). An alternative to the Born approximation is the so-called Rytov approximation; see Rytov, Kravtsov, and Tatarskii (1987) and Samuelides (1998) among others. A good reference to both the Born and Rytov approximation is Beydoun and Tarantola (1988).

The scattering Born integrals (2.6.18) and (2.6.21) can be expressed in many alternative forms. They can also be generalized in several ways. We shall briefly outline two such generalizations.

1. Scattering integral equations. The major inaccuracies in the first-order Born approximation (2.6.18) are caused by neglecting the last two terms in (2.6.14). These terms contain Δu_i and their spatial derivatives. In fact, the whole procedure leading to (2.6.18) may be performed even without neglecting these two terms. Then the final equation, which is analogous to (2.6.18), would be exact. The integral, however, would contain Δu_i. Consequently, the equation would represent an *integral equation for* $\Delta u_n(\vec{x}, \omega)$. Similar exact integral equations can be derived even for $\Delta p(\vec{x}, \omega)$ in the acoustic case (Lippman-Schwinger equation). Various approaches have been proposed to solve the integral equations (for example, the *Born series method*). The leading term in the Born series corresponds to the Born approximation derived here.

2. Generalized Born scattering. If reference model $\mathcal{M}^0$ is inhomogeneous, the Green functions in $\mathcal{M}^0$ are usually not known exactly. They are often calculated by some asymptotic high-frequency method (for example, by the ray method). This introduces errors into the computations. In principle, it is possible to find expressions for the relevant error terms and introduce them into the treatment outlined in this section. The scattering integrals then include these error terms. Consequently, the scattering integrals represent both *scattering from medium perturbations* and *scattering from errors*. If models $\mathcal{M}^0$ and $\mathcal{M}$ are identical, the scattering integrals represent merely the scattering from errors. The scattering integrals give satisfactorily accurate results even in regions where the asymptotic method used is highly inaccurate, or fails completely. As an example, let us consider the shear waves in inhomogeneous anisotropic media in the vicinity of shear wave singularities. The penalty for this increase in accuracy is a considerably more extensive computation. The method was proposed by Coates and Chapman (1991). For a very detailed and comprehensive theoretical treatment of inhomogeneous anisotropic media, see Chapman and Coates (1994), where simple expressions for the error terms are also derived. See also Coates and Chapman (1990b) and Coates and Charrette (1993).

2.6.3   Line-Source Solutions

In 2-D models in which the elastic parameters and the density do not depend on one Cartesian coordinate, say x_2, it is very common to consider a line source parallel to the x_2-axis instead of the point source. Such line sources have been broadly used in the interpretation of profile measurements and in 2-D finite-difference computations. Consequently, briefly discussing even the ray-theory solutions for such a case would be useful. In this section, we shall solve a relevant canonical example corresponding to a line source in a homogeneous medium. The solution given here will be generalized for a line source in a laterally varying layered structure in Chapter 5. In Sections 5.1.12 and 5.2.15, a considerably more general line source, situated in a 3-D laterally varying layered structure, with an arbitrary curvature and torsion and with an arbitrary distribution of the initial travel times along it, will be also considered.

Several methods can be used to solve the canonical example. For example, we could solve the wave equation in cylindrical coordinates; see, for example, Bleistein (1984). Here, however, we shall use a different approach based on representation theorems because we wish to demonstrate the simplicity of applications of the representation theorem and 3-D Green functions.

We shall consider a pressure wavefield in fluid media in the frequency domain. The pressure wavefield satisfies the equation (2.5.13), where f^p does not depend on x_2. The pressure does not then depend on x_2 in the whole space either. Without loss of generality, we shall consider receivers situated in plane $x_2 = 0$.

We put

$$f^p(x_j, \omega) = \delta(x_1 - x_{01})\delta(x_3 - x_{03}).$$
(2.6.22)

The relevant solution $p(\vec{x}, \omega)$ of (2.5.13) then represents the *2-D acoustic Green function* $G^{2D}(\vec{x}, \vec{x}_0, \omega)$ (for $x_2 = x_{20} = 0$). We emphasize that $\delta(x_2 - x_{02})$ is not present in (2.6.22). The coordinates of the line source in plane $x_2 = 0$ are x_{01}, x_{03}.

To determine $G^{2D}(\vec{x}, \vec{x}_0, \omega)$, we shall use the representation theorem (2.6.9) for the unbounded medium. The integral over S then vanishes, and (2.6.9) yields

$$G^{2D}(\vec{x}, \vec{x}_0, \omega) = \int_V \delta(x_1' - x_{01})\delta(x_3' - x_{03})G(\vec{x}, \vec{x}', \omega)\mathrm{d}V(\vec{x}')$$

$$= \tfrac{1}{4}\rho\pi^{-1}\int_{-\infty}^{\infty} (l^2 + r_l^2)^{-1/2} \exp\!\left[i\omega(l^2 + r_l^2)^{1/2}\big/c\right]\mathrm{d}l.$$
(2.6.23)

Here we have used (2.5.28) for the 3-D Green function $G(\vec{x}, \vec{x}', \omega)$, with $r = (l^2 + r_l^2)^{1/2}$, where l is the arclength along the line source, and r_l is the distance of the receiver from the line source in plane $x_2 = 0$, $r_l = [(x_1 - x_{01})^2 + (x_3 - x_{03})^2]^{1/2}$. The volume integral in (2.6.23) reduces to a line integral due to the delta functions in the integrand.

For high ω, the integral can be simply calculated by the method of stationary phase. The stationary point is $l = 0$, and we can put $(l^2 + r_l^2)^{1/2} \doteq r_l + \frac{1}{2}l^2/r_l$ in the exponent and $(l^2 + r_l^2)^{-1/2} \doteq r_l^{-1}$ in the amplitude terms. We then apply the Poisson integral (2.5.70) to (2.6.23) and obtain

$$G^{2D}(\vec{x}, \vec{x}_0, \omega) \doteq \tfrac{1}{2}\rho(c/2\pi\omega r_l)^{1/2} \exp[i\omega r_l/c + i\pi/4]. \tag{2.6.24}$$

Integral (2.6.23) can also be computed exactly. It can be transformed to another integral representing the Hankel function of the first kind and zeroth order, $H_0^{(1)}$:

$$G^{2D}(\vec{x}, \vec{x}_0, \omega) = \tfrac{1}{4}i\rho H_0^{(1)}(\omega r_l/c); \tag{2.6.25}$$

see Abramowitz and Stegun (1970), Bleistein (1984), and DeSanto (1992). Equation (2.6.25) represents the exact solution. Thus, the representation theorem gives an exact solution. Using the well-known asymptotic expression of $H_0^{(1)}(\omega r_l/c)$ for $\omega r_l/c \to \infty$, we obtain (2.6.24) from (2.6.25). Note that (2.6.24) and (2.6.25) correspond to $\omega > 0$. If $\omega < 0$, complex conjugate quantities must be considered.

Using the Fourier transform, the frequency-domain 2-D Green function can be transformed into a time-domain 2-D acoustic Green function. The time-domain 2-D acoustic Green function $G^{2D}(\vec{x}, t; \vec{x}_0, t_0)$ is a solution of the acoustic wave equation (2.1.25) with $f^p(\vec{x}, t)$ given by the relation

$$f^p(\vec{x}, t) = \delta(t - t_0)\delta(x_1 - x_{01})\delta(x_3 - x_{03}). \tag{2.6.26}$$

Using the Fourier transform relations given in Appendix A.1, the asymptotic expression (2.6.24) yields

$$G^{2D}(\vec{x}, t; \vec{x}_0, t_0) \doteq \tfrac{1}{2}\rho\pi^{-1}(c/2r_l)^{1/2} H(t - t_0 - r_l/c)(t - t_0 - r_l/c)^{-1/2}, \tag{2.6.27}$$

and the exact expression (2.6.25) yields

$$G^{2D}(\vec{x}, t; \vec{x}_0, t_0) = \tfrac{1}{2}\rho\pi^{-1} H(t - t_0 - r_l/c)\big[(t - t_0)^2 - r_l^2/c^2\big]^{-1/2}. \tag{2.6.28}$$

Here $H(\zeta)$ is the Heaviside function. In the first-motion approximation $t - t_0 \doteq r_l/c$, the exact equation (2.6.28) yields (2.6.27), as $(t - t_0)^2 - r_l^2/c^2 = (t - t_0 - r_l/c)(t - t_0 + r_l/c) \doteq (2r_l/c)(t - t_0 - r_l/c)$.

If we compare the 3-D (point-source) acoustic Green functions $G(\vec{x}, \vec{x}_0, \omega)$ and $G(\vec{x}, t; \vec{x}_0, t_0)$ given by (2.5.28) and (2.5.29) with the analogous 2-D (line-source) acoustic Green functions (2.6.24) through (2.6.28), we see several basic differences. Let us first discuss the frequency-domain Green functions.

1. Whereas the exact 3-D Green function (2.5.28) is very simple, the exact 2-D Green function (2.6.25) is more complex and is expressed in terms of the Hankel function. The expression simplifies if we consider $\omega r_l/c \to \infty$ (far-field zone and/or high-frequency asymptotics); see (2.6.24). Thus, in 2-D with a line source, we need to distinguish between exact and asymptotic computations, even in homogeneous media.

2. The 3-D Green function (2.5.28) decreases with increasing distance r from the point source as r^{-1}. The 2-D asymptotic Green function (2.6.24), however, decreases with increasing distance r_l from the line source as $r_l^{-1/2}$.

3. The Fourier spectrum of the 3-D Green function (2.5.28) is constant, independent of frequency. The Fourier spectrum of the 2-D Green function (2.6.24), however, is not constant, but rather proportional to $|\omega|^{-1/2} \exp[i(\pi/4)\,\mathrm{sgn}(\omega)]$.

As we can see from (2.5.28) and (2.6.24), the amplitude of the 2-D (line-source) asymptotic time-harmonic Green function $G^{2D}(\vec{x}, \vec{x}_0, \omega)$ can be calculated from the 3-D (point-source) time-harmonic Green function in two steps:

a. Replace r by $(r_l/c)^{1/2}$.

b. Multiply it by filter $F(\omega)$ given by the relation:

$$F(\omega) = (2\pi/|\omega|)^{1/2} \exp\left[i\tfrac{\pi}{4}\,\mathrm{sgn}(\omega)\right]. \qquad (2.6.29)$$

We shall call $F(\omega)$ given by (2.6.29) the two-dimensional frequency filter.

In the time domain, we can also calculate the 2-D (line-source) Green function $G^{2D}(\vec{x}, t; \vec{x}_0, t_0)$ given by (2.6.27) from the 3-D (point-source) Green function $G(\vec{x}, t; \vec{x}_0, t_0)$ given by (2.5.29) in two steps. Step a is the same as before, but step b should be replaced by step b*.

b*. Perform the following replacement:

$$\delta(t - t_0 - r/c) \rightarrow \sqrt{2}H(t - t_0 - r_l/c)(t - t_0 - r_l/c)^{-1/2}. \qquad (2.6.30)$$

In Chapter 5, we shall show that the 2-D ray-theory Green function can be calculated from the 3-D ray-theory Green function in the same way as already shown for 2-D laterally varying layered structures. Step a is equivalent to replacing the relative geometrical spreading $\mathcal{L}$ by the in-plane relative geometrical spreading $\mathcal{L}^{\|}$. In a homogeneous medium, this is equivalent to step a. Steps b and b* remain valid even in this case without any change.

Regarding the time-domain expressions (2.6.27) and (2.6.28) for the 2-D acoustic Green functions, the most striking difference is that the 2-D Green functions do not contain the delta functions as in 3-D; see (2.5.29). The time-domain 2-D Green functions are infinite at the wavefront, but they also include a *long tail* that behaves with increasing time t as $[(t - t_0)^2 - r_l^2/c^2]^{-1/2}$. The long tail can be simply understood if we realize that the travel time from a point on the line source to the fixed receiver increases with the distance of the point from the plane $x_2 = 0$.

The procedure just outlined can also be used to find the 2-D elastodynamic Green functions for isotropic media. Similarly, as in the acoustic case, we need to distinguish between exact and asymptotic 2-D Green functions. Asymptotic high-frequency expressions for 2-D Green functions, however, can be constructed from 3-D Green functions in the same way as in the acoustic medium, see steps a, b, and b* and Section 5.2.15. We shall not repeat the whole procedure here; the interested reader is referred to Hudson (1980a) for exact expressions and to Section 5.2.15.

CHAPTER THREE

Seismic Rays and Travel Times

T he two most important concepts in the propagation of high-frequency seismic body waves in smoothly varying, layered and block structures are their *travel times* and *rays*. Both concepts are closely related. Many procedures to compute rays and travel times have been proposed. The selection of the appropriate procedure to compute seismic rays and the relevant travel times is greatly influenced by such factors as:

a. The dimensionality of the model under consideration (1-D, 2-D, 3-D).

b. The computer representation and the complexity of the model (for example, a smooth model with smooth interfaces, a grid model, or a cell model).

c. The source-receiver configuration (localization of earthquakes, surface profile measurements, VSP, cross-hole configuration, migration, and the like) and by the volume of the required computations. As an example, compare the volume of computations for a point-source 2-D surface profile configuration and for extensive 3-D migration/inversion grid computations.

d. The required accuracy of computations.

e. The required numerical efficiency of computations.

f. The type of computed travel times (first arrivals only, later arrivals, diffracted waves, and the like).

g. The required comprehensiveness of computations (travel times only, rays and travel times, or also Green function, synthetic seismograms, and particle ground motions).

h. The practical purpose of ray tracing and travel-time computation.

Some of these factors, of course, overlap and/or are mutually connected. Moreover, both travel times and rays have been defined and used in seismology and in seismic exploration in various conflicting ways and with different definitions. In this book, we shall introduce the rays and the travel times using mostly the high-frequency asymptotic methods applied to acoustic and elastodynamic wave equations. For exceptions, see Section 3.8.

In the *acoustic case*, the application of HF asymptotic methods to the acoustic wave equation immediately yields the eikonal equation $(\nabla T)^2 = 1/c^2$ as the basic equation to calculate travel time T and rays Ω; see Section 2.4.1. In an *isotropic elastic medium*, the derivation of similar eikonal equations is not quite straightforward. The application of HF asymptotic methods to the elastodynamic isotropic equation yields the separation of the wavefield into two waves, which may propagate separately: P and S waves. The relevant travel times and rays of these waves are, however, again controlled by similar eikonal equations: $(\nabla T)^2 = 1/\alpha^2$ for P waves and $(\nabla T)^2 = 1/\beta^2$ for S waves, where α and β are the local velocities of the P and S waves given by (2.4.23). See Section 2.4.2 for details.

Finally, in *anisotropic inhomogeneous media*, the application of HF asymptotic methods to the elastodynamic equation yields three different types of waves that may separately propagate in the medium: one qP and two qS (qS1 and qS2). The relevant eikonal equation for any of these three waves is $G_m(x_i, p_i) = 1, m = 1, 2, 3$, where G_m are eigenvalues of the Christoffel matrix; see Section 2.4.3.

Thus, if we exclude the anisotropic medium, we can formally use *the same eikonal equation* $(\nabla T)^2 = 1/V^2$, for both acoustic and elastic isotropic media. Here $V = c$ for the acoustic case, $V = \alpha$ for P waves in the elastic medium, and $V = \beta$ for S waves in the elastic medium. The eikonal equation $(\nabla T)^2 = 1/V^2$ is the basic equation of Chapter 3. The exception is only Section 3.6, where anisotropic inhomogeneous media are treated.

The eikonal equation $(\nabla T)^2 = 1/V^2(x_i)$ is a nonlinear partial differential equation of the first order. In mathematics, such equations are usually solved for T in terms of characteristics (see Bleistein 1984). The characteristics of the eikonal equation are some trajectories, described by a system of ordinary differential equations that can be solved easily by standard numerical procedures. The main advantage of characteristics of the eikonal equation is that the travel time T along them can be calculated by simple quadratures.

In the seismic ray theory, we define *rays as characteristics of the eikonal equation* and call the system of ordinary differential equations for the characteristics the *ray tracing system* or the *system of ray equations*. The ray tracing system may be expressed in various forms; see Section 3.1. In layered media, we need to supplement the definition of rays, based on the characteristics of the eikonal equation, with Snell's law at those points where the ray has contact with a structural interface; see Section 2.4.5.

Both the rays and travel times may also be introduced in different ways than explained here. For different definitions of seismic rays, see Section 3.1. Regarding the travel times, we distinguish between *two basic definitions*.

a. **Ray-theory travel times.** Ray-theory travel times are introduced here as the travel times of the individual elementary waves (such as direct, reflected, multiply reflected, and converted), calculated along the rays of these waves. Thus, for different elementary waves, we obtain different ray-theory travel times. For this reason, ray-theory travel times are also sometimes called elementary travel times. It is obvious that the term ray-theory travel time only has an approximate, asymptotic HF meaning.

b. **First-arrival travel times.** The first-arrival travel times have an exact, not asymptotic, meaning, even for inhomogeneous layered structures. They correspond to the exact solution of an elastodynamic equation in a given model and to the complete wavefield, which is not separated into individual elementary waves.

For more detailed definitions of these two types of travel times, their extensions, and an explanation of differences, see Section 3.8.1.

In the following text, unless otherwise stated, we shall refer to the ray-theory travel times as travel times. We remind the readers that any ray-theory travel time corresponds to a selected elementary wave. Only if the travel times could be confused, will their type be specified.

Several direct methods to calculate the travel times of seismic body waves and wavefronts, without computation of rays, have been used in seismology and seismic prospecting. They will be briefly discussed in Section 3.8. The rays, if they are needed, are computed afterwards, from known travel times. The computation of travel times by first-order perturbation methods is discussed in Section 3.9. Both isotropic and anisotropic layered structures are considered. The advantage of the perturbation methods is that they require ray tracing to be performed only in a simpler background medium. The travel time in a perturbed medium is then obtained by quadratures along the known ray in a background medium. The perturbation methods are very attractive in the solution of inverse kinematic structural problems (seismic tomography).

Mostly, however, ray tracing is a necessary prerequisite for the ray-theory travel-time computations. For this reason, considerable attention is devoted in Chapter 3 to ray tracing. There are two main types of the ray tracing: (a) initial-value ray tracing and (b) boundary-value ray tracing. In the *initial-value ray tracing*, the direction of the ray is known at some point of the ray (or, at least, it may be simply determined from some other known data, for example, at an initial surface). The position of the point and the direction of the ray at that point then constitute the complete system of *initial conditions* for the ray tracing system, and the ray trajectory can be calculated with the required accuracy using various methods. In the *boundary-value ray tracing*, the direction of the ray is not known at any of its points. Instead, some other conditions are known at the ray; for example, we are seeking the ray of a specified elementary wave that connects two given points. The direction of the ray at any of these two points is not known a priori. We then speak of *two-point ray tracing*. In yet another example, we know how to compute the initial direction of rays at all points on some initial surface and are seeking the ray that passes through a specified fixed point R outside the initial surface. We then speak of *initial surface-fixed point ray tracing*. For details, refer to Section 3.11.

In this chapter, we devote most of our attention to various approaches to solving of the initial-value ray tracing problem in relation to the computation of ray-theory travel times. In the following, we shall present a brief review of these approaches. We shall consider an arbitrary elementary wave propagating in general 3-D inhomogeneous layered and block structures. We shall mainly discuss approaches suitable for isotropic media. For anisotropic media, see Section 3.6.

There are four main methods of **initial-value ray tracing**.

1. The first method involves the *numerical solution of ray-tracing equations*, supplemented by Snell's law at those points where the ray contacts structural interfaces. The relevant ray-theory travel times are either obtained automatically by ray tracing (if the variable in the ray tracing system is T) or by quadratures along the known rays. The numerical solution of the ray tracing system can be performed, with controlled accuracy using various methods such as the Runge-Kutta method and Hamming's predictor-corrector method. For details, refer to Sections 3.2 and 3.3. In Section 3.2.3, various anomalous situations in numerical ray tracing (boundary rays, critical rays, diffracted rays, sliding rays, edge rays, and the like) are described.

2. The second method is based on *analytical solutions* of the ray tracing system. Analytical solutions, however, are available only exceptionally, for example, for models in which $V^{-n}(x_i)$ ($n = \pm 1, \pm 2, \ldots$) is a linear function of Cartesian coordinates. See Section 3.4, particularly Sections 3.4.1 through 3.4.5. The most general case for which the analytical solutions are known corresponds to $V^{-2}(x_i) = a + b_i x_i + c_{ij} x_i x_j$; see Körnig (1995) and Section 3.4.6.

3. The third method is based on *semianalytical solutions*. This method has two alternatives. *In the first alternative*, the model is divided into layers and/or blocks, separated by curved structural interfaces. The velocity distribution inside the individual layers and blocks is specified in such a way that it allows the analytical computation of rays. *In the second alternative*, the model is divided into cells (usually rectangular

or triangular in 2-D and rectangular or tetrahedral in 3-D) in which the velocity is again specified in such a way that it allows the analytical computation of rays. In both alternatives, the resulting rays in the whole model are then obtained as a chain of ray elements in the individual layers, blocks, or cells, computed analytically. Both alternatives may, of course, be combined. For details see Sections 3.4.7 and 3.4.8.

4. The fourth method is applicable only to 1-D models (vertically inhomogeneous with plane structural interfaces, radially symmetric with spherical interfaces). The ray tracing and ray-theory travel-time computations can then be performed by *standard quadratures of ray integrals*. This is a classical problem in seismology and is well described in the seismological literature. See Section 3.7 for the basic ray integrals.

Ray tracing systems can be expressed (and solved) in any *curvilinear coordinate system*, including nonorthogonal coordinates; see Section 3.5, which devotes considerable attention to spherical coordinates, due to their importance in global seismology. See Sections 3.5.4 and 3.5.5. The systems derived can, however, also be applied to ellipsoidal and other coordinates.

In anisotropic inhomogeneous media, the numerical solution of the ray tracing system is the most important; analytical solutions are quite exceptional. See Section 3.6.

The initial-value ray tracing is also widely used in the **boundary-value ray tracing problem**, particularly in the so-called shooting method. A classification and explanation of different boundary-value ray tracing problems can be found in Section 3.11. In addition to shooting methods, many other methods are also discussed there; these methods include bending methods and methods based on the perturbation theory and on the paraxial ray theory. Section 3.11 only gives the classification and a very brief explanation of various boundary-value ray tracing approaches and the relevant literature; the individual methods on which these approaches are based are explained in greater detail elsewhere in the book. For the shooting methods, see Sections 3.1 through 3.7. The solution of paraxial boundary-value ray tracing problems is discussed in Chapter 4, particularly in Section 4.9. For the application of perturbation methods in the two-point computation of travel times, see Section 3.9. Finally, Section 3.8 treats the computation of first-arrival travel times along 2-D and 3-D grids of points.

Section 3.10 is not devoted to single rays but to *orthonomic systems of rays*, corresponding to an arbitrary elementary seismic wave propagating in a 3-D layered structure. These systems of rays are also called simply *ray fields*. Many important concepts connected with ray fields, particularly ray parameters, ray coordinates, ray Jacobians, elementary ray tubes and their cross-sectional areas, geometrical spreading, caustics, and the KMAH index, are introduced. The properties of the ray Jacobian are discussed, and various methods of its computation are described. Section 3.10 also briefly explains how the transport equations can be simply solved along rays in terms of the ray Jacobian. Such solutions of the transport equation will be broadly used in Chapter 5 to study the ray amplitudes of seismic body waves. The derived equations are very general; many of them are applicable both to isotropic and anisotropic media.

3.1 Ray Tracing Systems in Inhomogeneous Isotropic Media

Rays play a basic role in various branches of physics. For this reason, it is not surprising that many different approaches can be used to define them and to derive ray tracing systems.

The most general approach to deriving seismic ray tracing systems is based on the asymptotic high-frequency solution of the elastodynamic equation. This approach yields a very important result, namely that the high-frequency seismic wave field *in a smoothly inhomogeneous isotropic medium* is approximately separated into two independent waves: the P and the S wave. The travel-time fields of these two independent waves satisfy the relevant eikonal equations. See Section 2.4 for an approximate derivation of the eikonal equations. Such an approach is formal, but it is also the most rigorous and straightforward and will be used to derive the ray tracing system in this section.

It will also be shown that the rays of P and S waves in smoothly inhomogeneous isotropic media, introduced as characteristics of the eikonal equation, satisfy the following important geometrical and physical properties.

- They are orthogonal to wavefronts.
- They are extremal curves of Fermat's functional.
- They satisfy locally Snell's law, modified for smooth inhomogeneous media.
- They represent trajectories along which the high-frequency part of the time-averaged energy flows.

Some of these properties can also be used alternatively to derive the ray tracing systems of P and S waves. Only the energy approach, however, offers the possibility of separating high-frequency P and S waves in smoothly inhomogeneous isotropic media. In the three remaining approaches, the separation is not derived, but it is assumed a priori.

3.1.1 Rays as Characteristics of the Eikonal Equation

The eikonal equation of high-frequency P and S waves propagating in smoothly inhomogeneous isotropic media $(\nabla T)^2 = 1/V^2$ was derived in Section 2.4.2. In Cartesian coordinates, it reads

$$p_i p_i = 1/V^2(x_i), \qquad \text{where } p_i = \partial T/\partial x_i. \tag{3.1.1}$$

Here $T = T(x_i)$ is the travel time (eikonal), p_i are components of the slowness vector, $\vec{p} = \nabla T$, $V = \alpha$ for P waves, and $V = \beta$ for S waves.

Equation (3.1.1) is a nonlinear partial differential equation of the first order for $T(x_i)$. It can be expressed in many alternative forms. In general, we shall write the eikonal equation in the following form:

$$\mathcal{H}(x_i, p_i) = 0, \tag{3.1.2}$$

where function $\mathcal{H}$ may be specified in different ways. For example, $\mathcal{H}(x_i, p_i) = p_i p_i - V^{-2}$, $\mathcal{H}(x_i, p_i) = \frac{1}{2}(V^2 p_i p_i - 1)$, $\mathcal{H}(x_i, p_i) = (p_i p_i)^{1/2} - 1/V$, and $\mathcal{H}(x_i, p_i) = \frac{1}{2}\ln(p_i p_i) + \ln V$.

The nonlinear partial differential equation (3.1.2) is usually solved in terms of *characteristics*. The characteristics of (3.1.2) are 3-D space trajectories $x_i = x_i(u)$ (u being some parameter along the trajectory), along which $\mathcal{H}(x_i, p_i) = 0$ is satisfied, and along which travel time $T(u)$ can be simply evaluated by quadratures. The characteristic curve is a solution of the so-called *characteristic system* of ordinary differential equations of the first order. The detailed derivation of the characteristic system can be found in many textbooks and monographs such as Smirnov (1953), Morse and Feshbach (1953), Kline and Kay (1965), Courant and Hilbert (1966), Bleistein (1984), and Babich, Buldyrev, and Molotkov (1985). In particular, the book by Bleistein (1984) offers a very detailed and tutorial treatment. Here we shall not derive the characteristic system of equations, but refer the reader to these papers and books. The characteristic system of the nonlinear partial differential equation (3.1.2) reads

$$\frac{dx_i}{du} = \frac{\partial \mathcal{H}}{\partial p_i}, \qquad \frac{dp_i}{du} = -\frac{\partial \mathcal{H}}{\partial x_i}, \qquad \frac{dT}{du} = p_k \frac{\partial \mathcal{H}}{\partial p_k}, \qquad i = 1, 2, 3.$$

$$(3.1.3)$$

In a 3-D medium, the system consists of seven equations. The six equations for $x_i(u)$ and $p_i(u)$ are, in general, coupled and must be solved together. The solution to these six equations is $x_i = x_i(u)$, the characteristic curve as a 3-D trajectory, and $p_i = p_i(u)$, the components of the slowness vector along the characteristic. The seventh equation for the travel time along the trajectory, $T = T(u)$, is not coupled with the other six equations and can be solved independently, as soon as the characteristic is known. The solution $T = T(u)$ is then obtained by simple quadratures. It is not difficult to see that $\mathcal{H}(x_i, p_i) = 0$ is satisfied along the characteristic, as soon as it is satisfied at one reference point of the characteristic.

Parameter u along the characteristic cannot be chosen arbitrarily. It depends on the specific form of function $\mathcal{H}$ in (3.1.2). Increment du along the characteristic is related to the increment of the travel time, dT, as

$$du = dT/(p_k \partial \mathcal{H}/\partial p_k). \qquad (3.1.4)$$

Since the rays have been defined as characteristic curves of the eikonal equation, the system of ordinary differential equations (3.1.3) can be used to determine the ray trajectory and the travel time along it. We call this the *system of ray equations*, or the *ray tracing system*.

In the seismological literature, it is common to give a mechanical interpretation to the eikonal equation, written in the form of (3.1.2), and to the ray tracing system (3.1.3). In classical mechanics, equations (3.1.3) represent the canonical equations of motion of a particle that moves in the field governed by the Hamiltonian function $\mathcal{H}(x_i, p_i)$ and has energy $\mathcal{H} = 0$. See Kline and Kay (1965) and Goldstein (1980). The Hamiltonian function $\mathcal{H}(x_i, p_i)$ is usually called just the Hamiltonian, quantities p_i are called the momenta, and (3.1.3) are called the Hamiltonian canonical equations. This terminology has recently been used in seismology; see, for example, Thomson and Chapman (1985) and Farra and Madariaga (1987). We shall also use it in this book; but we shall prefer the term *components of the slowness vector* in the case of p_i.

In the Hamiltonian formalism of classical mechanics, x_i and p_i are considered to be independent coordinates in a *six-dimensional phase space*. They are also often called *canonical coordinates*, and the 6×1 column matrix $(x_1, x_2, x_3, p_1, p_2, p_3)^T$ is called the *canonical vector*. Equation $\mathcal{H}(x_i, p_i) = 0$ then represents a hypersurface in 6-D phase space. On the hypersurface,

$$d\mathcal{H} = \frac{\partial \mathcal{H}}{\partial x_i} dx_i + \frac{\partial \mathcal{H}}{\partial p_i} dp_i = 0.$$

The relation is satisfied if we put

$$dx_i/(\partial \mathcal{H}/\partial p_i) = -dp_i/(\partial \mathcal{H}/\partial x_i), \qquad i = 1, 2, 3 \qquad (3.1.5)$$

(no summation over i). If we put these expressions equal to the differential of auxiliary variable du, we obtain the first six equations of the canonical Hamilton system (3.1.3). The additional equation for T is then obtained simply as

$$p_i \frac{\partial \mathcal{H}}{\partial p_i} = p_i \frac{dx_i}{du} = \frac{\partial T}{\partial x_i} \frac{dx_i}{du} = \frac{dT}{du}.$$

Thus, the solution of the Hamilton system of equations, where $x_i = x_i(u)$ and $p_i = p_i(u)$, can also be interpreted as the parametric equations of a curve in 6-D phase space x_i, p_i ($i = 1, 2, 3$) on hypersurface $\mathcal{H}(x_i, p_i) = 0$. This curve is also often called the (6-D) *characteristic curve* of (3.1.2). Ray $x_i = x_i(u)$ is then a 3-D projection of the 6-D characteristic curve into the x_1, x_2, x_3-space.

We shall now express the characteristic systems (3.1.3) corresponding to several different forms of eikonal equation (3.1.2). We shall first use a rather general form of the Hamiltonian, which includes many other options,

$$\mathcal{H}(x_i, p_i) = n^{-1}\{(p_i p_i)^{n/2} - 1/V^n\}, \tag{3.1.6}$$

where n is a real-valued quantity. In applications, we shall consider n to be an integer. We shall use (3.1.6) also in the limit for $n \to 0$. The l'Hospital rule then yields

$$\mathcal{H}(x_i, p_i) = \tfrac{1}{2}\ln(p_i p_i) + \ln V = \tfrac{1}{2}\ln(V^2 p_i p_i). \tag{3.1.7}$$

Factor $1/n$ in (3.1.6) is used to obtain a suitable parameter u along the characteristic.

The characteristic system of equations (3.1.3) corresponding to Hamiltonian (3.1.6) reads,

$$\frac{dx_i}{du} = (p_k p_k)^{n/2-1} p_i, \qquad \frac{dp_i}{du} = \frac{1}{n}\frac{\partial}{\partial x_i}\left(\frac{1}{V^n}\right),$$

$$\frac{dT}{du} = (p_k p_k)^{n/2} = V^{-n}; \tag{3.1.8}$$

see (3.1.3). Because we have identified the characteristics of the eikonal equations as rays, system (3.1.8) represents the ray tracing system.

It will be useful to write explicitly several forms of the ray tracing system for different n and, consequently, for different parameters u along the ray.

For $n = 0$, $dT/du = 1$. Thus, parameter u along the ray is directly equal to travel time T. The ray tracing system then reads

$$\frac{dx_i}{dT} = (p_k p_k)^{-1} p_i, \qquad \frac{dp_i}{dT} = -\frac{\partial \ln V}{\partial x_i}. \tag{3.1.9}$$

For $n = 1$, $dT/du = 1/V$. Parameter u along the ray is arclength s along the ray. The ray tracing system reads

$$\frac{dx_i}{ds} = (p_k p_k)^{-1/2} p_i, \qquad \frac{dp_i}{ds} = \frac{\partial}{\partial x_i}\left(\frac{1}{V}\right), \qquad \frac{dT}{ds} = \frac{1}{V}. \tag{3.1.10}$$

The simplest ray tracing system is obtained for $n = 2$. We now denote parameter u along the ray by σ:

$$\frac{dx_i}{d\sigma} = p_i, \qquad \frac{dp_i}{d\sigma} = \frac{1}{2}\frac{\partial}{\partial x_i}\left(\frac{1}{V^2}\right), \qquad \frac{dT}{d\sigma} = \frac{1}{V^2}. \tag{3.1.11}$$

Finally, we shall also use the ray tracing system for $n = -1$. We then denote parameter u along the ray by ζ, and the ray tracing system reads

$$\frac{dx_i}{d\zeta} = (p_k p_k)^{-3/2} p_i, \qquad \frac{dp_i}{d\zeta} = -\frac{\partial V}{\partial x_i}, \qquad \frac{dT}{d\zeta} = V. \tag{3.1.12}$$

All the ray tracing systems (3.1.8) through (3.1.12) are shown in the form of Hamilton canonical equations (3.1.3) for eikonal equation $\mathcal{H}(x_i, p_i) = 0$; see (3.1.2).

One of the great advantages of systems (3.1.8) through (3.1.12) is that the RHSs of equations for dx_i/du depend only on canonical coordinates p_i (not x_i), and the RHSs of equations for dp_i/du only on canonical coordinates x_i (not p_i). This allows closed-form analytical solutions of the ray tracing equations to be found in many useful cases.

Many other forms of the Hamiltonian $\mathcal{H}(x_i, p_i)$ also yield suitable ray tracing systems. For example, we can multiply (3.1.6) by V^n,

$$\mathcal{H}(x_i, p_i) = n^{-1}\{(V^2 p_k p_k)^{n/2} - 1\} = 0.$$

An interesting property of this Hamiltonian is that $dT/du = (V^2 p_k p_k)^n = 1$ for arbitrary n. Thus, the variable u along the ray equals travel time T for any n. The most common case is to consider $n = 2$. The Hamiltonian then reads

$$\mathcal{H}(x_i, p_i) = \tfrac{1}{2}(V^2 p_k p_k - 1) = 0, \tag{3.1.13}$$

and the relevant ray tracing system is

$$dx_i/dT = V^2 p_i, \qquad dp_i/dT = -\tfrac{1}{2} p_k p_k \partial V^2/\partial x_i. \tag{3.1.14}$$

Another suitable form of the Hamiltonian is

$$\mathcal{H}(x_i, p_i) = \tfrac{1}{2} V^{2-n}(p_k p_k - V^{-2}). \tag{3.1.15}$$

Because the Hamiltonian (3.1.15) satisfies the relation $p_i \partial \mathcal{H}/\partial p_i = V^{-n}$, it yields the same monotonic parameter u along the ray as (3.1.6) for the same n. For $n = 0$, (3.1.15) yields $u = T$, for $n = 1$ it yields $u = s$, and for $n = 2$ it yields $u = \sigma$. The Hamiltonian (3.1.15) has often been used for $n = 1$ (that is, for $u = s$). Then it reads

$$\mathcal{H}(x_i, p_i) = \tfrac{1}{2} V(p_k p_k - V^{-2}), \tag{3.1.16}$$

and the relevant ray tracing system is

$$dx_i/ds = V p_i, \qquad dp_i/ds = -\tfrac{1}{2} p_k p_k \partial V/\partial x_i + \tfrac{1}{2}\partial(1/V)/\partial x_i. \tag{3.1.17}$$

Even more general forms of Hamiltonians can be introduced. As an example, see (3.4.11). The next example is $\mathcal{H}(x_i, p_i) = \tfrac{1}{2} F(x_i)(p_k p_k - V^{-2})$, where $F(x_i)$ is an arbitrary continuous positive function. Then, $dT/du = F(x_i)/V^2(x_i)$.

Because the eikonal equation $\mathcal{H}(x_i, p_i) = 0$ is satisfied along the whole ray (once it is satisfied at one of its points), we can modify the ray tracing systems by inserting $p_k p_k = V^{-2}$. Equations (3.1.8) through (3.1.10) and (3.1.12) then yield

$$\frac{dx_i}{du} = V^{2-n} p_i, \qquad \frac{dp_i}{du} = \frac{1}{n}\frac{\partial}{\partial x_i}\left(\frac{1}{V^n}\right), \qquad \frac{dT}{du} = \frac{1}{V^n}, \tag{3.1.18}$$

$$\frac{dx_i}{dT} = V^2 p_i, \qquad \frac{dp_i}{dT} = -\frac{\partial \ln V}{\partial x_i}, \tag{3.1.19}$$

$$\frac{dx_i}{ds} = V p_i, \qquad \frac{dp_i}{ds} = \frac{\partial}{\partial x_i}\left(\frac{1}{V}\right), \qquad \frac{dT}{ds} = \frac{1}{V}, \tag{3.1.20}$$

$$\frac{dx_i}{d\zeta} = V^3 p_i, \qquad \frac{dp_i}{d\zeta} = -\frac{\partial V}{\partial x_i}, \qquad \frac{dT}{d\zeta} = V. \tag{3.1.21}$$

Similarly, (3.1.14) and (3.1.17) also yield (3.1.19) and (3.1.20). Equations (3.1.18) through (3.1.21) are not expressed in the Hamiltonian form (3.1.3). Nevertheless, they represent very useful ray tracing systems.

The simplest version of ray tracing system (3.1.11) corresponds to the Hamiltonian $\mathcal{H}(x_i, p_i) = \frac{1}{2}(p_k p_k - 1/V^2)$; see Burridge (1976). The relevant variable σ along the ray, connected with travel time T by relation $dT/d\sigma = 1/V^2$, is sometimes called the *natural variable along the ray*.

In a 3-D medium, the ray tracing system consists of six ordinary differential equations of the first order, with one additional equation for the travel time. The system of six ordinary differential equations of the first order can be expressed as a system of three ordinary differential equations of the second order. Let us demonstrate this on system (3.1.18):

$$\frac{d}{du}\left(V^{n-2}\frac{dx_i}{du}\right) = \frac{1}{n}\frac{\partial}{\partial x_i}\left(\frac{1}{V^n}\right), \qquad i = 1, 2, 3. \tag{3.1.22}$$

The simplest system of ordinary differential equations of the second order is again obtained for $n = 2$,

$$\frac{d^2 x_i}{d\sigma^2} = \frac{1}{2}\frac{\partial}{\partial x_i}\left(\frac{1}{V^2}\right), \qquad i = 1, 2, 3. \tag{3.1.23}$$

Instead of the components of the slowness vector p_i in the ray tracing systems, it would also be possible to use some other quantities specifying the slowness vector. For example, we can express p_i in terms of the three angles i_1, i_2, and i_3 as follows: $p_1 = V^{-1}\cos i_1$, $p_2 = V^{-1}\cos i_2$, and $p_3 = V^{-1}\cos i_3$. The ray tracing system for a 3-D medium then consists of six ordinary differential equations of the first order for x_k and i_k, $k = 1, 2, 3$. See Yeliseyevnin (1964) and Červený and Ravindra (1971, p. 25). Another possibility is to express p_i in terms of two angles i and ϕ as follows: $p_1 = V^{-1}\sin i \cos\phi$, $p_2 = V^{-1}\sin i \sin\phi$, and $p_3 = V^{-1}\cos i$. The ray tracing system for a 3-D medium then consists of five ordinary differential equations of the first order for x_1, x_2, x_3, i, and ϕ only. The eikonal equation is automatically satisfied along the ray. For a 2-D medium, the relevant system consists of three ordinary differential equations of the first order only. See Pšenčík (1972) and Červený, Molotkov, and Pšenčík (1977, pp. 59, 62). In general, however, the foregoing ray tracing systems in terms of p_i are numerically more efficient than those expressed in terms of some angular quantities. For this reason, we do not present and discuss the latter systems here but refer the reader to the previously given references.

In general, the number of ordinary differential equations of the first order in the ray tracing systems *can always be reduced from six to four*. One of the three equations for p_i can be replaced by the eikonal equation, and one of the three equations for x_i can be removed by taking u equal to coordinate x_i. Similarly, if we express the ray tracing system in terms of ordinary differential equations of the second order, the number of equations can always be reduced from three to two. This applies not only to ray tracing systems in Cartesian coordinates but also to any other orthogonal or nonorthogonal coordinates. Moreover, this also applies to anisotropic media. Many such "reduced" ray tracing systems are known from the seismological literature.

The general procedure of deriving the reduced ray tracing system is simple and is based on the so-called *reduced Hamiltonian*. Here we shall introduce the reduced Hamiltonian for Cartesian coordinates only; for curvilinear coordinates, see Section 3.5. We shall first solve the eikonal equation for p_3 and obtain

$$p_3 = -\mathcal{H}^R(x_1, x_2, x_3, p_1, p_2). \tag{3.1.24}$$

We shall call $\mathcal{H}^R(x_i, p_I)$ the reduced Hamiltonian. The minus sign $(-)$ in (3.1.24) is taken for convenience. Now we define the Hamiltonian $\mathcal{H}(x_i, p_i)$ as

$$\mathcal{H}(x_i, p_i) = p_3 + \mathcal{H}^R(x_1, x_2, x_3, p_1, p_2) = 0. \tag{3.1.25}$$

For this Hamiltonian, we can express the standard ray tracing system using (3.1.3). As a variable u along the ray, we use x_3. Consequently, we do not need to calculate dx_3/du because it equals unity. Moreover, we do not need to calculate dp_3/du because p_3 is given explicitly by (3.1.24). Thus, ray tracing system (3.1.3) reduces to four ordinary differential equations of the first order:

$$\frac{dx_I}{dx_3} = \frac{\partial \mathcal{H}^R}{\partial p_I}, \qquad \frac{dp_I}{dx_3} = -\frac{\partial \mathcal{H}^R}{\partial x_I}, \qquad I = 1, 2. \tag{3.1.26}$$

The additional relation for the travel-time calculation along the ray is

$$\frac{dT}{dx_3} = p_3 + p_I \frac{\partial \mathcal{H}^R}{\partial p_I}; \tag{3.1.27}$$

see (3.1.3). Alternative equations to (3.1.24) through (3.1.27) are obtained if the eikonal equation is solved for p_1 and the variable u along the ray is taken as $u = x_1$. Similarly, we can solve the eikonal equation for p_2 and take $u = x_2$.

As $p_i = \partial T/\partial x_i$, Equation (3.1.24) represents a nonlinear partial differential equation of the first order, which is known as the Hamilton-Jacobi equation. Note that the reduced Hamiltonian $\mathcal{H}^R(x_i, p_I)$ does not vanish along the ray but equals $-p_3$. Consequently, it is not constant along the ray.

We shall now express the reduced Hamiltonian $\mathcal{H}^R(x_i, p_I)$ in Cartesian coordinates:

$$\mathcal{H}^R(x_i, p_I) = -\left[1/V^2(x_1, x_2, x_3) - p_1^2 - p_2^2\right]^{1/2}. \tag{3.1.28}$$

The minus sign $(-)$ corresponds to propagation in the direction of increasing x_3 (positive p_3); see (3.1.24). For the opposite direction of propagation, it would be necessary to take the $+$ in (3.1.28). The reduced ray tracing system (3.1.26) is taken as

$$\frac{dx_I}{dx_3} = \frac{V p_I}{\left[1 - V^2(p_1^2 + p_2^2)\right]^{1/2}},$$
$$\frac{dp_I}{dx_3} = \frac{1}{\left[1 - V^2(p_1^2 + p_2^2)\right]^{1/2}} \frac{\partial}{\partial x_I}\left(\frac{1}{V}\right), \qquad I = 1, 2. \tag{3.1.29}$$

For the travel time, we obtain

$$dT/dx_3 = V^{-1}\left[1 - V^2(p_1^2 + p_2^2)\right]^{-1/2} = 1/V^2 p_3; \tag{3.1.30}$$

see (3.1.27).

Ray tracing system (3.1.29) consisting of four ordinary differential equations of the first order can also be expressed as a system of two ordinary differential equations of the second order. We merely express p_I in terms of $x'_I = dx_I/dx_3$ from the first equation of (3.1.29) and insert them into the second equation of (3.1.29). We obtain $p_I = V^{-1}x'_I[1 + x'^2_1 + x'^2_2]^{-1/2}$ and

$$\frac{d}{dx_3}\left[\frac{1}{V}\frac{x'_I}{\sqrt{1 + x'^2_1 + x'^2_2}}\right] - \frac{\partial}{\partial x_I}\left(\frac{1}{V}\right)\sqrt{1 + x'^2_1 + x'^2_2} = 0,$$

$$I = 1, 2. \tag{3.1.31}$$

Ray tracing system (3.1.29) is simple and consists of four ordinary differential equations of the first order only. The great disadvantage of the system is that parameter $u = x_3$ is not necessarily monotonic along the ray. Ray tracing system (3.1.29) fails at *ray turning points* at which the ray has a minimum with respect to coordinate x_3. In the following, we shall consider only the complete ray tracing systems consisting of six equations, unless otherwise stated. For more details, see Sections 3.3.1 and 3.3.2.

3.1.2 Relation of Rays to Wavefronts

In isotropic inhomogeneous media, the *rays are orthogonal trajectories to wavefronts*. This is a well-known fact from general courses of physics. However, in certain types of media (for example, in anisotropic media), the rays are not orthogonal to wavefronts. Thus, wavefronts and rays are not orthogonal in general. We are not justified in assuming this, even in isotropic media a priori; we must prove it.

In this section, wavefronts $T = T(x_i)$, satisfying eikonal equation $\nabla T \cdot \nabla T = 1/V^2$, are assumed to be known, and we wish to derive the system of differential equations for the orthogonal trajectories to wavefronts. We shall see that the system is identical with the ray tracing system.

We specify the orthogonal trajectory to the wavefronts by parameteric equations $x_i = x_i(s)$, where x_i are the Cartesian coordinates of points along the orthogonal trajectory and s is the arclength. The unit vector tangent to the orthogonal trajectory is $\vec{t} = d\vec{x}/ds$. The vector perpendicular to the wavefront is denoted by $\vec{p} = \nabla T$. At any point on the orthogonal trajectory, $\vec{t}$ must be of the same direction as $\vec{p}$. Thus, $t_i = \lambda p_i$, where λ is some constant. This constant is simply obtained from the eikonal equation, $\lambda = V$. Consequently, $t_i = dx_i/ds = V p_i$. Comparing this equation with (3.1.20), we can conclude that the rays in isotropic inhomogeneous media are perpendicular to wavefronts.

To derive the complete set of differential equations for the orthogonal trajectories, we must still find the equations for dp_i/ds. Taking into account that $p_i = \partial T/\partial x_i$ and $p_i p_i = 1/V^2$,

$$\frac{dp_i}{ds} = \frac{\partial p_i}{\partial x_j}\frac{dx_j}{ds} = V p_j \frac{\partial}{\partial x_j}\left(\frac{\partial T}{\partial x_i}\right) = V p_j \frac{\partial}{\partial x_i}\left(\frac{\partial T}{\partial x_j}\right)$$

$$= V p_j \frac{\partial p_j}{\partial x_i} = \frac{1}{2}V\frac{\partial}{\partial x_i}(p_j p_j) = \frac{1}{2}V\frac{\partial}{\partial x_i}\left(\frac{1}{V^2}\right) = \frac{\partial}{\partial x_i}\left(\frac{1}{V}\right).$$

Together with $dx_i/ds = V p_i$, this result yields ray tracing system (3.1.20).

The rays are orthogonal to the wavefronts in isotropic media irrespective of the coordinate system used. Thus, we can use this concept to derive the ray tracing systems for isotropic media in any curvilinear coordinate system.

3.1.3 Rays as Extremals of Fermat's Functional

Rays are often introduced using variational principles, namely using Fermat's principle. The weakness of this approach in the case of seismic waves was mentioned earlier. Fermat's principle is applied to P and S waves as if they were two independent waves. But as we know, this separation cannot be performed generally; the elastic wavefield is fully separated into P and S waves only in homogeneous media. This separation can be proved only by asymptotic methods.

Let us denote the propagation velocity of the wave under consideration (either P or S) by V and assume that $V(x_i)$ with its first derivatives are continuous functions of Cartesian coordinates x_i. Consider the following integral:

$$J = \int_S^R \mathrm{d}T = \int_S^R \frac{\mathrm{d}s}{V}, \tag{3.1.32}$$

where S and R denote two arbitrary but fixed points. The integration is performed along some curve connecting S and R, where s is the arclength along this curve. The value of J denotes the travel time from S to R. It, of course, depends on the curve along which the integral is taken. Integral (3.1.32) is called Fermat's functional.

Fermat's principle establishes along which curve the signal propagates from S to R. The signal propagates from point S to point R along a curve that renders Fermat's functional (3.1.32) stationary. The curve for which Fermat's functional is stationary is called the *extremal curve* or the *extremal of Fermat's functional*.

We shall now prove that the extremal of Fermat's functional satisfies the same system of ordinary differential equations as the characteristic of the eikonal equation. Recall that the statement that the functional is stationary along a curve means that the first variation of the functional vanishes along that curve,

$$\delta J = \delta \int_S^R \mathrm{d}T = \delta \int_S^R \frac{\mathrm{d}s}{V} = 0. \tag{3.1.33}$$

In the calculus of variations, the functional is often expressed in the following form:

$$J = \int_{x_0}^{x_1} \mathcal{L}(x, y, z, y', z') \mathrm{d}x, \tag{3.1.34}$$

where $y = y(x)$, $z = z(x)$, $y' = \mathrm{d}y/\mathrm{d}x$, and $z' = \mathrm{d}z/\mathrm{d}x$. Points $[x_0, y(x_0), z(x_0)]$ and $[x_1, y(x_1), z(x_1)]$ are fixed. The extremal curve $y = y(x)$, $z = z(x)$ of the preceding functional J satisfies *Euler's equations*,

$$\frac{\mathrm{d}}{\mathrm{d}x}\left(\frac{\partial \mathcal{L}}{\partial y'}\right) - \frac{\partial \mathcal{L}}{\partial y} = 0, \qquad \frac{\mathrm{d}}{\mathrm{d}x}\left(\frac{\partial \mathcal{L}}{\partial z'}\right) - \frac{\partial \mathcal{L}}{\partial z} = 0, \tag{3.1.35}$$

with the relevant boundary conditions at x_0 and x_1.

Function $\mathcal{L}(x, y, z, x', y')$ is often called the Lagrangian. (3.1.34) represents Fermat's functional if we put

$$\mathcal{L}(x, y, z, x', y') = (1 + y'^2 + z'^2)^{1/2} / V(x, y, z). \tag{3.1.36}$$

Euler's equations (3.1.35) then read

$$\frac{\mathrm{d}}{\mathrm{d}x}\left(\frac{1}{V}\frac{y'}{(1 + y'^2 + z'^2)^{1/2}}\right) = \frac{\partial}{\partial y}\left(\frac{1}{V}\right)(1 + y'^2 + z'^2)^{1/2},$$

$$\frac{\mathrm{d}}{\mathrm{d}x}\left(\frac{1}{V}\frac{z'}{(1 + y'^2 + z'^2)^{1/2}}\right) = \frac{\partial}{\partial z}\left(\frac{1}{V}\right)(1 + y'^2 + z'^2)^{1/2}. \tag{3.1.37}$$

These equations are fully equivalent to ray tracing equations (3.1.31) (in the notation $x = x_3, y = x_1, z = x_2$). Thus, we have proved that *the rays may be interpreted as extremals of Fermat's functional.*

Functional (3.1.34) and Euler's equations (3.1.35) are suitable mainly if the extremal has no turning points with respect to the x-coordinate. We, however, prefer to use a monotonic parameter along the ray instead of the x-coordinate. It is then more convenient to consider the variational problem *in parameteric form*. We shall consider any monotonic parameter

u along the curves. The functional then reads

$$J = \int_S^R \mathcal{L}(x_i, x_i')\mathrm{d}u, \qquad i = 1, 2, \ldots, n, \tag{3.1.38}$$

where $x_i = x_i(u), x_i' = \mathrm{d}x_i/\mathrm{d}u$. Points S and R are fixed, but parameter u need not be fixed at these points. We assume that the same function $\mathcal{L}(x_i, x_i')$ can be used for an arbitrary choice of parameter u. In other words, we require that the same extremal curve be obtained for any parameter u. This implies that $\mathcal{L}(x_i, x_i')$ is *a homogeneous function of the first degree in x_i'*. Then again, the extremal curve $x_i = x_i(u)$ of functional (3.1.38) satisfies Euler's equations,

$$\frac{\mathrm{d}}{\mathrm{d}u}\left(\frac{\partial \mathcal{L}}{\partial x_i'}\right) - \frac{\partial \mathcal{L}}{\partial x_i} = 0, \qquad i = 1, 2, \ldots, n. \tag{3.1.39}$$

For details, see Smirnov (1953), Jeffreys and Jeffreys (1966), Courant and Hilbert (1966), Babich and Buldyrev (1972), and Goldstein (1980).

We shall refer to (3.1.39) as *Euler's equations in parameteric form*. Note that (3.1.39) are invariant with respect to the choice of parameter u.

We shall now seek Euler's equations in parameteric form for Fermat's functional. The infinitesimal length element $\mathrm{d}s$ along the curve can be expressed in terms of x_i' as

$$\mathrm{d}s = \left(x_1'^2 + x_2'^2 + x_3'^2\right)^{1/2}\mathrm{d}u, \qquad x_i' = \mathrm{d}x_i/\mathrm{d}u.$$

Fermat's functional (3.1.38) then reads

$$J = \int_S^R V^{-1}(x_i)\left(x_1'^2 + x_2'^2 + x_3'^2\right)^{1/2}\mathrm{d}u$$

so that $\mathcal{L} = V^{-1}(x_i)(x_1'^2 + x_2'^2 + x_3'^2)^{1/2}$ is a homogeneous function of the first order in x_i'. Euler's equations in this particular case read

$$\frac{\mathrm{d}}{\mathrm{d}u}\left(\frac{1}{V}\frac{x_i'}{\left(x_1'^2 + x_2'^2 + x_3'^2\right)^{1/2}}\right) - \frac{\partial}{\partial x_i}\left(\frac{1}{V}\right)\left(x_1'^2 + x_2'^2 + x_3'^2\right)^{1/2} = 0. \tag{3.1.40}$$

Let us now consider a special parameter u along the extremal, which is related to arclength s as $\mathrm{d}u = V^{n-1}\mathrm{d}s$. Note that n and u now have the same meaning as they have in Section 3.1.1. Therefore, $(x_1'^2 + x_2'^2 + x_3'^2)^{1/2} = \mathrm{d}s/\mathrm{d}u = V^{1-n}$, and (3.1.40) reads

$$\frac{\mathrm{d}}{\mathrm{d}u}\left(V^{n-2}\frac{\mathrm{d}x_i}{\mathrm{d}u}\right) = \frac{1}{n}\frac{\partial}{\partial x_i}\left(\frac{1}{V^n}\right), \qquad i = 1, 2, 3.$$

This equation is fully equivalent to (3.1.22). We have thus again proved that the rays may be interpreted as extremals of Fermat's functional, even if we consider Fermat's functional in the parameteric form.

Fermat's principle, in its original version, states that the rays represent the paths that require the least time for a signal to travel from S to R. Thus, it speaks of the minimum of Fermat's functional, not of its stationary value. Consequently, we shall speak of *Fermat's minimum-time principle*. Fermat's minimum-time principle may still play an important role in seismology, if we are interested in the computations of the *first arrival times* only; see Section 3.8. We know, however, that the rays do not necessarily correspond to the least time. In seismology, situations in which the travel-time curve of a certain wave has a loop are common, even if the velocity distribution without interfaces is quite smooth. For a selected elementary wave, the first arrival corresponds to an absolute minimum of Fermat's functional, and the other merely corresponds to the stationary values. The original version of Fermat's principle does not explain these later arrivals. Note that rays never correspond to the absolute maximum of Fermat's functional. They may correspond to the maximum of some selected system of curves, but Fermat's functional is actually stationary in this case.

Thus, the same ray tracing systems can be derived from either the Hamiltonian function $\mathcal{H}(x_i, p_i)$ (related to the eikonal equation) or the Lagrangian function $\mathcal{L}(x_i, x_i')$ (related to Fermat's principle). The ray tracing systems derived from the Hamiltonian function $\mathcal{H}(x_i, p_i)$ consist of six ordinary differential equations of the first order for x_i and p_i (position and slowness vector components). We also speak of ray tracing systems in Hamiltonian form, or of Hamiltonian ray tracing systems. The ray tracing systems derived from the Lagrangian function $\mathcal{L}(x_i, x_i')$ consist of three ordinary differential equations of the second order, but they contain x_i and $x_i' = dx_i/du$ instead of x_i and p_i. We also speak of ray tracing systems in a Lagrangian form, or of Lagrangian ray tracing systems. Here we have derived the Lagrangian ray tracing systems (3.1.37) and (3.1.40) as extremals of Fermat's functional. It should be, however, emphasized that the Lagrangian ray tracing systems can also be derived directly from the Hamiltonian ray tracing systems, without involving Fermat's principle at all. See (3.1.22) and (3.1.23).

The Hamiltonian and Lagrangian ray tracing systems have a well-known analogy in classical mechanics. The canonical equations of motion (3.1.3) of a particle moving in a field governed by the Hamiltonian function $\mathcal{H}(x_i, p_i)$ can also be expressed in terms of the Lagrangian equation of motion, corresponding to Euler's equations (3.1.39). For this reason, it is also common to call (3.1.39) the Euler-Lagrange equations. Of course, this also applies to Equations (3.1.35). In this book, we consistently use the approach based on the asymptotic high-frequency solutions of the wave equations, and on the consequent eikonal and transport equations. As a result, we do not need Fermat's principle and the Lagrangian function at all. For more details on mutual relations between the Hamiltonian and Lagrangian functions and on the relevant Legendre transformation, see Kline and Kay (1965) and Goldstein (1980).

An alternative approach to the definition of rays is also based on variational principles. We shall prove that the *rays are geodesics* in a Riemannian space with a specially chosen metric tensor. Let us consider a Riemannian space with curvilinear coordinates x^i ($i = 1, 2, 3$) and with metric tensor g_{ij}. The square of the distance between two adjacent points is given by relation

$$ds^2 = g_{ij}dx^i dx^j. \tag{3.1.41}$$

The summation in (3.1.41) is over the same upper and lower indices. Distance s may have various physical meanings. The geodesic in a Riemannian space is a curve, whose length has a stationary value with respect to arbitrarily small variations of the curve and whose end points are fixed (see Synge and Schild 1952). Thus, the geodesic between points S and R satisfies the variational condition

$$\delta \int_S^R ds = \delta \int_S^R (g_{ij}dx^i dx^j)^{1/2} = 0. \tag{3.1.42}$$

Let us now assume that ds in (3.1.41) represents the infinitesimal travel time between two adjacent points. In this case, the definition of the geodesic is exactly the same as the definition of the ray; see (3.1.33).

We shall present a simple example of a Riemannian space in which the distance corresponds to the travel time. We shall consider Cartesian coordinates x^i and adopt the following metric tensor:

$$g_{ij} = V^{-2}(x^i)\delta_{ij}, \tag{3.1.43}$$

where V is the velocity and δ_{ij} is the Kronecker delta symbol. As a result, (3.1.41) yields $g_{ij} \mathrm{d}x^i \mathrm{d}x^j = V^{-2}[(\mathrm{d}x^1)^2 + (\mathrm{d}x^2)^2 + (\mathrm{d}x^3)^2] = \mathrm{d}T^2$. Thus, *rays are geodesics* in the Riemannian space with metric tensor g_{ij} given by (3.1.43).

3.1.4 Ray Tracing System from Snell's Law

In this section, we shall derive the ray tracing system by applying Snell's law (2.4.70) locally, without any additional assumption. The approach, however, is not as strict as other approaches, as we simulate a smooth medium by a system of thin homogeneous layers and then decrease the thicknesses of the layers to zero. The limiting process is performed without rigorous proof and is, more or less, intuitive.

We shall consider only unconverted transmitted waves, either P or S. We denote the velocity of the selected wave V. First, we derive an approximate version of Snell's law (2.4.70) for an interface with a small velocity contrast $\tilde{V} - V$. If $(\vec{p} \cdot \vec{n})^2 \gg |\tilde{V}^{-2} - V^{-2}|$, then, approximately,

$$\left[\tilde{V}^{-2} - V^{-2} + (\vec{p} \cdot \vec{n})^2\right]^{1/2} \doteq |\vec{p} \cdot \vec{n}| - V^{-3}(\tilde{V} - V)/|\vec{p} \cdot \vec{n}|.$$

Inserting this into Snell's law (2.4.70) and taking into account that $\epsilon|\vec{p} \cdot \vec{n}| = \vec{p} \cdot \vec{n}$, we obtain Snell's law in the following form:

$$\tilde{\vec{p}} \doteq \vec{p} - V^{-3}(\tilde{V} - V)\vec{n}/(\vec{p} \cdot \vec{n}). \tag{3.1.44}$$

We shall now simulate a smooth medium by using a system of thin homogeneous layers, with first-order interfaces along isovelocity surfaces; see Figure 3.1. (We understand isovelocity surface to be a surface along which the velocity is constant.) If the medium is smooth, the isovelocity surfaces are only slightly curved. We shall replace them locally by tangent planes at the point of intersection with the ray. Moreover, if we choose sufficiently thin layers, the velocity contrast across the individual fictitious interfaces will be small. We can then use (3.1.44). Unit normal $\vec{n}$ at any point of incidence is perpendicular to the isovelocity surface so that $\vec{n} = \pm\nabla V/|\nabla V|$. Equation (3.1.44) then yields

$$\Delta\vec{p} = -\frac{\Delta V}{V^3}\frac{\nabla V}{\vec{p} \cdot \nabla V}, \qquad \Delta\vec{p} = \tilde{\vec{p}} - \vec{p}, \qquad \Delta V = \tilde{V} - V.$$

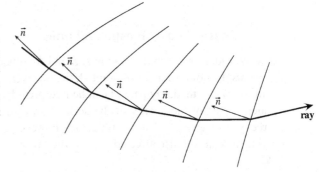

Figure 3.1. Derivation of the ray tracing system from Snell's law. The smooth velocity distribution is simulated by a system of thin homogeneous layers with interfaces along isovelocity surfaces (see thin lines). The arrows denote the unit normals to the isovelocity surfaces.

In the limit, for infinitesimally thin layers, we obtain

$$d\vec{p} = -\frac{1}{V^3}\frac{\nabla V}{\vec{p}\cdot\nabla V}dV.$$

We now denote the ray trajectory by the parametric equations $x_i = x_i(s)$, where s is the arclength along the ray. We also denote the unit vector to the ray $\vec{t}$, so that $d\vec{x}/ds = \vec{t}$. Because Snell's law (2.4.70) implicitly takes the slowness vector to be tangent to the ray, we can put $\vec{t} = V\vec{p}$ and $d\vec{x}/ds = V\vec{p}$. Moreover, we obtain $dV/ds = \vec{t}\cdot\nabla V = V\vec{p}\cdot\nabla V$, and then

$$\frac{d\vec{p}}{ds} = -\frac{1}{V^3}\frac{\nabla V}{\vec{p}\cdot\nabla V}\frac{dV}{ds} = \nabla\left(\frac{1}{V}\right).$$

Thus, the final ray tracing system is

$$d\vec{x}/ds = V\vec{p}, \qquad d\vec{p}/ds = \nabla(1/V).$$

This is fully equivalent to ray tracing system (3.1.20), derived from the eikonal equation by the method of characteristics.

3.1.5   Relation of Rays to the Energy Flux Trajectories

In this section, we shall show that the time-averaged energy of high-frequency elastic waves in smoothly inhomogeneous isotropic media flows along rays. If we use travel time T as a parameter along the ray, the ray tracing system in isotropic inhomogeneous media reads

$$dx_i/dT = V^2 p_i = V N_i, \qquad dp_i/dT = -\partial\ln V/\partial x_i; \qquad (3.1.45)$$

see (3.1.19). Here $V = \alpha$ for P waves and $V = \beta$ for S waves. Derivatives dx_i/dT represent components of vector $\vec{v}_r$, often called the *ray velocity vector*. Ray velocity vector $\vec{v}_r$ is tangent to the ray at any point of the ray. Moreover, ray velocity $v_r = (\vec{v}_r.\vec{v}_r)^{1/2}$ represents the propagation velocity of the wave along the ray trajectory.

The group velocity vector $\vec{U}$ of high-frequency elastic waves propagating in smooth inhomogeneous media was introduced in Section 2.4.4 as the velocity vector of the energy flux. For elastic inhomogeneous *isotropic* media, it is given by the relation $\vec{U} = V\vec{N}$, where $V = \alpha$ for P waves, $V = \beta$ for S waves, and $\vec{N}$ is the unit vector perpendicular to the wavefront; see (2.4.57) and (2.4.58). Comparing $\vec{U}$ with (3.1.45), we see that ray velocity vector $\vec{v}_r$ equals group velocity vector $\vec{U}$ in isotropic media. Thus, the high-frequency part of the elastic energy flows along rays in smoothly inhomogeneous isotropic media. In Section 3.6, we shall prove that the same conclusion applies to inhomogeneous anisotropic media.

3.1.6   Physical Rays. Fresnel Volumes

Let us consider ray Ω, and two points, S and R, situated on this ray. Assume that point S represents the point source and point R, the receiver.

In the ray method, ray path Ω can be interpreted as a trajectory along which the high-frequency part of the energy of the seismic wave under consideration propagates from source S to receiver R. The ray trajectory, however, is only mathematical fiction. In fact, the wavefield at R is also affected by the structure and velocity distribution in some vicinity of Ω.

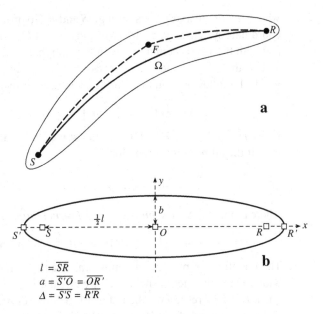

Figure 3.2. Fresnel volumes. (a) The Fresnel volume in an inhomogeneous medium, corresponding to the point source at S and receiver at R, for a selected frequency f. The bold line denotes the ray Ω from S to R. The Fresnel volume consists of points F satisfying the condition (3.1.46). (b) The Fresnel volume in a homogeneous medium corresponding to the point source at S and receiver at R is represented by an ellipsoid of revolution with foci at S and R.

$$l = \overline{SR}$$
$$a = \overline{S'O} = \overline{OR'}$$
$$\Delta = \overline{S'S} = \overline{R'R}$$

The region that actually affects the wavefield at R has been the subject of interest and investigation for a long time. As a result of numerous experiments, it is believed that the wavefield at R is affected by the structure in some vicinity of central ray Ω which is called the *Fresnel volume*. The Fresnel volume, of course, depends on the position of S and R. Therefore, we shall speak of the *Fresnel volume corresponding to a point source at S and a receiver at R*.

The Fresnel volume has been defined in different ways. Here we shall use a very simple definition proposed by Kravtsov and Orlov (1980). Let us consider an elementary harmonic wave propagating from the point source at S to the receiver at R. Denote the frequency of the harmonic wave by f and the ray connecting S and R by Ω. Let us consider an auxiliary point F in the vicinity of central ray Ω and construct the rays connecting F with S and R. Point F then belongs to the Fresnel volume corresponding to the point source at S and the receiver at R if, and only if,

$$|T(F, S) + T(F, R) - T(R, S)| < \tfrac{1}{2} f^{-1}. \tag{3.1.46}$$

Here $T(F, S)$ is the travel time from S to F, and similarly for $T(F, R)$ and $T(R, S)$. See Figure 3.2(a).

The physical meaning of condition (3.1.46) is obvious. Points F such that time difference $|T(F, S) + T(F, R) - T(R, S)|$ is larger than one half of the period do not affect the wavefield at R significantly. For these time differences, destructive interference plays an important role.

Fresnel volumes are closely related to Fresnel zones. Here we shall define the *Fresnel zone at point O of ray Ω* as a section of the Fresnel volume by a plane perpendicular to ray Ω at O. In a similar way, we can define the *interface Fresnel zone* at a point of incidence Q on interface Σ as a section of the Fresnel volume by interface Σ.

As a simple but very important example, we shall briefly discuss the Fresnel volumes of direct waves generated by a point source *in a homogeneous medium*. We denote the distance between S and R by l and the propagation velocity by V. It is not difficult to prove that the Fresnel volume of the direct wave in a homogeneous medium is an ellipsoid of revolution, whose rotational axis passes through S and R. To prove this, we introduce a local Cartesian coordinate system x, y, z with its origin at point O, situated in the middle between S and R, and with the x-axis coinciding with the straight line $\overline{SR}$. The x-coordinates of points S, O and R are $-\frac{1}{2}l$, 0, and $\frac{1}{2}l$, respectively. See Figure 3.2(b). In local coordinates, (3.1.46) yields the following equation for the boundary of the Fresnel volume:

$$\left[(x - \tfrac{1}{2}l)^2 + r^2\right]^{1/2} + \left[(x + \tfrac{1}{2}l)^2 + r^2\right]^{1/2} - l = \tfrac{1}{2}\lambda, \tag{3.1.47}$$

where $r^2 = y^2 + z^2$, $\lambda = Vf^{-1}$. After some simple algebra, this equation yields the equation of the ellipsoid of revolution,

$$\frac{x^2}{a^2} + \frac{y^2 + z^2}{b^2} = 1, \tag{3.1.48}$$

where a and b are the *semiaxes of the Fresnel ellipsoid*. They are given by relations

$$a = \tfrac{1}{2}l\big(1 + \tfrac{1}{2}\lambda/l\big), \qquad b = \tfrac{1}{2}\sqrt{\lambda l}\big(1 + \tfrac{1}{4}\lambda/l\big)^{1/2}. \tag{3.1.49}$$

The quantity b represents the most important parameter of the Fresnel volume: its half-width. Note that the Fresnel zones are circular in a homogeneous medium. Consequently, quantity b also represents the radius of the Fresnel zone at point O.

As we can see from Figure 3.2(b), the boundary of the Fresnel volume intersects the x-axis at points S' and R', outside S and R. We can introduce *overshooting distance* Δ by the relation

$$\Delta = \overline{S'S} = \overline{R'R} = a - \tfrac{1}{2}l = \tfrac{1}{4}\lambda. \tag{3.1.50}$$

The overshooting distance has an obvious physical meaning. It does not depend on the distance between the source S and the receiver R.

In the high-frequency approximation ($\lambda \ll l$), we can neglect $\frac{1}{2}\lambda/l$ and $\frac{1}{4}\lambda/l$ in the brackets of (3.1.49) with respect to unity. We then obtain the simple equations,

$$a \doteq \tfrac{1}{2}l, \qquad b \doteq \tfrac{1}{2}\sqrt{\lambda l} = \tfrac{1}{2}f^{-1/2}\sqrt{lV}, \qquad \Delta = \tfrac{1}{4}\lambda = \tfrac{1}{4}f^{-1}V. \tag{3.1.51}$$

For $f \to \infty$, the frequency dependence of the three quantities a, b, and Δ is different: $b \sim f^{-1/2}$, $a \sim f^0$, and $\Delta \sim f^{-1}$.

Thus, the half-width of the Fresnel volume b is proportional to $f^{-1/2}$. Consequently, for high frequencies, the Fresnel volume is closely concentrated to ray Ω connecting S and R. With increasing frequency, the width of the Fresnel volume decreases as $f^{-1/2}$.

The Fresnel volumes introduced by (3.1.46) correspond to the wavefield generated by a point source situated at S. The Fresnel volumes (and relevant Fresnel zones), however, may be introduced even for a wavefield generated at an arbitrary initial surface passing through S. Let us consider a simple case of a plane wave, with wavefront Σ^S perpendicular to ray Ω at S. Similarly, as in the case of a point source, it is not difficult to prove that the relevant Fresnel volume corresponding to a receiver at R in a homogeneous medium is *a paraboloid of revolution*, whose rotational axis is normal to Σ^S and passes through R. In the high-frequency approximation, the radius of the relevant Fresnel zone at Σ^S is given by the relation

$$b \sim f^{-1/2}\sqrt{lV} = \sqrt{\lambda l}. \tag{3.1.52}$$

Here l is again the distance between S and R.

The term Fresnel volume is due to Kravtsov and Orlov (1980). Fresnel volumes are also known as 3-D Fresnel zones and regions responsible for diffraction among others. They are also called physical rays, as opposed to the mathematical ray Ω. For a more detailed treatment of Fresnel volumes, including a historical review and many references, refer to Gelchinsky (1985), Červený and Soares (1992), and Kvasnička and Červený (1994, 1996).

Fresnel volumes play an important role in the solution of many wave propagation problems. For example, they can be used effectively to formulate the validity conditions of the ray method (see Section 5.9) and to study the resolution of seismic methods. For efficient calculations and a more detailed discussion of Fresnel volumes in laterally varying 3-D layered structures and for many other references refer to Section 4.11.

3.2 Rays in Laterally Varying Layered Structures

In Section 3.1, we derived the ray tracing systems to evaluate rays in general laterally varying isotropic media without interfaces. To compute rays in actual laterally varying layered structures, we must also discuss the problem of the continuation of rays across curved interfaces. Moreover, we must also specify the initial conditions for the ray tracing system.

3.2.1 Initial Conditions for a Single Ray

Each ray and the travel time along the ray are fully specified by the coordinates of the initial point S of the ray, by the initial components of slowness vector $\vec{p}_0$ at point S and by the initial travel time at S. We shall use the following notation:

$$\text{At } S: \qquad x_i = x_{i0}, \qquad p_i = p_{i0}, \qquad T = T_0. \tag{3.2.1}$$

Quantities p_{i0} determine the initial direction of the ray at S. They are not arbitrary; they must satisfy the eikonal equation at S,

$$p_{i0} p_{i0} = 1/V_0^2, \qquad V_0 = V(x_{i0}). \tag{3.2.2}$$

Instead of the p_{i0}, which must satisfy (3.2.2), the initial direction of the ray may be defined by two take-off angles at S, i_0 and ϕ_0. These can be defined as spherical polar coordinates at S:

$$p_{10} = V_0^{-1} \sin i_0 \cos \phi_0, \qquad p_{20} = V_0^{-1} \sin i_0 \sin \phi_0,$$
$$p_{30} = V_0^{-1} \cos i_0, \tag{3.2.3}$$

with $0 \le i_0 \le \pi, 0 \le \phi_0 \le 2\pi$; see Figure 3.3. Such p_{i0} satisfy (3.2.2) automatically.

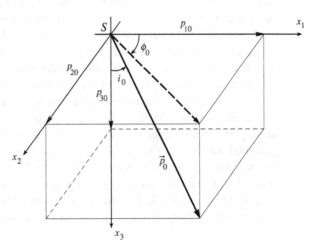

Figure 3.3. The definition of initial take-off angles i_0 and ϕ_0 at initial point S. The bold continuous line denotes initial slowness vector $\vec{p}_0$; the bold dashed line represents the projection of $\vec{p}_0$ into the $x_3 = 0$ plane.

Note that spherical polar coordinates r, i_0, and ϕ_0, where r is the distance from S, form a right-handed orthogonal coordinate system with its origin at S. This system is common in studying wavefields generated by a point source at S. Cartesian coordinate system $x_i(\equiv x, y, z)$ is defined in the following way. The origin of the coordinate system is situated at the epicenter, axis $x_1(\equiv x)$ is tangent to the Earth's surface and points to the North, axis $x_2(\equiv y)$ is tangent to the surface and points to the East, and axis $x_3(\equiv z)$ points downward. See Aki and Richards (1980, pp. 114–15).

It is also possible to use some alternative angles to define the direction of the ray at S. For example, we can use $\theta_0 = \frac{1}{2}\pi - i_0$, $-\frac{1}{2}\pi \leq \theta_0 \leq \frac{1}{2}\pi$. Then

$$p_{10} = V_0^{-1} \cos\theta_0 \cos\phi_0, \qquad p_{20} = V_0^{-1} \cos\theta_0 \sin\phi_0,$$
$$p_{30} = V_0^{-1} \sin\theta_0, \tag{3.2.4}$$

It is simple to see that angle θ_0 is now measured from the equator, not from the polar point as i_0. A right-hand rectangular coordinate system is then formed by r, θ_0, and $-\phi_0$, with its origin at S.

We do not present here the initial conditions for the reduced ray tracing system (3.1.29) and for equation (3.1.30) for the travel time. They follow immediately from (3.2.1) and (3.2.2).

We have so far been interested only in the initial conditions for a single ray. The problem of determining the initial conditions for a single ray, if the initial travel-time field is known along an initial surface or along an initial line, will be discussed in Section 4.5, together with the initial conditions for the dynamic ray tracing system.

3.2.2   Rays in Layered and Block Structures. Ray Codes

Let us now consider a 3-D laterally varying model divided by structural interfaces into layers and blocks, in which the distribution of model parameters is smooth. Assume that all layers (blocks) in the model are suitably numbered and also that the individual structural interfaces between blocks and layers are numbered. The numbering may, of course, be performed in many ways.

We now wish to compute ray Ω in such a model, with the initial point situated at S and with a specified initial direction at S; see Section 3.2.1. To start the ray tracing, we need to know whether the first element of the ray at S is P or S. Then the computation is unique, but only until ray Ω strikes a structural interface. Then we must decide which of the generated waves should be chosen (reflected, transmitted, P, or S). See Figure 3.4. A similar decision must be made at every other point of incidence. The relevant information on the selection of generated waves at structural interfaces must be available a priori, in the input data (similarly as x_{i0} and p_{i0} at S). This a priori information is known as the *ray code*. The ray code is also often called the *ray signature*. There are many ways of constructing ray codes. Usually, the ray codes are formed by a chain of characters (mostly integers and letters) and spaces. The ray code may successively specify the numbers of interfaces at which points of incidence $Q_1, Q_2, \ldots, Q_N$ are situated, and also the types of the selected generated waves at $S, Q_1, Q_2, \ldots, Q_N$; see Figure 3.5. Alternatively, the ray code may follow individual elements of the ray ($\overline{SQ_1}, \overline{Q_1Q_2}, \ldots$), specifying the numbers of layers (blocks) in which the elements are situated and the types of waves along these elements (P or S). Many variants and modifications of ray codes are possible. They depend, in large measure, on the method used to construct the model (model building). For example, the

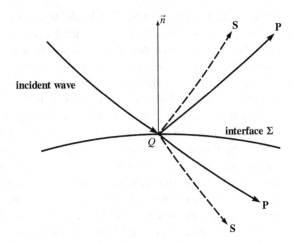

Figure 3.4. Reflection/transmission of a seismic body wave at a curved interface Σ between two inhomogeneous isotropic media. In a high-frequency approximation, the problem is locally reduced to the problem of R/T of a plane wave at a plane interface between two homogeneous media in the vicinity of the point of incidence Q. Four seismic body waves are generated: Two reflected (P and S) and two transmitted waves (P and S).

3-D ray tracing program Complete Ray Tracing (CRT), described in Červený, Klimeš, and Pšenčík (1988b), uses optionally five alternative ray codes constructed in different ways. Each of them may be suitable in certain computations.

Let us now consider rays with the initial point at S and with a specified ray code. The initial directions of rays at S may be arbitrary. We call the wave propagating along these rays, described by the ray code, the *elementary wave*. Consequently, we can also speak of *ray codes of elementary waves*. We may have numerical codes of elementary waves (Červený, Molotkov, and Pšenčík 1977, p. 88) and alphanumerical codes of elementary waves among others. As the ray codes may be constructed in different ways, also the decomposition of the complete wavefield into elementary waves depends on the coding of waves used. One elementary wave in one code may include two or more elementary waves in another code. For examples, see Červený, Molotkov, and Pšenčík (1977, pp. 88, 89).

Thus, the computation of ray Ω, starting at points S with known initial condition (3.2.1), and specified by a ray code, is as follows. We start the ray tracing at S, using any of the ray tracing systems derived in Section 3.1, for the type of wave specified by the ray code (P or S). As soon as the computed ray strikes an interface, we check whether the number of the interface corresponds to the ray code. If not, we stop the computation. If it does, we select the proper generated wave specified by the ray code and try to determine the appropriate initial conditions for the slowness vector of the selected generated wave using (2.4.70). If (2.4.70) cannot be applied (for example, at edges in interfaces or for a generated

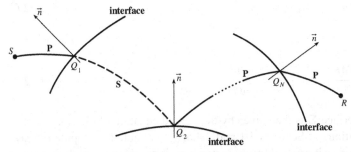

Figure 3.5. Ray Ω of a multiply reflected/transmitted wave in a laterally varying layered and block model. The points of incidence at the interfaces are successively denoted by $Q_1, Q_2, \ldots, Q_N$. Between the individual interfaces, the wave may propagate as either P or S.

inhomogeneous wave, see Section 3.2.3), we also must stop the computation. If (2.4.70) can be applied, we start the new ray tracing using the appropriate ray tracing system. The procedure is successively repeated, interface by interface or layer by layer. The ray may also strike the bottom or side boundary of the model; we then also need to stop the computation. We use the term *successful ray* to refer to the ray that terminates on the specified reference surface Σ^R, along which the receivers are distributed.

It is useful to assign to each computed ray (even to unsuccessful rays) the so-called *history function*. The history function specifies completely the numbers of interfaces crossed by the ray, the numbers of layers (blocks) in which the individual elements of the rays are situated, the types of waves along the individual elements (P or S), the caustics encountered, the positions of the termination points, and the reason for termination. Let us emphasize the difference between the ray code and the ray history. The ray code is an a priori information, which should be known before starting computations. The ray history is known only when its computation has been terminated. Ray histories play an important role in boundary value ray tracing by shooting methods; see Section 3.11.2.

In the discussion so far, we have considered only the elementary waves generated by point source S. Later on, however, we shall also consider the elementary waves generated at initial surfaces or at initial lines. The concepts of ray codes and ray histories also apply to these cases.

The next new problem that appears in ray tracing in layered structures involves the *determination of the intersection points* $Q_1, Q_2, \ldots, Q_N$ of ray Ω with structural interfaces. In some simple situations, the intersection point may be found analytically; in other situations it should be determined numerically. Various numerical algorithms may be used to find this intersection point, but we shall not discuss them here.

To apply Snell's law (2.4.70), we must know the slowness vector of incident wave $\vec{p}(Q)$, the velocities $V(Q)$ and $\tilde{V}(Q)$ of the incident wave and of the selected reflected/transmitted wave, and the unit normal $\vec{n}(Q)$ to the interface. Assume that the interface is described by equation $\Sigma(x_i) = 0$. We can then compute $\vec{n}$ at point Q of interface Σ,

$$\vec{n} = \frac{\epsilon^* \nabla \Sigma}{(\nabla \Sigma \cdot \nabla \Sigma)^{1/2}}, \quad \text{that is,} \quad n_k = \epsilon^* \frac{\partial \Sigma}{\partial x_k} \bigg/ \left(\frac{\partial \Sigma}{\partial x_n} \frac{\partial \Sigma}{\partial x_n} \right)^{1/2}. \quad (3.2.5)$$

Here ϵ^* equals either 1 or -1; which we choose is quite arbitrary. The second equation of (3.2.5) is expressed in Cartesian coordinates. For curvilinear coordinates, see Section 3.5.

The interface is often described by relation $x_3 = f(x_1, x_2)$. Then

$$\Sigma(x_i) = x_3 - f(x_1, x_2) = 0, \quad (3.2.6)$$

and, consequently,

$$\frac{\partial \Sigma}{\partial x_1} = -\frac{\partial f}{\partial x_1}, \quad \frac{\partial \Sigma}{\partial x_2} = -\frac{\partial f}{\partial x_2}, \quad \frac{\partial \Sigma}{\partial x_3} = 1. \quad (3.2.7)$$

Unit normal $\vec{n}(Q)$ to interface Σ is then given by the second equation of (3.2.5).

Thus, using the equations of Sections 3.1 and 2.4.5, we are able to compute the rays and relevant travel times of arbitrary multiply reflected (possibly converted) high-frequency body waves in a 3-D laterally varying layered and block structure.

3.2.3 Anomalous Rays in Layered Structures

In certain situations, ray tracing in layered and block structures fails. It may fail for several reasons. For example, Snell's law (2.4.70) may yield complex-valued components $\tilde{p}_i(Q)$ of the slowness vector for the reflected or transmitted wave we wish to compute. Additionally, if the interface is not smooth at the point of incidence, the vector $\vec{n}$ normal to the interface is not defined there. Consequently, Snell's law cannot be applied.

In some other situations, a ray can be calculated by standard ray tracing, but it may still be anomalous in some way. As examples, we can name the boundary rays, which separate an illuminated region from a shadow zone, critical rays, and the like. Although the computation of such anomalous rays does not cause any difficulty, the amplitudes of seismic waves propagating along the anomalous rays cannot be computed by standard ray procedures described in Section 2.4, based on the transport equation, Snell's law, and R/T coefficients. Usually, the amplitudes along anomalous rays and in their vicinity are frequency-dependent.

All such situations will be discussed in Section 5.9. Extensions of the ray method to compute the wavefield in anomalous regions will be discussed there. Here we shall only very briefly describe the most important anomalous rays and anomalous regions, typical in routine 2-D and 3-D ray tracing in laterally varying, layered and block structures. In many cases, we shall refer to other sections of the book, where the individual anomalous regions are treated in more detail. To simplify the explanations in this section, we shall primarily discuss anomalous rays and anomalous regions for unconverted waves only.

a. Postcritical incidence and inhomogeneous waves. The angle of reflection or transmission $\tilde{i}$ is related to the angle of incidence i by the standard Snell's law, $\sin \tilde{i} = (\tilde{V}/V) \sin i$. For $\sin i > V/\tilde{V}$, angle $\tilde{i}$, the relevant eikonal and components of slowness vector $\tilde{p}_i$ become complex-valued. The generated waves with the complex-valued eikonal are called inhomogeneous or evanescent; see Section 2.2.10. They physically exist but cannot be computed by the standard ray method. Note that the angle of incidence $i = i^*$, for which $\sin i^* = V/\tilde{V}$, is called the *critical angle of incidence*, angles $i > i^*$ are called *postcritical* (as well as *overcritical* or *supercritical*), and angles $i < i^*$ are called *subcritical*. See Figure 3.6. This means that inhomogeneous waves are generated for postcritical angles of incidence only. The ray tracing must be stopped at the interface if the ray code

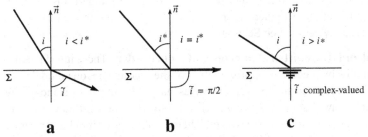

Figure 3.6. Subcritical (a), critical (b), and postcritical (c) angles of incidence i. For the subcritical angle of incidence i, the R/T angle $\tilde{i}$ and the corresponding ray are real-valued; see (a). For the critical angle of incidence $i = i^*$, for which $\sin i^* = V/\tilde{V}$, R/T angle $\tilde{i}$ equals $\frac{1}{2}\pi$, and the ray of the R/T wave is parallel to the interface; see (b). For the postcritical (also called overcritical) angles of incidence i, R/T angle $\tilde{i}$ and the relevant ray of the R/T wave are complex-valued; see (c). The R/T wave is then inhomogeneous.

would continue with a complex eikonal. For more details on inhomogeneous waves refer to Section 5.9.3. The ray of the reflected wave corresponding to the critical angle of incidence is called the critical ray. The critical ray and rays close to the critical ray are real-valued and may be simply evaluated by standard ray tracing, but the wavefield in the critical region is singular. See Section 5.9.2, part 2.

b. Edges and vertexes in interfaces. If the ray of an incident wave is incident at an edge or a vertex in the interface, vector $\vec{n}$ normal to the interface is not defined, and the initial conditions for reflected/transmitted rays cannot be computed. The ray tracing must be stopped at the edge or at the vertex. The wavefield of waves generated at the edges and vertexes becomes more complicated. There is a one-parameteric system of normals at any point on the edge and a relevant one-parameteric system of rays. They correspond to the so-called *edge wave*. Similarly, there is a two-parameteric system of normals at the vertex and a relevant two-parameteric system of rays. They correspond to the so-called *vertex wave* or *tip wave*. For more details on edge and vertex waves, see Section 5.9.2, part 3, and Section 4.5.

c. Rays tangent to an interface. In laterally varying layered structures, rays locally tangent to an interface (or to the boundary of a smooth body) are quite common. A ray locally tangent to a smooth interface can be simply computed by standard ray tracing, including adjacent rays not crossing the interface. The wavefield in the vicinity of this ray is, however, singular beyond the tangent point. A shadow zone is usually formed beyond the tangent point, in the region between the ray and the interface. Thus, beyond the tangent point, the ray locally tangent to the interface becomes a boundary ray between the shadow zone and the illuminated region. Various waves of diffractive nature penetrate into the shadow zone. We speak of smooth interface diffraction or smooth body diffraction. The computation of diffracted waves again requires special treatment. See Section 5.9.2, part 3.

In some simple situations (for example, in the case of a plane interface between two homogeneous media), a ray may even be globally tangent to the interface (grazing ray). This applies, for example, to the ray of a transmitted wave if the angle of incidence is critical; see Figure 3.6. The relevant transmitted wave generates a head wave propagating to the other side of the interface. The rays of head waves satisfy Snell's law and may be calculated by standard ray tracing, with initial ray points distributed along the interface. The wavefield of head waves, however, cannot be computed by the standard ray method. Head waves belong to the class of higher order waves and can only be treated using the higher-order terms of the ray series. See Sections 5.6.7 and 5.7.10.

d. Rays reflected and transmitted at interfaces of higher order. These interfaces are often introduced formally, by piecewise approximation of the inhomogeneous model. The rays and travel times of all reflected and transmitted waves may be calculated by standard ray tracing, including Snell's law (2.4.70). The reflected and converted transmitted waves are, however, higher order waves, similar to head waves. Their amplitudes cannot be computed by standard ray equations. See Sections 5.6.4 and 5.7.6.

e. Caustics. Caustic surfaces are envelopes of rays. At any point situated close to the caustic surface on its illuminated side, there are always at least two intersecting rays, one approaching the caustics and the second leaving it. On the other side of the caustic surface, a caustic shadow is formed. No regular rays, but only complex-valued rays, penetrate into the caustic shadow. The rays tangent to caustics on its illuminated side can be calculated simply; standard ray tracing may be used. The ray amplitudes, however, become infinite at caustic points (at points where the ray is tangent to a caustic). The wavefield is also anomalous in some vicinity of the caustic point, but not along the whole ray. At larger distances from the caustic point, the wavefield is again regular. See Section 5.9.2, part 1.

f. Regions of high velocity gradient. In these regions, the validity conditions of the ray method are not satisfied. Although the rays passing through such a region are formally simple to evaluate, the accuracy of the wavefield computations will be low.

g. Large curvature of interface. The rays of reflected/transmitted waves are formally simple to calculate, even if the curvature of the interface is large at the point of incidence. The ray calculations of the wavefield will, however, be very inaccurate.

h. Strong interference effects. Any situation in which the wavefield displays a strong interference character and is formed by a superposition of a large number of elementary waves can hardly be treated by the ray method, or, at least, the application of the ray method becomes cumbersome. The best conditions for applying the ray method are those in which the wavefield is formed by several (not too many) noninterfering waves. See also Section 5.9.4.

i. Chaotic behavior of rays. The chaotic behavior of rays is characterized by the exponential sensitivity of rays to initial conditions. The chaotic rays diverge exponentially, although they are initially close. The consequence is that the rays exceeding in length some "predictability horizon" cannot be traced backward to recover their initial conditions, and the two-point ray tracing problem cannot be solved. The chaotic propagation may be quantified by *Lyapunov exponents*. The rays may behave chaotically both in deterministic and stochastic (random) media. Multiple scattering by irregularities of the model (rough structural interfaces, smooth obstacles, and the like) are most likely to cause the chaotic behavior of rays. For more details, see Section 4.10.7.

3.2.4 Curvature and Torsion of the Ray

Rays are general 3-D curves. In certain applications, it is useful to know the curvature K and the torsion T of the ray. These quantities may be determined simply from the ray tracing system. We denote $\vec{t}$, $\vec{n}$, and $\vec{b}$ as the *unit tangent, unit normal*, and *unit binormal* to the ray, respectively, and use *Frenet's formulae*:

$$\mathrm{d}\vec{t}/\mathrm{d}s = K\vec{n}, \qquad \mathrm{d}\vec{n}/\mathrm{d}s = T\vec{b} - K\vec{t}, \qquad \mathrm{d}\vec{b}/\mathrm{d}s = -T\vec{n}, \qquad (3.2.8)$$

where s denotes the arclength along the ray.

Using ray tracing system (3.1.10) in vector form and by inserting $\vec{p} = V^{-1}\vec{t}$, we arrive at

$$\mathrm{d}(V^{-1}\vec{t})/\mathrm{d}s = \nabla(V^{-1}).$$

Performing the derivatives and using the first of Frenet's formulae yields

$$KV^{-1}\vec{n} + \vec{t}\,\mathrm{d}(V^{-1})/\mathrm{d}s = \nabla(V^{-1}). \qquad (3.2.9)$$

This equation shows that vectors $\vec{n}$, $\vec{t}$, and $\nabla(V^{-1})$ are coplanar and that binormal $\vec{b}$ is always perpendicular to $\nabla(V^{-1})$ and $\vec{t}$.

Curvature K is obtained immediately from (3.2.9) if multiplied by $\vec{n}$,

$$K = V\vec{n} \cdot \nabla V^{-1} = -V^{-1}\vec{n} \cdot \nabla V. \tag{3.2.10}$$

To determine torsion T, we first differentiate (3.2.9) with respect to s,

$$\vec{n}\frac{\mathrm{d}}{\mathrm{d}s}(KV^{-1}) + KV^{-1}\frac{\mathrm{d}\vec{n}}{\mathrm{d}s} + \vec{t}\frac{\mathrm{d}^2}{\mathrm{d}s^2}V^{-1} + \frac{\mathrm{d}\vec{t}}{\mathrm{d}s}\frac{\mathrm{d}}{\mathrm{d}s}V^{-1} = \frac{\mathrm{d}}{\mathrm{d}s}\nabla V^{-1}.$$

Multiplying this equation by $\vec{b}$ yields

$$KV^{-1}\frac{\mathrm{d}\vec{n}}{\mathrm{d}s} \cdot \vec{b} + \frac{\mathrm{d}\vec{t}}{\mathrm{d}s} \cdot \vec{b}\frac{\mathrm{d}}{\mathrm{d}s}(V^{-1}) = \vec{b} \cdot \frac{\mathrm{d}}{\mathrm{d}s}\nabla\left(\frac{1}{V}\right).$$

Using Frenet's formulae, we obtain the expression for T,

$$T = VK^{-1}\vec{b} \cdot \mathrm{d}\nabla V^{-1}/\mathrm{d}s. \tag{3.2.11}$$

For completeness, we also give the equations for $\vec{n}$ and $\vec{b}$,

$$\vec{b} = \frac{\vec{t} \times \nabla(V^{-1})}{|\vec{t} \times \nabla(V^{-1})|}, \qquad \vec{n} = \vec{b} \times \vec{t} = \frac{(\vec{t} \times \nabla(V^{-1})) \times \vec{t}}{|\vec{t} \times \nabla(V^{-1})|}. \tag{3.2.12}$$

3.3 Ray Tracing

This section is devoted mainly to initial-value ray tracing; for boundary-value ray tracing, see Section 3.11. Ray tracing systems in laterally varying structures can be solved in four ways: numerically, analytically, semianalytically, or by cell ray tracing approaches. The most general approach is based on the numerical solution of the ray tracing system; see Section 3.3.1. The simplest and fastest solution of the ray tracing system, however, is based on the analytical solution, wherever the complexity of the model allows such solutions. It is, however, very difficult to describe the velocity distribution in the whole model by a simple velocity function that would allow the analytical solution of the ray tracing system. Therefore, the whole model (or the whole layer or the whole block) is often divided into suitable cells, in which the velocity distribution can be approximated by simple functions that permit analytical ray computations. The ray is then obtained as a chain of analytically computed segments. Each of these ray tracing approaches has its advantages and disadvantages. One approach may be suitable in some applications, whereas another approach is more appropriate in other applications. In this section, we shall devote our attention mainly to numerical ray tracing; for analytical, semianalytical, and cell ray tracing approaches, refer to the next section.

3.3.1 Numerical Ray Tracing

The ray tracing systems presented in Section 3.1 are mostly systems of ordinary differential equations of the first order, with appropriate initial conditions. The numerical procedures for the solution of a system of ordinary differential equations of the first order with specified initial conditions are well known. Various standard numerical techniques, which give the solution of such systems with the required accuracy, are described in detail in textbooks on numerical mathematics. The routines designed for numerical integration of systems of ordinary differential equations of the first order can be found in many subroutine packages. Among the most popular are the Runge-Kutta method and Hamming's predictor-corrector method. We shall not go into details of these procedures from the mathematical point of view; we refer the interested reader to any textbook on numerical methods.

To perform numerical ray tracing, we must first select the proper ray tracing system; in other words, we need to select the parameter u along the ray we wish to use (arclength s, travel time T, parameter σ, and so on). The selection of the integration parameter will be discussed in the next section. We then need to select regular step Δu in the parameter along the ray, the accuracy of computations of the evaluated quantities required in one step, and the maximum number of regular steps.

The standard algorithms for numerical ray tracing can keep the required accuracy of the computation below some limit. For example, the computation for each step Δu is performed in two ways: (a) with step Δu and (b) twice with step $\Delta u/2$. If the differences between both these computations satisfy the accuracy conditions, the computation continues. If not, the step is halved, and the computations are repeated. If the accuracy conditions are not satisfied even if the step is halved, the step is halved again, and so on. The maximum number of halvings in standard procedures is optional, but it is usually 5 to 10. After several halvings, the required accuracy is usually achieved. In standard subroutines, a message is given if the required accuracy has not been reached, even with the maximum number of halvings. This is, however, an exceptional case. Thus, we can see that the choice of step Δu in the integration is not particularly critical. If the medium is sufficiently smooth and the layers thick, the choice of a larger step Δu can often increase the numerical efficiency of the computations, without detriment to accuracy. The accuracy is kept below the required limit by halving the intervals wherever necessary.

An independent test of the accuracy of computations is also the eikonal equation $\mathcal{H}(x_i, p_i) = 0$, which must be satisfied along the whole ray. Thus, we can require $|p_i p_i - 1/V^2| < \delta$, where δ is some specified small quantity. The accuracy and stability of the ray tracing is increased if $p_i p_i$ is normalized to $1/V^2$ at any step of the numerical ray tracing. Actually, this normalization is always very useful. Although the constraint (eikonal) equation should be theoretically satisfied along the whole ray, the numerical noise could influence its validity.

The maximum number of steps in the computation of the ray has a more or less formal meaning; it stops the ray computation if the ray becomes extremely long for some reason (for example, incorrect input data).

At any step of the numerical ray tracing, it is necessary to perform a check for crossing interfaces, boundaries of the model, and possible other surfaces at which we wish to know or to store the results. If the check is positive, it is necessary to find the point of intersection of the ray with the surface under consideration. The point of intersection can be determined in many ways. A two-point iterative method is usually used. The method of halving of intervals is safe, but very slow. This method may be combined with some faster algorithms.

If we compute a ray in a layered/block structure, the new initial slowness vector components $\tilde{p}_i(Q)$ of the relevant R/T wave must be determined at each point Q at which the ray crosses an interface; see Sections 3.2.2 and 3.2.3. We also need to remember that point Q does not correspond, as a rule, to a regular grid of points computed along the ray with step Δu. Thus, we must find a new point on the ray of the R/T wave that corresponds to the regular grid even across the interface. Only then can we start standard ray tracing at regular steps Δu.

For a detailed description of the algorithms of numerical initial-value ray tracing in 3-D laterally varying inhomogeneous layered and block structures, refer to Červený, Klimeš, and Pšenčík (1988b).

3.3.2 Choice of the Integration Parameter Along the Ray

In practical ray tracing, the most popular parameters along a ray are travel time T and arclength s. We can, however, also choose many other parameters, such as σ ($d\sigma = V^2 dT$) or ζ ($d\zeta = dT/V$). From a numerical point of view, none of these parameters has a distinct advantage in ray tracing. If we wish to evaluate not only rays but also wavefronts, it may perhaps be useful to use travel time T as the variable along the ray. The wavefronts are then obtained automatically as a by-product of the ray tracing.

All these parameters vary monotonically along the ray. This is a great advantage because we have no problems with the turning points. In certain applications, it is also common to use one of the Cartesian coordinates as the parameter along the ray and the relevant reduced Hamiltonian; see (3.1.28) through (3.1.30).

System (3.1.29) is suitable for computing rays corresponding to waves propagating roughly in the x_3-direction. In this case, the sign in the expression for p_3 can be determined uniquely. The ray tracing system fails completely for rays nearly perpendicular to the x_3-axis. Its application to rays that may have some turning points with respect to the x_3-axis will be problematic.

In some applications in geophysics, rays propagate roughly in the direction of one of the coordinate axes. Examples are common in reflection seismology if the rays do not deviate considerably from the vertical axis. Another example is the propagation of acoustic waves in an ocean waveguide.

Let us discuss a very simple example of simplifying system (3.1.37), which is analogous to (3.1.29) if the wave propagates roughly in the x-direction. Assume that V depends on z only, that the initial conditions are $y_0 = p_{y0} = 0$, and that the ray deviates only slightly from the x-direction so that $z'^2 \ll 1$. In this case, $y = 0$, $p_y = 0$ along the whole ray, and (3.1.37) reduces to the second equation only. Since $1 + z'^2 \doteq 1$,

$$\frac{\mathrm{d}}{\mathrm{d}x}\left(\frac{1}{V}z'\right) = \frac{\partial}{\partial z}\left(\frac{1}{V}\right).$$

Because $\mathrm{d}(V^{-1}z')/\mathrm{d}x = z'^2 \partial V^{-1}/\partial z + V^{-1}z''$, the ray tracing equation becomes

$$\mathrm{d}^2 z/\mathrm{d}x^2 = -V^{-1}\partial V/\partial z. \tag{3.3.1}$$

This ray tracing equation is used broadly in the investigation of long-range sound transmission in oceans, see Flatté et al. (1979).

3.3.3 Travel-Time Computation Along a Ray

The basic quantity we wish to determine by ray tracing is the travel time. The travel time along the ray can be computed in several different ways.

The first is to use the travel time directly as parameter u along the ray. This leads to ray tracing systems (3.1.9), (3.1.14), and (3.1.19). As discussed in Section 3.3.2, this approach also allows us to plot the wavefronts simply.

If we use some other parameter u along the ray, we must add one ordinary differential equation of the first order for travel time T to the system of equations for rays, $dT/du = p_i \partial\mathcal{H}/\partial p_i$; see (3.1.3) or (3.1.27) for the reduced Hamiltonian $\mathcal{H}^R$. We can solve this equation either together with the equations for rays or after the ray trajectory has been computed.

There are several reasons for computing the travel time together with the ray, using a system consisting of seven equations. In this case, it is simple to keep the accuracy of the travel-time computation (not only of the ray) below some limit. If the accuracy is not sufficient, integration step Δu is halved automatically. If we evaluate travel time T by quadratures along a known ray, the integration step cannot be decreased unless we perform a new ray tracing.

3.3.4 Ray Tracing in Simpler Types of Media

In Section 3.1.1, we derived several forms of the ray tracing system for arbitrary 3-D laterally varying structures. In this section we shall start with the ray tracing systems consisting of six ordinary differential equations of the first order for x_i and p_i and reformulate the systems for certain simpler situations, mainly for situations in which the velocity model has lower dimensions. We shall not try to seek analytical solutions of these systems; this will be done in the next section. We shall only seek to separate the equations or to decrease the number of equations in the system.

As a starting point, we shall mainly use the simplest ray tracing system (3.1.11) with variable σ along the ray ($dT = d\sigma / V^2$). Ray tracing system (3.1.11) corresponds to the general ray tracing system (3.1.8) for $n = 2$. Ray tracing system (3.1.11) provides useful simplifications even if the general ray tracing system cannot be simplified at all. We shall use the initial conditions specified by Equations (3.2.1) with (3.2.2).

1. PARTIAL SEPARATION OF VARIABLES

Let us assume that the square of the slowness distribution is described by the following relation:

$$1/V^2(x_1, x_2, x_3) = A(x_1, x_3) + B(x_2). \tag{3.3.2}$$

Ray tracing system (3.1.11) then separates into two independent systems, coupled only by the initial conditions. The first system reads

$$\frac{dx_1}{d\sigma} = p_1, \qquad \frac{dx_3}{d\sigma} = p_3, \qquad \frac{dp_1}{d\sigma} = \frac{1}{2}\frac{\partial A}{\partial x_1}, \qquad \frac{dp_3}{d\sigma} = \frac{1}{2}\frac{\partial A}{\partial x_3}, \tag{3.3.3}$$

and the second reads

$$\frac{dx_2}{d\sigma} = p_2, \qquad \frac{dp_2}{d\sigma} = \frac{1}{2}\frac{\partial B}{\partial x_2}. \tag{3.3.4}$$

The initial conditions are again given by (3.2.1), but (3.2.2) takes the form

$$p_{10}^2 + p_{20}^2 + p_{30}^2 = A(x_{10}, x_{30}) + B(x_{20}). \tag{3.3.5}$$

Travel time T is given by the relation

$$T(\sigma) = T(\sigma_0) + \int_{\sigma_0}^{\sigma} (A + B)d\sigma$$

$$= T(\sigma_0) + \int_{\sigma_0}^{\sigma} (p_1^2 + p_3^2)d\sigma + \int_{\sigma_0}^{\sigma} p_2^2 d\sigma. \tag{3.3.6}$$

For certain simple functions $B(x_2)$, system (3.3.4) can be solved analytically. The system of six ordinary differential equations then reduces to four equations (3.3.3). Equations (3.3.4) can be useful if we wish to extend the 2-D computations to simple 3-D models.

Note that the general ray tracing system (3.1.8) yields this separation only for $n = 2$, but for no other n.

128 SEISMIC RAYS AND TRAVEL TIMES

2. 3-D COMPUTATIONS IN 2-D MODELS

We shall now consider a 2-D model. We understand a 2-D model to be a model in which the velocity does not vary along a selected direction. Without loss of generality, we shall consider this direction to be along the x_2-axis. Then the coordinate x_2 does not appear explicitly in the eikonal equation. In mechanics, such a coordinate is usually called *cyclic* or *implicit coordinate*. The computations in 2-D models are usually performed only in plane x_1-x_3, perpendicular to the x_2-axis. These computations will be considered later. Here, however, we shall consider general 3-D computations in a 2-D model, with arbitrary orientation of the initial slowness vector. Such computations are also sometimes called $2\frac{1}{2}$-dimensional computations; see Brokešová (1994).

We shall obtain our ray tracing systems in a straightforward way from (3.3.3) and (3.3.4), by putting $B = 0$ in (3.3.2). Thus, $A = 1/V^2$. System (3.3.4) can be solved analytically to yield

$$x_2(\sigma) = x_{20} + p_{20}(\sigma - \sigma_0), \qquad p_2(\sigma) = p_{20}. \tag{3.3.7}$$

System (3.3.3) remains valid even now, only $A = 1/V^2$,

$$\frac{dx_1}{d\sigma} = p_1, \qquad \frac{dp_1}{d\sigma} = \frac{1}{2}\frac{\partial}{\partial x_1}\left(\frac{1}{V^2}\right),$$
$$\frac{dx_3}{d\sigma} = p_3, \qquad \frac{dp_3}{d\sigma} = \frac{1}{2}\frac{\partial}{\partial x_3}\left(\frac{1}{V^2}\right). \tag{3.3.8}$$

The initial conditions for the ray tracing system are given by (3.2.1), where p_{10} and p_{30} satisfy the relation

$$p_{10}^2 + p_{30}^2 = V_0^{-2} - p_{20}^2. \tag{3.3.9}$$

Thus, p_{20} also affects the solution of system (3.3.8) due to initial conditions (3.3.9).

Please note, however, that the projection of the ray into horizontal plane x_1-x_2 is not a straight line. This means that the rays in the 2-D models are, in general, 3-D spatial curves, with nonzero torsion.

Travel time $T(\sigma)$ is given by a simple integral, namely,

$$T(\sigma) = T(\sigma_0) + \int_{\sigma_0}^{\sigma} V^{-2}d\sigma$$
$$= T(\sigma_0) + p_{20}^2(\sigma - \sigma_0) + \int_{\sigma_0}^{\sigma} \left(p_1^2 + p_3^2\right)d\sigma. \tag{3.3.10}$$

Ray tracing system (3.1.8) can also be simplified if V^{-n} depends only on x_1 and x_3, but not on x_2. Then

$$x_2(u) = x_{20} + p_{20}\int_{u_0}^{u} (p_k p_k)^{n/2-1}du = x_{20} + p_{20}\int_{u_0}^{u} V^{2-n}du,$$
$$p_2(u) = p_{20}. \tag{3.3.11}$$

In plane x_1-x_3, the ray tracing system remains the same as (3.1.8), but consists of four, not six equations. The travel-time equation is not changed. For more details, see Brokešová (1994).

3. 2-D COMPUTATIONS IN 2-D MODELS

We have so far considered quite general initial conditions p_{10}, p_{20}, and p_{30}, which satisfy (3.2.2). For the first time in this chapter, we shall consider special orientation of the initial slowness vector. We assume that

$$p_{20} = 0. \tag{3.3.12}$$

This means that the initial slowness vector is perpendicular to the x_2-axis. Along the whole ray, the slowness vector is situated in the x_1-x_3 plane, with $p_2 = 0$. Ray tracing system (3.1.8) then yields the following equations:

$$\frac{dx_1}{du} = A^{n/2-1} p_1, \qquad \frac{dp_1}{du} = \frac{1}{n}\frac{\partial}{\partial x_1}\left(\frac{1}{V^n}\right),$$

$$\frac{dx_3}{du} = A^{n/2-1} p_3, \qquad \frac{dp_3}{du} = \frac{1}{n}\frac{\partial}{\partial x_3}\left(\frac{1}{V^n}\right), \tag{3.3.13}$$

$$\frac{dT}{du} = A^{n/2} = V^{-n},$$

with $A = p_1^2 + p_3^2 = V^{-2}$.

System (3.3.13) is usually called the *2-D ray tracing system*. The initial conditions for the system at initial point O are again given by (3.2.1) with $p_2 = p_{20} = 0$, $x_2 = x_{20}$. The components of the slowness vector, p_{10} and p_{30}, satisfy relation $p_{10}^2 + p_{30}^2 = 1/V_0^2$, where $V_0 = V(x_{i0})$,

The 2-D ray tracing system (3.3.13) can again be presented in many alternative forms, similar to the 3-D ray tracing system (3.1.8); see Section 3.1. If $n = 0$, parameter u along the ray equals T. The RHS of (3.3.13) becomes

$$\frac{1}{n}\frac{\partial}{\partial x_i}\left(\frac{1}{V^n}\right) = -\frac{\partial}{\partial x_i}\ln V.$$

See also (3.1.14) and (3.1.19).

4. 1-D MODELS

Ray tracing systems can be simplified even more and solved in terms of closed-form integrals in 1-D models in which V^{-n} depends on one coordinate only. In this case, two coordinates are cyclic. For details refer to Section 3.7

3.4 Analytical Ray Tracing

For certain simple types of models, the ray tracing system can be solved analytically. In this section, we shall give several examples, suitable mainly for 2-D and 3-D computations. Some additional analytical solutions of the ray tracing system for 1-D structures will be given in Section 3.7.

The simplest analytical expressions for rays and travel times exist for homogeneous media and for media with a constant gradient of the square of slowness, $1/V^2$. Consequently, we shall deal with them first. We shall then go on to general analytical solutions for media with a constant gradient of V^{-n} and a constant gradient of $\ln V$. We shall also discuss models in which the rays and travel times are polynomials in some conveniently chosen parameter.

Section 3.3.1 showed that ray tracing systems could be solved numerically without any problem. We also know that the real structure is often rather complicated and can hardly be described by simple velocity laws that would allow analytical solution of the ray tracing system. We are thus faced with the problem whether there is any sense in seeking analytical solutions at all. Analytical solutions are important for at least three reasons. First, in many applications (such as seismic prospecting), the velocity distributions within blocks and layers may be effectively described by simple velocity laws. Second, the analytical solutions are valuable in the cell approach, in which the whole block (layer) is subdivided into a system of cells with simple velocity distributions within each cell. For more details on cell ray tracing, refer to Section 3.4.7. Third, the analytical solutions are usually computationally more efficient and flexible than standard numerical ray tracing.

To come up with analytical solutions of the ray tracing system, we shall mostly use the ray tracing systems presented in Section 3.1 and the initial conditions given by (3.2.1). Velocity V at the initial point will again be denoted by V_0 and must satisfy the eikonal equation at that point so that $V_0 = (p_{i0}p_{i0})^{-1/2}$.

3.4.1 Homogeneous Media

The ray tracing system for a homogeneous medium yields very simple solutions for any parameter u along the ray. The most useful system is obtained with arclength s as the parameter; see (3.1.10) or (3.1.20). The constant velocity in the model is denoted V_0. Under standard initial conditions (3.2.1), the solution now reads

$$
\begin{aligned}
&x_i(s) = x_{i0} + V_0 p_{i0}(s - s_0), \qquad p_i(s) = p_{i0}, \\
&T(s) = T_0 + (s - s_0)/V_0.
\end{aligned}
\tag{3.4.1}
$$

Thus, the ray is a straight line.

The great advantage of analytical solution (3.4.1) is not only in the simplicity of the ray tracing itself but also in the simple determination of the intersection of the straight lines with the interfaces.

3.4.2 Constant Gradient of the Square of Slowness, V^{-2}

Among inhomogeneous media, the simplest analytical solutions of the ray tracing system are obtained if the gradient of V^{-2} is constant. Assume that V^{-2} is given by

$$
V^{-2}(x_i) = A_0 + A_1 x_1 + A_2 x_2 + A_3 x_3.
\tag{3.4.2}
$$

We shall consider only that part of the space in which $V^{-2}(x_i) = A_0 + A_i x_i > 0$ and assume that initial point x_{i0} is also situated in that part so that $V_0^{-2} = A_0 + A_i x_{i0} > 0$.

The analytical solution of ray tracing system (3.1.11) is then

$$
\begin{aligned}
&x_i(\sigma) = x_{i0} + p_{i0}(\sigma - \sigma_0) + \tfrac{1}{4} A_i(\sigma - \sigma_0)^2, \\
&p_i(\sigma) = p_{i0} + \tfrac{1}{2} A_i(\sigma - \sigma_0), \\
&T(\sigma) = T(\sigma_0) + V_0^{-2}(\sigma - \sigma_0) + \tfrac{1}{2} A_i p_{i0}(\sigma - \sigma_0)^2 + \tfrac{1}{12} A_i A_i(\sigma - \sigma_0)^3.
\end{aligned}
\tag{3.4.3}
$$

Here the parameter σ along the ray is related to travel time T and to arclength s by relation $d\sigma = V^2 dT = V ds$. Thus, the ray is a quadratic parabola; the components of slowness vector p_i are linear in $(\sigma - \sigma_0)$, and the travel time is a cubic polynomial in $(\sigma - \sigma_0)$.

Although the velocity distribution considered here is a special case of that considered in the next section, we have given the relevant analytical solutions explicitly because they are very simple, and important.

As we can see, system (3.4.3) reduces to (3.4.1) for $A_i A_i \to 0$ ($i = 1, 2, 3$), if we take into account that $\sigma - \sigma_0 = V_0(s - s_0)$ in a homogeneous medium. This is the great advantage of (3.4.3) in comparison with similar relations for models with a constant gradient of V^{-n} ($n \neq 2$), where the final relations are usually indefinite for $A_i A_i \to 0$, and alternative equations must usually be used if $A_i A_i$ is very small.

3.4.3 Constant Gradient of the nth Power of Slowness, V^{-n}

We shall consider the velocity distribution given by

$$V^{-n} = A_0 + A_1 x_1 + A_2 x_2 + A_3 x_3. \tag{3.4.4}$$

Here n is an arbitrary nonvanishing integer, positive or negative. We shall again consider only the part of space in which $V^{-n} = A_0 + A_i x_i > 0$ and assume that initial point x_{i0} is situated in that part so that $V_0^{-n} = A_0 + A_i x_{i0} > 0$.

We shall seek analytical solutions of ray tracing system (3.1.8). A simple linear solution is obtained for the components of the slowness vector:

$$p_i(u) = p_{i0} + n^{-1} A_i (u - u_0). \tag{3.4.5}$$

Here u is a parameter along the ray related to the travel time as follows: $dT = V^{-n} du$. Ray tracing system (3.1.8) may then be solved in closed integral form:

$$
\begin{aligned}
x_i(u) &= x_{i0} + \int_0^{u-u_0} X^{n/2-1}(w) \left(p_{i0} + \frac{1}{n} A_i w \right) dw, \\
T(u) &= T_0 + \int_0^{u-u_0} X^{n/2}(w) dw,
\end{aligned}
\tag{3.4.6}
$$

where

$$
\begin{aligned}
&X(w) = p_i p_i = a w^2 + b w + c, \\
&a = n^{-2} A_i A_i, \qquad b = 2n^{-1} A_i p_{i0}, \qquad c = p_{i0} p_{i0} = V_0^{-2}.
\end{aligned}
\tag{3.4.7}
$$

Integrals (3.4.6) can be evaluated analytically for any $n \neq 0$. The relevant primitive functions can be found in mathematical handbooks.

3.4.4 Constant Gradient of Logarithmic Velocity, $\ln V$

Simple analytical ray tracing solutions are also obtained if the velocity distribution is given by

$$\ln V = A_0 + A_1 x_1 + A_2 x_2 + A_3 x_3. \tag{3.4.8}$$

It is then possible to use ray tracing system (3.1.9), obtained from (3.1.8) for $n = 0$. The procedure is fully analogous to that in Section 3.4.3. In this case, the most convenient parameter is travel time T. The ray tracing system yields the solutions

$$
\begin{aligned}
p_i(T) &= p_{i0} - A_i (T - T_0), \\
x_i(T) &= x_{i0} + \int_0^{T-T_0} X^{-1}(w)(p_{i0} - A_i w) dw,
\end{aligned}
\tag{3.4.9}
$$

where

$$X(w) = aw^2 + bw + c,$$
$$a = A_i A_i, \qquad b = -2A_i p_{i0}, \qquad c = p_{i0} p_{i0} = 1/V_0^2. \tag{3.4.10}$$

3.4.5 Polynomial Rays

As a rule, the simplest computational algorithms are based on polynomial functions. In Section 3.4.2, we arrived at a polynomial solution of the ray tracing system for the medium with a constant gradient of the square of slowness, V^{-2}. The ray trajectory is a quadratic polynomial in terms of $(\sigma - \sigma_0)$. Here we shall consider more complex velocity distributions that also yield polynomial rays.

Eikonal equation $p_i p_i = 1/V^2(x_i)$ may be expressed in an even more general Hamiltonian form,

$$\mathcal{H}(x_i, p_i) = \tfrac{1}{2}\{F[p_i p_i] - F[1/V^2]\} = 0, \tag{3.4.11}$$

where $F[q]$ is a continuous function of q with continuous first and second derivatives. We denote $F'[q] = dF[q]/dq$ and assume that $qF'[q] > 0$ in the whole region of our interest. The physical meaning of the condition $qF'[q] > 0$ will be explained later. Ray tracing system (3.1.3) then reads

$$\frac{dx_i}{du} = F'[p_k p_k] p_i, \qquad \frac{dp_i}{du} = \frac{1}{2} F'[V^{-2}] \frac{\partial}{\partial x_i}\left(\frac{1}{V^2}\right),$$
$$\frac{dT}{du} = (p_i p_i) F'[p_k p_k]. \tag{3.4.12}$$

Ray tracing system (3.4.12) is very general and yields many interesting analytical solutions. In previous sections, special cases of (3.4.12) with $F[q] = 2n^{-1}q^{\frac{n}{2}}$ and $F[q] = \ln q$ were considered. Very simple analytical solutions are obtained if the spatial gradient of $F[1/V^2]$ is constant,

$$F[1/V^2] = A_0 + A_1 x_1 + A_2 x_2 + A_3 x_3. \tag{3.4.13}$$

Then Equations (3.4.12) for dp_i/du can be simplified and read $dp_i/du = -\partial \mathcal{H}/\partial x_i = \tfrac{1}{2} A_i$. Ray tracing system (3.4.12) then has the following solution:

$$p_i(u) = p_{i0} + \tfrac{1}{2} A_i (u - u_0),$$
$$x_i(u) = x_{i0} + \int_0^{u-u_0} F'[X]\left(p_{i0} + \tfrac{1}{2} A_i w\right) dw, \tag{3.4.14}$$
$$T(u) = T_0 + \int_0^{u-u_0} F'[X] X \, dw,$$

where

$$X = aw^2 + bw + c, \qquad a = \tfrac{1}{4} A_i A_i, \qquad b = p_{i0} A_i, \qquad c = p_{i0} p_{i0}. \tag{3.4.15}$$

Here u is a parameter along the ray, which is related to the travel time as

$$dT = V^{-2} F'[V^{-2}] du. \tag{3.4.16}$$

Since $qF'[q] > 0$ in the region of $q = 1/V^2$ under consideration, parameter u along the ray is monotonic. For this reason, we assumed $qF'[q] > 0$.

We shall now consider polynomial function $F[q]$, satisfying condition $qF'[q] > 0$ in the region of $q = 1/V^2$ under consideration,

$$F[q] = q + \sum_{j=2}^{N} c_j q^j. \tag{3.4.17}$$

Then we obtain

$$F'[X] = 1 + \sum_{j=2}^{N} jc_j(aw^2 + bw + c)^{j-1} = \sum_{j=0}^{2N-2} C_j w^j. \tag{3.4.18}$$

Coefficients C_j can be obtained simply from c_j. For example, if $N = 2$,

$$C_0 = 1 + 2c_2 c, \qquad C_1 = 2c_2 b, \qquad C_2 = 2c_2 a.$$

Inserting (3.4.18) into (3.4.14) yields a polynomial of order $2N$ for $x_i(u)$ and a polynomial of order $2N + 1$ for $T(u)$, in terms of $(u - u_0)$.

Let us present the complete solution for $N = 2$. The velocity distribution is given by

$$V^{-2} + c_2 V^{-4} = A_0 + A_1 x_1 + A_2 x_2 + A_3 x_3, \tag{3.4.19}$$

where $c_2 \geq 0$. The solution of the ray tracing system is then

$$
\begin{aligned}
p_i(u) &= p_{i0} + \tfrac{1}{2} A_i (u - u_0), \\
x_i(u) &= x_{i0} + (1 + 2c_2 c)p_{i0}(u - u_0) + \left[c_2 b p_{i0} + \tfrac{1}{4} A_i(1 + 2c_2 c) \right](u - u_0)^2 \\
&\quad + \tfrac{1}{3}(2c_2 a p_{i0} + A_i c_2 b)(u - u_0)^3 + \tfrac{1}{4} A_i c_2 a (u - u_0)^4, \\
T(u) &= T_0 + c(1 + 2c_2 c)(u - u_0) + \tfrac{1}{2} b(1 + 4c_2 c)(u - u_0)^2 \\
&\quad + \tfrac{1}{3}(a + 4c_2 c a + 2c_2 b^2)(u - u_0)^3 + c_2 a b (u - u_0)^4 \\
&\quad + \tfrac{2}{5} c_2 a^2 (u - u_0)^5,
\end{aligned}
\tag{3.4.20}
$$

where a, b, and c are given by (3.4.15).

As expected, we have obtained a polynomial solution of the fourth order for $x_i(u)$ and of the fifth order for travel time $T(u)$ in $u - u_0$ for velocity distribution (3.4.17), with $N = 2$.

The velocity distribution given by (3.4.17) with (3.4.13) is not the only distribution for which we can find analytical solutions of the ray tracing system in the form of a power series in $u - u_0$. The advantage of the solutions presented here is that the power series is finite (a polynomial). We shall now give one very general and useful velocity distribution, which leads to analytical solutions of the ray tracing system in the form of an infinite power series in $(\sigma - \sigma_0)$. Assume that $1/V^2$ is a general polynomial in Cartesian coordinates x_i,

$$V^{-2} = A_0 + A_j x_j + A_{jk} x_j x_k + \cdots + A_{jk...n} x_j x_k \ldots x_n. \tag{3.4.21}$$

We can then assume the solution of ray tracing system $x_i(\sigma)$, $p_i(\sigma)$, $T(\sigma)$ in the form of a power series in $\sigma - \sigma_0$, where the coefficients of this series are as yet unknown. If these ansatz power series are inserted into the ray tracing system, a recurrent system of equations is obtained from which all coefficients can successively be determined for $j = 0, 1, 2, \ldots$.

A finite power series in $\sigma - \sigma_0$ is obtained only if $1/V^2$ is a polynomial of the first order in (3.4.21). The solution is then the same as (3.4.3). If $n \geq 2$, an infinite power series in $\sigma - \sigma_0$ is obtained. For relevant equations and more details, see Červený (1987b).

3.4.6 More General V^{-2} Models

The ray tracing system can also be solved for more complicated $1/V^2$ models. We shall give two examples.

1. Consider the $1/V^2$ distribution in the form

$$1/V^2 = A^{(1)}(x_1) + A^{(2)}(x_2) + A^{(3)}(x_3). \tag{3.4.22}$$

By inserting (3.4.22) into ray tracing system (3.1.23), we obtain three ordinary differential equations of the second order:

$$\frac{d^2 x_i}{d\sigma^2} - \frac{1}{2}\frac{\partial A^{(i)}}{\partial x_i} = 0, \qquad i = 1, 2, 3$$

(no summation over i). Assume now that $A^{(i)}(x_i)$ is a quadratic polynomial in x_i,

$$A^{(i)}(x_i) = A + Bx_i + Cx_i^2. \tag{3.4.23}$$

Then the ordinary differential equation for x_i reduces to

$$\frac{d^2 x_i}{d\sigma^2} - Cx_i = \frac{1}{2}B.$$

The solutions of this equation are familiar. They correspond to the so-called parabolic layer and can be expressed in terms of trigonometric or hyperbolic functions. See also Section 3.7.2 for the analytical solutions of (3.4.23) and for more details.

2. Now we shall consider the general quadratic distribution of $1/V^2$,

$$1/V^2 = A + B_i x_i + C_{ij} x_i x_j, \tag{3.4.24}$$

with $C_{ij} = C_{ji}$. The analytical solutions for this distribution were found by Körnig (1995) using the Laplace transform. The ray tracing system (3.1.23) for distribution (3.4.24) reads

$$\frac{d^2 x_i}{d\sigma^2} = \frac{1}{2}B_i + C_{ij}x_j, \tag{3.4.25}$$

and the initial conditions are $x_i = x_{i0}$ and $p_i = p_{i0}$ for $\sigma = 0$. We denote the Laplace variable corresponding to σ by s and the Laplace transform of $x_i(\sigma)$ by $X_i(s)$. Then (3.4.25) yields the system of three equations in the Laplace domain for s:

$$s^2 X_i(s) - s x_{i0} - p_{i0} = B_i/2s + C_{ij}X_j(s), \qquad i = 1, 2, 3.$$

If we express it in the following form:

$$X_j(s)(C_{ij} - s^2\delta_{ij}) = -\frac{B_i + 2s^2 x_{i0} + 2p_{i0}s}{2s},$$

the solution for $X_i(\sigma)$ comes out as

$$X_i(s) = -\frac{(B_j + 2s^2 x_{j0} + 2p_{j0}s)}{2s\,\det(C_{nk} - s^2\delta_{nk})}F_{ij}(s). \tag{3.4.26}$$

Here $F_{ij}(s)$ denotes the cofactor of $C_{ij} - s^2\delta_{ij}$. In general, (3.4.26) is the ratio of polynomials of the sixth and seventh order. It is suitable to express $\det(C_{ij} - s^2\delta_{ij})$ in terms of eigenvalues of C_{ij}, λ_1, λ_2, and λ_3, such that $\det(C_{ij} - s^2\delta_{ij}) = (\lambda_1 - s^2)(\lambda_2 - s^2)(\lambda_3 - s^2)$. The expressions for rays, $x_i(\sigma)$, are obtained from (3.4.26) by inverse Laplace transform of $X_i(s)$. These expressions can be found analytically, using partial fraction expansions. There are seven different forms of solution, depending on the mutual relations between λ_1, λ_2, and λ_3. All these possible forms are listed and discussed by Körnig (1995) in detail. In addition to a

polynomial of the second order in σ, these solutions contain the following trigonometric and hyperbolic functions:

$$C_m(\sigma) = \cosh\left(|\lambda_m|^{1/2}\sigma\right), \qquad S_m(\sigma) = \sinh\left(|\lambda_m|^{1/2}\sigma\right)/|\lambda_m|^{1/2}$$
$$\text{for } \lambda_m > 0,$$
$$= \cos\left(|\lambda_m|^{1/2}\sigma\right), \qquad = \sin\left(|\lambda_m|^{1/2}\sigma\right)/|\lambda_m|^{1/2}$$
$$\text{for } \lambda_m < 0.$$

Körnig (1995) also gives expressions for slowness vector $p_i(\sigma) = dx_i(\sigma)/d\sigma$ and
for the travel time:

$$T(\sigma) = T(0) + \int_0^\sigma p_i(\sigma')p_i(\sigma')\mathrm{d}\sigma'.$$

Note that distributions (3.4.2) and (3.4.23) represent special cases of Körnig's general
distribution (3.4.24).

3.4.7 Cell Ray Tracing

In the cell approximation, the whole model (or the whole layer or the whole block) is
subdivided into a network of cells. Velocity V, or some other function related to velocity
V, such as slowness $1/V$ or the square of slowness $1/V^2$, can be specified at grid points
of the network, alternatively in the centers of the cells. The velocity distribution within the
individual cells is then approximated by some simple velocity laws.

The simplest case is to consider a constant velocity within the individual cells. Fictitious
interfaces of the first order are then introduced at the boundaries of the cells, but the ray
tracing is extremely simple and the intersections of the ray with the interfaces can be
determined very easily. Rectangular box cells with a constant velocity within the cells have
been used in seismology for a long time, for example, in the method of Aki, Christoffersen,
and Husebye (1977); see also Koch (1985). Even though this velocity approximation is
rather crude, it has yielded a number of important results, both in seismology and seismic
prospecting (see Langan, Lerche, and Cutler 1985).

It is, however, not difficult to adopt velocity laws that do not introduce interfaces of
the first order at cell boundaries but that introduce only interfaces of the second order.
The velocity is then continuous across the boundaries of the cells, and only the velocity
gradient is discontinuous. We shall describe one such case, suitable for tetrahedral cells.
The results can easily be simplified for triangular cells in 2-D structures.

In tetrahedral cells, the velocity distribution can be described by analytical distribution
(3.4.4) or (3.4.8). The four constants A_0, A_1, A_2, and A_3 can then be determined from
the velocity values at the four apexes of the tetrahedron. The most popular in seismology
and seismic exploration is the case of the constant gradient of velocity V within the cells
($n = -1$). The ray in such a cell is a part of a circle, and the travel time along the circular ray
can be expressed in terms of inverse hyperbolic or logarithmic functions. See, for example,
Gebrande (1976), Will (1976), Whittal and Clowes (1979), Marks and Hron (1980), Cassel
(1982), Müller (1984), Chapman (1985), and Weber (1988). The simplest, polynomial
analytical solution for the ray and travel time, however, corresponds to the case of the
constant gradient of the square of slowness V^{-2} ($n = 2$); see (3.4.3). The exact equation of
the ray trajectory in such a tetrahedral cell is a quadratic polynomial in $(\sigma - \sigma_0)$ and a cubic
polynomial in $(\sigma - \sigma_0)$ for the travel time. The determination of the intersection of the ray
with the plane boundary of the cell leads to the solution of the quadratic equation. Thus, the

procedure of ray tracing in a tetrahedral cell with a constant gradient of V^{-2} requires the solution of four quadratic equations in σ and the computation of some simple polynomial expressions of a low order. The solution is exact; no computations of trigonometric or transcendental functions are required. The equations are very simple; they do not require special treatment for $A \to 0$ (very small gradient). See Červený (1987a) and Virieux, Farra, and Madariaga (1988).

The analytical ray tracing in a cell with a constant gradient of velocity V is similar, being slightly more complicated for programming and numerically less efficient. It requires the computation of transcendental functions.

In the tetrahedral cells with the analytical distribution (3.4.4) or (3.4.8), the velocity is continuous across the boundaries of the cells, but its gradient is not. Thus, the boundaries of cells represent *fictitious interfaces of the second order*. (They actually do not exist in the model, but are introduced by the approximation used.) The interfaces of the second order give, of course, a smoother approximation of the model than the interfaces of the first order; nevertheless, they cause numerous difficulties in the ray computations. The fictitious interfaces of the second order yield anomalies in the computation of the ray field in the vicinity of rays tangent to these interfaces. The ray field changes very drastically in these regions and often contains loops, caustics, and small shadow zones. Consequently, the cell computation of geometrical spreading and amplitudes may often be rather chaotic and may contain zeros and infinities. Sometimes, such computations do not allow the actual trend of the ray amplitudes to be followed. Thus, removing not only the interfaces of the first order, but also the fictitious interfaces of the second order would be useful.

In 1-D models, such fictitious interfaces of the second order may be removed using polynomial rays; see Section 3.7.3. In laterally varying media, however, the attempts to remove the fictitious interfaces of the second order would be considerably more complicated. The only known proposal that deals with removing the fictitious interfaces of the second order in the cell approach is by Körnig (1995). In his proposal, Körnig (1995) uses the general quadratic approximation of $1/V^2$ given by (3.4.24).

3.4.8 Semianalytical Ray Tracing in Layered and Block Structures

We consider a general 3-D laterally varying model consisting of thick layers and/or large blocks, separated by smoothly curved structural interfaces. We assume that the velocity distributions inside any individual layers and blocks are specified by simple velocity laws that allow the analytical computation of rays. Such models have been commonly used is seismic exploration. Layers/blocks of constant velocity, constant velocity gradient, or constant gradient of the square of slowness are primarily considered here. The rays in these three velocity distributions are straight lines, circles, or parabolas; see Section 3.7.2. The only actual problem of ray tracing consists of finding the intersections of straight lines (circles, parabolas) with the structural interfaces. At these intersections, Snell's law must, of course, be applied. It is very important to realize that there are no fictitious interfaces in the model; all interfaces have a structural meaning. This is the great advantage of this model in comparison with the cell model. The disadvantage is that the model allows only rather crude velocity variations inside the individual layers/blocks. See Yacoub, Scott, and McKeown (1970), Sorrells, Crowley, and Veith (1971), Shah (1973a, 1973b), Hubral and Krey (1980), Lee and Langston (1983a, 1983b), Langston and Lee (1983). For many other references see Hubral and Krey (1980).

3.4.9 Approximate Ray Tracing

In complex 3-D structures, the ray tracing and travel-time computation can be rather time consuming, particularly if the problem under study requires many rays to be computed. It may even prove useful to apply some approximate methods to the ray tracing and travel-time computations. These methods have been broadly used in boundary-value ray tracing, mainly as ray estimators for further applications of bending methods; see Section 3.11.3. For completeness, we shall also mention the main principles here.

Two basic approaches have been proposed for approximate ray tracing. The first consists of computing an exact ray in a laterally averaged structure. The 3-D structure is usually replaced by a 1-D (vertically inhomogeneous or radially symmetric) layered structure in which the velocity distribution within the individual layers is laterally averaged. The structure may also be averaged in many other ways. The second approach consists of approximate ray computations through an actual, nonaveraged structure. The rays under consideration, however, satisfy Fermat's principle only approximately. Various alternatives to these two methods are available. For a good review of approximate ray tracing approaches, see Thurber (1986), where many other references can also be found. See also Section 3.7 for ray tracing in 1-D models, Section 3.11 for boundary ray tracing, and Section 3.9 for perturbation methods.

3.5 Ray Tracing in Curvilinear Coordinates

In the previous sections, we have assumed that the model is specified in Cartesian coordinates. In some applications, however, it may be useful to consider models specified in some curvilinear coordinate systems, such as spherical, cylindrical, and ellipsoidal. In global seismology, it is common to use spherical coordinates to describe the model. For this reason, we shall discuss the ray tracing systems for models specified in various curvilinear coordinates.

3.5.1 Curvilinear Orthogonal Coordinates

Consider a right-handed curvilinear orthogonal coordinate system ξ_1, ξ_2, ξ_3 with the relevant unit basis vectors $\vec{e}_1, \vec{e}_2$, and $\vec{e}_3$, and denote the corresponding scale factors by h_1, h_2, and h_3. The square of the infinitesimal length element ds^2 in the given coordinate system is

$$ds^2 = h_1^2 d\xi_1^2 + h_2^2 d\xi_2^2 + h_3^2 d\xi_3^2. \tag{3.5.1}$$

We are familiar with expressing vectorial differential operators in the orthogonal curvilinear coordinate system ξ_i from vector calculus. In the following, we shall only need the expression for the gradient:

$$\nabla \Phi = \frac{1}{h_1} \frac{\partial \Phi}{\partial \xi_1} \vec{e}_1 + \frac{1}{h_2} \frac{\partial \Phi}{\partial \xi_2} \vec{e}_2 + \frac{1}{h_3} \frac{\partial \Phi}{\partial \xi_3} \vec{e}_3. \tag{3.5.2}$$

The slowness vector is defined as $\vec{p} = \nabla T$, so

$$\vec{p} = \frac{1}{h_1} \frac{\partial T}{\partial \xi_1} \vec{e}_1 + \frac{1}{h_2} \frac{\partial T}{\partial \xi_2} \vec{e}_2 + \frac{1}{h_3} \frac{\partial T}{\partial \xi_3} \vec{e}_3. \tag{3.5.3}$$

Thus, the components of slowness vector $\vec{p}$ in curvilinear orthogonal coordinates ξ_1, ξ_2, and ξ_3 read

$$p_1 = \frac{1}{h_1}\frac{\partial T}{\partial \xi_1}, \qquad p_2 = \frac{1}{h_2}\frac{\partial T}{\partial \xi_2}, \qquad p_3 = \frac{1}{h_3}\frac{\partial T}{\partial \xi_3}. \tag{3.5.4}$$

As we can see from (3.5.4), we must strictly distinguish between p_i and $\partial T/\partial \xi_i$. Because we shall often use partial derivatives $\partial T/\partial \xi_i$, we shall use an abbreviated symbol T_i for them,

$$T_i = \partial T/\partial \xi_i. \tag{3.5.5}$$

3.5.2 The Eikonal Equation in Curvilinear Orthogonal Coordinates

In vector form, the eikonal equation in isotropic media is expressed as $\nabla T \cdot \nabla T = 1/V^2$. In curvilinear orthogonal coordinates, the gradient is given by (3.5.2). The eikonal equation can thus be expressed in arbitrary curvilinear orthogonal coordinates as follows:

$$\frac{1}{h_1^2}\left(\frac{\partial T}{\partial \xi_1}\right)^2 + \frac{1}{h_2^2}\left(\frac{\partial T}{\partial \xi_2}\right)^2 + \frac{1}{h_3^2}\left(\frac{\partial T}{\partial \xi_3}\right)^2 = \frac{1}{V^2(\xi_1,\xi_2,\xi_3)}. \tag{3.5.6}$$

Using notation (3.5.5),

$$\frac{1}{h_1^2}T_1^2 + \frac{1}{h_2^2}T_2^2 + \frac{1}{h_3^2}T_3^2 = \frac{1}{V^2(\xi_1,\xi_2,\xi_3)}. \tag{3.5.7}$$

Eikonal equation (3.5.7) can be expressed as $\mathcal{H}(\xi_i, T_i) = 0$; see (3.1.2). We shall again consider a very general form:

$$\mathcal{H}(\xi_i, T_i) = \frac{1}{n}\left\{\left(\frac{1}{h_1^2}T_1^2 + \frac{1}{h_2^2}T_2^2 + \frac{1}{h_3^2}T_3^2\right)^{n/2} - \frac{1}{V^n}\right\} = 0. \tag{3.5.8}$$

This also includes the case of $n = 0$:

$$\mathcal{H}(\xi_i, T_i) = \frac{1}{2}\ln\left(\frac{1}{h_1^2}T_1^2 + \frac{1}{h_2^2}T_2^2 + \frac{1}{h_3^2}T_3^2\right) + \ln V. \tag{3.5.9}$$

Equations (3.5.8) and (3.5.9) for Hamiltonian $\mathcal{H}(\xi_i, T_i)$ may be expressed in many alternative forms. For example, if we consider $n = 2$ in (3.5.8) and multiply it by V^2, we obtain

$$\mathcal{H}(\xi_i, T_i) = \frac{1}{2}\left\{V^2\left(T_1^2/h_1^2 + T_2^2/h_2^2 + T_3^2/h_3^2\right) - 1\right\} = 0. \tag{3.5.10}$$

This Hamiltonian formally corresponds to (3.1.13) in Cartesian coordinates. The disadvantage of Hamiltonian (3.5.10) is that velocity variations $V(\xi_i)$ are mixed with the variations of scale factors $h_1(\xi_i)$, $h_2(\xi_i)$, and $h_3(\xi_i)$. For this reason, we shall mostly use (3.5.8) and (3.5.9) in the following discussion. The eikonal equation for the components of slowness vector p_i in standard form reads

$$p_1^2 + p_2^2 + p_3^2 = 1/V^2(\xi_1,\xi_2,\xi_3). \tag{3.5.11}$$

It is also possible to introduce the reduced Hamiltonian $\mathcal{H}^R(\xi_i, T_I)$ by solving the eikonal equation (3.5.7) for T_1, T_2, or T_3. Without any loss of generality, we shall solve it for T_3:

$$T_3 = -\mathcal{H}^R(\xi_i, T_I), \tag{3.5.12}$$

where

$$\mathcal{H}^R(\xi_i, T_I) = -h_3 \big[V^{-2}(\xi_1, \xi_2, \xi_3) - h_1^{-2} T_1^2 - h_2^{-2} T_2^2 \big]^{1/2}. \tag{3.5.13}$$

The Hamiltonian $\mathcal{H}(\xi_i, T_i)$ can then be defined as

$$\mathcal{H}(\xi_i, T_i) = T_3 + \mathcal{H}^R(\xi_i, T_I). \tag{3.5.14}$$

3.5.3 The Ray Tracing System in Curvilinear Orthogonal Coordinates

The ray tracing system in curvilinear orthogonal coordinates can be derived simply by the method of characteristics; see (3.1.3). The characteristic system for the general form of the eikonal equation (3.5.8) is as follows:

$$\frac{d\xi_i}{du} = A^{n/2-1} \frac{T_i}{h_i^2}, \qquad \frac{dT_i}{du} = \frac{1}{n} \frac{\partial}{\partial \xi_i} \left(\frac{1}{V} \right)^n + A^{n/2-1} \sum_{k=1}^{3} \frac{T_k^2}{h_k^3} \frac{\partial h_k}{\partial \xi_i},$$

$$\frac{dT}{du} = A^{n/2} = \frac{1}{V^n}, \tag{3.5.15}$$

where

$$A = \frac{1}{h_1^2} T_1^2 + \frac{1}{h_2^2} T_2^2 + \frac{1}{h_3^2} T_3^2 = V^{-2}.$$

In the limit for $n = 0$, parameter u along the ray equals travel time T, and the ray tracing system reads

$$\frac{d\xi_i}{dT} = A^{-1} \frac{T_i}{h_i^2}, \qquad \frac{dT_i}{dT} = -\frac{\partial \ln V}{\partial \xi_i} + A^{-1} \sum_{k=1}^{3} \frac{T_k^2}{h_k^3} \frac{\partial h_k}{\partial \xi_i}. \tag{3.5.16}$$

Note that (3.5.16) can also be directly obtained from the Hamiltonian (3.5.10). As in Cartesian coordinates, parameter u along the ray equals arclength s for $n = 1$ ($ds = V\,dT$), and $u = \sigma$ for $n = 2$ ($d\sigma = V^2\,dT$) in (3.5.15).

Ray tracing systems (3.5.15) and (3.5.16) are expressed in Hamiltonian form for eikonal equations (3.5.8) and (3.5.9). We can, however, insert $A = V^{-2}$ into (3.5.15) and (3.5.16). Although the ray tracing systems are no longer in Hamiltonian form, they are simpler for numerical ray tracing. The simplest ray tracing system is again obtained for $n = 2$.

The initial conditions for (3.5.15) and (3.5.16) at point S situated on the ray are

$$\xi_i = \xi_{i0}, \qquad T_i = T_{i0}, \qquad T = T_0. \tag{3.5.17}$$

Here T_{i0} must satisfy the condition

$$h_{10}^{-2} T_{10}^2 + h_{20}^{-2} T_{20}^2 + h_{30}^{-2} T_{30}^2 = V_0^{-2}, \tag{3.5.18}$$

where V_0 and h_{i0} are V and h_i at point S. We can again specify the initial direction of the ray by two take-off angles, i_0 and ϕ_0 at S, where $0 \le i_0 \le \pi$ and $0 \le \phi_0 \le 2\pi$. Let us consider the basis unit vectors $\vec{e}_1$, $\vec{e}_2$, and $\vec{e}_3$ at S. Then i_0 and ϕ_0 can be taken as in Section 3.2.1, but with respect to the triplet of unit vectors $\vec{e}_1$, $\vec{e}_2$, and $\vec{e}_3$. The components of slowness vector $p_{i0} = \vec{p}_0 \cdot \vec{e}_i$ can be expressed in terms of i_0 and ϕ_0. When we determine p_{10}, p_{20}, and p_{30}, we obtain the required expressions for T_{10}, T_{20}, and T_{30} as follows:

$$T_{10} = h_{10} p_{10}, \qquad T_{20} = h_{20} p_{20}, \qquad T_{30} = h_{30} p_{30}. \tag{3.5.19}$$

These initial conditions T_{i0} automatically satisfy eikonal equation (3.5.18) at S.

If the ray is incident at curved interface Σ, the initial conditions for the rays of reflected/transmitted waves must be determined. In Section 2.4.5, we presented the general form of Snell's law (2.4.70) for arbitrary curved interfaces. We shall now express Snell's law (2.4.70) in an arbitrary curvilinear orthogonal coordinate system in terms of T_i.

Assume that the interface is given by equation $\Sigma(\xi_1, \xi_2, \xi_3) = 0$. The unit normal $\vec{n}$ to Σ at the point of incidence Q is then given by vectorial equation (3.2.5). Due to its vector character, the equation is also valid in curvilinear orthogonal coordinates so that

$$n_i = \frac{\epsilon^*}{h_i} \frac{\partial \Sigma}{\partial \xi_i} \Big/ A, \qquad A = \left[\sum_{k=1}^{3} \frac{1}{h_k^2} \left(\frac{\partial \Sigma}{\partial \xi_k} \right)^2 \right]^{1/2} \tag{3.5.20}$$

(no summation over i). Here $\epsilon^* = \pm 1$ specifies the required orientation of unit normal $\vec{n}$; see (3.2.5). Then, (2.4.70) yields Snell's law in the following form:

$$\tilde{T}_i = T_i - h_i \{ B \mp \epsilon [\tilde{V}^{-2} - V^{-2} + B^2]^{1/2} \} n_i,$$

$$B = \epsilon^* \left(\sum_{k=1}^{3} \frac{1}{h_k^2} T_k \frac{\partial \Sigma}{\partial \xi_k} \right) \Big/ A \tag{3.5.21}$$

(no summation over i). Here ϵ is the orientation index so that $\epsilon = \operatorname{sgn} B$. The upper sign corresponds to transmitted waves, the lower sign refers to reflected waves. As we can see, Snell's law (3.5.21) is expressed fully in terms of T_i, not p_i. Quantities $\tilde{T}_i$ and $\tilde{V}$ correspond to the selected wave reflected/transmitted at Q.

The equations derived in this section are sufficient to perform ray tracing in 3-D inhomogeneous layered models in arbitrary curvilinear orthogonal coordinates. As soon as analytical expressions for the scale factors h_1, h_2, and h_3 in an orthogonal coordinate system are known, the application of the ray tracing system (3.5.16) is straightforward. The analytical expressions for the scale factors h_1, h_2, and h_3 for many coordinate systems can be found in many mathematical textbooks and handbooks; see, for example, Korn and Korn (1961). For this reason, we shall not present here specific ray tracing systems for different types of coordinates. The only exceptions are the spherical polar coordinates because they play a basic role in global seismological studies. See also Yan and Yen (1995) for the discussion of ray tracing in ellipsoidal coordinates.

Ray tracing systems (3.5.15) and (3.5.16) consist of six ordinary differential equations of the first order for ξ_i and T_i. The number of equations in the system can be reduced to four if we use the reduced Hamiltonian (3.5.13) and choose $u = \xi_3$. Then, the reduced ray tracing system reads:

$$\frac{d\xi_I}{d\xi_3} = \frac{1}{T_3} \left(\frac{h_3}{h_I} \right)^2 T_I,$$

$$\frac{dT_I}{d\xi_3} = \frac{T_3}{h_3} \frac{\partial h_3}{\partial \xi_I} + \frac{h_3^2}{2T_3} \left[\frac{\partial V^{-2}}{\partial \xi_I} - \frac{\partial h_1^{-2}}{\partial \xi_I} T_1^2 - \frac{\partial h_2^{-2}}{\partial \xi_I} T_2^2 \right], \tag{3.5.22}$$

with the equation for the travel-time computation along the ray:

$$dT/d\xi_3 = h_3 V^{-1} \left[1 - V^2 (h_1^{-2} T_1^2 + h_2^{-2} T_2^2) \right]^{-1/2} = h_3^2 V^{-2} T_3^{-1}. \tag{3.5.23}$$

Mutatis mutandis, Equations (3.5.12) through (3.5.14) and Equations (3.5.22) and (3.5.23) can be used even for $u = \xi_1$ or $u = \xi_2$ as a variable along the ray. Equations for initial conditions at a point S and at a structural interface, derived in this section, can easily be modified even for the reduced ray tracing system (3.5.22).

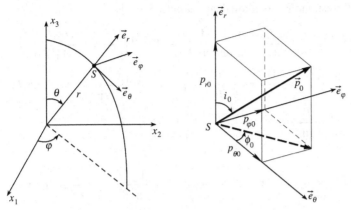

Figure 3.7. The definition of the initial take-off angles i_0 and ϕ_0 at point S in spherical coordinates r, θ, φ. The bold continuous line denotes the initial slowness vector $\vec{p}_0$; the dashed bold line represents the horizontal projection of the initial slowness vector into the plane specified by $\vec{e}_\theta$ and $\vec{e}_\varphi$.

3.5.4 Ray Tracing in Spherical Polar Coordinates

We denote the spherical polar coordinates r, θ, and φ such that r is the radial distance, θ is the colatitude, and φ is the longitude, where $0 \leq \theta \leq \pi$ and $0 \leq \varphi < 2\pi$; see Figure 3.7. The spherical coordinates are related to Cartesian coordinates x_i as follows:

$$x_1 = r \sin\theta \cos\varphi, \qquad x_2 = r \sin\theta \sin\varphi, \qquad x_3 = r \cos\theta. \qquad (3.5.24)$$

The square of the infinitesimal length element is

$$ds^2 = dr^2 + r^2 d\theta^2 + r^2 \sin^2\theta \, d\varphi^2. \qquad (3.5.25)$$

It follows from (3.5.25) that the scale factors are

$$h_r = 1, \qquad h_\theta = r, \qquad h_\varphi = r \sin\theta. \qquad (3.5.26)$$

The components of the slowness vector can be expressed as

$$p_r = T_r, \qquad p_\theta = r^{-1} T_\theta, \qquad p_\varphi = (r \sin\theta)^{-1} T_\varphi, \qquad (3.5.27)$$

with

$$T_r = \partial T / \partial r, \qquad T_\theta = \partial T / \partial \theta, \qquad T_\varphi = \partial T / \partial \varphi.$$

Eikonal equation (3.5.7) can then be expressed as

$$T_r^2 + \frac{1}{r^2} T_\theta^2 + \frac{1}{r^2 \sin^2\theta} T_\varphi^2 = \frac{1}{V^2(r, \theta, \varphi)} \qquad (3.5.28)$$

or, in a more general form, as

$$\mathcal{H}(r, \theta, \varphi, T_r, T_\theta, T_\varphi)$$
$$= \frac{1}{n}\left\{ \left(T_r^2 + \frac{1}{r^2} T_\theta^2 + \frac{1}{r^2 \sin^2\theta} T_\varphi^2 \right)^{n/2} - \frac{1}{V^n(r, \theta, \varphi)} \right\} = 0. \qquad (3.5.29)$$

For $n = 0$,

$$\mathcal{H}(r, \theta, \varphi, T_r, T_\theta, T_\varphi)$$
$$= \frac{1}{2} \ln\left(T_r^2 + \frac{1}{r^2} T_\theta^2 + \frac{1}{r^2 \sin^2\theta} T_\varphi^2 \right) + \ln V(r, \theta, \varphi) = 0. \qquad (3.5.30)$$

The ray tracing system corresponding to eikonal equation (3.5.29) is

$$\frac{dr}{du} = A^{n/2-1} T_r, \qquad \frac{dT_r}{du} = \frac{1}{n}\frac{\partial}{\partial r}\left(\frac{1}{V}\right)^n + A^{n/2-1}\left(\frac{T_\theta^2}{r^3} + \frac{T_\varphi^2}{r^3 \sin^2\theta}\right),$$

$$\frac{d\theta}{du} = A^{n/2-1}\frac{T_\theta}{r^2}, \qquad \frac{dT_\theta}{du} = \frac{1}{n}\frac{\partial}{\partial\theta}\left(\frac{1}{V}\right)^n + A^{n/2-1}\frac{T_\varphi^2 \cos\theta}{r^2 \sin^3\theta}, \qquad (3.5.31)$$

$$\frac{d\varphi}{du} = A^{n/2-1}\frac{T_\varphi}{r^2 \sin^2\theta}, \qquad \frac{dT_\varphi}{du} = \frac{1}{n}\frac{\partial}{\partial\varphi}\left(\frac{1}{V}\right)^n,$$

with

$$\frac{dT}{du} = A^{n/2} = \frac{1}{V^n}, \qquad A = T_r^2 + \frac{1}{r^2}T_\theta^2 + \frac{1}{r^2 \sin^2\theta}T_\varphi^2 = \frac{1}{V^2}. \qquad (3.5.32)$$

The RHS of the equation for dT_r/du can be simplified, as $T_\theta^2/r^3 + T_\varphi^2/(r^3 \sin^2\theta) = (V^{-2} - T_r^2)/r$, see (3.5.32).

Ray tracing system (3.5.31) can be used for any n. The simplest ray tracing system is obtained for $n = 2$ such that $A^{n/2-1} = 1$. Parameter u along the ray then equals σ such that $d\sigma = V^2 dT$. For $n = 0$, parameter u equals travel time T. If $n = 0$, expression $n^{-1}\partial(V^{-n})/\partial r$ reads $-\partial \ln V/\partial r$, and the two other relevant derivatives can be found similarly. For $n = 1$, parameter u equals arclength s along the ray.

The initial conditions at point S for ray tracing system (3.5.31) are

$$r = r_0, \qquad \theta = \theta_0, \qquad \varphi = \theta_0,$$
$$T_r = T_{r0}, \qquad T_\theta = T_{\theta 0}, \qquad T_\varphi = T_{\varphi 0}, \qquad T = T_0. \qquad (3.5.33)$$

Quantities T_{r0}, $T_{\theta 0}$, and $T_{\varphi 0}$ must satisfy the relation

$$T_{r0}^2 + \frac{1}{r_0^2}T_{\theta 0}^2 + \frac{1}{r_0^2 \sin^2\theta_0}T_{\varphi 0}^2 = \frac{1}{V_0^2}. \qquad (3.5.34)$$

We can again introduce take-off angles i_0 and ϕ_0 to specify the initial direction of the ray. In seismology, it is usual to consider initial angles i_0 and ϕ_0 in spherical coordinates as shown in Figure 3.7; see Aki and Richards (1980, p. 724). Then

$$p_{r0} = V_0^{-1}\cos i_0, \qquad p_{\theta 0} = V_0^{-1}\sin i_0 \cos\phi_0, \qquad p_{\varphi 0} = V_0^{-1}\sin i_0 \sin\phi_0.$$

Taking into account the scale factors,

$$T_{r0} = V_0^{-1}\cos i_0, \qquad T_{\theta 0} = V_0^{-1}r_0 \sin i_0 \cos\phi_0,$$
$$T_{\varphi 0} = V_0^{-1}r_0 \sin\theta_0 \sin i_0 \sin\phi_0. \qquad (3.5.35)$$

These initial values satisfy eikonal equation (3.5.34) at S automatically.

3.5.5 Modified Ray Tracing Systems in Spherical Polar Coordinates

The ray tracing systems in spherical polar coordinates may be expressed in many alternative forms. Particularly familiar in seismology is the ray tracing system proposed by Julian and Gubbins (1977), see also Aki and Richards (1980, pp. 724–5) and Dahlen and Tromp (1998, Chap. 15). Instead of T_r, T_θ, and T_φ, Julian and Gubbins use two polar angles, i and ξ, which are introduced at any point of the ray in much the same way that angles i_0 and ϕ_0, described in Section 3.5.4, were introduced: $T_r = V^{-1}\cos i$, $T_\theta = V^{-1}r \sin i \cos\xi$, and $T_\varphi = V^{-1}r \sin\theta \sin i \sin\xi$. The system consists of five equations only, but it is rather complicated (it contains trigonometric functions of three angles). The number of equations in the ray tracing system can even be reduced to four, if we use r or φ or θ as the variable u along the ray. For example, Liu and Tromp (1996) use $u = \varphi$. Various ray tracing systems consisting of four equations only may also be obtained by introducing the reduced Hamiltonian $\mathcal{H}^R$; see (3.5.13) and (3.5.22). The relevant reduced ray tracing systems follow immediately from (3.5.22) if we insert (3.5.26) into it. We do not present the final ray tracing systems here because their derivation is simple. Moreover, the reduced ray tracing systems do not offer distinct advantages when compared with the full ray tracing system (3.5.31). For other alternative systems, see Jacob (1970) and Comer (1984).

Here we shall present several particularly simple ray tracing systems, including 2-D ray tracing systems. We shall not give the initial conditions and Snell's law; they are straightforward.

Instead of spherical polar coordinates r, θ, and φ, we shall introduce new coordinates z, a, and b:

$$z = -R\ln(r/R), \qquad a = R\theta, \qquad b = R\varphi. \tag{3.5.36}$$

Here R is an arbitrary positive number (for example, the radius of the Earth). If $r = R$, then $z = 0$, and if $r \to 0$, then $z \to \infty$. Thus, z is a modified depth coordinate. Consequently,

$$r = R\exp(-z/R), \qquad dr/dz = -r/R,$$
$$T_z = -rR^{-1}T_r, \qquad T_a = R^{-1}T_\theta, \qquad T_b = R^{-1}T_\varphi.$$

Note that z, a, and b are still curvilinear orthogonal coordinates, with $h_z = r/R$, $h_a = r/R$, and $h_b = r\sin\theta/R$. The eikonal equation now becomes

$$T_z^2 + T_a^2 + \sin^{-2}\theta T_b^2 = r^2/R^2V^2. \tag{3.5.37}$$

Here r and θ should be expressed in terms of z and a. We shall again express the eikonal equation in a more general form:

$$\mathcal{H} = n^{-1}\left\{\left(T_z^2 + T_a^2 + \sin^{-2}\theta T_b^2\right)^{n/2} - \eta^n\right\} = 0, \qquad \eta = r/RV. \tag{3.5.38}$$

Using (3.1.3), we obtain the relevant ray tracing system,

$$\frac{dz}{du} = A^{n/2-1}T_z, \qquad \frac{dT_z}{du} = \frac{1}{n}\frac{\partial\eta^n}{\partial z},$$
$$\frac{da}{du} = A^{n/2-1}T_a, \qquad \frac{dT_a}{du} = \frac{1}{n}\frac{\partial\eta^n}{\partial a} + A^{n/2-1}\frac{T_b^2\cos\theta}{R\sin^3\theta}, \tag{3.5.39}$$
$$\frac{db}{du} = A^{n/2-1}\frac{T_b}{\sin^2\theta}, \qquad \frac{dT_b}{du} = \frac{1}{n}\frac{\partial\eta^n}{\partial b},$$

with

$$dT/du = A^{n/2} = \eta^n, \qquad A = T_z^2 + T_a^2 + \sin^{-2}\theta T_b^2 = \eta^2. \tag{3.5.40}$$

Ray tracing system (3.5.39) is particularly simple for $n = 2$, since $A^{n/2-1} = 1$. Parameter u along the ray is then given by the relation $du = \eta^2 dT$. If $n = 0$, parameter u along the ray equals travel time T. Derivative $n^{-1}\partial\eta^n/\partial z$ then reads $-\partial(\ln\eta)/\partial z$; the two other partial derivatives are specified similarly.

As we can see from (3.5.39), quantity $\eta^{-1} = VR/r$ is used instead of velocity V to describe the model. The relation between η and V is unique, and one quantity can be simply expressed in terms of the other.

System (3.5.39) is perhaps the simplest 3-D ray tracing system in the modified spherical coordinates, particularly for $n = 2$. Let us emphasize that ray tracing system (3.5.39) is quite general; no approximation has been made in deriving it.

Ray tracing systems (3.5.31) and (3.5.39) require an alternative treatment if the computed ray enters a region of θ close to 0 or to π (that is, for a close to 0 or to $R\pi$). These singularities have a formal meaning and depend on the definition of coordinates θ and φ. The local transformation of coordinates $\theta, \varphi \longrightarrow \theta', \varphi'$ eliminates the problem. For example, Liu and Tromp (1996) propose to rotate the Earth's model so that both the source and receiver are located on the equator. An alternative treatment is also needed for $r \to 0$.

In the general case, ray tracing system (3.5.39) must be solved as a whole because all the equations are coupled with one another. There are some exceptions. If $n = 2$ and we assume that $\eta(r, \theta, \varphi) = \eta_1(r) + \eta_2(\theta, \varphi)$, that is, $\eta(z, a, b) = \eta_1(z) + \eta_2(a, b)$, system (3.5.39) then splits into two subsystems, which are fully separated; they are coupled only by the initial conditions.

We shall now specify ray tracing system (3.5.39) for two 2-D situations, in which the system is particularly simple.

First, we consider η *independent of* φ (that is, η is independent of $b = R\varphi$). We also assume that the initial conditions are such that $p_{\varphi 0} = 0$, which implies that $T_{\varphi 0} = 0$. Then $b = b_0, T_b = T_{b0} = 0$ along the whole ray, and the ray trajectory is situated in plane $\varphi = \varphi_0$. Ray tracing system (3.5.39) then yields

$$\frac{dz}{du} = A^{n/2-1} T_z, \qquad \frac{dT_z}{du} = \frac{1}{n} \frac{\partial \eta^n}{\partial z},$$
$$\frac{da}{du} = A^{n/2-1} T_a, \qquad \frac{dT_a}{du} = \frac{1}{n} \frac{\partial \eta^n}{\partial a}, \tag{3.5.41}$$

with

$$dT/du = \eta^n = A^{n/2}, \qquad A = T_z^2 + T_a^2 = \eta^2. \tag{3.5.42}$$

Second, we consider η *independent of* r (that is, η is independent of z). We assume that the initial conditions are such that $p_{z0} = 0$, which implies that $T_{z0} = 0$. Then $z = z_0$ (or $r = r_0$) and $T_z = T_{z0} = 0$ along the whole ray. An example is the ray tracing along the spherical surface of the Earth. It would, of course, be possible to use (3.5.39) again, but we shall first rewrite the eikonal equation for our case. If we put $T_z = 0$ and multiply (3.5.38) by $\sin^n \theta = \sin^n(a/R)$, we obtain

$$\mathcal{H} = \frac{1}{n} \left\{ \left(T_a^2 \sin^2 \frac{a}{R} + T_b^2 \right)^{n/2} - \kappa^n \right\} = 0, \qquad \kappa = \frac{r_0 \sin(a/R)}{RV}. \tag{3.5.43}$$

Now we introduce a new coordinate c given by the relation (Mercator transformation)

$$c = R \ln \tan \frac{\theta}{2} = R \ln \tan \frac{a}{2R}, \qquad T_a = \frac{1}{\sin(a/R)} T_c,$$

instead of a. Coordinates c and b are still 2-D orthogonal coordinates, with the same scale factors:

$$h_c = h_b = \frac{r_0}{R} \sin \frac{a}{R} = \frac{2r_0}{R} \frac{\exp(c/R)}{1 + \exp^2(c/R)}. \tag{3.5.44}$$

Eikonal equation (3.5.43) then becomes

$$\mathcal{H} = n^{-1}\left\{\left(T_c^2 + T_b^2\right)^{n/2} - \kappa^n\right\} = 0, \qquad (3.5.45)$$

with the relevant ray-tracing system,

$$\frac{dc}{du} = A^{n/2-1} T_c, \qquad \frac{dT_c}{du} = \frac{1}{n}\frac{\partial \kappa^n}{\partial c},$$
$$\frac{db}{du} = A^{n/2-1} T_b, \qquad \frac{dT_b}{du} = \frac{1}{n}\frac{\partial \kappa^n}{\partial b}, \qquad (3.5.46)$$

with

$$\frac{dT}{du} = A^{n/2} = \kappa^n, \qquad A = T_c^2 + T_b^2 = \kappa^2. \qquad (3.5.47)$$

We have thus obtained two 2-D ray tracing systems, (3.5.41) and (3.5.46), which are fully equivalent to the 2-D ray tracing system in Cartesian coordinates.

In the first case, ray tracing system (3.5.41) corresponds to ray tracing in a plane of constant φ. The relevant transformation leading to this system is

$$z = R\ln(R/r), \qquad a = R\theta, \qquad \eta = r/R\,V(r, \theta, \varphi_0). \qquad (3.5.48)$$

We speak of a *2-D Earth flattening transformation* (2-D EFT).

In the second case, ray tracing system (3.5.46) corresponds to ray tracing along a spherical surface $r = r_0 = $ const. The transformation formulae are as follows:

$$c = R\ln\tan\left(\tfrac{1}{2}\theta\right), \qquad b = R\varphi, \qquad \kappa = r_0\sin\theta/R\,V(r_0, \theta, \varphi). \qquad (3.5.49)$$

We can speak of a *2-D Earth surface flattening transformation* (2-D ESFT), or of a Mercator transformation. This transformation was first proposed by Jobert and Jobert (1983) for ray tracing of surface waves. See also Jobert and Jobert (1987).

Both the 2-D EFT and 2-D ESFT transformations are very important, at least from the following two points of view.

- Any analytical solution of the 2-D ray tracing system in Cartesian coordinates can be used without change in spherical coordinates r, θ or θ, φ, and vice versa.
- Computer programs for 2-D ray tracing in Cartesian coordinates can be directly used for 2-D ray tracing in spherical coordinates r, θ or θ, φ by a simple modification of the input and output data.

3.5.6 Ray Tracing in Curvilinear Nonorthogonal Coordinates

For completeness, we shall give the eikonal equation and ray tracing systems in general curvilinear nonorthogonal coordinates. For more details, see Červený, Klimeš, and Pšenčík (1988b) and Hrabě (1994). In nonorthogonal coordinates, it is necessary to distinguish strictly between covariant and contravariant components of vectors and tensors; see, for example, Synge and Schild (1952). As usual, we shall write the covariant indices as subscripts and the contravariant indices as superscripts. As in Section 3.1.3, we denote the coordinates x^i and the covariant components of the metric tensor g_{ij}. The square of the infinitesimal distance ds^2 between two adjacent points is given by relation (3.1.41). The contravariant components of the metric tensor, g^{ij}, can be calculated from g_{ij} using the relations

$$g_{ij}g^{ik} = \delta^k_j, \tag{3.5.50}$$

where the mixed covariant and contravariant Kronecker delta δ^k_j equals 1 if $k = j$ and 0 otherwise. The summation in (3.5.50) and in all other equations in this section is over the same superscripts and subscripts, that is, over i in (3.5.50). Covariant components f_i of a vector $\vec{f}$ may be expressed in terms of contravariant components f^i as follows:

$$f_i = g_{ij}f^j, \qquad f^i = g^{ij}f_j. \tag{3.5.51}$$

The eikonal equation in the general coordinate system for P or S waves propagating in an inhomogeneous isotropic medium can be expressed in several alternative forms, using either covariant or contravariant components of slowness vector $\vec{p}$, p_i or p^i. Note that the covariant components of slowness vector p_i are $\partial T/\partial x^i$. The three forms of the eikonal equation are

$$g^{ij}p_ip_j = V^{-2}(x^i),$$
$$g_{ij}p^ip^j = V^{-2}(x^i), \qquad p_ip^i = V^{-2}(x^i). \tag{3.5.52}$$

Here V is either α or β depending on the type of wave. All three forms of the eikonal equation are fully equivalent; see (3.5.51). We shall primarily use the first form of the eikonal equation, expressed in terms of the covariant components of the slowness vector.

We shall again use a more general form of eikonal equation (3.5.52):

$$\mathcal{H}(x^i, p_i) = n^{-1}\left\{(g^{kj}p_kp_j)^{n/2} - V^{-n}(x^i)\right\} = 0, \tag{3.5.53}$$

where n is an arbitrary number.

The ray tracing system can be obtained from (3.5.53) in several alternative forms. Because $p_i = \partial T/\partial x^i$, we can again express the characteristic system of (3.5.53) in Hamiltonian canonical form (3.1.3):

$$\frac{dx^i}{du} = A^{n/2-1}g^{ik}p_k, \qquad \frac{dp_i}{du} = \frac{1}{n}\frac{\partial}{\partial x^i}\left(\frac{1}{V}\right)^n - \frac{1}{2}A^{n/2-1}p_kp_j\frac{\partial g^{kj}}{\partial x^i}, \tag{3.5.54}$$

with

$$\frac{dT}{du} = A^{n/2} = \left(\frac{1}{V}\right)^n, \qquad A = g^{kj}p_kp_j = \frac{1}{V^2(x^i)}. \tag{3.5.55}$$

Here u is a parameter along the ray, related to travel time T by the first equation of (3.5.55). Equations (3.5.54) represent the final form of the ray tracing system in general nonorthogonal coordinates.

System (3.5.54) contains the derivatives of the contravariant components of the metric tensor, $\partial g^{kj}/\partial x^i$. They can be replaced by the derivatives of the covariant components of the metric tensor using the relation

$$\partial g^{kj}/\partial x^i = -g^{km}g^{jn}\partial g_{mn}/\partial x^i.$$

Christoffel symbols of the second kind are frequently used instead of the derivatives of the metric tensor:

$$\Gamma_{ij}^{k} = \frac{1}{2}g^{kl}\left(\frac{\partial g_{il}}{\partial x^{j}} + \frac{\partial g_{jl}}{\partial x^{i}} - \frac{\partial g_{ij}}{\partial x^{l}}\right). \tag{3.5.56}$$

After some algebra, we obtain an alternative expression for dp_i/du in (3.5.54):

$$\frac{dp_i}{du} = \frac{1}{n}\frac{\partial}{\partial x^i}\left(\frac{1}{V(x^i)}\right)^n + A^{n/2-1}p_k p_j g^{km}\Gamma_{im}^{j}. \tag{3.5.57}$$

As in Sections 3.5.3 through 3.5.5, parameter u along the ray corresponds to travel time T if $n = 0$, to arclength s along the ray if $n = 1$, and to σ ($d\sigma = V^2 dT$) if $n = 2$. A particularly simple ray tracing system is obtained for $n = 2$, since $A^{n/2-1} = 1$. If $n = 0$, we need to put $n^{-1}\partial(V^{-n})/\partial x^i = -\partial \ln V/\partial x^i$ in (3.5.54) and (3.5.57).

Equation (3.5.57) can be formally simplified if we introduce the so-called absolute derivatives along curve $x^i = x^i(u)$. The absolute derivative Df_r/Du of the covariant component of a vector along curve $x^i = x^i(u)$ is defined as

$$\frac{Df_r}{Du} = \frac{df_r}{du} - \Gamma_{rn}^{m}f_m\frac{dx^n}{du}; \tag{3.5.58}$$

see Synge and Schild (1952). Ray tracing system (3.5.57) may then be expressed as

$$\frac{Dp_i}{Du} = \frac{1}{n}\frac{\partial}{\partial x^i}\left(\frac{1}{V}\right)^n. \tag{3.5.59}$$

Note that absolute derivative Dp_i/Du is also often denoted as $\delta p_i/\delta u$.

In the Riemannian geometry, the concept of *parallel transport of a vector along a curve* plays an important role. We say that vector f_r is propagated parallel along curve Ω if it satisfies the differential equation

$$\frac{Df_r}{Du} = \frac{df_r}{du} - \Gamma_{rn}^{m}f_m\frac{dx^n}{du} = 0 \tag{3.5.60}$$

along Ω. This definition also implies that the length of the vector, given by the expression $(g^{ik}f_i f_k)^{1/2}$, remains fixed along Ω.

3.5.7 Comments on Ray Tracing in Curvilinear Coordinates

Ray tracing in models specified in curvilinear coordinates may be particularly useful in the case of some symmetries (for example, in radially symmetric media). In such cases, it is often possible to find suitable analytical solutions; see Section 3.7.4.

In general, however, the ray tracing systems for models specified in curvilinear coordinates are more complex than those in Cartesian coordinates. Moreover, the ray tracing systems in curvilinear coordinates often fail in some regions. For example, the ray tracing system in spherical polar coordinates cannot be used for θ close to 0 or to π (polar regions) and for r close to 0 (close to the center of the Earth); see Section 3.5.5.

It is, however, possible to use an universal approach to ray tracing in curvilinear coordinates, which is quite safe and removes all the singularities introduced by the coordinate system under consideration. A model specified in curvilinear coordinates may be transformed into a model specified in Cartesian coordinates, using standard transformation relations, and the ray tracing may be performed in general Cartesian coordinates. After this, the results may again be transformed back to the curvilinear coordinates under consideration. If the model is specified in a 3-D mesh, the individual transformations are straightforward. This approach has some advantages.

1. It may be used for models specified in any coordinate system. The relevant ray tracing system need not be derived.
2. It removes all the singularities of ray tracing equations, which would be introduced by the relevant curvilinear coordinate system.
3. Most equations presented in this book may be applied without any change, including dynamic ray tracing in Cartesian or ray-centered coordinates. It is not necessary to derive new dynamic ray tracing systems for the relevant curvilinear coordinate system.

To apply this universal approach, we must add some transformation routines to the general programs for model building and ray tracing that are designed in Cartesian coordinates.

3.6 Ray Tracing in Inhomogeneous Anisotropic Media

In this section, we shall discuss ray tracing and travel-time computations of any of the three waves propagating in an inhomogeneous anisotropic medium. The ray tracing system is the same for all three waves propagating in an inhomogeneous anisotropic medium; the type of elementary wave we wish to compute is specified only by the initial conditions. See Sections 3.6.2 and 3.6.3.

We remind the reader that the basic role in the ray method in an inhomogeneous anisotropic medium is played by a 3×3 matrix $\Gamma_{ik}(x_i, p_i) = a_{ijkl} p_j p_l$, where $a_{ijkl} = c_{ijkl}/\rho$ are the density-normalized elastic parameters, and p_i are the Cartesian components of the slowness vector, $p_i = \partial T/\partial x_i$. We denote the three eigenvalues and eigenvectors of $\Gamma_{ik}(x_i, p_i)$ by $G_m(x_i, p_i)$ and $\vec{g}^{(m)}(x_i, p_i)$, $m = 1, 2, 3$.

For more details on ray tracing and travel-time computations in inhomogeneous anisotropic media and for numerical computations, see Babich (1961a), Červený (1972), Červený and Pšenčík (1972), Suchy (1972), Červený, Molotkov, and Pšenčík (1977), Jech (1983), Petrashen and Kashtan (1984), Červený and Firbas (1984), Mochizuki (1987), Gajewski and Pšenčík (1987a, 1990, 1992), Shearer and Chapman (1989), Farra and Le Bégat (1995), and Mensch and Farra (1999). Here we shall mainly follow Červený (1972).

3.6.1 Eikonal Equation

Let us consider a nondegenerate case of three different eigenvalues of matrix Γ_{ik},

$$G_1(x_i, p_i) \neq G_2(x_i, p_i) \neq G_3(x_i, p_i). \qquad (3.6.1)$$

The eikonal equation for any of the three waves then reads

$$G_m(x_i, p_i) = 1, \qquad m = 1, 2, 3; \qquad (3.6.2)$$

see Section 2.4.3. Equation (3.6.2) represents a nonlinear partial differential equation of the first order for $T(x_i)$ and describes the propagation of the wavefront $T(x_i) = $ const. of the wave under consideration.

We shall use eikonal equation (3.6.2) in Hamiltonian form:

$$\mathcal{H}(x_i, p_i) = \tfrac{1}{2}(G_m(x_i, p_i) - 1) = 0, \qquad m = 1, 2, 3. \tag{3.6.3}$$

This form of Hamiltonian corresponds to (3.1.13) for isotropic media, where $G_m = V^2 p_k p_k$.

The index m specifies the type of elementary wave under consideration. Unless otherwise stated, we shall use $m = 1$ for the qS1 wave, $m = 2$ for the qS2 wave, and $m = 3$ for the qP wave. Most of the equations we shall derive will, however, be applicable to any m.

Condition (3.6.1) plays an important role particularly in the investigation of qS waves. Many results derived in this section fail for $G_1 = G_2$ as well as for G_1 close to G_2. This is the case of the qS waves in weakly anisotropic medium (close to isotropic). G_1 may, however, be locally close to G_2 even in strongly anisotropic media (shear wave singularities). The qS1 and qS2 waves cannot be treated as two independent waves in these cases because they are mutually coupled. We speak of qS wave coupling. For a more detailed treatment of the qS waves if $G_1 \doteq G_2$, see Sections 3.9.4 and 5.4.6. In this section, we shall consider only the classical ray-theory situation of two well-separated qS waves, propagating in a region where G_1 is not close to G_2.

3.6.2 Ray Tracing System

The ray tracing system will be derived here from eikonal equation (3.6.3) in the standard way using the method of characteristics. We shall express the characteristics in Hamiltonian canonical form,

$$\frac{\mathrm{d}x_i}{\mathrm{d}u} = \frac{1}{2}\frac{\partial G_m}{\partial p_i}, \qquad \frac{\mathrm{d}p_i}{\mathrm{d}u} = -\frac{1}{2}\frac{\partial G_m}{\partial x_i}, \qquad \frac{\mathrm{d}T}{\mathrm{d}u} = \frac{1}{2}p_i\frac{\partial G_m}{\partial p_i};$$

see (3.1.3). Here T is the travel time and u is a parameter along the ray, specified by the last equation. Applying Euler's theorem and the eikonal equation, we obtain

$$\mathrm{d}T/\mathrm{d}u = \tfrac{1}{2}p_i\,\partial G_m/\partial p_i = G_m = 1.$$

Thus, the parameter along the ray, u, equals T, and the ray tracing system reads

$$\frac{\mathrm{d}x_i}{\mathrm{d}T} = \frac{1}{2}\frac{\partial G_m}{\partial p_i}, \qquad \frac{\mathrm{d}p_i}{\mathrm{d}T} = -\frac{1}{2}\frac{\partial G_m}{\partial x_i}. \tag{3.6.4}$$

Ray tracing system (3.6.4) is the same for all three waves propagating in inhomogeneous anisotropic media.

Instead of parameter T along the ray, we can also introduce such other parameters as arclength s along the ray. Since $\mathrm{d}s^2 = \mathrm{d}x_k\mathrm{d}x_k$, the first equation of (3.6.4) yields

$$\left(\frac{\mathrm{d}s}{\mathrm{d}T}\right)^2 = \frac{\mathrm{d}x_k}{\mathrm{d}T}\frac{\mathrm{d}x_k}{\mathrm{d}T} = \frac{1}{4}\frac{\partial G_m}{\partial p_k}\frac{\partial G_m}{\partial p_k}$$

(no summation over m). Ray tracing system (3.6.4) can then be expressed as

$$\frac{\mathrm{d}x_i}{\mathrm{d}s} = A\frac{\partial G_m}{\partial p_i}, \qquad \frac{\mathrm{d}p_i}{\mathrm{d}s} = -A\frac{\partial G_m}{\partial x_i}, \qquad \frac{\mathrm{d}T}{\mathrm{d}s} = 2A \tag{3.6.5}$$

with

$$A = \left(\frac{\partial G_m}{\partial p_k}\frac{\partial G_m}{\partial p_k}\right)^{-1/2}$$

(no summation over m). Obviously, system (3.6.5) is more complicated than (3.6.4). Consequently, we shall mainly use ray tracing system (3.6.4) in the following text. Only for some special types of anisotropic media, such as the factorized anisotropic inhomogeneous media described in Section 3.6.6, shall we also use some other parameters along the ray instead of T.

Eigenvalues $G_m(x_i, p_i)$ of matrix Γ_{ij} in ray tracing systems (3.6.4) and (3.6.5) are the solutions of cubic equation (2.2.28). Thus, the straightforward, but very cumbersome, approach to expressing the RHS of (3.6.4) or (3.6.5) for a particular anisotropic medium would be based on the analytical solution of the cubic equation in G_m and on the determination of partial derivatives $\partial G_m / \partial p_i$ and $\partial G_m / \partial x_i$ from these analytical solutions. This approach may be used efficiently only for very simple anisotropy symmetries. In a general case, this "hungry wolf" method is not efficient. Instead, however, we can use a considerably more efficient method. In the ray tracing systems, we do not require the eigenvalues G_m themselves to be known; we require knowledge of their partial derivatives only. These partial derivatives can be determined even without solving the cubic equation in G_m. The first option is to determine them using the theorem of implicit functions; see Červený (1972). The other option is to express the derivatives in terms of eigenvectors $\vec{g}^{(m)}$. Both options are equivalent and lead to the same ray tracing system. Here we shall use the second option. We express G_m in terms of matrix Γ_{ij} and the eigenvector components in the following form:

$$G_m = \Gamma_{jl} g_j^{(m)} g_l^{(m)};\tag{3.6.6}$$

see (2.2.34). We now take the derivative of G_m with respect to p_i,

$$
\begin{aligned}
\frac{\partial G_m}{\partial p_i} &= \frac{\partial \Gamma_{jl}}{\partial p_i} g_j^{(m)} g_l^{(m)} + 2\Gamma_{jl} \frac{\partial g_j^{(m)}}{\partial p_i} g_l^{(m)} \\
&= \frac{\partial \Gamma_{jl}}{\partial p_i} g_j^{(m)} g_l^{(m)} + 2 G_m \frac{\partial g_j^{(m)}}{\partial p_i} g_j^{(m)} \\
&= \frac{\partial \Gamma_{jl}}{\partial p_i} g_j^{(m)} g_l^{(m)} + G_m \frac{\partial}{\partial p_i} \left(g_j^{(m)} g_j^{(m)} \right) \\
&= \frac{\partial \Gamma_{jl}}{\partial p_i} g_j^{(m)} g_l^{(m)}.
\end{aligned}
$$

The same is true for $\partial G_m / \partial x_i$. On aggregate,

$$\frac{\partial G_m}{\partial p_i} = \frac{\partial \Gamma_{jl}}{\partial p_i} g_j^{(m)} g_l^{(m)}, \qquad \frac{\partial G_m}{\partial x_i} = \frac{\partial \Gamma_{jl}}{\partial x_i} g_j^{(m)} g_l^{(m)}.\tag{3.6.7}$$

The ray tracing system then reads

$$\frac{dx_i}{dT} = \frac{1}{2} \frac{\partial \Gamma_{jl}}{\partial p_i} g_j^{(m)} g_l^{(m)}, \qquad \frac{dp_i}{dT} = -\frac{1}{2} \frac{\partial \Gamma_{jl}}{\partial x_i} g_j^{(m)} g_l^{(m)}.\tag{3.6.8}$$

It is easy to determine the derivatives of $\Gamma_{jl} = a_{jkln} p_k p_n$:

$$\frac{\partial \Gamma_{jl}}{\partial p_i} = (a_{ijkl} + a_{jkli}) p_k, \qquad \frac{\partial \Gamma_{jl}}{\partial x_i} = \frac{\partial a_{jkln}}{\partial x_i} p_k p_n.\tag{3.6.9}$$

Inserting (3.6.9) into (3.6.8) yields the final form of the ray tracing system for general inhomogeneous anisotropic media:

$$\frac{dx_i}{dT} = a_{ijkl} p_l g_j^{(m)} g_k^{(m)}, \qquad \frac{dp_i}{dT} = -\frac{1}{2} \frac{\partial a_{jkln}}{\partial x_i} p_k p_n g_j^{(m)} g_l^{(m)}. \qquad (3.6.10)$$

To determine the RHS of ray tracing system (3.6.10), we need to know the density-normalized elastic parameters a_{ijkl} and their spatial derivatives, the components of slowness vector $\vec{p}$, and the components of eigenvector $\vec{g}^{(m)}$. The components of eigenvector $\vec{g}^{(m)}$ can be obtained as solutions of equations

$$(\Gamma_{jk} - G_m \delta_{jk}) g_k^{(m)} = 0, \qquad g_k^{(m)} g_k^{(m)} = 1 \qquad (3.6.11)$$

(no summation over m). Using these equations, components $g_1^{(m)}$, $g_2^{(m)}$, and $g_3^{(m)}$ can be calculated at any point of the ray numerically. We can also find the expressions for $g_i^{(m)}$ and their products analytically:

$$g_j^{(m)} g_k^{(m)} = D_{jk}/D, \qquad (3.6.12)$$

where D_{jk} and D are given by relations

$$\begin{aligned}
D_{11} &= (\Gamma_{22} - G_m)(\Gamma_{33} - G_m) - \Gamma_{23}^2, \\
D_{22} &= (\Gamma_{11} - G_m)(\Gamma_{33} - G_m) - \Gamma_{13}^2, \\
D_{33} &= (\Gamma_{11} - G_m)(\Gamma_{22} - G_m) - \Gamma_{12}^2, \\
D_{12} &= D_{21} = \Gamma_{13}\Gamma_{23} - \Gamma_{12}(\Gamma_{33} - G_m), \\
D_{13} &= D_{31} = \Gamma_{12}\Gamma_{23} - \Gamma_{13}(\Gamma_{22} - G_m), \\
D_{23} &= D_{32} = \Gamma_{12}\Gamma_{31} - \Gamma_{23}(\Gamma_{11} - G_m), \\
D &= D_{11} + D_{22} + D_{33}.
\end{aligned} \qquad (3.6.13)$$

In a more compact form,

$$D_{ij} = \tfrac{1}{2} \epsilon_{ikl} \epsilon_{jrs} (\Gamma_{kr} - G_m \delta_{kr})(\Gamma_{ls} - G_m \delta_{ls}), \qquad D = D_{ii} \qquad (3.6.14)$$

(no summation over m). Here ϵ_{ijk} is the Levi-Civitta symbol where

$$\epsilon_{123} = \epsilon_{312} = \epsilon_{231} = 1, \qquad \epsilon_{321} = \epsilon_{213} = \epsilon_{132} = -1,$$
$$\epsilon_{ijk} = 0 \qquad \text{otherwise.}$$

Thus, ray tracing system (3.6.10) may also alternatively be expressed as

$$\frac{dx_i}{dT} = a_{ijkl} p_l D_{jk}/D, \qquad \frac{dp_i}{dT} = -\tfrac{1}{2} \frac{\partial a_{jkln}}{\partial x_i} p_k p_n D_{jl}/D. \qquad (3.6.15)$$

Here D_{jk} and D are given by (3.6.13), in which we can put $G_m = 1$ because the eikonal equation $G_m = 1$ is satisfied along the ray.

Ray tracing systems (3.6.8), (3.6.10), and (3.6.15) may fail if the eigenvalue, G_m, corresponding to the wave whose ray is being calculated, is equal or close to one of the remaining eigenvalues. Condition (3.6.1) is not satisfied, and the relevant eigenvectors in (3.6.8) and (3.6.10) cannot be determined uniquely. This constraint only applies to quasi-shear qS1 and qS2 waves. Let us denote the eigenvalues of the two quasi-shear waves $G_1(x_i, p_i)$ and $G_2(x_i, p_i)$. Ray tracing systems (3.6.8) and (3.6.10) cannot be used if $G_1 = G_2$ because the relevant eigenvectors $\vec{g}^{(1)}$ and $\vec{g}^{(2)}$ cannot be determined uniquely.

In the same way, ray tracing system (3.6.15) fails, as in this case D_{jk}/D is an indefinite expression of the type $0/0$.

For the qP waves, the ray tracing systems derived here apply universally and can be used in both anisotropic and isotropic media. For the qS waves, however, the situation is more complicated. We shall describe here briefly the possible ways to overcome these complications.

    a. Ray tracing system (3.6.10) can be used even in isotropic media, if $\vec{g}^{(m)}$ is *chosen* in the plane perpendicular to known $\vec{g}^{(3)}$. The direction of $\vec{g}^{(m)}$ in this plane may be chosen arbitrarily.

    b. In anisotropic media, ray tracing system (3.6.10) can be used even when the ray under consideration passes through a singular direction. The ray tracing system (3.6.10), however, must be supplemented by an algorithm that controls the choice of the correct eigenvector after passing the singularity. See more details in Vavryčuk (2000).

In certain applications, it may be useful to compute a ray of the hypothetical qS wave, corresponding to an average eigenvalue G^{av} of both qS waves, $G^{av} = \frac{1}{2}(G_1 + G_2)$. Using (3.6.6), we obtain

$$G^{av} = \tfrac{1}{2}(G_1 + G_2) = \tfrac{1}{2}\Gamma_{jl}\big(g_j^{(1)}g_l^{(1)} + g_j^{(2)}g_l^{(2)}\big) = \tfrac{1}{2}\Gamma_{jl}\big(\delta_{jl} - g_j^{(3)}g_l^{(3)}\big).$$

Thus, the average eigenvalue G^{av} does not depend on $\vec{g}^{(1)}$ and $\vec{g}^{(2)}$, but only on the known $\vec{g}^{(3)}$. The ray of the hypothetical qS wave, corresponding to the average eigenvalue G^{av}, can be again computed using the ray tracing system (3.6.10), where $g_j^{(m)}g_l^{(m)}$ are replaced by $\frac{1}{2}(\delta_{jl} - g_j^{(3)}g_l^{(3)})$. The ray tracing system is then valid quite universally, both in anisotropic and isotropic media, including the shear wave singularities and their vicinity in anisotropic media.

As a result of ray tracing for a selected wave, we obtain the Cartesian coordinates x_i along the ray which define the ray trajectory. At each point on the ray, we determine the following other important quantities.

- The components p_i of slowness vector $\vec{p}$, perpendicular to the wavefront. Components p_i are obtained automatically at each point on the ray.
- The components of phase velocity vector $\mathcal{C}_i$,

$$\mathcal{C}_i = p_i/(p_k p_k). \tag{3.6.16}$$

- The phase velocity, $\mathcal{C} = (\mathcal{C}_k \mathcal{C}_k)^{1/2} = (p_k p_k)^{-1/2}$.
- The components of the unit normal perpendicular to the wavefront, $N_i = \mathcal{C}p_i$. Recall relation $\mathcal{C} = G_m(x_i, N_i)^{1/2}$.
- The components of polarization vector $g_i^{(m)}$ corresponding to the wave under consideration. These components are needed in ray tracing system (3.6.10) and must be computed at each step; see (3.6.11). If we use ray tracing system (3.6.15), we do not need them, but we can easily calculate them from (3.6.11). They define the direction of the displacement vector of the wave under consideration.
- The components of group velocity vector $\mathcal{U}_i$

$$\mathcal{U}_i = \mathrm{d}x_i/\mathrm{d}T = a_{ijkl}p_l g_j^{(m)}g_k^{(m)} = a_{ijkl}p_l D_{jk}/D. \tag{3.6.17}$$

- The group velocity, $\mathcal{U} = (\mathcal{U}_k \mathcal{U}_k)^{1/2}$.
- The components of the unit vector tangent to the ray, t_i,

$$t_i = \mathrm{d}x_i/\mathrm{d}s = \mathcal{U}_i/(\mathcal{U}_k \mathcal{U}_k)^{1/2}.$$

- Angle γ between the ray and the normal to the wavefront,

$$\cos\gamma = \vec{t} \cdot \vec{N} = (\mathcal{C}/\mathcal{U})(\vec{p} \cdot \vec{\mathcal{U}}) = \mathcal{C}/\mathcal{U}; \qquad (3.6.18)$$

  see (2.4.51).
- Angle ξ between the ray and polarization vector $\vec{g}^{(m)}$,

$$\cos\xi = \vec{t} \cdot \vec{g}^{(m)}.$$

- Angle η between the polarization vector and the normal to the wavefront,

$$\cos\eta = \vec{N} \cdot \vec{g}^{(m)}.$$

- The derivative of slowness vector $\vec{p}$ with respect to the travel time along the ray trajectory, $\mathrm{d}\vec{p}/\mathrm{d}T$.

3.6.3 Initial Conditions for a Single Ray in Anisotropic Inhomogeneous Media

For simplicity, we shall only consider one of the derived ray tracing systems, namely system (3.6.10).

Ray tracing system (3.6.10) is identical for all three types of waves that can propagate in an anisotropic smoothly inhomogeneous medium (i.e., for one quasi-P and two quasi-S waves). The type of wave whose ray is to be computed must be specified by initial conditions. Thus, the initial conditions for the ray tracing system play a more important role in anisotropic media than in isotropic media. They specify not only the initial point and the initial direction of the ray but also the type of wave that is to be computed.

The initial conditions for a single ray of one particular selected wave passing through point S can be most easily expressed by defining the initial direction of slowness vector $\vec{p}$ at S, and not the initial direction of the ray. (We know that, in an anisotropic medium, the slowness vector is perpendicular to the wavefront and is not tangent to the ray.)

The initial conditions for ray tracing system (3.6.10) may then be expressed as

$$\text{At } S: \quad x_i = x_{i0}, \qquad p_i = p_{i0}, \qquad (3.6.19)$$

where p_{i0} satisfy the eikonal equation at S,

$$G_m(x_{i0}, p_{i0}) = 1, \qquad (3.6.20)$$

corresponding to the particular wave we wish to compute ($m = 1, 2, 3$). When eikonal equation (3.6.20) is satisfied at S, ray tracing system (3.6.10) keeps the eikonal equation satisfied along the whole ray. This means that the type of wave does not change along the ray in a smooth anisotropic medium; however, it may change at interfaces. In a smooth anisotropic medium, the only exception is related to the rays of qS1 and qS2 waves passing through shear-wave singularities. See Section 5.4.6.

As in isotropic media, the components of slowness vector $\vec{p}_0$ (at S), which satisfy (3.6.20), can be expressed in terms of two take-off angles i_0 and ϕ_0 (see Figure 3.3):

$$p_{10} = N_{10}/\mathcal{C}_0, \qquad p_{20} = N_{20}/\mathcal{C}_0, \qquad p_{30} = N_{30}/\mathcal{C}_0, \qquad (3.6.21)$$

where

$$N_{10} = \sin i_0 \cos\phi_0, \qquad N_{20} = \sin i_0 \sin\phi_0, \qquad N_{30} = \cos i_0, \qquad (3.6.22)$$

and $\mathcal{C}_0$ denotes phase velocity $\mathcal{C}$ at point S for direction $\vec{N}_0$ and for the type of wave whose ray we wish to compute,

$$\mathcal{C}_0 = G_m(x_{i0}, N_{i0})^{1/2}. \tag{3.6.23}$$

We can easily prove that p_{i0}, given by (3.6.21) through (3.6.23), satisfy (3.6.20).

It should be reemphasized that angles i_0 and ϕ_0 do not determine the initial direction of the ray (that is, the direction of the group velocity vector); instead, they determine the direction of slowness vector $\vec{p}$ at S. If the slowness vector is known, the direction of the group velocity vector can be uniquely obtained from (3.6.17). In most applications, we do not need to know the initial direction of the ray in advance, before we start the ray tracing. A procedure that would use the take-off angles of the ray as initial parameters would be more cumbersome and, in fact, not necessary in most applications. It would require p_{10}, p_{20}, and p_{30} to be determined from the following set of three equations:

$$t_{i0} = \frac{\mathcal{U}_{i0}}{(\mathcal{U}_{k0}\mathcal{U}_{k0})^{1/2}},$$

$$\text{where } \mathcal{U}_{i0} = a_{ijkl}(S)p_{l0}(D_{jk}/D)_0, \qquad i = 1, 2, 3, \tag{3.6.24}$$

and where components t_{i0} of the unit vector tangent to the ray at S are assumed known; they may be expressed in terms of take-off angles i_0 and ϕ_0 by similar relations as N_{i0} in (3.6.22). The system of equations (3.6.24) can be efficiently solved only for exceptionally simple types of anisotropy. Note that D_{jk} and D (with $G_m = 1$) are polynomials of the fourth order in three variables p_{i0}, $i = 1, 2, 3$.

3.6.4 Rays in Layered and Block Anisotropic Structures

As in isotropic media, the process of reflection/transmission of high-frequency seismic waves at a curved interface between two inhomogeneous anisotropic media may be treated locally. It is reduced to the problem of reflection/transmission of plane waves at a plane interface between two homogeneous media.

To compute rays in a layered or block inhomogeneous anisotropic structure, we need to determine the initial slowness vectors of the relevant R/T waves generated at points at which the ray is incident at an interface. In isotropic media, the initial slowness vector of the R/T wave under consideration is given in explicit form by Snell's law. In anisotropic media, it must be computed numerically. The appropriate equations are given in Section 2.3.3.

3.6.5 Ray Tracing for Simpler Types of Anisotropic Media

In certain situations, the ray tracing systems simplify considerably, particularly if we consider a simpler anisotropy and/or if the ray is confined to certain planes of symmetry of the anisotropic medium. For example, considerably simpler ray tracing systems are obtained for transversely isotropic media.

We shall describe an example of such a simplification in general terms here. Consider a situation in which

$$\Gamma_{12} = \Gamma_{23} = 0 \tag{3.6.25}$$

is satisfied along the whole ray. This case is very important in practical applications. The equations for the eigenvalues and eigenvectors are then simplified. Eigenvalues G_m satisfy the equation

$$\det \begin{pmatrix} \Gamma_{11} - G_m & 0 & \Gamma_{13} \\ 0 & \Gamma_{22} - G_m & 0 \\ \Gamma_{13} & 0 & \Gamma_{33} - G_m \end{pmatrix}$$

$$= (\Gamma_{22} - G_m)\{G_m^2 - (\Gamma_{11} + \Gamma_{33})G_m + \Gamma_{11}\Gamma_{33} - \Gamma_{13}^2\} = 0. \quad (3.6.26)$$

As we can see, the eigenvalues are given by the relations

$$G_2 = \Gamma_{22}, \qquad G_m^2 - (\Gamma_{11} + \Gamma_{33})G_m + \Gamma_{11}\Gamma_{33} - \Gamma_{13}^2 = 0, \qquad m = 1, 3. \quad (3.6.27)$$

Thus, eigenvalue G_2 equals Γ_{22} and the two remaining eigenvalues, G_1 and G_3, can be obtained as the solution of a quadratic equation (not cubic). This simplifies the problem considerably.

For $m = 2$, Equations (3.6.4) and $G_2 = \Gamma_{22}$ yield a very simple ray tracing system:

$$\frac{dx_i}{dT} = \frac{1}{2}\frac{\partial \Gamma_{22}}{\partial p_i}, \qquad \frac{dp_i}{dT} = -\frac{1}{2}\frac{\partial \Gamma_{22}}{\partial x_i}. \quad (3.6.28)$$

Ray tracing system (3.6.28) is quite universal and remains valid even for a degenerate case of S waves in isotropic media.

For $m = 1$ and $m = 3$, we can again use (3.6.4). The partial derivatives of G_1 and G_3 can be simply obtained from quadratic equation (3.6.27) using the theorem on implicit functions

$$\frac{\partial G_m}{\partial p_i} = \left[G_m \frac{\partial}{\partial p_i}(\Gamma_{11} + \Gamma_{33}) - \frac{\partial}{\partial p_i}(\Gamma_{11}\Gamma_{33} - \Gamma_{13}^2) \right]$$

$$\Big/ [2G_m - (\Gamma_{11} + \Gamma_{33})]; \quad (3.6.29)$$

$\partial G_m/\partial x_i$ can be found similarly. Along the ray, we can put $G_m = 1$ in (3.6.29). The ray tracing system then reads:

$$\frac{dx_i}{dT} = \frac{1}{2}\left[\frac{\partial}{\partial p_i}\left(\Gamma_{11} + \Gamma_{33} - \Gamma_{11}\Gamma_{33} + \Gamma_{13}^2\right) \right]$$

$$\Big/ [2 - (\Gamma_{11} + \Gamma_{33})],$$

$$\frac{dp_i}{dT} = -\frac{1}{2}\left[\frac{\partial}{\partial x_i}\left(\Gamma_{11} + \Gamma_{33} - \Gamma_{11}\Gamma_{33} + \Gamma_{13}^2\right) \right] \quad (3.6.30)$$

$$\Big/ [2 - (\Gamma_{11} + \Gamma_{33})].$$

In ray tracing systems (3.6.28) and (3.6.30), the derivatives of Γ_{ij} can be calculated using (3.6.9). Ray tracing system (3.6.30) is again quite universal and remains valid even for a degenerate case of S waves in isotropic media.

Although ray tracing systems (3.6.28) and (3.6.30) are quite universal for $\Gamma_{12} = \Gamma_{23} = 0$, eigenvectors $\vec{g}^{(1)}$ and $\vec{g}^{(2)}$ cannot be determined uniquely for $G_2 = G_1$. Eigenvectors $\vec{g}^{(m)}$, $m = 1, 2, 3$, can be determined for $\Gamma_{12} = \Gamma_{23} = 0$ from the following system of equations:

$$(\Gamma_{11} - G_m)g_1^{(m)} + \Gamma_{13}g_3^{(m)} = 0,$$

$$(\Gamma_{33} - G_m)g_3^{(m)} + \Gamma_{13}g_1^{(m)} = 0, \quad (3.6.31)$$

$$(\Gamma_{22} - G_m)g_2^{(m)} = 0,$$

with the additional condition,

$$g_i^{(m)}g_i^{(m)} = 1 \quad (3.6.32)$$

(no summation over m). System (3.6.31) with (3.6.32) does not give a unique solution for eigenvectors $\vec{g}^{(1)}$ and $\vec{g}^{(2)}$ if $G_1 = G_2 = \Gamma_{22}$. Even in this case, however, ray tracing systems (3.6.28) and (3.6.30) work quite safely.

Equations (3.6.28) and (3.6.30) represent the final ray tracing systems for the three waves if $\Gamma_{12} = \Gamma_{23} = 0$ along the whole ray. We shall now discuss situations that lead to the relations $\Gamma_{12} = \Gamma_{23} = 0$. Assume a 2-D medium in which the elastic parameters do not depend on x_2 and also assume that $p_{20} = 0$. Then, in view of (3.6.10), $p_2 = 0$ along the whole ray. Because $\Gamma_{12} = a_{1i2j} p_i p_j$ and $\Gamma_{23} = a_{2i3j} p_i p_j$, we obtain $\Gamma_{12} = \Gamma_{23} = 0$, if

$$a_{1112} = a_{1123} = a_{1213} = a_{1233} = a_{1323} = a_{2333} = 0. \tag{3.6.33}$$

Many important anisotropic media satisfy relations (3.6.33). They include transversely isotropic media and hexagonal media. The rays computed by ray tracing system (3.6.28) *are still three-dimensional* because $dx_2/dT \neq 0$. The system simplifies even more if we make assumptions in addition to (3.6.33), namely,

$$a_{1222} = a_{2223} = 0;$$

quantity dx_2/dT in ray tracing system (3.6.28) vanishes so that the relevant rays are fully confined to plane $x_2 = x_{20}$.

Very much the same approach can be used if $\Gamma_{12} = \Gamma_{13} = 0$ or if $\Gamma_{23} = \Gamma_{13} = 0$.

We shall not give the explicit ray tracing systems for the individual types of anisotropic media here because they can be simply obtained from (3.6.28) and (3.6.30). Moreover, many of them are known from the literature, where numerical examples are also presented. For 2-D transversely isotropic media with a vertical axis of symmetry, see Červený, Molotkov, and Pšenčík (1977); for 2-D transversely isotropic media with an arbitrary orientation of the axis of symmetry in plane x_1-x_3, see Jech (1983); and for certain hexagonal systems, see Červený and Pšenčík (1972). See also discussions in Hanyga (1988) and Pratt and Chapman (1992) among others.

If equations $\Gamma_{12} = 0$ and $\Gamma_{23} = 0$ are satisfied and the elastic parameters depend on one coordinate only (for example, on depth), ray tracing systems (3.6.28) and (3.6.30) can be solved in terms of closed-form integrals. These integrals for inhomogeneous transversely isotropic media were first presented by Vlaar (1968) and in a slightly simpler form by Červený, Molotkov, and Pšenčík (1977).

In the computation of rays and travel times in inhomogeneous anisotropic media, perturbation methods have often been used; see Section 3.9. In the perturbation method, ray-theory computations are first performed in some simpler reference model that is close to the actual model. These computations are then used to obtain analogous solutions in the perturbed model.

The isotropic model has often been used as a reference model; see Section 3.9. In certain cases, it may be better to consider a reference medium with *ellipsoidal anisotropy* (with an ellipsoidal slowness surface). For example, the slowness surface of qP waves, propagating in an ellipsoidally anisotropic medium, is given by relation $G_m(x_i, p_i) = 1$, where

$$G_m(x_i, p_i) = a_{1111} p_1^2 + a_{2222} p_2^2 + a_{3333} p_3^2.$$

The ray tracing system for ellipsoidally anisotropic media may be obtained simply from (3.6.4). It is extremely simple and similar to isotropic media. For more details see Mensch and Farra (1999), where the ellipsoidal anisotropy reference model was used to compute rays, travel times, and slowness vectors of qP waves in orthorhombic media.

3.6.6 Ray Tracing in Factorized Anisotropic Media

We shall now introduce an important concept of a special type of anisotropic inhomogeneous medium, and call it the factorized anisotropic inhomogeneous (FAI) medium. In the FAI medium, the density-normalized elastic parameters a_{ijkl} depend on Cartesian coordinates x_i in the following way:

$$a_{ijkl}(x_i) = f^2(x_i)A_{ijkl},\tag{3.6.34}$$

where A_{ijkl} are constants, independent of Cartesian coordinates, satisfying symmetry relations $A_{ijkl} = A_{jikl} = A_{ijlk} = A_{klij}$; see Červený (1989c). Function $f(x_i)$ is an arbitrary positive continuous function of Cartesian coordinates. Thus, all the density-normalized elastic parameters $a_{ijkl}(x_i)$ in the FAI medium depend on Cartesian coordinates x_i in the same way; the relative spatial variations of all a_{ijkl} are equal. A special case of a factorized anisotropic medium (3.6.34) was first studied by Shearer and Chapman (1989). As we can see, relation (3.6.34) in a way separates anisotropy from inhomogeneity; a_{ijkl} are factorized. This factorization of the density-normalized elastic parameters automatically factorizes many other important expressions and relations (for example, matrix Γ_{ij} and its eigenvalues, the eikonal equation, and phase and group velocities). Phase velocity $\mathcal{C}$ and group velocity $\mathcal{U}$ in the FAI medium are given by relations

$$\mathcal{C}(x_i, N_i) = f(x_i)\mathcal{C}_0(N_i), \qquad \mathcal{U}(x_i, N_i) = f(x_i)\mathcal{U}_0(N_i),\tag{3.6.35}$$

where $\mathcal{C}_0(N_i)$ and $\mathcal{U}_0(N_i)$ are the position-independent phase and group velocities in a homogeneous anisotropic medium with elastic parameters A_{ijkl}.

In the following, we shall consider $f(x_i)$ to be a *dimensionless function* of Cartesian coordinates x_i. Then, both a_{ijkl} and A_{ijkl} have dimensions of squared velocity $(\mathrm{m^2s^{-2}})$. Similarly, $\mathcal{C}$, $\mathcal{C}_0$, $\mathcal{U}$, and $\mathcal{U}_0$ have the dimension of velocity. In a homogeneous anisotropic medium, we can put $f(x_i) = 1$, so that $a_{ijkl} = A_{ijkl}$. For inhomogeneous anisotropic media, it may be convenient to fix A_{ijkl} at some selected point x_{iA} as $A_{ijkl} = a_{ijkl}(x_{iA})$. Then also $f(x_{iA}) = 1$. The variations of $f(x_i)$ with coordinates then express the variations of phase and group velocities within the model; see (3.6.35).

It would also be possible to interpret (3.6.34) in a different way: to take parameters A_{ijkl} as dimensionless. Then, $f(x_i)$ would have the dimension of velocity $(\mathrm{km\,s^{-1}})$. See Červený (1989c) and Červený and Simões-Filho (1991) for examples of this interpretation. Here, however, we shall consider only dimensionless $f(x_i)$.

Using (3.6.34), matrix $\Gamma_{ik}(x_i, p_i)$ and its eigenvalues can be expressed as

$$\Gamma_{ik}(x_i, p_i) = f^2(x_i)\Gamma_{ik}^0, \qquad \text{where } \Gamma_{ik}^0 = A_{ijkl}p_j p_l,$$
$$G_m(x_i, p_i) = f^2(x_i)G_m^0(p_i).\tag{3.6.36}$$

Here $G_m^0(p_i)$ are position-independent eigenvalues of $\Gamma_{ik}^0(p_i)$.

In the FAI medium, eikonal equation (3.6.2) takes the form

$$f^2(x_i)G_m^0(p_i) = 1.\tag{3.6.37}$$

This may appear in a rather general Hamiltonian form,

$$\mathcal{H}(x_i, p_i) = n^{-1}\left\{\left[G_m^0(p_i)\right]^{n/2} - 1/f^n(x_i)\right\} = 0,\tag{3.6.38}$$

where n is an arbitrary real-valued number. Equation (3.6.38) can also be expressed for $n = 0$,

$$\mathcal{H}(x_i, p_i) = \tfrac{1}{2} \ln G_m^0(p_i) + \ln f(x_i) = 0. \tag{3.6.39}$$

The ray tracing system can be obtained from eikonal equations (3.6.38) or (3.6.39), using Hamiltonian canonical equations (3.1.3). For $n \neq 0$,

$$\frac{dx_i}{du} = \frac{1}{n} \frac{\partial}{\partial p_i} (G_m^0)^{n/2}, \qquad \frac{dp_i}{du} = \frac{1}{n} \frac{\partial f^{-n}}{\partial x_i}, \qquad \frac{dT}{du} = (G_m^0)^{n/2} = f^{-n}. \tag{3.6.40}$$

Alternatively,

$$\frac{dx_i}{du} = \frac{1}{2} (G_m^0)^{n/2-1} \frac{\partial G_m^0}{\partial p_i}, \qquad \frac{dp_i}{du} = \frac{1}{n} \frac{\partial f^{-n}}{\partial x_i}, \qquad \frac{dT}{du} = (G_m^0)^{n/2} = f^{-n}. \tag{3.6.41}$$

Similarly, for $n = 0$,

$$\frac{dx_i}{dT} = \frac{1}{2} \frac{\partial \ln G_m^0}{\partial p_i}, \qquad \frac{dp_i}{dT} = -\frac{\partial \ln f}{\partial x_i}. \tag{3.6.42}$$

Ray tracing systems (3.6.40) through (3.6.42), expressed in Hamiltonian canonical form, are suitable for finding various analytical or semianalytical solutions and for deriving the dynamic ray tracing system and ray perturbation equations. However, if we are interested in ray tracing and travel-time computations only, we can simplify ray tracing systems (3.6.40) through (3.6.41) slightly, using eikonal equation $G_m^0 = f^{-2}$:

$$\frac{dx_i}{du} = \frac{1}{2} f^{2-n} \frac{\partial G_m^0}{\partial p_i}, \qquad \frac{dp_i}{du} = \frac{1}{n} \frac{\partial f^{-n}}{\partial x_i}, \qquad \frac{dT}{du} = f^{-n}. \tag{3.6.43}$$

As in isotropic media, the simplest ray tracing system is obtained for $n = 2$. Denoting parameter u by σ in this case, we obtain

$$\frac{dx_i}{d\sigma} = \frac{1}{2} \frac{\partial G_m^0}{\partial p_i}, \qquad \frac{dp_i}{d\sigma} = \frac{1}{2} \frac{\partial f^{-2}}{\partial x_i}, \qquad \frac{dT}{d\sigma} = f^{-2}. \tag{3.6.44}$$

It is immediately evident that ray tracing systems (3.6.40) through (3.6.44) for the FAI medium are considerably simpler than for a general inhomogeneous anisotropic medium. The RHS of the equations for x_i do not depend explicitly on x_i (with the exception of (3.6.43)), and the RHS of the equations for p_i do not depend explicitly on p_i. Instead of 63 first partial spatial derivatives of elastic parameters a_{ijkl}, we only need to determine the three first partial derivatives of $f^{-n}(x_i)$ at each step of the ray tracing in the FAI medium.

In ray tracing systems (3.6.40) through (3.6.44), we must know $\partial G_m^0 / \partial p_i$. As in (3.6.7) and (3.6.9), we easily obtain

$$\partial G_m^0 / \partial p_i = 2 A_{ijkl} p_l g_j^{(m)} g_k^{(m)} = 2 A_{ijkl} p_l D_{jk}^0 / D^0. \tag{3.6.45}$$

Functions D_{jk}^0 and D^0 are again given by the same equations as D_{jk} and D; see (3.6.13) or (3.6.14). Only Γ_{ik} is replaced with Γ_{ik}^0, and G_m is replaced with G_m^0.

Using ray tracing systems (3.6.40) and (3.6.41), we can obtain some analytical or semianalytical solutions of the ray tracing systems for FAI models with a constant gradient of f^{-n} or $\ln f$.

Assume that the spatial distribution of f^{-n} (for $n \neq 0$) is given as

$$f^{-n}(x_i) = A_0 + A_1 x_1 + A_2 x_2 + A_3 x_3, \tag{3.6.46}$$

where A_0, A_1, A_2, and A_3 are constants. Ray tracing system (3.6.40) then yields analytical solutions for p_i,

$$p_i(u) = p_i(u_0) + n^{-1} A_i(u - u_0), \tag{3.6.47}$$

where u is a parameter along the ray, related to the travel time as $dT = f^{-n} du$. The remaining equations of the ray tracing system read

$$\frac{dx_i}{du} = \frac{1}{n} \frac{\partial}{\partial p_i} \left(G_m^0 \right)^{n/2}, \qquad \frac{dT}{du} = \left(G_m^0 \right)^{n/2}. \tag{3.6.48}$$

As we can see, the RHSs of (3.6.48) do not explicitly depend on x_i; they depend only on p_i ($i = 1, 2, 3$). From (3.6.47), we know that p_i depend explicitly only on u. Thus, Equations (3.6.48) may be solved by quadratures,

$$x_i(u) = x_i(u_0) + \frac{1}{n} \int_{u_0}^{u} \frac{\partial}{\partial p_i} \left(G_m^0 \right)^{n/2} du,$$
$$\tag{3.6.49}$$
$$T(u) = T(u_0) + \int_{u_0}^{u} \left(G_m^0 \right)^{n/2} du.$$

We now assume that function $f(x_i)$ is given by the relations

$$\ln f = A_0 + A_1 x_1 + A_2 x_2 + A_3 x_3. \tag{3.6.50}$$

This yields the solution for $n = 0$, with parameter $u = T$ along the ray,

$$p_i(T) = p_i(T_0) - A_i(T - T_0), \qquad x_i(T) = x_i(T_0) + \frac{1}{2} \int_{T_0}^{T} \frac{\partial \ln G_m^0}{\partial p_i} dT. \tag{3.6.51}$$

The simplest solutions are obtained for (3.6.46) with $n = 2$. Then, using (3.6.47), (3.6.49), and (3.6.45), we arrive at

$$p_i(\sigma) = p_i(\sigma_0) + \tfrac{1}{2} A_i(\sigma - \sigma_0),$$

$$x_i(\sigma) = x_i(\sigma_0) + A_{ijkl} \int_{\sigma_0}^{\sigma} p_l g_k^{(m)} g_j^{(m)} d\sigma, \tag{3.6.52}$$

$$T(\sigma) = T(\sigma_0) + A_{ijkl} \int_{\sigma_0}^{\sigma} p_i p_l g_k^{(m)} g_j^{(m)} d\sigma.$$

Parameter σ along the ray is also related to travel time T as follows: $dT = f^{-2} d\sigma$.

3.6.7 Energy Considerations

It was shown in Section 3.1.5 that the time-averaged energy of high-frequency elastic waves in slightly inhomogeneous isotropic media flows along rays. It is not difficult to generalize this fact even for anisotropic inhomogeneous media.

As we can see from (3.6.10), the ray velocity vector components $v_{ri} = dx_i/dT$ are given by relations $v_{ri} = a_{ijkl} p_l g_j^{(m)} g_k^{(m)}$, where m specifies the type of wave under consideration (qP, qS1, or qS2). The same expression $\mathcal{U}_i = a_{ijkl} p_l g_j^{(m)} g_k^{(m)}$ was also derived in Section 2.4.4 for the components of group velocity vector $\vec{\mathcal{U}}$; see (2.4.59). Consequently, $\vec{\mathcal{U}} = \vec{v}_r$

and the time-averaged energy of high-frequency elastic waves flows along rays, even in inhomogeneous anisotropic media.

3.7 Ray Tracing and Travel-Time Computations in 1-D Models

In this section, we shall consider one-dimensional media in which the velocity depends on one coordinate only. Thus, the two remaining coordinates are cyclic, and the ray tracing system for isotropic media, expressed in Cartesian coordinates, can be solved in terms of closed-form integrals. In certain cases (but not generally), similar closed-form integral solutions can also be found for one-dimensional media in orthogonal curvilinear coordinates.

We shall mainly discuss two seismologically important one-dimensional media. The first corresponds to the *vertically inhomogeneous medium*, specified in Cartesian coordinates. The velocity of propagation V depends only on depth in the vertically inhomogeneous medium. The second corresponds to the *radially symmetric medium*, specified in spherical polar coordinates. In the radially symmetric medium, velocity V depends on radial distance r only.

The computation of rays and travel times in these two types of 1-D models has been broadly discussed in the seismological literature. Practically any textbook on seismology and on seismic prospecting discusses these problems, at least for one of the 1-D models. Let us refer to Savarenskiy and Kirnos (1955), Puzyrev (1959), Bullen (1965), Jeffreys (1970), Aki and Richards (1980), and Bullen and Bolt (1985). The interested reader is also referred to general books on wave propagation problems; see, for example, Kline and Kay (1965), Felsen and Marcuvitz (1973), Pilant (1979), Kravtsov and Orlov (1980), and Hanyga, Lenartowicz, and Pajchel (1984). In view of these references, we shall be as brief as possible.

In general orthogonal curvilinear coordinates, the ray tracing system for 1-D media cannot be solved in terms of closed-form integrals. The reason is that the scale factors h_i also depend on coordinates; see Section 3.5. These solutions are available only in some special cases.

In 1-D anisotropic media, all density-normalized elastic parameters depend on one coordinate only. If we use Cartesian coordinates, the ray tracing system can always be solved in terms of closed-form integrals, as in isotropic media. The integrals, however, can be rather complicated, particularly if we consider complex anisotropy symmetries. The numerical evaluation of these integrals may be as time consuming as the direct numerical solution of the relevant differential equations for rays. For travel-time computations in anisotropic media, see Gassman (1964).

For 1-D isotropic media, the closed-form integral equations for rays and travel times can simply be obtained directly, without invoking a general 3-D ray tracing system. The most common approaches are based either on Fermat's principle or on Snell's law. The relevant simple derivations can be found in seismological textbooks. Here, however, we shall consider the 1-D medium as a special case of general 3-D inhomogeneous media and derive all equations from the general 3-D ray tracing system.

3.7.1 Vertically Inhomogeneous Media

We shall consider Cartesian coordinates x_1, x_2, and x_3 and denote them x, y, and z. Similarly, the Cartesian components of the slowness vector, p_1, p_2, and p_3, will be denoted p_x, p_y, and p_z. We assume that velocity V depends on z only, $V = V(z)$. We also assume that the z-axis is vertical and points downward. The origin of the Cartesian coordinate system is located at zero depth, so that z may be referred to as depth. Axes x and y are situated in the plane perpendicular to axis z at zero depth.

Let the initial point of the ray S be situated at point x_0, y_0, z_0, and the initial slowness vector be p_{x0}, p_{y0}, p_{z0}, satisfying relation $p_{x0}^2 + p_{y0}^2 + p_{z0}^2 = 1/V^2(S)$. Without loss of generality, we shall assume that

$$y_0 = 0, \qquad p_{y0} = 0. \tag{3.7.1}$$

The ray as a whole is then situated in plane x-z, and $y = 0$, $p_y = 0$ along the whole ray.

Ray tracing system (3.3.13) yields the ray tracing system in the 1-D medium ($V = V(z)$) in the following form:

$$\frac{dx}{du} = A^{n/2-1} p_x, \qquad \frac{dp_x}{du} = 0,$$

$$\frac{dz}{du} = A^{n/2-1} p_z, \qquad \frac{dp_z}{du} = \frac{1}{n}\frac{\partial}{\partial z}\left(\frac{1}{V^n}\right), \tag{3.7.2}$$

$$\frac{dT}{du} = A^{n/2} = V^{-n},$$

where $A = (p_x^2 + p_z^2) = V^{-2}$. System (3.7.2) yields $p_x = p_{x0} = $ const. along the whole ray. We shall call the constant horizontal component of slowness vector p_x the *parameter of the ray* and denote it p. This is a standard notation in seismology. If the acute angle between the ray and the vertical line is i,

$$p_x(u) = |\vec{p}| \sin i = \sin i / V. \tag{3.7.3}$$

We have thus arrived at the generalized Snell's law for a vertically inhomogeneous medium,

$$p = \sin i(z) / V(z). \tag{3.7.4}$$

The final system of differential equations then reads

$$\frac{dx}{du} = A^{n/2-1} p, \qquad \frac{dp_z}{du} = \frac{1}{n}\frac{\partial}{\partial z}\left(\frac{1}{V^n}\right),$$

$$\frac{dz}{du} = A^{n/2-1} p_z, \qquad \frac{dT}{du} = A^{n/2} = V^{-n}, \tag{3.7.5}$$

where $A = p^2 + p_z^2 = V^{-2}$.

Equations (3.7.5) can always be solved in terms of closed-form integrals. To find them, we must express p_z not from (3.7.5) but from the eikonal equation $p^2 + p_z^2 = V^{-2}$,

$$p_z = \pm(V^{-2} - p^2)^{1/2}. \tag{3.7.6}$$

The disadvantage of (3.7.6) is that it contains two signs. The plus sign ($+$) applies to the downgoing part of the ray (p_z positive), and the minus sign ($-$) refers to the upgoing part of the ray. As we can see from (3.7.6), p_z depends on z only. Using (3.7.6), we can eliminate parameter u from the ray tracing system and use depth z instead. For simplicity, we shall consider a wave that propagates in the direction of increasing x. (The modification of the resulting equations for waves propagating in the direction of decreasing x is straightforward.) We then get

$$\frac{dx}{dz} = \frac{p}{p_z} = \pm\frac{p}{(V^{-2} - p^2)^{1/2}}, \qquad \frac{dT}{dz} = \frac{A}{p_z} = \pm\frac{V^{-2}}{(V^{-2} - p^2)^{1/2}}. \tag{3.7.7}$$

The solution of (3.7.7) is

$$x(p, z) = x(p, z_0) \pm \int_{z_0}^{z} \frac{p\,dz}{(V^{-2} - p^2)^{1/2}},$$

$$T(p, z) = T(p, z_0) \pm \int_{z_0}^{z} \frac{V^{-2}dz}{(V^{-2} - p^2)^{1/2}}. \tag{3.7.8}$$

As we can see, the simplest solution of (3.7.8) will again be obtained for the square-of-slowness (V^{-2}) models. Of course, (3.7.8) may take many alternative forms. The most common is

$$x(p, z) = x(p, z_0) \pm \int_{z_0}^{z} \frac{pV\,dz}{(1 - p^2 V^2)^{1/2}},$$

$$T(p, z) = T(p, z_0) \pm \int_{z_0}^{z} \frac{dz}{V(1 - p^2 V^2)^{1/2}}. \tag{3.7.9}$$

We shall now consider a *vertically inhomogeneous layered structure*, containing horizontal parallel interfaces of the first or higher order. It is easy to generalize Equations (3.7.9) for this case. Assume that the ray under consideration has N points of reflection/transmission at these interfaces, at depths $z_1, z_2, \ldots, z_N$. The wave may propagate downward or upward along different segments of the ray, but it always propagates in the direction of increasing x. The equations, alternative to (3.7.9), then read

$$x(p, z) = x(p, z_0) + \sum_{k=1}^{N} x(p, z_{k-1}, z_k) + x(p, z_N, z),$$

$$T(p, z) = T(p, z_0) + \sum_{k=1}^{N} T(p, z_{k-1}, z_k) + T(p, z_N, z), \tag{3.7.10}$$

where

$$x(p, z_i, z_{i+1}) = \pm \int_{z_i}^{z_{i+1}} \frac{pV\,dz}{(1 - p^2 V^2)^{1/2}},$$

$$T(p, z_i, z_{i+1}) = \pm \int_{z_i}^{z_{i+1}} \frac{dz}{V(1 - p^2 V^2)^{1/2}}. \tag{3.7.11}$$

Here the plus sign is used for $z_{i+1} > z_i$, the minus sign is used for $z_{i+1} < z_i$. Thus, $x(p, z_{i-1}, z_i)$ and $T(p, z_{i-1}, z_i)$ are always positive. (It would be possible to use absolute values in (3.7.11) instead of $\pm$.) Obviously,

$$x(p, z_i, z_{i+1}) = x(p, z_{i+1}, z_i),$$

$$T(p, z_i, z_{i+1}) = T(p, z_{i+1}, z_i). \tag{3.7.12}$$

Equations (3.7.10) are very general and can be used for arbitrary multiply reflected waves, including converted waves. In the individual layers, V may be either α or β.

If the ray passes through a turning point at depth z_M, Equations (3.7.10) must be modified. The element of the ray with the turning point at z_M must be formally divided into two elements: one going downward and the other going upward. We can, however, again use (3.7.10), if we formally take z_M as one of the interfaces. The depth of turning point z_M can be determined from (3.7.4) with $i(z_M) = \pi/2$,

$$1/V(z_M) = p. \tag{3.7.13}$$

For example, let us consider a refracted wave, with the initial and end points of the ray at depth $z = 0$ and with $x_0 = T_0 = 0$. Then,

$$x(p) = x(p, 0, z_M) + x(p, z_M, 0) = 2 \int_0^{z_M} \frac{pV dz}{(1 - p^2 V^2)^{1/2}},$$

$$T(p) = T(p, 0, z_M) + T(p, z_M, 0) = 2 \int_0^{z_M} \frac{dz}{V(1 - p^2 V^2)^{1/2}}.$$

(3.7.14)

These two equations represent the parametric form of the travel-time curve of the refracted wave, with parameter p. As in (3.7.10), expressions (3.7.14) can be simply generalized for any ray multiply reflected/transmitted or refracted in a vertically inhomogeneous layered structure.

As we can see, the integrands of (3.7.11) grow above all limits for $z_i = z_M$ or $z_{i+1} = z_M$. This may cause some complications, particularly if we wish to determine derivative $dx(p)/dp$ or $dT(p)/dp$. These derivatives are needed in evaluating geometrical spreading and, consequently, in computing amplitudes. It is, however, not difficult to change the form of the integrals using the integration-by-parts method so that the singularity of the integrand at $z = z_M$ is removed.

Function $T(p) - px(p)$ plays an important role in seismology. It is commonly referred to as the *delay time* and denoted by $\tau(p)$. The integral expression for $\tau(p, z_i, z_{i+1})$ reads

$$\tau(p, z_i, z_{i+1}) = T(p, z_i, z_{i+1}) - px(p, z_i, z_{i+1})$$

$$= \pm \int_{z_i}^{z_{i+1}} (V^{-2} - p^2)^{1/2} dz.$$

(3.7.15)

Delay time $\tau(p)$ is used in the WKBJ (G. Wentzel, H. A. Kramers, L. Brillouin, and H. Jeffreys) method and in many other applications. It has no singular behavior at turning point $z = z_M$. Note that the delay time is related to x in the following way:

$$d\tau(p, z_i, z_{i+1})/dp = -x(p, z_i, z_{i+1}).$$

(3.7.16)

The closed-form integral solutions given in this section can be computed either numerically or, in simple cases, analytically. Suitable *numerical procedures*, based on a spline approximation of the velocity-depth distribution, were proposed by Chapman (1971). The spline approximation removes all fictitious interfaces of first, second, and third orders and the relevant anomalies in the amplitude-distance curves. In the same way as in the cell approach, an alternative option is to divide the whole structure (or the whole layer) into formal fictitious sublayers and to specify a simple velocity-depth distribution within each sublayer. An example is the piecewise linear velocity approximation. The velocity-depth distribution in each sublayer is chosen to provide simple analytical expressions for $x(p, z_i, z_{i+1})$, $T(p, z_i, z_{i+1})$, and $\tau(p, z_i, z_{i+1})$. For a more detailed discussion refer to the next section.

3.7.2 Analytical Solutions for Vertically Inhomogeneous Media

For simple velocity-depth distributions $V = V(z)$, the rays and travel times can be expressed analytically. There are two types of analytical solutions for vertically inhomogeneous media. The first involves a monotonic parameter along the ray (arclength s, travel time T, parameter σ, and so on). The turning points do not cause any problem in these equations; the monotonic parameter passes quite smoothly through them. The analytical solutions of the second type do not involve a monotonic parameter along the ray, but are expressed in terms of depth. As shown in Section 3.7.1, very simple closed-form integrals can be used in this case for any velocity distribution. The analytical solutions of these integrals exist in certain simple cases. We must be careful to take the proper signs for the descending and ascending parts of the ray. Some complications in these solutions may be caused by turning points.

We again use Cartesian coordinates x, y, and z, with the initial point of the ray situated at point $S(x_0, y_0, z_0)$, and with the initial slowness vector, p_{x0}, p_{y0}, and p_{z0}, satisfying the relation $p_{x0}^2 + p_{y0}^2 + p_{z0}^2 = 1/V^2(S)$. We assume that (3.7.1) holds so that

$$y = 0, \qquad p_y = 0 \tag{3.7.17}$$

is satisfied along the whole ray, and the whole ray is situated in the plane $y = 0$. Slowness vector component p_x is constant along the whole ray, $p_x = p$, so that the eikonal equation reads $p^2 + p_z^2 = 1/V^2$.

We shall consider the models in which the vertical gradient of $V^{-n}(z)$ or $\ln V(z)$ is constant:

$$V^{-n}(z) = a + bz, \qquad \ln V(z) = a + bz. \tag{3.7.18}$$

Here n may be any integer, $n \neq 0$. Velocity-depth distributions (3.7.18) are special cases of the 3-D velocity distributions discussed in Sections 3.4.2 through 3.4.4; therefore, we can use the solutions derived there. We only need to put $A_1 = A_2 = 0$, $A_0 = a$, and $A_3 = b$; see (3.4.4) and (3.4.8). All these solutions are expressed in terms of a monotonic parameter along the ray (s, T, σ, and so on).

The expressions in terms of depth z can be obtained directly from (3.7.8) or (3.7.9). Alternatively, they can also be obtained from the monotonic parameter expressions, if we use the eikonal equation to determine monotonic parameter u and to eliminate it from other expressions. We can use (3.4.5), (3.4.9), and (3.7.6) to obtain

$$u - u_0 = \pm n b^{-1} \left[(V^{-2} - p^2)^{1/2} - \left(V_0^{-2} - p^2 \right)^{1/2} \right] \qquad \text{for } n \neq 0,$$

$$T - T_0 = \pm b^{-1} \left[(V^{-2} - p^2)^{1/2} - \left(V_0^{-2} - p^2 \right)^{1/2} \right] \qquad \text{for } n = 0.$$

Here the sign must be such that $u - u_0$ (or $T - T_0$) is positive in the direction of wave propagation.

The disadvantage of the models with a constant vertical gradient of $V^{-n}(z)$ or $\ln V(z)$ is that they cannot properly simulate smooth maxima or minima of velocity, low-velocity channels, and the like, without introducing interfaces of first or higher orders. For this reason, we shall also briefly discuss the parabolic velocity-depth distribution, which overcomes these difficulties.

We shall present the analytical solutions explicitly for five special velocity-depth distributions:

1. A homogeneous medium $V = $ const.
2. A constant gradient of the square of slowness $V^{-2}(z)$
3. A constant gradient of the logarithmic velocity, $\ln V(z)$
4. A constant gradient of velocity $V(z)$
5. A quadratic depth distribution of $V^{-2}(z)$, $V^{-2}(z) = a + bz + cz^2$.

The simplest expressions are again obtained for homogeneous media and for the constant gradient of the square of slowness, $V^{-2}(z)$. In all cases, with the exception of the first and the last case, we shall give three sets of expressions: (a) in terms of a monotonic parameter along the ray; (b) in terms of depth z, for the ray segment without a turning point; and (c) in terms of depth, for the ray segment with a turning point. In the last case, $z_{i+1} = z_i$.

To simplify the equations expressed in terms of depth z, we shall use the following notations:

$$V_i = V(z_i), \qquad V_{i+1} = V(z_{i+1}),$$
$$w_i = \left[V_i^{-2} - p^2\right]^{1/2}, \qquad w_{i+1} = \left[V_{i+1}^{-2} - p^2\right]^{1/2},$$
$$c_i = V_i w_i = \left[1 - p^2 V_i^2\right]^{1/2}, \qquad c_{i+1} = V_{i+1} w_{i+1} = \left[1 - p^2 V_{i+1}^2\right]^{1/2}.$$

The sign convention remains the same as in Section 3.7.1: the upper sign corresponds to the downgoing part of the ray, and the lower sign refers to the upgoing part of the ray. Note that c_k can also be expressed as $\cos i(z_k)$, where $i(z_k)$ is the acute angle between the ray and the vertical at depth z_k.

1. HOMOGENEOUS LAYER, $V(z) = V_i = $ CONST.

We know that the ray in a homogeneous medium is a straight line and that the travel time is a linear function of arclength s along the ray, or alternatively, of depth z. We shall only give the relevant expressions in terms of depth z. In this case, equations (3.7.11) and (3.7.15) yield

$$x(p, z_i, z_{i+1}) = pV_i d/c_i, \qquad T(p, z_i, z_{i+1}) = d/V_i c_i,$$
$$\tau(p, z_i, z_{i+1}) = w_i d, \qquad d = |z_{i+1} - z_i|$$

(no summation over i). These equations have a very simple geometric interpretation, since $pV_i = \sin i$ and $c_i = [1 - p^2 V_i^2]^{1/2} = \cos i$, where i is the acute angle between the ray and the vertical in the layer under consideration.

2. CONSTANT GRADIENT OF THE SQUARE OF SLOWNESS, $V^{-2}(z)$

This is the second simplest case, corresponding to $n = 2$. We assume the velocity distribution as follows:

$$1/V^2(z) = a + bz, \qquad \text{that is,} \qquad V(z) = 1/[a + bz]^{1/2}.$$

The suitable monotonic parameter along the ray is σ such that $d\sigma = V^2 dT$. Simple polynomial equations are obtained:

$$p_z(\sigma) = p_z(\sigma_0) + \tfrac{1}{2}b(\sigma - \sigma_0),$$
$$z(\sigma) = z(\sigma_0) + p_z(\sigma_0)(\sigma - \sigma_0) + \tfrac{1}{4}b(\sigma - \sigma_0)^2,$$
$$x(\sigma) = x(\sigma_0) + p(\sigma - \sigma_0),$$
$$T(\sigma) = T(\sigma_0) + (a + bz(\sigma_0))(\sigma - \sigma_0)$$
$$\qquad + \tfrac{1}{2}bp_z(\sigma_0)(\sigma - \sigma_0)^2 + \tfrac{1}{12}b^2(\sigma - \sigma_0)^3.$$

Alternative expressions in terms of depth z, for the ray segment without a turning point, are

$$x(p, z_i, z_{i+1}) = \pm 2pb^{-1}(w_{i+1} - w_i),$$
$$T(p, z_i, z_{i+1}) = \pm \left[2p^2 b^{-1}(w_{i+1} - w_i) + \tfrac{2}{3}b^{-1}\left(w_{i+1}^3 - w_i^3\right)\right],$$
$$\tau(p, z_i, z_{i+1}) = \pm \tfrac{2}{3}b^{-1}\left(w_{i+1}^3 - w_i^3\right).$$

For the ray segment with a turning point, we obtain

$$x(p, z_i, z_{i+1} = z_i) = 4pw_i/|b|,$$

$$T(p, z_i, z_{i+1} = z_i) = 4w_i\left(p^2 + \tfrac{1}{3}w_i^2\right)/|b|,$$

$$\tau(p, z_i, z_{i+1} = z_i) = \tfrac{4}{3}w_i^3/|b|.$$

It can easily be shown that the ray is parabolic in the model considered. The equations for $z(\sigma)$ and $x(\sigma)$ yield the following expression for the ray trajectory:

$$z = z_S + p_{zS}p^{-1}(x - x_S) - \tfrac{1}{4}Bp^{-2}(x - x_S)^2.$$

Here x_S and z_S are coordinates of the initial point S, $B = -b$, and $p_{zS} = \pm w_S = \pm[V^{-2}(z_S) - p^2]^{1/2}$ is the z-component of the initial slowness vector at S. An alternative expression for the ray trajectory is

$$B(x - x_S - 2B^{-1}p_{zS}p)^2 = 4p^2(z_S + B^{-1}p_{zS}^2 - z).$$

For $p_{zS} > 0$, the ray has a minimum at point $M[x_M, z_M]$ with coordinates $x_M = x_S + 2p_{zS}pB^{-1}$ and $z_M = z_S + p_{zS}^2B^{-1}$.

The preceding ray equation can be solved for p. We shall consider two points, $S[x_S, z_S]$ and $R[x_R, z_R]$, situated in half-space $-\infty < z < a/B$. (For $z = a/B$, the velocity is infinite, and for $z > a/B$, it is complex-valued.) We also denote $u_S = 1/V(S)$ and $u_R = 1/V(R)$. Then the ray parameter p of the ray passing through S and R is

$$p_{1,2}^2 = \tfrac{1}{4}x^2r^{-2}\left[u_R^2 + u_S^2 \pm \sqrt{\left(u_R^2 + u_S^2\right)^2 - B^2r^2}\right],$$

where $x = x_R - x_S$, $r = [x^2 + (z_R - z_S)^2]^{1/2}$. Thus, we have obtained two ray parameters p_1 and p_2 and two rays Ω_1 and Ω_2 connecting points S and R, assuming that $Br < (u_R^2 + u_S^2)$.

The result is very interesting. No ray arrives from S at point R for which $r > B^{-1}(u_R^2 + u_S^2)$. Boundary surface $r = B^{-1}(u_R^2 + u_S^2)$ can also be expressed in the following form:

$$B^2x^2 = 4(a - Bz)(a - Bz_S).$$

The boundary surface represents a *caustic surface* and has the form of a paraboloid of revolution. If point R is situated inside the caustic surface, two rays Ω_1 and Ω_2 connect it with S.

The multiplicity of rays is unpleasant in practical computations. We can, however, remove it by decreasing the size of the model. Assume that both S and R are situated at the same depth z_S. We can then consider the model $-\infty < z < z_S + \tfrac{1}{2}B^{-1}u_S^2 = \tfrac{1}{2}(z_S + aB^{-1})$. Because one of the two rays arriving at R always has a minimum at depth $z_M > z_S + \tfrac{1}{2}B^{-1}u_S^2$, the multiplicity of rays is fully removed.

The two-point travel times $T(R, S)$ can easily be calculated for any points S and R situated inside the caustic paraboloid of revolution. We can use any of the two expressions given above; for example,

$$T(R, S) = p^{-1}|x|\left(u_S^2 + \tfrac{1}{12}p^{-2}B^2x^2\right) - \tfrac{1}{2}Bp_{zS}p^{-2}x^2.$$

In general, we obtain two travel times. One arrival may again be eliminated by decreasing the size of the model.

In applications, it may be suitable to replace half-space $-\infty < z < 0$ by a homogeneous half-space, in which velocity $V = 1/\sqrt{a}$ is constant. Even though the velocity is continuous across plane $z = 0$, the form of the caustics becomes considerably complicated. For a detailed discussion, see Kravtsov and Orlov (1980, Section 3.3.5).

3. CONSTANT GRADIENT OF LOGARITHMIC VELOCITY, ln V

The ray expressions for this model (corresponding to $n = 0$) contain a logarithmic function and an inverse trigonometric function, arctan. The assumed velocity distribution is

$$\ln V(z) = a + bz, \qquad \text{that is,} \qquad V(z) = \exp(a + bz).$$

The suitable monotonic parameter along the ray is travel time T; hence,

$$p_z(T) = p_z(T_0) - b(T - T_0),$$
$$z(T) = z(T_0) - \tfrac{1}{2}b^{-1}\ln\left(V_0^2 X\right),$$
$$x(T) = x(T_0) + \frac{1}{b}\left(\arctan\frac{b(T - T_0) - p_{z0}}{p} - \arctan\left(-\frac{p_{z0}}{p}\right)\right),$$

where

$$p_{z0} = p_z(T_0),$$
$$X = b^2(T - T_0)^2 - 2bp_{z0}(T - T_0) + V_0^{-2}, \qquad V_0 = V(T_0).$$

Because travel time T is the parameter along the ray, we do not require an equation for it. Alternative expressions in terms of depth z, for the ray segment without a turning point, are

$$x(p, z_i, z_{i+1}) = \mp\frac{1}{b}\left(\arctan\frac{w_{i+1}}{p} - \arctan\frac{w_i}{p}\right),$$

$$T(p, z_i, z_{i+1}) = \mp\frac{1}{b}(w_{i+1} - w_i),$$

$$\tau(p, z_i, z_{i+1}) = \mp\frac{1}{b}\left\{w_{i+1} - w_i - p\left(\arctan\frac{w_{i+1}}{p} - \arctan\frac{w_i}{p}\right)\right\}.$$

For the ray segment with a turning point, we obtain

$$x(p, z_i, z_{i+1} = z_i) = 2\arctan(w_i/p)/|b|,$$
$$T(p, z_i, z_{i+1} = z_i) = 2w_i/|b|,$$
$$\tau(p, z_i, z_{i+1} = z_i) = 2[w_i - p\arctan(w_i/p)]/|b|.$$

4. CONSTANT GRADIENT OF VELOCITY, V(z)

This is a model often used in seismology. It corresponds to $n = -1$. The velocity distribution is assumed in the following form:

$$V(z) = a + bz.$$

The suitable parameter along the ray is ξ such that $d\xi = V^{-1}dT$; hence,

$$p_z(\xi) = p_z(\xi_0) - b(\xi - \xi_0),$$

$$z(\xi) = z(\xi_0) + \frac{1}{b}(X^{-1/2} - V_0),$$

$$x(\xi) = x(\xi_0) + \frac{1}{pX^{1/2}}\left(\xi - \xi_0 - \frac{1}{b}p_{z0}\right) + \frac{V_0 p_{z0}}{bp},$$

$$T(\xi) = T(\xi_0) + \frac{1}{|b|} \ln \frac{\epsilon X^{1/2} - b(\xi - \xi_0) + p_{z0}}{\epsilon V_0^{-1} + p_{z0}},$$

where $\epsilon = \text{sgn}(-b)$ and

$$p_{z0} = p_z(\xi_0),$$

$$X = b^2(\xi - \xi_0)^2 - 2bp_{z0}(\xi - \xi_0) + V_0^{-2}, \qquad V_0 = V(\xi_0).$$

Alternative expressions in terms of depth z for the ray segment without a turning point are

$$x(p, z_i, z_{i+1}) = \mp \frac{1}{pb}(c_{i+1} - c_i),$$

$$T(p, z_i, z_{i+1}) = \pm \frac{1}{b} \ln \frac{V_{i+1}(1 + c_i)}{V_i(1 + c_{i+1})},$$

$$\tau(p, z_i, z_{i+1}) = \pm \frac{1}{b}\left(c_{i+1} - c_i + \ln \frac{V_{i+1}(1 + c_i)}{V_i(1 + c_{i+1})}\right).$$

For the ray segment with a turning point, we obtain

$$x(p, z_i, z_{i+1} = z_i) = 2c_i/p|b|,$$

$$T(p, z_i, z_{i+1} = z_i) = 2\ln[(1 + c_i)/pV_i]/|b|,$$

$$\tau(p, z_i, z_{i+1} = z_i) = 2[\ln((1 + c_i)/pV_i) - c_i]/|b|.$$

It is not difficult to prove that the ray trajectory in the model with a constant velocity gradient is circular. Eliminating $(\xi - \xi_0)$ from the relations for $z(\xi)$ and $x(\xi)$ yields

$$\left[x - x_0 - \frac{p_{z0}V_0}{bp}\right]^2 + \left[z - z_0 + \frac{V_0}{b}\right]^2 = \frac{1}{p^2b^2}.$$

Here x_0 and z_0 are the coordinates of the initial point, $V_0 = V(z_0)$, $p_{z0} = (V_0^{-2} - p^2)^{1/2}$. Thus, the radius of the circle is $(pb)^{-1}$ and the center of the circle is situated at point

$$\left[x_0 + (bp)^{-1}p_{z0}V_0; z_0 - V_0/b\right].$$

Similarly, eliminating $(\xi - \xi_0)$ and p from the expressions for $T(\xi)$, $x(\xi)$, and $z(\xi)$, we can also prove that the wavefront for a point source at (x_0, z_0) is a circle:

$$[x - x_0]^2 + [z - z_0 + V_0 b^{-1}(1 - \cosh(b(T - T_0)))]^2$$

$$= V_0^2 b^{-2} \sinh^2(b(T - T_0)).$$

For fixed T, the radius of the wavefront circle is $b^{-1}V_0 \sinh(b(T - T_0))$, and the coordinates of the center are $x_0; z_0 - V_0 b^{-1}(1 - \cosh(b(T - T_0)))$. Circular rays and wavefronts in the model with constant velocity gradient have often been used in seismological applications.

The preceding equation for the wavefront can be used to derive simple equations for the two-point travel time $T(R, S)$ from any point S to any other point R in the model with the constant gradient of velocity b. Denote by r the distance between S and R. Then

$$T(R, S) = |b^{-1} \text{arccosh}[1 + b^2 r^2 / 2 V_R V_S]|$$
$$= b^{-1} \text{arcsinh}[br(V_R V_S)^{-1/2}(1 + b^2 r^2 / 4 V_S V_R)^{1/2}].$$

Here V_S is the velocity at S, and V_R is the velocity at R. The expressions for $T(R, S)$ given here can take many other alternative forms.

5. PARABOLIC LAYER

The distribution of the square of slowness $V^{-2}(z)$ is assumed to be quadratic in z,

$$V^{-2}(z) = a + bz + cz^2.$$

This velocity distribution is very suitable for investigating wave propagation in smooth low-velocity channels and in models with smooth velocity maxima and minima. For a more detailed discussion of ray fields, turning points, and caustics and of various waveguide and barrier effects in a parabolic layer, see Kravtsov and Orlov (1980) and other references therein.

We shall again use σ such that $d\sigma = V^2 dT$ as the suitable monotonic parameter along the ray. Ray tracing system (3.7.2) then reads

$$\frac{dx}{d\sigma} = p, \qquad \frac{dp_z}{d\sigma} = \tfrac{1}{2}b + cz,$$
$$\frac{dz}{d\sigma} = p_z, \qquad \frac{dT}{d\sigma} = A = V^{-2},$$

where $A = (p^2 + p_z^2)$. The two ordinary differential equations of the first order in z and p_z can be combined into one ordinary differential equation of the second order:

$$d^2 z / d\sigma^2 - cz = \tfrac{1}{2}b.$$

The two linearly independent solutions of this differential equation are $\sin(\sqrt{-c}\,(\sigma - \sigma_0))$ and $\cos(\sqrt{-c}(\sigma - \sigma_0))$ for $c < 0$ and $\exp(\sqrt{c}(\sigma - \sigma_0))$ and $\exp(-\sqrt{c}\,(\sigma - \sigma_0))$ for $c > 0$. Taking into account the proper initial conditions, the solution is

$$z(\sigma) = \tfrac{1}{2}(A + B)\exp[\sqrt{c}(\sigma - \sigma_0)] + \tfrac{1}{2}(A - B)\exp[-\sqrt{c}(\sigma - \sigma_0)]$$
$$- \tfrac{1}{2}c^{-1}b, \qquad \text{for } c > 0,$$
$$z(\sigma) = A\,\cos[\sqrt{-c}(\sigma - \sigma_0)] + C\sin[\sqrt{-c}(\sigma - \sigma_0)]$$
$$- \tfrac{1}{2}c^{-1}b, \qquad \text{for } c < 0.$$

Here $z_0 = z(\sigma_0)$, $p_{z0} = p_z(\sigma_0)$, $A = z_0 + \tfrac{1}{2}bc^{-1}$, $B = p_{z0}/\sqrt{c}$, and $C = p_{z0}/\sqrt{-c}$. The solution for $x(\sigma)$ is simple: $x(\sigma) = x_0 + p(\sigma - \sigma_0)$. As we can see, $x(\sigma)$ is a monotonic function of σ so that $\sigma - \sigma_0$ may be replaced by $p^{-1}(x - x_0)$ in the expressions for $z(\sigma)$. Hence,

$$z(x) = \tfrac{1}{2}(A + B)\exp[\sqrt{c}p^{-1}(x - x_0)]$$
$$+ \tfrac{1}{2}(A - B)\exp[-\sqrt{c}p^{-1}(x - x_0)] - \tfrac{1}{2}c^{-1}b, \qquad \text{for } c > 0,$$
$$z(x) = A\cos[\sqrt{-c}p^{-1}(x - x_0)]$$
$$+ C\sin[\sqrt{-c}p^{-1}(x - x_0)] - \tfrac{1}{2}c^{-1}b, \qquad \text{for } c < 0.$$

Here we have used the notation $x_0 = x(\sigma_0)$.

For $c < 0$, the velocity distribution represents a smooth low-velocity layer (wave-guide), and rays $z = z(x)$ oscillate within the layer. In the opposite case of $c > 0$, the velocity distribution represents a smooth high-velocity layer (antiwaveguide), with a smooth velocity maximum at some depth. The rays then have an exponential character.

It is simple to derive the expressions for $p_z(\sigma) = dz/d\sigma$ from those for $z(\sigma)$. Travel time $T(\sigma)$ can then be determined by integrating $(p^2 + p_z^2)$ with respect to σ, or, alternatively, by integrating $V^{-2}(z) = a + bz + cz^2$. The relevant equations are left to the reader as homework.

6. OTHER SIMPLE VELOCITY DISTRIBUTIONS

Analytical solutions for the ray trajectory and for the travel time can be computed for various other velocity-depth distributions. Many such analytical solutions can be found in the seismological literature. Suitable analytical solutions can be found even for various velocity-depth distributions specified by relations $z = F[1/V^2]$, where F is some simple function. Particularly simple analytical expressions for rays are obtained if function F is a polynomial; see Section 3.7.3. Moreover, even the velocity-depth distribution $z = a + bV + cV^2 + \cdots$ yields suitable analytical solutions (although not as simple as the velocity-depth distribution $z = a + b/V^2 + c/V^4 + \ldots$). These analytical solutions can be found in Puzyrev (1959) and Červený and Pretlová (1977).

7. TIME-DEPTH RELATIONSHIPS

In seismic exploration, considerable attention has been devoted to the so-called time-depth relationships (that is, to the analytical relations between depth and travel time, measured along the vertical in a vertically inhomogeneous medium). As an example, consider a linear velocity-depth distribution $V(z) = V_0 + kz$. Then, $dT/dz = (V_0 + kz)^{-1}$ and $T(z) = k^{-1} \ln(1 + kz/V_0)$. This gives the well-known time-depth relation $z = k^{-1} V_0(\exp[kT] - 1)$. Similar analytical time-depth relations are known for many other velocity-depth distributions. Let us name several of them: $V^{-n}(z) = a + bz$, $V(z) = a + bz^{1/n}$ (where n is an arbitrary integer), $V(z) = a \exp(bz)$, $V(z) = a - b \exp(-cz)$, $V(z) = c \tanh(a + bz)$, $V(z) = a - b/(c + zd)$, $V^{-2}(z) = a - b/(c + zd)$, $V(z) = 1/(a_0 + a_1 z + \cdots + a_m z^m)$, and $V(z) = a(1 + b^2 z^2)^{1/2}$ among others. For a detailed discussion of time-depth relations for these and other velocity-depth distributions see, for example, Kaufman (1953), Puzyrev (1959), and Al-Chalabi (1997a, 1997b). The last reference, Al-Chalabi (1997b), also offers time-depth relations for vertically inhomogeneous layered structures.

3.7.3 Polynomial Rays in Vertically Inhomogeneous Media

It was shown in Section 3.4.5 that the ray tracing system yields polynomial rays if the velocity distribution is specified by the relation $F[1/V^2] = A_0 + A_i x_i$, where F is a polynomial function in $1/V^2$. These polynomial rays also play an important role in vertically inhomogeneous media, where $F[1/V^2] = a + bz$. Here we shall use the velocity-depth distribution in a slightly different form:

$$z = f[1/V^2]. \tag{3.7.19}$$

We shall first consider the general function f in (3.7.19) and only discuss the polynomial functions f later on. In all these cases, however, we only consider functions f that are monotonic in the range of $1/V^2$ being considered. In other words, there is a unique correspondence between z and $1/V^2$.

From the eikonal equation, we obtain (3.7.6), so that ray tracing system (3.7.2) for $n = 2$ yields

$$dz/d\sigma = \pm(V^{-2} - p^2)^{1/2}. \tag{3.7.20}$$

Then

$$d\sigma = \pm(V^{-2} - p^2)^{-1/2}dz = \pm(V^{-2} - p^2)^{-1/2}f'[V^{-2}]d(V^{-2}), \tag{3.7.21}$$

where $f'[V^{-2}] = df[q]/dq$ with $q = V^{-2}$. This equation can simply be integrated to give

$$\sigma - \sigma_0 = \pm \int_{(1/V_0)^2}^{(1/V)^2} f'[V^{-2}](V^{-2} - p^2)^{-1/2}dV^{-2}.$$

If we change the variable under the integral by substituting

$$w^2 = V^{-2} - p^2,$$

we get

$$\sigma - \sigma_0 = \pm 2 \int_{w_0}^{w} f'[w^2 + p^2]dw, \tag{3.7.22}$$

where

$$w = (V^{-2} - p^2)^{1/2}, \qquad w_0 = (V_0^{-2} - p^2)^{1/2}. \tag{3.7.23}$$

Finally, using (3.7.5) with $u = \sigma(n = 2)$ for x and T, we arrive at

$$x = x_0 + 2p \int_{w_0}^{w} f'[w^2 + p^2]dw,$$
$$T = T_0 + 2 \int_{w_0}^{w} (w^2 + p^2)f'[w^2 + p^2]dw. \tag{3.7.24}$$

In (3.7.22) and (3.7.24), expression $f'[w^2 + p^2]$ denotes $df[q]/dq$ for $q = w^2 + p^2$.

Integrals (3.7.22) and (3.7.24) are very simple indeed. They offer a large amount of simple analytical solutions. We shall only discuss one particularly simple solution, corresponding to polynomial function $f[q]$, in more detail. Let us assume that function $f[q]$ is given by the polynomial relation

$$f[q] = \sum_{n=0}^{N} a_n q^n, \qquad \text{that is,} \qquad z = \sum_{n=0}^{N} a_n V^{-2n}. \tag{3.7.25}$$

Hence,

$$f'[p^2 + w^2] = \sum_{n=1}^{N} n a_n (w^2 + p^2)^{n-1}.$$

The final expressions for x and T then read

$$x(w) = x(w_0) + 2p \int_{w_0}^{w} \left(\sum_{n=1}^{N} n a_n (w^2 + p^2)^{n-1} \right) dw,$$
$$T(w) = T(w_0) + 2 \int_{w_0}^{w} (w^2 + p^2) \left(\sum_{n=1}^{N} n a_n (w^2 + p^2)^{n-1} \right) dw. \tag{3.7.26}$$

As we can see, the integrands are polynomials in w so that $x(w)$ and $T(w)$ are also polynomials in w.

Note that expressions (3.7.26) with $N = 3$ were used to compile very efficient algorithms and fast computer programs for ray tracing and travel-time computations in an arbitrary, 1-D, vertically inhomogeneous, layered model; see Červený (1980) and Červený and Janský (1985). The advantage of these algorithms is that they do not introduce fictitious interfaces of the second and third order. Such interfaces could produce fictitious anomalies in the amplitude-distance curves. We shall describe the algorithm briefly.

The velocity in the model is specified at n grid points z_i, $i = 1, 2, \ldots, n$. Grid point $z = z_1 = 0$ corresponds to the surface of the Earth; $z = z_n$ corresponds to the bottom of the model. At any grid point, the velocity is either discontinuous (interface of the first order), continuous with a discontinuous first derivative (interface of the second order), or continuous. If the velocity is discontinuous at $z = z_i$, two velocities must be specified at that point, one just above the interface and the second immediately below it. We shall formally call the region between the two consecutive grid points z_i and z_{i+1} the ith subinterval of depths. The standard term layer will be used to denote the region between two physical interfaces of the first order. Thus, any layer may be composed of several subintervals of depths.

The velocity distribution between the individual grid points is approximated by the function

$$z = a_i + b_i V^{-2} + c_i V^{-4} + d_i V^{-6}, \tag{3.7.27}$$

where a_i, b_i, c_i, and d_i are constants. These constants can be determined for the whole group of subintervals between the individual interfaces (of the first and second order) by a smoothed spline algorithm (see, for example, Reinsch 1967; Pretlová 1976) or by some other spline algorithms (see, for example, Cline 1981). The velocity-depth distribution between the two interfaces is then smooth together with the first and second derivatives of velocity, and function $V = V(z)$ does not oscillate there. The smoothed spline algorithm might slightly change quantities z_i, but this does not cause any complications. The depths z_i, corresponding to physical interfaces, are, of course, fixed. The degree of smoothing of the velocity-depth distribution can be controlled by a special parameter. Even a slight smoothing, which does not change the velocity-depth function visually, increases the stability of the results considerably.

Let us now assume that the source is situated at depth z_s and the receiver is at z_r. Assume that the ray of the wave under consideration is composed of $N + 1$ elements, each of which is completely within one subinterval of depths, in which the velocity is specified by Equation (3.7.27). The end points of the elements are situated either on the boundaries between the subintervals, at the source, at the receiver, or at the turning points of the ray. We denote the end points of the elements successively such that $Q_0 \equiv S$ (source) $Q_1, Q_2, \ldots, Q_{N+1} \equiv R$ (receiver). Note that the same subinterval of depth may be encountered several times, depending on the number of times the wave passes through it.

The equations for the total epicentral distance $x(p, z_s, z_r)$, total travel time $T(p, z_s, z_r)$, and total delay time $\tau(p, z_s, z_r)$ can simply be obtained from (3.7.26):

$$x(p, z_s, z_r) = \sum_{i=1}^{N+1} x_i(p, z(Q_{i-1}), z(Q_i)),$$

$$T(p, z_s, z_r) = \sum_{i=1}^{N+1} T_i(p, z(Q_{i-1}), z(Q_i)), \tag{3.7.28}$$

$$\tau(p, z_s, z_r) = \sum_{i=1}^{N+1} \tau_i(p, z(Q_{i-1}), z(Q_i)),$$

where $x_i(p, z(Q_{i-1}), z(Q_i))$, $T_i(p, z(Q_{i-1}), z(Q_i))$ and $\tau_i(p, z(Q_{i-1}), z(Q_i))$ are given by the relations

$$
\begin{aligned}
x_i(p, z(Q_{i-1}), z(Q_i)) = \pm\{ & (2b_i p + 4c_i p^3 + 6d_i p^5)(w_i - w_{i-1}) \\
& + \left(\tfrac{4}{3}c_i p + 4d_i p^3\right)\left(w_i^3 - w_{i-1}^3\right) \\
& + \tfrac{6}{5}d_i p\left(w_i^5 - w_{i-1}^5\right)\}, \\
T_i(p, z(Q_{i-1}), z(Q_i)) = \pm\{ & (2b_i p^2 + 4c_i p^4 + 6d_i p^6)(w_i - w_{i-1}) \\
& + \left(\tfrac{2}{3}b_i + \tfrac{8}{3}c_i p^2 + 6d_i p^4\right)\left(w_i^3 - w_{i-1}^3\right) \\
& + \left(\tfrac{4}{5}c_i + \tfrac{18}{5}d_i p^2\right)\left(w_i^5 - w_{i-1}^5\right) \\
& + \tfrac{6}{7}d_i\left(w_i^7 - w_{i-1}^7\right)\}, \\
\tau_i(p, z(Q_{i-1}), z(Q_i)) = \pm\{ & \left(\tfrac{2}{3}b_i + \tfrac{4}{3}c_i p^2 + 2d_i p^4\right)\left(w_i^3 - w_{i-1}^3\right) \\
& + \left(\tfrac{4}{5}c_i + \tfrac{12}{5}d_i p^2\right)\left(w_i^5 - w_{i-1}^5\right) \\
& + \tfrac{6}{7}d_i\left(w_i^7 - w_{i-1}^7\right)\}.
\end{aligned}
$$

Quantities w_i have a standard meaning, $w_i = (V^{-2}(Q_i) - p^2)^{1/2}$. The upper signs correspond to the descending part of the ray ($z(Q_i) > z(Q_{i-1})$), and the lower signs correspond to the ascending part of the ray ($z(Q_i) < z(Q_{i-1})$). For the ray element with a turning point, we insert $w_i = 0$ and multiply the result by two.

As we can see, the algorithm only requires one square root w_i and some simple polynomials to be computed for each subinterval of depth; transcendental functions are not needed at all.

The proposed algorithm has two limitations:

a. It cannot simulate a velocity distribution with smooth local maxima and minima in the velocity-depth distribution. Interfaces of the second order must be allowed at the points of maxima and minima.

b. If the gradient of velocity changes abruptly in some region with a slowly varying velocity, the depth-velocity relation $z = z[V^{-2}]$ given by (3.7.27) may oscillate at the relevant depths. The oscillations have no physical meaning and must be removed. This can be done, for example, by introducing an artificial interface of the second order. Such oscillations, however, should appear only exceptionally.

If we allow the existence of second-order interfaces, we can just put $c_i = d_i = 0$ in (3.7.27) and determine a_i and b_i from the velocities at grid points Q_{i-1} and Q_i. The following relations are then obtained for x_i, T_i and τ_i

$$
\begin{aligned}
x_i(p, z(Q_{i-1}), z(Q_i)) &= \pm 2b_i p(w_i - w_{i-1}), \\
T_i(p, z(Q_{i-1}), z(Q_i)) &= \pm\left[2b_i p^2(w_i - w_{i-1}) + \tfrac{2}{3}b_i\left(w_i^3 - w_{i-1}^3\right)\right], \\
\tau_i(p, z(Q_{i-1}), z(Q_i)) &= \pm\tfrac{2}{3}b_i\left(w_i^3 - w_{i-1}^3\right),
\end{aligned}
$$

with $b_i = (z(Q_i) - z(Q_{i-1}))/(V(Q_i)^{-2} - V(Q_{i-1})^{-2})$. For the ray element with a turning point, we obtain $x_i = 4|b_i| p w_{i-1}$, $T_i = 4|b_i| w_{i-1}(p^2 + \tfrac{1}{3}w_{i-1}^2)$, $\tau_i = \tfrac{4}{3}|b_i| w_{i-1}^3$. These equations are more efficient in computation and simpler to program than a piecewise linear approximation of the velocity-depth function. They are also optionally used in the present algorithms of the WKBJ method; see Chapman, Chu, and Lyness (1988).

3.7.4 Radially Symmetric Media

The basic equations of the ray method for radially symmetric media can be derived in several ways. The first option is to derive them directly, by applying Fermat's principle. This approach has mostly been used in the seismological literature; see Savarenskiy and Kirnos (1955), Bullen (1965), Pilant (1979), Aki and Richards (1980), and Bullen and Bolt (1985) among others. The second option is to start with general ray tracing systems derived in spherical coordinates (see Section 3.5.4) and to specify them for radially symmetric media. A similar approach for vertically inhomogeneous media was used in Section 3.7.1. Finally, it is possible to use the Earth flattening approximation (EFA) derived and discussed for 2-D media in Section 3.5.5. Using EFA, we can transform any result derived for vertically inhomogeneous media to that appropriate for radially symmetric media and vice versa. Note that the Earth flattening approximation has also been successfully used in the reflectivity method; see Müller (1985). The synthetic seismograms for radially symmetric media can be computed by the application of EFA to the reflectivity synthetic seismograms computed for the corresponding vertically inhomogeneous model.

Because we have derived many useful analytical solutions for vertically inhomogeneous media in Sections 3.7.1–3.7.3, we shall only describe the way in which these results can be transformed for radially symmetric media. However, we shall first give several general equations for the radially symmetric media.

We shall use spherical coordinates r, θ, and φ and assume that the velocity does not depend on θ and φ. Without loss of generality, we may adopt initial conditions (3.5.33) with

$$\varphi = \varphi_0, \qquad T_\varphi = T_{\varphi 0} = 0. \tag{3.7.29}$$

A similar situation was discussed in Section 3.5.5. The ray as a whole is then situated in plane $\varphi = \varphi_0$, and $p_\varphi = T_\varphi = 0$ along the whole ray. We can now express ray tracing system (3.5.31) in the following form:

$$\frac{dr}{du} = A^{n/2-1} T_r, \qquad \frac{dT_r}{du} = \frac{1}{n}\frac{\partial}{\partial r}\left(\frac{1}{V}\right)^n + A^{n/2-1} T_\theta^2 r^{-3},$$
$$\frac{d\theta}{du} = A^{n/2-1} T_\theta r^{-2}, \qquad \frac{dT_\theta}{du} = 0, \tag{3.7.30}$$

with

$$\frac{dT}{du} = A^{n/2} = V^{-n}, \qquad A = T_r^2 + r^{-2} T_\theta^2 = V^{-2}.$$

Thus, T_θ is constant along the whole ray. If the acute angle between the ray and the vertical line is denoted $i(r)$,

$$T_\theta(r) = r p_\theta(r) = \frac{r \sin i(r)}{V(r)} = \text{const.}$$

In seismology, this constant is usually called the *ray parameter* and denoted p, as in vertically inhomogeneous models. Hence,

$$\frac{r \sin i(r)}{V(r)} = p. \tag{3.7.31}$$

This is the *generalized Snell's law* for a radially symmetric medium.

Equations (3.7.30) can always be solved in terms of closed-form integrals. Quantity T_r can be expressed from the eikonal equation, $T_r^2 + r^{-2} T_\theta^2 = V^{-2}$,

$$T_r = \pm(V^{-2} - r^{-2} p^2)^{1/2}. \tag{3.7.32}$$

Here the plus sign corresponds to the upgoing part of the ray, and the minus sign corresponds to the downgoing part of the ray (decreasing r). Equation (3.7.30) then yields

$$\frac{d\theta}{dr} = \frac{T_\theta}{r^2 T_r} = \pm\frac{pV}{r(r^2 - V^2 p^2)^{1/2}}, \qquad \frac{dT}{dr} = \frac{A}{T_r} = \pm\frac{r}{V(r^2 - V^2 p^2)^{1/2}}. \tag{3.7.33}$$

The solution of (3.7.33) is

$$\theta(p, r) = \theta(p, r_0) \pm \int_{r_0}^{r} \frac{pV \, dr}{r(r^2 - V^2 p^2)^{1/2}},$$
$$T(p, r) = T(p, r_0) \pm \int_{r_0}^{r} \frac{r \, dr}{V(r^2 - V^2 p^2)^{1/2}}. \tag{3.7.34}$$

These equations can be expressed in several alternative forms; see Bullen and Bolt (1985).

For a radially symmetric medium containing structural interfaces of the first order along spherical surfaces $r = $ const., the equations for θ and T may again be expressed as sums of elements corresponding to the individual layers, as in (3.7.10). The layer contributions are given by relations

$$\theta(p, r_i, r_{i+1}) = \pm \int_{r_i}^{r_{i+1}} \frac{pV \, dr}{r(r^2 - V^2 p^2)^{1/2}},$$
$$T(p, r_i, r_{i+1}) = \pm \int_{r_i}^{r_{i+1}} \frac{r \, dr}{V(r^2 - V^2 p^2)^{1/2}}. \tag{3.7.35}$$

The plus sign refers to $r_{i+1} > r_i$, and the minus sign refers to $r_{i+1} < r_i$.

As in vertically inhomogeneous media, the element of the ray passing through a turning point must be formally divided into two elements, one going downward and the other going upward. We denote the coordinate r of the turning point r_M. It satisfies the relation

$$r_M / V(r_M) = p; \tag{3.7.36}$$

see (3.7.31) with $i(r) = \frac{1}{2}\pi$. For example, if a direct (refracted) wave with a source and receiver close to the Earth's surface is involved, with $\theta_0 = T_0 = 0$,

$$\theta(p) = 2p \int_{r_M}^{R} \frac{V \, dr}{r(r^2 - V^2 p^2)^{1/2}},$$
$$T(p) = 2 \int_{r_M}^{R} \frac{r \, dr}{V(r^2 - V^2 p^2)^{1/2}}. \tag{3.7.37}$$

At turning point $r = r_M$, the integrands of (3.7.37) are infinite. This fact causes some complications in calculating integrals (3.7.37), particularly if we wish to compute $d\theta/dp$ or dT/dp. As in vertically inhomogeneous media, however, the problem can be solved using the integration-by-parts method. The relevant final equations can be found in Bullen and Bolt (1985).

In radially symmetric media, delay time $\tau(p, r_i, r_{i+1})$ is defined in a way similar to that in (3.7.15):

$$\tau(p, r_i, r_{i+1}) = T(p, r_i, r_{i+1}) - p\theta(p, r_i, r_{i+1})$$
$$= \pm \int_{r_i}^{r_{i+1}} \frac{(r^2 - V^2 p^2)^{1/2}}{rV} dr. \tag{3.7.38}$$

Function $\tau(p, r_i, r_{i+1})$ has broadly been used in various seismological applications. It satisfies the relation

$$d\tau(p, r_i, r_{i+1})/dp = -\theta(p, r_i, r_{i+1}). \tag{3.7.39}$$

The ray integrals presented in this section can be calculated *numerically* using methods similar to those used in Section 3.7.1 for vertically inhomogeneous media.

Analytical solutions for radially symmetric media can be obtained directly from (3.7.35) and (3.7.38). They can, however, also be obtained in a simpler way from the analytical solutions for vertically inhomogeneous media, using the EFA. Because we do not wish to cause confusion, we shall denote, in the remaining part of this section, the velocity distribution and the ray parameter in the radially symmetric media V_R and p_R, respectively. For vertically inhomogeneous media, we shall still use the standard notation V and p. Any analytical solution for a velocity-depth distribution $V(z)$ in a vertically inhomogeneous medium can then be transformed to a radially symmetric model by making the following substitution:

$$V \longrightarrow RV_R/r, \qquad z \longrightarrow R\ln(R/r). \tag{3.7.40}$$

In the final expressions, we only have to put

$$x \longrightarrow R\theta, \qquad\qquad p \longrightarrow R^{-1}p_R. \tag{3.7.41}$$

Any analytical solution of the ray tracing system, presented in Section 3.7.2 or 3.7.3, can thus be modified to satisfy a radially symmetric medium.

We shall give two examples that play an important role in seismological applications. The first example presents the simplest solution for an inhomogeneous radially symmetric medium, and the second corresponds to the classical Mohorovičić velocity law. We shall follow this with a brief discussion of polynomial rays. In all these cases, we shall only give the expressions in terms of radius r as a parameter. The upper sign corresponds to the upgoing part of the ray; the lower sign refers to the downgoing part of the ray.

1. SIMPLEST SOLUTIONS FOR INHOMOGENEOUS RADIALLY SYMMETRIC MEDIA

It is not surprising that the simplest analytical solutions are again obtained for the velocity distribution corresponding to $V^{-2}(z) = a + bz$. The relevant velocity distribution in the radially symmetric model is as follows:

$$\left(\frac{RV_R}{r}\right)^{-2} = a + bR\ln\frac{R}{r}, \qquad \text{that is,}$$
$$V_R(r) = \frac{r}{R}\left(a + bR\ln\frac{R}{r}\right)^{-1/2}. \tag{3.7.42}$$

Very simple analytical solutions can be obtained for this velocity distribution; see Section 3.7.2, §2. If we use (3.7.41),

$$\theta(p_R, r_i, r_{i+1}) = \mp \frac{2p_R}{bR^3}(w_{i+1} - w_i),$$

$$T(p_R, r_i, r_{i+1}) = \mp \frac{2p_R^2}{bR^3}(w_{i+1} - w_i) \mp \frac{2}{3bR^3}(w_{i+1}^3 - w_i^3), \qquad (3.7.43)$$

$$\tau(p_R, r_i, r_{i+1}) = \mp \frac{2}{3bR^3}(w_{i+1}^3 - w_i^3),$$

where

$$w_k = \left[(r_k/V_R(r_k))^2 - p_R^2\right]^{1/2}.$$

Thus, the computation of rays and travel times requires the calculation of simple square roots only; no transcendental functions are necessary. For the ray element with a turning point, (3.7.43) yields $\theta = 4p_R w_i / |b| R^3$, $T = 4w_i(p_R^2 + \frac{1}{3}w_i^2)/|b|R^3$, and $\tau = \frac{4}{3}w_i^3/|b|R^3$.

2. MOHOROVIČIĆ VELOCITY DISTRIBUTION

We shall now use EFT to transform the analytical solutions for the constant gradient of the logarithmic velocity, $\ln V(z) = a + bz$, from a vertically inhomogeneous to a radially symmetric model. The corresponding velocity is

$$V_R(r) = A(r/R)^{1-k}, \qquad \text{with } A = \exp(a), \qquad k = bR. \qquad (3.7.44)$$

Velocity distribution (3.7.44) is known as the Mohorovičić velocity law (see Bullen and Bolt 1985) or also as Bullen's velocity law. The analytical solutions are again obtained simply from those given in Section 3.7.2, §3, if we insert $x = R\theta$, $p = R^{-1}p_R$, $b = k/R$:

$$\theta(p_R, r_i, r_{i+1}) = \pm \frac{1}{k}\left(\arctan \frac{w_{i+1}}{p_R} - \arctan \frac{w_i}{p_R}\right),$$

$$T(p_R, r_i, r_{i+1}) = \pm \frac{1}{k}(w_{i+1} - w_i), \qquad (3.7.45)$$

$$\tau(p_R, r_i, r_{i+1}) = \pm \frac{1}{k}\left\{w_{i+1} - w_i - p_R\left(\arctan \frac{w_{i+1}}{p_R} - \arctan \frac{w_i}{p_R}\right)\right\},$$

where

$$w_k = \left[(r_k/V_R(r_k))^2 - p_R^2\right]^{1/2}.$$

For the ray element with a turning point, (3.7.45) yields $\theta = 2\arctan(w_i/p_R)/|k|$, $T = 2w_i/|k|$, and $\tau = 2(w_i - p_R \arctan(w_i/p_R))/|k|$.

3. POLYNOMIAL RAYS IN RADIALLY SYMMETRIC MEDIA

For vertically inhomogeneous media, simple polynomial analytical solutions of the ray tracing systems were found in Section 3.7.3. These polynomial solutions can also be found for radially symmetric media. In fact, solution (3.7.43) is a special simple example of these solutions. We shall apply the Earth flattening approximation to the velocity-depth distribution (3.7.25). The corresponding velocity distribution for a radially symmetric model is

$$R \ln \frac{R}{r} = \sum_{n=0}^{N} a_n \left(\frac{r}{RV_R}\right)^{2n}. \qquad (3.7.46)$$

The analytical solutions are then obtained simply from (3.7.26), or for $N = 3$ from (3.7.28) and subsequent equations. If constants a_n in (3.7.46) are slightly modified, we can also discuss the velocity distribution

$$\ln r = \sum_{n=0}^{N} b_n \, (r/V_R)^{2n} \, . \tag{3.7.47}$$

Complete analytical solutions for velocity distribution (3.7.47) can be found in Červený and Janský (1983), with examples of computations in Janský and Červený (1981) and Zedník, Janský, and Červený (1993).

Many other analytical solutions for radially symmetric media can be obtained from those for vertically inhomogeneous media presented in Sections 3.7.2 and 3.7.3. We shall not give them here, although some of them may be of interest. For example, velocity distribution $V(r) = a - br^2$ yields circular ray trajectories; see Bullen and Bolt (1985).

3.8 Direct Computation of Travel Times and/or Wavefronts

In the previous sections of this chapter, we have discussed the computation of rays and ray-theory travel times of selected elementary waves. The ray-theory travel times are calculated as a by-product of ray tracing. Using several nearby rays, it is also simple to construct wavefronts, with a specified travel-time step ΔT. See Section 3.3.3.

The travel times are usually a more important result of ray tracing than the rays themselves. Consequently, it is not surprising that a great effort in seismological applications has also been devoted to the direct computation of travel times and wavefronts, without invoking ray tracing at all.

Analytical computation of *two-point travel times* is possible only exceptionally, for very simple models. They include homogeneous models, models with a constant gradient of velocity and models with a constant gradient of the square of slowness. See Section 3.7.2 for the relevant equations.

Classical methods of computing wavefronts without invoking rays are based on the local application of the *Huygens principle*. The current wavefront for travel time $T = T_0 + k \Delta T$ is calculated from the previous wavefront $T = T_0 + (k - 1) \Delta T$ as an *envelope surface* of spheres, with their centers distributed along the previous wavefront and with radii $V \Delta T$, where V is the local velocity at the center of the sphere. Velocity V may vary laterally along the wavefront but must be smooth; the model is considered to be locally homogeneous in the region of each sphere. The method has found broad applications in the solution of direct and inverse seismic structural problems in 2-D laterally varying layered isotropic structures. It has been based mainly on a *graphical construction* of Huygens circles using a pair of compasses. Various alternatives of this method have been developed, for example the *method of wavefronts* (Thornburgh 1930; Rockwell 1967), and the *method of time fields* (Riznichenko 1946, 1985). The method is very stable, but its accuracy depends strongly on the accuracy of used graphical construction, on travel-time step ΔT, and on the smoothness of the medium.

The preceding methods are very simple in 2-D models, if the graphical construction is used. Unfortunately, the computer realization of the method is rather complicated, particularly in 3-D layered structures. Recent computer methods of *wavefront construction* are based on different principles: on a hybrid combination with standard ray tracing. The current wavefront in the wavefront construction method is computed from the previous wavefront by ray tracing of short ray elements. For more details, refer to Section 3.8.5.

Several other important methods have been proposed to compute first arrival travel times in rectangular, 2-D, and 3-D grid models. See Section 3.8.3, which is devoted to *network shortest-path ray tracing*, and Section 3.8.4, which is devoted to *finite-difference methods*. The accuracy of the computation of the first-arrival travel times in grid models depends, as a rule, on the *grid step, h*.

In all travel-time computations, it is very important whether we evaluate the ray-theory travel times or first-arrival travel times. The ray-theory travel times and first-arrival travel times concepts were briefly explained in the introduction to Chapter 3. Because these concepts play a very important role both in the ray theory and in seismological applications, we shall summarize the main properties of both types of travel times and the differences between them in Section 3.8.1.

Note that the wavefronts, or some surfaces close to them, are also the "corner stones" of some asymptotic methods of computing seismic wavefields in complex laterally varying structures. See the *phase-front method* by Haines (1983, 1984a, 1984b) and other references given there.

The methods to compute travel times and/or wavefronts of seismic body waves in 2-D and 3-D models are developing very fast. In the future, we can expect the appearance of new powerful methods that will surpass the methods discussed here in speed, accuracy, and stability of computations. Recently, a new method of computing the first-arrival travel times in 3-D models has been proposed by Sethian and Popovici (1999); they call it the *fast marching method*. In the fast marching method, the problem of computing first-arrival travel times is treated as the problem of tracking evolving interfaces, the solution of which was developed by J. A. Sethian in a number of publications and used by him and by others in various problems. The method can be applied to models with arbitrarily large gradients of velocity. For more details and references, see Sethian and Popovici (1999).

Finally, it should be mentioned that the high-energy travel times can also be measured from complete synthetic seismograms, which are computed, for example, by finite differences. The complete synthetic seismograms allow us to pick up the high-energy travel times in the frequency band we need in interpretations. See Loewenthal and Hu (1991). This procedure is, however, time consuming because it requires full wavefield modeling. An alternative, considerably faster approach to calculating high-frequency travel times in the frequency band under consideration was proposed by Nichols (1996). In the method, it is sufficient to solve the Helmholtz equation for a very small number of frequencies in the frequency band under consideration. From these computations, it is possible to appreciate the travel times and relevant Green functions in the specified frequency band. See also Audebert et al. (1997).

3.8.1 Ray Theory Travel Times and First-Arrival Travel Times

In this section, we shall define the ray-theory travel times and first-arrival travel times and explain the main differences between them.

1. RAY-THEORY TRAVEL TIMES

Ray-theory travel times are being introduced as the travel times of individual elementary waves, calculated along the rays of these waves. This definition is closely related to the high-frequency asymptotic solutions of the elastodynamic equation and to the eikonal equation. The calculation of ray-theory travel times along the relevant ray may be numerical, analytical, semianalytical, and the like.

Let us now discuss the preceding definition and the properties of the ray-theory travel times in greater detail.

a. The ray-theory travel times are defined separately for the individual elementary waves. Thus, the ray-theory travel time is not a global property of the wavefield, but the property of a selected elementary wave. For example, we have ray-theory travel times of direct waves, reflected waves, converted waves, and multiply-reflected waves. The ray-theory travel time is not only a function of position but also of the ray code of the elementary seismic body wave.

b. The ray-theory travel time of a selected elementary wave is, in general, a multivalued function of coordinates of the receiver, even in a smooth medium without interfaces. This is due to multipathing.

c. In certain regions of the model, the ray-theory travel times of the elementary waves under consideration are not defined. Such regions are partly due to the specification of the ray code (for example, reflected waves exist only on one side of the interface) and partly due to the shadow zones of the elementary wave under consideration. For different elementary waves, the shadow zones are, in general, situated in different regions of the model.

d. The ray-theory travel times may correspond to the first-arrival travel times. Mostly, however, they correspond to later arrivals and may carry a considerable amount of energy.

e. The definition of the ray-theory travel times is fully based on ray concepts. They do not have an exact, but only asymptotic (high-frequency) meaning.

f. In addition to computing ray-theory travel times, it is usually possible to calculate the relevant ray amplitudes of the elementary wave under consideration also.

A very important subclass of the ray-theory travel times are *zeroth-order ray-theory travel times*. They correspond to the elementary waves of the zeroth-order ray approximation; see Sections 2.4, 5.6, and 5.7. In recent applications of the ray method in seismology and seismic exploration, mostly the elementary waves corresponding to the zeroth-order ray approximation (zeroth-order elementary waves) have been used. The definition of the zeroth-order ray-theory travel times does not include such elementary waves as higher-order elementary waves (head waves) and diffracted waves (edge waves, tip waves, sliding waves) among others.

The zeroth-order ray-theory travel times satisfy all the properties listed earlier. Regarding item c, even if we consider all possible zeroth-order elementary waves for a given source, there may be some regions where *no zeroth-order elementary wave* arrives for a specified position of the source. We shall call them the *absolute shadow zones* (in the zeroth-order ray approximation of the ray method).

2. FIRST-ARRIVAL TRAVEL TIMES

They are related to the exact solution of the elastodynamic equation, with proper initial and boundary conditions. The first-arrival travel time corresponds to the first arrival of the complete wavefield at a specified receiver position. The surface of constant first-arrival travel time t_0 separates the illuminated and nonilluminated regions of the model at $t = t_0$. Two suitable methods of calculating first-arrival travel times are briefly outlined in Sections 3.8.3 and 3.8.4.

Let us now discuss certain properties of the first-arrival travel times.

a. The first-arrival travel times are not related to a (somewhat ambiguous) decomposition of the complete wavefield into elementary waves; see Section 3.2.2. They are properties of the complete wavefield.

b. The first-arrival travel time is a function of position only, not of the type of wave to arrive first.

c. The first-arrival travel time is a *unique* function of position. It is defined at any point of the model. There are no shadow zones. Moreover, the first-arrival travel time is a continuous function of coordinates. The first spatial derivatives of the first-arrival travel time, however, may be discontinuous. They may be discontinuous even at points where the velocities are continuous. (Example of such discontinuity include the intersection of the wavefronts of direct and head waves.)

d. The first-arrival travel times have an exact meaning because they are based on the solution of the elastodynamic equation. They have no direct connection with ray concepts, but the Fermat's minimum time principle may be applied.

e. The concept of the first-arrival travel times is not related, in any way, to the amplitudes of the wavefield. It is not a simple theoretical problem to assign an amplitude to the evaluated first-arrival travel times. Only in those regions where the first-arrival travel time coincides with the zeroth-order ray-theory travel times, may the amplitudes be calculated by standard ray concepts (transport equation and geometrical spreading).

The concept of first-arrival travel time may also be *extended to certain later arrivals* imposing some constraints on the algorithms used in the computations of first-arrival travel times. This is the way the first-arrival travel times of waves reflected at structural interfaces can be introduced. These first-arrival travel times of reflected waves, however, again have a different meaning from the ray-theory travel times of reflected waves. They are a unique function of position, without possible triplications and shadows. At postcritical distances, they may correspond to elementary travel times of head waves, or to zeroth-order ray-theory travel times of postcritical reflection, depending on the constraints imposed. If the structural interface has edges, the first-arrival travel times of reflected waves may correspond to travel times of diffracted (edge) waves in certain regions.

3. SEVERAL REMARKS ON BOTH CONCEPTS

It is obvious that both ray-theory travel times and first-arrival travel times have the same meaning in certain situations. This applies, for example, to the homogeneous medium with a point source of P waves. The ray-theory travel time of the direct P wave then corresponds exactly to the first-arrival travel time. The same also applies to a halfspace in which the velocity increases linearly with depth.

In models with structural interfaces, however, both terms often have a different meaning. Due to velocity variations, shadow zones and caustics may be formed. The ray-theory travel times of direct waves in shadow zones are not defined, but the first-arrival travel times are well defined at any point of the shadow zone. On the contrary, the ray-theory travel time of a direct wave beyond the caustics is multivalued, but the first-arrival travel time is single-valued even there. It corresponds to the ray-theory travel time of the fastest branch of the direct wave.

In seismic applications, it may be very important to distinguish carefully between ray-theory travel times and first-arrival travel times. For example, a wave connected with the first arrival may be very weak, whereas a considerable amount of energy may be carried by an elementary wave. A seismologist picking up first-arrival travel times may prefer the first-arrival travel times, whereas, in seismic prospecting, the migrated image could be distorted if the first-arrival travel times were used to back-propagate the energy in recorded seismic sections; see Geoltrain and Brac (1993) and Gray and May (1994).

3.8.2 Solution of the Eikonal Equation by Separation of Variables

The method of separation of variables can be used to solve directly the eikonal equation for certain simple classes of models, including laterally varying ones. We shall not discuss here the method of separation of variables from a general point of view, but only explain it on two simple but important examples. For a more general treatment, see Kravtsov and Orlov (1980), where many other references can be found.

Let us consider Cartesian coordinates and assume that the velocity distribution in the model is specified by the relation:

$$1/V^2(x_i) = \epsilon_1(x_1) + \epsilon_2(x_2) + \epsilon_3(x_3). \tag{3.8.1}$$

We shall seek the solution of the eikonal equation in the form $T(x_1, x_2, x_3) = T_1(x_1) + T_2(x_2) + T_3(x_3)$. Inserting the expressions for $T(x_i)$ and $1/V^2(x_i)$ into the eikonal equation (3.1.1), we obtain

$$\left[\left(\frac{\partial T_1}{\partial x_1} \right)^2 - \epsilon_1(x_1) \right] + \left[\left(\frac{\partial T_2}{\partial x_2} \right)^2 - \epsilon_2(x_2) \right] + \left[\left(\frac{\partial T_3}{\partial x_3} \right)^2 - \epsilon_3(x_3) \right] = 0. \tag{3.8.2}$$

The equation (3.8.2) must be satisfied identically for any x_1, x_2, and x_3. Because the first term in (3.8.2) is a function of x_1 only, and the two other terms are functions of x_2 and x_3 only, the first term must be constant. The same is valid even for the second and third terms. Consequently, we obtain three separated equations:

$$(\partial T_1/\partial x_1)^2 - \epsilon_1(x_1) = \alpha_1, \qquad (\partial T_2/\partial x_2)^2 - \epsilon_2(x_2) = \alpha_2,$$
$$(\partial T_3/\partial x_3)^2 - \epsilon_3(x_3) = -\alpha_1 - \alpha_2. \tag{3.8.3}$$

Here α_1 and α_2 are arbitrary constants. All three equations can be solved by quadratures:

$$T_1(x_1) = \pm \int_{x_{10}}^{x_1} [\epsilon_1(x_1) + \alpha_1]^{1/2} dx_1,$$

$$T_2(x_2) = \pm \int_{x_{20}}^{x_2} [\epsilon_2(x_2) + \alpha_2]^{1/2} dx_2,$$

$$T_3(x_3) = \pm \int_{x_{30}}^{x_3} [\epsilon_3(x_3) - \alpha_1 - \alpha_2]^{1/2} dx_3.$$

The *complete solution of the eikonal equation* (3.1.1) for the velocity distribution (3.8.1) in Cartesian coordinates x_i is then

$$T_1(x_1, x_2, x_3) = \pm \int_{x_{10}}^{x_1} [\epsilon_1(x_1) + \alpha_1]^{1/2} dx_1 \pm \int_{x_{20}}^{x_2} [\epsilon_2(x_2) + \alpha_2]^{1/2} dx_2$$
$$\pm \int_{x_{30}}^{x_3} [\epsilon_3(x_3) - \alpha_1 - \alpha_2]^{1/2} dx_3 + \alpha_3. \tag{3.8.4}$$

Because the eikonal equation contains only partial derivatives of $T_i(x_i)$, not $T_i(x_i)$ themselves, we have added an additional constant α_3.

A similar procedure may be used even to find a complete solution of the eikonal equation for various curvilinear coordinate systems. Consider eikonal equation (3.5.6) in orthogonal curvilinear coordinates ξ_i, with scale factors h_i, and assume that the velocity distribution is specified by the relation

$$1/V^2(\xi_i) = h_1^{-2}\epsilon_1(\xi_1) + h_2^{-2}\epsilon_2(\xi_2) + h_3^{-2}\epsilon_3(\xi_3). \tag{3.8.5}$$

We shall seek the solution of the eikonal equation (3.5.6) in the form $T(\xi_i) = T_1(\xi_1) + T_2(\xi_2) + T_3(\xi_3)$. Then, (3.5.6) with (3.8.5) yields

$$h_1^{-2}[(\partial T_1/\partial \xi_1)^2 - \epsilon_1] + h_2^{-2}[(\partial T_2/\partial \xi_2)^2 - \epsilon_2]$$
$$+ h_3^{-2}[(\partial T_3/\partial \xi_3)^2 - \epsilon_3] = 0. \tag{3.8.6}$$

This equation, however, cannot be always directly separated because h_1, h_2, and h_3 are, in general, functions of all three coordinates ξ_i. In many important curvilinear orthogonal coordinate systems, however, (3.8.6) can be simply modified to yield a separable equation. This may be achieved, for example, by suitable multiplications.

As an example, we shall consider spherical polar coordinates $\xi_1 = r$, $\xi_2 = \theta$, and $\xi_3 = \varphi$; see Section 3.5.4. Then, $h_1 = 1$, $h_2 = r$, and $h_3 = r\sin\theta$. Expressing (3.8.6) in spherical polar coordinates and multiplying it by r^2, we obtain

$$r^2\left[\left(\frac{\partial T_r}{\partial r}\right)^2 - \epsilon_1(r)\right] + \left[\left(\frac{\partial T_\theta}{\partial \theta}\right)^2 - \epsilon_2(\theta)\right]$$
$$+ \frac{1}{\sin^2\theta}\left[\left(\frac{\partial T_\varphi}{\partial \varphi}\right)^2 - \epsilon_3(\varphi)\right] = 0. \tag{3.8.7}$$

The first term is a function of r only, and the two next terms are functions of θ and φ only. Thus,

$$r^2\left[\left(\frac{\partial T_r}{\partial r}\right)^2 - \epsilon_1(r)\right] = -\alpha_1,$$

$$\left[\left(\frac{\partial T_\theta}{\partial \theta}\right)^2 - \epsilon_2(\theta)\right] + \frac{1}{\sin^2\theta}\left[\left(\frac{\partial T_\varphi}{\partial \varphi}\right)^2 - \epsilon_3(\varphi)\right] = \alpha_1.$$

Here α_1 is an arbitrary constant. Multiplying the second equation by $\sin^2\theta$, we obtain three separated equations:

$$\left(\frac{\partial T_r}{\partial r}\right)^2 - \epsilon_1(r) = -\frac{\alpha_1}{r^2}, \qquad \left(\frac{\partial T_\varphi}{\partial \varphi}\right)^2 - \epsilon_3(\varphi) = \alpha_2,$$

$$\left(\frac{\partial T_\theta}{\partial \theta}\right)^2 - \epsilon_2(\theta) = \alpha_1 - \frac{\alpha_2}{\sin^2\theta}.$$

The complete solution of the eikonal equation in spherical polar coordinates r, θ, and φ is then as follows:

$$T(r, \theta, \varphi) = \pm\int_{r_0}^{r}[\epsilon_1(r) - \alpha_1/r^2]^{1/2}dr \pm \int_{\varphi_0}^{\varphi}[\epsilon_3(\varphi) + \alpha_2]^{1/2}d\varphi$$
$$\pm\int_{\theta_0}^{\theta}[\epsilon_2(\theta) + \alpha_1 - \alpha_2/\sin^2\theta]^{1/2}d\theta + \alpha_3. \tag{3.8.8}$$

In a similar way, we can construct the complete solutions of the eikonal equation in many other orthogonal curvilinear coordinate systems ξ_i, for which the velocity distribution is specified by (3.8.5).

As soon as the complete solution $T(\xi_i)$ in the curvilinear orthogonal coordinates ξ_i is known in the form analogous to (3.8.8), we can compute the components of the slowness vector by equations (3.5.4). Similarly, we can obtain the ray equations from the complete solution of the eikonal equation. Without a derivation, we shall present here the final ray equations:

$$\partial T(\xi_1, \xi_2, \xi_3, \alpha_1, \alpha_2)/\partial \alpha_1 = \beta_1,$$
$$\partial T(\xi_1, \xi_2, \xi_3, \alpha_1, \alpha_2)/\partial \alpha_2 = \beta_2. \tag{3.8.9}$$

The ray equations (3.8.9) contain two new constants, β_1 and β_2. If (3.8.9) are resolved for ξ_2 and ξ_3, the ray equations can be expressed in explicit form:

$$\xi_2 = \xi_2(\xi_1, \alpha_1, \alpha_2, \beta_1, \beta_2), \qquad \xi_3 = \xi_3(\xi_1, \alpha_1, \alpha_2, \beta_1, \beta_2). \tag{3.8.10}$$

There are four free parameters in ray equations (3.8.10), so that the complete system of rays is four-parameteric. See more details on systems of rays in Section 3.10.

Assume now that one coordinate, say x_2, is cyclic, so that $\epsilon_2(x_2) = 0$. Then, $T_2(x_2) = (x_2 - x_{20})p_2$, where $p_2 = \pm\sqrt{\alpha_2}$. From (3.8.4), we obtain

$$T(x_1, x_2, x_3) = p_2 x_2 \pm \int_{x_{10}}^{x_1} [\epsilon_1(x_1) + \alpha_1]^{1/2} dx_1$$
$$\pm \int_{x_{30}}^{x_3} [\epsilon_3(x_3) - \alpha_1 - p_2^2]^{1/2} dx_3 + \alpha_3. \tag{3.8.11}$$

Similarly, for two cyclic coordinates, x_1 and x_2, we have $\epsilon_1(x_1) = \epsilon_2(x_2) = 0$, and the complete solution of the eikonal equation is

$$T(x_1, x_2, x_3) = p_1 x_1 + p_2 x_2 \pm \int_{x_{30}}^{x_3} [\epsilon_3(x_3) - p_1^2 - p_2^2]^{1/2} dx_3 + \alpha_3. \tag{3.8.12}$$

Let us now consider the simplest case of a 1-D, vertically inhomogeneous medium. If we put $p_2 = 0$, and $\alpha_3 = 0$, (3.8.12) exactly coincides with (3.7.15), derived in a different way. Note that the integral on the RHS of (3.8.12) represents the delay time τ in this case. Consequently, the delay times for various coordinate systems for one or two cyclic coordinates can be simply constructed from the complete solution of the eikonal equation.

Instead of the complete separation of variables, we can also perform an *incomplete separation*. The incomplete separation of variables reduces the number of independent variables in the eikonal equation. This may be useful particularly if one coordinate, say x_2, is cyclic. Then, the eikonal equation may be applied to the travel-time function $T(x_1, x_2, x_3)$ given by

$$T(x_1, x_2, x_3) = p_2 x_2 + \tilde{T}(x_1, x_3). \tag{3.8.13}$$

An analogous equation can often be used even in curvilinear coordinates.

3.8.3 Network Shortest-Path Ray Tracing

Let us consider point source S and receiver R and construct various curves connecting S and R. According to Fermat's minimum-time principle, the first-arrival travel time is the minimum time over all possible paths connecting S and R.

Efficient ways of finding the minimum travel time from S to R and the relevant trajectories on a discrete grid of points are based on the *theory of graphs*. The trajectory corresponding to the minimum time is usually called the *shortest path*, where "shortest" means the minimum travel time. For this reason, the methods described in this section are also called the *shortest-path methods*.

The model is represented by a discrete grid of points at which the velocities are specified. By *graph*, we understand the mathematical object, composed of *grid points* (representing *nodes*) and their *connections* (called *edges* or *arcs*). The graph becomes a *network* when weights are assigned to the connections. In our case, the weights are taken to be equal to the travel times between two connected points. The shortest path in the network may then be interpreted as an approximation to the *seismic ray* due to Fermat's principle. We also speak of *network rays* and call the whole procedure *network ray tracing*.

Each node of the network may be connected only with a limited number of nodes in the neighborhood, but not with the nodes that lie farther away. To specify these connections, a very important concept of the forward star has been introduced. The *forward star corresponding to an arbitrarily selected node* of the network is the set of other nodes with which the node is connected. The forward star may be constructed in various ways, depending on the problem under consideration and on other conditions. The forward stars corresponding to different nodes may be different. They should be small in regions of fast changes of velocities, and in regions close to interfaces. In regions of smooth velocity changes, they may be larger.

In the theory of graphs, a very efficient algorithm to determine the shortest path in networks was proposed by Dijkstra (1959). Various modifications of this algorithm have been used broadly in network ray tracing.

The first references related to network ray tracing are probably those of Nakanishi and Yamaguchi (1986), Moser (1991, 1992), and Saito (1989). The proposed algorithms have been further developed and generalized and/or modified, for example, in papers by Mandal (1992), Asakawa and Kawanaka (1993), Cao and Greenhalgh (1993), Fischer and Lees (1993), Klimeš and Kvasnička (1994), and Cheng and House (1996).

A very detailed and tutorial treatment can be found in Moser (1992), which is recommended for further reading. The same reference also discusses, in great detail, many applications of network ray tracing in seismology and in seismic exploration. See also Nolet and Moser (1993) for applications.

In large 3-D models, the network ray tracing may be rather time-consuming. It would be suitable to decrease the size of the model. If the first estimate of the ray Ω connecting points S and R is known, it may be useful to consider a Fresnel volume-like model connected with the ray estimate and to perform network ray tracing computations only in it. The exact size and shape of the model is not too decisive; it is only required that the model be chosen sufficiently broad to include the ray update.

In network ray tracing, it is very important to determine *estimates of the maximum error* in computing the first-arrival travel times. A detailed analysis of this error is given in Klimeš and Kvasnička (1994). The authors propose a suitable way to estimate the maximum error of computations in network ray tracing and to minimize the error by optimizing the sizes of the forward star. Their computer programs also yield, in addition to first-arrival travel times and network rays, estimates of these errors.

Note that the method of network ray tracing can also be generalized for 3-D inhomogeneous *anisotropic* media.

3.8.4 Finite-Difference Method

In fact, the finite-difference method described here has nothing in common with the finite-difference method of solving linear partial differential equations, such as the acoustic wave equation or the elastodynamic equation. It is usually specified in the seismological literature as the *finite-difference solver of the eikonal equation*. This, however, does not mean that the classical finite-difference method is applied to the eikonal equation. We must remember that the eikonal equation is a *nonlinear* equation, and that its direct finite-difference solution is not simple. The finite-difference method discussed here uses certain important consequences of the eikonal equation, as well as of the consequences of some other concepts (such as the Huygens principle). The direct numerical solution of the eikonal equation has been used only exceptionally; see Pilipenko (1979, 1983) for details, a description of the algorithm, and numerical examples of applications.

The method of finite differences for computing the first-arrival travel times along an *expanding square* (in 2-D) and along an *expanding cube* (in 3-D) was proposed by Vidale (1988, 1989, 1990). In the method, the first-arrival travel times at points situated on the expanded square (cube) are computed from the first-arrival travel times known at points situated on the original squares (cubes). The expansion equations are different, depending on the position of the point on the square, and on its distance from the source. At large distances from the source, the expansion is performed by a local plane-wave approximation.

The main principles of the finite-difference computation of first-arrival travel times can be very simply demonstrated on the case of an *expanding halfspace*. Reshef and Kosloff (1986) were the first to propose this algorithm, mainly for applications in seismic exploration reflection methods. They consider only *one-way propagation* (forward continuation). In 2-D, the eikonal equation $(T_{,x})^2 + (T_{,z})^2 = V^{-2}$ is expressed in explicit form:

$$T_{,z} = +\left[1/V^2 - (T_{,x})^2\right]^{1/2}. \tag{3.8.14}$$

The plus sign corresponds to the forward continuation, so that the algorithm yields first-arrival travel times. The algorithm to solve (3.8.14) follows:

   a. The derivative $T_{,x} = \partial T(x, z)/\partial x$ along level $z = $ const. is evaluated by finite differences, assuming $T(x, z)$ along this level is known.
   b. The depth continuation (along z) is realized from one level to another by the Runge-Kutta method of the fourth order.

The initial values of T for $z = 0$ have the form $T(x, z = 0) = T^0(x)$ (the initial travel-time curves or the source). Function $T^0(x)$ may be obtained by standard ray tracing in the known model.

The same algorithm can also be used for the backward continuation (decreasing time). In this case, it would be necessary to use the minus sign in (3.8.14). The method was extended to 3-D by Reshef (1991). The assumption of one-way propagation automatically eliminates the waves propagating in the opposite direction.

The next "finite-difference (FD)" method, mentioned earlier, was proposed by Vidale (1988, 1989, 1990) for 2-D and 3-D models with an arbitrarily situated point source. The first-arrival travel times are computed along a discrete *expanding square* (in 2-D) or along an *expanding cube* (in 3-D). For FD computations along expanding circles or spheres (in polar coordinates), see Schneider (1995).

The expanding-square procedure has certain *drawbacks*. For example, it does not give sufficiently accurate results in regions of higher velocity contrast. It does not yield any estimate of the error of computations. The approximations used are not sufficiently accurate in situations in which the wavefront is strongly curved (for example, in the vicinity of a point source). To eliminate these problems, the algorithms proposed by Vidale have recently been generalized and extended in many ways. Similar algorithms have also been proposed for anisotropic media. We shall not give any details but only refer to published papers. See Qin et al. (1992, 1993), van Trier and Symes (1991), Podvin and Lecomte (1991), Matsuoka and Ezaka (1992), Schneider et al. (1992), Lecomte (1993), Eaton (1993), Li and Ulrych (1993a, 1993b), Cao and Greenhalgh (1994), Faria and Stoffa (1994), Riahi and Juhlin (1994), Schneider (1995), Hole and Zelt (1995), and Klimeš (1996). Among all these extensions and modifications, the algorithms proposed by Podvin and Lecomte (1991) should be particularly emphasized. The algorithms work well even in regions containing large velocity contrasts on structural interfaces of arbitrary shape. Instead of finite differences, the finite elements also were used; see Daley, Marfurt, and McCarron (1999).

In the method described, only the first-arrival travel times, not the rays, are calculated. If the rays are needed for some other purpose, they must be calculated a posteriori from known travel times. It should, however, be mentioned that the method yields rays of different elementary waves in different regions of the model.

The main drawback of the FD eikonal solver is that only the first-arrival times are computed. Recently, some new hybrid extensions of the FD solver have been proposed; they can also be used in multivalued travel-time computations. These extensions combine global standard ray tracing with the local FD solution of the eikonal equation. See the "big ray tracing" of Benamou (1996) and Abgrall and Benamou (1999).

In conclusion, we can say that the finite-difference method is the fastest method of computing the first-arrival travel times. The accuracy of most present algorithms, however, is lower than the accuracy of recent versions of the network ray tracing algorithm, particularly in 3-D structures. See Klimeš and Kvasnička (1994).

3.8.5 Wavefront Construction Method

The purpose of the wavefront construction method is to compute successively the wavefronts of an elementary wave under consideration for travel times $T = T_0 + k\Delta T$, $k = 1, 2, \ldots$, starting from the initial wavefront $T = T_0$. The current kth wavefront $T = T_0 + k\Delta T$ is constructed from the previous $(k-1)$th wavefront $T = T_0 + (k-1)\Delta T$ using short elements of rays calculated by ray tracing. The number of rays in the method is not fixed but is adjusted at each wavefront (for each k). The new rays at a current wavefront are introduced as soon as some imposed criteria are not satisfied. The criteria may include a too large distance between neighboring rays, or a large difference in the directions of two neighboring rays. The initial conditions for new rays at the wavefront are determined from neighboring rays using some sort of interpolation. The wavefront construction method has been proposed and successfully applied to 3-D laterally varying layered structures with a smooth velocity distribution in individual layers by Vinje et al. (1992, 1993, 1996), Coultrip

(1993), Ettrich and Gajewski (1996), Lucio, Lambaré, and Hanyga (1996), and Lambaré, Lucio, and Hanyga (1996). It may, however, be easily applied even to grid models. The travel times at grid points in the individual small cells between succeeding wavefronts and neighboring rays are determined by interpolation.

There is a basic difference between the wavefront construction method, described here, and the classical time-field method, described at the beginning of Section 3.8. In the wavefront construction method, the current wavefront is obtained from the previous wavefront by ray tracing short ray elements. In the time-field method, the current wavefront is obtained from the previous one by using Huygens principle, that is, by constructing the envelope surface to spheres with the centers distributed along the previous wavefront. The time-field method can easily be realized graphically in 2-D models, using a pair of compasses. Its computer realization, however, is difficult, particularly in 3-D models. Consequently, the wavefront construction method is considerably more suitable for computer treatment than the time-field method.

Thus, the wavefront construction method is based on a computation of rays and wavefronts. It differs both from standard ray tracing and from direct travel-time computation. Consequently, it is to some extent questionable whether it should be included in the section devoted to the direct computation of travel times because it uses short ray segments to construct the wavefronts successively. See also a brief discussion and comparison of the wavefront construction method with the controlled shooting method in Section 3.11.2.

3.8.6 Concluding Remarks

In Sections 3.8.3 and 3.8.4, we have described the *grid computations* of first-arrival travel times. It is, however, necessary to emphasize that even the *ray-theory travel times* of selected elementary waves can be evaluated in grid models. As shown in Section 3.8.5, the wavefront construction method is suitable for this purpose. Several other methods suitable for such computations are described in Section 3.11.2. Let us name, among others, the controlled ray tracing supplemented by ray paraxial approximation or by weighting of ray paraxial approximations.

As discussed in Sections 3.8.3 and 3.8.4, the methods of finite differences and of shortest-path ray tracing are suitable for computing first-arrival travel times in grid models. They are, however, not as suitable if we also wish to compute geometrical spreading and amplitudes; see Section 3.10. These quantities are usually computed along rays, but the rays are not known in the first-arrival travel-time grid computations. There are two options in treating this problem:

a. To compute the rays a posteriori from travel times and to calculate the geometrical spreading and ray amplitudes along these rays.
b. To calculate the geometrical spreading and ray amplitudes directly from the travel-time field.

Theoretically, it is possible to calculate the geometrical spreading from the travel-time field of a specified elementary wave. Two such methods are described in Section 4.10.5; another is discussed in Vidale and Houston (1990). It is also possible to solve the transport equation by finite differences; see Buske (1996). All these direct methods, however, require the knowledge of the *second spatial derivatives of the travel-time field*, which must be determined numerically from known travel times at grid points. Because the accuracy of the travel times is usually not high, the numerical determination of the second derivatives of travel times may be rather problematic, particularly in 3-D models. Moreover, there may be additional problems with reflection/transmission coefficients at structural interfaces in case of layered and block models.

There is, however, another great danger in calculating amplitudes from grid computations of first-arrival travel times. Usually, the zeroth-order ray approximation is used for calculation, but the waves arriving first are not necessarily zeroth-order elementary waves, at least in certain regions. Simple examples are head waves and diffracted waves penetrating into shadow zones. Thus, we would apply the zeroth-order approximation equations to compute amplitudes of waves that are not zeroth-order elementary waves. This, of course, could yield unpredictable errors.

The preceding problems play an important role only in first-arrival travel time computations such as finite differences and network ray tracing. In the method of wavefront construction (see Section 3.8.5), the geometrical spreading and ray amplitudes can be calculated quite safely along the rays of the elementary wave under consideration.

3.9 Perturbation Methods for Travel Times

The methods of ray tracing and travel-time computations described in this chapter are simple and straightforward in principle and can be used to solve direct kinematic problems in any laterally inhomogeneous, three-dimensional, isotropic or anisotropic, layered and block structures. It would, however, be rather time-consuming and cumbersome to use the methods to solve inverse kinematic problems by numerical modeling.

A simple procedure to solve both direct and inverse kinematic problems in inhomogeneous, isotropic or anisotropic structures is based on perturbation theory. Perturbation methods can be used to solve even more complex problems of seismic wave fields, not just the problem of computing travel times; see Sections 2.6.2 and 4.7.4. In this section, however, we shall discuss only the first-order travel-time perturbations. The application of perturbation methods to travel times is particularly attractive in the solution of the kinematic inverse problem.

There are three main approaches to deriving first-order perturbation equations for travel times. *The first approach*, most common in the seismological literature, is based on Fermat's principle. It exploits the fact that the ray is a curve that renders Fermat's functional stationary. Thus, in the first-order perturbation theory for travel times, the ray-path changes can be ignored because they are of the second order. See, for example, Aki and Richards (1980, p. 797) and Nolet (1987) for isotropic media and Chapman and Pratt (1992) for anisotropic media. *The second approach* is based directly on the eikonal equation and does not exploit Fermat's principle at all. See Romanov (1972, 1978) for isotropic media and Červený (1982a), Červený and Jech (1982), Hanyga (1982b), and Jech and Pšenčík (1989) for anisotropic media. The procedure becomes particularly simple if the eikonal equation is expressed in Hamiltonian form, and the problem is solved in terms of canonical coordinates x_i, p_i in 6-D phase space. See Farra and Madariaga (1987), Nowack and Lutter (1988), Farra, Virieux, and Madariaga (1989), Farra and Le Bégat (1995), and Farra (1999). *The third approach* is based on a Lagrangian formulation. As a starting point, it uses the Euler-Lagrange equation of rays, (3.1.37) or (3.1.40), expressed in terms of x_i and $x_i' = \mathrm{d}x_i/\mathrm{d}u$. For a good description, see Snieder and Sambridge (1992), Sambridge and Snieder (1993), Snieder and Spencer (1993), Snieder and Aldridge (1995), and Snieder and Lomax (1996). Note that the second and third approaches can be used more broadly in the ray perturbation theory, not just in the derivation of first-order perturbation equations for travel times; see Section 4.7.4.

The derivation of the first-order perturbation equations for travel times presented in this section is based mainly on the eikonal equation, expressed in Hamiltonian form

$\mathcal{H}(x_i, p_i) = 0$. The results may be applied both to isotropic and anisotropic media, including the singular directions in anisotropic media. They also consider nonfixed end points of the ray and perturbations of structural interfaces. The derivation, close to that given by Farra and Le Bégat (1995), is very general, objective, and straightforward.

3.9.1 First-Order Perturbation Equations for Travel Times in Smooth Media

Let us consider an elementary wave propagating in a smooth reference medium $\mathcal{M}^0$, characterized by Hamiltonian $\mathcal{H}^0$. Similarly, as in Section 2.6.2, reference medium $\mathcal{M}^0$ is also called the background or nonperturbed medium. The medium $\mathcal{M}^0$ may be isotropic or anisotropic, and the elementary wave may be P or S in an isotropic medium or qP, qS1, or qS2 in an anisotropic medium. The results will be generalized for layered media containing structural interfaces in Section 3.9.5.

We further consider reference ray Ω^0 in $\mathcal{M}^0$, connecting two points S and R. We introduce monotonic parameter u along Ω^0 and denote by $x_i^0(u)$ and $p_i^0(u)$ the Cartesian coordinates of points and Cartesian components of the slowness vector along reference ray Ω^0. Monotonic parameter u along Ω^0 cannot be chosen arbitrarily; it is determined by the form of the Hamiltonian $\mathcal{H}^0$ under consideration. At points S and R, monotonic parameter u takes the values u_S and u_R. Then travel time $T^0(x_i^0(u_R), x_i^0(u_S))$ from S to R along Ω^0 in reference medium $\mathcal{M}^0$ is given by the integral

$$T^0\left(x_i^0(u_R), x_i^0(u_S)\right) = \int_{u_S}^{u_R} p_i^0(\partial\mathcal{H}^0/\partial p_i^0)\mathrm{d}u = \int_{u_S}^{u_R} p_i^0\dot{x}_i^0\mathrm{d}u; \qquad (3.9.1)$$

see (3.1.3). Here $\dot{x}_i^0 = \mathrm{d}x_i^0/\mathrm{d}u$.

Now we shall consider perturbed model $\mathcal{M}$, which differs only slightly from reference model $\mathcal{M}^0$, and denote the relevant Hamiltonian $\mathcal{H}$. We introduce the model perturbations $\Delta\mathcal{H}$ of the Hamiltonian by relation $\mathcal{H} = \mathcal{H}^0 + \Delta\mathcal{H}$. Consider ray Ω in perturbed medium $\mathcal{M}$, which deviates only slightly from reference ray Ω^0 in background medium $\mathcal{M}^0$. See Figure 3.8. We define ray Ω by parameteric equation $x_i(u) = x_i^0(u) + \Delta x_i(u)$, where u is

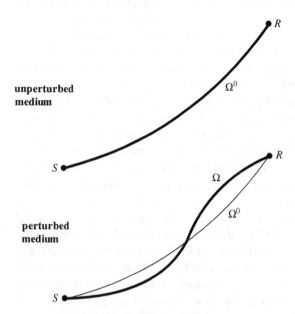

Figure 3.8. Ray Ω^0 from S to R in the unperturbed (background) medium, and the relevant ray Ω from S to R in the perturbed medium. The first-order travel-time perturbations $\Delta T(R, S)$ due to perturbations of Hamiltonian $\Delta\mathcal{H}$ can be calculated in terms of quadratures of $-\Delta\mathcal{H}$ along the reference ray Ω^0 from S to R. It is not necessary to determine ray Ω.

the monotonic parameter introduced along reference ray Ω^0. We allow $\Delta x_i(u_S) \neq 0$ and $\Delta x_i(u_R) \neq 0$, so that end points S' and R' of perturbed ray Ω may be different from end points S and R of reference ray Ω^0. Similarly, the components of the slowness vector along Ω are given by relations $p_i(u) = p_i^0(u) + \Delta p_i(u)$. We also introduce $\Delta \dot{x}_i(u)$ with relation $\dot{x}_i(u) = \dot{x}_i^0(u) + \Delta \dot{x}_i(u)$. Considering only the first-order perturbations, we can express travel time $T(x_i(u_R), x_i(u_S))$ along perturbed ray Ω in perturbed medium $\mathcal{M}$ as follows:

$$T(x_i(u_R), x_i(u_S)) = \int_{u_S}^{u_R} p_i \dot{x}_i \, du$$

$$\doteq \int_{u_S}^{u_R} \left(p_i^0 \dot{x}_i^0 + p_i^0 \Delta \dot{x}_i + \dot{x}_i^0 \Delta p_i \right) du. \tag{3.9.2}$$

We shall now express $\dot{x}_i^0 \Delta p_i$ in (3.9.2) in terms of $\Delta \mathcal{H}(x_i^0, p_i^0)$. We use the expansion

$$\mathcal{H}(x_i, p_i) = \mathcal{H}^0(x_i, p_i) + \Delta \mathcal{H}(x_i, p_i)$$

$$= \mathcal{H}^0(x_i^0, p_i^0) + (\partial \mathcal{H}^0 / \partial x_i^0) \Delta x_i$$

$$+ (\partial \mathcal{H}^0 / \partial p_i^0) \Delta p_i + \Delta \mathcal{H}(x_i^0, p_i^0)$$

$$= \mathcal{H}^0(x_i^0, p_i^0) - \dot{p}_i^0 \Delta x_i + \dot{x}_i^0 \Delta p_i + \Delta \mathcal{H}(x_i^0, p_i^0).$$

Here we have used $\Delta \mathcal{H}(x_i, p_i) \doteq \Delta \mathcal{H}(x_i^0, p_i^0)$ because we are considering only first-order perturbations. Also, $\dot{p}_i^0(u) = \mathrm{d}p_i^0/\mathrm{d}u$. If we insert $\mathcal{H}(x_i, p_i) = 0$ and $\mathcal{H}^0(x_i^0, p_i^0) = 0$, we obtain

$$\dot{x}_i^0 \Delta p_i = \dot{p}_i^0 \Delta x_i - \Delta \mathcal{H}(x_i^0, p_i^0),$$

$$p_i \dot{x}_i = p_i^0 \dot{x}_i^0 - \Delta \mathcal{H}(x_i^0, p_i^0) + \mathrm{d}(p_i^0 \Delta x_i)/\mathrm{d}u.$$

Using these relations in (3.9.2) yields

$$T(x_i(u_R), x_i(u_S)) = T^0(x_i^0(u_R), x_i^0(u_S))$$

$$- \int_{u_S}^{u_R} \Delta \mathcal{H}(x_i^0, p_i^0) \, du + \Delta T^e(u_R, u_S), \tag{3.9.3}$$

where $\Delta T^e(u_R, u_S) = \Delta T^e(R, S)$ is the *end-point travel-time contribution*, given by the relation

$$\Delta T^e(R, S) = p_i^0(R) \Delta x_i(R) - p_i^0(S) \Delta x_i(S). \tag{3.9.4}$$

For fixed end points S and R, we have $\Delta x_i(R) = 0$ and $\Delta x_i(S) = 0$, and $\Delta T^e(R, S)$ vanishes. $\Delta T^e(R, S)$ will be discussed in more detail later.

For fixed points S and R, travel-time perturbation $\Delta T(R, S)$ is introduced as follows:

$$\Delta T(R, S) = T(x_i(u_R), x_i(u_S)) - T^0(x_i^0(u_R), x_i^0(u_S)). \tag{3.9.5}$$

In the first-order perturbation theory, $\Delta T(R, S)$ is given by the relation

$$\Delta T(R, S) = - \int_{\Omega^0(R,S)} \Delta \mathcal{H}(x_i^0, p_i^0) \, du; \tag{3.9.6}$$

see (3.9.3). This is the final relation, valid both for isotropic and anisotropic media. The integration is taken along reference ray $\Omega^0(R, S)$ in nonperturbed medium $\mathcal{M}^0$, from point S to R. The relation remains valid even in layered media with fixed, nonperturbed structural interfaces; see Section 3.9.5. If the structural interfaces are also perturbed, it is necessary to add an additional term to (3.9.6). See Section 3.9.5 for its derivation.

One thing should be emphasized at this point. In the derivation of $\Delta T(R, S)$, we have used perturbations Δx_i and Δp_i, specifying perturbed ray Ω and slowness vector $\vec{p}$ along it. The final result (3.9.6), however, does not depend at all on Δx_i and Δp_i; the integration is performed along reference ray Ω^0 in unperturbed medium $\mathcal{M}^0$. To calculate $\Delta T(R, S)$, the only quantity that should be known along Ω^0 is the structural perturbation of Hamiltonian $\Delta \mathcal{H}(x_i^0, p_i^0)$, for x_i^0 and p_i^0 fixed along Ω^0. The derivation of Δx_i and Δp_i themselves is more complex and requires dynamic ray tracing. See Section 4.7.4.

Let us now briefly explain the end-point travel-time contribution $\Delta T^e(R, S)$, given by (3.9.4). It does not depend on medium perturbations and merely expresses the change of the travel time if the end points of ray Ω^0 are slightly shifted. Consider two points, S' and R', situated close to S and R. Expression $\Delta T^e(R, S)$ can then be used to calculate the travel time from S' to R', if the travel time from S to R is known:

$$
\begin{aligned}
T(R', S') &\doteq T(R, S) + \Delta T^e(R, S) \\
&= T(R, S) + p_i^0(R)\Delta x_i(R) - p_i^0(S)\Delta x_i(S).
\end{aligned}
\tag{3.9.7}
$$

Here $\Delta x_i(R)$ and $\Delta x_i(S)$ express the shifts from R to R' and from S to S', respectively. Consequently, it is not necessary to perform new ray tracing from S' to R' if we wish to determine $T(R', S')$. Equation (3.9.7) is, however, valid only up to linear terms $\Delta x_i(S)$ and $\Delta x_i(R)$. A more general expression for $T(R', S')$, quadratic in $\Delta x_i(R)$ and $\Delta x_i(S)$, will be derived in Section 4.9.2. It will, however, require dynamic ray tracing along Ω^0 and computing the ray propagator matrix.

Equation (3.9.7) can be used both for isotropic and anisotropic media. We shall not discuss it anymore and consider only fixed end points S and R of Ω^0 in the next sections.

3.9.2 Smooth Isotropic Medium

In isotropic media, it is common to consider the arclength s as an independent variable u along reference ray Ω^0 in background model $\mathcal{M}^0$. The relevant Hamiltonian is given by the relation $\mathcal{H}(x_i, p_i) = \sqrt{p_i p_i} - 1/V$; see (3.1.6) for $n = 1$. The perturbation $\Delta \mathcal{H}(x_i^0, p_i^0)$ of Hamiltonian is then given by the relation

$$
\Delta \mathcal{H}(x_i^0, p_i^0) = -\Delta(1/V).
\tag{3.9.8}
$$

The first-order travel-time perturbation is as follows:

$$
\Delta T(R, S) = \int_{\Omega^0(R,S)} \Delta(1/V)\mathrm{d}s;
\tag{3.9.9}
$$

see (3.9.6). This is a famous relation, well known from most seismological textbooks. It is valid both for P waves (with $V = \alpha$) and for S waves ($V = \beta$). Its alternative derivation, based on Fermat's principle, is elementary. Considering $T(R, S) = \int_S^R (1/V)\mathrm{d}s$, where the integration is performed along the ray Ω^0, and taking into account that the ray-path changes can be ignored, we immediately obtain (3.9.9) for $\Delta T(R, S)$.

Relation (3.9.9) is valid even in layered isotropic media, when the position of the interfaces is fixed. See Firbas (1984) and Section 3.9.5. An additional term to (3.9.9) for layered medium with perturbed interfaces will be derived in Section 3.9.5.

Equation (3.9.9) plays a very important role in the solution of inverse kinematic problems of isotropic laterally varying media, particularly in tomographic methods. Its attractive feature is that it gives a linear relationship between ΔT and $\Delta(1/V)$. Consequently, it can be used to determine the slowness perturbations $\Delta(1/V)$ from the measured travel-time perturbations. We must, however, remember that (3.9.9) is only approximate, based on first-order perturbation theory. Therefore, it should be used iteratively in practical applications. Actually, the relation between $\Delta T(R, S)$ and $\Delta(1/V)$ is nonlinear. For this reason, (3.9.9) is also often called the *linearized travel-time equation*.

3.9.3 Smooth Anisotropic Medium

We shall consider the anisotropic Hamiltonian (3.6.3)

$$\mathcal{H}(x_i, p_i) = \tfrac{1}{2}[G_m(x_i, p_i) - 1]. \tag{3.9.10}$$

Here $G_m(x_i, p_i)$ is an eigenvalue of the Christoffel matrix $\Gamma_{ik} = a_{ijkl}p_j p_l$, and $\vec{g}^{(m)}$ is the relevant eigenvector. The index m specifies the type of the elementary wave under consideration: $m = 1$ for qS1 wave, $m = 2$ for qS2 wave, and $m = 3$ for qP wave. We assume that the eigenvalue $G_m(x_i, p_i)$ under consideration is not equal or close to any other eigenvalue $G_k(x_i, p_i), k \neq m$, at any point of the ray Ω^0. Hamiltonian (3.9.10) corresponds to monotonic parameter $u = T$ along ray Ω^0. Equation (3.9.10) yields

$$\Delta\mathcal{H}(x_i^0, p_i^0) = \tfrac{1}{2}\Delta G_m(x_i^0, p_i^0). \tag{3.9.11}$$

To determine $\Delta G_m(x_i^0, p_i^0)$, we shall exploit equation $(\Gamma_{jk} - G_m \delta_{jk})g_k^{(m)} = 0$. Taking first-order perturbation, we obtain the basic equation for ΔG_m:

$$(\Gamma_{jk} - G_m\delta_{jk})\Delta g_k^{(m)} + (\Delta\Gamma_{jk} - \Delta G_m\delta_{jk})g_k^{(m)} = 0. \tag{3.9.12}$$

To simplify the notation, we shall not use zeros in superscripts to emphasize that the quantities are taken at Ω^0. We must, however, remember that all quantities in (3.9.12) and in the following equations are taken at Ω^0. Consequently, we also have

$$\Delta\Gamma_{jk} = \Delta a_{ijkl}p_j p_l. \tag{3.9.13}$$

Multiplying (3.9.12) by $g_j^{(m)}$ yields

$$(\Delta\Gamma_{jk} - \Delta G_m\delta_{jk})g_k^{(m)}g_j^{(m)} = 0,$$

as $(\Gamma_{jk} - G_m\delta_{jk})g_j^{(m)} = 0$. Since $\vec{g}^{(m)}$ is a unit vector, $g_k^{(m)}g_k^{(m)} = 1$; hence,

$$\Delta G_m(x_i, p_i) = \Delta\Gamma_{jk}g_k^{(m)}g_j^{(m)} = \Delta a_{ijkl}p_i p_l g_j^{(m)}g_k^{(m)}. \tag{3.9.14}$$

Consequently,

$$\Delta T(R, S) = -\tfrac{1}{2}\int_{\Omega^0(R,S)} \Delta a_{ijkl}p_i p_l g_j^{(m)}g_k^{(m)}\,dT. \tag{3.9.15}$$

The integration is performed along reference ray Ω^0 in $\mathcal{M}^0$ from S to R, and all quantities in (3.9.15) are taken along Ω^0. Integration variable dT can be expressed in terms of ds, $dT = ds/\mathcal{U}$, where s is the arclength along Ω^0 and $\mathcal{U}$ is the group velocity. Thus, the alternative form of (3.9.15) is as follows:

$$\Delta T(R, S) = -\tfrac{1}{2}\int_{\Omega^0(R,S)} \Delta a_{ijkl}p_i p_l g_j^{(m)}g_k^{(m)}\mathcal{U}^{-1}\,ds. \tag{3.9.16}$$

Equation (3.9.15) was first derived (in a slightly different form than given here) from the eikonal equation by Červený (1982a), using Romanov's ideas (Romanov 1972, 1978). An alternative derivation was given by Hanyga (1982b). For a very detailed discussion of (3.9.15) and for its specification for many special anisotropic situations, see Červený and Jech (1982). Numerical examples can be found in Červený and Firbas (1984) and in Firbas (1984). An independent derivation, based on Fermat's principle, was given by Chapman and Pratt (1992). It will be also derived in Section 5.4.6 using the quasi-isotropic approximation.

It is straightforward to derive from (3.9.15) and (3.9.16) expressions for the perturbations of the phase and group velocities at any point of the ray. Note that the expressions for the perturbation of phase velocity in a homogeneous anisotropic medium were first derived as early as in 1960s in a classical paper by Backus (1965).

In (3.9.15) and (3.9.16), both reference medium $\mathcal{M}^0$ and perturbed medium $\mathcal{M}$ are, in general, anisotropic. For qP waves ($m = 3$), the equations may be used quite universally, even for reference medium $\mathcal{M}^0$ isotropic and perturbed medium $\mathcal{M}$ anisotropic (weak anisotropy). For both $\mathcal{M}^0$ and $\mathcal{M}$ isotropic, (3.9.16) yields (3.9.9) for qP waves. For qS waves, however, equations (3.9.15) and (3.9.16) fail if the two eigenvalues G_1 and G_2 of the two qS waves are equal or very close to each other along some part of reference ray Ω^0 or along the whole ray Ω^0 in unperturbed medium $\mathcal{M}^0$. The reason is that eigenvectors $\vec{g}^{(1)}$ and $\vec{g}^{(2)}$ cannot be defined uniquely for $G_1 = G_2$. See the detailed treatment of this case in the next section.

As we can see in (3.9.15) and (3.9.16), the relation between travel-time perturbation $\Delta T(R, S)$ and perturbations of density-normalized elastic parameters Δa_{ijkl} is linear even in anisotropic media. For this reason, (3.9.15) and (3.9.16) again represent linearized travel-time equations. The equations can be suitably used in the solution of inverse kinematic problems of anisotropic media, particularly in tomographic studies. See Chapman and Pratt (1992), Jech and Pšenčík (1992), and Pratt and Chapman (1992).

3.9.4 Degenerate Case of qS Waves in Anisotropic Media

If the eigenvalues G_1 and G_2 of the two qS waves are equal or very close to each other along the whole ray Ω^0, or along some part of it in unperturbed medium $\mathcal{M}^0$, the linearized travel-time equations for $\Delta T(R, S)$, derived in Section 3.9.3, fail. There are two important situations when such difficulties appear in practical applications. Globally, G_1 equals G_2 along the whole ray Ω^0 if reference medium $\mathcal{M}^0$ is isotropic. We then speak of the *quasi-isotropic case*. Locally, G_1 equals G_2, or is close to it when the direction of slowness vector $\vec{p}$ at some part of Ω^0 is close to the shear wave singular direction. We then speak of *the quasi-degenerate case*. See Section 2.2.8 and 2.2.9 for more details. This terminology was introduced by Kravtsov and Orlov (1980). We shall distinguish both cases only if necessary; otherwise, we shall speak of the *degenerate case of qS waves*.

The equations for the travel-time perturbations in the degenerate case of qS waves were first derived and discussed by Jech and Pšenčík (1989); see also Nowack and Pšenčík (1991). For the solution of similar problems in quantum physics, see Landau and Lifschitz (1974), and for the solution of similar problems in the theory of electromagnetic waves, see Kravtsov (1968), Kravtsov and Orlov (1980). The last reference also gives an extensive bibliography on this subject. See also Section 5.4.6.

We shall consider reference ray Ω^0 situated in unperturbed medium $\mathcal{M}^0$, with $p_i(u)$ known along Ω^0. Consider a point $u = u_0$ of reference ray Ω^0 at which $G_1(u_0) = G_2(u_0)$, with $G_3(u_0) \neq G_1(u_0)$, for the relevant slowness vector $\vec{p}(u_0)$. In the following, we shall consider point $u = u_0$ of Ω^0, but we shall not write u_0 as the argument of the individual quantities. Only eigenvector $\vec{g}^{(3)}$, not $\vec{g}^{(1)}$ and $\vec{g}^{(2)}$, can be uniquely determined at that point. We only know that $\vec{g}^{(1)}$ and $\vec{g}^{(2)}$ are mutually perpendicular and that they are situated in a plane perpendicular to $\vec{g}^{(3)}$. We select two arbitrary, mutually perpendicular unit vectors $\vec{e}^{(1)}$ and $\vec{e}^{(2)}$ in a plane perpendicular to $\vec{g}^{(3)}$. We number unit vectors $\vec{e}^{(1)}$ and $\vec{e}^{(2)}$ so that triplet $\vec{e}^{(1)}$, $\vec{e}^{(2)}$, $\vec{g}^{(3)}$ is right-handed. We define two new, mutually perpendicular, unit vectors $\vec{g}^{(1)}$, $\vec{g}^{(2)}$ in a plane perpendicular to $\vec{g}^{(3)}$ as follows:

$$g_k^{(M)} = a_J^{(M)} e_k^{(J)}, \tag{3.9.17}$$

with the summation over $J = 1, 2$. We choose them so that triplet $\vec{g}^{(1)}$, $\vec{g}^{(2)}$, $\vec{g}^{(3)}$ is right-handed. Then $a_1^{(2)} = -a_2^{(1)}$, $a_2^{(2)} = a_1^{(1)}$, $a_1^{(1)2} + a_2^{(1)2} = 1$. We can also denote $a_1^{(1)} = \cos\varphi$, $a_2^{(1)} = \sin\varphi$ where φ is the angle between $\vec{g}^{(1)}$ and $\vec{e}^{(1)}$, $\cos\varphi = \vec{g}^{(1)} \cdot \vec{e}^{(1)}$, and $\sin\varphi = \vec{g}^{(1)} \cdot \vec{e}^{(2)}$.

Using (3.9.12) for $m = M$ (with $M = 1, 2$), we obtain

$$(\Gamma_{jk} - G_M\delta_{jk})\Delta g_k^{(M)} + (\Delta\Gamma_{jk} - \Delta G_M\delta_{jk})g_k^{(M)} = 0 \tag{3.9.18}$$

(no summation over M). Inserting (3.9.17) into (3.9.18) and multiplying the result by $e_j^{(M)}$, we obtain

$$(B_{IK} - \Delta G_M\delta_{IK})a_I^{(M)} = 0, \tag{3.9.19}$$

where

$$B_{IK} = \Delta\Gamma_{jk}e_j^{(I)}e_k^{(K)} = \Delta a_{ijkl} p_i p_l e_j^{(I)} e_k^{(K)}. \tag{3.9.20}$$

The 2×2 matrix **B** is usually called the *weak-anisotropy matrix*. In the derivation of (3.9.19) from (3.9.18), we have taken into account that $(\Gamma_{jk} - G_M\delta_{jk})e_j^{(M)} = 0$ for any vector $\vec{e}^{(M)}$ perpendicular to $\vec{g}^{(3)}$. We have also used $e_j^{(I)}e_j^{(K)} = \delta_{IK}$.

Thus, the problem of determining ΔG_M is reduced to the solution of eigenvalue problem (3.9.19) for a 2×2 weak-anisotropy matrix **B** given by (3.9.20). It is easy to find eigenvalues ΔG_M and the relevant eigenvectors $\vec{a}^{(M)}$ ($M = 1, 2$):

$$\Delta G_{1,2} = \tfrac{1}{2}[(B_{11} + B_{22}) \pm D], \tag{3.9.21}$$

$$a_1^{(1)} = a_2^{(2)} = \frac{1}{\sqrt{2}}\left(1 + \frac{B_{11} - B_{22}}{D}\right)^{1/2},$$

$$a_2^{(1)} = -a_1^{(2)} = \frac{\operatorname{sgn} B_{12}}{\sqrt{2}}\left(1 - \frac{B_{11} - B_{22}}{D}\right)^{1/2}, \tag{3.9.22}$$

with

$$D = \left[(B_{11} - B_{22})^2 + 4B_{12}^2\right]^{1/2}. \tag{3.9.23}$$

Note that the 2-D eigenvectors $\vec{a}^{(1)}$ and $\vec{a}^{(2)}$ of the 2×2 weak-anisotropy matrix **B** are specified with respect to the frame given by $\vec{e}^{(1)}$ and $\vec{e}^{(2)}$ in a plane perpendicular to $\vec{g}^{(3)}$; see (3.9.17). To calculate actual 3-D vectors $\vec{g}^{(1)}$ and $\vec{g}^{(2)}$ from $\vec{a}^{(1)}$ and $\vec{a}^{(2)}$ given by (3.9.22), we must use (3.9.17).

The result (3.9.22) is very interesting. The two eigenvectors $\vec{g}^{(1)}$ and $\vec{g}^{(2)}$, given by (3.9.17) with (3.9.22), can be determined uniquely even if the background model is degenerate (for example, isotropic medium). Eigenvectors $\vec{g}^{(1)}$ and $\vec{g}^{(2)}$, however, depend on the properties of the perturbed medium, specifically on perturbations Δa_{ijkl}. For different

Δa_{ijkl}, different eigenvectors $\vec{g}^{(1)}$ and $\vec{g}^{(2)}$ are obtained. Thus, the *perturbation eliminates degeneration* (Landau and Lifschitz 1974; Jech and Pšenčík, 1989). In the degenerate case of qS waves, $\vec{g}^{(1)}$ and $\vec{g}^{(2)}$ have a double function:

1. In 3-D, they represent the eigenvectors of the 3×3 Christoffel matrix Γ_{ik}, corresponding to eigenvalues $G_1 = G_2$.
2. In the plane perpendicular to $\vec{g}^{(3)}$, they represent the eigenvectors of the 2×2 matrix B_{IK}, corresponding to eigenvalues ΔG_1 and ΔG_2.

Now we shall discuss the travel-time perturbations for qS waves in the weakly anisotropic media, with the isotropic background medium $\mathcal{M}^0$. In this case, $\vec{g}^{(3)}$ is tangent to the reference ray, and $\vec{e}^{(1)}$ and $\vec{e}^{(2)}$ are perpendicular to it. Along reference ray Ω^0, we choose unit vectors $\vec{e}^{(1)}$ and $\vec{e}^{(2)}$, which vary smoothly (but arbitrarily) along Ω^0. They may represent unit normal $\vec{n}$ and unit binormal $\vec{b}$, the basis vectors of the ray-centered coordinate system $\vec{e}_1$, $\vec{e}_2$ (see Section 4.1.1), and so on. Inserting (3.9.21) into (3.9.11) and (3.9.6), we immediately obtain expressions for the travel-time perturbations $\Delta T_{1,2}(R, S)$:

$$\Delta T_{1,2}(R, S) = -\tfrac{1}{4} \int_{\Omega^0(R,S)} [(B_{11} + B_{22}) \pm D] \mathrm{d}T. \qquad (3.9.24)$$

All quantities in (3.9.24) are taken along reference ray Ω^0 in the background isotropic medium $\mathcal{M}^0$. We can also use $\mathrm{d}T = \mathrm{d}s/\beta$, where s is the arclength along Ω^0. The two signs in (3.9.24) correspond to the two qS waves polarized along $\vec{g}^{(1)}$ and $\vec{g}^{(2)}$. The plus sign stands for the faster qS wave, and the minus sign stands for the slower qS wave. As we can see from (3.9.24), (3.9.23), and (3.9.20), the relation between $\Delta T_{1,2}(R, S)$ and Δa_{ijkl} is nonlinear in the degenerate case of qS waves.

Unit vectors $\vec{e}^{(1)}$ and $\vec{e}^{(2)}$ may also represent the eigenvectors $\vec{g}^{(1)}$ and $\vec{g}^{(2)}$, given by (3.9.17) with (3.9.22). We denote the weak-anisotropy matrix $\mathbf{B}$ corresponding to $\vec{e}^{(1)} = \vec{g}^{(1)}$ and $\vec{e}^{(2)} = \vec{g}^{(2)}$ by $\mathbf{B}^g$,

$$B_{IJ}^g = \Delta a_{ijkl} p_i p_l g_j^{(I)} g_k^{(J)}. \qquad (3.9.25)$$

It is simple to see that $\mathbf{B}^g$ is a diagonal 2×2 matrix, related to $\mathbf{B}$ as follows:

$$\mathbf{B}^g = \mathbf{A}^T \mathbf{B} \mathbf{A}, \quad \mathbf{A} = \begin{pmatrix} a_1^{(1)} & -a_2^{(1)} \\ a_2^{(1)} & a_1^{(1)} \end{pmatrix}. \qquad (3.9.26)$$

Consequently,

$$\begin{aligned} B_{11}^g &= \tfrac{1}{2}[(B_{11} + B_{22}) + D] = \Delta G_1, \\ B_{22}^g &= \tfrac{1}{2}[(B_{11} + B_{22}) - D] = \Delta G_2, \qquad B_{12}^g = 0; \end{aligned} \qquad (3.9.27)$$

see (3.9.21). Let us emphasize that B_{IJ}^g, given by (3.9.25), are not linear functions of Δa_{ijkl} because $\vec{a}^{(1)}$ and $\vec{a}^{(2)}$ also depend on Δa_{ijkl}; see (3.9.22).

Let us now consider three consequences of perturbation equation (3.9.24).

a. Average qS-wave travel-time perturbation. This is given by the relation

$$\Delta T^a = \tfrac{1}{2}[\Delta T_1(R, S) + \Delta T_2(R, S)] = -\tfrac{1}{4} \int_{\Omega^0(R,S)} (B_{11} + B_{22}) \mathrm{d}T. $$

$$(3.9.28)$$

The average qS-wave travel-time perturbation ΔT^a is linear in Δa_{ijkl}. Relation (3.9.28) was first derived by Červený and Jech (1982).

b. The time delay between the two split qS waves. This is given by the relation

$$\Delta T^s = \Delta T_2(R, S) - \Delta T_1(R, S) = \tfrac{1}{2} \int_{\Omega^0(R,S)} D \, dT, \tag{3.9.29}$$

where D is given by (3.9.23). As we can see, ΔT^s is again nonlinear in Δa_{ijkl}.

c. Separation of isotropic and anisotropic perturbations. The perturbations of the density-normalized elastic parameters Δa_{ijkl} include both isotropic and anisotropic perturbations: We shall express $\Delta a_{ijkl}(x_i)$ as follows:

$$\Delta a_{ijkl}(x_i) = \Delta A^0_{ijkl}(x_i) + \Delta A_{ijkl}(x_i), \tag{3.9.30}$$

where $\Delta A^0_{ijkl}(x_i)$ represents the *isotropic perturbation*, and $\Delta A_{ijkl}(x_i)$ represents the *anisotropic perturbation*. The weak-anisotropy matrix $\mathbf{B}$ then reads

$$B_{IJ} = p_l p_j e_i^{(I)} e_k^{(J)} \left(\Delta A^0_{ijkl} + \Delta A_{ijkl} \right) = 2\beta^{-1} \Delta \beta \delta_{IJ} + C_{IJ}, \tag{3.9.31}$$

where

$$C_{IJ} = p_j p_l e_i^{(I)} e_k^{(J)} \Delta A_{ijkl}; \tag{3.9.32}$$

see (3.9.20). Inserting (3.9.31) into (3.9.24) yields

$$\Delta T_{1,2} = - \int_{\Omega^0(R,S)} \beta^{-1} \Delta \beta \, dT - \tfrac{1}{4} \int_{\Omega^0(R,S)} (C_{11} + C_{22}) \, dT$$

$$\mp \tfrac{1}{4} \int_{\Omega^0(R,S)} \left[(C_{11} - C_{22})^2 + 4C_{12}^2 \right]^{1/2} dT. \tag{3.9.33}$$

Consequently, the average qS-wave travel-time perturbation ΔT^a is

$$\Delta T^a = - \int_{\Omega^0(R,S)} \beta^{-1} \Delta \beta \, dT - \tfrac{1}{4} \int_{\Omega^0(R,S)} (C_{11} + C_{22}) \, dT, \tag{3.9.34}$$

and the time delay ΔT^s between the two split qS waves is

$$\Delta T^s = \tfrac{1}{2} \int_{\Omega^0(R,S)} \left[(C_{11} - C_{22})^2 + 4C_{12}^2 \right]^{1/2} dT. \tag{3.9.35}$$

As we can see, the time delay ΔT^s does not depend on isotropic perturbations $\Delta \beta$, but *depends only on anisotropy perturbations* ΔA_{ijkl}. This makes the inversion of the time delay between two split qS waves very attractive. As expected, the relation between ΔT^s and ΔA_{ijkl} is nonlinear.

The foregoing relation for ΔT^s can also be obtained considering factorized anisotropic media; then, ΔA_{ijkl} are constant in the region under interest. For more details and numerical examples related to ΔT^s in factorized anisotropic media, see Červený and Simões-Filho (1991).

The results of this section can be generalized considering a reference common ray $\Omega^0(R, S)$ different from the ray of S wave in the background isotropic medium. For example, it may be useful to consider the ray of the hypothetical qS wave corresponding to the average eigenvalue G^{av} of both qS waves as the reference ray. Such a ray can be safely computed even in weakly anisotropic media; see Section 3.6.2. In this way, we can obtain more accurate expressions even for ΔT^a and ΔT^s.

3.9.5 Travel-Time Perturbations in Layered Media

In this section, we shall mostly follow the approach proposed by Farra and Le Bégat (1995). Consider the reference ray $\Omega^0(R, S)$ of an arbitrary elementary multiply reflected (possibly converted) wave propagating in a 3-D laterally varying layered structure $\mathcal{M}^0$, passing through fixed points S and R. We further consider N points of reflection/transmission on Ω^0, between S and R, and denote successively the points of incidence on structural interfaces $\Sigma_1, \Sigma_2, \ldots, \Sigma_N$ by $Q_1, Q_2, \ldots, Q_N$; see Figure 3.5. We can then apply successively (3.9.3) with (3.9.4) to each segment of the ray between two R/T points and obtain

$$
T(x_i(u_R); x_i(u_S)) = \int_{u_S}^{u_R} p_i^0 \dot{x}_i^0 du
$$
$$
- \int_{u_S}^{u_R} \Delta\mathcal{H}(x_i^0, p_i^0) du + \Delta T^i(R, S), \tag{3.9.36}
$$

where $\Delta T^i(R, S)$ is given by the relation

$$
\Delta T^i(R, S) = - \sum_{k=1}^{N} \left[\tilde{p}_i^0(Q_k) - p_i^0(Q_k) \right] \Delta x_i(Q_k). \tag{3.9.37}
$$

Here $p_i^0(Q_k)$ are components of the slowness vector corresponding to the incident wave at Q_k in $\mathcal{M}^0$, and $\tilde{p}_i^0(Q_k)$ correspond to the R/T wave at Q_k in $\mathcal{M}^0$. Because Ω^0 must be continuous across the interface, $\Delta x_i(Q_k)$ should be the same for incident and R/T segments at Q_k.

Equation (3.9.36) implies, for fixed end points S and R,

$$
\Delta T(R, S) = - \int_{\Omega^0(R,S)} \Delta\mathcal{H}(x_i^0, p_i^0) du + \Delta T^i(R, S). \tag{3.9.38}
$$

We shall now specify $\Delta T^i(R, S)$ for the case of perturbed structural interfaces. Let us consider interface Σ_k and the point of incidence Q_k. Because the procedure is the same for any interface, we shall leave out subscript k in the symbol for the interface Σ_k and speak simply of interface Σ. Assume that the interface is specified by equation $\Sigma(x_i) = 0$ in the perturbed model and by $\Sigma^0(x_i) = 0$ in the background model. We introduce the perturbations of interface $\Delta\Sigma(x_i)$ by the relation $\Sigma(x_i) = \Sigma^0(x_i) + \Delta\Sigma(x_i)$. Then $\Delta x_i(Q_k)$ in (3.9.37) should be taken so as to represent the shift between the point of incidence of Ω^0 on Σ^0 and the point of incidence of Ω on Σ. Consequently, if point $x_i(Q_k)$ represents the point of incidence of Ω^0 on Σ^0, $x_i(Q_k) + \Delta x_i(Q_k)$ represents the point of incidence of Ω on Σ. Hence,

$$
\Sigma(x_i + \Delta x_i) = \Sigma^0(x_i + \Delta x_i) + \Delta\Sigma(x_i + \Delta x_i)
$$
$$
\doteq \Sigma^0(x_i) + (\partial\Sigma^0/\partial x_k)\Delta x_k + \Delta\Sigma(x_i).
$$

Here we have considered only the first-order perturbations, so that $\Delta\Sigma(x_i + \Delta x_i) \doteq \Delta\Sigma(x_i)$. Because point x_i is situated on Σ^0 and point $x_i + \Delta x_i$ falls on Σ, we also have $\Sigma^0(x_i) = 0$, $\Sigma(x_i + \Delta x_i) = 0$. This yields

$$
\Sigma_{,i}^0(Q_k)\Delta x_i(Q_k) = -\Delta\Sigma(Q_k). \tag{3.9.39}
$$

Here $\Sigma_{,i}^0 = \partial\Sigma^0/\partial x_i$.

Now we shall modify $\tilde{p}_i^0(Q_k) - p_i^0(Q_k)$ in (3.9.37). As the tangential components of $\vec{\tilde{p}}^0(Q_k)$ and $\vec{p}^0(Q_k)$ are the same, vector $\vec{\tilde{p}}^0(Q_k) - \vec{p}^0(Q_k)$ has the direction of the normal to Σ^0 at Q_k. If we use (3.2.5) with $\epsilon^* = 1$ for the normal, we obtain

$$\tilde{p}_i^0(Q_k) - p_i^0(Q_k) = \left[\left(\tilde{p}_l^0(Q_k) - p_l^0(Q_k) \right) \Sigma_{,l}^0(Q_k) \right]$$
$$\times \Sigma_{,i}^0(Q_k) / \left(\Sigma_{,n}^0(Q_k) \Sigma_{,n}^0(Q_k) \right).$$

Taking into account this relation with (3.9.39) in (3.9.37) finally yields

$$\Delta T^i(R, S) = \sum_{k=1}^{N} \frac{\left[\left(\tilde{p}_l^0(Q_k) - p_l^0(Q_k) \right) \Sigma_{,l}^0(Q_k) \right]}{\Sigma_{,n}^0(Q_k) \Sigma_{,n}^0(Q_k)} \Delta \Sigma(Q_k). \tag{3.9.40}$$

This is the final expression for the travel-time perturbation due to perturbations of structural interfaces. It is valid both for isotropic and anisotropic media. For isotropic media, it was first derived by Farra, Virieux, and Madariaga (1989). Later on, Farra and Le Bégat (1995) proved that (3.9.40) is also valid for anisotropic media. Note that expressions $\Sigma_{,l}^0(Q_k)$, $p_l^0(Q_k)$, and $\tilde{p}_l^0(Q_k)$ in (3.9.40) are known from ray tracing of reference ray Ω^0 so that the numerical realization of (3.9.40) is elementary and fast. Equation (3.9.40) also implies that $\Delta T^i(R, S) = 0$ if interfaces $\Sigma_1, \Sigma_2, \ldots, \Sigma_N$ are fixed. This was proved earlier by Firbas (1984).

3.10 Ray Fields

In the preceding sections, we mostly considered an individual single ray, specified by the proper initial conditions. The exception was Section 3.8, where the computation of travel times was discussed. In this section, we shall consider the whole system of rays, corresponding to a system of wavefronts of a selected wave, propagating in the model under consideration. We call this system of rays that corresponds to the system of wavefronts of the wave under consideration *the orthonomic system of rays* or the *normal congruency of rays*.

In this section, we shall briefly discuss certain important concepts and properties of the ray field, such as ray coordinates, the Jacobians, the ray tube, geometrical spreading, caustics, and shadow zones. Some properties of the ray field, expressed in terms of the ray Jacobian or geometrical spreading, are very important in solving the transport equation and in computing amplitudes.

3.10.1 Ray Parameters. Ray Coordinates

In 3-D media, each ray of the orthonomic system of rays can be specified by two parameters. We shall denote them γ_1 and γ_2 and call them *ray parameters*. We shall present several examples.

a. RAY PARAMETERS FOR A POINT SOURCE

If the point source is situated at S, the ray parameters can be introduced in several ways. The most common way is to introduce them as two take-off angles i_0 and ϕ_0 at the source. They are also called radiation angles. These angles specify the direction of the initial slowness vector $\vec{p}_0$ at the point source, see Figure 3.3.

Instead of the take-off angles, it would be possible to consider any other two parameters that specify the initial direction of the ray as the ray parameters. For example, it is possible to consider two components of the slowness vector at the source, say p_{01} and p_{02}. The third component of the slowness vector, p_{03}, can then be calculated from p_{01} and p_{02} using the existence condition $p_{01}^2 + p_{02}^2 + p_{03}^2 = 1/V_0^2$, where V_0 is the relevant velocity at the source. It is, however, necessary to specify the sign of p_{03}.

In isotropic media, the initial slowness vector $\vec{p}_0$ also specifies the initial direction of the ray because the slowness vector is tangent to the ray. Thus, in this case, i_0 and ϕ_0 represent the *initial direction of the ray*. In anisotropic media, the initial direction of the ray is different from the direction of the initial slowness vector $\vec{p}_0$. The initial direction of the ray is given by group velocity $\vec{U}$. If, however, the initial slowness vector $\vec{p}_0$ and the elastic tensor at S are known, group velocity vector $\vec{U}$ can be uniquely determined from them. Thus, i_0 and ϕ_0 can be used as ray parameters even in anisotropic media, although they *do not represent the initial direction of the ray*, but rather the directions of the initial slowness vector.

Take-off angles i_0 and ϕ_0 can also represent the ray parameters in other cases, not just in the case of a point source. For example, the ray field of *rays diffracted at a vertex* can also be parametrized by two take-off angles at the vertex.

b. RAY PARAMETERS AT A WAVEFRONT. RAY FIELD OF NORMAL RAYS

Let us consider wavefront $T(x_i) = T^0$ and introduce curvilinear coordinates ξ_1 and ξ_2 along it. In isotropic media, the rays are perpendicular to the wavefront. Ray parameters γ_1 and γ_2 of any ray can then be chosen as the coordinates ξ_1 and ξ_2 of the initial point of the ray on the wavefront.

Similarly, the parameters of the normal rays generated at initial surface Σ^0 can be chosen in the same way. For example, this applies to the ray parameters of rays with the initial points along an "exploding reflector." See Figure 3.9(a).

In anisotropic media, the rays are not perpendicular to wavefronts. Nevertheless, the curvilinear coordinates ξ_1 and ξ_2 of the initial point of the ray on the wavefront may be taken as the ray parameters of the ray under consideration.

c. RAY PARAMETERS ALONG AN INITIAL SURFACE

At the initial surface Σ^0, along which the distribution of initial time T^0 is not constant, the rays are not perpendicular to the wavefront, even in isotropic media. Notwithstanding, the angle between surface Σ^0 and the ray can be calculated at each point of Σ^0 from the

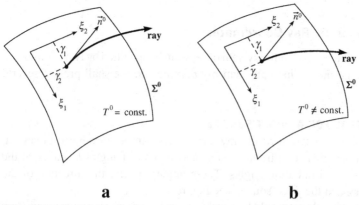

a   **b**

Figure 3.9. The initial parameters of a ray at initial surface Σ^0. Ray parameters γ_1 and γ_2 can be chosen as curvilinear coordinates ξ_1 and ξ_2 introduced along initial surface Σ^0. (a) If initial travel time T^0 is constant along Σ^0, the rays are perpendicular to Σ^0. (b) If initial travel time T^0 varies along Σ^0, the rays are not perpendicular to Σ^0.

known geometry of Σ^0, from the distribution of initial time T^0 along Σ^0, and from the elastic parameters in the medium surrounding Σ^0. The curvilinear coordinates along initial surface Σ^0 can then again be regarded as ray parameters γ_1 and γ_2. See Figure 3.9(b).

d. RAY PARAMETERS FOR A LINE SOURCE

First assume a line source C^0, along which the distribution of the initial travel time T^0 is constant. In isotropic media, the rays are perpendicular to line C^0, and any ray generated by a line source can be specified by two parameters:

1. By γ_2, specifying the position of initial point S of the ray on line C^0 (for example, the arclength of initial point S from some reference point on line C^0).
2. By $\gamma_1 = i_0$, the initial radiation angle in the plane perpendicular to C^0 at initial point S.

Now consider a line source, situated in an isotropic medium, along which the distribution of initial travel time T^0 is not constant. At any point S of C^0, the tangents to the generated rays form a radiation cone. The apex of the radiation cone is situated at S, and the axis of the radiation cone is tangent to C^0 at S. The apex angle can be determined from the local derivative of the initial travel time T^0 along C^0 at S, from the geometry of C^0 at S, and from the elastic parameters in the medium surrounding C^0 at S. Thus, any ray generated by line source C^0 can be specified by two ray parameters: By γ_2, similarly as in the case of $T^0 = $ const. along C^0, and by $\gamma_1 = i_0$, the angle that determines the position of the tangent of the selected ray at S on the radiation cone. In an anisotropic medium, the situation is similar, but we must again distinguish between the initial direction of rays and the direction of slowness vectors.

We now introduce the *ray coordinates*,

$$(\gamma_1, \gamma_2, \gamma_3), \tag{3.10.1}$$

where γ_1 and γ_2 are the ray parameters that specify the ray and $\gamma_3 = u$ is a monotonic parameter along the ray. The ray coordinates simplify the solution of many problems in the ray theory considerably. The following monotonic parameters $\gamma_3 = u$ are most common: arclength s along the ray, travel time T along the ray, or parameter σ, related to the travel time by the relation $dT = d\sigma/V^2$.

Let us now consider the parameteric equation,

$$\vec{x} = \vec{x}(\gamma_1, \gamma_2, T). \tag{3.10.2}$$

If γ_1 and γ_2 are fixed and T varies, Equation (3.10.2) represents the parameteric equation of the ray specified by ray parameters γ_1 and γ_2. For fixed travel time T and γ_1 and γ_2 varying, this equation becomes the equation of the wavefront.

Similarly, we can write the parameteric equations,

$$\vec{x} = \vec{x}(\gamma_1, \gamma_2, s), \qquad \vec{x} = \vec{x}(\gamma_1, \gamma_2, \sigma), \qquad \vec{x} = \vec{x}(\gamma_1, \gamma_2, u). \tag{3.10.3}$$

For fixed s, σ, or u, Equations (3.10.3) are the parameteric equations of some surfaces. In general, these equations do not correspond to the wavefronts, they differ from them both in shape and position. Formally, the surface

$$\vec{x} = \vec{x}(\gamma_1, \gamma_2, \gamma_3), \tag{3.10.4}$$

with $\gamma_3 = \gamma_{30} = $ const., will be called the *surface of constant γ_3*. It corresponds to the wavefront if $\gamma_3 = T$.

Ray coordinates γ_1, γ_2, and γ_3 are general curvilinear coordinates. Only in some special situations may they be orthogonal, but in inhomogeneous media they are, as a rule, nonorthogonal. It would be possible to introduce the relevant metric tensor and to apply the well-developed methods of Riemannian geometry; see Section 3.5.6. This would really lead to shorter derivations and simpler final expressions in certain situations. We shall not, however, use the methods of Riemannian geometry here. We believe that our explanations and derivations will be quite simple and straightforward even without the application of Riemannian geometry.

Let us now introduce the 3×3 transformation matrix $\hat{\mathbf{Q}}^{(x)}$ from ray coordinates γ_1, γ_2, and γ_3 to general Cartesian coordinates x_1, x_2, and x_3 and denote its elements $Q_{ij}^{(x)}$, $i, j = 1, 2, 3$,

$$Q_{ij}^{(x)} = \partial x_i / \partial \gamma_j. \tag{3.10.5}$$

We are using symbol $\hat{\mathbf{Q}}^{(x)}$ because we shall later introduce another alternative transformation matrix from ray coordinates to ray-centered coordinates and denote it simply $\hat{\mathbf{Q}}$. Thus, superscript (x) specifies that the transformation is performed from ray coordinates to general Cartesian coordinates x_i.

The knowledge of the transformation matrix $\hat{\mathbf{Q}}^{(x)}$ is very useful in various applications. We can write

$$dx_i = (\partial x_i / \partial \gamma_j) d\gamma_j = Q_{ij}^{(x)} d\gamma_j, \tag{3.10.6}$$

If we introduce 3×1 column matrices $d\hat{\mathbf{x}}$ and $d\hat{\gamma}$, we can express (3.10.6) in matrix form,

$$d\hat{\mathbf{x}} = \hat{\mathbf{Q}}^{(x)} d\hat{\gamma}, \qquad d\hat{\gamma} = (\hat{\mathbf{Q}}^{(x)})^{-1} d\hat{\mathbf{x}}. \tag{3.10.7}$$

The importance of these relations is obvious.

Transformation matrix $\hat{\mathbf{Q}}^{(x)}$ can be calculated along a known ray using the procedure called *dynamic ray tracing*; see Chapter 4, particularly Sections 4.2 and 4.7. It, of course, depends on the choice of ray coordinate γ_3, which may equal any monotonic parameter along the ray such as s, T, or σ.

3.10.2 Jacobians of Transformations

In some regions, the behavior of the ray field may be complicated. There are some regions into which the rays do not penetrate at all (shadow zones). On the contrary, two or more rays may pass through each point in other regions. Such situations are well known in seismology, particularly the shadow zones and the regions of loops in the travel-time curves. The transformation from ray coordinates to general Cartesian coordinates is not regular in such cases. The fundamental role in the investigation of such situations is played by the Jacobian $J^{(u)}$ (the Jacobian of the transformation from ray coordinates γ_1, γ_2, $\gamma_3 \equiv u$ to general Cartesian coordinates x_1, x_2, x_3)

$$J^{(u)} = \partial(x_1, x_2, x_3) / \partial(\gamma_1, \gamma_2, u) = \det \hat{\mathbf{Q}}^{(x)}. \tag{3.10.8}$$

Here transformation matrix $\hat{\mathbf{Q}}^{(x)}$ is taken for $\gamma_3 = u$.

If Jacobian $J^{(u)}$ is defined and does not vanish at any point of region D, the ray field is called regular in the region. On the other hand, the ray field is called singular at any point where $J^{(u)}$ is not defined or where it vanishes.

Jacobian $J^{(u)}$ does not depend solely on ray parameters γ_1 and γ_2; it also relies on the third ray coordinate $\gamma_3 = u$, which represents a monotonic parameter along the ray; see

(3.10.1). We can use any monotonic parameter along the ray to define the appropriate Jacobian of transformation ($u = T, s, \sigma$). We denote the relevant Jacobians of transformation $J^{(T)}$, $J^{(s)}$, and $J^{(\sigma)}$. Thus,

$$J^{(s)} = \partial(x_1, x_2, x_3)/\partial(\gamma_1, \gamma_2, s),$$
$$J^{(T)} = \partial(x_1, x_2, x_3)/\partial(\gamma_1, \gamma_2, T), \qquad (3.10.9)$$
$$J^{(\sigma)} = \partial(x_1, x_2, x_3)/\partial(\gamma_1, \gamma_2, \sigma).$$

We shall mostly consider Jacobians $J^{(T)}$ or $J^{(s)}$. The relations between various Jacobians of transformation (3.10.9) will be given in the next section.

The last column of transformation matrix $\hat{\mathbf{Q}}^{(x)}$ represents the Cartesian components of the following vector:

$$(\partial\vec{x}/\partial u)_{\gamma_1,\gamma_2} = (\mathrm{d}\vec{x}/\mathrm{d}u)_{\text{along the ray}} = g_u^{-1}\vec{t},$$

where g_u is given by the relation

$$g_u = (\mathrm{d}u/\mathrm{d}s)_{\text{along the ray}}, \qquad (3.10.10)$$

and $\vec{t}$ is the unit vector tangent to the ray. We can now give a useful expression for Jacobian $J^{(u)}$,

$$J^{(u)} = \frac{1}{g_u}\det\begin{pmatrix} (\partial x_1/\partial\gamma_1)_u & (\partial x_1/\partial\gamma_2)_u & t_1 \\ (\partial x_2/\partial\gamma_1)_u & (\partial x_2/\partial\gamma_2)_u & t_2 \\ (\partial x_3/\partial\gamma_1)_u & (\partial x_3/\partial\gamma_2)_u & t_3 \end{pmatrix} = \frac{1}{g_u}\vec{\Omega}^{(u)}\cdot\vec{t}, \quad (3.10.11)$$

where

$$\vec{\Omega}^{(u)} = (\partial\vec{x}/\partial\gamma_1 \times \partial\vec{x}/\partial\gamma_2)_u. \qquad (3.10.12)$$

Subscripts u again emphasize the fact that the expressions are taken for constant u; the symbol $\times$ denotes the cross product.

3.10.3 Elementary Ray Tube. Geometrical Spreading

Jacobians $J^{(u)}$ are closely connected with certain geometrical properties of the ray field, particularly with the density of the ray field. The density of the ray field can be expressed in terms of the cross-sectional area of the ray tube. In this section, we shall introduce the relevant terminology related to the ray tube and discuss its relation to Jacobians. The relations derived in this section are valid both for isotropic and anisotropic media.

By the *elementary ray tube*, we understand the family of rays, the parameters of which are within the limits ($\gamma_1; \gamma_1 + \mathrm{d}\gamma_1$) and ($\gamma_2; \gamma_2 + \mathrm{d}\gamma_2$); see Figure 3.10. The elementary ray tube is also called simply the *ray tube*.

An important role in the ray method is played by the vectorial surface element $\mathrm{d}\vec{\Omega}^{(u)}$, which is cut out of the surface of constant $\gamma_3 = u$ by the ray tube. The vectorial surface element $\mathrm{d}\vec{\Omega}^{(u)}$ is given by the relation, well known from the vector calculus,

$$\mathrm{d}\vec{\Omega}^{(u)} = (\partial\vec{x}/\partial\gamma_1 \times \partial\vec{x}/\partial\gamma_2)_u\,\mathrm{d}\gamma_1\,\mathrm{d}\gamma_2 = \vec{\Omega}^{(u)}\mathrm{d}\gamma_1\mathrm{d}\gamma_2, \qquad (3.10.13)$$

where $\times$ denotes the cross product and $\vec{\Omega}^{(u)}$ is given by (3.10.12); see Jeffreys and Jeffreys (1966). Because vectors $(\partial\vec{x}/\partial\gamma_1)_u$ and $(\partial\vec{x}/\partial\gamma_2)_u$ are tangent to the surface of constant $\gamma_3 = u$, the vectorial surface element $\mathrm{d}\vec{\Omega}^{(u)}$ is always perpendicular to the surface of constant u. It is obvious that surface elements $\mathrm{d}\vec{\Omega}^{(s)}$, $\mathrm{d}\vec{\Omega}^{(T)}$, and $\mathrm{d}\vec{\Omega}^{(\sigma)}$ do not, in general, have the same direction. See Figure 3.11.

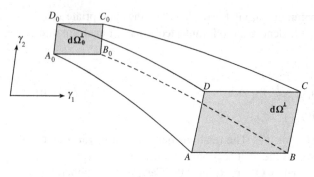

Figure 3.10. Elementary ray tube. Ray $A_0 A$ corresponds to ray parameters γ_1 and γ_2, ray $B_0 B$ corresponds to $\gamma_1 + d\gamma_1$ and γ_2, ray $C_0 C$ corresponds to $\gamma_1 + d\gamma_1$ and $\gamma_2 + d\gamma_2$, and ray $D_0 D$ corresponds to ray parameters γ_1 and $\gamma_2 + d\gamma_2$. Quantities $d\Omega_0^{\perp}$ and $d\Omega^{\perp}$ denote the cross-sectional areas of the ray tube. In isotropic media, $d\Omega^{\perp}$ equals $d\Omega^T$, an area cut out by the ray tube from the wavefront. In anisotropic media, the rays are not perpendicular to the wavefront and $d\Omega^{\perp} \neq d\Omega^{(T)}$.

The vectorial surface element $d\vec{\Omega}^{(u)}$ has a simple and clear physical meaning for $u = T$. In this case, $d\vec{\Omega}^{(T)}$ represents the *vectorial surface element cut out of the wavefront by the ray tube*. $d\vec{\Omega}^{(T)}$ has the direction of $\pm\vec{N}$, where $\vec{N}$ is the unit vector normal to the wavefront, oriented in the direction of the propagation of the wavefront. We shall now introduce $d\Omega^{(T)}$, the *scalar surface element cut out of the wavefront of the ray tube*, by any of the two alternative relations:

$$d\vec{\Omega}^{(T)} = d\Omega^{(T)} \vec{N}, \qquad d\Omega^{(T)} = d\vec{\Omega}^{(T)} \cdot \vec{N}. \tag{3.10.14}$$

The scalar surface element $d\Omega^{(T)}$ may be positive, negative, or zero.

Quantities $d\vec{\Omega}^{(T)}$ and $d\Omega^{(T)}$ are infinitesimal because they contain factor $d\gamma_1 d\gamma_2$. We can, however, express them in terms of noninfinitesimal quantities $\vec{\Omega}^{(T)}$ and $\Omega^{(T)}$ as follows:

$$\vec{\Omega}^{(T)} = d\vec{\Omega}^{(T)}/d\gamma_1 d\gamma_2,$$
$$\Omega^{(T)} = \vec{\Omega}^{(T)} \cdot \vec{N} = d\Omega^{(T)}/d\gamma_1 d\gamma_2 = (d\vec{\Omega}^{(T)} \cdot \vec{N})/d\gamma_1 d\gamma_2. \tag{3.10.15}$$

see (3.10.13) and (3.10.14). Thus, $\vec{\Omega}^{(T)}$ and $\Omega^{(T)}$ represent the vectorial and scalar surface elements cut out of the wavefront by the ray tube, *normalized with respect to* $d\gamma_1 d\gamma_2$. They are usually more suitable for analytical treatment, applications, and computation than infinitesimal quantities $d\vec{\Omega}^{(T)}$ and $d\Omega^{(T)}$.

Let us again return to the general parameter u along the ray. We shall briefly discuss the scalar product of $\vec{\Omega}^{(u)}$ with $\vec{t}$, the unit vector tangent to the ray, oriented positively in the direction of propagation of the wave under consideration. We introduce quantity J by

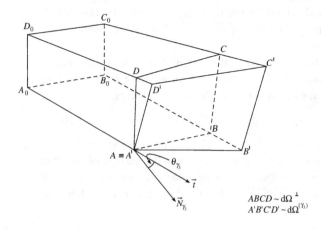

Figure 3.11. Scalar surface element $d\Omega^{(\gamma_3)}$ represents the area $A'B'C'D'$ cut out from the surface of constant γ_3 by the ray tube. In general, the scalar surface element $d\Omega^{(\gamma_3)}$ differs from the cross-sectional area $d\Omega^{\perp}$ of the ray tube, $ABCD$. In anisotropic media, for $\gamma_3 = T$, surface $A'B'C'D'$ corresponds to the wavefront, but $d\Omega^{(T)}$ differs from $d\Omega^{\perp}$. Only in isotropic media for $\gamma_3 = T$ is the surface of constant $\gamma_3 = T$ (wavefront) perpendicular to rays and $A'B'C'D' \equiv ABCD$. Unit vector $\vec{t}$ is perpendicular to $ABCD$; unit vector $\vec{N}_{\gamma_3}$ is perpendicular to $A'B'C'D'$.

the relation

$$J = (\partial \vec{x}/\partial \gamma_1 \times \partial \vec{x}/\partial \gamma_2)_u \cdot \vec{t}. \tag{3.10.16}$$

Quantity J vanishes when the cross product vanishes. Such points are called caustic points; see Section 3.10.5. Otherwise, J may be either positive or negative, depending on the mutual orientation of both vectors. Using (3.10.11) and (3.10.13), we can express J in three different ways:

$$J = \vec{t} \cdot \vec{\Omega}^{(u)} = \vec{t} \cdot d\vec{\Omega}^{(u)}/d\gamma_1 d\gamma_2 = g_u J^{(u)}. \tag{3.10.17}$$

Mixed product $\vec{t} \cdot d\vec{\Omega}^{(u)}$ represents the projection of vectorial surface element $d\vec{\Omega}^{(u)}$ into a plane perpendicular to the ray. We introduce $d\Omega^{\perp}$ by the relation

$$d\Omega^{\perp} = \vec{t} \cdot d\vec{\Omega}^{(u)} = J d\gamma_1 d\gamma_2 = J^{(u)} g_u d\gamma_1 d\gamma_2. \tag{3.10.18}$$

The quantity $d\Omega^{\perp}$ represents the *cross-sectional area of the ray tube*, that is, the area cut out from the plane perpendicular to the ray by the ray tube. See Figures 3.10 and 3.11. Equation (3.10.18) shows that J also represents the cross-sectional area of the ray tube, *normalized with respect to* $d\gamma_1 d\gamma_2$. It also yields

$$J^{(u)} = d\Omega^{\perp}/g_u d\gamma_1 d\gamma_2. \tag{3.10.19}$$

Equation (3.10.18) also indicates that $g_u J^{(u)}$ is the same for arbitrary parameter u so that

$$J = d\Omega^{\perp}/d\gamma_1 d\gamma_2 = g_u J^{(u)} = J^{(s)} = g_T J^{(T)} = g_\sigma J^{(\sigma)} = \cdots. \tag{3.10.20}$$

Here we have used the obvious relation $g_s = 1$; see (3.10.10).

As we can see from (3.10.18), quantity J given by (3.10.16), related to the cross-sectional area of ray tube $d\Omega^{\perp}$ by (3.10.18), equals $J^{(s)}$, the Jacobian of transformation from ray coordinates $\gamma_1, \gamma_2, \gamma_3 = s$ to Cartesian coordinates x_1, x_2, x_3. Thus, $J^{(s)}$ in some way plays an exceptional role among the other Jacobians $J^{(u)}$. For this reason, we shall call it the *ray Jacobian*.

Equation (3.10.20) can be used to recalculate mutually the individual Jacobians $J \equiv J^{(s)}, J^{(\sigma)}, J^{(T)}$, and so on. The very important Jacobian $J^{(T)}$ related to $\gamma_3 = T$ can be expressed in terms of J as follows:

$$J^{(T)} = \mathcal{U}J, \tag{3.10.21}$$

where $\mathcal{U}$ is the group velocity ($g_T = 1/\mathcal{U}$). This relation is also valid in anisotropic media.

Quantities J and $J^{(T)}$ can also be suitably expressed in terms of $\Omega^{(T)}$; see (3.10.15). Using (3.10.14), (3.10.15), (3.10.17), and (3.6.18), we obtain

$$J = (\vec{t} \cdot d\vec{\Omega}^{(T)})/d\gamma_1 d\gamma_2 = (\vec{t} \cdot \vec{N}) d\Omega^{(T)}/d\gamma_1 d\gamma_2 = \mathcal{C}\mathcal{U}^{-1}\Omega^{(T)}. \tag{3.10.22}$$

Thus, the final relations between J, $J^{(T)}$ and $\Omega^{(T)}$ are

$$J^{(T)} = \mathcal{U}J = \mathcal{C}\Omega^{(T)}. \tag{3.10.23}$$

All three quantities $J^{(T)}$, J, and $\Omega^{(T)}$ have been broadly used in the seismic ray method, particularly in computing amplitudes of high-frequency seismic body waves propagating in complex structures. All three quantities $J^{(T)}$, J, and $\Omega^{(T)}$ are mutually different in anisotropic medium, where $\mathcal{U} \neq \mathcal{C}$. We summarize the physical meaning of $J^{(T)}$, J, and $\Omega^{(T)}$. $J^{(T)}$ represents the Jacobian of transformation from ray coordinates $\gamma_1, \gamma_2, \gamma_3 = T$ to general Cartesian coordinates x_1, x_2, x_3. $J \equiv J^{(s)}$ is the Jacobian of transformation from ray coordinates $\gamma_1, \gamma_2, \gamma_3 = s$ to general Cartesian coordinates x_1, x_2, x_3. At the same time,

J represents the cross-sectional area of ray tube $d\Omega^{\perp}$ (perpendicular to the ray), normalized with respect to $d\gamma_1 d\gamma_2$. Finally, $\Omega^{(T)}$ represents the scalar surface element cut out of the wavefront by the ray tube and normalized with respect to $d\gamma_1 d\gamma_2$. In an isotropic medium, J equals $\Omega^{(T)}$ because $\mathcal{U} = \mathcal{C}$ there. Thus, $J = \Omega^{(T)} = J^{(T)}/\mathcal{C}$.

Ray Jacobian J, other related Jacobians $J^{(u)}$, and $\Omega^{(T)}$ play a fundamental role in the calculation of amplitudes; see Section 3.10.6. The amplitudes are inversely proportional to $|J|^{1/2}$. Thus, the amplitudes are high in regions in which the density of rays is high (small $d\Omega^{\perp}$ and, consequently, small J). In regions in which the density of rays is small (high $d\Omega^{\perp}$, and, consequently, high J) these amplitudes are low. This implies that the ray diagrams, together with the travel-time curves, provide the possibility of estimating not only the kinematic properties but also the ray amplitudes of seismic body waves under consideration.

Function $|J|^{1/2}$ is often called *geometrical spreading* in the literature devoted to the seismic ray method. However, the terminology is not uniform in the seismological literature.

3.10.4 Properties and Computation of the Ray Jacobian J

Ray Jacobian J represents the Jacobian of transformation $J^{(s)}$ from ray coordinates $\gamma_1, \gamma_2,$ $\gamma_3 = s$ to Cartesian coordinates x_1, x_2, x_3. It measures the cross-sectional area of ray tube $d\Omega^{\perp}$; see (3.10.18). Relation (3.10.18) is valid not only in isotropic media but also in anisotropic media. In this section, we shall discuss several other properties of ray Jacobian J and outline some possibilities of its computation. All the derived equations may be simply expressed in terms of any other Jacobian $J^{(u)}$, using relations $J = g_u J^{(u)}$; see (3.10.20).

1. RELATION OF J TO THE DIVERGENCE OF THE GROUP VELOCITY VECTOR, $\nabla \cdot \vec{\mathcal{U}}$

We shall derive the relation between $\nabla \cdot \vec{\mathcal{U}}$ and J, which plays an important role in the solution of the transport equations; see Section 3.10.6. The relation is valid both for isotropic and anisotropic media.

We shall consider an elementary ray tube $(\gamma_1, \gamma_1 + d\gamma_1)$ and $(\gamma_2, \gamma_2 + d\gamma_2)$ and denote by Ω the ray specified by ray parameters γ_1 and γ_2. We construct two wavefronts at the points of Ω corresponding to travel times T and $T + dT$; see Figure 3.12. We shall discuss the body cut out from the elementary tube by these two wavefronts. The lateral walls of the body are formed by rays, the front and back surfaces by the wavefronts. We denote the volume of the body by A, its surface by S, and any inner point of volume A, situated on Ω, by M. We can then use the well-known expression for the divergence,

$$\nabla \cdot \vec{\mathcal{U}} = \lim_{A \to M} \frac{1}{A} \iint_S \vec{\mathcal{U}} \cdot d\vec{S},$$

where $d\vec{S}$ denotes an elementary vectorial element of surface S. As regards volume A,

$$A = J^{(T)} d\gamma_1 d\gamma_2 dT, \qquad J^{(T)} = \partial(x_1, x_2, x_3)/\partial(\gamma_1, \gamma_2, T).$$

Here $J^{(T)}$ is the Jacobian of the transformation from ray coordinates γ_1, γ_2, T to Cartesian coordinates x_1, x_2, x_3.

Because the lateral walls of the body are formed by rays and $\vec{\mathcal{U}}$ is parallel to the rays, $\vec{\mathcal{U}} \cdot d\vec{S} = 0$ along these walls. Using (3.10.18) along the wavefront sections, we can also

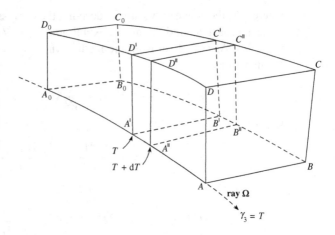

Figure 3.12. Computation of the divergence of the group velocity vector $\nabla \cdot \vec{\mathcal{U}}$. The figure shows body A cut out from the elementary ray tube by two wavefronts, corresponding to travel times $\gamma_3 = T$ (see $A'B'C'D'$) and $\gamma_3 = T + dT$ (see $A''B''C''D''$).

write $\vec{\mathcal{U}} \cdot d\vec{S} = \mathcal{U}\vec{t} \cdot d\vec{\Omega}^{(T)} = \mathcal{U}d\Omega^{\perp} = \mathcal{U}J d\gamma_1 d\gamma_2 = J^{(T)}d\gamma_1 d\gamma_2$. Thus,

$$\nabla \cdot \vec{\mathcal{U}} = \lim_{A \to M} \frac{1}{J^{(T)}d\gamma_1 d\gamma_2 dT} \{(d\Omega^{\perp}\mathcal{U})_{T+dT} - (d\Omega^{\perp}\mathcal{U})_T\}$$

$$= \lim_{A \to M} \frac{1}{J^{(T)}dT} \{(J^{(T)})_{T+dT} - (J^{(T)})_T\}.$$

This yields the final equation,

$$\nabla \cdot \vec{\mathcal{U}} = \frac{1}{J^{(T)}} \frac{dJ^{(T)}}{dT}. \tag{3.10.24}$$

Using (3.10.21), we express $\nabla \cdot \vec{\mathcal{U}}$ in terms of J and $\Omega^{(T)}$,

$$\nabla \cdot \vec{\mathcal{U}} = \frac{1}{J} \frac{d}{ds}(\mathcal{U}J) = \frac{1}{C\Omega^{(T)}} \frac{d}{dT}(C\Omega^{(T)}). \tag{3.10.25}$$

Similarly, for general $\gamma_3 = u$, we obtain

$$\nabla \cdot \vec{\mathcal{U}} = \frac{1}{J^{(u)}} \frac{d}{du}(g_u \mathcal{U}J^{(u)}). \tag{3.10.26}$$

Equations (3.10.24) through (3.10.26) can be expressed in many alternative forms. They are valid both for isotropic and anisotropic media.

   Relation (3.10.24) can also be derived directly from the ray tracing system using *Smirnov's lemma* (Smirnov 1964; Thomson and Chapman 1985). Assume that we have a system of differential equations $d\vec{x}/dT = \vec{F}(\vec{x})$. Smirnov's lemma yields a differential equation for the Jacobian of transformation $J^{(T)} = D(x_1, x_2, x_3)/D(\gamma_1, \gamma_2, T)$:

$$d(\ln J^{(T)})/dT = \nabla \cdot \vec{F}. \tag{3.10.27}$$

We can readily see that (3.10.27) immediately yields (3.10.24), as in our case $\vec{F} = \vec{\mathcal{U}}$.

2. RELATION OF J TO $\nabla^2 T$

   In isotropic media, the transport equation is expressed in terms of $\nabla^2 T$. We shall derive an expression for $\nabla^2 T$ in terms of Jacobian J, valid in isotropic inhomogeneous media, using the relation

$$\nabla^2 T = \nabla \cdot \nabla T = \nabla \cdot \vec{p}, \tag{3.10.28}$$

where $\vec{p}$ is the slowness vector. In deriving the expression, it would be possible to use the same approach as in deriving the expression for $\nabla \cdot \vec{\mathcal{U}}$. It is, however, simpler to employ (3.10.24). As $\mathcal{U} = V$ in isotropic media, where $V = c, \alpha,$ or β, we can put $\vec{p} = V^{-2}\vec{\mathcal{U}}$. We then obtain

$$\nabla^2 T = -\frac{2}{V^3}\nabla V \cdot \vec{\mathcal{U}} + \frac{1}{V^2}\nabla \cdot \vec{\mathcal{U}} = -\frac{2}{V^2}\vec{t} \cdot \nabla V + \frac{1}{V^2}\frac{\mathrm{d}\ln J^{(T)}}{\mathrm{d}T}.$$

Since

$$-\frac{2}{V^2}\vec{t} \cdot \nabla V = -\frac{2}{V^3}\frac{\mathrm{d}V}{\mathrm{d}T} = \frac{1}{V^2}\frac{\mathrm{d}\ln V^{-2}}{\mathrm{d}T},$$

we finally obtain

$$\nabla^2 T = \frac{1}{V^2}\frac{\mathrm{d}}{\mathrm{d}T}\ln\left(V^{-2}J^{(T)}\right) = \frac{1}{J^{(T)}}\frac{\mathrm{d}}{\mathrm{d}T}\left(\frac{J^{(T)}}{V^2}\right). \tag{3.10.29}$$

There are again several alternative forms of (3.10.29); for example,

$$\nabla^2 T = \frac{1}{VJ}\frac{\mathrm{d}}{\mathrm{d}T}\left(\frac{J}{V}\right) = \frac{1}{J}\frac{\mathrm{d}}{\mathrm{d}s}\left(\frac{J}{V}\right) = \frac{1}{J^{(\sigma)}}\frac{\mathrm{d}J^{(\sigma)}}{\mathrm{d}\sigma}. \tag{3.10.30}$$

3. APPROXIMATE FINITE-DIFFERENCE COMPUTATION OF RAY JACOBIAN J ALONG THE RAY

The ray Jacobian can be computed approximately by direct numerical measurement of the cross-sectional area of the ray tube $\mathrm{d}\Omega^\perp$, $J = \mathrm{d}\Omega^\perp/\mathrm{d}\gamma_1\mathrm{d}\gamma_2$. If we replace the differentials by finite differences, we obtain

$$J \doteq \Delta\Omega^\perp/\Delta\gamma_1\Delta\gamma_2. \tag{3.10.31}$$

This relation is valid both for isotropic and anisotropic media.

In anisotropic media, the wavefront is not perpendicular to the ray. It is then more usual to use *scalar surface element* $\mathrm{d}\Omega^{(T)}$ cut out of wavefront $T = $ const. by the ray tube; see (3.10.14). Using (3.10.23) and (3.10.15), we finally obtain

$$J = \mathrm{d}\Omega^\perp/\mathrm{d}\gamma_1\mathrm{d}\gamma_2 = \mathcal{C}\mathrm{d}\Omega^{(T)}/\mathcal{U}\mathrm{d}\gamma_1\mathrm{d}\gamma_2. \tag{3.10.32}$$

If we replace the differentials by finite differences, we obtain

$$J \doteq \frac{\mathcal{C}\Delta\Omega^{(T)}}{\mathcal{U}\Delta\gamma_1\Delta\gamma_2}. \tag{3.10.33}$$

The procedure is as follows; see Figure 3.11. We compute the three rays specified by close ray parameters. For example, we compute one "central" ray A_0A specified by two parameters γ_1 and γ_2 and two "supporting" rays B_0B and D_0D, specified by ray parameters $\gamma_1 + \Delta\gamma_1, \gamma_2$ and $\gamma_1, \gamma_2 + \Delta\gamma_2$. We shall use travel time T as the third ray coordinate γ_3 along these rays. We now wish to compute ray Jacobian J at a given point of the ray, $A \equiv A'$; see Figure 3.11. The coordinates of points A', B', D' are known from ray tracing, so that we can simply compute the area S of triangle $A'B'D'$. The scalar surface element $\Delta\Omega^{(T)}$ is then given by relation $\Delta\Omega^{(T)} \doteq 2S$. Because $\Delta\gamma_1, \Delta\gamma_2, \mathcal{U}$, and $\mathcal{C}$ are known, J can be determined from (3.10.33). The procedure can be used even for anisotropic media. For isotropic media, of course, $B' \equiv B, C' \equiv C, D' \equiv D$, and $\mathcal{U} = \mathcal{C}$.

We can also compute the fourth ray C_0C specified by ray parameters $\gamma_1 + \Delta\gamma_1$ and $\gamma_2 + \Delta\gamma_2$. This provides the possibility of estimating the accuracy of the computations.

This procedure determines the absolute value of J, but not its sign. The sign of J can be determined using (3.10.16). We construct two vectors $\vec{A'B'}$ and $\vec{A'D'}$ and cross product $\vec{A'B'} \times \vec{A'D'}$. If $(\vec{A'B'} \times \vec{A'D'}) \cdot \vec{t} > 0$, the sign of J is '+'. If $(\vec{A'B'} \times \vec{A'D'}) \cdot \vec{t} < 0$, the sign of J is '−'. Note that a caustic point is situated at A' if $(\vec{A'B'} \times \vec{A'D'}) \cdot \vec{t} = 0$.

In 2-D media, the procedure reduces to the calculation of the length of one straight line element, connecting two points situated on two close rays. The way to determine the sign of J is then obvious.

4. COMPUTATION OF THE RAY JACOBIAN BY DYNAMIC RAY TRACING

The procedure of computing the ray Jacobian outlined in Section 3.10.4.3 requires the calculation of at least three near rays in 3-D media, or two near rays in 2-D media. The ray Jacobian, however, can be computed along one ray only using the procedure that is now usually called dynamic ray tracing. It involves solving an additional system of linear ordinary differential equations of the first or second order. In the system, quantities $\partial x_i / \partial \gamma_1$ and $\partial x_i / \partial \gamma_2$ are calculated ($i = 1, 2, 3$). The additional system of ordinary differential equations can be solved simultaneously with the ray tracing system or after it, along a known ray. The results of ray tracing, supplemented by the results of the dynamic ray tracing, allow us to compute the ray Jacobian and the geometrical spreading analytically. Dynamic ray tracing will be described in more detail in Chapter 4. It plays a very important role not only in computing the ray Jacobian and geometrical spreading but also in many other seismological applications. In Chapter 4, a special ray-centered coordinate system will be used. That makes the dynamic ray tracing particularly simple. Dynamic ray tracing, however, can also be performed in general Cartesian coordinates in which the required partial derivatives are computed directly. Dynamic ray tracing in Cartesian coordinates can be useful in certain applications. For more details on dynamic ray tracing in Cartesian coordinates, see Sections 4.2 and 4.7.

5. RAY JACOBIAN ACROSS A STRUCTURAL INTERFACE

Let us consider ray Ω of a reflected/transmitted wave and denote the point of incidence Q. We also denote $J(Q)$ and $\tilde{J}(Q)$ the ray Jacobians of the incident and selected reflected/transmitted wave at Q. If we interpret J in terms of the cross-sectional area of ray tube $d\Omega^{\perp}$, see (3.10.18), we can easily prove that $J(Q)$ and $\tilde{J}(Q)$ are related as follows:

$$J(Q)/\tilde{J}(Q) = t_i(Q)n_i(Q)/\tilde{t}_k(Q)n_k(Q) = \pm \cos i(Q)/\cos \tilde{i}(Q). \quad (3.10.34)$$

The individual symbols in (3.10.34) have a standard meaning: $\vec{t}$ is the unit vector tangent to Ω, $\vec{n}$ the unit normal to the interface, and the tilde corresponds to the selected R/T wave. Consequently, $i(Q)$ is the angle of incidence, and $\tilde{i}(Q)$ is the angle of reflection/transmission; the plus sign corresponds to the transmitted wave, and the minus sign corresponds to the reflected wave. Equation (3.10.34) is valid both for isotropic and anisotropic media.

6. RAY JACOBIAN IN ORTHOGONAL CURVILINEAR COORDINATES ξ_1, ξ_2, ξ_3

Using (3.10.9), we obtain

$$J^{(T)} = \frac{\partial(x_1, x_2, x_3)}{\partial(\gamma_1, \gamma_2, T)} = \frac{\partial(x_1, x_2, x_3)}{\partial(\xi_1, \xi_2, \xi_3)} \frac{\partial(\xi_1, \xi_2, \xi_3)}{\partial(\gamma_1, \gamma_2, T)}.$$

In an isotropic medium, the last column in the determinant $D(\xi_1, \xi_2, \xi_3)/D(\gamma_1, \gamma_2, T)$ can be expressed using the ray tracing system (3.5.16):

$$(\partial \xi_i / \partial T)_{\xi_l} = (d\xi_i / dT)_{\text{along the ray}} = V^2 p_i / h_i = V t_i / h_i,$$

where h_i are scale factors and $p_i = h_i^{-1} \partial T / \partial \xi_i$ (no summation over i); see (3.5.4). If we use relations $\partial(x_1, x_2, x_3)/\partial(\xi_1, \xi_2, \xi_3) = h_1 h_2 h_3$ and $J^{(T)} = VJ$, we obtain the final expression for the ray Jacobian in orthogonal curvilinear coordinates:

$$J = \det \begin{pmatrix} h_1(\partial \xi_1 / \partial \gamma_1)_T & h_1(\partial \xi_1 / \partial \gamma_2)_T & t_1 \\ h_2(\partial \xi_2 / \partial \gamma_1)_T & h_2(\partial \xi_2 / \partial \gamma_2)_T & t_2 \\ h_3(\partial \xi_3 / \partial \gamma_1)_T & h_3(\partial \xi_3 / \partial \gamma_2)_T & t_3 \end{pmatrix}. \tag{3.10.35}$$

This equation is simple to understand from the geometrical point of view.

7. ANALYTICAL COMPUTATION OF THE RAY JACOBIAN IN HOMOGENEOUS MEDIA

In homogeneous media, the rays are straight lines, and the Jacobians can be calculated analytically. We shall give three examples for the *central ray field* (point source ray field).

- *Isotropic medium*, the ray parameters correspond to take-off angles: $\gamma_1 = i_0$, $\gamma_2 = \phi_0$. Then

$$x_1 = l \sin i_0 \cos \phi_0, \qquad x_2 = l \sin i_0 \sin \phi_0, \qquad x_3 = l \cos i_0,$$

where l is the distance from the source. Taking derivatives $\partial x_i / \partial \gamma_J$ and computing the determinant, we obtain

$$J = l^2 \sin i_0. \tag{3.10.36}$$

- *Isotropic medium*, the ray parameters correspond to slowness vector components p_{10} and p_{20} at the source: $\gamma_1 = p_{10}$, $\gamma_2 = p_{20}$. Then

$$x_1 = V p_{10} l, \qquad x_2 = V p_{20} l, \qquad x_3 = \left[1 - V^2 \left(p_{10}^2 + p_{20}^2\right)\right]^{1/2} l,$$

where l is again the distance from the source. Taking derivatives $\partial x_i / \partial \gamma_J$, determinant J yields

$$J = V^2 l^2 / \left[1 - V^2 \left(p_{10}^2 + p_{20}^2\right)\right]^{1/2}. \tag{3.10.37}$$

Here V is acoustic velocity c, P wave velocity α or S wave velocity β, depending on the wave under consideration.

- *Anisotropic medium*, arbitrary ray parameters γ_1 and γ_2. We shall use travel time T as the monotonic parameter along the ray, and compute $J^{(T)}$. We can put

$$x_1 = T \mathcal{U}_1, \qquad x_2 = T \mathcal{U}_2, \qquad x_3 = T \mathcal{U}_3,$$

where $\mathcal{U}_i$ are components of the group velocity, $T = r/\mathcal{U}$. We then obtain

$$J^{(T)} = T^2 \det \begin{pmatrix} \partial \mathcal{U}_1 / \partial \gamma_1 & \partial \mathcal{U}_1 / \partial \gamma_2 & \mathcal{U}_1 \\ \partial \mathcal{U}_2 / \partial \gamma_1 & \partial \mathcal{U}_2 / \partial \gamma_2 & \mathcal{U}_2 \\ \partial \mathcal{U}_3 / \partial \gamma_1 & \partial \mathcal{U}_3 / \partial \gamma_2 & \mathcal{U}_3 \end{pmatrix}. \tag{3.10.38}$$

Jacobian $J^{(T)}$ can be expressed in various alternative forms (for example, in terms of the curvature of the slowness surface). Analytically, the elements of Jacobian $J^{(T)}$ can be computed using relation

$$\mathcal{U}_i = \frac{1}{2} \frac{\partial G_m}{\partial p_i}, \qquad \frac{\partial \mathcal{U}_i}{\partial \gamma_J} = \frac{1}{2} \frac{\partial^2 G_m}{\partial p_i \partial p_k} \frac{\partial p_k}{\partial \gamma_J}. \tag{3.10.39}$$

8. RAY JACOBIAN IN ONE-DIMENSIONAL ISOTROPIC MEDIA

We shall consider a point source situated at S in a model specified in orthogonal curvilinear coordinates ξ_1, ξ_2, ξ_3 with scale factors h_1, h_2, h_3. We make three assumptions:

a. Velocity V depends on ξ_3 only, $V = V(\xi_3)$.

b. Scale factors h_1 and h_3 depend on ξ_3 only, $h_1 = h_1(\xi_3)$, $h_3 = h_3(\xi_3)$. Scale factor h_2 may also depend on ξ_1, $h_2 = h_2(\xi_1, \xi_3)$, but not on ξ_2.

c. Component p_2 of slowness vector $\vec{p}$ vanishes at initial point S of ray Ω, $p_{20} = 0$. Ray tracing system (3.5.16) then shows that $p_2 = 0$ along the whole ray Ω and that ray Ω is completely situated on surface $\xi_2 = \xi_{20}$.

We choose ray parameters γ_1 and γ_2 as follows: $\gamma_2 = \xi_{20}$, $\gamma_1 = i_0$, where i_0 is the angle between ray Ω and coordinate line ξ_3 at S. Then $(\partial \xi_2/\partial \gamma_1)_T = 0$, $(\partial \xi_2/\partial \gamma_2)_T = 1$, and (3.10.35) yields

$$J = h_2 \det \begin{pmatrix} h_1(\partial \xi_1/\partial i_0)_T & t_1 \\ h_3(\partial \xi_3/\partial i_0)_T & t_3 \end{pmatrix}. \tag{3.10.40}$$

Alternative forms of (3.10.40) can be obtained if we replace the derivatives taken along wavefront $T = $ const. by derivatives taken for $\xi_3 = $ const. or for $\xi_1 = $ const. Geometrical considerations yield $\det(\cdots) = h_1(\cos i)(\partial \xi_1/\partial i_0)_{\xi_3} = -h_3(\sin i)(\partial \xi_3/\partial i_0)_{\xi_1}$, and

$$J = h_1 h_2(\cos i)(\partial \xi_1/\partial i_0)_{\xi_3} = -h_2 h_3(\sin i)(\partial \xi_3/\partial i_0)_{\xi_1}. \tag{3.10.41}$$

Instead of $(\partial \xi_1/\partial i_0)_{\xi_3}$ and $(d\xi_3/\partial i_0)_{\xi_1}$, we can also consider derivatives of the travel-time curve $T = T(\xi_i)$ along lines $\xi_3 = $ const.,

$$J = V h_2(\cot an\, i)(\partial T/\partial i_0)_{\xi_3}. \tag{3.10.42}$$

Expressions (3.10.41) and (3.10.42) can also be expressed in several alternative forms using other ray parameters γ_1 than i_0. We shall employ a particularly useful parameter γ_1 broadly used in seismological applications. Ray tracing system (3.5.16) shows that the quantity $h_1 p_1$ is constant along the whole ray Ω. We denote this constant by p:

$$p = h_1 p_1 = h_1(\xi_3)(\sin i(\xi_3))/V(\xi_3) = h_{10}(\sin i_0)/V_0. \tag{3.10.43}$$

Equation $p = $ const. along the ray represents the generalized Snell's law for a 1-D isotropic medium in orthogonal curvilinear coordinates ξ_1, ξ_2, and ξ_3. Equation (3.10.43) implies

$$(\partial \xi_1/\partial i_0)_{\xi_3} = h_{10} V_0^{-1}(\cos i_0)(\partial \xi_1/\partial p)_{\xi_3}.$$

The relevant relations for J computed along profiles $\xi_3 = $ const. are

$$\begin{aligned} J &= h_1 h_2 h_{10} V_0^{-1}(\cos i_0)(\cos i)(\partial \xi_1/\partial p)_{\xi_3}, \\ J &= V h_2 h_{10} V_0^{-1}(\cos i_0)(\cot an\, i)(\partial T/\partial p)_{\xi_3}, \end{aligned} \tag{3.10.44}$$

where p is given by (3.10.43). The relations computed along profiles $\xi_1 = $ const. are analogous. Let us emphasize again that J, given by (3.10.41), (3.10.42) and (3.10.44), corresponds to the ray parameters $\gamma_1 = i_0$ and $\gamma_2 = \xi_{20}$, even though we have also used the parameter p given by (3.10.43) in (3.10.44).

We shall now specify expressions (3.10.40) through (3.10.44) in two important orthogonal curvilinear coordinate systems.

a. Cylindrical coordinates. This system of coordinates has often been used to study the point-source solutions in *vertically inhomogeneous media*. We choose $\xi_1 = \eta$, $\xi_2 = \varphi$, and $\xi_3 = z$, where z is the depth and η is the distance from the cylindrical axis. Then, $h_1 = h_3 = 1$ and $h_2 = \eta$. Quantity $p = (\sin i_0)/V_0 = (\sin i(z))/V(z)$ represents the standard seismological ray parameter p for vertically inhomogeneous media, see (3.7.4). Equations (3.10.41), (3.10.42), and (3.10.44) then yield four alternative expressions for J, calculated for $\gamma_1 = i_0$ and $\gamma_2 = \varphi_0$ along horizontal profiles $z = $ const.:

$$
\begin{aligned}
J &= \eta \,(\cos i)(\partial\eta/\partial i_0)_z \\
&= V_0^{-1}\eta \,(\cos i_0)(\cos i)(\partial\eta/\partial p)_z \\
&= V(z)\eta \,(\cotan i)(\partial T/\partial i_0)_z \\
&= V_0^{-1} V(z)\eta \,(\cos i_0)(\cotan i)(\partial T/\partial p)_z.
\end{aligned}
\tag{3.10.45}
$$

Expressions (3.10.45) can be applied to analytical ray equations $x = x(p)$ and $T = T(p)$ derived in Sections 3.7.1 through 3.7.3 for vertically inhomogeneous media. (Note that $x(p)$ is equivalent to $\eta(p)$ in our notation.) The first equation of (3.10.45) has been well known in seismology for a long time. Its simple geometrical derivation can be found in Červený and Ravindra (1971). The same reference also gives many examples of its application. Similar equations can also be easily obtained along profiles $\eta = $ const. (VSP, cross-hole): $J = -\eta \,(\sin i)\,(\partial z/\partial i_0)_\eta$.

b. Spherical coordinates. Spherical coordinates are suitable for studying seismic wave propagation in radially symmetric media. We choose $\xi_1 = \theta$, $\xi_2 = \varphi$, $\xi_3 = r$, where r, φ, and θ have the same meaning as in Section 3.7.4. The scale factors are $h_1 = r$, $h_2 = r \sin\theta$, and $h_3 = 1$. Quantity $p = r_0(\sin i_0)/V_0 = r(\sin i(r))/V(r)$ represents the standard seismological ray parameter p for a radially symmetric medium; see (3.7.31). Four alternative expressions for J in radially symmetric media are

$$
\begin{aligned}
J &= r^2(\sin\theta)(\cos i)(\partial\theta/\partial i_0)_r \\
&= r^2 r_0 V_0^{-1}(\sin\theta)(\cos i_0)(\cos i)(\partial\theta/\partial p)_r \\
&= r V(r)(\sin\theta)(\cotan i)(\partial T/\partial i_0)_r \\
&= r V(r) r_0 V_0^{-1}(\sin\theta)(\cos i_0)(\cotan i)(\partial T/\partial p)_r.
\end{aligned}
\tag{3.10.46}
$$

The relations (3.10.46) can be applied to analytical ray equations $\theta = \theta(p)$ and $T = T(p)$ derived in Section 3.7.4 for radially symmetric media. They correspond to the ray parameters $\gamma_1 = i_0$ and $\gamma_2 = \varphi_0$.

9. RAY JACOBIAN IN TERMS OF THE CURVATURE OF THE WAVEFRONT

A popular method of computing ray Jacobian J is based on curvatures of the wavefront. For general inhomogeneous media, the appropriate relations will be derived in detail in Section 4.10.3; see (4.10.30). Here we shall only present an important classical equation, valid for *homogeneous isotropic media*. In this equation, ray Jacobian J is expressed in terms of the *Gaussian curvature of wavefront K*:

$$
J(R)/J(S) = K(S)/K(R).
\tag{3.10.47}
$$

As we can see, the larger the Gaussian curvature of the wavefront is the smaller the ray Jacobian becomes. The relation is intuitively simple to understand, if we realize that the rays are straight lines in homogeneous media and that the main curvature directions do not rotate about a planar ray in a homogeneous medium.

The simple relation (3.10.47) can, of course, be expressed in many other alternative forms; for example, it can be expressed in terms of the two main curvatures or main radii of curvature of the wavefront.

The method based on curvatures of the wavefront has also been used in 2-D and 3-D models composed of homogeneous layers separated by curved structural interfaces. The rays can be computed semianalytically in this case; see Section 3.4.8. However, (3.10.47) can be applied only to elements of the ray between the individual interfaces, but not across the interfaces. The procedure must be supplemented by other equations for the change of curvature of the wavefront across a structural interface. Such an equation for a curved interface between two *inhomogeneous* media was first derived by Gel'chinskiy (1961); see also Červený and Ravindra (1971). Special cases of the general Gel'chinskiy equation, valid for a curved interface between two homogeneous media were well known even earlier; see references in Červený and Ravindra (1971). The general equations will also be derived in detail in Section 4.6.3, see particularly (4.6.21).

For many other details and applications, and for specific equations related to layered models, see, for example, Alekseyev and Gel'chinskiy (1959), Červený and Ravindra (1971), Deschamps (1972), Shah (1973b), Hubral (1979, 1980), Goldin (1979), Hubral and Krey (1980), Červený and Hron (1980), Ursin (1982a, 1982b), Lee and Langston (1983a), and Gjøystdal, Reinhardsen, and Ursin (1984).

3.10.5 Caustics. Classification of Caustics

The points of the ray, at which the ray Jacobian vanishes ($J = 0$) are called *caustic points*. At these points, the cross-sectional area of the ray tube shrinks to zero.

Obviously the determinant of a 3×3 matrix $\hat{\mathbf{W}}$ vanishes if the rank of $\hat{\mathbf{W}}$ is less than 3. Thus, we have two types of caustic points along the ray. We shall call them caustic points of the first and second order.

A *caustic point of the first order* is a point on the ray at which the following relation holds:

$$\mathrm{rank}\big(\hat{\mathbf{Q}}^{(x)}\big) = 2. \tag{3.10.48}$$

At a caustic point of the first order, the ray tube shrinks to an elementary arc, perpendicular to the direction of propagation. See Figure 3.13(a).

A *caustic point of the second order*, also called the *focus point*, is a point on the ray, at which the following relation holds:

$$\mathrm{rank}\big(\hat{\mathbf{Q}}^{(x)}\big) = 1. \tag{3.10.49}$$

At a caustic point of the second order, the ray tube shrinks to a point. See Figure 3.13b.

In computing the displacement vector along the ray, we need to know ray Jacobian $J = \det \hat{\mathbf{Q}}^{(x)}$ because the displacement vector is proportional to $J^{-1/2}$. In passing through the caustic point of the first order, ray Jacobian J changes sign, and the argument of $J^{1/2}$ takes the phase term $\pm\pi/2$. Similarly, in passing through the caustic point of the second order, the phase term is $\pm\pi$. The phase shift at caustic points plays an important role in computing the vectorial complex-valued amplitudes of seismic body waves.

The *phase shift due to caustics* is cummulative. If we pass through several caustic points along the ray, the total phase shift is the sum of the individual phase shifts. Consider ray Ω

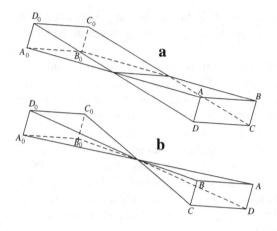

Figure 3.13. Two types of caustic points along the ray. (a) At a caustic point of the first order, the ray tube shrinks into an elementary arc, perpendicular to the direction of propagation. (b) At a caustic point of the second order (also called the focus or the double caustic point), the ray tube shrinks to a point.

from S to R. The phase shift due to caustics along Ω from S to R is then

$$T^c(R, S) = \pm \tfrac{1}{2}\pi k(R, S). \tag{3.10.50}$$

Here $k(S, R)$ is called the *index of ray trajectory Ω from S to R*, or also the *KMAH index*. In isotropic media, it equals the number of caustic points along ray trajectory Ω from S to R, caustic points of the second order being counted twice. The choice of sign in (3.10.50) will be discussed in Sections 3.10.6, 4.12, and 5.8.8. For anisotropic media, see Sections 4.14.13 and 5.8.8.

The term "index of the ray trajectory" for $k(R, S)$ is used, for example, by Kravtsov and Orlov (1980). The alternative term, the KMAH index, was introduced by Ziolkowski and Deschamps (1980), acknowledging the work by Keller (1958), Maslov (1965), Arnold (1967), and Hörmander (1971) on this problem. See also the discussion in Chapman and Drummond (1982).

In space, caustic points of the first order are not isolated, they form *caustic surfaces*. From a physical point of view, these surfaces are envelopes of rays. There are many possible forms of caustic surfaces. They can, however, be grouped into several basic types. A detailed classification of caustic surfaces can be found in Kravtsov and Orlov (1980). The classification is very close and takes its terminology from the theory of catastrophes; see, for example, Thom (1972), Arnold (1974), Gilmore (1981), Brown and Tappert (1987). In Gilmore (1981), the interested reader will find a whole chapter devoted to the catastrophe theory in relation to the eikonal equation.

3.10.6   Solution of the Transport Equation in Terms of the Ray Jacobian

Transport equations can be suitably solved *along rays* in terms of the ray Jacobian. Let us first consider the *acoustic case* with variable density ρ. The transport equation is then given by (2.4.11). Along the ray, $\nabla T = c^{-1}\vec{t}$, where c is the acoustic velocity and $\vec{t}$ is the unit vector tangent to the ray. We can put $\vec{t} \cdot \nabla(P/\sqrt{\rho}) = \mathrm{d}(P/\sqrt{\rho})/\mathrm{d}s$. The transport equation then reads

$$\frac{\mathrm{d}}{\mathrm{d}s}\left(\frac{P}{\sqrt{\rho}}\right) + \frac{c}{2}\frac{P}{\sqrt{\rho}}\nabla^2 T = 0. \tag{3.10.51}$$

Inserting (3.10.30) for $\nabla^2 T$, we obtain

$$\frac{d}{ds} \ln\left\{ P(s)\sqrt{\frac{J(s)}{\rho(s)c(s)}} \right\} = 0.$$

The solution of this equation is

$$P(s) = \Psi(\gamma_1, \gamma_2)(\rho(s)c(s)/J(s))^{1/2}. \tag{3.10.52}$$

Here $\Psi(\gamma_1, \gamma_2)$ is constant along the ray. It may, of course, vary from one ray to another. The solution can also be given another form:

$$P(s) = \left[\frac{\rho(s)c(s)J(s_0)}{\rho(s_0)c(s_0)J(s)} \right]^{1/2} P(s_0). \tag{3.10.53}$$

Equation (3.10.53) can be used to determine the amplitude $P(s)$ along the whole ray when $P(s_0)$ is known at some reference point $s = s_0$ of the ray.

Equations (3.10.52) and (3.10.53) represent two final forms of the solution of the acoustic transport equation. They can, of course, be expressed in many alternative forms. For example, instead of the ray Jacobian, we use general Jacobian $J^{(u)}$ and relation $J = g_u J^{(u)}$. For example, if $u = T$, we have $g_u = c^{-1}$ and (3.10.53) becomes

$$P(T) = \left[\frac{\rho(T)c^2(T)J^{(T)}(T_0)}{\rho(T_0)c^2(T_0)J^{(T)}(T)} \right]^{1/2} P(T_0). \tag{3.10.54}$$

If we use $u = \sigma$, $g_u = c$ and the velocity factors are eliminated from (3.10.53).

In a similar way, we can solve the transport equations for *elastic P and S waves. For P waves*, the transport equation for amplitude factor A is given by (2.4.32), and its solution reads

$$A(s) = \frac{\Psi(\gamma_1, \gamma_2)}{[\rho(s)\alpha(s)J(s)]^{1/2}} \tag{3.10.55}$$

or, alternatively,

$$A(s) = \left[\frac{\rho(s_0)\alpha(s_0)J(s_0)}{\rho(s)\alpha(s)J(s)} \right]^{1/2} A(s_0). \tag{3.10.56}$$

If we prefer some other Jacobian, we can again use $J = g_u J^{(u)}$.

For S waves, the form of the transport equations depends on the choice of unit vectors $\vec{e}_1$ and $\vec{e}_2$. If they satisfy (2.4.36), the transport equations for $B(s)$ and $C(s)$ are decoupled and have exactly the same form as the transport equation for P waves, only velocity α is replaced by β. Their solutions are

$$B(s) = \frac{\Psi_1(\gamma_1, \gamma_2)}{[\rho(s)\beta(s)J(s)]^{1/2}}, \qquad C(s) = \frac{\Psi_2(\gamma_1, \gamma_2)}{[\rho(s)\beta(s)J(s)]^{1/2}} \tag{3.10.57}$$

or, alternatively,

$$B(s) = \left[\frac{\rho(s_0)\beta(s_0)J(s_0)}{\rho(s)\beta(s)J(s)} \right]^{1/2} B(s_0),$$

$$C(s) = \left[\frac{\rho(s_0)\beta(s_0)J(s_0)}{\rho(s)\beta(s)J(s)} \right]^{1/2} C(s_0). \tag{3.10.58}$$

If (2.4.36) is not satisfied, $\vec{e}_1$ and $\vec{e}_2$ do not represent polarization vectors of S waves, and the transport equations for B and C are coupled. It is not difficult to find the solution of these transport equations, but we do not present them here. For $\vec{e}_1 = \vec{n}$ and $\vec{e}_2 = \vec{b}$, where $\vec{n}$ is the unit normal and $\vec{b}$ is the unit binormal to the ray, the solutions of the coupled transport equations may be found, for example, in Červený and Ravindra (1971).

Finally, for *anisotropic media*, the transport equation is given by (2.4.49). Along the ray,

$$\frac{\mathrm{d}}{\mathrm{d}s}(\sqrt{\rho}A) + \frac{1}{2\mathcal{U}}\sqrt{\rho}A\nabla \cdot \vec{\mathcal{U}} = 0, \tag{3.10.59}$$

where $\vec{\mathcal{U}}$ is the group velocity vector. Taking into account relation (3.10.25) for $\nabla \cdot \vec{\mathcal{U}}$, we obtain

$$A(s) = \frac{\Psi(\gamma_1, \gamma_2)}{[\rho(s)\mathcal{U}(s)J(s)]^{1/2}}. \tag{3.10.60}$$

An alternative form reads

$$A(s) = \left[\frac{\rho(s_0)\mathcal{U}(s_0)J(s_0)}{\rho(s)\mathcal{U}(s)J(s)}\right]^{1/2} A(s_0). \tag{3.10.61}$$

Instead of $J(s)$, we can again use $J = g_u J^{(u)}$. In anisotropic media, it is most usual to use $u = T$, where T is the travel time along the ray. As $g_T = 1/\mathcal{U}$, (3.10.61) and (3.10.23) yield

$$A(T) = \left[\frac{\rho(T_0)J^{(T)}(T_0)}{\rho(T)J^{(T)}(T)}\right]^{1/2} A(T_0) = \left[\frac{\rho(T_0)\mathcal{C}(T_0)\Omega^{(T)}(T_0)}{\rho(T)\mathcal{C}(T)\Omega^{(T)}(T)}\right]^{1/2} A(T_0)$$

$$= \left[\frac{\rho(T_0)\mathcal{U}(T_0)J(T_0)}{\rho(T)\mathcal{U}(T)J(T)}\right]^{1/2} A(T_0). \tag{3.10.62}$$

This equation also holds for P and S waves in isotropic media; see (3.10.56) and (3.10.58) for $u = T$, with $J = J^{(T)}/\alpha$ or $J = J^{(T)}/\beta$. However, it does not hold for the pressure amplitudes in acoustic waves; see (3.10.54).

Thus, we have obtained three different expressions for amplitudes of seismic body waves in anisotropic media. They contain different quantities J, $J^{(T)}$, and $\Omega^{(T)}$, describing the divergence of rays. These three expressions also depend in a different way on phase and group velocities. One of them contains the group velocity $\mathcal{U}$, one phase velocity $\mathcal{C}$, and one does not contain a velocity at all; see (3.10.62). These differences have been a source of confusion in the computation of amplitudes of seismic body waves in anisotropic media. Consequently, we need to specify precisely which of the three quantities (J, $J^{(T)}$, or $\Omega^{(T)}$) we wish to use to describe the divergence of rays, if we use expressions for amplitudes of seismic body waves in anisotropic media. According to the chosen quantity J, $J^{(T)}$, or $\Omega^{(T)}$, we have to use the proper equation.

All the equations derived in this section have a very simple physical meaning. As we know, the energy of HF seismic body waves flows along rays. There is no energy flux through the walls of the ray tube according to the zeroth-order approximation of the ray method. In other words, the energy flux through $\mathrm{d}\Omega^{\perp}$ must be the same along the whole ray, independently of s. We denote the energy flux through $\mathrm{d}\Omega^{\perp}$ by $E^{\perp}$. For an inhomogeneous anisotropic medium, we then obtain

$$E^\perp = \rho \mathcal{U} A A^* f_c d\Omega^\perp = \rho \mathcal{U} J A A^* f_c d\gamma_1 d\gamma_2 = \text{const.};$$

see Section 2.4.4. Because $f_c d\gamma_1 d\gamma_2$ is constant along Ω, we obtain $\rho \mathcal{U} J A A^* = \text{const.}$ along the whole ray. This is in full agreement with (3.10.61).

Because the ray Jacobian may be negative, it is useful to determine strictly the sign of $J^{1/2}$ in the equation derived. This can be done using the *phase shift due to caustics* $T^c(s_0, s)$, which was discussed in the previous section. We shall consider only one of the derived equations, (3.10.61), corresponding to general anisotropic media. This equation can take the following form:

$$A(s) = \left[\frac{\rho(s_0)\mathcal{U}(s_0)|J(s_0)|}{\rho(s)\mathcal{U}(s)|J(s)|} \right]^{1/2} A(s_0) \exp[iT^c(s, s_0)], \qquad (3.10.63)$$

where the phase shift due to caustics is given by (3.10.50).

The correct sign of the phase shift due to caustics $T^c(s, s_0)$ in (3.10.50) has been discussed in the wave propagation literature for a long time. For the analytical signal given by (2.2.9), with the plus sign in the definition equation $F(\zeta) = x(\zeta) + ig(\zeta)$, the correct sign in (3.10.50) is minus ($-$) so that

$$T^c(s, s_0) = -\tfrac{1}{2}\pi k(s, s_0), \qquad (3.10.64)$$

where $k(s_0, s)$ is the KMAH index. Equation (3.10.64) is, of course, also valid for time-harmonic waves with time factor $\exp[-i\omega(t - T)]$. For a proof, see Section 5.8.8.

The analytical signal with the minus sign in the definition equation, $F(\zeta) = x(\zeta) - ig(\zeta)$, is also commonly used. In this case, the sign in (3.10.64) would be plus ($+$) instead of minus ($-$). This also applies to time-harmonic waves with time factor $\exp[+i\omega(t - T)]$.

3.11 Boundary-Value Ray Tracing

Boundary-value ray tracing plays a considerably more important role in seismology and in seismic exploration than initial-value ray tracing. On the other hand, it is also more complex and time-consuming.

Many different procedures of boundary-value ray tracing have been proposed. It is not simple to classify these procedures and to separate them into certain groups. This classification may always be applied in different ways. The classification given here is far from complete. Moreover, we shall consider only the ray tracing methods suitable for computing ray-theory travel times. The grid methods of computing the first-arrival travel times are described in Section 3.8. They represent automatically the solution of the boundary-value travel-time problem.

Before we discuss the methods of solving the boundary-value ray tracing problems in Sections 3.11.2 through 3.11.4, we shall briefly review various types of these problems and demonstrate them on simple examples. For completeness, we shall also include initial-value ray tracing problems.

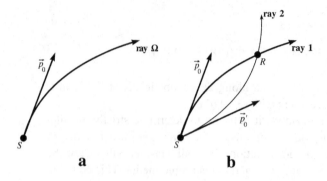

Figure 3.14. Initial-value ray tracing. (a) Ray Ω is specified by the coordinates of initial point S and by the initial direction of the ray at S (the direction of initial slowness vector $\vec{p}_0$). (b) There may be two or more rays leaving S in different directions but that pass through the same point R (multiple rays from S to R).

3.11.1 Initial-Value and Boundary-Value Ray Tracing: A Review

1. INITIAL-VALUE RAY TRACING

The initial conditions for a single ray were discussed in Section 3.2.1. At an initial point, the initial direction may be specified by two take-off angles i_0 and ϕ_0; see Figures 3.14(a) and 3.3. The components of the slowness vector at the initial point are then obtained from (3.2.3). Once the initial conditions are known, the ray can be easily computed by solving the relevant ray tracing system. When we start computing the ray, we do not know through which points the ray will pass.

If we also wish to determine the travel time along the ray, we must specify the initial travel time, that is, the travel time at the initial point of the ray.

In a smooth medium without interfaces, the ray is uniquely defined by its initial conditions. In layered and block structures, the ray is defined not only by the initial conditions but also by the ray code of the elementary wave under consideration.

Note that various rays of the same elementary wave may intersect so that there may be two or more rays leaving point S in different directions and passing through the same point $R \neq S$. See Figure 3.14(b). We shall call the rays of the same elementary wave that intersect at point R *multiple rays* at R. The multiplicity of rays depends, of course, on the position of point R. Often, there will be only one ray; in other cases, there will be two or more rays that connect S with R. If several multiple rays connect S and R, the travel times are also multivalued.

In certain applications, the initial direction of the ray at point S is not given, but it can be calculated from some other data. This applies mainly to such problems in which the initial travel time is given along the initial surface, Σ^0. The initial direction of the ray can then be calculated at any point S of initial surface Σ^0, assuming that the geometry of Σ^0, the distribution of the initial travel time along Σ^0, and the propagation velocity are known in the vicinity of point S. More details will be given in Section 4.5.

The simplest initial-surface ray tracing problems correspond to the ray tracing of normal rays. If the travel-time distribution along initial surface Σ^0 is constant, all rays are perpendicular to Σ^0; see Figure 3.9a. Initial surface Σ^0 then behaves like a wavefront. The initial direction of any ray can be simply determined using (3.2.5) as it corresponds to the direction of the unit normal to Σ^0. For this reason, we often speak of normal rays. The initial components of slowness vector $p_k(S)$ are obtained from $n_k(S)$ determined by (3.2.5) using relation $p_k = n_k/V$, where V is the relevant propagation velocity at S.

Ray tracing of normal rays has important applications in seismic prospecting. In the "exploding reflector" algorithm in seismic reflection methods, it is assumed that the reflector

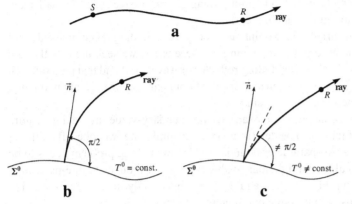

Figure 3.15. Boundary-value ray tracing. (a) In two-point ray tracing, we seek the ray that passes through two fixed points S and R. (b, c) In initial surface-fixed point ray tracing, we seek the rays that leave initial surface Σ_0 and pass through point R. The ray is perpendicular to Σ^0 if the initial time T^0 is constant along Σ^0; see (b). The position of initial point S on initial surface Σ^0 is not known a priori.

explodes at some specified time. Thus, the reflector behaves like a wavefront and all rays are perpendicular to it.

If the initial travel time is not constant along Σ^0, the rays are not perpendicular to Σ^0, and the initial surface does not behave like a wavefront; see Figure 3.9b. Nevertheless, the initial direction of rays may simply be computed at any point of Σ^0. The procedure of determining the initial direction of the ray in this case is described in Section 4.5.

2. BOUNDARY-VALUE RAY TRACING

As mentioned earlier, the ray is not specified by the initial conditions, but by some other conditions, related to different points of the ray in boundary-value ray tracing.

The most important case of boundary-value ray tracing is *two-point ray tracing*. In two-point ray tracing, we seek the ray that connects two fixed points S and R; see Figure 3.15(a). The initial directions of the ray at S and R are not known in this case. In fact, the problem of two-point ray tracing corresponds to the definition of the ray by Fermat's principle.

As we can see in Figure 3.14(b), the solution of two-point ray tracing is not necessarily unique; sometimes we can find two or more multiple rays that connect S with R. The travel time along one of them corresponds to the absolute minimum. The travel times calculated along other rays are larger, but they still render Fermat's functional stationary. Such situations are very common in applications, especially in the case of refracted waves, waves reflected from interfaces of the syncline form, and the waveguide propagation.

There are numerous examples in seismology and in seismic prospecting where two-point ray tracing is needed. Let us name, among others, the location of earthquake hypocenters, construction of synthetic seismograms, particle ground-motion diagrams, and solution of many inverse problems (including tomography and migration).

The next example of boundary-value ray tracing is *initial surface-fixed point ray tracing*. In this case, the distribution of the initial travel-time field T^0 is given along initial surface Σ^0, and we wish to find the relevant ray passing through a fixed point R; see Figure 3.15(c). We know neither the initial direction of the ray at R nor the position of the initial point S of the ray at Σ^0. As in two-point ray tracing, initial surface-fixed point ray tracing is not unique; several rays starting from different points along Σ^0 may arrive at R. We shall again speak of multiple rays and multivalued travel times.

Special cases of initial surface-fixed point ray tracing are exploding reflector-fixed point ray tracing and wavefront-fixed point ray tracing. In these cases, we seek the rays that are perpendicular to initial surface Σ^0 (exploding reflector or wavefront) and that pass through fixed point R; see Figure 3.15(b). We can also speak of boundary-value normal ray tracing, as opposed to initial-value normal ray tracing.

In three-dimensional media, all these examples of boundary-value ray tracing require two initial parameters of the ray to be determined (for example, the two take-off angles i_0 and ϕ_0 at S or R) so that a two-parameter boundary-value ray tracing problem needs to be solved. In two-dimensional media, boundary-value ray tracing reduces to one-parameter boundary-value ray tracing. Figures 3.14 and 3.15 are shown only in 2-D, but imagining the relevant 3-D situations with two free parameters is simple.

Literature devoted to boundary-value ray tracing is very extensive. Let us list only a few important references: Wesson (1970, 1971), Chander (1975), Julian and Gubbins (1977), Pereyra, Lee, and Keller (1980), Thurber and Ellsworth (1980), Lee and Stewart (1981), Keller and Perozzi (1983), Docherty (1985), Um and Thurber (1987), Waltham (1988), Hanyga (1988), Obolentseva and Grechka (1988), Pereyra (1988, 1992, 1996), Prothero, Taylor, and Eickemeyer (1988), Virieux, Farra, and Madariaga (1988), Farra, Virieux, and Madariaga (1989), Sambridge and Kennett (1990), Virieux and Farra (1991), Farra (1992), Moser, Nolet, and Snieder (1992), Hanyga and Pajchel (1995), Passier and Snieder (1995), Wang and Houseman (1995), Bulant (1996, 1999), Grechka and McMechan (1996), Guiziou, Mallet, and Madariaga (1996), Clarke and Jannaud (1996), Hanyga (1996b), Liu and Tromp (1996), Mao and Stuart (1997), and Koketsu and Sekine (1998). For a point-to-curve ray tracing, see Hanyga and Pajchel (1995) and Hanyga (1996b).

3.11.2 Shooting Methods

The shooting method is an iterative procedure that uses standard initial-value ray tracing to solve a boundary-value ray tracing problem.

Initial-value ray tracing, however, may be applied in different ways. We shall describe two of them.

1. Standard shooting method. In the standard shooting method, the initial-value ray tracing procedure is put within an iterative loop to find the ray passing through receiver R. Standard shooting is usually applied to the selected elementary wave under consideration, but it may also be applied to a group of elementary waves, the travel times of which are in some way connected. For simplicity, we shall consider one elementary wave in the following. The loop, however, should be organized very carefully because the ray field corresponding to the elementary wave may be very complicated. There may be shadow regions where no rays of the elementary wave under consideration arrive. Similarly, there may be regions of overlapping, where two or more rays of the same elementary wave arrive at the same receiver point R (multipathing). It is useful to find boundaries between the individual regions of the same ray history of the elementary wave under consideration. The greatest danger in the standard shooting method is represented by elementary waves, the ray field of which corresponds to a very narrow range of ray parameters but that covers a large part of the region of the model of interest. Typical representatives of such waves are slightly refracted waves, similar to head waves, in refraction studies.

Figure 3.16. Solution of boundary-value ray tracing by the shooting method. (a) Two-point ray tracing. We wish to find the ray from S that passes through R. Three trial rays are shown. (b) Initial surface-fixed point ray tracing. We wish to find the ray from initial surface Σ^0 that passes through R. The ray must satisfy the relevant conditions at S on Σ^0. Three trial rays are shown. The loops may also be organized in the opposite directions, by shooting from R.

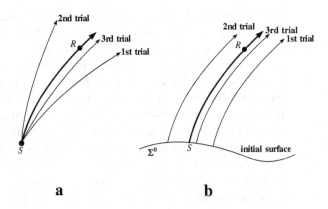

a   **b**

We shall now present several simple examples. We shall first consider a two-point ray tracing problem. We wish to find the ray that connects points S and R in Figure 3.16(a). We take one of the two points as the initial point, for example, point S. We then shoot the rays from S under different take-off angles i_0 and ϕ_0 in the direction of point R. Each ray is computed by standard initial-value ray tracing. An iterative loop is used to find the ray that passes through the other point, R. In this way, the shooting method resembles a gun firing from S at a target at R. See Figure 3.16(a) for a 2-D situation.

Various methods can be used to accelerate the convergence of the iterations. For example, considerable acceleration can be achieved by applying one of the recent sophisticated methods, such as the method of paraxial ray approximation. See Section 4.9.

An alternative shooting procedure can also be applied in initial-surface to fixed-point ray tracing. Here we wish to find a ray shot from initial surface Σ^0 that passes through point R; see Figure 3.16(b). We successively shoot rays from different points S_i on Σ^0. At any point S_i, the initial direction of the ray is fully determined by the geometry of Σ^0, the distribution of the initial travel time along Σ^0, and the propagation velocity at S_i. Thus, the initial direction of the ray at any point S_i is fixed and cannot be changed, but the ray from S_i does not, in general, pass through R. An iterative loop must be used to find point S on Σ^0 from which the ray passes through R. Alternatively, we can also shoot rays from R to Σ^0 and minimize iteratively the differences between the directions of these and computed initial directions along Σ^0.

The shooting method has been successfully used mainly in 2-D models, in situations in which we need to find rays shot from a point source to a series of receivers distributed regularly or irregularly in some region along the surface of the Earth. We start shooting rays that hit the Earth's surface outside the region with receivers. We then regularly vary the take-off angle to come closer to the receiver region. As soon as we overshoot a receiver, we return and determine the ray passing through the point using standard numerical interpolation techniques. Traditionally, the regula falsi, halving of intervals, or one of their combinations has been used. The method of halving of intervals is slower, but safer. We then continue to shoot new rays at regular steps in the take-off angles. In the *modification of this procedure*, special care is devoted to the boundaries of ray fields with different ray histories, particularly to the boundaries of shadow zones, to critical rays, to multiple arrivals of the individual waves, and to the search for waves that may easily be overlooked (such as slightly refracted waves). If the loop is properly organized, the shooting method safely determines most of the rays of all the elementary waves under consideration arriving at the receiver positions. If an elementary wave has several branches, the modification of the shooting method determines all relevant multiple arrivals of the elementary waves. This modified shooting method has been found useful in the construction of 2-D ray synthetic seismograms; see Červený and Pšenčík (1984b).

A similar approach can be used if the sources and/or receivers are distributed in a borehole (VSP and/or cross-hole profiles) in a 2-D structure. It can also be used in other types of 2-D boundary-value ray tracing, such as boundary normal ray tracing. Boundary normal ray tracing has found applications in the construction of synthetic time sections in seismic explorations using the "normal ray" algorithm.

Let us now briefly discuss the main principles of the shooting method in a *general 3-D layered and block model*. For simplicity, we shall discuss only the two-point ray tracing problem. The algorithm described here is, however, applicable even to other boundary-value ray tracing problems. We shall consider a point source at S, situated arbitrarily in the model, and a system of receivers R_i, $i = 1, 2, \ldots, n$, distributed along some reference surface Σ^R inside the model or on its boundary. Surface Σ^R may represent the Earth's surface, a structural interface, or merely a formal surface inside the model, which may, for example, contain borehole(s) with receivers. Reference surface Σ^R may be finite. We shall seek the rays of a selected elementary wave generated by a point source situated at S, specified by a proper ray code. We call any ray of the elementary wave under consideration, with the initial point at S and with the terminal point at Σ^R, the *successful ray*. The ray parameters of rays of the elementary wave under consideration with the initial point at S form a 2-D *ray-parameter domain*. The size of the ray-parameter domain may be specified by input data, but it must cover ray parameters of all two-point rays connecting S with R_i, $i = 1, 2, \ldots, n$. Because these rays are not known a priori, it is useful to consider greater ray-parameter domains. Not all ray parameters in the ray-parameter domain represent successful rays; some of the rays may be terminated inside the model or on its boundaries. Moreover, the ray histories of the individual successful rays may be different. The basic problem of two-point ray tracing resides in the accurate decomposition of the 2-D ray-parameter domain into *homogeneous subdomains*, containing the ray parameters of rays with the same histories, and in the determination of demarcation belts between homogeneous subdomains. Two-point ray tracing may be done by means of sophisticated triangularization of the ray domain. A proposal of the algorithm of such triangularization is described in detail by Bulant (1996, 1999), where many numerical examples can also be found.

The triangularization of any homogenous subdomain corresponding to successful rays in the ray-parameter domain is mapped by rays into a triangularization of some region of reference surface Σ^R. As soon as the receiver R_k is situated in any homogeneous triangle on Σ^R, the relevant two-point ray from S to R_k can be found by routine interpolation methods; see Bulant and Klimeš (1999). Alternatively, the paraxial ray approximation from the closest apex point, or weighting of the paraxial ray approximation from all three apex points, may also be used. See Section 4.9.

It should be emphasized that the decomposition of the ray-parameter domain into homogeneous subdomains is the most important step of the procedure. The routine interpolation methods and/or the paraxial ray approximation methods can be used only inside homogeneous subdomains. They usually fail if they are applied across boundaries of the homogeneous subdomains because the rays in different homogeneous subdomains have different histories. Moreover, the decomposition is also a crucial step in the search for all multiple rays of the elementary wave under consideration arriving at the receiver.

2. Controlled initial-value ray tracing. The controlled initial-value ray tracing does not require rays passing through specified receivers to be found, but it does require that the model (or some target region of the model) be covered by a sufficiently dense system of rays of some elementary wave specified by the ray code. The decomposition of the ray-parameter domain into homogeneous subdomains, and the triangularization of homogeneous subdomains corresponding to rays of equal ray history again represents the key part of the algorithm. We shall introduce a ray tube corresponding to a homogeneous triangle of the ray-parameter domain and call it the *homogeneous ray tube*. The homogeneous ray tube can then be decomposed into *ray cells*, separated by consecutive wavefronts $T = T_1, T_2, \ldots, T_m$. Inside each ray cell, the travel times, Green function amplitudes, and other ray-theory quantities may be calculated by routine interpolation, the paraxial ray method, or weighting of paraxial ray approximations. Similarly as in the case of a two-point ray tracing described earlier, this interpolation would be questionable between rays of different ray histories. For this reason, it is necessary to interpolate only within homogeneous ray tubes. See Bulant (1999) and Bulant and Klimeš (1999).

The controlled ray tracing may be very efficient if we need to calculate the ray-theory quantities at grid points in a target region of a model represented by a discrete grid of points. For example, it may play an important role in the migration of seismic data in seismic exploration for oil.

It should be noted that certain extensions of the ray method (like the Maslov-Chapman method or the method of summation of Gaussian beams) do not require that the ray passing exactly through the receiver to be known to compute the wavefield at the receiver; it is sufficient to know the rays at some neighboring points. Using these methods, it would be sufficient to cover a target region by a sufficiently dense system of ray points to compute the wavefield at any point of the target region. And this may be performed by controlled ray tracing.

The main principles of the controlled ray tracing method are very similar to the wavefront construction method described in Section 3.8.4. There are, however, several important differences. The main difference follows. In the controlled ray tracing method, all rays are computed directly from the source (or from the initial surface). Sufficient density of the ray field in the target region is ensured by triangularization of the ray-parameter domain. In the wavefront construction method, the rays are used to compute the wavefronts successively. At each wavefront, the number of rays is adjusted according to the *local behavior of the ray field*. Consequently, the ray field in the wavefront construction method is always sufficiently dense, and the succeeding interpolations are sufficiently accurate. Moreover, the computed rays are shorter because they are not computed directly from the source but rather from the individual wavefronts.

The preceding two shooting methods may also be used in anisotropic inhomogeneous media.

3.11.3 Bending Methods

The next method used to solve boundary-value ray tracing problems is called the bending method. The method does not use standard initial-value ray tracing. In the bending method, an initial ray path is guessed and then perturbed iteratively so as to find the relevant boundary-value ray. Note that the guessed trajectory need not, in general, correspond to any actual ray, it may just be an auxiliary reference curve connecting points S and R;

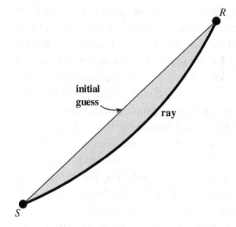

Figure 3.17. The solution of two-point ray tracing by the bending method. The initial guess ray path is guessed and then perturbed iteratively. The final ray is shown as the bold line.

see Figure 3.17. For example, in a layered medium, it does not need to satisfy Fermat's principle in smoothly varying medium and Snell's law at structural interfaces. If the initial guess is far from the actual ray connecting points S and R, the method may diverge.

Thus, bending methods do not represent a complete solution of the boundary-value ray tracing problem and of the subsequent determination of ray-theory travel times. First, an independent algorithm to estimate the guessed trajectories must be used. We can call this algorithm the *ray estimator*. After the ray estimator has been applied and has generated preliminary ray trajectory guesses for all sources and receivers, the bending method may be applied as a *postprocessor*, correcting the preliminary trajectories. This correction procedure is also known as bending the rays. However, the number and nature of the finally determined boundary-value rays depends on the ray estimator used.

On the one hand, the bending method is usually faster than the shooting method. On the other hand, it requires reasonable ray estimators; otherwise, many rays and relevant travel times may be lost.

The bending method may be very useful in combination with other methods. For example, it may be useful in combination with the shooting method, where it may be used to determine rays difficult for shooting such as slightly refracted waves. Indeed, these rays may easily be computed by bending. However, the shooting method may be used to find the guess reference curve to initiate the bending procedure. The reference curve need not be calculated with high accuracy because it serves only as a first guess (Pereyra, 1996). It may also be useful to combine the bending method with shortest-path ray tracing (see Section 3.8.3), taking the network rays as preliminary ray estimates. The postprocessing by the bending method may increase the accuracy of network ray tracing considerably. However, it would also be more time-consuming. See Moser, Nolet, and Snieder (1992).

We shall discuss only the calculation of one selected ray, corresponding to specified positions of the source and receiver, assuming that the guessed reference curve is known. In practical applications in seismic exploration for oil, a whole system of closely spaced receivers is often considered. In this case, the *receiver continuation strategy* (Pereyra, 1996) can be used to accelerate the computations for the whole system of receivers. In this strategy, a known ray for one source and receiver position may be used as a reference guess curve to initiate the bending procedure for the same source and neighboring receiver positions. The same, of course, also applies to the *source continuation strategy* for closely spaced point sources. Note that paraxial ray methods can also be successfully applied in this case; see Section 4.9. In paraxial ray methods, the ray passing through points S' and R' situated close to S and R can be approximately found analytically, if the ray propagator matrix along the reference ray from S to R is known. This matrix can be simply determined by dynamic ray tracing along a reference ray from S to R.

In the following, we shall discuss the three most common groups of bending post-processing methods. In all three groups, the structure of the model is assumed to be fixed. For the methods based on structural perturbations, see the Section 3.11.4.

1. METHODS BASED ON FITTING RAY TRACING EQUATIONS

The basic idea regarding the solution of the two-point ray tracing problem using these methods consists in determining such a curve connecting fixed points S and R, along which the ray tracing system is satisfied. It is assumed that a guessed reference curve Ω_0 connecting points S and R is known. To demonstrate the method, we shall consider the ray tracing system (3.1.22), consisting of three ordinary differential equations of the second order for $n = 1$,

$$\mathrm{d}(V^{-1}\mathrm{d}x_i/\mathrm{d}s)/\mathrm{d}s - \partial(1/V)/\partial x_i = 0, \qquad i = 1, 2, 3, \qquad (3.11.1)$$

where s is the arclength along the ray. If the reference curve Ω_0 represents a ray, the LHSs of (3.11.1) vanish at all points along Ω_0. If, however, it differs from a ray, the LHSs of (3.11.1) are, in general, different from zero, at least at some points along Ω_0. The bending procedure involves determining the curve Ω in the vicinity of Ω_0, with vanishing LHSs of (3.11.1) at all points along Ω. For this purpose, (3.11.1) is discretized, and the derivatives at nodal points are replaced by finite differences. The resulting system of nonlinear algebraic equations is then linearized and solved iteratively for displacements of nodal points, for which (3.11.1) is satisfied. The size of the resulting linear algebraic system is related to the number of finite difference nodes and may be rather large. See, for example, Julian and Gubbins (1977), Pereyra, Lee, and Keller (1980), Lee and Stewart (1981), and Thomson and Gubbins (1982). The approach may be applied to ray tracing systems expressed in both Lagrangian and Hamiltonian forms. The Lagrangian approach uses the Euler-Lagrange ordinary differential equations of the second order in x_i and $x_i' = \mathrm{d}x_i/\mathrm{d}s$; see, for example, (3.11.1). A detailed and tutorial treatment can be found in Snieder and Spencer (1993); see also Snieder and Sambridge (1992) and Pulliam and Snieder (1996). In the Hamiltonian approach, the standard ray tracing system of ordinary differential equations of the first order in x_i and p_i is applied; see, for example, Pereyra (1996). Pereyra (1996) also shows how the structural discontinuities are handled in the procedure and how the additional computations such as 3-D geometrical spreading, sensitivity studies, and travel-time inversion may be included.

2. METHODS BASED ON PARAXIAL RAY APPROXIMATION

If the reference curve represents a true ray, the paraxial ray approximation can be used to compute any paraxial ray in the quadratic vicinity of the true ray and to solve an arbitrary boundary-value problem for the paraxial rays. An important role in these computations is played by the ray propagator matrix, which can be determined by dynamic ray tracing along the reference ray. The computations may be performed in ray-centered coordinates (see Sections 4.3.1 through 4.3.6 for 4×4 ray propagator matrices), or in Cartesian coordinates (see Sections 4.3.7 and 4.7.2 for 6×6 ray propagator matrices). Once the ray propagator matrix is known along the reference ray, the paraxial rays and slowness vectors along these rays may be calculated analytically.

The paraxial ray approximation yields only approximate results. Thus, the computation should be iterative. The complication is that paraxial rays are not exact so that they must be considered only as guess reference curves in iterations (not rays). This creates a new problem, namely the *construction of rays in the vicinity of an arbitrary reference curve*, which does not represent a ray.

In the Hamiltonian formalism, such a problem was first solved by Farra (1992). The formalism leads to dynamic ray tracing along the reference curve. The dynamic ray tracing system, however, is not homogeneous in this case but contains a "source term." (The source term vanishes if the reference curve is a true ray.) Using the propagator technique, the solution of the dynamic ray tracing system with a source term can be expressed analytically in terms of the propagator matrix of the homogeneous dynamic ray tracing system. Only one additional integration along the reference curve is required.

A complete treatment of the Hamiltonian approach to the construction of rays situated close to an arbitrary reference curve in a laterally varying layered structure (including interfaces) is given by Farra (1992). The treatment is applicable to any Hamiltonian so that it can be applied both to isotropic and anisotropic media. Note that the solution of the same problem in anisotropic media using Lagrangian formulation would be considerably more complicated. Farra (1992) also solves the relevant boundary-value problem and shows that it may be reduced to several linear algebraic equations. Thus, it is not necessary to solve a large system of linear equations. Note that the Hamiltonian approach to the bending method proposed by Farra (1992) may also be easily combined with structural perturbations; see Section 3.11.4.

3. METHODS BASED ON MINIMIZING THE TRAVEL TIME

Let us consider an elementary wave propagating in a layered/block model and estimate a guessed trajectory Ω_0 connecting the source and receiver and satisfying the ray code of the elementary wave. The optimization procedure is then used to find the curve with the minimum travel time among the curves situated in the vicinity of the guessed trajectory and corresponding to the same ray code. The guessed reference curve Ω_0 is discretized and solved iteratively for such displacements of the nodal points, for which the travel time along the curve would be minimized. The method was proposed by Um and Thurber (1987); see also Prothero, Taylor, and Eickemeyer (1988), Moser, Nolet, and Snieder (1992), and Koketsu and Sekine (1998). The method is also often called the *pseudo-bending* method. Various methods of linear programming (such as the simplex method, see Prothero et al., 1988) or even nonlinear programming have been applied (Schneider et al. 1992). The method has been broadly used in seismic exploration for models consisting of homogeneous layers and blocks. In this case, the ray elements between interfaces are straight lines, and the ray of any elementary wave may be fully specified by the coordinates of points of contact with the individual structural interfaces. In the optimization procedures, the coordinates of the points of contact of the ray trajectory with structural interfaces are sought. The guessed ray trajectories do not need to satisfy Snell's law at structural interfaces, but the optimization principles yield it. In other words, the final rays obtained by bending postprocessing satisfy Snell's law at any point of contact of the ray with a structural interface. Note that the procedure does not seek the general first-arrival travel times and corresponding rays but seeks instead the *ray-theory travel times corresponding to selected elementary waves*. If multiplicity occurs, however, the procedure is not safe; it may yield any of the arrivals corresponding to the selected elementary wave. Often, only one arrival of the elementary wave is obtained; the others are lost. The method is very popular in seismic exploration because it is very fast. It has also been applied to more complex structures than described here; see Guiziou and Haas (1988), Guiziou (1989), and Vesnaver (1996a, 1996b).

3.11.4 Methods Based on Structural Perturbations

In both the shooting and bending methods, the structure of the medium is fixed. Other methods of boundary-value ray tracing are based on structural perturbations.

We shall first describe the method, which has been called *the continuation method*, and which was proposed by Keller and Perozzi (1983). In the continuation method, the structure is gradually deformed. The method has mostly been applied to ray tracing in layered 2-D models with constant velocity within the individual layers, a buried source, and the receivers distributed along the surface. In the procedure, a model (also called the background model), which is simpler than the actual model being considered, is used first. For example, the method starts with a horizontally layered model only. In the background model, the ray from the source to the receiver, situated on the surface exactly above it, is merely a vertical straight line. The interfaces are then gradually deformed until the desired model is achieved. At each deformation step, the ray equations are solved, and the ray from the source to the receiver, situated vertically above it, is found. Source continuation is then applied by moving the source in a grid within the region of interest. At each source position, the rays to other receivers are found using the continuation of the receiver location. See Keller and Perozzi (1983) and Docherty (1985). The continuation method can also be efficiently combined with the shooting method and/or with the bending method; see Hanyga (1988).

Let us now consider the basic problem of the *ray perturbation theory* for an arbitrary 3-D laterally varying layered and block structure. We assume that the ray Ω^0 connecting two points S and R in a reference background medium $\mathcal{M}^0$ is known, and we wish to determine the ray Ω situated in a perturbed medium $\mathcal{M}$, connecting the same points S and R. We also assume that perturbed medium $\mathcal{M}$ differs only weakly from background medium $\mathcal{M}^0$. An analogous problem for travel-time perturbations was solved in Section 3.9; here we are interested in the perturbation of the ray itself. The problem of the ray perturbation will be treated in more detail in Section 4.7.4; here we shall only qualitatively explain its application to two-point ray tracing. Note that ray Ω^0 plays the role of the ray estimate in the bending method. Often, ray estimate Ω^0 may be found analytically, if model $\mathcal{M}^0$ is simple enough, or it may at least be found in a simpler way than ray Ω.

Similarly as in Section 3.9, we can again use the Hamiltonian or the Lagrangian formalism in the ray perturbation theory.

The Hamiltonian approach is based on the Hamiltonian formalism and solves the problem in the x_i-p_i-phase space. The Hamiltonian approach was used in the seismic ray perturbation problem in a series of papers by Farra, Madariaga, and Virieux; see, for example, Farra and Madariaga (1987), Farra, Virieux, and Madariaga (1989), Virieux (1989, 1991), Virieux, Farra, and Madariaga (1988), Virieux and Farra (1991), Farra (1989, 1990, 1992, 1993), Farra and Le Bégat (1995), and Farra (1999). The procedure is analogous to the construction of a true ray in the vicinity of a reference curve, described briefly in Section 3.11.3.2. It requires the inhomogeneous dynamic ray tracing system with a source term to be solved along Ω^0. The source term depends on the spatial derivatives of structural perturbations $\Delta \mathcal{H}$ of Hamiltonian $\mathcal{H}$ in this case. The solution can be expressed in terms of the propagator matrix of the homogeneous dynamic ray tracing system; see Section 4.3.8. Only one additional integration along reference ray Ω^0 is required. The approach may be applied to any form of Hamiltonian, to a layered medium, to isotropic and anisotropic media. The positions of endpoints S and R can be perturbed simultaneously with the structure. The solution of any boundary value ray tracing problem for ray Ω is not difficult; it only requires the solution of a few linear algebraic equations, containing the minors of the ray propagator matrix. Consequently, the Hamiltonian approach can also be applied if we wish to find ray Ω in $\mathcal{M}$ connecting points S' and R', where S' and R' do not coincide with S and R but are close to them. In a similar way, we can also solve any other boundary-value ray tracing problem for ray Ω.

The Lagrangian approach exploits the Euler-Lagrange ray tracing equations, expressed in terms of x_i and dx_i/du. For a detailed description with many references, see Snieder and Sambridge (1992, 1993), Snieder and Spencer (1993), and Pulliam and Snieder (1996). The Lagrangian approach leads to a system of linear algebraic equations that are tridiagonal. Using this approach, the second-order perturbation of travel time can also be determined by integration along reference ray Ω^0 in the background medium. The Lagrangian approach has so far been applied only to isotropic media; its application to general anisotropic media would be more involved.

In both Hamiltonian and Lagrangian approaches, the methods based on structural perturbations do not yield ray Ω in $\mathcal{M}$ exactly, but only approximately. If we wish to increase the accuracy of Ω, it is possible to apply standard bending methods in the perturbed medium $\mathcal{M}$ such as the methods based on the paraxial ray approximation, described in Section 3.11.3.2. In fact, in any of these two approaches, we can use a unified formulation that includes ray bending, structural perturbations, and paraxial ray tracing.

3.12 Surface-Wave Ray Tracing

In seismology, ray tracing has been applied not only to high-frequency seismic body waves propagating in laterally varying layered structures but also to surface waves propagating along a surface of a laterally varying elastic layered structure. Although this book is devoted primarily to the ray method for high-frequency seismic body waves, we shall also briefly describe the main principles of surface-wave ray tracing. The most important difference between high-frequency seismic body waves and surface waves is that the phase velocities of surface waves depend distinctly on frequency ω (even in nondissipative media), whereas the phase velocities of high-frequency seismic body waves in such media do not depend on ω. The rays of surface waves along a surface of the model are computed for a fixed frequency. For different frequencies, different rays are obtained.

The theory of seismic surface waves propagating along a flat surface of a one-dimensional (vertically inhomogeneous) layered structure has been well described in many seismological textbooks and papers. Two surface waves can propagate in such models: the slower *Rayleigh wave*, polarized in a vertical plane containing the direction of propagation, and the faster *Love wave*, polarized perpendicularly to this plane. The amplitudes of both waves are effectively concentrated close to the surface, in a layer the effective thickness of which is greater for lower frequencies ω. Both Rayleigh and Love waves consist of modes ($m = 0, 1, \ldots$). The most important are the lower modes, particularly the fundamental mode ($m = 0$). The higher modes are restricted to higher frequencies. The phase velocities of surface waves depend on frequency ω. The velocity dispersion is controlled by the so-called *dispersion relation*. The dispersion relations of different modes are different.

The dispersion relations can be expressed in various forms. For isotropic models, it is usual to specify them by the relation $\mathcal{C} = \mathcal{C}(\omega)$, where $\mathcal{C}$ is the phase velocity along the surface. An alternative form of dispersion relation, which will be used here, is $\omega = \omega(k)$, where k is the wavenumber, $k = \omega/\mathcal{C}(\omega)$. The determination of dispersion relations for a vertically inhomogeneous layered structure with a flat surface Σ is now a well-understood seismological problem. Propagator techniques for 1-D models and various matrix methods are mostly used to determine them. These methods are described in many seismological textbooks and papers; see, for example, Aki and Richards (1980, Chap. 7), where many other references can be found. At present, safe procedures and computer programs to calculate the dispersion relation for 1-D *anisotropic* layered structures with a flat surface are also well known; see Thomson (1996a, 1996b, 1997b) and Martin and Thomson (1997). Also in anisotropic models, the dispersion relations can be derived and computed by applying propagator techniques and matrix methods. In this case, the dispersion relation reads $\omega = \Omega(k_1, k_2)$, where k_1 and k_2 are two components of the wave vector at the surface. In both the isotropic and anisotropic case, the dispersion relations can be computed for arbitrary local realistic models of vertically inhomogeneous layered structures, with material parameters strongly varying with depth.

3.12.1 Surface Waves Along a Surface of a Laterally Varying Structure

Now we shall assume that the structure described here varies slowly laterally. Also the surface may be smoothly curved. Thus, the material parameters may vary strongly with depth, but the lateral variations should be slight. Often such models are called the "almost layered models." Because we shall consider a curved surface Σ of the model, it is useful to introduce 2-D curvilinear coordinates x^I ($I = 1, 2$) along the surface. They may be represented by Gaussian coordinates, for example. Coordinates x^1 and x^2 may be nonorthogonal. We denote the covariant components of the relevant 2-D metric tensor by g_{IJ}; see Section 3.5.6.

The main idea of the investigation of surface wave propagation in almost layered models of the Earth is a different treatment of the surface wave wavefield in the "horizontal direction" (along the surface Σ), and in the "vertical direction" (along normal to Σ). In a vertical direction, the wavefield is expressed locally by the normal mode theory, while the propagation along a surface is treated approximately, in much the same way as in the ray method. The theoretical treatment involves the *stretching* of horizontal coordinates and time, using a small parameter ϵ. In ocean acoustics, the method is called the *method of two-scale expansion* or the *method of horizontal rays and vertical modes*; see Burridge and Weinberg (1977) and Brekhovskikh and Godin (1989).

We shall select arbitrarily one mode of a surface wave and describe it by the ansatz relation:

$$\vec{u}(x^I, n, t) = \vec{A}(x^I, n, t) \exp[i\theta(x^I, t)], \qquad (3.12.1)$$

where x^1 and x^2 are the coordinates introduced along the surface Σ, n is the distance from Σ, measured along normal $\vec{n}$ to surface Σ, and t is time. The vectorial amplitude $\vec{A}$ is assumed to depend both on x^I and n, and the phase function θ is assumed to depend on x^I, but not on n. Moreover, it is assumed that $\vec{A}$ is a slowly varying function and θ a rapidly varying function of x^I and t. Thus, (3.12.1) represents some sort of space-time ray theory ansatz; see Section 2.4.6, especially Equation (2.4.72).

We define frequency ω and the covariant components of the wave vector k_I by the relations

$$\omega = -\partial\theta/\partial t, \qquad k_I = \partial\theta/\partial x^I. \qquad (3.12.2)$$

Quantities $\vec{A}$, k_I, and ω and the structure may vary smoothly laterally. (The variation of the structure in the normal direction may, however, be strong.) It is then possible to prove that

$\vec{A}(x^I, n, t)$ must be an eigenfunction of the local eigenvalue problem for the 1-D vertically inhomogeneous layered structure, for the relevant x^I. At different x^I, the eigenfunctions will be different.

In the following, we shall not try to solve completely the problem of surface waves propagating along surface Σ of a laterally varying structure. We shall not discuss the determination of $\vec{A}(x^I, n, t)$, but only the surface-wave ray tracing along surface Σ of the model and the computation of θ along these rays. For a more detailed treatment, see Woodhouse (1974, 1996), Gjevik (1973, 1974), Gregersen (1974), Babich, Chikhachev, and Yanovskaya (1976), Jobert and Jobert (1983, 1987), Levshin et al. (1987), Yomogida (1988), Virieux (1989), Virieux and Ekström (1991), Martin and Thomson (1997), Thomson (1997b), and Dahlen and Tromp (1998). For an analogous treatment of waves propagating in slowly varying waveguides, see Bretherton (1968), Burridge and Weinberg (1977), and Brekhovskikh and Godin (1989), among others. Various aspects of surface waves propagating in laterally varying media are also discussed by Woodhouse and Wong (1986), Yomogida (1985), Yomogida and Aki (1985), Tanimoto (1987), Kennett (1995), Wang and Dahlen (1995), Montagner (1996), Snieder (1996), and Ben-Hador and Buchen (1999).

3.12.2 Dispersion Relations and Surface-Wave Ray Tracing

For a generally curved surface and laterally varying structure, the local dispersion relation is different at different points x^I of the surface. Thus, we must add x^I to the arguments of the dispersion relation

$$\omega = \Omega(k_I, x^I). \tag{3.12.3}$$

Actually, an analogous dispersion relation may also be used for moving media; it would only be necessary to add t to the arguments of the dispersion relation. These dispersion relations play an important role in the propagation of waves in fluid media, but not in elastic models. For this reason, we shall use the dispersion relation in the time-independent form (3.12.3). Dispersion relation (3.12.3) is applicable both to isotropic and anisotropic media.

Inserting (3.12.2) into (3.12.3), we obtain

$$\partial\theta/\partial t + \Omega(\partial\theta/\partial x^I, x^I) = 0. \tag{3.12.4}$$

This is a nonlinear partial differential equation of the first order in phase θ. It belongs to the class of Hamilton-Jacobi equations and may be solved in terms of characteristics. Thus, in the theory of surface-wave ray tracing, the dispersion relation plays the same role as the eikonal equation in seismic body wave ray tracing. The four Hamilton's canonical equations of (3.12.4), representing the surface-wave ray tracing system, read

$$\mathrm{d}x^I/\mathrm{d}t = \partial\Omega/\partial k_I, \qquad \mathrm{d}k_I/\mathrm{d}t = -\partial\Omega/\partial x^I; \tag{3.12.5}$$

see (3.1.26). In addition, we also obtain equations for the phase θ (see (3.1.27)) and frequency ω:

$$\mathrm{d}\theta/\mathrm{d}t = -\omega + k_I \partial x^I/\mathrm{d}t, \tag{3.12.6}$$

$$\mathrm{d}\omega/\mathrm{d}t = (\partial\Omega/\partial x^I)(\mathrm{d}x^I/\mathrm{d}t) + (\partial\Omega/\partial k_I)(\mathrm{d}k_I/\mathrm{d}t) = 0. \tag{3.12.7}$$

System (3.12.5) represents the surface-wave ray tracing system, corresponding to the dispersion relation (3.12.4). Equation (3.12.7) shows that the frequency ω is constant along the whole ray. Note that the first equation of (3.12.5) defines the contravariant components of the group velocity vector along the ray,

$$\mathcal{U}^I = dx^I/dt = \partial\Omega/\partial k_I. \tag{3.12.8}$$

The group velocity $\mathcal{U}$ is then given by the expression

$$\mathcal{U} = (g_{IJ}\mathcal{U}^I\mathcal{U}^J)^{1/2}. \tag{3.12.9}$$

Finally, Equation (3.12.6) can be solved along the ray to give phase θ:

$$\theta(t) = -\omega t + \int k_I(dx^I/dt)dt = -\omega t + \int k_I\mathcal{U}^I dt. \tag{3.12.10}$$

The integral is taken along the ray.

The initial conditions for the surface-wave ray tracing system (3.12.5) for a specified frequency ω are

$$\text{At } t = 0: \quad x^I = x_0^I, \quad k_I = k_{I0}. \tag{3.12.11}$$

The quantities k_{I0}, however, cannot be chosen arbitrarily at x_0^I. They must satisfy the local dispersion relation at $x^I = x_0^I$, for the frequency ω under consideration:

$$\omega = \Omega(k_{I0}, x_0^I). \tag{3.12.12}$$

The surface-wave ray tracing system (3.12.5), with (3.12.6) through (3.12.12), is valid quite universally. It may be used for both isotropic and anisotropic media as well as for any smooth surface Σ. The curved surface Σ may be specified in arbitrary curvilinear coordinates, including nonorthogonal. To perform the computations, we need to know the local dispersion relations $\omega = \Omega(k_I, x^I)$ along Σ. These local dispersion relations can be obtained for locally 1-D media using propagator techniques and matrix methods.

For anisotropic media, a very detailed treatment of the surface-wave ray theory can be found in Martin and Thomson (1997), including the computation of amplitudes and complete wave forms. The authors also present useful relations to alternative approaches, and many references. They discuss the computation of local dispersion relations and give numerical examples of surface-wave rays along a flat surface of laterally varying anisotropic structures. Considerable attention is devoted to numerical problems encountered in surface-wave ray tracing.

3.12.3 Surface-Wave Ray Tracing Along a Surface of an Isotropic Structure

In this section, we shall simplify the surface-wave ray tracing system (3.12.5) for the isotropic medium. Consider an arbitrary, smoothly curved surface Σ, with the position-dependent metric tensor g_{IJ}. The material parameters (λ, μ, ρ) may vary strongly with distance from Σ, but only smoothly laterally.

In isotropic models, it is useful to introduce the wavenumber k by the relation $k^2 = g^{IJ}k_Ik_J$ along Σ, where g^{IJ} are the contravariant components of the metric tensor. This yields useful relations:

$$\partial k/\partial k_M = k^{-1}g^{MJ}k_J = k^{-1}k^M, \tag{3.12.13}$$
$$\partial k/\partial x^M|_{k_N} = (2k)^{-1}(\partial g^{IJ}/\partial x^M)k_Ik_J. \tag{3.12.14}$$

The symbol $|_{k_N}$ indicates that the derivative is taken for constant k_1 and k_2. We further modify the general dispersion relation (3.12.3) for isotropic media as follows:

$$\omega = \Omega(k_I, x^I) = \bar{\omega}(k, x^I).$$ (3.12.15)

Thus, ω depends on k only, not on k_1 and k_2. Then

$$\partial \Omega / \partial k_I = (\partial \bar{\omega} / \partial k)(\partial k / \partial k_I) = (\partial \bar{\omega} / \partial k)(k^I / k).$$ (3.12.16)

Quantity $\partial \bar{\omega} / \partial k$ has an important seismological interpretation. To explain it, we introduce ds, the arclength element along the ray. Using the first equation of (3.12.5) and (3.12.16), we obtain

$$\begin{aligned}
ds^2 &= g_{IJ}dx^I dx^J = \left(g_{IJ}\frac{dx^I}{dt}\frac{dx^J}{dt} \right)dt^2 \\
&= \left(g_{IJ}k^I k^J \right)k^{-2}\left(\frac{d\bar{\omega}}{dk} \right)^2 dt^2 = \left(\frac{d\bar{\omega}}{dk} \right)^2 dt^2.
\end{aligned}$$

This yields

$$ds/dt = d\bar{\omega}/dk = \mathcal{U},$$ (3.12.17)

where $\mathcal{U}$ is the group velocity. This corresponds to the well-known definition of group velocity in isotropic media. Finally, we find an important relation for $\partial \bar{\omega} / \partial x^I$. Because ω is constant along the ray, we obtain

$$\partial \bar{\omega} / \partial x^I = -(\partial \bar{\omega} / \partial k)(\partial k / \partial x^I).$$ (3.12.18)

Using relations (3.12.13) through (3.12.18), it is easy to simplify the surface-wave ray tracing system (3.12.5) for isotropic media. Equation (3.12.16) can be used to express the first equation of (3.12.5) as follows:

$$dx^M / dt = \partial \Omega / \partial k_I = (\partial \bar{\omega} / \partial k)k^M / k.$$

Using (3.12.8), (3.12.14), and (3.12.15) in the second equation of (3.12.5) yields

$$\begin{aligned}
\frac{dk_M}{dt} &= -\left.\frac{\partial \Omega}{\partial x^M}\right|_{k_N} = -\frac{\partial \bar{\omega}}{\partial x^M} - \left.\frac{\partial \bar{\omega}}{\partial k}\frac{\partial k}{\partial x^M}\right|_{k_N} \\
&= \frac{\partial \bar{\omega}}{\partial k}\left(\frac{\partial k}{\partial x^M} - \frac{1}{2k}\frac{\partial g^{IJ}}{\partial x^M}k_I k_J \right).
\end{aligned}$$

Now we use (3.12.17) and obtain the final form of the surface-wave ray tracing system for isotropic media as follows:

$$\frac{dx^M}{ds} = \frac{g^{MJ}k_J}{k}, \qquad \frac{dk_M}{ds} = \frac{\partial k}{\partial x^M} - \frac{1}{2k}\frac{\partial g^{IJ}}{\partial x^M}k_I k_J.$$ (3.12.19)

The surface-wave ray tracing system (3.12.19) does not use the dispersion relation $\omega = \bar{\omega}(k, x^I)$ explicitly, but rather uses the alternative dispersion relation $k = k(\omega, x^I)$. System (3.12.19) can be used for any smoothly curved surface Σ, with a position-dependent metric tensor g_{IJ}. It can, of course, also be applied to orthogonal curvilinear coordinates (spherical, ellipsoidal, and the like).

We can compute phase θ along the ray using (3.12.10) and (3.12.19),

$$\theta(t) = -\omega t + \int k_I \frac{dx^I}{ds}ds = -\omega t + \int k\,ds = -\omega \left(t - \int ds/C \right).$$ (3.12.20)

Here we have used $k_I g^{IJ} k_J = k^2$ and $k = \omega/C$. The integral is taken along the ray.

The initial conditions for a specified frequency ω at point $S(x_0^I)$ situated on Σ for system (3.12.19) are as follows:

$$\text{At } S: \qquad x^I = x_0^I, \qquad k_I = k_{I0}. \tag{3.12.21}$$

Here $k_{I0} = k_I(\omega, x_0^I)$ must satisfy the local dispersion relation for the frequency ω under consideration.

Indeed, system (3.12.19) is fully analogous to the ray tracing system (3.5.54), derived for seismic body waves propagating in a model specified by nonorthogonal coordinates x^i. We specify surface Σ by relation $x^3 = $ const. and use arclength s as the variable along the ray. Then $n = 1$, $A^{n/2-1} = V$, and system (3.5.54) reads

$$\mathrm{d}x^I/\mathrm{d}s = V g^{IK} p_K, \qquad \mathrm{d}p_I/\mathrm{d}s = \partial(1/V)/\partial x^I - \tfrac{1}{2} V p_K p_J \partial g^{KJ}/\partial x^I. \tag{3.12.22}$$

System (3.12.22) is alternative to (3.12.19). This is simple to see if we use $k = \omega/C(\omega)$, $k_M = \omega p_M(\omega)$, and $V = C(\omega)$ in (3.12.19), for $\omega = $ const. Thus, standard ray tracing systems derived for seismic body waves can also be used in surface-wave ray tracing. However, we must remember that phase velocity C and p_1 and p_2 depend on frequency. The frequency itself remains fixed along the whole ray, and the rays computed for different frequencies are different.

As an important example, we shall present the surface-wave ray tracing system along spherical surface Σ given by relation $r = R$ in spherical polar coordinates r, ϑ, and φ. (Here we use ϑ instead of standard θ because we have already used the symbol θ for the phase function.) In this case, $x^1 = \vartheta$, $x^2 = \varphi$, $g^{11} = R^{-2}$, $g^{22} = R^{-2}(\sin \vartheta)^{-2}$, and $g^{12} = g^{21} = 0$. Inserting this into (3.12.19), we obtain the surface-wave ray tracing system in the following form:

$$\frac{\mathrm{d}\vartheta}{\mathrm{d}s} = \frac{1}{kR^2} k_\vartheta, \qquad \frac{\mathrm{d}k_\vartheta}{\mathrm{d}s} = \frac{\partial k}{\partial \vartheta} + \frac{k_\varphi^2 \cos \vartheta}{kR^2 \sin^3 \vartheta},$$
$$\frac{\mathrm{d}\varphi}{\mathrm{d}s} = \frac{1}{kR^2 \sin^2 \vartheta} k_\varphi, \qquad \frac{\mathrm{d}k_\varphi}{\mathrm{d}s} = \frac{\partial k}{\partial \varphi}, \tag{3.12.23}$$

where $k_\vartheta = \partial \theta/\partial \vartheta$ and $k_\varphi = \partial \theta/\partial \varphi$. If we put $k = \omega/C$, $k_\vartheta = \omega T_\vartheta$, and $k_\varphi = \omega T_\varphi$, we can again see that (3.12.23) is fully equivalent to (3.5.31) with $n = 1$, $A^{n/2-1} = V$, $V = C$, and $r = R$.

We remind the reader that the spherical surface in the ray tracing system (3.5.31) can be transformed into a flat surface using the Earth's surface flattening transformation (Mercator transformation); see (3.5.49). The application of the Mercator transformation in surface-wave ray tracing along a spherical surface was first proposed by Jobert and Jobert (1983, 1987). Consequently, we can also apply standard 2-D ray tracing computer routines in Cartesian coordinates to surface-wave ray tracing along a spherical surface of the Earth. It is only necessary to modify the input and output data slightly.

For surface-wave ray tracing systems along an ellipsoidal surface Σ, see Jobert (1976) and Mochizuki (1989).

CHAPTER FOUR

Dynamic Ray Tracing. Paraxial Ray Methods

D ynamic ray tracing is a powerful procedure that has recently found broad applications in the evaluation of high-frequency seismic wavefields in laterally inhomogeneous layered structures and in the solution of inverse seismic problems. It consists of solving a system of several ordinary differential equations along a known ray Ω and yields the first derivatives of phase space coordinates of points on Ω (position, slowness vector components) with respect to initial phase space coordinates or ray parameters. Although the dynamic ray tracing system is very simple and can be integrated without any larger additional numerical effort, in contrast with standard ray tracing, it extends the possibilities of the standard ray method considerably.

The dynamic ray tracing system can be expressed in many forms and in various coordinate systems. Certain versions of the system have been known for a long time. For example, dynamic ray tracing was used by Belonosova, Tadzhimukhamedova, and Alekseyev (1967) to calculate geometrical spreading in *2-D laterally varying isotropic structures*. The dynamic ray tracing system for 3-D laterally varying *anisotropic media* in general Cartesian coordinates was first proposed and applied to the computation of geometrical spreading by Červený (1972). A similar procedure of calculating geometrical spreading in a 3-D laterally varying *isotropic layered structure* was discussed in detail by Červený, Langer, and Pšenčík (1974). The reference also gives the relations for dynamic ray tracing across a structural interface. For dynamic ray tracing in general *nonorthogonal coordinates* in isotropic layered and blocked structures, see Červený, Klimeš, and Pšenčík (1988b). For spherical coordinates, see Liu and Tromp (1996) and Dahlen and Tromp (1998).

The simplest form of the dynamic ray tracing system in isotropic media is obtained in ray-centered coordinates connected with ray Ω. The ray-centered coordinate system q_1, q_2, and q_3 connected with ray Ω is a curvilinear orthogonal coordinate system, introduced in such a way that the ray Ω represents the q_3-axis of the system. Coordinate lines q_1 and q_2 for any fixed point q_3 on Ω are formed by two mutually perpendicular straight lines intersecting at Ω, situated in a plane perpendicular to Ω at q_3. Thus, the coordinate plane $q_3 = $ const. is tangent to the wavefront, and the ray Ω is specified by equations $q_1 = q_2 = 0$. For more details on the ray-centered coordinate system and on the computation of its basis vectors, see Sections 4.1.1 through 4.1.3.

In ray-centered coordinates, the eikonal equation can be used to derive a simple approximate system of linear ordinary differential equations of the first order for rays, situated in the vicinity of central ray Ω. Such rays are called the *paraxial rays*, and the relevant system is called the *paraxial ray tracing system*; see Červený, Klimeš, and Pšenčík (1984) and Beydoun and Keho (1987). The term *paraxial* is taken from optics; it represents the vicinity of the axis of the optical system. In our case, it represents the vicinity of the central ray Ω (that is, the vicinity of the q_3-axis). The paraxial rays represent curves in a four-dimensional phase space, with phase-space coordinates q_1, q_2, $p_1^{(q)} = \partial T / \partial q_1$, and $p_2^{(q)} = \partial T / \partial q_2$. Note that the paraxial rays computed in this way are not exact outside ray Ω; they are only approximate. Their accuracy decreases with the increasing distance from Ω. See Section 4.1.6. The paraxial ray tracing system also represents the *dynamic ray tracing system* for partial derivatives $\partial q_I / \partial \gamma$ and $\partial p_I^{(q)} / \partial \gamma$ along Ω, where γ is an arbitrarily selected initial parameter of the system. The dynamic ray tracing system again consists of four linear ordinary differential equations of the first order. In fact, the system matrices of both systems are the same, only the computed quantities have a different physical meaning. The paraxial ray tracing system computes approximately the phase-space coordinates q_I and $p_I^{(q)}$ along paraxial rays, and the dynamic ray tracing system computes exactly the partial derivatives $\partial q_I / \partial \gamma$ and $\partial p_I^{(q)} / \partial \gamma$ along the central ray Ω. See Section 4.1.7.

If we consider a two-parameteric orthonomic system of rays, specified by ray parameters γ_1 and γ_2, we can use the dynamic ray tracing system to compute the 2×2 matrices $\mathbf{Q}$ and $\mathbf{P}$, with elements $Q_{IJ} = \partial q_I / \partial \gamma_J$ and $P_{IJ} = \partial p_I^{(q)} / \partial \gamma_J$ along Ω. The dynamic ray tracing system then consists of two matrix equations for $\mathbf{Q}$ and $\mathbf{P}$. Note that matrix $\mathbf{Q}$ represents the transformation matrix from ray coordinates γ_1 and γ_2 to the ray-centered coordinates q_1 and q_2 and can be used to compute geometrical spreading. Matrices $\mathbf{Q}$ and $\mathbf{P}$ can also be used to compute the 2×2 matrix $\mathbf{M}$ of the second derivative of the travel-time field with respect to q_1 and q_2, $\mathbf{M} = \mathbf{PQ}^{-1}$. The knowledge of $\mathbf{M}$ is quite sufficient to determine the distribution of *paraxial travel times*, which are quadratic in q_1 and q_2. It can also be used to compute paraxial wavefronts, the surfaces along which the paraxial travel times are constant. Note that the paraxial rays can be also defined as orthogonal trajectories to the system of paraxial wavefronts. Matrix $\mathbf{M}$ itself satisfies a nonlinear ordinary differential equation of the first order of the Riccati type. See Sections 4.1.7 and 4.1.8.

An alternative approach to the derivation of the dynamic ray tracing system is possible. It consists of the derivation of the Riccati equation for $\mathbf{M}$ from the eikonal equation in ray-centered coordinates and the transformation of the Riccati equation into the dynamic ray tracing system for $\mathbf{Q}$ and $\mathbf{P}$.

The dynamic ray tracing system in ray-centered coordinates for 2×2 matrices $\mathbf{Q}$ and $\mathbf{P}$ was first proposed for computing geometrical spreading along the central ray by Popov and Pšenčík (1978a, 1978b); see also Červený and Pšenčík (1979) and Pšenčík (1979). The derivation was based on the transformation of the Riccati equation for $\mathbf{M}$ into the dynamic ray tracing system for $\mathbf{Q}$ and $\mathbf{P}$. Later, it was shown how the system can be used to compute curvatures of the wavefront along the central ray; see Hubral (1979, 1980), Hubral and Krey (1980), and Červený and Hron (1980). Because of the importance of geometrical spreading and of the curvatures of wavefronts in evaluating the dynamic properties of seismic waves (ray amplitudes), Červený and Hron (1980) suggested that the procedure be called dynamic ray tracing. We shall continue to call the procedure dynamic ray tracing, although its applications are now much broader than just calculating the geometrical spreading and amplitudes along ray Ω.

Because the dynamic ray tracing system in ray-centered coordinates consists of four scalar linear ordinary differential equations of the first order, it has four linearly independent solutions. The 4×4 fundamental matrix, which is an identity matrix at some point S on ray Ω, is called the *propagator matrix of the dynamic ray tracing system from S*. We shall also speak of the *ray propagator matrix from S*. After the ray propagator matrix from S has been found along Ω, the solution of the dynamic ray tracing system for any initial conditions at S is obtained merely by multiplying the ray propagator matrix by the matrix of the initial conditions, without any additional dynamic ray tracing. Moreover, when the propagator matrix is known from S to R, it is also simple to find analytically the inverse matrix, which represents the ray propagator matrix from R to S. The ray propagator matrix is symplectic and may be chained along central ray Ω. It has also other useful properties that can be conveniently used in various seismological applications.

For isotropic media, the dynamic ray tracing system in ray-centered coordinates, derived in Section 4.1, and the relevant 4×4 ray propagator matrix will be used as the basic system of equations in the whole of Chapter 4. In Section 4.4, it will be generalized to isotropic layered media containing curved structural interfaces. The initial conditions for the dynamic ray tracing system in ray-centered coordinates, corresponding to a smooth initial curved surface, to a smooth initial line, and to a point source, situated in an isotropic medium, are derived in Section 4.5. Paraxial travel-time fields and slowness vectors and the matrices of curvature of the wavefront are discussed in detail in Section 4.6. Section 4.8 presents some analytical solutions of the dynamic ray tracing system for simple structures. In Section 4.9, it is shown that the boundary-value problems for paraxial rays can be solved analytically, once the relevant propagator matrix is known. This applies particularly to two-point paraxial ray tracing and to the computation of the two-point eikonal. The important problem of determining geometrical spreading along the central ray in a layered medium is treated in Section 4.10. The application of dynamic ray tracing in ray-centered coordinates to the computation of Fresnel volumes and Fresnel zones is described in Section 4.11, and it is applied to the computation of the KMAH index along Ω in Section 4.12. All previous results are specified for planar rays and for 2-D models in Section 4.13. It should be noted that dynamic ray tracing is a basic procedure in many extensions of the ray method such as the Maslov-Chapman method and the Gaussian beam summation procedures (see Section 5.8), in ray perturbation methods, and in the investigation of chaotic behavior of rays, among others.

General dynamic ray tracing systems can also be derived using *the Hamiltonian formalism*, without specifying the actual specific form of the Hamiltonian; see Sections 4.2 and 4.7. Consequently, the results can be used for both isotropic and anisotropic media and for an arbitrary curvilinear coordinate system (including the Cartesian coordinate system). The 3-D dynamic ray tracing system derived in this way consists of six scalar linear ordinary differential equations of the first order. The system, however, must satisfy one constraint relation, which follows from the eikonal equation. In actual computations, the initial conditions for the dynamic ray tracing system cannot be chosen arbitrarily but must satisfy the constraint relation at the initial point. After the constraint relation is satisfied at the initial point, it is satisfied along the whole ray Ω. The numerical noise, however, may cause deviations from the constraint relation and decrease the stability of computations. It is useful to normalize the results at any step of the ray so that the constraint relation is satisfied. The situation is analogous to the standard ray tracing system, where the initial components of the slowness vector must satisfy the eikonal equation $p_i p_i - 1/V^2 = 0$ at the initial point; see (3.2.2). Because the general dynamic ray tracing system consists of six linear ordinary differential equations of the first order, it has six linearly independent solutions. It will be shown in Section 4.2 that two of these linearly independent solutions, here referred to as the ray-tangent solutions and the noneikonal solutions, can be found analytically along a known ray. Consequently, the number of equations of the dynamic ray tracing system can always be reduced from six to four. See examples of such reduction in Sections 4.2.2 and 4.2.4.1. The RHSs of the system consisting of four equations become, however, usually more complex than the RHSs of the system consisting of six equations.

The 6×6 propagator matrices corresponding to dynamic ray tracing systems consisting of six equations are introduced in Section 4.3.7. These 6×6 propagator matrices *always contain* the ray-tangent and noneikonal solutions, which *are automatically removed in the 4×4 propagator matrices*. Consequently, the application of the 4×4 propagator matrices in the solution of various seismological problems related to orthonomic systems of rays is usually more transparent and physically more appealing than the application of 6×6 propagator matrices.

The 6×6 propagator matrices have been broadly applied in the ray perturbation theory; see Farra and Le Bégat (1995) and Farra (1999) where many other references can be found. For this reason, Sections 4.7.4 and 4.7.5 give only a brief exposition of the ray perturbation theory in terms of 6×6 propagator matrices. The basic equations of these sections are very general, valid for both isotropic and anisotropic media and for any coordinate system. Of course, it is possible to find alternative equations for isotropic media and for ray-centered coordinates.

In Section 4.2.4.2, it is shown how the dynamic ray tracing system consisting of six equations can be transformed from one coordinate system to another. A similar transformation of the 6×6 propagator matrix is described in Section 4.7.3. Consequently, the dynamic ray tracing can be performed in any coordinate system such as the Cartesian and transformed to any other coordinate system, without changing the Hamiltonian.

The Hamiltonian formalism, used in Sections 4.2, 4.3.7, and 4.7, is applicable both to isotropic and anisotropic media. The whole of Section 4.14, however, is devoted exclusively to the anisotropic inhomogeneous media. In anisotropic media, the Hamiltonian has the simplest form in Cartesian coordinates. The relevant dynamic ray tracing system in Cartesian coordinates, consisting of six linear ordinary differential equations of the first order, is discussed in detail in Section 4.14.1. The dynamic ray tracing system in Cartesian coordinates can, however, be transformed to any other coordinate system, keeping the Hamiltonian expressed in Cartesian coordinates. As an alternative to the ray-centered coordinates used in isotropic media, it is convenient to introduce the so-called wavefront orthonormal coordinates y_1, y_2, and y_3 along the ray in anisotropic media. In this system, the y_3-axis is oriented along the slowness vector, not along the ray. If anisotropy vanishes, the wavefront orthonormal coordinates yield the local Cartesian ray-centered coordinates; see Section 4.14.2. We can also construct the 4×4 propagator matrix in these coordinates (see Section 4.14.3) and use it in applications analogous to the 4×4 ray-centered propagator matrix.

4.1 Dynamic Ray Tracing in Ray-Centered Coordinates

In principle, it is not necessary to introduce ray-centered coordinates if we wish to perform dynamic ray tracing. Dynamic ray tracing may be performed in general Cartesian coordinates, or in any other orthogonal or nonorthogonal coordinates. However, the simplest version of dynamic ray tracing in isotropic media is obtained in ray-centered coordinates. We shall, therefore, explain the ray-centered coordinate system in detail in this section. In Section 4.2, we shall show how to derive alternative versions of dynamic ray tracing in Cartesian and other coordinate systems and how to transform one version into another.

4.1.1 Ray-Centered Coordinates: Definition, Orthogonality

For any selected ray Ω, we shall introduce a ray-centered coordinate system q_1, q_2, and q_3, connected with Ω in the following way. One coordinate, say q_3, corresponds to any

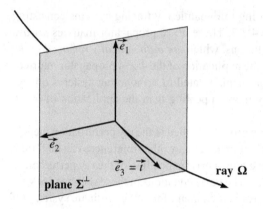

Figure 4.1. Basis vectors $\vec{e}_1$, $\vec{e}_2$, and $\vec{e}_3$ of the ray-centered coordinate system q_i connected with ray Ω. Ray Ω is the q_3-axis of the system. At any point on the ray (q_3 fixed), unit vector $\vec{e}_3$ equals $\vec{t}$, the unit tangent to Ω. Unit vectors $\vec{e}_1$ and $\vec{e}_2$ are situated in plane $\Sigma^\perp$, perpendicular to Ω at a given q_3, and are mutually perpendicular. The triplet $\vec{e}_1$, $\vec{e}_2$, $\vec{e}_3$ is right-handed.

monotonic parameter u along the ray. For simplicity, we shall mostly consider $u = s$, where s is the arclength of ray Ω, specified at an arbitrary reference point by $s = s_0$, where s_0 is given. Coordinates q_1 and q_2 form a 2-D Cartesian coordinate system in plane $\Sigma^\perp$ perpendicular to Ω at $q_3 = s$, with the origin at Ω. Thus, *ray Ω is one of the coordinate axes* of the ray-centered coordinate system. Coordinates q_1 and q_2 in $\Sigma^\perp$ may be chosen in many ways. We shall choose them so that the ray-centered coordinate system is *orthogonal*. This condition determines the ray-centered coordinate system q_1 and q_2 uniquely along the whole ray Ω, when the 2-D Cartesian system q_1 and q_2 has been specified at any reference point of the ray. The vector basis of the ray-centered coordinate system connected with Ω is formed at an arbitrary point corresponding to the arclength $q_3 = s$ of ray Ω by a right-handed triplet of unit vectors $\vec{e}_1(s)$, $\vec{e}_2(s)$, and $\vec{e}_3(s)$, where $\vec{e}_3(s) = \vec{t}(s)$ is the unit tangent to ray Ω and vectors $\vec{e}_1(s)$, $\vec{e}_2(s)$ are perpendicular to Ω; see Fig. 4.1.

We shall now show that the unit vectors $\vec{e}_I(s)$ can be computed along ray Ω by solving the vectorial differential equations of the first order:

$$d\vec{e}_I(s)/ds = a_I(s)\vec{p}(s), \qquad I = 1, 2. \tag{4.1.1}$$

Here $\vec{p}(s)$ is the slowness vector, known from ray tracing, and $a_I(s)$ are some continuous functions of s along Ω, which are not yet determined. We shall determine them in such a way as to keep $\vec{e}_I(s)$ perpendicular to Ω, $\vec{e}_I(s) \cdot \vec{p}(s) = 0$. Taking the derivative of $\vec{e}_I(s) \cdot \vec{p}(s) = 0$ with respect to s along Ω, we obtain

$$d\vec{e}_I(s)/ds \cdot \vec{p}(s) + \vec{e}_I(s) \cdot d\vec{p}(s)/ds = 0, \qquad I = 1, 2.$$

Now we multiply (4.1.1) by $\vec{p}(s)$ and use the previous equation. This yields

$$a_I(s) = (\vec{p} \cdot \vec{p})^{-1} d\vec{e}_I/ds \cdot \vec{p} = -(\vec{p} \cdot \vec{p})^{-1}\vec{e}_I(s) \cdot d\vec{p}/ds, \qquad I = 1, 2. \tag{4.1.2}$$

Equations (4.1.1) with (4.1.2) yield

$$d\vec{e}_I/ds = -(\vec{p} \cdot \vec{p})^{-1}(\vec{e}_I \cdot d\vec{p}/ds)\vec{p}, \qquad I = 1, 2. \tag{4.1.3}$$

We can also insert the eikonal equation $\vec{p} \cdot \vec{p} = 1/V^2$ such that

$$d\vec{e}_I/ds = -V^2(\vec{e}_I \cdot d\vec{p}/ds)\vec{p}, \qquad I = 1, 2. \tag{4.1.4}$$

Finally, we can insert $d\vec{p}/ds = \nabla(1/V) = -V^{-2}\nabla V$ and obtain

$$d\vec{e}_I/ds = (\vec{e}_I \cdot \nabla V)\vec{p}, \qquad I = 1, 2; \tag{4.1.5}$$

see (3.1.10). Any of the three systems (4.1.3) through (4.1.5) can be used alternatively to compute $\vec{e}_1(s)$ and $\vec{e}_2(s)$ along ray Ω. In the following text, we shall consider mainly (4.1.5).

Assume that $\vec{e}_1(s_0)$ and $\vec{e}_2(s_0)$ satisfy the following three conditions at an initial point $s = s_0$ of ray Ω:

a. $\vec{e}_I(s_0) \cdot \vec{p}(s_0) = 0, I = 1, 2;$
b. $\vec{e}_1(s_0) \cdot \vec{e}_2(s_0) = 0;$
c. $\vec{e}_1(s_0) \cdot \vec{e}_1(s_0) = 1, \vec{e}_2(s_0) \cdot \vec{e}_2(s_0) = 1.$

It is not difficult to prove that the solutions of (4.1.5) then satisfy the same conditions along the whole ray Ω (for any s). More specifically, $\vec{e}_1(s)$ and $\vec{e}_2(s)$ satisfy the following three conditions:

a. $\vec{e}_I(s) \cdot \vec{p}(s) = 0, I = 1, 2$ (both $\vec{e}_1(s)$ and $\vec{e}_2(s)$ are perpendicular to $\vec{p}(s)$, that is, they are perpendicular to ray Ω).
b. $\vec{e}_1(s) \cdot \vec{e}_2(s) = 0$ ($\vec{e}_1(s)$ is perpendicular to $\vec{e}_2(s)$).
c. $\vec{e}_1(s) \cdot \vec{e}_1(s) = 1, \vec{e}_2(s) \cdot \vec{e}_2(s) = 1$ ($\vec{e}_1(s)$ and $\vec{e}_2(s)$ are unit vectors).

In other words, when $\vec{e}_1(s_0), \vec{e}_2(s_0)$, and $\vec{e}_3(s_0) = \vec{t}(s_0)$ form a mutually perpendicular triplet of unit vector at an initial point $s = s_0$ of ray Ω, with $\vec{t}(s_0)$ tangent to Ω, then $\vec{e}_1(s), \vec{e}_2(s)$, and $\vec{e}_3(s) = \vec{t}(s)$ form a mutually perpendicular triplet of unit vectors at any point s of ray Ω, with $\vec{t}(s)$ tangent to Ω.

The arclength s in (4.1.3) through (4.1.5) can be easily replaced by any other monotonic parameter u along Ω, for example, by travel time T, $ds = V dT$.

We shall now describe the determination of the ray-centered coordinates q_1, q_2, and $q_3 = s$ of any point R' situated close to ray Ω, $R' = R'[q_1, q_2, q_3]$, in greater detail. First, we construct plane $\Sigma^\perp$ perpendicular to ray Ω and passing through R'. We then find the point of intersection of $\Sigma^\perp$ with Ω and denote it by R. Plane $\Sigma^\perp$ is tangent to the wavefront at R. It represents the q_1q_2-plane. See Figure 4.2. Because point R is situated at Ω, its ray-centered coordinates are $q_1 = q_2 = 0, q_3 = s$, so that $R = R[0, 0, s]$. Note that the q_3 coordinates of points R and R' are the same so that this construction also yields the $q_3 = s$ coordinate of point R'. Coordinates $q_1(R')$ and $q_2(R')$ are then easily obtained in plane $\Sigma^\perp$ using the known basis vectors $\vec{e}_1(s)$ and $\vec{e}_2(s)$. Radius vector $\vec{r}$ of point R' can be expressed in ray-centered coordinates as follows:

$$\vec{r}(q_1, q_2, s) = \vec{r}(0, 0, s) + q_1 \vec{e}_1(s) + q_2 \vec{e}_2(s). \tag{4.1.6}$$

Figure 4.2. Ray-centered coordinates q_1, q_2, and q_3 of point R' situated in the vicinity of ray Ω. Point R' is situated in plane $\Sigma^\perp$ perpendicular to Ω and crossing Ω at point R. The position of point R determines $q_3(R')$ because $q_3(R') = q_3(R)$. Then, $q_1(R')$ and $q_2(R')$ are determined as Cartesian coordinates of R' in plane $\Sigma^\perp$, with basis vectors $\vec{e}_1$ and $\vec{e}_2$.

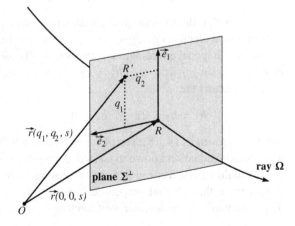

Consequently, $\vec{r}(0, 0, s)$ is the radius-vector of point R situated on Ω. The equation $\vec{r} = \vec{r}(0, 0, s)$ is a parameteric equation of ray Ω, with parameter s. Equation (4.1.6) defines the ray-centered coordinates q_1, q_2, and s of point R', assuming that $\vec{r}(0, 0, s)$, $\vec{e}_1(s)$, and $\vec{e}_2(s)$ are known.

The ray-centered coordinates q_1, q_2, and s of point R' connected with ray Ω are introduced uniquely if only one plane $\Sigma^\perp$ perpendicular to ray Ω passing through R' can be constructed. For points R' situated far from ray Ω, this requirement is sometimes not satisfied. We shall concentrate our attention only on points R', which are not situated far from Ω and at which the ray-centered coordinates connected with ray Ω may be introduced uniquely. Such *region of validity* of the ray-centered coordinate system actually depends on the curvature of ray Ω. The region of validity is broad along slightly curved rays, but it may be narrow along rays with a high curvature.

It is not difficult to show that the ray-centered coordinate system introduced here is *orthogonal*. Remember that length element $\mathrm{d}l$ in a general coordinate system q_i satisfies relation $(\mathrm{d}l)^2 = \mathrm{d}\vec{r} \cdot \mathrm{d}\vec{r} = g_{ik} \mathrm{d}q_i \mathrm{d}q_k$, where g_{ik} are elements of the relevant metric tensor. *The coordinate system q_i is called orthogonal if metric tensor g_{ij} is diagonal, that is, if $g_{ij} = 0$ for $i \neq j$.*

In our case, (4.1.6) and (4.1.5) yield

$$
\begin{aligned}
\mathrm{d}\vec{r} &= [\mathrm{d}\vec{r}(0, 0, s)/\mathrm{d}s + q_1 \mathrm{d}\vec{e}_1/\mathrm{d}s + q_2 \mathrm{d}\vec{e}_2/\mathrm{d}s]\mathrm{d}s + \vec{e}_1 \mathrm{d}q_1 + \vec{e}_2 \mathrm{d}q_2 \\
&= [1 + V^{-1}(\vec{e}_1 \cdot \nabla V)q_1 + V^{-1}(\vec{e}_2 \cdot \nabla V)q_2]\vec{t}\,\mathrm{d}s + \vec{e}_1 \mathrm{d}q_1 + \vec{e}_2 \mathrm{d}q_2 \\
&= [1 + (V^{-1}\partial V/\partial q_1)_{q_1=q_2=0}q_1 \\
&\quad + (V^{-1}\partial V/\partial q_2)_{q_1=q_2=0}q_2]\vec{t}\,\mathrm{d}s + \vec{e}_1 \mathrm{d}q_1 + \vec{e}_2 \mathrm{d}q_2 \\
&= h\vec{t}\,\mathrm{d}s + \vec{e}_1 \mathrm{d}q_1 + \vec{e}_2 \mathrm{d}q_2,
\end{aligned}
\tag{4.1.7}
$$

where

$$
h = 1 + (V^{-1}\partial V/\partial q_I)_{q_1=q_2=0}q_I.
\tag{4.1.8}
$$

This yields

$$
(\mathrm{d}l)^2 = \mathrm{d}\vec{r} \cdot \mathrm{d}\vec{r} = \mathrm{d}q_1^2 + \mathrm{d}q_2^2 + h^2 \mathrm{d}s^2.
\tag{4.1.9}
$$

From (4.1.9), we can see that the components of metric tensor g_{ij} of the ray-centered coordinate system are given by relations

$$
g_{11} = g_{22} = 1, \qquad g_{33} = h^2, \qquad g_{ij} = 0 \qquad \text{for } i \neq j.
\tag{4.1.10}
$$

This shows that the ray-centered coordinate system q_1, q_2, and q_3 is *orthogonal* because only diagonal elements g_{11}, g_{22}, and g_{33} are nonvanishing.

In an orthogonal coordinate system, we usually use *scale factors* h_1, h_2, and h_3 instead of metric tensor g_{ij}; $h_1^2 = g_{11}, h_2^2 = g_{22}, h_3^2 = g_{33}$. In the ray-centered coordinate system, the scale factors are

$$
h_1 = h_2 = 1, \qquad h_3 = h,
\tag{4.1.11}
$$

where h is given by (4.1.8); see (4.1.10). Scale factors (4.1.11) are sufficient to transform any vectorial equation known in Cartesian coordinates into ray-centered coordinates.

It should be emphasized that the orthogonality of the ray-centered coordinate system means more than mutual perpendicularity of $\vec{e}_1, \vec{e}_2$, and $\vec{e}_3 \equiv \vec{t}$ at any point of ray Ω. Three mutually perpendicular unit vectors $\vec{e}_1, \vec{e}_2$, and $\vec{e}_3$ may be constructed in many ways at different points of ray Ω, but the relevant coordinate systems are not, in general, orthogonal. It is not difficult to see that an orthogonal coordinate system is obtained only if the expression for $d\vec{r}$ (4.1.7) consists only of three terms with factors $\vec{t}\,ds$, $\vec{e}_1 dq_1$, and $\vec{e}_2 dq_2$ in the individual terms, but with no mixed factors $\vec{t}\,dq_1$, $\vec{t}\,dq_2$, $\vec{e}_1 ds$, $\vec{e}_1 dq_2$, $\vec{e}_2 dq_1$, and $\vec{e}_2 ds$. This requirement is satisfied only if $d\vec{e}_1/ds$ and $d\vec{e}_2/ds$ are parallel to $\vec{t}$; see (4.1.1). For this reason, we have defined $\vec{e}_1$ and $\vec{e}_2$ using (4.1.1).

We shall number unit vectors $\vec{e}_1(s)$ and $\vec{e}_2(s)$ to render the triplet $\vec{e}_1$, $\vec{e}_2$, $\vec{t}$ *right-handed*:

$$\vec{e}_1 \times \vec{e}_2 = \vec{t}, \qquad \vec{e}_2 \times \vec{t} = \vec{e}_1, \qquad \vec{t} \times \vec{e}_1 = \vec{e}_2. \tag{4.1.12}$$

Equations (4.1.12) have an important consequence. As $\vec{t}$ is known from ray tracing, $\vec{t} = V\vec{p}$, it is sufficient to compute only one of the two vectors $\vec{e}_1$ or $\vec{e}_2$ using the differential equations derived earlier. The second vector can be calculated using (4.1.12).

Let us emphasize the difference between basis vectors $\vec{e}_1$, $\vec{e}_2$, and $\vec{e}_3 \equiv \vec{t}$, defined by (4.1.5), and vectors $\vec{n}$, $\vec{b}$, and $\vec{t}$, where $\vec{n}$ is the unit normal, $\vec{b}$ is the unit binormal, and $\vec{t}$ is the unit tangent to Ω. Only $\vec{t}$ are the same in both triplets, but $\vec{e}_1$ and $\vec{e}_2$ are, in general, different from $\vec{n}$ and $\vec{b}$. Unit vectors $\vec{n}$, $\vec{b}$, and $\vec{t}$ can be determined uniquely at any point R of curve Ω (with the exception of $\vec{n}$ and $\vec{b}$ along a straight line) from the local geometrical properties of curve Ω in the vicinity of R; see Section 3.2.4. They depend only on these local properties at R, and not on the behavior of curve Ω at points distant from R. No initial conditions are required to determine $\vec{n}(s)$, $\vec{b}(s)$, and $\vec{t}(s)$. Basis unit vectors $\vec{e}_1(s)$ and $\vec{e}_2(s)$, however, depend on the initial conditions $\vec{e}_1(s_0)$ and $\vec{e}_2(s_0)$. At the initial point s_0, $\vec{t}(s_0)$ must be tangent to Ω, and $\vec{e}_1(s_0)$, $\vec{e}_2(s_0)$, and $\vec{t}(s_0)$ must form a right-handed system of unit vectors; see (4.1.12). Otherwise, $\vec{e}_1(s_0)$ and $\vec{e}_2(s_0)$ may be chosen arbitrarily; for example, $\vec{e}_1(s_0) = \vec{n}(s_0)$ and $\vec{e}_2(s_0) = \vec{b}(s_0)$. Differential equations (4.1.5) then yield uniquely the ray-centered basis vectors $\vec{e}_1(s)$ and $\vec{e}_2(s)$ along the whole ray Ω, which are, in general, different from $\vec{n}(s)$ and $\vec{b}(s)$. See Figure 4.3.

The next difference between $\vec{e}_1$ and $\vec{e}_2$ and $\vec{n}$ and $\vec{b}$ is that unit vectors $\vec{n}$ and $\vec{b}$ do not form an orthogonal coordinate system connected with curve Ω. The derivatives $d\vec{n}/ds$, $d\vec{b}/ds$, and $d\vec{t}/ds$ along Ω are given by well-known Frenet's formulas (3.2.8). As we can see from (3.2.8), $d\vec{n}/ds$ and $d\vec{b}/ds$ are not parallel to $\vec{t}$. The consequence is that $\vec{n}$, $\vec{b}$, and $\vec{t}$ do not

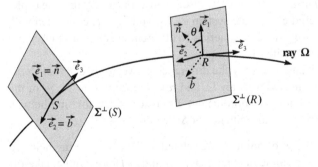

Figure 4.3. Difference between basis vectors $\vec{e}_1$, $\vec{e}_2$, and $\vec{e}_3 = \vec{t}$ and unit vectors $\vec{n}$, $\vec{b}$, and $\vec{t}$, where $\vec{n}$ is the unit normal, $\vec{b}$ is the unit binormal, and $\vec{t}$ is the unit tangent to ray Ω. Unit vectors $\vec{n}(R)$ and $\vec{b}(R)$ depend only on the local properties of ray Ω at R and not on $\vec{n}(S)$ and $\vec{b}(S)$. Unit vectors $\vec{e}_1(R)$ and $\vec{e}_2(R)$ depend on initial conditions $\vec{e}_1(S)$ and $\vec{e}_2(S)$. Thus, even for $\vec{e}_1(S) = \vec{n}(S)$ and $\vec{e}_2(S) = \vec{b}(S)$, $\vec{e}_1(R)$ and $\vec{e}_2(R)$ may be different from $\vec{n}(R)$ and $\vec{b}(R)$. The ray-centered coordinate system connected with basis vectors $\vec{e}_1$, $\vec{e}_2$, and $\vec{e}_3$ is orthogonal along the whole ray Ω, but the coordinate system connected with $\vec{n}$, $\vec{b}$, and $\vec{t}$ is not, in general, orthogonal.

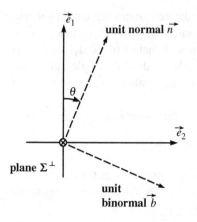

Figure 4.4. Definition of the Rytov angle θ in a plane $\Sigma^{\perp}$ perpendicular to ray Ω.

form the basis vectors of an orthogonal coordinate system, but rather form the basis vectors of a coordinate system which is, in general, nonorthogonal.

It would, however, be possible to use the known $\vec{n}$ and $\vec{b}$ along Ω and to compute $\vec{e}_1(s)$ and $\vec{e}_2(s)$ in terms of them. Following the suggestions of Popov and Pšenčík (1978a, 1978b), we can use equations

$$\vec{e}_1 = \vec{n}\cos\theta - \vec{b}\sin\theta, \qquad \vec{e}_2 = \vec{n}\sin\theta + \vec{b}\cos\theta; \qquad (4.1.13)$$

see Fig. 4.4. Angle θ varies along curve Ω and is known as the *Rytov angle*. It can be calculated along Ω using the equation

$$d\theta/ds = T(s), \qquad (4.1.14)$$

where $T(s)$ is the torsion of Ω. For details see Popov and Pšenčík (1978a, 1978b). This approach of calculating basis vectors $\vec{e}_1$ and $\vec{e}_2$ of the orthogonal coordinate system connected with curve Ω can be used for any 3-D curve Ω; its use is not limited to a ray. In this case, the scale factors are $h_1 = h_2 = 1$ and $h_3 = 1 - K(q_1\cos\theta + q_2\sin\theta)$, where K is the curvature of the curve. For rays, it yields the same results as (4.1.3), (4.1.4), (4.1.5), and (4.1.8).

If $\vec{e}_1(s)$ and $\vec{e}_2(s)$ satisfy (4.1.13) and (4.1.14) along curve Ω, we say that the vectors $\vec{e}_1(s)$ and $\vec{e}_2(s)$ are *transported parallelly* along Ω. Consequently, we can say that the triplet of basis vectors of the ray-centered coordinate system $\vec{e}_1$, $\vec{e}_2$, and $\vec{e}_3 = \vec{t}$, connected with ray Ω, is *transported parallelly along ray* Ω. Numerically, differential equations (4.1.5) are usually more efficient than (4.1.13) with (4.1.14).

It is very important to understand correctly the difference between *ray-centered coordinates* q_1, q_2, q_3 of point R', connected with ray Ω, and *ray coordinates* $\gamma_1, \gamma_2, \gamma_3$ of the same point R' (see Section 3.10.1 for ray coordinates). Let us consider, for simplicity, a point-source ray field, with the point source situated at S; see Figure 4.5.

a. The *ray coordinates* $\gamma_1, \gamma_2, \gamma_3$ of point R' are related to the complete ray field. To find them, we must first determine ray parameters γ_1 and γ_2 (for example, take-off angles i_0 and ϕ_0) of ray Ω', passing through S and R' (two-point ray tracing). See Figure 4.5(a) in 2-D (for $\gamma_2 = 0$). Then, we determine $\gamma_3 = s$, the arclength from S to R' along Ω'. Note that ray coordinates $\gamma_1, \gamma_2, \gamma_3$ are *mostly nonorthogonal*.

b. The *ray-centered coordinates* q_1, q_2, q_3 of the same point R' are connected with *a specified reference ray* passing through S, say Ω. Ray Ω' passing through S and R' does not play any important role in determining q_1, q_2, q_3. The ray-centered

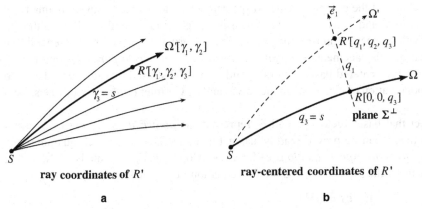

Figure 4.5. Differences between ray coordinates γ_1, γ_2, γ_3 and ray-centered coordinates q_1, q_2, q_3. The central ray field from point S is considered. (a) To determine ray coordinates γ_i of point R', it is necessary to find the ray Ω' passing through S and R'. Ray parameters γ_1 and γ_2 of Ω' then represent ray coordinates γ_1 and γ_2 of R', and γ_3 is the arclength along Ω' from S to R' (or any other alternative parameter along Ω' from S to R'). (b) To determine the ray-centered coordinates q_i of point R', connected with an arbitrarily chosen ray Ω, it is not necessary to know ray Ω' passing through S and R'. We construct plane $\Sigma^\perp$, passing through R' and perpendicular to Ω. The intersection of $\Sigma^\perp$ with Ω determines point R, and q_3 is the arclength along Ω from S to R (or any other alternative parameter along Ω from S to R). The remaining coordinates q_1 and q_2 are determined as Cartesian coordinates of R' in $\Sigma^\perp$, in a frame specified by basis vectors $\vec{e}_1$ and $\vec{e}_2$.

coordinates q_1, q_2, q_3 of point R' depend on the position of R' and on the specified reference ray Ω. For different rays Ω, different ray-centered coordinates are obtained. For a given ray Ω, $q_1(R')$, $q_2(R')$, $q_3(R')$ are constructed as follows. First, we construct plane $\Sigma^\perp$, passing through R' and perpendicular to Ω, and determine point R, at which $\Sigma^\perp$ intersects Ω. See Figure 4.5(b), again in 2-D (for $q_2 = 0$). Then we determine $q_3(R)$, the arclength from S to R along Ω (not along ray Ω' passing through R'). By definition, $q_3(R') \equiv q_3(R)$. Coordinates $q_1(R')$ and $q_2(R')$ determine the Cartesian coordinates of R' in plane $\Sigma^\perp$, with the origin at R and with axes oriented along $\vec{e}_1$ and $\vec{e}_2$. The reference ray Ω represents the coordinate axis in the ray-centered coordinate system. Note that $q_3(R')$ and $\gamma_3(R')$ are different, even if both measure arclength. $\gamma_3(R')$ measures arclength SR' along Ω' (passing through S and R'), and $q_3(R') \equiv q_3(R)$ measures arclength SR along the reference ray Ω. Note that the ray-centered coordinates q_1, q_2, q_3 are *always orthogonal*.

4.1.2 Ray-Centered Basis Vectors as Polarization Vectors

Unit vectors $\vec{e}_i(s)$ play an important role in all high-frequency asymptotic methods used to investigate seismic wavefields in inhomogeneous isotropic media. Among others, they determine the direction of the displacement vectors of high-frequency seismic body waves propagating in laterally varying isotropic structures. Unit vector $\vec{e}_3 = \vec{t}$ determines the direction of the displacement vector of P waves, which is always linearly polarized. Especially important are unit vectors $\vec{e}_1$ and $\vec{e}_2$ because they determine the *polarization of S waves*. In a smooth medium without interfaces, the displacement vector of an S wave propagating along Ω is polarized in the direction of $\vec{e}_1(R)$ at point R of Ω if it is polarized along $\vec{e}_1(S)$ at any reference point S of Ω. The same applies to the S wave polarized along

$\vec{e}_2$. The direction of the complex-valued displacement vector of the S wave remains fixed with respect to $\vec{e}_1$ and $\vec{e}_2$ as the wave progresses along Ω in a smooth medium, although it rotates with respect to unit normal $\vec{n}$ and unit binormal $\vec{b}$ to the ray. In general, we can say that the displacement vector of the S wave is *transported parallel* along ray Ω. Note that the ray-centered basis vectors $\vec{e}_1$ and $\vec{e}_2$ were first used by Luneburg (1964) to study the polarization of electromagnetic waves and by Cormier (1984) to study seismic S waves.

The fact that basis vectors $\vec{e}_1$ and $\vec{e}_2$ represent the *polarization vectors of S waves* in smooth media can be proved readily, using Equations. (2.4.36), which guarantee the *decoupling of the transport equations* of S waves. Along ray $T_{,i}e_{1j,i}$ equals $V^{-1}\mathrm{d}e_{1j}/\mathrm{d}s$, where T denotes the travel time, and the first equation of (2.4.36) reads

$$V^{-1}\mathrm{d}\vec{e}_1/\mathrm{d}s = a\nabla T.$$

Multiplying this equation by $\vec{p}$ and taking into account that $\vec{p} \cdot \vec{p} = 1/V^2$ and $\vec{e}_1 \cdot \vec{p} = 0$, we obtain $a = V\vec{p} \cdot \mathrm{d}\vec{e}_1/\mathrm{d}s = -V\vec{e}_1 \cdot \mathrm{d}\vec{p}/\mathrm{d}s$, in a manner similar to that in (4.1.2). This yields

$$\mathrm{d}\vec{e}_1/\mathrm{d}s = -V^2(\vec{e}_1 \cdot \mathrm{d}\vec{p}/\mathrm{d}s)\,\vec{p},$$

and the same equation for $\vec{e}_2$. But these equations are exactly the same as (4.1.4). We conclude that *the transport equations of S waves are decoupled* if unit vectors $\vec{e}_1$ and $\vec{e}_2$ are chosen to satisfy (4.1.4).

It is not difficult to explain simply and objectively (and to derive directly) Equations (4.1.4) *using Snell's law*. We shall use the same approach as in Section 3.1.4, where we derived the ray tracing systems from Snell's law. Let us consider a plane S wave incident at a plane interface between two homogeneous media. As we know from Section 2.3, the displacement vector of the transmitted S wave is polarized in the plane of incidence if the displacement vector of the incident S wave is also polarized in the plane of incidence. Similarly, the displacement vector of the transmitted S wave is perpendicular to the plane of incidence if the displacement vector of the incident S wave is perpendicular to the plane of incidence. This also applies approximately to high-frequency S waves incident at a slightly curved interface; see Section 2.4.5. We can exploit this fact to find the differential equation for the polarization vector of an S wave propagating in a smoothly inhomogeneous medium. As in Section 3.1.4, we shall simulate a smooth medium by a system of thin homogeneous layers, with interfaces along isovelocity surfaces (that is, along surfaces of constant velocity). Normal $\vec{n}$ to the interface at any point of incidence is perpendicular to the isovelocity surface so that $\vec{n} = \pm\nabla V/|\nabla V|$. See Figure 3.1. We wish to keep polarization vector $\vec{e}_1$ in the planes of incidence along the whole ray Ω. Given that the polarization vectors of the incident and transmitted S wave are situated in the plane of incidence, vector $\mathrm{d}\vec{e}_1/\mathrm{d}s$ must also be situated in the plane of incidence close to the point of incidence. The plane of incidence is specified by vectors $\vec{p}$ and $\vec{n}$ so that $\mathrm{d}\vec{e}_1/\mathrm{d}s = A\vec{p} + B\vec{n}$, where A and B remain to be determined. Multiplying this relation by $\vec{e}_1$, we obtain $\vec{e}_1 \cdot \mathrm{d}\vec{e}_1/\mathrm{d}s = A\vec{p} \cdot \vec{e}_1 + B\vec{n} \cdot \vec{e}_1$. Because $\vec{e}_1 \cdot \vec{e}_1 = 1$, we have $\vec{e}_1 \cdot \mathrm{d}\vec{e}_1/\mathrm{d}s = 0$. Moreover, $\vec{p} \cdot \vec{e}_1 = 0$ because $\vec{e}_1$ is perpendicular to the ray. This yields $B = 0$ and, thus, $\mathrm{d}\vec{e}_1/\mathrm{d}s = A\vec{p}$. From this we obtain the ordinary differential equation (4.1.4) in the same way as determined earlier.

As we know, the basis vectors of the ray-centered coordinate system $\vec{e}_1$, $\vec{e}_2$ can be chosen arbitrarily at the initial point S of the ray. Similarly, if ray Ω is incident at an interface, the basis vectors of the ray-centered coordinate system of reflected/transmitted waves can be taken arbitrarily at the point of reflection/transmission. Some options may, however, be more convenient. In this book, we systematically use the *standard option*, as given by Equation (2.3.45). In general, however, the basis vectors of the ray-centered coordinate system of reflected/transmitted waves at the reflection/transmission point do not necessarily represent the polarization vectors of reflected/transmitted waves at that point. The reason is that the actual polarization of reflected/transmitted S waves is also affected by the reflection/transmission coefficients at the interface. An incident S wave, linearly polarized at the point of incidence, can generate quasi-elliptically polarized reflected/transmitted S waves at the point of reflection/transmission. Even in this case, however, basis vectors $\vec{e}_1$ and $\vec{e}_2$ represent a suitable frame in which the polarization properties of reflected/transmitted S waves can be expressed. In Section 6.4, we shall give the exact rules for calculating the polarization properties of S waves propagating in a layered medium in terms of basis vectors $\vec{e}_1$ and $\vec{e}_2$. For this reason, we shall continue to refer to vectors $\vec{e}_1$ and $\vec{e}_2$ alternatively as basis and polarization vectors, even in a layered medium.

4.1.3 Computation of Ray-Centered Basis Vectors Along Ray Ω

This section is devoted to the computation of ray-centered basis vectors $\vec{e}_1(s)$, $\vec{e}_2(s)$, and $\vec{e}_3(s) = \vec{t}(s)$ along ray Ω. As $\vec{t}(s) = V\vec{p}(s)$ is known from ray tracing, it is sufficient to compute $\vec{e}_1(s)$ and $\vec{e}_2(s)$ only. Moreover, due to (4.1.12), we can only compute $\vec{e}_2(s)$ and determine $\vec{e}_1(s)$ using the relation $\vec{e}_1(s) = \vec{e}_2(s) \times \vec{t}(s)$ (or, alternatively, only compute $\vec{e}_1(s)$ and use $\vec{e}_2(s) = \vec{t}(s) \times \vec{e}_1(s)$). For simplicity, we shall discuss the computation of $\vec{e}_2(s)$ only. After simple modifications, the conclusions also apply to $\vec{e}_1(s)$.

The ray-centered unit vector $\vec{e}_2(s)$ can be calculated along ray Ω in four ways:

1. Direct numerical solution of differential equation (4.1.5) for $\vec{e}_2$. The method is quite general and may be used along any 3-D ray Ω. The relevant differential equation for $\vec{e}_2$ may be solved together with ray tracing, or after it (along a known ray Ω). Vector $\vec{e}_2(s_0)$ at the initial point s_0 must be chosen to satisfy relations $\vec{e}_2(s_0) \cdot \vec{e}_2(s_0) = 1$ and $\vec{e}_2(s_0) \cdot \vec{p}(s_0) = 0$. This guarantees that $\vec{e}_2(s_0)$ is a unit vector perpendicular to ray Ω at $s = s_0$. A consequence of the differential equation used is that $\vec{e}_2(s) \cdot \vec{e}_2(s) = 1$ and $\vec{e}_2(s) \cdot \vec{p}(s) = 0$ along the whole ray, that is, $\vec{e}_2(s)$ is a unit vector perpendicular to the ray along the whole ray Ω, and that $\vec{e}_1(s)$, $\vec{e}_2(s)$, and $\vec{e}_3(s) = \vec{t}(s)$ form a right-handed triplet along the whole ray Ω. Equations $\vec{e}_2(s) \cdot \vec{e}_2(s) = 1$ and $\vec{e}_2(s) \cdot \vec{p}(s) = 0$ can be used to check the accuracy of computations if $\vec{p}(s)$ and $\vec{e}_2(s)$ are determined numerically.

2. Computation in terms of $\vec{n}$ and $\vec{b}$, where $\vec{n}$ and $\vec{b}$ are the unit normal and unit binormal to Ω; see (4.1.13). This method requires not only the computation of $\vec{n}$ and $\vec{b}$, but also the computation of the curvature and torsion of ray Ω; see Section 3.2.4. By computing torsion $T(s)$ and the quadratures of (4.1.14) along ray Ω, we obtain the Rytov angle $\theta(s)$ and can use (4.1.13). The method is usually numerically less efficient than the previous method.

3. Computation in terms of an auxiliary vector $\vec{A}$. Let us consider an auxiliary unit vector $\vec{A}$, which may be constant or variable in space. We can then express $\vec{e}_2$ in terms of $\vec{A}$ and $\vec{t}$ as follows:

$$\vec{e}_2 = [(\vec{t} \times \vec{A}) \sin \Phi + (\vec{t} \times \vec{t} \times \vec{A}) \cos \Phi] / |\vec{t} \times \vec{A}|. \qquad (4.1.15)$$

The differential equations (4.1.3), (4.1.4), or (4.1.5) for $\vec{e}_2$ then yield a closed-form integral for angle Φ along ray Ω:

$$\Phi(s) = \Phi(s_0) + \int_{s_0}^{s} \frac{(\vec{t} \times \vec{A}) \cdot [V^{-1}(\vec{t} \cdot \vec{A})\nabla V + d\vec{A}/ds]}{(\vec{t} \times \vec{A})^2} \, ds. \qquad (4.1.16)$$

Vector $\vec{A}$ may be chosen in different ways, for example, as a constant arbitrarily oriented unit vector or as a unit vector oriented locally along a gradient of velocity, $\nabla V/V$. The derivation of Equations (4.1.15) and (4.1.16) and a more detailed discussion of various choices of $\vec{A}$ can be found in Popov and Pšenčík (1978b), Pšenčík (1979), Červený and Hron (1980), and Červený (1987b). The disadvantage of the method is that (4.1.16) fails if the direction of Ω is close to $\vec{A}$, that is, for small $|\vec{t} \times \vec{A}|^2$. It is then necessary to jump to another vector $\vec{A}$.

4. Analytic solution. In certain simple situations, $\vec{e}_2(s)$ may be calculated analytically along the ray. As a very important example, which has a number of applications, we can name a planar ray Ω. If ray Ω is situated in plane $\Sigma^{\|}$, we can choose $\vec{e}_2(s_0)$ perpendicular to $\Sigma^{\|}$ at the initial point s_0. The second equation of (4.1.3) then guarantees that $\vec{e}_2(s)$ equals $\vec{e}_2(s_0)$ along the whole ray Ω. The complete triplet $\vec{e}_1(u)$, $\vec{e}_2(u)$, $\vec{e}_3(u)$, where u is any monotonic parameter along ray Ω, is then given analytically by

$$\vec{e}_1(u) = -V(u)(\vec{p}(u) \times \vec{e}_2(u_0)), \qquad \vec{e}_2(u) = \vec{e}_2(u_0),$$
$$\vec{e}_3(u) = \vec{t}(u) = V(u)\vec{p}(u). \qquad (4.1.17)$$

Slowness vector $\vec{p}(u)$ is known from ray tracing.

As an example of the velocity distribution that yields the analytical solution for $\vec{e}_1(u)$, $\vec{e}_2(u)$, and $\vec{t}(u)$, we shall consider a model in which the nth power of slowness, V^{-n}, is a linear function of coordinates,

$$V^{-n} = A_0 + A_1 x_1 + A_2 x_2 + A_3 x_3. \qquad (4.1.18)$$

In this case, we obtain an analytical expression for $\vec{p}(u)$

$$\vec{p}(u) = \vec{p}(u_0) + \vec{A}(u - u_0)/n,$$

where $\vec{A}$ is a vector with components A_1, A_2, and A_3; see Section 3.4.3. The whole ray is planar, with plane $\Sigma^{\|}$ specified by $\vec{p}(u_0)$ and $\vec{A}$. We again choose $\vec{e}_2(u_0)$ perpendicular to $\Sigma^{\|}$ that is, $\vec{e}_2(u_0) = (\vec{p}(u_0) \times \vec{A})/|\vec{p}(u_0) \times \vec{A}|$ and obtain

$$\vec{e}_2(u) = \vec{e}_2(u_0), \qquad \vec{e}_1(u) = -V(u)[(\vec{p}(u_0) + \vec{A}(u - u_0)/n) \times \vec{e}_2(u_0)].$$
$$(4.1.19)$$

from (4.1.17). Here $V(u) = (A_0 + A_i x_i(u))^{-1/n}$, where $x_i(u)$ are given by (3.4.6). The relations for $x_i(u)$ reduce to a quadratic polynomial for $n = 2$, that is, for $u = \sigma$; see (3.4.3).

Similar analytical expressions for $\vec{e}_1(u)$ and $\vec{e}_2(u)$ can also be found for a model in which velocity V is given by relation $\ln V = A_0 + A_i x_i$, as well as for some other models.

4.1.4 Local Ray-Centered Cartesian Coordinate System

It is often useful to introduce a local Cartesian coordinate system with its origin at a specified point R on ray Ω and with basis vectors $\vec{j}_1$, $\vec{j}_2$, and $\vec{j}_3$, given by the relations

$$\vec{j}_1 = \vec{e}_1(R), \qquad \vec{j}_2 = \vec{e}_2(R), \qquad \vec{j}_3 = \vec{e}_3(R) = \vec{t}(R). \qquad (4.1.20)$$

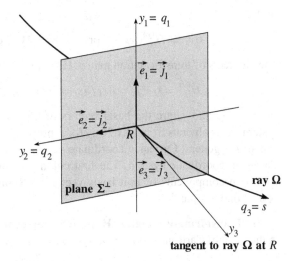

Figure 4.6. Local ray-centered Cartesian coordinate system y_1, y_2, y_3 at R on ray Ω. The basis vectors $\vec{j}_1, \vec{j}_2$, and $\vec{j}_3$ of the ray-centered Cartesian coordinate system at R coincide with $\vec{e}_1(R), \vec{e}_2(R)$, and $\vec{e}_3(R)$. The q_3-axis of the ray-centered coordinate system coincides with ray Ω, but the y_3-axis of the local Cartesian coordinate system constructed at R coincides with the tangent to ray Ω at R.

Thus, the basis vectors $\vec{j}_1, \vec{j}_2$, and $\vec{j}_3$ of the local ray-centered Cartesian coordinate system at R coincide with the basis vectors of the ray-centered coordinate system at the same point. Away from point R, however, the basis vectors of both systems differ. Basis vectors $\vec{j}_1, \vec{j}_2$, and $\vec{j}_3$ are constant throughout the space (the coordinate system is Cartesian), but basis vectors $\vec{e}_1, \vec{e}_2$, and $\vec{e}_3$ vary from place to place.

We denote the local Cartesian ray-centered coordinate by y_1, y_2, and y_3. A simple sketch to compare both types of coordinates in the 2-D case is shown in Figure 4.6.

Note the difference between the ray-centered coordinate $q_3 = s$ and the local Cartesian ray-centered coordinate y_3. Coordinate q_3 is measured along ray Ω, whereas coordinate y_3 is measured along the tangent to Ω at R; see Figure 4.6. Hence,

$$ds = h^{-1}dy_3, \tag{4.1.21}$$

where h is the scale factor given by (4.1.8). Approximately, in the vicinity of R,

$$s - s_0 \doteq h^{-1}y_3. \tag{4.1.22}$$

4.1.5 Transformation Matrices

In this section, we shall introduce several important 3×3 transformation matrices that are related to the ray-centered coordinate system and some others that will be needed in this chapter. In the whole section, we shall denote the general Cartesian coordinates as x_1, x_2, x_3 and the basis vectors of the general Cartesian coordinate system as $\vec{i}_1, \vec{i}_2$, and $\vec{i}_3$.

a. Transformation matrix $\hat{\mathbf{H}}^{(y)}$ from the local ray-centered Cartesian coordinate system y_1, y_2, y_3 to the general Cartesian coordinate system x_1, x_2, x_3.

Let us consider a local ray-centered Cartesian coordinate system y_1, y_2, y_3, with its origin at a selected point R of ray Ω. The transformation relations are

$$dx_k = H_{kl}^{(y)}dy_l \quad \text{or} \quad d\hat{\mathbf{x}} = \hat{\mathbf{H}}^{(y)}d\hat{\mathbf{y}}, \tag{4.1.23}$$

where $d\hat{\mathbf{x}} = (dx_1, dx_2, dx_3)^T$ and $d\hat{\mathbf{y}} = (dy_1, dy_2, dy_3)^T$. Matrix $\hat{\mathbf{H}}^{(y)}$ is orthonormal,

$$\hat{\mathbf{H}}^{(y)-1} = \hat{\mathbf{H}}^{(y)^T}, \qquad \det \hat{\mathbf{H}}^{(y)} = 1, \tag{4.1.24}$$

so that

$$dy_k = H_{lk}^{(y)}dx_l \qquad \text{or} \qquad d\hat{\mathbf{y}} = \hat{\mathbf{H}}^{(y)T}d\hat{\mathbf{x}}. \tag{4.1.25}$$

The elements of transformation matrix $\hat{\mathbf{H}}^{(y)}$ are

$$H_{kl}^{(y)} = \vec{i}_k \cdot \vec{j}_l = \partial x_k/\partial y_l = \partial y_l/\partial x_k. \tag{4.1.26}$$

Here $\vec{j}_1, \vec{j}_2,$ and $\vec{j}_3$ are the basis vectors of the local ray-centered Cartesian coordinate system. As is obvious from (4.1.26), $H_{kl}^{(y)}$ represents the cosine of the angle between the kth axis of the general Cartesian coordinate system and the lth axis of the local ray-centered Cartesian coordinate system. The first column of matrix $\hat{\mathbf{H}}^{(y)}$ is formed by the general Cartesian components of unit basis vector $\vec{j}_1$. Similarly, the second and third columns correspond to $\vec{j}_2$ and $\vec{j}_3$.

b. Transformation matrix $\hat{\mathbf{H}}$ from the ray-centered coordinate system $q_1, q_2, q_3 = s$ to the general Cartesian coordinate system x_1, x_2, x_3. The transformation relations are

$$dx_k = H_{kl}dq_l \qquad \text{or} \qquad d\hat{\mathbf{x}} = \hat{\mathbf{H}}d\hat{\mathbf{q}}. \tag{4.1.27}$$

Because the ray-centered coordinates are not Cartesian, matrix $\hat{\mathbf{H}}$ is not as simple as transformation matrix $\hat{\mathbf{H}}^{(y)}$. In general, it is orthogonal but not orthonormal.

We will mostly work with transformation matrix $\hat{\mathbf{H}}$ on central ray Ω, that is, for $q_I = 0$. Along central ray Ω, the properties of $\hat{\mathbf{H}}$ are simple. We will give several of these properties. Let us emphasize, however, that these relations are valid only along central ray Ω. Let us select point R on ray Ω and use (4.1.20). Then

$$\hat{\mathbf{H}}(R) = \hat{\mathbf{H}}^{(y)}(R). \tag{4.1.28}$$

Thus, matrix $\hat{\mathbf{H}}(R)$ is orthonormal along Ω,

$$\hat{\mathbf{H}}^{-1}(R) = \hat{\mathbf{H}}^T(R), \qquad \det\hat{\mathbf{H}}(R) = 1. \tag{4.1.29}$$

The elements of matrix $\hat{\mathbf{H}}(R)$ are given by relations

$$H_{kl}(R) = \vec{i}_k \cdot \vec{e}_l(R) = (\partial x_k/\partial q_l)_R = (\partial q_l/\partial x_k)_R. \tag{4.1.30}$$

The first column of $\hat{\mathbf{H}}$ represents the Cartesian components of basis vector $\vec{e}_1$, the second column represents the Cartesian components of $\vec{e}_2$, and the third column represents the Cartesian components of $\vec{e}_3 \equiv \vec{t}$. Hence,

$$d\hat{\mathbf{x}} = \hat{\mathbf{H}}(R)d\hat{\mathbf{q}}(R), \qquad d\hat{\mathbf{q}}(R) = \hat{\mathbf{H}}^T(R)d\hat{\mathbf{x}}, \tag{4.1.31}$$

where $d\hat{\mathbf{q}}(R) = (dq_1(R), dq_2(R), dq_3(R))^T$. Because the ith column of $\hat{\mathbf{H}}$ is represented by the Cartesian components of $\vec{e}_i$ along Ω, the determination of $\hat{\mathbf{H}}$ along Ω is equivalent to the determination of $\vec{e}_1, \vec{e}_2,$ and $\vec{e}_3$ along Ω.

c. Transformation matrix $\hat{\mathbf{Q}}$ from ray coordinates $\gamma_1, \gamma_2, \gamma_3$ to ray-centered coordinates q_1, q_2, q_3. The transformation relations are

$$dq_k = Q_{kl}d\gamma_l \qquad \text{or} \qquad d\hat{\mathbf{q}} = \hat{\mathbf{Q}}d\hat{\gamma}. \tag{4.1.32}$$

where $d\hat{\gamma} = (d\gamma_1, d\gamma_2, d\gamma_3)^T$. The properties of $\hat{\mathbf{Q}}$ away from ray Ω are not simple; we shall again consider $\hat{\mathbf{Q}}$ only along central ray Ω. Let us select an arbitrary point R on ray Ω. Then

$$Q_{ij}(R) = (\partial q_i/\partial \gamma_j)_R. \tag{4.1.33}$$

Along ray Ω, $q_1 = q_2 = 0$. This yields

$$Q_{13}(R) = (\partial q_1/\partial \gamma_3)_R = 0, \qquad Q_{23}(R) = (\partial q_2/\partial \gamma_3)_R = 0,$$

so that

$$\hat{\mathbf{Q}}(R) = \begin{pmatrix} Q_{11}(R) & Q_{12}(R) & 0 \\ Q_{21}(R) & Q_{22}(R) & 0 \\ Q_{31}(R) & Q_{32}(R) & Q_{33}(R) \end{pmatrix}. \tag{4.1.34}$$

Equation (4.1.34) immediately yields

$$\det \hat{\mathbf{Q}}(R) = Q_{33}(R) \det \mathbf{Q}(R) = (\partial q_3/\partial \gamma_3)_R \det \mathbf{Q}(R). \tag{4.1.35}$$

If we take $q_3 = \gamma_3$ along Ω, we obtain

$$\det \hat{\mathbf{Q}}(R) = \det \mathbf{Q}(R). \tag{4.1.36}$$

d. Transformation matrix $\hat{\mathbf{Q}}^{(x)}$ from ray coordinates $\gamma_1, \gamma_2, \gamma_3$ to Cartesian coordinates x_1, x_2, x_3. Matrix $\hat{\mathbf{Q}}^{(x)}$ was introduced in Section 3.10.1. It is defined as

$$dx_k = Q_{kl}^{(x)} d\gamma_l, \qquad d\hat{\mathbf{x}} = \hat{\mathbf{Q}}^{(x)} d\hat{\boldsymbol{\gamma}}. \tag{4.1.37}$$

We shall again consider matrix $\hat{\mathbf{Q}}^{(x)}$ only along central ray Ω. At point R, the elements of matrix $\hat{\mathbf{Q}}^{(x)}$ are given by relations

$$Q_{kl}^{(x)}(R) = (\partial x_k/\partial \gamma_l)_R. \tag{4.1.38}$$

It is simple to see from (4.1.27), (4.1.32), and (4.1.37) that

$$\hat{\mathbf{Q}}^{(x)}(R) = \hat{\mathbf{H}}(R)\hat{\mathbf{Q}}(R). \tag{4.1.39}$$

e. Transformation matrix $\hat{\mathbf{P}}$ from ray coordinates $\gamma_1, \gamma_2, \gamma_3$ to covariant ray-centered components of slowness vector $p_1^{(q)}, p_2^{(q)}$, and $p_3^{(q)}$. We denote the covariant ray-centered components of slowness vector $\vec{p}$ by $p_i^{(q)}$, $i = 1, 2, 3$. They are given by relations

$$p_i^{(q)} = \partial T/\partial q_i, \qquad i = 1, 2, 3.$$

Let us emphasize that the covariant components $p_i^{(q)} = \partial T/\partial q_i$ are not physical components of the slowness vector. The difference resides in scale factors h_i. In ray-centered coordinates, the physical components of the slowness vector equal $\partial T/\partial q_1$, $\partial T/\partial q_2$, and $h^{-1}\partial T/\partial q_3$. See more details in Section 3.5.

In the following text, we shall call $p_i^{(q)} = \partial T/\partial q_i$ simply ray-centered components of the slowness vector. We shall use the word *covariant* only in case of possible misunderstanding. The transformation relations are

$$dp_k^{(q)} = P_{kl} d\gamma_l \qquad \text{or} \qquad d\hat{\mathbf{p}}^{(q)} = \hat{\mathbf{P}} d\hat{\boldsymbol{\gamma}}. \tag{4.1.40}$$

We shall use this matrix only along central ray Ω. At any point R of the central ray,

$$P_{kl}(R) = \left(\partial p_k^{(q)}/\partial \gamma_l\right)_R = \left(\partial^2 T/\partial \gamma_l \partial q_k\right)_R. \tag{4.1.41}$$

Then

$$d\hat{\mathbf{p}}^{(q)}(R) = \hat{\mathbf{P}}(R) d\hat{\boldsymbol{\gamma}}, \qquad d\hat{\boldsymbol{\gamma}} = \hat{\mathbf{P}}^{-1}(R) d\hat{\mathbf{p}}^{(q)}(R), \tag{4.1.42}$$

where $d\hat{\mathbf{p}}^{(q)}(R) \equiv \left(dp_1^{(q)}(R), dp_2^{(q)}(R), dp_3^{(q)}(R)\right)^T$.

f. Transformation matrix $\hat{\mathbf{P}}^{(x)}$ from ray coordinates γ_1, γ_2, γ_3 to Cartesian components of the slowness vector p_1, p_2, and p_3. Matrix $\hat{\mathbf{P}}^{(x)}$ is defined by the relation

$$\mathrm{d}p_k = P_{kl}^{(x)}\mathrm{d}\gamma_l \qquad \text{or} \qquad \mathrm{d}\hat{\mathbf{p}} = \hat{\mathbf{P}}^{(x)}\mathrm{d}\hat{\gamma}. \tag{4.1.43}$$

Here $p_k = \partial T/\partial x_k$ are the Cartesian components of the slowness vector. At any point R on ray Ω, $\hat{\mathbf{P}}^{(x)}(R)$ can be simply expressed in terms of matrices $\hat{\mathbf{H}}(R)$ and $\hat{\mathbf{P}}(R)$,

$$\hat{\mathbf{P}}^{(x)}(R) = \hat{\mathbf{H}}(R)\hat{\mathbf{P}}(R). \tag{4.1.44}$$

Transformation matrices $\hat{\mathbf{Q}}$, $\hat{\mathbf{Q}}^{(x)}$, $\hat{\mathbf{P}}$, and $\hat{\mathbf{P}}^{(x)}$ depend on the choice of ray coordinate γ_3, corresponding to the monotonic parameter along the ray.

4.1.6 Ray Tracing in Ray-Centered Coordinates. Paraxial Ray Tracing System

The basis vectors $\vec{e}_1$, $\vec{e}_2$, and $\vec{e}_3$ may be introduced along any 3-D curve C, not only along a ray Ω; see (4.1.13) and (4.1.14). Let us consider the orthogonal ray-centered coordinate system q_1, q_2, q_3 connected with a curve C, with scale factors $h_1 = h_2 = 1$ and $h_3 = h$ and basis vectors $\vec{e}_1$, $\vec{e}_2$, and $\vec{e}_3 = \vec{t}$. The gradient of any function $T(q_1, q_2, q_3)$ is then given by the relation

$$\nabla T = \frac{\partial T}{\partial q_1}\vec{e}_1 + \frac{\partial T}{\partial q_2}\vec{e}_2 + \frac{1}{h}\frac{\partial T}{\partial q_3}\vec{e}_3;$$

see (3.5.2). The eikonal equation $(\nabla T)^2 = V^{-2}$ reads

$$\left(\frac{\partial T}{\partial q_1}\right)^2 + \left(\frac{\partial T}{\partial q_2}\right)^2 + \frac{1}{h^2}\left(\frac{\partial T}{\partial q_3}\right)^2 = \frac{1}{V^2(q_1, q_2, q_3)}. \tag{4.1.45}$$

To express the ray-tracing system in ray-centered coordinates q_1, q_2, q_3, we shall use the general form of the ray-tracing systems in orthogonal coordinates (3.5.15). We denote

$$T_1 = \partial T/\partial q_1, \qquad T_2 = \partial T/\partial q_2, \qquad T_3 = \partial T/\partial q_3. \tag{4.1.46}$$

Let us emphasize that T_i are covariant ray-centered components of the slowness vector, $p_i^{(q)} = T_i$. The physical ray-centered components of the slowness vector are given by expressions T_1, T_2, and $h^{-1}T_3$.

Variable q_3 along curve C is chosen to increase monotonically with travel time T so that $\partial T/\partial q_3 > 0$ is automatically satisfied. In the following discussion, we shall consider a vicinity of C in which $\partial T/\partial q_3 > 0$ is also satisfied. In other words, we do not consider paraxial rays, which have a turning point with respect to q_3. We can then solve the eikonal equation (4.1.45) for $T_3 = \partial T/\partial q_3$ and express it in terms of the reduced Hamiltonian $\mathcal{H}^R(q_i, T_I)$ as follows:

$$T_3 = -\mathcal{H}^R(q_i, T_I), \qquad \mathcal{H}^R(q_i, T_I) = -h\left[V^{-2}(q_i) - T_1^2 - T_2^2\right]^{1/2}. \tag{4.1.47}$$

Using (3.5.22), we obtain from (4.1.47) the ray tracing system in ray-centered coordinates:

$$\frac{\mathrm{d}q_1}{\mathrm{d}q_3} = \frac{h^2}{T_3}T_1, \qquad \frac{\mathrm{d}T_1}{\mathrm{d}q_3} = \frac{h^2}{T_3}\left[\frac{1}{2}\frac{\partial}{\partial q_1}\left(\frac{1}{V^2}\right) + \frac{1}{h^3}T_3^2\frac{\partial h}{\partial q_1}\right],$$
$$\frac{\mathrm{d}q_2}{\mathrm{d}q_3} = \frac{h^2}{T_3}T_2, \qquad \frac{\mathrm{d}T_2}{\mathrm{d}q_3} = \frac{h^2}{T_3}\left[\frac{1}{2}\frac{\partial}{\partial q_2}\left(\frac{1}{V^2}\right) + \frac{1}{h^3}T_3^2\frac{\partial h}{\partial q_2}\right]. \tag{4.1.48}$$

Here T_3 is given by the relation following from (4.1.47)

$$T_3 = h[V^{-2} - T_1^2 - T_2^2]^{1/2}. \qquad (4.1.49)$$

The number of equations in the ray tracing system connected with the curve C (4.1.48) is reduced to four. The advantage of the system (4.1.48) is that there are no problems with turning points if the curve C is situated close to a ray Ω, and if the paraxial rays situated close to C are studied. Otherwise, the system has no distinct advantages over the ray tracing system expressed in general Cartesian coordinates. It is nonlinear and algebraically more complicated. We are interested mainly in ray tracing in the vicinity of central ray Ω for small q_1 and q_2. For $C \equiv \Omega$ and for $q_1, q_2 \to 0$, the RHSs of equations (4.1.48) for dT_I/dq vanish. Unfortunately, they are expressed as differences of two terms, which remain finite even on central ray Ω. This fact increases the inaccuracies in computations. It will be shown, however, that this complication may be removed in the quadratic vicinity of Ω. The system may be simplified approximately and becomes linear.

The ray tracing system in ray-centered coordinates connected with a ray Ω was first derived by Červený and Pšenčík (1979), in a slightly different form than here. To find a convenient approximation for eikonal equation (4.1.45) and for ray tracing system (4.1.48) in the close vicinity of central ray Ω, we shall use the Taylor expansions of the travel-time field, velocities, and the like on central ray Ω. To distinguish between the quantities defined in the whole space and only along central ray Ω, we shall use a special notation for certain quantities defined only along Ω. This notation will be particularly useful for the velocity and its derivatives:

$$
\begin{aligned}
v(s) &= [V(q_1, q_2, s)]_{q_1=q_2=0}, \\
v_{,i}(s) &= [\partial V(q_1, q_2, s)/\partial q_i]_{q_1=q_2=0}, \\
v_{,ij}(s) &= [\partial^2 V(q_1, q_2, s)/\partial q_i \partial q_j]_{q_1,q_2=0}.
\end{aligned}
\qquad (4.1.50)
$$

For example, the abbreviated notation for scale factor h is

$$h = 1 + v^{-1}v_{,I}q_I; \qquad (4.1.51)$$

see (4.1.8). We shall use this notation only should the standard notation with V cause confusion, or where it would make the equations too long.

We shall now find an approximate expression for the reduced Hamiltonian $\mathcal{H}^R(q_i, T_I)$ given by (4.1.47), in the paraxial vicinity of ray Ω, that is, for small q_1 and q_2. Note that we use $q_3 = s$, the arclength along Ω. For small q_1 and q_2, T_1 and T_2 are also small. Using the Taylor expansion up to the second-order terms in q_1 and q_2, we obtain $V \doteq v + v_{,I}q_I + \frac{1}{2}v_{,IJ}q_Iq_J$. Hence

$$h/V \doteq v^{-1}\left(1 - \tfrac{1}{2}v^{-1}v_{,KL}q_Kq_L\right). \qquad (4.1.52)$$

Here we have also used (4.1.51). The approximate expression for the reduced Hamiltonian in the paraxial vicinity of Ω is then

$$\mathcal{H}^R(q_1, q_2, s, T_1, T_2) \doteq -v^{-1}\left[1 - \tfrac{1}{2}v^{-1}v_{,KL}q_Kq_L - \tfrac{1}{2}v^2\left(T_1^2 + T_2^2\right)\right]. \qquad (4.1.53)$$

For small q_I and T_I, this yields

$$\partial\mathcal{H}^R/\partial T_I \doteq vT_I, \qquad \partial\mathcal{H}^R/\partial q_I \doteq v^{-2}v_{,IJ}q_J,$$

and the paraxial ray tracing system in ray-centered coordinates reads

$$\mathrm{d}q_I/\mathrm{d}s = vT_I, \qquad \mathrm{d}T_I/\mathrm{d}s = -v^{-2}v_{,IJ}q_J. \tag{4.1.54}$$

We saw in (4.1.46) that T_I represent the ray-centered covariant components of slowness vector $p_I^{(q)}$ (which are also the physical components). Thus, (4.1.54) can be expressed as

$$\mathrm{d}q_I/\mathrm{d}s = vp_I^{(q)}, \qquad \mathrm{d}p_I^{(q)}/\mathrm{d}s = -v^{-2}v_{,IJ}q_J. \tag{4.1.55}$$

This is the final form of the *paraxial ray tracing system in ray-centered coordinates*.

In the paraxial ray tracing system (4.1.55), the variable s along ray Ω represents the arclength. Instead of it, we can use any other monotonic variable along Ω such as travel time T. Taking into account $\mathrm{d}s = v\mathrm{d}T$, we obtain the paraxial ray tracing system (4.1.55) in the following form:

$$\mathrm{d}q_I/\mathrm{d}T = v^2 p_I^{(q)}, \qquad \mathrm{d}p_I^{(q)}/\mathrm{d}T = -v^{-1}v_{,IJ}q_J. \tag{4.1.56}$$

The paraxial ray tracing system (4.1.56) can be expressed in a more compact form. We introduce the 4×1 column matrix

$$\mathbf{W}(T) = \left(q_1, q_2, p_1^{(q)}, p_2^{(q)}\right)^T \tag{4.1.57}$$

and express (4.1.56) as

$$\mathrm{d}\mathbf{W}(T)/\mathrm{d}T = \mathbf{S}\mathbf{W}. \tag{4.1.58}$$

Here $\mathbf{S}$ is a 4×4 system matrix,

$$\mathbf{S} = \begin{pmatrix} \mathbf{0} & v^2\mathbf{I} \\ -v^{-1}\mathbf{V} & \mathbf{0} \end{pmatrix}, \tag{4.1.59}$$

where $\mathbf{0}$ is a 2×2 null matrix, $\mathbf{I}$ is a 2×2 identity matrix, and $\mathbf{V}$ is the 2×2 matrix of the second derivatives of velocity V with respect to q_I, whose elements are

$$V_{IJ} = v_{,IJ} = \left(\partial^2 V(q_1, q_2, s)/\partial q_I \partial q_J\right)_{q_1=q_2=0}. \tag{4.1.60}$$

The advantages of the paraxial ray tracing systems (4.1.56) or (4.1.58) over the standard ray tracing systems follow:

- The paraxial ray tracing system consists of four equations only. Remember that the standard ray tracing system consists of six equations.
- The paraxial ray tracing system is linear. This is a great advantage over the standard ray tracing system.
- Because the paraxial ray tracing system is linear, we can compute the propagator matrix of the system. This propagator matrix will be determined in Section 4.3, where its properties will also be studied.

Paraxial ray tracing systems (4.1.56) or (4.1.58) play a fundamental role in the paraxial seismic ray method. Many important applications of paraxial ray tracing will be described in this chapter. The disadvantage of paraxial ray tracing systems (4.1.56) and (4.1.58) is that they are only approximate. The systems can be used only if the paraxial ray does not deviate considerably from central ray Ω.

Let us now briefly discuss the initial conditions for the paraxial ray tracing system (4.1.56). Let us consider the central ray Ω, specified by initial conditions (3.2.1) at point S. The initial conditions give the Cartesian coordinates of point S, $x_i(S)$, and the Cartesian components of the initial slowness vector $\vec{p}(S)$ at S. Components $p_i(S)$ are not arbitrary;

they must satisfy the relation $p_1^2(S) + p_2^2(S) + p_3^2(S) = 1/V^2(S)$ at S. Along Ω, we construct polarization vectors $\vec{e}_1$ and $\vec{e}_2$ using (4.1.5). At initial point S, $\vec{e}_1(S)$ and $\vec{e}_2(S)$ may be chosen arbitrarily; the only requirement is that $\vec{e}_1(S)$, $\vec{e}_2(S)$, and $\vec{e}_3(S) = \vec{t}(S)$ form a triplet of orthogonal, mutually perpendicular unit vectors. Using $\vec{e}_1(S)$ and $\vec{e}_2(S)$, we construct the ray-centered coordinate system q_1, q_2 in plane $\Sigma^\perp$, perpendicular to Ω at S.

Consider point S' situated in plane $\Sigma^\perp$, close to S, and denote its ray-centered coordinates by $q_I(S')$. The slowness vector of any paraxial ray passing through S' may be specified by the ray-centered components of slowness vector $p_1^{(q)}(S')$ and $p_2^{(q)}(S')$. The third component is again not required; it can be determined using the eikonal equation.

Thus, the initial conditions for paraxial ray tracing system (4.1.56) at point S' situated in plane $\Sigma^\perp$, perpendicular to ray Ω at S, are

$$\text{At } S': \qquad q_I = q_I(S'), \qquad p_I^{(q)} = p_I^{(q)}(S'). \tag{4.1.61}$$

The *four-parameter system of rays* specified by initial conditions (4.1.61) *is complete*, including all possible rays situated close to Ω.

Let us emphasize that $v_{,I}$ and $v_{,IJ}$ denote the derivatives of the velocity with respect to the ray-centered coordinates q_1 and/or q_2 at the central ray. Using transformation matrix $\hat{\mathbf{H}}$, we can relate these derivatives to the derivatives of velocity with respect to Cartesian coordinates as

$$\begin{aligned}
v_{,I}(R) &= H_{kI}(R)(\partial V/\partial x_k)_R, \\
v_{,IJ}(R) &= H_{kI}(R)H_{lJ}(R)\big(\partial^2 V/\partial x_k \partial x_l\big)_R.
\end{aligned} \tag{4.1.62}$$

4.1.7 Dynamic Ray Tracing System in Ray-Centered Coordinates

The purpose of dynamic ray tracing is to determine the first partial derivatives of phase space coordinates (coordinates, slowness vector components) with respect to the initial parameters of ray Ω, along a known ray Ω. The initial parameters of ray Ω may represent the phase space coordinates at the initial point of Ω, ray parameters γ_1 and γ_2 introduced in Section 3.10.1, or any other parameters specifying the initial conditions of paraxial rays. Such partial derivatives are needed in many seismological applications, including the computation of transformation matrices $\mathbf{Q}$ and $\mathbf{P}$, the ray Jacobian J, and the paraxial travel times. Because we are considering ray-centered coordinates, we wish to find the partial derivatives of q_I and $p_I^{(q)}$ with respect to γ, where γ is an arbitrarily selected initial parameter of Ω. Partial derivatives $\partial q_I/\partial \gamma$ and $\partial p_I^{(q)}/\partial \gamma$ are taken on the central ray Ω for other initial parameter(s) fixed. As partial derivative $\partial/\partial \gamma$ commutes with d/ds, we obtain the dynamic ray tracing system from the paraxial ray tracing system (4.1.55):

$$\frac{d}{ds}\left(\frac{\partial q_I}{\partial \gamma}\right) = v\frac{\partial p_I^{(q)}}{\partial \gamma}, \qquad \frac{d}{ds}\left(\frac{\partial p_I^{(q)}}{\partial \gamma}\right) = -v^{-2}v_{,IJ}\frac{\partial q_J}{\partial \gamma}. \tag{4.1.63}$$

This system of four linear ordinary differential equations of the first order for $\partial q_I/\partial \gamma$ and $\partial p_I^{(q)}/\partial \gamma$ ($I = 1, 2$) is known as the *dynamic ray tracing system in ray-centered coordinates*.

If we compare paraxial ray tracing system (4.1.55) with dynamic ray tracing system (4.1.63), we can see that the systems are the same; only the computed quantities are different. Paraxial ray tracing system (4.1.55) computes approximately the phase space coordinates q_I and $p_I^{(q)}$ along paraxial rays, and dynamic ray tracing system (4.1.63) computes exactly the partial derivatives $\partial q_I/\partial \gamma$, $\partial p_I^{(q)}/\partial \gamma$ along the central ray. If only parameter γ varies and the other initial parameters are fixed, we obtain $dq_I = (\partial q_I/\partial \gamma)_\Omega d\gamma$, $dp_I^{(q)} = (\partial p_I^{(q)}/\partial \gamma)_\Omega d\gamma$. Thus, paraxial ray tracing system (4.1.55) is obtained from dynamic ray tracing system (4.1.63) by multiplying the latter by $d\gamma$. In the following discussion, we shall mostly consider the dynamic ray tracing system, but we shall remember that it also represents the paraxial ray tracing system.

In the seismological literature, there is no unity in the use of the terms *paraxial ray tracing* and *dynamic ray tracing*. Often, these two terms are not distinguished at all. Some authors prefer to speak of paraxial ray tracing, and some others of dynamic ray tracing, although they have the same in mind. There is, however, no danger in these terminological differences because both the paraxial and dynamic ray tracing systems are the same; only their physical interpretation is different.

Dynamic ray tracing system (4.1.63) can be solved four times, for $\gamma = q_{10}, q_{20}, p_{10}^{(q)}$, and $p_{20}^{(q)}$, representing the initial values of q_I and $p_I^{(q)}$; see (4.1.61). The 16 solutions then correspond to a complete system of paraxial rays. We shall, however, mostly consider only a two-parameter orthonomic system of rays, specified by ray parameters γ_1 and γ_2; see Section 3.10.1. Then, (4.1.63) can be used to compute the 2×2 transformation matrices $Q_{IJ} = (\partial q_I/\partial \gamma_J)_{q_1=q_2=0}$ and $P_{IJ} = (\partial p_I^{(q)}/\partial \gamma_J)_{q_1=q_2=0}$; see Sections 4.1.5.c and 4.1.5.e. Using (4.1.63), we obtain the following dynamic ray tracing system for matrices $\mathbf{Q}$ and $\mathbf{P}$:

$$d\mathbf{Q}/ds = v\mathbf{P}, \qquad d\mathbf{P}/ds = -v^{-2}\mathbf{V}\mathbf{Q}. \qquad (4.1.64)$$

This system is one of the most important forms of dynamic ray tracing systems used in seismological applications. It was first derived and used to compute geometrical spreading (related to $\mathbf{Q}$) by Popov and Pšenčík (1978a, 1978b).

In dynamic ray tracing systems (4.1.63) and (4.1.64), variable s along ray Ω represents the arclength. Similarly as in paraxial ray tracing system (4.1.56), we can use travel time T instead of arclength s, $ds = vdT$. Dynamic ray tracing system (4.1.64) then reads

$$d\mathbf{Q}/dT = v^2\mathbf{P}, \qquad d\mathbf{P}/dT = -v^{-1}\mathbf{V}\mathbf{Q}. \qquad (4.1.65)$$

Dynamic ray tracing system (4.1.65) can be expressed in a more compact form. If we introduce a 4×1 column matrix $\mathbf{W}$ given by relation

$$\mathbf{W} = \left(\partial q_1/\partial \gamma, \partial q_2/\partial \gamma, \partial p_1^{(q)}/\partial \gamma, \partial p_2^{(q)}/\partial \gamma\right)^T, \qquad (4.1.66)$$

dynamic ray tracing system (4.1.65) becomes

$$\frac{d\mathbf{W}}{dT} = \mathbf{S}\mathbf{W}, \qquad \text{where } \mathbf{S} = \begin{pmatrix} \mathbf{0} & v^2\mathbf{I} \\ -v^{-1}\mathbf{V} & \mathbf{0} \end{pmatrix}. \qquad (4.1.67)$$

The 4×4 matrix $\mathbf{S}$ will be referred to as the *system matrix of the dynamic ray tracing system* in isotropic media in ray-centered coordinates. It is the same as the system matrix of the paraxial ray tracing system; see (4.1.59). A more compact form of the dynamic ray tracing system (4.1.65) is

$$d\mathbf{X}/dT = \mathbf{S}\mathbf{X}, \qquad \text{where } \mathbf{X} = \begin{pmatrix} \mathbf{Q} \\ \mathbf{P} \end{pmatrix}, \qquad (4.1.68)$$

and where $\mathbf{S}$ is given by (4.1.67). An important property of matrix $\mathbf{S}$ is that $\text{tr } \mathbf{S} = 0$. Equation (4.1.68) is equivalent to (4.1.67), but (4.1.67) should be solved twice with different initial condition to obtain all eight elements of matrix $\mathbf{X}$. Alternatively, we can say that system (4.1.68) consists of eight equations (Popov and Pšenčík 1978a, 1978b).

Dynamic ray tracing system (4.1.64) can also be expressed in the form of one linear ordinary differential equation of the second order for $\mathbf{Q}$. Taking the derivative of the first equation in (4.1.64) with respect to s and inserting the second equation yields

$$v\frac{d^2\mathbf{Q}}{ds^2} - \frac{\partial v}{\partial s}\frac{d\mathbf{Q}}{ds} + \mathbf{VQ} = 0. \tag{4.1.69}$$

If we use σ instead of s as the variable along Ω such that $d\sigma = v\,ds$, we obtain the simplest dynamic ray tracing equation

$$d^2\mathbf{Q}/d\sigma^2 + v^{-3}\mathbf{VQ} = 0. \tag{4.1.70}$$

Similarly, we obtain dynamic ray tracing systems for any other variable u along Ω, for example, for $u = T$.

Using $\mathbf{Q}$ and $\mathbf{P}$, many other important quantities may also be computed, and the relevant ordinary differential equations for these quantities may be derived. This applies to the 2×2 matrix of the curvature of wavefront $\mathbf{K}(s)$, to the 2×2 matrix of radii of the curvature of wavefront $\mathbf{R}(s)$, to the ray Jacobian $J(s)$ and geometrical spreading $\mathcal{L}(s)$, and to the 2×2 matrix $\mathbf{M}(s)$ of the second derivatives of the travel-time fields with respect to q_I, among others. Here we shall discuss briefly matrix $\mathbf{M}(s)$ with elements

$$M_{IJ}(s) = (\partial^2 T/\partial q_I \partial q_J)_{q_1=q_2=0}. \tag{4.1.71}$$

Because $\partial^2 T/\partial q_I \partial q_J = (\partial^2 T/\partial q_I \partial \gamma_K)(\partial \gamma_K/\partial q_J)$, we obtain

$$\mathbf{M} = \mathbf{PQ}^{-1}. \tag{4.1.72}$$

Consequently, matrix $\mathbf{M}$ can be calculated along Ω by dynamic ray tracing (4.1.64). A differential equation for matrix $\mathbf{M}$ itself can also be derived:

$$\frac{d\mathbf{M}}{ds} = \frac{d\mathbf{P}}{ds}\mathbf{Q}^{-1} + \mathbf{P}\frac{d\mathbf{Q}^{-1}}{ds} = \frac{d\mathbf{P}}{ds}\mathbf{Q}^{-1} - \mathbf{PQ}^{-1}\frac{d\mathbf{Q}}{ds}\mathbf{Q}^{-1},$$

because $d\mathbf{Q}^{-1}/ds = -\mathbf{Q}^{-1}(d\mathbf{Q}/ds)\mathbf{Q}^{-1}$. This yields

$$d\mathbf{M}/ds + v\mathbf{M}^2 + v^{-2}\mathbf{V} = 0. \tag{4.1.73}$$

This is a nonlinear ordinary differential equation of the first order of the Riccati type, expressed in matrix form. In general, this equation cannot be solved by elementary analytical methods. Equations of the Riccati type similar to (4.1.73) have been known for some time in the literature devoted to wave propagation problems. The matrix Riccati equation for $\mathbf{M}$ in the simple form of (4.1.73) was first derived by Červený and Hron (1980).

Dynamic ray tracing system (4.1.64) for $\mathbf{Q}$ and $\mathbf{P}$ can be solved only if the initial conditions for $\mathbf{Q}$ and $\mathbf{P}$ are specified at some initial point, say, at point S of ray Ω. Let us denote by $\mathbf{Q}(R)$ and $\mathbf{P}(R)$ the solutions of dynamic ray tracing system (4.1.64) at point R of ray Ω, corresponding to the initial conditions $\mathbf{Q}(S)$ and $\mathbf{P}(S)$. For a fixed point R, $\mathbf{Q}(R)$ and $\mathbf{P}(R)$ do not depend on the integration parameter u used to solve the dynamic ray tracing system along the ray Ω from S to R (T, s, and the like). If $\mathbf{Q}(S)$ and $\mathbf{P}(S)$ are chosen so that $\mathbf{P}(S)\mathbf{Q}^{-1}(S) = \mathbf{M}(S)$ is the matrix of second derivatives of the travel-time fields with respect to ray-centered coordinates q_1 and q_2, then $\mathbf{M}(R) = \mathbf{P}(R)\mathbf{Q}^{-1}(R)$ also has the meaning of the matrix of second derivatives of the travel-time field with respect to q_1 and q_2 at R. Moreover, if $\mathbf{Q}(S)$ and $\mathbf{P}(S)$ are chosen to represent the transformation matrices from ray to ray-centered coordinates and from ray to the ray-centered slowness vector components at S, they then also represent the same at point R.

Note that the parameterization of the ray fields by ray parameters γ_1 and γ_2 affects both $\mathbf{Q}$ and $\mathbf{P}$, but not $\mathbf{M}$. Just for the purpose of this section, we will introduce new 2×2 matrices $\bar{\mathbf{Q}}$ and $\bar{\mathbf{P}}$, related to matrices $\mathbf{Q}$ and $\mathbf{P}$ as

$$\bar{\mathbf{Q}} = \mathbf{QC}, \qquad \bar{\mathbf{P}} = \mathbf{PC}. \tag{4.1.74}$$

Here $\mathbf{P}$ and $\mathbf{Q}$ satisfy dynamic ray tracing system (4.1.64), and $\mathbf{C}$ is an arbitrary nonsingular constant 2×2 matrix. It is simple to see that $\bar{\mathbf{P}}\bar{\mathbf{Q}}^{-1} = \mathbf{P}\mathbf{Q}^{-1}$ so that matrix $\mathbf{M}$ is not influenced by $\mathbf{C}$.

Matrix $\mathbf{C}$ is related to the parametrization of the ray field. If $\mathbf{Q}$ and $\mathbf{P}$ correspond to ray parameters γ_1 and γ_2, and if $\bar{\mathbf{Q}}$ and $\bar{\mathbf{P}}$ correspond to ray parameters $\bar{\gamma}_1$ and $\bar{\gamma}_2$, elements C_{IJ} of matrix $\mathbf{C}$ are then given by relation

$$C_{IJ} = \partial \gamma_I / \partial \bar{\gamma}_J. \tag{4.1.75}$$

It is easy to derive the dynamic ray tracing system for $\bar{\mathbf{P}}$ and $\bar{\mathbf{Q}}$, using (4.1.64) which reads

$$\mathrm{d}\bar{\mathbf{Q}}/\mathrm{d}s = v\bar{\mathbf{P}}, \qquad \mathrm{d}\bar{\mathbf{P}}/\mathrm{d}s = -v^{-2}\mathbf{V}\bar{\mathbf{Q}}. \tag{4.1.76}$$

The dynamic ray tracing system (4.1.76) for $\bar{\mathbf{Q}}$ and $\bar{\mathbf{P}}$ remains exactly the same as the dynamic ray tracing system (4.1.64) for $\mathbf{Q}$ and $\mathbf{P}$.

The *conclusions* follow. Assume that initial conditions $\mathbf{Q}(S)$ and $\mathbf{P}(S)$ correspond to some *intrinsic choice* of the parametrization of ray field γ_1 and γ_2 at S. For this intrinsic choice of initial conditions, the solution of the dynamic ray tracing system at point R of the ray is $\mathbf{Q}(R)$ and $\mathbf{P}(R)$. For a different *user's choice* of the parameterization of ray field $\bar{\gamma}_1$ and $\bar{\gamma}_2$ at S, the solutions at point R are $\bar{\mathbf{Q}}(R) = \mathbf{Q}(R)\mathbf{C}$ and $\bar{\mathbf{P}}(R) = \mathbf{P}(R)\mathbf{C}$, where matrix $\mathbf{C}$ is given by (4.1.75).

4.1.8 Paraxial Travel Times

In Section 4.1.7, we explained how the 2×2 matrix $\mathbf{M}(s)$ can be computed in terms of 2×2 matrices $\mathbf{Q}(s)$ and $\mathbf{P}(s)$ and how matrices $\mathbf{Q}(s)$ and $\mathbf{P}(s)$ can be calculated by dynamic ray tracing. We shall now assume that matrix $\mathbf{M}(s)$ is known along Ω. A simple quadratic expansion of travel time $T(q_1, q_2, s)$ in ray-centered coordinates q_1 and q_2 is then

$$T(q_1, q_2, s) = T(s) + \tfrac{1}{2}\mathbf{q}^T \mathbf{M}(s)\mathbf{q}, \qquad \mathbf{q} = (q_1, q_2)^T. \tag{4.1.77}$$

Here $T(s) = T(0, 0, s)$. Note that the linear terms are missing in (4.1.77) because the wavefront is perpendicular to Ω. Equation (4.1.77) determines the *paraxial travel times*.

A terminological note. In the paraxial vicinity of central ray Ω, we can, of course, calculate the travel times, wavefronts, and rays either exactly (by standard ray tracing) or approximately (using the Taylor expansion). For this reason, it would be convenient to speak of exact and approximate paraxial travel times, exact and approximate paraxial rays, and exact and approximate paraxial wavefronts. This section is, however, devoted mostly to approximate (Taylor expansion) computations. To simplify the terminology, we shall call the approximate paraxial travel times (rays, wavefronts) simply paraxial travel times (rays, wavefronts). Thus, paraxial travel times *are not exact away from ray* Ω, they only approximately simulate the actual travel times in the vicinity of Ω. We shall, however, treat them fully as actual travel times. We can obtain paraxial wavefronts as surfaces of constant paraxial travel times, and paraxial rays as orthogonal trajectories to the family of paraxial wavefronts.

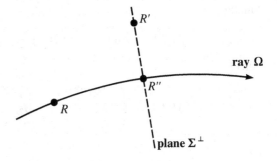

Figure 4.7. Definition of points R, R', and R''. Point R is situated arbitrarily on ray Ω, R' is situated close to R, possibly outside Ω. Point R'' is situated at the intersection of ray Ω with plane $\Sigma^{\perp}$, passing through R' and perpendicular to Ω.

Equation (4.1.77) for the paraxial travel times can be modified in two ways.

1. A considerably more flexible and efficient expression would be obtained if we found a quadratic expansion of $T(q_1, q_2, q_3)$ not only in q_1 and q_2 but also in q_3.
2. It would be useful to express expansion (4.1.77) also in local ray-centered Cartesian coordinates y_i (see Section 4.1.4) and in general Cartesian coordinates x_i.

In the following discussion, we will discuss both modifications.

Let us consider point R situated on central ray Ω and point R' situated in a close vicinity of R, but not necessarily in a plane perpendicular to Ω at R; see Figure 4.7. We introduce point R'' at which plane $\Sigma^{\perp}$, perpendicular to Ω and passing through R', intersects ray Ω. The ray-centered coordinates of these three points follow: $R \equiv [0, 0, s_0]$, $R'' \equiv [0, 0, s]$, and $R' \equiv [q_1, q_2, s]$. We can then express the Taylor expansion for $T(R'')$, up to the second-order terms in $(s - s_0)$, as

$$T(R'') \doteq T(R) + (\partial T/\partial s)_{s=s_0}(s - s_0) + \tfrac{1}{2}(\partial^2 T/\partial s^2)_{s=s_0}(s - s_0)^2.$$

The first and second derivatives of T with respect to s along Ω can be calculated simply,

$$(\mathrm{d}T/\mathrm{d}s)_{s=s_0} = v^{-1}(R), \qquad (\partial^2 T/\partial s^2)_{s=s_0} = -(v^{-2}\partial v/\partial s)_R.$$

Inserting these expressions into (4.1.77), we obtain the final Taylor expansion for the paraxial travel time $T(q_1, q_2, q_3 = s)$ in ray-centered coordinates $q_1, q_2, q_3 = s$, up to the second-order terms in q_1, q_2, and $s - s_0$,

$$T(R') \doteq T(R) + v^{-1}(R)(s - s_0)$$
$$- \tfrac{1}{2}v^{-2}(R)(\partial v/\partial s)_R(s - s_0)^2 + \tfrac{1}{2}q_I q_J M_{IJ}(R). \qquad (4.1.78)$$

Now we shall express Equation (4.1.78) in the *local ray-centered Cartesian coordinate system* y_1, y_2, y_3, with its origin at point R; see Section 4.1.4. We remind the reader that the y_1- and y_2-axes are fully equivalent to the q_1- and q_2-axes at R. In local Cartesian ray-centered coordinates, $R \equiv R(0, 0, 0)$ and $R' \equiv R'(y_1, y_2, y_3)$. Because $\mathrm{d}s = h^{-1}\mathrm{d}y_3$,

$$s - s_0 \doteq h^{-1}y_3 \doteq y_3(1 - v^{-1}v_{,I}y_I), \qquad (s - s_0)^2 \doteq y_3^2;$$

see (4.1.22) and (4.1.51). Inserting these relations into (4.1.78) yields

$$T(R') \doteq T(R) + v^{-1}(R)y_3 - \tfrac{1}{2}v^{-2}(R)v_{,3}(R)y_3^2$$
$$- v^{-2}(R)v_{,I}(R)y_I y_3 + \tfrac{1}{2}y_I y_J M_{IJ}(R). \qquad (4.1.79)$$

Finally,

$$T(R') \doteq T(R) + v^{-1}(R)y_3 + \tfrac{1}{2}y_iy_jM_{ij}(R), \tag{4.1.80}$$

where

$$M_{ij}(R) = \begin{pmatrix} M_{11}(R) & M_{12}(R) & -(v^{-2}v_{,1})_R \\ M_{12}(R) & M_{22}(R) & -(v^{-2}v_{,2})_R \\ -(v^{-2}v_{,1})_R & -(v^{-2}v_{,2})_R & -(v^{-2}v_{,3})_R \end{pmatrix}, \tag{4.1.81}$$

and

$$v_{,I}(R) = (\partial V/\partial y_I)_R = (\partial V/\partial q_I)_R, \qquad v_{,3}(R) = (\partial V/\partial y_3)_R = (\partial V/\partial s)_R; \tag{4.1.82}$$

see (4.1.50). In matrix form, (4.1.80) reads

$$T(R') \doteq T(R) + \hat{\mathbf{y}}^T \hat{\mathbf{p}}^{(y)}(R) + \tfrac{1}{2}\hat{\mathbf{y}}^T \hat{\mathbf{M}}(R)\hat{\mathbf{y}}. \tag{4.1.83}$$

Here

$$\hat{\mathbf{y}} = \begin{pmatrix} y_1 \\ y_2 \\ y_3 \end{pmatrix}, \qquad \hat{\mathbf{p}}^{(y)}(R) = \begin{pmatrix} p_1^{(y)}(R) \\ p_2^{(y)}(R) \\ p_3^{(y)}(R) \end{pmatrix} = \begin{pmatrix} 0 \\ 0 \\ 1/v(R) \end{pmatrix}. \tag{4.1.84}$$

$p_i^{(y)}(R)$ denote the components of the slowness vector in local Cartesian ray-centered coordinates y_i at point R. At R, only one of these components is nonvanishing.

Equations (4.1.80) and (4.1.83) require the construction of the local ray-centered Cartesian coordinate system at point R and the determination of coordinates y_1, y_2, y_3 of point R' in that system. It would be convenient to specify both points R and R' in the *general Cartesian coordinate system.*

We denote the general Cartesian coordinates of R and R' by $x_i(R)$ and $x_i(R')$. We now introduce the 3×1 column matrix

$$\hat{\mathbf{x}}(R', R) = \begin{pmatrix} x_1(R', R) \\ x_2(R', R) \\ x_3(R', R) \end{pmatrix} = \begin{pmatrix} x_1(R') - x_1(R) \\ x_2(R') - x_2(R) \\ x_3(R') - x_3(R) \end{pmatrix}. \tag{4.1.85}$$

Of course, we assume that elements $x_i(R', R)$ are small.

Equation (4.1.83) can be simply transformed from the local Cartesian ray-centered coordinates y_i to general Cartesian coordinates x_i using transformation matrix $\hat{\mathbf{H}}^{(y)}$; see Section 4.1.5. Because we apply this matrix only along central ray Ω, we can use $\hat{\mathbf{H}}$ instead of $\hat{\mathbf{H}}^{(y)}$; see (4.1.28). For small $\hat{\mathbf{y}}(R')$ and $\hat{\mathbf{x}}(R', R)$,

$$\hat{\mathbf{y}}(R') = \hat{\mathbf{H}}^T(R)\hat{\mathbf{x}}(R', R), \qquad \hat{\mathbf{p}}^{(y)}(R) = \hat{\mathbf{H}}^T(R)\hat{\mathbf{p}}^{(x)}(R),$$

where $\hat{\mathbf{p}}^{(x)} = (p_1^{(x)}, p_2^{(x)}, p_3^{(x)})^T$, $p_i^{(x)}$ being the general Cartesian components of the slowness vector. This yields $\hat{\mathbf{y}}^T(R')\hat{\mathbf{p}}^{(y)}(R) = \hat{\mathbf{x}}^T(R', R)\hat{\mathbf{H}}(R)\hat{\mathbf{H}}^T(R)\hat{\mathbf{p}}^{(x)}(R)$, and Equation (4.1.83) yields

$$T(R') \doteq T(R) + \hat{\mathbf{x}}^T(R', R)\hat{\mathbf{p}}^{(x)}(R) + \tfrac{1}{2}\hat{\mathbf{x}}^T(R', R)\hat{\mathbf{H}}(R)\hat{\mathbf{M}}(R)\hat{\mathbf{H}}^T(R)\hat{\mathbf{x}}(R', R). \tag{4.1.86}$$

Figure 4.8. Validity of paraxial travel-time expansions in ray-centered coordinates along Ω and in Cartesian coordinates at point R on Ω. (a) The paraxial expansion in ray-centered coordinates is valid along the whole ray Ω, and (b) the paraxial expansion in Cartesian coordinates is valid only in the vicinity of point R on Ω.

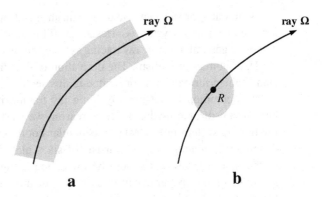

a   **b**

We now introduce the 3×3 matrix

$$\hat{\mathbf{M}}^{(x)}(R) = \hat{\mathbf{H}}(R)\hat{\mathbf{M}}(R)\hat{\mathbf{H}}^T(R). \tag{4.1.87}$$

It is simple to see that the elements of matrix $\hat{\mathbf{M}}^{(x)}$, $M_{ij}^{(x)}$, represent the second derivatives of the travel-time field with respect to the general Cartesian coordinates,

$$M_{ij}^{(x)}(R) = (\partial^2 T(x_1, x_2, x_3)/\partial x_i \partial x_j)_R. \tag{4.1.88}$$

Matrix $\hat{\mathbf{M}}^{(x)}$ is symmetric,

$$M_{ij}^{(x)}(R) = M_{ji}^{(x)}(R). \tag{4.1.89}$$

Using notation (4.1.87), Equation (4.1.86) can be expressed in a simpler form,

$$T(R') \doteq T(R) + \hat{\mathbf{x}}^T(R', R)\hat{\mathbf{p}}^{(x)}(R) + \tfrac{1}{2}\hat{\mathbf{x}}^T(R', R)\hat{\mathbf{M}}^{(x)}(R)\hat{\mathbf{x}}(R', R), \tag{4.1.90}$$

or, in component form,

$$T(R') \doteq T(R) + x_i(R', R)p_i^{(x)}(R) + \tfrac{1}{2}x_i(R', R)x_j(R', R)M_{ij}^{(x)}(R). \tag{4.1.91}$$

Let us again emphasize the difference between expansions (4.1.77) and (4.1.90). Paraxial expansion (4.1.77) is expressed in ray-centered coordinates and can be used at any point situated in the paraxial vicinity of ray Ω; see Figure 4.8(a). On the contrary, paraxial expansion (4.1.90) has a local character and is valid only in the close vicinity of point R situated on ray Ω; see Fig. 4.8(b). It is, however, very flexible and numerically very efficient because both points R and R' are specified in general Cartesian coordinates. The travel time given by (4.1.77) is exact along the whole ray Ω, but expansion (4.1.90) is exact only at one point – point R on ray Ω. At other points on ray Ω, expansion (4.1.90) is only approximate. Even for points R' situated along ray Ω, the accuracy of (4.1.90) decreases with increasing distance of R' from R.

4.2 Hamiltonian Approach to Dynamic Ray Tracing

In Section 4.1, we derived the dynamic ray tracing system for isotropic inhomogeneous media in ray-centered coordinates. The dynamic ray tracing system in ray-centered coordinates consists of four scalar linear ordinary differential equations of the first order only, which should be integrated along the central ray Ω.

A similar approach can also be used in an anisotropic inhomogeneous medium and in an arbitrary coordinate system (Cartesian, curvilinear orthogonal, curvilinear nonorthogonal). To derive general dynamic ray tracing systems applicable to these cases, it is useful to apply the Hamiltonian formalism. The derivation of dynamic ray tracing systems, based on the Hamiltonian formalism, is presented in this section.

It is common to perform dynamic ray tracing in Cartesian coordinates x_i, particularly in anisotropic media. For this reason, we start with the derivation of the dynamic ray tracing system in Cartesian rectangular coordinates x_i in Section 4.2.1. The derived dynamic ray tracing system consists of six scalar linear ordinary differential equations of the first order, which should be integrated along central ray Ω. It is shown that the solutions of the dynamic ray tracing system must satisfy one constraint equation, following from the eikonal equation. In the actual computations, the initial conditions must be chosen in such a way to satisfy the constraint equation. When the constraint equation is satisfied at the initial point S of ray Ω, it is satisfied along the whole ray Ω. Because the dynamic ray tracing system in Cartesian coordinates consists of six linear equations, it has six linearly independent solutions. It is shown that two such solutions can be found analytically. The first will be referred to as the *ray-tangent solution* and corresponds to central ray Ω. The second analytical solution will be referred to as the *noneikonal solution* because it does not satisfy the constraint relation. In most applications, the noneikonal solution should be eliminated by a proper choice of the initial conditions. The noneikonal solution is, however, useful in the construction of the 6×6 propagator matrix of the dynamic ray tracing system because it represents one of the linearly independent solutions.

The *wavefront orthonormal coordinates* y_i are introduced in Section 4.2.2. The origin of this coordinate system moves along central ray Ω with the propagating wavefront. Coordinate axis y_3 is oriented along local slowness vector $\vec{p}$, and the y_1- and y_2-axes are confined to the plane tangent to the wavefront. In isotropic media, the wavefront orthonormal coordinate system is equivalent to the local Cartesian ray-centered coordinate system introduced in Section 4.1.4. In anisotropic media, however, both systems differ because the slowness vector is not tangent to ray Ω. Actually, the ray-centered coordinate system in anisotropic media is always nonorthogonal. Because we prefer to work in Cartesian rectangular coordinate systems, we shall use the wavefront orthonormal system. In wavefront orthonormal coordinates, the number of equations in the dynamic ray tracing system can be reduced from six to four. All remaining solutions can be computed analytically. It is also shown how the solutions of the dynamic ray tracing system in Cartesian coordinates x_i can be transformed into solutions of the dynamic ray tracing in wavefront orthonormal coordinates y_i, and vice versa.

The orthonomic system of rays is considered in Section 4.2.3. It is shown there how the results of dynamic ray tracing can be used to determine the matrices of the second spatial derivatives of the travel-time field, geometrical spreading, and Jacobian $J^{(T)}$ along central ray Ω. The quadratic expansions for the paraxial travel times in the vicinity of central ray Ω, both in Cartesian and wavefront orthonormal coordinates, are derived. Mutual relations are found, both for isotropic and anisotropic media.

Dynamic ray tracing systems in arbitrary curvilinear, orthogonal or nonorthogonal, coordinates are derived in Section 4.2.4. The systems are again applicable both to isotropic and anisotropic media. A particularly simple dynamic ray tracing system is obtained in special nonorthogonal ray-centered coordinates, in which the 6×6 system matrix has 19 zeros, 1 unity, and only 16 elements influenced by the actual form of the Hamiltonian.

It clearly shows the relations between the dynamic ray tracing systems consisting of six and four equations.

4.2.1   Cartesian Rectangular Coordinates

In ray computations, it is very common to use Cartesian rectangular coordinates x_i. In these coordinates, travel time $T(x_i)$ satisfies eikonal equation (3.1.1) in isotropic medium and (3.6.2) in an anisotropic medium. The eikonal equation in both cases can be expressed in Hamiltonian form $\mathcal{H}^{(x)}(x_i, p_i^{(x)}) = 0$, where $p_i^{(x)} = \partial T/\partial x_i$ denote the Cartesian components of the slowness vector. See (3.1.2) for the isotropic medium and (3.6.3) for the anisotropic medium. In six-dimensional x_i-$p_i^{(x)}$-phase space, equation $\mathcal{H}^{(x)}(x_i, p_i^{(x)}) = 0$ represents a hypersurface.

We shall consider Hamiltonian $\mathcal{H}^{(x)}(x_i, p_i^{(x)})$, which satisfies the relation

$$p_i^{(x)}\partial\mathcal{H}^{(x)}/\partial p_i^{(x)} = 1. \tag{4.2.1}$$

The ray tracing system in Cartesian rectangular coordinate system x_i, corresponding to the eikonal equation $\mathcal{H}^{(x)}(x_i, p_i^{(x)}) = 0$, then reads

$$dx_i/dT = \partial\mathcal{H}^{(x)}/\partial p_i^{(x)}, \qquad dp_i^{(x)}/dT = -\partial\mathcal{H}^{(x)}/\partial x_i. \tag{4.2.2}$$

where the variable T along the ray represents the travel time and is determined by the form of Hamiltonian $\mathcal{H}^{(x)}(x_i, p_i^{(x)})$ satisfying (4.2.1).

We consider point $S(x_{i0}, p_{i0}^{(x)})$ in the x_i-$p_i^{(x)}$-phase space, situated on hypersurface $\mathcal{H}^{(x)}(x_i, p_i^{(x)}) = 0$, so that $\mathcal{H}^{(x)}(x_{i0}, p_{i0}^{(x)}) = 0$. We further consider ray Ω with its initial point at S, satisfying ray tracing system (4.2.2). Ray Ω is then *completely situated* on the hypersurface $\mathcal{H}^{(x)}(x_i, p_i^{(x)}) = 0$ passing through S. In other words, equation $\mathcal{H}^{(x)}(x_i, p_i^{(x)}) = 0$ is satisfied along the whole ray Ω, once it is satisfied at initial point S. In the phase space, we describe ray Ω by equations $x_i(T) = x_i^\Omega(T)$, $p_i^{(x)}(T) = p_i^{(x)\Omega}(T)$, with $x_i^\Omega(T_0) = x_{i0}$, $p_i^{(x)\Omega}(T_0) = p_{i0}^{(x)}$.

We now wish to derive the equations for computing the first partial derivatives of x_i and $p_i^{(x)}$ with respect to initial conditions x_{i0}, $p_{i0}^{(x)}$ along ray Ω from (4.2.2). We choose an arbitrary initial parameter and denote it γ. Initial parameter γ may represent any parameter from the set of initial parameters x_{i0}, $p_{i0}^{(x)}$, or from some other set of independent parameters specifying x_{i0}, $p_{i0}^{(x)}$. We denote

$$Q_i^{(x)} = (\partial x_i/\partial\gamma)_{T=\text{const.}}, \qquad P_i^{(x)} = (\partial p_i^{(x)}/\partial\gamma)_{T=\text{const.}} \tag{4.2.3}$$

Taking the partial derivative of ray tracing system (4.2.2) with respect to γ, and bearing in mind that $\partial/\partial\gamma$ commutes with d/dT, we obtain a system of six linear ordinary differential equations of the first order for $Q_i^{(x)}$ and $P_i^{(x)}$. It can be expressed in the following form:

$$dQ_i^{(x)}/dT = A_{ij}^{(x)}Q_j^{(x)} + B_{ij}^{(x)}P_j^{(x)}, \qquad dP_i^{(x)}/dT = -C_{ij}^{(x)}Q_j^{(x)} - D_{ij}^{(x)}P_j^{(x)}. \tag{4.2.4}$$

Here

$$A_{ij}^{(x)} = \partial^2\mathcal{H}^{(x)}/\partial p_i^{(x)}\partial x_j, \qquad B_{ij}^{(x)} = \partial^2\mathcal{H}^{(x)}/\partial p_i^{(x)}\partial p_j^{(x)},$$
$$C_{ij}^{(x)} = \partial^2\mathcal{H}^{(x)}/\partial x_i\partial x_j, \qquad D_{ij}^{(x)} = \partial^2\mathcal{H}^{(x)}/\partial x_i\partial p_j^{(x)}. \tag{4.2.5}$$

The system of six linear ordinary differential equations of the first order (4.2.4) for $Q_i^{(x)}$

and $P_i^{(x)}$ is called the *dynamic ray tracing system in Cartesian rectangular coordinates*. As we can see from (4.2.5), $A_{ij}^{(x)}$, $B_{ij}^{(x)}$, $C_{ij}^{(x)}$ and $D_{ij}^{(x)}$ satisfy the following three *symmetry relations*:

$$B_{ij}^{(x)} = B_{ji}^{(x)}, \qquad C_{ij}^{(x)} = C_{ji}^{(x)}, \qquad D_{ij}^{(x)} = A_{ji}^{(x)}. \tag{4.2.6}$$

If we take the partial derivatives with respect to γ, we wish to stay on the hypersurface $\mathcal{H}^{(x)}(x_i, p_i^{(x)}) = 0$, representing the eikonal equation. Consequently, we require that $\partial \mathcal{H}^{(x)}(x_i, p_i^{(x)})/\partial \gamma = 0$. Quantities $Q_i^{(x)}$, $P_i^{(x)}$ then satisfy the following *constraint relation*:

$$(\partial \mathcal{H}^{(x)}/\partial x_k) Q_k^{(x)} + (\partial \mathcal{H}^{(x)}/\partial p_k^{(x)}) P_k^{(x)} = 0. \tag{4.2.7}$$

It is not difficult to prove that constraint relation (4.2.7) is satisfied along the whole ray Ω after it is satisfied at the initial point S of Ω. Consequently, one of the six equations of dynamic ray tracing system (4.2.4) is redundant and may be replaced by (4.2.7).

To simplify the following equations, we shall use the notation along ray Ω:

$$\mathcal{U}_i^{(x)} = dx_i/dT = \partial \mathcal{H}^{(x)}/\partial p_i^{(x)}, \qquad \eta_i^{(x)} = dp_i^{(x)}/dT = -\partial \mathcal{H}^{(x)}/\partial x_i. \tag{4.2.8}$$

Here $\mathcal{U}_i^{(x)}$ are Cartesian rectangular components of the group velocity vector. Constraint equation (4.2.7) then reads:

$$\mathcal{U}_i^{(x)} P_i^{(x)} - \eta_i^{(x)} Q_i^{(x)} = 0. \tag{4.2.9}$$

Although the constraint equation (4.2.9) is theoretically satisfied along the whole ray Ω after it is satisfied at the initial point S of Ω, the numerical noise may cause deviations from it and decrease the stability of dynamic ray tracing. It is then useful to normalize $P_i^{(x)}$ and $Q_i^{(x)}$ at any step of the ray so that the constraint relation (4.2.9) is satisfied. Note that the dynamic ray tracing system (4.1.63) in ray-centered coordinates, consisting of four equations, does not suffer by these difficulties because *the constraints are built in*.

Dynamic ray tracing system (4.2.4) is also called *the paraxial ray tracing system* for rays situated in the vicinity of central ray Ω. Let us consider paraxial ray Ω' described by equations:

$$x_i(T) = x_i^\Omega(T) + \delta x_i(T), \qquad p_i^{(x)}(T) = p_i^{(x)\Omega}(T) + \delta p_i^{(x)}(T), \tag{4.2.10}$$

where $x_i(T) = x_i^\Omega(T)$ and $p_i^{(x)}(T) = p_i^{(x)\Omega}(T)$ are equations of central ray Ω, and $\delta x_i(T)$ and $\delta p_i^{(x)}(T)$ specify the linear approximations of the deviations from the central ray. Assume that the initial point $S'[x_{i0} + \delta x_{i0}, p_{i0}^{(x)} + \delta p_{i0}^{(x)}]$ of paraxial ray Ω' is situated close to the initial point $S[x_{i0}, p_{i0}^{(x)}]$ of central ray Ω, where $\delta x_{i0} = \delta x_i(T_0)$ and $\delta p_{i0}^{(x)} = \delta p_i^{(x)}(T_0)$ are small. Because $\delta x_i = (\partial x_i/\partial \gamma)\delta \gamma$ and $\delta p_i^{(x)} = (\partial p_i^{(x)}/\partial \gamma)\delta \gamma$, the paraxial ray tracing system for δx_i and $\delta p_i^{(x)}$ is obtained from dynamic ray tracing system (4.2.4) by multiplying it with $\delta \gamma$:

$$d\delta x_i/dT = A_{ij}^{(x)} \delta x_j + B_{ij}^{(x)} \delta p_j^{(x)}, \qquad d\delta p_i^{(x)}/dT = -C_{ij}^{(x)} \delta x_j - D_{ij}^{(x)} \delta p_j^{(x)}. \tag{4.2.11}$$

Here $A_{ij}^{(x)}$, $B_{ij}^{(x)}$, $C_{ij}^{(x)}$, and $D_{ij}^{(x)}$ are again given by (4.2.5) and satisfy symmetry relations (4.2.6). Because the eikonal equation must be satisfied along the paraxial ray Ω', we obtain a constraint relation analogous to (4.2.9)

$$\mathcal{U}_i^{(x)} \delta p_i^{(x)} - \eta_i^{(x)} \delta x_i = 0. \tag{4.2.12}$$

There are six initial conditions δx_{i0} and $\delta p_{i0}^{(x)}$ for paraxial ray tracing system (4.2.11). The complete ray field of paraxial rays, however, does not have six parameters, but only four; see Section 4.1.6. Consequently, two initial conditions and the relevant paraxial rays are redundant. *First*, δx_{i0} and $\delta p_{i0}^{(x)}$ should satisfy constraint relation (4.2.12); otherwise, the solution of (4.2.11) does not satisfy the eikonal equation. Such solutions of (4.2.11) not satisfying (4.2.12) will be referred to as the *noneikonal solutions*. *Second*, any ray Ω' of the remaining five-parameter system of paraxial rays is identical with a one-parameter system of paraxial rays, with initial points and initial slowness vectors distributed along Ω'. Consequently, the resulting system of paraxial rays is a four-parameter system. If we wish to retain only one ray of any one-parameter set of identical rays, we can specify the initial points S' along some surface Σ^0 passing through S. In certain applications, however, it may be suitable to consider a five-parameter system of rays, with S' distributed arbitrarily in the vicinity of S, not just along the surface Σ^0 passing through S.

Among the solutions of the paraxial ray tracing system (4.2.11) for nontrivial initial conditions δx_{i0} and $\delta p_{i0}^{(x)}$, there are also solutions that coincide with central ray Ω. Such initial conditions will be referred to as the *ray-tangent initial conditions*, and the corresponding solutions will be called the *ray-tangent solutions*.

Thus, the nontrivial initial conditions δx_{i0} and $\delta p_{i0}^{(x)}$ may be divided into three groups.

a. Ray-tangent initial conditions. These yield solutions coinciding with central ray Ω.
b. Noneikonal initial conditions. These do not yield paraxial rays satisfying the eikonal equation under consideration.
c. Standard paraxial initial conditions. These yield paraxial rays not coinciding with Ω, and satisfying the eikonal equation under consideration.

The discussion of the initial conditions for the dynamic ray tracing system (4.2.4) is analogous to that given earlier for the paraxial ray tracing system (4.2.11). The nontrivial initial conditions can again be divided into three groups:

a. Ray-tangent initial conditions.
b. Noneikonal initial conditions.
c. Standard paraxial initial conditions.

The ray-tangent and noneikonal solutions of dynamic ray tracing system (4.2.4) can be found analytically.

a. Ray tangent solutions. We shall prove that the solutions

$$Q_i^{(x)} = \mathcal{U}_i^{(x)}, \qquad P_i^{(x)} = \eta_i^{(x)} \qquad (4.2.13)$$

satisfy dynamic ray tracing system (4.2.4). Actually, we have

$$
\begin{aligned}
\mathrm{d}\mathcal{U}_i^{(x)}/\mathrm{d}T &= \mathrm{d}(\partial \mathcal{H}^{(x)}/\partial p_i^{(x)})/\mathrm{d}T = A_{ij}^{(x)}\mathcal{U}_j^{(x)} + B_{ij}^{(x)}\eta_j^{(x)}, \\
\mathrm{d}\eta_i^{(x)}/\mathrm{d}T &= -\mathrm{d}(\partial \mathcal{H}^{(x)}/\partial x_i)/\mathrm{d}T = -C_{ij}^{(x)}\mathcal{U}_j^{(x)} - D_{ij}^{(x)}\eta_j^{(x)}.
\end{aligned}
\qquad (4.2.14)
$$

This is equivalent to dynamic ray tracing system (4.2.4). Note that $\mathcal{U}_i^{(x)}$ and $\eta_i^{(x)}$ are known from ray tracing; see (4.2.2) and (4.2.8). It is simple to see that the ray-tangent solutions satisfy constraint equation (4.2.9) because

$$\mathcal{U}_i^{(x)} P_i^{(x)} - \eta_i^{(x)} Q_i^{(x)} = \mathcal{U}_i^{(x)} \eta_i^{(x)} - \eta_i^{(x)} \mathcal{U}_i^{(x)} = 0.$$

b. Noneikonal solutions. We shall consider Hamiltonian $\mathcal{H}^{(x)}(x_i, p_i^{(x)})$ for which first derivative $\partial \mathcal{H}^{(x)}/\partial x_i$ is a homogeneous function of the second degree in $p_i^{(x)}$. This assumption is satisfied, for example, for Hamiltonian $\mathcal{H}^{(x)}(x_i, p_i^{(x)}) = \frac{1}{2}(G(x_i, p_i^{(x)}) - 1)$, where G is an eigenvalue of the Christoffel matrix. This Hamiltonian is commonly considered in anisotropic media, see (3.6.3), but also in isotropic media, see (3.1.13). Using Euler's theorem for homogeneous functions (2.2.24) and (4.2.1), we then obtain

$$p_i^{(x)} \frac{\partial^2 \mathcal{H}^{(x)}}{\partial p_i^{(x)} \partial x_j} = 2 \frac{\partial \mathcal{H}^{(x)}}{\partial x_j} = -2\eta_j^{(x)}, \qquad p_i^{(x)} \frac{\partial^2 \mathcal{H}^{(x)}}{\partial p_i^{(x)} \partial p_j^{(x)}} = \frac{\partial \mathcal{H}^{(x)}}{\partial p_j^{(x)}} = \mathcal{U}_j^{(x)}.$$

$$(4.2.15)$$

Now we shall prove that $Q_i^{(x)}$ and $P_i^{(x)}$, given by relations

$$Q_i^{(x)} = (T - T_0)\mathcal{U}_i^{(x)}, \qquad P_i^{(x)} = p_i^{(x)} + (T - T_0)\eta_i^{(x)}, \qquad (4.2.16)$$

are solutions of dynamic ray tracing system (4.2.4), if Hamiltonian $\mathcal{H}^{(x)}(x_i, p_i^{(x)})$ satisfies (4.2.15). Using (4.2.14) and (4.2.15), we obtain

$$\mathrm{d}[(T - T_0)\mathcal{U}_i^{(x)}]/\mathrm{d}T = \mathcal{U}_i^{(x)} + (T - T_0)\mathrm{d}\mathcal{U}_i^{(x)}/\mathrm{d}T$$

$$= A_{ij}^{(x)}(T - T_0)\mathcal{U}_j^{(x)} + B_{ij}^{(x)}\big(p_j^{(x)} + (T - T_0)\eta_j^{(x)}\big),$$

$$\mathrm{d}[p_i^{(x)} + (T - T_0)\eta_i^{(x)}]/\mathrm{d}T = 2\eta_i^{(x)} - (T - T_0)\mathrm{d}\eta_i^{(x)}/\mathrm{d}T$$

$$= -C_{ij}^{(x)}(T - T_0)\mathcal{U}_j^{(x)} - D_{ij}^{(x)}\big(p_j^{(x)} + (T - T_0)\eta_j^{(x)}\big).$$

This is equivalent to dynamic ray tracing system (4.2.4) applied to (4.2.16). Consequently, (4.2.16) are solutions of (4.2.4). As in ray-tangent solutions (4.2.13), $\mathcal{U}_i^{(x)}$ and $\eta_i^{(x)}$ in (4.2.16) are known from ray tracing. It is not difficult to prove that solutions (4.2.16) do not satisfy constraint relation (4.2.9). Indeed,

$$\mathcal{U}_i^{(x)} P_i^{(x)} - \eta_i^{(x)} Q_i^{(x)} = \mathcal{U}_i^{(x)} p_i^{(x)} + (T - T_0)\big(\mathcal{U}_i^{(x)}\eta_i^{(x)} - \eta_i^{(x)}\mathcal{U}_i^{(x)}\big)$$

$$= \mathcal{U}_i^{(x)} p_i^{(x)} = 1.$$

Thus, $\mathcal{U}_i^{(x)} P_i^{(x)} - \eta_i^{(x)} Q_i^{(x)} \neq 0$, and (4.2.16) represents a noneikonal solution of dynamic ray tracing system (4.2.4).

Note that all equations of this section may be used both for isotropic and anisotropic media.

4.2.2 Wavefront Orthonormal Coordinates

Consider fixed point R on ray Ω and introduce wavefront orthonormal coordinates y_i whose origin is at R. The y_3-axis is oriented along slowness vector $\vec{p}$ at R, and axes y_1 and y_2 are confined to the plane tangent to the wavefront at R and are mutually perpendicular. In an isotropic medium, the wavefront orthonormal coordinate system is analogous to the local Cartesian ray-centered coordinate system, see Section 4.1.4. In an anisotropic medium, however, slowness vector $\vec{p}$ is not tangent to ray Ω at R, so that the y_3-axis is not tangent to Ω at R. Choosing the y_3-axis to be tangent to ray Ω would make the system nonorthogonal. We, however, prefer to work in orthonormal coordinate systems.

We now introduce the basis unit vectors $\vec{e}_1(R)$, $\vec{e}_2(R)$, and $\vec{e}_3(R)$ of the wavefront orthonormal coordinate system whose origin is at point R on ray Ω. Origin R moves along Ω with the wavefront under consideration; in other words, it is situated at the point of intersection of the wavefront with central ray Ω. Basis vector $\vec{e}_3(R)$ is given by a simple relation, $\vec{e}_3(R) = \mathcal{C}(R)\vec{p}(R)$, where $\mathcal{C}$ is the phase velocity. Basis vectors $\vec{e}_1(R)$ and $\vec{e}_2(R)$ are tangent to the wavefront at R and are mutually perpendicular. We shall specify the variations of $\vec{e}_1$ and $\vec{e}_2$ along Ω by a differential equation analogous to (4.1.3) for isotropic media,

$$\mathrm{d}\vec{e}_I/\mathrm{d}T = -(\vec{p} \cdot \vec{p})^{-1}(\vec{e}_I \cdot \vec{\eta})\vec{p}. \tag{4.2.17}$$

Here $\vec{\eta} = \mathrm{d}\vec{p}/\mathrm{d}T$. It is not difficult to prove that $\vec{e}_1(R)$ and $\vec{e}_2(R)$ determined from (4.2.17) form a right-handed mutually perpendicular triplet of unit vectors with $\vec{e}_3(R) = \mathcal{C}(R)\vec{p}(R)$ at any point R on Ω, if they form such a triplet at initial point S on Ω.

In contrast to the ray-centered coordinate system, ray Ω is not a coordinate line in the wavefront orthonormal coordinate system y_i; only the origin R of the system is situated on Ω. Moreover, in anisotropic media, basis vector $\vec{e}_3(R)$ is not tangent to ray Ω but is perpendicular to the wavefront. The wavefront orthonormal coordinate systems introduced at different points R on Ω do not coincide. There is an infinite number of wavefront orthonormal coordinate systems y_i connected with ray Ω, whose origins are at different points R on Ω. Coordinate y_3 of any point R' situated in the paraxial vicinity of ray Ω depends on the selection of the origin point R on Ω. It is always possible to choose point R so that point R' is situated in a plane tangent to the wavefront at R. The y_3-coordinate of R' is then zero. Consequently, if we consider $y_3 = 0$, the position of R' is fully specified by travel time T and by coordinates y_1 and y_2.

We now introduce the transformation matrices from wavefront orthonormal coordinate system y_1, y_2, and y_3 to the global Cartesian coordinate system x_k,

$$H_{kl} = \partial x_k/\partial y_l = \partial y_l/\partial x_k. \tag{4.2.18}$$

The 3×3 matrix $\hat{\mathbf{H}}$ with elements H_{kl} is unitary and is related to $\vec{e}_i$ as follows:

$$\hat{\mathbf{H}}^{-1} = \hat{\mathbf{H}}^T, \qquad H_{iK} = e_{Ki}^{(x)}, \qquad H_{i3} = e_{3i}^{(x)} = \mathcal{C}p_i^{(x)}. \tag{4.2.19}$$

Superscript (x) is used to emphasize that the components correspond to Cartesian coordinates x_i. Similarly, (y) will be used to emphasize wavefront orthonormal coordinates y_i. We also denote the Hamiltonian in the y_i-coordinates by $\mathcal{H}^{(y)}(y_i, p_i^{(y)})$, where $p_i^{(y)} = \partial T/\partial y_i$. Hamiltonian $\mathcal{H}^{(y)}(y_i, p_i^{(y)})$ again corresponds to variable T along ray Ω, and pressumably satisfies equations analogous to (4.2.1) and (4.2.15). We denote

$$\mathcal{U}_i^{(y)} = \frac{\partial \mathcal{H}^{(y)}}{\partial p_i^{(y)}}, \qquad \eta_i^{(y)} = -\frac{\partial \mathcal{H}^{(y)}}{\partial y_i}, \qquad Q_i^{(y)} = \frac{\partial y_i}{\partial \gamma}, \qquad P_i^{(y)} = \frac{\partial p_i^{(y)}}{\partial \gamma}. \tag{4.2.20}$$

Here γ is an arbitrary initial parameter and has the same meaning as in (4.2.3). The derivatives with respect to γ are taken for constant T. The transformation relations read:

$$\begin{aligned}
\mathcal{U}_n^{(y)} &= H_{in}\mathcal{U}_i^{(x)}, & p_n^{(y)} &= H_{in}p_i^{(x)}, & \eta_n^{(y)} &= H_{in}\eta_i^{(x)}, \\
Q_n^{(y)} &= H_{in}Q_i^{(x)}, & P_n^{(y)} &= H_{in}P_i^{(x)}, & & \\
\mathcal{U}_i^{(x)} &= H_{in}\mathcal{U}_n^{(y)}, & p_i^{(x)} &= H_{in}p_n^{(y)}, & \eta_i^{(x)} &= H_{in}\eta_n^{(y)}, \\
Q_i^{(x)} &= H_{in}Q_n^{(y)}, & P_i^{(x)} &= H_{in}P_n^{(y)}. & &
\end{aligned} \tag{4.2.21}$$

Note that

$$p_I^{(y)} = 0, \qquad p_3^{(y)} = 1/\mathcal{C}, \qquad \mathcal{U}_I^{(y)} \neq 0, \qquad \mathcal{U}_3^{(y)} = \mathcal{C}. \tag{4.2.22}$$

In an isotropic medium, $\mathcal{U}_I^{(y)} = 0$ also, but in an anisotropic medium, $\mathcal{U}_I^{(y)}$ do not vanish. Due to (4.2.19), Equation (4.2.17) can be used to determine the variations of H_{jI} along Ω:

$$\mathrm{d}H_{jI}/\mathrm{d}T = -\mathcal{C}^2\left(H_{kI}\eta_k^{(x)}\right)p_j^{(x)} = -\mathcal{C}^2\eta_I^{(y)}p_j^{(x)}. \tag{4.2.23}$$

Now we shall derive the equations for $Q_i^{(y)}$ and $P_i^{(y)}$ by transforming the dynamic ray tracing system in Cartesian coordinates x_i into wavefront orthonormal coordinates y_i. Consequently, even in the y_i-coordinates, we shall use Hamiltonian $\mathcal{H}^{(x)}(x_i, p_i^{(x)})$ and its derivatives with respect to x_i and $p_i^{(x)}$. This is particularly useful in anisotropic inhomogeneous media, as the elastic parameters may be specified in global Cartesian coordinate system x_i. In determining $Q_i^{(y)}$ and $P_i^{(y)}$, we shall proceed as follows. First, we determine the ray-tangent and noneikonal solutions analytically. Second, we determine $Q_3^{(y)}$ and $P_3^{(y)}$ for standard paraxial initial conditions analytically. Third, we find the dynamic ray tracing system, consisting of four equations in wavefront orthonormal coordinates for $Q_I^{(y)}$ and $P_I^{(y)}$.

The *ray-tangent solutions* are obtained from (4.2.13) using (4.2.21):

$$Q_i^{(y)} = \mathcal{U}_i^{(y)}, \qquad P_i^{(y)} = \eta_i^{(y)}. \tag{4.2.24}$$

Similarly, the *noneikonal solutions* are obtained from (4.2.16), using (4.2.21):

$$Q_i^{(y)} = (T - T_0)\mathcal{U}_i^{(y)}, \qquad P_i^{(y)} = p_i^{(y)} + (T - T_0)\eta_i^{(y)}. \tag{4.2.25}$$

For *standard paraxial conditions*, we can determine $Q_3^{(y)}$ and $P_3^{(y)}$ analytically in terms of $Q_I^{(y)}$ and $P_I^{(y)}$. Because partial derivatives $Q_i^{(y)} = \partial y_i/\partial \gamma$ are taken along a plane tangent to the wavefront at R on Ω, we obtain $p_i^{(y)}\partial y_i/\partial \gamma = 0$. If we take into account $p_I^{(y)} = 0$ and constraint relation (4.2.9) expressed in y_i-coordinates, we obtain

$$Q_3^{(y)} = 0, \qquad P_3^{(y)} = \mathcal{C}^{-1}\left(\eta_I^{(y)}Q_I^{(y)} - \mathcal{U}_I^{(y)}P_I^{(y)}\right). \tag{4.2.26}$$

We shall now derive the dynamic ray tracing system for $Q_I^{(y)}$ and $P_I^{(y)}$, consisting of four equations. We use (4.2.4), insert $Q_i^{(x)} = H_{in}Q_n^{(y)}$ and $P_i^{(x)} = H_{in}P_n^{(y)}$ (see (4.2.21)), and multiply the resulting equations by H_{im}:

$$\begin{aligned}
H_{im}\mathrm{d}\left(H_{in}Q_n^{(y)}\right)/\mathrm{d}T &= \bar{A}_{mn}Q_n^{(y)} + \bar{B}_{mn}P_n^{(y)}, \\
H_{im}\mathrm{d}\left(H_{in}P_n^{(y)}\right)/\mathrm{d}T &= -\bar{C}_{mn}Q_n^{(y)} - \bar{D}_{mn}P_n^{(y)},
\end{aligned} \tag{4.2.27}$$

where

$$\begin{aligned}
\bar{A}_{mn} &= H_{im}H_{jn}\partial^2\mathcal{H}^{(x)}/\partial p_i^{(x)}\partial x_j, & \bar{B}_{mn} &= H_{im}H_{jn}\partial^2\mathcal{H}^{(x)}/\partial p_i^{(x)}\partial p_j^{(x)}, \\
\bar{C}_{mn} &= H_{im}H_{jn}\partial^2\mathcal{H}^{(x)}/\partial x_i\partial x_j, & \bar{D}_{mn} &= H_{im}H_{jn}\partial^2\mathcal{H}^{(x)}/\partial x_i\partial p_j^{(x)}.
\end{aligned} \tag{4.2.28}$$

Instead of $m = 1, 2, 3$, we shall only use $M = 1, 2$. The left-hand sides (LHSs) of (4.2.27) can then be expressed as follows:

$$\begin{aligned}
H_{iM}\mathrm{d}\left(H_{in}Q_n^{(y)}\right)/\mathrm{d}T &= \mathrm{d}Q_M^{(y)}/\mathrm{d}T + \mathcal{C}\eta_M^{(y)}Q_3^{(y)}, \\
H_{iM}\mathrm{d}\left(H_{in}P_n^{(y)}\right)/\mathrm{d}T &= \mathrm{d}P_M^{(y)}/\mathrm{d}T + \mathcal{C}\eta_M^{(y)}P_3^{(y)}.
\end{aligned} \tag{4.2.29}$$

Using (4.2.15), (4.2.21), and (4.2.28), we also obtain these simple expressions for $\bar{B}_{M3}$ and $\bar{D}_{M3}$:

$$\bar{B}_{M3} = \mathcal{C}\mathcal{U}_M^{(y)}, \qquad \bar{D}_{M3} = -2\mathcal{C}\eta_M^{(y)}. \tag{4.2.30}$$

Inserting (4.2.26), (4.2.29), and (4.2.30) into (4.2.27) yields the final system of linear ordinary differential equations of the first order for $Q_M^{(y)}$ and $P_M^{(y)}$:

$$
\begin{aligned}
\mathrm{d}Q_M^{(y)}/\mathrm{d}T &= A_{MN}^{(y)}Q_N^{(y)} + B_{MN}^{(y)}P_N^{(y)} \\
\mathrm{d}P_M^{(y)}/\mathrm{d}T &= -C_{MN}^{(y)}Q_N^{(y)} - D_{MN}^{(y)}P_N^{(y)},
\end{aligned}
\tag{4.2.31}
$$

where $A_{MN}^{(y)}$, $B_{MN}^{(y)}$, $C_{MN}^{(y)}$, and $D_{MN}^{(y)}$ are given by the relations

$$
\begin{aligned}
A_{MN}^{(y)} &= H_{iM}H_{jN}\left[\partial^2\mathcal{H}^{(x)}/\partial p_i^{(x)}\partial x_j + \mathcal{U}_i^{(x)}\eta_j^{(x)}\right], \\
B_{MN}^{(y)} &= H_{iM}H_{jN}\left[\partial^2\mathcal{H}^{(x)}/\partial p_i^{(x)}\partial p_j^{(x)} - \mathcal{U}_i^{(x)}\mathcal{U}_j^{(x)}\right], \\
C_{MN}^{(y)} &= H_{iM}H_{jN}\left[\partial^2\mathcal{H}^{(x)}/\partial x_i\partial x_j - \eta_i^{(x)}\eta_j^{(x)}\right], \\
D_{MN}^{(y)} &= H_{iM}H_{jN}\left[\partial^2\mathcal{H}^{(x)}/\partial x_i\partial p_j^{(x)} + \eta_i^{(x)}\mathcal{U}_j^{(x)}\right],
\end{aligned}
\tag{4.2.32}
$$

and satisfy the following symmetry relations:

$$
B_{MN}^{(y)} = B_{NM}^{(y)}, \qquad C_{MN}^{(y)} = C_{NM}^{(y)}, \qquad D_{MN}^{(y)} = A_{NM}^{(y)}.
\tag{4.2.33}
$$

Equations (4.2.31) with (4.2.32) represent the final dynamic ray tracing system, consisting of four equations in wavefront orthonormal coordinates y_i. Together with (4.2.24) through (4.2.26), they represent the complete solution of the problem in the y_i-coordinates. All these equations can be used both for isotropic and anisotropic media. In the derivation, we have used the assumptions that $\partial\mathcal{H}^{(x)}/\partial p_i^{(x)}$ is a homogeneous function of the first degree, and that $\partial\mathcal{H}^{(x)}/\partial x_i$ is a homogeneous function of the second degree in $p_i^{(x)}$; see (4.2.15). Thus, we can use Hamiltonian $\mathcal{H}^{(x)}$ in the common form $\mathcal{H}^{(x)}(x_i, p_i^{(x)}) = \frac{1}{2}(G(x_i, p_i^{(x)}) - 1)$, where G is an eigenvalue of the Christoffel matrix. For isotropic media, the relevant Hamiltonian reads $\mathcal{H}^{(x)}(x_i, p_i^{(x)}) = \frac{1}{2}(V^2 p_i^{(x)}p_i^{(x)} - 1)$; see (3.1.13). For more details on isotropic media, see Section 4.7; anisotropic media are covered more completely in Section 4.14.

From the theoretical point of view, it does not matter very much whether we compute $Q_i^{(x)}$ and $P_i^{(x)}$ using (4.2.4) or $Q_M^{(y)}$ and $P_M^{(y)}$ using (4.2.31). Both sets are mutually related by transformation equations (4.2.21). At any point of ray Ω, we can use (4.2.21) and (4.2.26) to determine $Q_i^{(x)}$ and $P_i^{(x)}$ from $Q_M^{(y)}$ and $P_M^{(y)}$, and vice versa.

Dynamic ray tracing system (4.2.31) also represents the paraxial ray tracing system for rays situated in the vicinity of Ω:

$$
\mathrm{d}y_M/\mathrm{d}T = A_{MN}^{(y)}y_N + B_{MN}^{(y)}p_N^{(y)}, \qquad \mathrm{d}p_M^{(y)}/\mathrm{d}T = -C_{MN}^{(y)}y_N^{(y)} - D_{MN}^{(y)}p_N^{(y)}.
\tag{4.2.34}
$$

Note that y_M are zero along Ω, so that the paraxial ray tracing system is expressed directly in terms of y_I and $p_I^{(y)}$ (not in terms of δy_M and $\delta p_N^{(y)}$, as in (4.2.11)). Quantities y_I and $p_I^{(y)}$ represent the canonical coordinates in the four-dimensional y_I-$p_I^{(y)}$-phase space. The initial values y_{I0} and $p_{I0}^{(y)}$ at initial point S represent the wavefront orthonormal coordinates y_I of initial point S' and the relevant y_I-coordinates of the initial slowness vector. The initial point S' of paraxial ray Ω' is situated in a plane tangent to the wavefront at S on Ω. The solutions y_N and $p_N^{(y)}$ of paraxial ray tracing system (4.2.34) then represent the y_N-coordinates and relevant components of the slowness vector $p_N^{(y)}$ of paraxial ray Ω'. The system of paraxial rays is clearly four-parameter; the noneikonal and coinciding paraxial rays are automatically eliminated.

4.2.3 Orthonomic System of Rays

The system of paraxial rays Ω' situated close to central ray Ω is four-parameter. In this section, we shall consider a two-parameter subsystem, corresponding to the *normal congruency of rays* (orthonomic system of rays). The orthonomic system of rays is parameterized by two ray parameters γ_1 and γ_2; see Section 3.10.1. This system of rays is uniquely related to the system of wavefronts. In addition, we shall also consider ray coordinates γ_1, γ_2, and γ_3, where γ_3 is some variable along central ray Ω.

In the following text, we shall consider wavefront orthonormal coordinate system y_i, which can be simply used for both isotropic and anisotropic media. For isotropic media, it reduces to the local Cartesian ray-centered coordinate system. We introduce the 2×2 matrices $\mathbf{Q}^{(y)}$ and $\mathbf{P}^{(y)}$, with elements

$$Q_{IJ}^{(y)} = (\partial y_I / \partial \gamma_J)_{T=\text{const.}}, \qquad P_{IJ}^{(y)} = \left(\partial p_I^{(y)} / \partial \gamma_J\right)_{T=\text{const.}}. \qquad (4.2.35)$$

Dynamic ray tracing system (4.2.31) can then be expressed in simple matrix form:

$$d\mathbf{Q}^{(y)}/dT = \mathbf{A}^{(y)}\mathbf{Q}^{(y)} + \mathbf{B}^{(y)}\mathbf{P}^{(y)}, \qquad d\mathbf{P}^{(y)}/dT = -\mathbf{C}^{(y)}\mathbf{Q}^{(y)} - \mathbf{D}^{(y)}\mathbf{P}^{(y)}.$$
$$(4.2.36)$$

The elements of the 2×2 matrices $\mathbf{A}^{(y)}$, $\mathbf{B}^{(y)}$, $\mathbf{C}^{(y)}$, and $\mathbf{D}^{(y)}$ are given by (4.2.32) and satisfy the symmetry relations (4.2.33). Dynamic ray tracing system (4.2.36) is a generalization of dynamic ray tracing system (4.1.65), derived for the isotropic medium.

As in Sections 4.1.7 and 4.1.8, we obtain the following relation for the 2×2 matrix $\mathbf{M}^{(y)}$ of second derivatives of the travel-time field T with respect to y_1 and y_2:

$$M_{IJ}^{(y)} = \partial^2 T / \partial y_I \partial y_J, \qquad \mathbf{M}^{(y)} = \mathbf{P}^{(y)}\mathbf{Q}^{(y)-1}. \qquad (4.2.37)$$

Dynamic ray tracing system (4.2.36) consists of two matrix equations for the 2×2 matrices $\mathbf{Q}^{(y)}$ and $\mathbf{P}^{(y)}$. They correspond to four scalar equations, which should be solved twice. Thus, if we wish to compute $\mathbf{Q}^{(y)}$ and $\mathbf{P}^{(y)}$, we must solve eight scalar equations.

Equations (4.2.36) also represent the paraxial ray tracing system for the orthonomic system of rays in matrix form. Let us introduce the 2×1 paraxial column matrices $\mathbf{y} = \mathbf{Q}^{(y)}d\boldsymbol{\gamma}$ and $\mathbf{p}^{(y)} = \mathbf{P}^{(y)}d\boldsymbol{\gamma}$, where $d\boldsymbol{\gamma} = (d\gamma_1, d\gamma_2)^T$, $\mathbf{y} = (y_1, y_2)^T$, and $\mathbf{p}^{(y)} = (p_1^{(y)}, p_2^{(y)})^T$. Then, (4.2.36) yields the paraxial ray tracing system:

$$d\mathbf{y}/dT = \mathbf{A}^{(y)}\mathbf{y} + \mathbf{B}^{(y)}\mathbf{p}^{(y)}, \qquad d\mathbf{p}^{(y)}/dT = -\mathbf{C}^{(y)}\mathbf{y} - \mathbf{D}^{(y)}\mathbf{p}^{(y)}. \qquad (4.2.38)$$

Let us now briefly discuss the initial conditions for dynamic ray tracing system (4.2.36) at the initial point S of ray Ω, $\mathbf{Q}^{(y)}(S)$, and $\mathbf{P}^{(y)}(S)$. Matrices $\mathbf{Q}^{(y)}(S)$ and $\mathbf{P}^{(y)}(S)$ may, in principle, be arbitrary; no constraint is imposed on them. There are two important linearly independent matrix initial conditions for (4.2.36):

a. **Point-source initial conditions:**

$$\mathbf{Q}^{(y)}(S) = \mathbf{0}, \qquad \mathbf{P}^{(y)}(S) \neq \mathbf{0}, \qquad \text{rank } \mathbf{P}^{(y)}(S) = 2. \qquad (4.2.39)$$

b. **Plane-wavefront initial conditions:**

$$\mathbf{Q}^{(y)}(S) \neq \mathbf{0}, \qquad \mathbf{P}^{(y)}(S) = \mathbf{0}, \qquad \text{rank } \mathbf{Q}^{(y)}(S) = 2. \qquad (4.2.40)$$

The physical meaning of both conditions is obvious. See also Section 4.3.1 and Figure 4.9. Alternatively, it would be possible to introduce linearly independent solutions in terms of line sources.

Instead of the initial values of $\mathbf{Q}^{(y)}(S)$ and $\mathbf{P}^{(y)}(S)$, it may be convenient to specify the initial conditions by $\mathbf{Q}^{(y)}(S)$ and $\mathbf{M}^{(y)}(S)$. $\mathbf{P}^{(y)}(S)$ is then obtained using relation $\mathbf{P}^{(y)}(S) =$

$\mathbf{M}^{(y)}(S)\mathbf{Q}^{(y)}(S)$. Alternatively, it is also possible to use the 2×2 matrix of the curvature of wavefront $\mathbf{K}(S)$ or the 2×2 matrix of the radii of curvature of wavefront $\mathbf{R}(S)$ instead of $\mathbf{M}(S)$; see Section 4.6.3 for isotropic media and Section 4.14.6 for anisotropic media.

From the 2×2 matrices $\mathbf{Q}^{(y)}$ and $\mathbf{P}^{(y)}$, we can also construct the 3×3 matrices $\hat{\mathbf{Q}}^{(y)}$ and $\hat{\mathbf{P}}^{(y)}$, given by relations

$$Q_{ij}^{(y)} = \partial y_i/\partial \gamma_j, \qquad P_{ij}^{(y)} = \partial p_i^{(y)}/\partial \gamma_j. \tag{4.2.41}$$

If we take $\gamma_3 = T$, we can use the analytical expressions for ray-tangent solutions (4.2.24), and the analytical expressions for $Q_{3K}^{(y)}$ and $P_{3K}^{(y)}$, given by (4.2.26), where we insert $Q_{IK}^{(y)}$ for $Q_I^{(y)}$ and $P_{IK}^{(y)}$ for $P_I^{(y)}$. Then, we obtain

$$\hat{\mathbf{Q}}^{(y)} = \begin{pmatrix} \mathbf{Q}^{(y)} & \mathcal{U}_1^{(y)} \\ & \mathcal{U}_2^{(y)} \\ 0 \quad 0 & \mathcal{U}_3^{(y)} \end{pmatrix}, \qquad \hat{\mathbf{P}}^{(y)} = \begin{pmatrix} \mathbf{P}^{(y)} & \eta_1^{(y)} \\ & \eta_2^{(y)} \\ P_{31}^{(y)} \quad P_{32}^{(y)} & \eta_3^{(y)} \end{pmatrix}. \tag{4.2.42}$$

The 3×3 matrix $\hat{\mathbf{M}}^{(y)}$ of the second derivatives of travel-time field T with respect to the y_i-coordinates is given by relations

$$M_{ij}^{(y)} = \partial^2 T/\partial y_i \partial y_j, \qquad \hat{\mathbf{M}}^{(y)} = \hat{\mathbf{P}}^{(y)}\hat{\mathbf{Q}}^{(y)-1}. \tag{4.2.43}$$

This yields

$$\hat{\mathbf{M}}^{(y)} = \begin{pmatrix} \mathbf{M}^{(y)} & M_{13}^{(y)} \\ & M_{23}^{(y)} \\ M_{13}^{(y)} \quad M_{23}^{(y)} & M_{33}^{(y)} \end{pmatrix}, \tag{4.2.44}$$

where

$$\begin{aligned} \mathbf{M}^{(y)} &= \mathbf{P}^{(y)}\mathbf{Q}^{(y)-1}, \\ M_{I3}^{(y)} &= \mathcal{C}^{-1}\eta_I^{(y)} - \mathcal{C}^{-1}\mathcal{U}_J^{(y)}M_{IJ}^{(y)}, \\ M_{33}^{(y)} &= \mathcal{C}^{-1}\eta_3^{(y)} - \mathcal{C}^{-2}\eta_I^{(y)}\mathcal{U}_I^{(y)} + \mathcal{C}^{-2}\mathcal{U}_I^{(y)}\mathcal{U}_J^{(y)}M_{IJ}^{(y)}. \end{aligned} \tag{4.2.45}$$

Equations (4.2.44) and (4.2.45) are valid for both isotropic and anisotropic media. For isotropic media, (4.2.44) simplifies because $\mathcal{U}_I^{(y)} = 0$ and $\eta_i^{(y)} = -v^{-1}v_{,i}$.

As in Section 4.1.8, we can use $\hat{\mathbf{M}}^{(y)}$ given by (4.2.44) and (4.2.45) in simple expressions for paraxial travel times, which are valid even in anisotropic media. Travel time $T(R')$ at point R', situated close to R on ray Ω, is given by a relation analogous to (4.1.83):

$$T(R') = T(R) + \hat{\mathbf{y}}^T\hat{\mathbf{p}}^{(y)}(R) + \tfrac{1}{2}\hat{\mathbf{y}}^T\hat{\mathbf{M}}^{(y)}(R)\hat{\mathbf{y}}, \tag{4.2.46}$$

where $\hat{\mathbf{p}}^{(y)}(R) = (0, 0, 1/\mathcal{C}(R))^T$, $\hat{\mathbf{M}}^{(y)}(R)$ is given by (4.2.44), and $\hat{\mathbf{y}}$ is given by (4.1.84).

In general Cartesian coordinates x_i, we can compute $\hat{\mathbf{Q}}^{(x)}$ and $\hat{\mathbf{P}}^{(x)}$ from $\hat{\mathbf{Q}}^{(y)}$ and $\hat{\mathbf{P}}^{(y)}$ using (4.2.21),

$$\hat{\mathbf{Q}}^{(x)} = \hat{\mathbf{H}}\hat{\mathbf{Q}}^{(y)}, \qquad \hat{\mathbf{P}}^{(x)} = \hat{\mathbf{H}}\hat{\mathbf{P}}^{(y)}, \qquad \hat{\mathbf{M}}^{(x)} = \hat{\mathbf{H}}\hat{\mathbf{M}}^{(y)}\hat{\mathbf{H}}^T. \tag{4.2.47}$$

The 3×3 transformation matrix $\hat{\mathbf{H}}$ has elements H_{ij} given by (4.1.18); see also (4.1.19) and (4.2.23). The quadratic expansion for paraxial travel time $T(R')$ in Cartesian coordinates x_i then reads:

$$T(R') = T(R) + \hat{\mathbf{x}}^T\hat{\mathbf{p}}^{(x)}(R) + \tfrac{1}{2}\hat{\mathbf{x}}^T\hat{\mathbf{M}}^{(x)}(R)\hat{\mathbf{x}}, \tag{4.2.48}$$

where $\hat{\mathbf{x}} = \hat{\mathbf{x}}(R', R)$ is given by (4.1.85) and $\hat{\mathbf{p}}^{(x)} = (p_1^{(x)}, p_2^{(x)}, p_3^{(x)})^T$.

As we can see from (4.2.47), matrices $\hat{\mathbf{Q}}^{(x)}$ and $\hat{\mathbf{P}}^{(x)}$ for an orthonomic system of rays can be determined from $\hat{\mathbf{Q}}^{(y)}$ and $\hat{\mathbf{P}}^{(y)}$. Thus, we can solve the dynamic ray tracing system (4.2.31) (eight scalar equations) for $Q_{MN}^{(y)}$ and $P_{MN}^{(y)}$ numerically and supplement them by analytical solutions (4.2.24) and (4.2.26). Matrices $\hat{\mathbf{Q}}^{(x)}$ and $\hat{\mathbf{P}}^{(x)}$, however, can also be determined directly from dynamic ray tracing system (4.2.4) in Cartesian coordinates. In matrix form, (4.2.4) reads

$$d\hat{\mathbf{Q}}^{(x)}/dT = \hat{\mathbf{A}}^{(x)}\hat{\mathbf{Q}}^{(x)} + \hat{\mathbf{B}}^{(x)}\hat{\mathbf{P}}^{(x)}, \qquad d\hat{\mathbf{P}}^{(x)}/dT = -\hat{\mathbf{C}}^{(x)}\hat{\mathbf{Q}}^{(x)} - \hat{\mathbf{D}}^{(x)}\hat{\mathbf{P}}^{(x)}.$$

(4.2.49)

The elements $A_{ij}^{(x)}$, $B_{ij}^{(x)}$, $C_{ij}^{(x)}$, and $D_{ij}^{(x)}$ of the 3×3 matrices $\hat{\mathbf{A}}^{(x)}$, $\hat{\mathbf{B}}^{(x)}$, $\hat{\mathbf{C}}^{(x)}$, and $\hat{\mathbf{D}}^{(x)}$ are given by (4.2.5). System (4.2.49) consists of two matrix equations for the 3×3 matrices $\hat{\mathbf{Q}}^{(x)}$ and $\hat{\mathbf{P}}^{(x)}$. This is equivalent to 18 scalar equations. Without lose of generality, the number of equations *can be reduced from 18 to 12*. One column in each matrix $\hat{\mathbf{Q}}^{(x)}$ and $\hat{\mathbf{P}}^{(x)}$, corresponding to the ray-tangent solution, is known from ray tracing. The paraxial initial conditions for $Q_{iN}^{(x)}$ and $P_{iN}^{(x)}$, however, should be taken appropriately at S. They can be expressed in terms of $Q_{JK}^{(y)}(S)$ and $P_{JK}^{(y)}(S)$ as follows:

$$
\begin{aligned}
Q_{iN}^{(x)} &= H_{iJ} Q_{JN}^{(y)}, \\
P_{iN}^{(x)} &= H_{jI} p_i^{(x)} \eta_j^{(x)} Q_{IN}^{(y)} + \left(H_{iI} - H_{jI} p_i^{(x)} \mathcal{U}_j^{(x)} \right) P_{IN}^{(y)};
\end{aligned}
$$

(4.2.50)

see (4.2.21) and (4.2.26) . For *point-source initial conditions*, we insert $Q_{IN}^{(y)} = 0$:

$$Q_{iN}^{(x)} = 0, \qquad P_{iN}^{(x)} = \left(H_{iI} - H_{jI} p_i^{(x)} \mathcal{U}_j^{(x)} \right) P_{IN}^{(y)};$$

(4.2.51)

see (4.2.39). For *plane-wavefront initial conditions*, we insert $P_{IN}^{(y)} = 0$:

$$Q_{iN}^{(x)} = H_{iJ} Q_{JN}^{(y)}, \qquad P_{iN}^{(x)} = H_{jI} p_i^{(x)} \eta_j^{(x)} Q_{IN}^{(y)};$$

(4.2.52)

see (4.2.40). Thus, if we wish to calculate $Q_{iN}^{(x)}$ and $P_{iN}^{(x)}$ along ray Ω using dynamic ray tracing system (4.2.4) consisting of six equations, we can solve it only twice ($N = 1, 2$) with initial conditions (4.2.51) and (4.2.52). The remaining ray-tangent solutions are known from ray tracing; $Q_{i3}^{(x)} = \mathcal{U}_i^{(x)}$ and $P_{i3}^{(x)} = \eta_i^{(x)}$. Consequently, it is not necessary to solve 18 but only 12 scalar equations if we wish to determine all 18 elements of matrices $\hat{\mathbf{Q}}^{(x)}$ and $\hat{\mathbf{P}}^{(x)}$ for an orthonomic system of rays.

Note that $\mathbf{Q}^{(y)}$ and $\hat{\mathbf{Q}}^{(x)}$ can also be used to determine Jacobian $J^{(T)}$ and geometrical spreading; see (3.10.9). In fact, the computation of Jacobian $J^{(T)}$ and of geometrical spreading was historically the first application of dynamic ray tracing. We obtain

$$J^{(T)} = \det \hat{\mathbf{Q}}^{(x)} = \det \hat{\mathbf{Q}}^{(y)} = \mathcal{C} \det \mathbf{Q}^{(y)},$$

(4.2.53)

as $\det \hat{\mathbf{H}} = 1$, and $\mathcal{U}_3^{(y)} = \mathcal{C}$; see (4.2.42) and (4.2.47). Thus, having solved system (4.2.36) (8 scalar equations), we can calculate $\det \mathbf{Q}^{(y)}$ and $J^{(T)}$, even in an anisotropic medium. Alternatively, we can also solve 12 scalar equations in Cartesian coordinates, supplement them with (4.2.13), and determine $\det \hat{\mathbf{Q}}^{(x)}$.

4.2.4 Curvilinear Coordinates

The dynamic ray tracing system in arbitrary curvilinear, orthogonal, or nonorthogonal coordinates ξ_i, may be obtained directly from that in the Cartesian rectangular coordinates, which is presented in Section 4.2.1, especially (4.2.4). For the reader's convenience, its simple and brief derivation will be given. The derivation applies both to isotropic and anisotropic media.

We denote the covariant components of the slowness vector by $p_i^{(\xi)} = \partial T/\partial \xi_i$, and the Hamiltonian under consideration by $\mathcal{H}^{(\xi)}(\xi_i, p_i^{(\xi)})$. The eikonal equation in Hamiltonian form reads $\mathcal{H}^{(\xi)}(\xi_i, p_i^{(\xi)}) = 0$. To simplify the notation, all indices will be subscripts, and the summation will be performed over equal subscripts, even though we are considering curvilinear, possibly nonorthogonal coordinates. In the equations we shall present, all quantities are well defined so that there is no possibility of misunderstanding.

The ray tracing system in curvilinear nonorthogonal coordinates ξ_i, corresponding to eikonal equation $\mathcal{H}^{(\xi)}(\xi_i, p_i^{(\xi)}) = 0$, then reads

$$\mathrm{d}\xi_i/\mathrm{d}T = \partial\mathcal{H}^{(\xi)}\big/\partial p_i^{(\xi)}, \qquad \mathrm{d}p_i^{(\xi)}\big/\mathrm{d}T = -\partial\mathcal{H}^{(\xi)}/\partial\xi_i. \qquad (4.2.54)$$

It is assumed that $p_k^{(\xi)}\partial\mathcal{H}^{(\xi)}/\partial p_k^{(\xi)} = 1$; hence, the variable along the ray again represents travel time T.

Consider point $S(\xi_{i0}, p_{i0}^{(\xi)})$ in the six-dimensional ξ_i-$p_i^{(\xi)}$-phase space, situated on the hypersurface $\mathcal{H}^{(\xi)}(\xi_i, p_i^{(\xi)}) = 0$. Also consider ray Ω with initial point at S, satisfying ray tracing system (4.2.54). Equation $\mathcal{H}^{(\xi)}(\xi_i, p_i^{(\xi)}) = 0$ is then satisfied along the whole ray Ω.

We shall now use ray tracing system (4.2.54) to derive the equations for the first derivatives of ξ_i and $p_i^{(\xi)}$ with respect to initial parameter γ. We denote

$$Q_i^{(\xi)} = (\partial\xi_i/\partial\gamma)_{T=\mathrm{const.}}, \qquad P_i^{(\xi)} = \big(\partial p_i^{(\xi)}/\partial\gamma\big)_{T=\mathrm{const.}} \qquad (4.2.55)$$

Taking the partial derivatives of ray tracing system (4.2.54) with respect to γ, and bearing in mind that $\partial/\partial\gamma$ commutes with $\mathrm{d}/\mathrm{d}T$, we obtain the dynamic ray tracing system, consisting of six linear ordinary differential equations of the first order for $Q_i^{(\xi)}$ and $P_i^{(\xi)}$:

$$\mathrm{d}Q_i^{(\xi)}\big/\mathrm{d}T = A_{ij}^{(\xi)}Q_j^{(\xi)} + B_{ij}^{(\xi)}P_j^{(\xi)}, \qquad \mathrm{d}P_i^{(\xi)}\big/\mathrm{d}T = -C_{ij}^{(\xi)}Q_j^{(\xi)} - D_{ij}^{(\xi)}P_j^{(\xi)}. \qquad (4.2.56)$$

Here

$$A_{ij}^{(\xi)} = \partial^2\mathcal{H}^{(\xi)}\big/\partial p_i^{(\xi)}\partial\xi_j, \qquad B_{ij}^{(\xi)} = \partial^2\mathcal{H}^{(\xi)}\big/\partial p_i^{(\xi)}\partial p_j^{(\xi)},$$
$$C_{ij}^{(\xi)} = \partial^2\mathcal{H}^{(\xi)}\big/\partial\xi_i\partial\xi_j, \qquad D_{ij}^{(\xi)} = \partial^2\mathcal{H}^{(\xi)}\big/\partial\xi_i\partial p_j^{(\xi)}. \qquad (4.2.57)$$

They satisfy the symmetry relations

$$B_{ij}^{(\xi)} = B_{ji}^{(\xi)}, \qquad C_{ij}^{(\xi)} = C_{ji}^{(\xi)}, \qquad D_{ij}^{(\xi)} = A_{ji}^{(\xi)}. \qquad (4.2.58)$$

Quantities $Q_i^{(\xi)}$ and $P_i^{(\xi)}$ satisfy the following constraint relation:

$$\big(\partial\mathcal{H}^{(\xi)}\big/\partial\xi_k\big)Q_k^{(\xi)} + \big(\partial\mathcal{H}^{(\xi)}\big/\partial p_k^{(\xi)}\big)P_k^{(\xi)} = 0. \qquad (4.2.59)$$

Constraint relation (4.2.59) is satisfied along the whole ray Ω once it is satisfied at the initial point S on Ω.

Dynamic ray tracing system (4.2.56) is also called the paraxial ray tracing system for paraxial rays Ω' situated close to central ray Ω:

$$\mathrm{d}\delta\xi_i/\mathrm{d}T = A_{ij}^{(\xi)}\delta\xi_j + B_{ij}^{(\xi)}\delta p_j^{(\xi)}, \qquad \mathrm{d}\delta p_i^{(\xi)}\big/\mathrm{d}T = -C_{ij}^{(\xi)}\delta\xi_j - D_{ij}^{(\xi)}\delta p_j^{(\xi)}, \qquad (4.2.60)$$

where $A_{ij}^{(\xi)}$, $B_{ij}^{(\xi)}$, $C_{ij}^{(\xi)}$, and $D_{ij}^{(\xi)}$ are again given by (4.2.57) and satisfy symmetry relations (4.2.58). As in Cartesian coordinates, quantities $\delta\xi_i$ and $\delta p_i^{(\xi)}$ represent the linear approximations of the deviation of paraxial ray Ω' from central ray Ω and of the deviations of the relevant slowness vector covariant components $p_i^{(\xi)}$. For more details, refer to Section 4.2.1. The constraint relation for paraxial rays reads

$$\left(\partial\mathcal{H}^{(\xi)}/\partial\xi_k\right)\delta\xi_k + \left(\partial\mathcal{H}^{(\xi)}/\partial p_k^{(\xi)}\right)\delta p_k^{(\xi)} = 0.$$

Similarly as in Cartesian coordinates, two solutions of dynamic ray tracing system (4.2.56) can be found analytically:

a. Ray-tangent solutions:

$$Q_i^{(\xi)} = \partial\mathcal{H}^{(\xi)}/\partial p_i^{(\xi)}, \qquad P_i^{(\xi)} = -\partial\mathcal{H}^{(\xi)}/\partial\xi_i. \tag{4.2.61}$$

b. The noneikonal solutions:

$$Q_i^{(\xi)} = (T - T_0)\partial\mathcal{H}^{(\xi)}/\partial p_i^{(\xi)}, \qquad P_i^{(\xi)} = p_i^{(\xi)} + (T - T_0)\partial\mathcal{H}^{(\xi)}/\partial\xi_i. \tag{4.2.62}$$

Derivatives $\partial\mathcal{H}^{(\xi)}/\partial\xi_i$ and $\partial\mathcal{H}^{(\xi)}/\partial p_i^{(\xi)}$ are known from ray tracing (4.2.54). The ray-tangent solution satisfies constraint relation (4.2.59), but the noneikonal solution does not satisfy it.

The dynamic ray tracing system (4.2.56) presented here in general curvilinear coordinates ξ_i can be modified in many ways. Next we shall discuss several such modifications. We shall also discuss determining paraxial travel times and geometrical spreading in curvilinear coordinates for orthonomic systems of rays.

1. APPLICATION OF THE REDUCED HAMILTONIAN

We shall now prove that the dynamic ray tracing system in arbitrary curvilinear coordinates can always be reduced to four equations, for both isotropic and anisotropic media. The price we pay for this reduction of the number of equation is that we need to take one coordinate of the system as a variable along central ray Ω. This approach is not convenient if central ray Ω has a turning point with respect to the selected coordinate in the region of interest.

We shall use coordinates ξ_i and assume that ray Ω does not have a turning point with respect to the ξ_3-coordinate in the region of interest. We introduce the reduced Hamiltonian

$$\mathcal{H}^{(\xi)}\left(\xi_i, p_i^{(\xi)}\right) = p_3^{(\xi)} + \mathcal{H}^R\left(\xi_i, p_I^{(\xi)}\right) \tag{4.2.63}$$

(see (3.1.25)), and take ξ_3 as the variable along ray Ω. The ray tracing system then consists of four equations only:

$$d\xi_I/d\xi_3 = \partial\mathcal{H}^R/\partial p_I^{(\xi)}, \qquad dp_I^{(\xi)}/d\xi_3 = -\partial\mathcal{H}^R/\partial\xi_I. \tag{4.2.64}$$

As in other sections, we introduce $Q_I^{(R)}$ and $P_I^{(R)}$ by relations $Q_I^{(R)} = (\partial\xi_I/\partial\gamma)_{\xi_3=\text{const.}}$ and $P_I^{(R)} = (\partial p_I^{(\xi)}/\partial\gamma)_{\xi_3=\text{const.}}$, where γ is some initial parameter, that is, $\xi_{10}, \xi_{20}, p_{10}^{(\xi)}$, or $p_{20}^{(\xi)}$. The derivatives are taken *for constant* ξ_3 and for other constant initial parameters. The dynamic ray tracing system for $Q_I^{(R)}$ and $P_I^{(R)}$ then consists of four equations only and reads

$$dQ_I^{(R)}/d\xi_3 = A_{IJ}^{(R)}Q_J^{(R)} + B_{IJ}^{(R)}P_J^{(R)}, \qquad dP_I^{(R)}/d\xi_3 = -C_{IJ}^{(R)}Q_J^{(R)} - D_{IJ}^{(R)}P_J^{(R)}, \tag{4.2.65}$$

where

$$A_{IJ}^{(R)} = \partial^2 \mathcal{H}^R / \partial p_I^{(\xi)} \partial \xi_J, \qquad B_{IJ}^{(R)} = \partial^2 \mathcal{H}^R / \partial p_I^{(\xi)} \partial p_J^{(\xi)},$$
$$C_{IJ}^{(R)} = \partial^2 \mathcal{H}^R / \partial \xi_I \partial \xi_J, \qquad D_{IJ}^{(R)} = \partial^2 \mathcal{H}^R / \partial \xi_I \partial p_J^{(\xi)}. \qquad (4.2.66)$$

Equations (4.2.65) represent the *"reduced" dynamic ray tracing system*. We can also use $\delta\xi_I$ and $\delta p_I^{(\xi)}$ instead of $Q_I^{(R)}$ and $P_I^{(R)}$; then, (4.2.65) represents the *"reduced" paraxial ray tracing system*. The reduced paraxial ray tracing system fails if the paraxial ray has a turning point with respect to the ξ_3-coordinate.

As in previous sections, the solutions of the dynamic ray tracing system (4.2.65), $Q_I^{(R)}$ and $P_I^{(R)}$, can also be used to construct the complete solutions $Q_i^{(\xi)}$ and $P_i^{(\xi)}$ of the dynamic ray tracing system with variable $u = \xi_3$ along Ω and with $\mathcal{H}^{(\xi)} = p_3^{(\xi)} + \mathcal{H}^R(\xi_i, p_I^{(\xi)})$. We take into account that $\partial \mathcal{H}^{(\xi)} / \partial p_3^{(\xi)} = 1$, and use constraint relation (4.2.59). In addition, the ray-tangent and noneikonal solutions can be constructed from the solutions of ray tracing system (4.2.64).

Dynamic ray tracing system (4.2.65) is valid in an arbitrary coordinate system, both in isotropic and anisotropic media. It may also be useful in Cartesian rectangular coordinates (see (3.1.25)) and in orthogonal curvilinear coordinates, see (3.5.13).

2. TRANSFORMATIONS OF DYNAMIC RAY TRACING SYSTEMS

It may be convenient to transform the dynamic ray tracing system from one coordinate system to another, with the Hamiltonian and its derivatives specified in one (presumably the simpler) coordinate system. For example, we can specify the Hamiltonian and its derivatives in the Cartesian coordinate system, in which their expressions are simple, even in anisotropic media, and perform the computations in some curvilinear coordinate system.

We shall transform the dynamic ray tracing system from curvilinear coordinates x_i to curvilinear coordinates ξ_i. We emphasize that x_i represent any curvilinear nonorthogonal coordinates. The Hamiltonian in x_i-coordinates reads $\mathcal{H}^{(x)}(x_i, p_i^{(x)})$, where $p_i^{(x)} = \partial T / \partial x_i$ are covariant components of the slowness vector in the x_i-coordinates. We assume that both Hamiltonians $\mathcal{H}^{(x)}$ and $\mathcal{H}^{(\xi)}$ satisfy relations analogous to (4.2.1) and (4.2.15), so that variable u along the ray equals T. We denote the transformation matrices from ξ_i to x_k and back by $H_{im}^{(\xi)} = \partial x_i / \partial \xi_m$ and $\bar{H}_{mi}^{(\xi)} = \partial \xi_m / \partial x_i$. They satisfy relations $H_{im}^{(\xi)} \bar{H}_{mj}^{(\xi)} = \delta_{ij}$ and $\bar{H}_{mi}^{(\xi)} H_{in}^{(\xi)} = \delta_{mn}$. The ray tracing system is transformed from one coordinate system to another as follows:

$$\frac{dx_i}{dT} = H_{ij}^{(\xi)} \frac{d\xi_j}{dT}, \qquad \frac{dp_i^{(x)}}{dT} = \frac{d}{dT}\left(\frac{\partial T}{\partial \xi_j}\frac{\partial \xi_j}{\partial x_i}\right) = \bar{H}_{ji}^{(\xi)} \frac{dp_j^{(\xi)}}{dT} + p_j^{(\xi)} \frac{d}{dT} \bar{H}_{ji}^{(\xi)}. \qquad (4.2.67)$$

As in preceeding sections, we use the notation $Q_m^{(x)} = \partial x_m / \partial \gamma$, $P_m^{(x)} = \partial p_m^{(x)} / \partial \gamma$, $Q_m^{(\xi)} = \partial \xi_m / \partial \gamma$, and $P_m^{(\xi)} = \partial p_m^{(\xi)} / \partial \gamma$, where γ is some initial parameter. The derivatives with respect to γ are taken for constant T. The transformation relations between $Q_m^{(\xi)}$, $P_m^{(\xi)}$ and $Q_i^{(x)}$, $P_i^{(x)}$, are

$$Q_m^{(\xi)} = \bar{H}_{mi}^{(\xi)} Q_i^{(x)}, \qquad P_m^{(\xi)} = H_{im}^{(\xi)} P_i^{(x)} + F_{mn} Q_n^{(\xi)}, \qquad (4.2.68)$$

where

$$F_{mn} = p_i^{(x)} \partial^2 x_i / \partial \xi_m \partial \xi_n. \qquad (4.2.69)$$

The inversion of (4.2.68) is

$$Q_i^{(x)} = H_{im}^{(\xi)} Q_m^{(\xi)}, \qquad P_i^{(x)} = \bar{H}_{mi}^{(\xi)}\big(P_m^{(\xi)} - F_{mn} Q_n^{(\xi)}\big). \tag{4.2.70}$$

Now we shall transform the dynamic ray tracing system from the x_i-coordinates to the ξ_k-coordinates. The dynamic ray tracing system in curvilinear coordinates x_i is given by equations,

$$\mathrm{d}Q_i^{(x)}/\mathrm{d}T = A_{ij}^{(x)} Q_j^{(x)} + B_{ij}^{(x)} P_j^{(x)}, \qquad \mathrm{d}P_i^{(x)}/\mathrm{d}T = -C_{ij}^{(x)} Q_j^{(x)} - D_{ij}^{(x)} P_j^{(x)}, \tag{4.2.71}$$

where $A_{ij}^{(x)}$, $B_{ij}^{(x)}$, $C_{ij}^{(x)}$, and $D_{ij}^{(x)}$ are given by (4.2.57), ξ being replaced by x; see (4.2.56). We shall now use the same approach as in Section 4.2.2 to transform (4.2.71) to coordinates ξ_i. We insert (4.2.70) into (4.2.71), multiply the equation for $Q_i^{(x)}$ by $\bar{H}_{ni}^{(\xi)}$, and multiply the equation for $P_i^{(x)}$ by $H_{in}^{(\xi)}$. After some algebra, we obtain the *transformed dynamic ray tracing system* for $Q_n^{(\xi)}$ and $P_n^{(\xi)}$ in the same form as (4.2.56), where $A_{ij}^{(\xi)}$, $B_{ij}^{(\xi)}$, and $C_{ij}^{(\xi)}$, and $D_{ij}^{(\xi)}$ are given by relations

$$
\begin{aligned}
A_{ij}^{(\xi)} &= \bar{A}_{ij}^{(\xi)} - \bar{B}_{im}^{(\xi)} F_{mj} - \bar{H}_{im}^{(\xi)}\mathrm{d}\big(H_{mj}^{(\xi)}\big)/\mathrm{d}T, \\
B_{ij}^{(\xi)} &= \bar{B}_{ij}^{(\xi)}, \\
C_{ij}^{(\xi)} &= \bar{C}_{ij}^{(\xi)} - \bar{D}_{im}^{(\xi)} F_{mj} - F_{im} A_{mj}^{(\xi)} - H_{mi}^{(\xi)}\mathrm{d}\big(\bar{H}_{nm}^{(\xi)} F_{nj}\big)/\mathrm{d}T, \\
D_{ij}^{(\xi)} &= \bar{D}_{ij}^{(\xi)} - F_{im} \bar{B}_{mj}^{(\xi)} + H_{mi}^{(\xi)}\mathrm{d}\big(\bar{H}_{jm}^{(\xi)}\big)/\mathrm{d}T.
\end{aligned}
\tag{4.2.72}
$$

Here

$$
\begin{aligned}
\bar{A}_{ij}^{(\xi)} &= \bar{H}_{in}^{(\xi)} H_{mj}^{(\xi)} \partial^2 \mathcal{H}^{(x)}/\partial p_n^{(x)} \partial x_m, & \bar{B}_{ij}^{(\xi)} &= \bar{H}_{in}^{(\xi)} \bar{H}_{jm}^{(\xi)} \partial^2 \mathcal{H}^{(x)}/\partial p_n^{(x)} \partial p_m^{(x)}, \\
\bar{C}_{ij}^{(\xi)} &= H_{ni}^{(\xi)} H_{mj}^{(\xi)} \partial^2 \mathcal{H}^{(x)}/\partial x_n \partial x_m, & \bar{D}_{ij}^{(\xi)} &= H_{ni}^{(\xi)} \bar{H}_{jm}^{(\xi)} \partial^2 \mathcal{H}^{(x)}/\partial x_n \partial p_m^{(x)}.
\end{aligned}
\tag{4.2.73}
$$

Note that symmetry relations (4.2.58) are again satisfied. Constraint relation (4.2.59) can be expressed in terms of Hamiltonian $\mathcal{H}^{(x)}$ as

$$Q_l^{(\xi)}\left(\frac{\partial \mathcal{H}^{(x)}}{\partial x_k} H_{kl}^{(\xi)} - \frac{\partial \mathcal{H}^{(x)}}{\partial p_k^{(x)}} \bar{H}_{mk}^{(\xi)} F_{ml}\right) + P_l^{(\xi)} \frac{\partial \mathcal{H}^{(x)}}{\partial p_k^{(x)}} \bar{H}_{lk}^{(\xi)} = 0. \tag{4.2.74}$$

Thus, we can specify Hamiltonian $\mathcal{H}^{(x)}$ and its derivatives in one curvilinear coordinate system and perform dynamic ray tracing in another curvilinear coordinate system. The transformed dynamic ray tracing system is particularly useful if coordinates x_i correspond to the global Cartesian coordinate system.

3. NONORTHOGONAL RAY-CENTERED COORDINATES

We shall now use transformation equations (4.2.72) to derive the dynamic ray tracing system in nonorthogonal ray-centered coordinates ζ_i. The nonorthogonal ray-centered coordinate system is introduced here so that it can also be applied to anisotropic media. As in isotropic media, we choose *central ray Ω to be the ζ_3-coordinate axis* and the ζ_1- and ζ_2-axes to be straight lines tangent to the wavefront, intersecting at central ray Ω. Coordinate ζ_3 represents travel time T. Obviously, the ray-centered coordinate system ζ_i in anisotropic media is nonorthogonal because the ray is not perpendicular to the wavefront.

The mutual relation between global Cartesian rectangular coordinates x_i and nonorthogonal ray-centered coordinates ζ_i reads

$$x_i(\zeta_j) = x_i^{\Omega}(\zeta_3) + H_{iN}^{(\zeta)}(\zeta_3)\zeta_N. \qquad (4.2.75)$$

Here $x_i = x_i^{\Omega}(\zeta_3)$ represent the equations of central ray Ω, along which $\zeta_1 = \zeta_2 = 0$. The transformation matrices from Cartesian coordinates x_i to nonorthogonal ray-centered coordinates ζ_m and back are again denoted by $H_{im}^{(\zeta)}$ and $\bar{H}_{mi}^{(\zeta)}$ and satisfy the relations $H_{im}^{(\zeta)}\bar{H}_{mj}^{(\zeta)} = \delta_{ij}$ and $\bar{H}_{mi}^{(\zeta)}H_{in}^{(\zeta)} = \delta_{mn}$. Note that $H_{ij}^{(\zeta)}$ and $\bar{H}_{ij}^{(\zeta)}$ are always different from H_{ij} introduced in Section 4.1.5 even in isotropic media because $\zeta_3 = T$, but $q_3 = s$. We also introduce standard notations $\mathcal{U}_i^{(\zeta)} = d\zeta_i/dT$ and $\eta_i^{(\zeta)} = \partial p_i^{(\zeta)}/dT$. We realize that in isotropic media $\mathcal{U}_I^{(\zeta)} = 0$ and $p_I^{(\zeta)} = 0$, but in anisotropic media $\mathcal{U}_I^{(\zeta)} = 0$ and $p_I^{(\zeta)} \neq 0$. Note that the situation is different in the wavefront orthonormal coordinate system, where $\mathcal{U}_I^{(y)} \neq 0$, $p_I^{(y)} = 0$. We also obtain:

$$\bar{H}_{3i}^{(\zeta)} = p_i^{(x)}, \qquad H_{i3}^{(\zeta)} = \mathcal{U}_i^{(x)}, \qquad p_i^{(x)}H_{in}^{(\zeta)} = \delta_{n3}, \qquad \bar{H}_{ni}^{(\zeta)}\mathcal{U}_i^{(x)} = \delta_{3n},$$

$$F_{MN} = 0, \qquad F_{m3} = F_{3m} = p_i^{(x)}dH_{im}^{(\zeta)}/dT = -H_{im}^{(\zeta)}\eta_i^{(x)}, \qquad (4.2.76)$$

$$dF_{m3}/dT = -\eta_i^{(x)}dH_{im}^{(\zeta)}/dT + H_{im}^{(\zeta)}(C_{ij}^{(x)}\mathcal{U}_j^{(x)} + D_{ij}^{(x)}\eta_j^{(x)}).$$

Using (4.2.76) in (4.2.72), we obtain the *dynamic ray tracing system in nonorthogonal ray-centered coordinates* ζ_i as follows:

$$dQ_i^{(\zeta)}/dT = A_{ij}^{(\zeta)}Q_j^{(\zeta)} + B_{ij}^{(\zeta)}P_j^{(\zeta)}, \qquad dP_i^{(\zeta)}/dT = -C_{ij}^{(\zeta)}Q_j^{(\zeta)} - D_{ij}^{(\zeta)}P_j^{(\zeta)}. \qquad (4.2.77)$$

The expressions for $A_{ij}^{(\zeta)}$, $B_{ij}^{(\zeta)}$, $C_{ij}^{(\zeta)}$, and $D_{ij}^{(\zeta)}$ are surprisingly simple. For $i = I$ and $j = J$, we obtain

$$A_{IJ}^{(\zeta)} = \bar{A}_{IJ}^{(\zeta)} - \bar{H}_{Ii}^{(\zeta)}dH_{iJ}^{(\zeta)}/dT, \qquad B_{IJ}^{(\zeta)} = \bar{B}_{IJ}^{(\zeta)},$$

$$C_{IJ}^{(\zeta)} = \bar{C}_{IJ}^{(\zeta)} - H_{iI}^{(\zeta)}H_{jJ}^{(\zeta)}\eta_i^{(x)}\eta_j^{(x)}, \qquad D_{IJ}^{(\zeta)} = \bar{D}_{IJ}^{(\zeta)} - \bar{H}_{Ji}^{(\zeta)}dH_{iI}^{(\zeta)}/dT. \qquad (4.2.78)$$

Here $\bar{A}_{IJ}^{(\zeta)}$, $\bar{B}_{IJ}^{(\zeta)}$, $\bar{C}_{IJ}^{(\zeta)}$ and $\bar{D}_{IJ}^{(\zeta)}$ are given by (4.2.73) in which curvilinear coordinates ξ_i are replaced by nonorthogonal ray-centered coordinates ζ_i. All other expressions $A_{nk}^{(\zeta)}$, $B_{nk}^{(\zeta)}$, $C_{nk}^{(\zeta)}$, and $D_{nk}^{(\zeta)}$, for $n = 3$ and/or $k = 3$, vanish, with the exception of $B_{33}^{(\zeta)}$:

$$A_{i3}^{(\zeta)} = A_{3i}^{(\zeta)} = 0, \qquad C_{i3}^{(\zeta)} = C_{3i}^{(\zeta)} = 0, \qquad D_{i3}^{(\zeta)} = D_{3i}^{(\zeta)} = 0,$$

$$B_{I3}^{(\zeta)} = B_{3I}^{(\zeta)} = 0, \qquad B_{33}^{(\zeta)} = 1. \qquad (4.2.79)$$

As we can see from (4.2.78) and (4.2.79), $A_{ij}^{(\zeta)}$, $B_{ij}^{(\zeta)}$, $C_{ij}^{(\zeta)}$, and $D_{ij}^{(\zeta)}$ again satisfy symmetry relations (4.2.58). The symmetry relations are also satisfied for $i = I$ and $j = J$. Constraint relation (4.2.74) takes a very simple form in the nonorthogonal ray-centered coordinates

$$P_3^{(\zeta)} = 0. \qquad (4.2.80)$$

The dynamic ray tracing system (4.2.77) in nonorthogonal ray-centered coordinates, consisting of six equations, can be decomposed into two subsystems. The first subsystem consists of four equations:

$$dQ_I^{(\zeta)}/dT = A_{IJ}^{(\zeta)}Q_J^{(\zeta)} + B_{IJ}^{(\zeta)}P_J^{(\zeta)}, \qquad dP_I^{(\zeta)}/dT = -C_{IJ}^{(\zeta)}Q_J^{(\zeta)} - D_{IJ}^{(\zeta)}P_J^{(\zeta)}, \qquad (4.2.81)$$

$I = 1, 2$, and the second subsystem consists of two equations for $Q_3^{(\zeta)}$ and $P_3^{(\zeta)}$:

$$dQ_3^{(\zeta)}/dT = P_3^{(\zeta)}, \qquad dP_3^{(\zeta)}/dT = 0. \tag{4.2.82}$$

The second subsystem can be solved analytically. For *standard paraxial initial conditions* (with $Q_{30}^{(\zeta)} = P_{30}^{(\zeta)} = 0$), we obtain the following relations valid along the whole ray:

$$Q_3^{(\zeta)}(T) = 0, \qquad P_3^{(\zeta)}(T) = 0. \tag{4.2.83}$$

For *ray-tangent initial conditions* (with $Q_{30}^{(\zeta)} = 1$, but $P_{i0}^{(\zeta)} = 0$ and $Q_{I0}^{(\zeta)} = 0$), we obtain the complete linearly independent solution:

$$Q_3^{(\zeta)}(T) = Q_{30}^{(\zeta)} = 1, \qquad Q_I^{(\zeta)}(T) = 0, \qquad P_i^{(\zeta)}(T) = 0. \tag{4.2.84}$$

Finally, for the *noneikonal initial conditions* (with $P_{30}^{(\zeta)} \neq 0$, $Q_{i0}^{(\zeta)} = 0$, and $P_{I0}^{(\zeta)} = 0$), the linearly independent solution is

$$\begin{aligned} P_3^{(\zeta)}(T) &= P_{30}^{(\zeta)}, & Q_3^{(\zeta)}(T) &= (T - T_0)P_{30}^{(\zeta)}, \\ P_I^{(\zeta)}(T) &= 0, & Q_I^{(\zeta)}(T) &= 0. \end{aligned} \tag{4.2.85}$$

This solution does not satisfy constraint relation (4.2.80) and does not correspond to the eikonal equation under consideration. Nevertheless, it may be useful in constructing the 6×6 propagator matrix of the dynamic ray tracing system.

Dynamic ray tracing system (4.2.77) with (4.2.78) in nonorthogonal ray-centered coordinates ζ_i depends on the specification of $H_{ik}^{(\zeta)}$ and $\bar{H}_{nj}^{(\zeta)}$. As we can see in (4.2.76), $\bar{H}_{3i}^{(\zeta)}$ and $H_{i3}^{(\zeta)}$ are known along Ω, $\bar{H}_{3i}^{(\zeta)} = p_i^{(x)}$, and $H_{i3}^{(\zeta)} = \mathcal{U}_i^{(x)}$. It remains to specify $H_{iK}^{(\zeta)}$ and $\bar{H}_{Ni}^{(\zeta)}$, which determine the actual orientation of axes ζ_1 and ζ_2 in the plane tangent to the wavefront at any point of ray Ω. Dynamic ray tracing system (4.2.77) with (4.2.78) is valid for arbitrarily chosen $H_{iK}^{(\zeta)}$ and $\bar{H}_{Ni}^{(\zeta)}$ along the ray, which satisfy $H_{im}^{(\zeta)}\bar{H}_{mj}^{(\zeta)} = \delta_{ij}$ and $\bar{H}_{mi}^{(\zeta)}H_{in}^{(\zeta)} = \delta_{mn}$. For different options, see Hanyga (1982a), Kendall, Guest, and Thomson (1992), and Klimeš (1994). Here we shall specify $H_{iK}^{(\zeta)}$ and $\bar{H}_{Ni}^{(\zeta)}$ along Ω in a simple way:

$$dH_{iK}^{(\zeta)}/dT = -\big(H_{kK}^{(\zeta)}\eta_k^{(x)}\big)\mathcal{U}_i^{(x)}, \qquad d\bar{H}_{Ni}^{(\zeta)}/dT = -\big(\bar{H}_{Nk}^{(\zeta)}d\mathcal{U}_k^{(x)}/dT\big)p_i^{(x)}. \tag{4.2.86}$$

This specification simplifies the expressions for $A_{IJ}^{(\zeta)}$, $B_{IJ}^{(\zeta)}$, $C_{IJ}^{(\zeta)}$, and $D_{IJ}^{(\zeta)}$ considerably:

$$\begin{aligned} A_{IJ}^{(\zeta)} &= \bar{H}_{In}^{(\zeta)}H_{mJ}^{(\zeta)}\partial^2\mathcal{H}^{(x)}/\partial p_n^{(x)}\partial x_m, \\ B_{IJ}^{(\zeta)} &= \bar{H}_{In}^{(\zeta)}\bar{H}_{Jm}^{(\zeta)}\partial^2\mathcal{H}^{(x)}/\partial p_n^{(x)}\partial p_m^{(x)}, \\ C_{IJ}^{(\zeta)} &= H_{nI}^{(\zeta)}H_{mJ}^{(\zeta)}\big(\partial^2\mathcal{H}^{(x)}/\partial x_n\partial x_m - \eta_n^{(x)}\eta_m^{(x)}\big), \\ D_{IJ}^{(\zeta)} &= H_{nI}^{(\zeta)}\bar{H}_{Jm}^{(\zeta)}\partial^2\mathcal{H}^{(x)}/\partial x_n\partial p_m^{(x)}. \end{aligned} \tag{4.2.87}$$

Expressions (4.2.79) also apply to option (4.2.86).

Let us return to (4.2.86). If we solve (4.2.86) and add $\bar{H}_{3i}^{(\zeta)} = p_i^{(x)}$ and $H_{i3}^{(\zeta)} = \mathcal{U}_i^{(x)}$, we obtain $H_{ik}^{(\zeta)}$ and $\bar{H}_{ki}^{(\zeta)}$ in full. Assume that $H_{iK}^{(\zeta)}(S)$ and $\bar{H}_{Ni}^{(\zeta)}(S)$, with $H_{i3}^{(\zeta)}(S) = \mathcal{U}_i^{(x)}(S)$ and $\bar{H}_{3i}^{(\zeta)}(S) = p_i^{(x)}(S)$, are specified at the initial point S of ray Ω so that $\bar{H}_{mi}^{(\zeta)}(S)H_{in}^{(\zeta)}(S) = \delta_{mn}$. Relation $\bar{H}_{mi}^{(\zeta)}H_{in}^{(\zeta)} = \delta_{mn}$ is then satisfied along the whole ray Ω. Consequently, it is sufficient to solve numerically only one of the two equations of (4.2.86), preferably the first, and to calculate the second transformation matrix as the inverse of the first.

Note that the dynamic ray tracing system in wavefront orthonormal coordinates y_i, discussed in Section 4.2.2, can be obtained from the dynamic ray tracing system in nonorthogonal ray-centered coordinates ζ_i using the transformations

$$h_{im} = \partial y_i / \partial \zeta_m, \qquad \bar{h}_{mi} = \partial \zeta_m / \partial y_i. \tag{4.2.88}$$

Transformation matrices h_{im} and $\bar{h}_{mi}$ can be expressed in terms of $\mathcal{U}_i^{(y)}$ as follows:

$$h_{iM} = \bar{h}_{iM} = \delta_{iM}, \qquad h_{i3} = \mathcal{U}_i^{(y)}, \qquad \bar{h}_{I3} = -\mathcal{C}^{-1}\mathcal{U}_I^{(y)}, \qquad \bar{h}_{33} = \mathcal{C}^{-1}. \tag{4.2.89}$$

They satisfy relations $h_{im}\bar{h}_{mj} = \delta_{ij}$ and $\bar{h}_{mi}h_{in} = \delta_{mn}$. Analogously to (4.2.68) and (4.2.70), we can then use the transformations

$$\begin{aligned} Q_i^{(y)} &= h_{im} Q_m^{(\zeta)}, & P_i^{(y)} &= \bar{h}_{mi}\big(P_m^{(\zeta)} - f_{mk}Q_k^{(\zeta)}\big), \\ Q_m^{(\zeta)} &= \bar{h}_{mi} Q_i^{(y)}, & P_m^{(\zeta)} &= h_{im} P_i^{(y)} + f_{mn}\bar{h}_{ni} Q_i^{(y)}, \end{aligned} \tag{4.2.90}$$

where

$$f_{MN} = 0, \qquad f_{M3} = f_{3M} = -\eta_M^{(y)}, \qquad f_{33} = -\mathcal{U}_i^{(y)}\eta_i^{(y)}. \tag{4.2.91}$$

4. ORTHONOMIC SYSTEM OF RAYS

We shall consider a curvilinear, orthogonal, or nonorthogonal coordinate system ξ_i, and introduce 3×3 matrices $\hat{\mathbf{Q}}^{(\xi)}$ and $\hat{\mathbf{P}}^{(\xi)}$, with elements $Q_{ij}^{(\xi)} = \partial\xi_i / \partial\gamma_j$ and $P_{ij}^{(\xi)} = \partial p_i^{(\xi)} / \partial\gamma_j$, where γ_j are ray coordinates. The dynamic ray tracing system in matrix form, for 3×3 matrices $\hat{\mathbf{Q}}^{(\xi)}$ and $\hat{\mathbf{P}}^{(\xi)}$, then reads

$$d\hat{\mathbf{Q}}^{(\xi)}/dT = \hat{\mathbf{A}}^{(\xi)}\hat{\mathbf{Q}}^{(\xi)} + \hat{\mathbf{B}}^{(\xi)}\hat{\mathbf{P}}^{(\xi)}, \qquad d\hat{\mathbf{P}}^{(\xi)}/dT = -\hat{\mathbf{C}}^{(\xi)}\hat{\mathbf{Q}}^{(\xi)} - \hat{\mathbf{D}}^{(\xi)}\hat{\mathbf{P}}^{(\xi)}. \tag{4.2.92}$$

The elements of $\hat{\mathbf{A}}^{(\xi)}$, $\hat{\mathbf{B}}^{(\xi)}$, $\hat{\mathbf{C}}^{(\xi)}$, and $\hat{\mathbf{D}}^{(\xi)}$ are given either by (4.2.57) or by (4.2.72). As in Cartesian rectangular coordinates, (4.2.92) is equivalent to 18 scalar equations for $Q_{ij}^{(\xi)}$ and $P_{ij}^{(\xi)}$. For orthonomic system of rays, the number of equations can be reduced from 18 to 12 because the third column in each matrix $\hat{\mathbf{Q}}^{(\xi)}$ and $\hat{\mathbf{P}}^{(\xi)}$ represents the ray-tangent solution and can be calculated analytically:

$$Q_{m3}^{(\xi)} = \mathcal{U}_m^{(\xi)} = \partial\mathcal{H}^{(\xi)}/\partial p_m^{(\xi)}, \qquad P_{m3}^{(q)} = \eta_m^{(\xi)} = -\partial\mathcal{H}^{(\xi)}/\partial\xi_m;$$

see (4.2.61). We can introduce the 3×3 matrix $\hat{\mathbf{M}}^{(\xi)}$ of second derivatives of the travel-time field with respect to ξ_i,

$$M_{ij}^{(\xi)} = \partial^2 T/\partial\xi_i\partial\xi_j, \qquad \hat{\mathbf{M}}^{(\xi)} = \hat{\mathbf{P}}^{(\xi)}\hat{\mathbf{Q}}^{(\xi)-1}. \tag{4.2.93}$$

Then the expression for the paraxial travel time at point R', situated in the vicinity of point R, in curvilinear coordinates ξ_i, is

$$T(R') = T(R) + \hat{\boldsymbol{\xi}}^T \hat{\mathbf{p}}^{(\xi)}(R) + \tfrac{1}{2}\hat{\boldsymbol{\xi}}^T \hat{\mathbf{M}}^{(\xi)}(R)\hat{\boldsymbol{\xi}}, \tag{4.2.94}$$

where $\hat{\boldsymbol{\xi}} = (\xi_1(R') - \xi_1(R), \xi_2(R') - \xi_2(R), \xi_3(R') - \xi_3(R))^T$, and $\hat{\mathbf{p}}^{(\xi)}(R) = (p_1^{(\xi)}(R), p_2^{(\xi)}(R), p_3^{(\xi)}(R))^T$. Similarly, we can find the expression for Jacobian $J^{(T)}$, at any point R on Ω, valid in curvilinear coordinates ξ_i:

$$J^{(T)} = \det\hat{\mathbf{H}}^{(\xi)} \det\hat{\mathbf{Q}}^{(\xi)}, \tag{4.2.95}$$

where $\det\hat{\mathbf{H}}^{(\xi)} = \partial(x_1, x_2, x_3)/\partial(\xi_1, \xi_2, \xi_3)$ and $\det\hat{\mathbf{Q}}^{(\xi)} = \partial(\xi_1, \xi_2, \xi_3)/\partial(\gamma_1, \gamma_2, T)$.

As a special case of curvilinear coordinates ξ_i, we shall consider nonorthogonal ray-centered coordinates ζ_i. In these coordinates, all expressions simplify because $P_{3i}^{(\zeta)} = P_{i3}^{(\zeta)} = 0$, $Q_{13}^{(\zeta)} = Q_{31}^{(\zeta)} = 0$, and $Q_{33}^{(\zeta)} = 1$. This yields $M_{i3}^{(\zeta)} = M_{3i}^{(\zeta)} = 0$. Equation (4.2.94) for the paraxial travel time at point R' in nonorthogonal ray-centered coordinates ζ_i then reads

$$T(R') = T(R) + \zeta^T \mathbf{p}^{(\zeta)}(R) + \tfrac{1}{2}\zeta^T \mathbf{M}^{(\zeta)}(R)\zeta, \tag{4.2.96}$$

where $\zeta = (\zeta_1(R'), \zeta_2(R'))^T$ and $\mathbf{p}^{(\zeta)}(R) = (p_1^{(\zeta)}(R), p_2^{(\zeta)}(R))^T$. In isotropic media, $\mathbf{p}^{(\zeta)}(R) = \mathbf{0}$ and (4.2.96) yields (4.1.77). The expression (4.2.95) for Jacobian $J^{(T)}$ also simplifies in nonorthogonal ray-centered coordinates. Bearing in mind that $Q_{33}^{(\zeta)} = 1$, we obtain $\det \hat{\mathbf{Q}}^{(\zeta)} = \det \mathbf{Q}^{(\zeta)}$. We also use $H_{i3}^{(\zeta)} = \mathcal{U}_i^{(x)}$ and obtain $\det(H_{in}^{(\zeta)}) = \mathcal{U} \cos \gamma = \mathcal{C}$, where γ is the acute angle between $\vec{p}$ and $\mathcal{U}$, $\cos \gamma = \mathcal{C}/\mathcal{U}$. This yields

$$J^{(T)} = \det \hat{\mathbf{H}}^{(\zeta)} \det \hat{\mathbf{Q}}^{(\zeta)} = \mathcal{C} \det \mathbf{Q}^{(\zeta)}. \tag{4.2.97}$$

Because $\det \mathbf{Q}^{(\zeta)} = \det \mathbf{Q}^{(y)}$, $J^{(T)}$ in nonorthogonal ray-centered coordinates and in wavefront orthonormal coordinates is given by the same expression; see (4.2.53).

4.3 Propagator Matrices of Dynamic Ray Tracing Systems

The dynamic ray tracing systems derived in Sections 4.1 and 4.2 consist of four or six linear ordinary differential equations of the first order. Because the systems are linear, it is possible and useful to introduce 4×4 and 6×6 propagator matrices for these systems. For simplicity, we shall discuss in greater detail only the 4×4 *propagator matrices* for the dynamic ray tracing systems consisting of four scalar equations. The results derived for 4×4 propagator matrices, however, remain valid also for the 6×6 *propagator matrices*, applicable to dynamic ray tracing systems consisting of six scalar equations. See the brief treatment of the 6×6 propagator matrices in Section 4.3.7. We shall use travel time T as the variable along ray Ω, as in Equations (4.1.65) through (4.1.68) in Section 4.1.7 and in the whole of Section 4.2. The propagator matrices discussed in this section are, of course, also applicable to paraxial ray tracing systems.

We shall now consider the dynamic ray tracing system consisting of four linear ordinary differential equations of the first order in the following general form:

$$d\mathbf{W}/dT = \mathbf{S}\mathbf{W}. \tag{4.3.1}$$

Here $\mathbf{W}$ is a 4×1 column matrix, $\mathbf{S}$ is the 4×4 system matrix of the dynamic ray tracing system, and T is the travel time along ray Ω. (4.3.1) represents several dynamic ray tracing systems derived in the previous sections. The most important is dynamic ray tracing system (4.1.67) for isotropic media, expressed in ray-centered coordinates. The next is dynamic ray tracing system (4.2.31) in wavefront orthonormal coordinates, valid in both isotropic and anisotropic media. Moreover, paraxial ray tracing systems (4.1.58) in ray-centered coordinates and (4.2.34) in wavefront orthonormal coordinates can also be expressed in the form of (4.3.1). Finally, if the 4×1 matrix $\mathbf{W}$ in (4.3.1) is replaced by the 4×2 matrix $\mathbf{X}$ (see (4.1.68)), dynamic ray tracing system (4.3.1) can also be used to compute the 2×2 matrices $\mathbf{Q}$ and $\mathbf{P}$ in ray-centered coordinates in isotropic media (see (4.1.68)), or the 2×2 matrices $\mathbf{Q}^{(y)}$ and $\mathbf{P}^{(y)}$ in wavefront orthonormal coordinates (see (4.2.36)). Hereinafter, we shall assume that all elements of system matrix $\mathbf{S} = \mathbf{S}(T)$ are continuous functions of T along Ω. The points at which the elements of $\mathbf{S}$ are discontinuous will be considered later.

The 4×4 matrix $\mathbf{A}(T)$ is called the *integral matrix* of (4.3.1) if it satisfies the relation

$$\mathrm{d}\mathbf{A}/\mathrm{d}T = \mathbf{SA}. \tag{4.3.2}$$

Thus, each column of integral matrix $\mathbf{A}$ satisfies equation (4.3.1).

We shall consider two special cases of integral matrices of (4.3.1): the fundamental matrix of (4.3.1) and the propagator matrix of (4.3.1). The integral matrix of (4.3.1) is called *fundamental matrix* of (4.3.1) if it is nonsingular for every T in its domain of definition. In other words, the fundamental matrix of (4.3.1) is formed by four linearly independent solutions of (4.3.1). The integral matrix of (4.3.1) is called the *propagator matrix of (4.3.1) from T_0* if it is equal to the 4×4 identity matrix at $T = T_0$.

In this section, we shall derive and discuss in detail the solutions of dynamic ray tracing system (4.3.1). The solutions of systems of linear ordinary differential equations of the first order have been broadly investigated in the mathematical literature; see, for example, Coddington and Levinson (1955) and Kamke (1959). In the seismological literature, considerable attention has been devoted to such systems in connection with the computation of complete seismic wave fields in a 1-D stratified medium; see, for example, Gilbert and Backus (1966) and Ursin (1983). The form of system matrix $\mathbf{S}$ in these seismological applications is, of course, different from our system matrix $\mathbf{S}$, and the number of equations is also different. Even though certain general properties of the solutions of (4.3.1) do not depend on the form of matrix $\mathbf{S}$ and may be adopted directly from the foregoing seismological references, certain other properties are related only to the specific form of $\mathbf{S}$ we use. For this reason, we shall now derive, in a simple and objective way, all the properties we shall need in the following text. Without proof, we shall only present the *uniqueness theorem*: If $S_{ij}(T)$ $(i, j = 1, 2, 3, 4)$ are continuous functions of T, then, for any 4×1 column matrix $\mathbf{W}_0$ and any T_0, there is but one solution $\mathbf{W}(T)$ of (4.3.1) such that $\mathbf{W}(T_0) = \mathbf{W}_0$.

4.3.1 Definition of the Propagator Matrix

We introduce the *propagator matrix from T_0* $\mathbf{\Pi}(T, T_0)$ as the 4×4 integral matrix of (4.3.1)

$$\mathrm{d}\mathbf{\Pi}/\mathrm{d}T = \mathbf{S}\mathbf{\Pi}, \tag{4.3.3}$$

which satisfies the following initial conditions at $T = T_0$,

$$\mathbf{\Pi}(T_0, T_0) = \mathbf{I}, \tag{4.3.4}$$

where $\mathbf{I}$ is the 4×4 identity matrix. Note that we do not *assume* that the propagator matrix $\mathbf{\Pi}(T, T_0)$ is the fundamental matrix of (4.3.1); this property is a consequence of (4.3.4) and will be proved in Section 4.3.3.

We shall now consider ray Ω and two points S and R situated on Ω corresponding to travel times T_0 and T and introduce the *following notation*:

$$\mathbf{\Pi}(R, S) = \begin{pmatrix} \mathbf{Q}_1(R, S) & \mathbf{Q}_2(R, S) \\ \mathbf{P}_1(R, S) & \mathbf{P}_2(R, S) \end{pmatrix}. \tag{4.3.5}$$

Here $\mathbf{Q}_1, \mathbf{Q}_2, \mathbf{P}_1$, and $\mathbf{P}_2$ are 2×2 matrices. They have a very simple physical meaning for the orthonomic system of rays, if the dynamic ray tracing system is considered in the form of (4.1.68) or (4.2.36).

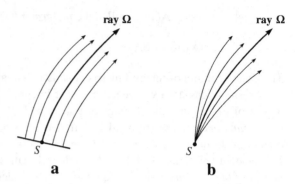

Figure 4.9. Explanation of normalized plane wavefront (a) and normalized point source (b) initial conditions for dynamic ray tracing along ray Ω from S.

- $\mathbf{Q}_1$ and $\mathbf{P}_1$ are solutions of dynamic ray tracing system (4.1.54) for the initial conditions

$$\mathbf{Q}(S) = \mathbf{I}, \qquad \mathbf{P}(S) = \mathbf{0}, \tag{4.3.6}$$

where $\mathbf{I}$ is a 2×2 identity matrix and $\mathbf{0}$ is a 2×2 null matrix; see also (4.2.40). We shall call the initial conditions (4.3.6) "*normalized telescopic point*" initial conditions, or "*normalized plane wavefront*" initial conditions. Matrix $\mathbf{M}$ of the second derivatives of the travel-time field with respect to q_1 and q_2 vanishes at initial point S, $\mathbf{M}(S) = \mathbf{P}(S)\mathbf{Q}^{-1}(S) = \mathbf{0}$. Thus, the wavefront is locally planar at S and the initial slowness vectors are parallel in the vicinity of S. See Figure 4.9(a).

- $\mathbf{Q}_2$ and $\mathbf{P}_2$ are solutions of the dynamic ray tracing system for the initial conditions

$$\mathbf{Q}(S) = \mathbf{0}, \qquad \mathbf{P}(S) = \mathbf{I}; \tag{4.3.7}$$

see also (4.2.39). We shall call these initial conditions the *normalized point-source initial conditions*. Matrix $\mathbf{M}$ of the second derivatives of the travel-time field with respect to q_1 and q_2 is infinite at initial point S because $\mathbf{Q}(S) = \mathbf{0}$. This corresponds to the point-source solution. See Figure 4.9(b).

We shall also use the 4×2 matrices $\mathbf{\Pi}_1$ and $\mathbf{\Pi}_2$ defined in the following way:

$$\mathbf{\Pi}_1(R, S) = \begin{pmatrix} \mathbf{Q}_1(R, S) \\ \mathbf{P}_1(R, S) \end{pmatrix}, \qquad \mathbf{\Pi}_2(R, S) = \begin{pmatrix} \mathbf{Q}_2(R, S) \\ \mathbf{P}_2(R, S) \end{pmatrix}. \tag{4.3.8}$$

Thus, $\mathbf{\Pi}_1$ corresponds to the normalized telescopic initial conditions at S, and $\mathbf{\Pi}_2$ corresponds to the normalized point-source initial conditions at S. A similar notation can also be introduced for other 4×4 dynamic ray tracing systems, particularly for dynamic ray tracing system (4.2.36) in wavefront orthonormal coordinates.

It would also be possible to arrange the columns of the propagator matrix in a different order and to construct two 4×2 solutions different from $\mathbf{\Pi}_1$ and $\mathbf{\Pi}_2$. Such solutions would correspond to two line source solutions. The physical meaning of individual elements of the 4×4 propagator matrix $\mathbf{\Pi}(R, S)$ will also be discussed in Section 4.3.6.

To simplify the terminology, we shall also use the simpler term *ray propagator matrix* instead of the term *propagator matrix of the dynamic ray tracing system*, where this simpler terminology can cause no misunderstanding.

An important note. For fixed points S and R on ray Ω, the 4×4 propagator matrix $\mathbf{\Pi}(R, S)$, corresponding to the dynamic ray tracing system in ray-centered coordinates, does not depend on the integration parameter u used to solve the dynamic ray tracing system along ray Ω from S to R (travel time T, arclength s, and so on).

4.3.2 Symplectic Properties

The 4×4 matrix $\mathbf{A}$ is called symplectic if it satisfies the matrix equation

$$\mathbf{A}^T \mathbf{J} \mathbf{A} = \mathbf{J}, \tag{4.3.9}$$

where $\mathbf{J}$ is the 4×4 matrix

$$\mathbf{J} = \begin{pmatrix} \mathbf{0} & \mathbf{I} \\ -\mathbf{I} & \mathbf{0} \end{pmatrix}. \tag{4.3.10}$$

Here $\mathbf{I}$ is the 2×2 identity matrix and $\mathbf{0}$ is the 2×2 null matrix. We shall now study whether the propagator matrix $\mathbf{\Pi}$ of (4.3.1) is symplectic. Because $d\mathbf{\Pi}/dT = \mathbf{S}\mathbf{\Pi}$, we obtain

$$\begin{aligned} d(\mathbf{\Pi}^T \mathbf{J} \mathbf{\Pi})/dT &= (d\mathbf{\Pi}^T/dT)\mathbf{J}\mathbf{\Pi} + \mathbf{\Pi}^T \mathbf{J}(d\mathbf{\Pi}/dT) \\ &= \mathbf{\Pi}^T \mathbf{S}^T \mathbf{J}\mathbf{\Pi} + \mathbf{\Pi}^T \mathbf{J}\mathbf{S}\mathbf{\Pi} = \mathbf{\Pi}^T (\mathbf{S}^T \mathbf{J} + \mathbf{J}\mathbf{S})\mathbf{\Pi}. \end{aligned}$$

Thus, $\mathbf{\Pi}^T \mathbf{J}\mathbf{\Pi}$ is constant along Ω if $\mathbf{S}^T \mathbf{J} + \mathbf{J}\mathbf{S} = \mathbf{0}$ along Ω. We now introduce the 2×2 minors of the 4×4 system matrix $\mathbf{S}$ as

$$\mathbf{S} = \begin{pmatrix} \mathbf{S}_{11} & \mathbf{S}_{12} \\ -\mathbf{S}_{21} & -\mathbf{S}_{22} \end{pmatrix}. \tag{4.3.11}$$

Condition $\mathbf{S}^T \mathbf{J} + \mathbf{J}\mathbf{S} = \mathbf{0}$ can then be expressed in terms of $\mathbf{S}_{IJ}$ as follows:

$$\mathbf{S}_{12} = \mathbf{S}_{12}^T, \qquad \mathbf{S}_{21} = \mathbf{S}_{21}^T, \qquad \mathbf{S}_{22} = \mathbf{S}_{11}^T. \tag{4.3.12}$$

We now take into account the fact that $\mathbf{\Pi} = \mathbf{I}$ at the initial point of the ray, see (4.3.4), so that $\mathbf{\Pi}^T \mathbf{J}\mathbf{\Pi} = \mathbf{J}$ at that point. This proves that the propagator matrix satisfies the symplecticity relation

$$\mathbf{\Pi}^T \mathbf{J}\mathbf{\Pi} = \mathbf{J} \tag{4.3.13}$$

along the whole ray Ω, if the minors of the system matrix of the dynamic ray tracing system satisfy relations (4.3.12). Relations (4.3.12), however, represent the *symmetry conditions*, satisfied by all dynamic ray tracing systems we have studied, see, for example, (4.2.33) for the wavefront orthonormal coordinates. The conclusion is that the *propagator matrices of the dynamic ray tracing systems are symplectic.*

If we use notation (4.3.5), symplectic relation (4.3.13) leads to the following relations for the 2×2 matrices $\mathbf{P}_1$, $\mathbf{Q}_1$, $\mathbf{P}_2$, and $\mathbf{Q}_2$:

$$\begin{pmatrix} \mathbf{Q}_1^T & \mathbf{P}_1^T \\ \mathbf{Q}_2^T & \mathbf{P}_2^T \end{pmatrix} \begin{pmatrix} \mathbf{0} & \mathbf{I} \\ -\mathbf{I} & \mathbf{0} \end{pmatrix} \begin{pmatrix} \mathbf{Q}_1 & \mathbf{Q}_2 \\ \mathbf{P}_1 & \mathbf{P}_2 \end{pmatrix} = \begin{pmatrix} \mathbf{0} & \mathbf{I} \\ -\mathbf{I} & \mathbf{0} \end{pmatrix}. \tag{4.3.14}$$

Equation (4.3.14) yields four invariants, which remain constant along ray Ω,

$$\begin{aligned} \mathbf{Q}_1^T \mathbf{P}_1 - \mathbf{P}_1^T \mathbf{Q}_1 = \mathbf{0}, \qquad & \mathbf{P}_2^T \mathbf{Q}_1 - \mathbf{Q}_2^T \mathbf{P}_1 = \mathbf{I}, \\ \mathbf{Q}_2^T \mathbf{P}_2 - \mathbf{P}_2^T \mathbf{Q}_2 = \mathbf{0}, \qquad & \mathbf{Q}_1^T \mathbf{P}_2 - \mathbf{P}_1^T \mathbf{Q}_2 = \mathbf{I}. \end{aligned} \tag{4.3.15}$$

These invariants generate many other convenient relations, that are valid along the ray.

Without deriving them, we shall present a few:

$$\mathbf{P}_1\mathbf{Q}_1^{-1} - \left(\mathbf{P}_1\mathbf{Q}_1^{-1}\right)^T = \mathbf{0}, \qquad \mathbf{P}_2\mathbf{Q}_2^{-1} - \left(\mathbf{P}_2\mathbf{Q}_2^{-1}\right)^T = \mathbf{0},$$
$$\mathbf{P}_2\mathbf{Q}_2^{-1} - \mathbf{P}_1\mathbf{Q}_1^{-1} = \mathbf{Q}_2^{-1\,T}\mathbf{Q}_1^{-1} = \mathbf{Q}_1^{-1\,T}\mathbf{Q}_2^{-1},$$
$$\mathbf{P}_2\mathbf{Q}_1^T - \mathbf{P}_1\mathbf{Q}_2^T = \mathbf{I}, \qquad \mathbf{Q}_1\mathbf{P}_2^T - \mathbf{Q}_2\mathbf{P}_1^T = \mathbf{I}, \tag{4.3.16}$$
$$\mathbf{Q}_1\mathbf{Q}_2^T - \mathbf{Q}_2\mathbf{Q}_1^T = \mathbf{0}, \qquad \mathbf{P}_1\mathbf{P}_2^T - \mathbf{P}_2\mathbf{P}_1^T = \mathbf{0},$$

etc.

4.3.3 Determinant of the Propagator Matrix. Liouville's Theorem

We take the determinant of both sides of Equation (4.3.13). Because $\det \mathbf{J} = +1$, $\det(\mathbf{\Pi}^T)\det\mathbf{\Pi} = (\det\mathbf{\Pi})^2 = 1$. This yields

$$\det\mathbf{\Pi}(R, S) = 1. \tag{4.3.17}$$

(The possibility of $\det\mathbf{\Pi}(R, S) = -1$ is excluded by initial conditions (4.3.4).)

Thus, the *ray propagator matrix is nonsingular along the whole ray* Ω, and the ray propagator matrix is also the fundamental matrix of (4.3.1).

Property (4.3.17) also follows from Liouville's theorem, known from the theory of linear ordinary differential equations of the first order (see Kamke 1959; Coddington and Levinson 1955). According to Liouville's theorem, the determinant of any integral matrix $\mathbf{A}(T)$ of (4.3.1), $\det\mathbf{A}(T)$, satisfies the following equations along the ray:

$$\mathrm{d}(\det\mathbf{A}(T))/\mathrm{d}T = \mathrm{tr}\,\mathbf{S}(T)\det\mathbf{A}(T).$$

This equation can be solved to yield

$$\det\mathbf{A}(T) = \det\mathbf{A}(T_0)\exp\left[\int_{T_0}^{T}\mathrm{tr}\,\mathbf{S}(T')\mathrm{d}T'\right].$$

Parameter T' in the integral represents the travel time along ray Ω. As a consequence of (4.3.12), $\mathrm{tr}\,\mathbf{S}(T) = 0$ in the dynamic ray tracing system, and the foregoing equation immediately yields (4.3.17) for propagator matrix $\mathbf{\Pi}(R, S)$.

For a detailed discussion, see Gilbert and Backus (1966). In the following text, we shall refer to (4.3.17) as Liouville's theorem.

Ray propagator matrix $\mathbf{\Pi}(R, S)$ has four eigenvalues. Liouville's theorem (4.3.17) implies that all eigenvalues are different from zero along the whole ray Ω. The symplectic properties of the 4×4 ray propagator matrix $\mathbf{\Pi}$ imply that the inverse $\mathbf{\Pi}^{-1}$ has the same eigenvalues as $\mathbf{\Pi}$. Let us denote by μ_1, μ_2, μ_3, and μ_4 the four eigenvalues of $\mathbf{\Pi}(R, S)$ and sort them according their absolute values,

$$|\mu_1| \geq |\mu_2| \geq |\mu_3| \geq |\mu_4|. \tag{4.3.18}$$

Because $\mathbf{\Pi}$ and $\mathbf{\Pi}^{-1}$ have the same eigenvalues, the four eigenvalues of $\mathbf{\Pi}$, μ_1, μ_2, μ_3, and μ_4, form two reciprocal pairs:

$$\mu_1\mu_4 = 1, \qquad \mu_2\mu_3 = 1. \tag{4.3.19}$$

Note that the eigenvalues of the propagator matrix play an important role in the investigation of the chaotic behavior of rays and in the computation of Lyapunov exponents; see Section 4.10.7.

4.3.4 Chain Rule

In a smooth medium, if $S_{ij}(T)$ are continuous functions of T, the ray propagator matrix satisfies the following chain rule:

$$\mathbf{\Pi}(T, T_0) = \mathbf{\Pi}(T, T_1)\mathbf{\Pi}(T_1, T_0). \tag{4.3.20}$$

Here T_1 corresponds to an arbitrary reference point on ray Ω, not necessarily situated between T_0 and T.

The proof of relation (4.3.20) can be found, for example, in Gilbert and Backus (1966). It follows from the uniqueness theorem. Solutions $\mathbf{\Pi}(T, T_0)$ and $\mathbf{\Pi}(T, T_1)\mathbf{\Pi}(T_1, T_0)$ are both solutions of (4.3.3), and they are equal if $T = T_1$. Indeed,

$$\begin{aligned}
\mathrm{d}[\mathbf{\Pi}(T, T_1)\mathbf{\Pi}(T_1, T_0)]/\mathrm{d}T &- [\mathbf{S}\mathbf{\Pi}(T, T_1)\mathbf{\Pi}(T_1, T_0)] \\
&= [\mathrm{d}\mathbf{\Pi}(T, T_1)/\mathrm{d}T - \mathbf{S}\mathbf{\Pi}(T, T_1)]\mathbf{\Pi}(T_1, T_0) = \mathbf{0}, \\
[\mathbf{\Pi}(T, T_1)\mathbf{\Pi}(T_1, T_0)]_{T=T_1} &= \mathbf{\Pi}(T_1, T_1)\mathbf{\Pi}(T_1, T_0) = \mathbf{\Pi}(T_1, T_0).
\end{aligned}$$

Thus, the uniqueness theorem proves that (4.3.20) is valid for any T.

This equation can be used to connect the propagator matrices calculated independently along different segments of the ray. For example, the equation can be used when ray tracing and dynamic ray tracing are being performed analytically in certain parts of the model and numerically in other parts.

Equation (4.3.20) can be generalized simply. Let us consider points $S_0, S_1, \ldots, S_n$ along ray Ω, corresponding to travel times $T_0, T_1, \ldots, T_n$, respectively. It is not required that $T_{i+1} > T_i$. Then

$$\mathbf{\Pi}(S_n, S_0) = \mathbf{\Pi}(S_n, S_{n-1})\mathbf{\Pi}(S_{n-1}, S_{n-2}) \cdots \mathbf{\Pi}(S_1, S_0) = \prod_{i=n}^{1} \mathbf{\Pi}(S_i, S_{i-1}). \tag{4.3.21}$$

Another consequence of (4.3.20) is related to the inverse of the propagator matrix. Let us consider $T = T_0$ in (4.3.20). Then

$$\mathbf{\Pi}(T_0, T_0) = \mathbf{I} = \mathbf{\Pi}(T_0, T_1)\mathbf{\Pi}(T_1, T_0). \tag{4.3.22}$$

It follows from (4.3.22) that

$$\mathbf{\Pi}^{-1}(T_1, T_0) = \mathbf{\Pi}(T_0, T_1). \tag{4.3.23}$$

4.3.5 Inverse of the Ray Propagator Matrix

We multiply (4.3.13) by $\mathbf{J}^T$ from the LHS and obtain $\mathbf{J}^T\mathbf{\Pi}^T\mathbf{J}\mathbf{\Pi} = \mathbf{I}$; consequently,

$$\mathbf{\Pi}^{-1} = \mathbf{J}^T\mathbf{\Pi}^T\mathbf{J}. \tag{4.3.24}$$

This immediately yields

$$\mathbf{\Pi}^{-1}(R, S) = \begin{pmatrix} \mathbf{P}_2^T(R, S) & -\mathbf{Q}_2^T(R, S) \\ -\mathbf{P}_1^T(R, S) & \mathbf{Q}_1^T(R, S) \end{pmatrix}; \tag{4.3.25}$$

see (4.3.5). Using also relation (4.3.23), we finally obtain

$$\mathbf{\Pi}(S,\,R) = \begin{pmatrix} \mathbf{Q}_1(S,\,R) & \mathbf{Q}_2(S,\,R) \\ \mathbf{P}_1(S,\,R) & \mathbf{P}_2(S,\,R) \end{pmatrix}$$

$$= \mathbf{\Pi}^{-1}(R,\,S) = \begin{pmatrix} \mathbf{P}_2^T(R,\,S) & -\mathbf{Q}_2^T(R,\,S) \\ -\mathbf{P}_1^T(R,\,S) & \mathbf{Q}_1^T(R,\,S) \end{pmatrix}. \qquad (4.3.26)$$

Hence,

$$\begin{aligned} \mathbf{Q}_1(S,\,R) &= \mathbf{P}_2^T(R,\,S), & \mathbf{P}_1(S,\,R) &= -\mathbf{P}_1^T(R,\,S), \\ \mathbf{Q}_2(S,\,R) &= -\mathbf{Q}_2^T(R,\,S), & \mathbf{P}_2(S,\,R) &= \mathbf{Q}_1^T(R,\,S). \end{aligned} \qquad (4.3.27)$$

Equation (4.3.25) is of great importance in various applications because it allows us to combine the boundary conditions for the ray given at different points of the ray. The equation may be used, for example, in boundary-value ray tracing of paraxial rays; see Section 4.9.

Note. Matrices $\mathbf{\Pi}(S,\,R)$ and $\mathbf{\Pi}^{-1}(R,\,S)$ discussed here correspond to the ray-centered coordinate system q_1, q_2, and q_3, introduced for the wave propagating along Ω from S to R. The used ray-centered coordinate system is connected with the right-handed triplet $\vec{e}_1, \vec{e}_2, \vec{e}_3 \equiv \vec{t}$, where $\vec{t}$ is tangent to the ray Ω and is oriented in the direction of propagation from S to R. For the backward propagator matrix $\mathbf{\Pi}^b(S,\,R)$ corresponding to the ray-centered coordinate system introduced for the wave propagating in the backward direction from R to S, see Section 4.4.9.

4.3.6 Solution of the Dynamic Ray Tracing System in Terms of the Propagator Matrix

Let us again consider ray Ω and two points S and R on Ω. Assume that propagator matrix $\mathbf{\Pi}(R,\,S)$ is known. The solution of dynamic ray tracing system (4.3.1) at R can then be expressed for any initial conditions at S in the following form:

$$\mathbf{W}(R) = \mathbf{\Pi}(R,\,S)\mathbf{W}(S). \qquad (4.3.28)$$

The 4×1 column matrices $\mathbf{W}(R)$ and $\mathbf{W}(S)$ have an obvious meaning. For example, if we consider the dynamic ray tracing system (4.1.67) in ray-centered coordinates, $\mathbf{W}$ is given by (4.1.66). For paraxial ray tracing system (4.1.58) in ray-centered coordinates, $\mathbf{W}$ is given by (4.1.57).

A similar relation can be obtained for any 4×2 matrix solution. Let us consider the 4×2 matrix $\mathbf{X}$ and the dynamic ray tracing system (4.1.68). Then,

$$\mathbf{X}(R) = \mathbf{\Pi}(R,\,S)\mathbf{X}(S). \qquad (4.3.29)$$

Equations (4.3.28) and (4.3.29) are very powerful and play a fundamental role in dynamic ray tracing. When the ray propagator matrix is known, we can find the solution of the dynamic ray tracing system analytically for any initial conditions, without repeating the dynamic ray tracing.

Moreover, we also obtain

$$\mathbf{W}(S) = \mathbf{\Pi}^{-1}(R,\,S)\mathbf{W}(R), \qquad \mathbf{X}(S) = \mathbf{\Pi}^{-1}(R,\,S)\mathbf{X}(R). \qquad (4.3.30)$$

The elements of matrix $\mathbf{\Pi}^{-1}(R,\,S)$ can be simply calculated from the known matrix $\mathbf{\Pi}(R,\,S)$ using (4.3.26).

Equation (4.3.28) offers a simple physical explanation of the elements of the 4×4 ray propagator matrix $\Pi_{\alpha\beta}(R, S)$ ($\alpha, \beta = 1, 2, 3, 4$): $\Pi_{\alpha\beta}(R, S) = \partial W_\alpha(R)/\partial W_\beta(S)$.

4.3.7 6×6 Propagator Matrices

All properties of the 4×4 propagator matrices, derived in Sections 4.3.2 through 4.3.6, also apply to the 6×6 propagator matrices of dynamic ray tracing systems consisting of six equations. Only the 2×2 minors of the 4×4 propagator matrices should be replaced by analogous 3×3 minors:

$$\Pi(R, S) = \begin{pmatrix} \hat{\Pi}_{11}(R, S) & \hat{\Pi}_{12}(R, S) \\ \hat{\Pi}_{21}(R, S) & \hat{\Pi}_{22}(R, S) \end{pmatrix}. \tag{4.3.31}$$

Consequently, the 2×2 minors $\mathbf{Q}_1$, $\mathbf{Q}_2$, $\mathbf{P}_1$, and $\mathbf{P}_2$, introduced by (4.3.5), should be replaced by the 3×3 minors $\hat{\Pi}_{11}$, $\hat{\Pi}_{12}$, $\hat{\Pi}_{21}$, and $\hat{\Pi}_{22}$ in all equations of the previous sections. The 6×6 propagator matrices of all dynamic ray tracing systems we have studied are symplectic because the relevant dynamic ray tracing systems satisfy symmetry relations (4.3.12). In other words, symplectic relation $\Pi^T \mathbf{J} \Pi = \mathbf{J}$ is satisfied along the whole ray Ω, where $\mathbf{J}$ is the 6×6 matrix given by (4.3.10), $\mathbf{I}$ is the 3×3 identity matrix, and $\mathbf{0}$ the 3×3 null matrix. The chain rule $\Pi(R, S) = \Pi(R, S_1)\Pi(S_1, S)$ and Liouville's theorem $\det \Pi(R, S) = 1$ are also satisfied, and the inverse of the 6×6 propagator matrix is given by the relations shown in Section 4.3.5. Finally, the equations analogous to (4.3.28) through (4.3.30) also remain valid.

We shall now briefly discuss several important special cases of the 6×6 propagator matrices. In all cases, Hamiltonians corresponding to the integration parameter T along Ω are considered in this section.

1. CARTESIAN RECTANGULAR COORDINATES x_i

We shall denote the 6×6 propagator matrix, corresponding to dynamic ray tracing system (4.2.4) in Cartesian rectangular coordinates, by $\Pi^{(x)}(R, S)$. The computation of all elements of the 6×6 propagator matrix requires the system of six equations to be solved numerically six times. In other words, we must solve numerically 36 equations. After all elements of $\Pi^{(x)}(R, S)$ are known, the solution of dynamic ray tracing system (4.2.4) at any point R on Ω can be expressed analytically for arbitrary initial conditions given at S. A similar conclusion is also valid for the solutions of paraxial ray tracing system (4.2.11):

$$\begin{pmatrix} \hat{\mathbf{Q}}^{(x)}(R) \\ \hat{\mathbf{P}}^{(x)}(R) \end{pmatrix} = \Pi^{(x)}(R, S) \begin{pmatrix} \hat{\mathbf{Q}}^{(x)}(S) \\ \hat{\mathbf{P}}^{(x)}(S) \end{pmatrix},$$

$$\begin{pmatrix} \delta\hat{\mathbf{x}}(R) \\ \delta\hat{\mathbf{p}}^{(x)}(R) \end{pmatrix} = \Pi^{(x)}(R, S) \begin{pmatrix} \delta\hat{\mathbf{x}}(S) \\ \delta\hat{\mathbf{p}}^{(x)}(S) \end{pmatrix}. \tag{4.3.32}$$

There is, however, one important difference between the 4×4 and 6×6 propagator matrices. If we are considering eikonal equation $\mathcal{H}^{(x)}(x_i, p_i^{(x)}) = 0$, the 6×6 propagator matrix $\Pi^{(x)}(R, S)$ also contains the solution that does not satisfy the eikonal equation (noneikonal solution). Constraint relation (4.2.9) or (4.2.12) should be applied at initial point S to eliminate this solution. After $\hat{\mathbf{Q}}^{(x)}(S)$ and $\hat{\mathbf{P}}^{(x)}(S)$ are chosen to satisfy constraint relation (4.2.9) at initial point S, constraint relation (4.2.9) is satisfied along the whole ray Ω. Similarly, if $\delta\hat{\mathbf{x}}(S)$ and $\delta\hat{\mathbf{p}}^{(x)}(S)$ satisfy constraint relation (4.2.12) at initial point S, the constraint relation is also satisfied by $\delta\hat{\mathbf{x}}(R)$ and $\delta\hat{\mathbf{p}}^{(x)}(R)$.

Note that one of the columns of the 6×6 propagator matrix $\mathbf{\Pi}^{(x)}(R, S)$ can be expressed analytically, if one of the coordinate axes of the Cartesian coordinate system is chosen in the direction of the slowness vector at the initial point S. For example, we can take the x_3-axis along $\vec{p}(S)$. Then $p_1^{(x)}(S) = p_2^{(x)}(S) = 0$ and $p_3^{(x)}(S) = 1/\mathcal{C}(S)$, where $\mathcal{C}(S)$ is the phase velocity along the x_3-axis at S. The sixth column of the 6×6 propagator matrix $\Pi_{\alpha\beta}^{(x)}(R, S)$ represents the noneikonal solution (4.2.16) and may be expressed analytically as

$$\Pi_{\alpha 6}^{(x)} = \mathcal{C}(S)(T - T_0)\mathcal{U}_\alpha^{(x)} \qquad \text{for } 1 \le \alpha \le 3,$$
$$\Pi_{\alpha 6}^{(x)} = \mathcal{C}(S)\big[p_{\alpha-3}^{(x)} + (T - T_0)\eta_{\alpha-3}^{(x)}\big] \qquad \text{for } 4 \le \alpha \le 6.$$

Indeed, these expressions yield $\Pi_{66} = 1$ and $\Pi_{\alpha 6} = 0$ for $\alpha \ne 6$ at the initial point $T = T_0$.

Equations (4.3.32) can also be used to find $\hat{\mathbf{Q}}^{(x)}(S)$ and $\hat{\mathbf{P}}^{(x)}(S)$ from known $\hat{\mathbf{Q}}^{(x)}(R)$ and $\hat{\mathbf{P}}^{(x)}(R)$, analogously to (4.3.30).

Finally, Equations (4.3.32) offer a simple physical explanation of the individual elements of $\mathbf{\Pi}^{(x)}(R, S)$. Denote $\mathbf{W} = (Q_1^{(x)}, Q_2^{(x)}, Q_3^{(x)}, P_1^{(x)}, P_2^{(x)}, P_3^{(x)})^T$. Then $\Pi_{\alpha\beta}^{(x)}$ (for $\alpha = 1, 2, \ldots, 6, \beta = 1, 2, \ldots, 6$) are given by relations

$$\Pi_{\alpha\beta}^{(x)} = \partial W_\alpha(R)/\partial W_\beta(S). \tag{4.3.33}$$

2. CURVILINEAR COORDINATES ξ_i

In the same way as for Cartesian rectangular coordinates, we can construct the 6×6 propagator matrix $\mathbf{\Pi}^{(\xi)}(R, S)$ for dynamic ray tracing system (4.2.56) and for paraxial ray tracing system (4.2.60), consisting of six equations. The continuation equations in curvilinear coordinates ξ_i are fully analogous to equations in Cartesian coordinates:

$$\begin{pmatrix} \hat{\mathbf{Q}}^{(\xi)}(R) \\ \hat{\mathbf{P}}^{(\xi)}(R) \end{pmatrix} = \mathbf{\Pi}^{(\xi)}(R, S) \begin{pmatrix} \hat{\mathbf{Q}}^{(\xi)}(S) \\ \hat{\mathbf{P}}^{(\xi)}(S) \end{pmatrix},$$
$$\begin{pmatrix} \delta\hat{\boldsymbol{\xi}}(R) \\ \delta\hat{\mathbf{p}}^{(\xi)}(R) \end{pmatrix} = \mathbf{\Pi}^{(\xi)}(R, S) \begin{pmatrix} \delta\hat{\boldsymbol{\xi}}(S) \\ \delta\hat{\mathbf{p}}^{(\xi)}(S) \end{pmatrix}. \tag{4.3.34}$$

Again, the initial conditions at point S must satisfy constraint relations (4.2.59). After the constraint relation is satisfied at S, it is satisfied along the whole ray Ω. The 6×6 propagator matrix $\mathbf{\Pi}^{(\xi)}(R, S)$ is again symplectic because the system matrix of the dynamic ray tracing system satisfies symmetry relations (4.2.58). It also satisfies Liouville's theorem $\det \mathbf{\Pi}^{(\xi)}(R, S) = 1$, chain rule (4.3.20), and the relations for its inverse (4.3.26).

3. TRANSFORMATIONS OF PROPAGATOR MATRICES

We shall consider the 6×6 propagator matrix $\mathbf{\Pi}^{(\xi)}(R, S)$ in curvilinear coordinates ξ_i, and the 6×6 propagator matrix $\mathbf{\Pi}^{(x)}(R, S)$. However, we shall assume that x_i may represent curvilinear coordinates. The transformation relations between $Q_m^{(\xi)}$, $P_m^{(\xi)}$ and $Q_k^{(x)}$, $P_k^{(x)}$ are given by (4.2.68). These relations can be used to find the equations connecting $\mathbf{\Pi}^{(\xi)}(R, S)$ with $\mathbf{\Pi}^{(x)}(R, S)$. In matrix form, we can express (4.2.68) and (4.2.70) as

$$\hat{\mathbf{Q}}^{(\xi)} = \hat{\bar{\mathbf{H}}}^{(\xi)}\hat{\mathbf{Q}}^{(x)}, \qquad \hat{\mathbf{P}}^{(\xi)} = \hat{\mathbf{H}}^{(\xi)T}\hat{\mathbf{P}}^{(x)} + \hat{\mathbf{F}}\hat{\bar{\mathbf{H}}}^{(\xi)}\hat{\mathbf{Q}}^{(x)},$$
$$\hat{\mathbf{Q}}^{(x)} = \hat{\mathbf{H}}^{(\xi)}\hat{\mathbf{Q}}^{(\xi)}, \qquad \hat{\mathbf{P}}^{(x)} = \hat{\bar{\mathbf{H}}}^{(\xi)T}\hat{\mathbf{P}}^{(\xi)} - \hat{\mathbf{H}}^{(\xi)T}\hat{\mathbf{F}}\hat{\mathbf{Q}}^{(\xi)}. \tag{4.3.35}$$

Here $\hat{\mathbf{Q}}^{(\xi)}$, $\hat{\mathbf{P}}^{(\xi)}$, $\hat{\mathbf{Q}}^{(x)}$, and $\hat{\mathbf{P}}^{(x)}$ represent 3×1 column matrices with components $Q_i^{(\xi)}$, $P_i^{(\xi)}$, $Q_i^{(x)}$, and $P_i^{(x)}$, $i = 1, 2, 3$. The elements of the 3×3 matrix $\hat{\mathbf{H}}^{(\xi)}$ are given by

$H_{im}^{(\xi)} = \partial x_i / \partial \xi_m$, and the elements of the 3 × 3 matrix $\hat{\bar{\mathbf{H}}}^{(\xi)}$ are given by $\bar{H}_{im}^{(\xi)} = \partial \xi_i / \partial x_m$. We also have $\hat{\mathbf{H}}^{(\xi)} \hat{\bar{\mathbf{H}}}^{(\xi)} = \mathbf{I}$. Finally, the elements of the 3 × 3 matrix $\hat{\mathbf{F}}$ are given by (4.2.69). Equations (4.3.35) can then be expressed as

$$
\begin{pmatrix} \hat{\mathbf{Q}}^{(x)} \\ \hat{\mathbf{P}}^{(x)} \end{pmatrix} = \begin{pmatrix} \hat{\mathbf{H}}^{(\xi)} & \hat{\mathbf{0}} \\ -\hat{\mathbf{H}}^{(\xi)T} \hat{\mathbf{F}} & \hat{\mathbf{H}}^{(\xi)T} \end{pmatrix} \begin{pmatrix} \hat{\mathbf{Q}}^{(\xi)} \\ \hat{\mathbf{P}}^{(\xi)} \end{pmatrix},
$$

$$
\begin{pmatrix} \hat{\mathbf{Q}}^{(\xi)} \\ \hat{\mathbf{P}}^{(\xi)} \end{pmatrix} = \begin{pmatrix} \hat{\bar{\mathbf{H}}}^{(\xi)} & \hat{\mathbf{0}} \\ \hat{\mathbf{F}} \hat{\bar{\mathbf{H}}}^{(\xi)} & \hat{\mathbf{H}}^{(\xi)T} \end{pmatrix} \begin{pmatrix} \hat{\mathbf{Q}}^{(x)} \\ \hat{\mathbf{P}}^{(x)} \end{pmatrix}.
$$

(4.3.36)

Equations (4.3.32) immediately yield

$$
\begin{pmatrix} \hat{\mathbf{Q}}^{(\xi)}(R) \\ \hat{\mathbf{P}}^{(\xi)}(R) \end{pmatrix} = \begin{pmatrix} \hat{\bar{\mathbf{H}}}^{(\xi)}(R) & \hat{\mathbf{0}} \\ \hat{\mathbf{F}}(R) \hat{\bar{\mathbf{H}}}^{(\xi)}(R) & \hat{\mathbf{H}}^{(\xi)T}(R) \end{pmatrix} \mathbf{\Pi}^{(x)}(R, S)
$$

$$
\times \begin{pmatrix} \hat{\mathbf{H}}^{(\xi)}(S) & \hat{\mathbf{0}} \\ -\hat{\mathbf{H}}^{(\xi)T}(S) \hat{\mathbf{F}}(S) & \hat{\mathbf{H}}^{(\xi)T}(S) \end{pmatrix} \begin{pmatrix} \hat{\mathbf{Q}}^{(\xi)}(S) \\ \hat{\mathbf{P}}^{(\xi)}(S) \end{pmatrix}.
$$

(4.3.37)

Consequently, the 6 × 6 propagator matrix $\mathbf{\Pi}^{(\xi)}(R, S)$ can be expressed in terms of the 6 × 6 propagator matrix $\mathbf{\Pi}^{(x)}(R, S)$ as

$$
\mathbf{\Pi}^{(\xi)}(R, S) = \begin{pmatrix} \hat{\bar{\mathbf{H}}}^{(\xi)}(R) & \hat{\mathbf{0}} \\ \hat{\mathbf{F}}(R) \hat{\bar{\mathbf{H}}}^{(\xi)}(R) & \hat{\mathbf{H}}^{(\xi)T}(R) \end{pmatrix}
$$

$$
\times \mathbf{\Pi}^{(x)}(R, S) \begin{pmatrix} \hat{\mathbf{H}}^{(\xi)}(S) & \hat{\mathbf{0}} \\ -\hat{\mathbf{H}}^{(\xi)T}(S) \hat{\mathbf{F}}(S) & \hat{\mathbf{H}}^{(\xi)T}(S) \end{pmatrix}.
$$

(4.3.38)

It is easy to check that (4.3.38) yields $\mathbf{\Pi}^{(\xi)}(S, S) = \mathbf{I}$.

Equation (4.3.38) has certain important consequences. Assume that we wish to compute $\mathbf{\Pi}^{(\xi)}(R, S)$. In this case, we can map coordinates ξ_i onto Cartesian coordinates and compute $\mathbf{\Pi}^{(x)}(R, S)$ in Cartesian coordinates. Propagator matrix $\mathbf{\Pi}^{(\xi)}(R, S)$ can then be obtained from $\mathbf{\Pi}^{(x)}(R, S)$ by two simple matrix multiplications (4.3.38) at the end points S and R of ray Ω. Note that the transformation matrices in (4.3.38) simplify considerably if both x_i and ξ_i coordinates are Cartesian; then $\hat{\mathbf{F}}(R) = \hat{\mathbf{0}}$ and $\hat{\mathbf{F}}(S) = \hat{\mathbf{0}}$.

4. NONORTHOGONAL RAY-CENTERED COORDINATES ζ_i

A very simple 6 × 6 propagator matrix $\mathbf{\Pi}^{(\zeta)}(R, S)$ is obtained in nonorthogonal ray-centered coordinates ζ_i, for dynamic ray tracing system (4.2.77). Let us denote the 36 elements of $\mathbf{\Pi}^{(\zeta)}(R, S)$ by $\Pi_{\alpha\beta}^{(\zeta)}(R, S)$, with $\alpha = 1, 2, \ldots, 6$ and $\beta = 1, 2, \ldots, 6$. Then, $\Pi_{3\beta}^{(\zeta)}(R, S) = \Pi_{\alpha3}^{(\zeta)}(R, S) = \Pi_{\alpha6}^{(\zeta)}(R, S) = \Pi_{6\beta}^{(\zeta)}(R, S) = 0$, with the exception of three elements $\Pi_{33}^{(\zeta)}(R, S)$, $\Pi_{36}^{(\zeta)}(R, S)$, and $\Pi_{66}^{(\zeta)}(R, S)$:

$$
\Pi_{33}^{(\zeta)}(R, S) = \Pi_{66}^{(\zeta)}(R, S) = 1, \qquad \Pi_{36}^{(\zeta)}(R, S) = T(R, S) = T(R) - T(S).
$$

(4.3.39)

See (4.2.84) and (4.2.85). Actually, only 16 elements of propagator matrix $\mathbf{\Pi}^{(\zeta)}(R, S)$ should be computed numerically using the dynamic ray tracing system (4.2.81), solving it for normalized plane-wavefront initial conditions and for normalized point-source initial conditions. Three other elements are given by (4.3.39), and the remaining 17 elements are zero. After the propagator matrix $\mathbf{\Pi}^{(\zeta)}(R, S)$ is known along Ω, we can again use the equations analogous to (4.3.32). The initial conditions must satisfy constraint relation $P_3^{(\zeta)}(S) = 0$; see (4.2.80). Then $P_3^{(\zeta)}(R) = 0$ for all points R along Ω. Thus, the noneikonal solutions are eliminated very simply in nonorthogonal ray-centered coordinates.

5. WAVEFRONT ORTHONORMAL COORDINATES y_i

In wavefront orthonormal coordinates y_i, the situation is similar to the nonorthogonal ray-centered coordinates. Only 16 elements of propagator matrix $\mathbf{\Pi}^{(y)}(R, S)$ should be computed numerically using dynamic ray tracing system (4.2.31). The system should be solved for normalized plane-wavefront initial conditions (4.2.40) and for normalized point-source initial conditions (4.2.39). The remaining elements and linearly independent solutions, however, are not as simple as in general ray-centered coordinates. Ray-tangent solution (4.2.24) cannot be chosen to render only its third element nonvanishing at initial point S. It would be necessary to construct a new linearly independent solution, combining the ray-tangent solution with other solutions. The relevant equations, however, are not given here because we shall not need them at all in this book. All equations we wish to derive in wavefront orthonormal coordinates y_i can be obtained directly from the 4×4 propagator matrix of dynamic ray tracing system (4.2.31). The construction and application of the 6×6 propagator matrix $\mathbf{\Pi}^{(y)}(R, S)$ would only represent an alternative approach to that presented here.

4.3.8 Inhomogeneous Dynamic Ray Tracing System

Inhomogeneous dynamic and paraxial ray tracing systems have been successfully used in certain applications, particularly in the ray perturbation method. For simplicity, we shall only discuss the inhomogeneous dynamic ray tracing system in Cartesian coordinates, consisting of six linear ordinary differential equations of the first order. The corresponding homogeneous dynamic ray tracing system is given by (4.2.4). In matrix form, the inhomogeneous dynamic ray tracing system reads

$$\frac{\mathrm{d}}{\mathrm{d}T}\begin{pmatrix}\hat{\mathbf{Q}}^{(x)} \\ \hat{\mathbf{P}}^{(x)}\end{pmatrix} = \begin{pmatrix}\hat{\mathbf{A}}^{(x)} & \hat{\mathbf{B}}^{(x)} \\ -\hat{\mathbf{C}}^{(x)} & -\hat{\mathbf{D}}^{(x)}\end{pmatrix}\begin{pmatrix}\hat{\mathbf{Q}}^{(x)} \\ \hat{\mathbf{P}}^{(x)}\end{pmatrix} + \begin{pmatrix}\hat{\mathbf{E}}^{(x)} \\ \hat{\mathbf{F}}^{(x)}\end{pmatrix}. \qquad (4.3.40)$$

Here $\hat{\mathbf{A}}^{(x)}$, $\hat{\mathbf{B}}^{(x)}$, $\hat{\mathbf{C}}^{(x)}$, and $\hat{\mathbf{D}}^{(x)}$ are 3×3 matrices with elements given by (4.2.5), and $\hat{\mathbf{Q}}^{(x)}$, $\hat{\mathbf{P}}^{(x)}$, $\hat{\mathbf{E}}^{(x)}$, and $\hat{\mathbf{F}}^{(x)}$ are 3×1 column matrices. For $\hat{\mathbf{E}}^{(x)} = \hat{\mathbf{0}}$ and $\hat{\mathbf{F}}^{(x)} = \hat{\mathbf{0}}$, inhomogeneous dynamic ray tracing system (4.3.40) reduces to the standard dynamic ray tracing system (4.2.4).

We denote by $\mathbf{\Pi}^{(x)}(T, T_0)$ the 6×6 propagator matrix of homogeneous dynamic ray tracing system (4.2.4), expressed as a function of travel time T along Ω, with $\mathbf{\Pi}^{(x)}(T_0, T_0) = \mathbf{I}$. The solution of inhomogeneous dynamic ray tracing system (4.3.40) can then be expressed in terms of propagator matrix $\mathbf{\Pi}^{(x)}(T, T_0)$ as follows:

$$\begin{pmatrix}\hat{\mathbf{Q}}^{(x)}(T) \\ \hat{\mathbf{P}}^{(x)}(T)\end{pmatrix} = \mathbf{\Pi}^{(x)}(T, T_0)\begin{pmatrix}\hat{\mathbf{Q}}^{(x)}(T_0) \\ \hat{\mathbf{P}}^{(x)}(T_0)\end{pmatrix} + \int_{T_0}^{T}\mathbf{\Pi}^{(x)}(T, T')\begin{pmatrix}\hat{\mathbf{E}}^{(x)}(T') \\ \hat{\mathbf{F}}^{(x)}(T')\end{pmatrix}\mathrm{d}T'.$$
$$(4.3.41)$$

Here the integral is taken along central ray Ω and integration variable T' represents the travel time. See Gilbert and Backus (1966) for more details. Solution (4.3.41) can be simply verified by direct inspection. Taking the derivatives of both of its sides with respect to T, (4.3.40) is obtained.

Thus, once the propagator matrix of the homogeneous dynamic ray tracing system is known, the solution of the relevant inhomogeneous dynamic ray tracing system can be obtained by quadratures along central ray Ω.

In an analogous way, it is possible to find the solutions of any 4×4 and 6×6 inhomogeneous dynamic ray tracing system, expressed in any coordinate system.

4.4 Dynamic Ray Tracing in Isotropic Layered Media

This section is devoted to dynamic ray tracing in ray-centered coordinates in isotropic layered media containing structural interfaces. We shall consider an orthonomic system of rays and the dynamic ray tracing system (4.1.64) or (4.1.65), consisting of four equations. First, we shall derive an equation for the travel-time distribution along any curved interface Σ crossing the central ray, in the vicinity of the point of incidence. Surface Σ may represent a structural interface. We then apply the *phase matching method*, which requires that the phase functions (travel times) of incident, reflected, and transmitted waves be equal along Σ; see Section 2.4.5. This will be sufficient to recompute the matrix $\mathbf{M}$ of the second derivatives of the travel-time field across interface Σ. After this, we shall take into account the continuity of rays across interface Σ and obtain equations for the transformation of matrices $\mathbf{Q}$ and $\mathbf{P}$ and for the transformation of the ray propagator matrix $\mathbf{\Pi}$ across Σ. For anisotropic media, see Section 4.14, and for dynamic ray tracing systems consisting of six equations see Section 4.7.

As a by-product of the investigation of dynamic ray tracing in a layered medium, we shall also obtain a special version of the dynamic ray tracing, which we shall call *surface-to-surface dynamic ray tracing*. In this version, the initial points of the central ray and of all paraxial rays are situated along some *anterior surface*, and the termination points of all these rays are situated along a *posterior surface*. The ray propagator matrices for surface-to-surface dynamic ray tracing can be simply obtained from the general ray propagator matrix $\mathbf{\Pi}$ merely by rearranging the terms. See Section 4.4.7.

4.4.1 Geometry of the Interface

Let us assume that surface Σ crossing the central ray Ω is described by equation

$$\Sigma(x_i) = 0. \tag{4.4.1}$$

Let ray Ω of the incident wave strike surface Σ at point Q. We assume that surface Σ is smooth at Q. More specifically, we assume that at least Σ, $\partial\Sigma/\partial x_i$, and $\partial^2\Sigma/\partial x_i\partial x_j$ $(i, j = 1, 2, 3)$ are continuous at Q.

Surface Σ may also be a structural interface. It is then very useful to introduce the reflection/transmission point $\tilde{Q}$. The point of incidence Q and the point of reflection/transmission $\tilde{Q}$ coincide, but point Q corresponds to the incident wave, and point $\tilde{Q}$ corresponds to the reflected/transmitted wave; see Figure 4.10. As we know, many quantities computed by ray tracing and dynamic ray tracing are discontinuous across Σ, for example $\vec{p}(Q) \neq \vec{p}(\tilde{Q})$, $\mathbf{M}(Q) \neq \mathbf{M}(\tilde{Q})$, $\mathbf{Q}(Q) \neq \mathbf{Q}(\tilde{Q})$, and $\mathbf{P}(Q) \neq \mathbf{P}(\tilde{Q})$. To simplify the notation, we shall also use the following convention. If arguments Q and $\tilde{Q}$ are omitted, we understand that $\vec{p}, \mathbf{M}, \mathbf{P}, \mathbf{Q}$, and so on correspond to point Q (incident wave), and that $\tilde{\vec{p}}, \tilde{\mathbf{M}}, \tilde{\mathbf{P}}, \tilde{\mathbf{Q}}$, and so on correspond to point $\tilde{Q}$ (reflected/transmitted wave).

To discuss the problem of the dynamic ray tracing across the interface, it is, in principle, not necessary to introduce a local Cartesian coordinate system on interface Σ at point Q;

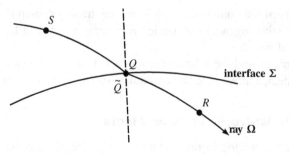

Figure 4.10. Definition of the point of incidence Q and of reflection/transmission point $\tilde{Q}$ on interface Σ. Both points coincide, but point Q corresponds to the incident wave, and point $\tilde{Q}$ corresponds to the selected reflected/transmitted wave.

it is merely necessary to know the derivatives $\partial \Sigma / \partial x_i$ and $\partial^2 \Sigma / \partial x_i \partial x_j$ at Q. The first derivatives $(\partial \Sigma / \partial x_i)_Q$ are related to the normal to Σ at Q. We remind the reader that the problem of reflection/transmission of HF elastic waves at a curved interface was also solved in Section 2.4.5 without introducing the local Cartesian coordinate system connected with interface Σ at Q.

Nevertheless, in certain applications, such as in the surface-to-surface dynamic ray tracing, it will be useful to consider local Cartesian coordinate systems connected with interfaces.

We introduce a local Cartesian coordinate system (z_1, z_2, z_3) with its origin at Q and basis vectors $\vec{i}_1^{(z)}, \vec{i}_2^{(z)}$, and $\vec{i}_3^{(z)}$. We specify unit vector $\vec{i}_3^{(z)}$ to coincide with the unit vector normal to interface Σ at Q, $\vec{n}(Q)$,

$$\vec{i}_3^{(z)} = \vec{n}(Q). \tag{4.4.2}$$

The two other basis vectors $\vec{i}_1^{(z)}$ and $\vec{i}_2^{(z)}$ are obviously situated in the plane tangent to Σ at Q. It is possible to specify them arbitrarily in the tangent plane. We only require triplet $\vec{i}_1^{(z)}, \vec{i}_2^{(z)}, \vec{i}_3^{(z)}$ to be mutually orthogonal and right-handed.

Unit vector $\vec{n}(Q)$ *normal to* Σ *at* Q is given by relation

$$\vec{n}(Q) = \frac{\epsilon^* \nabla \Sigma}{(\nabla \Sigma \cdot \nabla \Sigma)^{1/2}}. \tag{4.4.3}$$

Here ϵ^* equals either $+1$ or -1; the choice is optional. In the general Cartesian coordinate system, Equation (4.4.3) yields the following relations for the Cartesian components of unit normal $\vec{n}(Q)$,

$$n_k(Q) = \epsilon^* \frac{\partial \Sigma}{\partial x_k} \Big/ \left(\frac{\partial \Sigma}{\partial x_j} \frac{\partial \Sigma}{\partial x_j} \right)^{1/2}. \tag{4.4.4}$$

If surface Σ is specified by equation

$$x_3 = f(x_1, x_2), \tag{4.4.5}$$

we can put $\Sigma(x_i) = x_3 - f(x_1, x_2) = 0$ and obtain

$$\partial \Sigma / \partial x_1 = - \partial f / \partial x_1, \qquad \partial \Sigma / \partial x_2 = - \partial f / \partial x_2, \qquad \partial \Sigma / \partial x_3 = 1. \tag{4.4.6}$$

Inserting these expressions into (4.4.4) yields $n_k(Q)$.

We denote the slowness vector of the incident wave at Q by $\vec{p}(Q)$. We call the plane specified by $\vec{p}(Q)$ and $\vec{n}(Q)$ the *plane of incidence*. If $\vec{p}(Q)$ is parallel to $\vec{n}(Q)$ (normal incidence), the plane of incidence is not defined, but it may be chosen as an arbitrary plane containing $\vec{n}(Q)$.

In this way, the z_3-axis of the local Cartesian coordinate system connected with Σ at Q is strictly specified. We will specify the local Cartesian coordinate system z_1, z_2, z_3 completely by the 3×3 *transformation matrix* $\hat{\mathbf{Z}}$ from the local Cartesian coordinate system z_i to the general Cartesian coordinate system x_i,

$$\mathrm{d}x_k = Z_{kl}\mathrm{d}z_l \quad \text{or} \quad \mathrm{d}\hat{\mathbf{x}} = \hat{\mathbf{Z}}\mathrm{d}\hat{\mathbf{z}}, \tag{4.4.7}$$

where $\mathrm{d}\hat{\mathbf{z}} \equiv (\mathrm{d}z_1, \mathrm{d}z_2, \mathrm{d}z_3)^T$. Matrix $\hat{\mathbf{Z}}$ is orthonormal, so that $\hat{\mathbf{Z}}^{-1} = \hat{\mathbf{Z}}^T$, and $\det \hat{\mathbf{Z}} = 1$. Consequently,

$$\mathrm{d}z_k = Z_{lk}\mathrm{d}x_l \quad \text{or} \quad \mathrm{d}\hat{\mathbf{z}} = \hat{\mathbf{Z}}^T \mathrm{d}\hat{\mathbf{x}}. \tag{4.4.8}$$

The elements of transformation matrix $\hat{\mathbf{Z}}$ are given by any of the following relations:

$$Z_{kl} = \vec{i}_k \cdot \vec{i}_l^{(z)} = \partial x_k / \partial z_l = \partial z_l / \partial x_k. \tag{4.4.9}$$

The first column of matrix $\hat{\mathbf{Z}}$ is formed by the general Cartesian components of basis vector $\vec{i}_1^{(z)}$, the second column corresponds to $\vec{i}_2^{(z)}$, and the third column corresponds to $\vec{i}_3^{(z)} \equiv \vec{n}$. Thus, unit basis vectors $\vec{i}_1^{(z)}, \vec{i}_2^{(z)}$, and $\vec{i}_3^{(z)}$ are fully specified by matrix $\hat{\mathbf{Z}}$.

In addition to matrix $\hat{\mathbf{Z}}$, we also introduce the 3×3 *transformation matrix* $\hat{\mathbf{G}}(Q)$ from ray-centered coordinates $q_1, q_2, q_3 = s$ to the local Cartesian coordinates z_1, z_2, z_3 at the point of incidence Q,

$$\mathrm{d}z_k = G_{kl}(Q)\mathrm{d}q_l(Q) \quad \text{or} \quad \mathrm{d}\hat{\mathbf{z}} = \hat{\mathbf{G}}(Q)\mathrm{d}\hat{\mathbf{q}}(Q), \tag{4.4.10}$$

where $\mathrm{d}\hat{\mathbf{q}}(Q) \equiv (\mathrm{d}q_1(Q), \mathrm{d}q_2(Q), \mathrm{d}q_3(Q))^T$. Matrix $\hat{\mathbf{G}}(Q)$ is orthonormal, and $\hat{\mathbf{G}}^{-1}(Q) = \mathbf{G}^T(Q)$, $\det(\hat{\mathbf{G}}(Q)) = 1$. Thus,

$$\mathrm{d}q_k(Q) = G_{lk}(Q)\mathrm{d}z_l(Q) \quad \text{or} \quad \mathrm{d}\hat{\mathbf{q}}(Q) = \hat{\mathbf{G}}^T(Q)\mathrm{d}\hat{\mathbf{z}}(Q). \tag{4.4.11}$$

The elements of transformation matrix $\hat{\mathbf{G}}(Q)$ are given by equation

$$G_{kl}(Q) = \vec{i}_k^{(z)} \cdot \vec{e}_l(Q) = (\partial z_k / \partial q_l)_Q = (\partial q_l / \partial z_k)_Q, \tag{4.4.12}$$

where $\vec{e}_l(Q)$, $l = 1, 2, 3$, are the basis vectors of the ray-centered coordinate system of the incident wave at Q. Thus, the first column of $\hat{\mathbf{G}}(Q)$ represents the basis vector $\vec{e}_1$, the second column represents the basis vector $\vec{e}_2$, and the third column represents the basis vector $\vec{e}_3$; they are all expressed in their z_i-components ($i = 1, 2, 3$).

It is simple to prove that matrices $\hat{\mathbf{Z}}, \hat{\mathbf{G}}(Q)$, and $\hat{\mathbf{H}}(Q)$ are mutually related as follows:

$$\hat{\mathbf{G}}(Q) = \hat{\mathbf{Z}}^T \hat{\mathbf{H}}(Q), \qquad \hat{\mathbf{H}}(Q) = \hat{\mathbf{Z}}\hat{\mathbf{G}}(Q). \tag{4.4.13}$$

Matrix $\hat{\mathbf{Z}}$ is, of course, the same for the wave incident at Q and for the waves reflected/transmitted at $\tilde{Q}$. Matrices $\hat{\mathbf{G}}(\tilde{Q})$ and $\hat{\mathbf{H}}(\tilde{Q})$ are, however, different from $\hat{\mathbf{G}}(Q)$ and $\hat{\mathbf{H}}(Q)$. The mutual relations between $\hat{\mathbf{G}}(\tilde{Q})$ and $\hat{\mathbf{G}}(Q)$ and between $\hat{\mathbf{H}}(\tilde{Q})$ and $\hat{\mathbf{H}}(Q)$ will be derived later on.

In the dynamic ray tracing across interface Σ, an important role is played by the *curvature of interface* Σ at point Q. We denote the 2×2 matrix of the curvature of interface Σ at Q by $\mathbf{D}(Q)$, with elements $D_{IJ}(Q)$. We define $\mathbf{D}(Q)$ in the standard way. We describe interface Σ in the vicinity of point Q by equation

$$z_3 \doteq -\tfrac{1}{2}z_I z_J D_{IJ}(Q). \tag{4.4.14}$$

To find explicit formulae for D_{IJ}, we expand $\Sigma(z_i)$ in the vicinity of Q up to the second-order terms in z_i,

$$\Sigma(Q) + (\partial \Sigma / \partial z_i)_Q z_i + \tfrac{1}{2}(\partial^2 \Sigma / \partial z_i \partial z_j)_Q z_i z_j + \cdots = 0;$$

see (4.4.1). As the interface passes through point Q, $\Sigma(Q) = 0$. Then

$$(\partial \Sigma/\partial z_I)_Q z_I + (\partial \Sigma/\partial z_3)_Q z_3 + \tfrac{1}{2}(\partial^2 \Sigma/\partial z_I \partial z_J)_Q z_I z_J + \cdots = 0.$$

Terms with $z_I z_3$ and z_3^2 have been omitted because they are of higher order. Since $(\partial \Sigma/\partial z_I)_Q = 0$, we finally obtain

$$(\partial \Sigma/\partial z_3)_Q z_3 + \tfrac{1}{2}(\partial^2 \Sigma/\partial z_I \partial z_J)_Q z_I z_J + \cdots = 0.$$

This yields

$$D_{IJ}(Q) = (\partial^2 \Sigma/\partial z_I \partial z_J)_Q / (\partial \Sigma/\partial z_3)_Q. \qquad (4.4.15)$$

We can now replace the local Cartesian coordinates z_i by the general Cartesian coordinates x_i using transformation relations

$$(\partial \Sigma/\partial z_3)_Q = Z_{i3}(\partial \Sigma/\partial x_i)_Q,$$
$$(\partial^2 \Sigma/\partial z_I \partial z_J)_Q = Z_{iI} Z_{kJ}(\partial^2 \Sigma/\partial x_i \partial x_k)_Q.$$

We then obtain equation

$$D_{IJ} = Z_{iI} Z_{kJ}(\partial^2 \Sigma/\partial x_i \partial x_k)_Q / [Z_{j3}(\partial \Sigma/\partial x_j)_Q]. \qquad (4.4.16)$$

We remind the reader that Z_{j3} are components of the unit normal to interface Σ at Q. Using (4.4.4), we can express the denominator of the preceding equation in the form

$$Z_{j3}(\partial \Sigma/\partial x_j)_Q = \epsilon^*(\partial \Sigma/\partial x_j)_Q(\partial \Sigma/\partial x_j)_Q / [(\partial \Sigma/\partial x_k)_Q(\partial \Sigma/\partial x_k)_Q]^{1/2}$$
$$= \epsilon^*[(\partial \Sigma/\partial x_j)_Q(\partial \Sigma/\partial x_j)_Q]^{1/2}.$$

This yields the final equation for curvature matrix $\mathbf{D}(Q)$,

$$D_{IJ}(Q) = \epsilon^* Z_{iI} Z_{kJ} D_{ik}^{(x)}(Q), \qquad (4.4.17)$$

where

$$D_{ik}^{(x)}(Q) = (\partial^2 \Sigma/\partial x_i \partial x_k)_Q / [(\partial \Sigma/\partial x_j)_Q(\partial \Sigma/\partial x_j)_Q]^{1/2}. \qquad (4.4.18)$$

We emphasize that $D_{ik}^{(x)}(Q)$ are independent of the local Cartesian coordinate system z_i.

It will also be useful to know the components of the unit vector $\vec{n}$ normal to interface Σ in the vicinity of point Q, in local Cartesian coordinate system z_i. Assuming that the interface is given by Equation (4.4.14), and using (4.4.5) with (4.4.4) and (4.4.6), we obtain

$$n_I = \epsilon^* D_{IJ} z_J, \qquad n_3 = \epsilon^*. \qquad (4.4.19)$$

These equations are valid with an accuracy up to the linear terms in z_I.

Let us return to the *selection of the basis vectors* $\vec{i}_1^{(z)}$ and $\vec{i}_2^{(z)}$ of the local Cartesian coordinate system at Q. They can be chosen in various ways. We shall describe one simple option. We will consider any vector $\vec{v}$ that is not parallel to $\vec{n}(Q)$. We can then put

$$\vec{i}_2^{(z)} = \frac{\vec{n} \times \vec{v}}{|\vec{n} \times \vec{v}|}, \qquad \vec{i}_1^{(z)} = \frac{\vec{n} \times (\vec{v} \times \vec{n})}{|\vec{n} \times \vec{v} \times \vec{n}|} = \frac{\vec{v} - \vec{n}(\vec{v} \cdot \vec{n})}{[\vec{v} \cdot \vec{v} - (\vec{v} \cdot \vec{n})^2]^{1/2}}. \qquad (4.4.20)$$

As a special case of this choice, we can take $\vec{v} = \vec{p}(Q)$, the slowness vector of the incident wave at Q. This option has been traditionally considered in the seismological literature. Together with (4.4.2), we obtain

$$\vec{i}_3^{(z)} = \vec{n}, \qquad \vec{i}_2^{(z)} = \frac{\vec{n} \times \vec{p}}{|\vec{n} \times \vec{p}|}, \qquad \vec{i}_1^{(z)} = \frac{\vec{p} - \vec{n}(\vec{p} \cdot \vec{n})}{[\vec{p} \cdot \vec{p} - (\vec{p} \cdot \vec{n})^2]^{1/2}}. \qquad (4.4.21)$$

In option (4.4.21), unit vector $\vec{i}_2^{(z)}$ is perpendicular to the plane of incidence, and $\vec{i}_1^{(z)}$ is situated along the intersection of the plane of incidence with the plane tangent to Σ at Q. Since $\vec{p}(Q) \cdot \vec{i}_1^{(z)} \geq 0$, the positive orientation of the $\vec{i}_1^{(z)}$ axis is "along the slowness vector of the incident wave." In the following text, we shall call the option based on (4.4.21) the *standard option* of the local Cartesian coordinate system z_i at Q.

Option $\vec{v} = \vec{p}$ in (4.4.20) fails for normal incidence because $\vec{v}$ is then parallel to $\vec{n}(Q)$. However, we can take $\vec{i}_2^{(z)} = \vec{e}_2(Q)$.

For the readers' convenience, we shall express the 3×3 matrix $\hat{\mathbf{G}}(Q)$, corresponding to the standard option of the local Cartesian coordinate system z_1, z_2, z_3 at Q, specified by (4.4.21), explicitly. Using (4.4.12), we obtain

$$\hat{\mathbf{G}}(Q) = \begin{pmatrix} \epsilon \cos i_S \cos \kappa & -\epsilon \cos i_S \sin \kappa & \sin i_S \\ \sin \kappa & \cos \kappa & 0 \\ -\sin i_S \cos \kappa & \sin i_S \sin \kappa & \epsilon \cos i_S \end{pmatrix} = \hat{\mathbf{G}}^{\parallel}(Q)\hat{\mathbf{G}}^{\perp}(Q),$$

(4.4.22)

where $\hat{\mathbf{G}}^{\parallel}(Q)$ and $\hat{\mathbf{G}}^{\perp}(Q)$ are rotational matrices, given by relations

$$\hat{\mathbf{G}}^{\parallel}(Q) = \begin{pmatrix} \epsilon \cos i_S & 0 & \sin i_S \\ 0 & 1 & 0 \\ -\sin i_S & 0 & \epsilon \cos i_S \end{pmatrix}, \qquad \hat{\mathbf{G}}^{\perp}(Q) = \begin{pmatrix} \cos \kappa & -\sin \kappa & 0 \\ \sin \kappa & \cos \kappa & 0 \\ 0 & 0 & 1 \end{pmatrix}.$$

(4.4.23)

The meaning of the individual symbols follows: i_S is the acute angle of incidence, $0 \leq i_S \leq \frac{1}{2}\pi$, ϵ is the orientation index introduced by (2.4.71), that is $\epsilon = \text{sign}(\vec{p}(Q) \cdot \vec{n})$. Finally, κ is the angle between $\vec{e}_2$ and $\vec{i}_2^{(z)}$, $0 \leq \kappa \leq 2\pi$, specified by relations $\cos \kappa = \vec{e}_2 \cdot \vec{i}_2^{(z)}$ and $\sin \kappa = \vec{e}_1 \cdot \vec{i}_2^{(z)}$. Matrix $\hat{\mathbf{G}}^{\perp}$ performs the rotation in the plane perpendicular to ray Ω at the point of incidence Q so as to shift $\vec{e}_2$ into $\vec{i}_2^{(z)}$. Similarly, matrix $\hat{\mathbf{G}}^{\parallel}$ performs the rotation in the plane of incidence so as to shift the unit vector $\vec{N}$, perpendicular to the wavefront of the incident wave at Q, into $\vec{i}_3^{(z)}$.

Matrix $\hat{\mathbf{G}}(\tilde{Q})$, corresponding to reflected/transmitted waves, can be expressed explicitly in a similar way. We shall consider the same local Cartesian coordinate system z_i as in (4.4.22), corresponding to standard option (4.4.21). In addition, we shall consider the *standard option* for $\vec{e}_1(\tilde{Q})$ and $\vec{e}_2(\tilde{Q})$ given by (2.3.45). Hence,

$$\hat{\mathbf{G}}(\tilde{Q}) = \begin{pmatrix} \pm\epsilon \cos i_R \cos \kappa & \mp\epsilon \cos i_R \sin \kappa & \sin i_R \\ \sin \kappa & \cos \kappa & 0 \\ -\sin i_R \cos \kappa & \sin i_R \sin \kappa & \pm\epsilon \cos i_R \end{pmatrix} = \hat{\mathbf{G}}^{\parallel}(\tilde{Q})\hat{\mathbf{G}}^{\perp}(\tilde{Q}),$$

(4.4.24)

where $\hat{\mathbf{G}}^{\parallel}(\tilde{Q})$ and $\hat{\mathbf{G}}^{\perp}(\tilde{Q})$ are given by relations

$$\hat{\mathbf{G}}^{\parallel}(\tilde{Q}) = \begin{pmatrix} \pm\epsilon \cos i_R & 0 & \sin i_R \\ 0 & 1 & 0 \\ -\sin i_R & 0 & \pm\epsilon \cos i_R \end{pmatrix}, \quad \hat{\mathbf{G}}^{\perp}(\tilde{Q}) = \hat{\mathbf{G}}^{\perp}(Q). \quad (4.4.25)$$

Here i_R is the acute angle of reflection/transmission, the upper sign corresponds to the transmitted wave, the lower sign corresponds to the reflected wave, and κ is the same as in (4.4.22). Relation $\hat{\mathbf{G}}^{\perp}(\tilde{Q}) = \hat{\mathbf{G}}^{\perp}(Q)$ guarantees standard option (2.3.45). Consequently, the angles between $\vec{e}_2$ and $\vec{i}_2^{(z)}$ are the same for incident and R/T waves,

$\vec{e}_2(Q) \cdot \vec{i}_2^{(z)} = \vec{e}_2(\tilde{Q}) \cdot \vec{i}_2^{(z)}$. The same is, of course, also valid for the angles between $\vec{e}_1$ and $\vec{i}_2^{(z)}$ at Q and $\tilde{Q}$.

To determine $\hat{\mathbf{G}}(\tilde{Q})$ from $\hat{\mathbf{G}}(Q)$, it is not necessary to compute angles i_S and i_R. Let us assume that the elements $G_{21}(Q)$ and $G_{22}(Q)$, corresponding to the incident wave at Q, are known. Then, assuming standard option (4.4.21) and standard option (2.3.45), we simply obtain

$$\hat{\mathbf{G}}(\tilde{Q}) = \begin{pmatrix} G_{22}(Q)G_{33}(\tilde{Q}) & -G_{21}(Q)G_{33}(\tilde{Q}) & G_{13}(\tilde{Q}) \\ G_{21}(Q) & G_{22}(Q) & 0 \\ -G_{22}(Q)G_{13}(\tilde{Q}) & G_{21}(Q)G_{13}(\tilde{Q}) & G_{33}(\tilde{Q}) \end{pmatrix}. \qquad (4.4.26)$$

Here $G_{13}(\tilde{Q}) = \vec{i}_1^{(z)} \cdot \vec{e}_3(\tilde{Q}) = V(\tilde{Q})\vec{i}_1^{(z)} \cdot \vec{p}(\tilde{Q})$ and $G_{33}(\tilde{Q}) = \vec{i}_3^{(z)} \cdot \vec{e}_3(\tilde{Q}) = V(\tilde{Q})\vec{i}_3^{(z)} \cdot \vec{p}(\tilde{Q})$. See Snell's law (2.4.70) for $\vec{p}(\tilde{Q})$. It is interesting to note that matrix $\hat{\mathbf{G}}(\tilde{Q})$ is fully specified by four quantities $G_{21}(Q)$, $G_{22}(Q)$, $G_{13}(\tilde{Q})$, and $G_{33}(\tilde{Q})$. Using $\hat{\mathbf{G}}(\tilde{Q})$, we can compute matrix $\hat{\mathbf{H}}(\tilde{Q})$,

$$\hat{\mathbf{H}}(\tilde{Q}) = \hat{\mathbf{Z}}\hat{\mathbf{G}}(\tilde{Q}), \qquad (4.4.27)$$

which gives the Cartesian components of basis vectors $\vec{e}_1(\tilde{Q})$, $\vec{e}_2(\tilde{Q})$, and $\vec{e}_3(\tilde{Q})$.

Matrix $\hat{\mathbf{H}}(\tilde{Q})$ can also be determined directly, without introducing the local Cartesian coordinate system z_i. It is sufficient to know unit vector $\vec{n}(Q)$ normal to Σ at Q, and slowness vectors $\vec{p}(Q) = \vec{N}(Q)/V(Q)$ and $\vec{p}(\tilde{Q}) = \vec{N}(\tilde{Q})/V(\tilde{Q})$. Standard option (2.3.45) yields

$$\vec{e}_2(\tilde{Q}) = \frac{(\vec{e}_2(Q) \cdot \vec{n})[(\vec{N}(Q) \cdot \vec{N}(\tilde{Q}))\vec{n} - (\vec{n} \cdot \vec{N}(\tilde{Q}))\vec{N}(Q)] - (\vec{n} \cdot \vec{e}_1(Q))(\vec{n} \times \vec{N}(Q))}{|\vec{n} \times \vec{N}(Q)|^2},$$

$$\vec{e}_1(\tilde{Q}) = \vec{e}_2(\tilde{Q}) \times \vec{N}(\tilde{Q}).$$

$$(4.4.28)$$

4.4.2 Matrix M Across the Interface

We shall now derive the equations for the distribution of the *travel time along interface* Σ, in the vicinity of Q. First, however, we shall write a general equation for the distribution of the travel time in the vicinity of Q. At point Q', situated close to Q, see (4.1.86),

$$T(Q') = T(Q) + \hat{\mathbf{x}}^T(Q', Q)\hat{\mathbf{p}}^{(x)}(Q)$$
$$+ \tfrac{1}{2}\hat{\mathbf{x}}^T(Q', Q)\hat{\mathbf{H}}(Q)\hat{\mathbf{M}}(Q)\hat{\mathbf{H}}^T(Q)\hat{\mathbf{x}}(Q', Q). \qquad (4.4.29)$$

We remind the reader that $\hat{\mathbf{x}}(Q', Q) = \hat{\mathbf{x}}(Q') - \hat{\mathbf{x}}(Q)$; see (4.1.85). Using transformation matrix $\hat{\mathbf{Z}}$, we can easily express (4.4.29) in the local Cartesian coordinate system z_i:

$$\hat{\mathbf{p}}^{(x)}(Q) = \hat{\mathbf{Z}}(Q)\hat{\mathbf{p}}^{(z)}(Q), \qquad \hat{\mathbf{x}}(Q', Q) = \hat{\mathbf{Z}}(Q)\hat{\mathbf{z}},$$
$$\hat{\mathbf{x}}^T(Q', Q)\hat{\mathbf{p}}^{(x)}(Q) = \hat{\mathbf{z}}^T\hat{\mathbf{Z}}^T(Q)\hat{\mathbf{Z}}(Q)\hat{\mathbf{p}}^{(z)}(Q) = \hat{\mathbf{z}}^T\hat{\mathbf{p}}^{(z)}(Q).$$

Here $\hat{\mathbf{z}} \equiv (z_1(Q'), z_2(Q'), z_3(Q'))^T$, $z_i(Q')$ are z_i-coordinates of point Q', and $\hat{\mathbf{p}}^{(z)}(Q)$ represents a column matrix of the components of slowness vector $\vec{p}$ in the local Cartesian coordinate system z_i. Expansion (4.4.29) then becomes

$$T(Q') = T(Q) + \hat{\mathbf{z}}^T\hat{\mathbf{p}}^{(z)}(Q) + \tfrac{1}{2}\hat{\mathbf{z}}^T\hat{\mathbf{M}}^{(z)}(Q)\hat{\mathbf{z}}, \qquad (4.4.30)$$

where

$$\hat{\mathbf{M}}^{(z)}(Q) = \hat{\mathbf{Z}}^T(Q)\hat{\mathbf{H}}(Q)\hat{\mathbf{M}}(Q)\hat{\mathbf{H}}^T(Q)\hat{\mathbf{Z}}(Q) = \hat{\mathbf{G}}(Q)\hat{\mathbf{M}}(Q)\hat{\mathbf{G}}^T(Q).$$

$$(4.4.31)$$

In component form, (4.4.30) reads

$$T(Q') = T(Q) + z_i p_i^{(z)}(Q) + \tfrac{1}{2} z_i z_j M_{ij}^{(z)}(Q). \tag{4.4.32}$$

We remind the reader that z_i are the local Cartesian coordinates of point Q'. The general expansions (4.4.30) and (4.4.32) are valid in the whole medium surrounding Σ from the side of the incident wave, in the vicinity of point Q.

We shall now specify general expression (4.4.32) for points Q' situated along interface Σ, in the vicinity of Q. Coordinates z_i of point Q', situated on interface Σ, in the vicinity of Q, are as follows: $Q' \equiv [z_1, z_2, z_3 = -\tfrac{1}{2} z_I z_J D_{IJ}(Q)]$. Thus, the position of point Q' on Σ is fully specified by coordinates z_1, z_2 and by curvature matrix $\mathbf{D}(Q)$. If we insert $z_3 = -\tfrac{1}{2} z_I z_J D_{IJ}(Q)$ into (4.4.32), we obtain travel time $T(Q')$ as a function of two variables only, z_1, and z_2. For this reason, we shall use the notation $T(Q') \equiv T^\Sigma(z_1, z_2)$. In the expression for $T^\Sigma(z_1, z_2)$, we shall only retain the terms up to the second order in z_1 and z_2. First, we express (4.4.32) as

$$\begin{aligned} T^\Sigma(z_1, z_2) = {}& T(Q) + z_I p_I^{(z)}(Q) + z_3 p_3^{(z)}(Q) \\ & + \tfrac{1}{2} z_I z_J G_{Ik}(Q) G_{Jm}(Q) M_{km}(Q). \end{aligned}$$

All the other terms are of a higher order, due to (4.4.14). We can put

$$\begin{aligned} G_{Ik} G_{Jm} M_{km} = {}& G_{IK} G_{JM} M_{KM} + G_{I3} G_{JM} M_{3M} \\ & + G_{IK} G_{J3} M_{K3} + G_{I3} G_{J3} M_{33}. \end{aligned}$$

We denote

$$\begin{aligned} E_{IJ}(Q) = {}& G_{I3}(Q) G_{JM}(Q) M_{3M}(Q) + G_{IK}(Q) G_{J3}(Q) M_{K3}(Q) \\ & + G_{I3}(Q) G_{J3}(Q) M_{33}(Q). \end{aligned} \tag{4.4.33}$$

We can then write

$$T^\Sigma(z_1, z_2) = T(Q) + z_I p_I^{(z)}(Q) + \tfrac{1}{2} z_I z_J F_{IJ}(Q), \tag{4.4.34}$$

where

$$F_{IJ}(Q) = G_{IK}(Q) G_{JM}(Q) M_{KM}(Q) + E_{IJ}(Q) - p_3^{(z)}(Q) D_{IJ}(Q). \tag{4.4.35}$$

Equation (4.4.34), with (4.4.33) and (4.4.35), represents the *final solution of our problem*, in component form. It gives the *distribution of the travel time T along any surface Σ crossing the central ray Ω at point Q*. In matrix form,

$$T^\Sigma(z_1, z_2) = T(Q) + \mathbf{z}^T \mathbf{p}^{(z)}(Q) + \tfrac{1}{2} \mathbf{z}^T \mathbf{F}(Q) \mathbf{z}, \tag{4.4.36}$$

with

$$\mathbf{F}(Q) = \mathbf{G}(Q) \mathbf{M}(Q) \mathbf{G}^T(Q) + \mathbf{E}(Q) - p_3^{(z)}(Q) \mathbf{D}(Q). \tag{4.4.37}$$

In actual computations, we shall usually work with transformation matrices $\hat{\mathbf{H}}$, $\hat{\mathbf{G}}$, and $\hat{\mathbf{Z}}$. For this reason, we also express $p_1^{(z)}(Q)$, $p_2^{(z)}(Q)$, and $p_3^{(z)}(Q)$ in (4.4.34) through (4.4.37) in terms of these quantities. Hence,

$$p_i^{(z)}(Q) = Z_{ki}(Q) p_k^{(x)}(Q) = V^{-1}(Q) Z_{ki}(Q) H_{k3}(Q) = V^{-1}(Q) G_{i3}(Q). \tag{4.4.38}$$

Here $V(Q)$ denotes the propagation velocity of the incident wave at Q.

Expansions similar to (4.4.36) can be obtained *for any R/T wave* in the vicinity of point $\tilde{Q}$. We denote the travel time of the R/T wave along surface Σ in the vicinity of $\tilde{Q}$ by $\tilde{T}^\Sigma(z_1, z_2)$. Then

$$\tilde{T}^\Sigma(z_1, z_2) = T(\tilde{Q}) + \mathbf{z}^T \mathbf{p}^{(z)}(\tilde{Q}) + \tfrac{1}{2}\mathbf{z}^T \mathbf{F}(\tilde{Q})\mathbf{z}, \tag{4.4.39}$$

where

$$\mathbf{F}(\tilde{Q}) = \mathbf{G}(\tilde{Q})\mathbf{M}(\tilde{Q})\mathbf{G}^T(\tilde{Q}) + \mathbf{E}(\tilde{Q}) - p_3^{(z)}(\tilde{Q})\mathbf{D}(Q). \tag{4.4.40}$$

Here $\mathbf{E}(\tilde{Q})$ is again given by (4.4.33), where Q has been is replaced by $\tilde{Q}$. We also have $\mathbf{D}(\tilde{Q}) = \mathbf{D}(Q)$.

We now apply *phase matching along* Σ, in the vicinity of Q. We remind the reader that the phase matching follows from the boundary conditions at interface Σ; see Section 2.4.5. The phase matching implies that $\tilde{T}^\Sigma(z_1, z_2) = T^\Sigma(z_1, z_2)$. Thus,

$$T(Q) + \mathbf{z}^T \mathbf{p}^{(z)}(Q) + \tfrac{1}{2}\mathbf{z}^T \mathbf{F}(Q)\mathbf{z} = T(\tilde{Q}) + \mathbf{z}^T \mathbf{p}^{(z)}(\tilde{Q}) + \tfrac{1}{2}\mathbf{z}^T \mathbf{F}(\tilde{Q})\mathbf{z}. \tag{4.4.41}$$

Equation (4.4.41) yields three equations:

$$T(Q) = T(\tilde{Q}), \qquad \mathbf{p}^{(z)}(Q) = \mathbf{p}^{(z)}(\tilde{Q}), \qquad \mathbf{F}(Q) = \mathbf{F}(\tilde{Q}). \tag{4.4.42}$$

The meaning of the first equation, $T(Q) = T(\tilde{Q})$, is obvious. The second equation indicates that the tangential components of the slowness vectors of the incident and R/T waves are equal. In fact, it implies Snell's law. The main result of this section is given by the relation $\mathbf{F}(Q) = \mathbf{F}(\tilde{Q})$. We can express it as

$$\mathbf{G}(Q)\mathbf{M}(Q)\mathbf{G}^T(Q) + \mathbf{E}(Q) - p_3^{(z)}(Q)\mathbf{D}$$
$$= \mathbf{G}(\tilde{Q})\mathbf{M}(\tilde{Q})\mathbf{G}^T(\tilde{Q}) + \mathbf{E}(\tilde{Q}) - p_3^{(z)}(\tilde{Q})\mathbf{D}. \tag{4.4.43}$$

Equation (4.4.43) yields an important relation for the *transformation of matrix* $\mathbf{M}$ *across interface* Σ,

$$\mathbf{M}(\tilde{Q}) = \mathbf{G}^{-1}(\tilde{Q})[\mathbf{G}(Q)\mathbf{M}(Q)\mathbf{G}^T(Q) + \mathbf{E}(Q) - \mathbf{E}(\tilde{Q}) - u\mathbf{D}]\mathbf{G}^{-1T}(\tilde{Q}), \tag{4.4.44}$$

where

$$u = p_3^{(z)}(Q) - p_3^{(z)}(\tilde{Q}) = V^{-1}(Q)G_{33}(Q) - V^{-1}(\tilde{Q})G_{33}(\tilde{Q}). \tag{4.4.45}$$

In abbreviated form, (4.4.44) reads

$$\tilde{\mathbf{M}} = \tilde{\mathbf{G}}^{-1}[\mathbf{G}\mathbf{M}\mathbf{G}^T + \mathbf{E} - \tilde{\mathbf{E}} - u\mathbf{D}]\tilde{\mathbf{G}}^{-1T}. \tag{4.4.46}$$

The physical interpretation of (4.4.44) or (4.4.46) is straightforward. The equations allow us to evaluate the matrix of the second derivatives of travel-time field $\mathbf{M}(\tilde{Q})$ corresponding to any wave reflected/transmitted at point $\tilde{Q}$, if the matrix of the second derivatives of travel-time field $\mathbf{M}(Q)$, corresponding to the incident wave, is known at the point of incidence Q. The first term in (4.4.46), $\tilde{\mathbf{G}}^{-1}\mathbf{G}\mathbf{M}\mathbf{G}^T\tilde{\mathbf{G}}^{-1T}$, represents the actual transformation of matrix $\mathbf{M}$ from the ray-centered coordinate system at Q to the ray-centered coordinate system at $\tilde{Q}$. The second term, $\tilde{\mathbf{G}}^{-1}(\mathbf{E} - \tilde{\mathbf{E}})\tilde{\mathbf{G}}^{-1T}$, characterizes the *effect of inhomogeneities* of the medium close to Q and $\tilde{Q}$. Finally, the third term, $-u\tilde{\mathbf{G}}^{-1}\mathbf{D}\tilde{\mathbf{G}}^{-1T}$, represents the *effect of the curvature* of interface Σ at Q.

Equation (4.4.33) for *inhomogeneity matrix* $\mathbf{E}$ can be expressed as

$$E_{IJ}(Q) = -V^{-2}(Q)\{[G_{I3}(Q)G_{JK}(Q) + G_{IK}(Q)G_{J3}(Q)](\partial V/\partial q_K)_Q$$
$$+ G_{I3}(Q)G_{J3}(Q)(\partial V/\partial q_3)_Q\}; \tag{4.4.47}$$

see (4.1.81).

For the readers' convenience, we shall explicitly express the individual quantities derived in this section, corresponding to the standard option of the local coordinate Cartesian system z_1, z_2, z_3 at Q specified by (4.4.21) and to the standard option of $\vec{e}_1(\tilde{Q})$ and $\vec{e}_2(\tilde{Q})$ given by (2.3.45).

From (4.4.22) through (4.4.25), we immediately obtain the expressions for the 2×2 matrices $\mathbf{G}(Q)$, $\mathbf{G}(\tilde{Q})$, $\mathbf{G}^{\|}(Q)$, $\mathbf{G}^{\perp}(Q)$, $\mathbf{G}^{\|}(\tilde{Q})$, and $\mathbf{G}^{\perp}(\tilde{Q})$:

$$\mathbf{G}(Q) = \mathbf{G}^{\|}(Q)\mathbf{G}^{\perp}(Q), \qquad \mathbf{G}(\tilde{Q}) = \mathbf{G}^{\|}(\tilde{Q})\mathbf{G}^{\perp}(\tilde{Q}), \tag{4.4.48}$$

$$\mathbf{G}^{\|}(Q) = \begin{pmatrix} \epsilon \cos i_S & 0 \\ 0 & 1 \end{pmatrix}, \qquad \mathbf{G}^{\perp}(Q) = \begin{pmatrix} \cos \kappa & -\sin \kappa \\ \sin \kappa & \cos \kappa \end{pmatrix},$$

$$\mathbf{G}^{\|}(\tilde{Q}) = \begin{pmatrix} \pm\epsilon \cos i_R & 0 \\ 0 & 1 \end{pmatrix}, \qquad \mathbf{G}^{\perp}(\tilde{Q}) = \begin{pmatrix} \cos \kappa & -\sin \kappa \\ \sin \kappa & \cos \kappa \end{pmatrix}. \tag{4.4.49}$$

As we can see from (4.4.22) through (4.4.25), (4.4.48), and (4.4.49), the determinants of the 3×3 matrices $\hat{\mathbf{G}}(Q)$, $\hat{\mathbf{G}}(\tilde{Q})$, $\hat{\mathbf{G}}^{\|}(Q)$, $\hat{\mathbf{G}}^{\perp}(Q)$, $\hat{\mathbf{G}}^{\|}(\tilde{Q})$, $\hat{\mathbf{G}}^{\perp}(\tilde{Q})$, $\mathbf{G}^{\perp}(Q)$, and $\mathbf{G}^{\perp}(\tilde{Q})$ equal unity. For $\mathbf{G}(Q)$, $\mathbf{G}(\tilde{Q})$, $\mathbf{G}^{\|}(Q)$, and $\mathbf{G}^{\|}(\tilde{Q})$, however, we have

$$\det \mathbf{G}(Q) = \det \mathbf{G}^{\|}(Q) = \epsilon \cos i_S,$$
$$\det \mathbf{G}(\tilde{Q}) = \det \mathbf{G}^{\|}(\tilde{Q}) = \pm\epsilon \cos i_R. \tag{4.4.50}$$

Now we shall discuss quantity u and matrices $\mathbf{E}$ and $\tilde{\mathbf{E}}$. As we can see from (4.4.22) and (4.4.24), quantity u given by (4.4.45) can be expressed as

$$u = \epsilon(V^{-1}(Q)\cos i_S \mp V^{-1}(\tilde{Q})\cos i_R), \tag{4.4.51}$$

where the upper sign corresponds to the transmitted wave, and the lower sign corresponds to the reflected wave. For an unconverted reflected wave, $V(Q) = V(\tilde{Q})$ and $i_S = i_R$. Then, (4.4.51) yields

$$u = 2\epsilon V^{-1}(Q)\cos i_S. \tag{4.4.52}$$

Note that u does not depend at all on the orientation of $\vec{e}_1$ and $\vec{e}_2$.

Finally, the expression for the 2×2 matrices $\mathbf{E}$ can also be written in a simpler form in the standard option of the local Cartesian coordinate system z_i at Q, specified by (4.4.21). We can consider (4.4.47) and insert $G_{23} = 0$; see (4.4.22). This yields $E_{22}(Q) = E_{22}(\tilde{Q}) = 0$. If we also introduce the derivatives of velocities with respect to coordinates z_i instead of q_i, we obtain for the incident wave

$$E_{11}(Q) = -\sin i_S \, V^{-2}(Q)[(1 + \cos^2 i_S)V_{,1}^{(z)} - \epsilon \cos i_S \sin i_S \, V_{,3}^{(z)}],$$
$$E_{12}(Q) = E_{21}(Q) = -\sin i_S \, V^{-2}(Q)V_{,2}^{(z)}, \tag{4.4.53}$$
$$E_{22}(Q) = 0.$$

Similarly, for R/T waves,

$$E_{11}(\tilde{Q}) = -\sin i_R V^{-2}(\tilde{Q})[(1 + \cos^2 i_R)\tilde{V}_{,1}^{(z)} \mp \epsilon \cos i_R \sin i_R \tilde{V}_{,3}^{(z)}],$$
$$E_{12}(\tilde{Q}) = E_{21}(\tilde{Q}) = -\sin i_R \, V^{-2}(\tilde{Q})\tilde{V}_{,2}^{(z)}, \tag{4.4.54}$$
$$E_{22}(\tilde{Q}) = 0.$$

Here we have used the notation

$$V_{,i}^{(z)} = (\partial V/\partial z_i)_Q = (\partial V/\partial x_j)_Q Z_{ji},$$
$$\tilde{V}_{,i}^{(z)} = (\partial \tilde{V}/\partial z_i)_{\tilde{Q}} = (\partial \tilde{V}/\partial x_j)_{\tilde{Q}} Z_{ji}. \tag{4.4.55}$$

It is remarkable that E_{IJ} and $\tilde{E}_{IJ}$ do not depend on the orientation of unit vectors $\vec{e}_1$ and $\vec{e}_2$ in the plane perpendicular to ray Ω.

For the unconverted reflected wave, we obtain a very simple expression for $\mathbf{E} - \tilde{\mathbf{E}}$:

$$E_{11} - \tilde{E}_{11} = 2\epsilon \cos i_S \sin^2 i_S \, V^{-2}(Q) V_{,3}^{(z)},$$
$$E_{12} - \tilde{E}_{12} = E_{21} - \tilde{E}_{21} = E_{22} - \tilde{E}_{22} = 0. \tag{4.4.56}$$

Thus, $\mathbf{E} - \tilde{\mathbf{E}}$ depends only on the derivatives of the velocity in the direction perpendicular to the interface Σ in this case.

4.4.3 Paraxial Slowness Vector

Because the general expression (4.4.32) for the travel-time field in the vicinity of point Q in local Cartesian coordinates z_1, z_2, z_3 is known, we can also easily compute expressions for the z_i-components of slowness vector $\vec{p}$, $p_i^{(z)}(Q') = (\partial T/\partial z_i)_{Q'}$. Since expression (4.4.32) for $T(Q')$ is valid up to the second-order terms in z_i, the expressions for $p_i^{(z)}(Q')$ are valid only *up to the first order in z_i*. From (4.4.32), we obtain

$$p_i^{(z)}(Q') = p_i^{(z)}(Q) + M_{ij}^{(z)}(Q)z_j, \tag{4.4.57}$$

where $M_{ij}^{(z)}(Q)$ is given by (4.4.31) and z_j are the coordinates of point Q'. Equation (4.4.57) is generally valid at any point Q' close to Q, on the side of interface Σ that contains the incident wave.

We will now use these general equations to write the relevant expressions for $\vec{p}^{(z)}(Q')$ *at points Q' situated directly on surface Σ*, $Q' \equiv [z_1, z_2, -\frac{1}{2}z_I z_J D_{IJ}(Q)]$. From (4.4.57), we obtain

$$p_i^{(z)}(Q') = p_i^{(z)}(Q) + M_{iJ}^{(z)}(Q)z_J, \qquad \text{for } Q' \in \Sigma. \tag{4.4.58}$$

We resolve $\vec{p}^{(z)}(Q')$ into two vectorial components, $\vec{p}^{(n)}(Q')$ and $\vec{p}^{(\Sigma)}(Q')$, normal and tangential to Σ at Q',

$$\vec{p}^{(z)}(Q') = \vec{p}^{(n)}(Q') + \vec{p}^{(\Sigma)}(Q'), \qquad \text{for } Q' \in \Sigma, \tag{4.4.59}$$

where

$$\vec{p}^{(n)}(Q') = (\vec{n}(Q') \cdot \vec{p}^{(z)}(Q'))\vec{n}(Q'),$$
$$\vec{p}^{(\Sigma)}(Q') = \vec{p}^{(z)}(Q') - (\vec{n}(Q') \cdot \vec{p}^{(z)}(Q'))\vec{n}(Q'). \tag{4.4.60}$$

Inserting (4.4.19) for $\vec{n}(Q')$ and (4.4.58) for $\vec{p}^{(z)}(Q')$ into (4.4.60), we obtain the following results:

- For the normal component of slowness vector $\vec{p}$ at $Q' \in \Sigma$,

$$p_I^{(n)}(Q') = p_3^{(z)}(Q)D_{IJ}(Q)z_J,$$
$$p_3^{(n)}(Q') = p_3^{(z)}(Q) + [p_I^{(z)}(Q)D_{IK}(Q) + M_{3K}^{(z)}(Q)]z_K. \tag{4.4.61}$$

- For the tangential component of the slowness vector at $Q' \in \Sigma$,

$$p_I^{(\Sigma)}(Q') = p_I^{(z)}(Q) + [M_{IJ}^{(z)}(Q) - p_3^{(z)}(Q)D_{IJ}(Q)]z_J$$
$$= p_I^{(z)}(Q) + F_{IJ}(Q)z_J, \qquad (4.4.62)$$
$$p_3^{(\Sigma)}(Q') = -p_I^{(z)}(Q)D_{IK}(Q)z_K.$$

See (4.4.35). Similar equations can also be immediately obtained for the waves reflected/transmitted at point $\tilde{Q}$. It is simple to check that the tangential components of slowness vector $\vec{p}^{(z)}$ remain continuous across interface Σ, $p_i^{(\Sigma)}(\tilde{Q}') = p_i^{(\Sigma)}(Q')$. Of course, the normal components of the slowness vector are not continuous across the interface, $\vec{p}^{(n)}(\tilde{Q}') \neq \vec{p}^{(n)}(Q')$.

4.4.4 Transformation of Matrices Q and P Across the Interface

We require the paraxial rays to be continuous across interface Σ. We denote the ray parameters of the central ray Ω passing through point Q by γ_I and the ray parameter of an arbitrarily selected paraxial ray by γ_I'. We assume that the paraxial ray is incident at interface Σ at point Q' close to point Q; $Q' \equiv [z_1, z_2, -\frac{1}{2}z_I z_J D_{IJ}(Q)]$, $\gamma_I' = \gamma_I(Q')$. Then

$$d\gamma_I = (\partial\gamma_I/\partial z_J)_Q dz_J = (\partial\gamma_I/\partial q_m)_Q(\partial q_m/\partial z_J)_Q dz_J,$$

where $d\gamma_I = \gamma_I' - \gamma_I$, $dz_I = z_I(Q')$. Since $(\partial\gamma_I/\partial q_3)_Q = 0$,

$$d\gamma_I = (\partial\gamma_I/\partial q_M)_Q(\partial q_M/\partial z_J)_Q dz_J.$$

The same relation can, of course, be applied to rays reflected/transmitted at point $\tilde{Q}$. Hence,

$$(\partial\gamma_I/\partial q_M)_Q(\partial q_M/\partial z_J)_Q = (\partial\gamma_I/\partial\tilde{q}_M)_{\tilde{Q}}(\partial\tilde{q}_M/\partial z_J)_{\tilde{Q}},$$

where $\tilde{q}_M$ are the ray-centered coordinates corresponding to the R/T waves at $\tilde{Q}$, q_M correspond to the incident wave at Q. It follows from (4.1.33) and (4.1.34) that $(\partial\gamma_I/\partial q_M)_Q = (\mathbf{Q}^{-1}(Q))_{IM}$, and from (4.4.12) that $(\partial q_M/\partial z_J)_Q = G_{JM}(Q)$. Thus,

$$(\mathbf{Q}^{-1}(Q))_{IM}G_{JM}(Q) = (\mathbf{Q}^{-1}(\tilde{Q}))_{IM}G_{JM}(\tilde{Q}).$$

In matrix form,

$$\mathbf{Q}^{-1}(Q)\mathbf{G}^T(Q) = \mathbf{Q}^{-1}(\tilde{Q})\mathbf{G}^T(\tilde{Q}). \qquad (4.4.63)$$

The final result is

$$\mathbf{Q}(\tilde{Q}) = \mathbf{G}^T(\tilde{Q})\mathbf{G}^{-1\,T}(Q)\mathbf{Q}(Q). \qquad (4.4.64)$$

Similarly,

$$d\boldsymbol{\gamma} = \mathbf{Q}^{-1}(Q)\mathbf{G}^T(Q)d\mathbf{z}, \qquad d\mathbf{z} = \mathbf{G}^{-1\,T}(Q)\mathbf{Q}(Q)d\boldsymbol{\gamma}, \qquad (4.4.65)$$

where $d\boldsymbol{\gamma} \equiv (d\gamma_1, d\gamma_2)^T$, $d\mathbf{z} \equiv (dz_1, dz_2)^T$, $d\gamma_I = \gamma_I' - \gamma_I = \gamma_I(Q') - \gamma_I(Q)$, and $dz_I = z_I(Q')$. Equation (4.4.64) also implies

$$\det \mathbf{Q}(\tilde{Q})/\det \mathbf{G}(\tilde{Q}) = \det \mathbf{Q}(Q)/\det \mathbf{G}(Q). \qquad (4.4.66)$$

The derivation of the transformation relations between $\mathbf{P}(\tilde{Q})$ and $\mathbf{P}(Q)$ is straightforward. Because $\mathbf{M}(\tilde{Q}) = \mathbf{P}(\tilde{Q})\mathbf{Q}^{-1}(\tilde{Q})$, we obtain $\mathbf{P}(\tilde{Q}) = \mathbf{M}(\tilde{Q})\mathbf{Q}(\tilde{Q})$. We can then use relations (4.4.46) for $\mathbf{M}(\tilde{Q})$ and (4.4.64) for $\mathbf{Q}(\tilde{Q})$. This yields

$$\tilde{\mathbf{P}} = \tilde{\mathbf{G}}^{-1}[\mathbf{GMG}^T - u\mathbf{D} + \mathbf{E} - \tilde{\mathbf{E}}]\tilde{\mathbf{G}}^{-1T}\tilde{\mathbf{G}}^T\mathbf{G}^{-1T}\mathbf{Q}.$$

Because $\mathbf{M} = \mathbf{PQ}^{-1}$,

$$\tilde{\mathbf{P}} = \tilde{\mathbf{G}}^{-1}[\mathbf{GP} + (\mathbf{E} - \tilde{\mathbf{E}} - u\mathbf{D})\mathbf{G}^{-1T}\mathbf{Q}]. \qquad (4.4.67)$$

This is the final relation for the transformation of matrix $\mathbf{P}$ across interface Σ.

We shall now look for the *projection of matrices* $\mathbf{Q}$ *and* $\mathbf{P}$ *on an arbitrary surface* Σ crossing ray Ω at point Q. Using (4.4.62), we can put

$$dp_I^{(\Sigma)}(Q') = p_I^{(\Sigma)}(Q') - p_I^{(\Sigma)}(Q) = F_{IJ}(Q)z_J.$$

Using (4.4.37) and (4.4.65), this yields

$$d\mathbf{p}^{(\Sigma)}(Q') = \left[\mathbf{GP} + (\mathbf{E} - p_3^{(z)}\mathbf{D})\mathbf{G}^{-1\,T}\mathbf{Q}\right]d\gamma. \qquad (4.4.68)$$

Combining Equations (4.4.65) for $d\mathbf{z}$ and (4.4.68) for $d\mathbf{p}^{(\Sigma)}(Q')$, we finally obtain

$$\begin{pmatrix} d\mathbf{z}(Q') \\ d\mathbf{p}^{(\Sigma)}(Q') \end{pmatrix} = \mathbf{Y}(Q)\begin{pmatrix} \mathbf{Q}d\gamma \\ \mathbf{P}d\gamma \end{pmatrix}, \qquad (4.4.69)$$

where $\mathbf{Q}d\gamma = \mathbf{Q}(Q)(\gamma(Q') - \gamma(Q))$, $\mathbf{P}d\gamma = \mathbf{P}(Q)(\gamma(Q') - \gamma(Q))$, and $\gamma = (\gamma_1, \gamma_2)^T$. $\mathbf{Y}(Q)$ is a 4×4 matrix

$$\mathbf{Y}(Q) = \begin{pmatrix} \mathbf{G}^{-1T}(Q) & \mathbf{0} \\ (\mathbf{E}(Q) - p_3^{(z)}(Q)\mathbf{D})\mathbf{G}^{-1T}(Q) & \mathbf{G}(Q) \end{pmatrix}. \qquad (4.4.70)$$

We also have

$$\begin{pmatrix} \mathbf{Q}d\gamma \\ \mathbf{P}d\gamma \end{pmatrix} = \mathbf{Y}^{-1}(Q)\begin{pmatrix} d\mathbf{z}(Q') \\ d\mathbf{p}^{(\Sigma)}(Q') \end{pmatrix} \qquad (4.4.71)$$

with

$$\mathbf{Y}^{-1}(Q) = \begin{pmatrix} \mathbf{G}^T(Q) & \mathbf{0} \\ -\mathbf{G}^{-1}(Q)(\mathbf{E}(Q) - p_3^{(z)}(Q)\mathbf{D}) & \mathbf{G}^{-1}(Q) \end{pmatrix}. \qquad (4.4.72)$$

We shall call $\mathbf{Y}(Q)$ the *projection matrix*.

Direct inspection proves that projection matrix $\mathbf{Y}(Q)$ and its inverse $\mathbf{Y}^{-1}(Q)$ are *symplectic*,

$$\mathbf{Y}^T(Q)\mathbf{J}\mathbf{Y}(Q) = \mathbf{J}, \qquad \mathbf{Y}^{-1\,T}(Q)\mathbf{J}\mathbf{Y}^{-1}(Q) = \mathbf{J}. \qquad (4.4.73)$$

Here $\mathbf{J}$ is given by relation (4.3.10). This also implies Liouville's relations $\det \mathbf{Y}(Q) = 1$ and $\det \mathbf{Y}^{-1}(Q) = 1$. Similarly, matrices $\mathbf{Y}(Q)$ and $\mathbf{Y}^{-1}(Q)$ satisfy important inverse relations valid for the inverse of the propagator matrix:

$$\text{if } \mathbf{Y}(Q) = \begin{pmatrix} \mathbf{a} & \mathbf{b} \\ \mathbf{c} & \mathbf{d} \end{pmatrix}, \qquad \text{then } \mathbf{Y}^{-1}(Q) = \begin{pmatrix} \mathbf{d}^T & -\mathbf{b}^T \\ -\mathbf{c}^T & \mathbf{a}^T \end{pmatrix}; \qquad (4.4.74)$$

see (4.3.25). This can again be proved by direct inspection from (4.4.70) and (4.4.72).

4.4.5 Ray Propagator Matrix Across a Curved Interface

Deriving expressions for the transformations of the ray propagator matrix across interface Σ at point Q is not difficult. Using Equations (4.4.64) and (4.4.67), we obtain

$$\begin{pmatrix} \mathbf{Q}(\tilde{Q}) \\ \mathbf{P}(\tilde{Q}) \end{pmatrix} = \mathbf{\Pi}(\tilde{Q}, Q)\begin{pmatrix} \mathbf{Q}(Q) \\ \mathbf{P}(Q) \end{pmatrix}, \qquad (4.4.75)$$

where the 4×4 matrix $\mathbf{\Pi}(\tilde{Q}, Q)$ is given as

$$\mathbf{\Pi}(\tilde{Q}, Q) = \begin{pmatrix} \mathbf{G}^T(\tilde{Q})\mathbf{G}^{-1T}(Q) & \mathbf{0} \\ \mathbf{G}^{-1}(\tilde{Q})(\mathbf{E}(Q) - \mathbf{E}(\tilde{Q}) - u\mathbf{D})\mathbf{G}^{-1\,T}(Q) & \mathbf{G}^{-1}(\tilde{Q})\mathbf{G}(Q) \end{pmatrix},$$

(4.4.76)

and u is given by (4.4.51). For $\mathbf{G}(Q)$ and $\mathbf{G}(\tilde{Q})$, refer to (4.4.48) and (4.4.49); for $\mathbf{E}(Q)$ and $\mathbf{E}(\tilde{Q})$, refer to (4.4.53) and (4.4.54); and for $\mathbf{D}$, refer to (4.4.15). Matrix $\mathbf{\Pi}(\tilde{Q}, Q)$ plays an important role in the investigation of the propagation of high-frequency seismic body waves in layered structures. We shall call it the *interface propagator matrix*.

We can see from (4.4.76) that the interface propagator matrix $\mathbf{\Pi}(\tilde{Q}, Q)$ can be expressed as a product of two matrices,

$$\mathbf{\Pi}(\tilde{Q}, Q) = \mathbf{Y}^{-1}(\tilde{Q})\mathbf{Y}(Q),$$

(4.4.77)

where $\mathbf{Y}$ and $\mathbf{Y}^{-1}$ are given by (4.4.70) and (4.4.72). Relation (4.4.75) also immediately follows from Equations (4.4.69) and (4.4.71) and from the continuity of the matrix

$$\begin{pmatrix} d\mathbf{z} \\ d\mathbf{p}^{(\Sigma)} \end{pmatrix}$$

across the interface. Hence,

$$\begin{pmatrix} \mathbf{Q}(\tilde{Q})d\gamma \\ \mathbf{P}(\tilde{Q})d\gamma \end{pmatrix} = \mathbf{Y}^{-1}(\tilde{Q}) \begin{pmatrix} d\mathbf{z} \\ d\mathbf{p}^{(\Sigma)} \end{pmatrix} = \mathbf{Y}^{-1}(\tilde{Q})\mathbf{Y}(Q) \begin{pmatrix} \mathbf{Q}(Q)d\gamma \\ \mathbf{P}(Q)d\gamma \end{pmatrix}.$$

(4.4.78)

This implies (4.4.75) with (4.4.77).

As in (4.3.5), we shall also use the standard notation for the 2×2 minors of the interface propagator matrix $\mathbf{\Pi}(\tilde{Q}, Q)$,

$$\mathbf{\Pi}(\tilde{Q}, Q) = \begin{pmatrix} \mathbf{Q}_1(\tilde{Q}, Q) & \mathbf{Q}_2(\tilde{Q}, Q) \\ \mathbf{P}_1(\tilde{Q}, Q) & \mathbf{P}_2(\tilde{Q}, Q) \end{pmatrix},$$

(4.4.79)

where

$$\begin{aligned} \mathbf{Q}_1(\tilde{Q}, Q) &= \mathbf{G}^T(\tilde{Q})\mathbf{G}^{-1T}(Q), \qquad \mathbf{Q}_2(\tilde{Q}, Q) = \mathbf{0}, \\ \mathbf{P}_1(\tilde{Q}, Q) &= \mathbf{G}^{-1}(\tilde{Q})[\mathbf{E}(Q) - \mathbf{E}(\tilde{Q}) - u\mathbf{D}]\mathbf{G}^{-1T}(Q), \\ \mathbf{P}_2(\tilde{Q}, Q) &= \mathbf{G}^{-1}(\tilde{Q})\mathbf{G}(Q). \end{aligned}$$

(4.4.80)

We shall now discuss the *chain property* of the ray propagator matrix along ray Ω crossing interface Σ at point Q; see Figure 4.10. Consider point S situated on an incident branch of ray Ω and point R situated on the reflected/transmitted branch of ray Ω. Assume that basis vectors $\vec{e}_1$ and $\vec{e}_2$ and the relevant ray propagator matrix $\mathbf{\Pi}(Q, S)$ are known at the point of incidence Q. Hence, for a wave arbitrarily reflected/transmitted at point $\tilde{Q}$,

$$\mathbf{\Pi}(\tilde{Q}, S) = \mathbf{\Pi}(\tilde{Q}, Q)\mathbf{\Pi}(Q, S),$$

(4.4.81)

where $\mathbf{\Pi}(\tilde{Q}, Q)$ is the interface propagator matrix given by (4.4.76) or (4.4.77). At point R situated on the ray of the selected reflected/transmitted wave,

$$\mathbf{\Pi}(R, S) = \mathbf{\Pi}(R, \tilde{Q})\mathbf{\Pi}(\tilde{Q}, Q)\mathbf{\Pi}(Q, S).$$

(4.4.82)

Thus, the chain property of the ray propagator matrix is satisfied even along rays Ω crossing interfaces; it is merely necessary to introduce the interface propagator matrix $\Pi(\tilde{Q}, Q)$ at the point of intersection of ray Ω with interface Σ. The point of intersection should be split into two points: the point of incidence Q and the R/T point $\tilde{Q}$.

It is not difficult to show that the product of two symplectic matrices is also symplectic. Equation (4.4.77) then indicates that the interface propagator matrix $\Pi(\tilde{Q}, Q)$ is symplectic,

$$\Pi^T(\tilde{Q}, Q)\mathbf{J}\Pi(\tilde{Q}, Q) = \mathbf{J}. \tag{4.4.83}$$

We also have $\det \Pi(\tilde{Q}, Q) = 1$. The inverse of the interface propagator matrix $\Pi^{-1}(\tilde{Q}, Q)$ is given by relation

$$
\begin{aligned}
\Pi^{-1}(\tilde{Q}, Q) = \Pi(Q, \tilde{Q}) &= \begin{pmatrix} \mathbf{Q}_1(Q, \tilde{Q}) & \mathbf{Q}_2(Q, \tilde{Q}) \\ \mathbf{P}_1(Q, \tilde{Q}) & \mathbf{P}_2(Q, \tilde{Q}) \end{pmatrix} \\
&= \begin{pmatrix} \mathbf{P}_2^T(\tilde{Q}, Q) & -\mathbf{Q}_2^T(\tilde{Q}, Q) \\ -\mathbf{P}_1^T(\tilde{Q}, Q) & \mathbf{Q}_1^T(\tilde{Q}, Q) \end{pmatrix} \\
&= \begin{pmatrix} \mathbf{G}^T(Q)\mathbf{G}^{-1\,T}(\tilde{Q}) & \mathbf{0} \\ -\mathbf{G}^{-1}(Q)[\mathbf{E}(Q) - \mathbf{E}(\tilde{Q}) - u\mathbf{D}]\mathbf{G}^{-1\,T}(\tilde{Q}) & \mathbf{G}^{-1}(Q)\mathbf{G}(\tilde{Q}) \end{pmatrix}.
\end{aligned}
\tag{4.4.84}
$$

4.4.6 Ray Propagator Matrix in a Layered Medium

We shall consider a laterally inhomogeneous medium containing curved interfaces of the first order $\Sigma_1, \Sigma_2, \ldots, \Sigma_k$. The surface of the model is considered to be one of these interfaces. Assume a ray of an *arbitrary multiply reflected (possibly converted) wave* and denote it again by Ω. We consider $N + 2$ points situated on ray Ω: the initial point S, the end point R, and N points of reflection/transmission on Ω between S and R; see Figure 4.11. At any point where ray Ω is incident at an interface, we shall again formally distinguish between the *point of incidence*, situated on the ray of the incident wave, and the *point of reflection/transmission*, situated on the ray of the reflected/transmitted wave. We denote the points of incidence (situated on the interfaces Σ_i, $i = 1, 2, \ldots, k$) consecutively $Q_1, Q_2, \ldots, Q_N$, and the corresponding points of reflection/transmission $\tilde{Q}_1, \tilde{Q}_2, \ldots, \tilde{Q}_N$. Point Q_i, of course, coincides with point $\tilde{Q}_i$, but the parameters of the medium and of the wave propagating along Ω may be different at Q_i and $\tilde{Q}_i$. Formally, we shall also use $\tilde{Q}_0 = S$. Interfaces Σ_i are assumed to be continuous together with at least the second derivatives in the vicinity of points Q_i. The segments of ray Ω in the individual layers may correspond either to a P or to an S wave. Thus, a completely general, multiply-reflected, converted high-frequency wave is considered.

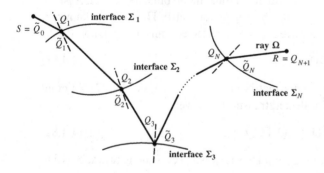

Figure 4.11. Ray Ω of an elementary multiply reflected/transmitted ray in a layered/blocked medium. Points Q_1, $Q_2, \ldots, Q_N$ represent points of incidence, and points $\tilde{Q}_1, \tilde{Q}_2, \ldots, \tilde{Q}_N$ represent the relevant reflection/transmission points.

By dynamic ray tracing along the individual segments of ray Ω (see (4.3.29)), and by applying the interface propagator matrix (see (4.4.75)), we finally obtain

$$\begin{pmatrix} \mathbf{Q}(R) \\ \mathbf{P}(R) \end{pmatrix} = \mathbf{\Pi}(R, \tilde{Q}_N) \prod_{i=N}^{1} [\mathbf{\Pi}(\tilde{Q}_i, Q_i)\mathbf{\Pi}(Q_i, \tilde{Q}_{i-1})] \begin{pmatrix} \mathbf{Q}(S) \\ \mathbf{P}(S) \end{pmatrix}. \quad (4.4.85)$$

The symbol $\prod_{i=N}^{1} \mathbf{A}_i$ denotes matrix product $\mathbf{A}_N \mathbf{A}_{N-1} \dots \mathbf{A}_1$.

It follows immediately from (4.4.85) that the ray propagator matrix $\mathbf{\Pi}(R, S)$ is given by relation

$$\mathbf{\Pi}(R, S) = \mathbf{\Pi}(R, \tilde{Q}_N) \prod_{i=N}^{1} [\mathbf{\Pi}(\tilde{Q}_i, Q_i)\mathbf{\Pi}(Q_i, \tilde{Q}_{i-1})]. \quad (4.4.86)$$

All symbols in (4.4.86) have the same meaning as in (4.4.85).

We shall now show that the ray propagator matrix $\mathbf{\Pi}(R, S)$ corresponding to an arbitrary multiply reflected (possibly converted) wave propagating in a *3-D laterally varying layered medium* satisfies the properties proved in Section 4.3 for the ray propagator matrix in a smooth medium.

a. **The symplectic property.** All matrices in the product on the RHS of (4.4.86) are symplectic. This implies that the final matrix $\mathbf{\Pi}(R, S)$ in (4.4.86) is also symplectic.
b. **Liouville's theorem.** The determinants of all matrices on the RHS of (4.4.86) equal 1. This implies that $\det \mathbf{\Pi}(R, S) = 1$, even in a laterally varying structure.
c. **Chain property.** Considering (4.3.20) for smooth parts of ray Ω and (4.4.82) across interfaces, we obtain a generally valid chain rule,

$$\mathbf{\Pi}(R, S) = \mathbf{\Pi}(R, O)\mathbf{\Pi}(O, S). \quad (4.4.87)$$

Here O is an arbitrary point situated on ray Ω of a wave multiply reflected in a layered medium. Point O may be situated even on an interface, but we must strictly distinguish between the points of incidence (Q_i) and the points of reflection/transmission ($\tilde{Q}_i$). Point O may, of course, be situated also outside points S and R, but on the ray Ω.
d. **The inverse of the ray propagator matrix.** If we express the ray propagator matrix $\mathbf{\Pi}(R, S)$ given by (4.4.86) as

$$\mathbf{\Pi}(R, S) = \begin{pmatrix} \mathbf{Q}_1(R, S) & \mathbf{Q}_2(R, S) \\ \mathbf{P}_1(R, S) & \mathbf{P}_2(R, S) \end{pmatrix}, \quad (4.4.88)$$

then the inverse of $\mathbf{\Pi}(R, S)$ is given by relation

$$\mathbf{\Pi}^{-1}(R, S) = \mathbf{\Pi}(S, R) = \begin{pmatrix} \mathbf{P}_2^T(R, S) & -\mathbf{Q}_2^T(R, S) \\ -\mathbf{P}_1^T(R, S) & \mathbf{Q}_1^T(R, S) \end{pmatrix}. \quad (4.4.89)$$

The proof of (4.4.89) follows immediately from the symplecticity and chain property of $\mathbf{\Pi}(R, S)$. It is also possible to prove (4.4.89) directly. We can simply show that the product of two matrices with inverses given by (4.4.89) has an inverse that is again given by (4.4.89). We then take into account that all matrices in the product on the RHS of (4.4.86) satisfy (4.4.89); see (4.3.26) and (4.4.84).

4.4.7 Surface-to-Surface Ray Propagator Matrix

We multiply (4.4.86) by $\mathbf{Y}^{-1}(S)$ from the right and by $\mathbf{Y}(R)$ from the left. We also introduce 4×4 matrix

$$\mathbf{T}(R, S) = \mathbf{Y}(R)\mathbf{\Pi}(R, S)\mathbf{Y}^{-1}(S). \tag{4.4.90}$$

We then take into account (4.4.77) and rewrite (4.4.86) as

$$\mathbf{T}(R, S) = \mathbf{T}(R, \tilde{Q}_N) \prod_{i=N}^{1} \mathbf{T}(Q_i, \tilde{Q}_{i-1}). \tag{4.4.91}$$

The 4×4 matrix $\mathbf{T}(R, S)$ will be called the *surface-to-surface ray propagator matrix*. We remind the reader that $\tilde{Q}_0 \equiv S$.

The surface-to-surface ray propagator matrix $\mathbf{T}(R, S)$ satisfies the same properties as ray propagator matrix $\mathbf{\Pi}(R, S)$. We can immediately see from (4.4.90) that $\mathbf{T}(S, S) = \mathbf{I}$. Matrix $\mathbf{T}(R, S)$ is symplectic because it is a product of symplectic matrices. Also, $\det \mathbf{T}(R, S) = 1$ because the determinants of $\mathbf{Y}(R)$, $\mathbf{Y}^{-1}(S)$, and $\mathbf{\Pi}(R, S)$ equal 1. If we introduce the 2×2 matrices $\mathbf{A}(R, S)$, $\mathbf{B}(R, S)$, $\mathbf{C}(R, S)$, and $\mathbf{D}(R, S)$ as

$$\mathbf{T}(R, S) = \begin{pmatrix} \mathbf{A}(R, S) & \mathbf{B}(R, S) \\ \mathbf{C}(R, S) & \mathbf{D}(R, S) \end{pmatrix}, \tag{4.4.92}$$

the inverse of $\mathbf{T}(R, S)$ is

$$\mathbf{T}^{-1}(R, S) = \mathbf{T}(S, R) = \begin{pmatrix} \mathbf{D}^{T}(R, S) & -\mathbf{B}^{T}(R, S) \\ -\mathbf{C}^{T}(R, S) & \mathbf{A}^{T}(R, S) \end{pmatrix}. \tag{4.4.93}$$

The proof of (4.4.93) is straightforward; any of the three matrices in (4.4.90) satisfies the same property. The chain property of $\mathbf{T}(R, S)$ follows from the chain property of $\mathbf{\Pi}(R, S)$.

We shall now discuss the physical meaning of the surface-to-surface ray propagator matrix. We can put

$$\mathbf{\Pi}(R, S) = \mathbf{Y}^{-1}(R)\mathbf{T}(R, S)\mathbf{Y}(S).$$

Inserting this into (4.3.29) yields

$$\mathbf{Y}(R) \begin{pmatrix} \mathbf{Q}(R) \\ \mathbf{P}(R) \end{pmatrix} = \mathbf{T}(R, S)\mathbf{Y}(S) \begin{pmatrix} \mathbf{Q}(S) \\ \mathbf{P}(S) \end{pmatrix}.$$

Using (4.4.69), we obtain

$$\begin{pmatrix} \mathrm{d}\mathbf{z}(R') \\ \mathrm{d}\mathbf{p}^{(\Sigma)}(R') \end{pmatrix} = \mathbf{T}(R, S) \begin{pmatrix} \mathrm{d}\mathbf{z}(S') \\ \mathrm{d}\mathbf{p}^{(\Sigma)}(S') \end{pmatrix}, \tag{4.4.94}$$

where R' denotes a point situated on the posterior surface passing through point R, and S' stands for a point situated on the anterior surface passing through point S. It is assumed that R' is situated close to R, and S' close to S. Equation (4.4.94) represents the final *continuation equation* for the surface-to-surface dynamic ray tracing.

We shall now explain in greater detail the physical meaning of 4×1 matrices

$$\begin{pmatrix} \mathrm{d}\mathbf{z}(R') \\ \mathrm{d}\mathbf{p}^{(\Sigma)}(R') \end{pmatrix} \quad \text{and} \quad \begin{pmatrix} \mathrm{d}\mathbf{z}(S') \\ \mathrm{d}\mathbf{p}^{(\Sigma)}(S') \end{pmatrix}.$$

Consider central ray Ω connecting point S situated on anterior surface Σ_a and point R situated on posterior surface Σ_p. We introduce local Cartesian coordinate systems z_1, z_2, z_3 at both surfaces, with origins at S and R. The z_3-axes of both systems correspond to unit normals to Σ_a and Σ_p at S and R. Axes z_1 and z_2 at S and R are chosen so that both systems z_1, z_2, z_3 form a right-handed Cartesian coordinate system; otherwise, they may be chosen arbitrarily.

Let us now select paraxial ray Ω', which passes through point S' situated on anterior surface Σ_a specified by coordinates $z_1(S')$ and $z_2(S')$. Let us emphasize that point S' is situated on surface Σ_a, not in the plane tangent to Σ_a at S. The z_3-coordinate of point S' is given by the relation $z_3 = -\frac{1}{2} z_I z_J D_{IJ}(S)$. If anterior surface Σ_a and its curvature matrix D_{IJ} at S are known, point S' is fully specified by $z_1(S')$ and $z_2(S')$; the z_3-coordinate of S' can be determined from them, if necessary. We also denote $\mathbf{dz}(S') \equiv (z_1(S'), z_2(S'))^T$.

The direction of the selected paraxial ray Ω' at S' is fully specified by $dp_1^{(\Sigma)}(S')$ and $dp_2^{(\Sigma)}(S')$, given by relations

$$dp_I^{(\Sigma)}(S') = p_I^{(\Sigma)}(S') - p_I^{(\Sigma)}(S), \qquad (4.4.95)$$

where $\vec{p}^{(\Sigma)}(S')$ denotes the vectorial component of the slowness vector, tangent to Σ_a at S', in the same way that $\vec{p}^{(\Sigma)}(S)$ denotes the vectorial component of the slowness vector, tangent to Σ_a at S. If $dp_I^{(\Sigma)}(S')$ are known at S', the complete slowness vector $\vec{p}(S')$ can be determined using Equations (4.4.59) through (4.4.62), and vice versa.

Equation (4.4.94) then determines the coordinates $z_1(R')$ and $z_2(R')$ of point R' at which paraxial ray Ω' intersects posterior surface Σ_p. In addition, it also determines components $dp_1^{(\Sigma)}(R') = p_1^{(\Sigma)}(R') - p_1^{(\Sigma)}(R)$ and $dp_2^{(\Sigma)}(R') = p_2^{(\Sigma)}(R') - p_2^{(\Sigma)}(R)$. All quantities have exactly the same meaning as they do on the anterior surface.

We shall emphasize an important point. The column 4×1 matrix $(z_1, z_2, p_1^{(\Sigma)}, p_2^{(\Sigma)})^T$ is continuous across any interface Σ. Thus, it is not necessary to distinguish between points Q_i and $\tilde{Q}_i$, and Equation (4.4.91) simplifies to

$$\mathbf{T}(Q_N, Q_0) = \prod_{i=N}^{1} \mathbf{T}(Q_i, Q_{i-1}), \qquad (4.4.96)$$

where Q_0 is situated on the anterior surface and Q_N is on the posterior surface.

Using (4.4.90) and (4.4.92), we can find the relations between $\mathbf{A}(R, S)$, $\mathbf{B}(R, S)$, $\mathbf{C}(R, S)$, and $\mathbf{D}(R, S)$ given by (4.4.92) and $\mathbf{Q}_1(R, S)$, $\mathbf{Q}_2(R, S)$, $\mathbf{P}_1(R, S)$, and $\mathbf{P}_2(R, S)$, given by (4.3.5). We shall present three important relations:

$$
\begin{aligned}
\mathbf{B} &= \mathbf{G}^{-1T}(R)\mathbf{Q}_2\mathbf{G}^{-1}(S) \\
\mathbf{DB}^{-1} &= \mathbf{G}(R)\mathbf{P}_2\mathbf{Q}_2^{-1}\mathbf{G}^T(R) + \mathbf{E}(R) - p_3^{(z)}(R)\mathbf{D}(R), \\
-\mathbf{B}^{-1}\mathbf{A} &= -\mathbf{G}(S)\mathbf{Q}_2^{-1}\mathbf{Q}_1\mathbf{G}^T(S) + \mathbf{E}(S) - p_3^{(z)}(S)\mathbf{D}(S).
\end{aligned}
\qquad (4.4.97)
$$

Thus, $\mathbf{B}$ is simply related to $\mathbf{Q}_2$, which plays an important role in the computation of geometrical spreading; see Section 4.10.2. The physical meaning of $\mathbf{DB}^{-1}$ is obvious from (4.4.37). It represents a 2×2 matrix of second derivatives of the travel-time field $T^\Sigma(z_1, z_2)$ with respect to the z_I-coordinates *along the posterior surface* Σ_p, for $\mathbf{M} = \mathbf{P}_2\mathbf{Q}_2^{-1}$. It will be shown in Section 4.6 that $\mathbf{M} = \mathbf{P}_2\mathbf{Q}_2^{-1}$ corresponds to a point source situated at S, that is, on the anterior surface Σ_a; see (4.6.8). The meaning of $-\mathbf{B}^{-1}\mathbf{A}$ is similar. It represents the matrix of second derivatives of the travel-time field $T^\Sigma(z_1, z_2)$ with respect to the z_I-coordinates *along the anterior surface* Σ_a, due to a point source situated at R, that is, on posterior surface Σ_p; see (4.6.9).

Surface-to-surface computations have been effectively used in seismic exploration for oil, particularly in models composed of homogeneous layers separated by curved interfaces; see Bortfeld (1989), Bortfeld and Kemper (1991), and Hubral, Schleicher, and Tygel (1992), among others. The complete theoretical treatment can be found in Bortfeld (1989), who speaks of the *theory of seismic systems*. See also Section 4.9.5.

4.4.8 Chain Rules for the Minors of the Ray Propagator Matrix. Fresnel Zone Matrix

One of the most important relations in the paraxial ray theory is the chain rule for the ray propagator matrix $\mathbf{\Pi}(R, S)$ in layered laterally varying media; see (4.4.87) as well as (4.4.81) or (4.4.82). A similar chain rule can also be written for 2×2 matrices $\mathbf{Q}_1(R, S)$, $\mathbf{Q}_2(R, S)$, $\mathbf{P}_1(R, S)$, and $\mathbf{P}_2(R, S)$, minors of the ray propagator matrix $\mathbf{\Pi}(R, S)$; see (4.3.5). The chain rules allow matrices $\mathbf{Q}_1$, $\mathbf{Q}_2$, $\mathbf{P}_1$, and $\mathbf{P}_2$ to be factorized into individual factors, corresponding to specified branches of ray Ω. These factors can be calculated independently, and in many cases analytically.

The chain rules for $\mathbf{Q}_1$, $\mathbf{Q}_2$, $\mathbf{P}_1$, and $\mathbf{P}_2$ have certain important applications. For example, the chain rule for matrix $\mathbf{Q}_2(R, S)$ can be used to factorize the geometrical spreading; see Section 4.10.4. It also provides a suitable tool for investigating the KMAH index and the phase shift due to caustics; see Section 4.12.2.

Let us consider ray Ω, connecting two points S and R. The point S may represent a point source, or it may merely represent a reference point on Ω. At point Q, situated on ray Ω between S and R, ray Ω is incident at interface Σ. The ray propagator matrix $\mathbf{\Pi}(R, S)$ can then be expressed in the following form (see (4.4.82)):

$$\mathbf{\Pi}(R, S) = \begin{pmatrix} \mathbf{Q}_1(R, \tilde{Q}) & \mathbf{Q}_2(R, \tilde{Q}) \\ \mathbf{P}_1(R, \tilde{Q}) & \mathbf{P}_2(R, \tilde{Q}) \end{pmatrix} \begin{pmatrix} \mathbf{Q}_1^I & \mathbf{0} \\ \mathbf{P}_1^I & \mathbf{P}_2^I \end{pmatrix} \begin{pmatrix} \mathbf{Q}_1(Q, S) & \mathbf{Q}_2(Q, S) \\ \mathbf{P}_1(Q, S) & \mathbf{P}_2(Q, S) \end{pmatrix},$$

$$(4.4.98)$$

where $\mathbf{Q}_1^I = \mathbf{Q}_1(\tilde{Q}, Q)$, $\mathbf{P}_1^I = \mathbf{P}_1(\tilde{Q}, Q)$, and $\mathbf{P}_2^I = \mathbf{P}_2(\tilde{Q}, Q)$ are 2×2 minors of the interface propagator matrix $\mathbf{\Pi}(\tilde{Q}, Q)$,

$$\mathbf{Q}_1^I = \mathbf{G}^T(\tilde{Q})\mathbf{G}^{-1T}(Q), \qquad \mathbf{P}_1^I = \mathbf{G}^{-1}(\tilde{Q})(\mathbf{W}(Q) - \mathbf{W}(\tilde{Q}))\mathbf{G}^{-1T}(Q),$$

$$\mathbf{P}_2^I = \mathbf{G}^{-1}(\tilde{Q})\mathbf{G}(Q), \qquad \mathbf{W}(Q) = \mathbf{E}(Q) - p_3^{(z)}(Q)\mathbf{D};$$

$$(4.4.99)$$

see (4.4.80). From (4.4.98), we then obtain

$$\mathbf{Q}_1(R, S) = \mathbf{Q}_1(R, \tilde{Q})\mathbf{U}^1\mathbf{Q}_1(Q, S),$$
$$\mathbf{Q}_2(R, S) = \mathbf{Q}_2(R, \tilde{Q})\mathbf{U}^2\mathbf{Q}_2(Q, S),$$
$$\mathbf{P}_1(R, S) = \mathbf{P}_1(R, \tilde{Q})\mathbf{U}^3\mathbf{P}_1(Q, S),$$
$$\mathbf{P}_2(R, S) = \mathbf{P}_2(R, \tilde{Q})\mathbf{U}^4\mathbf{P}_2(Q, S).$$

$$(4.4.100)$$

These equations represent the basic form of *the chain rules for minors of the ray propagator matrix*. The whole ray Ω is decomposed into two branches: from S to Q and from $\tilde{Q}$ to R. The 2×2 matrices $\mathbf{U}^1$, $\mathbf{U}^2$, $\mathbf{U}^3$, and $\mathbf{U}^4$ are given by relations

$$\mathbf{U}^1 = \mathbf{Q}_1^I + \mathbf{Q}_1^{-1}(R, \tilde{Q})\mathbf{Q}_2(R, \tilde{Q})\mathbf{P}_1^I$$
$$\qquad + \mathbf{Q}_1^{-1}(R, \tilde{Q})\mathbf{Q}_2(R, \tilde{Q})\mathbf{P}_2^I\mathbf{P}_1(Q, S)\mathbf{Q}_1^{-1}(Q, S),$$

$$\mathbf{U}^2 = \mathbf{P}_1^I + \mathbf{Q}_2^{-1}(R, \tilde{Q})\mathbf{Q}_1(R, \tilde{Q})\mathbf{Q}_1^I + \mathbf{P}_2^I\mathbf{P}_2(Q, S)\mathbf{Q}_2^{-1}(Q, S),$$

$$\mathbf{U}^3 = \mathbf{Q}_1^I\mathbf{Q}_1(Q, S)\mathbf{P}_1^{-1}(Q, S) + \mathbf{P}_1^{-1}(R, \tilde{Q})\mathbf{P}_2(R, \tilde{Q})\mathbf{P}_2^I$$
$$\qquad + \mathbf{P}_1^{-1}(R, \tilde{Q})\mathbf{P}_2(R, \tilde{Q})\mathbf{P}_1^I\mathbf{Q}_1(Q, S)\mathbf{P}_1^{-1}(Q, S),$$

$$\mathbf{U}^4 = \mathbf{P}_2^I + \mathbf{P}_1^I\mathbf{Q}_2(Q, S)\mathbf{P}_2^{-1}(Q, S)$$
$$\qquad + \mathbf{P}_2^{-1}(R, \tilde{Q})\mathbf{P}_1(R, \tilde{Q})\mathbf{Q}_1^I\mathbf{Q}_2(Q, S)\mathbf{P}_2^{-1}(Q, S).$$

The expressions for $\mathbf{U}^1, \mathbf{U}^2, \mathbf{U}^3$, and $\mathbf{U}^4$ are not simple and contain minors of ray propagator matrices $\mathbf{\Pi}(R, \tilde{Q})$ and $\mathbf{\Pi}(Q, S)$. All these minors, however, can be expressed in terms of the 2×2 matrices $\mathbf{M}$ of the second derivatives of the travel-time field at Q or $\tilde{Q}$, due to a point source or a telescopic source at S or R. Here we shall discuss in greater detail only the chain rule for the minor $\mathbf{Q}_2(R, S)$, which plays a very important role in many applications. We shall denote by $\mathbf{M}(Q, S)$ the matrix $\mathbf{M}$ of the second derivatives of the travel-time field at Q due to a point source situated at S, and by $\mathbf{M}(\tilde{Q}, R)$ the matrix $\mathbf{M}$ at $\tilde{Q}$ due to a point source situated at R. Applying (4.1.72) and (4.3.26) to the point sources at S and R, we obtain

$$\mathbf{M}(Q, S) = \mathbf{P}_2(Q, S)\mathbf{Q}_2^{-1}(Q, S), \qquad \mathbf{M}(\tilde{Q}, R) = -\mathbf{Q}_2^{-1}(R, \tilde{Q})\mathbf{Q}_1(R, \tilde{Q}).$$
(4.4.101)

For more details on (4.4.101), see Section 4.6.1. Using (4.4.101), the expression for $\mathbf{U}^2$ may be expressed in the following way:

$$\begin{aligned} \mathbf{U}^2 &= \mathbf{P}_1^I - \mathbf{M}(\tilde{Q}, R)\mathbf{Q}_1^I + \mathbf{P}_2^I \mathbf{M}(Q, S) \\ &= \mathbf{G}^{-1}(\tilde{Q})[\mathbf{F}(Q, S) - \mathbf{F}(\tilde{Q}, R)]\mathbf{G}^{-1T}(Q), \end{aligned}$$
(4.4.102)

where

$$\begin{aligned} \mathbf{F}(Q, S) &= \mathbf{G}(Q)\mathbf{M}(Q, S)\mathbf{G}^T(Q) + \mathbf{E}(Q) - p_3^{(z)}(Q)\mathbf{D}(Q), \\ \mathbf{F}(\tilde{Q}, R) &= \mathbf{G}(\tilde{Q})\mathbf{M}(\tilde{Q}, R)\mathbf{G}^T(\tilde{Q}) + \mathbf{E}(\tilde{Q}) - p_3^{(z)}(\tilde{Q})\mathbf{D}(Q). \end{aligned}$$
(4.4.103)

As we can see from (4.4.37) and (4.4.40), $\mathbf{F}(Q, S)$ represents the matrix of second derivatives of the travel-time field along surface Σ at point Q due to a point source situated at S. Matrix $\mathbf{F}(\tilde{Q}, R)$ has a fully analogous meaning, but the source is situated at R. Similarly, we can introduce $\mathbf{F}(\tilde{Q}, S)$ and $\mathbf{F}(Q, R)$. Matrices $\mathbf{F}$ are continuous across the interface Σ:

$$\mathbf{F}(Q, S) = \mathbf{F}(\tilde{Q}, S), \qquad \mathbf{F}(Q, R) = \mathbf{F}(\tilde{Q}, R).$$

We now introduce the 2×2 matrix $\mathbf{M}^F(Q; R, S)$ by the relation

$$\mathbf{U}^2 = \mathbf{G}^{-1}(\tilde{Q})\mathbf{M}^F(Q; R, S)\mathbf{G}^{-1T}(Q).$$
(4.4.104)

Using (4.4.102), we obtain

$$\begin{aligned} \mathbf{M}^F(Q; R, S) &= \mathbf{F}(Q, S) - \mathbf{F}(\tilde{Q}, R) \\ &= \mathbf{F}(Q, S) - \mathbf{F}(Q, R) \\ &= \mathbf{F}(\tilde{Q}, S) - \mathbf{F}(\tilde{Q}, R). \end{aligned}$$
(4.4.105)

Matrix $\mathbf{M}^F(Q; R, S)$ plays an important role in the computation of Fresnel volumes and Fresnel zones; see Section 4.11. For this reason, we call it the *Fresnel zone matrix*. The term Fresnel zone matrix was introduced by Hubral, Schleicher, and Tygel (1992), using the surface-to-surface ray propagator matrix. Fresnel zone matrices have also found applications in the calculation of relative geometrical spreading (see Section 4.10.4) and in the computation of the KMAH index (see Section 4.12.2).

The Fresnel zone matrix $\mathbf{M}^F(Q; R, S)$ is related to the selected ray Ω and to three points situated on it: point source S, receiver R, and the point Q at which the Fresnel zone matrix is evaluated. It is continuous across interface Σ,

$$\mathbf{M}^F(Q; R, S) = \mathbf{M}^F(\tilde{Q}; R, S);$$
(4.4.106)

see (4.4.105).

An important note. In our treatment, the travel time increases from S to R, and the introduced ray-centered coordinate system corresponds to the direction of propagation from S to R. Both $\mathbf{F}(Q, S)$ and $\mathbf{F}(\tilde{Q}, R)$ are expressed in the same ray-centered coordinate system. This implies that the numerical value of $\mathbf{F}(Q, S)$ will mostly have the opposite sign to the numerical value of $\mathbf{F}(\tilde{Q}, R)$. For example, in a homogeneous medium, $\mathbf{M}(Q, S)$ is always positive, and $\mathbf{M}(\tilde{Q}, R)$ is always negative.

As shown in (4.4.105), the Fresnel zone matrix can be expressed in several general forms. If we insert (4.4.103) into (4.4.105), we obtain more specific expressions:

$$
\begin{aligned}
\mathbf{M}^F(Q; R, S) &= \mathbf{G}(Q)[\mathbf{M}(Q, S) - \mathbf{M}(Q, R)]\mathbf{G}^T(Q) \\
&= \mathbf{G}(\tilde{Q})[\mathbf{M}(\tilde{Q}, S) - \mathbf{M}(\tilde{Q}, R)]\mathbf{G}^T(\tilde{Q}) \\
&= \mathbf{G}(\tilde{Q})\big[\mathbf{P}_1^I + \mathbf{P}_2^I \mathbf{M}(Q, S) - \mathbf{M}(\tilde{Q}, R)\mathbf{Q}_1^I\big]\mathbf{G}^T(Q).
\end{aligned}
\tag{4.4.107}
$$

Alternatively, other useful forms are

$$
\begin{aligned}
\mathbf{M}^F(Q; R, S) &= \mathbf{G}(Q)\mathbf{M}(Q, S)\mathbf{G}^T(Q) - \mathbf{G}(\tilde{Q})\mathbf{M}(\tilde{Q}, R)\mathbf{G}^T(\tilde{Q}) \\
&\quad + \mathbf{E}(Q) - \mathbf{E}(\tilde{Q}) - u\mathbf{D}(Q) \\
&= \mathbf{G}(Q)\mathbf{P}_2(Q, S)\mathbf{Q}_2^{-1}(Q, S)\mathbf{G}^T(Q) \\
&\quad + \mathbf{G}(\tilde{Q})\mathbf{Q}_2^{-1}(R, \tilde{Q})\mathbf{Q}_1(R, \tilde{Q})\mathbf{G}^T(\tilde{Q}) \\
&\quad + \mathbf{E}(Q) - \mathbf{E}(\tilde{Q}) - u\mathbf{D}(Q).
\end{aligned}
\tag{4.4.108}
$$

The expressions (4.4.108) for the Fresnel zone matrix $\mathbf{M}^F(Q; R, S)$ are very general. If we choose the standard option of the local Cartesian coordinate system z_i at point Q on Σ, given by (4.4.21), and a standard choice of $\vec{e}_1(\tilde{Q})$ and $\vec{e}_2(\tilde{Q})$ given by (2.3.45), the individual matrices in (4.4.108) simplify. We can insert expressions (4.4.48) with (4.4.49) for $\mathbf{G}(Q)$ and $\mathbf{G}(\tilde{Q})$, (4.4.51) for u, (4.4.53) for $E_{IJ}(Q)$, and (4.4.54) for $E_{IJ}(\tilde{Q})$. For other possible simplifications, see Section 4.8.4.

The final equation for the decomposition of $\mathbf{Q}_2(R, S)$ using the Fresnel zone matrix can be obtained from (4.4.100) and (4.4.104):

$$
\mathbf{Q}_2(R, S) = \mathbf{Q}_2(R, \tilde{Q})\mathbf{G}^{-1}(\tilde{Q})\mathbf{M}^F(Q; R, S)\mathbf{G}^{-1T}(Q)\mathbf{Q}_2(Q, S).
\tag{4.4.109}
$$

Relation (4.4.109) can, of course, be further chained, if we decompose $\mathbf{Q}_2(R, \tilde{Q})$ and/or $\mathbf{Q}_2(Q, S)$. It would also be possible to write a similar expression for N points of incidence, $Q_1, Q_2, \ldots, Q_N$, situated along Ω between S and R. See Hubral, Tygel, and Schleicher (1995).

Points Q and $\tilde{Q}$ in the preceding relations are situated on a curved interface that crosses ray Ω. Point Q corresponds to the point of incidence, and point $\tilde{Q}$ corresponds to the point of reflection/transmission. All the presented relations, of course, also remain valid for point Q, situated on the smooth part of the ray.

4.4.9 Backward Propagation

In certain applications, particularly in the investigation of various source-receiver reciprocities, it would be useful to compare the results of the forward computations (from S to R) with those of the backward computation (from R to S). In the backward propagation, we must consider an orientation of the slowness vector opposite to that in the forward propagation.

We shall first briefly discuss the forward dynamic ray tracing along Ω, from S to R. At the initial point S, we specify the right-handed triplet of unit vectors $\vec{e}_1(S)$, $\vec{e}_2(S)$, and $\vec{e}_3(S) \equiv \vec{t}(S)$, where $\vec{t}(S)$ is tangent to the ray Ω and oriented in the direction of propagation from S to R. The unit vectors $\vec{e}_1(S)$ and $\vec{e}_2(S)$ may be chosen arbitrarily in the plane perpendicular to Ω at S, only they must be mutually perpendicular and form a right-handed triplet with $\vec{t}(S)$. At any point of incidence on a structural interface, we use a standard option (4.4.21) to construct a local Cartesian coordinate system z_i, with the z_3-axis oriented along the normal $\vec{n}$ to the interface at that point. The orientation of the unit normal $\vec{n}$ may be chosen arbitrarily to any side of the interface. We also use the standard choice (2.3.45) to determine $\vec{e}_1$ and $\vec{e}_2$ for the generated R/T waves. See also (4.4.25). As a result, we obtain the 4×4 propagator matrix $\mathbf{\Pi}(R, S)$ and the relevant right-handed triplet $\vec{e}_1(R)$, $\vec{e}_2(R)$, and $\vec{e}_3(R) \equiv \vec{t}(R)$ at R. An important consequence of the standard choice of $\vec{e}_1$ and $\vec{e}_2$ at individual points of incidence follows. If the unit vectors $\vec{e}_1(S)$ and $\vec{e}_2(S)$ are rotated by an angle φ in the plane perpendicular to the ray Ω at S, the unit vectors $\vec{e}_1(R)$ and $\vec{e}_2(R)$ will be rotated by the same angle in the plane perpendicular to the ray Ω at R.

Now we shall specify the *backward propagation* and denote the quantities corresponding to the backward propagation by the superscript b. At R, we choose

$$\vec{e}_1^b(R) = -\vec{e}_1(R), \qquad \vec{e}_2^b(R) = \vec{e}_2(R), \qquad \vec{e}_3^b(R) = -\vec{e}_3(R). \qquad (4.4.110)$$

Here we have changed the orientation of $\vec{e}_3^b(R)$ because the slowness vector $\vec{p}^b(R)$ is opposite to $\vec{p}(R)$. Since we wish to use the right-handed triplet, we have also changed the orientation of $\vec{e}_1^b(R)$. It would also be possible to use different alternative choices, but here we shall consider (4.4.110) systematically in the backward propagation.

At all interfaces, we must relate $\vec{i}_3^{(z)} \equiv \vec{n}$ used in the forward propagation to $\vec{i}_3^{(z)b}$ in the backward propagation. We shall use the following relation at all points of incidence:

$$\vec{i}_3^{(z)b} = -\vec{i}_3^{(z)}. \qquad (4.4.111)$$

Thus, the unit vectors $\vec{i}_3^{(z)b}$ are taken opposite to $\vec{i}_3^{(z)}$ at all points of incidence. The other two unit basis vectors $\vec{i}_1^{(z)b}$ and $\vec{i}_2^{(z)b}$ are computed in a standard way, using the standard option (4.4.21). This gives $\vec{i}_1^{(z)b} = -\vec{i}_1^{(z)}$ and $\vec{i}_2^{(z)b} = \vec{i}_2^{(z)}$. Consequently, the local Cartesian coordinate systems are again right-handed at all points of incidence. (See Figure 5.9.) The standard choice should be used to determine $\vec{e}_1^b$, $\vec{e}_2^b$, and $\vec{e}_3^b$ at relevant R/T points.

Thus, the backward propagation from R to S is fully specified by conditions (4.4.110) and (4.4.111). By dynamic ray tracing in the backward direction, from R to S, we obtain at S the propagator matrix $\mathbf{\Pi}^b(S, R)$ and the basis vectors $\vec{e}_1^b(S)$, $\vec{e}_2^b(S)$, and $\vec{e}_3^b(S)$. The basis vectors $\vec{e}_i^b(S)$ are related to $\vec{e}_i(S)$ as follows:

$$\vec{e}_1^b(S) = -\vec{e}_1(S), \qquad \vec{e}_2^b(S) = \vec{e}_2(S), \qquad \vec{e}_3^b(S) = -\vec{e}_3(S), \qquad (4.4.112)$$

We also obtain a simple relation between the forward propagator matrix $\mathbf{\Pi}(R, S)$ (see (4.3.5)) and the backward propagator matrix $\mathbf{\Pi}^b(S, R)$:

$$\mathbf{\Pi}^b(S, R) = \begin{pmatrix} \mathbf{Q}_1^b(S, R) & \mathbf{Q}_2^b(S, R) \\ \mathbf{P}_1^b(S, R) & \mathbf{P}_2^b(S, R) \end{pmatrix} = \begin{pmatrix} \bar{\mathbf{Q}}_1(S, R) & -\bar{\mathbf{Q}}_2(S, R) \\ -\bar{\mathbf{P}}_1(S, R) & \bar{\mathbf{P}}_2(S, R) \end{pmatrix}$$

$$= \begin{pmatrix} \bar{\mathbf{P}}_2^T(R, S) & \bar{\mathbf{Q}}_2^T(R, S) \\ \bar{\mathbf{P}}_1^T(R, S) & \bar{\mathbf{Q}}_1^T(R, S) \end{pmatrix}. \qquad (4.4.113)$$

Here $\bar{\mathbf{Q}}_I$ and $\mathbf{Q}_I$ are the same; only their off-diagonal terms have opposite signs. The relations between $\bar{\mathbf{P}}_I$ and $\mathbf{P}_I$ are analogous.

Equations (4.4.112) and (4.4.113) can be used in various applications of forward and backward propagation. The most important consequence of (4.4.113) is

$$\det \mathbf{Q}_2^b(S, R) = \det \mathbf{Q}_2(S, R) = \det \mathbf{Q}_2(R, S). \qquad (4.4.114)$$

It will be shown in Section 4.10.2 that (4.4.114) implies the reciprocity of relative geometrical spreading in forward and backward propagation.

4.5 Initial Conditions for Dynamic Ray Tracing

To compute ray propagator matrix $\mathbf{\Pi}$ along ray Ω, we do not need to specify the initial conditions at the initial point S of ray Ω; they are given by the identity matrix. To determine the actual matrices $\mathbf{Q}$, $\mathbf{P}$, and $\mathbf{M}$ along Ω, however, we must know the initial values of these matrices at S (at least of two of them). The problem is that these matrices are not, as a rule, known at S; they must be computed from other known quantities at S. All three initial matrices $\mathbf{Q}(S)$, $\mathbf{P}(S)$, and $\mathbf{M}(S)$ depend on the selection of the ray-centered coordinate system $\vec{e}_1(S)$, $\vec{e}_2(S)$, and $\vec{e}_3(S)$. Matrices $\mathbf{Q}(S)$ and $\mathbf{P}(S)$ also depend on the parameterization of the ray field close to S. In addition, $\mathbf{M}(S)$ and $\mathbf{P}(S)$ depend on the travel-time field in the vicinity of S.

In this section, we shall first consider the case of a smooth curved initial surface Σ^0 situated in *an isotropic inhomogeneous medium*, along which the initial time field T^0 is specified. To perform ray tracing and dynamic ray tracing from Σ^0, it is necessary to solve two problems:

   a. To determine the initial slowness vectors for ray tracing.
   b. To determine $\mathbf{Q}(S)$ and $\mathbf{P}(S)$ for dynamic ray tracing in ray-centered coordinates.

The first problem will be solved in Section 4.5.1; the second will be solved in Section 4.5.2. In Section 4.5.3, the results are specified for ray parameters γ_I corresponding to local Cartesian coordinates z_I in a plane tangent to Σ^0 at S. In addition to a smooth initial surface Σ^0, we shall also consider two very important cases of a point source at S (see Section 4.5.4) and of an initial line C^0 (see Section 4.5.5). Finally, a general initial surface Σ^0 containing edges and vertexes will be discussed in Section 4.5.6.

There are many analogies between the problem of initial surface Σ^0 and the problem of reflection/transmission at a structural interface Σ. The main difference consists in the different functional descriptions of the initial surface Σ^0 and structural interface Σ we use. In the case of initial surface Σ^0, it is convenient to consider a *global parameteric description*, with two parameters γ_1 and γ_2, which may also be taken as the *ray parameters* of the orthonomic system of rays generated at Σ^0. The initial travel time T^0 along Σ^0 may also be simply introduced as a function of these two parameters. The implicit description $\Sigma(x_i) = 0$ (see (4.4.1)) or the explicit description $x_3 = f(x_I)$ (see (4.4.5)) used in the solution of the reflection/transmission problems are not as suitable in solving the initial surface problem because they do not assign initial conditions to given ray parameters. Because we shall use the parameteric description of Σ^0, we cannot directly apply the results of Section 4.4 here. Another difference with respect to Section 4.4 is that we shall not construct an auxiliary Cartesian coordinate system z_1, z_2, z_3 at point S of initial surface Σ^0. The parameteric description of Σ^0 does not require it. This will be done in Section 4.5.3 only as an example to the general approach.

Figure 4.12. Initial surface Σ^0, specified by parameteric equation $\vec{x} = \vec{x}(\gamma_1, \gamma_2)$, where γ_1 and γ_2 are two parameters such as the Gaussian coordinates along Σ^0. Isolines $\gamma_1 = $ const. and $\gamma_2 = $ const. and the unit normal $\vec{n}^0$ at point $S[\gamma_1, \gamma_2]$ on Σ^0 are shown.

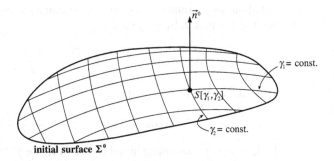

4.5.1 Initial Slowness Vector at a Smooth Initial Surface

We shall consider a smooth initial surface Σ^0. Surface Σ^0 may correspond to a structural interface, to a free surface, to an interface between a solid and fluid medium (for example, an ocean bottom), to a wavefront, to a hypothetical exploding reflector, or to an auxiliary surface situated in a smooth medium. Initial surface Σ^0 will be described by parameteric vectorial equation

$$\vec{x} = \vec{x}(\gamma_1, \gamma_2). \tag{4.5.1}$$

Here $\vec{x}$ is the radius-vector and γ_1 and γ_2 are two parameters (for example, Gaussian coordinates along Σ^0). Parameters γ_1 and γ_2 will also be taken as *ray parameters*, specifying the rays generated at the individual points of Σ^0. See Figure 4.12. Equation (4.5.1) is very general and can also be used to describe closed smooth surfaces, salt domes, and the like. We shall use the standard notation for first and second partial derivatives,

$$\vec{x}_{,I} = \partial\vec{x}/\partial\gamma_I, \qquad \vec{x}_{,IJ} = \partial^2\vec{x}/\partial\gamma_I\partial\gamma_J, \tag{4.5.2}$$

and denote

$$E = \vec{x}_{,1}\cdot\vec{x}_{,1}, \qquad F = \vec{x}_{,1}\cdot\vec{x}_{,2}, \qquad G = \vec{x}_{,2}\cdot\vec{x}_{,2}. \tag{4.5.3}$$

Assume that initial time field $T^0(\gamma_1, \gamma_2)$ is known along Σ^0. Initial time field $T^0(\gamma_1, \gamma_2)$ may be constant along Σ^0 (wavefront, exploding reflector) or may vary along it. We also assume that the first and second derivatives of the initial time field, $T_{,I}^0 = \partial T^0/\partial\gamma_I$ and $T_{,IJ}^0 = \partial^2 T^0/\partial\gamma_I\partial\gamma_J$, are known along Σ^0 (they can be determined from known $T^0(\gamma_1, \gamma_2)$).

Now we shall determine the tangential component $\vec{p}^{(\Sigma)}(S)$ of the initial slowness vector $\vec{p}^0(S)$ at any selected point S situated on Σ^0. Because $\vec{p}^{(\Sigma)}$ is tangential to Σ^0, we can put $\vec{p}^{(\Sigma)} = a\vec{x}_{,1} + b\vec{x}_{,2}$, where a and b are not yet known. Multiplying this equation successively by $\vec{x}_{,1}$ and $\vec{x}_{,2}$, we obtain two equations for a and b:

$$\vec{p}^{(\Sigma)}\cdot\vec{x}_{,1} = T_{,1}^0 = aE + bF, \qquad \vec{p}^{(\Sigma)}\cdot\vec{x}_{,2} = T_{,2}^0 = aF + bG.$$

Solving these equations for a and b, we obtain

$$\vec{p}^{(\Sigma)} = (EG - F^2)^{-1}\big[\big(GT_{,1}^0 - FT_{,2}^0\big)\vec{x}_{,1} + \big(-FT_{,1}^0 + ET_{,2}^0\big)\vec{x}_{,2}\big]. \tag{4.5.4}$$

This is the final expression for $\vec{p}^{(\Sigma)}$. Magnitude $p^{(\Sigma)}$ of $\vec{p}^{(\Sigma)}$ can be determined from the relation

$$p^{(\Sigma)2} = \vec{p}^{(\Sigma)}\cdot\vec{p}^{(\Sigma)} = (EG - F^2)^{-1}\big(GT_{,1}^{02} + ET_{,2}^{02} - 2FT_{,1}^0T_{,2}^0\big). \tag{4.5.5}$$

Equations (4.5.4) and (4.5.5) can also be expressed in a more compact form if we introduce the 2×2 matrix $\mathbf{A}$ by the relation

$$\mathbf{A} = (EG - F^2)^{-1} \begin{pmatrix} G & -F \\ -F & E \end{pmatrix}. \tag{4.5.6}$$

Then (4.5.4) and (4.5.5) read

$$\vec{p}^{(\Sigma)} = A_{IK} T^0_{,K} \vec{x}_{,I}, \qquad p^{(\Sigma)2} = A_{IK} T^0_{,I} T^0_{,K}. \tag{4.5.7}$$

To determine the complete initial slowness vector $\vec{p}^0(S)$, we also need to find its component normal to Σ^0. We denote the unit normal to initial surface Σ^0 at S by $\vec{n}^0$. If Σ^0 is smooth in the vicinity of S, $\vec{n}^0$ is given by relation

$$\vec{n}^0 = (\vec{x}_{,1} \times \vec{x}_{,2})/|\vec{x}_{,1} \times \vec{x}_{,2}| = (EG - F^2)^{-1/2}(\vec{x}_{,1} \times \vec{x}_{,2}). \tag{4.5.8}$$

For points S situated at the edges and vertices in Σ^0, see Sections 4.5.4 through 4.5.6.

Slowness vector $\vec{p}^0$ is situated in the *radiation plane*, specified by $\vec{n}^0$ and $\vec{p}^{(\Sigma)}$,

$$\vec{p}^0 = \sigma \vec{n}^0 + \vec{p}^{(\Sigma)}. \tag{4.5.9}$$

Quantity σ can be determined from the relevant eikonal equation. For isotropic media, the eikonal equation reads $\sigma^2 + p^{(\Sigma)2} = 1/V^2$, where V is the propagation velocity at S, but outside Σ^0. Consequently,

$$\sigma = \epsilon \left(V^{-2} - p^{(\Sigma)2} \right)^{1/2}. \tag{4.5.10}$$

Here $\epsilon = \pm 1$ is the given orientation index, $\epsilon = \text{sgn}(\vec{p}^0 \cdot \vec{n}^0)$, specifying to which side of Σ^0 the wave under consideration propagates. If Σ^0 is a structural interface, we have, in general, four possible velocities $V(S)$: P and S wave velocities on each side of Σ^0 at S. Consequently, we have *four generated waves* at S. The number of velocities and relevant generated waves is smaller in some special cases (for example, acoustic case or Σ^0 representing a formal surface).

The initial slowness vectors $\vec{p}^0(S)$ of all waves generated at S are situated in the same radiation plane so that they are *coplanar*. The radiation plane plays a role very similar to the role of the plane of incidence in the R/T problems. See Figure 4.13.

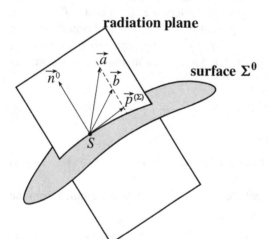

radiation plane

surface Σ^0

Figure 4.13. Radiation plane, constructed at point S of initial surface Σ^0. It is specified by unit normal $\vec{n}^0$ to Σ^0 at S, and by the tangential component $\vec{p}^{(\Sigma)}(S)$ of the slowness vector at S. The initial slowness vectors of all elementary waves generated on initial surface Σ^0 at S are situated in the radiation plane; see vectors $\vec{a}$ and $\vec{b}$.

The slowness vector $\vec{p}^0$ given by (4.5.9) corresponds to a homogeneous wave only if σ is real-valued, that is, if

$$p^{(\Sigma)2} \leq V^{-2}(S). \tag{4.5.11}$$

In the opposite case, the relevant slowness vector $\vec{p}^0$ and the corresponding ray are not real, but complex-valued. The corresponding wave is then *inhomogeneous* and cannot be computed by the standard ray method. It is obvious that certain of the possible four generated waves may be inhomogeneous (with a complex-valued initial slowness vector), and the other may represent standard homogeneous waves. There are several various combinations of generated homogeneous and inhomogeneous waves, as in the R/T problem. For the limiting case of $p^{(\Sigma)2} = V^{-2}(S)$, we obtain $\sigma = 0$. The relevant ray grazes initial surface Σ^0 at S.

The procedure described here may also be used if initial surface Σ^0 is situated in an inhomogeneous *anisotropic* medium. Relations (4.5.7) and (4.5.9) remain valid; only σ in (4.5.9) must be determined from the eikonal equation for an anisotropic medium. The method outlined in Section 2.3.3 can be used for this purpose. In general, three different values of σ, corresponding to one qP and two qS waves, may be obtained on each side of Σ^0 at S.

4.5.2 Initial Values of Q, P, and M at a Smooth Initial Surface

First, we need to determine the initial values of basis vectors $\vec{e}_1(S)$, $\vec{e}_2(S)$, and $\vec{e}_3(S)$ of the ray-centered coordinate system at S. Unit vector $\vec{e}_3(S)$ is tangent to ray Ω at S,

$$\vec{e}_3(S) = V(S)\vec{p}^0(S) = V(S)\left(\sigma \vec{n}^0 + \vec{p}^{(\Sigma)}(S)\right). \tag{4.5.12}$$

The other two unit vectors, $\vec{e}_1(S)$ and $\vec{e}_2(S)$, may be taken arbitrarily in the plane perpendicular to $\vec{e}_3(S)$; the only requirement is that $\vec{e}_1(S)$, $\vec{e}_2(S)$, and $\vec{e}_3(S)$ be mutually perpendicular and form a right-handed system. It may also be convenient to take either $\vec{e}_1(S)$ or $\vec{e}_2(S)$ perpendicular to the radiation plane.

Now we shall determine $\mathbf{Q}(S)$ and $\mathbf{P}(S)$. The determination of $\mathbf{Q}(S)$ is easy. We have

$$Q_{IJ} = \partial q_I / \partial \gamma_J = (\partial q_I / \partial x_k)(\partial x_k / \partial \gamma_J) = e_{Ik}(\partial x_k / \partial \gamma_J).$$

Consequently,

$$Q_{IJ}(S) = \vec{e}_I(S) \cdot \vec{x}_{,J}(S). \tag{4.5.13}$$

The determination of $P_{IJ}(S)$ is more involved. We need to determine the partial derivatives of slowness vector $\vec{p}$ with respect to γ_J along the wavefront (not along Σ^0). In addition to ray coordinates $\gamma_1, \gamma_2, \gamma_3 = T$ we also introduce auxiliary ray coordinates $\gamma_1' = \gamma_1$, $\gamma_2' = \gamma_2$, and $\gamma_3' = T - T^0(\gamma_1, \gamma_2)$. Whereas the coordinate surface $\gamma_3 = T$ represents the wavefront, coordinate surface $\gamma_3' = T - T^0 = 0$ is the initial surface, Σ^0. The transformation matrix $(\partial \gamma_i / \partial \gamma_j')$ from coordinates $\gamma_1, \gamma_2, \gamma_3$ to coordinates $\gamma_1', \gamma_2', \gamma_3'$ and its inverse are given by relations

$$(\partial \gamma_i / \partial \gamma_j') = \begin{pmatrix} 1 & 0 & 0 \\ 0 & 1 & 0 \\ T^0_{,1} & T^0_{,2} & 1 \end{pmatrix}, \quad (\partial \gamma_i' / \partial \gamma_j) = \begin{pmatrix} 1 & 0 & 0 \\ 0 & 1 & 0 \\ -T^0_{,1} & -T^0_{,2} & 1 \end{pmatrix}. \tag{4.5.14}$$

Now we shall discuss the partial derivatives of the slowness vector $\vec{p}$ with respect to ray parameters γ_I. We shall use the following notation:

$$\vec{p}_{,J} = (\partial \vec{p}/\partial \gamma_J)_{\text{wavefront}}, \qquad \vec{p}^{\,0}_{,J} = (\partial \vec{p}/\partial \gamma_J)_{\Sigma^0} = (\partial \vec{p}^{\,0}/\partial \gamma_J)_{\Sigma^0}. \quad (4.5.15)$$

Here $\vec{p}^{\,0}$ is given by (4.5.9). The relation between $\vec{p}_{,J}$ and $\vec{p}^{\,0}_{,J}$ follows from (4.5.14),

$$\vec{p}_{,J} = \vec{p}^{\,0}_{,J} - (\partial \vec{p}/\partial \gamma_3')T^0_{,J} = \vec{p}^{\,0}_{,J} - (\partial \vec{p}/\partial T)T^0_{,J} = \vec{p}^{\,0}_{,J} + V^{-1}\nabla V T^0_{,J}. \quad (4.5.16)$$

Derivatives $\vec{p}^{\,0}_{,J}$ are obtained from (4.5.9), and (4.5.16) yields

$$\vec{p}_{,J} = \sigma \vec{n}^0_{,J} + \sigma_{,J}\vec{n}^0 + \vec{p}^{(\Sigma)}_{,J} + V^{-1}\nabla V T^0_{,J}. \quad (4.5.17)$$

Derivatives $\vec{n}^0_{,J}$, $\sigma_{,J}$, $T^0_{,J}$, and $\vec{p}^{(\Sigma)}_{,J}$ are taken along Σ^0. Finally, for $P_{IJ} = \vec{e}_I \cdot \vec{p}_{,J}$, we get

$$P_{IJ} = \sigma\left(\vec{e}_I \cdot \vec{n}^0_{,J}\right) + \sigma_{,J}(\vec{e}_I \cdot \vec{n}^0) + \left(\vec{e}_I \cdot \vec{p}^{(\Sigma)}_{,J}\right) + V^{-1}T^0_{,J}(\vec{e}_I \cdot \nabla V). \quad (4.5.18)$$

Here σ, $\vec{n}^0$, $\vec{e}_1$, $\vec{e}_2$, $T^0_{,J}$, and ∇V are presumably known; we only need to find the expressions for $\vec{p}^{(\Sigma)}_{,J}$, $\sigma_{,J}$, and $\vec{n}^0_{,J}$.

For $\vec{p}^{(\Sigma)}_{,J}$, (4.5.7) yields

$$\vec{p}^{(\Sigma)}_{,J} = A_{NK,J}T^0_{,K}\vec{x}_{,N} + A_{NK}T^0_{,KJ}\vec{x}_{,N} + A_{NK}T^0_{,K}\vec{x}_{,NJ}. \quad (4.5.19)$$

To determine $\sigma_{,J}$, we use (4.5.10) and (4.5.7):

$$\sigma_{,J} = -\sigma^{-1}\left[V^{-3}(\nabla V \cdot \vec{x}_{,J}) + \tfrac{1}{2}A_{NK,J}T^0_{,N}T^0_{,K} + A_{NK}T^0_{,N}T^0_{,KJ}\right]. \quad (4.5.20)$$

Finally, the expressions for $\vec{n}^0_{,J}$ are known from the differential geometry of surfaces (Weingarten equations); see Korn and Korn (1961). We shall present a simple and objective derivation. We put $\vec{n}^0_{,J} = a_J\vec{x}_{,1} + b_J\vec{x}_{,2}$, and multiply it successively by $\vec{x}_{,1}$ and $\vec{x}_{,2}$. The resultant two equations read

$$a_J E + b_J F = -L_{1J}, \qquad a_J F + b_J G = -L_{2J}.$$

Here $L_{NJ} = -\vec{n}^0_{,J} \cdot \vec{x}_{,N}$. Because $\vec{n}^0 \cdot \vec{x}_{,N} = 0$, we also have $\vec{n}^0_{,J} \cdot \vec{x}_{,N} + \vec{n}^0 \cdot \vec{x}_{,NJ} = 0$. This yields

$$L_{NJ} = \vec{n}^0 \cdot \vec{x}_{,NJ}. \quad (4.5.21)$$

The solution of the two previous equations reads

$$\vec{n}^0_{,J} = -(EG - F^2)^{-1}[(L_{1J}G - L_{2J}F)\vec{x}_{,1} + (L_{2J}E - L_{1J}F)\vec{x}_{,2}]. \quad (4.5.22)$$

Using A_{IJ} given by (4.5.6), we can express (4.5.22) in compact form

$$\vec{n}^0_{,J} = -L_{JN}A_{NK}\vec{x}_{,K}. \quad (4.5.23)$$

Quantities L_{NJ} are closely related to the curvature of Σ^0 at S. In the differential geometry of surfaces, they are usually denoted as follows: $L_{11} = L$, $L_{12} = L_{21} = M$, and $L_{22} = N$.

Equations (4.5.13) and (4.5.18), with (4.5.19) through (4.5.23) give the final expressions for $Q_{IJ}(S)$ and $P_{IJ}(S)$. It would, of course, be possible to express these equations in many alternative forms. If $T^0(\gamma_1, \gamma_2)$ is constant along Σ^0, (4.5.18) simplifies considerably:

$$P_{IJ} = -\epsilon V^{-2}(\nabla V \cdot \vec{x}_{,J})(\vec{e}_I \cdot \vec{n}^0) - \epsilon V^{-1}L_{JK}A_{NK}(\vec{e}_I \cdot \vec{x}_{,N}). \quad (4.5.24)$$

If initial surface Σ^0 is planar, the second (curvature) term in (4.5.24) vanishes. If initial surface Σ^0 is situated in a homogeneous medium, the first (inhomogeneity) term vanishes. Using (4.5.13) and (4.5.18), we can determine the 2×2 matrix of the second derivatives of travel-time field $\mathbf{M}(S)$:

$$\mathbf{M}(S) = \mathbf{P}(S)\mathbf{Q}^{-1}(S). \tag{4.5.25}$$

It would also be possible to derive the equations for slowness vector $\vec{p}^0(S)$ and for matrices $\mathbf{Q}(S)$ and $\mathbf{P}(S)$ considering initial surface Σ^0 to be a 2-D Riemannian space, with curvilinear nonorthogonal coordinates γ_1 and γ_2. The covariant components of the relevant metric tensor g_{IJ} are given by relations $g_{11} = E$, $g_{12} = g_{21} = F$, and $g_{22} = G$, and its contravariant components are $g^{IJ} = A_{IJ}$; see (4.5.6). The derivatives $A_{NK,J}$ of the metric tensor are closely related to the Christoffel symbols; see (3.5.56).

4.5.3 Special Case: Local Cartesian Coordinates z_I as Ray Parameters

As an important example of the general approach described in Section 4.5.2, we shall specify the equation derived for one choice of ray parameters γ_1 and γ_2. We shall construct a local Cartesian coordinate system z_i at S, with its origin at S and with the z_3-axis along normal vector $\vec{n}^0$. Axes z_1 and z_2 are tangent to Σ^0 at S and may be chosen arbitrarily. Such a system was introduced in Section 4.4.1 so that we can use the notation introduced in that section. $\hat{\mathbf{Z}}$ denotes the 3×3 transformation matrix from the local Cartesian coordinate system z_1, z_2, z_3 to the general Cartesian coordinate system x_1, x_2, x_3, and $\vec{i}_1^{(z)}, \vec{i}_2^{(z)}$, and $\vec{i}_3^{(z)}$ are the basis vectors of the z_i-coordinate system, with $\vec{i}_3^{(z)} = \vec{n}^0$. Initial surface Σ^0 in the "quadratic" vicinity of S is approximated by equation $z_3 = -\frac{1}{2}z_I z_J D_{IJ}(S)$, where $D_{IJ}(S)$ is the matrix of curvature of Σ^0 at S.

We shall now specify the ray parameters γ_1, γ_2 to be equal to z_1, z_2, the local Cartesian coordinates in the plane tangent to Σ^0 at S. For given $D_{IJ}(S)$, two parameters $\gamma_1 = z_1$ and $\gamma_2 = z_2$ fully specify the relevant point S' on Σ^0, $S'[z_1, z_2, z_3 = -\frac{1}{2}z_I z_J D_{IJ}]$. It is important to keep in mind that the derivatives with respect to $\gamma_I (\equiv z_I)$ represent the derivatives along initial surface Σ^0, not along the plane tangent to Σ^0 at S.

For $x_{k,I}$ and $x_{k,IJ}$, we obtain

$$x_{k,I} = Z_{kI} - Z_{k3}z_J D_{IJ}, \qquad x_{k,IJ} = -Z_{k3}D_{IJ}. \tag{4.5.26}$$

Using these expressions, we can calculate other important quantities of Section 4.5.2. Specifying them for point S, we obtain

$$E = 1, \qquad F = 0, \qquad G = 1, \qquad A_{IJ} = \delta_{IJ},$$
$$A_{IJ,N} = 0, \qquad L_{NJ} = -D_{NJ}.$$

Equation (4.5.7) yields the following expressions for the tangential component of the slowness vector:

$$\vec{p}^{(\Sigma)} = T_{,I}^0 \vec{i}_I^{(z)}, \qquad p^{(\Sigma)2} = T_{,I}^0 T_{,I}^0. \tag{4.5.27}$$

Similarly, slowness vector $\vec{p}^0$ is given by (4.5.9) with σ given by (4.5.10). For $Q_{IJ}(S)$ and $P_{IJ}(S)$, we can again use (4.5.13) and (4.5.18); the individual expressions in (4.5.18), however, simplify as follows:

$$P_{IJ} = -\sigma^{-1}V^{-3}\left(\nabla V \cdot \vec{i}_J^{(z)}\right)G_{3I} - \sigma^{-1}T_{,N}^0 T_{,NJ}^0 G_{3I} + \sigma D_{KJ}G_{KI} + T_{,NJ}^0 G_{NI}$$
$$- T_{,N}^0 D_{NJ}G_{3I} + V^{-1}(\nabla V \cdot \vec{e}_I)T_{,J}^0. \tag{4.5.28}$$

Here we have used the notation $G_{IJ} = \vec{i}_I^{(z)} \cdot \vec{e}_J$ as in Section 4.4.1.

Now we need to determine $\vec{e}_1$, $\vec{e}_2$, and $\vec{e}_3$. We shall specify them in terms of the 3×3 matrix $\hat{\mathbf{G}}$:

$$\hat{\mathbf{G}} = \hat{\mathbf{G}}^{\parallel}\hat{\mathbf{G}}^{\perp}, \qquad \hat{\mathbf{G}}^{\parallel} = \begin{pmatrix} \sigma V g^{-1} T^0_{,1} & -g^{-1} T^0_{,2} & V T^0_{,1} \\ \sigma V g^{-1} T^0_{,2} & g^{-1} T^0_{,1} & V T^0_{,2} \\ -Vg & 0 & V\sigma \end{pmatrix}. \tag{4.5.29}$$

Here $g = T^0_{,I} T^0_{,I}$ and rotation matrix $\hat{\mathbf{G}}^{\perp}$ is given by (4.4.23). The third column of $\hat{\mathbf{G}}^{\parallel}$ represents the basis vector $\vec{e}_3$ tangent to the ray. The second column of $\hat{\mathbf{G}}^{\parallel}$ represents the basis vector $\vec{e}_2$ perpendicular to the radiation plane, and the first column of $\hat{\mathbf{G}}^{\parallel}$ represents basis vector $\vec{e}_1$, confined to the radiation plane. Matrix $\hat{\mathbf{G}}^{\perp}$ rotates $\vec{e}_1$ and $\vec{e}_2$ about $\vec{e}_3$ into any other position. Using (4.5.29), we also obtain equations

$$G_{MI}G_{NI} = \delta_{MN} - V^2 T^0_{,M} T^0_{,N}, \qquad G_{MI}G_{3I} = -\sigma V^2 T^0_{,M}. \tag{4.5.30}$$

It will be useful to compute $G_{MI}P_{IJ}$ from (4.5.28) instead of P_{IJ}. Multiplying (4.5.28) by G_{MI}, and using (4.5.30), we obtain

$$G_{MI}P_{IJ} = T^0_{,MJ} + \sigma D_{MJ} - E_{MJ}, \tag{4.5.31}$$

where E_{MJ} is the inhomogeneity matrix, given by the relation

$$E_{MJ} = \sigma^{-1} V^{-3}\big(\nabla V \cdot \vec{i}^{(z)}_J\big) G_{MI}G_{3I} - V^{-1}(\vec{e}_I \cdot \nabla V)G_{MI}T^0_{,J}.$$

Using (4.5.30) for $G_{MI}G_{3I}$, expressing $T^0_{,J} = V^{-1}G_{J3}$ (see (4.5.29)) and transforming $\nabla V \cdot \vec{i}^{(z)}_J = (\nabla V \cdot \vec{e}_i)G_{Ji}$, we obtain

$$E_{MJ} = -V^{-2}[(G_{M3}G_{JK} + G_{MK}G_{J3})(\partial V/\partial q_K) + G_{M3}G_{J3}(\partial V/\partial q_3)].$$

But this is exactly the same as the inhomogeneity matrix in Section 4.4.2; see (4.4.47). The final expressions for $\mathbf{Q}$, $\mathbf{P}$, and $\mathbf{M}$ are

$$\mathbf{Q} = \mathbf{G}^T, \qquad \mathbf{P} = \mathbf{G}^{-1}\big(\mathbf{T}^0 + p^{(z)}_3\mathbf{D} - \mathbf{E}\big),$$
$$\mathbf{M} = \mathbf{G}^{-1}\big(\mathbf{T}^0 + p^{(z)}_3\mathbf{D} - \mathbf{E}\big)\mathbf{G}^{-1T}. \tag{4.5.32}$$

Here $\mathbf{T}^0$ represents the 2×2 matrix with elements $T^0_{,IJ} = \partial^2 T^0/\partial z_I \partial z_J$. The derivatives should be calculated along Σ^0.

Notes.

1. Equations (4.5.32) for $\mathbf{Q}$, $\mathbf{P}$, and $\mathbf{M}$ were derived by direct application of the general equations of Sections 4.5.1 through 4.5.3 to specific ray parameters z_1 and z_2 corresponding to local Cartesian coordinates in the plane tangent to Σ^0 at S. The resultant equations for $\mathbf{P}$ and $\mathbf{M}$, however, can also be derived in a considerably simpler way, using the relations of Section 4.4. We use (4.4.37) and bear in mind that $\mathbf{F} = \mathbf{T}^0$. Then (4.4.37) immediately yields (4.5.32) for $\mathbf{M}$. Matrix $\mathbf{P}$ is obtained from the relation $\mathbf{P} = \mathbf{MQ}$. Consequently, (4.4.37) yields an independent test of the results.

2. Let us now mutually compare the general results of Section 4.5.2 with the results of this section; see (4.5.32). The results of this section look considerably simpler when compared with those in Section 4.5.2. They have, however, certain serious disadvantages. The equations of Section 4.5.2 have a *global character*; they can be used to study the whole ray field generated at initial surface Σ^0. Surface Σ^0 is specified quite exactly, using general parameteric equation (4.5.1) with parameters

γ_1 and γ_2, and initial time T^0 is expressed in terms of the same parameters γ_1, γ_2, so that $T^0 = T^0(\gamma_1, \gamma_2)$. Thus, it is easy to calculate $T^0_{,I}$ and $T^0_{,IJ}$, either analytically or numerically. On the other hand, (4.5.32) have a *local character*. At any point S on Σ^0, where we wish to compute $\mathbf{Q}$, $\mathbf{P}$, and/or $\mathbf{M}$, we must construct the local Cartesian coordinate system z_1, z_2, z_3. We also describe initial surface Σ^0 by the local quadratic approximation $z_3 = -\frac{1}{2}z_I z_J D_{IJ}(S)$. The local Cartesian coordinate system and the quadratic approximation of Σ^0 are different from one point of Σ^0 to another. Moreover, the determination of $T^0_{,IJ}$ needed in (4.5.32) and $T^0_{,I}$ needed in (4.5.29) will also be more complicated because initial time T^0 is not, as a rule, specified in terms of local Cartesian coordinates z_1 and z_2 (different from one point S to another), but in some global parameters γ_1 and γ_2.

3. The derived equations (4.5.32) are also closely related to surface-to-surface dynamic ray tracing; see Section 4.4.7 and relation (4.4.69). Consider (4.4.69), taken at point S' close to S on anterior surface Σ^0. The initial conditions for surface-to-surface dynamic ray tracing are $d\mathbf{z}(S') = \mathbf{z}(S')$ and $d\mathbf{p}^{(\Sigma)}(S') = \mathbf{T}^0(S)d\mathbf{z}(S')$. Thus, to start surface-to-surface dynamic ray tracing, we must know $\mathbf{T}^0(S)$. It can be proved that (4.4.69) immediately yields (4.5.32). We only need to insert $d\gamma = d\mathbf{z}$ and multiply (4.4.69) by $\mathbf{Y}^{-1}(S)$. This procedure provides an independent test of Equations (4.5.32). For an analogous derivation in anisotropic media, see Section 4.14.10.

4.5.4 Point Source

We shall now consider a point source situated at S. Because the "initial surface Σ^0" representing a point source is not smooth, we do not obtain one unit normal $\vec{n}^0$, as in (4.5.8), but a two-parameteric system of unit normals $\vec{n}^0$,

$$\vec{n}^0 = \vec{n}^0(\gamma_1, \gamma_2), \tag{4.5.33}$$

pointing to all sides of S. The relevant slowness vector $\vec{p}^0$ is then given by the relation $\vec{p}^0 = V^{-1}\vec{n}^0$; see (4.5.9) for $\vec{p}^{(\Sigma)} = 0$ and $\epsilon = 1$. Unit vector $\vec{e}_3$ of the ray-centered coordinate system coincides with the normal, $\vec{e}_3 = \vec{n}^0(\gamma_1, \gamma_2)$. The other two unit vectors, $\vec{e}_1$ and $\vec{e}_2$, can be taken arbitrarily in the plane perpendicular to $\vec{e}_3$; the only requirement is that $\vec{e}_1$, $\vec{e}_2$ and $\vec{e}_3$ are mutually perpendicular and form a right-handed system. In principle, it is sufficient to specify for given γ_1 and γ_2 only two of $\vec{e}_1$, $\vec{e}_2$, and $\vec{e}_3$, for example, $\vec{e}_1 = \vec{e}_1(\gamma_1, \gamma_2)$, $\vec{e}_2 = \vec{e}_2(\gamma_1, \gamma_2)$, and to evaluate the third one as a vector product, for example, $\vec{n}^0 = \vec{e}_3 = \vec{e}_1 \times \vec{e}_2$.

Now we shall determine $Q_{IJ}(S)$ and $P_{IJ}(S)$. Because $Q_{IJ} = \vec{e}_I \cdot \vec{x}_{,J}$ and $P_{IJ} = \vec{e}_I \cdot \vec{p}_{,J}$,

$$Q_{IJ} = 0, \qquad P_{IJ} = V^{-1}\vec{e}_I \cdot \vec{n}^0_{,J}. \tag{4.5.34}$$

These are the final expressions for $\mathbf{Q}$ and $\mathbf{P}$, assuming that $\vec{n}^0$ is given by (4.5.33). The ray parameters γ_1 and γ_2 in (4.5.33) may be taken arbitrarily.

As an example, we shall choose polar spherical coordinates as ray parameters, $\gamma_1 = i_0$ and $\gamma_2 = \phi_0$, as in Section 3.2.1. Then

$$\vec{n}^0 \equiv [\sin i_0 \cos \phi_0, \sin i_0 \sin \phi_0, \cos i_0]. \tag{4.5.35}$$

We choose $\vec{e}_3 = V\vec{p}^0 = \vec{n}^0$. The other two unit vectors, $\vec{e}_1$ and $\vec{e}_2$, may be taken as follows:

$$\vec{e}_1 \equiv [\cos i_0 \cos \phi_0, \cos i_0 \sin \phi_0, -\sin i_0],$$
$$\vec{e}_2 \equiv [-\sin \phi_0, \cos \phi_0, 0]. \tag{4.5.36}$$

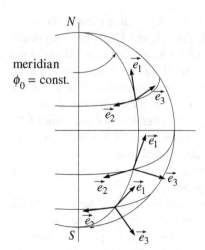

Figure 4.14. Possible choice of ray-centered basis vectors $\vec{e}_1$, $\vec{e}_2$, and $\vec{e}_3$ at a point source.

The direction of $\vec{e}_1$ and $\vec{e}_2$ is demonstrated in Figure 4.14 on a unit sphere with its center at S. $\vec{e}_1$ is oriented along the meridian (constant ϕ_0) and is positive northward; $\vec{e}_2$ is oriented along the parallel (constant i_0) and is positive westward.

Using (4.5.34) through (4.5.36), we obtain the final result:

$$\mathbf{Q}(S) = \mathbf{0}, \qquad \mathbf{P}(S) = V^{-1}(S) \begin{pmatrix} 1 & 0 \\ 0 & \sin i_0 \end{pmatrix}, \qquad \mathbf{M}^{-1}(S) = \mathbf{0}. \quad (4.5.37)$$

Results (4.5.34) and (4.5.37) can also be obtained by direct application of the equations derived in Sections 4.5.1 through 4.5.3. We only need to start with a smooth surface Σ^0. For example, we can consider a *spherical initial surface*, Σ^0, with its center at S and with radius a:

$$\vec{x} - \vec{x}(S) \equiv [a \sin i_0 \cos \phi_0, a \sin i_0 \sin \phi_0, a \cos i_0]. \quad (4.5.38)$$

We assume that the distribution of the initial time $T^0(\gamma_1, \gamma_2)$ along the spherical surface Σ^0 is constant and consider point S' on the spherical surface specified by parameters i_0 and ϕ_0. Using (4.5.38), we can compute $\vec{x}_{,I}$ and $\vec{x}_{,IJ}$ at S' and obtain $E = a^2$, $F = 0$, $G = a^2 \sin^2 i_0$, $A_{11} = a^{-2}$, $A_{12} = A_{21} = 0$, $A_{22} = (a \sin i_0)^{-2}$, $L_{11} = -a$, $L_{12} = L_{21} = 0$, and $L_{22} = -a \sin^2 i_0$. Again using $\vec{e}_1$ and $\vec{e}_2$ given by (4.5.36), we obtain the result

$$\mathbf{Q}(S') = a \begin{pmatrix} 1 & 0 \\ 0 & \sin i_0 \end{pmatrix}, \qquad \mathbf{P}(S') = \frac{1}{V(S')} \begin{pmatrix} 1 & 0 \\ 0 & \sin i_0 \end{pmatrix},$$

$$\mathbf{M}(S') = \frac{\mathbf{I}}{a V(S')}. \quad (4.5.39)$$

For $a \to 0$, point S' approaches S and (4.5.39) yields (4.5.37).

4.5.5 Initial Line

Let us consider a smooth 3-D initial line, C^0, situated in an inhomogeneous isotropic medium. Initial line C^0 may be described by a vectorial parameteric equation,

$$\vec{x} = \vec{x}(\gamma_2). \quad (4.5.40)$$

Here γ_2 is an arbitrary parameter along initial line C^0, for example, the arclength measured from some reference point on C^0. Thus, any point S situated on C^0 may be specified by γ_2.

We shall use the following notation:

$$\vec{x}_{,2} = \partial\vec{x}/\partial\gamma_2, \qquad \vec{x}_{,22} = \partial^2\vec{x}/\partial\gamma_2^2, \qquad G = \vec{x}_{,2} \cdot \vec{x}_{,2},$$
$$\vec{t} = \vec{x}_{,2}/|\vec{x}_{,2}| = G^{-1/2}\vec{x}_{,2}. \tag{4.5.41}$$

Here $\vec{t}$ denotes the unit tangent to initial line C^0, and G has the same meaning as in Section 4.5.1 (see (4.5.3)). If γ_2 represents the arclength along C^0, $G = 1$.

We consider an arbitrary distribution of initial travel time T^0 along C^0, $T^0 = T^0(\gamma_2)$. For $T^0 = $ const., we speak of an exploding line.

The tangential component of the initial slowness vector at S can be determined in the same way as in Section 4.5.1. We obtain

$$\vec{p}^{(\Sigma)} = G^{-1}T_{,2}^0\vec{x}_{,2}, \qquad p^{(\Sigma)2} = G^{-1}T_{,2}^{02}. \tag{4.5.42}$$

We now introduce unit normal $\vec{n}^0$, perpendicular to initial line C^0 at S. We do not, however, obtain only one normal $\vec{n}^0$ at S, as in (4.5.8). For fixed γ_2, we have a *one-parameteric system of unit normals* $\vec{n}^0$

$$\vec{n}^0 = \vec{n}^0(\gamma_1, \gamma_2), \tag{4.5.43}$$

parameterized by γ_1 and pointing to all sides from S in a plane perpendicular to C^0 at S. Vector $\vec{e}_1$ of the ray-centered coordinate system may then be defined as

$$\vec{e}_1 = \vec{t}(\gamma_2) \times \vec{n}^0(\gamma_1, \gamma_2). \tag{4.5.44}$$

Alternatively, we may have given, at each γ_2, a one-parameteric system of unit vectors $\vec{e}_1$ and corresponding system of normals $\vec{n}^0$:

$$\vec{e}_1 = \vec{e}_1(\gamma_1, \gamma_2), \qquad \vec{n}^0 = \vec{e}_1(\gamma_1, \gamma_2) \times \vec{t}(\gamma_2). \tag{4.5.45}$$

Unit vector $\vec{e}_1$ is parameterized by γ_1 and perpendicular to $\vec{t}(\gamma_2)$.

We shall now discuss initial slowness vector $\vec{p}^0$. We consider point S on C^0 and *the radiation plane at S*, specified by $\vec{p}^{(\Sigma)}(S)$ and $\vec{n}^0(S)$,

$$\vec{p}^0 = \sigma\vec{n}^0 + G^{-1/2}T_{,2}^0\vec{t}; \tag{4.5.46}$$

see (4.5.42) and (4.5.43). Quantity σ can be determined from the eikonal equation. For isotropic media,

$$\sigma = \left(V^{-2} - G^{-1}T_{,2}^{02}\right)^{1/2}. \tag{4.5.47}$$

Here we have taken the plus sign (+) because all generated waves propagate away from C^0. Two waves (P and S) may be generated at any point S of C^0. For $G^{-1}T_{,2}^{02} > V^{-2}$, the generated wave is inhomogeneous. For $G^{-1}T_{,2}^{02} = V^{-2}$, the generated rays graze the initial line at S.

As in Section 4.5.1, expression (4.5.46) for $\vec{p}^0$ remains valid even if initial line C^0 is situated in an inhomogeneous *anisotropic* medium. Only σ needs to be determined from the eikonal equation for the anisotropic medium. In this case, three waves may be generated at any point S of C^0 (qP, qS1, qS2).

For $T_{,2}^0 \neq 0$ along C^0, the initial slowness vectors of a selected generated wave with velocity V at S form a conical surface in isotropic media, with its axis oriented along $-\vec{t}(S)$ and with the apex angle $\arctan(\sigma\sqrt{G}/T_{,2}^0)$. The wavefront is also conical, with the axis along $\vec{t}(S)$ and with apex angle $\arctan(T_{,2}^0/\sigma\sqrt{G})$. For T^0 constant along C^0 (exploding line), the initial slowness vectors at S are perpendicular to C^0, and the wavefront is tubular.

Now we shall discuss the determination of P_{IJ} and Q_{IJ}. For this purpose, we need to specify unit vectors $\vec{e}_1$, $\vec{e}_2$, and $\vec{e}_3$. We shall do this in a standard way: $\vec{e}_3 = V\vec{p}^0$ and $\vec{e}_1$, $\vec{e}_2$ perpendicular to $\vec{e}_3$. More specific expressions will be discussed later. Matrices $\mathbf{Q}$ and $\mathbf{P}$ are then given by relations

$$Q_{IJ} = \vec{e}_I \cdot \vec{x}_{,J}, \qquad P_{IJ} = \vec{e}_I \cdot \vec{p}_{,J}. \tag{4.5.48}$$

Here $\vec{p}_{,J}$ represents the derivative $\partial\vec{p}/\partial\gamma_J$, taken along the wavefront, $\vec{p}_{,J} = \vec{p}^0_{,J} + \delta_{J2}V^{-1}T^0_{,2}\nabla V$; see (4.5.16). Expressions for Q_{IJ} are simple because $\vec{x}_{,1} = 0$, $\vec{x}_{,2} = \sqrt{G}\vec{t}$. The determination of P_{IJ} is more involved. As in (4.5.18), we obtain

$$P_{IJ} = \sigma(\vec{e}_I \cdot \vec{n}^0_{,J}) + \delta_{J2}[\sigma_{,2}(\vec{e}_I \cdot \vec{n}_0) + (\vec{e}_I \cdot \vec{p}^{(\Sigma)}_{,2}) + V^{-1}T^0_{,2}(\vec{e}_I \cdot \nabla V)]. \tag{4.5.49}$$

Here

$$\vec{p}^{(\Sigma)}_{,2} = -G^{-2}G_{,2}T^0_{,2}\vec{x}_{,2} + G^{-1}T^0_{,22}\vec{x}_{,2} + G^{-1}T^0_{,2}\vec{x}_{,22},$$
$$\sigma_{,2} = -\sigma^{-1}[V^{-3}(\nabla V \cdot \vec{x}_{,2}) - \tfrac{1}{2}G^{-2}G_{,2}T^{02}_{,2} + G^{-1}T^0_{,2}T^0_{,22}]. \tag{4.5.50}$$

The derivatives $\vec{n}^0_{,J}$ in (4.5.49) depend on a particular choice of $\vec{n}^0(\gamma_1, \gamma_2)$; see (4.5.43). An important choice of $\vec{n}^0$ will be discussed next. Otherwise, (4.5.48) through (4.5.50) represent the final solution of the problem of ray tracing and dynamic ray tracing from initial line C^0 with an arbitrary distribution of initial time T^0 along it.

For γ_2 fixed, that is, for a fixed point S on initial line C^0, $\vec{n}^0(\gamma_1, \gamma_2)$ represents a one-parameteric system of normals, with parameter γ_1. This one-parameteric system can be constructed in many ways. Here we shall discuss two possibilities.

a. SPECIFICATION OF $\vec{n}^0(\gamma_1, \gamma_2)$ IN TERMS OF AUXILIARY VECTOR $\vec{A}$

Consider vector $\vec{A}$, which may be constant along C^0 or may vary smoothly about it, $\vec{A} = \vec{A}(\gamma_2)$. We can then determine two auxiliary unit vectors $\vec{n}^A$ and $\vec{b}^A$ using the relations

$$\vec{b}^A = (\vec{x}_{,2} \times \vec{A})/|\vec{x}_{,2} \times \vec{A}|, \qquad \vec{n}^A = [(\vec{x}_{,2} \times \vec{A}) \times \vec{x}_{,2}]/|(\vec{x}_{,2} \times \vec{A}) \times \vec{x}_{,2}|. \tag{4.5.51}$$

Unit vectors $\vec{n}^A$ and $\vec{b}^A$ are perpendicular to C^0 and are mutually perpendicular. Together with $\vec{t}$, triplet $\vec{t}$, $\vec{n}^A$, $\vec{b}^A$ is right-handed and mutually orthogonal at any point S of C^0. Using these two unit vectors, we can define the system of $\vec{n}^0$ as

$$\vec{n}^0(\gamma_1, \gamma_2) = \vec{n}^A(\gamma_2)\cos i_0 + \vec{b}^A(\gamma_2)\sin i_0. \tag{4.5.52}$$

Here parameter $\gamma_1 = i_0$ specifies the selected radiation plane. Then

$$\vec{n}^0_{,1} = -\vec{n}^A \sin i_0 + \vec{b}^A \cos i_0, \qquad \vec{n}^0_{,2} = \vec{n}^A_{,2}\cos i_0 + \vec{b}^A_{,2}\sin i_0. \tag{4.5.53}$$

The determination of $\vec{n}^A_{,2}$ and $\vec{b}^A_{,2}$ from (4.5.51) is straightforward. Unit vectors $\vec{e}_1$, $\vec{e}_2$, and $\vec{e}_3$ may be taken as

$$\vec{e}_1 = \vec{b}^A \cos i_0 - \vec{n}^A \sin i_0,$$
$$\vec{e}_2 = V\sigma\vec{t} - VG^{-1/2}T^0_{,2}(\vec{n}^A \cos i_0 + \vec{b}^A \sin i_0),$$
$$\vec{e}_3 = VG^{-1/2}T^0_{,2}\vec{t} + V\sigma(\vec{n}^A \cos i_0 + \vec{b}^A \sin i_0). \tag{4.5.54}$$

Here $\vec{e}_1$ is perpendicular to the radiation plane, and $\vec{e}_2$ is perpendicular to the generated ray in the radiation plane. Inserting (4.5.51) through (4.5.54) into (4.5.48) and (4.5.49) yields the final expressions for Q_{IJ} and P_{IJ}, for an arbitrarily chosen auxiliary vector $\vec{A}(\gamma_2)$.

b. SPECIFICATION OF $\vec{n}^0(\gamma_1, \gamma_2)$ IN TERMS OF MAIN NORMAL $\vec{n}$ AND BINORMAL $\vec{b}$

One particular choice of auxiliary vector $\vec{A}$ plays a great role in differential geometry of curves: $\vec{A}(\gamma_2) = \vec{x}_{,22}$. From a computational point of view, this choice is not as convenient as some other choices (for example, $\vec{A} = \text{const.}$) because it also requires the computation of the third derivatives $\vec{x}_{,222}$ along C^0. On the other hand, it offers a simple geometrical interpretation of the results in terms of the main curvature K and torsion T of initial line C^0. For this reason, we shall present the complete solution of the initial line problem for this choice.

If $\vec{A} = \vec{x}_{,22}$, the unit vectors $\vec{n}^A$ and $\vec{b}^A$ given by (4.5.51) have the geometrical meaning of the main normal $\vec{n}$ and binormal $\vec{b}$ to initial line C^0. Taking the derivatives of $\vec{t}$, given by (4.5.41), and of $\vec{n}$ and $\vec{b}$, given by (4.5.51) with $\vec{A} = \vec{x}_{,22}$, we obtain

$$\vec{t}_{,2} = \sqrt{G}K\vec{n}, \qquad \vec{n}_{,2} = \sqrt{G}(T\vec{b} - K\vec{t}), \qquad \vec{b}_{,2} = -\sqrt{G}T\vec{n}. \quad (4.5.55)$$

Here K is the main curvature and T is the torsion of initial line C^0 at S. They are given by relations

$$K = G^{-3/2}|\vec{x}_{,2} \times \vec{x}_{,22}|, \qquad T = K^{-2}G^{-3}[\vec{x}_{,2} \cdot (\vec{x}_{,22} \times \vec{x}_{,222})]. \quad (4.5.56)$$

In fact, Equations (4.5.55) represent Frenet's formulae (3.2.8), modified for general parameter γ_2 along the initial line. For $\gamma_2 = s$, where s is the arclength along C^0 measured from some reference point, we obtain $G = 1$ and $\vec{x}_{,2} \cdot \vec{x}_{,22} = 0$. The expression (4.5.55) for main curvature K then simplifies considerably: $K = |\vec{x}_{,22}|$. As we can see in (4.5.56), the expression for T contains $\vec{x}_{,222}$. This is the disadvantage of choice $\vec{A} = \vec{x}_{,22}$.

Before we write the general expressions for Q_{IJ} and P_{IJ}, we shall present the relations for $\vec{p}_{,J} = \vec{p}^0_{,J} + \delta_{J2}V^{-1}T^0_{,2}\nabla V$:

$$\vec{p}_{,1} = \sigma(-\vec{n}\sin i_0 + \vec{b}\cos i_0),$$

$$\vec{p}_{,2} = \sigma_{,2}(\vec{n}\cos i_0 + \vec{b}\sin i_0) + \sigma\sqrt{G}T(\vec{b}\cos i_0 - \vec{n}\sin i_0)$$

$$- \sigma\sqrt{G}K\cos i_0\vec{t} + KT^0_{,2}\vec{n} + (G^{-1/2}T^0_{,2})_{,2}\vec{t} + V^{-1}T^0_{,2}\nabla V.$$

The final expressions for **Q**, **P**, and **M**$^{-1}$ are then

$$\mathbf{Q} = \begin{pmatrix} 0 & 0 \\ 0 & V\sigma\sqrt{G} \end{pmatrix}, \qquad \mathbf{P} = \begin{pmatrix} \sigma & P_{12} \\ 0 & P_{22} \end{pmatrix}, \qquad \mathbf{M}^{-1} = \begin{pmatrix} 0 & 0 \\ 0 & V\sigma\sqrt{G}/P_{22} \end{pmatrix}. \quad (4.5.57)$$

Here P_{12} and P_{22} are given by the relations

$$P_{12} = \sigma\sqrt{G}T - KT^0_{,2}\sin i_0 + V^{-1}T^0_{,2}(\vec{e}_1 \cdot \nabla V)$$

$$P_{22} = V^{-1}\sigma^{-1}\sqrt{G}[-\sigma K\cos i_0 + G^{-1}T^0_{,2}(V^{-1}(\nabla V \cdot \vec{x}_{,2}) \quad (4.5.58)$$

$$- G^{-1}(\vec{x}_{,2} \cdot \vec{x}_{,22})) + G^{-1}T^0_{,22}] + V^{-1}T^0_{,2}(\vec{e}_2 \cdot \nabla V).$$

For $\gamma_2 = s$, we can insert $G = 1$ and $\vec{x}_{,2} \cdot \vec{x}_{,22} = 0$. Equation (4.5.58) then yields

$$P_{22} = V^{-1}\sigma^{-1}[-\sigma K\cos i_0 + V^{-1}(\nabla V \cdot \vec{x}_{,2})T^0_{,2} + T^0_{,22}]$$

$$+ V^{-1}T^0_{,2}(\vec{e}_2 \cdot \nabla V). \quad (4.5.59)$$

For an *exploding initial line* C^0 and $\gamma_2 = s$, (4.5.57) and (4.5.59) yield

$$\mathbf{Q} = \begin{pmatrix} 0 & 0 \\ 0 & 1 \end{pmatrix}, \qquad \mathbf{P} = V^{-1} \begin{pmatrix} 1 & T \\ 0 & -K \cos i_0 \end{pmatrix},$$

$$\mathbf{M}^{-1} = \begin{pmatrix} 0 & 0 \\ 0 & -V/K \cos i_0 \end{pmatrix}. \tag{4.5.60}$$

For an *exploding planar initial line* C^0, we even have $T = 0$, and matrix $\mathbf{P}$ becomes diagonal. Finally, for an exploding straight line C^0, $K = T = 0$. Each of matrices $\mathbf{Q}, \mathbf{P}$, and $\mathbf{M}^{-1}$ has only one nonvanishing element in this case. The eigenvalues of matrix $\mathbf{M}^{-1}$ approach 0 and ∞ so that the generated wavefront is cylindrical, and the rays are perpendicular to C^0.

As in Section 4.5.4 for a point source, the equations for $\vec{p}^0$ and $\mathbf{Q}, \mathbf{P}$, and $\mathbf{M}^{-1}$ may be derived by direct application of the equations for the smooth initial surface Σ^0; see Sections 4.5.1 and 4.5.2. Instead of initial line C^0, we consider an initial tubular surface Σ^0 with its axis along initial line C^0 and with radius a, and afterwards we take the limit $a \to 0$. The final results are, of course, the same.

4.5.6 Initial Surface with Edges and Vertexes

The results of Sections 4.5.4 and 4.5.5 for a point source and a line source play an important role in various applications. They, however, also generalize the results obtained in Sections 4.5.1 through 4.5.3 for initial surface Σ^0. We can consider a general *initial surface Σ^0 with edges and vertices*, and with an arbitrary distribution of initial time T^0 along it. Let us discuss one arbitrary generated wave (P or S) at one side of Σ^0 only. If point S is situated on the smooth part of Σ^0, one ray is generated at S. The initial conditions for the relevant slowness vector $\vec{p}^0(S)$ are given by (4.5.9), and the initial conditions for $\mathbf{Q}(S)$ and $\mathbf{P}(S)$ are given by (4.5.13) and (4.5.18). If point S is situated on an edge, a one-parameteric system of rays is obtained at S. For any selected ray of the system, the initial conditions for slowness vector $\vec{p}^0(S)$ are given by (4.5.46), and for $\mathbf{Q}(S)$ and $\mathbf{P}(S)$, the initial conditions are given by (4.5.48) or (4.5.49). Finally, if point S is situated at a vertex, a two-parameteric system of rays is obtained. For any selected ray of the system, the initial conditions for slowness vector $\vec{p}^0(S)$ are given by $\vec{p}^0(S) = V^{-1}(S)\vec{n}^0(S)$, where $\vec{n}^0(S)$ is given by (4.5.35), and $\mathbf{P}(S)$ and $\mathbf{Q}(S)$ are given by (4.5.34) or (4.5.37). In general, up to four waves are generated at any point S of Σ^0 (P and S waves at both sides of Σ^0). The derived equations are valid for all these generated waves, only propagation velocity $V(S)$ should be properly specified.

4.6 Paraxial Travel-Time Field and Its Derivatives

In the standard ray method, travel times can be simply determined along the rays by numerical quadratures. In addition, standard ray tracing also yields components of the slowness vector along the ray. The components of the slowness vector represent the first spatial derivatives of the travel-time field or are directly related to them. Using the slowness vector, we can also determine the travel-time field in some vicinity of the ray under consideration, without new ray tracing. This simple approach, which uses the *first derivatives* of the travel-time field to determine the travel-time field in the vicinity of the ray, corresponds to a *local plane wave approximation*.

The paraxial ray method is considerably more powerful. It yields the matrix of the *second derivatives* of the travel-time field along the ray. This matrix can then be used to calculate the travel-time field in the vicinity of the ray under consideration with a considerably higher accuracy than the local plane wave approximation. In ray-centered coordinates q_I, the matrix of second derivatives of the travel-time field can be used to express the travel-time field in the vicinity of the ray in the form of a Taylor series in q_I, up to the quadratic terms. We speak of the *quadratic approximation* or the *curved wavefront approximation* of the travel-time field. Moreover, knowledge of the second derivatives of the travel-time field can also be used to determine the linear paraxial distribution of the slowness vector in the vicinity of the ray.

In this section, we shall first discuss the second derivatives of the travel-time field because they represent the cornerstones of the paraxial ray method. Since the matrices of the second derivatives of the travel-time field are closely related to the matrices of the curvature of the wavefront, we shall also introduce and briefly discuss these matrices. We shall then summarize the most important equations for the paraxial travel times and paraxial slowness vectors.

In the whole section, we treat only initial-value travel-time problems for isotropic media. The boundary-value problems for isotropic media, such as the problem of the determination of the two-point eikonal, will be treated in Section 4.9. For anisotropic inhomogeneous media, see Section 4.14.

We shall consider central ray Ω and two points, S and R, situated on it. We assume that transformation matrices $\hat{\mathbf{H}}(S)$ and $\hat{\mathbf{H}}(R)$ are known and that the 4×4 ray propagator matrix $\mathbf{\Pi}(R, S)$ has also been determined by dynamic ray tracing in ray-centered coordinates along ray Ω from S to R. We shall use the standard notation for the ray propagator matrix (4.3.5). Ray Ω may correspond to an arbitrary seismic body wave (P, S, multiply reflected, converted) propagating in an inhomogeneous isotropic layered medium.

4.6.1 Continuation Relations for Matrix M

In ray-centered coordinates q_I, the 2×2 matrix $\mathbf{M}$, with elements $M_{IJ} = \partial^2 T / \partial q_I \partial q_J$, can be calculated along ray Ω in several ways. First, it can be determined by solving the nonlinear matrix Riccati equation (4.1.73). Second, it can be expressed in terms of the 2×2 transformation matrices $\mathbf{Q}$ and $\mathbf{P}$ as

$$\mathbf{M} = \mathbf{PQ}^{-1},\tag{4.6.1}$$

and matrices $\mathbf{Q}$ and $\mathbf{P}$ can be calculated along ray Ω using the linear dynamic ray tracing system in ray-centered coordinates (4.1.64) or (4.1.65). Third, it is possible to perform the dynamic ray tracing in Cartesian coordinates, and to determine $\mathbf{M}$ from the computed quantities; see Sections 4.2.1 and 4.2.3. Here we shall discuss the second approach; for the third approach, see Section 4.2.3.

We now wish to find the continuation relations for matrix $\mathbf{M}$ in terms of the minors of ray propagator matrix (4.3.5). In other words, we wish to determine $\mathbf{M}(R)$ assuming that $\mathbf{M}(S)$ is known. Before we solve this problem, we shall present the continuation relations for matrices $\mathbf{Q}$ and $\mathbf{P}$. Using (4.3.29), we can express them in the following form:

$$\begin{aligned}
\mathbf{Q}(R) &= \mathbf{Q}_1(R, S)\mathbf{Q}(S) + \mathbf{Q}_2(R, S)\mathbf{P}(S), \\
\mathbf{P}(R) &= \mathbf{P}_1(R, S)\mathbf{Q}(S) + \mathbf{P}_2(R, S)\mathbf{P}(S).
\end{aligned}\tag{4.6.2}$$

Conversely, we can determine $\mathbf{Q}(S)$ and $\mathbf{P}(S)$ from $\mathbf{Q}(R)$ and $\mathbf{P}(R)$ (backward continuation):

$$\mathbf{Q}(S) = \mathbf{P}_2^T(R, S)\mathbf{Q}(R) - \mathbf{Q}_2^T(R, S)\mathbf{P}(R),$$
$$\mathbf{P}(S) = -\mathbf{P}_1^T(R, S)\mathbf{Q}(R) + \mathbf{Q}_1^T(R, S)\mathbf{P}(R). \tag{4.6.3}$$

We shall use the following notation, related to the important case of *point sources*. We denote the quantities $\mathbf{Q}(R)$ and $\mathbf{P}(R)$ corresponding to the *point source situated at S* by $\mathbf{Q}(R, S)$ and $\mathbf{P}(R, S)$. In this case, $\mathbf{Q}(S) = \mathbf{0}$, so that (4.6.2) yields

$$\mathbf{Q}(R, S) = \mathbf{Q}_2(R, S)\mathbf{P}(S), \qquad \mathbf{P}(R, S) = \mathbf{P}_2(R, S)\mathbf{P}(S). \tag{4.6.4}$$

Similarly, for a *point source situated at R*, we obtain the backward continuation equations,

$$\mathbf{Q}(S, R) = -\mathbf{Q}_2^T(R, S)\mathbf{P}(R), \qquad \mathbf{P}(S, R) = \mathbf{Q}_1^T(R, S)\mathbf{P}(R). \tag{4.6.5}$$

Matrices $\mathbf{Q}$ and $\mathbf{P}$ and the relevant continuation relations (4.6.2) and (4.6.3) play an important role in various applications such as, in the computation of paraxial rays, in the solution of two-point ray tracing for paraxial rays, and in the evaluation of geometrical spreading. In this section, we shall use them to find the continuation relations for matrix $\mathbf{M}$.

Using (4.6.1) through (4.6.3), we can write the continuation relations for matrix $\mathbf{M}$. Assume first that we know $\mathbf{M}(S)$ and wish to compute $\mathbf{M}(R)$. Then, in view of (4.6.1) and (4.6.2),

$$\mathbf{M}(R) = [\mathbf{P}_1(R, S) + \mathbf{P}_2(R, S)\mathbf{M}(S)][\mathbf{Q}_1(R, S) + \mathbf{Q}_2(R, S)\mathbf{M}(S)]^{-1}. \tag{4.6.6}$$

This is the *continuation relation for matrix $\mathbf{M}$*. Using (4.6.3), we can calculate $\mathbf{M}(S)$ from known $\mathbf{M}(R)$,

$$\mathbf{M}(S) = \left[-\mathbf{P}_1^T(R, S) + \mathbf{Q}_1^T(R, S)\mathbf{M}(R)\right]\left[\mathbf{P}_2^T(R, S) - \mathbf{Q}_2^T(R, S)\mathbf{M}(R)\right]^{-1}. \tag{4.6.7}$$

Equation (4.6.7) represents the *backward continuation*. In the same way that we introduced $\mathbf{P}(R, S)$ and $\mathbf{Q}(R, S)$ for a point source situated at S, we can also introduce $\mathbf{M}(R, S)$, representing $\mathbf{M}(R)$ for a point source situated at S. For the *point source at S*, we have $\mathbf{M}(S) \to \infty$, and (4.6.6) with (4.3.17) yields

$$\mathbf{M}(R, S) = \mathbf{P}_2(R, S)\mathbf{Q}_2^{-1}(R, S) = \mathbf{Q}_2^{-1T}(R, S)\mathbf{P}_2^T(R, S). \tag{4.6.8}$$

If the *point source is at R*, we obtain the backward formula

$$\mathbf{M}(S, R) = -\mathbf{Q}_1^T(R, S)\mathbf{Q}_2^{-1T}(R, S) = -\mathbf{Q}_2^{-1}(R, S)\mathbf{Q}_1(R, S). \tag{4.6.9}$$

As we can see from (4.6.8) and (4.6.9), matrices $\mathbf{M}(R, S)$ and $\mathbf{M}(S, R)$ are symmetric,

$$\mathbf{M}(R, S) = \mathbf{M}^T(R, S) \qquad \mathbf{M}(S, R) = \mathbf{M}^T(S, R).$$

Matrix $\mathbf{M}(R, S)$, however, differs from $\mathbf{M}(S, R)$. Thus, if the positions of the source and receiver are exchanged, the relevant matrices $\mathbf{M}(R, S)$ and $\mathbf{M}(S, R)$ are not reciprocal.

The 2×2 matrix $\mathbf{M}(R)$ can be simply extended to the 3×3 matrix $\hat{\mathbf{M}}(R)$, which represents the matrix of second derivatives of the travel-time field with respect to local Cartesian ray-centered coordinates y_1, y_2, y_3, with the origin at R; see (4.1.81). Using

$\hat{\mathbf{M}}(R)$, we can compute the 3×3 matrix $\mathbf{M}^{(x)}(R)$ of second derivatives of the travel-time field with respect to Cartesian coordinates x_i, $M_{ij}^{(x)}(R) = (\partial^2 T / \partial x_i \partial x_j)_R$; see (4.1.87).

The continuation relations for the 3×3 matrix $\hat{\mathbf{M}}^{(x)}$ are more involved than the continuation relations for the 2×2 matrix $\mathbf{M}$; see (4.6.6). Assume that $\hat{\mathbf{M}}^{(x)}(S)$ and $\hat{\mathbf{H}}(S)$ are known at S, and that $\hat{\mathbf{H}}(R)$ is known at R. $\hat{\mathbf{M}}^{(x)}(R)$ can then be determined from $\hat{\mathbf{M}}^{(x)}(S)$ in five steps:

  a. From $\hat{\mathbf{M}}^{(x)}(S)$, we determine $\hat{\mathbf{M}}(S) = \hat{\mathbf{H}}^T(S)\hat{\mathbf{M}}^{(x)}(S)\hat{\mathbf{H}}(S)$; see (4.1.87).
  b. From $\hat{\mathbf{M}}(S)$, we determine $\mathbf{M}(S)$; see (4.1.81).
  c. From $\mathbf{M}(S)$, we calculate $\mathbf{M}(R)$ using (4.6.6).
  d. We use (4.1.81) to determine $\hat{\mathbf{M}}(R)$ from $\mathbf{M}(R)$.
  e. We determine $\hat{\mathbf{M}}^{(x)}(R)$ from $\hat{\mathbf{M}}(R)$ using (4.1.87).

If the *point source* is situated at S, matrices $\hat{\mathbf{M}}(R, S)$ and $\hat{\mathbf{M}}^{(x)}(R, S)$ are given by relations

$$\hat{\mathbf{M}}(R, S) = \begin{pmatrix} & \mathbf{M}(R, S) & & -(v^{-2}v_{,1})_R \\ & & & -(v^{-2}v_{,2})_R \\ -(v^{-2}v_{,1})_R & -(v^{-2}v_{,2})_R & -(v^{-2}v_{,3})_R \end{pmatrix}, \qquad (4.6.10)$$

$$\hat{\mathbf{M}}^{(x)}(R, S) = \hat{\mathbf{H}}(R)\hat{\mathbf{M}}(R, S)\hat{\mathbf{H}}^T(R). \qquad (4.6.11)$$

4.6.2 Determination of Matrix M from Travel Times Known Along a Data Surface

Assume that the travel-time field is known along a *data surface*, Σ, in the vicinity of point R on Σ. We also assume that the travel-time field corresponds to ray-theory travel times of some elementary wave and that the propagation velocity V of this wave is known in the vicinity of data surface Σ, close to R. The travel times along Σ may be known from seismic measurements, or even from computations. The problem of determining matrix $\mathbf{M}(R)$ of the approaching elementary wave from travel times known along data surface Σ is then fully analogous to the problem of determining matrix $\mathbf{M}(S)$ of the wave generated at an initial surface Σ^0, along which the initial travel-time field is known. The latter problem was solved in Sections 4.5.1 through 4.5.3 so that we can merely take the results from that section. The only difference is in the signs of certain quantities, but this is simple to understand. In Sections 4.5.1 through 4.5.3, we treated the wavefield *generated* on Σ^0, and here we are interested in the wavefield *approaching* on Σ. We can use (4.5.9) to determine the slowness vector of the elementary wave under consideration. This equation also determines the direction of the approaching ray. After this, we can use (4.5.25) with (4.5.13) and (4.5.18) or, alternatively, (4.5.32), to determine the relevant matrix $\mathbf{M}(R)$.

The relevant equations contain the derivatives of the travel-time field along data surface Σ. To determine the slowness vector of the elementary wave under consideration at R, we need to know the *first derivatives* of the travel-time field along Σ. If we wish to determine matrix $\mathbf{M}(R)$, the situation is even worse: we must know the *second derivatives* of the travel-time field along Σ. Numerically, the determination of these derivatives is often unstable, particularly if we are treating empirical data from seismic measurements. Various alternatives of the popular $T^2 - X^2$ method can be also used to determine the second derivatives of the travel-time field along Σ. For this purpose, it is suitable to transform the paraxial travel-time field from the parabolic to the hyperbolic form; see Section 4.6.4.

All the matrices in (4.5.32) depend on the choice of basis vectors $\vec{e}_1(R)$ and $\vec{e}_2(R)$. Basis vectors $\vec{e}_1(R)$ and $\vec{e}_2(R)$ are perpendicular to unit vector $\vec{e}_3(R) \equiv \vec{t}(R)$, tangent to the ray, and $\vec{e}_1(R)$, $\vec{e}_2(R)$, and $\vec{e}_3(R)$ form a right-handed triplet of unit vectors. Otherwise, $\vec{e}_1(R)$ and $\vec{e}_2(R)$ may be chosen arbitrarily. Matrix $\mathbf{M}(R)$ determined by (4.5.32) is, of course, related to the choice of $\vec{e}_1(R)$ and $\vec{e}_2(R)$ but may, without loss of generality, be rotated to any other choice of $\vec{e}_1'(R)$ and $\vec{e}_2'(R)$.

4.6.3 Matrix of Curvature of the Wavefront

The 2×2 matrix of the curvature of wavefront $\mathbf{K}$ is related very simply to matrix $\mathbf{M}$ of the second derivatives of the travel-time field.

The general relations for the matrix of curvature $\mathbf{D}$ of surface Σ were derived in Section 4.4.1. Let us consider surface Σ which is continuous with its first and second tangential derivatives at point R. We construct a local Cartesian coordinate system z_1, z_2, z_3 with its origin at R and with the local z_3-axis situated along the normal to Σ at R. Thus, the z_1- and z_2-axes are tangent to the surface at point R. In the vicinity of R, the interface can be described by equation

$$z_3 = \pm \tfrac{1}{2} z_I z_J D_{IJ}. \tag{4.6.12}$$

Here $\mathbf{D}$ with elements D_{IJ} is a 2×2 matrix of the curvature of surface Σ at R. Its elements depend on the orientation of axes z_1 and z_2 in the plane tangent to Σ at R. The choice of the sign in (4.6.12) is arbitrary and may be specified by convention.

We shall now consider the wavefront at any point R of ray Ω. We shall choose the local ray-centered Cartesian coordinates y_1, y_2, y_3, instead of z_1, z_2, z_3 because axis y_3 coincides with the normal to the wavefront (it is tangent to ray Ω at R). The wavefront passing through R is given by the relation $T(y_1, y_2, y_3) = T(R) = \text{const}$. By convention, we shall consider the sign in Equation (4.6.12) to be a minus sign. Thus, we shall define matrix $\mathbf{K}$ of the curvature of the wavefront by relation

$$y_3 = -\tfrac{1}{2} y_I y_J K_{IJ}. \tag{4.6.13}$$

The relation between K_{IJ} and M_{IJ} can now be easily obtained from (4.1.80). In this equation, we can neglect the terms with y_3^2 and $y_I y_3$ because they are of higher order due to (4.6.13). If point R' is situated on the wavefront passing through point R, we have $T(R') = T(R)$. Then, (4.1.80) yields

$$v^{-1}(R)y_3 + \tfrac{1}{2} y_I y_J M_{IJ}(R) = 0.$$

This equation can be expressed as

$$y_3 = -\tfrac{1}{2} y_I y_J v(R) M_{IJ}(R). \tag{4.6.14}$$

Comparing (4.6.13) with (4.6.14) yields the final relation between the matrix of curvature of wavefront $\mathbf{K}(R)$ and the matrix of second derivatives of the travel-time field, $\mathbf{M}(R)$,

$$\mathbf{K}(R) = v(R)\mathbf{M}(R). \tag{4.6.15}$$

We can easily derive the ordinary differential equation for $\mathbf{K}$. By inserting $\mathbf{M} = v^{-1}\mathbf{K}$ into (4.1.73), we obtain

$$v d\mathbf{K}/ds - \partial v/\partial s \, \mathbf{K} + v\mathbf{K}^2 + \mathbf{V} = 0. \tag{4.6.16}$$

As we can see, the matrix of curvature of the wavefront is again controlled by an ordinary differential nonlinear first-order equation of the Riccati type. The equation is only slightly more complicated than the relevant equation for $\mathbf{M}$, due to the additional term $\mathbf{K}\partial v/\partial s$.

We can also define the 2×2 *matrix of the radii of curvature of the wavefront,*

$$\mathbf{R}(R) = \mathbf{K}^{-1}(R). \qquad (4.6.17)$$

Inserting $\mathbf{K} = \mathbf{R}^{-1}$ into (4.6.16) yields the ordinary nonlinear differential equation of the first order for $\mathbf{R}$:

$$v d\mathbf{R}/ds + \partial v/\partial s\, \mathbf{R} - \mathbf{R}\mathbf{V}\mathbf{R} - v\mathbf{I} = 0. \qquad (4.6.18)$$

The nonlinear equations (4.6.16) and (4.6.18) of the Riccati type for $\mathbf{K}$ and $\mathbf{R}$ are usually inconvenient for applications. It is more convenient to use the linear dynamic ray tracing system (4.1.64) or (4.1.65) for $\mathbf{P}$ and $\mathbf{Q}$. Using (4.6.6) and (4.6.15), we obtain the *continuation relation for curvature matrix* $\mathbf{K}$:

$$\begin{aligned}\mathbf{K}(R) &= v(R)[\mathbf{P}_1(R, S) + v^{-1}(S)\mathbf{P}_2(R, S)\mathbf{K}(S)]\\ &\times [\mathbf{Q}_1(R, S) + v^{-1}(S)\mathbf{Q}_2(R, S)\mathbf{K}(S)]^{-1}.\end{aligned} \qquad (4.6.19)$$

The backward continuation formula is obtained from (4.6.7),

$$\begin{aligned}\mathbf{K}(S) &= v(S)\left[-\mathbf{P}_1^T(R, S) + v^{-1}(R)\mathbf{Q}_1^T(R, S)\mathbf{K}(R)\right]\\ &\times \left[\mathbf{P}_2^T(R, S) - v^{-1}(R)\mathbf{Q}_2^T(R, S)\mathbf{K}(R)\right]^{-1}.\end{aligned} \qquad (4.6.20)$$

We shall also give an equation for the transformation of the curvature matrix of the wavefront across a curved interface between two inhomogeneous media. Using (4.6.15) and (4.4.46), we obtain

$$\tilde{\mathbf{K}} = \tilde{v}v^{-1}\tilde{\mathbf{G}}^{-1}[\mathbf{G}\mathbf{K}\mathbf{G}^T - vu\mathbf{D} + v(\mathbf{E} - \tilde{\mathbf{E}})]\tilde{\mathbf{G}}^{-1T}. \qquad (4.6.21)$$

All the symbols have the same meaning as in (4.4.46).

Because matrix $\mathbf{K}$ is symmetric, it has two real-valued eigenvalues. Let us denote these eigenvalues K_1 and K_2. They represent the principal curvatures of the wavefront at the point of ray Ω under consideration. The principal directions of the curvature of the wavefront are determined by the relevant eigenvectors $\vec{e}_1^K$ and $\vec{e}_2^K$. At any point R of the ray, the three unit vectors $\vec{e}_1^K(R)$, $\vec{e}_2^K(R)$, and $\vec{t}(R)$ are mutually orthogonal.

Instead of principal curvatures $K_1(R)$ and $K_2(R)$, we can also use the principal radii of the curvature of the wavefront on Ω at R,

$$R_1(R) = 1/K_1(R), \qquad R_2(R) = 1/K_2(R). \qquad (4.6.22)$$

Quantities $K_{1,2}(R)$ and $R_{1,2}(R)$ may take any real values, including 0 and ∞. For $K_{1,2}(R) \neq 0$ and $R_{1,2}(R) \neq 0$, the point R of the wavefront is called *elliptic* (if $K_1 K_2 > 0$) or *hyperbolic* (if $K_1 K_2 < 0$). For $K_1(R) = 0$ and $K_2(R) \neq 0$, or for $K_1(R) \neq 0$ and $K_2(R) = 0$, the point R is called *parabolic at R*. A special case of ellipsoidal wavefronts is the *spherical wavefront at R*, if $K_1(R) = K_2(R)$. For $K_1(R) = K_2(R) = 0$, the wavefront is *locally planar at R*. See Figure 4.15. For $K_1(R) = K_2(R)$, the point R is also called *ombilic*. Note that the wavefront in the vicinity of an ombilic point is either locally spherical ($K_1 = K_2 \neq 0$) or locally planar ($K_1 = K_2 = 0$).

Some other quantities are also often used to describe the curvatures of surfaces. We commonly use the *mean curvature* of the wavefront, $H = \frac{1}{2}(K_1 + K_2)$, and the *Gaussian*

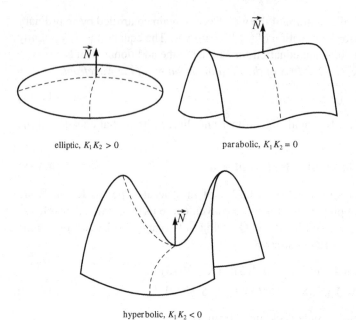

Figure 4.15. Elliptic, parabolic, and hyperbolic points on a wavefront. At an elliptic point, $K_1 K_2 > 0$; at a parabolic point, $K_1 K_2 = 0$; and at a hyperbolic point, $K_1 K_2 < 0$.

curvature of the wavefront, $K = K_1 K_2$. Consequently,

$$
\begin{aligned}
K(R) &= K_1(R)K_2(R) = \det \mathbf{K}(R) = v^2(R) \det \mathbf{M}(R), \\
H(R) &= \tfrac{1}{2}(K_1(R) + K_2(R)) = \tfrac{1}{2}\operatorname{tr}\mathbf{K}(R) = \tfrac{1}{2}v(R)\operatorname{tr}\mathbf{M}(R).
\end{aligned}
\tag{4.6.23}
$$

Here tr denotes the trace.

4.6.4 Paraxial Travel Times. Parabolic and Hyperbolic Travel Times

Using the first and second derivatives of the travel-time field, we obtain convenient approximate paraxial equations for the travel-time field in the vicinity of the ray under consideration. These equations were derived in Section 4.1.8; here we shall discuss them only very briefly.

Let us consider central ray Ω and point R situated on this ray. In addition, we shall consider point R', situated in the vicinity of point R. We can then express the travel-time field at R' by convenient quadratic equations. In the case of point R' situated in plane $\Sigma^\perp$ perpendicular to ray Ω at R, we can use ray-centered coordinates $q_I(R')$. For a generally situated point R', we can use either local Cartesian ray-centered coordinates $y_i(R', R)$ or general Cartesian coordinates $x_i(R', R)$. We obtain

$$
\begin{aligned}
T(R') &= T(R) + \tfrac{1}{2}\mathbf{q}^T(R')\mathbf{M}(R)\mathbf{q}(R') \qquad \text{for } R' \in \Sigma^\perp, \\
T(R') &= T(R) + \hat{\mathbf{y}}^T(R', R)\hat{\mathbf{p}}^{(y)}(R) + \tfrac{1}{2}\hat{\mathbf{y}}^T(R', R)\hat{\mathbf{M}}(R)\hat{\mathbf{y}}(R', R), \\
T(R') &= T(R) + \hat{\mathbf{x}}^T(R', R)\hat{\mathbf{p}}^{(x)}(R) + \tfrac{1}{2}\hat{\mathbf{x}}^T(R', R)\hat{\mathbf{M}}^{(x)}(R)\hat{\mathbf{x}}(R', R).
\end{aligned}
\tag{4.6.24}
$$

The meaning of the individual symbols is obvious; for details see Section 4.1.8.

Equations (4.6.24) can be used without any knowledge of the travel-time field outside R. The only quantities that must be explicitly known if we wish to apply (4.6.24) represent the travel time and its first and second derivatives at R. The travel-time field under consideration may correspond to a point source, or line source, or surface source. In addition, it is not necessary to know the position of the source.

Let us now discuss the distribution of travel times in a plane perpendicular to Ω at R. As we can see in the first equation of (4.6.24), the travel-time curves $T = T(q_1)$ (for q_2 fixed) and $T = T(q_2)$ (for q_1 fixed) represent *parabolas*. It is also common to speak of *parabolic approximation of travel times*. In certain applications, particularly in the seismic prospecting for oil, it is more common to use a *hyperbolic approximation of travel times*. For positive-definite matrix $M_{IJ}(R)$, we can easily transform the parabolic approximation into the hyperbolic approximation and back. Taking the square of the first equation of (4.6.24), and neglecting the terms higher than quadratic in q_I, we obtain a general expression for the hyperbolic approximation of travel times:

$$T^2(R') \doteq T^2(R) + T(R)q_I(R')q_J(R')M_{IJ}(R). \tag{4.6.25}$$

Equation (4.6.25) can be extended even to an arbitrary surface Σ passing through R. The relation (4.4.34) gives the parabolic travel-time approximation along Σ. For positive-definite matrix $F_{IJ}(R)$, the relevant hyperbolic approximation then reads

$$\left(T^\Sigma(z_1, z_2)\right)^2 \doteq \left[T(R) + z_I p_I^{(z)}(R)\right]^2 + T(R)z_I z_J F_{IJ}(R). \tag{4.6.26}$$

Here $F_{IJ}(R)$ are given by (4.4.37).

The hyperbolic approximation (4.6.26) is suitable if we wish to determine $F_{IJ}(R)$ from known travel-time data along Σ, using an alternative of the $T^2 - X^2$ method. Note that the simplest version of (4.6.26) is represented by the well-known reflection travel-time hyperbola $T^2 = T_0^2 + X^2/V_{NMO}^2$, where X is the offset and V_{NMO} is the so-called normal moveout velocity. Equation (4.6.26) generalizes this equation for a fully general case of an arbitrary elementary wave (multiply reflected, transmitted, converted) in general 3-D laterally varying layered and block structure and can be used to define some parameters alternative to the normal moveout velocities in terms of $F_{IJ}(R)$ for such general cases. On the contrary, if these parameters are determined from the travel-time data known along Σ, they can be used to compute $F_{IJ}(R)$.

In the next part of this section, we shall consider ray Ω and *two points S and R* on it. Moreover, we shall assume that a *point source is situated at S*. We denote by $T(R, S)$ the travel time from point source S to receiver R along Ω. This notation is introduced due to consistency with other similar symbols $\Pi(R, S)$, $\mathbf{M}(R, S)$, and the like. It is obvious that travel time $T(R, S)$ is reciprocal,

$$T(R, S) = T(S, R). \tag{4.6.27}$$

In an analogous way, we shall denote $T(R', S)$ the *paraxial travel times* from a point source situated at S to the receiver situated at R', close to R. Simple expressions for $T(R', S)$ can be obtained from (4.6.24):

$$T(R', S) = T(R, S) + \tfrac{1}{2}\mathbf{q}^T(R')\mathbf{M}(R, S)\mathbf{q}(R') \qquad \text{for } R' \in \Sigma^\perp,$$
$$T(R', S) = T(R, S) + \hat{\mathbf{y}}^T(R', R)\hat{\mathbf{p}}^{(y)}(R) + \tfrac{1}{2}\hat{\mathbf{y}}^T(R', R)\hat{\mathbf{M}}(R, S)\hat{\mathbf{y}}(R', R),$$
$$T(R', S) = T(R, S) + \hat{\mathbf{x}}^T(R', R)\hat{\mathbf{p}}^{(x)}(R) + \tfrac{1}{2}\hat{\mathbf{x}}^T(R', R)\hat{\mathbf{M}}^{(x)}(R, S)\hat{\mathbf{x}}(R', R).$$
$$\tag{4.6.28}$$

For $\mathbf{M}(R, S)$, $\hat{\mathbf{M}}(R, S)$, and $\hat{\mathbf{M}}^{(x)}(R, S)$, refer to (4.6.8), (4.6.10), and (4.6.11), respectively.

We need to emphasize the difference between $T(R, S)$ and $T(R', S)$. Presumably, all these quantities correspond to a point source at S on Ω. If both points S and R are situated on ray Ω, quantity $T(R, S)$ has a broader meaning: it also represents the travel time from S to R along Ω, where S is an arbitrary reference point on ray Ω (not a point source). Paraxial expressions (4.6.28) for $T(R', S)$, however, are valid only for a point source at S.

We shall add three remarks regarding Equations (4.6.28). To calculate $T(R', S)$, where S represents a point source situated on central ray Ω, the whole ray propagator matrix (4.3.5) does not need to be known. It is sufficient to calculate only minors $\mathbf{Q}_2(R, S)$ and $\mathbf{P}_2(R, S)$. Thus, the dynamic ray tracing needs to be performed only once, for the point-source initial conditions. As will be shown in Section 4.9, the whole propagator matrix (4.3.5) must be known to calculate $T(R', S')$, where both points S' and R' are situated outside central ray Ω (but close to S and R, respectively).

The second note is related to the *accuracy* of Equations (4.6.24) and (4.6.28) for paraxial travel times. Equations are only approximate, and their accuracy decreases with increasing distance of point R' from R. It is high if point R' is situated close to R and if the wavefront is smooth at R. It may, however, be rather low if the wavefront behaves anomalously in the vicinity of R (R close to a caustic point, to a structural interface, to boundary rays, and the like).

The third note is terminological. Paraxial travel time $T(R', S)$ is not computed by ray tracing from S to R' but is approximated by using the paraxial ray methods; see (4.6.28). For this reason, we can also call Equations (4.6.28) equations for a *paraxial two-point eikonal*. The general equations for a paraxial two-point eikonal $T(R', S')$ will be derived in Section 4.9.

4.6.5 Paraxial Slowness Vector

By paraxial slowness vector, we understand the gradient of the paraxial travel-time field, $\vec{p} = \nabla T$, in the vicinity of central ray Ω. Because the expression for the paraxial travel times are quadratic in terms of q_I (or y_i or x_i), the expressions for the paraxial slowness vector are linear only.

The derivatives of the paraxial travel times (4.6.24) yield

$$\mathbf{p}^{(q)}(R') = \mathbf{M}(R)\mathbf{q}(R') \qquad p_3^{(q)}(R') = v^{-1}(R), \qquad \text{for } R' \in \Sigma^{\perp},$$
$$\hat{\mathbf{p}}^{(y)}(R') = \hat{\mathbf{p}}^{(y)}(R) + \hat{\mathbf{M}}(R)\hat{\mathbf{y}}(R', R), \qquad (4.6.29)$$
$$\hat{\mathbf{p}}^{(x)}(R') = \hat{\mathbf{p}}^{(x)}(R) + \hat{\mathbf{M}}^{(x)}(R)\hat{\mathbf{x}}(R', R).$$

It should be emphasized that $\hat{\mathbf{p}}^{(y)}(R)$ has only one nonvanishing component: $p_3^{(y)}(R) = 1/v(R)$. To obtain the paraxial slowness vector at R' due to a point source situated at S, we merely replace $\mathbf{M}(R)$ by $\mathbf{M}(R, S)$, $\hat{\mathbf{M}}(R)$ by $\hat{\mathbf{M}}(R, S)$, and $\hat{\mathbf{M}}^{(x)}(R)$ by $\hat{\mathbf{M}}^{(x)}(R, S)$ in (4.6.29).

Using Equations (4.6.29) and (4.6.24), we can express the paraxial travel time at R' in terms of the slowness vectors at R and R',

$$T(R') = T(R) + \tfrac{1}{2}\mathbf{q}^{T}(R')\mathbf{p}^{(q)}(R') \qquad \text{for } R' \in \Sigma^{\perp},$$
$$T(R') = T(R) + \tfrac{1}{2}\hat{\mathbf{y}}^{T}(R', R)\big[\hat{\mathbf{p}}^{(y)}(R') + \hat{\mathbf{p}}^{(y)}(R)\big], \qquad (4.6.30)$$
$$T(R') = T(R) + \tfrac{1}{2}\hat{\mathbf{x}}^{T}(R', R)\big[\hat{\mathbf{p}}^{(x)}(R') + \hat{\mathbf{p}}^{(x)}(R)\big].$$

These equations will be used very conveniently to compute the paraxial travel times, if we first determine the paraxial slowness vector; see Section 4.9.

As in (4.6.24), the travel-time field under consideration in (4.6.29) and (4.6.30) may correspond to an arbitrarily situated point source, line source, or surface source.

4.7 Dynamic Ray Tracing in Cartesian Coordinates

The dynamic ray tracing system in ray-centered coordinates has played an important role in the seismic ray theory. It can be used to evaluate a simple 4×4 propagator matrix $\mathbf{\Pi}(R, S)$, which offers a large number of important applications. Nevertheless, it is of some interest to express the dynamic ray tracing systems also in other coordinates and to determine the relevant 6×6 propagator matrices. In certain situations, dynamic ray tracing in other coordinate systems may be numerically more efficient when compared with dynamic ray tracing in ray-centered coordinates. Particularly important is the dynamic ray tracing system in Cartesian coordinates. The number of equations in this dynamic ray tracing system is higher than in the ray-centered coordinates, but the system matrix may be simpler. Moreover, they do not require the computation of the basis vectors $\vec{e}_1$ and $\vec{e}_2$ of the ray-centered coordinate system along the ray. The dynamic ray tracing in Cartesian coordinates is also suitable in anisotropic media, as will be shown in Section 4.14.

The dynamic ray tracing system in Cartesian coordinates was derived and discussed in Section 4.2.1. Here we shall use the same notation as in that section. We shall first slightly generalize the system (4.2.4) for an arbitrary monotonic variable u along ray Ω and give several examples of dynamic ray tracing systems for isotropic media, using different u; see Section 4.7.1. In Section 4.7.2, we shall construct the 6×6 propagator matrix of the dynamic ray tracing system for an arbitrary monotonic variable u along Ω and then discuss the transformation of the dynamic ray tracing system across a structural interface. The relevant expressions for the interface propagator matrix will be given. These expression are, of course, also valid for $u = T$. The expressions for the 6×6 propagator matrix are extended to rays Ω situated in a layered medium. In Section 4.7.3, 6×6 propagator matrices in layered media in arbitrary curvilinear coordinates are discussed. The derived expressions for the 6×6 propagator matrices can be suitably used in the ray perturbation theory; see Sections 4.7.4 and 4.7.5. The main purpose of Section 4.7.4 is to derive convenient expressions for paraxial rays in a perturbed medium, situated close to a reference ray Ω^0, constructed in the unperturbed background medium. Finally, the computation of third and higher derivatives of the travel-time field along rays will be discussed in Section 4.7.6. It will be shown that "higher-order dynamic ray tracing systems" are not necessary to compute "higher derivatives" of the travel-time field. It is sufficient to use a standard dynamic ray tracing system and to supplement it by some quadratures along ray Ω.

4.7.1 Dynamic Ray Tracing System in Cartesian Coordinates

The general dynamic ray tracing system in Cartesian coordinates, valid for both isotropic and anisotropic media, was derived in Section 4.2.1; see (4.2.4) with (4.2.5). The monotonic paramater u along Ω in (4.2.4) represents travel time T. The analogous system, however, remains valid also for any other monotonic parameter u. We introduce

$$Q_i^{(x)}(u) = (\partial x_i / \partial \gamma)_{u=\text{const.}}, \qquad P_i^{(x)}(u) = \left(\partial p_i^{(x)} / \partial \gamma\right)_{u=\text{const.}}. \qquad (4.7.1)$$

We use the same notation for $Q_i^{(x)}$ and $P_i^{(x)}$ as in Section 4.2.1, where monotonic variable u represented travel time T. In this section, however, we shall use $Q_i^{(x)}$ and $P_i^{(x)}$ for arbitrary monotonic variable u. Of course, the variable u along ray Ω is related to the form of Hamiltonian $\mathcal{H}^{(x)}(x_i, p_i^{(x)})$ under consideration; see (3.1.4). Taking the partial derivative of ray tracing system (3.1.3) with respect to γ, and bearing in mind that $\partial/\partial\gamma$ commutes with d/du, we again obtain the dynamic ray tracing system in the following form:

$$dQ_i^{(x)}/du = A_{ij}^{(x)}Q_j^{(x)} + B_{ij}^{(x)}P_j^{(x)}, \qquad dP_i^{(x)}/du = -C_{ij}^{(x)}Q_j^{(x)} - D_{ij}^{(x)}P_j^{(x)}.$$

(4.7.2)

Here $A_{ij}^{(x)}$, $B_{ij}^{(x)}$, $C_{ij}^{(x)}$, and $D_{ij}^{(x)}$ are given by (4.2.5). The Hamiltonian $\mathcal{H}^{(x)}(x_i, p_i^{(x)})$ in (4.2.5), however, must correspond to the variable u along Ω. Quantities $A_{ij}^{(x)}$, $B_{ij}^{(x)}$, $C_{ij}^{(x)}$, and $D_{ij}^{(x)}$ satisfy symmetry relations (4.2.6) and constraint relation (4.2.7) for any u. Note that (4.7.2) also represents the paraxial ray tracing system, if we replace $Q_i^{(x)}$ by δx_i, and $P_i^{(x)}$ by $\delta p_i^{(x)}$; see (4.2.11).

As in Section 4.2.1, we can find the ray tangent solutions of dynamic ray tracing system (4.7.2) analytically,

$$Q_i^{(x)} = dx_i/du = \partial\mathcal{H}^{(x)}/\partial p_i^{(x)} = a\mathcal{U}_i^{(x)},$$
$$P_i^{(x)} = dp_i^{(x)}/du = -\partial\mathcal{H}^{(x)}/\partial x_i = a\eta_i^{(x)}.$$

(4.7.3)

Here $\mathcal{U}_i^{(x)}$ and $\eta_i^{(x)}$ are given by (4.2.8), and $\mathcal{U}_i^{(x)}$ represent the Cartesian components of the group velocity vector. Quantity $a(T)$ is given by the relation

$$a(T) = dT/du = p_k^{(x)}\partial\mathcal{H}^{(x)}/\partial p_k^{(x)}.$$

(4.7.4)

Of course, $a(T) = 1$ for $u = T$.

As an example of general dynamic ray tracing system (4.7.2), we shall specify it for isotropic media, for several different Hamiltonians $\mathcal{H}^{(x)}(x_i, p_i^{(x)})$ and for the relevant variables u along Ω.

Travel time T is most commonly considered to be a monotonic parameter, u, along the ray. In this case, we can take Hamiltonian $\mathcal{H}^{(x)}$ in the form $\mathcal{H}^{(x)} = \frac{1}{2}\ln(p_i^{(x)}p_i^{(x)}) + \ln V(x_i)$; see (3.1.7). The dynamic ray tracing for $u = T$ reads

$$\frac{d}{dT}Q_i^{(x)} = \frac{\delta_{ij}p_n^{(x)}p_n^{(x)} - 2p_i^{(x)}p_j^{(x)}}{(p_k^{(x)}p_k^{(x)})^2}P_j^{(x)}, \qquad \frac{d}{dT}P_i^{(x)} = -\frac{\partial^2\ln V}{\partial x_i\partial x_j}Q_j^{(x)}.$$

(4.7.5)

We note that $\partial^2\mathcal{H}^{(x)}/\partial p_i^{(x)}\partial x_j = \partial^2\mathcal{H}^{(x)}/\partial x_i\partial p_j^{(x)} = 0$ in our case.

An alternative form of the dynamic ray tracing system for $u = T$ can be obtained from Hamiltonian $\mathcal{H}^{(x)} = \frac{1}{2}(V^2 p_i^{(x)}p_i^{(x)} - 1)$:

$$\frac{d}{dT}Q_i^{(x)} = \frac{\partial V^2}{\partial x_j}p_i^{(x)}Q_j^{(x)} + V^2 P_i^{(x)},$$
$$\frac{d}{dT}P_i^{(x)} = -\frac{1}{2}\frac{\partial^2 V^2}{\partial x_i\partial x_j}p_k^{(x)}p_k^{(x)}Q_j^{(x)} - \frac{\partial V^2}{\partial x_i}p_j^{(x)}P_j^{(x)};$$

(4.7.6)

see (3.1.13). Here we can, of course, insert $p_k^{(x)}p_k^{(x)} = V^{-2}$, as the Hamiltonian vanishes along the whole ray. It may seem surprising that we have obtained two different dynamic ray tracing systems, (4.7.5) and (4.7.6), for the same parameter T along the ray. The explanation is simple. The dynamic ray tracing systems satisfy the constraint equation (4.2.7) along the whole ray. Applying this constraint to the initial conditions for (4.7.5) and (4.7.6) and taking into account that the constraint is satisfied along the whole ray, we find that both systems are equivalent. The meaning of $Q_i^{(x)}$ and $P_i^{(x)}$ in (4.7.5) and (4.7.6) is $Q_i^{(x)} = (\partial x_i / \partial \gamma)_{T=\text{const.}}$ and $P_i^{(x)} = (\partial p_i^{(x)} / \partial \gamma)_{T=\text{const.}}$.

If we assume *arclength s* to be the monotonic parameter u along the ray, we can take Hamiltonian $\mathcal{H}^{(x)}$ in the form $\mathcal{H}^{(x)} = (p_i^{(x)} p_i^{(x)})^{1/2} - V^{-1}(x_i)$. We then obtain the dynamic ray tracing system as follows:

$$\frac{d}{ds} Q_i^{(x)} = \frac{\delta_{ij}\left(p_n^{(x)} p_n^{(x)}\right) - p_i^{(x)} p_j^{(x)}}{\left(p_k^{(x)} p_k^{(x)}\right)^{3/2}} P_j^{(x)}, \qquad \frac{d}{ds} P_i^{(x)} = \frac{\partial^2}{\partial x_i \partial x_j}\left(\frac{1}{V}\right) Q_j^{(x)}.$$

$$(4.7.7)$$

As for $u = T$, we can also write for $u = s$ a dynamic ray tracing system alternative to (4.7.7). We use Hamiltonian $\mathcal{H} = \frac{1}{2} V (p_k^{(x)} p_k^{(x)} - V^{-2})$ (see (3.1.16)), and obtain

$$\frac{d}{ds} Q_i^{(x)} = \frac{\partial V}{\partial x_j} p_i^{(x)} Q_j^{(x)} + V P_i^{(x)},$$

$$\frac{d}{ds} P_i^{(x)} = -\frac{1}{V^2}\left(\frac{\partial^2 V}{\partial x_i \partial x_j} - \frac{1}{V}\frac{\partial V}{\partial x_i}\frac{\partial V}{\partial x_j}\right) Q_j^{(x)} - \frac{\partial V}{\partial x_i} p_j^{(x)} P_j^{(x)}.$$

$$(4.7.8)$$

The meaning of $Q_i^{(x)}$ and $P_i^{(x)}$ in (4.7.7) and (4.7.8) is $Q_i^{(x)} = (\partial x_i / \partial \gamma)_{s=\text{const.}}$ and $P_i^{(x)} = (\partial p_i^{(x)} / \partial \gamma)_{s=\text{const.}}$.

The simplest dynamic ray tracing system is obtained for *monotonic parameter* σ along the ray. Hamiltonian $\mathcal{H}^{(x)}$ is then given by the relation $\mathcal{H}^{(x)} = \frac{1}{2}(p_i^{(x)} p_i^{(x)} - 1/V^2)$, and the dynamic ray tracing system reads

$$\frac{d}{d\sigma} Q_i^{(x)} = P_i^{(x)}, \qquad \frac{d}{d\sigma} P_i^{(x)} = \frac{1}{2}\frac{\partial^2}{\partial x_i \partial x_j}\left(\frac{1}{V^2}\right) Q_j^{(x)}.$$

$$(4.7.9)$$

If $1/V^2$ is a linear function of coordinates x_i, dynamic ray tracing system (4.7.9) can be simply solved analytically:

$$P_i^{(x)}(\sigma) = P_i^{(x)}(\sigma_0), \qquad Q_i^{(x)}(\sigma) = Q_i^{(x)}(\sigma_0) + (\sigma - \sigma_0) P_i^{(x)}(\sigma_0).$$

All these dynamic ray tracing systems can be expressed in terms of ordinary differential equations of the second order. Particularly simple equations are obtained for $u = \sigma$,

$$\frac{d^2}{d\sigma^2} Q_i^{(x)} - \frac{1}{2}\frac{\partial^2}{\partial x_i \partial x_j}\left(\frac{1}{V^2}\right) Q_j^{(x)} = 0;$$

$$(4.7.10)$$

see (4.7.9). The meaning of $Q_i^{(x)}$ and $P_i^{(x)}$ in (4.7.9) and (4.7.10) is $Q_i^{(x)} = (\partial x_i / \partial \gamma)_{\sigma=\text{const.}}$ and $P_i^{(x)} = (\partial p_i^{(x)} / \partial \gamma)_{\sigma=\text{const.}}$.

4.7.2 6 × 6 Propagator Matrix in a Layered Medium

Dynamic ray tracing system (4.7.2) consists of six linear ordinary differential equations of the first order. Consequently, we can construct the 6 × 6 propagator matrix $\boldsymbol{\Pi}^{(x)}(u, u_0)$ of this dynamic ray tracing system. The 6 × 6 matrix $\boldsymbol{\Pi}^{(x)}(u, u_0)$ satisfies the initial condition $\boldsymbol{\Pi}^{(x)}(u_0, u_0) = \mathbf{I}$ at $u = u_0$, where $\mathbf{I}$ is the 6 × 6 identity matrix. See Section 4.3.7. Because the dynamic ray tracing system (4.7.2) satisfies symmetry relations (4.2.6), the 6 × 6 propagator matrix $\boldsymbol{\Pi}^{(x)}(u, u_0)$ is symplectic for any monotonic variable u along Ω. The 6 × 6 propagator matrix $\boldsymbol{\Pi}^{(x)}(u, u_0)$ also satisfies Liouville's theorem, chain property, the relations for its inverse, and so on. Consider two points, R and S, situated on ray Ω, where

S corresponds to variable u_0 and R corresponds to variable u. After propagator matrix $\mathbf{\Pi}^{(x)}(R, S)$ is known, (4.3.32) can be used to compute the solution of the dynamic ray tracing system (4.7.2) at any point R on Ω, assuming the initial conditions are known at point S.

To compute the 6×6 propagator matrix $\mathbf{\Pi}^{(x)}(R, S)$ along ray Ω crossing structural interface Σ, it is necessary to know the transformation of the propagator matrix across the interface. Let us denote the point of incidence of Ω on Σ by Q and the related R/T point by $\tilde{Q}$. As in ray-centered coordinates, the transformation of the 6×6 propagator matrix across interface Σ can be expressed in the following form:

$$\mathbf{\Pi}^{(x)}(R, S) = \mathbf{\Pi}^{(x)}(R, \tilde{Q})\mathbf{\Pi}^{(x)}(\tilde{Q}, Q)\mathbf{\Pi}^{(x)}(Q, S); \qquad (4.7.11)$$

see (4.4.82). Here $\mathbf{\Pi}^{(x)}(\tilde{Q}, Q)$ is the 6×6 *interface propagator matrix* in Cartesian coordinates. Interface propagator matrix $\mathbf{\Pi}^{(x)}(\tilde{Q}, Q)$ depends on the Hamiltonian $\mathcal{H}^{(x)}(x_i, p_i^{(x)})$ used so that it is different for different monotonic variables u.

A simple derivation of the interface propagator matrix, based on the continuity of paraxial rays across the interface, is given by Farra and Le Bégat (1995). We shall not repeat the derivation here; we only present the final results. We consider an interface described by the relation $\Sigma(x_1, x_2, x_3) = 0$; see (4.4.1). Partial derivatives $\Sigma_{,i} = \partial\Sigma/\partial x_i$ are closely related to the components of the vector normal to interface Σ, see (4.4.4), and $\Sigma_{,ij} = \partial^2\Sigma/\partial x_i\partial x_j$ are closely related to the curvature of the interface, see (4.4.16). Then

$$\mathbf{\Pi}^{(x)}(\tilde{Q}, Q) = \begin{pmatrix} \hat{\mathbf{\Pi}}_{11}^{(x)}(\tilde{Q}, Q) & \hat{\mathbf{0}} \\ \hat{\mathbf{\Pi}}_{21}^{(x)}(\tilde{Q}, Q) & \hat{\mathbf{\Pi}}_{22}^{(x)}(\tilde{Q}, Q) \end{pmatrix} = \mathbf{T\Pi} + \mathbf{\Delta}, \qquad (4.7.12)$$

where the 6×6 matrices $\mathbf{\Pi}$, $\mathbf{T}$, and $\mathbf{\Delta}$ are given by relations

$$\mathbf{\Pi} = \begin{pmatrix} \hat{\mathbf{\Pi}}_1 & \hat{\mathbf{0}} \\ \hat{\mathbf{\Pi}}_2 & \hat{\mathbf{I}} \end{pmatrix}, \qquad \mathbf{T} = \begin{pmatrix} \hat{\mathbf{I}} & \hat{\mathbf{0}} \\ \hat{\mathbf{T}}_1 & \hat{\mathbf{T}}_2 \end{pmatrix}, \qquad \mathbf{\Delta} = \begin{pmatrix} \hat{\mathbf{\Delta}}_1 & \hat{\mathbf{0}} \\ \hat{\mathbf{\Delta}}_2 & \hat{\mathbf{0}} \end{pmatrix}. \qquad (4.7.13)$$

The physical meaning of projection matrices $\mathbf{\Pi}$ and $\mathbf{T}$ and of matrix $\mathbf{\Delta}$ is explained in detail by Farra and Le Bégat (1995). In (4.7.13), $\hat{\mathbf{I}}$ and $\hat{\mathbf{0}}$ are 3×3 unity and null matrices, and 3×3 matrices $\hat{\mathbf{\Pi}}_1$, $\hat{\mathbf{\Pi}}_2$, $\hat{\mathbf{T}}_1$, $\hat{\mathbf{T}}_2$, $\hat{\mathbf{\Delta}}_1$, and $\hat{\mathbf{\Delta}}_2$ are as follows:

$$\begin{aligned} (\Pi_1)_{ij} &= \delta_{ij} - \Phi_{ij}/\Phi, & (\Pi_2)_{ij} &= \Psi_{ij}/\Phi, \\ (T_1)_{ij} &= (\Psi_{ji} - \tilde{\Psi}_{ji})/\tilde{\Phi} - A_1(\delta_{ik} - \tilde{\Phi}_{ki}/\tilde{\Phi})\Sigma_{,kj}, & & \\ (T_2)_{ij} &= \delta_{ij} - (\tilde{\Phi}_{ji} - \Phi_{ji})/\tilde{\Phi}, & (\Delta_1)_{ij} &= \tilde{\Phi}_{ij}/\Phi, & (\Delta_2)_{ij} &= -\tilde{\Psi}_{ij}/\Phi. \end{aligned}$$

$$(4.7.14)$$

Individual symbols in (4.7.14) have the following meaning:

$$\begin{aligned} \Phi_{ik} &= (\partial\mathcal{H}^{(x)}/\partial p_i^{(x)})\Sigma_{,k}, & \Phi &= \Phi_{ii}, \\ \Psi_{ik} &= (\partial\mathcal{H}^{(x)}/\partial x_i)\Sigma_{,k}, & A_1 &= (p_i^{(x)} - \tilde{p}_i^{(x)})\Sigma_{,i}/\Sigma_{,k}\Sigma_{,k}. \end{aligned} \qquad (4.7.15)$$

Using (4.7.12) through (4.7.15), we obtain the final expressions for the 3×3 minors $\hat{\mathbf{\Pi}}_{11}^{(x)}(\tilde{Q}, Q)$, $\hat{\mathbf{\Pi}}_{21}^{(x)}(\tilde{Q}, Q)$, and $\hat{\mathbf{\Pi}}_{22}^{(x)}(\tilde{Q}, Q)$ of the 6×6 interface propagator matrix $\mathbf{\Pi}^{(x)}(\tilde{Q}, Q)$,

$$\left(\Pi_{11}^{(x)}(\tilde{Q}, Q)\right)_{ij} = \delta_{ij} - (\Phi_{ij} - \tilde{\Phi}_{ij})/\Phi,$$

$$\left(\Pi_{22}^{(x)}(\tilde{Q}, Q)\right)_{ij} = \delta_{ij} - (\tilde{\Phi}_{ji} - \Phi_{ji})/\tilde{\Phi},$$

$$\left(\Pi_{21}^{(x)}(\tilde{Q}, Q)\right)_{ij} = (\Psi_{ij} - \tilde{\Psi}_{ij})/\Phi - (\tilde{\Psi}_{ji} - \Psi_{ji})/\tilde{\Phi} - A_2\Sigma_{,i}\Sigma_{,j}/\Phi\tilde{\Phi}$$

$$- A_1(\delta_{ik} - \tilde{\Phi}_{ki}/\tilde{\Phi})\Sigma_{,kn}(\delta_{nj} - \Phi_{nj}/\Phi).$$

$$(4.7.16)$$

Here A_2 is given by the relation:

$$A_2 = \left(\partial\tilde{\mathcal{H}}^{(x)}/\partial p_i^{(x)}\right)\left(\partial\mathcal{H}^{(x)}/\partial x_i\right) - \left(\partial\mathcal{H}^{(x)}/\partial p_i^{(x)}\right)\left(\partial\tilde{\mathcal{H}}^{(x)}/\partial x_i\right).$$

All quantities with a tilde correspond to the R/T point $\tilde{Q}$ on Ω, and all quantities without a tilde correspond to the point of incidence Q on Ω. Computation of $\mathbf{\Pi}^{(x)}(\tilde{Q}, Q)$ does not require any local coordinate system to be introduced at Q. Equation (4.7.12) is valid for any form of Hamiltonian, including anisotropic media. It is possible to prove that $\mathbf{\Pi}^{(x)}(\tilde{Q}, Q)$ is symplectic and that $\det \mathbf{\Pi}^{(x)}(\tilde{Q}, Q) = 1$. Farra and Le Bégat (1995) also noticed that certain previously published interface matrices analogous to (4.7.12) were not symplectic. Using relation (4.7.12), it can also be proved that $Q_i^{(x)}(\tilde{Q})$ and $P_i^{(x)}(\tilde{Q})$ (corresponding to the R/T wave) satisfy constraint relation (4.2.7) if $Q_i^{(x)}(Q)$ and $P_i^{(x)}(Q)$ (corresponding to the incident wave) satisfy them.

The relations presented in this section can be used to calculate the 6×6 propagator matrix $\mathbf{\Pi}^{(x)}(u, u_0)$ in Cartesian coordinates, in any layered isotropic or anisotropic medium. Monotonic variable u along Ω may be arbitrary. To obtain the complete 6×6 ray propagator matrix, the system of six equations (4.7.2) should be solved six times. In other words, we need to solve a system consisting of 36 equations along ray Ω from S to R to determine $\mathbf{\Pi}^{(x)}(R, S)$.

Consider a 3-D laterally varying isotropic or anisotropic structure containing curved interfaces of the first order $\Sigma_1, \Sigma_2, \ldots, \Sigma_k$. Assume ray Ω of an arbitrary multiply reflected (possibly converted) elementary wave. Consider N points of reflection/transmission on Ω between initial point S and end point R. We denote the points of incidence by $Q_1, Q_2, \ldots, Q_N$, and the relevant R/T points by $\tilde{Q}_1, \tilde{Q}_2, \ldots, \tilde{Q}_N$. The final 6×6 propagator matrix of (4.7.2) in Cartesian coordinates is then given by the relations:

$$\mathbf{\Pi}^{(x)}(R, S) = \mathbf{\Pi}^{(x)}(R, \tilde{Q}) \prod_{i=N}^{1} \left[\mathbf{\Pi}^{(x)}(\tilde{Q}_i, Q_i)\mathbf{\Pi}^{(x)}(Q_i, \tilde{Q}_{i-1})\right]. \qquad (4.7.17)$$

The 6×6 propagator matrix $\mathbf{\Pi}^{(x)}(R, S)$ in a general 3-D layered medium satisfies all the properties of the ray propagator matrix $\mathbf{\Pi}^{(x)}(R, S)$ in a smooth medium at any point on ray Ω (including the interfaces): the symplectic property, Liouville's theorem, the chain property, and the like. Moreover, relation (4.7.17) can be used to prove that $Q_i^{(x)}(R)$ and $P_i^{(x)}(R)$ satisfy constraint relation (4.2.7) if $Q_i^{(x)}(S)$ and $P_i^{(x)}(S)$ satisfy it.

4.7.3 Transformation of the Interface Propagator Matrix

The dynamic ray tracing system in arbitrary curvilinear coordinates were derived and discussed in Section 4.2.4, and the relevant 6×6 propagator matrices were derived and discussed in Sections 4.3.7.2 and 4.3.7.3. Although the monotonic variable along ray Ω in Sections 4.2.4 and 4.3.7 was travel time T, the systems may easily be modified for an arbitrary monotonic variable u. Consequently, dynamic ray tracing in curvilinear coordinates in smooth anisotropic and anisotropic media does not cause any problem from a theoretical point of view. If we wish to perform dynamic ray tracing in curvilinear coordinates in a layered medium, however, we must know the 6×6 interface propagator matrix in these coordinates. In this section, we shall derive a simple expression for the 6×6 interface propagator matrix in curvilinear coordinates ξ_i.

We denote the interface propagator matrix in ξ_i coordinates by $\mathbf{\Pi}^{(\xi)}(\tilde{Q}, Q)$. We assume that this propagator matrix corresponds to the variable u along the ray. We can then transform expression (4.7.12) for the interface propagator matrix $\boldsymbol{\Pi}^{(x)}(\tilde{Q}, Q)$ from Cartesian coordinates to curvilinear coordinates ξ_i using relation (4.3.38),

$$\boldsymbol{\Pi}^{(\xi)}(\tilde{Q}, Q) = \begin{pmatrix} \hat{\bar{\mathbf{H}}}^{(\xi)}(\tilde{Q}) & \hat{\mathbf{0}} \\ \hat{\mathbf{F}}(\tilde{Q})\hat{\bar{\mathbf{H}}}^{(\xi)}(\tilde{Q}) & \hat{\bar{\mathbf{H}}}^{(\xi)T}(\tilde{Q}) \end{pmatrix}$$

$$\times \boldsymbol{\Pi}^{(x)}(\tilde{Q}, Q) \begin{pmatrix} \hat{\mathbf{H}}^{(\xi)}(Q) & \hat{\mathbf{0}} \\ -\hat{\mathbf{H}}^{(\xi)T}(Q)\hat{\mathbf{F}}(Q) & \hat{\mathbf{H}}^{(\xi)T}(Q) \end{pmatrix}. \qquad (4.7.18)$$

All the symbols have the same meaning as in (4.3.38). The 3×3 matrices $\hat{\mathbf{H}}^{(\xi)}(Q), \hat{\bar{\mathbf{H}}}^{(\xi)}(Q)$, and $\hat{\mathbf{F}}(Q)$ may be different from $\hat{\mathbf{H}}^{(\xi)}(\tilde{Q}), \hat{\bar{\mathbf{H}}}^{(\xi)}(\tilde{Q})$, and $\hat{\mathbf{F}}(\tilde{Q})$, for example, for coordinate system ξ_i connected with central ray Ω. This applies to $\xi_i = y_i$ (wavefront orthonormal coordinates), to $\xi_i = \zeta_i$ (nonorthogonal ray-centered coordinates), and the like. Equation (4.7.18) can be expressed in many alternative forms.

4.7.4 Ray Perturbation Theory

In this section, we shall show that the Hamiltonian approach to dynamic ray tracing can be conveniently used in the ray perturbation theory. We shall use the notation of Section 3.9, where the first-order travel-time perturbations were studied. Here we are, however, interested in the perturbations of seismic rays. We consider the background medium $\mathcal{M}^0$, characterized by Hamiltonian $\mathcal{H}^{(x)0}$, and reference ray Ω^0 in the background medium $\mathcal{M}^0$, parameterized by monotonic parameter u. We also consider two points on Ω^0, S, and R, and denote $u_S = u(S)$ and $u_R = u(R)$. Ray Ω^0 in phase space is specified by parametric equation $x_i = x_i^0(u)$ and $p_i^{(x)} = p_i^{(x)0}(u)$. In the perturbed medium $\mathcal{M}$, the Hamiltonian is denoted by $\mathcal{H}^{(x)}$. We introduce $\Delta\mathcal{H}^{(x)}$ by relation $\mathcal{H}^{(x)}(x_i, p_i^{(x)}) = \mathcal{H}^{(x)0}(x_i, p_i^{(x)}) + \Delta\mathcal{H}^{(x)}(x_i, p_i^{(x)})$ and assume that $\Delta\mathcal{H}^{(x)}$ is small. Consider ray Ω in the perturbed medium $\mathcal{M}$, situated close to Ω^0. We describe ray Ω by the parametric relation:

$$x_i(u) = x_i^0(u) + \Delta x_i(u), \qquad p_i^{(x)}(u) = p_i^{(x)0}(u) + \Delta p_i^{(x)}(u), \qquad (4.7.19)$$

as in Section 3.9. We then obtain

$$\frac{dx_i}{du} = \frac{dx_i^0}{du} + \frac{d\Delta x_i}{du} = \frac{\partial \mathcal{H}^{(x)}}{\partial p_i^{(x)}} = \frac{\partial\left(\mathcal{H}^{(x)0} + \Delta\mathcal{H}^{(x)}\right)}{\partial\left(p_i^{(x)0} + \Delta p_i^{(x)}\right)}$$

$$\doteq \frac{\partial \mathcal{H}^{(x)0}}{\partial p_i^{(x)0}} + \frac{\partial^2 \mathcal{H}^{(x)0}}{\partial p_i^{(x)0}\partial x_j^0}\Delta x_j + \frac{\partial^2 \mathcal{H}^{(x)0}}{\partial p_i^{(x)0}\partial p_j^{(x)0}}\Delta p_j^{(x)} + \frac{\partial \Delta\mathcal{H}^{(x)}}{\partial p_i^{(x)0}},$$

$$\frac{dp_i^{(x)}}{du} = \frac{dp_i^{(x)0}}{du} + \frac{d\Delta p_i^{(x)}}{du} = -\frac{\partial \mathcal{H}^{(x)}}{\partial x_i} = -\frac{\partial\left(\mathcal{H}^{(x)0} + \Delta\mathcal{H}^{(x)}\right)}{\partial\left(x_i^0 + \Delta x_i\right)}$$

$$\doteq -\frac{\partial \mathcal{H}^{(x)0}}{\partial x_i^0} - \frac{\partial^2 \mathcal{H}^{(x)0}}{\partial x_i^0\partial x_j^0}\Delta x_j - \frac{\partial^2 \mathcal{H}^{(x)0}}{\partial x_i^0\partial p_j^{(x)0}}\Delta p_j^{(x)} - \frac{\partial \Delta\mathcal{H}^{(x)}}{\partial x_i^0}.$$

Because $dx_i^0/du = \partial\mathcal{H}^{(x)0}/\partial p_i^{(x)0}$ and $dp_i^{(x)0}/du = -\partial\mathcal{H}^{(x)0}/\partial x_i^0$, we obtain

$$d\Delta x_i/du = A_{ij}^{(x)0}\Delta x_j + B_{ij}^{(x)0}\Delta p_j^{(x)} + \partial\Delta\mathcal{H}^{(x)}/\partial p_i^{(x)0},$$

$$d\Delta p_i^{(x)}/du = -C_{ij}^{(x)0}\Delta x_j - D_{ij}^{(x)0}\Delta p_j^{(x)} - \partial\Delta\mathcal{H}^{(x)}/\partial x_i^0. \qquad (4.7.20)$$

Here $A_{ij}^{(x)0}$, $B_{ij}^{(x)0}$, $C_{ij}^{(x)0}$, and $D_{ij}^{(x)0}$ are given by (4.2.5) and are computed in the background, unperturbed medium. Like the solutions of standard paraxial ray tracing systems, the solutions of the inhomogeneous paraxial ray tracing system (4.7.20) must satisfy *the constraint relation*:

$$\left(\partial \mathcal{H}^{(x)0}/\partial x_i^0\right)\Delta x_i + \left(\partial \mathcal{H}^{(x)0}/\partial p_i^{(x)0}\right)\Delta p_i^{(x)} + \Delta \mathcal{H}^{(x)} = 0. \tag{4.7.21}$$

For $\Delta \mathcal{H}^{(x)} = 0$, (4.7.20) and (4.7.21) represent the standard paraxial ray tracing system (4.2.11) with (4.2.12) in the background medium. For $\Delta \mathcal{H}^{(x)} \neq 0$, however, (4.7.20) represents the inhomogeneous paraxial ray tracing system (4.3.40), and its solution is given by (4.3.41). We denote by $\mathbf{\Pi}^{(x)0}(u, u_0)$ the ray propagator matrix computed along Ω^0 in the unperturbed medium. Then

$$\begin{pmatrix} \Delta \hat{\mathbf{x}}(u) \\ \Delta \hat{\mathbf{p}}^{(x)}(u) \end{pmatrix} = \mathbf{\Pi}^{(x)0}(u, u_0) \begin{pmatrix} \Delta \hat{\mathbf{x}}(u_0) \\ \Delta \hat{\mathbf{p}}^{(x)}(u_0) \end{pmatrix} + \int_{u_0}^{u} \mathbf{\Pi}^{(x)0}(u, u') \begin{pmatrix} \hat{\mathbf{E}}^{(x)}(u') \\ \hat{\mathbf{F}}^{(x)}(u') \end{pmatrix} \mathrm{d}u'. \tag{4.7.22}$$

The integral in (4.7.22) is taken along ray Ω^0 in unperturbed medium $\mathcal{M}^0$, and $\hat{\mathbf{E}}^{(x)}$ and $\hat{\mathbf{F}}^{(x)}$ are 3×1 column matrices with components given by relations:

$$E_i^{(x)} = \partial \Delta \mathcal{H}^{(x)}/\partial p_i^{(x)}, \qquad F_i^{(x)} = -\partial \Delta \mathcal{H}^{(x)}/\partial x_i. \tag{4.7.23}$$

Equation (4.7.22) can be used to compute paraxial rays situated in the vicinity of ray Ω^0 in the perturbed medium. To use them, it is necessary to determine the derivatives of $\Delta \mathcal{H}^{(x)}$ with respect to x_i and $p_i^{(x)}$. This is not difficult because the expressions for $\Delta \mathcal{H}^{(x)}$ are known from Section 3.9, for both isotropic and anisotropic media; see (3.9.8) and (3.9.14).

Equations similar to (4.7.20) can also be used to construct *true rays situated in the vicinity of any reference curve C^0*. This problem found important applications in connection with the bending method of two-point ray tracing; see Section 3.11.3.2. For a detailed derivation and discussion, valid both for isotropic and anisotropic media, see Farra (1992).

4.7.5 Second-Order Travel-Time Perturbation

We again consider the background medium $\mathcal{M}^0$, characterized by Hamiltonian $\mathcal{H}^{(x)0}$, and perturbed medium $\mathcal{M}$, with Hamiltonian $\mathcal{H}^{(x)}$. We also consider reference ray Ω^0 in the background medium $\mathcal{M}^0$, parameterized by monotonic parameter u, and two points S and R situated on Ω^0. It was shown in Section 3.9 that the first-order travel-time perturbation can be computed by a quadrature of the first-order perturbation of the Hamiltonian along Ω^0, in the background medium. The knowledge of the perturbed ray is not required in these computations.

In the foregoing section, we derived the inhomogeneous paraxial ray tracing system (4.7.20), which can be used to compute the first-order ray perturbations in the paraxial vicinity of reference ray Ω^0. Intuitively, we expect (4.7.20) to be sufficient to determine the second-order travel-time perturbation. In this section, we shall derive the expressions for the second-order travel-time perturbation and prove that the knowledge of the first-order ray perturbation is really sufficient to determine them.

In the derivation of the second-order travel-time perturbation, we shall follow Farra (1999). We introduce small parameter ϵ and express formally $\mathcal{H}^{(x)}$, x_i, and $p_i^{(x)}$ in the perturbation series in powers of ϵ, up to quadratic terms:

$$\mathcal{H}^{(x)} = \mathcal{H}^{(x)0} + \epsilon \mathcal{H}^{(x)1} + \epsilon^2 \mathcal{H}^{(x)2}, \qquad x_i = x_i^0 + \epsilon x_i^1 + \epsilon^2 x_i^2,$$
$$p_i^{(x)} = p_i^{(x)0} + \epsilon p_i^{(x)1} + \epsilon^2 p_i^{(x)2}. \tag{4.7.24}$$

Let us emphasize that $x_i^1, x_i^2, p_i^{(x)1}$, and $p_i^{(x)2}$ are not powers, but first-order and second-order perturbations. Using (4.7.24), we easily obtain

$$p_i^{(x)} \dot{x}_i = p_i^{(x)0} \dot{x}_i^0 + \epsilon \big(p_i^{(x)0} \dot{x}_i^1 + p_i^{(x)1} \dot{x}_i^0 \big) + \epsilon^2 \big(p_i^{(x)0} \dot{x}_i^2 + p_i^{(x)1} \dot{x}_i^1 + p_i^{(x)2} \dot{x}_i^0 \big), \tag{4.7.25}$$

$$\begin{aligned}
\mathcal{H}^{(x)}\big(x_i, p_i^{(x)}\big) = {}& \mathcal{H}^{(x)0}\big(x_i^0, p_i^{(x)0}\big) + \epsilon \big(\mathcal{H}^{(x)1} - \dot{p}_i^{(x)0} x_i^1 + p_i^{(x)1} \dot{x}_i^0 \big) \\
& + \epsilon^2 \bigg(\mathcal{H}^{(x)2} - \dot{p}_i^{(x)0} x_i^2 + p_i^{(x)2} \dot{x}_i^0 + \frac{\partial \mathcal{H}^{(x)1}}{\partial x_i^0} x_i^1 \\
& + \frac{\partial \mathcal{H}^{(x)1}}{\partial p_i^{(x)0}} p_i^{(x)1} + A \bigg),
\end{aligned} \tag{4.7.26}$$

where

$$A = \tfrac{1}{2}\big[A_{ij}^{(x)0} p_i^{(x)1} x_j^1 + B_{ij}^{(x)0} p_i^{(x)1} p_j^{(x)1} + C_{ij}^{(x)0} x_i^1 x_j^1 + D_{ij}^{(x)0} x_i^1 p_j^{(x)1} \big];$$

see (4.2.5) for $A_{ij}^{(x)0}, B_{ij}^{(x)0}, C_{ij}^{(x)0}$, and $D_{ij}^{(x)0}$. Superscript 0 is used to emphasize that the quantities are computed along reference ray Ω^0 in the background medium. Using the paraxial ray tracing system (4.7.20), where we replace Δx_i and $\Delta p_i^{(x)}$ by x_i^1 and $p_i^{(x)1}$, we obtain

$$A = \tfrac{1}{2} x_i^1 \big(-\dot{p}_i^{(x)1} - \partial \mathcal{H}^{(x)1}/\partial x_i^0 \big) + \tfrac{1}{2} p_i^{(x)1} \big(\dot{x}_i^1 - \partial \mathcal{H}^{(x)1}/\partial p_i^{(x)0} \big). \tag{4.7.27}$$

We now combine expressions (4.7.25) and (4.7.26) to exclude $p_i^{(x)2}$. Taking into account $\mathcal{H}^{(x)}(x_i, p_i^{(x)}) = 0$ and $\mathcal{H}^{(x)0}(x_i^0, p_i^{(x)0}) = 0$, we obtain the final expression for $p_i^{(x)} \dot{x}_i$:

$$\begin{aligned}
p_i^{(x)} \dot{x}_i = {}& p_i^{(x)0} \dot{x}_i^0 - \epsilon \mathcal{H}^{(x)1} \\
& - \epsilon^2 \big[\mathcal{H}^{(x)2} + \tfrac{1}{2} x_i^1 \partial \mathcal{H}^{(x)1}/\partial x_i^0 + \tfrac{1}{2} p_i^{(x)1} \partial \mathcal{H}^{(x)1}/\partial p_i^{(x)0} \big] \\
& + \epsilon \big(p_i^{(x)0} x_i^1 \big)^{\cdot} + \epsilon^2 \big(p_i^{(x)0} x_i^2 \big)^{\cdot} + \tfrac{1}{2} \epsilon^2 \big(x_i^1 p_i^{(x)1} \big)^{\cdot}.
\end{aligned} \tag{4.7.28}$$

Here $(\ )^{\cdot}$ denotes $\mathrm{d}(\ )/\mathrm{d}u$. Using (4.7.28), we can obtain the expression for the travel time in the perturbed medium,

$$\begin{aligned}
T(x_i(u_R), x_i(u_S)) = {}& \int_{u_S}^{u_R} p_i^{(x)} \dot{x}_i \, \mathrm{d}u = \int_{u_S}^{u_R} p_i^{(x)0} \dot{x}_i^0 \, \mathrm{d}u - \int_{u_S}^{u_R} \bigg[\epsilon \mathcal{H}^{(x)1} \\
& + \epsilon^2 \bigg(\mathcal{H}^{(x)2} + \frac{1}{2} \frac{\partial \mathcal{H}^{(x)1}}{\partial x_i^0} x_i^1 + \frac{1}{2} \frac{\partial \mathcal{H}^{(x)1}}{\partial p_i^{(x)0}} p_i^{(x)1} \bigg) \bigg] \mathrm{d}u \\
& + \epsilon \big[p_i^{(x)0} x_i^1 \big]_{u_S}^{u_R} + \epsilon^2 \big[p_i^{(x)0} x_i^2 \big]_{u_S}^{u_R} + \frac{1}{2} \epsilon^2 \big[x_i^1 p_i^{(x)1} \big]_{u_S}^{u_R};
\end{aligned} \tag{4.7.29}$$

see Section 3.9.1. We shall now collect the terms with ϵ and ϵ^2 in (4.7.29). Quantity ϵ only has a formal meaning, and we can put $\epsilon = 1$. Equation (4.7.29) for the travel time in the perturbed medium can then be expressed in the following form:

$$T(x_i(u_S), x_i(u_R)) = T^0 + T^1 + T^2. \tag{4.7.30}$$

Here T^0 is the travel time from S to R along Ω^0 in the background medium, T^1 is the first-order travel-time perturbation, and T^2 the second-order travel-time perturbation. They are given by the following relations:

$$T^0 = \int_{u_S}^{u_R} p_i^{(x)0} \dot{x}_i^0 \mathrm{d}u,$$

$$T^1 = -\int_{u_S}^{u_R} \mathcal{H}^{(x)1} \mathrm{d}u + \left[p_i^{(x)0} x_i^1 \right]_{u_S}^{u_R},$$

$$T^2 = -\int_{u_S}^{u_R} \left[\mathcal{H}^{(x)2} + \frac{1}{2} \frac{\partial \mathcal{H}^{(x)1}}{\partial x_i^0} x_i^1 + \frac{1}{2} \frac{\partial \mathcal{H}^{(x)1}}{\partial p_i^{(x)0}} p_i^{(x)1} \right] \mathrm{d}u \tag{4.7.31}$$

$$+ \left[p_i^{(x)0} x_i^2 \right]_{u_S}^{u_R} + \tfrac{1}{2} \left[p_i^{(x)1} x_i^1 \right]_{u_S}^{u_R}.$$

In (4.7.31), all integrals are taken along reference ray Ω^0. For first-order travel-time perturbation T^1, the result (4.7.31) is fully equivalent to (3.9.3).

Expression (4.7.30) and (4.7.31) can be used to find the travel time in the perturbed medium from the fixed point S' situated close to S to the fixed point R' situated close to R. We specify the coordinates of points S' and R' by relations $x_i(S') = x_i(S) + x_{Si}^1$, $x_i(R') = x_i(R) + x_{Ri}^1$, and assume that x_{Si}^1 and x_{Ri}^1 are small. We use (4.7.30) and (4.7.31), where we put

$$x_i^1(u_S) = x_{Si}^1, \qquad x_i^1(u_R) = x_{Ri}^1, \qquad x_i^2(u_S) = x_i^2(u_R) = 0. \tag{4.7.32}$$

Then the term $[p_i^{(x)0} x_i^2]_{u_S}^{u_R}$ in the expression for T^2 vanishes, and term $\frac{1}{2}[p_i^{(x)1} x_i^1]_{u_S}^{u_R}$ reads

$$\tfrac{1}{2} \left[p_i^{(x)1} x_i^1 \right]_{u_S}^{u_R} = \tfrac{1}{2} \left(x_{Ri}^1 p_i^{(x)1}(u_R) - x_{Si}^1 p_i^{(x)1}(u_S) \right). \tag{4.7.33}$$

Consequently, second-order travel-time perturbation T^2 does not depend on the second-order ray perturbations x_i^2 or on the second-order slowness vector perturbations $p_i^{(x)2}$.

4.7.6 Higher Derivatives of the Travel-Time Field

The first derivatives of the travel-time field, represented by the slowness vector $\vec{p} = \nabla T$, can be determined along the ray by standard ray tracing. Similarly, the second derivatives of the travel-time field can be computed by dynamic ray tracing. To compute the first derivatives of the travel-time field, the first derivatives of the propagation velocity must be known. Similarly, the dynamic ray tracing system uses the second derivatives of the propagation velocity.

The question is whether it is also possible to compute the third and higher derivatives of the travel-time field along the ray. It seems natural to think that some "higher-order dynamic ray tracing systems" will be necessary to perform such computations. Fortunately, the calculation of higher derivatives is conceptually much simpler. After the 3×3 matrix $\hat{\mathbf{Q}}^{(x)}$ is known along the central ray, the higher derivatives of the travel-time field can be determined by simple quadratures along the ray. However, to calculate the nth derivative of the travel-time field, we need to know the nth derivatives of the propagation velocity along the ray.

The theory of calculation of higher derivatives of the travel-time field is due to Babich, Buldyrev, and Molotkov (1985). We shall present a modified derivation, given by Klimeš (1997b). To simplify the mathematical treatment, we shall work with parameter σ along the ray, related to arclength s and travel time T along the ray as follows: $\mathrm{d}\sigma = V \mathrm{d}s = V^2 \mathrm{d}T$.

The dynamic ray tracing system for 3×3 matrices $Q_{ik}^{(x)}$ and $P_{ik}^{(x)}$ in Cartesian coordinates then reads

$$\frac{d}{d\sigma} Q_{ik}^{(x)} = P_{ik}^{(x)}, \qquad \frac{d}{d\sigma} P_{ik}^{(x)} = \frac{1}{2} \frac{\partial^2}{\partial x_i \partial x_j} \left(\frac{1}{V^2} \right) Q_{jk}^{(x)}; \tag{4.7.34}$$

see (4.7.9). We assume that dynamic ray tracing system (4.7.34) has been solved along ray Ω, so that the 3×3 matrices $Q_{ik}^{(x)}$ and $P_{ik}^{(x)}$ are known at any point of the ray.

We now introduce the Nth derivative of the travel-time field $T_{,ijk...mn}$ as

$$T_{,ijk...mn} = \frac{\partial^N T(x_1, x_2, x_3)}{\partial x_i \partial x_j \partial x_k \dots \partial x_m \partial x_n}. \tag{4.7.35}$$

The total number of indices in $T_{,ijk...mn}$ representing the Nth derivative is N. The first derivative is $T_{,i} = p_i$, where p_i are components of the slowness vector. Similarly, the second derivatives are $T_{,ij} = M_{ij}^{(x)}$, where $M_{ij}^{(x)}$ are the elements of matrix $\hat{\mathbf{M}}^{(x)}$ given by relation $\hat{\mathbf{M}}^{(x)} = \hat{\mathbf{P}}^{(x)} \hat{\mathbf{Q}}^{(x)-1}$. Using this relation, we obtain a useful equation: $P_{ij}^{(x)} = M_{ik}^{(x)} Q_{kj}^{(x)} = T_{,ik} Q_{kj}^{(x)}$.

We shall now compute the following derivative along ray Ω:

$$\frac{d}{d\sigma} \left(T_{,ijk...mn} Q_{ia}^{(x)} Q_{jb}^{(x)} Q_{kc}^{(x)} \dots Q_{me}^{(x)} Q_{nf}^{(x)} \right)$$

$$= \frac{dT_{,ijk...mn}}{d\sigma} Q_{ia}^{(x)} Q_{jb}^{(x)} Q_{kc}^{(x)} \dots Q_{me}^{(x)} Q_{nf}^{(x)}$$

$$+ T_{,ijk...mn} \frac{dQ_{ia}^{(x)}}{d\sigma} Q_{jb}^{(x)} Q_{kc}^{(x)} \dots Q_{me}^{(x)} Q_{nf}^{(x)}$$

$$+ T_{,ijk...mn} Q_{ia}^{(x)} \frac{dQ_{jb}^{(x)}}{d\sigma} Q_{kc}^{(x)} \dots Q_{me}^{(x)} Q_{nf}^{(x)}$$

$$+ \cdots$$

$$+ T_{,ijk...mn} Q_{ia}^{(x)} Q_{jb}^{(x)} Q_{kc}^{(x)} \dots Q_{me}^{(x)} \frac{dQ_{nf}^{(x)}}{d\sigma}.$$

We insert

$$\frac{d}{d\sigma} T_{,ijk...mn} = \frac{\partial T_{,ijk...mn}}{\partial x_r} \frac{dx_r}{d\sigma} = T_{,r} T_{,rijk...mn},$$

$$dQ_{ia}^{(x)}/d\sigma = P_{ia}^{(x)} = T_{,ir} Q_{ra}^{(x)},$$

and the like and obtain

$$\frac{d}{d\sigma} \left(T_{,ijk...mn} Q_{ia}^{(x)} Q_{jb}^{(x)} Q_{kc}^{(x)} \dots Q_{me}^{(x)} Q_{nf}^{(x)} \right)$$

$$= K_{ijk...mn} Q_{ia}^{(x)} Q_{jb}^{(x)} Q_{kc}^{(x)} \dots Q_{me}^{(x)} Q_{nf}^{(x)}. \tag{4.7.36}$$

Here $K_{ijk...mn}$ is given by the relation

$$K_{ijk...mn} = T_{,r} T_{,rijk...mn} + T_{,ri} T_{,rjk...mn}$$
$$+ T_{,rj} T_{,irk...mn} + T_{,rk} T_{,ijr...mn} + \cdots$$
$$+ T_{,rm} T_{,ijk...rn} + T_{,rn} T_{,ijk...mr}. \tag{4.7.37}$$

If $T_{,ijk...mn}$ represents the Nth derivative of the travel-time field, the expression for $K_{ijk...mn}$ given by (4.7.37) contains the $(N + 1)$st derivatives; see $T_{,rijk...mn}$. Thus, Equations (4.7.36) with (4.7.37) cannot be directly used for the successive calculation of derivatives. The highest derivatives in (4.7.37), however, can be removed using the eikonal equation $T_{,r}T_{,r} = 1/V^2$:

$$\tfrac{1}{2}(1/V^2)_{,ijk...mn} - \tfrac{1}{2}(T_{,r}T_{,r})_{,ijk...mn} = 0. \tag{4.7.38}$$

By applying (4.7.38), the $(N + 1)$st derivative of the travel-time field can be replaced by the Nth derivative of the square of slowness. Adding (4.7.38) to (4.7.37) yields a new expression for $K_{ijk...mn}$:

$$\begin{aligned} K_{ijk...mn} = {} & \left(\tfrac{1}{2}V^{-2}\right)_{,ijk...mn} - \tfrac{1}{2}(T_{,r}T_{,r})_{,ijk...mn} \\ & + T_{,r}T_{,rijk...mn} + T_{,ri}T_{,rjk...mn} + T_{,rj}T_{,irk...mn} + \cdots \\ & + T_{,rm}T_{,ijk...rn} + T_{,rn}T_{,ijk...mr}. \end{aligned} \tag{4.7.39}$$

It is not difficult to see that the highest term $T_{,r}T_{,rijk...mn}$ is canceled by the highest derivative term of $\tfrac{1}{2}(T_{,r}T_{,r})_{,ijk...mn}$. For $N \geq 3$, the next term is also canceled. Consequently, the highest derivative of T in the expression (4.7.39) for $K_{ijk...mn}$ is of the $(N-1)$st order. We thus replace the highest derivatives of the travel-time field by the derivatives of the square of slowness.

Equation (4.7.36) with (4.7.39) represents the final form of the equations for computing the Nth derivatives of the travel-time field along ray Ω. Equation (4.7.36) can be simply integrated along the ray. Because $Q_{ij}^{(x)}$ are assumed to be known along the ray, the results of the integration, $T_{,ijk...mn}Q_{ia}^{(x)}Q_{jb}^{(x)}Q_{kc}^{(x)}\cdots Q_{me}^{(x)}Q_{nf}^{(x)}$, can be used to compute $T_{,ijk...mn}$.

The expressions for $K_{ijk...mn}$ given by (4.7.39) are straightforward, but for high N, they may be cumbersome. For low N, however, they are very simple. We shall give the expression for $N = 3$ and $N = 4$,

$$K_{ijk} = \tfrac{1}{2}(V^{-2})_{,ijk}, \tag{4.7.40}$$

$$K_{ijkl} = \tfrac{1}{2}(V^{-2})_{,ijkl} - T_{,rij}T_{,rkl} - T_{,rik}T_{,rjl} - T_{,ril}T_{,rjk}. \tag{4.7.41}$$

As indicated by (4.7.40), the third derivatives of the propagation velocity must be known along ray Ω if we wish to calculate the third derivatives of the travel-time field. No derivatives of the travel-time field appear in the expression for K_{ijk}. To calculate the fourth derivatives of the travel-time field, T_{ijkl}, we need to know the fourth derivatives of the square of slowness, and the third derivatives of the travel-time field.

4.8 Special Cases. Analytical Dynamic Ray Tracing

In certain simple situations, dynamic ray tracing systems in isotropic media can be solved analytically. Dynamic ray tracing then reduces to step-by-step computations, passing from one interface to another, and to applying appropriate interface matrices at structural interfaces. In this section, we shall briefly discuss several such situations.

4.8.1 Homogeneous Layers Separated by Curved Interfaces

We first consider dynamic ray tracing system (4.1.64) in ray-centered coordinates. In a homogeneous layer, velocity $V = \text{const.}$, and elements V_{IJ} of matrix $\mathbf{V}$ vanish. The dynamic ray tracing system is then as follows:

$$d\mathbf{Q}/ds = V\mathbf{P}, \qquad d\mathbf{P}/ds = \mathbf{0}; \tag{4.8.1}$$

see (4.1.64). The solution reads

$$\mathbf{Q}(R) = \mathbf{Q}(S) + Vl(R, S)\mathbf{P}(S), \qquad \mathbf{P}(R) = \mathbf{P}(S), \tag{4.8.2}$$

where $l(R, S)$ is the length of the straight-line segment from S to R.

Ray propagator matrix $\mathbf{\Pi}(R, S)$ along the segment of ray Ω from S to R in a smooth medium then yields

$$\mathbf{\Pi}(R, S) = \begin{pmatrix} \mathbf{I} & Vl(R, S)\mathbf{I} \\ \mathbf{0} & \mathbf{I} \end{pmatrix}. \tag{4.8.3}$$

Interface matrix $\mathbf{Y}(Q)$ and its inverse $\mathbf{Y}^{-1}(Q)$ also simplify because $\mathbf{E}(Q) = \mathbf{0}$. Equations (4.4.70) and (4.4.72) yield

$$\mathbf{Y}(Q) = \begin{pmatrix} \mathbf{G}^{-1\,T}(Q) & \mathbf{0} \\ -p_3^{(z)}(Q)\mathbf{D}(Q)\mathbf{G}^{-1\,T}(Q) & \mathbf{G}(Q) \end{pmatrix},$$
$$\mathbf{Y}^{-1}(Q) = \begin{pmatrix} \mathbf{G}^{T}(Q) & \mathbf{0} \\ p_3^{(z)}(Q)\mathbf{G}^{-1}(Q)\mathbf{D}(Q) & \mathbf{G}^{-1}(Q) \end{pmatrix}. \tag{4.8.4}$$

Interface propagator matrix $\mathbf{\Pi}(\tilde{Q}, Q)$ is then given by the following relation:

$$\mathbf{\Pi}(\tilde{Q}, Q) = \mathbf{Y}^{-1}(\tilde{Q})\mathbf{Y}(Q) = \begin{pmatrix} \mathbf{G}^{T}(\tilde{Q})\mathbf{G}^{-1\,T}(Q) & \mathbf{0} \\ -u\mathbf{G}^{-1}(\tilde{Q})\mathbf{D}(Q)\mathbf{G}^{-1\,T}(Q) & \mathbf{G}^{-1}(\tilde{Q})\mathbf{G}(Q) \end{pmatrix}, \tag{4.8.5}$$

where u is given by (4.4.45) or (4.4.51); see (4.4.76). Finally, we shall present an expression for the surface-to-surface ray propagator matrix, $\mathbf{T}(R, S) = \mathbf{Y}(R)\mathbf{\Pi}(R, S)\mathbf{Y}^{-1}(S)$,

$$\mathbf{T}(R, S) = \mathbf{Y}(R)\begin{pmatrix} \mathbf{I} & Vl(R, S)\mathbf{I} \\ \mathbf{0} & \mathbf{I} \end{pmatrix}\mathbf{Y}^{-1}(S). \tag{4.8.6}$$

Here $\mathbf{Y}(R)$ and $\mathbf{Y}^{-1}(S)$ are given by (4.8.4).

Let us now consider the solutions of dynamic ray tracing system (4.7.6) in Cartesian coordinates for a homogeneous medium. For a homogeneous medium, the system reads

$$dQ_i^{(x)}/dT = V^2 P_i^{(x)}, \qquad dP_i^{(x)}/dT = 0. \tag{4.8.7}$$

Assume that the initial values of $P_i^{(x)}$ and $Q_i^{(x)}$ are known at point S of ray Ω. The solution of (4.8.7) at point R of ray Ω is then

$$P_i^{(x)}(R) = P_i^{(x)}(S),$$
$$Q_i^{(x)}(R) = Q_i^{(x)}(S) + V^2 T(R, S)P_i^{(x)}(S). \tag{4.8.8}$$

Here $T(R, S)$ is the travel time from S to R, that is, $T(R, S) = l(R, S)/V$.

4.8.2 Homogeneous Layers Separated by Plane Interfaces

In this case, the curvatures of the interfaces at all points of incidence vanish, and matrices (4.8.4) and (4.8.5) simplify. We obtain

$$\mathbf{Y}(Q) = \begin{pmatrix} \mathbf{G}^{-1\,T}(Q) & \mathbf{0} \\ \mathbf{0} & \mathbf{G}(Q) \end{pmatrix}, \qquad \mathbf{Y}^{-1}(Q) = \begin{pmatrix} \mathbf{G}^{T}(Q) & \mathbf{0} \\ \mathbf{0} & \mathbf{G}^{-1}(Q) \end{pmatrix}, \tag{4.8.9}$$

and, similarly,

$$\mathbf{\Pi}(\tilde{Q}, Q) = \begin{pmatrix} \mathbf{G}^T(\tilde{Q})\mathbf{G}^{-1T}(Q) & \mathbf{0} \\ \mathbf{0} & \mathbf{G}^{-1}(\tilde{Q})\mathbf{G}(Q) \end{pmatrix}. \tag{4.8.10}$$

Otherwise, the equations remain the same.

4.8.3 Layers with a Constant Gradient of Velocity

We shall consider a model in which the velocity of propagation V is a linear function of Cartesian coordinates, $V(x_i) = V_0 + A_i x_i$. Using (4.1.63), we find that the matrix $\mathbf{V}$ of second derivatives of velocity with respect to q_1 and q_2 vanishes. The dynamic ray tracing in ray-centered coordinates can again be expressed in the form of (4.8.1). The solution, however, is a little more complex than (4.8.2) because velocity V is variable along ray Ω:

$$\mathbf{Q}(R) = \mathbf{Q}(S) + \mathbf{P}(S)\int_S^R V \, ds, \qquad \mathbf{P}(R) = \mathbf{P}(S). \tag{4.8.11}$$

We denote

$$\sigma(R, S) = \int_S^R V \, ds = \int_S^R V^2 \, dT \tag{4.8.12}$$

and obtain

$$\mathbf{Q}(R) = \mathbf{Q}(S) + \mathbf{P}(S)\sigma(R, S), \qquad \mathbf{P}(R) = \mathbf{P}(S). \tag{4.8.13}$$

It is obvious that σ in (4.8.12) and (4.8.13) is a monotonic parameter along the ray introduced in Section 3.1.1; see (3.1.11). The ray propagator matrix $\mathbf{\Pi}(R, S)$ along the segment of ray Ω from S to R then reads

$$\mathbf{\Pi}(R, S) = \begin{pmatrix} \mathbf{I} & \mathbf{I}\sigma(R, S) \\ \mathbf{0} & \mathbf{I} \end{pmatrix}. \tag{4.8.14}$$

Interface matrices $\mathbf{Y}(Q)$, $\mathbf{Y}^{-1}(Q)$, and $\mathbf{\Pi}(\tilde{Q}, Q)$, however, do not simplify in this case; general relations (4.4.70), (4.4.72), and (4.4.76) need to be used. The foregoing equations, together with expressions for the interface propagator matrix $\mathbf{\Pi}(\tilde{Q}, Q)$, can be also suitably used for analytical dynamic ray tracing in models consisting of tetrahedral cells with *constant gradients of velocity* inside the cells and in layered and block models with constant gradient of velocity inside individual layers and blocks. See also Lafond and Levander (1990).

For the surface-to-surface ray propagator matrix, we obtain

$$\mathbf{T}(R, S) = \mathbf{Y}(R)\begin{pmatrix} \mathbf{I} & \mathbf{I}\sigma(R, S) \\ \mathbf{0} & \mathbf{I} \end{pmatrix}\mathbf{Y}^{-1}(S). \tag{4.8.15}$$

4.8.4 Analytical Dynamic Ray Tracing in Cartesian Coordinates

The simplest solution of the dynamic ray tracing system in Cartesian coordinates is obtained from (4.7.9) for the model with a constant gradient of the square of slowness $V^{-2}(x_i)$, that is $V^{-2}(x_i) = A_0 + A_i x_i$; see Section 3.4.2. The relevant Hamiltonian in Cartesian coordinates is given by the relation $\mathcal{H}^{(x)} = \frac{1}{2}(p_i^{(x)}p_i^{(x)} - 1/V^2)$, and the monotonic parameter along the ray is σ, where $d\sigma = V^2 dT = V ds$. The solution of (4.7.9) reads

$$P_i^{(x)}(R) = P_i^{(x)}(S), \qquad Q_i^{(x)}(R) = Q_i^{(x)}(S) + \sigma(R, S)P_i^{(x)}(S). \tag{4.8.16}$$

Here $\sigma(R, S)$ is given by (4.8.12), $Q_i^{(x)} = (\partial x_i/\partial\gamma)_{\sigma=\text{const.}}$ and $P_i^{(x)} = (\partial p_i^{(x)}/\partial\gamma)_{\sigma=\text{const.}}$. The 6×6 propagator matrix $\mathbf{\Pi}^{(x)}(R, S)$ is then given by the relation

$$\mathbf{\Pi}^{(x)}(R, S) = \begin{pmatrix} \hat{\mathbf{I}} & \sigma(R, S)\hat{\mathbf{I}} \\ \hat{\mathbf{0}} & \hat{\mathbf{I}} \end{pmatrix}. \tag{4.8.17}$$

Expression (4.8.17) practically agrees with (4.8.3). The minors of (4.8.17), however, represent 3×3 matrices, but the minors of (4.8.3) represent 2×2 matrices. If we combine (4.8.17) with the interface ray propagator matrices (4.7.12) and use (4.7.11), we obtain the analytical expression for the 6×6 propagator matrix in general 3-D laterally varying layered and block structures, with a constant gradient of $1/V^2$ in the individual layers and in models consisting of tetrahedral cells with a constant gradient of $1/V^2$ in the individual cells.

If the gradient of $\ln V(x_i)$ or $1/V(x_i)$ is constant, it is possible to use dynamic ray tracing system (4.7.5) (with monotonic variable T) or (4.7.7) (with monotonic variable s). We again obtain constant $P_i^{(x)}$ along the whole ray, $P_i^{(x)}(R) = P_i^{(x)}(S)$. The equations for $Q_i^{(x)}$ are, however, more complicated. We can insert the proper analytical solutions for $p_i^{(x)}$ and use $P_i^{(x)} = \text{const.}$ $Q_i^{(x)}$ can then be obtained by quadratures along central ray Ω.

Analytical expression for the propagator matrix in terms of closed-form integrals can also be derived for *vertically inhomogeneous media*. In this case, it is suitable to use the reduced Hamiltonian $\mathcal{H}^R(x_i, p_I)$ introduced by (3.1.28). If we assume that the velocity depends on the Cartesian coordinate x_3 only, the reduced Hamiltonian is given by the relation $\mathcal{H}^R = -S(x_3, p_1, p_2)$, with $S = [1/V^2(x_3) - p_1^2 - p_2^2]^{1/2}$. This yields $\partial\mathcal{H}^R/\partial x_I = 0$ and $\partial\mathcal{H}^R/\partial p_I = S^{-1}p_I$. The ray tracing system then shows that p_1 and p_2 are constant along the whole ray. It is usual to take p_1 and p_2 as ray parameters γ_1 and γ_2. The 4×4 dynamic ray tracing system corresponding to the reduced Hamiltonian under consideration is then given by (4.2.65), with

$$A_{IJ}^R = 0, \qquad B_{IJ}^R = S^{-1}\delta_{IJ} + S^{-3}p_I p_J, \qquad C_{IJ}^R = 0, \qquad D_{IJ}^R = 0, \tag{4.8.18}$$

and with $Q_I^R = (\partial x_I/\partial\gamma)_{x_3=\text{const.}}$ and $P_I^R = (\partial p_I^{(x)}/\partial\gamma)_{x_3=\text{const.}}$. Consequently, dynamic ray tracing (4.2.65) is very simple and reads

$$d\mathbf{Q}^R/dx_3 = \mathbf{B}^R\mathbf{P}^R, \qquad d\mathbf{P}^R/dx_3 = \mathbf{0}. \tag{4.8.19}$$

Because $\mathbf{B}^R$ depends on x_3 only, (4.8.19) has the solution

$$\mathbf{Q}^R(x_3) = \mathbf{Q}^R(x_{30}) + \left(\int_{x_{30}}^{x_3} \mathbf{B}^R(x_3)dx_3\right)\mathbf{P}^R(x_{30}), \qquad \mathbf{P}^R(x_3) = \mathbf{P}^R(x_{30}). \tag{4.8.20}$$

Here we have assumed that $x_3 > x_{30}$ and that there is no turning point between x_{30} and x_3. Otherwise, it would be necessary to divide the ray at all turning points into upgoing and downgoing segments. Equation (4.8.20) shows that the reduced 4×4 ray propagator matrix $\mathbf{\Pi}^R(x_3, x_{30})$ is given by the relation

$$\mathbf{\Pi}^R(x_3, x_{30}) = \begin{pmatrix} \mathbf{I} & \int_{x_{30}}^{x_3} \mathbf{B}^R(x_3)dx_3 \\ \mathbf{0} & \mathbf{I} \end{pmatrix}. \tag{4.8.21}$$

Thus, the computation of the propagator matrix $\mathbf{\Pi}^R(x_3, x_{30})$ only requires the quadrature of matrix $\mathbf{B}^R(x_3)$ along the x_3 axis, from x_{30} to x_3.

Note that $\det \mathbf{Q}^R$ has a simple geometrical meaning. It represents the cross section of the ray tube by a plane $x_3 = $ const. It is interesting that this cross section remains constant along the ray, if plane-wavefront initial conditions are considered; see (4.8.21). The waves propagating in a vertically inhomogeneous medium and corresponding to plane wavefront initial conditions are often called *Snell's waves*.

It is not difficult to find relations between the 4×4 propagator matrix $\mathbf{\Pi}^R(R, S)$, given by (4.8.21), and the relevant 4×4 propagator matrix $\mathbf{\Pi}(R, S)$, corresponding to ray-centered coordinates (4.3.5). Actually, the reduced propagator matrix $\mathbf{\Pi}^R(R, S)$ can be interpreted as the surface-to-surface ray propagator matrix $\mathbf{T}(R, S)$, which was introduced in Section 4.4.7. In this case, all surfaces are plane and parallel. Using (4.4.90), we obtain

$$\mathbf{\Pi}(R, S) = \mathbf{Y}^{-1}(R)\mathbf{\Pi}^R(R, S)\mathbf{Y}(S). \tag{4.8.22}$$

Here the 2×2 matrices $\mathbf{Y}$ and $\mathbf{Y}^{-1}$ are given by (4.4.70) and (4.4.72), where $\mathbf{D} = \mathbf{0}$, $\mathbf{E}$ has only one nonvanishing term E_{11} (as $V_{,1}^{(z)} = V_{,2}^{(z)} = 0$), and $\mathbf{G}$ is given by (4.4.48) and (4.4.49). Notice that $\mathbf{\Pi}^R(R, S)$ is continuous across structural interfaces but that $\mathbf{\Pi}(R, S)$ is not continuous.

Without any loss of generality, we can consider only planar rays situated in plane $\Sigma^{\|}$, parallel to the x_3-axis and passing through S and R. We shift the x_3-axis parallel to pass through point S and rotate the Cartesian coordinate system about the x_3-axis in such a way that $\Sigma^{\|}$ represents plane $x_2 = 0$. We choose $p_2(S) = 0$ and $e_{22}(S) = 1$. Then $p_2 = 0$ and $e_{22} = 1$ along the whole ray. Consequently, polarization vector $\vec{e}_2$ is perpendicular to $\Sigma^{\|}$, and $\vec{e}_1$ and $\vec{e}_3$ are situated in plane $\Sigma^{\|}$. As usual, we denote $p_1 = p$. Because $p_2 = 0$ along the whole ray, we obtain $B_{11}^R = S^{-3}V^{-2}$, $B_{22}^R = S^{-1}$ and $B_{12}^R = B_{21}^R = 0$, where $S = [1/V^2(x_3) - p^2]^{1/2}$. Equations (4.8.21) and (3.7.9) then yield

$$\mathbf{Q}_2^R(x_3, x_{30}) = \int_{x_{30}}^{x_3} \mathbf{B}^R(x_3) \mathrm{d}x_3 = \begin{pmatrix} (\mathrm{d}x_1/\mathrm{d}p)_{x_3} & 0 \\ 0 & x_1/p \end{pmatrix}. \tag{4.8.23}$$

The derivative $(\mathrm{d}x_1/\mathrm{d}p)_{x_3}$ is taken along line $x_3 = $ const. It is straightforward to find the relation between $\mathbf{\Pi}^R(R, S)$ and $\mathbf{\Pi}(R, S)$ using (4.8.22). We use $\mathbf{G}^{\perp} = \mathbf{I}$ and $\mathbf{G} = \mathbf{G}^{\|}$, where $\mathbf{G}^{\|}$ has diagonal elements $G_{11}^{\|} = \cos i(S)$, $G_{22}^{\|} = 1$; see (4.4.48) and (4.4.49). Using (4.8.22) and (4.8.23) also, we obtain

$$\mathbf{Q}_2(R, S) = \mathbf{G}^T(R)\mathbf{Q}_2^R(R, S)\mathbf{G}(S)$$
$$= \begin{pmatrix} \cos i(S) \cos i(R)(\mathrm{d}x_1/\mathrm{d}p)_{x_3} & 0 \\ 0 & x_1/p \end{pmatrix}. \tag{4.8.24}$$

Using (4.8.24), we obtain an important relation for $\det \mathbf{Q}_2(R, S)$ for vertically inhomogeneous media, which is very useful in the computation of relative geometrical spreading,

$$\det \mathbf{Q}_2(R, S) = p^{-1} \cos i(S) \cos i(R) x_1(p)(\mathrm{d}x_1(p)/\mathrm{d}p)_{x_3}; \tag{4.8.25}$$

see Section 4.10.2. For a more detailed discussion of (4.8.25), see Section 4.10.2.5. If a line source parallel to x_2-axis is considered instead of the point source, Equations (4.8.23) through (4.8.25) remain valid, but x_1/p is replaced by 1.

Equation (4.8.25) remains valid for any ray parameter γ_1 (not necessarily p). Matrices $\mathbf{Q}$ and $\mathbf{P}$, however, depend on ray parameters. For a point source at S, relation $\det \mathbf{Q}(R) = \det \mathbf{Q}_2(R, S) \det \mathbf{P}(S)$ can be used to compute $J(R) = \det \mathbf{Q}(R)$. For ray parameters $\gamma_1 = i(S)$, $\gamma_2 = \phi(S)$, we obtain $\det \mathbf{P}(S) = V^{-1}(S)p$; see (4.5.37). Then $J(R) = V^{-1}(S) \cos i(S) \cos i(R) x_1(p)(\mathrm{d}x_1/\mathrm{d}p)_{x_3}$. This is fully equivalent to the second equation of

(3.10.45), derived in a quite different way. For ray parameters $\gamma_1 = p_1 = p$ and $\gamma_2 = p_2$, we obtain $\det \mathbf{P}(S) = 1/\cos i(S)$; see (4.10.16). Then $J(R) = p^{-1} \cos i(R) x_1(p)(dx_1/dp)_{x_3}$.

We shall now use (4.8.22) to derive the equations for $\mathbf{P}_2(R, S)$:

$$\mathbf{P}_2(R, S) = \mathbf{G}^{-1}(R)\mathbf{G}(S) - \mathbf{G}^{-1}(R)\mathbf{E}(R)\mathbf{Q}_2^R(R, S)\mathbf{G}(S). \tag{4.8.26}$$

If the gradient of velocity vanishes at R, $\mathbf{E}(R) = \mathbf{0}$, and (4.8.26) simplifies:

$$\mathbf{P}_2(R, S) = \mathbf{G}^{-1}(R)\mathbf{G}(S) = \begin{pmatrix} \cos i(S)/\cos i(R) & 0 \\ 0 & 1 \end{pmatrix}. \tag{4.8.27}$$

This is actually an expected result, following also from simple geometrical considerations.

The reduced 4×4 propagator matrices similar to (4.8.21) can also be derived for some other one-dimensional models, for example, for radially symmetric models. It is, however, necessary to be careful because the scale factors also depend on coordinates.

4.8.5 Reflection/Transmission at a Curved Interface

The general procedures for computing the ray propagator matrix $\mathbf{\Pi}(R, S)$ of a wave reflected/transmitted at the curved interface Σ between two laterally inhomogeneous media were presented in Sections 4.4.5 and 4.4.8. In this section, we shall discuss these equations in greater detail, considering *a wave generated by a point source* situated at S. The derivation would be similar for all four submatrices $\mathbf{Q}_1(R, S)$, $\mathbf{Q}_2(R, S)$, $\mathbf{P}_1(R, S)$, and $\mathbf{P}_2(R, S)$ of ray propagator matrix $\mathbf{\Pi}(R, S)$. Hence, we shall discuss only the 2×2 matrix $\mathbf{Q}_2(R, S)$, which plays a basic role in computing the amplitudes of the waves generated by the point source. As a by-product, we shall also obtain convenient equations for the Fresnel zone matrix $\mathbf{M}^F$.

We consider a ray of an arbitrary reflected or transmitted (possibly converted) wave. The ray has two segments: incident and reflected/transmitted. The point source is situated at the point S of the incident segment, and the receiver at the point R of the R/T segment. As usual, the point of incidence at interface Σ is denoted by Q, and the point of R/T is denoted by $\tilde{Q}$. Otherwise, we shall use the same notation as in Sections 4.4.5 and 4.4.8; see Figure 4.10.

We shall introduce the standard local Cartesian coordinate system z_i at the point of incidence Q using relations (4.4.21). We remind the reader that orientation index ϵ is given by the relation $\epsilon = \mathrm{sign}(\vec{p}(Q) \cdot \vec{n})$, where $\vec{p}(Q)$ is the slowness vector of the incident wave at Q. Consequently, we can use (4.4.48) and (4.4.49) for $\mathbf{G}(Q)$ and $\mathbf{G}(\tilde{Q})$. Unit vectors $\vec{e}_1$ and $\vec{e}_2$ may be specified at any point of ray Ω. We shall specify them at Q and $\tilde{Q}$ in such a way that $\vec{e}_2(Q) = \vec{e}_2(\tilde{Q}) = \vec{i}_2^{(z)}(Q)$. Thus, $\vec{e}_2(Q)$ and $\vec{e}_2(\tilde{Q})$ are perpendicular to the plane of incidence. If we wish to know $\vec{e}_1(S)$ and $\vec{e}_2(S)$, we need to recalculate $\vec{e}_2$ from Q back to S. As a consequence of this choice of $\vec{e}_2(Q)$ and $\vec{e}_2(\tilde{Q})$, we have $\mathbf{G}^\perp(Q) = \mathbf{G}^\perp(\tilde{Q}) = \mathbf{I}$, $\mathbf{G}(Q) = \mathbf{G}^\parallel(Q)$, and $\mathbf{G}(\tilde{Q}) = \mathbf{G}^\parallel(\tilde{Q})$. Equations (4.4.108) and (4.4.109) then yield

$$\mathbf{Q}_2(R, S)$$
$$= \mathbf{Q}_2(R, \tilde{Q}) \begin{pmatrix} \pm\epsilon/\cos i_R & 0 \\ 0 & 1 \end{pmatrix} \mathbf{M}^F(Q; R, S) \begin{pmatrix} \epsilon/\cos i_S & 0 \\ 0 & 1 \end{pmatrix} \mathbf{Q}_2(Q, S),$$
$$\tag{4.8.28}$$

$$\mathbf{M}^F(Q; R, S)$$
$$= \mathbf{G}^\parallel(Q)\mathbf{P}_2(Q, S)\mathbf{Q}_2^{-1}(Q, S)\mathbf{G}^\parallel(Q)$$
$$+ \mathbf{G}^\parallel(\tilde{Q})\mathbf{Q}_2^{-1}(R, \tilde{Q})\mathbf{Q}_1(R, \tilde{Q})\mathbf{G}^\parallel(\tilde{Q}) + \mathbf{E}(Q) - \mathbf{E}(\tilde{Q}) - u\mathbf{D}(Q).$$
$$\tag{4.8.29}$$

Here i_S is the acute angle of incidence, i_R the acute angle of R/T, $\mathbf{G}^{\parallel}(Q)$ and $\mathbf{G}^{\parallel}(\tilde{Q})$ are given by (4.4.49), $E_{IJ}(Q)$ and $E_{IJ}(\tilde{Q})$ are given by (4.4.53) and (4.4.54), and u is given by (4.4.51). The upper sign in (4.8.28) corresponds to the transmitted wave, the lower sign corresponds to the reflected wave. (4.8.28) also immediately yields an important relation for $\det \mathbf{Q}_2(R, S)$:

$$\det \mathbf{Q}_2(R, S) = \pm \frac{1}{\cos i_S \cos i_R} \det \mathbf{Q}_2(R, \tilde{Q}) \det \mathbf{M}^F(Q; R, S) \det \mathbf{Q}_2(Q, S).$$

$$(4.8.30)$$

The upper sign again corresponds to the transmitted wave; the lower sign corresponds to the reflected wave.

We shall now specify (4.8.28) through (4.8.30) for two simple velocity distributions.

a. INTERFACE BETWEEN MEDIA WITH CONSTANT VELOCITY GRADIENTS

We assume that the velocity $V(x_i)$ corresponding to the incident wave depends on coordinates x_i as $V(x_i) = V_0 + A_i x_i$, and the velocity $\tilde{V}(x_i)$ corresponding to the R/T wave depends on coordinates x_i as $\tilde{V}(x_i) = \tilde{V}_0 + \tilde{A}_i x_i$. Then we can use (4.8.14) and obtain $\mathbf{Q}_2(Q, S) = \sigma_S \mathbf{I}$ and $\mathbf{Q}_2(R, \tilde{Q}) = \sigma_R \mathbf{I}$, where $\sigma_S = \sigma(Q, S)$ and $\sigma_R = \sigma(R, \tilde{Q})$ are given by (4.8.12). Consequently,

$$\mathbf{Q}_2(R, S) = \sigma_S \sigma_R \begin{pmatrix} \pm\epsilon/\cos i_R & 0 \\ 0 & 1 \end{pmatrix} \mathbf{M}^F(Q; R, S) \begin{pmatrix} \epsilon/\cos i_S & 0 \\ 0 & 1 \end{pmatrix},$$

$$(4.8.31)$$

$$\det \mathbf{Q}_2(R, S) = \pm \frac{\sigma_S^2 \sigma_R^2}{\cos i_S \cos i_R} \det \mathbf{M}^F(Q; R, S). \qquad (4.8.32)$$

The elements of the Fresnel zone matrix $\mathbf{M}^F(Q; R, S)$ are given by the relations

$$\begin{aligned}
M_{11}^F(Q; R, S) &= (\cos^2 i_S)/\sigma_S + (\cos^2 i_R)/\sigma_R \\
&\quad + E_{11}(Q) - E_{11}(\tilde{Q}) - u(Q)D_{11}, \\
M_{12}^F(Q; R, S) &= M_{21}^F(Q; R, S) = E_{12}(Q) - E_{12}(\tilde{Q}) - uD_{12}, \\
M_{22}^F(Q; R, S) &= 1/\sigma_S + 1/\sigma_R - uD_{22}.
\end{aligned} \qquad (4.8.33)$$

Here $E_{IJ}(Q)$ and $E_{IJ}(\tilde{Q})$ are given by (4.4.53) and (4.4.54), where $V_{,i}^{(z)} = A_j Z_{ji}$ and $\tilde{V}_{,i}^{(z)} = \tilde{A}_j Z_{ji}$. Quantity $u(Q)$ is given by (4.4.51).

For an *unconverted reflected wave*, expressions (4.8.33) for the elements of the Fresnel zone matrix simplify. We use $i_S = i_R$, (4.4.52) for u, and (4.4.56) for $E_{IJ}(Q) - E_{IJ}(\tilde{Q})$,

$$\begin{aligned}
M_{11}^F(Q; R, S) &= \cos i_S \big[(1/\sigma_S + 1/\sigma_R) \cos i_S \\
&\quad + 2\epsilon \sin^2 i_S V^{-2} V_{,3}^{(z)} - 2\epsilon V^{-1} D_{11} \big], \\
M_{12}^F(Q; R, S) &= M_{21}^F(Q; R, S) = -2\epsilon V^{-1} \cos i_S D_{12}, \\
M_{22}^F(Q; R, S) &= 1/\sigma_S + 1/\sigma_R - 2\epsilon V^{-1} \cos i_S D_{22}.
\end{aligned} \qquad (4.8.34)$$

All quantities are taken at Q. Let us briefly discuss the signs of the inhomogeneity and curvature terms in (4.8.34). All these terms contain the factor ϵ. Instead of $V_{,3}^{(z)}$, we can introduce $V_3^+ = \epsilon V_{,3}^{(z)}$. Quantity V_3^+ is *positive* if velocity V increases toward interface Σ at Q. Similarly, instead of curvature matrix $\mathbf{D}$, we can introduce matrix $\mathbf{D}^+ = \epsilon \mathbf{D}$. Both eigenvalues of matrix $\mathbf{D}^+$ are positive if interface Σ is concave for the observer situated on the incident segment of ray Ω.

b. INTERFACE SEPARATING HOMOGENEOUS MEDIA

We denote the velocity corresponding to the incident wave by V_S and the length of the incident segment of the ray by l_S, $l_S = \overline{SQ}$. For the R/T segment of the ray, we use V_R and l_R, with $l_R = \overline{\tilde{Q}R}$. We then obtain $\sigma_S = V_S l_S$, $\sigma_R = V_R l_R$, and $E_{IJ}(Q) = E_{IJ}(\tilde{Q}) = 0$. Equations (4.8.31) and (4.8.32) remain the same, but we insert $\sigma_S = V_S l_S$ and $\sigma_R = V_R l_R$. Equations (4.8.33), however, yield

$$M_{11}^F(Q; R, S) = (\cos^2 i_S)/V_S l_S + (\cos^2 i_R)/V_R l_R - u D_{11},$$
$$M_{12}^F(Q; R, S) = M_{21}^F(Q; R, S) = -u D_{12}, \tag{4.8.35}$$
$$M_{22}^F(Q; R, S) = 1/V_S l_S + 1/V_R l_R - u D_{22}.$$

This gives a useful equation for $\det \mathbf{M}^F(Q; R, S)$:

$$\det \mathbf{M}^F(Q; R, S)$$
$$= \left[\frac{\cos^2 i_S}{V_S l_S} + \frac{\cos^2 i_R}{V_R l_R} - u D_{11}\right]\left[\frac{1}{V_S l_S} + \frac{1}{V_R l_R} - u D_{22}\right] - u^2 D_{12}^2. \tag{4.8.36}$$

For an unconverted reflected wave, $i_S = i_R$, $V_S = V_R$, and (4.8.35) yields

$$M_{11}^F(Q; R, S) = V_S^{-1}\cos i_S[(1/l_S + 1/l_R)\cos i_S - 2\epsilon D_{11}],$$
$$M_{12}^F(Q; R, S) = M_{21}^F(Q; R, S) = -2\epsilon \cos i_S V_S^{-1} D_{12}, \tag{4.8.37}$$
$$M_{22}^F(Q; R, S) = V_S^{-1}[1/l_S + 1/l_R - 2\epsilon \cos i_S D_{22}].$$

Similarly, (4.8.36) yields

$$\det \mathbf{M}^F(Q; R, S) = \frac{\cos i_S}{V_S^2}\left\{\left[\left(\frac{1}{l_S} + \frac{1}{l_R}\right)\cos i_S - 2\epsilon D_{11}\right]\right.$$
$$\left. \times \left[\frac{1}{l_S} + \frac{1}{l_R} - 2\epsilon \cos i_S D_{22}\right] - 4\cos i_S D_{12}^2\right\}. \tag{4.8.38}$$

4.9 Boundary-Value Ray Tracing for Paraxial Rays

Ray propagator matrix $\mathbf{\Pi}$ connects the properties of the ray field and of the travel-time field at different points of ray Ω. This property of the ray propagator matrix can be effectively used to solve analytically various boundary-value ray tracing problems for paraxial rays. In this section, considerable attention will be devoted to the analytical solution of the *two-point ray tracing problem* for paraxial rays and to the calculation of the *two-point eikonal*. In a similar way, it would be possible to solve analytically other boundary-value ray tracing problems for paraxial rays such as the initial surface-fixed point ray tracing. See Section 3.11 for the discussion of various boundary-value ray tracing problems.

We shall consider central ray Ω and two points, S and R, situated on Ω. Ray Ω may be situated in any 3-D laterally varying layered isotropic structure and may correspond to an arbitrary high-frequency (HF) seismic body wave, including multiply reflected and converted waves.

We assume that the following quantities are known.

a. Propagation velocity V at S and R, $V(S)$, and $V(R)$. Each of them corresponds to the velocity of either the P or S wave, depending on the type of elementary wave under consideration. The acoustic case may, of course, also be considered.

b. The gradients of velocity at S and R, $(\nabla V)_S$ and $(\nabla V)_R$. These gradients again correspond to the relevant elementary wave.

c. The transformation matrices from ray-centered coordinates q_i to Cartesian coordinates x_i, $\hat{\mathbf{H}}(S)$, and $\hat{\mathbf{H}}(R)$. These matrices specify the local directions of the basis unit vectors of the ray-centered coordinate systems at S and R, $\vec{e}_i(S)$ and $\vec{e}_i(R)$. We remind the reader that the slowness vectors at S and R are given by relations $\vec{p}(S) = V^{-1}(S)\vec{e}_3(S)$ and $\vec{p}(R) = V^{-1}(R)\vec{e}_3(R)$.

d. The travel time from S to R, $T(R, S)$, along central ray Ω, of the elementary wave under consideration.

e. Ray propagator matrix $\mathbf{\Pi}(R, S)$, computed by dynamic ray tracing in ray-centered coordinates. As usual, we shall denote the 2×2 minors of $\mathbf{\Pi}(R, S)$ by $\mathbf{Q}_1(R, S)$, $\mathbf{Q}_2(R, S)$, $\mathbf{P}_1(R, S)$, and $\mathbf{P}_2(R, S)$; see (4.3.5).

We shall also consider two additional points S' and R', point S' situated close to S and R' close to R. We shall solve various boundary-value ray tracing problems for these two points. In particular, we are interested in finding an expression for the *two-point eikonal*, $T(R', S')$, that is, the travel time from S' to R'.

4.9.1 Paraxial Two-Point Ray Tracing in Ray-Centered Coordinates

In this section, we shall assume that point S' is situated in a plane perpendicular to Ω at S, and point R' in a plane perpendicular to Ω at R. We assume that the ray-centered coordinates q_1 and q_2 of these two points are known and denote them $q_I(S')$ and $q_I(R')$. We wish to find the travel time from S' to R', $T(R', S')$, and the ray-centered components of slowness vectors $\vec{p}(S')$ and $\vec{p}(R')$, $p_I^{(q)}(S')$ and $p_I^{(q)}(R')$, corresponding to ray $\Omega'(R', S')$ connecting points S' and R'.

The ray-centered components of the slowness vectors at S' and R' can be found simply. The paraxial ray tracing system yields the continuation relations

$$\begin{pmatrix} \mathbf{q}(R') \\ \mathbf{p}^{(q)}(R') \end{pmatrix} = \mathbf{\Pi}(R, S) \begin{pmatrix} \mathbf{q}(S') \\ \mathbf{p}^{(q)}(S') \end{pmatrix}; \tag{4.9.1}$$

see (4.3.28). Here we have used the notation

$$\mathbf{q}(R') = \begin{pmatrix} q_1(R') \\ q_2(R') \end{pmatrix}, \qquad \mathbf{q}(S') = \begin{pmatrix} q_1(S') \\ q_2(S') \end{pmatrix},$$

$$\mathbf{p}^{(q)}(R') = \begin{pmatrix} p_1^{(q)}(R') \\ p_2^{(q)}(R') \end{pmatrix}, \qquad \mathbf{p}^{(q)}(S') = \begin{pmatrix} p_1^{(q)}(S') \\ p_2^{(q)}(S') \end{pmatrix}; \tag{4.9.2}$$

see (4.1.57). Equation (4.9.1) yields

$$\mathbf{q}(R') = \mathbf{Q}_1(R, S)\mathbf{q}(S') + \mathbf{Q}_2(R, S)\mathbf{p}^{(q)}(S'),$$
$$\mathbf{p}^{(q)}(R') = \mathbf{P}_1(R, S)\mathbf{q}(S') + \mathbf{P}_2(R, S)\mathbf{p}^{(q)}(S'). \tag{4.9.3}$$

The solution of (4.9.3) for $\mathbf{p}^{(q)}(S')$ and $\mathbf{p}^{(q)}(R')$ is

$$\mathbf{p}^{(q)}(S') = \mathbf{Q}_2^{-1}\mathbf{q}(R') - \mathbf{Q}_2^{-1}\mathbf{Q}_1\mathbf{q}(S'),$$
$$\mathbf{p}^{(q)}(R') = \left(\mathbf{P}_1 - \mathbf{P}_2\mathbf{Q}_2^{-1}\mathbf{Q}_1\right)\mathbf{q}(S') + \mathbf{P}_2\mathbf{Q}_2^{-1}\mathbf{q}(R'). \tag{4.9.4}$$

Here and in the following equation, we omit the arguments of the 2×2 matrices $\mathbf{Q}_1$, $\mathbf{Q}_2$, $\mathbf{P}_1$, and $\mathbf{P}_2$; we understand these arguments to be $\mathbf{Q}_1(R, S)$, $\mathbf{Q}_2(R, S)$, $\mathbf{P}_1(R, S)$, and $\mathbf{P}_2(R, S)$.

Taking into account that

$$\mathbf{P}_1 - \mathbf{P}_2\mathbf{Q}_2^{-1}\mathbf{Q}_1 = (\mathbf{P}_1\mathbf{Q}_1^{-1} - \mathbf{P}_2\mathbf{Q}_2^{-1})\mathbf{Q}_1 = -\mathbf{Q}_2^{-1T}\mathbf{Q}_1^{-1}\mathbf{Q}_1 = -\mathbf{Q}_2^{-1T}$$

(see (4.3.16)), we can express (4.9.4) in its final form,

$$
\begin{aligned}
\mathbf{p}^{(q)}(S') &= -\mathbf{Q}_2^{-1}(R, S)\mathbf{Q}_1(R, S)\mathbf{q}(S') + \mathbf{Q}_2^{-1}(R, S)\mathbf{q}(R'), \\
\mathbf{p}^{(q)}(R') &= -\mathbf{Q}_2^{-1T}(R, S)\mathbf{q}(S') + \mathbf{P}_2(R, S)\mathbf{Q}_2^{-1}(R, S)\mathbf{q}(R').
\end{aligned}
\tag{4.9.5}
$$

The paraxial slowness vectors at S' and R' have thus been determined analytically.

We shall use $\mathbf{M}(R, S) = \mathbf{P}_2(R, S)\mathbf{Q}_2^{-1}(R, S)$ to denote the matrix of the second derivatives of travel-time field $\mathbf{M}$ at R, due to the point source situated at S. Similarly, we use $\mathbf{M}(S, R) = -\mathbf{Q}_2^{-1}(R, S)\mathbf{Q}_1(R, S)$, which represents the matrix of the second derivatives of the travel-time field at S, due to the point source situated at R. See (4.6.8) and (4.6.9). An alternative form of (4.9.5) is then

$$
\begin{aligned}
\mathbf{p}^{(q)}(S') &= \mathbf{M}(S, R)\mathbf{q}(S') + \mathbf{Q}_2^{-1}(R, S)\mathbf{q}(R'), \\
\mathbf{p}^{(q)}(R') &= -\mathbf{Q}_2^{-1T}(R, S)\mathbf{q}(S') + \mathbf{M}(R, S)\mathbf{q}(R').
\end{aligned}
\tag{4.9.6}
$$

Assuming that the propagator matrix Π is known along the whole central ray Ω, and realizing that $\mathbf{q}(S')$ and $\mathbf{p}^{(q)}(S')$ are known, we can use the continuation relation (4.9.1) to determine completely the paraxial ray and the paraxial slowness vector at any point of the paraxial ray. Consequently, Equations (4.9.1) with (4.9.6) solve completely the problem of two-point ray tracing for paraxial rays in ray-centered coordinates.

The next problem is to determine the travel time from S' to R' along paraxial ray $\Omega'(R', S')$. We can put

$$T(R') = T(R) + \tfrac{1}{2}\mathbf{q}^T(R')\mathbf{p}^{(q)}(R'), \qquad T(S') = T(S) + \tfrac{1}{2}\mathbf{q}^T(S')\mathbf{p}^{(q)}(S');$$
$$\tag{4.9.7}$$

see (4.6.30). Subtracting these two equations, we arrive at

$$T(R', S') = T(R, S) + \tfrac{1}{2}\mathbf{q}^T(R')\mathbf{p}^{(q)}(R') - \tfrac{1}{2}\mathbf{q}^T(S')\mathbf{p}^{(q)}(S'). \tag{4.9.8}$$

As $\mathbf{p}^{(q)}(R')$ and $\mathbf{p}^{(q)}(S')$ are given by (4.9.6), we can use (4.9.8) to determine $T(R', S')$:

$$
\begin{aligned}
T(R', S') = T(R, S) &+ \tfrac{1}{2}\mathbf{q}^T(R')\left[\mathbf{M}(R, S)\mathbf{q}(R') - \mathbf{Q}_2^{-1T}(R, S)\mathbf{q}(S')\right] \\
&- \tfrac{1}{2}\mathbf{q}^T(S')\left[\mathbf{M}(S, R)\mathbf{q}(S') + \mathbf{Q}_2^{-1}(R, S)\mathbf{q}(R')\right].
\end{aligned}
$$

If we take into account that

$$\mathbf{q}^T(R')\mathbf{Q}_2^{-1T}(R, S)\mathbf{q}(S') = \mathbf{q}^T(S')\mathbf{Q}_2^{-1}(R, S)\mathbf{q}(R'),$$

we obtain the final expression for the *two-point eikonal in ray-centered coordinates*:

$$
\begin{aligned}
T(R', S') = T(R, S) &+ \tfrac{1}{2}\mathbf{q}^T(R')\mathbf{M}(R, S)\mathbf{q}(R') - \tfrac{1}{2}\mathbf{q}^T(S')\mathbf{M}(S, R)\mathbf{q}(S') \\
&- \mathbf{q}^T(S')\mathbf{Q}_2^{-1}(R, S)\mathbf{q}(R').
\end{aligned}
\tag{4.9.9}
$$

4.9.2 Paraxial Two-Point Ray Tracing in Cartesian Coordinates

In this section, we shall consider arbitrary positions of point S' close to S and point R' close to R. Points S' and R' need not be situated in planes perpendicular to Ω at S and R, respectively.

We shall first solve the problem in the local Cartesian ray-centered coordinate systems at points S and R and only then rewrite the results in general Cartesian coordinates x_i.

At both points S and R, we construct local Cartesian ray-centered coordinate systems y_i; see Section 4.1.4. We denote the local Cartesian ray-centered coordinates of point S' by $y_i(S', S)$ and the local Cartesian ray-centered coordinates of point R' by $y_i(R', R)$. The second argument in this notation emphasizes the origins of the relevant coordinate systems at S and R.

We denote the projection of S' into the plane $\Sigma^\perp(S)$ perpendicular to Ω at S by $S^\perp$ and the projection of R' into the plane $\Sigma^\perp(R)$ perpendicular to Ω at R by $R^\perp$. Then

$$y_I(S', S) = q_I(S^\perp), \qquad y_I(R', R) = q_I(R^\perp). \tag{4.9.10}$$

We introduce the 3×1 column matrices $\hat{\mathbf{y}}(S', S)$ and $\hat{\mathbf{y}}(R', R)$ by the following relations:

$$\hat{\mathbf{y}}(S', S) = \begin{pmatrix} y_1(S', S) \\ y_2(S', S) \\ y_3(S', S) \end{pmatrix} = \begin{pmatrix} q_1(S^\perp) \\ q_2(S^\perp) \\ y_3(S', S) \end{pmatrix},$$

$$\hat{\mathbf{y}}(R', R) = \begin{pmatrix} y_1(R', R) \\ y_2(R', R) \\ y_3(R', R) \end{pmatrix} = \begin{pmatrix} q_1(R^\perp) \\ q_2(R^\perp) \\ y_3(R', R) \end{pmatrix}. \tag{4.9.11}$$

We now wish to find the travel time from S' to R', $T(R', S')$, and the local Cartesian components of slowness vectors $p_i^{(y)}(S')$ and $p_i^{(y)}(R')$, corresponding to paraxial ray $\Omega'(R', S')$.

We shall first determine the components of slowness vectors $p_i^{(y)}(R')$ and $p_i^{(y)}(S')$, corresponding to paraxial ray $\Omega'(R', S')$. Using (4.6.29), we obtain

$$p_i^{(y)}(R') = p_i^{(y)}(R) + M_{ij}(R)y_j(R', R).$$

This yields

$$p_i^{(y)}(R') = p_i^{(y)}(R) + \delta_{iK} M_{KJ}(R)y_J(R', R) + M_{ij}^+(R)y_j(R', R). \tag{4.9.12}$$

Here $M_{IJ}^+ = 0$, $M_{3j}^+ = M_{3j}$, and $M_{i3}^+ = M_{i3}$. Symbol δ_{iK} $(i = 1, 2, 3; K = 1, 2)$ has the standard meaning: $\delta_{iK} = 1$ for $i = K$ and $\delta_{iK} = 0$ for $i \neq K$. Due to (4.6.29) and (4.9.10), we obtain

$$M_{KJ}(R)y_J(R', R) = M_{KJ}(R)q_J(R^\perp) = p_K^{(q)}(R^\perp).$$

Because points $R^\perp$ and $S^\perp$ are situated in planes perpendicular to ray Ω at R and S, we can use (4.9.6) for $p^{(q)}(R^\perp)$:

$$p_K^{(q)}(R^\perp) = M_{KJ}(R, S)q_J(R^\perp) - \left(\mathbf{Q}_2^{-1}(R, S)\right)_{JK}q_J(S^\perp). \tag{4.9.13}$$

Inserting (4.9.13) into (4.9.12) and considering (4.9.10) yields

$$p_i^{(y)}(R') = p_i^{(y)}(R) + M_{ij}^+(R)y_j(R', R)$$
$$+ \delta_{iK}\left\{M_{KJ}(R, S)y_J(R', R) - \left(\mathbf{Q}_2^{-1}(R, S)\right)_{JK}y_J(S', S)\right\}.$$

Again, taking into account that $M_{ij}^+(R) = M_{ij}^+(R, S)$ and combining the second and last terms yields the final relation

$$p_i^{(y)}(R') = p_i^{(y)}(R) + M_{ij}(R, S)y_j(R', R) - \delta_{iK}\left(\mathbf{Q}_2^{-1}(R, S)\right)_{JK}y_J(S', S). \tag{4.9.14}$$

To express (4.9.14) in matrix form, it is useful to introduce two matrices

$$\hat{\mathbf{I}}^M = \begin{pmatrix} 1 & 0 \\ 0 & 1 \\ 0 & 0 \end{pmatrix}, \qquad \hat{\mathbf{A}}(R, S) = \hat{\mathbf{I}}^M \mathbf{Q}_2^{-1}(R, S)\hat{\mathbf{I}}^{MT} = \begin{pmatrix} \mathbf{Q}_2^{-1}(R, S) & 0 \\ & & 0 \\ 0 & 0 & 0 \end{pmatrix}.$$
(4.9.15)

We also take into account that $\mathbf{y}(S', S) = \hat{\mathbf{I}}^{MT}\hat{\mathbf{y}}(S', S)$ and obtain

$$\begin{aligned} \hat{\mathbf{p}}^{(y)}(R') &= \hat{\mathbf{p}}^{(y)}(R) + \hat{\mathbf{M}}(R, S)\hat{\mathbf{y}}(R', R) - \hat{\mathbf{I}}^M \mathbf{Q}_2^{-1T}(R, S)\mathbf{y}(S', S) \\ &= \hat{\mathbf{p}}^{(y)}(R) + \hat{\mathbf{M}}(R, S)\hat{\mathbf{y}}(R', R) - \hat{\mathbf{A}}^T(R, S)\hat{\mathbf{y}}(S', S). \end{aligned}$$
(4.9.16)

In a similar way, we obtain the relations for $p_i^{(y)}(S')$,

$$p_i^{(y)}(S') = p_i^{(y)}(S) + M_{ij}(S, R)y_j(S', S) + \delta_{iK}\left(\mathbf{Q}_2^{-1}(R, S)\right)_{KJ}y_J(R', R),$$
(4.9.17)

or, in matrix form,

$$\begin{aligned} \hat{\mathbf{p}}^{(y)}(S') &= \hat{\mathbf{p}}^{(y)}(S) + \hat{\mathbf{M}}(S, R)\hat{\mathbf{y}}(S', S) + \hat{\mathbf{I}}^M \mathbf{Q}_2^{-1}(R, S)\mathbf{y}(R', R) \\ &= \hat{\mathbf{p}}^{(y)}(S) + \hat{\mathbf{M}}(S, R)\hat{\mathbf{y}}(S', S) + \hat{\mathbf{A}}(R, S)\hat{\mathbf{y}}(R', R). \end{aligned}$$
(4.9.18)

The next problem is to determine the travel time from S' to R' along paraxial ray $\Omega'(R', S')$, $T(R', S')$. We use (4.6.30),

$$\begin{aligned} T(R') &= T(R) + \tfrac{1}{2}\hat{\mathbf{y}}^T(R', R)\left[\hat{\mathbf{p}}^{(y)}(R') + \hat{\mathbf{p}}^{(y)}(R)\right], \\ T(S') &= T(S) + \tfrac{1}{2}\hat{\mathbf{y}}^T(S', S)\left[\hat{\mathbf{p}}^{(y)}(S') + \hat{\mathbf{p}}^{(y)}(S)\right]. \end{aligned}$$

For $T(R', S') = T(R') - T(S')$, we then obtain

$$\begin{aligned} T(R', S') &= T(R, S) + \tfrac{1}{2}\hat{\mathbf{y}}^T(R', R)\left(\hat{\mathbf{p}}^{(y)}(R') + \hat{\mathbf{p}}^{(y)}(R)\right) \\ &\quad - \tfrac{1}{2}\hat{\mathbf{y}}^T(S', S)\left(\hat{\mathbf{p}}^{(y)}(S') + \hat{\mathbf{p}}^{(y)}(S)\right). \end{aligned}$$

Inserting (4.9.16) and (4.9.18) into the preceding equation yields

$$\begin{aligned} T(R', S') &= T(R, S) + \hat{\mathbf{y}}^T(R', R)\hat{\mathbf{p}}^{(y)}(R) - \hat{\mathbf{y}}^T(S', S)\hat{\mathbf{p}}^{(y)}(S) \\ &\quad + \tfrac{1}{2}\hat{\mathbf{y}}^T(R', R)\hat{\mathbf{M}}(R, S)\hat{\mathbf{y}}(R', R) - \tfrac{1}{2}\hat{\mathbf{y}}^T(S', S)\hat{\mathbf{M}}(S, R)\hat{\mathbf{y}}(S', S) \\ &\quad - \tfrac{1}{2}\hat{\mathbf{y}}^T(R', R)\hat{\mathbf{A}}^T(R, S)\hat{\mathbf{y}}(S', S) - \tfrac{1}{2}\hat{\mathbf{y}}^T(S', S)\hat{\mathbf{A}}(R, S)\hat{\mathbf{y}}(R', R). \end{aligned}$$

It is not difficult to see that the last two terms are equal. Thus, the final form for the travel time from S' to R' is

$$\begin{aligned} T(R', S') &= T(R, S) + \hat{\mathbf{y}}^T(R', R)\hat{\mathbf{p}}^{(y)}(R) - \hat{\mathbf{y}}^T(S', S)\hat{\mathbf{p}}^{(y)}(S) \\ &\quad + \tfrac{1}{2}\hat{\mathbf{y}}^T(R', R)\hat{\mathbf{M}}(R, S)\hat{\mathbf{y}}(R', R) - \tfrac{1}{2}\hat{\mathbf{y}}^T(S', S)\hat{\mathbf{M}}(S, R)\hat{\mathbf{y}}(S', S) \\ &\quad - \hat{\mathbf{y}}^T(S', S)\hat{\mathbf{A}}(R, S)\hat{\mathbf{y}}(R', R). \end{aligned}$$
(4.9.19)

In all these equations, we can also use symplectic relations $\mathbf{Q}_2(R, S) = -\mathbf{Q}_2^T(S, R)$ and $\mathbf{A}(R, S) = -\mathbf{A}^T(S, R)$. The last term in (4.9.19) can also be expressed in the form $\hat{\mathbf{y}}^T(S', S)\hat{\mathbf{A}}^T(S, R)\hat{\mathbf{y}}(R', R)$.

All equations for the two-point ray tracing in local Cartesian ray-centered coordinates y_i can easily be expressed in *general Cartesian coordinates* x_i or in any local Cartesian coordinate system x_i introduced at S and R. The local Cartesian coordinate systems at S and R may be different. We specify the Cartesian coordinate systems at S and R by transformation matrices $\hat{\mathbf{H}}(S)$ and $\hat{\mathbf{H}}(R)$; see Section 4.1.5.

We denote the Cartesian coordinates of points under consideration by $x_i(R')$, $x_i(R)$, $x_i(S')$, and $x_i(S)$. We introduce the notations

$$x_i(R', R) = x_i(R') - x_i(R), \qquad x_i(S', S) = x_i(S') - x_i(S). \qquad (4.9.20)$$

Similarly, we introduce the 3×1 column matrices $\hat{\mathbf{x}}(R', R)$ and $\hat{\mathbf{x}}(S', S)$ using (4.1.85). The transformation relations are then

$$\hat{\mathbf{x}}(R', R) = \hat{\mathbf{H}}(R)\hat{\mathbf{y}}(R', R), \qquad \hat{\mathbf{y}}(R', R) = \hat{\mathbf{H}}^T(R)\hat{\mathbf{x}}(R', R),$$
$$\hat{\mathbf{x}}(S', S) = \hat{\mathbf{H}}(S)\hat{\mathbf{y}}(S', S), \qquad \hat{\mathbf{y}}(S', S) = \hat{\mathbf{H}}^T(S)\hat{\mathbf{x}}(S', S). \qquad (4.9.21)$$

We can also transform the slowness vectors in a similar manner.

For slowness vectors $\hat{\mathbf{p}}^{(x)}(S')$ and $\hat{\mathbf{p}}^{(x)}(R')$, (4.9.16) and (4.9.18) yield

$$\hat{\mathbf{p}}^{(x)}(R') = \hat{\mathbf{p}}^{(x)}(R) + \hat{\mathbf{M}}^{(x)}(R, S)\hat{\mathbf{x}}(R', R) - \hat{\mathbf{A}}^{(x)T}(R, S)\hat{\mathbf{x}}(S', S),$$
$$\hat{\mathbf{p}}^{(x)}(S') = \hat{\mathbf{p}}^{(x)}(S) + \hat{\mathbf{M}}^{(x)}(S, R)\hat{\mathbf{x}}(S', S) + \hat{\mathbf{A}}^{(x)}(R, S)\hat{\mathbf{x}}(R', R). \qquad (4.9.22)$$

These are the final equations for the two-point paraxial slowness vectors, $\hat{\mathbf{p}}^{(x)}(R')$ and $\hat{\mathbf{p}}^{(x)}(S')$. Matrices $\hat{\mathbf{M}}^{(x)}$ and $\hat{\mathbf{A}}^{(x)}$ are given by relations

$$\hat{\mathbf{M}}^{(x)}(R, S) = \hat{\mathbf{H}}(R)\hat{\mathbf{M}}(R, S)\hat{\mathbf{H}}^T(R),$$
$$\hat{\mathbf{M}}^{(x)}(S, R) = \hat{\mathbf{H}}(S)\hat{\mathbf{M}}(S, R)\hat{\mathbf{H}}^T(S), \qquad (4.9.23)$$
$$\hat{\mathbf{A}}^{(x)}(R, S) = \hat{\mathbf{H}}(S)\hat{\mathbf{A}}(R, S)\hat{\mathbf{H}}^T(R).$$

In the last equation of (4.9.23), we can also insert $\hat{\mathbf{A}}^{(x)}(R, S) = -\hat{\mathbf{A}}^{(x)T}(S, R)$.

As in ray-centered coordinates, Equations (4.9.22) solve completely the two-point ray tracing problem if the points S' and R' are situated arbitrarily in the vicinity of S and R and specified in Cartesian coordinates. We have two possibilities for determining the paraxial ray $\Omega'(R', S')$, which connects points S' and R'. *First*, we can use $x_i(S', S)$ and $p_i^{(x)}(S')$, determined by (4.9.22), and perform standard ray tracing from S'. Consequently, the problem of a two-point ray tracing is reduced to a problem of an initial-value ray tracing. Alternatively, ray tracing can also be started at R', using known $x_i(R', R)$ and $p_i^{(x)}(R')$. *Second*, we can exploit the known propagator matrix $\mathbf{\Pi}(R, S)$ and compute the paraxial ray Ω' analytically, without a new ray tracing. To do this, we need to know $q_I(S^\perp)$ and $p_I^{(q)}(S^\perp)$ at point $S^\perp$, where the paraxial ray $\Omega'(R', S')$ intersects plane $\Sigma^\perp(S)$, perpendicular to Ω at S. Actually, relevant projection equations follow immediately from those given earlier:

$$q_K(S^\perp) = y_K(S', S), \qquad y_i(S', S) = H_{ji}(S)x_j(S', S),$$
$$p_K^{(q)}(S^\perp) = H_{iK}(S)(p_i^{(x)}(S') - p_i^{(x)}(S)) - M_{K3}(S)y_3(S', S).$$

For isotropic media, $M_{K3}(S) = -(v^{-2}v_{,K})_S$; see (4.1.81). As soon as $q_K(S^\perp)$ and $p_K^{(q)}(S^\perp)$ are known, (4.9.1) can be used to compute the paraxial ray Ω' analytically. (Note that points S' and R' in (4.9.1) are situated in planes perpendicular to Ω at S and R so that they correspond to $S^\perp$ and $R^\perp$.) Analogous equations can be also obtained for $q_K(R^\perp)$ and $p_K^{(q)}(R^\perp)$.

For travel time $T(R', S')$ from S' to R' (4.9.19) yields

$$
\begin{aligned}
T(R', S') = {} & T(R, S) + \hat{\mathbf{x}}^T(R', R)\hat{\mathbf{p}}^{(x)}(R) - \hat{\mathbf{x}}^T(S', S)\hat{\mathbf{p}}^{(x)}(S) \\
& + \tfrac{1}{2}\hat{\mathbf{x}}^T(R', R)\hat{\mathbf{M}}^{(x)}(R, S)\hat{\mathbf{x}}(R', R) \\
& - \tfrac{1}{2}\hat{\mathbf{x}}^T(S', S)\hat{\mathbf{M}}^{(x)}(S, R)\hat{\mathbf{x}}(S', S) - \hat{\mathbf{x}}^T(S', S)\hat{\mathbf{A}}^{(x)}(R, S)\hat{\mathbf{x}}(R', R).
\end{aligned}
$$

(4.9.24)

4.9.3 Paraxial Two-Point Eikonal

Equation (4.9.24) represents one of the most important equations of the paraxial ray method. It allows travel time $T(R', S')$ from any point S' to any point R' in a laterally varying layered structure to be computed analytically, without any ray tracing. Ray $\Omega(R', S')$ connecting points S' and R' need not be computed. The requirement is that ray $\Omega(R', S')$ be a *paraxial ray* to central ray $\Omega(R, S)$, along which travel time $T(R, S)$, ray propagator matrix $\Pi(R, S)$, and some other quantities are known. Points S' and R' should be situated close to S and R, respectively, but otherwise their positions may be arbitrary.

The positions of all points S', S, R', and R are specified in general Cartesian coordinates so that Equation (4.9.24) is easy to use. Moreover, Equation (4.9.24) remains valid even if the Cartesian coordinate systems at points S and R are different. The local Cartesian coordinate systems under consideration at S and R are specified by matrices $\hat{\mathbf{H}}(R)$ and $\hat{\mathbf{H}}(S)$ in (4.9.23).

Equation (4.9.24) has a very simple structure that is easy to understand. The first term, $T(R, S)$, corresponds to the travel time along the central ray; all the remaining terms represent corrections to $T(R, S)$. Two corrections are linear in coordinates $x_i(R', R)$ and $x_i(S', S)$ and contain only the first derivatives of the travel-time field, represented by the relevant slowness vectors. The next three terms are quadratic in the coordinates. The first of them corresponds to a point source at S (see $\hat{\mathbf{M}}^{(x)}(R, S)$); the second corresponds to a point source at R (see $\hat{\mathbf{M}}^{(x)}(S, R)$). Finally, the last term is of a *mixed character*.

Equation (4.9.24) assumes that the ray tracing and dynamic ray tracing was performed from S to R. For this reason, the signs of the terms with $\hat{\mathbf{p}}^{(x)}(S)$ and $\hat{\mathbf{M}}^{(x)}(S, R)$ are minus.

4.9.4 Mixed Second Derivatives of the Travel-Time Field

Equation (4.9.24) can also be used to determine the mixed second derivatives of travel-time field $T(R, S)$ with respect to the Cartesian coordinates of source S and receiver R. We introduce the 3×3 matrix $\hat{\mathbf{M}}^{mix}(R, S)$ with elements M_{ij}^{mix} given by relations

$$
M_{ij}^{mix}(R, S) = (\partial^2 T(R', S')/\partial x_i(S', S)\partial x_j(R', R))_{R'=R, S'=S}.
$$

(4.9.25)

Matrix $M_{ij}^{mix}(R, S)$ can also be considered for different local Cartesian coordinate systems introduced at points S and R.

It is simple to determine matrix $\hat{\mathbf{M}}^{mix}$ of the mixed second derivatives of the travel-time field from (4.9.24):

$$
\hat{\mathbf{M}}^{mix}(R, S) = -\hat{\mathbf{A}}^{(x)}(R, S) = -\hat{\mathbf{H}}(S)\begin{pmatrix} \mathbf{Q}_2^{-1}(R, S) & & 0 \\ & & 0 \\ 0 & 0 & 0 \end{pmatrix}\hat{\mathbf{H}}^T(R);
$$

(4.9.26)

see (4.9.15) and (4.9.23).

The mixed derivatives of the travel-time field play an important role in the paraxial ray method. They can be used to determine geometrical spreading from travel-time data. See Section 4.10.5.

4.9.5 Boundary-Value Problems for Surface-to-Surface Ray Tracing

Boundary-value problems for surface-to-surface ray tracing are fully analogous to the boundary-value problems in ray-centered coordinates; see Section 4.9.1. We use the notation of Section 4.4.7 and make use of Equation (4.4.94). We remind the reader that the initial point S of central ray Ω and the initial point S' of paraxial ray Ω' are situated on anterior surface Σ_a, close to each other. Similarly, end points R and R' of Ω and Ω' are situated on posterior surface Σ_p. Assuming $d\mathbf{z}(S')$ and $d\mathbf{z}(R')$ are known, (4.4.94) can be solved for $d\mathbf{p}^{(\Sigma)}(S')$ and $d\mathbf{p}^{(\Sigma)}(R')$:

$$
\begin{aligned}
d\mathbf{p}^{(\Sigma)}(S') &= -\mathbf{B}^{-1}\mathbf{A}d\mathbf{z}(S') + \mathbf{B}^{-1}d\mathbf{z}(R'), \\
d\mathbf{p}^{(\Sigma)}(R') &= -\mathbf{B}^{-1T}d\mathbf{z}(S') + \mathbf{D}\mathbf{B}^{-1}d\mathbf{z}(R').
\end{aligned}
\tag{4.9.27}
$$

Here $\mathbf{A} = \mathbf{A}(R, S)$, $\mathbf{B} = \mathbf{B}(R, S)$, $\mathbf{C} = \mathbf{C}(R, S)$, and $\mathbf{D} = \mathbf{D}(R, S)$ are 2×2 minors of the surface-to-surface propagator matrix $\mathbf{T}(R, S)$; see (4.4.92).

Equations (4.9.27) can be used in a number of applications. Here we shall use them to derive an expression for a two-point eikonal, $T^{\Sigma}(R', S')$, from point S' on anterior surface Σ_a to point R' on posterior surface Σ_p. As in ray-centered coordinates, we obtain

$$
\begin{aligned}
T^{\Sigma}(R', S') &= T^{\Sigma}(R, S) + d\mathbf{z}^T(R')\mathbf{p}^{(z)}(R) - d\mathbf{z}^T(S')\mathbf{p}^{(z)}(S) \\
&\quad + \tfrac{1}{2}d\mathbf{z}^T(R')\mathbf{D}\mathbf{B}^{-1}d\mathbf{z}(R') + \tfrac{1}{2}d\mathbf{z}^T(S')\mathbf{B}^{-1}\mathbf{A}d\mathbf{z}(S') \\
&\quad - d\mathbf{z}^T(S')\mathbf{B}^{-1}d\mathbf{z}(R').
\end{aligned}
\tag{4.9.28}
$$

The positions of point S' on the anterior surface Σ_a and of point R' on posterior surface Σ may be arbitrary. It is only required that S' be close to S and R' be close to R. See also Bortfeld (1989) and Hubral, Schleicher, and Tygel (1992).

Expression (4.9.28) can also be generalized to consider points S' and R', which are situated outside Σ_a and Σ_p and not directly on them. In fact, in this case (4.9.28) again yields the earlier derived Equation (4.9.24). For this reason, we shall not give a detailed derivation here but only a brief outline. We take into account (4.4.97), (4.4.32), and (4.4.36) and observe that (4.4.36) is a special case of (4.4.32) for Q' situated on surface Σ, that is, for $z_3 = -\tfrac{1}{2}z_I z_J D_{IJ}$. Consequently, (4.4.36) should be replaced by (4.4.32) for Q' situated outside Σ. We do this for both points R' and S' and obtain (4.9.24).

As in (4.9.25) and (4.9.26), we can introduce mixed second derivatives of the travel-time field (4.9.28) with respect to $z_I(S', S) = dz_I(S')$ and $z_J(R', R) = dz_J(R')$ and express them in terms of the 2×2 matrix $\mathbf{B}$:

$$
M_{IJ}^{\Sigma mix}(R, S) = (\partial^2 T^{\Sigma}(R', S')/\partial z_I(S', S)\partial z_J(R', R))_{R'=R, S'=S},
\tag{4.9.29}
$$

$$
\mathbf{M}^{\Sigma mix}(R, S) = -\mathbf{B}^{-1}(R, S).
\tag{4.9.30}
$$

In $M_{IJ}^{\Sigma mix}$, one derivative is taken along anterior surface Σ_a, and the second is taken along posterior surface Σ_p. Equation (4.9.30) can be used to compute $\mathbf{Q}_2(R, S)$ from $\mathbf{M}^{\Sigma mix}(R, S)$; see (4.4.97) and Section 4.10.5.b.

4.9.6 Concluding Remarks

The solution of the two-point ray tracing problem, outlined in Section 4.9, is based on paraxial rays. Consequently, it is only *approximate, not exact*. Let us consider central ray Ω, and two points S and R situated on Ω. We further consider paraxial ray Ω', connecting two points S' (close to S) and R' (close to R). Assume that the 4×4 propagator matrix $\mathbf{\Pi}(R, S)$ corresponding to ray-centered coordinates is known. We can then compute analytically the paraxial slowness vectors $\vec{p}(R')$ and $\vec{p}(S')$, corresponding to paraxial ray Ω'. The actual trajectory of Ω' can be obtained analytically using (4.3.28); see Section 4.9.2. The solution is, however, only approximate; its accuracy decreases with increasing distances $\overline{S'S}$ and $\overline{R'R}$. If the accuracy of the solution is not sufficient in the actual problem under study, it is necessary to solve the problem iteratively. In the first iteration, we perform *exact ray tracing*, starting from S', with $\vec{p}(S')$ taken as initial data. Alternatively, we can start from R' with $\vec{p}(R')$, or perform ray computations from both sides. Along the ray(s), we determine new propagator matrix (matrices). The procedure should be repeated until the required accuracy is achieved. As we know from Section 3.11.1, the procedure is not necessarily unique due to the multiplicity of rays.

There are also some other possibilities for treating the two-point ray tracing problem. Moreover, there are also some other important boundary-value ray tracing problems that are not discussed here. A detailed treatment of all other boundary-value ray tracing problems would increase the length of this section inadmissibly. For this reason, we shall only make a few brief comments.

a. Regarding the boundary-value ray tracing problems connected with initial surfaces and initial lines, some of them may be easily solved in terms of the surface-to-surface propagator matrix; see (4.9.27) and (4.9.28). This is left as an exercise for the reader.

b. Regarding the application of 6×6 propagator matrices, instead of 4×4 propagator matrices, both approaches are alternative. We do not discuss the solution of the boundary-value ray tracing problems in terms of 6×6 propagator matrices. For a detailed treatment, see Farra and Le Bégat (1995) and Farra (1999). These references also give the solution of the two-point ray tracing problem, considering the structural perturbations of the model.

c. Regarding the inhomogeneous *anisotropic* medium, for more details refer to Section 4.14.12.

4.10 Geometrical Spreading in a Layered Medium

The concept of geometrical spreading plays a very important role in the computation of amplitudes of seismic body waves propagating in inhomogeneous, isotropic, or anisotropic layered structures. Usually, geometrical spreading is introduced with respect to the cross-sectional area of the ray tube or in some relation to the ray Jacobian. Unfortunately, the definition of geometrical spreading in the seismological literature is not unique. The general physical meaning of geometrical spreading as introduced by different authors is usually very similar, but the individual definitions differ in detail. In certain definitions, it is tacitly assumed that the wavefront is generated by a point source. For details and many other references, see Gel'chinskiy (1961), Wesson (1970), Červený and Ravindra (1971), Chen and Ludwig (1973), Červený, Langer, and Pšenčík (1974), Green (1976), Popov (1977), Červený, Molotkov, and Pšenčík (1977), Popov and Pšenčík (1978a, 1978b), Sharafutdinov

(1979), Goldin (1979, 1986), Pšenčík (1979), Červený and Pšenčík (1979), Grinfeld (1980), Červený and Hron (1980), Popov and Tyurikov (1981), Thomson and Chapman (1985), Červený (1985a, 1985b, 1987a, 1987b), Červený, Klimeš, and Pšenčík (1988a, 1988b), Kendall and Thomson (1989), Ursin (1990), Hubral, Schleicher, and Tygel (1992), Tygel, Schleicher, and Hubral (1992), Najmi (1996), and Snieder and Chapman (1998), among others.

Here we shall define geometrical spreading in a simple way. Let us consider an orthonomic system of rays, parameterized by ray parameters γ_1, and γ_2, and select arbitrarily a ray Ω and a point R on the ray. We then define geometrical spreading $\mathcal{L}(R)$ by the following alternative equations:

$$\mathcal{L}(R) = |J(R)|^{1/2} = \left|\mathcal{U}^{-1}(R)J^{(T)}(R)\right|^{1/2} = \left|\mathcal{U}^{-1}(R)\mathcal{C}(R)\Omega^{(T)}(R)\right|^{1/2}.$$
(4.10.1)

Here J, $J^{(T)}$, and $\Omega^{(T)}$ have the same meaning as introduced in Sections 3.10.2 and 3.10.3; see (3.10.23). Quantity $J = J^{(s)}$ is the Jacobian of transformation from ray coordinates $\gamma_1, \gamma_2, \gamma_3 = s$ to general Cartesian coordinates x_1, x_2, x_3, $J^{(T)}$ is the Jacobian of transformation from ray coordinates $\gamma_1, \gamma_2, \gamma_3 = T$ to general Cartesian coordinates x_1, x_2, x_3, $\Omega^{(T)}$ represents the scalar surface element cut out of the wavefront by the ray tube and normalized with respect to $d\gamma_1 d\gamma_2$, $\mathcal{U}$ is the group velocity, and $\mathcal{C}$ is the phase velocity along Ω. For acoustic waves and for elastic waves in isotropic media, group velocity $\mathcal{U}$ and phase velocity $\mathcal{C}$ can be replaced by the propagation velocity (c or α or β, depending on the type of wave). We remind the reader that $J = J^{(s)}$ is also called the ray Jacobian. Instead of $J^{(s)}$ and $J^{(T)}$, we can also use $J^{(u)}$, where u is an arbitrary monotonic parameter along ray Ω, and $J^{(u)}$ is the relevant Jacobian of transformation from ray coordinates $\gamma_1, \gamma_2, \gamma_3 = u$ to x_1, x_2, x_3. The relation between $J^{(s)}$ and $J^{(u)}$ is $J^{(s)} = g_u J^{(u)}$, with $g_u = du/ds$. Ray Jacobian $J = J^{(s)}$ also measures the cross-sectional area of the ray tube, $d\Omega^{\perp}$, perpendicular to the ray, normalized with respect to $d\gamma_1 d\gamma_2$. More details are given in Section 3.10.

Several useful relations for $|J(R)|$ were derived in Section 3.10.4. All these relations can be used to evaluate geometrical spreading $\mathcal{L}(R)$. In this section, we shall discuss the computation of geometrical spreading $\mathcal{L}(R)$ by dynamic ray tracing and derive some general properties of $J(R)$ and $\mathcal{L}(R)$. We shall also introduce a new quantity, the relative geometrical spreading (see Section 4.10.2) and discuss the possibilities of factorizing the geometrical spreading. Finally, we shall propose methods of determining the relative geometrical spreading from travel-time data known along a surface.

4.10.1 Geometrical Spreading in Terms of Matrices Q and Q̂$^{(x)}$

It was shown in Sections 4.2.3 and 4.2.4.4 that $J^{(T)}$ can be computed by dynamic ray tracing in any curvilinear coordinate system ξ_i in terms of the 3×3 matrix $\hat{\mathbf{Q}}^{(\xi)}$ (with elements $\partial\xi_i/\partial\gamma_j$); see (4.2.95). Using (4.10.1), geometrical spreading is obtained in the following form:

$$\mathcal{L}(R) = \left|\mathcal{U}^{-1}(R)\det\left(\hat{\mathbf{H}}^{(\xi)}\right)\det\left(\hat{\mathbf{Q}}^{(\xi)}\right)\right|^{1/2}.$$
(4.10.2)

Here $\hat{\mathbf{H}}^{(\xi)}$ is a 3×3 transformation matrix from curvilinear coordinates ξ_i to Cartesian coordinates x_i, $H_{ij}^{(\xi)} = \partial x_i/\partial\xi_j$, $\det\hat{\mathbf{Q}}^{(\xi)} = \partial(\xi_1, \xi_2, \xi_3)/\partial(\gamma_1, \gamma_2, T)$, and $\mathcal{U}$ is the group velocity. Equation (4.10.2) is valid for both isotropic and anisotropic media and in any coordinate system ξ_i.

Here we are mainly interested in rectangular Cartesian coordinates x_i and in wavefront orthonormal coordinates y_i. In isotropic media, the wavefront orthonormal coordinates y_i coincide with the local ray-centered coordinates. Consequently, we wish to express $\mathcal{L}(R)$ in terms of $\hat{\mathbf{Q}}^{(x)}$ and $\hat{\mathbf{Q}}^{(y)}$. Instead of $\hat{\mathbf{Q}}^{(y)}$, we shall also use the simpler notation $\hat{\mathbf{Q}}$, as in Section 4.1.5.c for isotropic media. In both cases, (4.10.2) simplifies as $|\det \hat{\mathbf{H}}^{(y)}| = 1$. Using (4.2.42), we obtain

$$
\mathcal{L}(R) = \left| \det \begin{pmatrix} Q_{11}^{(x)} & Q_{12}^{(x)} & t_1^{(x)} \\ Q_{21}^{(x)} & Q_{22}^{(x)} & t_2^{(x)} \\ Q_{31}^{(x)} & Q_{32}^{(x)} & t_3^{(x)} \end{pmatrix} \right|^{1/2} = \left| \det \begin{pmatrix} Q_{11}^{(y)} & Q_{12}^{(y)} & t_1^{(y)} \\ Q_{21}^{(y)} & Q_{22}^{(y)} & t_2^{(y)} \\ 0 & 0 & t_3^{(y)} \end{pmatrix} \right|^{1/2}.
$$

$$(4.10.3)$$

Here $t_i^{(x)}$ and $t_i^{(y)}$ are Cartesian and wavefront orthonormal components of the unit vector tangent to the ray, $t_i^{(x)} = \mathcal{U}_i^{(x)}/\mathcal{U}$ and $t_i^{(y)} = \mathcal{U}_i^{(y)}/\mathcal{U}$. Because $t_3^{(y)} = \mathcal{C}/\mathcal{U}$, we also obtain an important expression:

$$
\mathcal{L}(R) = \left| \mathcal{C}(R)\mathcal{U}^{-1}(R) \det \mathbf{Q}^{(y)}(R) \right|^{1/2} = \left| \mathcal{C}(R)\mathcal{U}^{-1}(R) \det \mathbf{Q}(R) \right|^{1/2}.
$$

$$(4.10.4)$$

Equations (4.10.1) through (4.10.4) are valid both for isotropic and anisotropic media. In the remaining part of Section 4.10, we shall consider only *isotropic media*, unless otherwise stated. Then $\mathcal{C} = \mathcal{U}$, $t_1^{(y)} = t_2^{(y)} = 0$, and $t_3^{(y)} = 1$. Equation (4.10.4) simplifies

$$
\mathcal{L}(R) = \left| \det \mathbf{Q}^{(y)}(R) \right|^{1/2} = \left| \det \mathbf{Q}(R) \right|^{1/2}.
$$

$$(4.10.5)$$

We shall now derive the *continuation relations* for geometrical spreading along a ray. We consider ray Ω and two points S and R situated on this ray. In view of relations (4.6.2),

$$
\mathcal{L}(R) = \left| \det[\mathbf{Q}_1(R, S) + \mathbf{Q}_2(R, S)\mathbf{M}(S)] \right|^{1/2} \mathcal{L}(S).
$$

$$(4.10.6)$$

Alternatively, we can use (4.6.3),

$$
\mathcal{L}(R) = \left| \det[\mathbf{P}_2^T(R, S) - \mathbf{Q}_2^T(R, S)\mathbf{M}(R)] \right|^{-1/2} \mathcal{L}(S).
$$

$$(4.10.7)$$

Here $\mathbf{Q}_1$, $\mathbf{Q}_2$, $\mathbf{P}_1$, and $\mathbf{P}_2$ are 2×2 minors of the ray propagator matrix (4.3.5). It can be proved that relations (4.10.6) and (4.10.7) are equivalent.

For a point source situated at point S, $\mathbf{Q}(S) = \mathbf{0}$, and (4.6.2) yields

$$
\mathcal{L}(R) = \left| \det \mathbf{Q}_2(R, S) \det \mathbf{P}(S) \right|^{1/2}.
$$

$$(4.10.8)$$

Similarly, for a point source situated at point R, $\mathbf{Q}(R) = \mathbf{0}$, and using Equation (4.6.3),

$$
\mathcal{L}(S) = \left| \det \mathbf{Q}_2(R, S) \det \mathbf{P}(R) \right|^{1/2},
$$

$$(4.10.9)$$

since $\det(-\mathbf{Q}_2^T(R, S)) = \det \mathbf{Q}_2(R, S)$. We can now write the general relation

$$
\mathcal{L}(R) = \left| \det \mathbf{P}(S) / \det \mathbf{P}(R) \right|^{1/2} \mathcal{L}(S).
$$

$$(4.10.10)$$

In general, $\det \mathbf{P}(S) \neq \det \mathbf{P}(R)$. Thus, Equation (4.10.10) implies that *geometrical spreading $\mathcal{L}$ is not reciprocal*.

4.10.2 Relative Geometrical Spreading

Let us consider ray Ω connecting point source S with the receiver situated at R. We define the *relative geometrical spreading* $\mathcal{L}(R, S)$ *due to a point source at S* as

$$\mathcal{L}(R, S) = \mathcal{L}(R)/|\det \mathbf{P}(S)|^{1/2} = |\det \mathbf{Q}_2(R, S)|^{1/2}. \qquad (4.10.11)$$

The relative geometrical spreading at S due to a point source situated at R

$$\mathcal{L}(S, R) = \mathcal{L}(S)/|\det \mathbf{P}(R)|^{1/2} = |\det \mathbf{Q}_2(R, S)|^{1/2} = \mathcal{L}(R, S). \qquad (4.10.12)$$

We have arrived at this important conclusion: *Relative geometrical spreading is reciprocal.*

The relative geometrical spreading $\mathcal{L}(R, S)$ plays a basic role in the computation of ray amplitudes of seismic body waves generated by a point source, particularly in the computation of ray-theory Green functions; see Chapter 5. For this reason, we shall now briefly summarize various methods of determining $\mathcal{L}(R, S)$ in different types of media. We shall also present several analytical expressions for simpler structures. Because $\mathcal{L}(R, S) = \mathcal{L}(S, R)$, we shall only discuss $\mathcal{L}(R, S)$ given by (4.10.11). We emphasize that we are considering an orthonomic central system of rays in an isotropic medium, generated by a point source situated at S. For fixed points S and R situated on ray Ω, the relative geometrical spreading $\mathcal{L}(R, S)$ does not depend on the integration parameter u used to solve the dynamic ray tracing system from S to R along ray Ω (travel time T, arclength s, and the like).

1. Dynamic ray tracing in ray-centered coordinates. $\mathbf{Q}_2(R, S)$ is then determined at any point R on Ω, and (4.10.11) can be used to calculate $\mathcal{L}(R, S)$. In simpler structures, the 4×4 propagator matrix $\mathbf{\Pi}(R, S)$ can be determined analytically; see Section 4.8. This applies particularly to a 3-D layered structure in which the layers with a constant velocity or with a constant gradient of velocity are separated by curved structural interfaces. The propagator matrix is then given by the analytic expression

$$
\mathbf{\Pi}(R, S) = \begin{pmatrix} \mathbf{I} & \mathbf{I}\sigma(R, \tilde{Q}_N) \\ \mathbf{0} & \mathbf{I} \end{pmatrix}
$$
$$
\times \prod_{i=N}^{1} \begin{pmatrix} \mathbf{Q}_1(\tilde{Q}_i, Q_i) & \mathbf{Q}_1(\tilde{Q}_i, Q_i)\sigma(Q_i, \tilde{Q}_{i-1}) \\ \mathbf{P}_1(\tilde{Q}_i, Q_i) & \mathbf{P}_1(\tilde{Q}_i, Q_i)\sigma(Q_i, \tilde{Q}_{i-1}) + \mathbf{P}_2(\tilde{Q}_i, Q_i) \end{pmatrix};
$$
$$(4.10.13)$$

see Section 4.4.6 and Fig. 4.11. Here $\sigma(Q_i, \tilde{Q}_{i-1})$ is given by (4.8.12), and $\mathbf{Q}_1(\tilde{Q}_i, Q_i)$, $\mathbf{P}_1(\tilde{Q}_i, Q_i)$, and $\mathbf{P}_2(\tilde{Q}_i, Q_i)$ are given by (4.4.80). $\mathbf{Q}_2(R, S)$ is then obtained as the upper right-hand 2×2 minor of $\mathbf{\Pi}(R, S)$; see (4.3.5). For homogeneous layers, $\sigma(Q_i, \tilde{Q}_{i-1}) = Vl(Q_i, \tilde{Q}_{i-1})$, where V is the velocity in the layer and $l(Q_i, \tilde{Q}_{i-1})$ is the distance between Q_i and $\tilde{Q}_{i-1}$. Equation (4.10.13) yields many other simpler expressions, for example, for plane dipping interfaces, plane-parallel interfaces, 2-D structures, and the like.

2. Dynamic ray tracing in Cartesian coordinates. Here we shall assume that the dynamic ray tracing along Ω is performed in Cartesian coordinates. We then can calculate J or $J^{(T)}$ along Ω as shown in Section 4.2.3 and 4.2.4.4. Using (4.10.1), we obtain

$$\mathcal{L}(R, S) = |J(R)/\det \mathbf{P}(S)|^{1/2} = V^{-1}(R)|J^{(T)}(R)/\det \mathbf{P}(S)|^{1/2}$$
$$= |\Omega^{(T)}(R)/\det \mathbf{P}(S)|^{1/2}. \qquad (4.10.14)$$

Thus, to determine $\mathcal{L}(R, S)$, we also need to know $\det \mathbf{P}(S)$ for a point source at S. This can be determined analytically. We can use $P_{IJ}(S) = (\vec{e}_I \cdot \partial \vec{p}/\partial \gamma_J)_S$, where γ_1 and γ_2 are ray parameters. We shall present two examples.

a. For $\gamma_1 = i_0$ and $\gamma_2 = \phi_0$, where i_0 and ϕ_0 are take-off angles at S, defined in Section 3.2.1 and used in Section 4.5.4, we can use (4.5.37) and obtain

$$\det \mathbf{P}(S) = V^{-2}(S) \sin i_0. \tag{4.10.15}$$

b. For $\gamma_1 = p_1(S)$ and $\gamma_2 = p_2(S)$, where p_1, p_2 are the components of the slowness vector, we obtain

$$\det \mathbf{P}(S) = 1/\cos \phi_0 = 1/V(S)p_3(S) = 1/\left[1 - V^2(S)\left(p_1^2(S) + p_2^2(S)\right)\right]^{1/2}. \tag{4.10.16}$$

If $J(R)$, $J^{(T)}(R)$, or $\Omega^{(T)}(R)$ is known, the relative geometrical spreading $\mathcal{L}(R, S)$ is obtained by inserting (4.10.15) or (4.10.16) into (4.10.14) (depending on the ray parameters γ_1 and γ_2 chosen). An analogous approach can be applied even if the dynamic ray tracing is performed in curvilinear coordinates.

3. Surface-to-surface ray tracing. To obtain $\mathcal{L}(R, S)$ from the results of surface-to-surface ray tracing, we shall use (4.4.50) and the first equation of (4.4.97):

$$
\begin{aligned}
\mathcal{L}(R, S) &= |\det \mathbf{G}(R) \det \mathbf{G}(S) \det \mathbf{B}(R, S)|^{1/2} \\
&= |\cos i_S \cos i_R \det \mathbf{B}(R, S)|^{1/2}.
\end{aligned} \tag{4.10.17}
$$

Here i_S is the acute angle between the direction of the normal $\vec{n}$ to anterior surface Σ_a at S, and i_R has the same meaning on posterior surface Σ_p at R. The 2×2 matrix $\mathbf{B}(R, S)$ is known from surface-to-surface ray tracing; see (4.4.92). Note that $\det \mathbf{B}$ is also reciprocal, $\det \mathbf{B}(R, S) = \det \mathbf{B}(S, R)$.

4. Homogeneous medium. Equation (4.10.13) yields a simple expression for a homogeneous medium without interfaces:

$$\mathcal{L}(R, S) = Vl(R, S), \tag{4.10.18}$$

where V is the velocity and $l(R, S)$ is the distance between S and R. (4.10.18) also immediately follows from (4.10.14), if we insert (3.10.36) into it for J and (4.10.15) for $\det \mathbf{P}(S)$, or (3.10.37) for J and (4.10.16) for $\det \mathbf{P}(S)$.

5. Vertically inhomogeneous media. We then can use (4.10.14) and insert there general relations (3.10.45) for J and (4.10.15) for $\det \mathbf{P}(S)$:

$$
\begin{aligned}
\mathcal{L}(R, S) &= |p^{-1} \cos i_0 \cos i \eta(p)(\partial \eta/\partial p)_z|^{1/2} \\
&= V(S)|\cos i \eta(i_0)(\partial \eta/\partial i_0)_z/\sin i_0|^{1/2} \\
&= |p^{-2} \cos i_0 \cos i \eta(p)(\partial T/\partial p)_z|^{1/2} \\
&= V(S)|p^{-1} \cos i \eta(i_0)(\partial T/\partial i_0)_z/\sin i_0|^{1/2}.
\end{aligned} \tag{4.10.19}
$$

It is assumed that source S is situated at depth z_0 and that the receiver is at depth z. $i_0 = i(S)$ and $i = i(R)$ represent the acute angles between the vertical axis and the ray at S and R. Quantity p represents the seismological ray parameter for a vertically inhomogeneous medium and is constant along the whole ray, $p = (\sin i_0)/V(S) = (\sin i(z))/V(z)$; see (3.7.4). Note also that $\cos i_0 = [1 - V^2(S)p^2]^{1/2}$ and $\cos i = [1 - V^2(R)p^2]^{1/2}$. Function $\eta = \eta(p)$ represents the horizontal distance between S and R, and $T = T(p)$ represents the relevant travel time. If the ray is downgoing or upgoing monotonically between S and R, we can use the notation of Section 3.7.1: $\eta(p) = x(p, z, z_0)$ and $T(p) = T(p, z, z_0)$; see (3.7.11).

In (4.10.19), most quantities are known from ray tracing and travel-time computations. In addition to these quantities, we only need to determine the derivatives $(\partial \eta / \partial p)_z$ (or $(\partial \eta / \partial i_0)_z$ or $(\partial T / \partial p)_z$ or $(\partial T / \partial i_0)_z$). For fixed z_0 and z, the expressions for the derivatives with respect to p can be found in terms of closed-form integrals for any velocity-depth distribution $V(z)$ using (3.7.11). We obtain

$$
\begin{aligned}
\frac{\partial x(p, z_0, z)}{\partial p} &= \pm \int_{z_0}^{z} \frac{V\,dz}{(1 - p^2 V^2)^{3/2}}, \\
\frac{\partial T(p, z_0, z)}{\partial p} &= \pm \int_{z_0}^{z} \frac{p V\,dz}{(1 - p^2 V^2)^{3/2}}.
\end{aligned}
\tag{4.10.20}
$$

The plus sign $(+)$ refers to $z > z_0$ (downgoing ray), and the minus sign $(-)$ to $z < z_0$ (upgoing ray). The integrals can be evaluated numerically for any velocity-depth distribution $V(z)$. Rays with turning points have to be divided into upgoing and downgoing segments. The only complication is for z_0 or z close to the depth of the turning point of the ray z_M, at which $p = 1/V(z_M)$. The integrands of (4.10.20) grow above all limits for $z_0 = z_M$ and/or $z = z_M$. The singularity, however, can be removed using the integration-by-parts method.

For simpler velocity-depth distributions, analytical expressions for $\eta(p)$ and $T(p)$ are known; see Section 3.7.2. Then the determination of the derivatives is straightforward, and the computation of $\mathcal{L}(R, S)$ using (4.10.19) is elementary, even for ray segments that contain the turning point or are directly at the turning point.

Notice that the first and third equations of (4.10.19) clearly display the reciprocity of $\mathcal{L}(R, S)$.

6. Vertically inhomogeneous layered structure. Equations (4.10.19) also remain valid in this case. We merely use

$$
\eta(p) = \sum_{k=1}^{N+1} x(p, z_{k-1}, z_k), \qquad T(p) = \sum_{k=1}^{N+1} T(p, z_{k-1}, z_k).
\tag{4.10.21}
$$

Depths z_i correspond to structural interfaces or simply to grid points of the velocity-depth approximation, z_0 and z_{N+1} represent the positions of the source and the receiver.

As the simplest example, we consider a model consisting of homogeneous plane-parallel layers, with the z-axis perpendicular to the interfaces. Then

$$
\eta(p) = p \sum_{k=1}^{N+1} V_k d_k / c_k, \qquad (\partial \eta / \partial p)_z = \sum_{k=1}^{N+1} V_k d_k / c_k^3.
$$

Here $d_k = |z_k - z_{k-1}|$ and $c_k = [1 - p^2 V_k^2]^{1/2}$. The final expression for $\mathcal{L}(R, S)$ is obtained from the first equation of (4.10.19):

$$
\begin{aligned}
\mathcal{L}(R, S) &= \left[\cos i_0 \cos i\, p^{-1} \eta(p) \sum_{k=1}^{N+1} V_k d_k / c_k^3 \right]^{1/2} \\
&= \left[\cos i_0 \cos i \left(\sum_{k=1}^{N+1} V_k d_k / c_k \right) \left(\sum_{k=1}^{N+1} V_k d_k / c_k^3 \right) \right]^{1/2}.
\end{aligned}
\tag{4.10.22}
$$

The same expression is obtained even if we use other equations of (4.10.19).

Expressions analogous to (4.10.22) can also be obtained for vertically inhomogeneous layered structures with some other simple velocity-depth distributions within the layers; see Section 3.7.2. This applies to layers with a constant vertical gradient of $V(z)$, $V^{-2}(z)$, $\ln V(z)$, and the like. One must only be careful to use the proper expressions for the ray elements with a turning point. Such expressions can be found in Section 3.7.2 for all the aforementioned velocity distributions.

7. Radially symmetric models. We use (3.10.46), (4.10.14), and (4.10.15) and obtain

$$
\begin{aligned}
\mathcal{L}(R, S) &= r\, r_0 |p^{-1} \sin\theta \cos i_0 \cos i\, (\partial\theta/\partial p)_r|^{1/2} \\
&= r\, V(S)|\sin\theta \cos i\, (\partial\theta/\partial i_0)_r / \sin i_0|^{1/2} \\
&= r\, r_0 p^{-1}|\sin\theta \cos i_0 \cos i\, (\partial T/\partial p)_r|^{1/2} \\
&= V(S)|r\, V(R) \sin\theta \cot i\, (\partial T/\partial i_0)_r / \sin i_0|^{1/2}.
\end{aligned}
\tag{4.10.23}
$$

Here $i(r)$ is the acute angle between the ray and the vertical line at r, $i_0 = i(S)$, and $i = i(R)$. Quantity p represents the seismological ray parameter for a radially symmetric medium and is constant along the whole ray, that is, $p = r \sin i(r)/V(r) = r_0 \sin i_0/V(S)$; see (3.7.31). We denote $r_0 = r(S)$, $r = r(R)$, so that $\cos i_0 = [1 - V^2(S)p^2 r_0^{-2}]^{1/2}$, $\cos i = [1 - V^2(R)p^2 r^{-2}]^{1/2}$. The other symbols in (4.10.23) have the same meaning as in (3.10.46). It is assumed that $\theta = 0$ at source S so that colatitude θ measures the angular distance between S and R, and is reciprocal. The first and third equations of (4.10.23) again clearly show that $\mathcal{L}(R, S)$ is reciprocal.

As in vertically inhomogeneous media, we introduce functions $\theta(p, r_i, r_{i+1})$ and $T(p, r_i, r_{i+1})$; see (3.7.35). For fixed r_i and r_{i+1}, we then obtain

$$
\begin{aligned}
\frac{\partial\theta(p, r_i, r_{i+1})}{\partial p} &= \pm \int_{r_i}^{r_{i+1}} \frac{r\, V\, dr}{(r^2 - V^2 p^2)^{3/2}}, \\
\frac{\partial T(p, r_i, r_{i+1})}{\partial p} &= \pm \int_{r_i}^{r_{i+1}} \frac{p\, V\, r\, dr}{(r^2 - V^2 p^2)^{3/2}}.
\end{aligned}
\tag{4.10.24}
$$

The numerical computation of (4.10.24) is easy for any velocity distribution $V(r)$. The complication is again connected with the turning point $r = r_M$, for which $r_M/V(r_M) = p$; see (3.7.36). This complication may be overcome by the integration-by-parts method.

Equations (4.10.23) remain valid also for radially symmetric media containing structural interfaces of the first order along spherical surfaces $r = r_k = \text{const}$. As for vertically inhomogeneous media [see(4.10.21)], we express $\theta(p)$ and $T(p)$ as the sum of elements corresponding to the individual layers:

$$
\theta(p) = \sum_{k=1}^{N+1} \theta(p, r_{k-1}, r_k), \qquad T(p) = \sum_{k=1}^{N+1} T(p, r_{k-1}, r_k).
\tag{4.10.25}
$$

The analytical expressions for $\theta(p, r_{k-1}, r_k)$ and $T(p, r_{k-1}, r_k)$ for several simple velocity distributions $V(r)$ can be found in Section 3.7.4. The computation of $\mathcal{L}(R, S)$ for layered radially symmetric media with these velocity distribution within the individual layers is straightforward.

8. Other analytical cases. Analytical expressions for relative geometrical spreading $\mathcal{L}(R, S)$ can be found for many other one-dimensional velocity distributions. Also, these expression may be simply modified to vertical profiles (VSP, cross-hole). The relevant expressions for Jacobian J in vertically inhomogeneous media, with the derivatives taken along a vertical profile, can be found in Section 3.10.4.8a.

It is also simple to find $\mathcal{L}(R, S)$ for polynomial rays in both vertically inhomogeneous and radially symmetric media. For vertically inhomogeneous media, we merely use (4.10.19) and (4.10.21) and the expressions for $x(p, z_{k-1}, z_k)$ given in Section 3.7.3. If the velocity-depth distribution between the individual grid points is approximated by $z = a_i + b_i V^{-2} + c_i V^{-4} + d_i V^{-6}$, the analytical expressions for $\mathcal{L}(R, S)$ are immediately obtained from (3.7.28). Analogous equations are also obtained for polynomial rays in radially symmetric media; see (3.7.47). The relevant derivatives with respect to p for polynomial rays, required in (4.10.19) and (4.10.23), can be found in Červený (1980) for vertically inhomogeneous media and in Červený and Janský (1983) for radially symmetric media.

The relative geometrical spreading $\mathcal{L}(R, S)$, as introduced here, is close to the geometrical spreading factors introduced in other publications. Unfortunately, the definitions differ in details. Moreover, most of the other definitions do not exhibit reciprocity $\mathcal{L}(R, S) = \mathcal{L}(S, R)$. For example, the "spreading function" L, introduced by Equation (2.108) in Červený and Ravindra (1971), is related to $\mathcal{L}(R, S)$ in media without interfaces as follows: $\mathcal{L}(R, S) = V(S)L$. In media with interfaces, the "interface factor" $\prod_{j=1}^{k}(d\sigma(O_j)/d\sigma'(O_j))^{1/2}$ should also be eliminated from (2.108). The reason is that we are connecting the interface factor with the R/T coefficients, not with geometrical spreading; see Chapter 5. Consequently, many analytical expressions for L for various types of media, published by Červený and Ravindra (1971), can be used to determine $\mathcal{L}(R, S)$. We only multiply L by $V(S)$ and exclude the interface factor. The same also applies to geometrical spreading L introduced by Červený, Molotkov, and Pšenčík (1977).

4.10.3 Relation of Geometrical Spreading to Matrices M and K

In an orthogonal ray-centered coordinate system, $q_1, q_2, q_3 = s$, with scale factors $h_1 = h_2 = 1$ and $h_3 = h$ (see (4.1.11)), the well-known expression for $\nabla^2 T$ reads

$$
\nabla^2 T = \frac{1}{h_1 h_2 h_3} \left[\frac{\partial}{\partial q_1} \left(\frac{h_2 h_3}{h_1} \frac{\partial T}{\partial q_1} \right) + \frac{\partial}{\partial q_2} \left(\frac{h_1 h_3}{h_2} \frac{\partial T}{\partial q_2} \right) + \frac{\partial}{\partial q_3} \left(\frac{h_1 h_2}{h_3} \frac{\partial T}{\partial q_3} \right) \right]
$$
$$
= \frac{1}{h} \left[\frac{\partial}{\partial q_1} \left(h \frac{\partial T}{\partial q_1} \right) + \frac{\partial}{\partial q_2} \left(h \frac{\partial T}{\partial q_2} \right) + \frac{\partial}{\partial q_3} \left(\frac{1}{h} \frac{\partial T}{\partial q_3} \right) \right].
$$

On the central ray Ω, $h = 1$, $\partial T / \partial q_I = 0$, and $\partial h / \partial q_3 = 0$. The expression for $\nabla^2 T$ simplifies to

$$
\nabla^2 T = \frac{\partial^2 T}{\partial q_1^2} + \frac{\partial^2 T}{\partial q_2^2} + \frac{\partial^2 T}{\partial q_3^2} = \text{tr}\,\mathbf{M} + \frac{d}{ds}\left(\frac{1}{v}\right).
$$

In view of (3.10.30),

$$
\text{tr}\,\mathbf{M} + \frac{d}{ds}\left(\frac{1}{v}\right) = \frac{1}{J}\frac{d}{ds}\left(\frac{J}{v}\right).
$$

This yields the final relation between J and $\mathbf{M}$:

$$
\text{tr}\,\mathbf{M} = (vJ)^{-1}dJ/ds = 2v^{-1}\mathcal{L}^{-1}d\mathcal{L}/ds. \tag{4.10.26}
$$

Relation (4.10.26) can be expressed in terms of the 2×2 matrix of the curvature of the wavefront. Taking into account $\mathbf{K} = v\mathbf{M}$, (4.10.26) yields

$$\operatorname{tr} \mathbf{K} = J^{-1} dJ/ds = 2\mathcal{L}^{-1} d\mathcal{L}/ds. \tag{4.10.27}$$

We can also use the mean curvature of the wavefront $H = \frac{1}{2}(K_1 + K_2)$, $\operatorname{tr} \mathbf{K} = 2H$. Hence, (4.10.27) yields this simple relation between H and J:

$$H = \frac{1}{2} J^{-1} dJ/ds = \mathcal{L}^{-1} d\mathcal{L}/ds. \tag{4.10.28}$$

All three equations (4.10.26) through (4.10.28) can be solved for J or $\mathcal{L}$. For example, the solution of (4.10.28) is

$$J(R) = J(S) \exp\left(2 \int_S^R H \, ds\right), \qquad \mathcal{L}(R) = \mathcal{L}(S) \exp\left(\int_S^R H \, ds\right). \tag{4.10.29}$$

In a *homogeneous medium*, (4.10.29) can be simply integrated. Because $K_I(s) = 1/R_I(s) = 1/[R_I(s_0) + s - s_0]$, we obtain $\int_S^R H \, ds = \frac{1}{2} \ln[K(S)/K(R)]$, where $K = K_1 K_2$ is the Gaussian curvature of the wavefront. Then (4.10.29) yields

$$J(R) = J(S)K(S)/K(R), \qquad \mathcal{L}(R) = \mathcal{L}(S)[K(S)/K(R)]^{1/2}. \tag{4.10.30}$$

Alternatively, we can insert

$$K(S)/K(R) = (R_1(S) + l)(R_2(S) + l)/R_1(S)R_2(S), \tag{4.10.31}$$

where l is the distance between S and R, $l = \overline{SR}$, and $R_{1,2}(S)$ are the main radii of curvature of the wavefront at S.

4.10.4 Factorization of Geometrical Spreading

It was shown in Section 4.4.8 that the 2×2 matrix $\mathbf{Q}_2(R, S)$ can be factorized. This capability immediately implies that relative geometrical spreading $\mathcal{L}(R, S) = |\det \mathbf{Q}_2(R, S)|^{1/2}$ can also be factorized (Goldin 1991).

Let us consider ray Ω in a layered laterally varying medium, connecting two points: S (point source) and R (receiver). In addition, let us consider point Q on Ω, at which ray Ω is incident at surface Σ. Surface Σ may be curved and may be arbitrarily inclined with respect to the incident ray Ω. It may also represent a structural interface. In this case, we also consider point $\tilde{Q}$, situated at the same position as Q on Σ, but from the side of the generated wave under consideration (reflected, transmitted). Using (4.4.100), we obtain

$$
\begin{aligned}
\mathcal{L}(R, S) &= |\det \mathbf{Q}_2(R, S)|^{1/2} \\
&= |\det \mathbf{Q}_2(R, \tilde{Q})|^{1/2} |\det \mathbf{U}^2|^{1/2} |\det \mathbf{Q}_2(Q, S)|^{1/2} \\
&= \mathcal{L}(R, \tilde{Q})\mathcal{L}(\tilde{Q}, Q)\mathcal{L}(Q, S).
\end{aligned} \tag{4.10.32}
$$

Here we have employed the following notation:

$$\mathcal{L}(\tilde{Q}, Q) = |\det \mathbf{U}^2|^{1/2}, \tag{4.10.33}$$

where $\mathbf{U}^2$ can be expressed in terms of the Fresnel zone matrix $\mathbf{M}^F(Q; R, S)$ using relation (4.4.104). We shall refer to factor $\mathcal{L}(\tilde{Q}, Q)$ as the *interface spreading factor*.

The interface spreading factor can be expressed in several alternative forms. We take into account that $\det \mathbf{G}(Q) = t_i(Q)n_i(Q)$, where n_i are components of the unit normal to

Σ at Q. From a physical point of view, $t_i(Q)n_i(Q) = \pm \cos i(Q)$, where $i(Q)$ is the angle of incidence (that is, the acute angle between the tangent to the incident ray and normal to interface Σ at Q). Similarly, we can put $\det \mathbf{G}(\tilde{Q}) = t_i(\tilde{Q})n_i(Q)$. Components $t_i(\tilde{Q})$ correspond to the reflected/transmitted ray. In view of (4.4.104),

$$\mathcal{L}(\tilde{Q}, Q) = |t_i(\tilde{Q})n_i(Q)t_k(Q)n_k(Q)|^{-1/2}|\det \mathbf{M}^F(Q; S, R)|^{1/2}, \quad (4.10.34)$$

where $\mathbf{M}^F$ is the Fresnel zone matrix introduced by (4.4.107). For *unconverted reflected waves* such as acoustic waves, purely P waves, and purely S waves, $t_i(\tilde{Q})n_i(Q) = -t_i(Q)n_i(Q)$, and Equation (4.10.34) yields

$$\mathcal{L}(\tilde{Q}, Q) = |\det \mathbf{M}^F(Q; S, R)|^{1/2}/|t_i(Q)n_i(Q)|. \quad (4.10.35)$$

Thus, the interface spreading factor can be expressed simply in terms of the Fresnel zone matrix.

4.10.5 Determination of the Relative Geometrical Spreading from Travel-Time Data

In certain seismological applications such as studies of the absorption of seismic body waves and true-amplitude studies in seismic prospecting for oil, it is very useful to eliminate the relative geometrical spreading from the measured amplitudes of seismic body waves. The question whether the relative geometrical spreading can be determined from purely kinematic measurements of the travel-time field, without any knowledge of the structure or of the actual type of the body wave under investigation, is one of the basic problems in the ray theory and is widely discussed in the literature.

This problem has recently also found new applications in theoretical studies and in the numerical modeling of seismic wave fields. As described in Section 3.8, the travel-time field can also be computed directly, without ray tracing. This capability applies mainly to travel-time fields of the first arrivals, but the method can be extended even to some more complex waves (for example, reflected and multiply reflected waves). As a result of these computations, the travel times are determined in a regularly spaced grid of points covering the model. It would add considerably to the possibilities of such computations if they were supplemented by the evaluation of geometrical spreading (and, consequently, of amplitudes). The problem in this case is the same as in the seismological applications described earlier: to determine the geometrical spreading from known travel times.

The problem of determining geometrical spreading from travel-time data, without any knowledge of the structure and of the type of wave and its ray field, is not straightforward. In general, it is not sufficient to know the travel-time field generated by a *single point source*; the travel-time fields generated by several points sources should be known. Also, the travel time generated by a source sufficiently different from the point source (for example, a plane wave) may be considered.

Let us consider an arbitrary multiply reflected, possibly converted, seismic body wave propagating in a general 3-D laterally varying structure. Consider ray Ω corresponding to this wave and two points, R and S, situated on this ray. The relative geometrical spreading can then be determined from the travel-time data in several ways.

a. The first option is to determine the geometrical spreading from several experiments, exploiting the measured matrix $\mathbf{M}$ of second derivatives of the travel-time field at points S and R. The method of determining matrix $\mathbf{M}$ from travel-time data known along a data surface is described in Section 4.6.2. For this purpose, velocities and their first derivatives and curvatures of data surfaces at S and R have to be also known. In general, at least three experiments must be performed. In some special source–receiver configurations (for example, if the source and receiver coincide, $S \equiv R$), the number of experiments is reduced to two.

b. The second option is to determine the relative geometrical spreading from mixed second derivatives of the travel-time field. The velocities at S and R must also be known.

c. The third option, convenient mainly in the case of grid computation of the travel-time field, was proposed by Vidale and Houston. The interested reader is referred to Vidale and Houston (1990).

We shall now briefly describe the first two possibilities.

a. DETERMINATION OF RELATIVE GEOMETRICAL SPREADING $\mathcal{L}(R, S)$ FROM MATRICES M

We shall consider the following three experiments; see Figure 4.16.

- For *the point source at S*, we determine the matrix of second derivatives of the travel-time field at R, $\mathbf{M}(R, S)$, from the travel-time data close to R.
- For *the point source at R*, we determine the matrix of second derivatives of the travel-time field at S, $\mathbf{M}(S, R)$, from the travel-time data close to S.
- In the third experiment, the source may be situated either at S or at R, but it must be different from a point source; in other words, the wavefront generated by the source must be sufficiently different from the wavefront generated by the point source. We can consider, for example, a locally plane wavefront. This source may, of course, be simulated by an array of point sources. Assume that such a source is situated at S. Denote the matrix of second derivatives of the travel-time field corresponding to this source at S by $\mathbf{M}(S)$, with $\det \mathbf{M}(S) \neq \infty$, and assume that $\mathbf{M}(S)$ is known. We then determine the matrix of second derivatives of the travel-time field at R, $\mathbf{M}(R)$, from the travel-time data close to R.

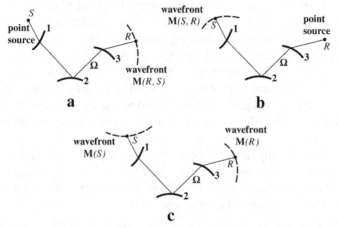

Figure 4.16. Determination of relative geometrical spreading from travel-time data. Description of three experiments to determine $\mathbf{M}(R, S)$, $\mathbf{M}(S, R)$, $\mathbf{M}(S)$, and $\mathbf{M}(R)$. (a) First experiment; (b) second experiment; (c) third experiment. For more details see text.

Thus, these three experiments yield four matrices: (a) $\mathbf{M}(R, S)$, (b) $\mathbf{M}(S, R)$, and (c) $\mathbf{M}(S)$ and $\mathbf{M}(R)$. We shall now derive equations that will allow us to determine $\mathcal{L}(R, S)$ from these four matrices. To shorten the derivation, we shall simply write $\mathbf{Q}_1, \mathbf{Q}_2, \mathbf{P}_1$, and $\mathbf{P}_2$ instead of $\mathbf{Q}_1(R, S), \mathbf{Q}_2(R, S), \mathbf{P}_1(R, S)$, and $\mathbf{P}_2(R, S)$.

We can put

$$\mathbf{M}(R, S) = \mathbf{P}_2\mathbf{Q}_2^{-1}, \qquad \mathbf{M}(S, R) = -\mathbf{Q}_2^{-1}\mathbf{Q}_1,$$
$$\mathbf{M}(R) = (\mathbf{P}_1 + \mathbf{P}_2\mathbf{M}(S))(\mathbf{Q}_1 + \mathbf{Q}_2\mathbf{M}(S))^{-1}; \tag{4.10.36}$$

see (4.6.6), (4.6.8), and (4.6.9). Equations (4.10.36) yield

$$\mathbf{M}(R, S)\mathbf{Q}_2 = \mathbf{P}_2, \qquad \mathbf{Q}_2\mathbf{M}(S, R) = -\mathbf{Q}_1,$$
$$\mathbf{M}(R)(\mathbf{Q}_1 + \mathbf{Q}_2\mathbf{M}(S)) = \mathbf{P}_1 + \mathbf{P}_2\mathbf{M}(S). \tag{4.10.37}$$

We now take into account the symplectic relation $\mathbf{P}_2\mathbf{Q}_1^T - \mathbf{P}_1\mathbf{Q}_2^T = \mathbf{I}$ (see (4.3.16)) and write

$$\mathbf{P}_1 = (\mathbf{P}_2\mathbf{Q}_1^T - \mathbf{I})\mathbf{Q}_2^{-1T} = [\mathbf{M}(R, S)\mathbf{Q}_2\mathbf{Q}_1^T - \mathbf{I}]\mathbf{Q}_2^{-1T}. \tag{4.10.38}$$

Inserting (4.10.38) into (4.10.37) and eliminating $\mathbf{P}_2$ from (4.10.37) yields

$$\mathbf{Q}_2\mathbf{M}(S, R) = -\mathbf{Q}_1,$$
$$\mathbf{M}(R)(\mathbf{Q}_1 + \mathbf{Q}_2\mathbf{M}(S)) = [\mathbf{M}(R, S)\mathbf{Q}_2\mathbf{Q}_1^T - \mathbf{I}]\mathbf{Q}_2^{-1T} + \mathbf{M}(R, S)\mathbf{Q}_2\mathbf{M}(S). \tag{4.10.39}$$

This can be expressed as

$$\mathbf{Q}_2\mathbf{M}(S, R) = -\mathbf{Q}_1,$$
$$\mathbf{M}(R)(\mathbf{Q}_1 + \mathbf{Q}_2\mathbf{M}(S))\mathbf{Q}_2^T = \mathbf{M}(R, S)[\mathbf{Q}_2\mathbf{Q}_1^T + \mathbf{Q}_2\mathbf{M}(S)\mathbf{Q}_2^T] - \mathbf{I}. \tag{4.10.40}$$

If we eliminate $\mathbf{Q}_1$,

$$\mathbf{M}(R)\mathbf{Q}_2(-\mathbf{M}(S, R) + \mathbf{M}(S))\mathbf{Q}_2^T$$
$$= \mathbf{M}(R, S)\mathbf{Q}_2(-\mathbf{M}(S, R) + \mathbf{M}(S))\mathbf{Q}_2^T - \mathbf{I}. \tag{4.10.41}$$

This yields

$$\mathbf{Q}_2(\mathbf{M}(S, R) - \mathbf{M}(S))\mathbf{Q}_2^T = (\mathbf{M}(R) - \mathbf{M}(R, S))^{-1}. \tag{4.10.42}$$

Taking the determinant of (4.10.42), we obtain

$$\mathcal{L}(R, S) = |\det(\mathbf{M}(R) - \mathbf{M}(R, S))\det(\mathbf{M}(S, R) - \mathbf{M}(S))|^{-1/4}. \tag{4.10.43}$$

This is the final equation we have been looking for. The equation was first published by Červený, Klimeš, and Pšenčík (1988a).

Some special cases of this equation are known from the literature (see, for example, Hubral 1983). Let us consider, as an example, the special case of waves reflected from an interface, with the positions of the source and receiver coinciding (zero-offset configuration). Two experiments are sufficient in this case because $\mathbf{M}(R, S) = \mathbf{M}(S, R)$. We still have, however, $\mathbf{M}(S) \neq \mathbf{M}(R)$. $\mathbf{M}(S)$ corresponds to the generated wavefront, and $\mathbf{M}(R)$ corresponds to the measured wavefront in the third experiment.

b. DETERMINATION OF RELATIVE GEOMETRICAL SPREADING $\mathcal{L}(R, S)$
FROM MIXED TRAVEL-TIME DERIVATIVES

It was proposed by Gritsenko (1984) to determine geometrical spreading from mixed second derivatives of the travel-time field. For a detailed discussion, see Goldin (1986).

We shall start the derivation with (4.9.26). We again consider ray Ω, connecting a point source at S and receiver at R, and assume that velocities $V(S)$ and $V(R)$ are known. We introduce two data surfaces: the source data surface Σ_1 at S and the receiver data surface Σ_2 at R. The surfaces may be curved and arbitrarily inclined with respect to the central ray Ω. We only exclude the case of ray Ω tangent to Σ_1 at S and/or to Σ_2 at R. At both surfaces, Σ_1 and Σ_2, we introduce standard local Cartesian coordinate systems $z_i(S)$ and $z_i(R)$ as described in Section 4.4.1, with the relevant 3×3 transformation matrices $\hat{\mathbf{Z}}(S)$ and $\hat{\mathbf{Z}}(R)$. We use (4.4.13) to obtain $\hat{\mathbf{H}}(S) = \hat{\mathbf{Z}}(S)\hat{\mathbf{G}}(S)$ and $\hat{\mathbf{H}}(R) = \hat{\mathbf{Z}}(R)\hat{\mathbf{G}}(R)$. Inserting these relations into (4.9.26) yields

$$\hat{\mathbf{M}}^{mix}(R, S) = -\hat{\mathbf{Z}}(S)\hat{\mathbf{G}}(S) \begin{pmatrix} \mathbf{Q}_2^{-1}(R, S) & 0 \\ 0 & 0 \\ 0 & 0 & 0 \end{pmatrix} \hat{\mathbf{G}}^T(R)\hat{\mathbf{Z}}^T(R).$$

This implies

$$-\hat{\mathbf{G}}(S) \begin{pmatrix} \mathbf{Q}_2^{-1}(R, S) & 0 \\ 0 & 0 \\ 0 & 0 & 0 \end{pmatrix} \hat{\mathbf{G}}^T(R) = \hat{\mathbf{M}}^{\Sigma mix}(R, S), \tag{4.10.44}$$

where $\hat{\mathbf{M}}^{\Sigma mix}$ is a 3×3 matrix of second mixed derivatives of the travel-time field with respect to the positions of the source and receiver, $z_i(S)$ and $z_i(R)$, respectively:

$$\hat{\mathbf{M}}^{\Sigma mix}(R, S) = \hat{\mathbf{Z}}^T(S)\hat{\mathbf{M}}^{mix}(R, S)\hat{\mathbf{Z}}(R). \tag{4.10.45}$$

The elements of matrix $\hat{\mathbf{M}}^{\Sigma mix}$ are given by relation

$$M_{ij}^{\Sigma mix}(R, S) = \left[\frac{\partial^2 T(R', S')}{\partial z_i(S')\partial z_j(R')} \right]_{R'=R, S'=S}. \tag{4.10.46}$$

In the following, we shall consider only the 2×2 upper-left-hand minors of both sides of (4.10.45),

$$-\mathbf{G}(S)\mathbf{Q}_2^{-1}(R, S)\mathbf{G}^T(R) = \mathbf{M}^{\Sigma mix}(R, S). \tag{4.10.47}$$

The 2×2 matrix $\mathbf{M}^{\Sigma mix}(R, S)$ only contains four mixed second derivatives of the travel-time field along Σ_1 at S and along Σ_2 at R. Relation (4.10.47) yields

$$\mathbf{Q}_2(R, S) = -\mathbf{G}^T(R)(\mathbf{M}^{\Sigma mix}(R, S))^{-1}\mathbf{G}(S). \tag{4.10.48}$$

This equation also immediately follows from the surface-to-surface ray tracing; see (4.9.30) and (4.4.97). The surface Σ_1, corresponds to the anterior surface Σ_a, and surface Σ_2 corresponds to the posterior surface Σ_p in this case. (4.10.48) can be used to determine all four elements of $\mathbf{Q}_2(R, S)$ only if basis vectors $\vec{e}_1$ and $\vec{e}_2$ are known at S and R because these vectors determine matrices $\mathbf{G}(S)$ and $\mathbf{G}(R)$. The determination of the relative geometrical spreading, however, is simpler because it does not require matrices $\mathbf{G}(S)$ and $\mathbf{G}(R)$, but only their determinants, to be known:

$$\det \mathbf{G}(S) = t_i(S)n_i(S), \qquad \det \mathbf{G}(R) = t_i(R)n_i(R). \tag{4.10.49}$$

Here $n_i(S)$ and $n_i(R)$ are unit normals to Σ_1 at S and to Σ_2 at R, and $t_i(S)$ and $t_i(R)$ are unit vectors tangent to Ω at S and R. Note that the positive orientation of ray Ω is from S to R and that unit tangents $t_i(S)$, $t_i(R)$ are taken positive along the positive orientations of ray Ω. The final equations for the relative geometrical spreading are

$$
\mathcal{L}(R, S) = |\det \mathbf{Q}_2(R, S)|^{1/2} = \left| \frac{(t_i(S)n_i(S))(t_k(R)n_k(R))}{\det \mathbf{M}^{\Sigma mix}(R, S)} \right|^{1/2}
$$
$$
= \left| \frac{\cos i(S) \cos i(R)}{\det \mathbf{M}^{\Sigma mix}(R, S)} \right|^{1/2}. \tag{4.10.50}
$$

Equation (4.10.50) can be conveniently used in *various seismological applications*. The source data surface, Σ_1, and the receiver data surface, Σ_2, may be different, but they may also coincide and form one common data surface. As a very important example, we can consider the surface of the Earth on which both the sources and receivers are distributed. Similar equations were also derived by Goldin (1986) and Hubral, Schleicher, and Tygel (1992), among others.

To determine $\mathcal{L}(R, S)$ using (4.10.50), we need to find four first derivatives of the travel-time field on surfaces Σ_1, Σ_2, $(\partial T/\partial z_I)_S$, and $(\partial T/\partial z_I)_R$ and four mixed second derivatives of the travel-time field $\partial^2 T/\partial z_I(S)\partial z_J(R)$. The second mixed derivatives form matrix $\mathbf{M}^{\Sigma mix}$. The first derivatives can serve to determine $t_i(S)n_i(S)$ and $t_k(R)n_k(R)$ using eikonal equation

$$
t_i(S)n_i(S) = V(S)(\partial T/\partial z_3)_S = \pm\{1 - V^2(S)[(\partial T/\partial z_1)_S^2 + (\partial T/\partial z_2)_S^2]\}^{1/2}.
$$

A similar equation can also be written for $t_i(R)n_i(R)$. The sign is obvious, but it has no effect on $\mathcal{L}(R, S)$ at all; see (4.10.50).

Equation (4.10.50) is convenient for computing the relative geometrical spreading in the numerical modeling of travel-time fields by direct methods, without ray tracing (see Section 3.8). Let us assume that such direct methods yield travel times in a regularly spaced grid of points, covering the model under consideration, due to a point source situated at a selected grid point. We then choose surfaces Σ_1 and Σ_2 on the coordinate surfaces, corresponding to the individual grid planes. Thus, we have nine different possible combinations of source and receiver data planes. Both data planes may, of course, coincide. The travel-time computations should be performed for *at least three source points* situated in source plane Σ_1.

We shall add one remark regarding Equation (4.10.50). Matrix $\mathbf{M}^{\Sigma mix}$ of second derivatives of the travel-time field in (4.10.50) can be approximately expressed in terms of first derivatives of components of the slowness vector, computed for at least three point sources. Consequently, the computation of second derivatives (of travel times) can be replaced by the computation of first derivatives (of the slowness vector), assuming that the slowness vectors are known at grid points, instead of the travel time, or in addition to it. The appropriate equations are straightforward.

4.10.6 Determination of the 4 × 4 Propagator Matrix from Travel-Time Data

The simplest derivation of the complete 4×4 propagator matrix from travel-time data is based on the formalism of the surface-to-surface ray tracing. Consider a ray Ω with the initial point S situated on the anterior surface Σ_a and the end point R situated on the posterior surface Σ_p. We wish to determine the 4×4 surface-to-surface propagator matrix $\mathbf{T}(R, S)$, given by (4.4.92); in other words, we wish to determine four 2×2 matrices $\mathbf{A}(R, S)$, $\mathbf{B}(R, S)$, $\mathbf{C}(R, S)$, and $\mathbf{D}(R, S)$. As we know from Section 4.9.5, matrix $\mathbf{B}(R, S)$ can be determined from the 2×2 matrix of mixed second derivatives of the travel-time field $\mathbf{M}^{\Sigma mix}(R, S)$; see (4.9.30) and (4.9.29). Thus, it remains to determine $\mathbf{A}(R, S)$, $\mathbf{C}(R, S)$, and $\mathbf{D}(R, S)$. Using (4.9.28), we obtain simple expressions for $\mathbf{DB}^{-1}$ and $\mathbf{B}^{-1}\mathbf{A}$:

$$
\begin{aligned}
(\mathbf{DB}^{-1})_{IJ} &= [\partial^2 T^\Sigma / \partial z_I(R') \partial z_J(R')]_{R'=R, S'=S}, \\
(\mathbf{B}^{-1}\mathbf{A})_{IJ} &= [\partial^2 T^\Sigma / \partial z_I(S') \partial z_J(S')]_{R'=R, S'=S}.
\end{aligned}
\tag{4.10.51}
$$

It follows from (4.10.51) that the components of $\mathbf{DB}^{-1}$ represent the second derivatives of the travel-time field along the posterior surface Σ_p, due to a point source situated on the anterior surface at S, and the components of $\mathbf{B}^{-1}\mathbf{A}$ represent the second derivatives of the travel-time field along the anterior surface Σ_a, due to a point source situated on the posterior surface at R. Because $\mathbf{B}(R, S)$ is known, (4.10.51) can be used to determine $\mathbf{A}(R, S)$ and $\mathbf{D}(R, S)$: $\mathbf{A} = \mathbf{B}(\mathbf{B}^{-1}\mathbf{A})$ and $\mathbf{D} = (\mathbf{DB}^{-1})\mathbf{B}$. Finally, $\mathbf{C}$ can be determined from the symplectic relation $\mathbf{D}^T\mathbf{A} - \mathbf{B}^T\mathbf{C} = \mathbf{I}$,

$$
\mathbf{C} = \mathbf{B}^{-1T}(\mathbf{D}^T\mathbf{A} - \mathbf{I}).
\tag{4.10.52}
$$

The derived relations for $\mathbf{A}$, $\mathbf{B}$, $\mathbf{C}$ and $\mathbf{D}$ represent the complete solution of the problem.

From the surface-to-surface propagator matrix $\mathbf{T}(R, S)$, represented by $\mathbf{A}(R, S)$, $\mathbf{B}(R, S)$, $\mathbf{C}(R, S)$, and $\mathbf{D}(R, S)$, it is possible to determine the 4×4 ray-centered propagator matrix $\mathbf{\Pi}(R, S)$, represented by $\mathbf{Q}_1(R, S)$, $\mathbf{Q}_2(R, S)$, $\mathbf{P}_1(R, S)$, and $\mathbf{P}_2(R, S)$. The matrices $\mathbf{Q}_2(R, S)$, $\mathbf{P}_2(R, S)$, and $\mathbf{Q}_1(R, S)$ are obtained from (4.4.97), and $\mathbf{P}_1(R, S)$ is obtained from the symplectic relation $\mathbf{P}_1 = \mathbf{Q}_2^{-1T}(\mathbf{P}_2^T\mathbf{Q}_1 - \mathbf{I})$.

Thus, the 4×4 propagator matrices can be determined from three 2×2 matrices of second derivatives of the travel-time field: one along the anterior surface, one along the posterior surface, and one mixed. The matrices of second derivatives of the travel-time field can be found from observed data. First, it is possible to determine them by numerical differentiation of observed travel-times. Second, the parabolic travel-time approximation can be replaced by the hyperbolic travel-time approximation (see (4.6.26)), and an alternative of the $T^2 - X^2$ method can be applied. Note that the hyperbolic approximation for the travel-time $T^\Sigma(R', S')$, given by (4.9.28), in surface-to-surface ray tracing reads

$$
\begin{aligned}
(T^\Sigma(R', S'))^2 &\doteq \left[T^\Sigma(R, S) + d\mathbf{z}^T(R')\mathbf{p}^{(z)}(R) - d\mathbf{z}^T(S')\mathbf{p}^{(z)}(S)\right]^2 \\
&\quad + T^\Sigma(R, S)[d\mathbf{z}^T(R')\mathbf{DB}^{-1}d\mathbf{z}(R') + d\mathbf{z}^T(S')\mathbf{B}^{-1}\mathbf{A}d\mathbf{z}(S') \\
&\quad - 2d\mathbf{z}^T(S')\mathbf{B}^{-1}d\mathbf{z}(R')].
\end{aligned}
\tag{4.10.53}
$$

4.10.7 Exponentially Increasing Geometrical Spreading. Chaotic Behavior of Rays

In a homogeneous medium, the geometrical spreading increases linearly with the increasing length of the ray l; see (4.10.18). In inhomogeneous media, the behavior of the geometrical spreading along the ray is more complex. In certain cases, the average geometrical spreading increases exponentially with increasing length of the ray. This increase occurs mainly in laterally varying 2-D and 3-D structures in which the heterogeneities exceed a certain degree. Such rays often exhibit chaotic behavior.

The chaotic behavior of rays is mostly characterized by their *extremely strong sensitivity* to the initial ray conditions, for example, to the ray parameters. The chaotic rays, with the same initial point and with a very small difference in ray parameters, tend to *diverge exponentially from each other* with the increasing length of the ray.

The chaotic behavior of rays introduces certain fundamental limitations on the feasibility of ray-theory computations. Let us assume that computers with a fixed length of the word are used. Then, the two-point ray tracing cannot be performed with a required accuracy if the length of rays under consideration exceeds some limit (predictability horizon). Similarly, sufficiently accurate interpolations within ray tubes cannot be performed if the rays are long because the ray tubes are extremely broad. The next disadvantage is that the number of multiple rays, corresponding to one elementary wave, arriving at the receiver, increases strongly with increasing distance between the source and receiver, measured along rays. Consequently, the complete system of multiple two-point rays corresponding to the elementary wave under consideration cannot be efficiently calculated at the receiver.

There are many approaches to the investigation of the chaotic behavior of rays. The ideas of these approaches have been mostly taken from the investigation of chaos in dynamical systems in physics. The exponential divergence of nearby rays in the phase space has been often quantified by the so-called *Lyapunov exponents*. The Lyapunov exponents can be defined in various ways. One of definitions is based on the results of dynamic ray tracing along the ray, particularly on the 4×4 propagator matrix $\mathbf{\Pi}(T, T_0)$; see Section 4.3.3. The two *positive Lyapunov exponents* l_I $(I = 1, 2)$ are then defined by the relation

$$l_I = \limsup_{T \to \infty}[(T - T_0)^{-1} \ln |\mu_I(T, T_0)|] \tag{4.10.54}$$

(Klimeš 1999a; Matyska 1999). Here the limit is computed along the ray, T being the time variable along the ray. Further, μ_1 and μ_2 represent the eigenvalues of the 4×4 propagator matrix $\mathbf{\Pi}$, with the largest absolute values. It is sufficient to study only the positive Lyapunov exponents l_1 and l_2 corresponding to the two largest eigenvalues μ_1 and μ_2 because μ_3 and μ_4 would yield the negative Lyapunov exponents of the same absolute value as l_2 and l_1. This result follows immediately from the property (4.3.19) of the eigenvalues of the propagator matrix $\mathbf{\Pi}$. Equation (4.10.54) can be also modified for any other monotonic variable along Ω.

The definition (4.10.54) is very general and can be applied to any ray Ω, situated in an arbitrary 3-D laterally varying layered structure. If $\mu_I(T)$ increases with T only slowly, (4.10.54) yields $l_I = 0$ (nonchaotic ray). Only if $\mu_I(T)$ depends exponentially on T are l_I nonvanishing. Consequently, l_I represent measures of the exponential deviation of rays. The main problem in the application of Equation (4.10.54) in realistic models of a finite size is that the limit $T \to \infty$ is not attainable. Actually, in realistic models, it would be necessary to consider some average values of $\ln |\mu_I|$ and perform some extrapolations.

Chaotic behavior of rays has played a very important role in long-range acoustic wave propagation in underwater acoustics. The main subject of interest has been the behavior of acoustic wavefields in ocean waveguides with smooth lateral heterogeneities at long-range distances. The basic principles of the ray chaos in underwater acoustics are described in detail in Palmer et al. (1988), Brown, Tappert, and Goni (1991), Brown et al. (1991), Smith, Brown, and Tappert (1992), and Abdullaev (1993). These references may serve as a good introduction to the subject. See also Tappert and Tang (1996), Mazur and Gilbert (1997), Jiang, Pitts, and Greenleaf (1997), and other references given there.

In seismology, the investigation of chaotic behavior of seismic rays has started recently. Keers, Dahlen, and Nolet (1997) investigated a chaotic ray behavior in models with laterally varying interfaces, particularly in the Earth's crust with the undulating Mohorovičić discontinuity. They found that the chaotic behavior due to laterally varying interfaces is very pronounced at large epicentral distances. For Lyapunov exponents in randomly layered media, see Scales and Van Vleck (1997). It is obvious that the chaotic behavior of seismic rays will also play an important role in seismic exploration for oil. An attempt to quantify the exponential divergence of rays with respect to the complexity of the model and to formulate explicit criteria enabling to construct models suitable for ray tracing was made by Klimeš (1999a). This area is open for further research.

The rays may exhibit chaotic behavior in both deterministic and stochastic (random) environments. In both of these types of media, the exponential increase of geometrical spreading due to chaotic behavior of rays competes with a large spreading due to diffractions and scattering. In general, the waves arriving as first arrivals in media, in which heterogeneities exceed certain degree, may be of diffractive character and need not be obtained by regular ray tracing (Wielandt 1987). For a discussion of wave propagation and ray tracing in random media, see Chernov (1960), Wu and Aki (1985), Ojo and Mereu (1986), Rytov, Kravtsov, and Tatarskii (1987), Müller, Roth, and Korn (1992), Roth, Müller, and Snieder (1993), Ryzhik, Papanicolaou, and Keller (1996), Witte, Roth, and Müller (1996), and Samuelides (1998), among others. Similarly, as in deterministic models, the subject of ray tracing and travel-time computation in random media requires further investigation.

4.11 Fresnel Volumes

Fresnel volumes were introduced in Section 3.1.6; see also (3.1.46) and Figure 3.2. Let us consider a point source of a monochromatic wave (of frequency f) situated at S, a receiver at R, and ray Ω connecting S and R. From the physical point of view, the Fresnel volume (3.1.46) represents the frequency-dependent spatial vicinity of ray Ω, which actually influences the wavefield at receiver R. The cross section of the Fresnel volume by a plane perpendicular to Ω is called the *Fresnel zone*, and the section of the Fresnel volume by interface Σ is called the *interface Fresnel zone*.

Equation (3.1.46) defines the monochromatic Fresnel volume. The seismic wavefield, however, is not monochromatic, but transient. Seismic signals are usually represented by short wavelets, the Fourier spectrum of which contains many frequencies. For narrow-band signals, the monochromatic Fresnel volumes, constructed for the prevailing frequency of the signal, may represent a good approximation of the Fresnel volume relevant to the signal. It would otherwise be necessary to consider the Fresnel volumes for several frequencies. See also Knapp (1991) and Brühl, Vermeer, and Kiehn (1996).

Recently, Fresnel volumes and Fresnel zones have found many applications in seismology and in seismic exploration. Traditionally, they have played an important role in the investigation of the resolution of seismic methods (Sheriff and Geldard 1982; Sheriff 1989; Lindsey 1989; Thore and Juliard 1999). Fresnel volumes have also been used to study the accuracy of the ray method (Kravtsov and Orlov 1980; Ben-Menahem and Beydoun 1985; Beydoun and Ben-Menahem 1985; Kravtsov 1988). They have been applied in tomographic studies and in other methods of inversion of seismic data (Yomogida 1992; Vasco, Peterson, and Majer 1995; Gudmundsson 1996; Schleicher et al. 1997; Pulliam and Snieder 1998). It can be expected that Fresnel volumes will find many other applications in both forward and inverse seismic methods.

Two methods have been proposed to compute Fresnel volumes in complex 2-D and 3-D layered structures.

The first method, called *Fresnel volume ray tracing*, proposed by Červený and Soares (1992), is based on the paraxial ray approximation. It consists in standard numerical ray tracing, supplemented by dynamic ray tracing. The dynamic ray tracing is used to compute the ray propagator matrix, which is needed to evaluate the Fresnel volume. Because dynamic ray tracing and computing the ray propagator matrix are standard procedures in most ray tracing packages, Fresnel volume ray tracing is simple to program and is numerically very efficient. See Section 4.11.2.

The second method is based on network ray tracing or on any other method of grid travel-time computation; see Kvasnička and Červený (1994) and Section 4.11.3.

Both of these methods have certain advantages and disadvantages. They cannot be applied universally, but only in certain situations. For a detailed comparison, see Section 4.11.4.

In some simple cases, the Fresnel volumes may be calculated analytically. Such analytical results are useful in various applications. Moreover, the analytical results offer a deeper insight into the properties of Fresnel volumes and Fresnel zones. See Section 4.11.1.

Fresnel volumes can be constructed not only for a point source, but also for waves generated at an initial surface. As an example, see (3.1.52) for the radius of the Fresnel zone corresponding to a plane wave (not a point source) at S. To shorten the treatment, this case is not discussed here. The interested reader can find the relevant equations in Kravtsov and Orlov (1980) and Červený (1987b).

Although this chapter is devoted to paraxial ray methods and to dynamic ray tracing, we describe here even computation of Fresnel volumes based on other methods. See, for example, Section 4.11.3 devoted to network ray tracing computations. This is done for completeness.

4.11.1 Analytical Expressions for Fresnel Volumes and Fresnel Zones

For simple structures such as a plane structural interface between two homogeneous half-spaces, the expressions for Fresnel volumes and Fresnel zones of certain important elementary waves can be found analytically. In some cases, we are even able to find exact expressions for Fresnel volumes defined by (3.1.46). We shall present here, without detailed derivation, several important relations that may be useful in applications. For a detailed derivation of all equations, many examples, and the physical discussion of the results, see Kvasnička and Červený (1996).

1. UNCONVERTED REFLECTED WAVES

Let us consider an unconverted reflected wave at a plane interface Σ between two homogeneous halfspaces. Let source S and receiver R be situated at distances h_S and h_R from Σ. Denote the velocity in the first halfspace (containing S and R) by V_1 and in the second halfspace by V_2. We also denote by i_S the angle of incidence, $g = \tan i_S$, and by Q the point of incidence of ray Ω on Σ. The intersection of the Fresnel volume of the reflected wave with interface Σ is referred to as the *interface Fresnel zone*. In our case, the interface Fresnel zone is an ellipse, with half-axes $r^{\parallel}$ and $r^{\perp}$. The *in-plane half-axis $r^{\parallel}$* corresponds to the plane of incidence, and the *transverse half-axis $r^{\perp}$* corresponds to the direction perpendicular to the plane of incidence.

A simple geometrical treatment yields the following expressions for $r^\parallel$ and $r^\perp$:

$$r^\parallel = b \frac{\sqrt{1+g^2}}{1+g^2 v^2} \left(1 - \frac{x_0^2}{a^2} + g^2 v^2 \right)^{1/2},$$

$$r^\perp = b \frac{1}{\sqrt{1+g^2 v^2}} \left(1 - \frac{x_0^2}{a^2} + g^2 v^2 \right)^{1/2};$$

$$(4.11.1)$$

see Kvasnička and Červený (1996). Here l is the length of the ray of the wave reflected from S to R, $l = \overline{SQ} + \overline{QR}$. The quantities a and b represent the half-axes of the rotational ellipsoidal Fresnel volume of the direct wave in a homogeneous medium with velocity V_1, assuming the distance between the source and receiver is l. They are given by expressions (3.1.49). In a high-frequency approximation $a = \frac{1}{2} l$ and $b = \frac{1}{2} f^{-1/2} (V_1 l)^{1/2}$. The quantity v denotes "the fatness ratio," $v = b/a$. In a high-frequency approximation, $v = f^{-1/2} (V_1/l)^{1/2}$. Finally, x_0 is the distance of the point of incidence Q from point O, situated on ray Ω in the middle between S and R (measured along Ω). For $h_S = h_R, x_0 = 0$. It may be useful to express l and x_0 in terms of more practical quantities h_S, h_R, and g:

$$x_0 = \frac{1}{2}(h_R - h_S)\sqrt{1+g^2}, \qquad l = (h_R + h_S)\sqrt{1+g^2}. \qquad (4.11.2)$$

Note that $\sqrt{1+g^2} = 1/\cos i_S$.

The accuracy of Equations (4.11.1) is very high. For $V_1 > V_2$, (4.11.1) are quite exact. For $V_1 < V_2$, they may lose some accuracy in the critical and overcritical regions, where the reflected waves are contaminated by head waves. Equations (4.11.1) are very general and can be simplified in many ways, for example, for $h_S = h_R$ (so that $x_0 = 0$), for normal incidence $i_S = 0$ (so that $g = 0$), and for the high-frequency approximation. The simplest and most useful is the high-frequency approximation, valid for $|x_0| \ll l$. The terms with v can be neglected in this case. If we also use $b = \frac{1}{2} f^{-1/2}(V_1 l)^{1/2}$ and (4.11.2) for l, we obtain

$$r^\parallel \doteq f^{-1/2} \left(\frac{V_1 h_R h_S}{(h_R + h_S)\cos^3 i_S} \right)^{1/2}, \qquad r^\perp \doteq f^{-1/2} \left(\frac{V_1 h_R h_S}{(h_R + h_S)\cos i_S} \right)^{1/2}.$$

$$(4.11.3)$$

The Fresnel volume of the reflected wave also penetrates into the second medium, below Σ. Let us denote the *maximum penetration distance* D (see Figure 4.17). Geometrical considerations yield an approximate equation, valid for modest angles of incidence:

$$D = \frac{1}{4} f^{-1} V_2 \left(1 - V_2^2 \sin^2 i_S / V_1^2 \right)^{-1/2}. \qquad (4.11.4)$$

In general, the center of the interface Fresnel ellipse is not situated at the point of incidence Q but is shifted outside the ray along the intersection of interface Σ with the plane of incidence. This *off-ray shift* d is given by the relation

$$d = |x_0 g| v^2 \sqrt{1+g^2} / (1+g^2 v^2). \qquad (4.11.5)$$

The off-ray shift d vanishes for $h_S = h_R$ (so that $x_0 = 0$), for $i_S = 0$ (so that $g = 0$), and in the high-frequency approximation (so that $v \sim f^{-1/2}$ and $d \sim f^{-1}$). The off-ray shift can be simply understood geometrically, considering the example of the unconverted reflected wave at a plane interface between two homogeneous halfspaces. The interface Fresnel zone is then represented by the intersection of the interface with the Fresnel ellipsoid, with foci situated at the receiver and at the image source. The intersection of an ellipsoid with a plane is always an ellipse, but the center of the ellipse is not necessarily situated on the axis of the ellipsoid (representing the ray in our case).

2. TRANSMITTED WAVES

We shall now consider receiver R situated in the second medium, again at distance h_R from Σ, and denote the angle of transmission i_R. In this case, the exact expressions for the half-axes of the interface Fresnel zone cannot be found. We shall present only the high-frequency approximation, valid for modest angles of incidence:

$$r^{\|} = f^{-1/2} \left(\frac{\cos^3 i_S}{V_S h_S} + \frac{\cos^3 i_R}{V_R h_R} \right)^{-1/2}, \qquad r^{\perp} = f^{-1/2} \left(\frac{\cos i_S}{V_S h_S} + \frac{\cos i_R}{V_R h_R} \right)^{-1/2}.$$

$$(4.11.6)$$

Here $V_S = V_1$ and $V_R = V_2$. For $V_S = V_R$, (4.11.6) yields (4.11.3). If $V_R > V_S$, (4.11.6) fails for angles of incidence i_S close to the critical angle of incidence i_S^*, given by relation $i_S^* = \arcsin(V_S/V_R)$. The interface Fresnel zone corresponding to i_S close to i_S^* may be very extensive. The relevant approximate expressions can be found in Kvasnička and Červený (1996). This reference also gives a detailed treatment of the Fresnel volumes of head waves.

3. CONVERTED REFLECTED AND TRANSMITTED WAVES

Relations (4.11.6) can again be used. Velocities V_S and V_R should be taken according to the type of converted wave under consideration: V_S correspond to the velocity at the source and V_R to the velocity at the receiver.

For the interface Fresnel zones of converted waves, see also Eaton, Stewart, and Harrison (1991) and Hubral et al. (1993).

4.11.2 Paraxial Fresnel Volumes. Fresnel Volume Ray Tracing

A simple and efficient algorithm for evaluating paraxial Fresnel volumes along the ray in a 3-D laterally varying layered structure, called *Fresnel volume ray tracing*, was proposed by Červený and Soares (1992). It is based on dynamic ray tracing and computing the ray propagator matrix along the ray. We shall consider ray Ω and two points S and R on Ω. Assume that point S represents a point source and point R, the receiver. Ray Ω corresponds to an arbitrary multiply reflected and converted elementary wave propagating in a 3-D layered laterally varying structure. Let us select an arbitrary point F situated in the paraxial vicinity of Ω. Point F then belongs to the Fresnel volume for frequency f if it satisfies (3.1.46). See also Figure 3.2.

We introduce plane $\Sigma_F^{\perp}$ passing through F, perpendicular to Ω and intersecting ray Ω at point O_F. Then, using (4.6.28) and (3.1.46) yields the equation of the boundary of the paraxial Fresnel volume on plane $\Sigma_F^{\perp}$ for frequency f:

$$|\mathbf{q}^T(F)[\mathbf{M}(O_F, S) - \mathbf{M}(O_F, R)]\mathbf{q}(F)| = f^{-1}. \qquad (4.11.7)$$

Here $\mathbf{q}(F) = (q_1(F), q_2(F))^T$, where $q_I(F)$ are ray-centered coordinates of point F. Equation (4.11.7) also represents the boundary of the *paraxial Fresnel zone* at point O_F of Ω.

Let us emphasize that $\mathbf{M}(O_F, S)$ corresponds to a point source situated at S, and $\mathbf{M}(O_F, R)$ corresponds to a point source situated at R. Thus, the computation of (4.11.7) would require two dynamic ray tracings: one from S and the other from R. It is, however, possible to obtain all the quantities in (4.11.7) by just one dynamic ray tracing, from S to R.

If we use the chain rule (4.3.20) and the equation for the inverse of the ray propagator matrix (4.3.26), we obtain

$$
\mathbf{M}(O_F, S) = \mathbf{P}_2(O_F, S)\mathbf{Q}_2^{-1}(O_F, S),
$$
$$
\mathbf{M}(O_F, R) = \left[-\mathbf{P}_1(O_F, S)\mathbf{Q}_2^T(R, S) + \mathbf{P}_2(O_F, S)\mathbf{Q}_1^T(R, S) \right]
$$
$$
\times \left[-\mathbf{Q}_1(O_F, S)\mathbf{Q}_2^T(R, S) + \mathbf{Q}_2(O_F, S)\mathbf{Q}_1^T(R, S) \right]^{-1}.
$$

These equations contain only the submatrices of ray propagator matrices $\mathbf{\Pi}(R, S)$ and $\mathbf{\Pi}(O_F, S)$. In the plane $\Sigma_F^\perp$, the 2×2 matrix

$$
\mathbf{M}^F(O_F; R, S) = \mathbf{M}(O_F, S) - \mathbf{M}(O_F, R) \tag{4.11.8}
$$

represents the *Fresnel zone matrix* (see (4.4.107) for $\mathbf{G} = \mathbf{I}$). We denote the eigenvalues of the Fresnel zone matrix by $M_1(O_F)$ and $M_2(O_F)$.

It is simple to see that (4.11.7) represents a quadratic curve in plane $\Sigma_F^\perp$. For $M_1(O_F) > 0$ and $M_2(O_F) > 0$, the curve is a *Fresnel ellipse* and for $M_1(O_F)M_2(O_F) < 0$ the *Fresnel hyperbola*. We shall discuss here the Fresnel ellipses and comment only briefly on the Fresnel hyperbolas later. The half-axes of the Fresnel ellipse $r_1(O_F)$ and $r_2(O_F)$ are given as

$$
r_1(O_F) = f^{-1/2}[M_1(O_F)]^{-1/2}, \qquad r_2(O_F) = f^{-1/2}[M_2(O_F)]^{-1/2}.
$$
$$
\tag{4.11.9}
$$

For a detailed description of the algorithm of Fresnel volume ray tracing and many numerical examples of paraxial Fresnel volumes, see Červený and Soares (1992). For alternative approaches, see Gelchinsky (1985), Hubral et al. (1993), and Pulliam and Snieder (1998).

The Fresnel zone, as introduced here, specifies the intersection of the Fresnel volume with plane $\Sigma_F^\perp$ perpendicular to ray Ω. It is often valuable to study the intersection of the Fresnel volume with some general surface Σ crossed by ray Ω or with structural interface Σ. We then speak of the *interface Fresnel zone* of the elementary wave under consideration. The interface Fresnel zone can be obtained by projecting the standard Fresnel zone onto interface Σ, taking into account relations (4.4.40). As in Section 4.4.2, we introduce the local Cartesian coordinate system z_1, z_2, z_3, with its origin at the point of incidence Q on Σ, and with the z_3-axis coinciding with the normal to Σ at Q. Let us now consider any point F situated on structural interface Σ in the vicinity of Q. Point F is situated on the *boundary of the interface Fresnel zone* if its coordinates $z_1(F)$ and $z_2(F)$ satisfy the relation

$$
|\mathbf{z}^T(F)\mathbf{M}^F(Q; R, S)\mathbf{z}(F)| = f^{-1}. \tag{4.11.10}
$$

Here $\mathbf{z}(F) = (z_1(F), z_2(F))^T$, and $\mathbf{M}^F(Q; R, S)$ is the *interface Fresnel zone matrix*. It is given by the relation

$$
\mathbf{M}^F(Q; R, S) = \mathbf{G}(Q)[\mathbf{M}(Q, S) - \mathbf{M}(Q, R)]\mathbf{G}^T(Q). \tag{4.11.11}
$$

The meaning of $\mathbf{G}(Q)$ is the same as in Section 4.4.2; see (4.4.107).

Equation (4.11.11) for the interface Fresnel zone matrix $\mathbf{M}^F(Q; R, S)$ generalizes (4.11.8). If we use a plane surface Σ perpendicular to ray Ω in a smooth medium and choose local Cartesian coordinates $z_1 = q_1$ and $z_2 = q_2$, (4.11.11) reduces to (4.11.8). For this reason, we call the general form of the matrix $\mathbf{M}^F(Q; R, S)$ given by (4.11.11) the *Fresnel zone matrix* and do not emphasize the word *interface*.

Note that the Fresnel zone matrix $\mathbf{M}^F(Q; R, S)$ is continuous across the interface Σ from the point of incidence Q to the R/T point $\tilde{Q}$, $\mathbf{M}^F(\tilde{Q}; R, S)$. The half-axes $r_1(Q)$ and

$r_2(Q)$ of the interface Fresnel zone are given by relations (4.11.9), where $M_1(O_F)$ and $M_2(O_F)$ are substituted by the eigenvalues $M_1(Q)$ and $M_2(Q)$ of the Fresnel zone matrix $\mathbf{M}^F(Q; R, S)$ given by (4.11.11).

It is interesting to realize that the Fresnel zone matrix $\mathbf{M}^F(Q; R, S)$ given by (4.11.11) corresponds fully to the analogous matrix obtained by factorization of the 2×2 matrix $\mathbf{Q}_2(R, S)$ in Section 4.4.8; see (4.4.105), (4.4.107), and (4.4.108). Section 4.4.8 gives several alternative general forms of the Fresnel zone matrix. All these forms can be used to calculate the dimensions of the interface Fresnel zones. It is merely necessary to find the eigenvalues of $\mathbf{M}^F(Q; R, S)$ and to insert them into (4.11.9). The Fresnel zone matrix is also discussed in great detail in Section 4.8.5, where explicit expressions are found for an interface Σ between two media with a constant velocity gradient and for an interface between two homogeneous media. For a plane interface $\Sigma(D_{IJ} = 0)$, Equations (4.11.9) with (4.8.35) immediately yield (4.11.6) for the interface Fresnel zone of transmitted (possibly converted) waves. This equation further yields (4.11.3) for unconverted reflected waves. Because $|\det \mathbf{Q}_2(R, S)|^{1/2}$ equals the relative geometrical spreading $\mathcal{L}(R, S)$ (see (4.10.11)), the Fresnel zone matrix is also directly related to the relative geometrical spreading. This relation was also studied by Sun (1996).

Equations (4.11.7) through (4.11.11) are related to elliptical Fresnel zones. The *hyperbolic Fresnel zones* are more complex, with long hyperbolic tails extending to infinity. Asatryan and Kravtsov (1988) showed that these tails do not influence the wavefield at the receiver significantly. The Fresnel zones are again satisfactorily described by (4.11.9), where $M_1(O_F)$ and $M_2(O_F)$ should be taken in absolute values, and f should be substituted by $2f$. For applications of hyperbolic Fresnel zones, see Neele and Snieder (1992).

4.11.3 Fresnel Volumes of First Arriving Waves

Fresnel volumes can also be computed by network ray tracing (see Kvasnička and Červený 1994) or by any other method of grid travel time computation. The method is fast and efficient. It can, however, be applied only to waves arriving at the receiver as the first arrivals. In addition, it may also be applied to reflected waves from structural interface Σ because the direct wave may be artificially removed from the computations. The reflected waves calculated by network ray tracing, however, are considered in a broader sense than usual. They correspond not only to reflections from the structural interface Σ itself but also from the whole second medium below Σ. Such reflected waves also include head waves (if they exist), waves refracted in the second medium and returning into the first medium, and edge waves generated at the edges of interface Σ, among others. In case of multiple arrivals of the reflected wave, only the travel times of the first arrivals are considered.

The algorithms for the computation of Fresnel volumes of direct waves and of reflected waves, based on network ray tracing, are different. The former requires two network ray tracings, and the latter need four network ray tracings.

a. Let us first consider the **direct wave** from point source S to receiver R. By the direct wave we understand the wave that is not reflected at any interface but that may be transmitted any number of times. We perform two network ray tracings and compute $T(F, S)$ from point S, and $T(F, R)$ from point R for all points F in the grid model. The boundary of the Fresnel volume for frequency f is then formed by points F satisfying the relation

$$T(F, S) + T(F, R) - T(S, R) = \tfrac{1}{2} f^{-1}. \tag{4.11.12}$$

Thus, the boundary of the Fresnel volume of the direct wave can be constructed as an isosurface, along which $T(F, S) + T(F, R) - T(S, R)$ is constant and equals $\frac{1}{2}f$.

b. Let us now consider the **reflected wave in a broader sense**, from interface Σ. We perform the computations in two steps. In the first step, we compute $T(F, S)$ and $T(F, R)$, corresponding to the direct wave, at all points F of the model, including interface Σ. In the second step, interface Σ is assumed to be a secondary source surface, and two network ray tracings are performed, starting from the interface travel times along Σ. The relevant travel times at points F are denoted by $T(F, \Sigma, S)$ and $T(F, \Sigma, R)$. The boundary of the Fresnel volume of reflected waves for frequency f is then formed by points F satisfying the relation

$$\min[T(F, S) + T(F, \Sigma, R)$$
$$- T(S, R); T(F, R) + T(F, \Sigma, S) - T(S, R)] = \tfrac{1}{2}f^{-1}. \quad (4.11.13)$$

Thus, the Fresnel volumes of reflected waves are constructed as an isosurface along which the quantity on the LHS of (4.11.13) is constant and equals $\frac{1}{2}f^{-1}$.

Note that the network ray tracing computation of Fresnel volumes is fully based on the computation of travel times, and the ray does not need to be known. In fact, the method may be used even for the computation of Fresnel volumes of first arriving waves for which standard ray tracing fails (head waves, waves diffracted from edges, diffractions behind smooth objects, and the like). For examples of such computations, see Kvasnička and Červený (1994).

4.11.4 Comparison of Different Methods of Calculating Fresnel Volumes and Fresnel Zones

In 3-D laterally varying layered structures, analytic methods can rarely be used to calculate Fresnel volumes, Fresnel zones, and interface Fresnel zones. At present, there are two basic methods of performing such computations (see Sections 4.11.2 and 4.11.3): (a) Fresnel volume ray tracing and (b) a method based on network ray tracing. There are several important differences between these two methods.

1. Both methods apply to different waves. Fresnel volume ray tracing is applicable only to zero-order ray theory elementary waves, but not to higher-order waves (such as head waves) and to diffracted waves. On the contrary, the method based on network ray tracing is applicable only to waves arriving in the first arrivals, even if these waves do not belong to the category of zero-order ray theory waves; see Section 3.8. The category of first arriving waves, however, may be extended even to reflected waves in a broader sense; see Section 4.11.3.

2. The sensitivity of the method to the existence of structural interfaces situated close to ray Ω, but not touching it, is another difference. Fresnel volume ray tracing is not sensitive to these interfaces at all because dynamic ray tracing is controlled only by the first and second derivatives of velocity *directly along ray* Ω and not by the structure in the vicinity of Ω. Thus, Fresnel volume ray tracing does not yield the correct width of the Fresnel volume in this case. The method based on network ray tracing, however, takes into account the structural interfaces in the vicinity of Ω automatically. See examples in Kvasnička and Červený (1994).

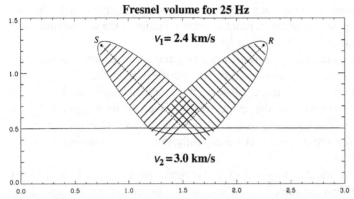

Figure 4.17. The Fresnel volume of a wave reflected at a plane interface between two homogeneous media. Only a section, perpendicular to the interface and passing through source S and receiver R, is displayed. Comparison of three different methods to compute the Fresnel volumes: (a) Analytic computations. (b) Network ray tracing. (c) Fresnel volume ray tracing. The continuous line, representing the boundary of the Fresnel volume, was computed by methods (a) and (b). Both methods gave the same result in this case. The bold lines perpendicular to the ray were computed by method (c). As we can see, Fresnel volume ray tracing mostly yields sufficiently accurate results. For exceptions and for a more detailed discussion, see the text.

3. The final difference involves the accuracy of both methods. The Fresnel volume ray tracing is a high-frequency method and yields sufficiently accurate results only for high-frequencies, for which the whole Fresnel volume is situated in the paraxial vicinity of ray Ω. The accuracy of network ray tracing is independent of frequency (except for numerical errors). Thus, in cases when both methods can be applied, the accuracy of Fresnel volume ray tracing is in general lower. Moreover, Fresnel volume ray tracing yields only quantities that are of the order of $f^{-1/2}$ (half-axes of Fresnel zones and of interface Fresnel zones) and does not yield the quantities that are of the order f^{-1} (the overshooting distance behind the source and receiver, penetration distance D of the Fresnel volume of a reflected wave below a structural interface, off-ray shift d of the center of the interface Fresnel zone). The methods based on the network ray tracing yield both the effects of $f^{-1/2}$ and f^{-1} correctly.

A simple example of the Fresnel volume for frequency $f = 25$ Hz of an unconverted wave reflected from plane interface Σ between two homogeneous halfspaces with velocities $V_1 = 2.4$ km/s and $V_2 = 3$ km/s and for $h_S = h_R = 0.75$ km is shown in Figure 4.17. The distance between S and R equals 1.5 km, and the angle of incidence $i_S = 45^0$. In this simple case, both methods can be used to calculate the Fresnel volume. Moreover, analytic methods can also be used to compute all quantities. The continuous thin line shows the boundary of the Fresnel volume computed by network ray tracing. The bold lines perpendicular to the ray show the half-axes $r^{\parallel}$ of the Fresnel ellipses computed by Fresnel volume ray tracing. In this case, the network ray tracing yields exact results so that it may serve to appreciate the accuracy of the Fresnel volume ray tracing.

As we can see in Figure 4.17, Fresnel volume ray tracing mostly yields sufficiently accurate results, with several exceptions.

a. Fresnel volume ray tracing does not give the overshooting distance Δ at S and R, which equals 24 m in our case; see (3.1.50). Close to S and R, slightly smaller Fresnel zones are obtained in comparison with exact values. Similar effects would be observed at caustic points. The paraxial Fresnel volume, computed by Fresnel volume ray tracing, degenerates to a point at a caustic point of the second order and to a line at a caustic point of the first order.

b. The Fresnel zones computed by Fresnel volume ray tracing in planes perpendicular to Ω are not sensitive to the existence of the interface Σ in the vicinity of Ω. Formally, they intersect interface Σ close to the point of incidence Q. This is, however, only a formal problem that can be easily solved by simple modifications in plotting routines. The boundary of the Fresnel volume and its intersection with the interface Σ is displayed correctly. Consequently, the size of the interface Fresnel zone is quite obvious.

c. Fresnel volume ray tracing does not give the actual penetration of the Fresnel volume into the second medium. This is the most serious problem in the application of Fresnel volume ray tracing. The maximum penetration distance D computed by network ray tracing is close to 54 m. This is slightly higher than the 48 m obtained from approximate formula (4.11.4). We must, however, take into account that (4.11.4) is valid only for modest angles of incidence.

Note that both methods yield the same value of the in-plane half-axis $r^{\|}$ of the interface Fresnel zone, which also agrees with the approximate formula (4.11.6). In all cases, $r^{\|} \doteq$ 320 m is obtained. Because $h_S = h_R$, the off-ray shift d in Figure 4.17 vanishes.

4.12 Phase Shift Due to Caustics. KMAH Index

The KMAH index $k(B, A)$ of ray trajectory Ω from A to B and the relevant phase shift due to caustics $T^c(B, A)$,

$$T^c(B, A) = -\tfrac{1}{2}\pi k(B, A), \tag{4.12.1}$$

are introduced in Sections 3.10.5 and 3.10.6 in terms of the 3×3 transformation matrix $\hat{\mathbf{Q}}^{(x)}$ from ray coordinates $\gamma_1, \gamma_2, \gamma_3$ to Cartesian coordinates x_1, x_2, x_3. By *caustic points*, we mean the points along Ω at which $\det \hat{\mathbf{Q}}^{(x)} = 0$. In isotropic media, we distinguish between the *caustic points of the first order*, at which rank $(\hat{\mathbf{Q}}^{(x)}) = 2$, and *caustic points of the second order*, at which rank $(\hat{\mathbf{Q}}^{(x)}) = 1$; see (3.10.48) and (3.10.49). The KMAH index of the ray trajectory Ω from A to B in isotropic media, $k(B, A)$, then equals the number of caustic points along Ω from A to B, the caustic points of the second order being considered twice. Alternatively, we can say that the KMAH index increases by 1 when the wave passes through a caustic point of the first order and by 2 when it passes through a caustic point of the second order. For a detailed derivation, both for isotropic and anisotropic media, see Section 5.8.8.

Instead of the 3×3 matrix $\hat{\mathbf{Q}}^{(x)}$, introduced in Section 3.10 and related to Cartesian coordinates, also the matrices $\hat{\mathbf{Q}}$ and $\mathbf{Q}$, introduced in Section 4.1 and related to ray-centered coordinates may be used to define the KMAH index.

If we use the 3×3 transformation matrix $\hat{\mathbf{Q}}$ from ray coordinates $\gamma_1, \gamma_2, \gamma_3$ to ray-centered coordinates q_1, q_2, q_3, the specification of the caustic points of the first and the second order remains the same as for $\hat{\mathbf{Q}}^{(x)}$; we only replace $\hat{\mathbf{Q}}^{(x)}$ by $\hat{\mathbf{Q}}$. This is simple to see from (4.1.39) and (4.1.29), because $\hat{\mathbf{Q}}^{(x)} = \hat{\mathbf{H}}\hat{\mathbf{Q}}$ and $\det \hat{\mathbf{H}} = 1$.

In ray-centered coordinates, it is more common to consider the 2×2 transformation matrix $\mathbf{Q}$ from ray coordinates γ_1, γ_2 to ray-centered coordinates q_1, q_2, instead of $\hat{\mathbf{Q}}$. The

2×2 matrix $\mathbf{Q}$ is computed by dynamic ray tracing along Ω. We can again specify the position of the caustic points along the ray by $\det \mathbf{Q} = 0$. At the caustic point of the first order,

$$\text{rank}(\mathbf{Q}) = 1. \tag{4.12.2}$$

Similarly, at the caustic point of the second order,

$$\text{rank}(\mathbf{Q}) = 0. \tag{4.12.3}$$

The KMAH index $k(B, A)$ depends on the initial values of matrices $\mathbf{Q}$ and $\mathbf{P}$ at point A. It is quite obvious that the number and position of the caustic points along ray Ω between A and B may be quite different for the normalized point source initial conditions ($\mathbf{Q}(A) = \mathbf{0}$, $\mathbf{P}(A) = \mathbf{I}$) and for the normalized plane wave initial conditions ($\mathbf{Q}(A) = \mathbf{I}$, $\mathbf{P}(A) = \mathbf{0}$). In practical applications, the most important role is played by the point source initial condition because they are required in Green function computations. For this reason, we shall discuss them in greater detail.

Let us consider a *point source at S* on Ω and two points A and B situated on the same ray Ω. Assume that the point source is not situated inside interval A, B. We introduce the KMAH index $k(B, A; S)$ of the ray trajectory Ω from A to B, for the point source situated at S. As in the previous cases, $k(B, A; S)$ equals the number of caustic points along Ω from A to B, with the caustic points of the second order being considered twice. If we take into account the relation $\mathbf{Q}(R) = \mathbf{Q}_2(R, S)\mathbf{P}(S)$, valid for the point source situated at S (see (4.6.4)), we can see that KMAH index $k(B, A; S)$ may also be determined by looking for zeros of the 2×2 matrix $\mathbf{Q}_2(R, S)$ along ray Ω.

It is possible to see that KMAH index $k(R, S; S)$ is reciprocal in the following sense:

$$k(R, S; S) = k(R, S; R). \tag{4.12.4}$$

The reciprocity relation (4.12.4) was proved by Goldin (1991) and by Klimeš (1997c). It can be understood if we take into account the properties of the inverse of the ray propagator matrix (4.3.26), particularly the property $\mathbf{Q}_2(R, S) = -\mathbf{Q}_2^T(S, R)$ (see (4.3.27)). Reciprocity relation (4.12.4) does not imply that the caustic points are situated at the same points of ray Ω if the source is situated at S and at R. The caustic points are, in general, situated at different points of the ray in both cases, but (4.12.4) remains valid. The reciprocity relation (4.12.4) remains valid even in anisotropic media; see Klimeš (1997c).

We remind the reader that the minus sign in (4.12.1) corresponds to the plus sign in the expression for the analytical signal $F(\zeta)$ under consideration, $F(\zeta) = x(\zeta) + \mathrm{i}g(\zeta)$. For time-harmonic waves, it corresponds to the exponential time factor $\exp[-\mathrm{i}\omega(t - T)]$ with positive ω. If we use $F(\zeta) = x(\zeta) - \mathrm{i}g(\zeta)$, the minus sign in (4.12.1) should be replaced by the plus sign. For time-harmonic waves, the minus sign in (4.12.1) should be changed to a plus sign if we use exponential time factor $\exp[+\mathrm{i}\omega(t - T)]$.

The phase shift due to caustics plays an important role in the computation of ray synthetic seismograms corresponding to the individual elementary waves. When the wave passes through a caustic point of the first order, the shape of the signal changes to its Hilbert transform. Similarly, if it passes through the caustic point of the second order, the seismic signal changes its sign. Thus, it is necessary to discuss the procedures of determining the phase shift due to the caustic (or, alternatively, the KMAH index) along ray Ω, starting from S to R. This is the subject of Sections 4.12.1 and 4.12.2.

4.12.1 Determination of the KMAH Index by Dynamic Ray Tracing

Let us consider ray Ω determined by solving the ray tracing system numerically. We assume that dynamic ray tracing has also been performed along Ω from S to R and that the 2×2 matrix $\mathbf{Q}$ is known at all points of Ω. To locate the caustic points of the first and second order, it is necessary to find the points at which $\det \mathbf{Q} = 0$, satisfying (4.12.2) or (4.12.3).

We shall now consider two consecutive points O^1 and O^2 on ray Ω, at which the 2×2 matrix $\mathbf{Q}$ takes values $\mathbf{Q}^1$ and $\mathbf{Q}^2$, so that $\mathbf{Q}^1 = \mathbf{Q}(O^1)$ and $\mathbf{Q}^2 = \mathbf{Q}(O^2)$. We wish to determine whether there is a caustic point on Ω between O^1 and O^2. Assume that $\det \mathbf{Q}^1 \neq 0$ and $\det \mathbf{Q}^2 \neq 0$ and that the number of caustic points between O^1 and O^2 does not exceed one. We can then use the following two criteria.

a. If

$$\det \mathbf{Q}^1 \det \mathbf{Q}^2 < 0, \tag{4.12.5}$$

there is a *caustic point of the first order* (line caustic) between O^1 and O^2.

b. Otherwise, if

$$\operatorname{tr}[\mathbf{Q}^1 (\mathbf{Q}^2)^{-1}] \det \mathbf{Q}^1 \det \mathbf{Q}^2 < 0, \tag{4.12.6}$$

there is a *caustic point of the second order* (focus) between O^1 and O^2. Criterion (4.12.6) can also be written in the following form, which is more useful in programming:

$$\left(Q_{11}^1 Q_{22}^2 - Q_{12}^1 Q_{21}^2 + Q_{22}^1 Q_{11}^2 - Q_{21}^1 Q_{12}^2 \right) \det \mathbf{Q}^1 < 0. \tag{4.12.7}$$

Note that there is no caustic point between O^1 and O^2 if $\det \mathbf{Q}^1 \det \mathbf{Q}^2 > 0$ and $\operatorname{tr}[\mathbf{Q}^1 (\mathbf{Q}^2)^{-1}] \det \mathbf{Q}^1 \det \mathbf{Q}^2 > 0$. Criterion (4.12.7) was proposed by L. Klimeš (see Červený, Klimeš, and Pšenčík 1988b).

For $\det \mathbf{Q}^1 = 0$ and/or $\det \mathbf{Q}^2 = 0$, the criteria should be modified. The caustic point is then situated directly at O^1 and/or at O^2.

Because the step along ray $\overline{O^1 O^2}$ is always finite in numerical ray tracing, the above discrete algorithms (4.12.5) and (4.12.6) cannot be, in principle, quite safe and may fail in exceptional cases due to numerical reasons and/or due to a greater number of caustic points between O^1 and O^2. It may be useful to decrease the computation step in dangerous regions where the values of $\det \mathbf{Q}^1$ and $\det \mathbf{Q}^2$ are very small.

In a layered medium, the KMAH index is calculated along ray segments between structural interfaces. Let us consider ray Ω in a layered medium, with N R/T points between S and R. Ray Ω consists of $N + 1$ segments. The points of incidence are denoted $Q_1, Q_2, \ldots, Q_N$, and the relevant R/T points, $\tilde{Q}_1, \tilde{Q}_2, \ldots, \tilde{Q}_N$. Then

$$k(R, S) = \sum_{k=1}^{N+1} k(Q_k, \tilde{Q}_{k-1}), \tag{4.12.8}$$

where $k(Q_k, \tilde{Q}_{k-1})$ is the number of caustic points corresponding to the elementary wave under consideration on ray Ω between $\tilde{Q}_{k-1}$ and Q_k (kth segment of the ray). In (4.12.8), we have used $S \equiv \tilde{Q}_0$ and $R \equiv Q_{N+1}$. Similarly, for the phase shift due to caustics, we obtain from (4.12.8) and (4.12.1),

$$T^c(R, S) = \sum_{k=1}^{N+1} T^c(Q_k, \tilde{Q}_{k-1}), \tag{4.12.9}$$

where

$$T^c(Q_k, \tilde{Q}_{k-1}) = -\tfrac{1}{2}\pi k(Q_k, \tilde{Q}_{k-1}). \qquad (4.12.10)$$

When a caustic point is situated directly at an interface at point Q_k, it should be taken only once in sum (4.12.8), either in $k(Q_k, \tilde{Q}_{k-1})$ or in $k(Q_{k+1}, \tilde{Q}_k)$.

4.12.2 Decomposition of the KMAH Index

As we know from Section 4.4.8, the 2×2 matrix $\mathbf{Q}_2(R, S)$ can be factorized; see (4.4.109). The factorization of $\mathbf{Q}_2(R, S)$ also implies the factorization of the relative geometrical spreading; see Section 4.10.4. The equations of Section 4.4.8 may also be used to decompose the KMAH index and the phase shift due to caustics. The decomposition equations play an important role mainly in investigating the wavefields generated by point sources.

We shall consider ray Ω of a wave reflected/transmitted from interface Σ in a laterally varying 3-D structure and two points S (point source) and R (receiver) situated on Ω. In addition, we shall also consider the point Q on Ω at which ray Ω is incident at interface Σ and the point $\tilde{Q}$, situated at the same position as Q, but corresponding to the relevant generated wave. We shall call the segment of the ray between S and Q the incident segment and the segment between $\tilde{Q}$ and R the R/T segment. Surface Σ may represent a structural interface as well as an arbitrary surface crossing the ray Ω in a smooth medium.

Equation (4.4.109) for the factorization of $\mathbf{Q}_2(R, S)$ yields

$$k(R, S; S) = k_1 + k_2 + k^F(Q). \qquad (4.12.11)$$

Here k_1 and k_2 have a standard meaning: $k_1 = k(Q, S; S)$ is the KMAH index corresponding to the incident segment of the ray, assuming a point source at S; $k_2 = k(R, \tilde{Q}; R)$ is the KMAH index corresponding to the R/T segment of the ray, *assuming a point source at* R. Let us emphasize that k_1 and k_2 correspond to point sources at S and R, respectively. Finally, the additional term $k^F(Q)$ is closely related to the interface Fresnel zone matrix $\mathbf{M}^F(Q; R, S)$ given by (4.4.107) or (4.4.108). Note that $k^F(Q) = k^F(\tilde{Q})$.

The relations for $k^F(Q)$ can be derived in different ways. For example, it is possible to derive the relation for $k^F(Q)$ by applying the method of stationary phase (for $\omega \to \infty$) to the Kirchhoff integral. This approach was used by Goldin (1991), Goldin and Piankov (1992), and Schleicher, Tygel, and Hubral (1993). It yields the relation

$$k^F(Q) = \tfrac{1}{2}(2 - \operatorname{Sgn} \mathbf{M}^F(Q; R, S)). \qquad (4.12.12)$$

Here $\operatorname{Sgn} \mathbf{M}^F$ denotes the signature of the 2×2 Fresnel zone matrix. It represents the number of positive eigenvalues of $\mathbf{M}^F$ minus the number of negative eigenvalues. Alternatively, we can also write

$$\operatorname{Sgn} \mathbf{M}^F = \operatorname{sgn} M_1 + \operatorname{sgn} M_2, \qquad (4.12.13)$$

where M_1 and M_2 represent the eigenvalues of the Fresnel zone matrix $\mathbf{M}^F$; see Section 4.11.2.

The decomposition equation (4.12.11) of the KMAH index plays an important role particularly if the geometrical spreading is not evaluated by dynamic ray tracing. Then the relations of Section 4.12.1 cannot be applied. As an example, let us consider models in which the ray tracing is performed analytically or semianalytically or in which the ray tracing is not needed at all (1-D models with an arbitrary velocity-depth distribution and

2-D models composed of homogeneous layers or of layers with constant gradients of velocity, slowness or quadratic slowness, and the like).

Let us explain in greater detail the difference between (4.12.8) and (4.12.11), which may seem to be in contradiction. Equation (4.12.8) does not consider any contribution from interfaces, but (4.12.11) does. To make the comparison simpler, we specify (4.12.8) for the point source at S and for only the two branches of the ray, considered in (4.12.11). (4.12.8) then yields

$$k(R, S; S) = k(R, \tilde{Q}; S) + k(Q, S; S). \qquad (4.12.14)$$

Both of the terms in (4.12.14) correspond to the point source at S, even the R/T segment. Thus, $k(R, \tilde{Q}; S)$ is influenced by the interface and may be nonvanishing even in a homogeneous medium. In (4.12.11), however, R/T segment $k_2 = k(R, \tilde{Q}; R)$ corresponds to the point source at R and is not affected by interface Σ at all. For a homogeneous medium, both $k_2 = k(R, \tilde{Q}; R)$ and $k_1 = k(Q, S; S)$ in (4.12.11) vanish, and (4.12.11) yields $k(R, S; S) = k^F(Q)$. In the same medium, (4.12.14) yields $k(Q, S; S) = 0$ and $k(R, S; S) = k(R, \tilde{Q}; S)$. Thus, in this case, $k(R, \tilde{Q}; S) = k^F(Q)$.

4.13 Dynamic Ray Tracing Along a Planar Ray. 2-D Models

Assume that Ω is a planar ray in an isotropic medium. For simplicity, we assume that the plane $\Sigma^{\|}$, in which ray Ω is situated, corresponds to the $x_1 x_3$-plane of the general Cartesian coordinate system and that the x_2-axis is perpendicular to that plane. It is obvious that the first derivatives of velocity with respect to x_2 must vanish along the whole ray Ω (otherwise ray Ω would leave plane $\Sigma^{\|}$). Similarly, the normals to the interfaces at all points Q_i are situated in the plane $\Sigma^{\|}$ (the x_2-components of all normals vanish).

The case described here also includes a *2-D model* in which the structural parameters do not depend on the x_2-coordinate. Ray Ω with its initial point and initial direction in plane $\Sigma^{\|}$ is then fully confined to plane $\Sigma^{\|}$ and does not deviate from it. Thus, in computing the rays and travel times, the foregoing assumptions are sufficient to obtain planar rays. To perform the dynamic ray tracing and to simplify it, we must make some other assumptions regarding the second derivatives of velocity and the curvature of interfaces. We shall do this later.

Let us now consider two points S and R situated on ray Ω and introduce the initial basis vectors of the ray-centered coordinate system $\vec{e}_1(S)$, $\vec{e}_2(S)$, and $\vec{e}_3(S) = \vec{t}(s)$. We assume that the initial slowness vector is fully situated in plane $\Sigma^{\|}$ so that $e_{32}(S) = 0$. We now take $\vec{e}_2(S)$ to have the direction of the x_2-axis so that $e_{21}(S) = e_{23}(S) = 0$ and $e_{22}(S) = 1$. Because system $\vec{e}_1, \vec{e}_2, \vec{e}_3$ is right-handed, we can put $\vec{e}_1(S) = \vec{e}_2(S) \times \vec{e}_3(S)$. According to Section 4.1.3, vector $\vec{e}_2$ is constant along the whole planar ray Ω in a smooth medium. If we denote any point on ray Ω by R, we can write

$$\begin{aligned} \vec{e}_2(R) &= \vec{e}_2(S), \qquad \vec{e}_3(R) = \vec{t}(R) = V(R)\vec{p}(R), \\ \vec{e}_1(R) &= \vec{e}_2(R) \times \vec{e}_3(R). \end{aligned} \qquad (4.13.1)$$

As usual, $\vec{t}$ denotes the unit vector tangent to the ray. These equations allow us to determine the transformation matrix $\hat{\mathbf{H}}$ from ray-centered to general Cartesian coordinates at any point R of the ray, $H_{ij} = e_{ji}$,

$$\hat{\mathbf{H}}(R) = \begin{pmatrix} H_{11}(R) & 0 & H_{13}(R) \\ 0 & 1 & 0 \\ H_{31}(R) & 0 & H_{33}(R) \end{pmatrix} = \begin{pmatrix} t_3(R) & 0 & t_1(R) \\ 0 & 1 & 0 \\ -t_1(R) & 0 & t_3(R) \end{pmatrix}. \qquad (4.13.2)$$

The last relation in (4.13.2) follows from the orthogonality of vectors $\vec{e}_1$ and $\vec{e}_3$. As we can see from (4.13.2), matrix $\hat{\mathbf{H}}(R)$ can be expressed completely in terms of Cartesian components of the unit vector tangent to the ray, $\vec{e}_3 = \vec{t}$. Basis vectors $\vec{e}_1$ and $\vec{e}_2$ do not need to be determined numerically.

Let us now consider a planar ray that is incident at interface Σ at point Q. We introduce the local Cartesian coordinate system (z_1, z_2, z_3) with its origin at Q and with basis vectors $\vec{i}_1^{(z)}$, $\vec{i}_2^{(z)}$, and $\vec{i}_3^{(z)}$. We wish to use the standard option (4.4.21) for the basis vectors $\vec{i}_1^{(z)}$, $\vec{i}_2^{(z)}$, and $\vec{i}_3^{(z)}$. In addition, we wish to choose $\vec{i}_2^{(z)}$ in such a way that it is perpendicular to the plane $\Sigma^{\|}$ and has the orientation of the x_2-axis. Consequently, we wish to choose $\vec{i}_2^{(z)} = \vec{e}_2(Q)$. For a given orientation of the unit normal $\vec{n}$, however, (4.4.21) would yield $\vec{i}_2^{(z)} = \pm \vec{e}_2(Q)$, not $\vec{i}_2^{(z)} = \vec{e}_2(Q)$. In a layered medium, this might yield jumps of the signs of $\vec{i}_2^{(z)}$ from one point of incidence to another. We can, however, easily remove these jumps by choosing proper orientation of the unit normals $\vec{n}$, taking $\epsilon^* = +1$ or $\epsilon^* = -1$ in (4.4.3). The appropriate choice of ϵ^* is then

$$\epsilon^* = \text{sgn}[\vec{e}_2(Q) \cdot (\nabla\Sigma \times \vec{p}(Q))] = \text{sgn}[p_1 \partial\Sigma/\partial x_3 - p_3 \partial\Sigma/\partial x_1]_Q. \qquad (4.13.3)$$

If we choose ϵ^* using (4.13.3) and compute $\vec{i}_1^{(z)}$, $\vec{i}_2^{(z)}$, and $\vec{i}_3^{(z)}$ using the standard option (4.4.21), the basis vector $\vec{i}_2^{(z)}$ has the same orientation as the x_2-axis at all points of incidence and equals $\vec{e}_2(Q)$. For a reflected/transmitted wave, we also take $\vec{e}_2(\tilde{Q}) = \vec{e}_2(Q)$. Consequently, the unit basis vector $\vec{e}_2$ is constant along the whole planar ray Ω, even in a layered medium. Note that $\vec{i}_1^{(z)}$ is given by (4.4.21) and does not depend on ϵ^*. It is always oriented "along the direction of the propagation of the wave" in the following sense: $\vec{p}(Q) \cdot \vec{i}_1^{(z)} \geq 0$.

Using the above standard choice of $\vec{i}_1^{(z)}$, $\vec{i}_2^{(z)}$, and $\vec{i}_3^{(z)}$ and (4.13.2) for $\hat{\mathbf{H}}(R)$, we obtain

$$\hat{\mathbf{G}}(Q) = \hat{\mathbf{G}}^{\|}(Q), \qquad \hat{\mathbf{G}}(\tilde{Q}) = \hat{\mathbf{G}}^{\|}(\tilde{Q}), \qquad (4.13.4)$$

where $\hat{\mathbf{G}}^{\|}(Q)$ and $\hat{\mathbf{G}}^{\|}(\tilde{Q})$ are given by (4.4.23) and (4.4.25). Consequently,

$$\mathbf{G}(Q) = \mathbf{G}^{\|}(Q), \qquad \mathbf{G}(\tilde{Q}) = \mathbf{G}^{\|}(\tilde{Q}), \qquad (4.13.5)$$

where $\mathbf{G}^{\|}(Q)$ and $\mathbf{G}^{\|}(\tilde{Q})$ are given by (4.4.49).

In a *2-D model*, independent of Cartesian x_2 coordinate, we have the following two conditions along the whole ray Ω:

$$V_{12} = V_{21} = V_{22} = 0, \qquad D_{12} = D_{21} = D_{22} = 0. \qquad (4.13.6)$$

Here V_{IJ} represents the second derivative of velocity with respect to q_I and q_J, and D_{IJ} are components of the curvature matrix of interfaces at the points of incidence. In our treatment, however, we shall consider a *more general situation*; we admit to $V_{22} \neq 0$ and $D_{22} \neq 0$ so that

$$V_{12} = V_{21} = 0, \qquad D_{12} = D_{21} = 0, \qquad V_{22} \neq 0, \qquad D_{22} \neq 0. \qquad (4.13.7)$$

The interpretation of the assumptions in (4.13.7) is obvious. Even though the first derivative of the velocity with respect to the x_2-coordinate vanishes, the second derivative of velocity with respect to x_2 may be nonzero. This means that plane $\Sigma^\|$ may represent the plane of symmetry of a low-velocity channel or, on the contrary, the plane of symmetry of a high-velocity layer. Similarly, the interfaces may be curved at the points of incidence along the x_2-axis, but the x_2-component of the normal to the interface must vanish at these points.

4.13.1 Transformation Matrices Q and P

Let us now express dynamic ray tracing system (4.1.64) explicitly for all eight components of the 2×2 matrices $\mathbf{Q}$ and $\mathbf{P}$,

$$
\begin{aligned}
&\frac{\mathrm{d}Q_{11}}{\mathrm{d}s} = v P_{11}, && \frac{\mathrm{d}P_{11}}{\mathrm{d}s} = -v^{-2} V_{11} Q_{11}, \\
&\frac{\mathrm{d}Q_{22}}{\mathrm{d}s} = v P_{22}, && \frac{\mathrm{d}P_{22}}{\mathrm{d}s} = -v^{-2} V_{22} Q_{22}, \\
&\frac{\mathrm{d}Q_{12}}{\mathrm{d}s} = v P_{12}, && \frac{\mathrm{d}P_{12}}{\mathrm{d}s} = -v^{-2} V_{11} Q_{12}, \\
&\frac{\mathrm{d}Q_{21}}{\mathrm{d}s} = v P_{21}, && \frac{\mathrm{d}P_{21}}{\mathrm{d}s} = -v^{-2} V_{22} Q_{21}.
\end{aligned}
\tag{4.13.8}
$$

Here we have assumed the validity of relations (4.13.7). We choose the initial condition for Q_{IJ} and P_{IJ} as follows:

$$
Q_{12}(S) = Q_{21}(S) = P_{12}(S) = P_{21}(S) = 0.
\tag{4.13.9}
$$

The other components of $\mathbf{Q}(S)$ and $\mathbf{P}(S)$ may be arbitrary. Thus, *we assume that the initial matrices $\mathbf{Q}(S)$ and $\mathbf{P}(S)$ are diagonal.* Due to initial conditions (4.13.9), (4.13.8) yields

$$
Q_{12}(R) = Q_{21}(R) = P_{12}(R) = P_{21}(R) = 0
\tag{4.13.10}
$$

for any point R situated on ray Ω.

The remaining part of (4.13.8) may then be divided into two independent systems. The first system is

$$
\mathrm{d}Q_{11}/\mathrm{d}s = v P_{11}, \qquad \mathrm{d}P_{11}/\mathrm{d}s = -v^{-2} V_{11} Q_{11},
\tag{4.13.11}
$$

and the second is

$$
\mathrm{d}Q_{22}/\mathrm{d}s = v P_{22}, \qquad \mathrm{d}P_{22}/\mathrm{d}s = -v^{-2} V_{22} Q_{22}.
\tag{4.13.12}
$$

System (4.13.11) (or (4.13.12)) is exactly the same as the dynamic ray tracing system (4.1.64), but it is in scalar, not matrix, form.

We now introduce the following notation:

$$
Q_{11} = Q^\|, \qquad P_{11} = P^\|, \qquad Q_{22} = Q^\perp, \qquad P_{22} = P^\perp.
\tag{4.13.13}
$$

We call quantities $Q^\|$ and $P^\|$ the *in-plane quantities* and $Q^\perp$ and $P^\perp$ the *transverse quantities*. Dynamic ray tracing systems (4.13.11) and (4.13.12) can then be expressed in more compact forms.

- The in-plane dynamic ray tracing system:

$$
\frac{\mathrm{d}\mathbf{X}^\|}{\mathrm{d}s} = \mathbf{S}^\| \mathbf{X}^\|, \qquad \mathbf{X}^\|(s) = \begin{pmatrix} Q^\|(s) \\ P^\|(s) \end{pmatrix}, \qquad \mathbf{S}^\| = \begin{pmatrix} 0 & v \\ -v^{-2} V_{11} & 0 \end{pmatrix}.
\tag{4.13.14}
$$

- The transverse dynamic ray tracing system:

$$\frac{d\mathbf{X}^\perp}{ds} = \mathbf{S}^\perp \mathbf{X}^\perp, \qquad \mathbf{X}^\perp(s) = \begin{pmatrix} Q^\perp(s) \\ P^\perp(s) \end{pmatrix}, \qquad \mathbf{S}^\perp = \begin{pmatrix} 0 & v \\ -v^{-2} V_{22} & 0 \end{pmatrix}.$$

(4.13.15)

4.13.2 In-Plane and Transverse Ray Propagator Matrices

We can construct the 2×2 ray propagator matrices $\mathbf{\Pi}^\parallel(R, S)$ and $\mathbf{\Pi}^\perp(R, S)$ corresponding to systems of equations (4.13.14) and (4.13.15) and satisfying conditions

$$\mathbf{\Pi}^\parallel(S, S) = \mathbf{I}, \qquad \mathbf{\Pi}^\perp(S, S) = \mathbf{I},$$

(4.13.16)

where $\mathbf{I}$ is a 2×2 identity matrix. We call $\mathbf{\Pi}^\parallel(R, S)$ the *in-plane ray propagator matrix* and $\mathbf{\Pi}^\perp(R, S)$ the *transverse ray propagator matrix*. Similarly, as in (4.3.5), we shall use the following notation:

$$\mathbf{\Pi}^\parallel(R, S) = \begin{pmatrix} Q_1^\parallel(R, S) & Q_2^\parallel(R, S) \\ P_1^\parallel(R, S) & P_2^\parallel(R, S) \end{pmatrix},$$

$$\mathbf{\Pi}^\perp(R, S) = \begin{pmatrix} Q_1^\perp(R, S) & Q_2^\perp(R, S) \\ P_1^\perp(R, S) & P_2^\perp(R, S) \end{pmatrix}.$$

(4.13.17)

The continuation relations (4.3.29) for planar ray Ω read

$$\mathbf{X}^\parallel(R) = \mathbf{\Pi}^\parallel(R, S)\mathbf{X}^\parallel(S), \qquad \mathbf{X}^\perp(R) = \mathbf{\Pi}^\perp(R, S)\mathbf{X}^\perp(S).$$

(4.13.18)

To produce the complete set of equations for the ray propagator matrix in a layered medium, we need to specify projection matrix $\mathbf{Y}(Q)$ for a planar ray. Let us consider planar ray Ω, which is incident at point Q on interface Σ. Hence,

$$\mathbf{D}(Q) = \begin{pmatrix} D^\parallel(Q) & 0 \\ 0 & D^\perp(Q) \end{pmatrix}, \qquad \mathbf{G}(Q) = \begin{pmatrix} \epsilon \cos i_S & 0 \\ 0 & 1 \end{pmatrix},$$

$$\mathbf{E}(Q) = \begin{pmatrix} E^\parallel(Q) & 0 \\ 0 & 0 \end{pmatrix}.$$

(4.13.19)

Here $D^\parallel = D_{11}$ and $D^\perp = D_{22}$ represent the in-plane and transverse elements of curvature matrix $\mathbf{D}$ of interface Σ at Q. In a 2-D model, $D^\perp = 0$. The expression for $\mathbf{G}(Q)$ follows from (4.13.5) and (4.4.49). $E^\parallel(Q)$ is the in-plane component of the inhomogeneity matrix; see (4.4.33). It may be expressed in several alternative forms,

$$\begin{aligned} E^\parallel(Q) &= -V^{-2} \sin i_S [2\epsilon \cos i_S \partial V/\partial q_1 + \sin i_S \partial V/\partial q_3] \\ &= -V^{-2} \sin i_S [(1 + \cos^2 i_S)\partial V/\partial z_1 - \epsilon \cos i_S \sin i_S \partial V/\partial z_3] \\ &= -V^{-1} \sin i_S [(p_1 \sin i_S + 2\epsilon p_3 \cos i_S)\partial V/\partial x_1 \\ &\quad + (p_3 \sin i_S - 2\epsilon p_1 \cos i_S)\partial V/\partial x_3]; \end{aligned}$$

(4.13.20)

see (4.4.53). Here p_1 and p_3 are Cartesian components of the slowness vector $\vec{p}$, i_S is the acute angle of incidence, and ϵ is the orientation index, that is $\epsilon = \text{sgn}(\vec{p} \cdot \vec{n})$. All quantities in (4.13.20) are taken at point Q. Using relations (4.13.19), we can express the in-plane and transverse projection matrices $\mathbf{Y}^{\parallel}(Q)$ and $\mathbf{Y}^{\perp}(Q)$,

$$
\begin{aligned}
\mathbf{Y}^{\parallel}(Q) &= \epsilon \begin{pmatrix} 1/\cos i_S & 0 \\ (E^{\parallel} - \epsilon V^{-1} \cos i_S D^{\parallel})/\cos i_S & \cos i_S \end{pmatrix}, \\
\mathbf{Y}^{\perp}(Q) &= \begin{pmatrix} 1 & 0 \\ -\epsilon V^{-1} \cos i_S D^{\perp} & 1 \end{pmatrix};
\end{aligned}
\tag{4.13.21}
$$

see (4.4.70). Similarly, we obtain

$$
\begin{aligned}
\mathbf{Y}^{\parallel -1}(Q) &= \epsilon \begin{pmatrix} \cos i_S & 0 \\ -(E^{\parallel} - \epsilon V^{-1} \cos i_S D^{\parallel})/\cos i_S & 1/\cos i_S \end{pmatrix}, \\
\mathbf{Y}^{\perp -1}(Q) &= \begin{pmatrix} 1 & 0 \\ \epsilon V^{-1} \cos i_S D^{\perp} & 1 \end{pmatrix}.
\end{aligned}
\tag{4.13.22}
$$

Expressions similar to (4.13.19) through (4.13.22) can be written even for reflected/transmitted waves at the point $\tilde{Q}$. We only replace V by $\tilde{V}$, $\sin i_S$ by $\sin i_R$, and $\cos i_S$ by $\pm \cos i_R$. The upper sign corresponds to transmitted waves, and the lower sign corresponds to reflected waves. Then, we also obtain expressions for the in-plane and transverse interface propagator matrices $\mathbf{\Pi}^{\parallel}(\tilde{Q}, Q) = \mathbf{Y}^{\parallel -1}(\tilde{Q})\mathbf{Y}^{\parallel}(Q)$ and $\mathbf{\Pi}^{\perp}(\tilde{Q}, Q) = \mathbf{Y}^{\perp -1}(\tilde{Q})\mathbf{Y}^{\perp}(Q)$:

$$
\begin{aligned}
\mathbf{\Pi}^{\parallel}(\tilde{Q}, Q) &= \pm \begin{pmatrix} \cos i_R/\cos i_S & 0 \\ [(E^{\parallel} - \tilde{E}^{\parallel}) - u D^{\parallel}]/(\cos i_R \cos i_S) & \cos i_S/\cos i_R \end{pmatrix}, \\
\mathbf{\Pi}^{\perp}(\tilde{Q}, Q) &= \begin{pmatrix} 1 & 0 \\ -u D^{\perp} & 1 \end{pmatrix}.
\end{aligned}
\tag{4.13.23}
$$

Here u is given by (4.4.51). The final equations for the in-plane and transverse ray propagator matrices $\mathbf{\Pi}^{\parallel}(R, S)$ and $\mathbf{\Pi}^{\perp}(R, S)$ in a layered medium are

$$
\begin{aligned}
\mathbf{\Pi}^{\parallel}(R, S) &= \mathbf{\Pi}^{\parallel}(R, \tilde{Q}_N) \prod_{i=N}^{1} (\mathbf{\Pi}^{\parallel}(\tilde{Q}_i, Q_i)\mathbf{\Pi}^{\parallel}(Q_i, \tilde{Q}_{i-1})), \\
\mathbf{\Pi}^{\perp}(R, S) &= \mathbf{\Pi}^{\perp}(R, \tilde{Q}_N) \prod_{i=N}^{1} (\mathbf{\Pi}^{\perp}(\tilde{Q}_i, Q_i)\mathbf{\Pi}^{\perp}(Q_i, \tilde{Q}_{i-1}));
\end{aligned}
\tag{4.13.24}
$$

see (4.4.86). The notation in (4.13.24) is the same as in (4.4.86), and $\tilde{Q}_0 \equiv S$.

In a similar way, we can construct the in-plane and transverse surface-to-surface ray propagator matrices $\mathbf{T}^{\parallel}(R, S)$ and $\mathbf{T}^{\perp}(R, S)$:

$$
\begin{aligned}
\mathbf{T}^{\parallel}(R, S) &= \mathbf{Y}^{\parallel}(R)\mathbf{\Pi}^{\parallel}(R, S)\mathbf{Y}^{\parallel -1}(S), \\
\mathbf{T}^{\perp}(R, S) &= \mathbf{Y}^{\perp}(R)\mathbf{\Pi}^{\perp}(R, S)\mathbf{Y}^{\perp -1}(S).
\end{aligned}
\tag{4.13.25}
$$

In a layered medium, $\mathbf{T}^{\parallel}(R, S)$ and $\mathbf{T}^{\perp}(R, S)$ are given by relations analogous to (4.4.96):

$$
\mathbf{T}^{\parallel}(R, S) = \prod_{i=N}^{1} \mathbf{T}^{\parallel}(Q_i, Q_{i-1}), \qquad \mathbf{T}^{\perp}(R, S) = \prod_{i=N}^{1} \mathbf{T}^{\perp}(Q_i, Q_{i-1}).
\tag{4.13.26}
$$

Here $S \equiv Q_0$ is situated on the anterior surface, and $R \equiv Q_N$ is on the posterior surface. The in-plane and transverse ray propagator matrices satisfy the symplectic properties

$$
(\mathbf{\Pi}^{\parallel})^T \mathbf{J} \mathbf{\Pi}^{\parallel} = \mathbf{J}, \qquad (\mathbf{\Pi}^{\perp})^T \mathbf{J} \mathbf{\Pi}^{\perp} = \mathbf{J}, \qquad \mathbf{J} = \begin{pmatrix} 0 & 1 \\ -1 & 0 \end{pmatrix}.
\tag{4.13.27}
$$

Similarly, they satisfy the relations for the inverse of the ray propagator matrix. If we use notation (4.13.17), we obtain

$$
\Pi^{\parallel-1}(R, S) = \Pi^{\parallel}(S, R) = \begin{pmatrix} P_2^{\parallel}(R, S) & -Q_2^{\parallel}(R, S) \\ -P_1^{\parallel}(R, S) & Q_1^{\parallel}(R, S) \end{pmatrix},
$$

$$
\Pi^{\perp-1}(R, S) = \Pi^{\perp}(S, R) = \begin{pmatrix} P_2^{\perp}(R, S) & -Q_2^{\perp}(R, S) \\ -P_1^{\perp}(R, S) & Q_1^{\perp}(R, S) \end{pmatrix}. \tag{4.13.28}
$$

Equations (4.13.28) also imply that

$$
\begin{aligned}
Q_1^{\parallel}(S, R) &= P_2^{\parallel}(R, S), & Q_1^{\perp}(S, R) &= P_2^{\perp}(R, S), \\
Q_2^{\parallel}(S, R) &= -Q_2^{\parallel}(R, S), & Q_2^{\perp}(S, R) &= -Q_2^{\perp}(R, S), \\
P_1^{\parallel}(S, R) &= -P_1^{\parallel}(R, S), & P_1^{\perp}(S, R) &= -P_1^{\perp}(R, S), \\
P_2^{\parallel}(S, R) &= Q_1^{\parallel}(R, S), & P_2^{\perp}(S, R) &= Q_1^{\perp}(R, S).
\end{aligned} \tag{4.13.29}
$$

Relations (4.13.27) through (4.13.29) are also satisfied by matrices $\mathbf{T}^{\parallel}$, $\mathbf{T}^{\perp}$, $\mathbf{Y}^{\parallel}$, and $\mathbf{Y}^{\perp}$. All the matrices also satisfy the chain property (4.4.87).

All the relations in this section are valid for planar ray Ω in a general layered model with $V_{22} \neq 0$ and $D_{22} \neq 0$; see (4.13.7). In a standard 2-D model, in which $V_{22} = D_{22} = 0$ (see (4.13.6)), the transverse matrices simplify considerably. The dynamic ray tracing system (4.13.12) for $P_{22}(s)$ and $Q_{22}(s)$ can be solved analytically to yield

$$
P_{22}(s) = P_{22}(s_0), \qquad Q_{22}(s) = Q_{22}(s_0) + P_{22}(s_0) \int_{s_0}^{s} v \, ds.
$$

The transverse ray propagator matrix, $\Pi^{\perp}(Q_i, \tilde{Q}_{i-1})$, is then given by relation

$$
\Pi^{\perp}(Q_i, \tilde{Q}_{i-1}) = \begin{pmatrix} 1 & \sigma(Q_i, \tilde{Q}_{i-1}) \\ 0 & 1 \end{pmatrix}. \tag{4.13.30}
$$

Here $\sigma(Q_i, \tilde{Q}_{i-1})$ is given by relation

$$
\sigma(Q_i, \tilde{Q}_{i-1}) = \int_{\tilde{Q}_{i-1}}^{Q_i} v \, ds, \tag{4.13.31}
$$

where the integration is taken over ray Ω. As $D_{22} = 0$, we also have

$$
\mathbf{Y}^{\perp}(Q_i) = \mathbf{I}, \qquad \mathbf{Y}^{\perp}(\tilde{Q}_i) = \mathbf{I}, \qquad \Pi^{\perp}(\tilde{Q}_i, Q_i) = \mathbf{I}.
$$

Then (4.13.24) yields

$$
\Pi^{\perp}(R, S) = \begin{pmatrix} 1 & \sigma(R, S) \\ 0 & 1 \end{pmatrix}. \tag{4.13.32}
$$

Relation (4.13.32) is the final expression for the transverse ray propagator matrix $\Pi^{\perp}(R, S)$ along planar ray Ω situated in the 2-D layered medium. In (4.13.32), $\sigma(R, S)$ is again given by (4.13.31), but it is taken along the whole ray from S to R, even across the interfaces.

Thus, the transverse ray propagator matrix $\Pi^{\perp}(R, S)$ in a 2-D layered medium is very simple and can be computed merely by one simple integration along the ray. If quantity σ is used as a monotonic parameter along the ray, the transverse ray propagator matrix is obtained immediately, without any additional computation.

4.13.3 Matrices M and K

We again consider a planar ray Ω connecting two points S and R and transformation matrices $\mathbf{Q}$ and $\mathbf{P}$ satisfying conditions (4.13.9) at point S. Matrices $\mathbf{Q}$ and $\mathbf{P}$ are then diagonal along the whole ray Ω; see (4.13.10). This configuration implies that matrix $\mathbf{M} = \mathbf{PQ}^{-1}$ of the second derivatives of the travel-time field with respect to ray-centered coordinates q_I is also diagonal. We introduce the in-plane and transverse second derivatives of the travel-time field, $M^{\|} = \partial^2 T/\partial q_1^2$ and $M^{\perp} = \partial^2 T/\partial q_2^2$, as follows:

$$\mathbf{M}(R) = \begin{pmatrix} M^{\|}(R) & 0 \\ 0 & M^{\perp}(R) \end{pmatrix}, \tag{4.13.33}$$

where

$$M^{\|}(R) = P^{\|}(R)/Q^{\|}(R), \qquad M^{\perp}(R) = P^{\perp}(R)/Q^{\perp}(R). \tag{4.13.34}$$

Assume that we know $M^{\|}(R)$ (or $M^{\perp}(S)$) at initial point S of Ω and wish to compute $M^{\|}(R)$ (or $M^{\perp}(R)$) at any point R of ray Ω. If we use (4.13.34), continuation relations (4.13.18), and the notation (4.13.17), we obtain

$$M^{\|}(R) = \left[P_1^{\|}(R, S) + P_2^{\|}(R, S)M^{\|}(S)\right]/\left[Q_1^{\|}(R, S) + Q_2^{\|}(R, S)M^{\|}(S)\right],$$
$$M^{\perp}(R) = \left[P_1^{\perp}(R, S) + P_2^{\perp}(R, S)M^{\perp}(S)\right]/\left[Q_1^{\perp}(R, S) + Q_2^{\perp}(R, S)M^{\perp}(S)\right].$$
$$\tag{4.13.35}$$

This can be expressed in alternative form using (4.13.29). We can also express $M^{\|}(S)$ in terms of $M^{\|}(R)$ by interchanging S and R in (4.13.35).

For a point source situated at point S, $M^{\|}(S) \to \infty$ and $M^{\perp}(S) \to \infty$. We again denote by $M^{\|}(R, S)$ and $M^{\perp}(R, S)$ the in-plane and transverse second derivatives of the travel-time field, $M^{\|}(R)$ and $M(R)$, corresponding to the point source at S. Hence,

$$M^{\|}(R, S) = P_2^{\|}(R, S)/Q_2^{\|}(R, S) = -Q_1^{\|}(S, R)/Q_2^{\|}(S, R),$$
$$M^{\perp}(R, S) = P_2^{\perp}(R, S)/Q_2^{\perp}(R, S) = -Q_1^{\perp}(S, R)/Q_2^{\perp}(S, R).$$
$$\tag{4.13.36}$$

Analogous expressions for $M^{\|}(S, R)$ and $M^{\perp}(S, R)$, corresponding to a point source at R, are obtained from (4.13.36) by interchanging R and S.

In local ray-centered coordinates y_1, y_2, y_3, the 3×3 matrix of the travel-time field $\hat{\mathbf{M}}(R)$ at point R of ray Ω is given by relation

$$\hat{\mathbf{M}}(R) = \begin{pmatrix} M^{\|}(R) & 0 & -(v^{-2}\partial v/\partial q_1)_R \\ 0 & M^{\perp}(R) & 0 \\ -(v^{-2}\partial v/\partial q_1)_R & 0 & -(v^{-2}\partial v/\partial q_3)_R \end{pmatrix}; \tag{4.13.37}$$

see (4.1.81). Finally, in general Cartesian coordinates x_1, x_2, x_3, the matrix of the second derivatives of the travel-time field $\hat{\mathbf{M}}^{(x)}(R)$, with components $M_{ij}^{(x)} = \partial^2 T/\partial x_i \partial x_j$, is given by relation

$$\hat{\mathbf{M}}^{(x)}(R) = \hat{\mathbf{H}}(R)\hat{\mathbf{M}}(R)\hat{\mathbf{H}}^T(R)$$
$$= \begin{pmatrix} t_3 & 0 & t_1 \\ 0 & 1 & 0 \\ -t_1 & 0 & t_3 \end{pmatrix} \begin{pmatrix} M^{\|} & 0 & -v^{-2}\partial v/\partial q_1 \\ 0 & M^{\perp} & 0 \\ -v^{-2}\partial v/\partial q_1 & 0 & -v^{-2}\partial v/\partial q_3 \end{pmatrix} \begin{pmatrix} t_3 & 0 & -t_1 \\ 0 & 1 & 0 \\ t_1 & 0 & t_3 \end{pmatrix};$$
$$\tag{4.13.38}$$

see (4.1.87) and (4.13.2). In (4.13.38), t_i are Cartesian components of unit vector $\vec{t}$ tangent to the ray, that is $\vec{t} = \vec{e}_3$. All quantities in (4.13.38) are taken at point R. We can again introduce the in-plane and transverse parts of $\hat{\mathbf{M}}^{(x)}(R)$,

$$
\mathbf{M}^{(x)\parallel}(R) = \begin{pmatrix} M_{11}^{(x)} & M_{13}^{(x)} \\ M_{31}^{(x)} & M_{33}^{(x)} \end{pmatrix}
$$
$$
= \begin{pmatrix} t_3 & t_1 \\ -t_1 & t_3 \end{pmatrix} \begin{pmatrix} M^{\parallel} & -v^{-2}\partial v/\partial q_1 \\ -v^{-2}\partial v/\partial q_1 & -v^{-2}\partial v/\partial q_3 \end{pmatrix} \begin{pmatrix} t_3 & -t_1 \\ t_1 & t_3 \end{pmatrix},
$$
(4.13.39)

and

$$
M^{(x)\perp}(R) = M_{22}^{(x)}(R) = M^{\perp}(R).
$$
(4.13.40)

Components $M_{12}^{(x)}$, $M_{21}^{(x)}$, $M_{23}^{(x)}$, and $M_{32}^{(x)}$ of matrix $\hat{\mathbf{M}}^{(x)}$ vanish. Note that $\partial v/\partial q_i$ in (4.13.38) and (4.13.39) can be expressed in terms of $\partial v/\partial x_1$ and $\partial v/\partial x_3$ as

$$
\partial v/\partial q_1 = t_3 \partial v/\partial x_1 - t_1 \partial v/\partial x_3, \qquad \partial v/\partial q_3 = t_1 \partial v/\partial x_1 + t_3 \partial v/\partial x_3;
$$
(4.13.41)

see (4.1.62).

Matrix $\mathbf{M}$ is closely related to the *matrix of the curvature of the wavefront* $\mathbf{K} = v\mathbf{M}$:

$$
\mathbf{K}(R) = \begin{pmatrix} K^{\parallel}(R) & 0 \\ 0 & K^{\perp}(R) \end{pmatrix},
$$
(4.13.42)

where $K^{\parallel} = vM^{\parallel}$ and $K^{\perp} = vM^{\perp}$. Similarly, we can express the continuation relations for $K^{\parallel}$ and $K^{\perp}$ using (4.13.35),

$$
\begin{aligned}
K^{\parallel}(R) &= v(R)\big[v(S)P_1^{\parallel}(R,S) + P_2^{\parallel}(R,S)K^{\parallel}(S)\big] \\
&\quad \big/ \big[v(S)Q_1^{\parallel}(R,S) + Q_2^{\parallel}(R,S)K^{\parallel}(S)\big], \\
K^{\perp}(R) &= v(R)\big[v(S)P_1^{\perp}(R,S) + P_2^{\perp}(R,S)K^{\perp}(S)\big] \\
&\quad \big/ \big[v(S)Q_1^{\perp}(R,S) + Q_2^{\perp}(R,S)K^{\perp}(S)\big].
\end{aligned}
$$
(4.13.43)

For a point source at S, we obtain

$$
\begin{aligned}
K^{\parallel}(R,S) &= v(R)P_2^{\parallel}(R,S)\big/Q_2^{\parallel}(R,S) = -v(R)Q_1^{\parallel}(S,R)\big/Q_2^{\parallel}(S,R), \\
K^{\perp}(R,S) &= v(R)P_2^{\perp}(R,S)\big/Q_2^{\perp}(R,S) = -v(R)Q_1^{\perp}(S,R)\big/Q_2^{\perp}(S,R).
\end{aligned}
$$
(4.13.44)

In Equations (4.13.43) and (4.13.44), we can again interchange S and R.

Instead of the in-plane and transverse curvatures of the wavefront, $K^{\parallel}$ and $K^{\perp}$, we can also introduce the *in-plane* and *transverse radii of curvatures of the wavefront* $R^{\parallel} = 1/K^{\parallel}$ and $R^{\perp} = 1/K^{\perp}$. The continuation equations for $R^{\parallel}$ and $R^{\perp}$ are obtained from (4.13.43):

$$
\begin{aligned}
R^{\parallel}(R) &= v^{-1}(R)\big[Q_2^{\parallel}(R,S) + v(S)Q_1^{\parallel}(R,S)R^{\parallel}(S)\big] \\
&\quad \big/ \big[P_2^{\parallel}(R,S) + v(S)P_1^{\parallel}(R,S)R^{\parallel}(S)\big], \\
R^{\perp}(R) &= v^{-1}(R)\big[Q_2^{\perp}(R,S) + v(S)Q_1^{\perp}(R,S)R^{\perp}(S)\big] \\
&\quad \big/ \big[P_2^{\perp}(R,S) + v(S)P_1^{\perp}(R,S)R^{\perp}(S)\big].
\end{aligned}
$$
(4.13.45)

Equations (4.13.35), (4.13.43), and (4.13.45) can also be used to find the transformations of $M^{\parallel}$, $M^{\perp}$, $K^{\parallel}$, $K^{\perp}$, $R^{\parallel}$, and $R^{\perp}$ across the interface at the point of incidence Q. We merely use $S = Q$ and $R = \tilde{Q}$ and insert the appropriate elements of the interface propagator matrices $\mathbf{\Pi}^{\parallel}(\tilde{Q}, Q)$ and $\mathbf{\Pi}^{\perp}(\tilde{Q}, Q)$ given by (4.13.23). Equations (4.13.35), (4.13.43), and (4.13.45) simplify considerably in this case as $Q_2^{\parallel}(\tilde{Q}, Q) = 0$ and $Q_2^{\perp}(\tilde{Q}, Q) = 0$; see (4.13.23).

The application of the radii of curvature of the wavefront has a long tradition in the seismic ray theory. Alekseyev and Gel'chinskiy (1958) proposed a method of calculating successively $R^{\parallel}$ and $R^{\perp}$ along planar ray Ω in a medium composed of *homogeneous layers* separated by curved interfaces and used the computed quantities $R^{\parallel}$ and $R^{\perp}$ to determine the geometrical spreading and ray amplitude. The algorithms proposed by Alekseyev and Gel'chinskiy, of course, follow from the equations presented in this section, which generalize them in several ways. See also the detailed description of the algorithm in Červený and Ravindra (1971, Eqs. (2.163)–(2.165)).

4.13.4   In-Plane and Transverse Geometrical Spreading

Along planar ray Ω, $t_2 = 0$, $Q_{21} = 0$, and $Q_{21}^{(x)} = 0$. We can then use Equations (4.10.3) and (4.10.4) and write the following equation for geometrical spreading $\mathcal{L}(R)$ at any point R situated on Ω:

$$\mathcal{L}(R) = \mathcal{L}^{\parallel}(R)\mathcal{L}^{\perp}(R). \tag{4.13.46}$$

Here $\mathcal{L}^{\parallel}(R)$ is the *in-plane geometrical spreading* and $\mathcal{L}^{\perp}(R)$ the *transverse geometrical spreading*. $\mathcal{L}^{\parallel}(R)$ and $\mathcal{L}^{\perp}(R)$ can be expressed in several alternative ways:

$$\mathcal{L}^{\parallel}(R) = |J^{\parallel}(R)|^{1/2} = \left| \det \begin{pmatrix} Q_{11}^{(x)}(R) & t_1(R) \\ Q_{31}^{(x)}(R) & t_3(R) \end{pmatrix} \right|^{1/2} = |Q^{\parallel}(R)|^{1/2},$$
$$\mathcal{L}^{\perp}(R) = |J^{\perp}(R)|^{1/2} = \left| Q_{22}^{(x)}(R) \right|^{1/2} = |Q^{\perp}(R)|^{1/2}.$$

$$\tag{4.13.47}$$

We remind the reader that we consider only isotropic media in this section so that $\mathcal{U} = C$. Quantities $Q_{11}^{(x)}(R)$ and $Q_{31}^{(x)}(R)$ in the expression for $\mathcal{L}^{\parallel}(R)$ are elements of matrix $\hat{\mathbf{Q}}^{(x)}$ and can be calculated by dynamic ray tracing in Cartesian coordinates.

From a physical point of view, $\mathcal{L}^{\parallel}(R)$ expresses the geometrical spreading of the ray tube in the plane of the ray $\Sigma^{\parallel}$, and $\mathcal{L}^{\perp}(R)$ expresses the geometrical spreading of the ray tube perpendicular to that plane.

Using ray propagator matrices (4.13.17) and continuation relations (4.13.18), we can write

$$Q^{\parallel}(R) = Q_1^{\parallel}(R, S)Q^{\parallel}(S) + Q_2^{\parallel}(R, S)P^{\parallel}(S), \tag{4.13.48}$$
$$Q^{\perp}(R) = Q_1^{\perp}(R, S)Q^{\perp}(S) + Q_2^{\perp}(R, S)P^{\perp}(S). \tag{4.13.49}$$

This also immediately follows from (4.6.2), if we take into account that the individual matrices in (4.6.2) are diagonalized for planar rays.

Equations (4.13.48) and (4.13.49) yield the continuation relations for $\mathcal{L}^{\parallel}(R)$ and $\mathcal{L}^{\perp}(R)$:

$$\mathcal{L}^{\parallel}(R) = \left| Q_1^{\parallel}(R, S) + Q_2^{\parallel}(R, S)M^{\parallel}(S) \right|^{1/2} \mathcal{L}^{\parallel}(S),$$
$$\mathcal{L}^{\perp}(R) = \left| Q_1^{\perp}(R, S) + Q_2^{\perp}(R, S)M^{\perp}(S) \right|^{1/2} \mathcal{L}^{\perp}(S).$$

$$\tag{4.13.50}$$

Now we wish to specify Equations (4.13.48), (4.13.49), and (4.13.50) for a point source situated at S. In 2-D models, however, we often consider yet another simple type of source: the line source, perpendicular to the plane $\Sigma^{\parallel}$ of ray Ω. We shall give the equations for the geometrical spreading for both types of sources.

a. Point source at S. $Q^{\parallel}(S) = Q^{\perp}(S) = 0$, and Equations (4.13.48) and (4.13.49) yield

$$\mathcal{L}^{\parallel}(R) = \left| Q_2^{\parallel}(R, S) P^{\parallel}(S) \right|^{1/2}, \qquad \mathcal{L}^{\perp}(R) = \left| Q_2^{\perp}(R, S) P^{\perp}(S) \right|^{1/2}.$$

$$(4.13.51)$$

As in 3-D structures, we can introduce *the relative in-plane and transverse geometrical spreading* for the *point source* situated at S. We denote them by $\mathcal{L}^{\parallel}(R, S)$ and $\mathcal{L}^{\perp}(R, S)$:

$$\mathcal{L}(R, S) = \mathcal{L}^{\parallel}(R, S)\mathcal{L}^{\perp}(R, S), \qquad\qquad (4.13.52)$$

$$\mathcal{L}^{\parallel}(R, S) = \left| Q_2^{\parallel}(R, S) \right|^{1/2}, \qquad \mathcal{L}^{\perp}(R, S) = \left| Q_2^{\perp}(R, S) \right|^{1/2}. \qquad (4.13.53)$$

Both $\mathcal{L}^{\parallel}(R, S)$ and $\mathcal{L}^{\perp}(R, S)$ are reciprocal, $\mathcal{L}^{\parallel}(R, S) = \mathcal{L}^{\parallel}(S, R)$ and $\mathcal{L}^{\perp}(R, S) = \mathcal{L}^{\perp}(S, R)$.

Equations (4.13.51) through (4.13.53) also apply to the general case of $V_{22} \neq 0$ and $D_{22} \neq 0$. We shall now consider a *standard 2-D model* in which $V_{22} = D_{22} = 0$. In a 2-D model, ray propagator matrix $\boldsymbol{\Pi}^{\perp}(R, S)$ is given by (4.13.32), which implies

$$Q_2^{\perp}(R, S) = \sigma(R, S) = \int_S^R v \, ds.$$

This yields simpler relations for transverse geometrical spreadings $\mathcal{L}^{\perp}(R)$ and $\mathcal{L}^{\perp}(R, S)$:

$$\mathcal{L}^{\perp}(R) = |\sigma(R, S) P^{\perp}(S)|^{1/2}, \qquad \mathcal{L}^{\perp}(R, S) = |\sigma(R, S)|^{1/2}. \qquad (4.13.54)$$

The relations for $\mathcal{L}^{\parallel}(R, S)$ and $\mathcal{L}^{\parallel}(R)$, (4.13.51) through (4.13.53), remain the same even in a standard 2-D model; they do not simplify. Both the relative in-plane and transverse geometrical spreading, $\mathcal{L}^{\parallel}(R, S)$ and $\mathcal{L}^{\perp}(R, S)$, are reciprocal in this case.

b. Line source at S. We consider a line source parallel to the x_2-axis, intersecting the plane $\Sigma^{\parallel}$ of the ray (plane $x_1 x_3$) at point S. We choose ray parameter γ_2 so that it is equal to the distance along the line source. We then obtain $Q^{\parallel}(S) = 0$, $Q^{\perp}(S) = 1$, $P^{\parallel}(S) \neq 0$, and $P^{\perp}(S) = 0$. Hence, Equations (4.13.48) and (4.13.49) yield

$$\mathcal{L}^{\parallel}(R) = \left| Q_2^{\parallel}(R, S) P^{\parallel}(S) \right|^{1/2}, \qquad \mathcal{L}^{\perp}(R) = \left| Q_1^{\perp}(R, S) \right|^{1/2}. \qquad (4.13.55)$$

The relative geometrical spreading for a point source at S was introduced by (4.10.11). Definition (4.10.11) cannot be applied to a line source because $\mathbf{P}(S) = \mathbf{P}^{\parallel}(S)\mathbf{P}^{\perp}(S) = \mathbf{0}$ in this case. Still, however, we can define the *relative in-plane geometrical spreading* $\mathcal{L}^{\parallel}(R, S)$ for a line source, in the same way as for a point source; see (4.13.53). *Relative transverse geometrical spreading* does not have any meaning for a line source.

In a standard 2-D model ($V_{22} = D_{22} = 0$), $Q_1^{\perp}(R, S) = 1$; see (4.13.32). Consequently,

$$\mathcal{L}^{\perp}(R) = 1. \qquad\qquad (4.13.56)$$

Let us add a terminological note to the computation of seismic wavefields in standard 2-D models (with $V_{22} = 0$ and $D_{22} = 0$). If we consider a line source parallel to the x_2-axis (that is, perpendicular to the plane $\Sigma^{\parallel}$) we speak of *standard 2-D computations*, or *standard 2-D case*. Standard 2-D computations are common mainly in the investigation of the complete wavefield by finite differences. In practical application, however, it is more important to consider a point source situated in the plane $\Sigma^{\parallel}$. Then we often speak of *2.5-D computations* (two-and-half-dimensional computations) or of a *2.5-D case*. See, for example, Bleistein, Cohen, and Hagin (1987). In the ray method, the only difference between a standard 2-D case and a 2.5-D case is the transverse geometrical spreading. In the standard 2-D case, there is no transverse spreading; see (4.13.56). In a 2.5-D case, the relative transverse geometrical spreading $\mathcal{L}^{\perp}(R, S)$ is given by (4.13.54). From a numerical point of view, the difference between standard 2-D and 2.5-D cases consists only in the computation of $\sigma(R, S)$ by numerical quadratures along the ray Ω. Thus, we can easily implement both options into any 2-D ray theory computer code. In finite-difference computations of complete wavefields, the transition from a line source to a point source is considerably more complicated.

There is, however, no uniqueness in the term *2.5-D computations*. The term has also been used frequently in a considerably more general meaning: for *general 3-D computations in standard 2-D models*; see Brokešová (1994). In this case, the rays are not confined to the plane $x_2 = 0$, but they are arbitrarily inclined with respect to it. The inclination depends on the component p_{20} of the slowness vector at the initial point. Only if $p_{20} = 0$ do we obtain standard 2-D computations with rays confined to the plane $x_2 = 0$. For more details, see Section 3.3.4.2.

4.13.5 Paraxial Travel Times

We shall consider planar ray Ω, points S and R situated on Ω, and point R' situated close to R. Point R' may also be situated outside the plane $\Sigma^{\parallel}$ so that $x_2(R') \neq 0$.

In ray-centered coordinates, we can use relations (4.13.33) and (4.6.24) to obtain

$$T(R') = T(R) + \tfrac{1}{2}q_1^2(R')M^{\parallel}(R) + \tfrac{1}{2}q_2^2(R')M^{\perp}(R). \qquad (4.13.57)$$

This relation is valid for points R' situated in a plane $\Sigma^{\perp}$ perpendicular to Ω at R. A similar relation in general Cartesian coordinates reads

$$T(R') = T(R) + \mathbf{x}^{\parallel T}(R', R)\mathbf{p}^{(x)\parallel}(R) + \tfrac{1}{2}\mathbf{x}^{\parallel T}(R', R)\mathbf{M}^{(x)\parallel}(R)\mathbf{x}^{\parallel}(R', R)$$
$$+ \tfrac{1}{2}x_2^2(R', R)M^{\perp}(R). \qquad (4.13.58)$$

Here we have used the notation

$$\mathbf{x}^{\parallel}(R', R) = \begin{pmatrix} x_1(R', R) \\ x_3(R', R) \end{pmatrix}, \qquad \mathbf{p}^{(x)\parallel}(R) = \begin{pmatrix} p_1^{(x)}(R) \\ p_3^{(x)}(R) \end{pmatrix}. \qquad (4.13.59)$$

The 2×2 matrix $\mathbf{M}^{(x)\parallel}(R)$ is given by relation (4.13.39). Point R' may be situated arbitrarily in the vicinity of R in this case.

For a point source situated at point S, (4.13.57) and (4.13.58) yield

$$T(R', S) = T(R, S) + \tfrac{1}{2}q_1^2(R')M^{\parallel}(R, S) + \tfrac{1}{2}q_2^2(R')M^{\perp}(R, S), \quad (4.13.60)$$

$$T(R', S) = T(R, S) + \mathbf{x}^{\parallel T}(R', R)\mathbf{p}^{(x)\parallel}(R)$$
$$+ \tfrac{1}{2}\mathbf{x}^{\parallel T}(R', R)\mathbf{M}^{(x)\parallel}(R, S)\mathbf{x}^{\parallel}(R', R) + \tfrac{1}{2}x_2^2(R', R)M^{\perp}(R, S).$$
$$(4.13.61)$$

Here $M^{\perp}(R, S)$ and $M^{\parallel}(R, S)$ are given by (4.13.36) and $\mathbf{M}^{(x)\parallel}(R, S)$ follows immediately from (4.13.39), specified for a point source at S.

4.13.6 Paraxial Rays Close to a Planar Central Ray

Let us consider a planar central ray Ω situated in a plane $\Sigma^{\|}$ and assume that $V_{12} = V_{21} = 0$. We, however, do not require $V_{22} = 0$ and $D_{22} = 0$; see (4.13.7). The paraxial ray tracing system then separates into two fully independent systems. The first system is for q_1, $p_1^{(q)}$:

$$dq_1/ds = vp_1^{(q)}, \qquad dp_1^{(q)}/ds = -v^{-2}v_{,11}q_1, \tag{4.13.62}$$

and the second is for q_2 and $p_2^{(q)}$:

$$dq_2/ds = vp_2^{(q)}, \qquad dp_2^{(q)}/ds = -v^{-2}v_{,22}q_2. \tag{4.13.63}$$

Both systems are fully equivalent to the relevant dynamic ray tracing systems along the plane central ray Ω; see (4.13.11) and (4.13.12). Thus, they can be solved using the 2×2 ray propagator matrices $\mathbf{\Pi}^{\|}(R, S)$ and $\mathbf{\Pi}^{\perp}(R, S)$, introduced in Section 4.13.2. All the equations are straightforward.

We shall now pay more attention only to system (4.13.63), which describes the ray deviations of the paraxial ray Ω' from plane $\Sigma^{\|}$ in which the central ray is situated. The deviations of paraxial ray Ω' from plane $\Sigma^{\|}$ are controlled mainly by the initial value of $p_2^{(q)}(s_0)$ and by the distribution of $v_{,22}$ along central ray Ω. To demonstrate the effect of these two quantities on the paraxial ray, we shall consider three simple possibilities: (a) $v_{,22} = 0$, (b) $v^{-3}v_{,22} = k^2 > 0$, and (c) $v^{-3}v_{,22} = -k^2 < 0$, where k is some real-valued constant.

To solve (4.13.63), it is convenient to introduce a new variable σ along central ray Ω, such that $d\sigma = vds$. Then (4.13.63) yields

$$dq_2/d\sigma = p_2^{(q)}, \qquad dp_2^{(q)}/d\sigma = -v^{-3}v_{,22}q_2. \tag{4.13.64}$$

This system can be combined into one ordinary differential equation of the second order,

$$d^2q_2/d\sigma^2 + v^{-3}v_{,22}q_2 = 0. \tag{4.13.65}$$

The initial conditions for this differential equation at $\sigma = \sigma_0$ are

$$q_2(\sigma) = q_2(\sigma_0), \qquad q_2'(\sigma) = p_2^{(q)}(\sigma_0). \tag{4.13.66}$$

After $q_2(\sigma)$ is determined from (4.13.65), $p_2^{(q)}(\sigma)$ is obtained from (4.13.64) as a derivative: $p_2^{(q)}(\sigma) = dq_2(\sigma)/d\sigma$.

We shall now consider the three cases of $v_{,22}$:

- $v_{,22} = 0$. Then $d^2q_2/d\sigma^2 = 0$, and $q_2(\sigma)$ is a linear function of σ. Taking into account initial conditions (4.13.66), we obtain

$$q_2(\sigma) = q_2(\sigma_0) + p_2^{(q)}(\sigma_0)(\sigma - \sigma_0), \qquad p_2^{(q)}(\sigma) = p_2^{(q)}(\sigma_0). \tag{4.13.67}$$

- $v^{-3}v_{,22} = k^2$. Then

$$d^2q_2/d\sigma^2 + k^2q_2 = 0. \tag{4.13.68}$$

The two linearly independent solutions of (4.13.68) are trigonometric functions $\sin k(\sigma - \sigma_0)$ and $\cos k(\sigma - \sigma_0)$. In view of initial conditions (4.13.66)

$$\begin{aligned} q_2(\sigma) &= q_2(\sigma_0)\cos k(\sigma - \sigma_0) + k^{-1}p_2^{(q)}(\sigma_0)\sin k(\sigma - \sigma_0), \\ p_2^{(q)}(\sigma) &= p_2^{(q)}(\sigma_0)\cos k(\sigma - \sigma_0) - kq_2(\sigma_0)\sin k(\sigma - \sigma_0). \end{aligned} \tag{4.13.69}$$

- $v^{-3}v_{,22} = -k^2$. Equation (4.13.65) then yields

$$d^2q_2/d\sigma^2 - k^2q_2 = 0. \tag{4.13.70}$$

The two linearly independent solutions of (4.13.70) are exponential functions $\exp(k(\sigma - \sigma_0))$ and $\exp(-k(\sigma - \sigma_0))$. Taking into account initial conditions (4.13.66), we obtain

$$q_2(\sigma) = \tfrac{1}{2}\big(q_2(\sigma_0) + k^{-1}p_2^{(q)}(\sigma_0)\big)e^{k(\sigma - \sigma_0)}$$
$$+ \tfrac{1}{2}\big(q_2(\sigma_0) - k^{-1}p_2^{(q)}(\sigma_0)\big)e^{-k(\sigma - \sigma_0)}, \tag{4.13.71}$$
$$p_2^{(q)}(\sigma) = \tfrac{1}{2}\big(kq_2(\sigma_0) + p_2^{(q)}(\sigma_0)\big)e^{k(\sigma - \sigma_0)} - \tfrac{1}{2}\big(kq_2(\sigma_0) - p_2^{(q)}(\sigma_0)\big)e^{-k(\sigma - \sigma_0)}.$$

The derived equations (4.13.67), (4.13.69), and (4.13.71) have a very interesting seismological interpretation. If $p_2^{(q)}(s_0) \neq 0$, paraxial rays Ω' behave in different ways for $v_{,22} = 0$, $v_{,22} > 0$, and $v_{,22} < 0$.

The case of $v_{,22} = 0$ corresponds to a standard 2-D model in which the velocity does not depend on x_2. Paraxial rays Ω' *deviate linearly* from $\Sigma^{\parallel}$ in a transverse direction with increasing σ.

For $v_{,22} > 0$, $\Sigma^{\parallel}$ represents the plane of symmetry of a low-velocity channel. It is not surprising that the paraxial rays Ω' have a typical waveguide character in this case. In the transverse direction, they *oscillate around the central ray*. The period of oscillation depends on $v_{,22}$.

For $v_{,22} < 0$, $\Sigma^{\parallel}$ represents the plane of symmetry of a high-velocity layer. In this case, paraxial rays Ω' *deviate exponentially* from $\Sigma^{\parallel}$ in the transverse direction with increasing σ.

Because the equations for paraxial rays (4.13.64) are fully equivalent to the equations for Q_{22} (see (4.13.12)), similar conclusions also apply to the relative transverse geometrical spreading $\mathcal{L}^{\perp}(R, S) = [Q^{\perp}(R, S)]^{1/2}$. The most interesting is the waveguide behavior of $\mathcal{L}^{\perp}(R, S)$ for $v_{,22} > 0$. The relative transverse geometrical spreading $\mathcal{L}^{\perp}(R, S)$ oscillates along central ray Ω and vanishes at a regularly distributed system of points along it. Consequently, caustics are formed at these points, and the relevant amplitudes blow up there.

4.13.7 Paraxial Boundary-Value Ray Tracing in the Vicinity of a Planar Ray. Two-Point Eikonal

We assume that the central ray Ω with points S and R is situated in plane x_1-x_3 and that points S' and R' are situated close to S and R, respectively. Points S' and R', however, *need not be situated in plane x_1-x_3*.

We shall again use the notations of Sections 4.9 and 4.13 and the initial conditions (4.13.9). As in Section 4.13.5, we introduce the notation (4.13.59) for $\mathbf{x}^{\parallel}(R', R)$, $\mathbf{x}^{\parallel}(S', S)$, $\mathbf{p}^{(x)\parallel}(R)$, and $\mathbf{p}^{(x)\parallel}(S)$. We shall also use the notations for $\mathbf{M}^{(x)\parallel}(R)$, and $M^{(x)\perp}(R)$ given by (4.13.39) and (4.13.40).

First, we determine the Cartesian components of slowness vectors $\vec{p}(R')$ and $\vec{p}(S')$, corresponding to the paraxial ray $\Omega'(R', S')$ connecting points S' and R'. Using general relations (4.9.22), we obtain

$$\mathbf{p}^{(x)\parallel}(R') = \mathbf{p}^{(x)\parallel}(R) + \mathbf{M}^{(x)\parallel}(R, S)\mathbf{x}^{\parallel}(R', R)$$
$$- Q_2^{\parallel-1}(R, S)\mathbf{\Gamma}^T(R, S)\mathbf{x}^{\parallel}(S', S), \tag{4.13.72}$$
$$p^{(x)\perp}(R') = p^{(x)\perp}(R) + M^{(x)\perp}(R, S)x_2(R', R) - Q_2^{\perp-1}(R, S)x_2(S', S).$$

Here we have used notation

$$\mathbf{\Gamma}(R, S) = \begin{pmatrix} t_3(S)t_3(R) & -t_3(S)t_1(R) \\ -t_1(S)t_3(R) & t_1(S)t_1(R) \end{pmatrix}. \tag{4.13.73}$$

In a similar way, we obtain, for $p^{(x)\parallel}(S')$ and $p^{(x)\perp}(S')$,

$$
\begin{aligned}
\mathbf{p}^{(x)\parallel}(S') = {} & \mathbf{p}^{(x)\parallel}(S) + \mathbf{M}^{(x)\parallel}(S, R)\mathbf{x}^{\parallel}(S', S) \\
& + Q_2^{\parallel-1}(R, S)\mathbf{\Gamma}(R, S)\mathbf{x}^{\parallel}(R', R), \qquad\qquad (4.13.74)
\end{aligned}
$$
$$
p^{(x)\perp}(S') = p^{(x)\perp}(S) + M^{(x)\perp}(S, R)x_2(S', S) + Q_2^{\perp-1}(R, S)x_2(R', R).
$$

Components $p_2^{(x)} = p^{(x)\perp}$ do not vanish only if at least one of the points R' or S' is situated outside plane x_1-x_3. For the travel time from S' to R' along paraxial ray $\Omega'(R', S')$ (4.9.24) immediately yields

$$
\begin{aligned}
T(R', S') = {} & T(R, S) + \mathbf{x}^{\parallel T}(R', R)\mathbf{p}^{(x)\parallel}(R) - \mathbf{x}^{\parallel T}(S', S)\mathbf{p}^{(x)\parallel}(S) \\
& + \tfrac{1}{2}\mathbf{x}^{\parallel T}(R', R)\mathbf{M}^{(x)\parallel}(R, S)\mathbf{x}^{\parallel}(R', R) \\
& - \tfrac{1}{2}\mathbf{x}^{\parallel T}(S', S)\mathbf{M}^{(x)\parallel}(S, R)\mathbf{x}^{\parallel}(S', S) \\
& - \mathbf{x}^{\parallel T}(S', S)\mathbf{\Gamma}(R, S)\mathbf{x}^{\parallel}(R', R)Q_2^{\parallel-1}(R, S) + x_2(R', R)p_2^{(x)}(R) \\
& - x_2(S', S)p_2^{(x)}(S) + \tfrac{1}{2}x_2^2(R', R)M_{22}^{(x)}(R, S) \\
& - \tfrac{1}{2}x_2^2(S', S)M_{22}^{(x)}(S, R) - x_2(S', S)x_2(R', R)Q_2^{\perp-1}(R, S).
\end{aligned}
$$
$$
(4.13.75)
$$

The last five terms represent the contribution corresponding to the deviations of paraxial ray $\Omega'(R', S')$ from plane x_1-x_3.

4.13.8 Determination of Geometrical Spreading from the Travel-Time Data in 2-D Media

In this section, we shall consider a planar ray Ω, fully situated in plane $\Sigma^{\parallel}$ (described by equation $x_2 = 0$), and two points S and R situated on Ω. Ray Ω may correspond to any multiply reflected, possibly converted, wave. The geometrical spreading $\mathcal{L}(R)$ can then be expressed as a product of the in-plane geometrical spreading $\mathcal{L}^{\parallel}(R)$ and the transverse geometrical spreading $\mathcal{L}^{\perp}(R)$; see (4.13.46). Regarding the source, we shall consider two options: (i) a point source situated at S and (ii) a line source, perpendicular to $\Sigma^{\parallel}$ and intersecting $\Sigma^{\parallel}$ at S. In both cases, the in-plane geometrical spreading $\mathcal{L}^{\parallel}(R)$ can be factorized further,

$$
\mathcal{L}^{\parallel}(R) = \left|Q_2^{\parallel}(R, S)P^{\parallel}(S)\right|^{1/2} = \mathcal{L}^{\parallel}(R, S)|P^{\parallel}(S)|^{1/2}; \qquad (4.13.76)
$$

see (4.13.51) and (4.13.53). Here $\mathcal{L}^{\parallel}(R, S) = |Q_2^{\parallel}(R, S)|^{1/2}$ is the *relative in-plane geometrical spreading*. In this section, we shall discuss possibilities of determination $\mathcal{L}^{\parallel}(R, S)$ from travel-time data known in the vicinity of S and R. We shall show that $\mathcal{L}^{\parallel}(R, S)$ can be completely determined from the travel-time data known *in the plane* $\Sigma^{\parallel}$ in the vicinity of S and R.

It is not surprising that the transverse geometrical spreading $\mathcal{L}^{\perp}(R)$ cannot be determined from the travel time data in plane $\Sigma^{\parallel}$. It would be necessary to know also the travel times in a vicinity of $\Sigma^{\parallel}$ or to proceed in some other way. If we, however, consider a standard 2-D model ($V_{22} = 0$ and $D_{22} = 0$) with a line source perpendicular to $\Sigma^{\parallel}$ at S, we obtain $\mathcal{L}^{\perp}(R) = 1$, and the determination of $\mathcal{L}^{\parallel}(R, S)$ solves the problem completely. See a more detailed discussion in Note 2 at the end of this section.

The equations for $\mathcal{L}^{\parallel}(R, S)$ we shall derive are applicable even for structural models with $V_{22} \neq 0$ and $D_{22} \neq 0$; see (4.13.7). The quantities V_{22} and D_{22} do not influence $\mathcal{L}^{\parallel}(R, S)$ at all. They influence only the transverse geometrical spreading $\mathcal{L}^{\perp}(R)$.

In Section 4.10.5, we used two methods to determine the relative geometrical spreading $\mathcal{L}(R, S)$ in 3-D layered structures from travel-time data known in the vicinity of S and R. Two analogous methods will be used here to determine the relative in-plane geometrical spreading $\mathcal{L}^\parallel(R, S)$.

a. The first method is based on the first and second derivatives of the travel-time fields along certain curves C_1 and C_2 situated in the plane $\Sigma^\parallel$ and passing through S and R. Exactly in the same way as in Section 4.10.5, we obtain the following expression for $\mathcal{L}^\parallel(R, S)$:

$$\mathcal{L}^\parallel(R, S) = |M^\parallel(R) - M^\parallel(R, S)|^{-1/4} |M^\parallel(S) - M^\parallel(S, R)|^{-1/4}. \quad (4.13.77)$$

Here $M^\parallel$ denote the second derivatives of the travel-time field *with respect to* q_1. All these derivatives can be determined from travel-time data known along the plane $\Sigma^\parallel$. They correspond to the three experiments described in Section 4.10.5. In the first experiment, the point (or line) source is situated at S, and $M^\parallel(R, S)$ denotes the second derivative of the travel-time field at R. In the second experiment, the point (or line) source is situated at R, and $M^\parallel(S, R)$ denotes the second derivative of the travel-time field at S. In the third experiment, the source may be situated either at S or at R, but the generated wavefront in the plane $x_2 = 0$ must be different from those considered in the two previous experiments. It may, for example, correspond to a straight line passing through S or R. Quantities $M^\parallel(R)$ and $M^\parallel(S)$ then correspond to the second derivatives of the relevant travel-time fields at R and S. See Figure 4.16.

The second derivatives of the travel-time field with respect to q_1 are not suitable for measurements. They can, however, be expressed in terms of the second derivatives of the travel-time field along any curves C_1 and C_2 passing through points S and R. Using Equation (4.5.32) and the notations of Section 4.13.2, we obtain

$$M^\parallel = V^{-2}(\vec{p} \cdot \vec{n})^{-2}[T^\parallel - E^\parallel + (\vec{p} \cdot \vec{n})D^\parallel]. \quad (4.13.78)$$

Here $T^\parallel$ represents the second derivative of the travel-time field along the relevant curve, C_1 or C_2, with respect to z_1 (tangent to the curve). It is interesting that the terms with $E^\parallel$ and $D^\parallel$ vanish if (4.13.78) is inserted into (4.13.77). We obtain

$$\mathcal{L}^\parallel(R, S) = |\cos i(S) \cos i(R)|^{1/2} |T^\parallel(R) - T^\parallel(R, S)|^{-1/4}$$
$$\times |T^\parallel(S, R) - T^\parallel(S)|^{-1/4}. \quad (4.13.79)$$

Here $T^\parallel(R)$, $T^\parallel(R, S)$, $T^\parallel(S, R)$, and $T^\parallel(S)$ have the same meaning as $M^\parallel(R)$, $M^\parallel(R, S)$, $M^\parallel(S, R)$, and $M^\parallel(S)$, with the exception that the second derivatives of the travel-time field are taken along curves C_1 and C_2 in the plane $\Sigma^\parallel$. The quantity $i(S)$ is the acute angle between the normal to the curve C_1 and slowness vector $\vec{p}$ at the point S. Analogously, $i(R)$ has the same meaning at R.

Curves C_1 and C_2 may be arbitrary, with the exception of curves tangent to ray Ω, for which $\vec{p} \cdot \vec{n} = 0$. Factor $|\cos i(S) \cos i(R)|^{1/2}$ can also be expressed in a more specific form,

$$|\cos i(S) \cos i(R)|^{1/2} = (V(S)V(R))^{1/2}[(\vec{p}(S) \cdot \vec{n}(S))(\vec{p}(R) \cdot \vec{n}(R))]^{1/2}$$
$$= (V(S)V(R))^{1/2}[(\partial T/\partial z_3)_S(\partial T/\partial z_3)_R]^{1/2}.$$

$$(4.13.80)$$

Thus, in addition to the second derivatives of the travel-time field along curves C_1 and

C_2, we also need to determine the first derivatives of the travel-time field perpendicular to curves C_1 and C_2. Moreover, we need to know $V(S)$ and $V(R)$.

In principle, the first derivatives of the travel-time field perpendicular to curves C_1 and C_2 need not be determined, it is sufficient to determine the first derivatives of the travel-time field on curves C_1 and C_2. Using the eikonal equation, we obtain $(\partial T/\partial z_3)^2 = 1/V^2 - (\partial T/\partial z_1)^2$. Thus, the alternative relation to (4.13.80) is

$$|\cos i(S) \cos i(R)|^{1/2} = \left|1 - V^2(S)(\partial T/\partial z_1)_S^2\right|^{1/4}\left|1 - V^2(R)(\partial T/\partial z_1)_R^2\right|^{1/4}.$$
(4.13.81)

Assuming that the velocities $V(S)$ and $V(R)$ are known, Equations (4.13.79) with (4.13.81) require only the *tangential* first and second derivatives of the travel-time field on C_1 and C_2 to be determined.

It may be useful to take curves C_1 and C_2 in the plane x_1x_3 in a special way: along coordinate lines $x_1 = $ const. (vertical line) and/or $x_3 = $ const. (horizontal line). As an example, let us consider the sources and receivers distributed along two vertical lines $x_1 = $ const. (cross-hole configuration). The final equation then reads

$$\mathcal{L}^{\|}(R, S) = \left|1 - V^2(S)(\partial T/\partial x_3)_S^2\right|^{1/4}\left|1 - V^2(R)(\partial T/\partial x_3)_R^2\right|^{1/4}$$
$$\times \left|T^{\|}(R) - T^{\|}(R, S)\right|^{-1/4}\left|T^{\|}(S, R) - T^{\|}(S)\right|^{-1/4};$$ (4.13.82)

see (4.13.79) and (4.13.81). The second derivatives $T^{\|}$ in (4.13.82) are taken with respect to x_3.

b. In the second method, the in-plane relative geometrical spreading $\mathcal{L}^{\|}(R, S)$ is obtained from the *second mixed* derivatives of the travel-time field. Using (4.10.50), we obtain

$$\mathcal{L}^{\|}(R, S) = \left|Q_2^{\|}(R, S)\right|^{1/2} = |\cos i(S) \cos i(R)|^{1/2}\left|M_{11}^{\Sigma mix}(R, S)\right|^{-1/2}.$$
(4.13.83)

Here $M_{11}^{\Sigma mix}$ is given by (4.10.46). It represents the mixed second derivative of the travel-time field with respect to $z_1(S)$ and $z_1(R)$, taken along arbitrary curves C_1 and C_2 passing through points S and R. To express $|\cos i(S) \cos i(R)|^{1/2}$ in terms of the first derivatives of the travel-time field and velocities, either Equation (4.13.80) or Equation (4.13.81) can be used.

It may again be convenient to consider the coordinate lines $x_1 = $ const. and/or $x_3 = $ const. of a general Cartesian coordinate system as C_1 and C_2. We have four such lines and four relevant equations for $\mathcal{L}^{\|}(R, S) = |Q_2^{\|}(R, S)|^{1/2}$:

$$\mathcal{L}^{\|}(R, S) = (V(S)V(R))^{1/2}|p_3(S)p_3(R)|^{1/2}\left|M_{11}^{mix}(R, S)\right|^{-1/2},$$
$$\mathcal{L}^{\|}(R, S) = (V(S)V(R))^{1/2}|p_3(S)p_1(R)|^{1/2}\left|M_{13}^{mix}(R, S)\right|^{-1/2},$$
$$\mathcal{L}^{\|}(R, S) = (V(S)V(R))^{1/2}|p_1(S)p_3(R)|^{1/2}\left|M_{31}^{mix}(R, S)\right|^{-1/2},$$
$$\mathcal{L}^{\|}(R, S) = (V(S)V(R))^{1/2}|p_1(S)p_1(R)|^{1/2}\left|M_{33}^{mix}(R, S)\right|^{-1/2}.$$
(4.13.84)

Here M_{ij}^{mix} represent the relevant mixed second derivatives of the travel-time field with respect to general Cartesian coordinates given by (4.9.25). Relations (4.13.84), of course, also follow from (4.13.83). For example, if both curves C_1 and C_2 are taken along x_1 (so that they are horizontal), (4.13.83) immediately yields the first equation of (4.13.84). Similarly, if C_1 and C_2 are vertical (cross-hole confuguration), (4.13.83) yields the last equation of (4.13.84).

As in the first method, the Cartesian component of slowness vector p_1 can be expressed in terms of p_3 using the eikonal equation, and vice versa. As an example, we shall again consider the sources and receivers distributed along two vertical lines $x_1 = $ const. (cross-hole configuration), as in (4.13.82). We then take the last equation of (4.13.84) and obtain

$$\mathcal{L}^{\parallel}(R, S) = \left|1 - V^2(S)(\partial T/\partial x_3)_S^2\right|^{1/4} \left|1 - V^2(R)(\partial T/\partial x_3)_R^2\right|^{1/4}$$
$$\times \left|\partial^2 T/\partial x_3(S)\partial x_3(R)\right|^{-1/2}. \tag{4.13.85}$$

Equations (4.13.84) can be expressed in many other alternative forms, depending on the source-receiver configurations under consideration.

Let us add three notes related to the determination of relative geometrical spreading from travel-time data in 2-D models.

Note 1. The determination of the first derivatives, and particularly of the second derivatives, of the travel-time field is, in general, a very unstable procedure. It is obvious that standard numerical methods may fail and that the travel-time data will require some smoothing.

Note 2. In practical applications, we are more interested in the geometrical spreading due to *point sources* situated in plane $x_2 = 0$ (the so-called 2.5-D case) than in line sources. In this case, the in-plane geometrical spreading $\mathcal{L}^{\parallel}(R, S)$ remains the same as for line sources, but the transverse geometrical spreading is changed. Unfortunately, $\mathcal{L}^{\perp}(R, S)$ cannot be determined from the travel-time data in plane $x_2 = 0$; the data outside this plane would also be required. Moreover, it would also be necessary to place several sources outside this plane. In fact, a complete 3-D system would be required. In certain applications, however, it would be possible to determine $\mathcal{L}^{\perp}(R, S)$ by simple quadratures along the ray,

$$\mathcal{L}^{\perp}(R, S) = |\sigma(R, S)|^{1/2} = \left|\int_S^R V \, ds\right|^{1/2}; \tag{4.13.86}$$

see (4.13.54). Here the integral is taken along ray Ω. The computation of $\mathcal{L}^{\perp}(R, S)$ using the foregoing integral is very robust and simple, but ray Ω must be known. The problem is that ray Ω is not known. Ray Ω, however, may be known approximately, and this could be sufficient to provide a good estimate of $\mathcal{L}^{\perp}(R, S)$. As an example, let us consider the direct computation of travel times at the grid points of a network, based on finite differences or the network ray tracing; see Section 3.8. In such direct computations, the rays are not used but may be approximately estimated. This approximate estimate of the ray may be applied in computing $\mathcal{L}^{\perp}(R, S)$ using (4.13.86) with good accuracy.

Note 3. The mixed second derivatives of the travel-time field in (4.13.84) can be approximately expressed in terms of first derivatives of the components of the slowness vector, computed for at least two point sources. The resulting equations may be suitable if the slowness vectors are known along the coordinate lines, instead of the travel time or in addition to it. See also Section 4.10.5 and Klimeš (2000).

4.14 Dynamic Ray Tracing in Inhomogeneous Anisotropic Media

Ray tracing in an inhomogeneous anisotropic medium is, in principle, simple, although tedious; see Section 3.6. If we express the eikonal equation in Hamiltonian form $\mathcal{H}^{(x)}(x_i, p_i^{(x)}) = 0$, we can express the ray tracing system in the well-known form (4.2.2). In a similar way, we can also express the dynamic ray tracing system; see (4.2.4).

The dynamic ray tracing systems for anisotropic inhomogeneous media can be expressed in various coordinate systems. The most important are the Cartesian coordinate system, the nonorthogonal ray-centered coordinate system, and the wavefront orthonormal coordinate system. The system in Cartesian coordinates consists of six linear ordinary differential equations of the first order. The system consisting of six linear ordinary equations does not require the basis vectors $\vec{e}_1$ and $\vec{e}_2$ to be computed along the ray. The systems in the nonorthogonal ray-centered coordinate system and in the wavefront orthonormal coordinates can be simply reduced to four equations. In this section, we shall only use the Cartesian coordinates and the wavefront orthonormal coordinates because we prefer to work with orthonormal coordinate systems. For more details on dynamic ray tracing in inhomogeneous anisotropic media, see Červený (1972), Hanyga (1982a), Gajewski and Pšenčík (1987a), Kendall and Thomson (1989), Gibson, Sena, and Toksöz (1991), Kendall, Guest, and Thomson (1992), Klimeš (1994), Farra and Le Bégat (1995), Bakker (1996), and Farra (1999). For some applications in seismic exploration, see Grechka, Tsvankin, and Cohen (1999).

Remember that basis vectors $\vec{e}_1$ and $\vec{e}_2$ in isotropic media play an important role not only from the computational point of view but also from the *seismological point of view*: $\vec{e}_1$ and $\vec{e}_2$ represent the polarization vectors of S waves. This is not the case in anisotropic inhomogeneous media, where the polarization of the individual wave propagating along the ray is represented by the relevant eigenvectors $\vec{g}^{(m)}$ of the Christoffel matrix $\hat{\Gamma}$. Thus, in anisotropic inhomogeneous media, basis vectors $\vec{e}_1$ and $\vec{e}_2$ represent merely auxiliary vectors, without any distinct importance in polarization studies.

4.14.1 Dynamic Ray Tracing in Cartesian Coordinates

In this section, we shall consider Hamiltonian $\mathcal{H}^{(x)}(x_i, p_i^{(x)})$ in the form of (3.6.3),

$$\mathcal{H}^{(x)}(x_i, p_i^{(x)}) = \tfrac{1}{2}\left(G_m(x_i, p_i^{(x)}) - 1\right), \tag{4.14.1}$$

where $G_m, m = 1, 2, 3$, are eigenvalues of the 3×3 Christoffel matrix $\hat{\Gamma}$ with components $\Gamma_{ik} = a_{ijkl} p_j^{(x)} p_l^{(x)}$. Eigenvalues G_m can be conveniently expressed as

$$G_m = \Gamma_{ik} g_i^{(m)} g_k^{(m)} = a_{ijkl} p_j^{(x)} p_l^{(x)} g_i^{(m)} g_k^{(m)}; \tag{4.14.2}$$

see (2.2.34). The monotonic parameter u along the ray corresponding to Hamiltonian (4.14.1) is travel time, T. The ray tracing system then reads

$$\frac{dx_n}{dT} = \frac{\partial \mathcal{H}^{(x)}}{\partial p_n^{(x)}}, \qquad \frac{dp_n^{(x)}}{dT} = -\frac{\partial \mathcal{H}^{(x)}}{\partial x_n}, \qquad n = 1, 2, 3. \tag{4.14.3}$$

We now put

$$Q_n^{(x)} = (\partial x_n / \partial \gamma)_{T=\text{const.}}, \qquad P_n^{(x)} = (\partial p_n^{(x)} / \partial \gamma)_{T=\text{const.}}, \tag{4.14.4}$$

where γ is some initial parameter such as the ray parameter, and $n = 1, 2, 3$. The *dynamic ray tracing system* then reads

$$dQ_n^{(x)}/dT = A_{nq}^{(x)} Q_q^{(x)} + B_{nq}^{(x)} P_q^{(x)}, \qquad dP_n^{(x)}/dT = -C_{nq}^{(x)} Q_q^{(x)} - D_{nq}^{(x)} P_q^{(x)}. \tag{4.14.5}$$

Here $A_{nq}^{(x)}$, $B_{nq}^{(x)}$, $C_{nq}^{(x)}$, and $D_{nq}^{(x)}$ are given by (4.2.5); see also (4.14.7). Indices n and q take the values $n = 1, 2, 3$ and $q = 1, 2, 3$. System (4.14.5) consists of six ordinary differential equations of the first order in six unknown quantities $Q_n^{(x)}$ and $P_n^{(x)}$, $n = 1, 2, 3$. The system is linear. We can also replace $\mathcal{H}^{(x)}$ by $\tfrac{1}{2} G_m$ in the whole system.

To express the RHSs of systems (4.14.3) and (4.14.5) in practical terms, we need to determine the first and second derivatives of the Hamiltonian. For the first derivatives, we obtain

$$\frac{\partial \mathcal{H}^{(x)}}{\partial p_n^{(x)}} = \frac{1}{2}\frac{\partial \Gamma_{ik}}{\partial p_n^{(x)}}g_i^{(m)}g_k^{(m)}, \qquad \frac{\partial \mathcal{H}^{(x)}}{\partial x_n} = \frac{1}{2}\frac{\partial \Gamma_{ik}}{\partial x_n}g_i^{(m)}g_k^{(m)}; \qquad (4.14.6)$$

see (3.6.7) and (4.14.1). Similarly, the second derivatives are given by equations

$$A_{nq}^{(x)} = \frac{\partial^2 \mathcal{H}^{(x)}}{\partial p_n^{(x)} \partial x_q} = \frac{1}{2}\frac{\partial^2 \Gamma_{ik}}{\partial p_n^{(x)} \partial x_q}g_i^{(m)}g_k^{(m)} + \frac{\partial \Gamma_{ik}}{\partial p_n^{(x)}}g_i^{(m)}\frac{\partial g_k^{(m)}}{\partial x_q},$$

$$B_{nq}^{(x)} = \frac{\partial^2 \mathcal{H}^{(x)}}{\partial p_n^{(x)} \partial p_q^{(x)}} = \frac{1}{2}\frac{\partial^2 \Gamma_{ik}}{\partial p_n^{(x)} \partial p_q^{(x)}}g_i^{(m)}g_k^{(m)} + \frac{\partial \Gamma_{ik}}{\partial p_n^{(x)}}g_i^{(m)}\frac{\partial g_k^{(m)}}{\partial p_q^{(x)}},$$

$$C_{nq}^{(x)} = \frac{\partial^2 \mathcal{H}^{(x)}}{\partial x_n \partial x_q} = \frac{1}{2}\frac{\partial^2 \Gamma_{ik}}{\partial x_n \partial x_q}g_i^{(m)}g_k^{(m)} + \frac{\partial \Gamma_{ik}}{\partial x_n}g_i^{(m)}\frac{\partial g_k^{(m)}}{\partial x_q}, \qquad (4.14.7)$$

$$D_{nq}^{(x)} = \frac{\partial^2 \mathcal{H}^{(x)}}{\partial x_n \partial p_q^{(x)}} = \frac{1}{2}\frac{\partial^2 \Gamma_{ik}}{\partial x_n \partial p_q^{(x)}}g_i^{(m)}g_k^{(m)} + \frac{\partial \Gamma_{ik}}{\partial x_n}g_i^{(m)}\frac{\partial g_k^{(m)}}{\partial p_q^{(x)}}.$$

We now need to express the first- and second-order derivatives of Γ_{ik} and the first-order derivatives of the components of eigenvector $\vec{g}^{(m)}$.

For the derivatives of Γ_{ik}, we obtain

$$\frac{\partial \Gamma_{ik}}{\partial p_n^{(x)}} = (a_{inkl} + a_{ilkn})p_l^{(x)}, \qquad \frac{\partial \Gamma_{ik}}{\partial x_n} = \frac{\partial a_{ijkl}}{\partial x_n}p_j^{(x)}p_l^{(x)},$$

$$\frac{\partial^2 \Gamma_{ik}}{\partial p_n^{(x)} \partial p_q^{(x)}} = a_{inkq} + a_{iqkn}, \qquad \frac{\partial^2 \Gamma_{ik}}{\partial x_n \partial p_q^{(x)}} = \left(\frac{\partial a_{iqkl}}{\partial x_n} + \frac{\partial a_{ilkq}}{\partial x_n}\right)p_l^{(x)},$$

$$\frac{\partial^2 \Gamma_{ik}}{\partial p_n^{(x)} \partial x_q} = \left(\frac{\partial a_{inkl}}{\partial x_q} + \frac{\partial a_{ilkn}}{\partial x_q}\right)p_l^{(x)}, \qquad \frac{\partial^2 \Gamma_{ik}}{\partial x_n \partial x_q} = \frac{\partial^2 a_{ijkl}}{\partial x_n \partial x_q}p_j^{(x)}p_l^{(x)}; \qquad (4.14.8)$$

see also (3.6.9). The *derivatives of the components of eigenvector* $\vec{g}^{(m)}$ can be expressed in terms of the two remaining eigenvectors $\vec{g}^{(r)}$, where $r \neq m$. Assume, for a while, that $m = 1$. We wish to determine $\partial g_k^{(1)}/\partial a$, where a may be x_q, $p_q^{(x)}$, or some other quantity. Since $g_k^{(1)}g_k^{(1)} = 1$, $g_k^{(1)}\partial g_k^{(1)}/\partial a = 0$. This means that $\partial \vec{g}^{(1)}/\partial a$ is perpendicular to $\vec{g}^{(1)}$ and

$$\partial g_k^{(1)}/\partial a = A_2 g_k^{(2)} + A_3 g_k^{(3)}. \qquad (4.14.9)$$

Taking the derivative of (2.2.32) with respect to a yields

$$(\partial \Gamma_{ik}/\partial a - \delta_{ik}\partial G_1/\partial a)g_k^{(1)} + (\Gamma_{ik} - G_1\delta_{ik})\big(A_2 g_k^{(2)} + A_3 g_k^{(3)}\big) = 0.$$

We now take into account that $\Gamma_{ik}g_k^{(2)} = G_2 g_i^{(2)}$ and $\Gamma_{ik}g_k^{(3)} = G_3 g_i^{(3)}$. Then

$$(\partial \Gamma_{ik}/\partial a - \delta_{ik}\partial G_1/\partial a)g_k^{(1)} + (G_2 - G_1)A_2 g_i^{(2)} + (G_3 - G_1)A_3 g_i^{(3)} = 0.$$

If we multiply this equation by $g_i^{(2)}$ (or $g_i^{(3)}$) and take into account that $(\partial G_1/\partial a)g_i^{(1)}g_i^{(2)} = 0$ (and $(\partial G_1/\partial a)g_i^{(1)}g_i^{(3)} = 0$), we obtain the final equations for A_2 (and A_3):

$$A_2 = \frac{1}{G_1 - G_2}\frac{\partial \Gamma_{ik}}{\partial a}g_k^{(1)}g_i^{(2)}, \qquad A_3 = \frac{1}{G_1 - G_3}\frac{\partial \Gamma_{ik}}{\partial a}g_k^{(1)}g_i^{(3)}. \quad (4.14.10)$$

Along the ray under consideration, $G_1 = 1$. This implies $G_2 = (\mathcal{C}^{(2)}/\mathcal{C}^{(1)})^2$ and $G_3 = (\mathcal{C}^{(3)}/\mathcal{C}^{(1)})^2$, where $\mathcal{C}^{(i)}$ are the phase velocities. Equations (4.14.10) then yield

$$A_2 = \frac{\mathcal{C}^{(1)2}}{\mathcal{C}^{(1)2} - \mathcal{C}^{(2)2}} \frac{\partial \Gamma_{ik}}{\partial a} g_k^{(1)} g_i^{(2)}, \qquad A_3 = \frac{\mathcal{C}^{(1)2}}{\mathcal{C}^{(1)2} - \mathcal{C}^{(3)2}} \frac{\partial \Gamma_{ik}}{\partial a} g_k^{(1)} g_i^{(3)}.$$

$$(4.14.11)$$

Inserting (4.14.11) into (4.14.9) yields the final expression for $\partial g_k^{(1)}/\partial a$. If $a = x_q$ or $a = p_q^{(x)}$, the relevant expressions for $\partial \Gamma_{ik}/\partial a$ are given by (4.14.8). The procedure fails for $\mathcal{C}^{(1)} = \mathcal{C}^{(2)}$ and for $\mathcal{C}^{(1)} = \mathcal{C}^{(3)}$.

Mutatis mutandis, we also obtain the relations for $\partial g_k^{(2)}/\partial a$ and $\partial g_k^{(3)}/\partial a$ from (4.14.9) and (4.14.11).

The relations we shall use contain group velocity vector $\vec{\mathcal{U}} = \mathrm{d}\vec{x}/\mathrm{d}T$ and vector $\vec{\eta} = \mathrm{d}\vec{p}/\mathrm{d}T$. In anisotropic media, the Cartesian components of $\vec{\mathcal{U}}$ and of $\vec{\eta}$ are given by relations

$$\mathcal{U}_i^{(x)} = \mathrm{d}x_i/\mathrm{d}T = \partial \mathcal{H}^{(x)}/\partial p_i^{(x)} = a_{ijkl} p_l^{(x)} g_j^{(m)} g_k^{(m)},$$

$$\eta_i^{(x)} = \mathrm{d}p_i^{(x)}/\mathrm{d}T = -\partial \mathcal{H}^{(x)}/\partial x_i \qquad (4.14.12)$$

$$= -\tfrac{1}{2}(\partial a_{jkln}/\partial x_i) p_k^{(x)} p_n^{(x)} g_j^{(m)} g_l^{(m)}.$$

4.14.2 Dynamic Ray Tracing in Wavefront Orthonormal Coordinates

Similarly as in Section 4.2.2, we now consider wavefront orthonormal coordinates y_i. We introduce basis vectors $\vec{e}_1, \vec{e}_2$, and $\vec{e}_3$ along ray Ω so that $\vec{e}_3 = \mathcal{C}\vec{p}$ and $\vec{e}_1$ and $\vec{e}_2$ are computed using (4.2.17). The 3×3 unitary transformation matrix $\hat{\mathbf{H}}$ from the wavefront orthonormal coordinates y_i to the global Cartesian coordinates x_k is defined by (4.2.18) and (4.2.19). Even in the dynamic ray tracing system in wavefront orthonormal coordinates y_i, we shall consider Hamiltonian $\mathcal{H}^{(x)}(x_i, p_i^{(x)})$; see (4.14.1). The solutions of this system for ray-tangent and noneikonal initial conditions can be found analytically; see (4.2.24) and (4.2.25). For standard paraxial initial conditions, the number of equations of the dynamic ray tracing system reduces from six to four:

$$\mathrm{d}Q_M^{(y)}/\mathrm{d}T = A_{MN}^{(y)} Q_N^{(y)} + B_{MN}^{(y)} P_N^{(y)}, \qquad \mathrm{d}P_M^{(y)}/\mathrm{d}T = -C_{MN}^{(y)} Q_N^{(y)} - D_{MN}^{(y)} P_N^{(y)}.$$

$$(4.14.13)$$

Here $Q_I^{(y)} = (\mathrm{d}y_I/\partial \gamma)_{T=const.}$ and $P_I^{(y)} = (\mathrm{d}p_I^{(y)}/\mathrm{d}\gamma)_{T=const.}$; see (4.2.20). Additionally, $A_{MN}^{(y)}$, $B_{MN}^{(y)}$, $C_{MN}^{(y)}$, and $D_{MN}^{(y)}$ are given by (4.2.32) and satisfy symmetry relations (4.2.33). The paraxial ray tracing system for y_M and $p_M^{(y)}$ can be expressed in the same form as (4.14.13), where $Q_M^{(y)}$ and $P_M^{(y)}$ are replaced by y_M and $p_M^{(y)}$.

Let us now consider an orthonomic system of rays, parameterized by two ray parameters γ_1 and γ_2. We introduce

$$Q_{nK}^{(x)} = (\partial x_n/\partial \gamma_K)_{T=const.}, \qquad P_{nK}^{(x)} = (\partial p_n^{(x)}/\partial \gamma_K)_{T=const.}$$

$$Q_{nK}^{(y)} = (\partial y_n/\partial \gamma_K)_{T=const.}, \qquad P_{nK}^{(y)} = (\partial p_n^{(y)}/\partial \gamma_K)_{T=const.} \qquad (4.14.14)$$

To find $Q_{nK}^{(x)}$ and $P_{nK}^{(x)}$ by dynamic ray tracing in Cartesian coordinates, dynamic ray tracing system (4.14.5) should be solved twice so that we must solve *12 equations*. Similarly, to find $Q_{NK}^{(y)}$ and $P_{NK}^{(y)}$ by dynamic ray tracing in wavefront orthonormal coordinates, dynamic ray tracing system (4.14.13) should be solved twice, so that we must solve *8 equations*.

After $Q_{nK}^{(x)}$ and $P_{nK}^{(x)}$ are known, we can find the complete 3×3 matrices $\hat{\mathbf{Q}}^{(x)}$ and $\hat{\mathbf{P}}^{(x)}$. We merely supplement $Q_{nK}^{(x)}$ and $P_{nK}^{(x)}$ by ray tangent solutions $Q_{n3}^{(x)} = \mathcal{U}_n^{(x)}$ and $P_{n3}^{(x)} = \eta_n^{(x)}$. We can, however, also find complete 3×3 matrices $\hat{\mathbf{Q}}^{(y)}$ and $\hat{\mathbf{P}}^{(y)}$. At the point of consideration, we can choose $\vec{e}_1$ and $\vec{e}_2$ arbitrarily; they only need to form a right-handed orthogonal system with $\vec{e}_3 = \mathcal{C}\vec{p}^{(x)}$. We can then use transformation relations $Q_{nk}^{(y)} = H_{in}Q_{ik}^{(x)}$ and $P_{nk}^{(y)} = H_{in}P_{ik}^{(x)}$. Thus, the computation of $\vec{e}_1$ and $\vec{e}_2$ along Ω is not required in this case.

The opposite situation is similar. After $Q_{NK}^{(y)}$ and $P_{NK}^{(y)}$ are determined from (4.14.13), the complete 3×3 matrices $\hat{\mathbf{Q}}^{(y)}$ and $\hat{\mathbf{P}}^{(y)}$ can be determined using (4.2.42). To determine $\hat{\mathbf{Q}}^{(x)}$ and $\hat{\mathbf{P}}^{(x)}$ from $\mathbf{Q}^{(y)}$ and $\mathbf{P}^{(y)}$, we can use (4.2.50) and supplement it by the ray-tangent solution (known from ray tracing). In this case, the computation of unit vectors $\vec{e}_1$ and $\vec{e}_2$ along the ray is required.

Both approaches are fully alternative. Deciding whether to perform dynamic ray tracing in Cartesian coordinates (12 equations) or in wavefront orthonormal coordinates (8 equations) depends mostly on the numerical efficiency of the systems. Most computer time in both systems is spent on computing second partial derivatives of the Hamiltonian. The disadvantage of the system in Cartesian coordinates is that it consists of 12 equations, whereas in wavefront orthonormal coordinates the number of equations is only 8. On the contrary, dynamic ray tracing system (4.14.13) requires H_{iN} to be determined along Ω; see (4.2.17). Moreover, the RHSs of the dynamic ray tracing (4.14.13) require many additional multiplications with H_{iN}; see (4.2.32).

In general, the dynamic ray tracing system (4.14.5) in Cartesian coordinates may be numerically more efficient in anisotropic media. Nevertheless, the construction of $\hat{\mathbf{Q}}^{(y)}$ and $\hat{\mathbf{P}}^{(y)}$, from computed $Q_{nK}^{(x)}$ and $P_{nK}^{(x)}$, may be very useful. For example, they can be used to construct the 4×4 propagator matrix $\mathbf{\Pi}$.

4.14.3 The 4×4 Ray Propagator Matrix in Anisotropic Inhomogeneous Media

Let us consider ray Ω and two points, S and R, situated on this ray. Because dynamic ray tracing system (4.14.13) is linear, we can introduce the 4×4 ray propagator matrix $\mathbf{\Pi}(R, S)$. It consists of four linearly independent solutions of (4.14.13), satisfying initial conditions $\mathbf{\Pi}(S, S) = \mathbf{I}$ at point S, where $\mathbf{I}$ is the 4×4 identity matrix. As in isotropic media, we introduce four 2×2 matrices $\mathbf{Q}_1(R, S)$, $\mathbf{Q}_2(R, S)$, $\mathbf{P}_1(R, S)$, and $\mathbf{P}_2(R, S)$ using (4.3.5). Matrices $\mathbf{Q}_1(R, S)$ and $\mathbf{P}_1(R, S)$ correspond to the *normalized plane wavefront initial conditions* at S: $\mathbf{Q}_1(S, S) = \mathbf{I}$, $\mathbf{P}_1(S, S) = \mathbf{0}$. Similarly, matrices $\mathbf{Q}_2(R, S)$ and $\mathbf{P}_2(R, S)$ correspond to the *normalized point source initial conditions*: $\mathbf{Q}_2(R, S) = \mathbf{0}$ and $\mathbf{P}_2(R, S) = \mathbf{I}$. The 4×4 ray propagator matrix $\mathbf{\Pi}(R, S)$ can be used to continue the $\mathbf{Q}^{(y)}$ and $\mathbf{P}^{(y)}$ along ray Ω:

$$\begin{pmatrix} \mathbf{Q}^{(y)}(R) \\ \mathbf{P}^{(y)}(R) \end{pmatrix} = \mathbf{\Pi}(R, S) \begin{pmatrix} \mathbf{Q}^{(y)}(S) \\ \mathbf{P}^{(y)}(S) \end{pmatrix}. \tag{4.14.15}$$

Similarly, paraxial ray Ω' is described by the relation

$$\begin{pmatrix} \mathbf{y}(R) \\ \mathbf{p}^{(y)}(R) \end{pmatrix} = \mathbf{\Pi}(R, S) \begin{pmatrix} \mathbf{y}(S) \\ \mathbf{p}^{(y)}(S) \end{pmatrix}. \tag{4.14.16}$$

Because symmetry relations (4.2.33) are satisfied along the whole ray, ray propagator matrix $\mathbf{\Pi}(R, S)$ is symplectic even in anisotropic inhomogeneous media. Moreover, $\mathbf{\Pi}(R, S)$ satisfies chain rule (4.3.20) and relations (4.3.26) for its inverse. Finally, $\det \mathbf{\Pi}(R, S) = 1$ along the whole ray Ω.

Note that the elements of the ray propagator matrix depend on the choice of $\vec{e}_1(S)$ and $\vec{e}_2(S)$. They, however, do not depend on the ray parameters, although $\mathbf{Q}^{(y)}(S)$ and $\mathbf{P}^{(y)}(S)$ depend on them. Equations (4.14.16) also offer *a simple physical interpretation of the elements of the ray propagator matrix* $\Pi_{\alpha\beta}(R, S)$, where $\alpha, \beta = 1, 2, 3, 4$. Let $(W_1, W_2, W_3, W_4)^T = (y_1, y_2, p_1^{(y)}, p_2^{(y)})^T$ be the four-dimensional phase-space coordinates, then $\Pi_{\alpha\beta}(R, S) = \partial W_\alpha(R)/\partial W_\beta(S)$.

It should again be emphasized that the 4×4 propagator matrix $\Pi(R, S)$ can also be calculated by solving dynamic ray tracing system (4.14.5) in Cartesian coordinates. We only need to use strictly defined initial conditions. For the normalized plane wavefront initial conditions ($Q_{IN}^{(y)} = \delta_{IN}$, $P_{IN}^{(y)} = 0$), we use

$$Q_{iN}^{(x)} = H_{iN}, \qquad P_{iN}^{(x)} = H_{jN}p_i^{(x)}\eta_j^{(x)}; \tag{4.14.17}$$

see (4.2.52). Similarly, for the normalized point-source initial conditions ($Q_{IN}^{(y)} = 0$, $P_{IN}^{(y)} = \delta_{IN}$), we use

$$Q_{iN}^{(x)} = 0, \qquad P_{iN}^{(x)} = H_{iN} - H_{jN}p_i^{(x)}\mathcal{U}_j^{(x)}; \tag{4.14.18}$$

see (4.2.51). Solving (4.14.5) for initial conditions (4.14.17) and (4.14.18), we obtain $Q_{iN}^{(x)}$ and $P_{iN}^{(x)}$ along ray Ω. They can be transformed to $Q_{IN}^{(y)}$ and $P_{IN}^{(y)}$ using relations $Q_{IN}^{(y)} = H_{il}Q_{iN}^{(x)}$ and $P_{IN}^{(y)} = H_{il}P_{iN}^{(x)}$. In this way, we obtain $\mathbf{Q}_1(R, S)$ and $\mathbf{P}_1(R, S)$ for initial conditions (4.14.17) and $\mathbf{Q}_2(R, S)$, $\mathbf{P}_2(R, S)$ for initial conditions (4.14.18). The computation of $\mathbf{Q}_1$, $\mathbf{Q}_2$, $\mathbf{P}_1$, and $\mathbf{P}_2$ requires (4.14.5) to be solved four times, as $N = 1, 2$. Consequently, we must solve only 24 equations, not the complete set of 36 equations that would be required in the computation of the 6×6 propagator matrix $\Pi^{(x)}(R, S)$.

4.14.4 The 4×4 Ray Propagator Matrix in Anisotropic Homogeneous Media

In this section, we shall solve dynamic ray tracing system (4.14.13) with (4.2.32) and find the 4×4 ray propagator matrix for a homogeneous anisotropic medium. We consider Hamiltonian $\mathcal{H}^{(x)}(x_i, p_i^{(x)}) = \frac{1}{2}(G_m(x_i, p_i^{(x)}) - 1)$, where G_m is an eigenvalue of the Christoffel matrix $\hat{\Gamma}$; see (4.14.1). In a homogeneous medium, we obtain $A_{MN}^{(y)} = C_{MN}^{(y)} = D_{MN}^{(y)} = 0$; see (4.2.32). Dynamic ray tracing system (4.14.13) then reads

$$dQ_{MK}^{(y)}/dT = B_{MN}^{(y)}P_{NK}^{(y)}, \qquad dP_{MK}^{(y)}/dT = 0.$$

Here $B_{MN}^{(y)}$ is given by (4.2.32). In a homogeneous medium, $B_{MN}^{(y)}$ is constant along the ray. The solution is

$$Q_{MK}^{(y)}(T) = Q_{MK}^{(y)}(T_0) + (T - T_0)B_{MN}^{(y)}P_{NK}^{(y)}, \qquad P_{MK}^{(y)}(T) = P_{MK}^{(y)}(T_0). \tag{4.14.19}$$

We shall consider an arbitrary straight-line ray Ω situated in a homogeneous anisotropic medium and two points, S and R, situated on it. Slowness vector $\vec{p}$ and group velocity vector $\vec{\mathcal{U}}$ are constant along Ω. We assume that both are known. We shall construct the 4×4 ray propagator matrix $\Pi(R, S)$ along Ω from S to R. For normalized plane-wave initial conditions at S ($Q_{MK}^{(y)}(S) = \delta_{MK}$, $P_{MK}^{(y)}(S) = 0$), the solution is

$$\mathbf{Q}_1(R, S) = \mathbf{I}, \qquad \mathbf{P}_1(R, S) = \mathbf{0}; \tag{4.14.20}$$

see (4.14.19). For normalized point-source initial conditions at S ($Q_{MK}^{(y)}(S) = 0$, $P_{MK}^{(y)}(S) = \delta_{MK}$), Equation (4.14.19) yields

$$\mathbf{Q}_2(R, S) = T(R, S)\mathbf{B}^{(y)}, \qquad \mathbf{P}_2(R, S) = \mathbf{I}. \qquad (4.14.21)$$

Here $T(R, S)$ is the travel time along Ω from S to R, $T(R, S) = l(R, S)/\mathcal{U}$, and $l(R, S)$ is the distance between S and R. The 2×2 matrix $\mathbf{B}^{(y)}$ with elements $B_{IJ}^{(y)}$ is given by (4.2.32). Consequently, the 4×4 ray propagator matrix $\mathbf{\Pi}(R, S)$ is given by relation:

$$\mathbf{\Pi}(R, S) = \begin{pmatrix} \mathbf{I} & T(R, S)\mathbf{B}^{(y)} \\ \mathbf{0} & \mathbf{I} \end{pmatrix}. \qquad (4.14.22)$$

As we can see from (4.14.22), matrix $\mathbf{B}^{(y)}$ plays a basic role in the propagation of seismic body waves in anisotropic homogeneous media. For this reason, we shall offer several alternative relations for it. We shall express $\mathbf{B}^{(y)}$ in terms of the 2×2 curvature matrices $\mathbf{D}^S$, $\mathbf{K}^G$, and $\mathbf{K}(R)$, corresponding to the slowness surface, group velocity surface, and wavefront at R, respectively. We shall also relate $\mathbf{B}^{(y)}$ to matrices $\mathbf{Q}_2(R, S)$ and $\mathbf{M}^{(y)}(R)$.

First, we shall find a very important relation between $\mathbf{B}^{(y)}$ and the 2×2 curvature matrix $\mathbf{D}^S$ of the slowness surface. Using (4.2.32) for $B_{MN}^{(y)}$, multiplying it by $H_{nM}H_{mN}$, we obtain

$$H_{nM}H_{mN}B_{MN}^{(y)} = H_{nM}H_{iM}H_{jN}H_{mN}B_{ij}^+, \qquad (4.14.23)$$

where B_{ij}^+ is given by the relation

$$B_{ij}^+ = \partial^2\mathcal{H}^{(x)}/\partial p_i^{(x)}\partial p_j^{(x)} - (\partial\mathcal{H}^{(x)}/\partial p_i^{(x)})(\partial\mathcal{H}^{(x)}/\partial p_j^{(x)}).$$

Before we discuss (4.14.23), we shall prove that $p_i^{(x)}B_{ij}^+ = 0$ and $p_j^{(x)}B_{ij}^+ = 0$. As we consider Hamiltonian $\mathcal{H}^{(x)}(x_i, p_i^{(x)})$ for which $\partial\mathcal{H}^{(x)}/\partial p_i^{(x)}$ is a homogeneous function of the first degree in $p_i^{(x)}$, we obtain

$$p_i^{(x)}B_{ij}^+ = p_i^{(x)}\partial^2\mathcal{H}^{(x)}/\partial p_i^{(x)}\partial p_j^{(x)} - p_i^{(x)}(\partial\mathcal{H}^{(x)}/\partial p_i^{(x)})(\partial\mathcal{H}^{(x)}/\partial p_j^{(x)}) = 0; \qquad (4.14.24)$$

see (4.2.1). Here we have used Euler's theorem (2.2.24) for homogeneous functions of the first degree $\partial\mathcal{H}^{(x)}/\partial p_j^{(x)}$ in $p_i^{(x)}$, $p_i^{(x)}\partial^2\mathcal{H}^{(x)}/\partial p_i^{(x)}\partial p_j^{(x)} = \partial\mathcal{H}^{(x)}/\partial p_j^{(x)}$, and the relation $p_i^{(x)}\mathcal{U}_i^{(x)} = 1$; see (2.4.51).

Now we return to (4.14.23). Using (2.5.60), we obtain $H_{nM}H_{iM} = \delta_{ni} - N_nN_i = \delta_{ni} - \mathcal{C}^2 p_n^{(x)}p_i^{(x)}$. Then (4.14.23) yields

$$H_{nM}H_{mN}B_{MN}^{(y)} = (\delta_{ni} - \mathcal{C}^2 p_n^{(x)}p_i^{(x)})(\delta_{mj} - \mathcal{C}^2 p_m^{(x)}p_j^{(x)})B_{ij}^+ = B_{nm}^+, \qquad (4.14.25)$$

due to (4.14.24).

In addition to orthonormal triplet $\vec{e}_1(S)$, $\vec{e}_2(S)$, and $\vec{e}_3(S) = \vec{N}(S)$, we shall also introduce an alternative orthonormal triplet $\vec{e}_1^*(S)$, $\vec{e}_2^*(S)$, and $\vec{e}_3^*(S) = \vec{t}(S)$. Whereas $\vec{e}_3(S) = \vec{N}(S)$ is perpendicular to the wavefront, $\vec{e}_3^*(S)$ is tangent to ray Ω. Consequently, $\vec{e}_1(S)$ and $\vec{e}_2(S)$ are tangent to the wavefront, but $\vec{e}_1^*(S)$ and $\vec{e}_2^*(S)$ are tangent to the slowness surface. We also introduce $H_{iN}^* = e_{Ni}^*$ and take into account that H_{iN}^* is constant along Ω in a homogeneous medium. Multiplying (4.14.25) by $H_{nL}^*H_{mK}^*$, we obtain

$$A_{ML}A_{NK}B_{MN}^{(y)} = H_{nL}^*H_{mK}^*B_{nm}^+ = H_{nL}^*H_{mK}^*\partial^2\mathcal{H}^{(x)}/\partial p_n^{(x)}\partial p_m^{(x)}. \qquad (4.14.26)$$

Here we have introduced a 2×2 matrix $\mathbf{A}$ with components A_{ML}:

$$A_{ML} = H_{nM}H_{nL}^*, \qquad \mathbf{A} = \begin{pmatrix} \vec{e}_1 \cdot \vec{e}_1^* & \vec{e}_1 \cdot \vec{e}_2^* \\ \vec{e}_2 \cdot \vec{e}_1^* & \vec{e}_2 \cdot \vec{e}_2^* \end{pmatrix}. \tag{4.14.27}$$

Because group velocity vector $\vec{\mathcal{U}}$ is perpendicular to $\vec{e}_1^*$ and $\vec{e}_2^*$, the second term of B_{nm}^+, containing $(\partial \mathcal{H}^{(x)}/\partial p_n^{(x)})\,(\partial \mathcal{H}^{(x)}/\partial p_m^{(x)})$, vanishes in (4.14.26).

It is not difficult to see that the RHS of (4.14.26) is closely related to the 2×2 curvature matrix $\mathbf{D}^S$ of the slowness surface. Using (4.4.17) and (4.4.18) (with $\epsilon^* = 1$), we obtain

$$D_{LN}^S = H_{nL}^* H_{mN}^* \big(\partial^2 \mathcal{H}^{(x)}/\partial p_n^{(x)} \partial p_m^{(x)}\big) \big/ \big[\big(\partial \mathcal{H}^{(x)}/\partial p_i^{(x)}\big)\big(\partial \mathcal{H}^{(x)}/\partial p_i^{(x)}\big)\big]^{1/2}. \tag{4.14.28}$$

Because $[(\partial \mathcal{H}^{(x)}/\partial p_i^{(x)})(\partial \mathcal{H}^{(x)}/\partial p_i^{(x)})]^{1/2} = \mathcal{U}$, we obtain the final result

$$\mathbf{B}^{(y)} = \mathcal{U}\mathbf{A}^{-1T}\mathbf{D}^S\mathbf{A}^{-1} \tag{4.14.29}$$

and

$$\mathbf{Q}_2(R, S) = \mathcal{U}T(R, S)\mathbf{A}^{-1T}\mathbf{D}^S\mathbf{A}^{-1} = l(R, S)\mathbf{A}^{-1T}\mathbf{D}^S\mathbf{A}^{-1}. \tag{4.14.30}$$

Note that (4.14.29) and (4.14.30) are valid for arbitrary orthonormal triplets $\vec{e}_1$, $\vec{e}_2$, and $\vec{e}_3$ and $\vec{e}_1^*$, $\vec{e}_2^*$, and $\vec{e}_3^*$, with $\vec{e}_3 = \vec{N}$ and $\vec{e}_3^* = \vec{t}$.

Equations (4.14.29) and (4.14.30) can also be used to find relations between the 2×2 matrix of the second derivatives of the travel-time field $\mathbf{M}^{(y)}(R)$ and the 2×2 curvature matrix of slowness surface $\mathbf{D}^S$ for a homogeneous medium. The position of point R may be arbitrary. For a general treatment of matrix $\mathbf{M}^{(y)}$ in an inhomogeneous anisotropic medium, see Section 4.14.6. Here we shall concentrate only on homogeneous anisotropic media for which we can find simple analytical relations. We use the continuation relation (4.6.6), which remains valid even in anisotropic media, and insert (4.14.20) and (4.14.21) for $\mathbf{P}_1, \mathbf{Q}_1, \mathbf{P}_2$, and $\mathbf{Q}_2$. This yields

$$\mathbf{M}^{(y)}(R) = \big[\mathbf{M}^{(y)-1}(S) + \mathbf{Q}_2(R, S)\big]^{-1}. \tag{4.14.31}$$

This is the general continuation relation for $\mathbf{M}^{(y)}$ for the anisotropic homogeneous medium. It is valid both for a point source at S ($\mathbf{M}^{(y)-1}(S) = \mathbf{0}$) and for a smooth wavefront at S ($\mathbf{M}^{(y)-1}(S) \neq \mathbf{0}$). If we insert (4.14.30), we obtain the final relation between $\mathbf{M}^{(y)}(R)$ and $\mathbf{D}^S$:

$$\mathbf{M}^{(y)}(R) = \big[\mathbf{M}^{(y)-1}(S) + l(R, S)\mathbf{A}^{-1T}\mathbf{D}^S\mathbf{A}^{-1}\big]^{-1}. \tag{4.14.32}$$

Similarly, we can find relations for the 2×2 curvature matrix of wavefront $\mathbf{K}(R)$ in terms of $\mathbf{D}^S$. We use $\mathbf{K}(R) = \mathcal{C}\mathbf{M}^{(y)}(R)$, and (4.14.32) yields

$$\mathbf{K}(R) = [\mathbf{K}^{-1}(S) + \mathcal{C}^{-1}l(R, S)\mathbf{A}^{-1T}\mathbf{D}^S\mathbf{A}^{-1}]^{-1}. \tag{4.14.33}$$

For a point source at S, relations (4.14.32) and (4.14.33) simplify, as $\mathbf{M}^{(y)-1}(S) = \mathbf{K}^{-1}(S) = \mathbf{0}$:

$$\mathbf{M}^{(y)}(R) = \mathbf{A}\mathbf{D}^{S-1}\mathbf{A}^T/l(R, S), \qquad \mathbf{K}(R) = \mathcal{C}\mathbf{A}\mathbf{D}^{S-1}\mathbf{A}^T/l(R, S). \tag{4.14.34}$$

Finally, the 2×2 curvature matrix $\mathbf{K}^G$ of the group velocity surface equals the 2×2 curvature matrix $\mathbf{K}$ of the wavefront taken for $T(R, S) = l(R, S)/\mathcal{U} = 1$, that is, for $l(R, S) = \mathcal{U}$.

Then (4.14.34) yields

$$\mathbf{K}^G = (\mathcal{C}/\mathcal{U})\mathbf{A}\mathbf{D}^{S-1}\mathbf{A}^T. \tag{4.14.35}$$

In the computation of ray amplitudes, it is useful to know the determinants of the 2×2 matrices derived earlier. The determinants of curvature matrices $\mathbf{D}^S$, $\mathbf{K}^G$, and $\mathbf{K}(R)$ represent the relevant Gaussian curvatures K^S, K^G, and $K(R)$ of the slowness surface, group velocity surface, and wavefront, respectively: $K^S = \det \mathbf{D}^S$, $K^G = \det \mathbf{K}^G$, $K(R) = \det \mathbf{K}(R)$. To find these relations, we need to know $\det \mathbf{A}$. The relation for $\det \mathbf{A}$ can be obtained by rotating $\vec{e}_1^*$ and $\vec{e}_2^*$ about $\vec{\mathcal{U}}$ and $\vec{e}_1$ and $\vec{e}_2$ about $\vec{p}$ to obtain $\vec{e}_2$ and $\vec{e}_2^*$ perpendicular to the plane specified by $\vec{\mathcal{U}}$ and $\vec{p}$. This is always possible. The determinants of the rotation matrices equal unity, and the elements of $\mathbf{A}$ are then $A_{11} = \vec{e}_1 \cdot \vec{e}_1^* = \cos \gamma = \mathcal{C}/\mathcal{U}$, $A_{22} = \vec{e}_2 \cdot \vec{e}_2^* = 1$, and $A_{12} = A_{21} = 0$. This yields

$$\det \mathbf{A} = \cos \gamma = \mathcal{C}/\mathcal{U}. \tag{4.14.36}$$

Here γ is the angle between $\vec{\mathcal{U}}$ and $\vec{p}$. Using (4.14.29), (4.14.30), and (4.14.34) through (4.14.36), we obtain

$$\det \mathbf{B}^{(y)} = \mathcal{U}^4 K^S/\mathcal{C}^2, \qquad K^G = \mathcal{C}^4/\mathcal{U}^4 K^S, \tag{4.14.37}$$

and

$$\det \mathbf{Q}_2(R, S) = \mathcal{U}^2 l^2 K^S/\mathcal{C}^2, \qquad \det \mathbf{M}^{(y)}(R, S) = \mathcal{C}^2/\mathcal{U}^2 l^2 K^S,$$
$$K(R) = \mathcal{C}^4/\mathcal{U}^2 l^2 K^S. \tag{4.14.38}$$

Here $l = l(R, S)$. Relations (4.14.37) do not depend on distance $l(R, S)$, but relations (4.14.38) do. Note that relation $K^G = \mathcal{C}^4/\mathcal{U}^4 K^S$ was also derived independently by Vavryčuk and Yomogida (1996).

All these relations, of course, remain valid even for isotropic media. We take the Hamiltonian $\mathcal{H}^{(x)} = \frac{1}{2}(V^2 p_k^{(x)} p_k^{(x)} - 1)$ and choose $\vec{e}_1^* = \vec{e}_1$ and $\vec{e}_2^* = \vec{e}_2$. Then $\mathbf{A} = \mathbf{I}$, $\mathbf{B}^{(y)} = V^2\mathbf{I}$, and $\mathbf{D}^S = V\mathbf{I}$. All other relations are straightforward.

4.14.5 Ray Jacobian and Geometrical Spreading

We shall again consider an orthonomic system of rays and wavefront orthonormal coordinates y_i. By dynamic ray tracing (4.14.13), we obtain the 2×2 matrices $\mathbf{Q}^{(y)}$ and $\mathbf{P}^{(y)}$. From the 2×2 matrices $\mathbf{Q}^{(y)}$ and $\mathbf{P}^{(y)}$, we can compute the complete 3×3 matrices $\hat{\mathbf{Q}}^{(y)}$ and $\hat{\mathbf{P}}^{(y)}$ (see (4.2.42)) and $\hat{\mathbf{Q}}^{(x)}$ and $\hat{\mathbf{P}}^{(x)}$ (see (4.2.47)). Consequently, Jacobians J and $J^{(T)}$ and quantity $\Omega^{(T)}$ can be computed in several alternative ways:

$$J^{(T)} = \det \hat{\mathbf{Q}}^{(x)} = \det \hat{\mathbf{Q}}^{(y)} = \mathcal{C} \det \mathbf{Q}^{(y)},$$
$$J = J^{(T)}/\mathcal{U} = \mathcal{U}^{-1} \det \hat{\mathbf{Q}}^{(x)} = \mathcal{U}^{-1} \det \hat{\mathbf{Q}}^{(y)} = (\mathcal{C}/\mathcal{U}) \det \mathbf{Q}^{(y)}, \tag{4.14.39}$$
$$\Omega^{(T)} = J^{(T)}/\mathcal{C} = \mathcal{C}^{-1} \det \hat{\mathbf{Q}}^{(x)} = \mathcal{C}^{-1} \det \hat{\mathbf{Q}}^{(y)} = \det \mathbf{Q}^{(y)};$$

see (3.10.23) and (4.2.42). Here $\Omega^{(T)}$ represents the surface scalar element cut out of the wavefront by the ray tube, normalized with respect to γ_1 and γ_2; see (3.10.15).

Using (4.14.15), we can express the continuation relation for $\mathbf{Q}^{(y)}$ along ray Ω as

$$\mathbf{Q}^{(y)}(R) = \mathbf{Q}_1(R, S)\mathbf{Q}^{(y)}(S) + \mathbf{Q}_2(R, S)\mathbf{P}^{(y)}(S). \tag{4.14.40}$$

For a point source situated at point S, we have $\mathbf{Q}^{(y)}(S) = \mathbf{0}$ so that $\mathbf{Q}^{(y)}(R) = \mathbf{Q}_2(R, S)\mathbf{P}^{(y)}(S)$.

This yields

$$J(R) = (C(R)/U(R)) \det \mathbf{Q}_2(R, S) \det \mathbf{P}^{(y)}(S). \tag{4.14.41}$$

Similarly, for a point source situated at R, we obtain

$$J(S) = (C(S)/U(S)) \det \mathbf{Q}_2^T(R, S) \det \mathbf{P}^{(y)}(R). \tag{4.14.42}$$

It is obvious from (4.14.42) and (4.14.41) that the ray Jacobian is not reciprocal.

The *geometrical spreading*, $\mathcal{L} = |J|^{1/2}$, is obtained from the foregoing equations for J. As we can see, Equation (4.10.3) is valid generally, even for anisotropic media. Alternative expressions are

$$\mathcal{L}(R) = \left| U^{-1}(R) \det \hat{\mathbf{Q}}^{(x)}(R) \right|^{1/2} = \left| (C(R)/U(R)) \det \mathbf{Q}^{(y)}(R) \right|^{1/2}. \tag{4.14.43}$$

For a point source situated at point S, we obtain

$$\mathcal{L}(R) = \left| (C(R)/U(R)) \det \mathbf{P}^{(y)}(S) \det \mathbf{Q}_2(R, S) \right|^{1/2}. \tag{4.14.44}$$

As in isotropic media, we shall introduce the *relative geometrical spreading*, $\mathcal{L}(R, S)$, corresponding to a point source situated at point S, as

$$\mathcal{L}(R, S) = |\det \mathbf{Q}_2(R, S)|^{1/2}. \tag{4.14.45}$$

The relative geometrical spreading is *reciprocal* and does not depend on the actual initial conditions for matrix $\mathbf{P}^{(y)}(S)$. The relation between geometrical spreading $\mathcal{L}(R)$ and the relative geometrical spreading in anisotropic media is

$$\mathcal{L}(R) = \left| (C(R)/U(R)) \det \mathbf{P}^{(y)}(S) \right|^{1/2} \mathcal{L}(R, S). \tag{4.14.46}$$

4.14.6 Matrix of Second Derivatives of the Travel-Time Field

For orthonomic system of rays, the 3×3 matrix $\hat{\mathbf{M}}^{(x)}$ of the second derivatives of the travel-time field with respect to Cartesian coordinates x_i ($M_{ij}^{(x)} = \partial^2 T/\partial x_i \partial x_j$), is given by the relations $\hat{\mathbf{M}}^{(x)} = \hat{\mathbf{P}}^{(x)} \hat{\mathbf{Q}}^{(x)-1}$. Similarly, in wavefront orthonormal coordinates y_i, we obtain $\hat{\mathbf{M}}^{(y)} = \hat{\mathbf{P}}^{(y)} \hat{\mathbf{Q}}^{(y)-1}$:

$$\hat{\mathbf{M}}^{(x)} = \hat{\mathbf{P}}^{(x)} \hat{\mathbf{Q}}^{(x)-1}, \qquad \hat{\mathbf{M}}^{(y)} = \hat{\mathbf{P}}^{(y)} \hat{\mathbf{Q}}^{(y)-1}, \qquad \hat{\mathbf{M}}^{(x)} = \hat{\mathbf{H}} \hat{\mathbf{M}}^{(y)} \hat{\mathbf{H}}^T. \tag{4.14.47}$$

In Section 4.2.3, we have derived equations (4.2.44) and (4.2.45) for $\hat{\mathbf{M}}^{(y)}$, expressed in terms of $\mathbf{P}^{(y)}, \mathbf{Q}^{(y)}, U_N^{(y)}$, and $\eta_n^{(y)}$. Because $U_N^{(y)} = H_{in} U_i^{(x)}$ and $\eta_n^{(y)} = H_{in} \eta_i^{(x)}$ are known from ray tracing, it is sufficient to compute $\mathbf{Q}^{(y)}$ and $\mathbf{P}^{(y)}$ by dynamic ray tracing (4.14.13) if we wish to determine $\mathbf{M}^{(y)}, \hat{\mathbf{M}}^{(y)}$, and $\hat{\mathbf{M}}^{(x)}$ along the whole ray.

The matrix of the curvature of wavefront $\mathbf{K}$ is given by relation

$$\mathbf{K}(R) = C(R)\mathbf{M}^{(y)}(R) = C(R)\mathbf{P}^{(y)}(R)\mathbf{Q}^{(y)-1}(R).$$

We denote by $\mathbf{M}^{(y)}(R, S)$ the 2×2 matrix $\mathbf{M}^{(y)}(R)$ corresponding to a point source situated at point S on ray Ω. Because the continuation relations for $\mathbf{Q}^{(y)}$ and $\mathbf{P}^{(y)}$ along Ω, given by (4.6.2) and (4.6.3), also remain valid in anisotropic media, we obtain

$$\begin{aligned}
\mathbf{M}^{(y)}(R, S) &= \mathbf{P}_2(R, S)\mathbf{Q}_2^{-1}(R, S), \\
\mathbf{M}^{(y)}(S, R) &= -\mathbf{Q}_2^{-1}(R, S)\mathbf{Q}_1(R, S).
\end{aligned} \tag{4.14.48}$$

Here $\mathbf{Q}_1(R, S)$, $\mathbf{Q}_2(R, S)$, and $\mathbf{P}_2(R, S)$ are the 2×2 minors of ray propagator matrix $\mathbf{\Pi}(R, S)$ for an anisotropic medium; see Section 4.14.3.

Using the dynamic ray tracing system (4.14.13), we can derive a differential equation for the 2×2 matrix $\mathbf{M}^{(y)} = \mathbf{P}^{(y)}\mathbf{Q}^{(y)-1}$:

$$d\mathbf{M}^{(y)}/dT = -\mathbf{C}^{(y)} - \left(\mathbf{A}^{(y)T}\mathbf{M}^{(y)} + \mathbf{M}^{(y)}\mathbf{A}^{(y)}\right) - \mathbf{M}^{(y)}\mathbf{B}^{(y)}\mathbf{M}^{(y)}. \quad (4.14.49)$$

Here $\mathbf{A}^{(y)}$, $\mathbf{B}^{(y)}$, and $\mathbf{C}^{(y)}$ are 2×2 matrices with elements $A_{IJ}^{(y)}$, $B_{IJ}^{(y)}$, and $C_{IJ}^{(y)}$, given by (4.2.32). This is a nonlinear ordinary differential equation of the first order of the Riccati type. In isotropic media, we can consider Hamiltonian $\mathcal{H}^{(x)} = \frac{1}{2}(V^2 p_k^{(x)} p_k^{(x)} - 1)$ and obtain $A_{IJ}^{(y)} = 0$, $B_{IJ}^{(y)} = V^2 \delta_{IJ}$, and $C_{IJ}^{(y)} = V^{-1}\partial^2 V/\partial y_I \partial y_J$. Then (4.14.49) yields (4.1.73).

Using (4.14.49), we can easily derive Riccati equations for the 2×2 matrix $\mathbf{K}$ of curvature of the wavefront, $\mathbf{K} = \mathcal{C}\mathbf{M}^{(y)}$, and for the 2×2 matrix $\mathbf{R}$ of the radii of curvature of the wavefront, $\mathbf{R} = \mathbf{K}^{-1} = \mathcal{C}^{-1}\mathbf{M}^{(y)-1}$. We shall present only the Riccati equation for $\mathbf{M}^{(y)-1}$ we shall need later. Using the relation $d\mathbf{M}^{(y)-1}/dT = -\mathbf{M}^{(y)-1}(d\mathbf{M}^{(y)}/dT)\mathbf{M}^{(y)-1}$, (4.14.49) yields

$$d\mathbf{M}^{(y)-1}/dT = \mathbf{M}^{(y)-1}\mathbf{C}^{(y)}\mathbf{M}^{(y)-1} + \mathbf{M}^{(y)-1}\mathbf{A}^{(y)T} + \mathbf{A}^{(y)}\mathbf{M}^{(y)-1} + \mathbf{B}^{(y)}.$$

$$(4.14.50)$$

4.14.7 Paraxial Travel Times, Slowness Vectors, and Group Velocity Vectors

Let us again consider ray Ω, fixed point R situated on Ω, and point R' situated in the vicinity of R. Because the expressions for the 3×3 matrices of second derivatives of the travel-time field $\hat{\mathbf{M}}^{(y)}(R)$ and $\hat{\mathbf{M}}^{(x)}(R)$ are known, we can construct quadratic expansions for *paraxial travel time* $T(R')$ in terms of $y_i(R', R)$ and $x_i(R', R)$; see (4.2.46) and (4.2.48).

Linear equations for the *paraxial slowness vector* in terms of $y_i(R', R)$ and $x_i(R', R)$ follow immediately from (4.2.46) and (4.2.48). Linear expansions can also be derived for the *paraxial group velocity vector*. In the x_i-$p_i^{(x)}$- phase space, we can expand $\partial \mathcal{H}^{(x)}/\partial p_n^{(x)}$ in terms of x_i and $p_i^{(x)}$ in the vicinity of R. Transforming the expansion into y_i-$p_i^{(y)}$- phase space, we obtain

$$\mathcal{U}_n^{(y)}(R') = \mathcal{U}_n^{(y)}(R) + \bar{B}_{nm}(R)\left(p_m^{(y)}(R') - p_m^{(y)}(R)\right) + \bar{A}_{nm}(R)y_m(R', R).$$

$$(4.14.51)$$

Here $\bar{A}_{nm}(R)$ and $\bar{B}_{nm}(R)$ are given by (4.2.28). Alternatively, we can express $p_m^{(y)}(R') - p_m^{(y)}(R)$ in terms of $y_j(R', R)$. This yields

$$\mathcal{U}_n^{(y)}(R') = \mathcal{U}_n^{(y)}(R) + \left[\bar{B}_{nm}(R)M_{mj}^{(y)}(R) + \bar{A}_{nj}(R)\right]y_j(R', R). \quad (4.14.52)$$

Expansions (4.14.51) and (4.14.52) can be used to derive many other alternative or special relations, but we shall not do this here.

It can be proved that linear expansions for $p_n^{(y)}(R')$ and $\mathcal{U}_n^{(y)}(R')$ yield the expected result $p_n^{(y)}(R')\mathcal{U}_n^{(y)}(R') = p_n^{(y)}(R)\mathcal{U}_n^{(y)}(R) = 1$ with an accuracy up to the linear terms in y_i in the whole paraxial vicinity of R, because the linear terms with $y_i(R', R)$ cancel one another.

4.14.8 Dynamic Ray Tracing Across a Structural Interface

We shall now discuss the transformation of matrices $\mathbf{M}^{(y)}$, $\mathbf{Q}^{(y)}$, and $\mathbf{P}^{(y)}$ across structural interface Σ in anisotropic media. We shall use the same approach and notation as in Section 4.4. At the point of incidence Q, we introduce a local Cartesian coordinate system z_1, z_2, z_3 as in Section 4.4.1. Axis z_3 is perpendicular to Σ at Q and z_1 and z_2 are tangent to it. The plane of incidence is specified by the slowness vector $\vec{p}(Q)$ of the incident wave at Q and by axis $\vec{i}_3^{(z)}$ (normal to the interface Σ at Q). We shall consider the "standard option" of the local Cartesian coordinate system given by (4.4.21) in which $\vec{i}_2^{(z)}(Q)$ is perpendicular to the plane of incidence. Consequently, the component of slowness vector $\vec{p}(Q)$ of the incident wave into the z_2-axis vanishes that is $\vec{p} \cdot \vec{i}_2^{(z)} = p_2^{(z)} = 0$. Note that the tangent to ray Ω at Q may deviate from the plane of incidence in anisotropic media. In the "quadratic" vicinity of Q, interface Σ is described by relation $z_3 \doteq -\frac{1}{2} z_I z_J D_{IJ}(Q)$, where $D_{IJ}(Q)$ are elements of the curvature matrix $\mathbf{D}(Q)$ of interface Σ at Q; see (4.4.14). We shall now consider point $Q'[z_1, z_2, z_3 \doteq -\frac{1}{2} z_I z_J D_{IJ}(Q)]$, situated on interface Σ in the quadratic vicinity of Q, and denote the travel time $T(Q')$ at Q' on Σ by $T^{\Sigma}(z_1, z_2)$. We must emphasize that $T^{\Sigma}(z_1, z_2)$ represents the travel time directly at the interface Σ, not in plane $z_3 = 0$. Using (4.2.48), we obtain

$$T^{\Sigma}(z_1, z_2) = T(Q) + z_I p_I^{(z)}(Q) + \tfrac{1}{2} z_I z_J F_{IJ}(Q), \qquad (4.14.53)$$

where $F_{IJ}(Q)$ are elements of the 2×2 matrix $\mathbf{F}(Q)$ given by relation

$$\mathbf{F} = (\mathbf{G} - \mathbf{A}^{an})\mathbf{M}^{(y)}(\mathbf{G} - \mathbf{A}^{an})^T + \mathbf{E} - p_3^{(z)}\mathbf{D}. \qquad (4.14.54)$$

All quantities in (4.14.54) are taken at point Q. The derivation of (4.14.53) and (4.14.54) is fully analogous to the derivation of (4.4.34) and (4.4.35) for isotropic media, and the structure of (4.14.54) is very similar to the structure of (4.4.37). The 2×2 matrices $\mathbf{G}$ and $\mathbf{D}$ have exactly the same meaning as in Section 4.4.1, and $\mathbf{M}^{(y)} = \mathbf{P}^{(y)}\mathbf{Q}^{(y)-1}$ (see (4.2.45)) is computed by dynamic ray tracing. The *inhomogeneity matrix* $\mathbf{E}$ is given by relation

$$E_{IJ} = \mathcal{C}^{-1}\big[G_{I3}G_{JM}\eta_M^{(y)} + G_{IK}G_{J3}\eta_K^{(y)} + G_{I3}G_{J3}\big(\eta_3^{(y)} - \mathcal{C}^{-1}\eta_L^{(y)}\mathcal{U}_L^{(y)}\big)\big]. \qquad (4.14.55)$$

Matrix $\mathbf{E}$ vanishes in a homogeneous medium. In an isotropic medium, $\mathcal{U}_L^{(y)} = 0$ and the relevant term in (4.14.55) vanishes. Finally, *anisotropy matrix* $\mathbf{A}^{an}$ is

$$A_{IN}^{an} = \mathcal{C}^{-1} G_{I3}\mathcal{U}_N^{(y)} = p_I^{(z)}\mathcal{U}_N^{(y)}. \qquad (4.14.56)$$

It vanishes in an isotropic medium, as $\mathcal{U}_N^{(y)} = 0$ there. In the standard option of local Cartesian coordinates z_i, quantity $p_2^{(z)} = 0$. Then

$$A_{IN}^{an} = \delta_{I1} p_1^{(z)} \mathcal{U}_N^{(y)}. \qquad (4.14.57)$$

For any reflected/transmitted wave, we can express the relevant travel time $\tilde{T}^{\Sigma}$ at point $\tilde{Q}$ analogously to (4.14.53). Consequently, $p_I^{(z)}(\tilde{Q}) = p_I^{(z)}(Q)$ and $F_{IJ}(\tilde{Q}) = F_{IJ}(Q)$. Relation $p_I^{(z)}(\tilde{Q}) = p_I^{(z)}(Q)$ represents Snell's law for anisotropic media. The tangent to the ray of a reflected/transmitted wave at $\tilde{Q}$ may again deviate from the plane of incidence, whereas the slowness vector of the R/T wave at $\tilde{Q}$ is confined to it.

Relation $F_{IJ}(\tilde{Q}) = F_{IJ}(Q)$ implies an important equation for computing $\tilde{\mathbf{M}}^{(y)}$ from $\mathbf{M}^{(y)}$:

$$\begin{aligned}
\tilde{\mathbf{M}}^{(y)} = (\tilde{\mathbf{G}} - \tilde{\mathbf{A}}^{an})^{-1}&\big[(\mathbf{G} - \mathbf{A}^{an})\mathbf{M}^{(y)}(\mathbf{G} - \mathbf{A}^{an})^T \\
&+ \mathbf{E} - \tilde{\mathbf{E}} - u\mathbf{D}\big](\tilde{\mathbf{G}} - \tilde{\mathbf{A}}^{an})^{-1T}.
\end{aligned} \qquad (4.14.58)$$

Here $u = p_3^{(z)} - \tilde{p}_3^{(z)}$. All the quantities with a tilde correspond to the R/T point $\tilde{Q}$, and the quantities without a tilde correspond to the point of incidence Q. Equation (4.14.58), in a slightly different form, was first derived by Bakker (1996).

To derive the relation between $\tilde{\mathbf{Q}}^{(y)}$ and $\mathbf{Q}^{(y)}$, we require the ray to be continuous across interface Σ; see Section 4.4.4. Since $(\partial \gamma_I / \partial y_3)_Q \neq 0$ in anisotropic media, the result differs from (4.4.64). We obtain

$$\mathbf{Q}^{(y)-1}(\mathbf{G} - \mathbf{A}^{an})^T = \tilde{\mathbf{Q}}^{(y)-1}(\tilde{\mathbf{G}} - \tilde{\mathbf{A}}^{an})^T. \tag{4.14.59}$$

This result yields

$$\tilde{\mathbf{Q}}^{(y)} = (\tilde{\mathbf{G}} - \tilde{\mathbf{A}}^{an})^T (\mathbf{G} - \mathbf{A}^{an})^{-1T} \mathbf{Q}^{(y)}. \tag{4.14.60}$$

Matrix $\mathbf{P}^{(y)}$ is given by relation $\mathbf{P}^{(y)} = \mathbf{M}^{(y)} \mathbf{Q}^{(y)}$. Applying (4.14.58) and (4.14.59) yields

$$\tilde{\mathbf{P}}^{(y)} = (\tilde{\mathbf{G}} - \tilde{\mathbf{A}}^{an})^{-1} \left[(\mathbf{E} - \tilde{\mathbf{E}} - u\mathbf{D})(\mathbf{G} - \mathbf{A}^{an})^{-1T} \mathbf{Q}^{(y)} + (\mathbf{G} - \mathbf{A}^{an})\mathbf{P}^{(y)} \right]. \tag{4.14.61}$$

Note that the foregoing equations fail for rays tangent to the interface.

We shall now briefly discuss the meaning of matrix $\mathbf{A}^{an}$. We remind the reader that group velocity vector $\vec{\mathcal{U}}$ can be decomposed into unit vectors $\vec{e}_i$ as follows: $\vec{\mathcal{U}} = \mathcal{C}\vec{e}_3 + \mathcal{U}_I^{(y)}\vec{e}_I$ because $\mathcal{U}_3^{(y)} = \mathcal{C}$. We shall now introduce auxiliary unit vectors $\vec{e}_I^A$:

$$\vec{e}_I^A = \vec{e}_I - \mathcal{C}^{-1}\vec{e}_3 \mathcal{U}_I^{(y)}. \tag{4.14.62}$$

Vector $\vec{e}_I^A$ coincides with basis vector $\vec{e}_I$ in isotropic media but deviates from it in anisotropic media. If we compute vector product $\vec{e}_1^A \times \vec{e}_2^A$ and denote it $\vec{e}_3^A$, we obtain $\vec{e}_3^A = \vec{e}_1^A \times \vec{e}_2^A = \vec{e}_1 \times \vec{e}_2 + \mathcal{C}^{-1}\mathcal{U}_I^{(y)}\vec{e}_I$. This yields

$$\vec{\mathcal{U}} = \mathcal{C}\vec{e}_3^A = \mathcal{C}(\vec{e}_1^A \times \vec{e}_2^A). \tag{4.14.63}$$

Thus, $\vec{e}_I^A$ are projections of $\vec{e}_I$ from the plane perpendicular to slowness vector $\vec{p}$ to the *plane perpendicular to ray* Ω. Using (4.14.62), we obtain

$$(\mathbf{G} - \mathbf{A}^{an})_{IJ} = \vec{i}_I^{(z)} \cdot \vec{e}_J^A. \tag{4.14.64}$$

This is the same as G_{IJ} for isotropic media, see (4.4.12), only $\vec{e}_J$ is replaced by $\vec{e}_J^A$. In general, however, $\vec{e}_1^A$ and $\vec{e}_2^A$ are not unit vectors and are not mutually perpendicular.

Using (4.14.63), (4.14.64), and the Laplace identity;

$$(\vec{a} \times \vec{b}) \cdot (\vec{c} \times \vec{d}) = (\vec{a} \cdot \vec{c})(\vec{b} \cdot \vec{d}) - (\vec{b} \cdot \vec{c})(\vec{a} \cdot \vec{d});$$

we also obtain a useful expression:

$$\det(\mathbf{G} - \mathbf{A}^{an}) = \mathcal{C}^{-1}(\vec{\mathcal{U}} \cdot \vec{i}_3^{(z)}) = (\mathcal{U}/\mathcal{C})(\vec{t} \cdot \vec{i}_3^{(z)}),$$

where $\vec{t}$ is the unit vector tangent to the ray.

4.14.9 The 4 × 4 Ray Propagator Matrix in Layered Anisotropic Media

Using (4.14.60) and (4.14.61), we obtain

$$\begin{pmatrix} \tilde{\mathbf{Q}}^{(y)} \\ \tilde{\mathbf{P}}^{(y)} \end{pmatrix} = \Pi(\tilde{Q}, Q) \begin{pmatrix} \mathbf{Q}^{(y)} \\ \mathbf{P}^{(y)} \end{pmatrix}, \tag{4.14.65}$$

where $\Pi(\tilde{Q}, Q)$ is the *interface propagator matrix for anisotropic media*. It is given by the relation

$$\Pi(\tilde{Q}, Q)$$
$$= \begin{pmatrix} (\tilde{\mathbf{G}} - \tilde{\mathbf{A}}^{an})^T (\mathbf{G} - \mathbf{A}^{an})^{-1T} & \mathbf{0} \\ (\tilde{\mathbf{G}} - \tilde{\mathbf{A}}^{an})^{-1} (\mathbf{E} - \tilde{\mathbf{E}} - u\mathbf{D})(\mathbf{G} - \mathbf{A}^{an})^{-1T} & (\tilde{\mathbf{G}} - \tilde{\mathbf{A}}^{an})^{-1} (\mathbf{G} - \mathbf{A}^{an}) \end{pmatrix}.$$
(4.14.66)

It is not difficult to prove that the interface propagator matrix $\Pi(\tilde{Q}, Q)$, given by (4.14.66), is symplectic and that $\det \Pi(\tilde{Q}, Q) = 1$.

Consequently, the 4×4 ray propagator matrix corresponding to any multiply reflected/transmitted wave propagating in a laterally inhomogeneous layered anisotropic medium is given by (4.4.86). The individual ray propagator matrices in (4.4.86) should, of course, correspond to the anisotropic medium, see Section 4.14.3 and relation (4.14.66). All relations and conclusions of Section 4.4.6 also apply to anisotropic media. Let us emphasize that $\Pi(R, S)$ is symplectic along the whole ray Ω, including the interfaces. Moreover, the chain rule and the relations for the inverse of the ray propagator matrix can be used.

4.14.10 Surface-to-Surface Ray Propagator Matrix

Matrices $\mathbf{Q}^{(y)}$ and $\mathbf{P}^{(y)}$ can be projected on structural interface Σ at the point of incidence Q. We again consider a local Cartesian coordinate system z_i with its origin at Q and with $\vec{i}_3^{(z)} = \vec{n}(Q)$. Basis vectors $\vec{i}_1^{(z)}(Q)$ and $\vec{i}_2^{(z)}(Q)$ may, however, be chosen arbitrarily in the plane tangent to Σ at Q; they only have to form a right-handed triplet with $\vec{i}_3^{(z)}(Q)$. We consider point Q' situated on Σ in the "quadratic" vicinity of Q and denote $dz_I = z_I(Q') - z_I(Q)$ and $dp_I^{(\Sigma)}(Q) = p_I^{(\Sigma)}(Q') - p_I^{(\Sigma)}(Q)$, where $\vec{p}^{(\Sigma)}$ is the tangential component of the slowness vector on Σ. The definition of $dz_I(Q)$ and $dp_I^{(\Sigma)}(Q)$ is the same as in isotropic media; see Section 4.4. In the same way as in Section 4.4.4, we obtain

$$\begin{pmatrix} d\mathbf{z}(Q') \\ d\mathbf{p}^{(\Sigma)}(Q') \end{pmatrix} = \mathbf{Y}(Q) \begin{pmatrix} \mathbf{Q}^{(y)} d\gamma \\ \mathbf{P}^{(y)} d\gamma \end{pmatrix}. \tag{4.14.67}$$

In anisotropic media, the 4×4 projection matrix $\mathbf{Y}(Q)$ is given by the relation

$$\mathbf{Y}(Q) = \begin{pmatrix} (\mathbf{G} - \mathbf{A}^{an})^{-1T} & \mathbf{0} \\ (\mathbf{E} - p_3^{(z)} \mathbf{D})(\mathbf{G} - \mathbf{A}^{an})^{-1T} & (\mathbf{G} - \mathbf{A}^{an}) \end{pmatrix}. \tag{4.14.68}$$

The inverse of $\mathbf{Y}(Q)$ is

$$\mathbf{Y}^{-1}(Q) = \begin{pmatrix} (\mathbf{G} - \mathbf{A}^{an})^T & \mathbf{0} \\ -(\mathbf{G} - \mathbf{A}^{an})^{-1}(\mathbf{E} - p_3^{(z)} \mathbf{D}) & (\mathbf{G} - \mathbf{A}^{an})^{-1} \end{pmatrix}. \tag{4.14.69}$$

All quantities in (4.14.68) and (4.14.69) are taken at Q. It is simple to show that matrices $\mathbf{Y}(Q)$ and $\mathbf{Y}^{-1}(Q)$ are symplectic. Also, $\mathbf{Y}^{-1}(\tilde{Q})\mathbf{Y}(Q) = \Pi(\tilde{Q}, Q)$.

Similarly as in isotropic media, we can introduce the 4×4 surface-to-surface ray propagator matrix $\mathbf{T}(R, S)$ by the relation $\mathbf{T}(R, S) = \mathbf{Y}(R)\Pi(R, S)\mathbf{Y}^{-1}(S)$; see Section 4.4.7. Point S is situated on *anterior surface* Σ_a, and point R is on *posterior surface* Σ_p. The surface-to-surface ray propagator matrix connects the column matrices $(dz_1, dz_2, dp_1^{(\Sigma)}, dp_2^{(\Sigma)})^T$ at Σ_a and Σ_p; see (4.4.94). All relations and conclusions of Section 4.4.7 also apply to the anisotropic inhomogeneous layered medium, where the surface-to-surface ray propagator matrix is again given by (4.4.96). The surface-to-surface ray propagator matrix is symplectic along the whole ray Ω from S to R, including the interfaces.

Equations (4.14.67) and (4.14.69) can also be used to derive expressions for $\mathbf{Q}^{(y)}(S)$ and $\mathbf{P}^{(y)}(S)$ at point $S = Q$ on initial surface Σ^0, situated in the inhomogeneous anisotropic media. Taking $d\gamma = d\mathbf{z}$, multiplying (4.14.67) by $\mathbf{Y}^{-1}(Q)$ from the left, and inserting $d\mathbf{p}^{(\Sigma)}(Q') = \mathbf{T}^0(Q)d\mathbf{z}(Q')$, Equation (4.14.67) with (4.14.69) then yields

$$\mathbf{Q}^{(y)}(S) = (\mathbf{G} - \mathbf{A}^{an})^T, \qquad \mathbf{P}^{(y)}(S) = (\mathbf{G} - \mathbf{A}^{an})^{-1}\left(\mathbf{T}^0 + p_3^{(z)}\mathbf{D} - \mathbf{E}\right).$$

(4.14.70)

All quantities $(\mathbf{G}, \mathbf{A}^{an}, \mathbf{T}^0, \mathbf{D}, \mathbf{E}, p_3^{(z)})$ are taken at S on Σ^0. See the analogous expression for $\mathbf{Q}(S)$ and $\mathbf{P}(S)$ in isotropic media in (4.5.32).

4.14.11 Factorization of $\mathbf{Q}_2$. Fresnel Zone Matrix

As in Section 4.4.8, we can factorize the 2×2 minors of the 4×4 ray propagator matrix $\mathbf{\Pi}(R, S)$, particularly matrix $\mathbf{Q}_2(R, S)$. We consider ray Ω connecting a point source at S and receiver at R. The ray may be situated in a 3-D layered inhomogeneous anisotropic structure and may correspond to an arbitrary, possibly converted, elementary wave. We select any one of the interfaces and denote the point of incidence by Q and the R/T point by $\tilde{Q}$. In the same way as in Section 4.4.8, we obtain

$$\mathbf{Q}_2(R, S) = \mathbf{Q}_2(R, \tilde{Q})(\tilde{\mathbf{G}} - \tilde{\mathbf{A}}^{an})^{-1}\mathbf{M}^F(Q; R, S)(\mathbf{G} - \mathbf{A}^{an})^{-1T}\mathbf{Q}_2(Q, S).$$

(4.14.71)

In the derivation, we have used (4.14.66). The 2×2 matrix $\mathbf{Q}_2(Q, S)$ corresponds to the incident branch of Ω, $\mathbf{Q}_2(R, \tilde{Q})$ corresponds to the R/T branch, and $\mathbf{M}^F(Q; R, S)$ is the Fresnel zone matrix at Q. It is given by the relation

$$\mathbf{M}^F(Q; R, S) = \mathbf{F}(Q, S) - \mathbf{F}(\tilde{Q}, R),$$

(4.14.72)

where $\mathbf{F}(Q, S)$ corresponds to the incident branch and $\mathbf{F}(\tilde{Q}, R)$ corresponds to the R/T branch. $\mathbf{F}(Q, S)$ is given by (4.14.54), where $\mathbf{M}^{(y)} = \mathbf{M}^{(y)}(Q, S)$, and all other quantities are taken at Q. $\mathbf{F}(\tilde{Q}, R)$ is given by the same equation, where $\mathbf{M}^{(y)} = \mathbf{M}^{(y)}(\tilde{Q}, R)$, and all other quantities are taken at $\tilde{Q}$. For $\mathbf{M}^{(y)}(Q, S)$ and $\mathbf{M}^{(y)}(\tilde{Q}, R)$ see (4.14.48). Equation (4.14.72) can be expressed in many other forms, as in Section 4.4.8. For example,

$$\mathbf{M}^F(Q; R, S) = (\mathbf{G} - \mathbf{A}^{an})\mathbf{M}^{(y)}(Q, S)(\mathbf{G} - \mathbf{A}^{an})^T$$
$$- (\tilde{\mathbf{G}} - \tilde{\mathbf{A}}^{an})\mathbf{M}^{(y)}(\tilde{Q}, R)(\tilde{\mathbf{G}} - \tilde{\mathbf{A}}^{an})^T + \mathbf{E} - \tilde{\mathbf{E}} - u\mathbf{D}.$$

(4.14.73)

Here $\mathbf{E}$ is given by (4.14.55), taken at Q, and $\tilde{\mathbf{E}}$ is given by the same equation, taken at $\tilde{Q}$. For $\mathbf{A}^{an}$ and $\tilde{\mathbf{A}}^{an}$, see (4.14.56).

As in isotropic media, Fresnel zone matrix $\mathbf{M}^F(Q; R, S)$ is useful in many applications; see Sections 4.4.8, 4.8.5, 4.10.4, 4.11.2, and 4.12.2.

4.14.12 Boundary-Value Ray Tracing for Paraxial Rays in Anisotropic Media

Boundary-value ray tracing for paraxial rays in isotropic inhomogeneous layered structures was investigated in Section 4.9. Most equations of Section 4.9 also apply to anisotropic media. In particular, the basic equation (4.9.24) for the two-point eikonal $T(R', S')$ can also be used in anisotropic media. Of course, matrices $\hat{\mathbf{M}}^{(x)}(R, S)$ and $\hat{\mathbf{M}}^{(x)}(S, R)$ in (4.9.24) should be specified for anisotropic media; see (4.14.47), (4.14.48), and (4.2.44) with (4.2.45).

Matrix $\hat{\mathbf{A}}(R, S)$ is given by (4.9.15), where $\mathbf{Q}_2(R, S)$ corresponds to the anisotropic ray propagator matrix; see Sections 4.14.3 and 4.14.9.

Similarly, (4.10.43) for relative geometrical spreading can be used in anisotropic media as it is. Equation (4.10.48) for determining relative geometrical spreading from mixed second derivatives of the travel-time field along the anterior and posterior surfaces should be modified: $\mathbf{G}$ should be replaced by $\mathbf{G} - \mathbf{A}^{an}$.

4.14.13 Phase Shift Due to Caustics. KMAH Index

The problem of the phase shift due to caustics in anisotropic inhomogeneous media is more complicated than in isotropic inhomogeneous media. The general reason for these complications is that the local slowness surface need not be convex in anisotropic media. The behavior of the slowness surface at the caustic point, expressed in terms of the 2×2 matrix $\mathbf{B}^{(y)}$, given by (4.2.32), plays an important role in the investigation of the phase shift due to caustics.

As in isotropic media, we shall denote the phase shift due to caustics along ray Ω from S to R by $T^c(R, S)$ and compute it by means of the KMAH index $k(R, S)$ using Equation (4.12.1). The KMAH index $k(R, S)$ is represented by the summation of contributions corresponding to the individual caustic points along ray Ω from S to R.

We denote by Δk the contribution to the KMAH index $k(R, S)$ corresponding to one caustic point situated between S and R. There are two main differences in the computation of Δk and of the KMAH index in isotropic and anisotropic media.

  a. In isotropic media, $\Delta k = 1$ when the wave passes through a caustic point of the first order (det $\mathbf{Q} = 0$, tr $\mathbf{Q} \neq 0$) and $\Delta k = 2$ when the wave passes through a focus, that is, caustic point of the second order (det $\mathbf{Q} = 0$, tr $\mathbf{Q} = 0$). In anisotropic media, however, Δk may also be negative, $\Delta k = -1$ or $\Delta k = -2$. See Lewis (1965), Garmany (1988), Kravtsov and Orlov (1993), Chapman and Coates (1994), Klimeš (1997a), and Bakker (1998).
  b. In isotropic media, the initial value of the KMAH index is always zero at a point source, if we compute the elastodynamic Green function. In anisotropic media, however, the initial value of the KMAH index may be nonvanishing in this case, depending on the local behavior of the slowness surface at S. It corresponds to $-\sigma_0$ in (2.5.76); see also (2.5.74) and Burridge (1967).

We shall now discuss problem (a), that is, the determination of the KMAH index along ray Ω in an inhomogeneous anisotropic medium. We consider dynamic ray tracing system (4.2.31) in wavefront orthonormal coordinates, with the 2×2 matrices $\mathbf{A}^{(y)}$, $\mathbf{B}^{(y)}$, $\mathbf{C}^{(y)}$, and $\mathbf{D}^{(y)}$ given by (4.2.32). Only one of these four matrices, namely $\mathbf{B}^{(y)}$, plays an important role in determining the KMAH index. Solving dynamic ray tracing system (4.2.31) along Ω, we can determine the 2×2 matrices $\mathbf{Q}^{(y)}$ and $\mathbf{P}^{(y)}$ along the whole ray Ω, assuming they are known at some initial point. We can also determine the 2×2 matrix $\mathbf{M}^{(y)}$ of the second derivatives of the travel-time field with respect to the wavefront orthonormal coordinates y_1, y_2, $\mathbf{M}^{(y)} = \mathbf{P}^{(y)}\mathbf{Q}^{(y)-1}$; see (4.2.45). We shall now consider caustic point C on ray Ω, where ray Ω touches the caustic surface. Consequently, det $\mathbf{Q}^{(y)}(C) = 0$. As in isotropic media, we shall consider two types of caustic points; see Section 4.12, especially Equations (4.12.2) and (4.12.3).

  a. *Caustic point of the second order* (also called the point caustic or focus or double caustic point). At this point, $\mathbf{Q}^{(y)}(C) = \mathbf{0}$, so that rank $\mathbf{Q}^{(y)}(C) = 0$.

b. *Caustic point of the first order* (also called the line caustic or a single caustic point). At this point, rank $\mathbf{Q}^{(y)}(C) = 1$.

Now we wish to determine the contribution Δk to the KMAH index when the ray passes through caustic point C. We shall present and discuss here simple and general criteria, which will be derived in Section 5.8.8, using the approach by Bakker (1998):

1. If $\mathbf{B}^{(y)}(C)$ is positive definite, then

$$\Delta k = 2 \qquad \text{at a point caustic,}$$
$$\Delta k = 1 \qquad \text{at a line caustic.}$$
(4.14.74)

2. If $\mathbf{B}^{(y)}(C)$ is negative definite, then

$$\Delta k = -2 \qquad \text{at a point caustic,}$$
$$\Delta k = -1 \qquad \text{at a line caustic.}$$
(4.14.75)

3. If $\mathbf{B}^{(y)}(C)$ is neither positive definite, nor negative definite, then

$$\Delta k = 0 \qquad\qquad\qquad \text{at a point caustic,}$$
$$\Delta k = \mathrm{sgn}\big(\mathbf{m}^T(C)\mathbf{B}^{(y)}(C)\mathbf{m}(C)\big) \qquad \text{at a line caustic.}$$
(4.14.76)

Here $\mathbf{m}(C)$ is the eigenvector of matrix $\mathbf{M}^{(y)}(C)$, corresponding to the singular eigenvalue of $\mathbf{M}^{(y)}(C)$. For the derivation and a more detailed explanation, see Section 5.8.8.

Here are several notes on criteria (4.14.74)–(4.14.76).

a. In isotropic media and ray-centered coordinates, $\mathbf{B}^{(y)}(C) = V(C)\mathbf{I}$, if monotonic parameter u along ray Ω represents arclength s. Consequently, $\mathbf{B}^{(y)}(C)$ is *always* positive definite, and criterion (4.14.74) can be always used. This fully corresponds to the treatment of the KMAH index in Section 4.12.

b. It can be shown (see Section 5.8.8), that the eigenvalues of matrix $\mathbf{B}^{(y)}(C)$ have the same signs as the eigenvalues of the matrix $\mathbf{D}^S(C)$ of the curvature of the slowness surface at C. Consequently, matrix $\mathbf{B}^{(y)}(C)$ can be replaced by $\mathbf{D}^S(C)$ in all the three criteria (4.14.74) through (4.14.76). Then (4.14.74) corresponds to the convex slowness surface, (4.14.75) corresponds to the concave slowness surface, and (4.14.76) corresponds to the hyperboloidal form of the slowness surface at C. Parabolic points of the slowness surface are not considered here; they correspond to singular directions.

c. The cases $\Delta k = -2$ and $\Delta k = 2$ are fully equivalent because they yield the same phase shift due to the caustic, $\exp(-i\pi) = \exp(i\pi)$.

d. The slowness surface of the qP wave in anisotropic media is always convex for all slowness vector directions. Consequently, criterion (4.14.74) can always be used for qP waves in anisotropic media as in isotropic media. Thus, the criteria (4.14.75) and (4.14.76) play an important role only for qS waves. For this reason, the case of $\Delta k = -1$ may occur only for qS waves in anisotropic media. It is also often called the *anomalous phase shift*.

e. Analogous criteria for the central ray field (corresponding to a point source) were derived by Klimeš (1997a) in a different form. These criteria are expressed in terms of the 2×2 minors $\mathbf{P}_2$ and $\mathbf{Q}_2$ of the 4×4 propagator matrix. Bakker (1998) showed that both criteria are equivalent if the central ray field is considered.

f. Klimeš (1997c) also proved that the *KMAH index is reciprocal* along ray Ω from S to R. See more details in Section 5.4.5.

CHAPTER FIVE

Ray Amplitudes

The computation of ray amplitudes along a known ray Ω in a smooth medium is simpler than the computation of ray Ω itself. The variations of ray amplitudes along the ray are controlled by the transport equation, which can be solved analytically in terms of geometrical spreading; see Sections 3.10.6 and 4.10.2. The geometrical spreading can be calculated along the ray by dynamic ray tracing, as described in detail in Chapter 4.

The computation is particularly simple for acoustic waves in fluid media. In this case, the ray amplitude has a scalar character. Let us consider ray Ω situated in a smooth fluid medium and two points S and R situated on this ray. Then the ray amplitude at R equals the ray amplitude at S, multiplied by a factor containing velocities, densities, and geometrical spreadings at S and R. No additional computations along Ω are required; the relation between the ray amplitudes at R and S contains only the preceding three quantities at points S and R.

In elastic media, the ray amplitudes have a vectorial character. In addition to geometrical spreading, it is also necessary to know polarization vectors at S and R. In isotropic elastic medium, the polarization vector of P waves is simple: it is tangent to the ray. For S waves, we need to know the polarization vectors $\vec{e}_1$ and $\vec{e}_2$; see Section 4.1.1. If the dynamic ray tracing along Ω is performed in ray-centered coordinates, the polarization vectors $\vec{e}_1$ and $\vec{e}_2$ must also be computed. Consequently, no additional computation along the ray Ω is required if we wish to determine the ray amplitudes of S waves at R. In an anisotropic medium, the situation is similar. The polarization vectors of individual elementary waves are represented by eigenvectors $\vec{g}^{(m)}$ of the Christoffel matrix $\hat{\Gamma}$. These eigenvectors are needed even in ray tracing; see (3.6.10). Thus, no additional computations along Ω are required either.

In layered media, the equations for ray amplitudes are not considerably more complicated than those for smooth media. It is only necessary to consider relevant reflection/transmission coefficients at points at which the ray Ω strikes structural interfaces. In fact, if the geometrical spreading and polarization vectors are known along the ray, the computation of R/T coefficients at structural interfaces is the only additional work that should be done if we wish to compute ray amplitudes. Thus, the R/T coefficients at structural interfaces play a very important role in the computation of ray amplitudes of an arbitrary multiply-reflected wave in layered structures.

The computation of vectorial ray amplitudes of S and converted waves in 3-D layered isotropic elastic structures is slightly more complicated than the other cases. The vectorial amplitudes of S waves have, in general, two complex-valued mutually perpendicular vectorial components, polarized along unit vectors $\vec{e}_1$ and $\vec{e}_2$. The orientation of these polarization vectors at the point of incidence at a structural interface may be arbitrary. Thus, the traditional approach based on the decomposition of the S wave into SV and SH components at a point of incidence (SH perpendicular to the plane of incidence and SV in the plane of incidence) cannot be used in general in the computation of R/T coefficients in 3-D models, and both S wave components are coupled there. As we shall show, it is very suitable to treat this problem in the matrix form.

In this chapter, we shall discuss the ray amplitudes independently for acoustic waves in fluids (Section 5.1), elastic waves in isotropic solids (Section 5.2), and elastic waves in anisotropic solids (Section 5.4). In all these cases, we consider an arbitrary multiply-reflected (possibly converted) wave propagating in a general 3-D laterally varying layered structure. We also consider important cases of the source and/or receiver situated directly on the structural interface or on the Earth's surface. Due to a large importance of R/T coefficients in elastic isotropic media, all of Section 5.3 is devoted to them. In Section 5.4.6, we also briefly treat the elastic waves in weakly anisotropic media.

Great attention is also devoted to point-source solutions, to relevant radiation functions and directivity patterns of point sources, and to the ray-theory Green functions. General expressions for ray-theory Green functions are derived and specified for many special but important cases such as planar rays and the 2-D case.

In all the foregoing cases, we present compact analytical expressions for ray amplitudes. These compact expressions are easy to understand. They factorize individual effects, which influence the ray amplitudes. They are valuable in many applications. For example, they are useful in the investigation of the reciprocity of ray-theory Green functions. We shall show that the ray-theory Green functions (as they are introduced here) are reciprocal for any elementary wave propagating in a 3-D laterally varying layered structure (fluid, elastic isotropic, elastic anisotropic). In numerical computations, however, the compact analytical expressions are not quite necessary. The computation may be performed by a straightforward, step-by-step algorithm, passing along the ray Ω from one structural interface to another and applying relevant transformations at individual points of reflection/transmission. An example of such algorithm is the CRT (Complete Ray Tracing) algorithm, described in detail in Červený, Klimeš, and Pšenčík (1988b).

The expressions for ray amplitudes, derived in Sections 5.1 through 5.4, can be modified even for weakly dissipative media. The modification requires only one additional quadrature along the ray Ω to compute the quantity $t^*(R, S)$, which represents a global absorption factor along Ω from S to R. The quantity $t^*(R, S)$ can then be used to construct various dissipative filters representing different dissipation models. See Section 5.5 for details.

The ray amplitudes, described in Sections 5.1 through 5.5, correspond to the zeroth-order approximation of the ray method and are only approximate. They can be generalized using the ray series. In the frequency domain, the ray series represents an asymptotic series in inverse powers of frequency ω. The ray-series method is described in detail in Section 5.6 for the acoustic case and in Section 5.7 for the elastic case (both isotropic and anisotropic). Methods to compute higher-order ray approximations are proposed and discussed. A great deal of attention is devoted to higher-order waves, particularly to head waves (see Sections 5.6.7 and 5.7.10).

The accuracy of ray amplitudes is very limited in singular regions of the ray field, such as the caustic region, the critical region, and the transition zone between shadow and illuminated regions. In some situations, the ray amplitudes may be completely invalid. For example, in the shadow zone, the ray amplitudes are vanishing, but directly at the caustics, they are infinite. Moreover, certain elementary waves, such as diffracted waves, cannot be described by the standard ray method at all. Various extensions of the ray method have been proposed to treat such singular regions, diffracted waves, and the like. For validity conditions and extensions of the ray method, see a brief exposition in Section 5.9. Let us emphasize here that the main aim of this book is to give a complete treatment of the ray method itself, not of its extensions. Such attempts would increase the length of the book inadmissibly. For this reason, the extensions are mostly described only qualitatively, without a more detailed theoretical treatment. Even the references to the extensions of the ray method are too numerous to be given here; consequently, we present only some literature for further reading. Only three extensions of the seismic ray method are discussed here in a greater detail:

1. Quasi-isotropic ray theory for seismic waves propagating in weakly anisotropic media and the qS-wave coupling. See Section 5.4.6.
2. Kirchhoff integrals, generalizing the results of the standard ray method for waves generated at an initial surface and for waves reflected/transmitted at a curved interface. See Section 5.1.11 for fluid models and Section 5.4.8 for elastic anisotropic models.
3. Paraxial ray approximations and paraxial Gaussian beams, specifying the wavefield not only along the ray but also in its vicinity. The construction of more general integral solutions of the elastodynamic equation, based on the summation of paraxial ray approximations and on the summation of Gaussian beams. Note that the summation of paraxial ray approximations yields integrals close or equal to the Maslov-Chapman integrals. See Section 5.8.

The final problem that should be mentioned is the sensitivity of ray amplitudes to minor details in the approximation of the model, for example, to artificial interfaces of second order introduced by cell approaches, to fictitious small oscillations of the velocity introduced by improper approximation of velocity distribution, and to triangular description of structural interfaces. There are several ways to avoid these problems.

a. It is possible to use, globally or locally, a more sophisticated approximation of the model that is smooth enough to suppress the artificial effects. This may be, however, a difficult problem in certain cases.
b. It is possible to smooth the computed ray amplitudes.
c. It is possible to develop special methods that satisfactorily treat the improper approximations, for example, the triangular description of structural interfaces. See Hanyga (1996a).
d. It is possible to use some methods that automatically include some smoothing of amplitudes. Examples include the Maslov-Chapman method and the method of the summation of Gaussian beams or of Gaussian wave packets. See Section 5.8.

5.1 Acoustic Case

In this section, we shall derive equations for the amplitudes of high-frequency pressure body waves propagating in fluid models with variable propagation velocity $c(x_i)$ and density $\rho(x_i)$. The pressure wavefield $p(x_i, t)$ then satisfies acoustic wave equations (2.4.9) and the relevant boundary conditions at the individual interfaces. As in Section 2.4.1, we express the solution of the acoustic wave equation in the following form

$$p(x_i, t) = P(x_i)F(t - T(x_i)). \tag{5.1.1}$$

Here $F(\zeta)$ is a high-frequency analytical signal. Travel-time field $T(x_i)$ satisfies eikonal equation $(\nabla T)^2 = c^{-2}(x_i)$, see (2.4.6), and amplitude function $P(x_i)$ transport equation $2\nabla T \cdot \nabla(P/\sqrt{\rho}) + (P/\sqrt{\rho})\nabla^2 T = 0$, see (2.4.11). The computation of the travel-time field was studied in detail in Chapters 3 and 4. Here we shall concentrate mainly on computing amplitude function $P(x_i)$.

For the properties of the analytical signal, see Appendix A. We only remind the reader that the analytical signal is given by the expression $F(\zeta) = \exp[-\mathrm{i}2\pi f\zeta]$ for harmonic waves, where f is the frequency. Alternatively, we shall also use analytical signal $F(\zeta) = \delta^{(A)}(\zeta) = \delta(\zeta) - \mathrm{i}(\pi\zeta)^{-1}$, where $\delta^{(A)}(\zeta)$ is the analytical delta function; see (2.2.12).

5.1.1 Continuation of Amplitudes Along a Ray

It was shown in Section 3.10.6 that the transport equation for $P(x_i)$ can be solved along ray Ω in terms of J, the Jacobian of transformation from ray coordinates $\gamma_1, \gamma_2, \gamma_3 = s$ to general Cartesian coordinates x_1, x_2, x_3, that is, $J = J^{(s)} = D(x_1, x_2, x_3)/D(\gamma_1, \gamma_2, s)$. As in Chapters 3 and 4, we shall also call J the ray Jacobian. Let us consider ray Ω and two points S and R situated on Ω. The continuation formula (3.10.53) for amplitudes along ray Ω can then be expressed in the following form:

$$P(R) = \left[\frac{\rho(R)c(R)J(S)}{\rho(S)c(S)J(R)}\right]^{1/2} P(S). \tag{5.1.2}$$

After we know the amplitude at reference point S on ray Ω, we can calculate it at any other point R on ray Ω using (5.1.2). Alternatively,

$$P(R) = \left[\frac{\rho(R)c(R)}{\rho(S)c(S)}\right]^{1/2} \frac{\mathcal{L}(S)}{\mathcal{L}(R)} \exp[\mathrm{i}T^c(R, S)]P(S), \tag{5.1.3}$$

where $T^c(R, S)$ is the phase shift due to caustics (see (3.10.64)) and $\mathcal{L}$ is the geometrical spreading, $\mathcal{L} = |J|^{1/2}$ (see (4.10.1)).

These are the final equations for the continuation of amplitudes along a ray Ω situated in a smooth 3-D laterally varying structure. If the ray strikes an interface, the equations should be modified, as will be shown in Sections 5.1.3 through 5.1.5.

Inserting expressions (5.1.2) or (5.1.3) into (5.1.1), we obtain the final equations for the continuation of pressure wave field $p(x_i, t)$ along ray Ω. Assume that $p(S, t)$ is given by relation $p(S, t) = P(S)F(t - T(S))$, where $P(S)$, $T(S)$ and analytical signal $F(\zeta)$ are known. Pressure wavefield $p(R, t)$ at point R situated on ray Ω passing through S is then given by the relation

$$p(R, t) = \left[\frac{\rho(R)c(R)}{\rho(S)c(S)}\right]^{1/2} \frac{\mathcal{L}(S)}{\mathcal{L}(R)} \exp[\mathrm{i}T^c(R, S)]$$
$$\times P(S)F(t - T(S) - T(R, S)); \tag{5.1.4}$$

see (5.1.3). Here $T(R, S)$ denotes the travel time from S to R along ray Ω.

In (5.1.2) through (5.1.4), $P(S)$ represents the *initial amplitude* at reference point S of ray Ω. To be able to use (5.1.2) and (5.1.3), we need to know, in addition to $P(S)$, the following initial quantities at S: velocity $c(S)$, density $\rho(S)$, and the ray Jacobian $J(S)$, or velocity $c(S)$, density $\rho(S)$, and geometrical spreading $\mathcal{L}(S)$. In (5.1.4), we also need to know $T(S)$ and the analytical signal $F(\zeta)$.

5.1.2 Point-Source Solutions. Radiation Function

In certain important situations, geometrical spreading $\mathcal{L}(S)$ vanishes at initial point S on ray Ω. For example, this applies to the point source at S, to the line source at S, and to the caustic point at S. For simplicity, we shall consider a point source. For a line source, see Section 5.1.12. If $P(S)$ is finite, amplitudes $P(R)$ then vanish along the whole ray Ω because $\mathcal{L}(S)P(S) = 0$ in (5.1.3). To obtain nonvanishing amplitudes along ray Ω, we need to require $P(S)$ to be infinite at S so that

$$\lim_{S' \to S}\{\mathcal{L}(S')P(S')\} = P^0(S),\tag{5.1.5}$$

where $P^0(S)$ is finite. In (5.1.5), point S' is situated on ray Ω, and the limit is taken along ray Ω.

Using (5.1.3) and (5.1.5), we obtain

$$P(R) = \left[\frac{\rho(R)c(R)}{\rho(S)c(S)}\right]^{1/2}\frac{\exp[iT^c(R,S)]}{\mathcal{L}(R)}P^0(S).\tag{5.1.6}$$

Equation (5.1.6) with (5.1.5) is more general than (5.1.3) because it also includes the situation with $\mathcal{L}(S) = 0$.

For a point source at S, we can also use the relative geometrical spreading $\mathcal{L}(R, S)$ instead of $\mathcal{L}(R)$. If we use (4.10.11), we obtain $\mathcal{L}(R) = \mathcal{L}(R, S)|\det \mathbf{P}(S)|^{1/2}$, $\mathcal{L}(S') = \mathcal{L}(S', S)|\det \mathbf{P}(S)|^{1/2}$, where $\mathbf{P}(S)$ has the same meaning as in Section 4.10. Then (5.1.5) and (5.1.6) yield

$$P(R) = \left[\frac{\rho(R)c(R)}{\rho(S)c(S)}\right]^{1/2}\frac{\exp[iT^c(R,S)]}{\mathcal{L}(R,S)}\mathcal{G}(S;\gamma_1,\gamma_2),\tag{5.1.7}$$

where

$$\mathcal{G}(S;\gamma_1,\gamma_2) = \lim_{S' \to S}\{\mathcal{L}(S',S)P(S')\}.\tag{5.1.8}$$

Function $\mathcal{G}(S;\gamma_1,\gamma_2)$ will be called here the *radiation function* of the wave under consideration, generated by a point source situated at S. The limit in (5.1.8) is taken along ray Ω, specified by ray parameters γ_1 and γ_2. Because the limit may be different for different rays, we introduce ray parameters γ_1 and γ_2 among the arguments of $\mathcal{G}$.

The advantage of this definition of the radiation function consists in its universality. The same definition is applicable to homogeneous and inhomogeneous media, to acoustic waves in fluid media and elastic waves in isotropic and anisotropic media. See Section 5.2.3 and 5.4.2.

Expression (5.1.7) for the ray-theory amplitudes $P(R)$ simplifies in a *homogeneous medium*. Then $T^c(R, S) = 0$ and $\mathcal{L}(R, S) = cl(R, S)$, where $l(R, S)$ is the distance between R and S. Equation (5.1.7) reads

$$P(R) = \mathcal{G}(S;\gamma_1,\gamma_2)/\mathcal{L}(R,S) = \mathcal{G}(S;\gamma_1,\gamma_2)/cl(R,S).\tag{5.1.9}$$

Thus, in a homogeneous medium $\mathcal{G}(S;\gamma_1,\gamma_2)$ represents the angular distribution of amplitudes $P(R)$ along a sphere with its center at S and with radius $l(R, S) = 1/c(S)$. If $\mathcal{G}(S;\gamma_1,\gamma_2)$ is independent of γ_1 and γ_2, we speak of the *omnidirectional radiation function*.

In seismological literature, the term *directivity pattern* of the wave has also been widely used. However, the meaning of this term has not always been unique and consistent. Here we shall introduce the directivity pattern of a wave generated by a point source in a *standard ray-theory meaning.* We assume that the medium is locally homogeneous in the vicinity of S. The directivity pattern of the wave generated by a point source is then represented by the distribution of ray-theory amplitudes over a unit sphere, whose center is at the source S and radius is 1. We denote the directivity pattern of a selected wave generated by a point source $\mathcal{F}(S; \gamma_1, \gamma_2)$ and define it by the relation

$$\mathcal{F}(S; \gamma_1, \gamma_2) = (\mathcal{G}(S; \gamma_1, \gamma_2)/\mathcal{L}(R, S))_{l(R,S)=1}. \tag{5.1.10}$$

Here $l(R, S)$ is the distance between S and R. We shall use this definition of the directivity pattern also for elastic isotropic and anisotropic media. This ray-theory definition of the directivity pattern may differ from some other seismological definitions, introduced by other authors. In the most important cases, however, the difference consists only in a multiplicative constant, which does not depend on γ_1 and γ_2. Thus, the actual angular dependence of the generated amplitudes remains the same in all definitions.

For acoustic waves propagating in fluid media, the difference between radiation function $\mathcal{G}(S; \gamma_1, \gamma_2)$ and directivity pattern $\mathcal{F}(S; \gamma_1, \gamma_2)$ is only formal. Inserting $\mathcal{L}(R, S) = c(S)l(R, S)$ into (5.1.10) yields

$$\mathcal{F}(S; \gamma_1, \gamma_2) = \mathcal{G}(S; \gamma_1, \gamma_2)/c(S). \tag{5.1.11}$$

In anisotropic media, however, the directivity pattern is quite different from the radiation function because the relative geometrical spreading $\mathcal{L}(R, S)$ depends on ray parameters γ_1 and γ_2.

In the following text, we shall prefer the term *radiation function* in our theoretical treatment. The definition of radiation function (5.1.8) is quite universal and simple to understand. Directivity patterns will be used merely to *demonstrate the angular dependence* of the ray-theory amplitudes of waves generated by a point source.

Let us now discuss the changes in amplitudes if the positions of the point source and receiver are interchanged (reciprocity relations). As we know, the relative geometrical spreading and the phase shift due to caustics are reciprocal; $\mathcal{L}(R, S) = \mathcal{L}(S, R)$ and $T^c(R, S) = T^c(S, R)$. *The ray amplitudes, however, are in general not reciprocal.* They are reciprocal only if

$$\frac{\mathcal{G}(S; \gamma_1, \gamma_2)}{\rho(S)c(S)} = \frac{\mathcal{G}(R; \gamma_1', \gamma_2')}{\rho(R)c(R)}. \tag{5.1.12}$$

Here γ_1, γ_2 and γ_1', γ_2' are the relevant ray parameters corresponding to the same ray Ω, for sources at S and R.

5.1.3 Amplitudes Across an Interface

Assume that ray Ω strikes interface Σ at point Q situated between points S and R. In addition to Q, we introduce point $\tilde{Q}$, coinciding with Q, but corresponding to the reflected/transmitted branch of the ray. Thus, S and Q are situated on the incident branch of the ray, and $\tilde{Q}$ and R are located on the reflected/transmitted branch of the ray. We wish to derive the continuation relations of amplitudes across the interface, from S to R.

Along the incident branch of the ray, we can use (5.1.3):

$$P(Q) = \left[\frac{\rho(Q)c(Q)}{\rho(S)c(S)}\right]^{1/2} \frac{\mathcal{L}(S)}{\mathcal{L}(Q)} \exp[iT^c(Q, S)]P(S).$$

Across the interface, we use (2.4.68),

$$P(\tilde{Q}) = RP(Q),$$

where R is the appropriate pressure reflection/transmission coefficient. Finally, along the reflected/transmitted branch, we can again use (5.1.3):

$$P(R) = \left[\frac{\rho(R)c(R)}{\rho(\tilde{Q})c(\tilde{Q})} \right]^{1/2} \frac{\mathcal{L}(\tilde{Q})}{\mathcal{L}(R)} \exp[iT^c(R, \tilde{Q})]P(\tilde{Q}).$$

The three foregoing equations yield the amplitude continuation relation from S to R,

$$P(R) = \left[\frac{\rho(R)c(R)}{\rho(S)c(S)} \right]^{1/2} \frac{\mathcal{L}(S)}{\mathcal{L}(R)} \mathcal{R}(Q) \exp[iT^c(R, S)]P(S), \tag{5.1.13}$$

where

$$T^c(R, S) = T^c(R, \tilde{Q}) + T^c(Q, S) \tag{5.1.14}$$

and

$$\mathcal{R}(Q) = R(Q) \left[\frac{\rho(Q)c(Q)}{\rho(\tilde{Q})c(\tilde{Q})} \right]^{1/2} \frac{\mathcal{L}(\tilde{Q})}{\mathcal{L}(Q)}. \tag{5.1.15}$$

We shall call $\mathcal{R}(Q)$ the *normalized pressure reflection/transmission coefficient*. The normalized pressure R/T coefficients $\mathcal{R}(Q)$ represent the standard pressure coefficients $R(Q)$ normalized with respect to the energy flux across the interface. For a more detailed explanation of normalized R/T coefficients, see Section 5.3.3. We now take into account the relation

$$\mathcal{L}(\tilde{Q})/\mathcal{L}(Q) = |\det \mathbf{Q}(\tilde{Q})/\det \mathbf{Q}(Q)|^{1/2} = (\cos i(\tilde{Q})/\cos i(Q))^{1/2},$$

where $i(Q)$ is the acute angle of incidence and $i(\tilde{Q})$ is the acute angle of reflection/transmission; see (4.4.66) and (4.4.50). Equation (5.1.15) then yields

$$\mathcal{R}(Q) = R(Q) \left[\frac{\rho(Q)c(Q) \cos i(\tilde{Q})}{\rho(\tilde{Q})c(\tilde{Q}) \cos i(Q)} \right]^{1/2}. \tag{5.1.16}$$

For a real-valued ray Ω, $\cos i(Q)$ and $\cos i(\tilde{Q})$ are always real-valued and nonnegative. Note that the change of sign of $\det \mathbf{Q}$ across the interface for reflected waves does not influence the phase shift $T^c(R, S)$.

Equation (5.1.13) represents the final form of the continuation relation for the amplitudes of reflected/transmitted waves. As in Section 5.1.2, (5.1.13) can be modified for a point source situated at S,

$$P(R) = \left[\frac{\rho(R)c(R)}{\rho(S)c(S)} \right]^{1/2} \frac{\exp[iT^c(R, S)]}{\mathcal{L}(R, S)} \mathcal{R}(Q)\mathcal{G}(S; \gamma_1, \gamma_2). \tag{5.1.17}$$

Here $\mathcal{L}(R, S) = |\det \mathbf{Q}_2(R, S)|^{1/2}$. Many useful relations for the relative geometrical spreading $\mathcal{L}(R, S)$ can be found in Section 4.10.2.

5.1.4 Acoustic Pressure Reflection/Transmission Coefficients

We shall use a notation similar to that in Sections 2.3.1 and 2.4.5. We denote $\rho_1 = \rho(Q)$ and $c_1 = c(Q)$ so that ρ_1 and c_1 correspond to the point of incidence Q. The same parameters on the opposite side of the interface are denoted ρ_2 and c_2. Similarly, we denote the angle of incidence i_1 and the acute angle of transmission i_2. Both angles are related by Snell's law $(\sin i_1)/c_1 = (\sin i_2)/c_2$. For the point of incidence, $i(Q) = i_1$. For reflected waves, point $\tilde{Q}$ is situated on the same side of the interface as point Q, but for transmitted waves, it is situated on the opposite side of the interface. Thus, for the reflected wave $i(\tilde{Q}) = i_1, c(\tilde{Q}) = c_1$ and $\rho(\tilde{Q}) = \rho_1$, but for the transmitted waves $i(\tilde{Q}) = i_2, c(\tilde{Q}) = c_2$ and $\rho(\tilde{Q}) = \rho_2$. We introduce *ray parameter* p using the relation

$$p = (\sin i(Q))/c(Q) = (\sin i_1)/c_1. \tag{5.1.18}$$

The symbol p that we use for the ray parameter should not be confused with the symbol for pressure. Due to Snell's law, $p = (\sin i_2)/c_2$ also.

We shall also use the notation

$$P_k = \cos i_k = \left(1 - c_k^2 p^2\right)^{1/2}, \qquad k = 1, 2. \tag{5.1.19}$$

For $p > 1/c_k$, the square root (5.1.19) is imaginary positive,

$$P_k = +\mathrm{i}\left(c_k^2 p^2 - 1\right)^{1/2}. \tag{5.1.20}$$

This choice guarantees that the amplitudes of generated inhomogeneous waves decrease exponentially with the increasing distance from the interface. The plus sign in the expression for P_k in (5.1.20) corresponds to the plus sign in the expression (2.2.9) for the analytical signal; $F(\zeta) = x(\zeta) + \mathrm{i}g(\zeta)$. For the analytical signal defined by $F(\zeta) = x(\zeta) - \mathrm{i}g(\zeta)$, it would be necessary to write a minus sign instead of a plus sign in (5.1.20).

Using notation (5.1.19), we can express pressure reflection coefficients $R^r(Q)$ and pressure transmission coefficients $R^t(Q)$ as follows:

$$R^r(Q) = \frac{\rho_2 c_2 P_1 - \rho_1 c_1 P_2}{\rho_2 c_2 P_1 + \rho_1 c_1 P_2}, \qquad R^t(Q) = \frac{2\rho_2 c_2 P_1}{\rho_2 c_2 P_1 + \rho_1 c_1 P_2}; \tag{5.1.21}$$

see (2.3.26). The expressions for the normalized pressure reflection/transmission coefficients $\mathcal{R}^r(Q)$ and $\mathcal{R}^t(Q)$ then read

$$\mathcal{R}^r(Q) = R^r(Q) = \frac{\rho_2 c_2 P_1 - \rho_1 c_1 P_2}{\rho_2 c_2 P_1 + \rho_1 c_1 P_2}, \qquad \mathcal{R}^t(Q) = \frac{2(\rho_1 \rho_2 c_1 c_2 P_1 P_2)^{1/2}}{\rho_2 c_2 P_1 + \rho_1 c_1 P_2}; \tag{5.1.22}$$

see (5.1.16). Thus, the normalized reflection coefficient $\mathcal{R}^r(Q)$ equals the standard reflection coefficient $R^r(Q)$. The normalized transmission coefficient $\mathcal{R}^t(Q)$, however, differs from the standard transmission coefficient $R^t(Q)$.

Now we shall discuss the *reciprocity relations* for the R/T coefficients. We say that the selected R/T coefficient is reciprocal, if it is the same for the wave propagating along ray Ω from S to R and from R to S, that is, if $R(Q) = R(\tilde{Q})$. It is obvious that both the pressure reflection coefficients R^r and $\mathcal{R}^r$ are reciprocal. For the pressure transmission coefficients, this is not true. The standard plane-wave pressure transmission coefficient R^t is not reciprocal, but *the normalized pressure transmission coefficient $\mathcal{R}^t$ is reciprocal*. To summarize,

$$\begin{aligned}
R^r(Q) &= R^r(\tilde{Q}), & \mathcal{R}^r(Q) &= \mathcal{R}^r(\tilde{Q}), \\
R^t(Q) &\neq R^t(\tilde{Q}), & \mathcal{R}^t(Q) &= \mathcal{R}^t(\tilde{Q}).
\end{aligned} \tag{5.1.23}$$

The reciprocity of the normalized pressure R/T coefficients is caused by the multiplicative square root factor in (5.1.16). In the ray method, factor $[\cos i(\tilde{Q})/\cos i(Q)]^{1/2}$

has often been connected with geometrical spreading $\mathcal{L}(R, S)$, not with the R/T coefficients; see, for example, Červený, Molotkov, and Pšenčík (1977). For transmitted waves, however, this approach yields nonreciprocal geometrical spreading and a nonreciprocal transmission coefficient. Because the reciprocity plays an important role in many wave propagation problems, we shall further use the normalized (reciprocal) R/T coefficients given by (5.1.16) and (5.1.22) systematically. The additional advantage of the normalized R/T coefficients consists in higher simplicity and a more compact form of the final equations for amplitudes of acoustic waves. Note that the normalized R/T coefficients $\mathcal{R}$ were introduced by Červený (1987b), who referred to them as reciprocal R/T coefficients.

In the following text, we shall consider only *real-valued rays* Ω *from S to R*. This implies that both $\cos i(Q)$ and $\cos i(\tilde{Q})$ are real-valued and nonnegative. First, we shall discuss reflection coefficients $\mathcal{R}^r = R^r$ and then the coefficients of transmission $\mathcal{R}^t$ and R^t.

a. PRESSURE REFLECTION COEFFICIENTS

The normalized pressure reflection coefficient $\mathcal{R}^r$ is exactly the same as the standard pressure reflection coefficient R^r; see (5.1.22). Consequently, we shall only speak of reflection coefficients R^r; the result also will be automatically valid for $\mathcal{R}^r$.

For reflected waves, $i(Q) = i(\tilde{Q}) = i_1$, $\rho(Q) = \rho(\tilde{Q}) = \rho_1$, and $c(Q) = c(\tilde{Q}) = c_1$. The square root P_1 in (5.1.21) corresponds both to the incident and the reflected waves, $P_1 = \cos i(Q) = \cos i(\tilde{Q}) = \cos i_1$. It is always real-valued and nonnegative. Square root P_2, however, may be complex-valued. As shown in (5.1.20), it is complex-valued for $p > 1/c_2$, that is, for

$$\sin i_1 > c_1/c_2. \tag{5.1.24}$$

If $c_1 \geq c_2$, Equation (5.1.24) is not satisfied for any real-valued acute angle of incidence i_1 so that reflection coefficient R^r is then always real-valued. If $c_1 < c_2$, the situation is more complex. The range of i_1 given by (5.1.24) corresponds to some real-valued acute angles of incidence. The limiting angle of incidence

$$i_1^* = \arcsin(c_1/c_2). \tag{5.1.25}$$

Usually, i_1^* is called the *critical angle of incidence*. Reflection coefficient R^r is real-valued only for *subcritical and critical* angles of incidence $i_1 \leq i_1^*$. For *postcritical angles of incidence* $i_1 > i_1^*$, it is complex-valued. Note that the postcritical angles of incidence are also often called *overcritical* angles of incidence or *supercritical* angles of incidence.

It is simple to see that $|R^r| = 1$ in the postcritical region $i_1 > i_1^*$. For this reason, this case is often called the *case of total reflection*. In the postcritical region,

$$R^r = |R^r| \exp[i \arg R^r], \qquad |R^r| = 1,$$
$$\arg R^r = -2 \arctan(\rho_1 c_1 |P_2|/\rho_2 c_2 P_1). \tag{5.1.26}$$

For normal incidence, $i_1 = 0$, $P_1 = P_2 = 1$, and acoustic reflection coefficient R^r is given by the simple relation

$$R^r = (Z_2 - Z_1)/(Z_2 + Z_1), \tag{5.1.27}$$

where

$$Z = \rho c \tag{5.1.28}$$

is called the *wave impedance*. Wave impedances play a very important role in reflection methods in seismic prospecting for oil. The reflection coefficient for normal incidence is positive for $Z_2 > Z_1$, negative for $Z_2 < Z_1$, and vanishes for $Z_2 = Z_1$.

For grazing incidence, $i_1 = \frac{1}{2}\pi$, $P_1 = 0$, and the reflection coefficient equals -1, $R^r = \mathcal{R}^r = -1$, independently of the values of the medium parameters.

For certain nonvanishing angles of incidence, reflection coefficient R^r may vanish. This may happen only in the following two cases: (a) $Z_2 > Z_1, c_2 < c_1, \rho_2 > \rho_1$, and (b) $Z_2 < Z_1, c_2 > c_1, \rho_2 < \rho_1$. Such situations are not very common in realistic structures because the density jump across the interface must be large enough to cause the signs of $Z_2 - Z_1$ and $c_2 - c_1$ to be opposite. In the theory of electromagnetic waves, the angle of incidence for which the reflection coefficient vanishes is usually called the *Brewster angle of incidence*. Here we shall use the same terminology and call the angle of incidence for which the relevant R/T coefficient vanishes the Brewster angle of incidence. If we denote the Brewster angle of incidence i_1^B, (5.1.21) yields

$$\sin i_1^B = \frac{1}{c_2}\left[\frac{Z_2^2 - Z_1^2}{\rho_2^2 - \rho_1^2}\right]^{1/2}. \tag{5.1.29}$$

It is not difficult to find a simple relation for $\partial R^r/\partial p$, valid for real-valued P_1 and P_2:

$$\partial R^r/\partial p = 2p\rho_1\rho_2 c_1 c_2 \left(c_2^2 - c_1^2\right)/P_1 P_2(\rho_2 c_2 P_1 + \rho_1 c_1 P_2)^2. \tag{5.1.30}$$

As we can see, $\partial R^r/\partial p$ is always zero for $p = 0$ (normal incidence), nonnegative for $c_2 > c_1$, and nonpositive for $c_2 < c_1$. Often, we consider $|R^r|$ instead of R^r. The derivative of $|R^r|$ with respect to p may have the opposite sign to $\partial R^r/\partial p$ given by (5.1.30). With the exception of situations (a) and (b) mentioned earlier, containing the Brewster angles, $\partial|R^r|/\partial p$ is always nonnegative. Thus, $|R^r|$ *always increases with the increasing angle of incidence* i_1 (or is constant). Only in the case of the Brewster angle is the situation more complex. The Brewster angle divides the regions of positive and negative $\partial|R^r|/\partial p$. Derivative $\partial|R^r|/\partial p$ is always negative up to the Brewster angle of incidence ($i_1 < i_1^B$), and positive beyond the Brewster angle of incidence ($i_1 > i_1^B$).

Figure 5.1 shows the modulus $|R^r|$ and argument $\arg R^r$ of pressure reflection coefficient R^r for one typical model of an interface between two fluid halfspaces. The parameters of the model are $c_1 = 2{,}000$ m/s, $\rho_1 = 1{,}500$ kg/m^3, $c_2 = 2{,}500$ m/s, and $\rho_2 = 1{,}661$ kg/m^3. The refraction index $n = c_1/c_2$ equals 0.8, and the critical angle of incidence i_1^* is 53.13°. The pressure reflection coefficient R^r equals the normalized pressure reflection coefficient $\mathcal{R}^r$ so that we can discuss only R^r. For subcritical angles of incidence ($i_1 < i_1^* = 53.13°$), R^r is real-valued and increasing. For postcritical angles of incidence ($i_1 > i_1^* = 53.13°$), R^r is complex-valued, with $|R^r| = 1$ and $\arg R^r$ negative. The phase $\arg R^r$ vanishes at $i_1 = i_1^*$ and decreases with i_1 increasing. For $i_1 = \frac{1}{2}\pi$, $\arg R^r = -\pi$. The wave impedances for both media follow: $Z_1 = 3 \cdot 10^6$ kg/m^{2}s, $Z_2 = 4 \cdot 15 \, 10^6$ kg/m^{2}s. Consequently, $Z_2 > Z_1$ and $c_2 > c_1$, so that the Brewster angle does not exist. The normal incidence reflection coefficient R^r is positive and equals $R^r = 0.161$; see (5.1.27).

In seismic exploration for oil, particularly in the AVO (amplitude-versus-offset) studies, it is often very useful to use simple approximations for reflection coefficient R^r, sufficiently accurate for small angles of incidence. The standard expansion of R^r in terms of p does not yield sufficiently accurate results. A most convenient approximation is

$$R^r = \frac{Z_2 - Z_1 P_2/P_1}{Z_2 + Z_1 P_2/P_1} \doteq \frac{1}{2}\left(\frac{Z_2 - Z_1}{Z_1} + \frac{c_2 - c_1}{c_1}\text{tg}^2 i_1\right). \tag{5.1.31}$$

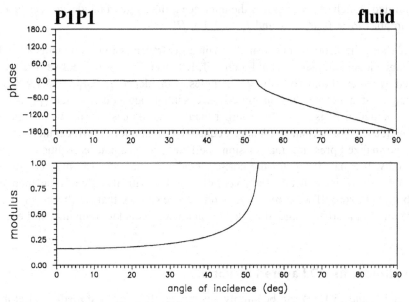

Figure 5.1. Pressure reflection coefficient at an interface between two fluid media. Model: $c_1 = 2,000$ m/s, $\rho_1 = 1,500$ kg/m^3, $c_2 = 2,500$ m/s, $\rho_2 = 1,661$ kg/m^3. Critical angle $i_1^* = 53.13°$. The curves also represent the normalized pressure reflection coefficient and the particle-velocity reflection coefficient; see Section 5.2.11.

To derive (5.1.31), we have taken into account the following relations:

$$\frac{P_2}{P_1} = \left[\frac{1 - \left(c_2^2/c_1^2\right)\sin^2 i_1}{\cos^2 i_1}\right]^{1/2} = \left[\frac{1 - \sin^2 i_1 + \left(1 - c_2^2/c_1^2\right)\sin^2 i_1}{\cos^2 i_1}\right]^{1/2}$$
$$= \left[1 + \left(1 - c_2^2/c_1^2\right)\mathrm{tg}^2 i_1\right]^{1/2} \doteq 1 + \tfrac{1}{2}\left(1 - c_2^2/c_1^2\right)\mathrm{tg}^2 i_1.$$

b. PRESSURE TRANSMISSION COEFFICIENTS

We shall discuss pressure transmission coefficients R^t and $\mathcal{R}^t$ for real-valued rays Ω. Thus, inhomogeneous waves are excluded from our treatment. Hence, $i(Q) = i_1$, $c(Q) = c_1$, $\rho(Q) = \rho_1$, and $i(\tilde{Q}) = i_2$, $c(\tilde{Q}) = c_2$, $\rho(\tilde{Q}) = \rho_2$. Because i_1 and i_2 are always real-valued acute angles for a real-valued ray Ω, square roots P_1 and P_2 are also real-valued and positive. This implies that transmission coefficients R^t and $\mathcal{R}^t$ are also real-valued and positive.

For *normal incidence*, $i_1 = 0$,

$$R^t = 2Z_2/(Z_1 + Z_2), \qquad \mathcal{R}^t = 2\sqrt{Z_1 Z_2}/(Z_1 + Z_2). \tag{5.1.32}$$

It is very simple to see in (5.1.32) that the normalized pressure transmission coefficient $\mathcal{R}^t$ is reciprocal, but that R^t is not.

In the following treatment, we shall distinguish two cases: $c_1 > c_2$ and $c_1 < c_2$.

a. If $c_1 > c_2$, the real-valued ray Ω of the transmitted wave exists for any acute angle of incidence, $0 \le i_1 \le \tfrac{1}{2}\pi$. For the grazing angle of incidence, and $i_1 = \tfrac{1}{2}\pi$, $P_1 = 0$, and $P_2 \ne 0$ so that $R^t = \mathcal{R}^t = 0$.

b. If $c_1 < c_2$, the real-valued ray Ω of the transmitted ray exists only for subcritical and critical angles of incidence i_1, $0 \le i_1 \le i_1^*$, where i_1^* is given by (5.1.25). Because $\sin i_2 = (c_2/c_1)\sin i_1$, $i_2 = \tfrac{1}{2}\pi$ for the critical angle of incidence. Thus, the transmitted branch of ray Ω grazes the interface if the angle of incidence is critical. This implies $P_1 \neq 0$, $P_2 = 0$, and $R^t = 2$, but $\mathcal{R}^t = 0$.

Figure 5.2 shows the standard and normalized pressure transmission coefficients R^t and $\mathcal{R}^t$, for the same model parameters as in Figure 5.1. Because $c_1 < c_2$, the real-valued rays of transmitted waves exist only for subcritical angles of incidence, $i_1 < i_1^* = 53.13°$. For postcritical angles of incidence, the transmitted wave is inhomogeneous. Standard pressure transmission coefficient R^t is real-valued, larger than 1, and increases in the whole range of angles of incidence. For normal incidence, it equals 1.161, and for the critical incidence, $R^t = 2$. The normalized pressure transmission coefficient $\mathcal{R}^t$ is again real-valued, but it is less than unity. It equals 0.986 for the normal incidence and changes only very little with increasing angle of incidence i_1. Only very close to the critical angle of incidence, it decreases abruptly to zero. If we compare R^t with $\mathcal{R}^t$, we can see that $\mathcal{R}^t$ (but not R^t) is very close to unity in a broad range of angles of incidence. This fact remains valid even for $c_2 < c_1$.

5.1.5 Amplitudes in 3-D Layered Structures

Equations (5.1.13) and (5.1.17) can be simply generalized for a layered medium. Let us consider ray Ω situated in a 3-D layered structure and two points S and R situated on this ray. We assume that ray Ω strikes various structural interfaces between S and R N-times. We denote the points of incidence Q_i, $i = 1, 2, \ldots, N$, and the corresponding points of reflection/transmission $\tilde{Q}_i$, $i = 1, 2, \ldots, N$. Point Q_i, of course, coincides with point $\tilde{Q}_i$ ($i = 1, 2, \ldots, N$), but they correspond to the incident and reflected/transmitted wave, respectively, and may be situated on different sides of the relevant interface.

The continuation relation for the amplitudes of acoustic waves then reads

$$P(R) = \left[\frac{\rho(R)c(R)}{\rho(S)c(S)} \right]^{1/2} \frac{\mathcal{L}(S)}{\mathcal{L}(R)} \mathcal{R}^C \exp[iT^c(R, S)]P(S), \qquad (5.1.33)$$

where

$$T^c(R, S) = \sum_{k=1}^{N+1} T^c(Q_k, \tilde{Q}_{k-1}), \qquad (5.1.34)$$

$$\mathcal{R}^C = \prod_{k=1}^{N} \mathcal{R}_k = \prod_{k=1}^{N} R_k \left[\frac{\rho(Q_k)c(Q_k)\cos i(\tilde{Q}_k)}{\rho(\tilde{Q}_k)c(\tilde{Q}_k)\cos i(Q_k)} \right]^{1/2}. \qquad (5.1.35)$$

In (5.1.34), we have denoted $S = \tilde{Q}_0$ and $R = Q_{N+1}$. Quantity R_k denotes the standard reflection/transmission coefficient at the point Q_k (see (5.1.21)), and $\mathcal{R}_k$ denotes the corresponding normalized reflection/transmission coefficient (see (5.1.22)). We shall call $\mathcal{R}^C$ the *complete reflection/transmission coefficient*. It equals the product of the normalized reflection/transmission coefficients at all points of incidence between S and R. It is obvious that the complete reflection/transmission coefficient $\mathcal{R}^C$ is reciprocal because the individual $\mathcal{R}_k$ in (5.1.35) are reciprocal.

For a point source situated at S,

$$P(R) = \left[\frac{\rho(R)c(R)}{\rho(S)c(S)} \right]^{1/2} \frac{\exp[iT^c(R, S)]}{\mathcal{L}(R, S)} \mathcal{R}^C \mathcal{G}(S; \gamma_1, \gamma_2). \qquad (5.1.36)$$

All the symbols have the same meaning as in (5.1.33) and in (5.1.17).

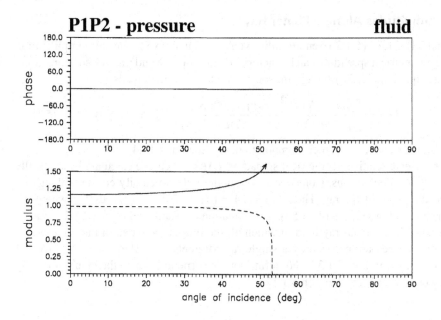

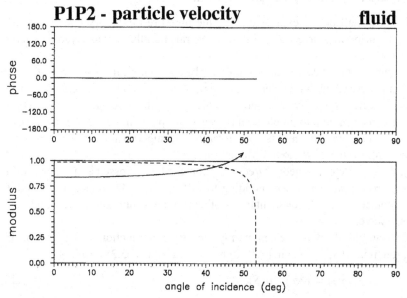

Figure 5.2. Transmission coefficients P1P2 at an interface between two fluid media. The model is the same as in Figure 5.1. Upper part: pressure transmission coefficients; lower part: particle velocity transmission coefficients (see Section 5.2.11). Continuous lines denote standard coefficients, and dashed lines represent the normalized coefficients. Note that the normalized transmission coefficients are the same for both pressure and particle velocity and are very close to unity for a broad range of angles of incidence. For $i_1 > 53.13°$, the transmitted wave is inhomogeneous, and the relevant coefficients are not shown.

5.1.6 Amplitudes Along a Planar Ray

Equations (5.1.33) and (5.1.36) remain valid even for a planar ray Ω. The only simplification is that the geometrical spreading can be factorized into in-plane and transverse geometrical spreading. Equation (5.1.36) for a point source situated at S then reads

$$P(R) = \left[\frac{\rho(R)c(R)}{\rho(S)c(S)}\right]^{1/2} \frac{\exp[iT^c(R, S)]}{\mathcal{L}^{\parallel}(R, S)\mathcal{L}^{\perp}(R, S)} \mathcal{R}^C \mathcal{G}(S; \gamma_1, \gamma_2). \qquad (5.1.37)$$

For details on $\mathcal{L}^{\parallel}$ and $\mathcal{L}^{\perp}$ and their computation, see Section 4.13.4.

The transverse relative geometrical spreading $\mathcal{L}^{\perp}(R, S)$ can be calculated analytically in many important situations. Consider a 2-D model with a velocity constant along any normal to the plane of the ray. Then $\mathcal{L}^{\perp}(R, S) = (\sigma(R, S))^{1/2} = (\int_S^R c\,ds)^{1/2}$, where the integral is taken along Ω, see (4.13.54). This situation is usually called the *2.5-D case*.

If we consider a planar ray field in which all the rays are situated in the same plane, the ray field can be parametrized by a single ray parameter γ_1. Then we write $\mathcal{G}(S; \gamma_1)$ instead of $\mathcal{G}(S; \gamma_1, \gamma_2)$ in (5.1.37). For a line source perpendicular to the plane of the ray and intersecting it at S, see Section 5.1.12.

5.1.7 Pressure Ray-Theory Green Function

In this section, we shall derive expressions for the pressure ray-theory Green function, corresponding to the point source situated at S. A general laterally varying layered structure is considered.

In inhomogeneous layered structures, there may be a finite or infinite number of rays Ω connecting point source S with receiver R. These rays correspond to different elementary waves (direct, reflected, multiply reflected, and the like). All the expressions derived in this section have been related to an elementary wave, propagating along the ray Ω being considered. The complete wavefield is then given by the superposition of all elementary waves, propagating from S to R along different rays.

Similarly, the complete ray-theory Green function can be expressed as a superposition of *elementary Green functions*, corresponding to different rays Ω connecting S and R. Here we shall consider only the elementary ray-theory Green function, corresponding to one elementary wave.

To obtain the amplitude of the elementary ray-theory Green function, we need to specify the radiation function $\mathcal{G}(S; \gamma_1, \gamma_2)$ in (5.1.36) properly. Using (2.5.28), we obtain

$$\mathcal{G}(S; \gamma_1, \gamma_2) = (4\pi)^{-1}\rho(S)c(S). \qquad (5.1.38)$$

Thus, the radiation function corresponding to the elementary acoustic ray-theory Green function is omnidirectional and satisfies reciprocity condition (5.1.12). Inserting (5.1.38) into (5.1.36) yields

$$P(R) = \frac{(\rho(S)\rho(R)c(S)c(R))^{1/2}}{4\pi \mathcal{L}(R, S)} \exp[iT^c(R, S)]\mathcal{R}^C. \qquad (5.1.39)$$

It is simple to see that the amplitude of the elementary ray-theory Green function is reciprocal. If we interchange the source and receiver, the amplitude remains the same.

For the reader's convenience, we shall give the final expressions for the elementary acoustic ray-theory Green function, corresponding to the selected ray Ω connecting the point source at S and the receiver at R.

In the frequency domain, the elementary Green function is expressed as

$$G(R, S, \omega) = \frac{(\rho(S)\rho(R)c(S)c(R))^{1/2}}{4\pi \mathcal{L}(R, S)} \mathcal{R}^C \exp[i\omega T(R, S) + iT^c(R, S)].$$

$$(5.1.40)$$

Similarly, in the time domain,

$$G(R, t; S, t_0) = \frac{(\rho(S)\rho(R)c(S)c(R))^{1/2}}{4\pi \mathcal{L}(R, S)} \mathcal{R}^C$$
$$\times \exp[iT^c(R, S)]\delta^{(A)}(t - t_0 - T(R, S)). \qquad (5.1.41)$$

It is obvious that both expressions are reciprocal in the following sense:

$$G(R, S, \omega) = G(S, R, \omega); \qquad G(R, t; S, t_0) = G(S, t; R, t_0). \qquad (5.1.42)$$

5.1.8 Receiver on an Interface

The equations derived for the pressure amplitudes are valid if the medium in the vicinity of source S and receiver R is smooth. If the source and/or receiver are situated on an interface, the derived equations need to be modified.

Let us first study the pressure amplitudes at the *receiver situated on interface* Σ^R passing through the point R. We shall use the following notation: if there is no interface at R, the pressure amplitude at R is $P^{SM}(R)$. We shall call $P^{SM}(R)$ the *smooth medium pressure amplitude at* R. This amplitude can be calculated using an equation derived in the previous sections. From $P^{SM}(R)$, we wish to determine $P(R)$ if receiver R is situated on structural interface Σ^R passing through R. We assume that point R is situated on Σ^R from the side of the incident wave, and introduce point R^+, situated on the opposite side of interface Σ^R. Otherwise, both points R and R^+ coincide. We also denote

$$c_1 = c(R), \qquad \rho_1 = \rho(R), \qquad c_2 = c(R^+), \qquad \rho_2 = \rho(R^+). \quad (5.1.43)$$

At interface Σ^R, two new waves are generated from the incident $P^{SM}(R)$: the reflected wave shown in Figure 5.3(a), and the transmitted wave shown in Figure 5.3(b). As usual, we denote the initial point of the generated wave being considered at interface Σ^R by $\tilde{R}$. It is obvious that $\tilde{R}$ coincides with R^+ for the transmitted wave and with R for the reflected wave. The travel times and analytical signals of all the three waves (incident, reflected, transmitted) coincide at R and $\tilde{R}$. At point R, two waves exist: the incident, smooth medium wave with amplitude $P^{SM}(R)$, and the reflected wave with amplitude $P^r(\tilde{R})$. At point R^+ situated on the opposite side of interface Σ^R, only one wave exists: the transmitted wave, with amplitude $P^t(\tilde{R})$. The amplitude of the total wavefield at interface Σ^R can be calculated in two ways

a. At the point R, as a superposition of amplitudes of incident and reflected waves, $P(R) = P^{SM}(R) + P^r(\tilde{R})$.

b. At the point R^+, as the amplitude of the transmitted wave, $P(R^+) = P^t(\tilde{R})$. Because pressure amplitude is continuous across the interface, it must hold that $P(R) = P(R^+)$.

We shall use both approaches and check that the results are the same.

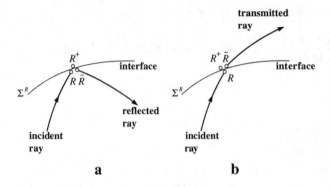

Figure 5.3. Explanation of symbols for the receiver situated on interface Σ^R. Point R corresponds to the wave incident on Σ^R; point $\tilde{R}$ corresponds to the generated R/T wave on Σ^R. Point R^+ is situated on the side of interface Σ^R opposite R.

At point R^+,

$$P(R^+) = P^t(\tilde{R}) = R^t(R)P^{SM}(R),\qquad(5.1.44)$$

where $R^t(R)$ is the transmission coefficient,

$$R^t(R) = \frac{2\rho_2 c_2 \cos i_1}{\rho_2 c_2 \cos i_1 + \rho_1 c_1 \cos i_2};\qquad(5.1.45)$$

see (5.1.21). Transmission coefficient $R^t(R)$ corresponds to the transmitted wave generated by a wave incident at interface Σ^R at point R. In (5.1.45), i_1 is the acute angle of incidence, and i_2 is the acute angle of transmission. Both angles i_1 and i_2 are connected by Snell's law. Note that $R^t(R) \neq R^t(\tilde{R})$, where $R^t(\tilde{R})$ is the transmission coefficient corresponding to the transmitted wave generated by the wave incident at interface Σ^R from the side of point $R^+ \equiv \tilde{R}$.

The second approach in deriving (5.1.44) takes into account the *incident and reflected waves at R*. At point R, the complete wavefield is given by the relation

$$P(R) = P^{SM}(R) + P^r(\tilde{R}) = (1 + R^r(R))P^{SM}(R).\qquad(5.1.46)$$

Here the reflection coefficient is

$$R^r(R) = \frac{\rho_2 c_2 \cos i_1 - \rho_1 c_1 \cos i_2}{\rho_2 c_2 \cos i_1 + \rho_1 c_1 \cos i_2}.\qquad(5.1.47)$$

Reflection coefficient $R^r(R)$ corresponds to the reflected wave generated by the wave incident at interface Σ^R at point R. It is simple to see that $1 + R^r(R) = R^t(R)$ (see (5.1.45)) and (5.1.47) so that both expressions (5.1.44) and (5.1.46) yield equivalent results.

From a physical point of view, it may seem strange that the amplitudes are discontinuous along the ray Ω under consideration at point R, where Ω strikes an interface Σ^R. At a point R, situated close to Σ^R, but not on it, the amplitude is $P^{SM}(R)$. Just at the interface, however, the amplitude jumps discontinuously and increases to $(1 + R^r(R))P^{SM}(R)$. The explanation is simple. Close to interface Σ^R, the complete wavefield is also obtained as a superposition of the incident wave and of the reflected wave, not by the incident wave only.

In the foregoing treatment, we have used the letter R to denote the point at which the receiver is situated as well as the R/T coefficients. They are not to be confused.

We shall now introduce the *pressure conversion coefficient* $\mathcal{D}(R)$ as

$$\mathcal{D}(R) = 1\qquad(5.1.48)$$

if point R is situated in a smooth medium, and

$$\mathcal{D}(R) = 1 + R^r(R), \qquad \mathcal{D}(R^+) = R^t(R) \tag{5.1.49}$$

if points R and R^+ are situated on interface Σ^R. Note again that $\mathcal{D}(R) = \mathcal{D}(R^+)$ in the case of pressure waves. The terminology "pressure conversion coefficients" is analogous to "elastic conversion coefficients," which will be introduced in Section 5.2.7.

The general expressions for amplitudes (5.1.33), (5.1.36), (5.1.37), and (5.1.39) can then be generalized by including the pressure conversion coefficients $\mathcal{D}(R)$ or $\mathcal{D}(R^+)$ given by (5.1.48) and (5.1.49). The expressions are then valid generally, both for the receiver situated in a smooth medium and on an interface. See Section 5.1.10 for the final equations for a point source.

5.1.9 Point Source at an Interface

We shall now discuss the pressure amplitudes at R due to a point source at S, situated on interface Σ^S.

First, we shall consider an auxiliary problem of a wave reflected/transmitted at interface Σ^S. This problem was discussed in Section 5.1.3. Here, however, we shall use a different notation, suitable for the purposes of this section. Consider a point source situated at point S_0 and a ray Ω_0 of a reflected/transmitted wave connecting S_0 with the receiver situated at R. We denote the point of incidence of ray Ω_0 at the interface $\tilde{S}$, and the point of reflection/transmission S. Thus, points $\tilde{S}$ and S_0 are situated on the incident branch of ray Ω_0, and points S and R are located on the reflected/transmitted branch of the ray. In addition to S, we also introduce point S^+, situated on the side of Σ^S opposite to S, and denote

$$c_1 = c(S), \qquad \rho_1 = \rho(S), \qquad c_2 = c(S^+), \qquad \rho_2 = \rho(S^+); \tag{5.1.50}$$

see Figure 5.4. Using (5.1.17), we obtain the general expression for $P(R)$:

$$P(R) = \left[\frac{\rho(R)c(R)}{\rho(S_0)c(S_0)}\right]^{1/2} \frac{\exp[iT^c(R, S_0)]}{\mathcal{L}(R, S_0)} \mathcal{R}(\tilde{S}) \mathcal{G}^{SM}(S_0; \gamma_1, \gamma_2). \tag{5.1.51}$$

Here $\mathcal{R}(\tilde{S})$ is the normalized R/T coefficient, and $\mathcal{G}^{SM}(S_0; \gamma_1, \gamma_2)$ is the "smooth medium" radiation function. As we know, the normalized R/T coefficient $\mathcal{R}(\tilde{S})$ is reciprocal so that

$$\mathcal{R}(\tilde{S}) = \mathcal{R}(S) = R(S)\left[\frac{\rho(S)c(S)\cos i(\tilde{S})}{\rho(\tilde{S})c(\tilde{S})\cos i(S)}\right]^{1/2},$$

where $R(S)$ is the standard R/T coefficient. Equation (5.1.51) then yields

$$P(R) = \left[\frac{\rho(R)c(R)\rho(S)c(S)\cos i(\tilde{S})}{\rho(S_0)c(S_0)\rho(\tilde{S})c(\tilde{S})\cos i(S)}\right]^{1/2}$$

$$\times \frac{\exp[iT^c(R, S_0)]}{\mathcal{L}(R, S_0)} R(S)\mathcal{G}^{SM}(S_0; \gamma_1, \gamma_2). \tag{5.1.52}$$

Now we shift point S_0 along ray Ω_0 to point $\tilde{S}$. Taking into account that $T^c(R, \tilde{S}) = T^c(R, S)$ and $\mathcal{L}(R, \tilde{S}) = \mathcal{L}(R, S)(\cos i(\tilde{S})/\cos i(S))^{1/2}$, (5.1.52) yields

$$P(R) = \left[\frac{\rho(R)c(R)}{\rho(S)c(S)}\right]^{1/2} \frac{\exp[iT^c(R, S)]}{\mathcal{L}(R, S)} R(S)\mathcal{G}^{SM}(\tilde{S}; \gamma_1, \gamma_2)\frac{\rho(S)c(S)}{\rho(\tilde{S})c(\tilde{S})}. \tag{5.1.53}$$

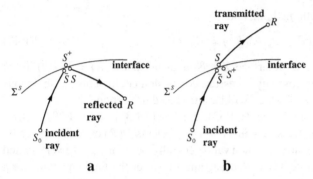

Figure 5.4. Explanation of symbols for the derivation of the pressure wavefield generated by a point source situated on interface Σ^S. The receiver is situated at R, an auxiliary point source is at S_0, the point of incidence is at $\tilde{S}$, and the point of reflection/transmission is at S. Point S^+ is situated on the opposite side of interface Σ^S than S. In the derivation, auxiliary source S_0 is shifted to $\tilde{S}$. (a) Auxiliary source S_0 and R are situated on the same side of Σ^S. (b) Auxiliary source S_0 and receiver R are situated on opposite sides of Σ^S.

This is the final expression of our auxiliary problem. We shall first specify it for the *transmitted wave*. Point $\tilde{S}$ is then situated at S^+, on the side of Σ^S opposite to S. Moreover, $R(S) = R^t(S)$, and is given by (5.1.45). It corresponds formally to the wave transmitted from S to S^+, *not from S^+ to S*. Angles i_1 and i_2 are given by relations $i_1 = i(S)$ and $i_2 = i(S^+)$ and are related by Snell's law $\sin i_1 = (c_1 \sin i_2)/c_2$.

Expression (5.1.53) can be rewritten in standard form as

$$P(R) = \left[\frac{\rho(R)c(R)}{\rho(S)c(S)} \right]^{1/2} \frac{\exp[iT^c(R, S)]}{\mathcal{L}(R, S)} \mathcal{G}(S^+; \gamma_1, \gamma_2), \tag{5.1.54}$$

where $\mathcal{G}(S^+; \gamma_1, \gamma_2)$ is the *generalized radiation function*, corresponding to the source situated on interface Σ^S:

$$\mathcal{G}(S^+; \gamma_1, \gamma_2) = R^t(S) \frac{\rho(S)c(S)}{\rho(S^+)c(S^+)} \mathcal{G}^{SM}(S^+; \gamma_1^t, \gamma_2^t). \tag{5.1.55}$$

Here $\mathcal{G}^{SM}(S^+; \gamma_1^t, \gamma_2^t)$ is the "smooth medium" radiation function, corresponding to the point S^+ situated on the side of Σ^S opposite to S. Ray parameters γ_1^t and γ_2^t correspond to the transmission from S^+ to S (along ray Ω_0). Similarly, γ_1 and γ_2 correspond to the ray Ω from S to R. Using Snell's law, we can relate γ_1^t and γ_2^t to γ_1 and γ_2.

Now we shall specify (5.1.53) for *reflected waves*. To obtain the complete wavefield at S, we need to add the reflected wave to the direct wave. The travel time and the analytical signals of the reflected and direct wave are the same in this case, so that we can add the amplitudes. This again yields

$$P(R) = \left[\frac{\rho(R)c(R)}{\rho(S)c(S)} \right]^{1/2} \frac{\exp[iT^c(R, S)]}{\mathcal{L}(R, S)} \mathcal{G}(S; \gamma_1, \gamma_2), \tag{5.1.56}$$

where the radiation function $\mathcal{G}(S; \gamma_1, \gamma_2)$ is given by the relation

$$\mathcal{G}(S; \gamma_1, \gamma_2) = \mathcal{G}^{SM}(S; \gamma_1, \gamma_2) + R^r(S)\mathcal{G}^{SM}(S; \gamma_1^r, \gamma_2^r). \tag{5.1.57}$$

Ray parameters γ_1^r and γ_2^r correspond to ray Ω_0 of the reflected wave.

As we can see from (5.1.55) and (5.1.57), the *generalized radiation function* of the source situated on interface Σ^S is different from the "smooth medium" radiation function of the source situated in a smooth medium. We even obtain different generalized radiation functions if the source is approaching the interface from different sides; see (5.1.55) for $\mathcal{G}(S^+; \gamma_1, \gamma_2)$ and (5.1.57) for $\mathcal{G}(S; \gamma_1, \gamma_2)$. This is not surprising given that a very general, possibly nonomnidirectional radiation function $\mathcal{G}^{SM}$ is being considered. In certain important situations, the generalized radiation functions simplifies. For example, for *omnidirectional radiation functions*, (5.1.57) yields

$$\mathcal{G}(S; \gamma_1, \gamma_2) = (1 + R'(S))\mathcal{G}^{SM}(S; \gamma_1, \gamma_2) = \mathcal{D}(S)\mathcal{G}^{SM}(S; \gamma_1, \gamma_2),$$
(5.1.58)

where $\mathcal{D}(S)$ is the pressure conversion coefficient, introduced by (5.1.48) and (5.1.49). In this case, Equation (5.1.55) yields

$$\mathcal{G}(S^+; \gamma_1, \gamma_2) = \frac{\rho(S)c(S)}{\rho(S^+)c(S^+)}\mathcal{D}(S^+)\mathcal{G}^{SM}(S^+; \gamma_1, \gamma_2).$$
(5.1.59)

Here $\mathcal{D}(S^+)$ is again given by (5.1.49), where R and R^+ are replaced by S and S^+. Equations (5.1.58) and (5.1.59) yield particularly simple results if we consider the smooth medium radiation functions corresponding to the acoustic ray-theory Green functions; see (5.1.38). Both (5.1.58) and (5.1.59) then yield the same, very simple result

$$\mathcal{G}(S; \gamma_1, \gamma_2) = \mathcal{G}(S^+; \gamma_1, \gamma_2) = \frac{\rho(S)c(S)}{4\pi}\mathcal{D}(S).$$
(5.1.60)

This result is valid both for the source situated in a smooth medium and on an interface. Moreover, if the source is situated on the interface Σ^S, it is irrelevant whether it is situated at S or at S^+. Even if it is situated at S^+, we take the parameters in (5.1.60) at S.

5.1.10 Final Equations for a Point Source

We shall now write final equations for the ray amplitudes of the pressure waves in 3-D layered structures, valid for any position of point source S and receiver R. Both points may be situated either in a smooth medium or on an interface. Taking into account (5.1.36) and the relations derived in Sections 5.1.8 and 5.1.9, we obtain

$$P(R) = \left[\frac{\rho(R)c(R)}{\rho(S)c(S)}\right]^{1/2} \frac{\exp[iT^c(R, S)]}{\mathcal{L}(R, S)} \mathcal{D}(R)\mathcal{R}^C\mathcal{G}(S; \gamma_1, \gamma_2).$$
(5.1.61)

Here $\mathcal{D}(R)$ is the pressure conversion coefficient, given by (5.1.48) and (5.1.49). If point S is situated in a smooth medium, $\mathcal{G}(S; \gamma_1, \gamma_2)$ represents the standard smooth medium radiation function. If S is situated on an interface, $\mathcal{G}(S; \gamma_1, \gamma_2)$ represents the generalized radiation function, given by (5.1.57). For omnidirectional smooth medium radiation functions, the generalized radiation functions are given by (5.1.58).

Equation (5.1.61) yields particularly simple results if we compute the amplitudes of an elementary pressure Green function. We can then use (5.1.60) to obtain

$$P(R) = \frac{[\rho(S)\rho(R)c(S)c(R)]^{1/2}}{4\pi \mathcal{L}(R, S)} \exp[iT^c(R, S)]\mathcal{D}(R)\mathcal{D}(S)\mathcal{R}^C.$$
(5.1.62)

Here $\mathcal{D}(R)$ and $\mathcal{D}(S)$ represent the pressure conversion coefficients, see (5.1.48) and (5.1.49).

For the reader's convenience, we shall give the final equations for the elementary pressure ray-theory Green function, corresponding to any elementary (multiply reflected) wave in a 3-D layered fluid medium. In the frequency domain,

$$G(R, S, \omega) = \frac{[\rho(S)\rho(R)c(S)c(R)]^{1/2}}{4\pi \mathcal{L}(R, S)} \mathcal{D}(R)\mathcal{D}(S)\mathcal{R}^C$$
$$\times \exp[i\omega T(R, S) + iT^c(R, S)]. \tag{5.1.63}$$

Similarly, in the time domain

$$G(R, t; S, t_0) = \frac{[\rho(S)\rho(R)c(S)c(R)]^{1/2}}{4\pi \mathcal{L}(R, S)} \mathcal{D}(R)\mathcal{D}(S)\mathcal{R}^C$$
$$\times \exp[iT^c(R, S)]\delta^{(A)}(t - t_0 - T(R, S)). \tag{5.1.64}$$

As we can see from (5.1.63) and (5.1.64), the elementary Green functions remain reciprocal even if the source and/or receiver are/is situated on an interface, in the sense of equations (5.1.42).

In equations (5.1.61) through (5.1.64), we can substitute R and/or S in the expressions for $\mathcal{D}(R)$, $\mathcal{D}(S)$, and $\mathcal{G}(S; \gamma_1, \gamma_2)$ by R^+ and/or S^+, if the receiver and/or source are situated on opposite sides of the relevant interfaces. In the case of pressure waves, this change may affect only Equation (5.1.61) because $\mathcal{D}(R^+) = \mathcal{D}(R)$ and $\mathcal{D}(S^+) = \mathcal{D}(S)$, but $\mathcal{G}(S; \gamma_1, \gamma_2)$ may be different from $\mathcal{G}(S^+; \gamma_1, \gamma_2)$.

5.1.11 Initial Ray-Theory Amplitudes at a Smooth Initial Surface. Acoustic Kirchhoff Integrals

We shall consider a smooth initial surface Σ^0. The initial surface may correspond to an interface between two fluid media, to a boundary of a fluid medium, to a wavefront, or to an auxiliary surface situated in a smooth medium. We assume that the initial time T^0 is specified along Σ^0. We can then determine the initial slowness vectors (see Section 4.5.1) and initial values of matrices $\mathbf{Q}$ and $\mathbf{P}$ of the waves generated at Σ^0 (see Sections 4.5.2 and 4.5.3). Consequently, we can start ray tracing and dynamic ray tracing at any point of Σ^0. To determine the ray amplitudes along the rays, however, we must also know the initial amplitudes along Σ^0, see $P(S)$ in (5.1.33). The determination of the initial ray-theory amplitudes is the main purpose of this section.

We shall start the treatment with the Kirchhoff integral (2.6.11) and use the same notation as in Section 2.6.1. The boundary surface S of volume V containing point $\vec{x}$ runs along initial surface Σ^0 and along a spherical surface C of an infinite radius. We assume that there are no sources in volume V so that $f^p = 0$. The volume integral in (2.6.11) then vanishes and the Kirchhoff integral in (2.6.11) over S reduces to the surface integral over Σ^0. The unit normal $\vec{n}$ is oriented *outward of V*.

As we can see in (2.6.11), we must know the distribution of pressure p and of the time derivative of the normal component of the particle velocity $\dot{v}^{(n)}$ along Σ^0 to be able to determine $p(\vec{x}, \omega)$. We shall now specify $p(\vec{x}', \omega)$ and $\dot{v}^{(n)}(\vec{x}', \omega)$ along Σ^0 in greater detail. We assume that the distribution of $p(\vec{x}', \omega)$ and $\dot{v}^{(n)}(\vec{x}', \omega)$ along Σ^0 is given by the relations

$$p(\vec{x}', \omega) = P^0(\vec{x}') \exp[i\omega T^0(\vec{x}')],$$
$$\dot{v}^{(n)}(\vec{x}', \omega) = -i\omega V^{(n)0}(\vec{x}') \exp[i\omega T^0(\vec{x}')]. \tag{5.1.65}$$

Here $P^0(\vec{x}')$ is the amplitude of pressure $p(\vec{x}', \omega)$ and $V^{(n)0}(\vec{x}')$ the amplitude of the normal component of particle velocity $v^{(n)}(\vec{x}', \omega)$ along initial surface Σ^0, and $T^0(\vec{x}')$ is the initial travel time along Σ^0. We assume that P^0, $V^{(n)0}$, and T^0 are independent of frequency and that the initial travel time $T^0(\vec{x}')$ is the same for $p(\vec{x}', \omega)$ and $v^{(n)}(\vec{x}', \omega)$. Inserting (5.1.65) into (2.6.11) yields

$$p(\vec{x}, \omega) = \int_{\Sigma^0} [P^0(\vec{x}')h(\vec{x}', \vec{x}, \omega) + i\omega V^{(n)0}(\vec{x}')G(\vec{x}', \vec{x}, \omega)]$$
$$\times \exp[i\omega T^0(\vec{x}')] d\Sigma^0(\vec{x}'). \qquad (5.1.66)$$

The Kirchhoff integral (5.1.66) is still exact, subject to assumption (5.1.65). In the following, however, we shall treat (5.1.66) asymptotically, for $\omega \to \infty$. We can then insert the asymptotic ray-theory expressions for the Green function $G(\vec{x}', \vec{x}, \omega)$ and for the relevant $h(\vec{x}', \vec{x}, \omega)$. As an approximation, only smooth-medium Green functions $G(\vec{x}', \vec{x}, \omega)$ and $h(\vec{x}', \vec{x}, \omega)$ will be used here. They correspond to the ray $\tilde{\Omega}(\vec{x}', \vec{x})$ connecting points $\vec{x}$ and $\vec{x}'$. Note that $\tilde{\Omega}(\vec{x}', \vec{x})$, $G(\vec{x}', \vec{x}, \omega)$, and $h(\vec{x}', \vec{x}, \omega)$ do not depend at all on the shape of Σ^0 and on $T^0(\vec{x}')$ at point $\vec{x}'$, but only on the position of points $\vec{x}'$ and $\vec{x}$. For any elementary ray-theory Green function $G(\vec{x}', \vec{x}, \omega)$ and relevant $h(\vec{x}', \vec{x}, \omega)$,

$$G(\vec{x}', \vec{x}, \omega) = G^A(\vec{x}', \vec{x}) \exp[i\omega T(\vec{x}', \vec{x})],$$
$$h(\vec{x}', \vec{x}, \omega) = -i\omega\tilde{\rho}^{-1}(\vec{x}')(\tilde{\vec{p}}(\vec{x}') \cdot \vec{n}(\vec{x}'))G^A(\vec{x}', \vec{x}) \exp[i\omega T(\vec{x}', \vec{x})].$$
$$(5.1.67)$$

Here $G^A(\vec{x}', \vec{x})$ is the amplitude of the elementary Green function under consideration, $T(\vec{x}', \vec{x})$ is its travel time, and $\tilde{\vec{p}}(\vec{x}')$ is its slowness vector at $\vec{x}'$. The tilde above the letters ρ and $\vec{p}$ is used to emphasize that these quantities correspond to the Green function $G(\vec{x}', \vec{x}, \omega)$ at point $\vec{x}'$ on the ray $\tilde{\Omega}(\vec{x}', \vec{x})$ connecting $\vec{x}'$ and $\vec{x}$. Density $\tilde{\rho}$ should always be taken at Σ^0 on the side of volume V. Inserting (5.1.67) into (5.1.66) yields

$$p(\vec{x}, \omega) = -i\omega \int_{\Sigma^0} a(\vec{x}')G^A(\vec{x}', \vec{x}) \exp[i\omega(T^0(\vec{x}') + T(\vec{x}', \vec{x}))] d\Sigma^0(\vec{x}'),$$
$$(5.1.68)$$

where

$$a(\vec{x}') = P^0(\vec{x}')\tilde{\rho}^{-1}(\vec{x}')(\tilde{\vec{p}}(\vec{x}') \cdot \vec{n}(\vec{x}')) - V^{(n)0}(\vec{x}'). \qquad (5.1.69)$$

This is the final form of the Kirchhoff integral we shall use here. The form of (5.1.68) is, of course, considerably influenced by assumption (5.1.65) regarding the form of $p(\vec{x}', \omega)$ and $\dot{v}^{(n)}(\vec{x}', \omega)$ along Σ^0. Note that G^A, T, and $\tilde{\vec{p}}$ depend on the position of points $\vec{x}$ and $\vec{x}'$, but not on the initial conditions along Σ^0. On the other hand, the quantities T^0, P^0, $V^{(n)0}$, $\tilde{\rho}$, and $\vec{n}$, depend on the position of $\vec{x}'$ only, not on $\vec{x}$.

As we have used the asymptotic expressions (5.1.67) for the Green function in (5.1.68), the Kirchhoff integral (5.1.68) represents only a high-frequency approximation. The two basic ways of treating the Kirchhoff integral (5.1.68) follow.

a. Numerical evaluation. Although (5.1.68) contains the ray-theory Green functions, the accuracy of the numerical treatment may be high. It, however, requires extensive two-point ray tracing from point $\vec{x}$ to all points $\vec{x}'$ along Σ^0. The family of rays $\tilde{\Omega}(\vec{x}', \vec{x})$ is two-parameteric. The relevant travel time $T(\vec{x}', \vec{x})$ and the Green function amplitude $G^A(\vec{x}', \vec{x})$ should be computed along each ray $\tilde{\Omega}$. In addition, slowness vector $\tilde{\vec{p}}(\vec{x}')$ should also be determined at $\vec{x}'$. Slowness vector $\tilde{\vec{p}}(\vec{x}')$ corresponds to the direction from $\vec{x}$ to $\vec{x}'$ along Ω.

b. Method of stationary phase. The high-frequency ($\omega \to \infty$) asymptotic methods can be further used to treat (5.1.68). The simplest is the method of stationary phase; see Bleistein (1984). It reduces the two-parameteric system of rays $\tilde{\Omega}$ to one (or a few) rays Ω^0. The price we pay for this is the lower accuracy of the results.

There are also some other methods of treating (5.1.68) and similar integrals, such as the expansion into Gaussian beams or Gaussian packets, the isochrone method, and the Maslov method. We shall not discuss these other methods here; for a description and many other references, see Spudich and Frazer (1984), Madariaga and Bernard (1985), Červený et al. (1987), and Huang, West, and Kendall (1998). See also Spencer, Chapman, and Kragh (1997) and Chapman (in press) for a very efficient approach to time-domain computations.

1. THE KIRCHHOFF INTEGRAL FOR A WAVE INCIDENT AT Σ^0

Equations (5.1.66) with (5.1.67) are very general and can be used for various initial conditions P^0 and $V^{(n)0}$ along Σ^0. We now shall consider this important special case: P^0 and $V^{(n)0}$ correspond to some elementary body wave incident at Σ^0. Both P^0 and $V^{(n)0}$ can then be expressed in terms of the pressure amplitudes P^{inc} of the incident wave along Σ^0. We shall consider a wave incident from the first medium, with medium parameters ρ_1 and c_1, and denote the medium parameters in the second medium by ρ_2 and c_2. All medium parameters may depend on coordinates. Quantities P^0 and $V^{(n)0}$ along Σ^0 are given by relations

$$P^0 = R^t P^{inc}, \qquad V^{(n)0} = \rho_2^{-1}(\vec{n} \cdot \vec{p}^{\,t})R^t P^{inc}, \tag{5.1.70}$$

or, alternatively,

$$P^0 = (1 + R^r)P^{inc}, \qquad V^{(n)0} = \rho_1^{-1}[(\vec{n} \cdot \vec{p}^{\,inc}) + (\vec{n} \cdot \vec{p}^{\,r})R^r]P^{inc}; \tag{5.1.71}$$

see (5.1.49) and (2.2.7). All quantities in (5.1.70) and (5.1.71) are taken at Σ^0, P^{inc} denotes the smooth medium pressure amplitude of the incident wave (not influenced by Σ^0), R^r and R^t are the pressure R/T coefficients given by (5.1.21), and $\vec{p}^{\,inc}$, $\vec{p}^{\,r}$, and $\vec{p}^{\,t}$ are the slowness vectors of the incident, reflected, and transmitted waves, respectively. Inserting (5.1.70) and (5.1.71) into (5.1.69) yields

$$a(\vec{x}') = \mathcal{A}(\vec{x}')P^{inc}(\vec{x}'), \tag{5.1.72}$$

where $\mathcal{A}(\vec{x}')$ is given by two alternative expressions:

$$\begin{aligned}
\mathcal{A}(\vec{x}') &= \left[\tilde{\rho}^{-1}(\vec{n} \cdot \tilde{\vec{p}}) - \rho_2^{-1}(\vec{n} \cdot \vec{p}^{\,t})\right]R^t \\
&= \tilde{\rho}^{-1}(\vec{n} \cdot \tilde{\vec{p}}) - \rho_1^{-1}(\vec{n} \cdot \vec{p}^{\,inc}) + (\tilde{\rho}^{-1}(\vec{n} \cdot \tilde{\vec{p}}) - \rho_1^{-1}(\vec{n} \cdot \vec{p}^{\,r}))R^r.
\end{aligned} \tag{5.1.73}$$

Again, all quantities in (5.1.73) are taken at $\vec{x}'$. The expressions in (5.1.73) are fully equivalent and can be used for both reflected and transmitted waves.

The Kirchhoff integral (5.1.68) corresponding to a wave incident at Σ^0 is then given by the expression

$$\begin{aligned}
p(\vec{x}, \omega) = -i\omega \int_{\Sigma^0} \mathcal{A}(\vec{x}')P^{inc}(\vec{x}')G^A(\vec{x}', \vec{x}) \\
\times \exp[i\omega(T^0(\vec{x}') + T(\vec{x}', \vec{x}))]d\Sigma^0(\vec{x}'),
\end{aligned} \tag{5.1.74}$$

where the weighting function $\mathcal{A}(\vec{x}')$ is given by any of the two expressions in (5.1.73). Thus, it is not necessary to store two quantities, P^0 and $V^{(n)0}$, along Σ^0, but only one quantity

P^{inc}. It should again be emphasized that the unit normal $\vec{n}$ to Σ^0 is oriented outward of the medium in which point $\vec{x}$ is situated.

Kirchhoff integral (5.1.74) is applicable to any wave incident at Σ^0 from the first medium. (The numbering of the media, however, is arbitrary.) Now we shall consider a special, but important case wherein the wave incident on Σ^0 is generated by a point source situated at $\vec{x}_0$ in the first medium, with the Green-function radiation function given by (5.1.38). The incident wave is then represented by the Green function $G(\vec{x}', \vec{x}_0, \omega)$, with $P^{inc}(\vec{x}') = G^A(\vec{x}', \vec{x}_0)$ and $T^0(\vec{x}') = T(\vec{x}', \vec{x}_0)$. Consequently, (5.1.74) yields the acoustic "Kirchhoff Green function" $G^K(\vec{x}, \vec{x}_0, \omega)$:

$$G^K(\vec{x}, \vec{x}_0, \omega) = -i\omega \int_{\Sigma^0} \mathcal{A}(\vec{x}')G(\vec{x}', \vec{x}_0, \omega)G(\vec{x}', \vec{x}, \omega)d\Sigma^0(\vec{x}'). \qquad (5.1.75)$$

Let us emphasize that both pressure ray-theory Green functions in the integral represent smooth medium Green functions, not influenced by Σ^0. Function $\mathcal{A}(\vec{x}')$ is given by any of the two expressions (5.1.73). All quantities in (5.1.73) are taken at $\vec{x}'$ on Σ^0.

The function $\mathcal{A}(\vec{x}')$ may be specified for the reflected waves, with both incident wave and point $\vec{x}$ situated in the first medium, and for the transmitted waves, with point $\vec{x}$ situated in the second medium, and the wave incidents at Σ^0 from the first medium.

For reflected waves (point $\vec{x}$ in the first medium), unit normal $\vec{n}$ is oriented into the second medium. We have $\tilde{\rho} = \rho_1$, $\tilde{c} = c_1$, $\vec{n} \cdot \vec{p}^{inc} = (\cos i_1)/c_1$, $\vec{n} \cdot \vec{p}^r = -(\cos i_1)/c_1$, $\vec{n} \cdot \vec{p}^t = (\cos i_2)/c_2$, and $\vec{n} \cdot \tilde{\vec{p}} = (\cos \tilde{i})/c_1$. Angles i_1, i_2, and $\tilde{i}$ have an obvious meaning, with $(\sin i_2)/c_2 = (\sin i_1)/c_1$. In general, angle $\tilde{i}$ is different from the specular angle i_1. Using any of the two expressions of (5.1.73), we obtain the *weighting function of the reflection Kirchhoff integral*:

$$\mathcal{A}(\vec{x}') = \frac{2 \cos i_1}{\rho_1 c_1} \frac{\rho_2 c_2 \cos \tilde{i} - \rho_1 c_1 \cos i_2}{\rho_2 c_2 \cos i_1 + \rho_1 c_1 \cos i_2}. \qquad (5.1.76)$$

The second factor in (5.1.76) resembles the pressure reflection coefficients R^r. However, it is only equal to it if $\tilde{i}$ equals the specular angle i_1. Otherwise, for $\tilde{i} \neq i_1$, the factor is different from R^r.

For transmitted waves (point $\vec{x}$ in the second medium), unit normal $\vec{n}$ is oriented into the first medium. We have $\tilde{\rho} = \rho_2$, $\tilde{c} = c_2$, $\vec{n} \cdot \vec{p}^{inc} = -(\cos i_1)/c_1$, $\vec{n} \cdot \vec{p}^r = (\cos i_1)/c_1$, $\vec{n} \cdot \vec{p}^t = -(\cos i_2)/c_2$, and $\vec{n} \cdot \tilde{\vec{p}} = (\cos \tilde{i})/c_2$. Any of the two expressions of (5.1.73) then yields the *weighting function of the transmission Kirchhoff integral*:

$$\mathcal{A}(\vec{x}') = \frac{2 \cos i_1(\cos \tilde{i} + \cos i_2)}{\rho_2 c_2 \cos i_1 + \rho_1 c_1 \cos i_2} = \frac{\cos \tilde{i} + \cos i_2}{\rho_2 c_2} R^t. \qquad (5.1.77)$$

Here R^t is the pressure transmission coefficient.

It may be of some interest to specify the Kirchhoff integrals (5.1.74) and (5.1.75) for a wave incident at an auxiliary surface Σ^0 situated in a smooth medium. The expression for $\mathcal{A}(\vec{x}')$, given by (5.1.73), simplifies because $\rho_1 = \rho_2 = \tilde{\rho}$, $c_1 = c_2 = \tilde{c}$, and $i_1 = i_2$. For the transmitted wave, we then obtain

$$\mathcal{A}(\vec{x}') = (\rho_1 c_1)^{-1}(\cos \tilde{i} + \cos i_1); \qquad (5.1.78)$$

see (5.1.77). If we wish to apply the Kirchhoff integral (5.1.74) or (5.1.75) to Σ^0 representing a wavefront in a smooth medium, we can use (5.1.78) with $i_1 = 0$ for the transmitted wave:

$$\mathcal{A}(\vec{x}') = (\rho_1 c_1)^{-1}(\cos \tilde{i} + 1). \qquad (5.1.79)$$

2. INITIAL RAY THEORY AMPLITUDES AT Σ^0

We shall now derive expressions for the initial ray-theory amplitudes at the initial surface Σ^0. To do this, we shall apply the method of stationary phase to the Kirchhoff integral (5.1.68). The distribution of $P^0(\vec{x}')$ and $V^{(n)0}(\vec{x}')$ along initial surface Σ^0 may be arbitrary (but smooth).

Let us first briefly discuss several important aspects of the problem. We denote by S any point on Σ^0. Using the methods of Section 4.5, we can determine the initial slowness vector $\vec{p}(S)$ and the initial values of 2×2 matrices $\mathbf{Q}(S)$ and $\mathbf{P}(S)$ at S and start ray tracing and dynamic ray tracing at S. Denote the relevant ray with the initial point at S on Σ^0 by Ω^0 and choose an arbitrary point $R \neq S$ on it. We wish to determine the initial ray-theory amplitude $P(S)$ at S, assuming $P^0(S)$ and $V^{(n)0}(S)$ are known. Thus, we are discussing the initial-value ray tracing problem; the position of R on Ω^0 plays no important part in it.

In the Kirchhoff integral computations, the situation is different because the position of receiver point $\vec{x}$ is fixed in advance. To calculate the Kirchhoff integral, it is necessary to determine rays $\tilde{\Omega}(\vec{x}', \vec{x})$ connecting receiver point R with all points $\vec{x}'$ along Σ^0. These rays should be calculated by two-point ray tracing, not by initial-value ray tracing from Σ^0. For high frequencies, however, the situation simplifies. We can use the stationary phase method, see Bleistein (1984, p. 88), which reduces the Kirchhoff integral computation to the computation of contributions from one (or several) stationary point(s). Any stationary point S of the Kirchhoff integral (5.1.68) is defined in such a way that the partial derivatives of $T^0(\vec{x}') + T(\vec{x}', \vec{x})$ along Σ^0 vanish at S. There may be several such stationary points along Σ^0, corresponding to several multiple rays from Σ^0 to R, as described in Section 3.10.1.c. The individual stationary points, however, can be treated independently. Here we shall consider only one of these stationary points. We shall again denote the position of the stationary point on Σ^0 by S and the relevant ray passing from S to R by Ω^0. It is obvious that the slowness vector $\vec{p}(S)$ of the elementary wave generated on Σ^0 is tangent to ray Ω^0 at S in a fluid medium and is related to $\tilde{\vec{p}}(S)$ as $\vec{p}(S) = -\tilde{\vec{p}}(S)$. Because the problem of determining the initial slowness vector $\vec{p}(S)$ and the 2×2 matrices $\mathbf{Q}(S)$ and $\mathbf{P}(S)$ at point S of initial surface Σ^0 was treated in great detail in Section 4.5, we shall not repeat the discussion here, but we shall concentrate only on determining the initial ray-theory amplitudes $P(S)$.

As in Section 4.4.1, we introduce the local Cartesian coordinate system z_i with its origin at S and with the z_3-axis perpendicular to Σ^0 at S. Coordinates z_1 and z_2 also play the role of ray parameters, $\gamma_1 = z_1$ and $\gamma_2 = z_2$. For more details and for the specification of the basis vectors $\vec{e}_1(S)$ and $\vec{e}_2(S)$, see Section 4.5.3.

The problem of determining the initial ray-theory amplitudes $P(S)$ has a *local character*, similar to the problem of determining the reflection/transmission amplitudes on an interface in the zeroth-order approximation of the ray method. The initial ray-theory amplitudes $P(S)$ do not depend on factors such as the curvature of Σ^0 at S, the gradients of medium parameters on both sides of Σ^0 at S, and the second tangential derivatives of the initial travel time $T^0(\vec{x}')$ at S. Consequently, we can consider a planar surface Σ^0, the homogeneous halfspaces on both sides of Σ^0, and $\partial^2 T^0/\partial z_I \partial z_J = 0$ at S. It should be emphasized that the foregoing factors influence the ray-theory amplitudes at R as a result of geometrical spreading. They also influence the initial values of the 2×2 matrices $\mathbf{Q}(S)$ and $\mathbf{P}(S)$ for dynamic ray tracing, and, consequently, $\mathbf{Q}(R)$ and $\mathbf{P}(R)$. They, however, do not influence the initial ray-theory amplitudes $P(S)$.

The stationary point contribution of the Kirchhoff integral (5.1.68) follows:

$$p(R, \omega) = -i\omega a(S) G^A(S, R) \exp[i\omega(T^0(S) + T(S, R))]$$
$$\times (2\pi/\omega)|\det \mathcal{M}(S, R)|^{-1/2} \exp\left[i\tfrac{\pi}{4} \operatorname{Sgn} \mathcal{M}(S, R)\right]; \quad (5.1.80)$$

see Bleistein (1984, Equation (2.8.23)). Here $G^A(S, R) = \tilde{\rho}(S)/4\pi r$, where r is the distance between S and R. $\mathcal{M}(S, R)$ is a 2×2 Hessian matrix with components

$$\mathcal{M}_{IJ}(S, R) = [\partial^2 (T^0 + T)/\partial z_I \partial z_J]_S.$$

Due to the local principle, $(\partial^2 T^0/\partial z_I \partial z_J)_S = 0$, and $\mathcal{M}_{IJ}(S, R)$ can be calculated using (4.4.37). We obtain

$$\mathcal{M}(S, R) = \mathbf{G}(S)\mathbf{M}(S, R)\mathbf{G}^T(S) = \frac{1}{\tilde{c}r}\begin{pmatrix} \cos^2 \tilde{i}(S) & 0 \\ 0 & 1 \end{pmatrix}.$$

This yields

$$|\det \mathcal{M}(S, R)|^{-1/2} = \tilde{c}r/\cos \tilde{i}(S), \qquad \text{Sgn}\, \mathcal{M}(S, R) = 2. \tag{5.1.81}$$

Inserting (5.1.81) into (5.1.80) yields

$$p(R, \omega) = (\tilde{\rho}(S)\tilde{c}(S)/2 \cos \tilde{i}(S))a(S) \exp[i\omega(T^0(S) + T(S, R)].\tag{5.1.82}$$

Because (5.1.82) actually represents a plane wave generated on Σ^0 in our simplified model, its amplitude is constant along the ray, $P(R) = P(S)$. Thus, the initial-value ray-theory amplitude is given by the simple relation

$$P(S) = \frac{\tilde{\rho}(S)\tilde{c}(S)a(S)}{2 \cos \tilde{i}(S)} = \frac{1}{2}\left[P^0(S) - \frac{\tilde{\rho}(S)\tilde{c}(S)}{\cos \tilde{i}(S)} V^{(n)0}(S) \right]. \tag{5.1.83}$$

This is the final result.

The minus sign in (5.1.83) may seem surprising because it would yield $P(S) = 0$ for Σ^0 situated in a smooth medium, if $V^{(n)0} = P^0(\cos \tilde{i})/\tilde{\rho}\tilde{c}$. We must, however, remember that the normal $\vec{n}$ to Σ^0 is oriented *outward of* the volume in which the receiver is situated. If $P^0(S)$ and $V^{(n)0}(S)$ correspond to an elementary wave propagating toward R, $V^{(n)0}(S) = -P^0(\cos \tilde{i})/\tilde{\rho}\tilde{c}$ and (5.1.83) yields $P(S) = P^0(S)$.

Although we have derived (5.1.83) using the local principle, it is valid generally, as well as for a curved Σ^0 at S, for nonvanishing gradients of medium parameters on both sides of Σ^0 at S, and for $[\partial^2 T^0/\partial z_I \partial z_J]_S \neq 0$. This can be verified by direct derivation, which is, however, more cumbersome than that presented here.

For completeness, we shall now present the final initial-value ray-theory expression for the pressure wavefield, generated at the initial surface Σ^0 situated in a laterally varying fluid medium containing interfaces. Consider point S on Σ^0, and construct ray Ω^0 from S, as described in Section 4.5. Also determine $\mathbf{Q}(S)$ and $\mathbf{P}(S)$ and perform dynamic ray tracing along Ω^0. We can then determine the geometrical spreading $\mathcal{L} = |\det \mathbf{Q}|^{1/2}$ along the whole ray Ω. The pressure wavefield at point R situated on Ω^0 is given by the relation

$$p(R, \omega) = P(R) \exp[i\omega(T^0(S) + T(R, S))]$$

$$P(R) = \left[\frac{\rho(R)c(R)}{\rho(S)c(S)} \right]^{1/2} \frac{\mathcal{L}(S)}{\mathcal{L}(R)} \mathcal{R}^C \exp[iT^c(R, S)]P(S). \tag{5.1.84}$$

Here $P(S)$ is given by (5.1.83). All symbols have the same meaning as in (5.1.33). For R situated on an interface, $P(R)$ should be multiplied by the conversion coefficient $\mathcal{D}(R)$; see Section 5.1.8.

Relations (5.1.83) and (5.1.84) can also be used if $P^0(S)$ and $V^{(n)0}(S)$ correspond to some elementary body wave incident at Σ^0. We again denote the angle of incidence at S by $i_1(S)$. Angle $\tilde{i}(S)$ satisfies Snell's law, $\tilde{c}^{-1}(S) \sin \tilde{i}(S) = c_1^{-1}(S) \sin i_1(S)$. For reflected waves, we have $\tilde{i}_1 = i_1$, $\tilde{c} = c_1$, and $\tilde{\rho} = \rho_1$. Similarly, for the transmitted wave $\tilde{i} = i_2$,

$\tilde{c} = c_2$, and $\tilde{\rho} = \rho_2$. We can then use (5.1.76) for reflected waves and obtain $\mathcal{A}(S) = (2 \cos i_1 / \rho_1 c_1) R^r(S)$. Similarly, for transmitted waves we use (5.1.77) and obtain $\mathcal{A}(S) = (2 \cos i_2 / \rho_2 c_2) R^t(S)$. Equations (5.1.72) and (5.1.83) yield very simple relations:

$$P(S) = R^r(S) P^{inc}(S) \qquad \text{for reflected waves,}$$
$$P(S) = R^t(S) P^{inc}(S) \qquad \text{for transmitted waves.}$$

$$(5.1.85)$$

Thus, we can use the ray-theory amplitude of the incident wave $P^{inc}(S)$, multiplied by the appropriate pressure R/T coefficient, to represent the initial ray-theory amplitude in (5.1.84). This result is not surprising; it fully corresponds to seismological intuition.

Equations (5.1.85) simplify further if initial surface Σ^0 is not an interface, but rather merely an auxiliary surface in a smooth medium. Then $c_1 = c_2$, $\rho_1 = \rho_2$, $R^r = 0$, and $R^t = 1$. Consequently, (5.1.85) yields

$$P(S) = 0 \qquad \text{for reflected waves,}$$
$$P(S) = P^{inc}(S) \qquad \text{for transmitted waves.}$$

$$(5.1.86)$$

Relations (5.1.86) have an obvious meaning. There is no reflection at auxiliary surface Σ^0, and the transmission amplitude equals the incident amplitude. Otherwise, all equations remain the same. If Σ^0 corresponds to a wavefront in a smooth medium, we can again use (5.1.86) for the initial amplitude $P(S)$. Reflections are not generated, and for transmissions $P(S) = P^{inc}(S)$. Even the solution of the kinematic part of the problem, described in Sections 4.5.1 through 4.5.3, simplifies considerably because $T^0(\gamma_1, \gamma_2)$ is constant along the wavefront. The rays are then normal to Σ^0.

Consequently, intermediate ray-theory solutions can be stored along auxiliary reference surfaces Σ^0 and used further for ray computations. We would proceed as follows. Consider an arbitrary elementary wave incident on Σ^0. We can parameterize Σ^0 by ray parameters γ_1 and γ_2 of the incident wave. Each point of incidence on Σ^0 then has parameters γ_1 and γ_2, corresponding to the parameters of the incident ray at that point. We can then store the travel time of incident wave $T^0(\gamma_1, \gamma_2)$ and the "smooth medium" pressure amplitude $P^{inc}(\gamma_1, \gamma_2)$. The values of $T^0(\gamma_1, \gamma_2)$ and $P^{inc}(\gamma_1, \gamma_2)$, stored along Σ^0, together with the information on which side of Σ^0 the wave is incident at Σ^0, are quite sufficient to recover completely the reflected and transmitted wavefields in the zeroth-order approximation of the ray method.

Because the computation of the initial slowness vectors and initial values of $\mathbf{Q}(S)$ and $\mathbf{P}(S)$ from $T^0(\gamma_1, \gamma_2)$, using the general methods described in Sections 4.5.1 through 4.5.3, may be time-consuming, it may be convenient to store along Σ^0 also some other quantities related to the incident wave (slowness vector, matrices $\mathbf{Q}$ and $\mathbf{P}$, and the like). For more details, refer to Section 5.5 of Červený, Klimeš, and Pšenčík (1988b).

3. REDUCTION TO THE RAY-THEORY SOLUTIONS – GENERAL CASE

In the previous derivation, we determined expressions for the initial ray-theory amplitudes on the initial surface Σ^0, using the principle of locality. Here we shall consider a quite general case of a curved surface Σ^0, laterally varying layered structures on both sides of Σ^0, and an arbitrary distribution of T^0 along Σ^0. We shall consider an elementary incident wave, generated by a point source situated at $\vec{x}_0$, and the receiver situated at $\vec{x}$. Both the incident and R/T waves may be arbitrarily multiply reflected/transmitted. We shall calculate the wavefield of a selected elementary wave, specified by a proper ray code, using two methods. First, we shall use the standard zeroth-order approximation of the ray method; see (5.1.36). Second, we shall apply the method of stationary phase to the Kirchhoff integral (5.1.74). We shall prove that both methods yield exactly the same results, even for the general model under consideration. In the derivation, it will be very convenient to use certain general properties of the Fresnel zone matrix.

We shall use the Kirchhoff integral (5.1.74), with $P^{inc}(\vec{x}') = P^{inc}(\vec{x}', \vec{x}_0)$ and $G^A(\vec{x}', \vec{x})$ given by the relations:

$$P^{inc}(\vec{x}', \vec{x}_0) = \left[\frac{\rho(\vec{x}')c(\vec{x}')}{\rho(\vec{x}_0)c(\vec{x}_0)} \right]^{1/2} \frac{\exp[iT^c(\vec{x}', \vec{x}_0)]}{\mathcal{L}(\vec{x}', \vec{x}_0)} \mathcal{R}^C(\vec{x}', \vec{x}_0) \mathcal{G}(\vec{x}_0; \gamma_1, \gamma_2),$$

$$G^A(\vec{x}', \vec{x}) = \frac{[\rho(\vec{x}')c(\vec{x}')\rho(\vec{x})c(\vec{x})]^{1/2}}{4\pi \mathcal{L}(\vec{x}', \vec{x})} \exp[iT^c(\vec{x}', \vec{x})]\mathcal{R}^C(\vec{x}', \vec{x}).$$

The individual symbols have the same meaning as in (5.1.36) and (5.1.40). $\mathcal{R}^C(\vec{x}', \vec{x}_0)$ denotes the complete R/T coefficient along the ray from $\vec{x}_0$ to $\vec{x}'$, and $\mathcal{R}^C(\vec{x}', \vec{x})$ denotes the complete R/T coefficient along the ray from $\vec{x}$ to $\vec{x}'$. Quantities $\rho(\vec{x}')$ and $c(\vec{x}')$ are taken at the appropriate side of Σ^0.

At the stationary point, the tangential components of the slowness vectors of the incident and generated R/T waves are equal. Consequently, Snell's law is satisfied there. We denote the position of the stationary point on Σ^0 by two symbols, Q and $\tilde{Q}$, corresponding to the point of incidence (Q), and to the relevant R/T point $(\tilde{Q})$. We also denote source point $\vec{x}_0$ by S and receiver point $\vec{x}$ by R. At the point of incidence Q, we introduce local Cartesian coordinates z_i using the standard option (4.4.21). The basis vectors $\vec{e}_1(\tilde{Q})$ and $\vec{e}_2(\tilde{Q})$ of the ray-centered coordinate system at the R/T point $\tilde{Q}$ are determined by the standard option (2.3.45).

The stationary point contribution of the Kirchhoff integral (5.1.74) is then as follows:

$$p(R, \omega) = -i\omega \mathcal{A}(\tilde{Q}) P^{inc}(Q, S) G^A(\tilde{Q}, R) \exp[i\omega(T^0(Q, S) + T(\tilde{Q}, R))]$$

$$\times (2\pi/\omega)|\det \mathcal{M}(Q; R, S)|^{-1/2} \exp\left[i\tfrac{\pi}{4} \operatorname{Sgn} \mathcal{M}(Q; R, S) \right].$$

$$(5.1.87)$$

see Bleistein (1984) and (5.1.80). Here $\mathcal{M}(Q; R, S) = \mathcal{M}(\tilde{Q}; R, S)$ is the 2×2 Hessian matrix with components

$$\mathcal{M}_{IJ}(Q; R, S) = [\partial^2(T^0(z_K) + T(z_K))/\partial z_I \partial z_J]_Q = M^F_{IJ}(Q; R, S).$$

$$(5.1.88)$$

Function $T^0(z_I)$ represents the distribution of the travel time of the incident wave (with a point source at S) along Σ^0, and $T(z_I)$ represents the distribution of the travel time along Σ^0 due to a point source at R. As shown in Section 4.4.8, $\mathcal{M}(Q; R, S)$ equals the Fresnel zone matrix $\mathbf{M}^F(Q; R, S)$; see (4.4.105). Sgn $\mathcal{M}$ denotes the signature of $\mathcal{M}$, that is, the number of positive eigenvalues of $\mathcal{M}$ minus the number of its negative eigenvalues.

Inserting expressions for $P^{inc}(Q, S)$ and $G^A(\tilde{Q}, R)$ into (5.1.87) yields

$$P(R, \omega) = -\tfrac{1}{2}i\mathcal{A}(\tilde{Q}) \left[\frac{\rho(R)c(R)}{\rho(S)c(S)} \right]^{1/2} \mathcal{G}(S; \gamma_1, \gamma_2)$$

$$\times \exp[i\omega(T^0(Q, S) + T(R, \tilde{Q}))]$$

$$\times [\rho(Q)c(Q)\rho(\tilde{Q})c(\tilde{Q})]^{1/2} \mathcal{R}^C(\tilde{Q}, R)\mathcal{R}^C(Q, S)$$

$$\times \frac{\exp\left[iT^c(Q, S) + iT^c(\tilde{Q}, R) + i\tfrac{\pi}{4} \operatorname{Sgn} \mathbf{M}^F(Q; R, S) \right]}{\mathcal{L}(Q, S)\mathcal{L}(\tilde{Q}, R)|\det \mathbf{M}^F(Q; R, S)|^{1/2}}.$$

$$(5.1.89)$$

The expression appears to be very cumbersome, but it can be simplified considerably. Using (4.8.30), we obtain

$$\mathcal{L}(Q, S)\mathcal{L}(\tilde{Q}, R)|\det \mathbf{M}^F(Q; R, S)|^{1/2} = (\cos i(Q) \cos i(\tilde{Q}))^{1/2}\mathcal{L}(R, S).$$

$$(5.1.90)$$

Here $i(Q)$ is the angle of incidence, and $i(\tilde{Q})$ is the angle of R/T. Equation (4.12.11) with (4.12.12) yields

$$T^c(Q, S) + T^c(\tilde{Q}, R) + \tfrac{\pi}{4} \operatorname{Sgn} \mathbf{M}^F(Q; R, S) = \tfrac{\pi}{2} + T^c(R, S). \qquad (5.1.91)$$

Both relations indicate the importance of the Fresnel zone matrix $\mathbf{M}^F(Q; R, S)$ in transforming the results obtained from the Kirchhoff integral by the stationary phase method to the ray-theory expressions. We take into account $\mathcal{A}(\tilde{Q}) = (2 \cos i(\tilde{Q})/\rho(\tilde{Q})c(\tilde{Q}))R(Q)$, where $R(Q)$ is the appropriate pressure R/T coefficient at Q on Σ^0(see (5.1.76) and (5.1.77)) and obtain

$$\mathcal{R}^C(R, \tilde{Q})\left[\frac{\rho(Q)c(Q) \cos i(\tilde{Q})}{\rho(\tilde{Q})c(\tilde{Q}) \cos i(Q)}\right]^{1/2} R(Q)\mathcal{R}^C(Q, S)$$

$$= \mathcal{R}^C(R, \tilde{Q})R(Q)\mathcal{R}^C(Q, S) = \mathcal{R}^C(R, S). \qquad (5.1.92)$$

Thus, the complete R/T coefficient $\mathcal{R}^C(R, S)$ now also includes the normalized R/T coefficient $\mathcal{R}(Q)$ at Q on Σ^0, which was not included in $\mathcal{R}^C(R, \tilde{Q})$ and $\mathcal{R}^C(Q, S)$. Inserting (5.1.90) through (5.1.92) into (5.1.89) yields the final, well-known ray-theory expressions:

$$p(R, \omega) = \left[\frac{\rho(R)c(R)}{\rho(S)c(S)}\right]^{1/2} \frac{\mathcal{R}^C(R, S)}{\mathcal{L}(R, S)}\mathcal{G}(S; \gamma_1, \gamma_2)$$

$$\times \exp[i\omega T(R, S) + iT^c(R, S)]; \qquad (5.1.93)$$

see (5.1.36) for comparison. Here we have used the standard notation $T(R, S) = T(R, \tilde{Q}) + T(Q, S)$. Consequently, the application of the stationary phase method to the Kirchhoff integral yields the same result as the zeroth-order ray method.

An important remark. The ray-theory solution (5.1.93) can also be obtained from (5.1.84), using (5.1.85) and an appropriate relation for $P^{inc}(S)$.

5.1.12 Initial Ray-Theory Amplitudes at a Smooth Initial Line in a Fluid Medium

The initial directions of rays generated at an initial line C^0 and relevant initial values of matrices $\mathbf{Q}$ and $\mathbf{P}$ were determined in Section 4.5.5. Thus, ray tracing and dynamic ray tracing from an initial line C^0 can be performed. To obtain complete ray-theory solutions, we must also specify the initial ray-theory amplitudes at an initial line C^0.

In the same way as for a point source, geometrical spreading vanishes at the initial line because $\det \mathbf{Q} = 0$ there; see (4.5.57). Thus, we obtain nonvanishing ray-theory amplitudes along the ray Ω from the initial line only if the initial ray-theory amplitude is infinite at the initial line. The product of initial ray-theory amplitude and geometrical spreading, however, must be finite there. As in the case of a point source, it will be useful to introduce some sort of *line source radiation functions*. We must, however, remember that $Q_{IJ} = 0$ for all I, J at a point source, but $Q_{22} \neq 0$ at the initial line; see (4.5.57). Consequently, the point-source radiation functions and the line-source radiation functions differ. The line-source radiation function at point S of the line source C^0 does not depend on the curvature and torsion of the initial line C^0 at S, on the inhomogeneity of the medium in the vicinity of S, and on the second derivatives of the initial travel time T^0 at S on C^0. All these factors are included in $\mathbf{Q}(S)$ and $\mathbf{P}(S)$ and influence the geometrical spreading but not the radiation function. Thus, in the computation of radiation function at S, we can consider a straight initial line C^0 situated in a homogeneous medium, with $T^{0''} = 0$, where T^0 is the initial travel time along C^0 and the primes denote the derivatives along C^0.

A similar problem was treated in Section 2.6.3, using the representation theorem. Here we shall consider a more general situation than in Section 2.6.3, considering initial line amplitudes P^0 and initial travel times T^0 varying along the initial line C^0 (with $T^{0''} = 0$). The relevant results will be derived using the representation theorem and the method of stationary phase. The final expressions for the line-source radiation functions will be obtained by matching these solutions with the ray-theory solutions for the same case. These radiation functions will be used to derive general ray-theory solutions for an arbitrary initial line C^0, situated in a general laterally varying 3-D layered structure, with nonvanishing curvature and torsion and with an arbitrary distribution of initial travel times along C^0.

1. REPRESENTATION THEOREM SOLUTIONS

We shall closely follow the treament of Section 2.6.3. The straight initial line C^0 is parallel to the x_2-axis in a homogeneous medium. We consider, however, a more general source term than (2.6.22):

$$f^p(x_j, \omega) = \delta(x_1 - x_{01})\delta(x_3 - x_{03})P^0(x_2)\exp[i\omega T^0(x_2)]. \tag{5.1.94}$$

We shall call $P^0(x_2)$ the *initial line amplitude* along C^0 and $T^0(x_2)$ the *initial travel time* along C^0. We shall consider a receiver R situated in the plane $x_2 = 0$, at a distance r_l from the initial line C^0. Then the exact solution for the pressure wavefield at point R is given by the representation theorem,

$$p(R, \omega) = \tfrac{1}{4}\rho\pi^{-1} \int_{-\infty}^{\infty} P^0(l)(l^2 + r_l^2)^{-1/2}$$
$$\times \exp\left[i\omega\left((l^2 + r_l^2)^{1/2}/c + T^0(l)\right)\right]dl; \tag{5.1.95}$$

see (2.6.7) and (2.6.23). Here we have used l instead of x_2 as the arclength along C^0. (The results will be modified for an arbitrary monotonous variable y_2 along C^0 at the end of this section.)

Now we shall apply the method of stationary phase to (5.1.95), assuming $\omega \to \infty$. The stationary point $l = l_0$ is defined by the relation $\phi'(l_0) = 0$, where $\phi(l) = (l^2 + r_l^2)^{1/2}/c + T^0(l)$:

$$\phi'(l_0) = l_0/cR_0 + T^{0'}(l_0) = 0, \qquad R_0 = (l_0^2 + r_l^2)^{1/2}. \tag{5.1.96}$$

We denote the stationary point $l = l_0$ situated on the initial line C^0 by S. For $T^{0'}(l_0) \neq 0$, the stationary point S is not situated in the plane $x_2 = 0$ as R but rather at a distance l_0 from it. The ray Ω generated at S and passing through R is inclined with respect to the plane $x_2 = 0$. The quantity R_0 denotes the distance between S and R. We introduce the acute angle φ between ray Ω and the plane perpendicular to C^0 at S. From (5.1.96), we obtain

$$\cos \varphi = r_l/R_0 = (1 - c^2 T^{0'2})^{1/2}. \tag{5.1.97}$$

For $T^{0'} = 0$ (exploding initial line C^0), we obtain $\varphi = 0$ and $R_0 = r_l$. The one-parameteric system of rays generated at S on C^0 then forms a central ray field in a plane perpendicular to C^0 at S. In all other cases, the one-parameteric system of rays generated at S on C^0 forms a conical surface, with the apex angle given by $\pi/2 - \varphi$ at S. Note that $|T^{0'}|$ is constant in our model and equals $1/b$, where b is the apparent velocity along C^0. For $c > b$, the generated waves would be inhomogeneous. Consequently, we consider only $c < b$.

As $T^{0''} = 0$ in our treatment, (5.1.96) yields immediately $\phi''(l_0) = r_l^2/c R_0^3$. The method of stationary phase then yields

$$p(R, \omega) \doteq \frac{\rho}{4\pi} \frac{P^0(S)}{\sqrt{R_0}} \frac{\sqrt{c}}{\cos \varphi} F(\omega) \exp[i\omega(R_0/c + T^0(S))]. \tag{5.1.98}$$

Here $F(\omega)$ is a two-dimensional frequency filter, given by (2.6.29) and R_0 is the distance between S and R. This is the final expression for the pressure wavefield generated by a straight line source C^0 situated in a homogeneous medium, assuming $T^{0''} = 0$. For $P^0(S) = 1$ and $T^0(S) = 0$, we obtain $\varphi = 0$ and $R_0 = r_l$ and (5.1.98) yields the 2-D acoustic Green function (2.6.24).

2. RAY-THEORY SOLUTIONS

Now we shall derive an alternative expression to (5.1.98) for the pressure wavefield $p(R, \omega)$ by the standard ray method. The solution contains the line-source radiation function $\mathcal{G}^L(S; \gamma_1)$, which cannot be determined by the ray method itself. Matching the ray-theory solution with the representation theorem solution for the same model, we shall express $\mathcal{G}^L(S; \gamma_1)$ in terms of initial line amplitudes $P^0(S)$.

We shall use the same approach as in Section 5.1.2 and introduce an auxiliary point S' on the ray Ω between S and R, close to S. Then,

$$P(R) = \left[\frac{\rho(R)c(R)}{\rho(S')c(S')} \right]^{1/2} \frac{\mathcal{L}(S')}{\mathcal{L}(R)} \exp[iT^c(R, S')]P(S'); \tag{5.1.99}$$

see (5.1.3). This is a general relation, valid even for inhomogeneous media. Now we assume that the medium close to S is locally homogeneous. Then we can use (4.6.2) and (4.8.3) to obtain

$$\mathcal{L}(S') = |\det \mathbf{Q}(S')|^{1/2} = |\det [\mathbf{Q}(S) + c R_0 \mathbf{P}(S)]|^{1/2}.$$

Here $R_0 = R_0(S', S)$ is the distance between S' and S, and $\mathbf{Q}(S)$ and $\mathbf{P}(S)$ are given by (4.5.57). Because we use the parameter $\gamma_2 = l$ (arclength) along the line source C^0, we can use $G = 1$ and $\sigma = c^{-1}\cos \varphi$ in (4.5.57). Then we obtain

$$\mathcal{L}(S') = \cos \varphi |R_0(S', S)(1 + R_0(S', S)\sigma^{-1} P_{22})|^{1/2}, \tag{5.1.100}$$

where P_{22} is given by (4.5.58). Inserting (5.1.100) into (5.1.99) and taking the limit $S' \to S$, we obtain

$$P(R) = \left[\frac{\rho(R)c(R)}{\rho(S)c(S)} \right]^{1/2} \frac{\exp[iT^c(R, S)]}{\mathcal{L}(R)} \mathcal{G}^L(S; \gamma_1), \tag{5.1.101}$$

where $\mathcal{G}^L(S; \gamma_1)$ is the line-source radiation function given by the relation

$$\mathcal{G}^L(S; \gamma_1) = \lim_{S' \to S} (\mathcal{L}(S')P(S')) = \lim_{S' \to S} (\cos \varphi(S')\sqrt{R_0(S', S)}P(S')).$$

$$(5.1.102)$$

The relation (5.1.101) is general and valid for initial lines C^0 with nonvanishing curvature and torsion, situated in an inhomogeneous medium, with $T^{0'}(S) \neq 0$ and $T^{0''}(S) \neq 0$. To match it with the representation theorem solution (5.1.98), we must specify it for a homogeneous medium, for a straight line source C^0 and for $T^{0''}(S) = 0$. It is simple to express $\mathcal{L}(R)$ for such a case because we can use (5.1.100), only we replace $R_0(S', S)$ by $R_0(R, S)$. In our model, we also have $P_{22} = 0$. Consequently,

$$P(R) = \mathcal{G}^L(S; \gamma_1)/\sqrt{R_0(R, S)} \cos \varphi(S),$$

$$(5.1.103)$$

and the pressure wavefield at point R is given by the relation

$$p(R, \omega) = P(R) \exp[i\omega T(R, S)], \qquad \text{with } T(R, S) = R_0(R, S)/c.$$

$$(5.1.104)$$

Equation (5.1.103) nicely shows the physical meaning of the line-source radiation function $\mathcal{G}^L(S; \gamma_1)$. Consider a one-parameteric system of rays generated at a point S situated on an initial line C^0, parameterized by the ray parameter γ_1. As we know, the system of rays forms a conical surface, with the apex angle $\pi/2 - \varphi$. Then the line-source radiation function $\mathcal{G}^L(S; \gamma_1)$ represents the amplitude at a distance $1/\cos^2\varphi(S)$ from S, measured along ray Ω generated at S on C^0, parameterized by the ray parameter γ_1. In other words, $\mathcal{G}^L(S; \gamma_1)$ represents a distribution of amplitudes along a circular intersection of the ray conical surface with a plane perpendicular to the cone axis, situated at a distance $1/\cos^2\varphi$ from S (measured along rays). If the initial travel time T^0 is constant along C^0, the line-source radiation function $\mathcal{G}^L(S; \gamma_1)$ represents the amplitude at a unit distance from C^0 in the plane perpendicular to C^0 at S. Note that the position of the point S may be parameterized by the ray parameter γ_2; see Sections 3.10.1 and 4.5.5. Consequently, $\mathcal{G}^L(S; \gamma_1)$ represents a two-parameteric radiation function.

3. GENERAL INITIAL LINE RAY-THEORY SOLUTIONS IN 3-D MODELS

Matching (5.1.104) with (5.1.98) offers three conclusions.

a. First, the ray-theory solutions should be multiplied by the two-dimensional frequency filter $F(\omega)$, given by (2.6.29).

b. The travel time $T(R, S)$ in the exponent of (5.1.104) should be supplemented by the initial travel time $T^0(S)$.

c. The initial-line radiation function $\mathcal{G}^L(S; \gamma_1)$ is related to the initial line amplitude $P^0(S)$ as follows:

$$\mathcal{G}^L(S; \gamma_1) = \tfrac{1}{4}\pi^{-1}\rho(S)\sqrt{c(S)}P^0(S).$$

$$(5.1.105)$$

Now we shall present final ray-theory expressions for the pressure wavefield generated at a line source C^0. We consider a smooth 3-D initial line C^0 with a (possibly) nonvanishing curvature and torsion. We shall specify the initial line by relations $\vec{x} = \vec{x}(\gamma_2)$, as in Section 4.5.5 (see (4.5.40)), where γ_2 is an arbitrary monotonous parameter along C^0 (not necessarily the arclength l). We also use $G = \partial\vec{x}/\partial\gamma_2 \cdot \partial\vec{x}/\partial\gamma_2$; see (4.5.41). The distribution of the initial travel time $T^0(\gamma_2)$ and the line-source amplitudes $P^0(\gamma_2)$ along C^0 may be arbitrary; we only assume that $P^0(\gamma_2)$ is smooth and that $T^{0'}(\gamma_2) = \partial T^0(\gamma_2)/\partial\gamma_2$

and $T^{0''}(\gamma_2) = \partial^2 T^0(\gamma_2)/\partial \gamma_2^2$ are continuous. Finally, we consider an arbitrary multiply reflected/transmitted elementary wave in a 3-D laterally varying layered structure. Then (5.1.101) and (5.1.105) yield

$$p(R, \omega) = P(R)F(\omega) \exp[i\omega(T(R, S) + T^0(S))], \qquad (5.1.106)$$

with

$$P(R) = \left[\frac{\rho(R)c(R)}{\rho(S)c(S)}\right]^{1/2} \frac{\exp[iT^c(R, S)]}{\mathcal{L}(R)} \mathcal{R}^C \mathcal{D}(R) \mathcal{G}^L(S; \gamma_1). \qquad (5.1.107)$$

Here $\mathcal{R}^C$ is the complete R/T coefficient (see (5.1.35)), $\mathcal{D}(R)$ is the pressure conversion coefficient (see (5.1.48) or (5.1.49)), $T^c(R, S)$ is the phase shift due to caustics (see (5.1.34)), and $\mathcal{G}^L(S; \gamma_1)$ is the line-source radiation function given by (5.1.105). The relation (5.1.97) for $\cos\varphi(S)$ should be modified here for a general parameter γ_2 along the initial line C^0:

$$\cos\varphi(S) = [1 - c^2(S)G^{-1}(S)T^{0'2}(S)]^{1/2}, \qquad (5.1.108)$$

with $G = \partial\vec{x}/\partial\gamma_2 \cdot \partial\vec{x}/\partial\gamma_2$. Geometrical spreading $\mathcal{L}(R)$ in (5.1.107) can be computed using standard relations (4.6.2),

$$\mathcal{L}(R) = |\det \mathbf{Q}(R)|^{1/2} = |\det (\mathbf{Q}_1(R, S)\mathbf{Q}(S) + \mathbf{Q}_2(R, S)\mathbf{P}(S))|^{1/2},$$

with $\mathbf{Q}(S)$ and $\mathbf{P}(S)$ given by (4.5.57).

4. 2-D COMPUTATIONS WITH A LINE SOURCE

Now we shall discuss a situation that has found useful applications in 2-D computations. Consider a 2-D laterally varying layered model in which the velocity and density are constant along any normal to the plane of rays $\Sigma^{\|}$. Moreover, let us consider a straight line source C^0 perpendicular to $\Sigma^{\|}$ and intersecting $\Sigma^{\|}$ at point S, with the initial travel time $T^0 = 0$ along C^0. Note that the computations would not be two-dimensional if we allowed $T^{0'} \neq 0$ because the rays generated at S would form a conical surface not confined to $\Sigma^{\|}$. We shall use the notation of Section 4.13 and specify the unit basis vectors $\vec{e}_1$, $\vec{e}_2$, and $\vec{e}_3$ as explained there. The rays in the plane $\Sigma^{\|}$ form a central ray field. To compute $P(R)$, we can then use $\mathcal{L}(R) = c^{-1/2}(S)\mathcal{L}^{\|}(R, S)$ where $\mathcal{L}^{\|}(R, S) = |Q_2^{\|}(R, S)|^{1/2}$. This relation follows from (4.13.46), (4.13.55), and (4.13.56), taking into account that $P^{\|}(S) = c^{-1}(S)$; see (4.5.60). Consequently,

$$\begin{aligned}
P(R) &= \left[\frac{\rho(R)c(R)}{\rho(S)}\right]^{1/2} \frac{\exp[iT^c(R, S)]}{\mathcal{L}^{\|}(R, S)} \mathcal{R}^C \mathcal{D}(R) \mathcal{G}^L(S; \gamma_1) \\
&= \frac{[\rho(R)c(R)c(S)\rho(S)]^{1/2}}{4\pi \mathcal{L}^{\|}(R, S)} \exp[iT^c(R, S)]\mathcal{R}^C \mathcal{D}(R)P^0(S).
\end{aligned}$$
$$(5.1.109)$$

5. 2-D PRESSURE RAY-THEORY GREEN FUNCTION

The amplitude of the 2-D pressure ray-theory Green function $G^{2D}(R, S, \omega)$ is obtained from (5.1.109), if we insert there $P^0(S) = 1$; see Section 2.6.3. Consequently, $G^{2D}(R, S, \omega)$ is given by the relation

$$\begin{aligned}
G^{2D}(R, S, \omega) = {} & \frac{[\rho(S)\rho(R)c(S)c(R)]^{1/2}}{4\pi \mathcal{L}^{\|}(R, S)} \mathcal{R}^C \mathcal{D}(R)F(\omega) \\
& \times \exp[i\omega T(R, S) + iT^c(R, S)];
\end{aligned}$$
$$(5.1.110)$$

see (5.1.106) and (5.1.109). In the time domain, we obtain

$$G^{2D}(R, t; S, t_0) = \frac{[\rho(S)\rho(R)c(S)c(R)]^{1/2}}{4\pi \mathcal{L}^{\parallel}(R, S)} \mathcal{R}^C \mathcal{D}(R)$$

$$\times \exp[iT^c(R, S)](\sqrt{2}H(\zeta)\zeta^{-1/2})^{(A)}, \qquad (5.1.111)$$

where $\zeta = t - t_0 - T(R, S)$. See (A.3.9) for $(H(\zeta)\zeta^{-1/2})^{(A)}$.

5.2 Elastic Isotropic Structures

In this section, we shall derive equations for the amplitudes and for the components of the displacement vector of high-frequency elastic body waves propagating in isotropic media with variable Lamé's elastic parameters $\lambda(x_i)$ and $\mu(x_i)$ and density $\rho(x_i)$. The displacement vector $\vec{u}(x_i, t)$ of an elastic wave propagating in the medium can be expressed in the following general form:

$$\vec{u}(x_i, t) = \vec{U}(x_i)F(t - T(x_i)); \qquad (5.2.1)$$

see (2.4.14). Here $F(\zeta)$ is a high-frequency analytical signal, $\vec{U}(x_i)$ is a *vectorial ray-theory complex-valued amplitude function*, and $T(x_i)$ is the travel time of the wave. For more details on analytical signal $F(\zeta)$, see the introduction to Section 5.1 and Appendix A.3.

As we know from Section 2.4.2, two high-frequency seismic body waves can propagate in a smoothly inhomogeneous isotropic medium.

a. **P waves**. The travel time $T(x_i)$ of the P wave satisfies the P-wave eikonal equation $\nabla T(x_i) \cdot \nabla T(x_i) = 1/\alpha^2(x_i)$, where $\alpha(x_i)$ is the position-dependent velocity of the P wave, $\alpha(x_i) = [(\lambda(x_i) + 2\mu(x_i))/\rho(x_i)]^{1/2}$. The vectorial amplitude function $\vec{U}(x_i)$ of P waves is given as

$$\vec{U}(x_i) = A(x_i)\vec{N}(x_i); \qquad (5.2.2)$$

see (2.4.26). Here $\vec{N}(x_i)$ denotes the unit vector perpendicular to the wavefront. It can be expressed in terms of slowness vector $\vec{p}(x_i) = \nabla T(x_i)$ using relation $\vec{N}(x_i) = \alpha(x_i)\vec{p}(x_i)$. Function $A(x_i)$ represents the *scalar ray-theory complex-valued amplitude function of P waves*. To simplify the terminology, we shall call it simply the *amplitude function of P waves*, or also the *amplitude of P waves*.

b. **S waves**. The travel time $T(x_i)$ of the S wave satisfies the S-wave eikonal equation $\nabla T(x_i) \cdot \nabla T(x_i) = 1/\beta^2(x_i)$, where $\beta(x_i)$ is the position-dependent velocity of the S wave, $\beta(x_i) = (\mu(x_i)/\rho(x_i))^{1/2}$. The vectorial complex-valued ray-theory amplitude function $\vec{U}(x_i)$ of S waves reads

$$\vec{U}(x_i) = B(x_i)\vec{e}_1(x_i) + C(x_i)\vec{e}_2(x_i); \qquad (5.2.3)$$

see (2.4.28). Here $\vec{e}_1$ and $\vec{e}_2$ are the ray-centered basis vectors (see Section 4.1.2), also called polarization vectors of S waves. The algorithms for their evaluation along rays are discussed in Section 4.1.3. Functions $B(x_i)$ and $C(x_i)$ represent *scalar complex-valued ray-centered components of the vectorial amplitude function $\vec{U}(x_i)$ of S waves*. We shall call $B(x_i)$ the S1 component and $C(x_i)$ the S2 component of the S wave. Thus, the S1 component represents the component of amplitude vector $\vec{U}$ along $\vec{e}_1$, and the S2 component represents the component of $\vec{U}$ along $\vec{e}_2$. We shall again simplify the terminology and refer to $B(x_i)$ as the *amplitude function (or amplitude) of the S1 wave* and to $C(x_i)$ as the *amplitude function (or* *amplitude) of the S2 wave.* We remind the reader, however, that S1 and S2 are not two independent waves; they merely represent two ray-centered components of the vectorial complex-valued amplitude function of the S wave.

5.2.1 Vectorial Complex-Valued Amplitude Functions of P and S Waves

We can also express (5.2.1) through (5.2.3) in a more general form, valid both for P and S waves:

$$\vec{u} = u_1^{(q)}\vec{e}_1 + u_2^{(q)}\vec{e}_2 + u_3^{(q)}\vec{e}_3,$$
$$\vec{U} = U_1^{(q)}\vec{e}_1 + U_2^{(q)}\vec{e}_2 + U_3^{(q)}\vec{e}_3. \tag{5.2.4}$$

Unit vectors $\vec{e}_1$, $\vec{e}_2$, and $\vec{e}_3 \equiv \vec{N}$ are the polarization vectors, introduced in Section 4.1.2, and form a triplet of basis vectors of the ray-centered coordinate system.

Quantities $u_1^{(q)}$, $u_2^{(q)}$, and $u_3^{(q)}$ represent the ray-centered components of displacement vector $\vec{u}$, and $U_1^{(q)}$, $U_2^{(q)}$, and $U_3^{(q)}$ represent the ray-centered components of vectorial amplitude function $\vec{U}$. Thus, Equation (5.2.1) can be expressed in ray-centered coordinate component form as

$$u_i^{(q)}(x_j, t) = U_i^{(q)}(x_j)F(t - T(x_j)). \tag{5.2.5}$$

It is obvious that $U_i^{(q)}$, $i = 1, 2, 3$, have different meanings for P and S waves:

a. For P waves,

$$U_1^{(q)} = U_2^{(q)} = 0, \qquad U_3^{(q)} = A. \tag{5.2.6}$$

b. For S waves,

$$U_1^{(q)} = B, \qquad U_2^{(q)} = C, \qquad U_3^{(q)} = 0. \tag{5.2.7}$$

In the following, we shall also use the matrix notation. We introduce two 3×1 matrices $\hat{\mathbf{u}}^{(q)}$ and $\hat{\mathbf{U}}^{(q)}$, related to the ray-centered components of the displacement vector and to the ray-centered components of the vectorial amplitude function:

$$\hat{\mathbf{u}}^{(q)} \equiv \left(u_1^{(q)},\ u_2^{(q)},\ u_3^{(q)}\right)^T, \qquad \hat{\mathbf{U}}^{(q)} \equiv \left(U_1^{(q)},\ U_2^{(q)},\ U_3^{(q)}\right)^T. \tag{5.2.8}$$

Equations (5.2.1) and (5.2.5) can then be expressed in matrix form,

$$\hat{\mathbf{u}}^{(q)} = \hat{\mathbf{U}}^{(q)}F(t - T). \tag{5.2.9}$$

We again emphasize the validity of relations (5.2.6) for P waves and of (5.2.7) for S waves. If we denote $\hat{\mathbf{U}}^{(q)} = \hat{\mathbf{U}}_P^{(q)}$ for P waves, and $\hat{\mathbf{U}}^{(q)} = \hat{\mathbf{U}}_S^{(q)}$ for S waves,

$$\hat{\mathbf{U}}_P^{(q)} = (0,\ 0,\ A)^T, \qquad \hat{\mathbf{U}}_S^{(q)} = (B,\ C,\ 0)^T. \tag{5.2.10}$$

As we have seen, it is very convenient to express vectorial amplitude function $\vec{U}$ in terms of ray-centered components. In the case of P waves, two components of $\vec{U}$ vanish, and in the case of S waves, one component of $\vec{U}$ vanishes. Nevertheless, it will prove convenient in many applications to express displacement vector $\vec{u}$ and vectorial amplitude function $\vec{U}$ in terms of components in other coordinate systems, mainly in general Cartesian components or in some local Cartesian components. If we use the matrix notation, it is necessary to distinguish between components in different coordinate systems by special notation. We shall do this by including appropriate coordinate system in brackets in the superscript in the same way that (q) has been used to emphasize the ray-centered coordinate system q_1, q_2, q_3 in (5.2.4).

Thus, in general Cartesian coordinate system x_1, x_2, x_3, we shall use the notation

$$\hat{\mathbf{u}}^{(x)} = \left(u_1^{(x)}, \; u_2^{(x)}, \; u_3^{(x)}\right)^T, \qquad \hat{\mathbf{U}}^{(x)} = \left(U_1^{(x)}, \; U_2^{(x)}, \; U_3^{(x)}\right)^T. \tag{5.2.11}$$

As in (5.2.9), we can write

$$\hat{\mathbf{u}}^{(x)} = \hat{\mathbf{U}}^{(x)} F(t - T). \tag{5.2.12}$$

In any other curvilinear orthogonal coordinate systems ξ_1, ξ_2, ξ_3,

$$\hat{\mathbf{u}}^{(\xi)} = \left(u_1^{(\xi)}, \; u_2^{(\xi)}, \; u_3^{(\xi)}\right)^T, \qquad \hat{\mathbf{U}}^{(\xi)} = \left(U_1^{(\xi)}, \; U_2^{(\xi)}, \; U_3^{(\xi)}\right)^T, \tag{5.2.13}$$

and

$$\hat{\mathbf{u}}^{(\xi)} = \hat{\mathbf{U}}^{(\xi)} F(t - T). \tag{5.2.14}$$

To transform mutually $\hat{\mathbf{U}}^{(q)}$, $\hat{\mathbf{U}}^{(x)}$, and $\hat{\mathbf{U}}^{(\xi)}$, we can use the 3×3 transformation matrices $\hat{\mathbf{H}}$ and $\hat{\mathbf{H}}^{(\xi)}$. Here $\hat{\mathbf{H}}$ is the transformation matrix from the ray-centered to the general Cartesian coordinate system; see Section 4.1.5. We remind the reader that $H_{kl} = e_{lk}$, where e_{lk} is the kth Cartesian component of unit basis vector $\vec{e}_l$. Similarly, $\hat{\mathbf{H}}^{(\xi)}$ is the transformation matrix from the curvilinear coordinate system ξ_1, ξ_2, ξ_3 to the general Cartesian coordinate system x_1, x_2, x_3. The transformation relations are

$$\begin{aligned}
\hat{\mathbf{U}}^{(x)} &= \hat{\mathbf{H}} \hat{\mathbf{U}}^{(q)}, & \hat{\mathbf{U}}^{(q)} &= \hat{\mathbf{H}}^T \hat{\mathbf{U}}^{(x)}, \\
\hat{\mathbf{U}}^{(x)} &= \hat{\mathbf{H}}^{(\xi)} \hat{\mathbf{U}}^{(\xi)}, & \hat{\mathbf{U}}^{(\xi)} &= \hat{\mathbf{H}}^{(\xi)-1} \hat{\mathbf{U}}^{(x)}, \\
\hat{\mathbf{U}}^{(q)} &= \hat{\mathbf{H}}^T \hat{\mathbf{H}}^{(\xi)} \hat{\mathbf{U}}^{(\xi)}, & \hat{\mathbf{U}}^{(\xi)} &= \hat{\mathbf{H}}^{(\xi)-1} \hat{\mathbf{H}} \hat{\mathbf{U}}^{(q)}.
\end{aligned} \tag{5.2.15}$$

Relations similar to (5.2.15) are, of course, valid also for the displacement vector components.

It is obvious that all the components of matrices $\hat{\mathbf{U}}^{(\xi)}$ and $\hat{\mathbf{U}}^{(x)}$ are, in general, nonvanishing, even though certain elements of matrix $\hat{\mathbf{U}}^{(q)}$ vanish; see (5.2.6) and (5.2.7).

Some local Cartesian coordinates z_1, z_2, and z_3 will often be used as ξ_1, ξ_2, and ξ_3. We denote by $\hat{\mathbf{Z}}$ the 3×3 transformation matrix from the local Cartesian coordinate system z_1, z_2, z_3 to the general Cartesian coordinate system x_1, x_2, x_3. A detailed specification of transformation matrices $\hat{\mathbf{Z}}$ for local Cartesian coordinate systems connected with interfaces can be found in Section 4.4.1. The transformation relations are then as follows:

$$\begin{aligned}
\hat{\mathbf{U}}^{(x)} &= \hat{\mathbf{Z}} \hat{\mathbf{U}}^{(z)}, & \hat{\mathbf{U}}^{(z)} &= \hat{\mathbf{Z}}^T \hat{\mathbf{U}}^{(x)}, \\
\hat{\mathbf{U}}^{(q)} &= \hat{\mathbf{G}}^T \hat{\mathbf{U}}^{(z)}, & \hat{\mathbf{U}}^{(z)} &= \hat{\mathbf{G}} \hat{\mathbf{U}}^{(q)},
\end{aligned} \tag{5.2.16}$$

where $\hat{\mathbf{G}}$ is a 3×3 transformation matrix given by the relation $\hat{\mathbf{G}} = \hat{\mathbf{Z}}^T \hat{\mathbf{H}}$.

5.2.2 Continuation of Amplitudes Along a Ray

The simplest continuation relations are obtained for the ray-centered components $U_1^{(q)}$, $U_2^{(q)}$, and $U_3^{(q)}$ of the vectorial amplitude function $\vec{U}$. Let us consider ray Ω and two points S and R situated on Ω. Using Equations (3.10.56) and (3.10.58), we obtain the general relation

$$\hat{\mathbf{U}}^{(q)}(R) = \left[\frac{V(S)\rho(S)J(S)}{V(R)\rho(R)J(R)} \right]^{1/2} \hat{\mathbf{U}}^{(q)}(S). \tag{5.2.17}$$

Here ρ is the density, V is the propagation velocity, and J is the ray Jacobian. For P waves, $V = \alpha$, and for S waves, $V = \beta$ in (5.2.17).

An alternative continuation relation uses geometrical spreading $\mathcal{L} = |J|^{1/2}$, and the phase shift due to caustics $T^c(R, S)$ instead of J,

$$\hat{\mathbf{U}}^{(q)}(R) = \left[\frac{V(S)\rho(S)}{V(R)\rho(R)} \right]^{1/2} \frac{\mathcal{L}(S)}{\mathcal{L}(R)} \exp[iT^c(R, S)]\hat{\mathbf{U}}^{(q)}(S). \qquad (5.2.18)$$

As in the acoustic case, we can write the continuation relation also for the ray-centered components $u_1^{(q)}$, $u_2^{(q)}$, and $u_3^{(q)}$ of displacement vector $\vec{u}^{(q)}$. Assume that $\hat{\mathbf{u}}^{(q)}$ is given by the following relation at the point S,

$$\hat{\mathbf{u}}^{(q)}(S, t) = \hat{\mathbf{U}}^{(q)}(S)F(t - T(S)),$$

where $F(\zeta)$, $\hat{\mathbf{U}}^{(q)}(S)$, and $T(S)$ are known. We then obtain a simple relation for $\hat{\mathbf{u}}^{(q)}(R, t)$,

$$\begin{aligned} \hat{\mathbf{u}}^{(q)}(R, t) = {} & \left[\frac{V(S)\rho(S)}{V(R)\rho(R)} \right]^{1/2} \frac{\mathcal{L}(S)}{\mathcal{L}(R)} \\ & \times \exp[iT^c(R, S)]\hat{\mathbf{U}}^{(q)}(S)F(t - T(S) - T(R, S)). \end{aligned} \qquad (5.2.19)$$

Here $T(R, S)$ is the travel time from S to R along ray Ω.

5.2.3 Point-Source Solutions. Radiation Matrices

As in the acoustic case (see Section 5.1.2), Equations (5.2.17) through (5.2.19) cannot be used if a point source is situated at point S because $\mathcal{L}(S)\hat{\mathbf{U}}^{(q)}(S) = 0$ for finite $\hat{\mathbf{U}}^{(q)}(S)$. At least one component of $\hat{\mathbf{U}}^{(q)}(S)$ has to be infinite so that

$$\lim_{S' \to S}\{\mathcal{L}(S')\hat{\mathbf{U}}^{(q)}(S')\} = \hat{\mathbf{U}}^{(q)0}(S), \qquad (5.2.20)$$

where $\hat{\mathbf{U}}^{(q)0}(S)$ is finite. Point S' is situated on Ω, and limit (5.2.20) is taken along Ω.

The modified equation (5.2.18), valid also for a point source situated at S, then reads

$$\hat{\mathbf{U}}^{(q)}(R) = \left[\frac{V(S)\rho(S)}{V(R)\rho(R)} \right]^{1/2} \frac{\exp[iT^c(R, S)]}{\mathcal{L}(R)} \hat{\mathbf{U}}^{(q)0}(S). \qquad (5.2.21)$$

It is now convenient to introduce the relative geometrical spreading $\mathcal{L}(R, S)$ instead of $\mathcal{L}(R)$; see Section 5.1.2. Then (5.2.20) and (5.2.21) yield

$$\hat{\mathbf{U}}^{(q)}(R) = \left[\frac{V(S)\rho(S)}{V(R)\rho(R)} \right]^{1/2} \frac{\exp[iT^c(R, S)]}{\mathcal{L}(R, S)} \hat{\mathcal{G}}^{(q)}(S; \gamma_1, \gamma_2). \qquad (5.2.22)$$

where

$$\hat{\mathcal{G}}^{(q)}(S; \gamma_1, \gamma_2) = \lim_{S' \to S}\{\mathcal{L}(S', S)\hat{\mathbf{U}}^{(q)}(S')\}. \qquad (5.2.23)$$

The limit in (5.2.23) is taken along ray Ω specified by ray parameters γ_1 and γ_2.

The 3×1 column matrix $\hat{\mathcal{G}}^{(q)}(S; \gamma_1, \gamma_2)$ will be referred to as the *ray-centered radiation matrix*, or simply the *radiation matrix*. Alternatively, we can also call it the *radiation vector*. The components of radiation matrix $\mathcal{G}_k^{(q)}(S; \gamma_1, \gamma_2)$ will be called *radiation functions*, as in the acoustic case. In fact, all the terminology and explanations related to the acoustic radiation function also apply to the elastic isotropic radiation functions; see Section 5.1.2. The only difference is that the radiation matrix has three components, corresponding to S1, S2, and P waves.

We shall now express (5.2.22) and (5.2.23) *for a homogeneous medium.* As for the acoustic medium in (5.1.9), we obtain

$$\hat{\mathbf{U}}^{(q)}(R) = \hat{\boldsymbol{\mathcal{G}}}^{(q)}(S; \gamma_1, \gamma_2)/V(S)l(R, S),$$

$$\hat{\boldsymbol{\mathcal{G}}}^{(q)}(S; \gamma_1, \gamma_2) = V(S)\left[\lim_{S' \to S}\{l(S', S)\hat{\mathbf{U}}^{(q)}(S')\}\right]_{\gamma_1, \gamma_2}.$$

Here $l(R, S)$ is the distance between S and R, and γ_1 and γ_2 are parameters of the ray under consideration. Thus, in a locally homogeneous medium, the ith component of the radiation matrix, $\mathcal{G}_i^{(q)}(S; \gamma_1, \gamma_2)$, represents the distribution of the ith ray-centered component $U_i^{(q)}$ of vectorial amplitude function $\vec{U}$ along a sphere whose center is at S and radius $l(R, S) = 1/V(S)$.

In addition to radiation functions $\mathcal{G}_k^{(q)}(S; \gamma_1, \gamma_2)$, we can again introduce *directivity patterns* $\mathcal{F}_k^{(q)}(S; \gamma_1, \gamma_2)$. In a locally homogeneous medium, $\mathcal{G}_k^{(q)}(S; \gamma_1, \gamma_2)$ and $\mathcal{F}_k^{(q)}(S; \gamma_1, \gamma_2)$ are related as follows:

$$\mathcal{F}_k^{(q)}(S; \gamma_1, \gamma_2) = \left(\mathcal{G}_k^{(q)}(S; \gamma_1, \gamma_2)/\mathcal{L}(R, S)\right)_{l(R,S)=1}. \tag{5.2.24}$$

See the relevant relation (5.1.10) and its discussion in the acoustic case. In the following text, we shall only use the radiation functions $\mathcal{G}_k^{(q)}(S; \gamma_1, \gamma_2)$, and not directivity patterns $\mathcal{F}_k^{(q)}(S; \gamma_1, \gamma_2)$. In isotropic media, the differences between both are only formal because $\mathcal{L}(R, S) = V(S)l(R, S)$. The differences will be important in anisotropic media; see Section 5.4.2.

Because the P and S waves generated by the point source must be investigated independently by the ray method, radiation matrix $\hat{\boldsymbol{\mathcal{G}}}^{(q)}$ also must be expressed independently for P and S waves, much like matrix $\hat{\mathbf{U}}^{(q)}$ in (5.2.10). Hence,

$$\hat{\boldsymbol{\mathcal{G}}}_P^{(q)} = \begin{pmatrix} 0, & 0, & \mathcal{G}_P^{(q)} \end{pmatrix}^T, \qquad \hat{\boldsymbol{\mathcal{G}}}_S^{(q)} = \begin{pmatrix} \mathcal{G}_{S1}^{(q)}, & \mathcal{G}_{S2}^{(q)}, & 0 \end{pmatrix}^T, \tag{5.2.25}$$

where

$$\mathcal{G}_P^{(q)}(S) = \lim_{S' \to S}\{\mathcal{L}(S', S)A(S')\},$$

$$\mathcal{G}_{S1}^{(q)}(S) = \lim_{S' \to S}\{\mathcal{L}(S', S)B(S')\}, \tag{5.2.26}$$

$$\mathcal{G}_{S2}^{(q)}(S) = \lim_{S' \to S}\{\mathcal{L}(S', S)C(S')\}.$$

The ray-centered radiation matrix can be transformed to different coordinate systems similar to $\hat{\mathbf{U}}^{(q)}$; see (5.2.15) and (5.2.16). If we denote the radiation matrix in the general coordinate system x_1, x_2, x_3 by $\hat{\boldsymbol{\mathcal{G}}}^{(x)}$ and the radiation matrix in the orthogonal coordinate system ξ_1, ξ_2, ξ_3 by $\hat{\boldsymbol{\mathcal{G}}}^{(\xi)}$, we can write

$$\begin{aligned} \hat{\boldsymbol{\mathcal{G}}}^{(x)} &= \hat{\mathbf{H}}\hat{\boldsymbol{\mathcal{G}}}^{(q)}, & \hat{\boldsymbol{\mathcal{G}}}^{(q)} &= \hat{\mathbf{H}}^T\hat{\boldsymbol{\mathcal{G}}}^{(x)}, \\ \hat{\boldsymbol{\mathcal{G}}}^{(x)} &= \hat{\mathbf{H}}^{(\xi)}\hat{\boldsymbol{\mathcal{G}}}^{(\xi)}, & \hat{\boldsymbol{\mathcal{G}}}^{(\xi)} &= \hat{\mathbf{H}}^{(\xi)-1}\hat{\boldsymbol{\mathcal{G}}}^{(x)}. \\ \hat{\boldsymbol{\mathcal{G}}}^{(q)} &= \hat{\mathbf{H}}^T\hat{\mathbf{H}}^{(\xi)}\hat{\boldsymbol{\mathcal{G}}}^{(\xi)}, & \hat{\boldsymbol{\mathcal{G}}}^{(\xi)} &= \hat{\mathbf{H}}^{(\xi)-1}\hat{\mathbf{H}}\hat{\boldsymbol{\mathcal{G}}}^{(q)}. \end{aligned} \tag{5.2.27}$$

Here the 3×3 transformation matrices $\hat{\mathbf{H}}$ and $\hat{\mathbf{H}}^{(\xi)}$ have the same meaning as in (5.2.15). Similarly, if we consider local Cartesian coordinates z_1, z_2, and z_3, we obtain

$$\begin{aligned} \hat{\boldsymbol{\mathcal{G}}}^{(x)} &= \hat{\mathbf{Z}}\hat{\boldsymbol{\mathcal{G}}}^{(z)}, & \hat{\boldsymbol{\mathcal{G}}}^{(z)} &= \hat{\mathbf{Z}}^T\hat{\boldsymbol{\mathcal{G}}}^{(x)}, \\ \hat{\boldsymbol{\mathcal{G}}}^{(q)} &= \hat{\mathbf{G}}^T\hat{\boldsymbol{\mathcal{G}}}^{(z)}, & \hat{\boldsymbol{\mathcal{G}}}^{(z)} &= \hat{\mathbf{G}}\hat{\boldsymbol{\mathcal{G}}}^{(q)}. \end{aligned} \tag{5.2.28}$$

Transformation matrices $\hat{\mathbf{Z}}$ and $\hat{\mathbf{G}}$ have the same meaning as in (5.2.16). It should be emphasized that $\mathcal{G}_1^{(q)} = \mathcal{G}_2^{(q)} = 0$ for P waves, and $\mathcal{G}_3^{(q)} = 0$ for S waves. In all other coordinate systems, the three components of the radiation matrix are, in general, nonvanishing.

Thus the final equation for the ray-centered components of the vectorial amplitude function $\vec{U}$ of the elastic wave generated by a point source at S is

$$\hat{\mathbf{U}}^{(q)}(R) = \left[\frac{\rho(S)V(S)}{\rho(R)V(R)} \right]^{1/2} \frac{\exp[iT^c(R, S)]}{\mathcal{L}(R, S)} \hat{\mathcal{G}}^{(q)}(S; \gamma_1\gamma_2). \tag{5.2.29}$$

Equation (5.2.29) is valid for any 3-D smooth elastic isotropic laterally varying medium. Point R is situated on the ray Ω passing through the point source situated at S; γ_1 and γ_2 denote the ray parameters of ray Ω.

Equation (5.2.29) can be simply written also for vectors $\hat{\mathbf{U}}^{(x)}$ and $\hat{\mathcal{G}}^{(x)}$. Using (5.2.15) and (5.2.27), we obtain

$$\hat{\mathbf{U}}^{(x)}(R) = \left[\frac{\rho(S)V(S)}{\rho(R)V(R)} \right]^{1/2} \frac{\exp[iT^c(R, S)]}{\mathcal{L}(R, S)} \hat{\mathbf{H}}(R)\hat{\mathbf{H}}^T(S)\hat{\mathcal{G}}^{(x)}(S; \gamma_1\gamma_2),$$

$$\tag{5.2.30}$$

where $\hat{\mathbf{H}}(R)$ and $\hat{\mathbf{H}}(S)$ are the relevant transformation matrices at R and S. Note that matrices $\hat{\mathbf{H}}(R)$ and $\hat{\mathbf{H}}(S)$ are different in inhomogeneous media because the rays are curved.

We shall now give the expressions for three seismologically important radiation matrices $\mathcal{G}_i^{(q)}(S; \gamma_1, \gamma_2)$.

a. OMNIDIRECTIONAL POINT SOURCE

The omnidirectional radiation matrix does not depend on ray parameters γ_1 and γ_2:

$$\mathcal{G}_i^{(q)}(S; \gamma_1, \gamma_2) = a_i(S), \tag{5.2.31}$$

where a_i are independent of γ_1 and γ_2.

b. SINGLE-FORCE POINT SOURCE

Assume a single force

$$\vec{f}(\vec{x}, t) = \delta(\vec{x} - \vec{x}(S))F(t)\vec{f}_0. \tag{5.2.32}$$

acting at point S. Here $F(t)$ is an analytical signal and δ the delta function. The radiation matrix then reads

$$\hat{\mathcal{G}}^{(q)}(S; \gamma_1, \gamma_2) = \frac{1}{4\pi\rho(S)V(S)} \hat{\mathbf{H}}^T(S)\hat{\mathbf{f}}_0^{(x)}(S), \tag{5.2.33}$$

where $\hat{\mathbf{f}}_0^{(x)}$ denotes the column matrix of the Cartesian components of $\vec{f}_0$,

$$\hat{\mathbf{f}}_0^{(x)}(S) = \left(f_{01}^{(x)}(S), \ f_{02}^{(x)}(S), \ f_{03}^{(x)}(S) \right)^T.$$

This expression for the radiation matrix may be obtained from relations (2.5.38) and (2.5.40). Relation (5.2.33) can also be expressed in an alternative form, separately for P and S waves,

$$\hat{\mathcal{G}}_P^{(q)}(S; \gamma_1, \gamma_2) = [4\pi\rho(S)\alpha(S)]^{-1} \left(0, 0, \vec{N}(S) \cdot \vec{f}_0^{(x)} \right)^T,$$

$$\hat{\mathcal{G}}_S^{(q)}(S; \gamma_1, \gamma_2) = [4\pi\rho(S)\beta(S)]^{-1} \left(\vec{e}_1(S) \cdot \vec{f}_0, \ \vec{e}_2(S) \cdot \vec{f}_0, \ 0 \right)^T.$$

Here $\vec{e}_1(S)$, $\vec{e}_2(S)$, and $\vec{e}_3(S) \equiv \vec{N}(S)$ are the polarization vectors at S, corresponding to the ray specified by the ray parameters γ_1 and γ_2.

c. MOMENT-TENSOR POINT SOURCE

Moment-tensor point sources play an important role in physics of earthquakes. They were investigated in detail by Aki and Richards (1980), Ben-Menahem and Singh (1981), and Kennett (1983), among others. Using the ray method, only high-frequency radiation of moment-tensor point sources can be investigated. In a high-frequency asymptotic solution, the general expressions for the radiation matrices of moment-tensor point sources simplify; see Červený, Pleinerová, Klimeš, and Pšenčík (1987). In the frequency domain, they can again be expressed in the form of (5.2.33); only $\hat{\mathbf{f}}_0^{(x)}(S)$ should be replaced by $-i\omega\hat{\mathbf{M}}_0^{(x)}(S)\hat{\mathbf{p}}_0^{(x)}(S)$. (This can be directly derived if we use the equivalent force system $f_i = -m_{ij,j}$; see Section 2.1.2 and Equation (2.5.41).) Here the 3×3 matrix $\hat{\mathbf{M}}_0^{(x)}$ represents the seismic moment tensor, and the 3×1 column matrix $\hat{\mathbf{p}}_0^{(x)}(S)$ represents the initial slowness vector; both are expressed in Cartesian coordinates. The factor $-i\omega$ may be connected with the spectrum of the analytical signal $F(\zeta)$, and yields a new analytical signal $\dot{F}(\zeta)$ in the time domain. Consequently, the moment-tensor radiation function is

$$\hat{\mathcal{G}}^{(q)}(S; \gamma_1, \gamma_2) = [4\pi\rho(S)V(S)]^{-1}\hat{\mathbf{H}}^T(S)\hat{\mathbf{M}}_0^{(x)}(S)\hat{\mathbf{p}}_0^{(x)}(S). \qquad (5.2.34)$$

The difference between the analytical signals $F(\zeta)$ and $\dot{F}(\zeta)$ does not play any role if the single-force and moment-tensor point sources are treated independently. If, however, both sources are treated together, it is actually necessary to consider the relevant analytical signals $F(\zeta)$ and $\dot{F}(\zeta)$ for both types of sources. For alternative expressions for moment-tensor point sources, see also Červený, Klimeš, and Pšenčík (1988b).

In (5.2.34), matrix $\hat{\mathbf{H}}^T(S)$ should again be treated separately for P and S waves, as in the case of the single-force radiation pattern.

For $(\mathbf{M}_0^{(x)})_{ij} = M\delta_{ij}$, the point source is called *the center of dilatation*. From a physical point of view, the center of dilatation consists of three couples without a torque moment acting along three mutually perpendicular axes. The quantity M is called the *scalar moment* and characterizes the strength of the source. The relevant radiation matrices are given by relations

$$\mathcal{G}_3^{(q)}(S; \gamma_1, \gamma_2) = M(S)/4\pi\rho(S)\alpha^2(S), \qquad \mathcal{G}_1^{(q)} = \mathcal{G}_2^{(q)} = 0. \qquad (5.2.35)$$

Thus, the center of dilatation generates only P waves (see $\mathcal{G}_3^{(q)}$) and no S waves (see $\mathcal{G}_1^{(q)} = \mathcal{G}_2^{(q)} = 0$). Moreover, the radiation function of P waves is omnidirectional. The dilatational source is also often called *the explosive source*. The relevant directivity pattern is given by the relation $\mathcal{F}_3^{(q)}(S; \gamma_1, \gamma_2) = M(S)/4\pi\rho(S)\alpha^3(S)$; see (5.2.24).

5.2.4 Amplitudes Across an Interface

Let us consider ray Ω, which strikes interface Σ at point Q. As in Section 5.1.3, we denote by $\tilde{Q}$ the point of reflection/transmission that is situated on Σ at Q but on the side of the selected R/T wave. Assume that point S is situated on the incidence branch of ray Ω. We can then use (5.2.18) to obtain

$$\hat{\mathbf{U}}^{(q)}(Q) = \left[\frac{V(S)\rho(S)}{V(Q)\rho(Q)}\right]^{1/2}\frac{\mathcal{L}(S)}{\mathcal{L}(Q)}\exp[iT^c(Q, S)]\hat{\mathbf{U}}^{(q)}(S). \qquad (5.2.36)$$

Across interface Σ, we can use the relation

$$\hat{\mathbf{U}}^{(q)}(\tilde{Q}) = \hat{\mathbf{R}}^T(Q)\hat{\mathbf{U}}^{(q)}(Q). \tag{5.2.37}$$

Here $\hat{\mathbf{R}}$ denotes the 3×3 matrix of the plane-wave reflection and transmission displacement coefficients R_{mn}, introduced in Section 2.3.2, and superscript T stands for the transpose.

Because $\hat{\mathbf{U}}^{(q)}(\tilde{Q})$ represents a selected reflected/transmitted wave, certain of its components vanish. For a P R/T wave, $U_1^{(q)}(\tilde{Q}) = U_2^{(q)}(\tilde{Q}) = 0$. Similarly, for an S R/T wave, $U_3^{(q)}(\tilde{Q}) = 0$. Relation (5.2.37), however, would give all components of $\hat{\mathbf{U}}^{(q)}(\tilde{Q})$ non-vanishing. Thus, if we apply (5.2.37) for selected elementary waves, we must put certain components of $\hat{\mathbf{U}}^{(q)}(\tilde{Q})$ or certain elements of $\hat{\mathbf{R}}(Q)$ equal to zero. The simplest way is to introduce four types of R/T matrices (P $\to$ P, P $\to$ S, S $\to$ P, and S $\to$ S) and to choose the proper matrix according to the alphanumerical code of the elementary wave. The relevant R/T matrices are given by relations

$$\hat{\mathbf{R}}_{P \to P}(Q) = \begin{pmatrix} 0 & 0 & 0 \\ 0 & 0 & 0 \\ 0 & 0 & R_{33}(Q) \end{pmatrix}, \quad \hat{\mathbf{R}}_{P \to S}(Q) = \begin{pmatrix} 0 & 0 & 0 \\ 0 & 0 & 0 \\ R_{31}(Q) & R_{32}(Q) & 0 \end{pmatrix},$$

$$\hat{\mathbf{R}}_{S \to P}(Q) = \begin{pmatrix} 0 & 0 & R_{13}(Q) \\ 0 & 0 & R_{23}(Q) \\ 0 & 0 & 0 \end{pmatrix}, \quad \hat{\mathbf{R}}_{S \to S}(Q) = \begin{pmatrix} R_{11}(Q) & R_{12}(Q) & 0 \\ R_{21}(Q) & R_{22}(Q) & 0 \\ 0 & 0 & 0 \end{pmatrix}. \tag{5.2.38}$$

The first index always indicates the incident wave; the second indicates the generated wave. These four matrices must be constructed both for reflected waves ($\hat{\mathbf{R}}_{P \to P}^r(Q)$, $\hat{\mathbf{R}}_{P \to S}^r(Q)$, $\hat{\mathbf{R}}_{S \to P}^r(Q)$, and $\hat{\mathbf{R}}_{S \to S}^r(Q)$) and for transmitted waves ($\hat{\mathbf{R}}_{P \to P}^t(Q)$, $\hat{\mathbf{R}}_{P \to S}^t(Q)$, $\hat{\mathbf{R}}_{S \to P}^t(Q)$, and $\hat{\mathbf{R}}_{S \to S}^t(Q)$).

Finally, at point R situated on the reflected/transmitted branch of ray Ω, we obtain

$$\hat{\mathbf{U}}^{(q)}(R) = \left[\frac{V(\tilde{Q})\rho(\tilde{Q})}{V(R)\rho(R)} \right]^{1/2} \frac{\mathcal{L}(\tilde{Q})}{\mathcal{L}(R)} \exp[iT^c(R, \tilde{Q})]\hat{\mathbf{U}}^{(q)}(\tilde{Q}). \tag{5.2.39}$$

The three relations (5.2.36), (5.2.37), and (5.2.39) yield the continuation relations for $\hat{\mathbf{U}}^{(q)}$ across the interface from S to R:

$$\hat{\mathbf{U}}^{(q)}(R) = \left[\frac{V(S)\rho(S)}{V(R)\rho(R)} \right]^{1/2} \frac{\mathcal{L}(S)}{\mathcal{L}(R)} \hat{\boldsymbol{\mathcal{R}}}^T(Q) \exp[iT^c(R, S)]\hat{\mathbf{U}}^{(q)}(S). \tag{5.2.40}$$

Here

$$T^c(R, S) = T^c(R, \tilde{Q}) + T^c(\tilde{Q}, S). \tag{5.2.41}$$

The 3×3 matrix $\hat{\boldsymbol{\mathcal{R}}}(Q)$ represents the matrix of *normalized reflection/transmission coefficients*. The elements of the matrix, $\mathcal{R}_{ij}(Q)$, satisfy the relation

$$\mathcal{R}_{ij}(Q) = R_{ij}(Q) \left[\frac{V(\tilde{Q})\rho(\tilde{Q})}{V(Q)\rho(Q)} \right]^{1/2} \frac{\mathcal{L}(\tilde{Q})}{\mathcal{L}(Q)}. \tag{5.2.42}$$

If we take into account the equation preceding (5.1.16), which is valid even for elastic waves, we obtain the final relation between standard displacement (R_{ij}) and normalized displacement $(\mathcal{R}_{ij})$ R/T coefficients,

$$\mathcal{R}_{ij}(Q) = R_{ij}(Q)\left[\frac{V(\tilde{Q})\rho(\tilde{Q})\,\cos i(\tilde{Q})}{V(Q)\rho(Q)\,\cos i(Q)}\right]^{1/2}. \qquad (5.2.43)$$

Here $i(Q)$ is the acute angle of incidence, and $i(\tilde{Q})$ is the acute angle of reflection/transmission, corresponding to the selected elementary wave. For the real-valued ray Ω, $\cos i(Q)$ and $\cos i(\tilde{Q})$ are always real-valued and nonnegative. Propagation velocities $V(\tilde{Q})$ and $V(Q)$ are taken according to the elementary wave under consideration: $V = \alpha$ for the P wave, and $V = \beta$ for the S wave.

Note one important point. The square-root modification factors for acoustic and elastic normalized reflection/transmission coefficients (see (5.1.16) and (5.2.43)) are different. This is due to the fact that the pressure R/T coefficients are considered in acoustics, and displacement R/T coefficients are considered in elastodynamics. Thus, we must be careful to use the correct modification factor to construct the normalized R/T coefficients. For a detailed discussion of normalized R/T coefficients, see Section 5.3.

Equation (5.2.40) represents the final relation for the continuation of the matrix of the ray-centered components of the vectorial amplitude factor across the interface. As in the acoustic case, the change of sign of $\det \mathbf{Q}$ across the interface for reflected waves does not influence the phase shift due to caustics $T^c(R, S)$.

The modification of (5.2.40) for a point source situated at S is simple. The final equation reads

$$\hat{\mathbf{U}}^{(q)}(R) = \left[\frac{V(S)\rho(S)}{V(R)\rho(R)}\right]^{1/2}\frac{\exp[iT^c(R, S)]}{\mathcal{L}(R, S)}\hat{\boldsymbol{\mathcal{R}}}^T(Q)\hat{\boldsymbol{\mathcal{G}}}^{(q)}(S; \gamma_1, \gamma_2).$$

$$(5.2.44)$$

Here $\mathcal{L}(R, S) = |\det \mathbf{Q}_2(R, S)|^{1/2}$ is the relative geometrical spreading. Many useful relations for $\det \mathbf{Q}_2(R, S)$ for the problem of reflection/transmission at a curved structural interface can be found in Section 4.8.5. See also Section 4.10.2.

5.2.5 Amplitudes in 3-D Layered Structures

Let us consider a ray Ω of an arbitrary multiply reflected, possibly converted, elementary elastic wave propagating in a 3-D isotropic layered structure. We also consider two points, S and R, situated on Ω and assume that ray Ω strikes various structural interfaces N times between S and R. The points of incidence are succesively denoted $Q_1, Q_2, \ldots, Q_N$, and the relevant points of R/T are denoted by $\tilde{Q}_1, \tilde{Q}_2, \ldots, \tilde{Q}_N$. We assume that the ray code of the elementary wave under consideration is strictly specified. In other words, the type of wave (P or S) along any element of the ray is known.

The continuation relations for the ray-centered amplitude matrix can be obtained by simple generalization of (5.2.40)

$$\hat{\mathbf{U}}^{(q)}(R) = \left[\frac{V(S)\rho(S)}{V(R)\rho(R)}\right]^{1/2}\frac{\mathcal{L}(S)}{\mathcal{L}(R)}\hat{\boldsymbol{\mathcal{R}}}^C \exp[iT^c(R, S)]\hat{\mathbf{U}}^{(q)}(S). \qquad (5.2.45)$$

where

$$T^c(R, S) = \sum_{i=1}^{N+1} T^c(Q_k, \tilde{Q}_{k-1}), \qquad \hat{\boldsymbol{\mathcal{R}}}^C = \prod_{k=N}^{1} \hat{\boldsymbol{\mathcal{R}}}^T(Q_k). \qquad (5.2.46)$$

Here $\prod_{k=N}^{1} \hat{\mathcal{R}}^{T}(Q_k) = \hat{\mathcal{R}}^{T}(Q_N)\hat{\mathcal{R}}^{T}(Q_{N-1}) \ldots \hat{\mathcal{R}}^{T}(Q_1)$. In (5.2.46), we have also used $\tilde{Q}_0 = S$ and $Q_{N+1} = R$. We shall call $\hat{\mathcal{R}}^{C}$ the complete matrix of normalized R/T coefficients along ray Ω between S and R.

For a point source at S, (5.2.45) yields

$$\hat{\mathbf{U}}^{(q)}(R) = \left[\frac{V(S)\rho(S)}{V(R)\rho(R)} \right]^{1/2} \frac{\exp[iT^c(R, S)]}{\mathcal{L}(R, S)} \hat{\mathcal{R}}^C \hat{\mathcal{G}}^{(q)}(S; \gamma_1, \gamma_2). \quad (5.2.47)$$

Here $\hat{\mathcal{G}}^{(q)}(S, \gamma_1, \gamma_2)$ is the ray-centered radiation matrix.

It is again necessary to emphasize an important point. Velocities V and the matrices of normalized R/T coefficients must be specified according to the ray code of the elementary wave under consideration. There are eight options for $\hat{\mathcal{R}}(Q_k)$ at any R/T point Q_k. It may equal $\hat{\mathcal{R}}^{r}(Q_k)$ (reflection) or $\hat{\mathcal{R}}^{t}(Q_k)$ (transmission) and may correspond to P → P, P → S, S → P, or S → S.

Equations (5.2.45) and (5.2.47) are very general, valid for any multiply reflected converted wave. In Sections 5.2.10 through 5.2.13, we shall discuss special cases of these equations, corresponding to unconverted P and S waves and to any multiply reflected, converted, elementary wave with a plane ray Ω.

5.2.6 Elastodynamic Ray-Theory Green Function

The complete elastodynamic ray-theory Green function can be expressed as the superposition of elementary ray-theory Green functions, corresponding to different rays connecting the source and receiver. Here we shall consider only one elementary ray-theory Green function.

The elastodynamic Green function $G_{in}(R, t; S, t_0)$ was defined in Section 2.5.4. We remind the reader that it represents the ith Cartesian component of the displacement vector at location R and time t, due to the point source situated at S, representing a single unit force oriented along the nth Cartesian axis, with the time dependence corresponding to an impulse delta function applied at time t_0.

To derive the expression for the elastodynamic ray-theory Green function corresponding to the ray Ω connecting S and R, we use (5.2.47). First, we transform it into Cartesian coordinates

$$\hat{\mathbf{U}}^{(x)}(R) = \left[\frac{V(S)\rho(S)}{V(R)\rho(R)} \right]^{1/2} \frac{\exp[iT^c(R, S)]}{\mathcal{L}(R, S)} \hat{\mathbf{H}}(R)\hat{\mathcal{R}}^C \hat{\mathcal{G}}^{(q)}(S; \gamma_1, \gamma_2).$$

Now we specify the radiation function $\hat{\mathcal{G}}^{(q)}(S; \gamma_1, \gamma_2)$ for a unit single-force source at S, oriented along the nth Cartesian axis. From (5.2.33), we obtain

$$\mathcal{G}_i^{(q)}(S; \gamma_1, \gamma_2) = \frac{1}{4\pi\rho(S)V(S)} H_{ni}(S). \quad (5.2.48)$$

As explained earlier, we put $i = 3$ for P waves and $i = 1, 2$ for S waves.

This finally yields the amplitude function of Green function $G_{in}(R, t; S, t_0)$,

$$U_i^{(x)}(R) = \frac{\exp[iT^c(R, S)]}{4\pi[V(S)V(R)\rho(S)\rho(R)]^{1/2}\mathcal{L}(R, S)} H_{ik}(R)\mathcal{R}_{kl}^C H_{nl}(S). \quad (5.2.49)$$

In this expression, the summation over k and l must be specified properly. If the first element of the ray (close to S) is P, we put $l = 3$, and if it is S, we put $l = L$, where $L = 1, 2$. Similarly, if the last element of the ray (close to R) is P, we put $k = 3$; however, if it is S, we put $k = K$ ($K = 1, 2$). The Einstein summation convention is applied to $K = 1, 2$ and $L = 1, 2$.

Thus, we have four specific alternatives for the amplitude function of Green function $G_{in}(R, t; S, t_0)$.

1. The first element of the ray is P (P wave source) and the last element is also P:

$$U_i^{(x)}(R) = \frac{\exp[iT^c(R, S)]}{4\pi [\alpha(S)\alpha(R)\rho(S)\rho(R)]^{1/2}\mathcal{L}(R, S)} H_{i3}(R)\mathcal{R}_{33}^C H_{n3}(S).$$

(5.2.50)

2. The first element of the ray is P (P wave source) and the last element is S:

$$U_i^{(x)}(R) = \frac{\exp[iT^c(R, S)]}{4\pi [\alpha(S)\beta(R)\rho(S)\rho(R)]^{1/2}\mathcal{L}(R, S)} H_{iK}(R)\mathcal{R}_{K3}^C H_{n3}(S).$$

(5.2.51)

3. The first element of the ray is S (S wave source) and the last element is P:

$$U_i^{(x)}(R) = \frac{\exp[iT^c(R, S)]}{4\pi [\beta(S)\alpha(R)\rho(S)\rho(R)]^{1/2}\mathcal{L}(R, S)} H_{i3}(R)\mathcal{R}_{3L}^C H_{nL}(S).$$

(5.2.52)

4. The first element of the ray is S (S wave source) and the last element is S:

$$U_i^{(x)}(R) = \frac{\exp[iT^c(R, S)]}{4\pi [\beta(S)\beta(R)\rho(S)\rho(R)]^{1/2}\mathcal{L}(R, S)} H_{iK}(R)\mathcal{R}_{KL}^C H_{nL}(S).$$

(5.2.53)

All these equations are valid for any multiply reflected, possibly converted wave in a 3-D layered structure. In all cases, we can put $H_{ij} = e_{ji}$, where e_{ji} represents the ith Cartesian component of polarization vector $\vec{e}_j$. We also remind the reader that $\vec{e}_3 \equiv \vec{N}$, the unit vector tangent to ray Ω. The summation for uppercase indices runs over 1 and 2.

For the reader's convenience, we shall present the final equations for the elementary ray-theory elastodynamic Green function $G_{in}(R, t; S, t_0)$, which is valid for any multiply reflected, possibly converted, elementary wave propagating in a 3-D isotropic layered structure.

In the frequency domain:

$$G_{in}(R, S, \omega) = \frac{1}{4\pi [\rho(S)\rho(R)V(S)V(R)]^{1/2}\mathcal{L}(R, S)} H_{ik}(R)\mathcal{R}_{kl}^C H_{nl}(S)$$
$$\times \exp[i\omega T(R, S) + iT^c(R, S)].$$

(5.2.54)

In the time domain:

$$G_{in}(R, t; S, t_0) = \frac{\exp[iT^c(R, S)]}{4\pi [\rho(S)\rho(R)V(S)V(R)]^{1/2}\mathcal{L}(R, S)} H_{ik}(R)\mathcal{R}_{kl}^C H_{nl}(S)$$
$$\times \delta^{(A)}(t - t_0 - T(R, S)).$$

(5.2.55)

The summation over k and l in (5.2.54) and (5.2.55) must be specified as shown in (5.2.50) through (5.2.53).

It will be shown in Section 5.3 that matrix $\hat{\mathcal{R}}^C$, computed along Ω from R to S, is a transpose of the same matrix, computed along Ω from S to R. This immediately yields the following relation:

$$G_{in}(R, t; S, t_0) = G_{ni}(S, t; R, t_0). \tag{5.2.56}$$

This is the famous *reciprocity relation of the elastodynamic Green function*. As we can see from (5.2.56), it is valid for the elementary ray-theory elastodynamic Green function, corresponding to any multiply reflected converted wave propagating in a 3-D laterally varying layered structure.

5.2.7 Receiver at an Interface. Conversion Coefficients

The foregoing equations for amplitude matrix $\hat{\mathbf{U}}^{(q)}(R)$ are valid only if the receiver (point R) is situated in a smooth medium. As in the acoustic case, the derived equations must be modified if the receiver is situated on an interface.

Let us consider interface Σ^R passing through point R. We shall use the following notation. If there is no interface at R, the amplitude matrix at R is denoted by $\hat{\mathbf{U}}^{(q)SM}(R)$ and called the *smooth medium amplitude matrix* (in ray-centered components) at R. It can be calculated by equations of the foregoing sections. We assume that point R is situated on Σ^R from the side of the incident wave. In addition to R, we introduce point R^+, situated on the opposite side of interface Σ^R. We denote

$$\begin{aligned} \lambda_1 &= \lambda(R), & \mu_1 &= \mu(R), & \rho_1 &= \rho(R), \\ \lambda_2 &= \lambda(R^+), & \mu_2 &= \mu(R^+), & \rho_2 &= \rho(R^+), \end{aligned} \tag{5.2.57}$$

In a similar way, we can introduce α_1, β_1 and α_2, β_2. When quantities (5.2.57) are known, we can construct the 3×3 matrices of the standard (nonnormalized) R/T displacement coefficients $\mathbf{R}^r(R)$ and $\mathbf{R}^t(R)$. Two reflected waves (P and S) are generated if the incident wave strikes interface Σ^R at point R. The complete wavefield at the point of incidence R is composed of the incident, reflected P, and reflected S waves. On the opposite side of interface Σ^R (at point R^+), the complete wavefield is composed of transmitted P and transmitted S waves. The amplitude matrix of the complete displacement wavefield is continuous across interface Σ^R. Thus, we have two options of evaluating the complete wavefield at interface Σ^R: either at R or at R^+.

Because we wish to compute the amplitudes of the complete wavefield at Σ^R, we need to express the individual elementary waves forming this wavefield in the same coordinate system. We shall use a general Cartesian coordinate system for this purpose. The transformation matrices $\hat{\mathbf{H}}$ from ray-centered to Cartesian coordinates are, of course, different for the individual elementary waves. We denote them by $\hat{\mathbf{H}}^{SM}(R)$ for the incident (smooth medium) wave, $\hat{\mathbf{H}}^{rP}(R)$ and $\hat{\mathbf{H}}^{rS}(R)$ for reflected P and S waves, $\hat{\mathbf{H}}^{tP}(R^+)$ and $\hat{\mathbf{H}}^{tS}(R^+)$ for transmitted P and S waves.

The amplitude matrix of the complete wavefield (in Cartesian coordinates) at the point of incidence R on Σ^R is then given by the relation

$$\hat{\mathbf{U}}^{(x)}(R) = [\hat{\mathbf{H}}^{SM}(R) + \hat{\mathbf{H}}^{rP}(R)(\hat{\mathbf{R}}^r(R))^T + \hat{\mathbf{H}}^{rS}(R)(\hat{\mathbf{R}}^r(R))^T]\hat{\mathbf{U}}^{(q)SM}(R). \tag{5.2.58}$$

Alternatively, at the point R^+ on the opposite side of Σ^R,

$$\hat{\mathbf{U}}^{(x)}(R^+) = [\hat{\mathbf{H}}^{tP}(R^+)(\hat{\mathbf{R}}^t(R))^T + \hat{\mathbf{H}}^{tS}(R^+)(\hat{\mathbf{R}}^t(R))^T]\hat{\mathbf{U}}^{(q)SM}(R). \tag{5.2.59}$$

In the symbols for the matrices of R/T coefficients, the superscript (r or t) indicates the reflected or transmitted wave. The appropriate matrices P $\rightarrow$ P, P $\rightarrow$ S, S $\rightarrow$ P, or S $\rightarrow$ S should be considered, see (5.2.38). Capital T stands for the transpose.

Expressions (5.2.58) and (5.2.59) can be written in the following form:

$$\hat{\mathbf{U}}^{(x)}(R) = \hat{\mathcal{D}}(R)\hat{\mathbf{U}}^{(q)SM}(R), \qquad \hat{\mathbf{U}}^{(x)}(R^+) = \hat{\mathcal{D}}(R^+)\hat{\mathbf{U}}^{(q)SM}(R), \quad (5.2.60)$$

or, alternatively, in component form,

$$U_i^{(x)}(R) = \mathcal{D}_{ik}(R)U_k^{(q)SM}(R), \qquad U_i^{(x)}(R^+) = \mathcal{D}_{ik}(R^+)U_k^{(q)SM}(R). \tag{5.2.61}$$

Here $\mathcal{D}_{ik}(R)$ and $\mathcal{D}_{ik}(R^+)$ are given by the following relations:

$$\mathcal{D}_{ik}(R) = H_{ik}^{SM}(R) + H_{i3}^{rP}(R)R_{k3}^r(R) + H_{iJ}^{rS}(R)R_{kJ}^r(R), \tag{5.2.62}$$

$$\mathcal{D}_{ik}(R^+) = H_{i3}^{tP}(R^+)R_{k3}^t(R) + H_{iJ}^{tS}(R^+)R_{kJ}^t(R); \tag{5.2.63}$$

see (5.2.58) and (5.2.59). Both the expressions (5.2.62) and (5.2.63) are equivalent because $\hat{\mathbf{U}}^{(x)}$ is continuous across interface Σ^R such that $\hat{\mathbf{U}}^{(x)}(R) = \hat{\mathbf{U}}^{(x)}(R^+)$.

In case of a smooth medium at R (without an interface Σ^R at R), Equations (5.2.60) and (5.2.61) remain valid, if we put

$$\mathcal{D}_{ik}(R) = H_{ik}^{SM}(R). \tag{5.2.64}$$

Thus, (5.2.60) and (5.2.61) can be used universally. We shall call the 3×3 matrix $\hat{\mathcal{D}}$ the *interface conversion matrix*, or simply the *conversion matrix*. The elements of the conversion matrix will be called the *interface conversion coefficients*, or simply *the conversion coefficients*.

We shall now specify conversion matrix $\hat{\mathcal{D}}$ for the incident (smooth medium) P wave $(U_1^{(q)SM}(R) = U_2^{(q)SM}(R) = 0)$ and for the incident (smooth medium) S wave $(U_3^{(q)SM}(R) = 0)$. For the incident P wave, we have

$$\mathcal{D}_{i1}(R) = \mathcal{D}_{i2}(R) = 0, \qquad \mathcal{D}_{i3}(R) \neq 0. \tag{5.2.65}$$

Similarly, for the incident S wave, we can use

$$\mathcal{D}_{i1}(R) \neq 0, \qquad \mathcal{D}_{i2}(R) \neq 0, \qquad \mathcal{D}_{i3}(R) = 0. \tag{5.2.66}$$

The same relations as (5.2.65) and (5.2.66) can also be written at the point R^+.

Equations (5.2.62) and (5.2.63) for the interface conversion coefficients can also be expressed in terms of polarization vectors $\vec{e}_1$, $\vec{e}_2$, and $\vec{e}_3 \equiv \vec{N}$, if we take into account $H_{ij} = e_{ji}$. Hence,

a. **For the incident P wave:**

$$\mathcal{D}_{i1}(R) = \mathcal{D}_{i2}(R) = \mathcal{D}_{i1}(R^+) = \mathcal{D}_{i2}(R^+) = 0,$$

$$\mathcal{D}_{i3}(R) = N_i^{SM}(R) + N_i^{rP}(R)R_{33}^r(R) + e_{1i}^{rS}(R)R_{31}^r(R) + e_{2i}^{rS}(R)R_{32}^r(R),$$

$$\mathcal{D}_{i3}(R^+) = N_i^{tP}(R^+)R_{33}^t(R) + e_{1i}^{tS}(R^+)R_{31}^t(R) + e_{2i}^{tS}(R^+)R_{32}^t(R). \tag{5.2.67}$$

b. **For the incident S wave:**

$$\mathcal{D}_{i3}(R) = \mathcal{D}_{i3}(R^+) = 0,$$

$$\mathcal{D}_{i1}(R) = e_{1i}^{SM}(R) + N_i^{rP}(R)R_{13}^r(R) + e_{1i}^{rS}(R)R_{11}^r(R) + e_{2i}^{rS}(R)R_{12}^r(R),$$

$$\mathcal{D}_{i1}(R^+) = N_i^{tP}(R^+)R_{13}^t(R) + e_{1i}^{tS}(R^+)R_{11}^t(R) + e_{2i}^{tS}(R^+)R_{12}^t(R),$$

$$\mathcal{D}_{i2}(R) = e_{2i}^{SM}(R) + N_i^{rP}(R)R_{23}^r(R) + e_{1i}^{rS}(R)R_{21}^r(R) + e_{2i}^{rS}(R)R_{22}^r(R),$$

$$\mathcal{D}_{i2}(R^+) = N_i^{tP}(R^+)R_{23}^t(R) + e_{1i}^{tS}(R^+)R_{21}^t(R) + e_{2i}^{tS}(R^+)R_{22}^t(R). \tag{5.2.68}$$

Let us briefly discuss the physical meaning of conversion matrix $\hat{\mathcal{D}}$. It performs two operations. First, it projects ray-centered amplitude matrix $\hat{\mathbf{U}}^{(q)}(R)$ into Cartesian amplitude matrix $\hat{\mathbf{U}}^{(x)}(R)$; see (5.2.60). Thus, it has a good physical meaning even in a smooth medium, where it equals transformation matrix $\hat{\mathbf{H}}^{SM}$; see (5.2.64). Second, it also takes into account the effects of the interface if the receiver is situated directly on this interface at point R or R^+; see (5.2.62) and (5.2.63). The most important application is for the receiver situated on the Earth's surface. The elements of the conversion matrix are then called *the free-surface conversion coefficients*, or simply *conversion coefficients*. See Červený and Ravindra (1971) and Červený, Molotkov, and Pšenčík (1977). The explicit formulae for $\mathcal{D}_{ij}$, both for the structural interface and for the Earth's surface, will be given in Section 5.3.8. Note that the "conversion coefficients" represent quite different quantities from "R/T coefficients of converted waves."

The general Cartesian coordinate system x_1, x_2, x_3 used in this section is in general different from the local Cartesian coordinate system z_1, z_2, z_3 connected with the interface. The results, however, can be transformed to the local Cartesian coordinate system. We merely use transformation matrix $\hat{\mathbf{Z}}$ and (5.2.16):

$$\hat{\mathbf{U}}^{(z)}(R) = \hat{\mathbf{Z}}^T(R)\hat{\mathcal{D}}(R)\hat{\mathbf{U}}^{(q)SM}(R),$$
$$\hat{\mathbf{U}}^{(z)}(R^+) = \hat{\mathbf{Z}}^T(R^+)\hat{\mathcal{D}}(R^+)\hat{\mathbf{U}}^{(q)SM}(R).$$

$$(5.2.69)$$

All equations for ray-centered amplitude matrix $\hat{\mathbf{U}}^{(q)}(R)$, derived in the previous sections, can be generalized for the case of a receiver situated on interface Σ^R passing through R. Cartesian amplitude matrix $\hat{\mathbf{U}}^{(x)}(R)$ is in this case obtained from $\hat{\mathbf{U}}^{(q)}(R)$ by multiplying it from the left by conversion matrix $\hat{\mathcal{D}}$. For the final equations, see Section 5.2.9.

5.2.8 Source at an Interface

In this section, we shall derive equations for the radiation matrix if the point source is situated on a structural interface Σ^S. For a more detailed treatment and many numerical examples see Jílek and Červený (1996). See also White (1983) for some special cases.

As in Section 5.1.9, we shall first consider the auxiliary problem of an elementary wave reflected/transmitted at interface Σ^S. This problem was treated in Section 5.2.4, but in this section we shall use slightly different notation that is more useful here. As in Section 5.1.9, we shall consider a point source situated at point S_0, and ray Ω_0 of a selected R/T wave connecting S_0 with a receiver situated at R. We denote the point of incidence of ray Ω_0 at Σ^S by $\tilde{S}$ and the point of R/T at Σ^S by S. (This notation is different from the notation of Section 5.2.4.) Thus, points S_0 and $\tilde{S}$ are situated on the same incident branch of ray Ω_0, and points S and R on the R/T branch of ray Ω_0. See Figure 5.4. As usual, we also denote the ray connecting S and R by Ω. Ray Ω_0 is then an extension of ray Ω outside point S. In addition to S, we also introduce point S^+ situated on Σ^S, but on the side opposite to S. Then we denote

$$\lambda_1 = \lambda(S), \qquad \mu_1 = \mu(S), \qquad \rho_1 = \rho(S),$$
$$\lambda_2 = \lambda(S^+), \qquad \mu_2 = \mu(S^+), \qquad \rho_2 = \rho(S^+).$$

$$(5.2.70)$$

The general relation (5.2.44) for the ray-centered amplitude matrix $\hat{\mathbf{U}}^{(q)}(R)$, related to the selected R/T wave, then reads

$$U_i^{(q)}(R) = \left[\frac{\rho(S_0)V(S_0)}{\rho(R)V(R)}\right]^{1/2} \frac{\exp[\mathrm{i}T^c(R, S_0)]}{\mathcal{L}(R, S_0)} \mathcal{R}_{ji}(\tilde{S})\mathcal{G}_j^{(q)SM}(S_0; \gamma_1, \gamma_2).$$

$$(5.2.71)$$

Here $\mathcal{G}_j^{(q)SM}$ is a "smooth medium" ray-centered radiation function that is not affected by the existence of interface Σ^S, and $\mathcal{R}_{ji}(\tilde{S})$ is the appropriate normalized R/T coefficient. Let us emphasize that the normalized R/T coefficient $\mathcal{R}_{ji}(\tilde{S})$ in (5.2.71) corresponds to the "positive" direction of propagation of the elementary wave under consideration along ray Ω_0 from S_0 through $\tilde{S}$ and S to R. Using (5.2.43), we can write

$$\mathcal{R}_{ji}(\tilde{S}) = R_{ji}(\tilde{S}) \left[\frac{V(S)\rho(S)\cos i(S)}{V(\tilde{S})\rho(\tilde{S})\cos i(\tilde{S})} \right]^{1/2}. \tag{5.2.72}$$

Here $R_{ji}(\tilde{S})$ is a standard displacement R/T coefficient. Now we insert (5.2.72) into (5.2.71), shift point S_0 along ray Ω_0 to point $\tilde{S}$, and take into account that $T^c(R, \tilde{S}) = T^c(R, S)$ and $\mathcal{L}(R, \tilde{S}) = (\cos i(\tilde{S})/\cos i(S))^{1/2}\mathcal{L}(R, S)$. Then (5.2.71) yields

$$U_i^{(q)}(R) = \left[\frac{\rho(S)V(S)}{\rho(R)V(R)} \right]^{1/2} \frac{\exp[iT^c(R, S)]}{\mathcal{L}(R, S)} \frac{\cos i(S)}{\cos i(\tilde{S})}$$

$$\times R_{ji}(\tilde{S})\mathcal{G}_j^{(q)SM}(\tilde{S}; \gamma_1, \gamma_2). \tag{5.2.73}$$

This is the final solution of the auxiliary problem, corresponding to the selected elementary wave.

We shall now compute the wavefield due to a point source situated at interface Σ^S. We assume that the type of the wave propagating from S to R (along ray Ω) is fixed. It may, of course, be either P or S. We have two options of computing the wavefield. In the *first option*, we shall consider the point source situated at point S^+, on the side of Σ^S opposite to S. See Figure 5.4(b). The source may generate both P and S waves. Both waves cross interface Σ^S as transmitted waves, from $\tilde{S}$ to S. The travel times of both waves from $\tilde{S}$ to R coincide and equal travel time $T(R, S)$. We also assume that the analytical signals of both generated waves are the same; otherwise, it would be necessary to treat both sources fully independently. We can then compute the ray-centered amplitude matrices of both "transmitted" waves. The final expression (see (5.2.73)) is

$$U_i^{(q)}(R) = \left[\frac{\rho(S)V(S)}{\rho(R)V(R)} \right]^{1/2} \frac{\exp(iT^c(R, S))}{\mathcal{L}(R, S)} \mathcal{G}_i^{(q)}(S^+; \gamma_1, \gamma_2), \tag{5.2.74}$$

where $\mathcal{G}_i^{(q)}(S^+; \gamma_1, \gamma_2)$ is the *generalized radiation function* corresponding to the point source situated on interface Σ^S:

$$\mathcal{G}_i^{(q)}(S^+; \gamma_1, \gamma_2) = (\cos i(S)/\cos i^P(S^+))R_{3i}^t(S^+)\mathcal{G}_3^{(q)SM}\left(S^+; \gamma_1^{tP}, \gamma_2^{tP}\right)$$

$$+ (\cos i(S)/\cos i^S(S^+))R_{ji}^t(S^+)\mathcal{G}_j^{(q)SM}\left(S^+; \gamma_1^{tS}, \gamma_2^{tS}\right). \tag{5.2.75}$$

Here $i^P(S^+)$ and $i^S(S^+)$ are angles of incidence of P and S waves at S^+. The ray-centered components $\mathcal{G}_i^{(q)}(S^+; \gamma_1, \gamma_2)$ of generalized radiation matrix $\hat{\mathcal{G}}^{(q)}(S^+, \gamma_1, \gamma_2)$ correspond to the polarization vectors $\vec{e}_1(S)$, $\vec{e}_2(S)$, and $\vec{e}_3(S) \equiv \vec{N}(S)$ of the elementary wave propagating from S to R. (5.2.75) yields the generalized radiation function of P waves for $i = 3$ and the generalized radiation pattern of S waves for $i = 1$ or 2. We also need to put $i(S) = i^P(S)$ for $i = 3$, and $i(S) = i^S(S)$ for $i = 1, 2$. The ray parameters $\gamma_1^{tP}, \gamma_2^{tP}$ and $\gamma_1^{tS}, \gamma_2^{tS}$ in the expressions for the smooth radiation functions correspond to the type of wave generated by the smooth-medium source. They are, of course, related to the ray parameters γ_1 and γ_2 of generalized radiation function $\mathcal{G}_i^{(q)}(S; \gamma_1, \gamma_2)$, and can be calculated from them using reflection/transmission laws (including Snell's law). Similarly, $i^P(S^+)$ and $i^S(S^+)$ should be calculated from $i(S)$.

In the *second option*, we shall consider the point source situated on the same side of interface Σ^S as S. See Figure 5.4(a). We then have to sum the direct wave and the two reflected waves. We again arrive at (5.2.74), where generalized radiation function $\mathcal{G}_i^{(q)}(S^+; \gamma_1, \gamma_2)$ is replaced by generalized radiation function $\mathcal{G}_i^{(q)}(S; \gamma_1, \gamma_2)$:

$$
\begin{aligned}
\mathcal{G}_i^{(q)}(S; \gamma_1, \gamma_2) = {} & \mathcal{G}_i^{(q)SM}(\tilde{S}; \gamma_1, \gamma_2) + (\cos i(S)/\cos i^P(\tilde{S})) \\
& \times R_{3i}^r(\tilde{S})\mathcal{G}_3^{(q)SM}(\tilde{S}; \gamma_1^{rP}, \gamma_2^{rP}) + (\cos i(S)/\cos i^S(\tilde{S})) \\
& \times R_{Ji}^r(\tilde{S})\mathcal{G}_J^{(q)SM}(\tilde{S}; \gamma_1^{rS}, \gamma_2^{rS}).
\end{aligned}
\tag{5.2.76}
$$

As in (5.2.75), $i^P(\tilde{S})$ and $i^S(\tilde{S})$ are angles of incidence of P and S waves at $\tilde{S}$. The generalized radiation function of P waves is obtained for $i = 3$; then also $i(S) = i^P(S)$. Similarly, the generalized radiation function of S waves is obtained for $i = 1, 2$; then also $i(S) = i^S(S)$. The ray parameters $\gamma_1^{rP}, \gamma_2^{rP}$ and $\gamma_1^{rS}, \gamma_2^{rS}$ in the expressions for smooth-medium radiation functions correspond to the type of wave generated by the smooth-medium source and may be calculated from γ_1 and γ_2 using the laws of reflection/transmission.

Equations (5.2.75) and (5.2.76) give two different relations for the generalized radiation matrices of a point source situated on an interface. If the point source is situated in a smooth medium, the generalized radiation function reduces to the smooth medium radiation function, corresponding to the first term in (5.2.76). The reflection coefficients in the second and third terms of (5.2.76) vanish for $\rho_1 = \rho_2$, $\lambda_1 = \lambda_2$, and $\mu_1 = \mu_2$.

It would be useful to add several remarks to the equations for the generalized radiation matrices (5.2.75) and (5.2.76). In general, these equations may give different results for sources of the same type approaching interface Σ^S from the two opposite sides. This is, of course, natural because smooth medium radiation functions depend, in general, on the medium parameters and ray parameters. Later on, however, we shall discuss a point source for which generalized radiation functions (5.2.75) and (5.2.76) are continuous across interface Σ^S. This point source corresponds to a single force, which plays an important role in many applications and in the definition of the elastodynamic Green function.

The ray-centered components of the smooth medium radiation matrices in expressions (5.2.75) and (5.2.76) depend, of course, on the choice of polarization vectors $\vec{e}_1, \vec{e}_2$, and $\vec{e}_3 \equiv \vec{N}$ at points $\tilde{S}$. The choice of $\vec{N}(\tilde{S})$ must correspond to the positive direction of propagation of the wave under consideration from S to R. Slowness vectors $\vec{p}(\tilde{S}) \equiv V(\tilde{S})\vec{N}(\tilde{S})$ and $\vec{p}(S) = V(S)\vec{N}(S)$ must satisfy the generalized Snell's law (2.4.70) for any of the elementary waves under consideration. Polarization vectors $\vec{e}_1(\tilde{S})$ and $\vec{e}_2(\tilde{S})$ must be mutually perpendicular, as well as perpendicular to $\vec{N}(\tilde{S})$. Otherwise, they can be chosen arbitrarily, but the R/T coefficients used must be consistent with this choice.

We shall now specify the generalized radiation matrix for a *single-force point source* situated at point S. We use (5.2.33) and obtain

$$
\begin{aligned}
\mathcal{G}_3^{(q)SM}(S; \gamma_1, \gamma_2) &= \frac{1}{4\pi\rho(S)\alpha(S)} H_{k3}(S) f_{0k}^{(x)}(S), \\
\mathcal{G}_J^{(q)SM}(S; \gamma_1, \gamma_2) &= \frac{1}{4\pi\rho(S)\beta(S)} H_{kJ}(S) f_{0k}^{(x)}(S).
\end{aligned}
\tag{5.2.77}
$$

Here $f_{0k}^{(x)}(S)$ are Cartesian components of the single force $\vec{f}_0(S)$, applied at point S. Exactly the same relations are obtained at point $\tilde{S}$.

For the single force $\vec{f}_0$ situated at point S, we insert (5.2.77) into (5.2.76) and obtain

$$\mathcal{G}_i^{(q)}(S; \gamma_1, \gamma_2) = \frac{1}{4\pi\rho(S)V(S)} \mathcal{D}_{ki}(S) f_{0k}^{(x)}(S), \tag{5.2.78}$$

with the 3×3 matrix $\mathcal{D}_{ki}(S)$ given by the relation,

$$\mathcal{D}_{ik}(S) = H_{ik}^{SM}(S) + [V(S)\cos i(S)/\alpha(\tilde{S})\cos i^P(\tilde{S})]H_{i3}^{rP}(\tilde{S})R_{3k}^r(\tilde{S})$$
$$+ [V(S)\cos i(S)/\beta(\tilde{S})\cos i^S(\tilde{S})]H_{iJ}^{rS}(\tilde{S})R_{Jk}^r(\tilde{S}). \tag{5.2.79}$$

Similarly, for the single force $\vec{f}_0$ situated at point S^+, we can insert (5.2.77) into (5.2.75) and obtain

$$\mathcal{G}_i^{(q)}(S^+; \gamma_1, \gamma_2) = \frac{1}{4\pi\rho(S)V(S)} \mathcal{D}_{ki}(S^+) f_{0k}^{(x)}(S^+), \tag{5.2.80}$$

with

$$\mathcal{D}_{ik}(S^+) = [\rho(S)V(S)\cos i(S)/\rho(S^+)\alpha(S^+)\cos i^P(S^+)]H_{i3}^{tP}(S^+)$$
$$\times R_{3k}^t(S^+) + [\rho(S)V(S)\cos i(S)/\rho(S^+)\beta(S^+)\cos i^S(S^+)]$$
$$\times H_{iJ}^{tS}(S^+)R_{Jk}^t(S^+). \tag{5.2.81}$$

The equations (5.2.79) and (5.2.81) can be expressed in many alternative forms. Suitable formulae for them will be derived in Section 5.3.8. It is possible to show that both expressions are equivalent. Thus, it does not matter whether we use (5.2.79) or (5.2.81). Moreover, we can use the reciprocity relations for the R/T coefficient (5.3.31) and find that they are also equivalent to the conversion matrices (5.2.62) and (5.2.63) and that $\mathcal{D}_{ik}$ represent the conversion coefficients. This plays an important role in the proof of reciprocity of the elastodynamic ray-theory Green function for a point source and/or receiver situated at structural interfaces.

In a matrix form, Equations (5.2.78) and (5.2.80) can be written as

$$\hat{\mathcal{G}}^{(q)}(S; \gamma_1, \gamma_2) = \frac{1}{4\pi\rho(S)V(S)} \hat{\mathcal{D}}^T(S)\hat{\mathbf{f}}_0^{(x)}(S). \tag{5.2.82}$$

5.2.9 Final Equations for Amplitude Matrices

In this section, we shall summarize the final equations for the amplitude matrices $\hat{\mathbf{U}}^{(x)}$, expressed in general Cartesian components. The elementary wave being considered may be any multiply-reflected, possibly converted, seismic body wave propagating in a general 3-D isotropic laterally varying layered structure. The source and/or receiver may be situated at any point of the medium, including the structural interfaces and the surface of the model.

We consider points S and R and the ray of the elementary wave involved connecting both these points. The receiver is situated at point R. The initial point S of ray Ω corresponds to a point source or merely to an arbitrarily selected point on the ray where the amplitude matrix is known.

In all the presented equations, we shall use the Einstein summation convention over $1, 2, 3$ for lowercase indices i, j, k, and over $1, 2$ for the uppercase indices I, J, K. We shall also use the following important *convention for the first and last elements of ray* Ω:

a. If the first element of ray Ω (starting at point S) is P, we put $j = 3$. If it is S, we put $j = J$.

b. If the last element of ray Ω (ending at point R) is P, we put $k = 3$. If it is S, we put $k = K$.

The final equations for $U_i^{(x)}(R)$ are then as follows.

a. The continuation relation:

$$U_i^{(x)}(R) = \left[\frac{V(S)\rho(S)}{V(R)\rho(R)} \right]^{1/2} \frac{\mathcal{L}(S)}{\mathcal{L}(R)} \, \exp[iT^c(R, S)] \mathcal{D}_{ik}(R) \mathcal{R}_{kj}^C U_j^{(q)}(S).$$

$$(5.2.83)$$

b. The point source at S:

$$U_i^{(x)}(R) = \left[\frac{V(S)\rho(S)}{V(R)\rho(R)} \right]^{1/2} \frac{\exp[iT^c(R, S)]}{\mathcal{L}(R, S)} \mathcal{D}_{ik}(R) \mathcal{R}_{kj}^C \mathcal{G}_j^{(q)}(S; \gamma_1, \gamma_2).$$

$$(5.2.84)$$

c. A single-force point source at S:

$$U_i^{(x)}(R) = \frac{\exp[iT^c(R, S)]}{4\pi [\rho(S)\rho(R)V(S)V(R)]^{1/2}} \frac{1}{\mathcal{L}(R, S)} \mathcal{D}_{ik}(R) \mathcal{R}_{kj}^C \mathcal{D}_{nj}(S) f_{0n}^{(x)}(S).$$

$$(5.2.85)$$

d. Elementary ray-theory Green function in the time domain:

$$G_{in}(R, t; S, t_0) = \frac{\exp[iT^c(R, S)]}{4\pi [\rho(S)\rho(R)V(S)V(R)]^{1/2}} \frac{1}{\mathcal{L}(R, S)} \mathcal{D}_{ik}(R) \mathcal{R}_{kj}^C \mathcal{D}_{nj}(S)$$
$$\times \delta^{(A)}(t - t_0 - T(R, S)). \qquad (5.2.86)$$

In (5.2.83) through (5.2.86), all the expressions have the same meaning as in the foregoing sections. Velocities $V(S)$ and $V(R)$ are specified as α or β according to the type of wave at S or R. $\mathcal{R}_{kj}^C$ are the elements of the complete matrix of the normalized R/T coefficients along the ray Ω from S to R. The complete matrix of normalized R/T coefficients is a product of the transposed matrices of normalized R/T coefficients along ray Ω from S to R. $\mathcal{D}_{ik}(R)$ and $\mathcal{D}_{ik}(S)$ are the conversion coefficients. If the source and/or receiver is situated in a smooth medium, the conversion coefficient $\mathcal{D}_{ik}$ reduces to H_{ik}. If the source and/or receiver is situated on an interface, the relevant equations of Sections 5.2.7 and 5.2.8 should be used. Expressions $\mathcal{D}_{ik}(R)$ and $\mathcal{D}_{ik}(S)$ in (5.2.83) through (5.2.85) should be replaced by $\mathcal{D}_{ik}(R^+)$ and $\mathcal{D}_{ik}(S^+)$ if the receiver and/or source are situated on the sides of the interfaces opposite to R and S. Finally, $\mathcal{G}_j^{(q)}(S; \gamma_1, \gamma_2)$ is the smooth medium radiation function if S is situated in a smooth medium. If the source is situated on the interface, $\mathcal{G}_j^{(q)}(S; \gamma_1, \gamma_2)$ is given by (5.2.76). It should again be replaced by $\mathcal{G}_j^{(q)}(S^+; \gamma_1, \gamma_2)$ if the source is situated on the side of the interface opposite to S; see (5.2.75).

Equation (5.2.86) determines the elementary elastodynamic ray-theory Green function in the time domain. The elementary elastodynamic ray-theory Green function $G_{in}(R, S, \omega)$ in the frequency domain is again given by (5.2.86); only the analytic delta function $\delta^{(A)}(t - t_0 - T(R, S))$ is replaced by $\exp(i\omega T(R, S))$.

5.2.10 Unconverted P Waves

General expressions (5.2.83) through (5.2.86) are simplified for unconverted P waves. In this case, we can put $j = k = 3$, and the only nonvanishing element of the matrix $\hat{\mathcal{R}}^C$

is $\mathcal{R}_{33}^C$, corresponding to the complete normalized P $\rightarrow$ P reflection/transmission coefficient. It represents a product of the individual normalized P $\rightarrow$ P R/T coefficients at points $Q_1, Q_2, \ldots, Q_N$. All these normalized P $\rightarrow$ P R/T coefficients and the complete normalized P $\rightarrow$ P R/T coefficient have a fully scalar character, as in the case of acoustic waves. To simplify the notation, we shall denote $\mathcal{R}_{33}^C$ by $\mathcal{R}^C$.

Equations (5.2.83) through (5.2.86) then simplify considerably and read as follows:

a. The continuation relation:

$$U_i^{(x)}(R) = \left[\frac{\alpha(S)\rho(S)}{\alpha(R)\rho(R)} \right]^{1/2} \frac{\mathcal{L}(S)}{\mathcal{L}(R)} \exp[iT^c(R, S)]\mathcal{D}_{i3}(R)\mathcal{R}^C U_3^{(q)}(S).$$

(5.2.87)

b. The point source at S:

$$U_i^{(x)}(R) = \left[\frac{\alpha(S)\rho(S)}{\alpha(R)\rho(R)} \right]^{1/2} \frac{\exp[iT^c(R, S)]}{\mathcal{L}(R, S)} \mathcal{D}_{i3}(R)\mathcal{R}^C \mathcal{G}_3^{(q)}(S; \gamma_1, \gamma_2).$$

(5.2.88)

c. The single-force point source at S:

$$U_i^{(x)}(R) = \frac{\exp[iT^c(R, S)]}{4\pi [\rho(S)\rho(R)\alpha(S)\alpha(R)]^{1/2}} \frac{1}{\mathcal{L}(R, S)} \mathcal{D}_{i3}(R)\mathcal{R}^C \mathcal{D}_{n3}(S) f_{0n}^{(x)}(S).$$

(5.2.89)

d. The elementary ray-theory P-wave Green function in the time domain:

$$G_{in}(R, t; S, t_0) = \frac{\exp[iT^c(R, S)]}{4\pi [\rho(S)\rho(R)\alpha(S)\alpha(R)]^{1/2}} \frac{1}{\mathcal{L}(R, S)} \mathcal{D}_{i3}(R)\mathcal{R}^C \mathcal{D}_{n3}(S)$$
$$\times \delta^{(A)}(t - t_0 - T(R, S)).$$

(5.2.90)

The reader is reminded that $\mathcal{D}_{i3}(R) = H_{i3}(R) = N_i(R)$ if the receiver is situated in a smooth medium, and an analogous relation is valid for $\mathcal{D}_{n3}(S)$. Here $N_i(R)$ is the Cartesian component of the unit vector tangent to the ray. Thus, in this case, the product $\mathcal{D}_{i3}(R)\mathcal{D}_{n3}(S)$ in (5.2.89) and (5.2.90) yields $N_i(R)N_n(S)$. If the receiver and/or source are situated on the interface, we can use (5.2.67). The meaning of all other symbols is the same as in the foregoing section.

5.2.11 P Waves in Fluid Media. Particle Velocity Amplitudes

Equations (5.2.87) through (5.2.90) remain valid even for P elastic waves propagating in fluid media. In all equations, we merely put $\beta = 0$. This substitution applies mainly to the expressions for $\mathcal{R}^C$, $\mathcal{D}_{i3}(R)$, and $\mathcal{D}_{n3}(S)$.

The results presented in this section are closely connected with those presented in Section 5.1 for *scalar pressure waves* $p(x_i, t)$. Here, however, we shall study the behavior of the *vectorial wavefield* $\vec{u}(x_i, t)$, which is related to the displacement vector. In the acoustic case, it is more common to consider *particle velocity vector* $\vec{v}(x_i, t) = \partial \vec{u}(x_i, t)/\partial t$ than displacement vector $\vec{u}(x_i, t)$. For this reason, we shall also give the relevant equations for the amplitudes of the particle velocity vector.

It would be more natural to include this paragraph in Section 5.1, "Acoustic Case." This would, however, require extending Section 5.1 considerably and deriving all the equations there related to the vectorial wavefield. At this point, we do not need to derive anything new, we merely insert $\beta = 0$ into the general equations for the vectorial elastic wavefields in solid media, derived in Sections 5.2.1 through 5.2.10.

First, we shall briefly discuss the complete R/T coefficient $\mathcal{R}^C$. Section 5.3.1 will show that the limiting process $\beta_1 \to 0$ and $\beta_2 \to 0$ in the expressions for the displacement R/T $P \to P$ coefficients yields the following relations:

$$R^r = \frac{\rho_2\alpha_2 P_1 - \rho_1\alpha_1 P_2}{\rho_2\alpha_2 P_1 + \rho_1\alpha_1 P_2}, \qquad R^t = \frac{2\rho_1\alpha_1 P_1}{\rho_2\alpha_2 P_1 + \rho_1\alpha_1 P_2}. \qquad (5.2.91)$$

Here we have used the following notation: $P_k = \cos i_k = (1 - \alpha_k^2 p^2)^{1/2}$, for $k = 1, 2$. If we compare (5.2.91) with (5.1.21), we can see that reflection coefficient (5.2.91) is exactly the same as the acoustic reflection coefficient but that the transmission coefficients are different. The explanation of this difference is simple: (5.2.91) represents the displacement R/T coefficients, but (5.1.21) the pressure R/T coefficients. We can, however, prove that the *normalized R/T coefficients* $\mathcal{R}$ are the same, both for displacement and pressure. If we use (5.2.43) and (5.2.91), we obtain the normalized displacement transmission coefficient as follows:

$$\mathcal{R}^t = \frac{2(\rho_1\rho_2\alpha_1\alpha_2 P_1 P_2)^{1/2}}{\rho_2\alpha_2 P_1 + \rho_1\alpha_1 P_2}. \qquad (5.2.92)$$

This exactly corresponds to the normalized pressure transmission coefficient; see (5.1.22). We conclude that, although the pressure and displacement transmission coefficients in fluid media differ, the normalized pressure and displacement transmission coefficients are the same.

Because the complete R/T coefficient $\mathcal{R}^C$ is a product of normalized R/T coefficients at the individual points of reflection and transmission between S and R, the complete displacement coefficient $\mathcal{R}^C$ in fluid media is the same as the complete pressure coefficient $\mathcal{R}^C$, introduced in Section 5.1.5; see (5.1.35).

Now we shall discuss the elements $\mathcal{D}_{i3}$ of conversion matrix $\hat{\mathbf{D}}$ in fluid media. We use (5.2.67) and consider only P waves. We then obtain $\mathcal{D}_{i3}(R) = N_i^{SM}(R) + N_i^{rP}(R)R_{33}^r(R)$, or, alternatively, $\mathcal{D}_{i3}(R^+) = N_i^{tP}(R^+)R_{33}^t(R)$. Here $N_i^{SM}(R)$ corresponds to the incident wave; $N_i^{rP}(R)$ and $N_i^{tP}(R)$ correspond to the reflected and transmitted waves. If we use (2.4.70), we arrive at

$$N_i^{rP}(R) = \alpha_1 p_i^{rP}(R) = \alpha_1 p_i(R) - 2P_1\epsilon n_i(R),$$
$$N_i^{tP}(R^+) = \alpha_2 p_i^{tP}(R^+) = \alpha_2 p_i(R) - \alpha_2\left(\alpha_1^{-1} P_1 - \alpha_2^{-1} P_2\right)\epsilon n_i(R).$$

Here P_1 and P_2 are given by (5.1.19). This finally yields the expressions for $\mathcal{D}_{i3}(R)$ and $\mathcal{D}_{i3}(R^+)$ in fluid media:

$$\mathcal{D}_{i3}(R) = \alpha_1 p_i(R) + (\alpha_1 p_i(R) - 2P_1\epsilon n_i(R))\frac{\rho_2\alpha_2 P_1 - \rho_1\alpha_1 P_2}{\rho_2\alpha_2 P_1 + \rho_1\alpha_1 P_2}, \qquad (5.2.93)$$

or, alternatively,

$$\mathcal{D}_{i3}(R^+) = [\alpha_1\alpha_2 p_i(R) - (\alpha_2 P_1 - \alpha_1 P_2)\epsilon n_i(R)]\frac{2\rho_1 P_1}{\rho_2\alpha_2 P_1 + \rho_1\alpha_1 P_2}. \qquad (5.2.94)$$

Equations (5.2.93) and (5.2.94) are valid for a receiver situated on two opposite sides of interface Σ^R with normal $\vec{n}$ at points R and R^+. If the receiver is situated *in a smooth medium*, (5.2.93) and (5.2.94) yield the same result:

$$\mathcal{D}_{i3}(R) = \mathcal{D}_{i3}(R^+) = \alpha_1 p_i(R) = N_i(R).$$

(5.2.95)

Assume now that the general Cartesian coordinate system is introduced so that the origin is situated at R and axis x_3 coincides with normal $\vec{n}$. Moreover, axis x_1 is situated in the plane of incidence. In this case, (5.2.93) yields

$$\mathcal{D}_{13}(R) = \frac{2\alpha_1\alpha_2\rho_2 p P_1}{\rho_2\alpha_2 P_1 + \rho_1\alpha_1 P_2}, \qquad \mathcal{D}_{23}(R) = 0,$$

$$\mathcal{D}_{33}(R) = \frac{2\rho_1\alpha_1 P_1 P_2 \epsilon}{\rho_2\alpha_2 P_1 + \rho_1\alpha_1 P_2}.$$

(5.2.96)

Alternatively, (5.2.94) yields

$$\mathcal{D}_{13}(R^+) = \frac{2\alpha_1\alpha_2\rho_1 p P_1}{\rho_2\alpha_2 P_1 + \rho_1\alpha_1 P_2}, \qquad \mathcal{D}_{23}(R^+) = 0,$$

$$\mathcal{D}_{33}(R^+) = \frac{2\rho_1\alpha_1 P_1 P_2 \epsilon}{\rho_2\alpha_2 P_1 + \rho_1\alpha_1 P_2}.$$

(5.2.97)

As we can see, the normal components $\mathcal{D}_{33}(R)$ and $\mathcal{D}_{33}(R^+)$, obtained from (5.2.93) and (5.2.94) are exactly the same. This was expected, of course, because the normal component of the displacement is continuous across the interface, even in fluid media. Tangential components $\mathcal{D}_{13}(R)$ and $\mathcal{D}_{13}(R^+)$, however, are not the same. In fluid media, the tangential component recorded at the top of interface Σ^R will be different from the tangential component recorded on the bottom of the interface, if $\rho_1 \neq \rho_2$.

Let us now consider values of $\mathcal{D}_{i3}$ at free surface Σ^R. If we put $\rho_2 \to 0$ and $\alpha_2 \to 0$, (5.2.96) or (5.2.97) yields

$$\mathcal{D}_{13}(R) = \mathcal{D}_{23}(R) = 0, \qquad \mathcal{D}_{33}(R) = \mathcal{D}_{33}(R^+) = 2\epsilon P_1.$$

(5.2.98)

Thus, the tangential components of the displacement vector vanish at a free surface. The normal component of the displacement is, however, nonvanishing. This is the great difference with respect to the pressure waves. The pressure wavefield vanishes at the free surface of a fluid medium, but the normal displacement component does not.

Finally, we should specify radiation functions $\mathcal{G}_3^{(q)}(S; \gamma_1, \gamma_2)$ and $\mathcal{D}_{n3}(S)$. The general relations for the radiation function $\mathcal{G}_3^{(q)}(S; \gamma_1, \gamma_2)$ are given by (5.2.76), which yield the following relations for fluid media:

$$\mathcal{G}_3^{(q)}(S, \gamma_1, \gamma_2) = \mathcal{G}_3^{(q)SM}(\tilde{S}; \gamma_1, \gamma_2) + R^r(S)\mathcal{G}_3^{(q)SM}(\tilde{S}; \gamma_1^{rP}, \gamma_2^{rP}),$$

(5.2.99)

where $R^r(S)$ is given by (5.2.91). Similarly, for a source situated at S^+, on the side of interface Σ^S opposite to S, (5.2.75) yields

$$\mathcal{G}_3^{(q)}(S^+; \gamma_1, \gamma_2) = (\rho_2 d_2/\rho_1 d_1)R^t(S)\mathcal{G}_3^{(q)SM}(S^+; \gamma_1^{tP}, \gamma_2^{tP}).$$

(5.2.100)

Here R^t is again given by (5.2.91). For a single-force point source situated at S or S^+, we obtain

$$\mathcal{G}_3^{(q)}(S; \gamma_1, \gamma_2) = (4\pi\rho_1\alpha_1)^{-1}\mathcal{D}_{k3}(S)f_{0k}^{(x)}(S),$$

$$\mathcal{G}_3^{(q)}(S^+; \gamma_1, \gamma_2) = (4\pi\rho_1\alpha_1)^{-1}\mathcal{D}_{k3}(S^+)f_{0k}^{(x)}(S^+).$$

(5.2.101)

Here $\mathcal{D}_{k3}(S)$ and $\mathcal{D}_{k3}(S^+)$ are given by equations (5.2.93) through (5.2.98); only R and R^+ should be replaced by S and S^+.

In fluid media, it is common to consider *particle velocity* $\vec{v}(x_j, t) = \dot{\vec{u}}(x_j, t)$, instead of $\vec{u}(x_j, t)$. From (5.2.1), we obtain

$$\vec{v}(x_j, t) = \dot{\vec{u}}(x_j, t) = \vec{U}(x_j)\dot{F}(t - T(x_j)). \tag{5.2.102}$$

Thus, $\vec{U}(x_j)$ again represents a vectorial amplitude function of the particle velocity wavefield, and all the relations derived for $\vec{U}(x_i)$ remain valid. Only the analytical signal for the particle velocity wavefield $\dot{F}(\zeta)$ is different from the analytical signal of displacement $F(\zeta)$. For $\vec{U}(x_j)$, we can write $\vec{U} = U_3^{(q)}\vec{N}$, where $\vec{N}$ is the unit vector perpendicular to the wavefront. The other two ray-centered components of $\vec{U}$, $U_1^{(q)}$ and $U_2^{(q)}$, vanish. Radiation matrix $\hat{\mathcal{G}}^{(q)}(S, \gamma_1, \gamma_2)$ also has only one nonvanishing component, $\mathcal{G}_3^{(q)}(S, \gamma_1, \gamma_2)$, defined in (5.2.26), where $A = U_3^{(q)}$. We need to modify, however, the single-force radiation function: instead of the single force $\vec{f}(x_j, t)$, given by (5.2.32), we must consider the *time derivative of the single force* $\partial \vec{f}(x_i, t)/\partial t$:

$$\partial \vec{f}(x_i, t)/\partial t = \delta(\vec{x} - \vec{x}(S))\dot{\vec{f}}_0 dF(t)/dt. \tag{5.2.103}$$

Here $\dot{\vec{f}}_0$ is a *constant* vector, specifying the direction and magnitude of $\partial f(x_i, t)/\partial t$. The dot above the letter is introduced to distinguish it from $\vec{f}_0$; see (5.2.32). Then, $f_{0k}^{(x)}$ in (5.2.101) should be replaced by $\dot{f}_{0k}^{(x)}$. Otherwise, all the equations remain the same.

5.2.12 Unconverted S Waves

In the case of unconverted S waves along the whole ray Ω, the only simplification is that the 3×3 and 3×1 matrices are reduced to 2×2 and 2×1 matrices. Equations (5.2.83) through (5.2.86) in this case read:

a. The continuation relation:

$$U_i^{(x)}(R) = \left[\frac{\beta(S)\rho(S)}{\beta(R)\rho(R)}\right]^{1/2} \frac{\mathcal{L}(S)}{\mathcal{L}(R)} \exp[iT^c(R, S)]\mathcal{D}_{iK}(R)\mathcal{R}_{KJ}^C U_J^{(q)}(S). \tag{5.2.104}$$

b. The point source at S:

$$U_i^{(x)}(R) = \left[\frac{\beta(S)\rho(S)}{\beta(R)\rho(R)}\right]^{1/2} \frac{\exp[iT^c(R, S)]}{\mathcal{L}(R, S)} \mathcal{D}_{iK}(R)\mathcal{R}_{KJ}^C \mathcal{G}_J^{(q)}(S; \gamma_1, \gamma_2). \tag{5.2.105}$$

c. A single-force point source at S:

$$U_i^{(x)}(R) = \frac{\exp[iT^c(R, S)]}{4\pi[\rho(S)\rho(R)\beta(S)\beta(R)]^{1/2}} \frac{1}{\mathcal{L}(R, S)} \mathcal{D}_{iK}(R)\mathcal{R}_{KJ}^C \mathcal{D}_{nJ}(S)f_{0n}^{(x)}(S). \tag{5.2.106}$$

d. Elementary ray-theory S-wave Green function:

$$G_{in}(R, t; S, t_0) = \frac{\exp[iT^c(R, S)]}{4\pi[\rho(S)\rho(R)\beta(S)\beta(R)]^{1/2}} \frac{1}{\mathcal{L}(R, S)} \mathcal{D}_{iK}(R)\mathcal{R}_{KJ}^C \mathcal{D}_{nJ}(S)$$
$$\times \delta^{(A)}(t - t_0 - T(R, S)). \tag{5.2.107}$$

Thus, we need to compute only the 2×2 complete matrix $\mathcal{R}^C$ of the normalized R/T coefficients. This represents a 2×2 upper-left-corner submatrix of the 3×3 matrix $\hat{\mathcal{R}}^C$.

The 2×2 matrix $\mathcal{R}^C$ is a product of the 2×2 matrices $\mathcal{R}^T$ at points $Q_1, Q_2, \ldots, Q_N$. If the receiver is situated in a smooth medium, then $\mathcal{D}_{iK}(R) = H_{iK}(R) = e_{Ki}(R)$. The same relation is valid for $\mathcal{D}_{nJ}(S)$. Thus, the product $\mathcal{D}_{iK}(R)\mathcal{D}_{nJ}(S)$ equals $e_{Ki}e_{Jn}(S)$. If the source and/or receiver are situated on an interface, we can use (5.2.68) to compute $\mathcal{D}_{iK}(R)$ and $\mathcal{D}_{nJ}(S)$. The meaning of all the other symbols is the same as in Section 5.2.9.

5.2.13 Amplitudes Along a Planar Ray. 2-D Case

The relations (5.2.83) through (5.2.86), valid for a general 3-D ray Ω, simplify for the planar ray Ω. We denote the plane in which the ray is situated $\Sigma^{\parallel}$. At the initial point S, we choose basis vector $\vec{e}_2$ perpendicular to $\Sigma^{\parallel}$. Then $\vec{e}_2$ is perpendicular to $\Sigma^{\parallel}$ along the whole ray Ω, and basis vectors $\vec{e}_1$ and $\vec{e}_3$ are confined to plane $\Sigma^{\parallel}$. For a detailed specification of the individual quantities in this case, see the introduction to Section 4.13.

Without loss of generality, we shall introduce general Cartesian coordinate system x_1, x_2, x_3 so that the x_2-axis is perpendicular to plane $\Sigma^{\parallel}$, which is specified by equation $x_2 = 0$. Conditions (2.3.45) are then satisfied at all R/T points. In this case, the matrix of R/T coefficients $\hat{\mathbf{R}}$ simplifies. R/T coefficients R_{12}, R_{21}, R_{23} and R_{32} vanish; see (2.3.44). We shall use special terminology for the five remaining R/T coefficients:

- R_{22} will be called the SH *R/T coefficients*.
- R_{11}, R_{13}, R_{31}, and R_{33} will be called the P-SV *R/T coefficients*.

Individually, R_{11} is the SV $\rightarrow$ SV coefficient, R_{13} the SV $\rightarrow$ P coefficient, R_{31} the P $\rightarrow$ SV coefficient, and R_{33} the P $\rightarrow$ P R/T coefficient. Detailed analytical expressions for SH and P-SV R/T coefficients will be given in Section 5.3.1.

From the general relations for amplitudes, we can see that component $U_2^{(q)}$ is no longer coupled with components $U_1^{(q)}$ and $U_3^{(q)}$ along planar ray Ω. It is common in seismology to call component $U_2^{(q)}$ the SH *component of the* S *wave* (S horizontal) and $U_1^{(q)}$ the SV *component of the* S *wave* (S vertical). This terminology corresponds to the seismological convention in which the x_3-axis of the general Cartesian coordinate system represents the vertical (depth) axis. Plane $\Sigma^{\parallel}$ is then vertical, and the SH component of the S wave, $U_2^{(q)}$, is horizontal. The terminology, however, may be confusing because the SV component of the S wave, $U_1^{(q)}$, is not necessarily vertical. It only means that the SV component is always *confined to the vertical plane* $\Sigma^{\parallel}$, which contains ray Ω and is specified by the x_1- and x_3-axes of the general Cartesian coordinate system. In our treatment in this section, we do not assume that plane $\Sigma^{\parallel}$ of ray Ω is vertical; nevertheless, we shall use the previously described standard seismological terminology. Because the SH and SV components are not coupled in our case of a planar ray Ω, we also often speak of the SV *and* SH *waves*, and not of the SV and SH components of the S wave.

We shall also take into account that geometrical spreading factor $\mathcal{L}(R, S)$ can be factorized into the *in-plane* and *transverse* factor along the planar ray Ω. The same is valid also for $\mathcal{L}(R)$ and $\mathcal{L}(S)$:

$$\mathcal{L}(R) = \mathcal{L}^{\parallel}(R)\mathcal{L}^{\perp}(R), \qquad \mathcal{L}(S) = \mathcal{L}^{\parallel}(S)\mathcal{L}^{\perp}(S),$$
$$\mathcal{L}(R, S) = \mathcal{L}^{\parallel}(R, S)\mathcal{L}^{\perp}(R, S).$$

For a detailed discussion and the relevant equations for the in-plane and transverse geometrical spreading factors, see Section 4.13.

1. SH WAVES

In the case of SH waves, we shall evaluate Cartesian component $U_2^{(x)}(R) = U_2^{(q)}(R)$. The SH wave is an unconverted S wave along the whole ray Ω so that we can use the equations of Section 5.2.12. The complete 2×2 matrix of normalized R/T coefficients $\mathcal{R}^C$ is diagonal in this case so that the resulting element $\mathcal{R}_{22}^C$ equals the product of the *scalar* normalized SH R/T coefficients. To emphasize this fact, we shall denote $\mathcal{R}_{22}^C$ by $\mathcal{R}_{SH}^C$ and call it the *complete normalized* SH *reflection/transmission coefficient* along ray Ω between S and R. The individual equations then read:

a. The continuation relation:

$$U_2^{(x)}(R) = \left[\frac{\beta(S)\rho(S)}{\beta(R)\rho(R)} \right]^{1/2} \frac{\mathcal{L}^{\parallel}(S)\mathcal{L}^{\perp}(S)}{\mathcal{L}^{\parallel}(R)\mathcal{L}^{\perp}(R)}$$
$$\times \exp[iT^c(R, S)]\mathcal{D}_{22}(R)\mathcal{R}_{SH}^C U_2^{(q)}(S). \qquad (5.2.108)$$

b. The point source at S:

$$U_2^{(x)}(R) = \left[\frac{\beta(S)\rho(S)}{\beta(R)\rho(R)} \right]^{1/2} \frac{\exp[iT^c(R, S)]}{\mathcal{L}^{\parallel}(R, S)\mathcal{L}^{\perp}(R, S)} \mathcal{D}_{22}(R)\mathcal{R}_{SH}^C \mathcal{G}_2^{(q)}(S; \gamma_1, \gamma_2). \qquad (5.2.109)$$

c. A single-force point source at S:

$$U_2^{(x)}(R) = \frac{\exp[iT^c(R, S)]}{4\pi [\rho(S)\rho(R)\beta(S)\beta(R)]^{1/2}}$$
$$\times \frac{1}{\mathcal{L}^{\parallel}(R, S)\mathcal{L}^{\perp}(R, S)} \mathcal{D}_{22}(R)\mathcal{D}_{22}(S)\mathcal{R}_{SH}^C f_{02}^{(x)}(S). \qquad (5.2.110)$$

d. Elementary ray-theory SH Green function:

$$G_{22}(R, t; S, t_0) = \frac{\exp[iT^c(R, S)]}{4\pi [\rho(S)\rho(R)\beta(S)\beta(R)]^{1/2}} \frac{1}{\mathcal{L}^{\parallel}(R, S)\mathcal{L}^{\perp}(R, S)}$$
$$\times \mathcal{D}_{22}(R)\mathcal{D}_{22}(S)\mathcal{R}_{SH}^C \delta^{(A)}(t - t_0 - T(R, S)). \qquad (5.2.111)$$

In all these expressions, components $\mathcal{D}_{22}(R)$ and $\mathcal{D}_{22}(S)$ of the conversion matrices $\hat{\mathcal{D}}(R)$ and $\hat{\mathcal{D}}(S)$ are given by the following relations:

- If point R is situated in a smooth medium:

$$\mathcal{D}_{22}(R) = 1.$$

- If point R is situated on an interface:

$$\mathcal{D}_{22}(R) = 1 + R_{22}^r(R), \qquad \mathcal{D}_{22}(R^+) = R_{22}^t(R).$$

The relation for component $\mathcal{G}_2^{(q)}(S; \gamma_1, \gamma_2)$ in (5.2.109) can also be simplified. From general equations (5.2.75) and (5.2.76), we obtain

$$\mathcal{G}_2^{(q)}(S^+; \gamma_1, \gamma_2) = (\cos i_1 / \cos i_2)R_{22}^t(S)\mathcal{G}_2^{(q)SM}(S^+; \gamma_1^{tS}, \gamma_2^{tS}),$$

$$\mathcal{G}_2^{(q)}(S; \gamma_1, \gamma_2) = \mathcal{G}_2^{(q)SM}(\tilde{S}; \gamma_1, \gamma_2) + R_{22}^r(S)\mathcal{G}_2^{(q)SM}(S; \gamma_1^{rS}, \gamma_2^{rS}).$$

Otherwise, all the symbols in (5.2.108) through (5.2.111) have their standard meaning.

2. P-SV WAVES

Here we shall consider an arbitrary *multiply reflected, possibly converted wave*, propagating along planar ray Ω. Because the SH component of the S wave is fully separated, we can consider only the SV component of the S wave, polarized in plane $\Sigma^{\parallel}$ of ray Ω.

In ray-centered coordinates, the SV component is represented by $U_1^{(q)}$, and the P wave is represented by $U_3^{(q)}$. We can again use general relations (5.2.83) through (5.2.86), but all the indices i, j, k, and n will only take values 1 or 3. *The convention for the first and last element* of ray Ω, formulated in Section 5.2.9, now simplifies:

    a. If the first element of ray Ω (starting at the point S) is P, we put $j = 3$. If it is S, we put $j = 1$.

    b. If the last element of ray Ω (ending at the point R) is P, we put $k = 3$. If it is S, we put $k = 1$.

As we can see from (5.2.83) through (5.2.86), this convention removes fully the summations due to the Einstein summation convention from these equations because indices k and j are fixed.

Let us now discuss element $\mathcal{R}_{kj}^C$ of the complete matrix of normalized R/T coefficients $\hat{\mathcal{R}}^C$. As we can check in Equations (5.2.38), each of the R/T matrices has only one non-vanishing element in our case, corresponding to the relevant P-SV R/T coefficient. Thus, matrix multiplication is not required to compute $\mathcal{R}_{kj}^C$; $\mathcal{R}_{kj}^C$ is merely a product of scalar normalized P-SV R/T coefficients. To emphasize this fact, we shall denote $\mathcal{R}_{kj}^C$ by $\mathcal{R}_{P,SV}^C$ and call it the *complete normalized P-SV R/T coefficient*. The product may contain P-SV reflection and transmission coefficients of four types: $\mathcal{R}_{11}$ (SV $\to$ SV), $\mathcal{R}_{13}$ (SV $\to$ P), $\mathcal{R}_{31}$ (P $\to$ SV), and $\mathcal{R}_{33}$ (P $\to$ P). The choice of the proper R/T coefficient must be consistent with the alphanumeric code of the wave.

The final equations are as follows:

    a. The continuation relation:

$$U_i^{(x)}(R) = \left[\frac{V(S)\rho(S)}{V(R)\rho(R)}\right]^{1/2} \frac{\mathcal{L}^{\parallel}(S)\mathcal{L}^{\perp}(S)}{\mathcal{L}^{\parallel}(R)\mathcal{L}^{\perp}(R)}$$
$$\times \exp[iT^c(R,S)]\mathcal{D}_{ik}(R)\mathcal{R}_{P,SV}^C U_j^{(q)}(S). \qquad (5.2.112)$$

    b. The point source at S:

$$U_i^{(x)}(R) = \left[\frac{V(S)\rho(S)}{V(R)\rho(R)}\right]^{1/2}$$
$$\times \frac{\exp[iT^c(R,S)]}{\mathcal{L}^{\parallel}(R,S)\mathcal{L}^{\perp}(R,S)}\mathcal{D}_{ik}(R)\mathcal{R}_{P,SV}^C \mathcal{G}_j^{(q)}(S;\gamma_1,\gamma_2). \qquad (5.2.113)$$

    c. A single-force point source at S:

$$U_i^{(x)}(R) = \frac{\exp[iT^c(R,S)]}{4\pi[\rho(S)\rho(R)V(S)V(R)]^{1/2}}$$
$$\times \frac{1}{\mathcal{L}^{\parallel}(R,S)\mathcal{L}^{\perp}(R,S)}\mathcal{D}_{ik}(R)\mathcal{D}_{nj}(S)\mathcal{R}_{P,SV}^C f_{0n}^{(x)}(S). \qquad (5.2.114)$$

    d. Elementary ray-theory P-SV Green function:

$$G_{in}(R,t;S,t_0) = \frac{\exp[iT^c(R,S)]}{4\pi[\rho(S)\rho(R)V(S)V(R)]^{1/2}} \frac{1}{\mathcal{L}^{\parallel}(R,S)\mathcal{L}^{\perp}(R,S)}$$
$$\times \mathcal{D}_{ik}(R)\mathcal{D}_{nj}(S)\mathcal{R}_{P,SV}^C \delta^{(A)}(t-t_0-T(R,S)). \qquad (5.2.115)$$

We again remind the reader that i, j, k, and n may take only values 1 or 3 and that k and j are strictly determined by the convention for the first and last element of ray Ω (3 for P, 1 for SV).

Equations (5.2.112) through (5.2.115) contain four elements of conversion matrix $\hat{\mathcal{D}}$: $\mathcal{D}_{11}$, $\mathcal{D}_{13}$, $\mathcal{D}_{31}$, and $\mathcal{D}_{33}$. The conversion coefficients $\mathcal{D}_{11}$, $\mathcal{D}_{13}$, $\mathcal{D}_{31}$, and $\mathcal{D}_{33}$ are given by simpler relations than (5.2.67) and (5.2.68) because certain R/T coefficients vanish. For the reader's convenience, we shall present these simplified equations for the conversion coefficients here:

$$
\begin{aligned}
\mathcal{D}_{11}(R) &= e_{11}^{SM}(R) + N_1^{rP}(R)R_{13}^r(R) + e_{11}^{rS}(R)R_{11}^r(R), \\
\mathcal{D}_{11}(R^+) &= N_1^{tP}(R^+)R_{13}^t(R) + e_{11}^{tS}(R^+)R_{11}^t(R), \\
\mathcal{D}_{13}(R) &= N_1^{SM}(R) + N_1^{rP}(R)R_{33}^r(R) + e_{11}^{rS}(R)R_{31}^r(R), \\
\mathcal{D}_{13}(R^+) &= N_1^{tP}(R^+)R_{33}^t(R) + e_{11}^{tS}(R^+)R_{31}^t(R), \\
\mathcal{D}_{31}(R) &= e_{13}^{SM}(R) + N_3^{rP}(R)R_{13}^r(R) + e_{13}^{rS}(R)R_{11}^r(R), \\
\mathcal{D}_{31}(R^+) &= N_3^{tP}(R^+)R_{13}^t(R) + e_{13}^{tS}(R^+)R_{11}^t(R), \\
\mathcal{D}_{33}(R) &= N_3^{SM}(R) + N_3^{rP}(R)R_{33}^r(R) + e_{13}^{rS}(R)R_{31}^r(R), \\
\mathcal{D}_{33}(R^+) &= N_3^{tP}(R^+)R_{33}^t(R) + e_{13}^{tS}(R^+)R_{31}^t(R).
\end{aligned}
\tag{5.2.116}
$$

Further specifications of $\mathcal{D}_{11}$, $\mathcal{D}_{13}$, $\mathcal{D}_{31}$, and $\mathcal{D}_{33}$ will be given in Section 5.3.8. The same equation can also be used for $\mathcal{D}_{ij}(S)$ and $\mathcal{D}_{ij}(S^+)$, if we replace R and R^+ by S and S^+ in (5.2.116).

The equation for the components of radiation functions $\mathcal{G}_j^{(q)}(S; \gamma_1, \gamma_2)$ (or $\mathcal{G}_j^{(q)}(S^+; \gamma_1, \gamma_2)$) in (5.2.113) can also be simplified. The general relations for the radiation matrix, if the source is situated on interface Σ^S, are given by (5.2.75) and (5.2.76), where J is fixed, $J = 1$.

Finally, in Equations (5.2.112) through (5.2.116), $V(S)$ and $V(R)$ are specified as α or β at S or R depending on the type of the wave at S or R.

5.2.14 Initial Ray-Theory Amplitudes at a Smooth Initial Surface in a Solid Medium

To solve the problem of initial ray-theory amplitudes at a smooth initial surface Σ^0, it would be natural to use the elastic Kirchhoff integral just as we used the acoustic Kirchhoff integral in Section 5.1.11. Moreover, the elastic Kirchhoff integral itself plays a very important role in various applications and extensions of the ray method. For anisotropic media, the Kirchhoff integral will be discussed in more detail in Section 5.4.8. All results that will be derived there are also immediately applicable to isotropic media. For this reason, we shall not discuss it here. We shall present only two important equations related to the initial ray-theory amplitudes on Σ^0, assuming that an elementary P or S wave is incident at it.

We shall consider a smooth initial surface Σ^0, situated inside or on the boundary of an elastic isotropic medium. The initial surface Σ^0 may represent, among others, a structural interface, a free surface, an auxiliary surface in a smooth medium, or a wavefront of a P or an S wave. The incident wave may be of any type (P or S) and may approach the surface Σ^0 from any side.

In Section 4.5, the problem of initial conditions along Σ^0 was solved from the kinematic point of view. It was shown there that P and S waves are generated at both sides of Σ^0. The initial values of slowness vector and of dynamic ray tracing matrices $\mathbf{Q}$ and $\mathbf{P}$ of these generated waves were determined from the distribution of travel time T^0 of the incident wave along Σ^0. To simplify the treatment, we shall formally consider two vectorial components of S waves independently, as S1 and S2 waves. Both components can be combined to give the complete S wave.

We denote the ray-centered displacement amplitude of the incident elementary wave along Σ^0 by $U_i^{(q)inc}(S)$, where $i = 1$ corresponds to the S1 wave, $i = 2$ to the S2 wave, and $i = 3$ to the P wave. Then the initial ray-centered displacement amplitude $U_k^{(q)}(S)$ of any generated wave is given by the relation

$$U_k^{(q)}(S) = R_{ik}(S)U_i^{(q)inc}(S) \tag{5.2.117}$$

(no summation over i). Here $R_{ik}(S)$ is the relevant displacement R/T coefficient. It corresponds to the reflection coefficient $R_{ik}^r(S)$ if the generated wave exists on the same side of Σ^0 as the incident wave and to the transmission coefficient $R_{ik}^t(S)$ if these waves exist on opposite sides of Σ^0. Thus, the determination of the initial ray-theory amplitudes on Σ^0 is simple. It is only necessary to multiply the smooth medium amplitude of the incident wave $U_i^{(q)inc}$ by the relevant displacement R/T coefficient.

Equation (5.2.117) is simplified if Σ^0 represents an auxiliary surface in a smooth medium (not a structural interface). Then only the transmission coefficient of the unconverted transmitted wave is nonvanishing and equals unity; all other R/T coefficients vanish. The initial amplitudes of the unconverted transmitted wave are given by a simple relation:

$$U_k^{(q)}(S) = U_k^{(q)inc}(S). \tag{5.2.118}$$

Consequently, the intermediate ray-theory solutions can be stored along arbitrary reference surfaces Σ^0 and used further for computations. The procedure is the same as in the acoustic case; see Section 5.1.11. Actually, it is sufficient to store the travel time $T^0(\gamma_1, \gamma_2)$ and the smooth medium amplitude $U_i^{(q)}(\gamma_1, \gamma_2)$ of the incident wave along the surface Σ^0. We also need to know the type of the incident wave (specified by index i = 1, 2, or 3) and the side from which the incident wave approaches Σ^0. Then, it is possible to recover all six generated elementary waves.

5.2.15 Initial Ray-Theory Amplitudes at a Smooth Initial Line in a Solid Medium

Here we shall derive general initial-line ray-theory solutions for 3-D elastic laterally varying structures. The curvature and the torsion of the initial line need not vanish. Similarly, the distribution of the initial travel time along C^0 may be arbitrary. The approach is very much the same as in Section 5.1.12, only vectorial ray-theory displacement amplitudes are considered instead of scalar ray-theory pressure amplitudes. For this reason, we shall be very brief.

First, we shall discuss the representation theorem solutions for a straight initial line C^0, situated in a homogeneous medium. In the volume integral (2.6.4), the source term $f_i(\vec{x}, \omega)$ is given by the relation,

$$f_i(\vec{x}, \omega) = \delta(x_1 - x_{01})\delta(x_3 - x_{03})f_{0i}^{(x)}(x_2)\exp[i\omega T^0(x_2)]. \tag{5.2.119}$$

Here $f_{0i}^{(x)}(x_2)$ represents the ith Cartesian component of a single force $\vec{f}_0(x_2)$ at the point x_2 of the initial line C^0. The elastodynamic ray-theory Green function $G_{in}(\vec{x}', \vec{x}, \omega)$ in (2.6.4) is given by (2.5.58). Consequently, we obtain two integrals, one for P waves and one for S waves. Alternatively, we can obtain three integrals for P, S1, and S2 waves, if we use $\delta_{in} - N_i N_n = e_{1i}e_{1n} + e_{2i}e_{2n}$; see (2.5.60). Using the ray-theory solution (5.2.18) for the same model as in the representation theorem solution and matching both these solutions, we obtain the final relation for the vectorial line-source radiation functions $\mathcal{G}_i^{(q)L}(S; \gamma_1)$:

$$\mathcal{G}_i^{(q)L}(S; \gamma_1) = \lim_{S' \to S} \left(\mathcal{L}(S') U_i^{(q)}(S') \right) = \frac{f_{0k}^{(x)}(S) e_{ik}(S)}{4\pi \rho(S) V^{3/2}(S)}. \qquad (5.2.120)$$

Here $\mathcal{G}_1^{(q)L}$ corresponds to S1 waves (with $V = \beta$ and with the polarization vector $\vec{e}_1(S)$), $\mathcal{G}_2^{(q)L}$ to S2 waves (with $V = \beta$ and $\vec{e}_2(S)$), and $\mathcal{G}_3^{(q)L}$ to P waves (with $V = \alpha$ and $\vec{e}_3(S) = \vec{N}(S)$). The final ray-theory solutions for the displacement vector of an arbitrary multiply-reflected wave generated at a point S of an initial line C^0 situated in a 3-D elastic laterally varying layered structure are then

$$u_i^{(x)}(R, \omega) = U_i^{(x)}(R) F(\omega) \exp[i\omega(T^0(S) + T(R, S))], \qquad (5.2.121)$$

with the vectorial ray-theory amplitude $U_i^{(x)}(R)$ given by the relation

$$U_i^{(x)}(R) = \left[\frac{V(S)\rho(S)}{V(R)\rho(R)} \right]^{1/2} \frac{\exp[iT^c(R, S)]}{\mathcal{L}(R)} \mathcal{D}_{ik}(R) \mathcal{R}_{kj}^C \mathcal{G}_j^{(q)L}(S; \gamma_1).$$

$$(5.2.122)$$

Here $F(\omega)$ is the two-dimensional frequency filter (2.6.29), and $\mathcal{G}_j^{(q)L}(S; \gamma_1)$ is the line-source radiation function. All other symbols have the same meaning as in a similar equation (5.2.84) valid for a point source. There are only two differences between the relation (5.2.84) for a point source and the relation (5.2.122) for a line source.

a. The radiation function $\mathcal{G}_j^{(q)}(S; \gamma_1, \gamma_2)$ in (5.2.84) corresponds to a point source situated at S, but the radiation function $\mathcal{G}_j^{(q)L}(S; \gamma_1)$ in (5.2.122) corresponds to a point S situated on a line source C^0.

b. Geometrical spreading $\mathcal{L}(R)$ in (5.2.122) is replaced by the relative geometrical spreading $\mathcal{L}(R, S)$ in (5.2.84).

We remind the reader the convention for the first and last elements of the ray, formulated in Section 5.2.9.

Because (5.2.122) has the same form as (5.2.84) (only $\mathcal{L}(R, S)$ is replaced by $\mathcal{L}(R)$ and $\mathcal{G}_j^{(q)}$ by $\mathcal{G}_j^{(q)L}$), we can even use, for a line source, other relations derived for a point source. This applies, for example, to (5.2.88) for unconverted P waves, to (5.2.105) for unconverted S waves, and so on.

Let us now consider *2-D computations* with a straight line source. As in (5.1.109), we obtain

$$U_i^{(x)}(R) = \left[\frac{V^2(S)\rho(S)}{V(R)\rho(R)} \right]^{1/2} \frac{\exp[iT^c(R, S)]}{\mathcal{L}^{\parallel}(R, S)} \mathcal{D}_{ik}(R) \mathcal{R}_{kj}^C \mathcal{G}_j^{(q)L}(S; \gamma_1),$$

$$(5.2.123)$$

or, alternatively,

$$U_i^{(x)}(R) = \frac{\exp[iT^c(R, S)]}{4\pi[\rho(S)\rho(R)V(S)V(R)]^{1/2}} \frac{\mathcal{D}_{ik}(R)\mathcal{R}_{kj}^C}{\mathcal{L}^{\parallel}(R, S)} f_{0l}^{(x)}(S) e_{jl}(S).$$

$$(5.2.124)$$

In (5.2.124), we can also use $H_{lj}(S)$ instead of $e_{jl}(S)$.

The relation (5.2.123) can also be generalized to consider the line source C^0 situated on a structural interface. Then, $\mathcal{G}_j^{(q)L}(S; \gamma_1)$ can be modified as explained in Section 5.2.8. For a single-force source, (5.2.124) can be used with $e_{jl}(S)$ replaced by $\mathcal{D}_{lj}(S)$.

We shall now specify (5.2.124) for SH and P-SV waves. As usual, we consider the x_2-axis of the Cartesian coordinate system perpendicular to the plane $\Sigma^\|$ of rays, and choose $\vec{e}_2$ parallel to x_2-axis. Then, (5.2.124) yields

a. SH waves, for $f_{01}^{(x)}(S) = f_{03}^{(x)}(S) = 0$:

$$U_2^{(x)}(R) = \frac{\exp[iT^c(R,S)]}{4\pi[\rho(S)\rho(R)\beta(S)\beta(R)]^{1/2}} \frac{\mathcal{D}_{22}(R)\mathcal{R}_{SH}^C \mathcal{D}_{22}(S)}{\mathcal{L}^\|(R,S)} f_{02}^{(x)}(S).$$

(5.2.125)

b. P-SV waves, for $f_{02}^{(x)}(S) = 0$:

$$U_i^{(x)}(R) = \frac{\exp[iT^c(R,S)]}{4\pi[\rho(S)\rho(R)V(S)V(R)]^{1/2}} \frac{\mathcal{D}_{ik}(R)\mathcal{R}_{kj}^C \mathcal{D}_{lj}(S)}{\mathcal{L}^\|(R,S)} f_{0l}^{(x)}(S).$$

(5.2.126)

In (5.2.126), i, j, k, and l are fixed and take values 1 or 3. As explained in Section 5.2.13, the matrix product $\mathcal{R}_{kj}^C$ reduces to a simple scalar multiplication of successive P-SV R/T coefficients.

It is now simple to give expressions for 2-D elementary ray-theory SH and P-SV Green functions $G_{in}^{2D}(R, S, \omega)$. We merely use $f_{0l}^{(x)}(S) = \delta_{ln}$ and obtain

a. For SH waves:

$$G_{22}^{2D}(R, S, \omega) = \frac{\exp[iT^c(R,S)]}{4\pi[\rho(S)\rho(R)\beta(S)\beta(R)]^{1/2}}$$
$$\times \frac{\mathcal{D}_{22}(R)\mathcal{D}_{22}(S)\mathcal{R}_{SH}^C}{\mathcal{L}^\|(R,S)} F(\omega)\exp[i\omega T(R,S)]. \quad (5.2.127)$$

b. For P-SV waves:

$$G_{in}^{2D}(R, S, \omega) = \frac{\exp[iT^c(R,S)]}{4\pi[\rho(S)\rho(R)V(S)V(R)]^{1/2}}$$
$$\times \frac{\mathcal{D}_{ik}(R)\mathcal{R}_{kj}^C \mathcal{D}_{nj}(S)}{\mathcal{L}^\|(R,S)} F(\omega)\exp[i\omega T(R,S)]. \quad (5.2.128)$$

It is not difficult to see that the 2-D elementary ray-theory Green functions are reciprocal in the following sense: $G_{in}^{2D}(R, S, \omega) = G_{ni}^{2D}(S, R, \omega)$. In the time domain, the 2-D elementary ray-theory Green functions $G_{22}^{2D}(R, t; S, t_0)$ and $G_{in}^{2D}(R, t; S, t_0)$ are again given by (5.2.127) and (5.2.128), only we replace $F(\omega)\exp[i\omega T(R,S)]$ by $[\sqrt{2}H(\zeta)\zeta^{-1/2}]^{(A)}$, where $\zeta = t - t_0 - T(R,S)$; see (A.3.9).

5.3 Reflection/Transmission Coefficients for Elastic Isotropic Media

The equations for the computation of the vectorial complex-valued displacement amplitudes of high-frequency elastic waves propagating in laterally varying nondissipative isotropic elastic layered and block structures were derived in Section 5.2. These equations contain displacement reflection/transmission coefficients R_{mn}. In this section, we shall discuss the displacement R/T coefficients in greater detail. In addition, we shall also discuss the normalized displacement R/T coefficients $\mathcal{R}_{mn}$.

The computation of R/T coefficients at a plane interface between two solids is a classical problem in seismology. Knott (1899) and Zöppritz (1919) were the first who published analytical expressions for R/T coefficients and gave some numerical examples. This is also the reason why the R/T coefficients are sometimes called the *Zöppritz coefficients* in the seismological literature. A large number of papers and books devoted to R/T coefficients have been published since. Let us name here only several of them: Muskat and Meres (1940), Nafe (1957), Vasil'yev (1959), Podyapol'skiy (1959), Bortfeld (1961), McCamy, Meyer, and Smith (1962), Koefoed (1962), Tooley, Spencer, and Sagoci (1965), Yanovskaya (1966), Červený and Ravindra (1971), Aki and Richards (1980), Schoenberg and Protázio (1992), and Borejko (1996). Even though the determination of the analytical expressions for R/T coefficients is not complicated, various conflicting expressions have been given. The problem is that the authors do not sometimes specify properly the conditions under which their computations of R/T coefficients have been performed. Moreover, many misprints and/or errors have appeared in the final expressions. Thus, the application of published equations, algorithms, and numerical results often becomes confusing.

In general, the R/T coefficients represent the ratios of certain quantities related to the amplitudes of generated waves to the amplitude of the incident wave. They may be introduced in many ways. The most common is to consider the *displacement R/T coefficients* introduced in Section 2.3 and used in Section 5.2. We remind the reader that the displacement R/T coefficients represent the ratios of ray-centered components of the displacement vector of generated R/T waves to the ray-centered components of the displacement vector of the incident wave. Alternatively, *potential R/T coefficients*, which are mostly based on the Lamé's potentials have also been used often. The potential R/T coefficients represent the ratios of the potential of any of the generated R/T waves to the potential of the incident wave. In the seismic ray method, we work mostly with displacements, not with potentials. For this reason, the potential R/T coefficients have not been used practically in the ray method and will not be used here at all either. Also the *energy R/T coefficients* have sometimes been used to demonstrate the partition of energy into individual generated waves at an interface. The energy R/T coefficients represent the ratios of the energy density of any of the generated R/T waves to the energy density of the incident wave. Even though the energy R/T coefficients offer a useful insight into the partition of the energy among the R/T waves generated at the interface, they have not found applications in the numerical modeling of seismic body wavefields. They do not yield any information on the argument of the complex-valued R/T coefficients. For this reason, they cannot be used in the computation of the complex-valued displacement vector along the ray.

In general, the displacement and potential R/T coefficients can simply be mutually recalculated, without loss of information. Similarly, the energy R/T coefficients may be evaluated simply from the displacement or potential R/T coefficients (but not vice versa).

As stated earlier, we shall use mainly the displacement R/T coefficients here. A disadvantage of the displacement R/T coefficients is that they are not in general reciprocal. (Note that the potential R/T coefficients are not reciprocal either.) In addition to the displacement R/T coefficients, we shall also use the *normalized displacement R/T coefficients* (or simply the *normalized R/T coefficients*). The normalized displacement R/T coefficients represent the displacement R/T coefficients normalized with respect to the energy flux across the interface; see Section 5.3.3. The normalized displacement R/T coefficients are reciprocal and are very useful in the ray method. They include the same phase information as the displacement R/T coefficients.

R/T coefficients have been broadly used in seismology in two different methods.

   a. In the ray method, to evaluate the wavefield of *selected elementary wave* propagating in complex laterally varying layered structures.

b. In matrix methods, to evaluate the *complete wavefield* in 1-D structures (for example, in the reflectivity method).

In ray applications, we usually apply individual R/T coefficients at any point of incidence, according to the ray code of the wave under consideration. In matrix methods, however, the R/T coefficients are grouped into R/T matrices. These matrices may be of different order in different applications (2×2, 4×4, 6×6, and so on). In the standard P-SV case, the 4×4 matrices are composed of four 2×2 matrices that correspond to reflection and transmission coefficients for "downgoing" and "upgoing" waves. Analogously, the 6×6 matrices consist of similar 3×3 reflection and transmission matrices, corresponding to the downgoing and upgoing waves. For a detailed treatment of R/T matrices in isotropic media, see Kennett (1983). For anisotropic media, see Section 5.4.7.

The general matrices, introduced in the reflectivity method, have, however, no direct application in the ray method because we do not need to consider all upgoing and downgoing waves for a selected elementary wave; the ray code for the elementary wave strictly specifies the proper R/T coefficient at any point of incidence. Without loss of generality, we can consider only downgoing R/T coefficients because the structure may be treated locally at the point of incidence; the "upper" medium coinciding with the medium with the incident wave.

Nevertheless, the 3×3 R/T *matrices* are very useful in the numerical modeling of seismic wavefields in 3-D isotropic elastic structures using the ray method, if we are also interested in S and converted waves. If the displacement vector of the incident S wave is polarized in the plane of incidence or perpendicular to it at all points of incidence, the systems of equations for the R/T coefficients are decomposed and the application of the 3×3 R/T matrices is not required. This includes all 1-D and 2-D ray tracing applications; see Section 5.2.13. The situation is, however, quite different *in 3-D models*. In general, the displacement vectors of the S wave are polarized arbitrarily at the individual points of incidence. The application of 3×3 R/T matrices is very useful in this case because it allows us to write *simple compact expressions* for the vectorial complex-valued amplitudes. These expressions become even simpler if we use the *normalized displacement R/T matrices* instead of standard displacement R/T matrices. See the more detailed treatment in Sections 5.2.4, 5.2.5, 5.3.5, and 5.3.6.

Let us add one remark. Displacement R/T coefficients depend, of course, on the orientation of the basis vectors of the ray-centered basis vectors $\vec{e}_1$, $\vec{e}_2$, and $\vec{e}_3 \equiv \vec{t}$, both for incident and R/T waves. Basis vectors $\vec{e}_1$ and $\vec{e}_2$, however, may be introduced in different ways; only unit vector $\vec{e}_3 \equiv \vec{t}$ is strictly specified for each wave.

In the computation of R/T coefficients, the unit vectors $\vec{e}_1$ and $\vec{e}_2$, corresponding to the incident wave and to the generated R/T waves, may be chosen arbitrarily. The resulting R/T coefficients, however, depend on this choice. This can be simply checked in system (2.3.37), which contains e^i_{Ij}, e^r_{Ij}, and e^t_{Ij} ($I = 1, 2; j = 1, 2, 3$). For any choice of e^i_{Ij}, e^r_{Ij}, and e^t_{Ij}, we obtain the relevant R/T coefficients. The same applies to the P-SV system (2.3.42), which contains e^i_{11}, e^i_{13}, e^r_{11}, e^r_{13}, e^t_{11}, and e^t_{13}. Corresponding R/T coefficients are obtained for any choice of these components of the basis vectors. To remove possible ambiguity, we shall specify the orientation of the unit vectors of $\vec{e}_1$ and $\vec{e}_2$ of all generated waves using the standard option (2.3.45).

Let us now return to the actual computation of the displacement R/T coefficients. The relevant systems of equations for the diplacement R/T coefficients are given in Section 2.3. In this section, we shall use the notations and equations of Section 2.3.2. We remind the reader that the velocities of P and S waves and the densities are denoted α_1, β_1, ρ_1 in the halfspace 1 and $\alpha_2, \beta_2, \rho_2$ in the halfspace 2. We also assume that the wave is incident at interface Σ from halfspace 1 and that normal $\vec{n}$ to interface Σ at the point of incidence Q is specified. It may be oriented to either side of interface Σ, that is, into halfspace 1 or into halfspace 2. We distinguish these two cases by *orientation index* ϵ,

$$\epsilon = \text{sgn}(\vec{p} \cdot \vec{n}), \tag{5.3.1}$$

where $\vec{p}$ is the slowness vector of the incident wave at Q; see (2.3.4). More details on the orientation index ϵ will be given in Section 5.3.2.

We shall first present the R/T coefficients for the case that the unit vector $\vec{e}_2$ of the incident wave is perpendicular to the plane of incidence; see Section 5.3.1. As is common in seismology, we shall call these R/T coefficients the P-SV and SH R/T coefficients. Only in Section 5.3.5, shall we consider the general case of $\vec{e}_2$ arbitrarily rotated along the ray of the incident wave.

As in Sections 2.3 and 5.2, we shall use the following notation for the displacement R/T coefficients. By R_{mn}^r we shall understand the reflection coefficient, where m specifies the type of incident wave and n specifies the type of reflected wave. Similarly, we denote the transmission coefficient by R_{mn}^t, where m specifies the type of the incident wave and n the type of transmitted wave. Index m is determined as follows:

> $m = 1$, S1 component of the incident S wave (polarized in the direction of $\vec{e}_1$)
> $m = 2$, S2 component of the incident S wave (polarized in the direction of $\vec{e}_2$)
> $m = 3$, P incident wave (polarized in the direction of $\vec{e}_3 \equiv \vec{t}$).

Index n is determined in the same way as m but corresponds to the selected reflected/transmitted wave.

5.3.1 P-SV and SH Reflection/Transmission Coefficients

In this section, we shall present the analytical expressions for the displacement R/T coefficients of P-SV and SH types. The basic assumption under which these "decomposed" R/T coefficients are derived is that the basis vector $\vec{e}_2$ corresponding to the incident wave is perpendicular to the plane of incidence at the point of incidence Q. The simplest choice for the basis vectors $\vec{e}_2$ of all generated R/T waves is to assume that they coincide with $\vec{e}_2$ for the incident wave. The system of six linear equations (2.3.37) then decomposes into two subsystems, (2.3.42) and (2.3.43). The first subsystem (2.3.42) consists of four linear equations for the P-SV R/T coefficients. The second system (2.3.43) consists of two linear equations for the SH R/T coefficients.

Before we give the equations for the P-SV and SH R/T coefficients, we shall make several remarks concerning the *local Cartesian coordinate system at the point of incidence* Q and the *orientation of $\vec{e}_1$ and $\vec{e}_2$ at Q*. In fact, we do not need to introduce the local Cartesian coordinate system at the point of incidence at all to evaluate the R/T coefficients, but it is often introduced in the ray method. Actually, we introduced the local Cartesian coordinate system at Q and the orientation of the basis vectors $\vec{e}_1$ and $\vec{e}_2$ at Q in Section 4.4.1, in connection with the dynamic ray tracing across an interface. To be consistent, we shall use the same options as in the dynamic ray tracing: the standard options (4.4.21) and (2.3.46). We remind the reader that the z_3-axis (specified by basis vector $\vec{i}_3^{(z)}$) is taken along normal $\vec{n}$ to Σ at Q. The normal $\vec{n}$ to Σ at Q may be oriented to either side of interface. Axis z_1 (basis vector $\vec{i}_1^{(z)}$) is taken along the intersection of the plane of incidence with the tangent plane to interface Σ at Q. The positive orientation of $\vec{i}_1^{(z)}$ is such that $\vec{p} \cdot \vec{i}_1^{(z)} > 0$, where $\vec{p}$ is the slowness vector of the incident wave. Thus, the positive z_1-axis points in the direction of propagation of the incident wave. Finally, the basis vectors $\vec{e}_2$ of all waves

Figure 5.5. The physical meaning of P-SV and SH reflection coefficients R_{33}, R_{31}, R_{13}, R_{11}, and R_{22}. All other R/T coefficients (R_{12}, R_{21}, R_{23}, and R_{32}) vanish. The orientation of the SV polarization vector $\vec{e}_1$ for $\epsilon = 1$ is also displayed by small arrows. For $\epsilon = -1$, the polarization vectors $\vec{e}_1$ would be opposite. (a) Reflection coefficients. (b) Transmission coefficients.

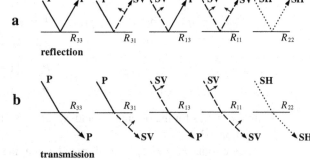

under consideration coincide with $\vec{i}_2^{(z)}$, and $\vec{e}_1 = \vec{e}_2 \times \vec{e}_3$. The calculated P-SV and SH R/T coefficients, however, do not depend on the local Cartesian coordinate system, they depend only on model parameters α_1, β_1, ρ_1, α_2, β_2, and ρ_2, on ray parameter $p = (\sin i)/V$ and on orientation index ϵ.

Figure 5.5 shows schematically the physical meaning of the individual nonvanishing R/T coefficients in the P-SV and SH case. We remind the reader that, in this case, $R_{12} = R_{21} = R_{23} = R_{32} = 0$ so that the only nonvanishing displacement R/T coefficients are $R_{11}(\equiv R_{SV \to SV})$, $R_{13}(\equiv R_{SV \to P})$, $R_{31}(\equiv R_{P \to SV})$, $R_{33}(\equiv R_{P \to P})$, and $R_{22}(\equiv R_{SH \to SH})$. Figure 5.5 also shows the orientation of the basis vectors $\vec{e}_1$ and $\vec{e}_3$ for $\epsilon = 1$. The unit vectors $\vec{e}_2$ of all waves are perpendicular to the plane of incidence and are pointing toward the reader. The unit vectors $\vec{e}_3$ have the same directions as the slowness vectors; see large arrows. Finally, the unit vectors $\vec{e}_1 = \vec{e}_2 \times \vec{e}_3$ are shown by small arrows. The basis vectors $\vec{e}_1$, $\vec{e}_2$, and $\vec{e}_3$ form a mutually perpendicular triplet of unit vectors for each wave.

The analytical expressions for R/T coefficients R_{11}, R_{13}, R_{31}, R_{33}, and R_{22} are as follows:

a. Displacement reflection coefficients:

$$R_{11} = D^{-1}[q^2 p^2 P_1 P_2 P_3 P_4 + \rho_1 \rho_2 (\alpha_1 \beta_2 P_2 P_3 - \beta_1 \alpha_2 P_1 P_4)$$
$$- \alpha_1 \beta_1 P_3 P_4 Y^2 + \alpha_2 \beta_2 P_1 P_2 X^2 - \alpha_1 \alpha_2 \beta_1 \beta_2 p^2 Z^2],$$
$$R_{13} = -2\epsilon \beta_1 p P_2 D^{-1}(q P_3 P_4 Y + \alpha_2 \beta_2 X Z),$$
$$R_{31} = 2\epsilon \alpha_1 p P_1 D^{-1}(q P_3 P_4 Y + \alpha_2 \beta_2 X Z), \qquad (5.3.2)$$
$$R_{33} = D^{-1}[q^2 p^2 P_1 P_2 P_3 P_4 + \rho_1 \rho_2 (\beta_1 \alpha_2 P_1 P_4 - \alpha_1 \beta_2 P_2 P_3)$$
$$- \alpha_1 \beta_1 P_3 P_4 Y^2 + \alpha_2 \beta_2 P_1 P_2 X^2 - \alpha_1 \alpha_2 \beta_1 \beta_2 p^2 Z^2],$$
$$R_{22} = \bar{D}^{-1}(\rho_1 \beta_1 P_2 - \rho_2 \beta_2 P_4).$$

b. Displacement transmission coefficients:

$$R_{11} = 2\beta_1 \rho_1 P_2 D^{-1}(\alpha_1 P_3 Y + \alpha_2 P_1 X),$$
$$R_{13} = 2\epsilon \beta_1 \rho_1 p P_2 D^{-1}(q P_1 P_4 - \alpha_1 \beta_2 Z),$$
$$R_{31} = -2\epsilon \alpha_1 \rho_1 p P_1 D^{-1}(q P_2 P_3 - \beta_1 \alpha_2 Z), \qquad (5.3.3)$$
$$R_{33} = 2\alpha_1 \rho_1 P_1 D^{-1}(\beta_2 P_2 X + \beta_1 P_4 Y),$$
$$R_{22} = 2\rho_1 \beta_1 P_2 \bar{D}^{-1}.$$

Here we have used the notation:

$$D = q^2 p^2 P_1 P_2 P_3 P_4 + \rho_1 \rho_2 (\beta_1 \alpha_2 P_1 P_4 + \alpha_1 \beta_2 P_2 P_3)$$
$$+ \alpha_1 \beta_1 P_3 P_4 Y^2 + \alpha_2 \beta_2 P_1 P_2 X^2 + \alpha_1 \alpha_2 \beta_1 \beta_2 p^2 Z^2, \qquad (5.3.4)$$
$$\bar{D} = \rho_1 \beta_1 P_2 + \rho_2 \beta_2 P_4,$$

and

$$q = 2(\rho_2\beta_2^2 - \rho_1\beta_1^2), \qquad X = \rho_2 - qp^2,$$
$$Y = \rho_1 + qp^2, \qquad Z = \rho_2 - \rho_1 - qp^2,$$
$$P_1 = (1 - \alpha_1^2 p^2)^{1/2}, \qquad P_2 = (1 - \beta_1^2 p^2)^{1/2}, \qquad (5.3.5)$$
$$P_3 = (1 - \alpha_2^2 p^2)^{1/2}, \qquad P_4 = (1 - \beta_2^2 p^2)^{1/2}.$$

Symbol ϵ stands for the orientation index; see (5.3.1). For a more detailed discussion of ϵ, see Section 5.3.2.

Square roots P_i, $i = 1$, 2, 3, 4, may be imaginary. The sign of the imaginary square root is taken positive, as in (5.1.20):

$$P_1 = i(\alpha_1^2 p^2 - 1)^{1/2} \qquad \text{for } p > 1/\alpha_1,$$
$$P_2 = i(\beta_1^2 p^2 - 1)^{1/2} \qquad \text{for } p > 1/\beta_1,$$
$$P_3 = i(\alpha_2^2 p^2 - 1)^{1/2} \qquad \text{for } p > 1/\alpha_2, \qquad (5.3.6)$$
$$P_4 = i(\beta_2^2 p^2 - 1)^{1/2} \qquad \text{for } p > 1/\beta_2.$$

The plus signs in the expressions for P_k in (5.3.6) again correspond to the plus sign in expression (2.2.9) for the analytical signal, $F(\zeta) = x(\zeta) + ig(\zeta)$. Had we defined the analytical signal as $F(\zeta) = x(\zeta) - ig(\zeta)$, it would have been necessary to replace i by $-i$ in (5.3.6). The choice (5.3.6) guarantees that the amplitudes of generated inhomogeneous waves decrease exponentially with the increasing distance from the interface.

Expressions (5.3.2) also apply to **displacement reflection coefficients from the Earth's surface**; we only need to put $\rho_2 = \alpha_2 = \beta_2 = 0$. The explicit relations for the displacement reflection coefficients from the Earth's surface are

$$R_{11} = D_1^{-1}[-(1 - 2\beta_1^2 p^2)^2 + 4p^2 P_1 P_2 \beta_1^3 \alpha_1^{-1}],$$
$$R_{13} = 4\epsilon p\beta_1^2 \alpha_1^{-1} P_2 D_1^{-1}(1 - 2\beta_1^2 p^2),$$
$$R_{31} = -4\epsilon p\beta_1 P_1 D_1^{-1}(1 - 2\beta_1^2 p^2), \qquad (5.3.7)$$
$$R_{33} = D_1^{-1}[-(1 - 2\beta_1^2 p^2)^2 + 4p^2 P_1 P_2 \beta_1^3 \alpha_1^{-1}],$$
$$R_{22} = 1.$$

In the same way, it is possible to obtain from (5.3.3) the **displacement transmission coefficients at the Earth's surface**. They have a formal meaning only but can be suitably used to evaluate the conversion coefficients at the Earth's surface; see Section 5.3.8. They are

$$R_{11} = 2P_2(1 - 2\beta_1^2 p^2)/D_1,$$
$$R_{13} = -4\epsilon\beta_1^2 p P_1 P_2/\alpha_1 D_1,$$
$$R_{31} = 4\epsilon\beta_1 p P_1 P_2/D_1, \qquad (5.3.8)$$
$$R_{33} = 2P_1(1 - 2\beta_1^2 p^2)/D_1,$$
$$R_{22} = 2.$$

In (5.3.7) and (5.3.8), D_1 is the so-called *Rayleigh function*,

$$D_1 = (1 - 2\beta_1^2 p^2)^2 + 4p^2 P_1 P_2 \beta_1^3 \alpha_1^{-1}. \qquad (5.3.9)$$

All other notations are the same as in (5.3.5).

Similarly, expressions (5.3.2) and (5.3.3) can also be used if one of the halfspaces is fluid. We only need to put $\beta_1 = 0$ or $\beta_2 = 0$ in the relevant fluid halfspace. Because S waves do not propagate in fluids, the corresponding R/T coefficients with S elements only have a formal meaning in the fluid halfspace. If both halfspaces are fluid, expressions (5.3.2) and (5.3.3) for R_{33} are indefinite, of the 0/0 type. The simple limiting process $\beta_1 = \beta_2 \to 0$, however, yields Equation (5.2.91), known from the acoustic case.

It should be emphasized that the choice of polarization vectors $\vec{e}_1$, $\vec{e}_2$, and $\vec{e}_3$ used here may be convenient in some applications. For example, it is fully consistent with the choice that has been used in dynamic ray tracing. However, any other choice is as good as ours and may be suitably used in some other computational systems. In fact, if the orientation of the unit vectors $\vec{e}_1$, $\vec{e}_2$, and $\vec{e}_3$ is arbitrarily changed, Equation (5.3.2) through (5.3.6) may be again used; it is only necessary to change the signs of relevant R/T coefficients appropriately.

5.3.2 Orientation Index ϵ

To calculate the P-SV R/T coefficients of converted waves SV $\to$ P and P $\to$ SV (R_{13} and R_{31}), we must know orientation index $\epsilon = \text{sgn}(\vec{p} \cdot \vec{n})$, where $\vec{p}$ is the slowness vector of the incident wave, and $\vec{n}$ is the unit normal to interface Σ at Q. This is not surprising because orientation index ϵ also affects the orientation of basis vectors $\vec{e}_1$ and $\vec{e}_2$. To prove this, we shall follow the construction of the local Cartesian coordinate system z_1, z_2, z_3 at Q and the specification of $\vec{e}_1$ and $\vec{e}_2$ as described in Section 5.3.1.

We have introduced the local Cartesian coordinate system at Q so that $\vec{i}_3^{(z)} \equiv \vec{n}$ and so that $\vec{i}_1^{(z)}$ is taken along the intersection of the plane of incidence with the plane tangent to Σ at Q, with $\vec{p} \cdot \vec{i}_1^{(z)} > 0$. The remaining basis vector $\vec{i}_2^{(z)}$ is then given by the relation $\vec{i}_2^{(z)} = \vec{i}_3^{(z)} \times \vec{i}_1^{(z)}$. Thus, the orientation of axis z_1 does not depend on ϵ, but $\vec{i}_2^{(z)}$ and $\vec{i}_3^{(z)}$ change their sign if ϵ changes its sign.

Because $\vec{e}_2 \equiv \vec{i}_2^{(z)}$ (by definition), both $\vec{e}_1$ and $\vec{e}_2$ also change their sign if ϵ changes its sign. See Figure 5.5, where the orientation of $\vec{e}_1$ is shown for $\epsilon = 1$. For $\epsilon = -1$, unit vectors $\vec{e}_1$ would point in the opposite direction.

It should again be emphasized that only the R/T coefficients of converted waves P $\to$ SV and SV $\to$ P depend on orientation index ϵ. No other R/T coefficients (R_{11}, R_{33}, R_{22}) depend on it.

5.3.3 Normalized Displacement P-SV and SH Reflection/Transmission Coefficients

The equations for amplitudes of seismic body waves propagating in layered structures are simplified if normalized displacement R/T coefficients $\mathcal{R}_{mn}$ are used instead of standard displacement R/T coefficients R_{mn}. Moreover, the normalized R/T coefficients $\mathcal{R}_{mn}$ have certain remarkable reciprocity properties. The normalized displacement P-SV and SH reflection/transmission coefficients are given by the relations

$$\mathcal{R}_{mn} = R_{mn} \left(\frac{V(\tilde{Q})\rho(\tilde{Q})P(\tilde{Q})}{V(Q)\rho(Q)P(Q)} \right)^{1/2}. \tag{5.3.10}$$

Here Q denotes the point of incidence; $\tilde{Q}$ denotes the point of reflection/transmission; $V(Q)$, $\rho(Q)$, and $P(Q)$ correspond to the incident wave; and $V(\tilde{Q})$, $\rho(\tilde{Q})$, and $P(\tilde{Q})$ correspond to the selected R/T wave. Moreover, ρ is density (ρ_1 or ρ_2), and V is velocity

(α_1, β_1, α_2, or β_2) depending on the type of the wave. Finally, $P(Q) = (1 - V^2(Q)p^2)^{1/2}$ and $P(\tilde{Q}) = (1 - V^2(\tilde{Q})p^2)^{1/2}$. Thus, $P(Q)$ and $P(\tilde{Q})$ may be any of P_1, P_2, P_3, or P_4, defined by (5.3.5). See also (5.2.43). Because we do not consider inhomogeneous waves here, $P(Q)$ and $P(\tilde{Q})$ are always real-valued and positive, and the whole normalization factor in (5.3.10) is always real-valued and positive. The R/T coefficients $\mathcal{R}_{mn}$ and R_{mn} themselves, however, may be complex-valued because they contain other square roots P_i (see (5.3.5)), and some of these square roots may be complex-valued. A typical example is the postcritically reflected wave. See Section 5.3.4.

To explain the physical meaning of the normalization factor, we shall consider the energy fluxes of incident and R/T plane waves across interface Σ at Q, perpendicular to Σ (along normal $\vec{n}$ to Σ). As in Section 2.2.7, we denote the Cartesian components of the energy flux $\hat{S}_i$. The energy flux along $\vec{n}$ is then $\hat{S}_i n_i$. Let us consider, for simplicity, that the SH component of the incident wave vanishes and assume that the amplitude of the incident wave (P or SV) equals unity. Then we obtain

$$|\hat{S}_i n_i|_{inc} = \rho(Q)V(Q)P(Q)f_c$$

for the incident wave (see (2.4.57) and (2.4.58)) and

$$|\hat{S}_i n_i|_{R/T} = \rho(\tilde{Q})V(\tilde{Q})P(\tilde{Q})RR^* f_c$$

for a selected R/T wave. Here R is the appropriate displacement R/T coefficient. The ratio of both energy fluxes is

$$\left| \frac{(\hat{S}_i n_i)_{R/T}}{(\hat{S}_i n_i)_{inc}} \right| = \frac{\rho(\tilde{Q})V(\tilde{Q})P(\tilde{Q})}{\rho(Q)V(Q)P(Q)} RR^* = \mathcal{R}\mathcal{R}^*. \qquad (5.3.11)$$

This yields

$$|\mathcal{R}| = |(\hat{S}_i n_i)_{R/T}/(\hat{S}_i n_i)_{inc}|^{1/2}. \qquad (5.3.12)$$

The physical meaning of $\mathcal{R}$ is obvious from (5.3.12). The *modulus of the normalized displacement R/T coefficient* $|\mathcal{R}|$ represents the square root of the absolute value of the ratio of the energy flux of the appropriate R/T wave to the energy flux of the incident wave, both taken along normal $\vec{n}$ to interface Σ at point Q. The argument of $\mathcal{R}$, however, is the same as the argument of the standard displacement R/T coefficient R.

For the reader's convenience, we shall give the expressions for the P-SV and SH normalized displacement R/T coefficients:

a. Reflection coefficients:

$$\mathcal{R}_{11} = R_{11}, \qquad \mathcal{R}_{22} = R_{22}, \qquad \mathcal{R}_{33} = R_{33},$$
$$\mathcal{R}_{13} = -2\epsilon p(\beta_1\alpha_1 P_1 P_2)^{1/2} D^{-1}(q P_3 P_4 Y + \beta_2\alpha_2 XZ), \qquad (5.3.13)$$
$$\mathcal{R}_{31} = 2\epsilon p(\beta_1\alpha_1 P_1 P_2)^{1/2} D^{-1}(q P_3 P_4 Y + \alpha_2\beta_2 XZ).$$

b. Transmission coefficients:

$$\mathcal{R}_{11} = 2(\beta_1\beta_2\rho_1\rho_2 P_2 P_4)^{1/2} D^{-1}(\alpha_1 P_3 Y + \alpha_2 P_1 X),$$
$$\mathcal{R}_{13} = 2\epsilon p(\beta_1\alpha_2\rho_1\rho_2 P_2 P_3)^{1/2} D^{-1}(q P_1 P_4 - \alpha_1\beta_2 Z),$$
$$\mathcal{R}_{31} = -2\epsilon p(\alpha_1\beta_2\rho_1\rho_2 P_1 P_4)^{1/2} D^{-1}(q P_2 P_3 - \beta_1\alpha_2 Z), \qquad (5.3.14)$$
$$\mathcal{R}_{33} = 2(\alpha_1\alpha_2\rho_1\rho_2 P_1 P_3)^{1/2} D^{-1}(\beta_2 P_2 X + \beta_1 P_4 Y),$$
$$\mathcal{R}_{22} = 2(\beta_1\beta_2\rho_1\rho_2 P_2 P_4)^{1/2} \bar{D}^{-1}.$$

All the symbols have the same meaning as in (5.3.2) through (5.3.6).

For a free surface, the expressions for the P-SV and SH normalized displacement reflection coefficients are

$$\mathcal{R}_{11} = R_{11}, \qquad \mathcal{R}_{22} = R_{22}, \qquad \mathcal{R}_{33} = R_{33},$$
$$\mathcal{R}_{13} = 4p\epsilon\beta_1(\beta_1 P_1 P_2/\alpha_1)^{1/2} D_1^{-1}(1 - 2\beta_1^2 p^2),$$
$$\mathcal{R}_{31} = -4p\epsilon\beta_1(\beta_1 P_1 P_2/\alpha_1)^{1/2} D_1^{-1}(1 - 2\beta_1^2 p^2).$$

$$(5.3.15)$$

D_1 is given by (5.3.9).

5.3.4 Displacement P-SV and SH R/T Coefficients: Discussion

It would not be simple to discuss all the P-SV and SH R/T coefficients for isotropic solid media in greater detail. The number of coefficients is rather high: five for reflections and five for transmissions. The coefficients are, in general, complex-valued quantities so that each coefficient is represented by two quantities: by the modulus and by the argument. The coefficients depend on six medium parameters $\alpha_1, \beta_1, \rho_1$ and $\alpha_2, \beta_2, \rho_2$, and on the angle of incidence i (or, alternatively, on the ray parameter p). Even though the number of medium parameters may be reduced to four by considering various ratios (for example, $\alpha_1/\alpha_2, \beta_1/\alpha_1, \beta_2/\alpha_2$, and ρ_1/ρ_2), the number of medium parameters still remains too high for a detailed parameteric study. Moreover, in many situations, the dependence of certain R/T coefficients on the angle of incidence is very complicated.

For this reason, we shall discuss the P-SV and SH R/T coefficients only very briefly. At present, good computer programs for computing the P-SV and SH R/T coefficients are available at most seismological institutions so that the readers may easily undertake this study themselves.

We shall be mainly interested in the behavior of P-SV and SH R/T coefficients for certain important situations. This applies mainly to the following situations: (1) normal incidence, (2) critical angles of incidence, and (3) Brewster angles of incidence. We shall also investigate the regions of angle of incidence in which the R/T coefficients are real-valued and in which they are complex-valued.

To obtain at least a rough idea of the behavior of P-SV and SH R/T coefficients, we shall present the moduli and arguments of these coefficients for two typical examples. The first example concerns a typical structural interface between two elastic halfspaces, with a weak velocity contrast (index of refraction = 0.8). See Figure 5.6. The second example concerns the free surface of an elastic halfspace such as the surface of the Earth. See Figure 5.7. In Figures 5.6 and 5.7, a classical alphanumerical seismological notation for P-SV R/T coefficients is used. The notation is self-explanatory and consists of a combination of letters P and S and numbers 1 (the first halfspace, with the incident wave) and 2 (the second halfspace). The relation of this notation to R_{ij} is as follows:

    a. For reflected waves, P1P1(R_{33}), P1S1(R_{31}), S1P1(R_{13}), and S1S1(R_{11}).
    b. For transmitted waves, P1P2(R_{33}), P1S2(R_{31}), S1P2(R_{13}), and S1S2(R_{11}).

In an analogous way, we shall also speak of reflected waves P1P1, P1S1, and the like.

    The first example corresponds to the following medium parameters: $\alpha_1 = 6{,}400$ m/s, $\beta_1 = \alpha_1/\sqrt{3} = 3{,}698$ m/s, $\rho_1 = 2{,}980$ kg/m^3, $\alpha_2 = 8{,}000$ m/s, $\beta_2 = \alpha_2/\sqrt{3} = 4{,}618$ m/s, and $\rho_2 = 3{,}300$ kg/m^3. Note that refraction index $n = \alpha_1/\alpha_2$ equals 0.8 in this case. Thus, we consider an interface with a positive, but only small increase of velocity ($\alpha_1/\alpha_2 = 0.8$). Such interfaces are very common in the Earth's interior, both in seismology

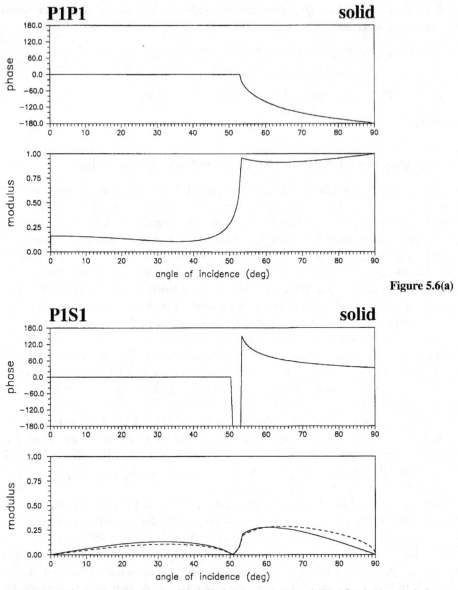

Figure 5.6. Displacement and normalized displacement P-SV and SH reflection/transmission coefficients at a plane interface between two homogeneous isotropic solid media. Model: $\alpha_1 = 6,400$ m/s, $\beta_1 = 3,698$ m/s, $\rho_1 = 2,980$ kg/m^3, $\alpha_2 = 8,000$ m/s, $\beta_2 = 4,618$ m/s, $\rho_2 = 3,300$ kg/m^3. Continuous lines: displacement R/T coefficients; dashed lines: normalized displacement R/T coefficients. Orientation index $\epsilon = 1$. (a) P1P1 (R_{33}^r) and P1S1 (R_{31}^r) reflection coefficients. (b) S1P1 (R_{13}^r) and S1S1 (R_{11}^r) reflection coefficients. (c) P1P2 (R_{33}^t) and P1S2 (R_{31}^t) transmission coefficients. (d) S1P2 (R_{13}^t) and S1S2 (R_{11}^t) transmission coefficients. (e) SH reflection (R_{22}^r) and transmission (R_{22}^t) coefficients. For $\epsilon = -1$, the signs of P1S1, S1P1, P1S2, and S1P2 coefficients would be opposite.

and in seismic exploration. In crustal seismology, this ratio corresponds roughly to the conditions at the Mohorovičić discontinuity. The same R/T coefficients would be obtained for many other values of α_1, β_1, ρ_1, α_2, β_2, and ρ_2 because the R/T coefficients depend on ratios α_1/α_2, β_1/α_1, β_2/α_2, and ρ_1/ρ_2. Thus, the medium parameters for which the

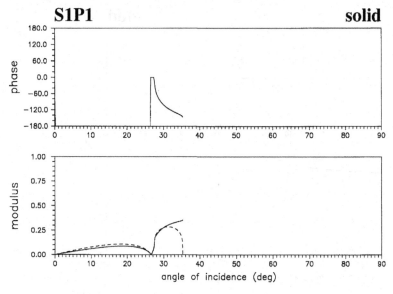

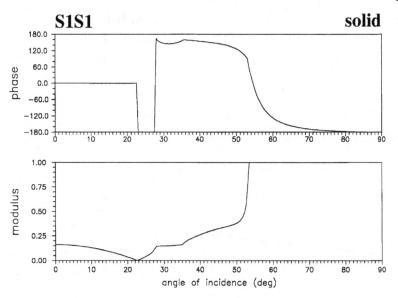

Figure 5.6(b)

R/T coefficients given in Figure 5.6 are computed may also be chosen, for example, as follows: $\alpha_1 = 2,000$ m/s, $\beta_1 = \alpha_1/\sqrt{3} = 1,155$ m/s, $\rho_1 = 1,500$ kg/m³, $\alpha_2 = 2,500$ m/s, $\beta_2 = \alpha_2/\sqrt{3} = 1,443$ m/s, and $\rho_2 = 1,661$ kg/m³. These medium parameters may be more typical for shallow structures in seismic exploration. Both the modulus and the phase of the relevant R/T coefficient are shown in all cases. The normalized displacement R/T coefficients $\mathcal{R}_{mn}$ are shown as dashed lines, if they differ from the standard displacement R/T coefficients R_{mn}. Because the inhomogeneous incident and/or R/T waves are not considered, only R/T coefficients for real-valued angles of incidence and real-valued angles of reflection/transmission of the wave under consideration are shown.

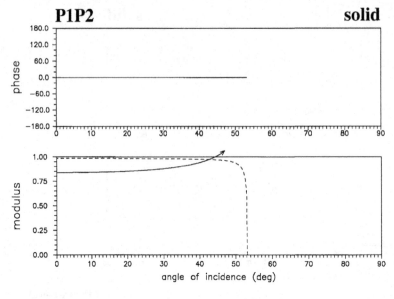

Figure 5.6(c)

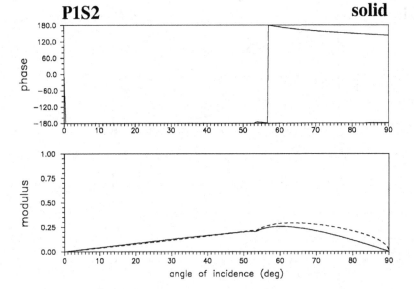

1. DISPLACEMENT R/T COEFFICIENTS FOR NORMAL INCIDENCE

For normal incidence (ray parameter $p = 0$), the R/T coefficients of converted waves (R_{13}, R_{31}) vanish, both for reflections and transmission. Only the R/T coefficients of unconverted waves are nonvanishing. To express them by simple equations, it is useful to introduce wave impedances Z^P and Z^S for P and S waves:

$$Z^P = \rho\alpha, \qquad Z^S = \rho\beta. \tag{5.3.16}$$

We also denote $Z_1^P = \rho_1\alpha_1$, $Z_2^P = \rho_2\alpha_2$, $Z_1^S = \rho_1\beta_1$, and $Z_2^S = \rho_2\beta_2$. The *reflection*

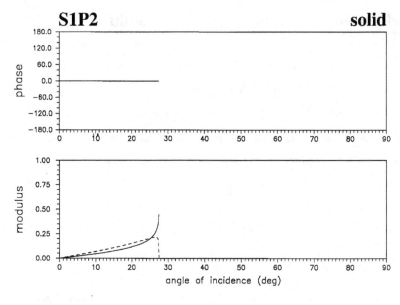

Figure 5.6(d)

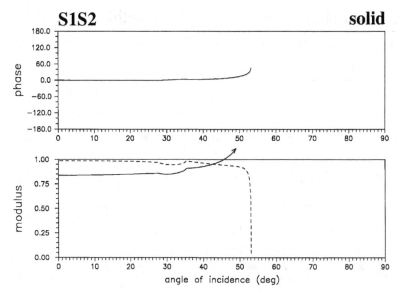

coefficients R_{11}, R_{22}, and R_{33} for normal incidence then read

$$R_{11} = -R_{22} = \left(Z_2^S - Z_1^S\right)/\left(Z_2^S + Z_1^S\right),$$
$$R_{33} = \left(Z_2^P - Z_1^P\right)/\left(Z_2^P + Z_1^P\right). \tag{5.3.17}$$

Thus, for normal incidence, the reflection coefficient R_{33} of the P1P1 reflected wave is given exactly by the same relation as the acoustic (pressure) reflection coefficient, only c must be replaced by α; see (5.1.27). The reflection coefficients for S waves are given by similar expressions, but the wave impedances for S waves must be considered. It may be surprising, to some extent, that the signs of the SH and SV reflection coefficients

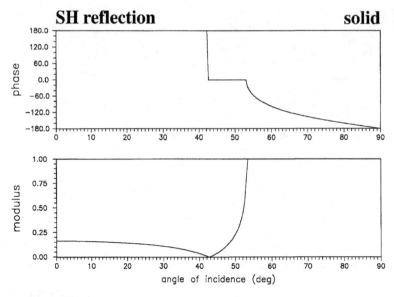

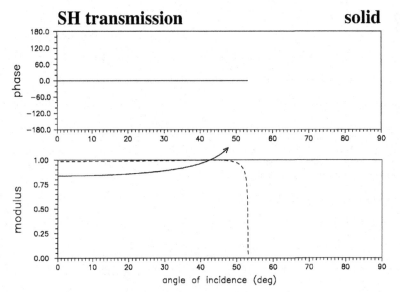

Figure 5.6(e)

are opposite. To explain this fact, we need to take into account our convention on the orientations of unit vectors $\vec{e}_1$ and $\vec{e}_2$. Basis vector $\vec{e}_2$, corresponding to SH waves, is the same for incident and reflected wave, but the signs of $\vec{e}_1$, corresponding to SV waves, are opposite for incident and reflected waves. This explains the difference in signs between R_{22} and R_{11} for normal incidence.

All the foregoing conclusions can be verified in Figure 5.6 in the numerical example under consideration. The normal incidence reflection coefficients $(R_{13},\ R_{31})$ vanish for converted waves (P1S1, S1P1). Unconverted reflection coefficients $(R_{11},\ R_{22},\ R_{33})$ take the values given by (5.3.17) for normal incidence: $R_{11} = R_{33} = -R_{22} = 0.161$. Thus, $R_{11} = R_{33}$ in our example. In general, however, R_{11} may differ from R_{33}. The reason why they are the same in Figure 5.6 is, that, in our example $\alpha_1/\alpha_2 = \beta_1/\beta_2$. Also note that the normalized displacement reflection coefficients $\mathcal{R}_{ij}$ do not differ from the standard displacement reflection coefficients R_{ij} for normal incidence.

In the investigation of R/T coefficients close to the normal incidence (e.g., in the AVO analysis), it may be suitable to use formally the polarization vectors $\vec{e}_1$ of reflected S waves in the direction opposite of that shown in Figure 5.5. In this case, the reflection coefficients R_{11} and R_{31} have opposite signs to those given in (5.3.2). All other R/T coefficients remain the same. Consequently, R_{11} equals R_{22} for the normal incidence. Such a choice of polarization vectors $\vec{e}_1$ of reflected S waves was used, for example, by Aki and Richards (1980).

The displacement *transmission coefficients* for normal incidence are given by the relation

$$R_{11} = R_{22} = 2Z_1^S/(Z_1^S + Z_2^S), \qquad R_{33} = 2Z_1^P/(Z_1^P + Z_2^P). \quad (5.3.18)$$

The expression for the displacement P1P2 transmission coefficient R_{33} is, in this case, different from the relevant expression for acoustic waves; see (5.1.32). This is not surprising given that the displacement transmission coefficients are considered in (5.3.18) but that the pressure transmission coefficients are given in (5.1.32). If we consider a particle velocity instead of pressure in the fluid medium (see (5.2.91)), we will obtain exactly the same expression as in (5.3.18).

It is obvious from (5.3.18) that the displacement transmission coefficients are not reciprocal for normal incidence. The normalized displacement transmission coefficients for normal incidence are

$$\mathcal{R}_{11} = \mathcal{R}_{22} = 2\sqrt{Z_1^S Z_2^S}/(Z_1^S + Z_2^S), \qquad \mathcal{R}_{33} = 2\sqrt{Z_1^P Z_2^P}/(Z_1^P + Z_2^P);$$
$$(5.3.19)$$

see (5.3.10). Evidently, the normalized R/T coefficients are reciprocal. Moreover, in this case, the normalized displacement coefficients equal the normalized pressure coefficients.

The differences between the standard displacement transmission coefficients R_{ij} and the normalized displacement transmission coefficients $\mathcal{R}_{ij}$ for normal incidence can be clearly seen in Figure 5.6. The numerical values of the individual transmission coefficients in the numerical example (Figure 5.6) are $R_{11} = R_{22} = R_{33} = 0.839$, $\mathcal{R}_{11} = \mathcal{R}_{22} = \mathcal{R}_{33} = 0.987$, and $R_{13} = R_{31} = \mathcal{R}_{13} = \mathcal{R}_{31} = 0$. Thus, the normalized displacement transmission coefficients $\mathcal{R}_{ij}$ are closer to unity under normal incidence than the standard displacement transmission coefficients.

Finally, we shall present the displacement reflection coefficient at a free surface for normal incidence:

$$R_{11} = -1, \qquad R_{22} = 1, \qquad R_{33} = -1. \quad (5.3.20)$$

2. CRITICAL ANGLES OF INCIDENCE

Critical angles of incidence are angles of incidence for which some square root P_i, $i = 1, 2, 3, 4$ in (5.3.5), vanishes. If we denote the velocity of the incident wave V and the velocity of the selected R/T wave $\tilde{V}$, one of the square roots P_i ($i = 1, 2, 3, 4$) vanishes for $\tilde{V}p = 1$, that is, for $(\tilde{V}/V)\sin i^* = 1$, where i^* is the critical angle of incidence. This yields the general definition of the critical angle of incidence,

$$i^* = \arcsin(V/\tilde{V}). \quad (5.3.21)$$

Clearly, in this case, the relevant angle of reflection/transmission equals $\frac{1}{2}\pi$ so that the ray of the relevant R/T wave is parallel to the interface. For angles of incidence i greater than critical angle i^*, the square root becomes positive imaginary (see (5.3.6)), and the R/T coefficients are complex-valued.

There may be several critical angles corresponding to different P_i. For the incident P wave, we may have two critical angles; for the incident SV wave, three critical angles; and for the incident SH wave, one critical angle. We shall treat these three cases separately.

a. Critical angles of incidence for the incident P wave. For the incident P wave, there exists no one critical angle or one critical angle or two critical angles. If $\alpha_2 < \alpha_1$, no critical angle exists. At least one critical angle exists for $\alpha_1 < \alpha_2$. The first critical angle is

$$i^* = \arcsin(\alpha_1/\alpha_2). \tag{5.3.22}$$

This angle always represents the *minimum critical angle* i^*_{min}. In addition, also the second critical angle i^{**} exists if $\beta_2 > \alpha_1$,

$$i^{**} = \arcsin(\alpha_1/\beta_2). \tag{5.3.23}$$

Thus, the second critical angle exists only at interfaces with a large velocity contrast.

In our numerical example in Figure 5.6(a), only the first critical angle exists for incident P waves, $i^* = \arcsin(\alpha_1/\alpha_2) = 53.13°$. This angle also represents the minimum critical angle, $i^*_{min} = 53.13°$. The second critical angle does not exist because $\beta_2 < \alpha_1$.

b. Critical angles of incidence for the incident SV wave. For the incident SV wave, one, two, or three critical angles of incidence exist. There is always *at least one critical angle* in this case. For $\alpha_2 > \beta_2 > \alpha_1 > \beta_1$,

$$i^* = \arcsin(\beta_1/\alpha_2), \qquad i^{**} = \arcsin(\beta_1/\beta_2), \qquad i^{***} = \arcsin(\beta_1/\alpha_1), \tag{5.3.24}$$

with $i^* < i^{**} < i^{***}$. In this case, angle i^* represents the minimum critical angle, $i^*_{min} = \arcsin(\beta_1/\alpha_2)$. For different relations between velocities α_1, β_1, α_2, and β_2, certain of the preceding critical angles shown in (5.3.24) do not exist, or their succession may be different. For example, for $\alpha_2 < \alpha_1$, only one critical angle of incidence exists, $i^* = i^*_{min} = \arcsin(\beta_1/\alpha_1)$. Note that this critical angle of incidence exists always because β_1 is always less than α_1.

In the numerical example shown in Figure 5.6(b), all three critical angles exist: $i^* = \arcsin(\beta_1/\alpha_2) = 27.53°$, $i^{**} = \arcsin(\beta_1/\alpha_1) = 35.30°$, and $i^{***} = \arcsin(\beta_1/\beta_2) = 53.13°$. Thus, the minimum critical angle $i^*_{min} = 27.53°$.

c. Critical angles of incidence for an incident SH wave. For the incident SH wave, no one critical angle exists for $\beta_1 > \beta_2$, and one critical angle exists for $\beta_1 < \beta_2$. It is given by the relation

$$i^* = \arcsin(\beta_1/\beta_2) \tag{5.3.25}$$

and also represents the minimum critical angle. Note that the minimum critical angles of incidence are different for SH and SV incident waves if $\alpha_2 > \alpha_1$.

In the numerical example under consideration, $i^* = \arcsin(\beta_1/\beta_2) = 53.13°$. This also represents the minimum critical angle.

d. Critical angles of incidence for waves reflected at the earth's surface. In this case, no critical angle exists for the incident P and SH waves, but one critical angle exists for the incident SV wave

$$i^* = \arcsin(\beta_1/\alpha_1). \tag{5.3.26}$$

This relation also defines the minimum critical angle.

Now we shall briefly discuss certain *properties of the displacement R/T coefficients* connected with the critical angles of incidence.

a. Maximum angles of incidence. Let us consider an R/T wave with velocity $\tilde{V}$, and an incident wave with velocity V. For the critical angle of incidence, $i^* = \arcsin(V/\tilde{V})$, and the angle of R/T is $\tilde{i} = \frac{1}{2}\pi$. For angles of incidence $i > i^*$, the relevant angle of R/T is complex-valued, and the generated wave is inhomogeneous. We are not considering inhomogeneous waves here so that critical angle $i^* = \arcsin(V/\tilde{V})$ is the *maximum angle of incidence* for which the relevant homogeneous R/T wave (characterized by velocity $\tilde{V}$) exists. Several examples can be seen in Figures 5.6. See the reflection coefficient R_{13} where the maximum angle of incidence $\arcsin(\beta_1/\alpha_1) = 35.30°$. See also transmission coefficient R_{13}, with the maximum angle of incidence $\arcsin(\beta_1/\alpha_2) = 27.53°$, and transmission coefficients R_{11}, R_{22}, and R_{33}, with the maximum angle of incidence $\arcsin(\alpha_1/\alpha_2) = \arcsin(\beta_1/\beta_2) = 53.13°$. Only transmission coefficient R_{31} corresponds to a homogeneous transmitted wave for an arbitrary angle of incidence, $0 \leq i \leq 90°$.

In Figure 5.6, it is interesting that the normalized R/T coefficients *always vanish* for the maximum angle of incidence, with the exception of unconverted reflected waves $(\mathcal{R}_{11}, \mathcal{R}_{22}, \mathcal{R}_{33})$. This is a great difference with respect to the standard displacement R/T coefficients, which may be rather high for angles of incidence close to the maximum angles of incidence. For unconverted transmitted waves, the standard displacement coefficients are even larger than unity. See transmission coefficients R_{11}, R_{22}, and R_{33} in Figure 5.6, where the values larger than unity are indicated by arrows. The normalized coefficients, however, vanish in all these cases. Note that the maximum angle of incidence of the relevant R/T wave *always* corresponds to a critical angle of incidence.

b. Maximum angles of reflection/transmission. As in (a), the range of angles of reflection/transmission is also sometimes limited, assuming that the angle of incidence i lies between 0 and $\frac{1}{2}\pi$. This occurs when $\tilde{V} < V$. Since $\sin \tilde{i} = (\tilde{V}/V)\sin i$, we can write $\tilde{i} = \arcsin(\tilde{V}/V)$ for the angle of incidence $i = 90°$. Because Figure 5.6 does not show the angles of reflection/transmission, we cannot identify the maximum angles of reflection/transmission in this figure. We can, however, easily calculate them. For example, for the P1S1 reflected wave, the maximum angle of reflection is $\tilde{i} = \arcsin(\beta_1/\alpha_1)$. In our case, $\tilde{i} = 35.30°$. The maximum angle of transmission exists only for one transmitted wave: for the P1S2 wave, that is, $\arcsin(\beta_2/\alpha_1) = 46.18°$.

The R/T waves, corresponding to angles of R/T larger than the maximum angles, physically exist; they are, however, generated by inhomogeneous incident waves with complex-valued angles of incidence. Examples are the so-called pseudospherical waves and various "star waves" such as the S^* wave.

c. Complex-valued R/T coefficients. All the P-SV and SH displacement R/T coefficients are always real-valued if the relevant angles of incidence are *smaller than the minimum critical angle*. Similarly, they are always complex-valued for angles of incidence *larger than the minimum critical angle*. Thus, the arguments of the R/T coefficients are zero or π for angles of incidence smaller than the critical angle, which can easily be verified in Figure 5.6. Consequently, the minimum critical angle plays a very important role. We call angles of incidence $i < i^*_{min}$ the *subcritical angles of incidence*, and angles of incidence $i > i^*_{min}$ the *postcritical angles of incidence*. The postcritical angles of incidence are also called *overcritical angles of incidence*. Note that the nonvanishing argument of the R/T coefficient affects the shape of the signal of the R/T waves; see Chapter 6.

Figure 5.6 also shows the phase shifts of individual R/T coefficients so that we can easily verify that the phase shifts vanish for angles of incidence less than the minimum critical angle of incidence and are nonvanishing for angles of incidence larger than the minimum critical angle (but less than the maximum angle of incidence). We must, however, inspect the phase shifts in Figure 5.6 carefully and distinguish them from change of sign (phase shift 180°).

Let us first consider the reflected waves. The minimum critical angles for reflection coefficients P1P1(R_{33}), P1S1(R_{31}), and SH → SH (R_{22}) equal 53.13°, and for reflection coefficients S1S1(R_{11}) and S1P1(R_{13}) they equal 27.53°. Thus, all the reflection coefficients are complex-valued if the angles of incidence are postcritical. For the P1P1, P1S1, and SH → SH reflected waves, the minimum critical angle is rather high (53.13°) so that the postcritical region is situated at great epicentral distances. For S1S1 and S1P1 reflection coefficients, the minimum critical angle is rather low (27.53°) so that the subcritical region is very narrow.

Transmission coefficients P1P2 (R_{33}), S1P2 (R_{13}), and SH → SH (R_{22}) are real-valued for all angles of incidence corresponding to homogeneous transmitted waves. Only the two transmission coefficients P1S2 and S1S2 are complex-valued for angles of incidence larger than the minimum critical angle. The minimum critical angle for the P1S2 transmitted wave $i^*_{min} = 53.13°$ and for the S1S2 transmitted wave $i^*_{min} = 27.53°$. In both cases, however, the arguments of these transmission coefficients are close to 0° or to ±180°. Thus, even certain transmission coefficients may be complex-valued.

d. Anomalous behavior of R/T coefficients near critical angles. The displacement R/T coefficients change very fast with respect to the angle of incidence i in the vicinity of critical angles. In most cases, derivatives $d|R_{mn}|/di$ and/or $d(\arg R_{mn})/di$ are infinite at critical angles of incidence. Moreover, the left-hand and right-hand derivatives are usually different there. For this reason, the modulus and/or argument of the R/T coefficient usually form an edge or an inflection point (with an infinite derivative) at a critical point. The fast changes of $|R_{mn}|$ and/or $\arg R_{mn}$ cause the ray method to be inapplicable in the critical region.

The most expressive changes of the R/T coefficients close to the critical angles can be observed for reflection coefficients P1P1(R_{33}), S1S1(R_{11}), and SH → SH (R_{22}).

e. Elliptic polarization of S waves for postcritical angles of incidence. As we can see from the equations for the SV and SH coefficient, or directly from Figure 5.6, the arguments of the SV → SV and SH → SH R/T coefficients are usually different. For example, for $\alpha_2 > \alpha_1 > \beta_2 > \beta_1$, the minimum critical angle for the S1S1 reflected wave (R_{11}) is $\arcsin(\beta_1/\alpha_2)$, but for the SH → SH reflected wave (R_{22}), it is $\arcsin(\beta_1/\beta_2)$. Assume now that both the SV and SH components of the incident wave are nonvanishing. The reflected S wave then has two mutually phase-shifted components. This property immediately implies that the reflected S wave is not polarized linearly, but rather elliptically. For more details, refer to Section 6.4.

3. BREWSTER ANGLES OF INCIDENCE

For a Brewster angle of incidence, the relevant R/T coefficient vanishes. The most typical and well-known example among the P-SV and SH R/T coefficients is the SH reflection coefficient; see Figure 5.6. For $\beta_2 > \beta_1$ and $\rho_2 > \rho_1$, we obtain $Z_2^S > Z_1^S$ so that R_{22} is negative under normal incidence. At the critical angle i^*, however, the reflection coefficient equals 1. Thus, the Brewster angle i_B is situated in the region $0 < i_B < i^*$, usually very close to the critical angle. This is the basic difference between the acoustic and SH reflection coefficients. For SH waves, the Brewster angle is very common, but for acoustic waves, it is exceptional. In our numerical example, the Brewster angle for reflected SH waves $i_B = 42.5°$.

Brewster angles are rather common even for other P-SV R/T coefficients, particularly for the SV and converted waves. Even the P1P1 reflection coefficient may vanish for certain angles of incidence in some cases, but this is rather unusual.

In our numerical example, several Brewster angles can be observed for reflected waves: For the P1S1 wave, $i_B \doteq 50.5°$; for the S1P1 wave, $i_B \doteq 26.5°$; and for the S1S1 wave, $i_B \doteq 22.5°$. Note that all these Brewster angles of incidence are situated close to minimum critical angles of incidence. Transmitted waves do not display Brewster angles.

The second example corresponds to the free surface of an elastic halfspace such as the Earth's surface. Figure 5.7 shows the reflection coefficients at the Earth's surface, for $\beta_1/\alpha_1 = 0.577$. Because the figures are self-explanatory, we shall be brief in the discussion.

Reflection coefficients R_{33} (P1P1) and R_{11} (S1S1) are given by the same analytical expressions, if they are expressed in terms of ray parameter p; see (5.3.7). If we present them in terms of the angles of incidence of P and S waves, they have an apparently different form because the coefficient R_{33} is stretched due to different angles of incidence. Both coefficients R_{11} and R_{33} display *two Brewster angles*. The reflection coefficient R_{33} is completely real-valued, but the S1S1 reflection coefficient R_{11} is complex-valued beyond the critical angle of incidence of S waves $\arcsin(\beta_1/\alpha_1) = 35.30°$. Note the proximity of one Brewster angle to the critical angle of incidence in the R_{11} reflection coefficient.

The reflection coefficients of converted waves R_{13} (S1P1) and R_{31} (P1S1) are both real-valued and reach values larger than unity in certain ranges of angles of incidence. The relevant normalized reflection coefficients $\mathcal{R}_{13}$ and $\mathcal{R}_{31}$, however, are both less than unity or equal to it. Moreover, both $\mathcal{R}_{13}$ and $\mathcal{R}_{31}$ are given by the same analytical expressions, if they are expressed in terms of ray parameter p (see (5.3.15)); they differ only in sign. As with the reflection coefficients R_{11} and R_{33}, the normalized reflection coefficients $\mathcal{R}_{13}$ and $\mathcal{R}_{31}$ are different in Figure 5.7 only as a result of different angles of incidence; $\mathcal{R}_{13}$ is stretched.

5.3.5 Displacement Reflection/Transmission Matrices

In this section, we shall consider an arbitrary orientation of polarization vectors $\vec{e}_1$ and $\vec{e}_2$ of incident and reflected/transmitted waves. The only requirement is that unit vectors $\vec{e}_1$, $\vec{e}_2$, and $\vec{e}_3 \equiv \vec{t}$ are mutually orthogonal and form a right-handed system for all the waves under consideration (incident, reflected, and transmitted). Thus, $\vec{e}_2$ need not be perpendicular to the plane of incidence and $\vec{e}_1$ need not be situated in the plane of incidence (as in the case of P-SV and SH R/T coefficients).

In this case, we obtain *nine reflection coefficients* R_{mn}^r ($m = 1, 2, 3; n = 1, 2, 3$). The convention for the choice of m and n is discussed at the beginning of Section 5.3; see also Figure 5.8. In the same way, we obtain *nine transmission coefficients* R_{mn}^t. We shall

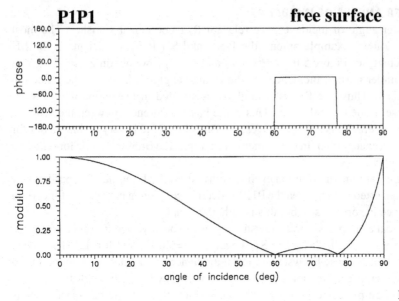

Figure 5.7(a)

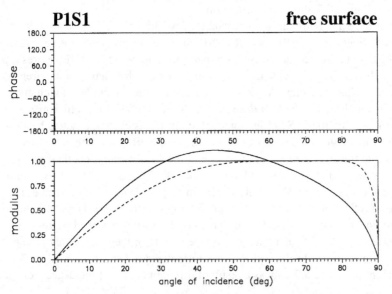

Figure 5.7. Displacement and normalized displacement P-SV reflection coefficients at a free surface of a solid halfspace. Model: $\alpha_1 = 6{,}400$ m/s, $\beta_1 = 3{,}698$ m/s, $\rho_1 = 2{,}980$ kg/m^3. Continuous lines: displacement coefficients; dashed lines: normalized displacement coefficients. Orientation index $\epsilon = 1$, see Figure 5.10. (a) P1P1 (R^r_{33}) and P1S1 (R^r_{31}) reflection coefficients. (b) S1S1 (R^r_{11}) and S1P1 (R^r_{13}) reflection coefficients. For $\epsilon = -1$, the signs of P1S1 and S1P1 reflection coefficients would be opposite.

use the superscripts r and t to specify the reflection and transmission coefficients only if symbol R_{mn} could cause an misunderstanding. Otherwise, superscripts r and t will not be used.

Reflection coefficients R^r_{mn} (m, $n = 1$, 2, 3) form a 3 × 3 displacement reflection matrix $\hat{\mathbf{R}}^r$. Similarly, R^t_{mn} form a 3 × 3 displacement transmission matrix $\hat{\mathbf{R}}^t$. These *full R/T matrices*, however, do not have any application in the ray method because the type of the R/T coefficient at any point of incidence is strictly specified by the ray code of the wave

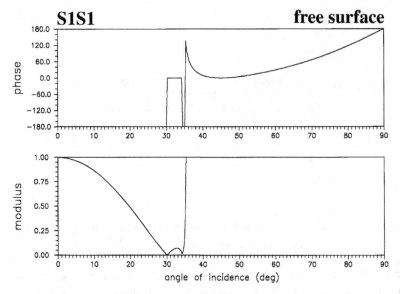

Figure 5.7(b)

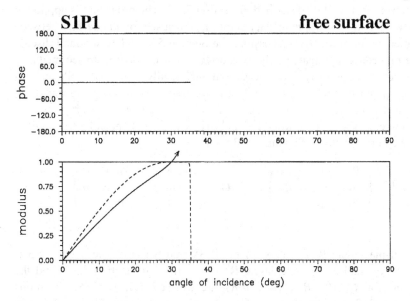

under consideration. The only exception is related to 3-D computations of S and converted waves. The S waves have, in general, two components, and the matrix notation allows for simple compact expressions of the vectorial amplitudes of S waves and converted waves propagating in 3-D layered structures. In this case, however, we need to introduce four types of displacement R/T matrices, corresponding to $P \rightarrow P$, $P \rightarrow S$, $S \rightarrow P$, and $S \rightarrow S$ reflection/transmission. The actual expressions for these matrices are given in (5.2.38). We remind the reader that the $P \rightarrow P$ R/T matrix has only one nonvanishing element R_{33}, the $P \rightarrow S$ R/T matrix has only two nonvanishing elements R_{31} and R_{32}, the $S \rightarrow P$ R/T matrix has also two nonvanishing elements R_{13} and R_{23}, and the $S \rightarrow S$ R/T matrix has four nonvanishing elements R_{11}, R_{12}, R_{21}, and R_{22}.

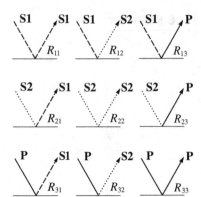

Figure 5.8. The physical meaning of elements R^r_{ij} of the 3×3 matrix $\hat{\mathbf{R}}^r$ of reflection coefficients. The physical meaning of the 3×3 matrix $\hat{\mathbf{R}}^t$ of transmission coefficients is analogous.

In the following, we shall discuss the *full R/T matrices* because the individual P $\to$ P, P $\to$ S, S $\to$ P, and S $\to$ S R/T matrices are obtained from the full matrices simply by putting some elements equal zero.

The 3×3 full R/T matrices $\hat{\mathbf{R}}^r$ and $\hat{\mathbf{R}}^t$ may be calculated in several ways. A very general procedure, based on the direct solution of the six linear algebraic equations (2.3.50), expressed in matrix form, will be described in Section 5.4.7. The procedure is applied to an interface between two anisotropic media but may also be used for an interface between two isotropic media. For a detailed discussion, see Section 5.4.7. Here we shall discuss another simple procedure, applicable only to interfaces between two isotropic media. It is based on the computation of R/T matrices from analytically computed P-SV and SH R/T coefficients, given by (5.3.2) and (5.3.3). In this case, it is necessary to transform the displacement ray-centered components $U^{(q)}_1$ and $U^{(q)}_2$ into a new rotated ray-centered coordinate system corresponding to the P-SV and SH case ($\vec{e}_2$ perpendicular to the plane of incidence). We introduce the following 3×3 matrices:

$$\hat{\mathbf{R}}^0 = \begin{pmatrix} R_{11} & 0 & R_{13} \\ 0 & R_{22} & 0 \\ R_{31} & 0 & R_{33} \end{pmatrix}, \qquad \hat{\mathbf{G}}^\perp = \begin{pmatrix} \cos \kappa & -\sin \kappa & 0 \\ \sin \kappa & \cos \kappa & 0 \\ 0 & 0 & 1 \end{pmatrix}.$$

$$(5.3.27)$$

Here $\hat{\mathbf{R}}^0$ is the displacement R/T matrix for the P-SV and SH case; see Section 5.3.1. The choice of $\vec{e}_1$, $\vec{e}_2$, and $\vec{e}_3$ for calculating $\hat{\mathbf{R}}^0$ is described in Section 5.3.1, and the analytical relations for R_{11}, R_{13}, R_{31}, R_{33}, and R_{22} are given by (5.3.2) and (5.3.3). In this case, the remaining R/T coefficients (R_{12}, R_{21}, R_{23}, and R_{32}) vanish. The 3×3 matrix $\hat{\mathbf{G}}^\perp$ is a rotation matrix that rotates unit vectors $\vec{e}_1$ and $\vec{e}_2$ by the angle κ about the ray. If we put $\cos \kappa = \vec{e}_2 \cdot \vec{i}^{(z)}_2$ and $\sin \kappa = \vec{e}_2 \cdot \vec{i}^{(z)}_1$, rotation matrix $\hat{\mathbf{G}}^\perp$ will shift unit vector $\vec{e}_2$ to $\vec{i}^{(z)}_2$. Thus, if $\hat{\mathbf{U}}^{(q)}(Q)$ correspond to arbitrarily chosen $\vec{e}_1$ and $\vec{e}_2$, then $\hat{\mathbf{G}}^\perp(Q)\hat{\mathbf{U}}^{(q)}(Q)$ corresponds to the P-SV and SH case, with $\vec{e}_2$ perpendicular to the plane of incidence, and $\vec{e}_1$ situated in the plane of incidence.

We shall now use (5.2.37), $\hat{\mathbf{U}}^{(q)}(\tilde{Q}) = \hat{\mathbf{R}}^T(Q)\hat{\mathbf{U}}^{(q)}(Q)$. If we use the rotated displacement matrices, the R/T matrix $\hat{\mathbf{R}}(Q)$ reduces to the P-SV and SH matrix $\hat{\mathbf{R}}^0$. Thus,

$$\hat{\mathbf{G}}^\perp(\tilde{Q})\hat{\mathbf{U}}^{(q)}(\tilde{Q}) = \hat{\mathbf{R}}^{0T}(Q)\hat{\mathbf{G}}^\perp(Q)\hat{\mathbf{U}}^{(q)}(Q).$$

This yields

$$\hat{\mathbf{U}}^{(q)}(\tilde{Q}) = \hat{\mathbf{G}}^{\perp T}(\tilde{Q})\hat{\mathbf{R}}^{0T}(Q)\hat{\mathbf{G}}^\perp(Q)\hat{\mathbf{U}}^{(q)}(Q), \qquad (5.3.28)$$

which can be expressed in the form of (5.2.37) if we put

$$\hat{\mathbf{R}}(Q) = \hat{\mathbf{G}}^{\perp T}(Q)\hat{\mathbf{R}}^0(Q)\hat{\mathbf{G}}^{\perp}(\tilde{Q}). \tag{5.3.29}$$

This is the final expression for the displacement R/T matrix corresponding to any orientation of unit vectors $\vec{e}_1$ and $\vec{e}_2$ corresponding to incident, reflected, and transmitted waves.

5.3.6 Normalized Displacement Reflection/Transmission Matrices

Instead of the displacement R/T coefficient matrices $\hat{\mathbf{R}}$, we can construct normalized displacement R/T matrices $\hat{\boldsymbol{\mathcal{R}}}$. Note that the general expressions for the amplitudes of seismic body waves propagating in 3-D layered structures contain the normalized displacement R/T matrices $\hat{\boldsymbol{\mathcal{R}}}$, not the standard displacement R/T matrices $\hat{\mathbf{R}}$; see (5.2.45). The 3×3 normalized displacement R/T matrix $\hat{\boldsymbol{\mathcal{R}}}$ has nine elements $\mathcal{R}_{mn}$ ($m, n = 1, 2, 3$), which are constructed from R_{mn} using relation (5.3.10). We remind the reader that the normalized displacement R/T coefficients $\mathcal{R}_{mn}$ represent the displacement R/T coefficients R_{mn} normalized with respect to the energy flux across the interface.

The normalized displacement R/T matrices $\hat{\boldsymbol{\mathcal{R}}}$ can be calculated similarly as the displacement R/T matrices $\hat{\mathbf{R}}$, but the individual elements R_{mn} must be reduced to $\mathcal{R}_{mn}$ using (5.3.10). We can, of course, also use the relation, alternative to (5.3.29),

$$\hat{\boldsymbol{\mathcal{R}}}(Q) = \hat{\mathbf{G}}^{\perp T}(Q)\hat{\boldsymbol{\mathcal{R}}}^0(Q)\hat{\mathbf{G}}^{\perp}(\tilde{Q}), \tag{5.3.30}$$

where $\hat{\boldsymbol{\mathcal{R}}}^0(Q)$ is the matrix of normalized P-SV and SH R/T coefficients, given by (5.3.13) through (5.3.15) (with $\mathcal{R}_{12} = \mathcal{R}_{21} = \mathcal{R}_{23} = \mathcal{R}_{32} = 0$).

5.3.7 Reciprocity of R/T Coefficients

The reciprocity of ray computations plays a basic role not only in theoretical considerations but also in various practical applications. The expressions for the ray amplitudes of seismic body waves propagating in layered media contain R/T coefficients. It is quite obvious that *displacement R/T coefficients are not, in general, reciprocal.* This observation, however, does not imply that the expressions for the ray amplitudes are not reciprocal. As we have shown in Section 5.2.5, displacement R/T coefficients can be combined with some other factors in the expressions for ray amplitudes to give the *normalized displacement R/T coefficients.* (The normalization of the displacement coefficients is performed with respect to the energy flux across the interface.) We shall show that the *normalized displacement R/T coefficients are reciprocal.* We shall first give the reciprocity relations for the P-SV and SH case and then for the case of arbitrary unit vectors $\vec{e}_1$ and $\vec{e}_2$. In both cases, we shall consider a ray Ω connecting two points S and R, and incident at point Q on interface Σ. As usual, we denote by $\tilde{Q}$ the point of R/T on Σ. Points Q and $\tilde{Q}$ coincide, but Q corresponds to the incident ray if the wave propagates from S to R, and $\tilde{Q}$ corresponds to the selected R/T ray. We consider only one interface; the result can be simply generalized for any number of interfaces.

1. RECIPROCITY RELATIONS FOR P-SV AND SH R/T COEFFICIENTS

We shall now consider an arbitrary seismic body wave propagating along Ω from S to R; see Figure 5.9(a). The arrows in Figure 5.9(a) show the direction of propagation (unit vector $\vec{e}_3 \equiv \vec{t}$) and the direction of unit vector $\vec{e}_1$. The backward propagation, from R to S, is specified by Equations (4.4.110) and (4.4.111); see Figure 5.9(b). We remind the

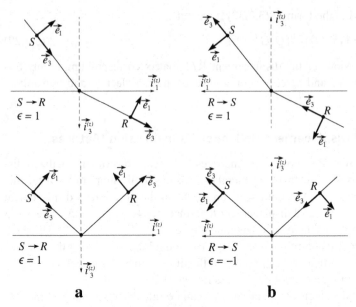

Figure 5.9. Choice of basis vector $\vec{e}_i$ in the forward (a) and backward (b) propagation along ray Ω. In the backward propagation, $\vec{e}_1$ and $\vec{e}_3$ are chosen in the opposite directions than in the forward propagation. The local Cartesian coordinate systems at the points of incidence on structural interfaces are also changed as follows: $\vec{i}_1^{(z)}$ and $\vec{i}_3^{(z)}$ are chosen in the opposite directions than in the forward propagation. Unit vectors $\vec{e}_2$ and $\vec{i}_2^{(z)}$ remain the same in both cases.

reader that the vectors $\vec{e}_1$ and $\vec{e}_3$ are taken opposite in forward and backward computations. Similarly, the unit vectors $\vec{i}_3^{(z)}$, perpendicular to the interfaces at all points of incidence, and the unit vectors $\vec{i}_1^{(z)}$ are taken opposite. Unit vectors $\vec{e}_2$ and $\vec{i}_2^{(z)}$ remain the same in both cases. Compare Figures 5.9(a) and 5.9(b).

The preceding relations also immediately imply the relation between the forward and backward orientation indices, ϵ and $\bar{\epsilon}$. In Figure 5.9(a), the unit normal $\vec{n} \equiv \vec{i}_3^{(z)}$ to the interface Σ was chosen downward so that $\epsilon = 1$. Because the unit vectors $\vec{i}_3^{(z)}$ in the backward computation are taken opposite, we obtain the following rule for the determination of $\bar{\epsilon}$:

$$\bar{\epsilon} = -\epsilon \qquad \text{for reflected waves,}$$
$$\bar{\epsilon} = \epsilon \qquad \text{for transmitted waves.}$$

see Figure 5.9(b). By direct inspection, we then obtain from (5.3.13) through (5.3.15) the following *reciprocity relation for the normalized R/T coefficients*:

$$\mathcal{R}_{mn}(Q) = \mathcal{R}_{nm}(\tilde{Q}). \tag{5.3.31}$$

We shall give two examples of reciprocity relations for the normalized transmission coefficients $\mathcal{R}_{11}^t$ and $\mathcal{R}_{31}^t$. We shall mark all quantities corresponding to the wave propagating from R to S at $\tilde{Q}$ with a bar above the symbols. The plain symbols, without bars, correspond to the wave propagating from S to R, at Q. Then

$$\bar{\alpha}_1 = \alpha_2, \qquad \bar{\beta}_1 = \beta_2, \qquad \bar{\rho}_1 = \rho_2, \qquad \bar{\alpha}_2 = \alpha_1,$$
$$\bar{\beta}_2 = \beta_1, \qquad \bar{\rho}_2 = \rho_1, \qquad \bar{P}_2 = P_4,$$
$$\bar{P}_4 = P_2, \qquad \bar{P}_1 = P_3, \qquad \bar{P}_3 = P_1, \qquad \bar{\epsilon} = \epsilon.$$

This yields

$$\bar{q} = -q, \qquad \bar{X} = Y, \qquad \bar{Y} = X, \qquad \bar{Z} = -Z, \qquad \bar{D} = D.$$

Using (5.3.14), we can write

$$
\begin{aligned}
\mathcal{R}^t_{11}(\tilde{Q}) &= 2(\bar{\beta}_1\bar{\beta}_2\bar{\rho}_1\bar{\rho}_2\bar{P}_2\bar{P}_4)^{1/2}\bar{D}^{-1}(\bar{\alpha}_1\bar{P}_3\bar{Y} + \bar{\alpha}_2\bar{P}_1\bar{X}) \\
&= 2(\beta_1\beta_2\rho_1\rho_2 P_2 P_4)^{1/2}D^{-1}(\alpha_2 P_1 X + \alpha_1 P_3 Y) = \mathcal{R}^t_{11}(Q), \\
\mathcal{R}^t_{13}(\tilde{Q}) &= 2\bar{\epsilon}p(\bar{\beta}_1\bar{\alpha}_2\bar{\rho}_1\bar{\rho}_2\bar{P}_2\bar{P}_3)^{1/2}\bar{D}^{-1}(\bar{q}\bar{P}_1\bar{P}_4 - \bar{\alpha}_1\bar{\beta}_2\bar{Z}) \\
&= 2\epsilon p(\alpha_1\beta_2\rho_1\rho_2 P_1 P_4)^{1/2}D^{-1}(-q P_2 P_3 + \alpha_2\beta_1 Z) = \mathcal{R}^t_{31}(Q).
\end{aligned}
$$

Similar results are also obtained for other types of normalized displacement P-SV and SH transmission coefficients. For reflection coefficients, the reciprocity relations (5.3.31) are immediately seen from (5.3.13).

2. RECIPROCITY RELATIONS IN THE GENERAL CASE

We shall now consider the general case of $\vec{e}_2$ not perpendicular to the plane of incidence. We again assume that the direction of unit vectors $\vec{e}_1$ and $\vec{e}_3$ for the reciprocal direction (from R to S) is opposite to the direction of unit vectors $\vec{e}_1$ and $\vec{e}_3$ from S to R. We can then use (5.3.30) and modify the matrices $\hat{G}^{\perp}$ properly for the reciprocal direction. As a result, we again obtain general relation (5.3.31). Thus, reciprocity relation (5.3.31) is valid generally.

Note that reciprocity relation (5.3.31) is an isotropic equivalent of the general reciprocity relation for normalized displacement R/T matrices, derived for anisotropic media by Chapman (1994); see also Section 5.4.7.

5.3.8 P-SV and SH Conversion Coefficients

In seismology and seismic exploration, receivers are most commonly situated on the Earth's surface (or very close to it). The surface of the Earth is a very distinct discontinuity and affects the seismic wavefield recorded by the receiver situated on it considerably. To express the effects of the Earth's surface on the incident wave quantitatively, the ray-centered amplitude matrix of the incident wave must be multiplied by the free-surface conversion matrix $\hat{D}$; see Section 5.2.7. The resulting products give the components of the displacement vector expressed in the general or a local Cartesian coordinate system at the point of incidence. We shall again call the elements of the free-surface conversion matrix the free-surface conversion coefficients, or simply the conversion coefficients.

It is not surprising that the free-surface conversion coefficients are very important in certain seismological applications, perhaps even more important than the R/T coefficients. Even if waves propagating in a homogeneous halfspace without interfaces are studied, free-surface conversion coefficients must be used for receivers situated on the Earth's surface. Thus, the application of the free-surface conversion coefficients in numerical modeling of seismic wavefields is nearly universal. For this reason, we shall give the explicit expressions for the free-surface conversion coefficients. We shall consider only the most common P-SV and SH case in which the unit vector $\vec{e}_2$ of the incident wave is perpendicular to the plane of incidence. In this case, $\mathcal{D}_{12} = \mathcal{D}_{21} = \mathcal{D}_{23} = \mathcal{D}_{32} = 0$, and conversion matrix $\hat{D}$ has only five nonvanishing elements $\mathcal{D}_{11}, \mathcal{D}_{13}, \mathcal{D}_{31}, \mathcal{D}_{33}$, and $\mathcal{D}_{22}$. The conversion coefficients should not be confused with the R/T coefficients of converted waves.

In the evaluation of conversion coefficients, we need to be very careful about their signs. Orientation index ϵ is not quite sufficient to determine the signs of the conversion coefficients $\mathcal{D}_{11}, \mathcal{D}_{13}, \mathcal{D}_{31}$, and $\mathcal{D}_{33}$ uniquely because the general Cartesian coordinate system can be chosen arbitrarily. For this reason, we shall use the local Cartesian coordinate system and fix it using the convention described in Section 5.3.1. In other words, we assume that the unit vector $\vec{e}_2$ of the ray-centered coordinate system coincides with the unit vector $\vec{i}_2$ of the local Cartesian coordinate system. We remind the reader that unit vector $\vec{e}_3$ is

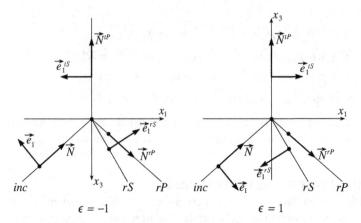

Figure 5.10. Orientation of polarization vectors $\vec{N}$ and $\vec{e}_1$ of incident, reflected, and transmitted P and S waves at a free surface of an isotropic solid halfspace. Left, orientation index $\epsilon = -1$; right, orientation index $\epsilon = 1$.

positive in the direction of propagation of the wave under consideration and that the positive orientation of $\vec{i}_1$ is such that $\vec{p} \cdot \vec{i}_1^{(z)} > 0$. (The local x_1-axis is tangent to the interface and is positive in the direction of propagation of the wave.) The local Cartesian coordinate system is then uniquely tied to the unit vectors $\vec{e}_1$, $\vec{e}_2$, and $\vec{e}_3$ of the incident wave, and is fully specified by orientation index ϵ. See Figure 5.10.

The derivation of analytical expressions for the conversion coefficients $\mathcal{D}_{11}$, $\mathcal{D}_{13}$, $\mathcal{D}_{31}$, $\mathcal{D}_{33}$, and $\mathcal{D}_{22}$ for the Earth's surface is simple:

1. We use the R/T displacement coefficients R_{11}, R_{13}, R_{31}, R_{33}, and R_{22} for the Earth's surface, given by (5.3.7) and (5.3.8).

2. If we use (5.2.116) to compute $\mathcal{D}_{ij}$, we need to specify the components of unit vectors $\vec{e}_1$ and $\vec{e}_3 \equiv \vec{N}$, corresponding to incidence, reflected and transmitted waves. They are given by the following relations:

$$
\begin{aligned}
N_1^{SM} &= \alpha_1 p, & N_3^{SM} &= \epsilon P_1, & e_{11}^{SM} &= \epsilon P_2, & e_{13}^{SM} &= -\beta_1 p, \\
N_1^{rP} &= \alpha_1 p, & N_3^{rP} &= -\epsilon P_1, & e_{11}^{rS} &= -\epsilon P_2, & e_{13}^{rS} &= -\beta_1 p, \\
N_1^{tP} &= 0, & N_3^{tP} &= \epsilon, & e_{11}^{tS} &= \epsilon, & e_{13}^{tS} &= 0.
\end{aligned}
$$

$$(5.3.32)$$

see Figure 5.10.

3. Inserting (5.3.7) or (5.3.8) and (5.3.32) into (5.2.116), we can compute $\mathcal{D}_{ij}(R)$ and $\mathcal{D}_{ij}(R^+)$. A considerably simpler approach is to compute $\mathcal{D}_{ij}(R^+)$, where only the transmission coefficients (5.3.8) (not the reflection coefficients (5.3.7)) are required. In fact, the free-surface conversion coefficients $\mathcal{D}_{ij}$ are directly equal the transmission coefficients; they should only be modified by a proper orientation index ϵ. The final equations then read:

$$
\begin{aligned}
\mathcal{D}_{11}(R^+) &= 2 P_2 \epsilon \left(1 - 2\beta_1^2 p^2\right)/D_1, & &\text{SV} \to x, \\
\mathcal{D}_{13}(R^+) &= 4\beta_1 p\, P_1 P_2 / D_1, & &\text{P} \to x, \\
\mathcal{D}_{31}(R^+) &= -4\beta_1^2 p\, P_1 P_2 / \alpha_1 D_1, & &\text{SV} \to z, \\
\mathcal{D}_{33}(R^+) &= 2 P_1 \epsilon \left(1 - 2\beta_1^2 p^2\right)/D_1, & &\text{P} \to z, \\
\mathcal{D}_{22}(R^+) &= 2, & &\text{SH} \to y.
\end{aligned}
$$

$$(5.3.33)$$

We shall briefly describe the physical meaning of the indices of $\mathcal{D}_{ij}$. The first index i specifies the component of the displacement vector in local Cartesian coordinate system x, y, z ($i = 1$ for the horizontal x-component, $i = 2$ for the horizontal y-component, $i = 3$ for the vertical z-component). The second index j denotes the type of incident wave ($j = 1$ for the incident SV-wave, $j = 2$ for the incident SH-wave, and $j = 3$ for the incident P wave). The relevant specification of $\mathcal{D}_{ij}$ is shown on each line in (5.3.33).

From (5.2.116) we can also determine $\mathcal{D}_{ij}(R)$, using reflection coefficients (5.3.7) and expressions (5.3.32). It is not surprising that $\mathcal{D}_{ij}(R) = \mathcal{D}_{ij}(R^+)$ because the displacement vector is continuous across structural interfaces (including the Earth's surface).

The conversion coefficients $\mathcal{D}_{11}$, $\mathcal{D}_{13}$, $\mathcal{D}_{31}$, and $\mathcal{D}_{33}$ for the Earth's surface depend only on the angle of incidence and on one parameter, $\gamma = \beta_1/\alpha_1$. Figure 5.11 displays the conversion coefficients $\mathcal{D}_{11}$, $\mathcal{D}_{13}$, $\mathcal{D}_{31}$, and $\mathcal{D}_{33}$ for the Earth's surface, assuming $\gamma = \beta_1/\alpha_1 = 0.577$ (that is, $\alpha_1/\beta_1 \doteq \sqrt{3}$).

Let us first discuss the conversion coefficients $\mathcal{D}_{13}$ (P $\rightarrow$ x) and $\mathcal{D}_{33}$ (P $\rightarrow$ z), corresponding to *the incident* P *wave*. They are real-valued and smooth, without singularities, for all angles of incidence $0 \le i_1 \le 90°$. Conversion coefficient $\mathcal{D}_{33}$ continuously decreases with increasing angle of incidence, starting from $\mathcal{D}_{33} = 2$ for normal incidence ($i = 0$). Coefficient $\mathcal{D}_{13}$ vanishes for angle of incidence $i = 0$ and for angle of incidence $i = 90°$. In between, it has a maximum for $i \doteq 63°$, where it reaches the value of about 1.75.

The conversion coefficients $\mathcal{D}_{11}$ (SV $\rightarrow$ x) and $\mathcal{D}_{31}$ (SV $\rightarrow$ z) corresponding to the incident SV wave are considerably more complex. They are real-valued only for the angle of incidence of the SV wave i less than the critical angle, $i^* = \arcsin(\beta_1/\alpha_1)$. The region $i < i^*$ of angles of incidence is usually called the *shear wave window*. The critical angle in our case equals 35.30°. Close to critical angle $i = i^*$, conversion coefficients $\mathcal{D}_{11}$ and $\mathcal{D}_{31}$ vary rapidly. For the normal angle of incidence $i = 0°$, $\mathcal{D}_{31} = 0$ and $\mathcal{D}_{11} = 2$. Thus, the vertical component vanishes, and the horizontal component doubles in this case. Conversion coefficient $\mathcal{D}_{11}$ vanishes also for $1 - 2\beta_1^2 p^2 = 0$ (that is, for $i = 45°$). In other words, the SV wave is purely vertical at the Earth's surface if it is incident at the Earth's surface under angle $i = 45°$. The complex behavior of $\mathcal{D}_{11}$ and $\mathcal{D}_{31}$ has serious consequences for the polarization of S waves at the Earth's surface; see Section 6.4.7.

Equations (5.3.33) correspond to the free-surface conversion coefficients. Similar equations as (5.3.33) can be derived even for a general structural interface:

$$\mathcal{D}_{11}(R^+) = 2\epsilon \rho_1 \beta_1 P_2 D^{-1}[\rho_2 \alpha_2 P_1 P_4 + \alpha_1 P_3 P_4 Y - \alpha_1 \alpha_2 \beta_2 p^2 Z],$$
$$\text{SV} \rightarrow x,$$

$$\mathcal{D}_{13}(R^+) = 2\rho_1 \alpha_1 p P_1 D^{-1}[\rho_2 \beta_1 \alpha_2 P_4 + \alpha_2 \beta_2 P_2 X - q P_2 P_3 P_4],$$
$$\text{P} \rightarrow x,$$

$$\mathcal{D}_{31}(R^+) = 2\rho_1 \beta_1 p P_2 D^{-1}[q P_1 P_3 P_4 - \alpha_2 \beta_2 P_1 X - \rho_2 \beta_2 \alpha_1 P_3],$$
$$\text{SV} \rightarrow z,$$

$$\mathcal{D}_{33}(R^+) = 2\epsilon \rho_1 \alpha_1 P_1 D^{-1}[\rho_2 \beta_2 P_2 P_3 + \beta_1 P_3 P_4 Y - \alpha_2 \beta_1 \beta_2 p^2 Z],$$
$$\text{P} \rightarrow z,$$

$$\mathcal{D}_{22}(R^+) = 2\rho_1 \beta_1 P_2 / \bar{D}, \qquad \text{SH} \rightarrow y.$$

$$(5.3.34)$$

All symbols have the same meaning as in (5.3.4) through (5.3.6).

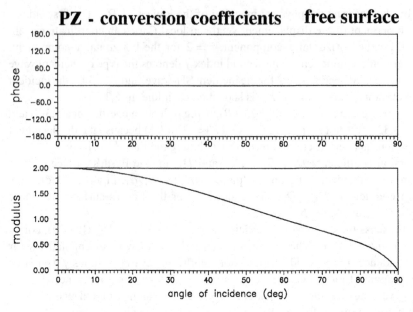

PZ - conversion coefficients   free surface

Figure 5.11(a)

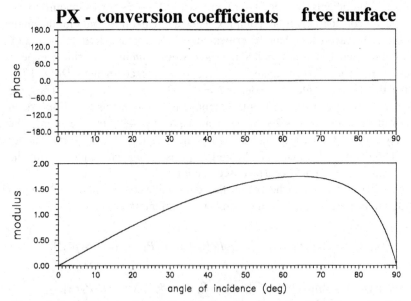

PX - conversion coefficients   free surface

Figure 5.11. P-SV conversion coefficients at a free surface of a solid halfspace. Model: $\alpha_1 = 6,400$ m/s, $\beta_1 = 3,698$ m/s, $\rho_1 = 2,980$ kg/m^3. Orientation index $\epsilon = 1$. (a) Conversion coefficients for the incident P wave, PZ($\mathcal{D}_{33}$) and PX($\mathcal{D}_{13}$). (b) Conversion coefficients for the incident S wave, SZ($\mathcal{D}_{31}$) and SX($\mathcal{D}_{11}$).

5.4 Elastic Anisotropic Structures

The expressions for amplitudes of elastic body waves propagating in inhomogeneous anisotropic layered structures are formally surprisingly simple. They are very similar to those for the amplitudes of pressure waves propagating in inhomogeneous fluid media, only the computation of the individual quantities in these expressions is considerably more involved.

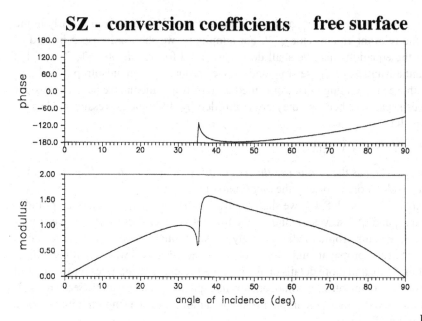

Figure 5.11(b)

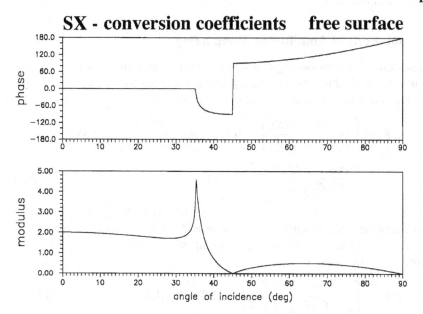

We shall again use the standard equations for the displacement vector $\vec{u}(x_j, t)$,

$$\vec{u}(x_j, t) = \vec{U}(x_j)F(t - T(x_j)). \tag{5.4.1}$$

Here $F(\zeta)$ is a high-frequency analytical signal, $\vec{U}(x_j)$ is a *vectorial ray theory complex-valued amplitude function*, and $T(x_j)$ is the travel time of the wave.

As shown in Section 2.4.3, three elastic body waves can propagate in a smooth inhomogeneous anisotropic medium: one qP and two qS (qS1 and qS2) waves. Each of these waves corresponds to one of the three eigenvalues G_m and eigenvectors $\vec{g}^{(m)}$, $m = 1, 2, 3$, of the Christoffel matrix $\hat{\Gamma}$; see (2.2.19). The travel time of the mth wave satisfies eikonal equation

$G_m(x_i, p_i) = 1$, with $p_i = \partial T/\partial x_i$, and amplitude function $\vec{U}$ is polarized linearly in the direction of the relevant eigenvector $\vec{g}^{(m)}$. For simplicity, we shall only use $\vec{g}$ instead of $\vec{g}^{(m)}$ because the equations that we shall derive are valid for any m ($m = 1, 2, 3$). This is an important advantage of the seismic body waves propagating in anisotropic media as compared to those propagating in isotropic media. In isotropic media, we need to consider P and S waves separately because they are controlled by different expressions. Thus, in anisotropic media,

$$\vec{U}(x_i) = A(x_i)\,\vec{g}(x_i). \qquad (5.4.2)$$

We shall refer to $A(x_i)$ as the *scalar ray-theory complex-valued amplitude function* of the wave under consideration, or briefly the *amplitude* of the wave.

In Sections 5.4.1 through 5.4.5, we shall briefly discuss the computation of amplitudes $A(x_i)$ of qP, qS1, and qS2 waves, including the multiply reflected converted waves propagating in a layered anisotropic media. We only consider situations in which these waves are fully separated and propagate independently. More details on qS wave coupling will be given in Section 5.4.6, and more details on the R/T coefficients and matrices on a structural interface between two anisotropic homogeneous halfspaces will be given in Section 5.4.7. Finally, Section 5.4.8 discusses the initial ray-theory amplitudes at a smooth initial surface and elastic Kirchhoff integrals.

5.4.1 Computations of Amplitudes Along a Ray

We shall consider ray Ω corresponding to a selected wave propagating in a smooth inhomogeneous anisotropic medium and two points S and R situated on Ω. The *continuation equations* for the amplitude function $A(x_i)$ of (5.4.2) then read

$$A(R) = \left[\frac{\rho(S)\mathcal{U}(S)J(S)}{\rho(R)\mathcal{U}(R)J(R)}\right]^{1/2} A(S) = \left[\frac{\rho(S)J^{(T)}(S)}{\rho(R)J^{(T)}(R)}\right]^{1/2} A(S)$$

$$= \left[\frac{\rho(S)\mathcal{C}(S)\Omega^{(T)}(S)}{\rho(R)\mathcal{C}(R)\Omega^{(T)}(R)}\right]^{1/2} A(S); \qquad (5.4.3)$$

see (3.10.62). Here $J = J^{(s)}$ and $J^{(T)}$ denote Jacobians (3.10.9), with $J^{(T)} = \mathcal{U}J$, and $\Omega^{(T)}$ represents the scalar surface element cut out of the wavefront by the ray tube, normalized with respect to $\mathrm{d}\gamma_1 \mathrm{d}\gamma_2$. $\mathcal{U}$ and $\mathcal{C}$ again denote the group and phase velocities.

Alternatively, we can also write,

$$A(R) = \left[\frac{\rho(S)\mathcal{U}(S)}{\rho(R)\mathcal{U}(R)}\right]^{1/2} \frac{\mathcal{L}(S)}{\mathcal{L}(R)} \exp[iT^c(R, S)]\,A(S), \qquad (5.4.4)$$

where $\mathcal{L}(R) = |J(R)|^{1/2}$ is the geometrical spreading, and $T^c(R, S)$ is the phase shift due to caustics.

5.4.2 Point-Source Solutions. Radiation Functions

Continuation equations (5.4.3) and (5.4.4) cannot be used if the point source is situated at S because $J(S) = J^{(T)}(S) = \Omega^{(T)}(S) = \mathcal{L}(S) = 0$ there. For finite $A(S)$, (5.4.3) and (5.4.4) would, in this case, yield $A(R) = 0$ for all points R along ray Ω. As in isotropic media, however, the continuation equations can be modified to include even this case, but we would then need to assume that $A(S) = \infty$.

We shall now choose a new point S' situated on Ω between S and R, close to S. We use (4.14.41) for $J(R)$ and modify (5.4.3) as follows:

$$A(R) = \left[\frac{\rho(S')\mathcal{C}(S') \det \mathbf{Q}_2(S', S)}{\rho(R)\mathcal{C}(R) \det \mathbf{Q}_2(R, S)} \right]^{1/2} A(S'). \tag{5.4.5}$$

Here $\mathbf{Q}_2$ is the element of ray propagator matrix $\mathbf{\Pi}(R, S)$ (see Section 4.14.3), and $\mathcal{C}$ is the phase velocity. Alternatively, we can also write

$$A(R) = \left[\frac{\rho(S')\mathcal{C}(S')}{\rho(R)\mathcal{C}(R)} \right]^{1/2} \frac{\mathcal{L}(S', S)}{\mathcal{L}(R, S)} \exp[iT^c(R, S')] A(S'), \tag{5.4.6}$$

where $\mathcal{L}(R, S)$ is the *relative geometrical spreading*, $\mathcal{L}(R, S) = |\det \mathbf{Q}_2(R, S)|^{1/2}$. As $\mathbf{Q}_2(R, S) = -\mathbf{Q}_2^T(S, R)$, the relative geometrical spreading *is reciprocal*, $\mathcal{L}(R, S) = \mathcal{L}(S, R)$.

Now we shall move point S' along Ω to point S. As we know, $\mathcal{L}(S', S) \to 0$ in this case. We shall, however, assume that product $\mathcal{L}(S', S)A(S')$ remains finite for $S' \to S$. If we introduce function

$$\mathcal{G}(S; \gamma_1, \gamma_2) = \lim_{S' \to S} \{\mathcal{L}(S', S)A(S')\}, \tag{5.4.7}$$

(5.4.6) will yield

$$A(R) = \left[\frac{\rho(S)\mathcal{C}(S)}{\rho(R)\mathcal{C}(R)} \right]^{1/2} \frac{\mathcal{G}(S; \gamma_1, \gamma_2)}{\mathcal{L}(R, S)} \exp[iT^c(R, S)]. \tag{5.4.8}$$

Radiation function $\mathcal{G}(S; \gamma_1, \gamma_2)$ has been introduced in much the same way as the radiation function $\mathcal{G}_k^{(q)}(S; \gamma_1, \gamma_2)$ for an isotropic medium. As the limit is taken along the ray and rays are parameterized by ray parameters γ_1 and γ_2, $\mathcal{G}$ is a function not only of the position of source S but also of ray parameters γ_1 and γ_2. The derivation of the radiation function of an arbitrarily oriented single-force point source will be given in Section 5.4.5.

In an isotropic homogeneous medium, the radiation function represents the distribution of amplitudes of the wave generated by a point source along a sphere whose center is at S. This is simple to see because $\mathcal{L}(R, S) = \mathcal{C}l(R, S)$ in isotropic media, where $l(R, S)$ is the distance between S and R. In anisotropic media, however, the situation is different. Relative geometrical spreading $\mathcal{L}(R, S)$ is a complicated function of γ_1 and γ_2 and affects the radiation properties of the source considerably. Thus, $\mathcal{G}(S; \gamma_1, \gamma_2)$ in anisotropic media does not completely represent the directional properties of the source but represents only one part of it. The second part is represented by the geometrical spreading.

In addition to the radiation function $\mathcal{G}(S; \gamma_1, \gamma_2)$ given by (5.4.7), we shall also introduce *directivity pattern* $\mathcal{F}(S; \gamma_1, \gamma_2)$. As in isotropic media, we introduce directivity pattern $\mathcal{F}(S; \gamma_1, \gamma_2)$ in a locally homogeneous medium in the vicinity of point S by the equation

$$\mathcal{F}(S; \gamma_1, \gamma_2) = (\mathcal{G}(S; \gamma_1, \gamma_2)/\mathcal{L}(R, S))_{l(R,S)=1}, \tag{5.4.9}$$

where $l(R, S)$ is the distance between R and S. Thus, in a locally homogeneous medium, the directivity pattern $\mathcal{F}(S; \gamma_1, \gamma_2)$ of the wave under consideration, generated by a point source situated at S, represents the *angular distribution* of ray theory amplitudes of the wave along a unit sphere whose center is at S. Note the basic difference between $\mathcal{F}$ and $\mathcal{G}$ in the anisotropic medium. Whereas the difference is only formal in the isotropic medium, it may be very distinct in the anisotropic media. For more details on point-source solutions in anisotropic media and on their radiation patterns, see Kawasaki and Tanimoto (1981),

Hanyga (1984), Ben-Menahem (1990), Ben-Menahem and Sena (1990), Gajewski (1993), Tsvankin (1995), and Pšenčík and Teles (1996). The last reference also presents interesting numerical examples showing nicely the influence of relative geometrical spreading $\mathcal{L}(R, S)$ on the directivity pattern $\mathcal{F}(S; \gamma_1, \gamma_2)$.

The computation of radiation function $\mathcal{G}(S; \gamma_1, \gamma_2)$ for a single-force point source will be described in Section 5.4.5. The solution will be obtained by matching (5.4.8) with the solutions for a homogeneous anisotropic medium given in Section 2.5.5. In this way, solutions for different types of sources can also be obtained.

For completeness, we shall also give the final equation for $\vec{u}(R, t)$, assuming a point source situated at point S. Using (5.4.1), (5.4.2), and (5.4.8), we obtain

$$\vec{u}(R, t) = \left[\frac{\rho(S)\mathcal{C}(S)}{\rho(R)\mathcal{C}(R)} \right]^{1/2} \frac{\mathcal{G}(S; \gamma_1, \gamma_2)}{\mathcal{L}(R, S)}$$
$$\times \exp[iT^c(R, S)]\vec{g}(R)F(t - T(R, S)).$$

5.4.3 Amplitudes Across an Interface

We shall now study the reflected/transmitted waves across structural interfaces. Consider ray Ω and two points S and R situated on Ω. We assume that ray Ω strikes interface Σ at point Q situated between S and R. In addition to Q, we also introduce point $\tilde{Q}$, coinciding with Q but situated on the reflected/transmitted branch of the ray. Thus, points S and Q are situated on the incident branch of the ray, and $\tilde{Q}$ and R are located on the R/T branch of the ray. We shall now derive the continuation relations for amplitudes valid across interface Σ from S to R.

Along the incident branch of the ray, we can use (5.4.4),

$$A(Q) = \left[\frac{\rho(S)\mathcal{U}(S)}{\rho(Q)\mathcal{U}(Q)} \right]^{1/2} \frac{\mathcal{L}(S)}{\mathcal{L}(Q)} \exp[iT^c(Q, S)] \, A(S).$$

Across the interface,

$$A(\tilde{Q}) = R \, A(Q),$$

where R is the appropriate displacement R/T coefficient. It may correspond either to an unconverted wave or to a converted wave, depending on the types of incident and R/T waves. We shall again use A for the amplitude function of the wave generated at $\tilde{Q}$, even though the R/T wave may be converted on Σ. Along the R/T branch of the ray, the continuation formula reads:

$$A(R) = \left[\frac{\rho(\tilde{Q})\mathcal{U}(\tilde{Q})}{\rho(R)\mathcal{U}(R)} \right]^{1/2} \frac{\mathcal{L}(\tilde{Q})}{\mathcal{L}(R)} \exp[iT^c(R, \tilde{Q})] \, A(\tilde{Q}).$$

Combining these three equations, we obtain the continuation formula from S to R in the following form:

$$A(R) = \left[\frac{\rho(S)\mathcal{U}(S)}{\rho(R)\mathcal{U}(R)} \right]^{1/2} \frac{\mathcal{L}(S)}{\mathcal{L}(R)} \mathcal{R}(Q) \exp[iT^c(R, S)] \, A(S), \tag{5.4.10}$$

where

$$\mathcal{R}(Q) = R(Q) \left[\frac{\rho(\tilde{Q})\mathcal{U}(\tilde{Q})}{\rho(Q)\mathcal{U}(Q)} \right]^{1/2} \frac{\mathcal{L}(\tilde{Q})}{\mathcal{L}(Q)}, \tag{5.4.11}$$

$$T^c(R, S) = T^c(R, \tilde{Q}) + T^c(Q, S). \tag{5.4.12}$$

$\mathcal{R}(Q)$ is referred to as the *normalized displacement R/T coefficient* to distinguish it from standard displacement R/T coefficient $R(Q)$.

We shall show that $\mathcal{R}(Q)$ represents the displacement R/T coefficient, normalized with respect to the energy flux in the direction perpendicular to the interface. Because geometrical spreading $\mathcal{L}(Q) = |J(Q)|^{1/2}$ represents the cross-sectional area of ray tube, we obtain $\mathcal{L}(\tilde{Q})/\mathcal{L}(Q) = (\cos i(\tilde{Q})/\cos i(Q))^{1/2}$, where $i(Q)$ is the angle of incidence and $i(\tilde{Q})$ the angle of reflection/transmission. Hence,

$$\mathcal{R}(Q) = R(Q)\left[\frac{\rho(\tilde{Q})\mathcal{U}(\tilde{Q})\cos i(\tilde{Q})}{\rho(Q)\mathcal{U}(Q)\cos i(Q)}\right]^{1/2} = R(Q)\left[\frac{\rho(\tilde{Q})\mathcal{U}_n(\tilde{Q})}{\rho(Q)\mathcal{U}_n(Q)}\right]^{1/2}.$$

$$(5.4.13)$$

Here $\mathcal{U}_n(Q)$ and $\mathcal{U}_n(\tilde{Q})$ are the normal components (perpendicular to the interface) of the group velocity vectors of incident and R/T waves, respectively. The physical explanation of normalization factor $\rho(\tilde{Q})\mathcal{U}_n(\tilde{Q})/\rho(Q)\mathcal{U}_n(Q)$ remains practically the same as for isotropic media; see Section 5.3.3. The modulus of the normalized displacement R/T coefficient $|\mathcal{R}(Q)|$ represents the square root of the absolute value of the ratio of the energy flux of the appropriate R/T wave to the energy flux of the incident wave, both of them considered along normal $\vec{n}$ to interface Σ at Q. The argument of normalized coefficient $\mathcal{R}(Q)$, however, is the same as the argument of standard displacement R/T coefficient $R(Q)$.

As we showed in Section 5.3.7, the normalized R/T coefficients for isotropic media are reciprocal, in the following sense: $\mathcal{R}_{mn}(Q) = \mathcal{R}_{nm}(\tilde{Q})$; see (5.3.31). The proof for isotropic media was straightforward, using explicit expressions for the R/T coefficients. For anisotropic media, the explicit expressions for R/T coefficients would be more complex. The interested reader is referred to Section 5.4.7 and to Chapman (1994) for the proof that reciprocity equation (5.3.31) remains valid even for anisotropic media.

Equation (5.4.10) can be simply modified to consider a point source at S:

$$A(R) = \left[\frac{\rho(S)\mathcal{C}(S)}{\rho(R)\mathcal{C}(R)}\right]^{1/2}\frac{\mathcal{G}(S;\gamma_1,\gamma_2)}{\mathcal{L}(R,S)}\mathcal{R}(Q)\exp[iT^c(R,S)]. \qquad (5.4.14)$$

A systematic parameteric investigation of the reflection/transmission coefficients of a plane wave on a plane interface between two homogeneous anisotropic halfspaces is not simple because these coefficients depend on a great number of parameters. In general, they depend on 2×21 elastic moduli, on two densities, and on two tangential components of the slowness vector of the incident wave. The number of parameters is reduced for some simpler anisotropy symmetries (for example, for transversely isotropic media), but still it remains prohibitively large for a systematic study. For many references to such investigations, see Section 2.3.3.

For more details on R/T coefficients and R/T matrices at an interface between two homogeneous anisotropic halfspaces, see Section 5.4.7.

5.4.4 Amplitudes in 3-D Layered Structures

It is simple to generalize (5.4.10) and (5.4.14) for any multiply reflected/transmitted, possibly converted, wave propagating in a 3-D laterally varying layered anisotropic structure. We again consider ray Ω and two points, S and R, situated on Ω. In addition, we assume that the ray strikes N times some structural interfaces between S and R. We denote the relevant points of incidence successively Q_i, $i = 1, 2, \ldots, N$, and the relevant points of R/T $\tilde{Q}_i$, $i = 1, 2, \ldots, N$. We also denote $\tilde{Q}_0 = S$ and $Q_{N+1} = R$.

The continuation formula (5.4.10) can be generalized to read

$$A(R) = \left[\frac{\rho(S)\mathcal{U}(S)}{\rho(R)\mathcal{U}(R)} \right]^{1/2} \frac{\mathcal{L}(S)}{\mathcal{L}(R)} \, \mathcal{R}^C \, \exp[iT^c(R, S)] \, A(S), \tag{5.4.15}$$

where

$$T^c(R, S) = \sum_{k=1}^{N+1} T^c(Q_k, \tilde{Q}_{k-1}), \tag{5.4.16}$$

$$\mathcal{R}^C = \prod_{k=1}^{N} \mathcal{R}(Q_k) = \prod_{k=1}^{N} R(Q_k) \left[\frac{\rho(\tilde{Q}_k)\mathcal{U}_n(\tilde{Q}_k)}{\rho(Q_k)\mathcal{U}_n(Q_k)} \right]^{1/2}. \tag{5.4.17}$$

Here $\mathcal{R}^C$ is the *complete reflection/transmission coefficient* along the ray Ω from S to R. It equals the product of the normalized displacement R/T coefficients $\mathcal{R}$ at all points of incidence Q_i, $i = 1, 2, \ldots, N$, between S and R.

For a point source situated at S, we can modify (5.4.15) to read

$$A(R) = \left[\frac{\rho(S)\mathcal{C}(S)}{\rho(R)\mathcal{C}(R)} \right]^{1/2} \frac{\mathcal{G}(S; \gamma_1, \gamma_2)}{\mathcal{L}(R, S)} \, \mathcal{R}^C \, \exp[iT^c(R, S)]. \tag{5.4.18}$$

The complete R/T coefficient $\mathcal{R}^C$ is reciprocal because it is the product of reciprocal coefficients.

The final equation for the displacement vector of an arbitrary multiply reflected wave propagating in a 3-D laterally varying anisotropic layered structure, generated by a point source situated at S, now reads

$$\vec{u}(R, t) = \left[\frac{\rho(S)\mathcal{C}(S)}{\rho(R)\mathcal{C}(R)} \right]^{1/2} \frac{\mathcal{G}(S; \gamma_1, \gamma_2)}{\mathcal{L}(R, S)} \, \mathcal{R}^C$$
$$\times \exp[iT^c(R, S)] \, \vec{g}(R) F(t - T(R, S)). \tag{5.4.19}$$

5.4.5 Ray-Theory Green Function

We shall now derive the general expressions for the ray-theory Green function corresponding to an arbitrary body wave propagating in an inhomogeneous anisotropic layered structure. In fact, we have practically derived it; see (5.4.18) and (5.4.19). The only thing that remains to be done is to determine the radiation function $\mathcal{G}(S; \gamma_1, \gamma_2)$ corresponding to the unit single-force point source. We shall determine it by matching (5.4.19) with expressions (2.5.75) for the high-frequency asymptotic Green function, derived for homogeneous anisotropic medium in Section 2.5. In the time domain,

$$G_{in}(R, t; S, 0) = \frac{g_i g_n \exp\left[i\frac{1}{2}\pi\sigma_0\right]}{4\pi\rho\mathcal{U}\sqrt{|K^S|r}} \, \delta^{(A)}(t - T(R, S)). \tag{5.4.20}$$

Here $\delta^{(A)}(\zeta)$ denotes the analytical delta function. Specifying (5.4.19) for a homogeneous medium and for $F(\zeta) = \delta^{(A)}(\zeta)$ yields

$$u_i(R) = \frac{\mathcal{G}(S; \gamma_1, \gamma_2)}{\mathcal{L}(R, S)} \, g_i(R) \delta^{(A)}(t - T(R, S)). \tag{5.4.21}$$

Taking into account the relation (4.14.38) between det $\mathbf{Q}_2(R, S)$ and K^S, we obtain a simple expression for the radiation function of an unit single force oriented along the x_n-axis in a homogeneous anisotropic medium,

$$\mathcal{G}(S; \gamma_1, \gamma_2) = \frac{g_n(S)}{4\pi\rho(S)\mathcal{C}(S)} \exp\left[\tfrac{1}{2}i\pi\sigma_0(S)\right]. \tag{5.4.22}$$

In certain applications, it is useful to know also a more general radiation function, corresponding to an *arbitrarily oriented single-force point source* $\vec{f}_0(S)$. It is given by the relation

$$\mathcal{G}(S; \gamma_1, \gamma_2) = \frac{g_k(S)f_{0k}^{(x)}(S)}{4\pi\rho(S)\mathcal{C}(S)} \exp\left[\tfrac{1}{2}i\pi\sigma_0(S)\right]. \tag{5.4.23}$$

Here $f_{0k}^{(x)}$ denotes the kth Cartesian components of $\vec{f}_0(S)$.

Considering (5.4.22) in (5.4.19) yields the final expression for the ray-theory Green function in an inhomogeneous anisotropic layered structure:

$$G_{in}(R, t; S, t_0) = \frac{g_n(S)g_i(R)\exp[iT^G(R, S)]}{4\pi[\rho(S)\rho(R)\mathcal{C}(S)\mathcal{C}(R)]^{1/2}\mathcal{L}(R, S)}$$
$$\times \mathcal{R}^C \, \delta^{(A)}(t - t_0 - T(R, S)). \tag{5.4.24}$$

Equation (5.4.24) can be used also for the Green function $G_{in}(R, S, \omega)$ in the frequency domain; we only replace $\delta^{(A)}(t - t_0 - T)$ by $\exp[i\omega T]$. The meanings of all the symbols in (5.4.24) have been explained earlier. Only the phase shift due to caustics $T^c(R, S)$ have been replaced by $T^G(R, S)$ to include also σ_0. We can call $T^G(R, S)$ the complete phase shift due to caustics of the ray-theory Green function in anisotropic media. It is given by the relation

$$T^G(R, S) = T^c(R, S) + \tfrac{1}{2}\pi\sigma_0(S) = -\tfrac{1}{2}\pi[k(R, S) - \sigma_0(S)]. \tag{5.4.25}$$

Here $k(R, S)$ is the KMAH index, corresponding to caustics situated on ray Ω *between S and R*. Its computation is discussed in detail in Section 4.14.13. $\sigma_0(S)$ corresponds to the point source; see Section 2.5.5. We can also introduce $k^G(R, S)$ using the relation

$$k^G(R, S) = k(R, S) - \sigma_0(S). \tag{5.4.26}$$

Then $k^G(R, S)$ represents the KMAH index of the ray-theory Green function in anisotropic media and even includes its initial value $-\sigma_0(S)$ at the point source S.

Klimeš (1997c) has proved that the complete phase shift of the ray-theory Green function due to caustics is reciprocal, $T^G(R, S) = T^G(S, R)$; see Section 4.14.13. Also the relative geometrical spreading and the complete normalized R/T coefficients are reciprocal: $\mathcal{L}(R, S) = \mathcal{L}(S, R)$ and $\mathcal{R}^C(R, S) = \mathcal{R}^C(S, R)$. Equation (5.4.24) then shows that the ray-theory Green function $G_{in}(R, t; S, t_0)$ is reciprocal in the following sense:

$$G_{in}(R, t; S, t_0) = G_{ni}(S, t; R, t_0). \tag{5.4.27}$$

The reciprocity relation (5.4.27) is valid for the ray-theory Green function corresponding to any multiply reflected (possibly converted) wave propagating in a 3-D laterally varying anisotropic layered structure. The same reciprocity relation is, of course, valid in isotropic media; see (5.2.56).

Expressions for the ray-theory Green function for anisotropic inhomogeneous media were derived independently by several authors. See Červený (1990), Ben-Menahem,

Gibson, and Sena (1991), Kendall, Guest, and Thomson (1992), and Pšenčík and Teles (1996).

5.4.6 Quasi-Isotropic Ray Theory. qS Wave Coupling

Ray theory for inhomogeneous anisotropic media, presented in Section 3.6, cannot be used for qS waves if the eigenvalues G_1 and G_2 of the Christoffel matrix, related to qS1 and qS2 waves, coincide or are close to each other. The qS1 and qS2 waves do not propagate independently in this case but are *mutually coupled*. We speak of *the qS wave coupling*. This may happen globally in an inhomogeneous weakly anisotropic medium (close to isotropic) or locally in the vicinity of shear wave singular directions, see Sections 2.2.8 and 2.2.9. In a limit of infinitely weak anisotropy, the zero-order ray theory for two independent qS waves, described in Section 3.6, does not yield the results known in isotropic inhomogeneous media. Thus, the "isotropic" and "anisotropic" ray methods are in conflict in this case. The methods to investigate the qS waves in such cases must take into account the coupling between the two qS waves (coupling ray theory). See also a brief discussion in Sections 3.6.1 and 3.9.4. The travel times of qS waves in situations with $G_1 \doteq G_2$ were derived and discussed in Section 3.9.4 using the degenerate perturbation method. Here we shall discuss the amplitudes of qS waves propagating in inhomogeneous weakly anisotropic media. The most important property of amplitudes of coupled qS waves propagating in inhomogeneous weakly anisotropic media is that they are frequency dependent. We shall use the quasi-isotropic approximation and the quasi-isotropic ray theory; see Section 3.9.4. For completeness, we shall also discuss the propagation of qP waves in inhomogeneous weakly anisotropic media.

Various methods have been used to investigate the qS wave coupling in inhomogeneous weakly anisotropic elastic media. The most comprehensive treatment, based on generalized Born approximation, was given by Coates and Chapman (1990b). See Section 2.6.2 for a brief explanation of generalized Born approximation. In the generalized Born approximation, the error terms produced by substituting a zeroth-order ray-theory Green's function into the elastodynamic equations are treated as source terms of scattered field. Using the perturbation and asymptotic methods, the volume scattering integral is simplified and reduced to quadratures along the ray. For quasi-isotropic and alternative approaches, see, for example, Kravtsov (1968), Kravtsov and Orlov (1980), Chapman and Shearer (1989), Guest, Thomson, and Kendall (1992), Thomson, Kendall, and Guest (1992), Kiselev (1994), Sharafutdinov (1994), Kravtsov, Naida, and Fuki (1996), Druzhinin (1996), Zillmer, Kashtan, and Gajewski (1998), and Pšenčík (1998). A broad literature related to the quasi-isotropic approximation in other branches of physics is given by Kravtsov and Orlov (1980). The accuracy of different approaches was numerically studied by Bulant, Klimeš, and Pšenčík (1999).

Here we do not intend to treat the problem of coupling of qS waves in weakly anisotropic media in a great detail. For this reason, we shall use only a simple derivation based on the quasi-isotropic approximation and on the formal Debye procedure, in which the perturbation Δa_{ijkl} of density-normalized elastic parameters a_{ijkl} from isotropic background to weakly anisotropic medium is formally considered to be of the order ω^{-1}. See Kravtsov and Orlov (1980) for electromagnetic waves and Pšenčík (1998) for elastic waves.

We shall consider an isotropic inhomogeneous background, described by velocities $\alpha(x_i)$ and $\beta(x_i)$ and the density $\rho(x_i)$, and a perturbed weakly anisotropic inhomogeneous medium, with density-normalized elastic parameters $a_{ijkl}(x_i)$. We define the perturbations

Δa_{ijkl} by the relation:

$$a_{ijkl}(x_i) = a_{ijkl}^0(x_i) + \Delta a_{ijkl}(x_i), \tag{5.4.28}$$

where

$$a_{ijkl}^0(x_i) = (\alpha^2 - 2\beta^2)\delta_{ij}\delta_{kl} + \beta^2(\delta_{ik}\delta_{jl} + \delta_{il}\delta_{jk}). \tag{5.4.29}$$

We shall now use (2.4.15) in the frequency domain in the weakly anisotropic medium and obtain

$$i\omega\bar{N}_i(\vec{U}) + \bar{M}_i(\vec{U}) = 0, \tag{5.4.30}$$

where $\bar{N}_i$ and $\bar{M}_i$ are density-normalized N_i and M_i, given by (2.4.41): $\bar{N}_i = N_i/\rho$ and $\bar{M}_i = M_i/\rho$. Now we insert (5.4.28) into (5.4.30) and take into account the Debye procedure, $\Delta a_{ijkl} \sim 1/\omega$. Neglecting terms of the order of $\sim\omega^{-1}$, we obtain

$$i\omega\bar{N}_i^0(\vec{U}) + \bar{M}_i^0(\vec{U}) + i\omega\Delta a_{ijkl}p_l p_j U_k = 0. \tag{5.4.31}$$

Here $\bar{N}_i^0$ and $\bar{M}_i^0$ are density-normalized N_i and M_i, corresponding to isotropic background, given by (2.4.16). Equation (5.4.31) yields two equations

$$\bar{N}_i^0(\vec{U}) = 0, \tag{5.4.32}$$
$$\bar{M}_i^0(\vec{U}) + i\omega\Delta a_{ijkl}p_l p_j U_k = 0. \tag{5.4.33}$$

As in Section 2.4.2, Equation (5.4.32) describes the kinematic properties and polarization of P and S waves propagating in background isotropic media. It yields eikonal equations $p_i p_i = 1/\alpha^2$ for P waves and $p_i p_i = 1/\beta^2$ for S waves and relevant ray tracing systems. It also yields the polarization of P and S waves:

$$\begin{aligned}\vec{U} &= A\,\vec{N} & &\text{for P waves,}\\ \vec{U} &= B\,\vec{e}^{(1)} + C\,\vec{e}^{(2)} & &\text{for S waves;}\end{aligned} \tag{5.4.34}$$

see (2.4.26) and (2.4.28). Here $\vec{N}$ is the unit normal to the wavefront in the isotropic background medium, and $\vec{e}^{(1)}$ and $\vec{e}^{(2)}$ are two mutually perpendicular unit vectors, perpendicular to $\vec{N}$.

Now we shall discuss (5.4.33). For $\Delta a_{ijkl} = 0$, (5.4.33) yields standard transport equations; see (2.4.30) for P waves and (2.4.34) for S waves. For $\Delta a_{ijkl} \neq 0$, the transport equations (2.4.30) and (2.4.34) have nonvanishing right-hand sides. We again emphasize that the second term in (5.4.33) is of the order of $\sim\omega^0$, due to the Debye procedure. We shall discuss (5.4.33) independently for qP and qS waves. For P waves in weakly anisotropic media, see also Sayers (1994).

1. qP WAVES
Inserting $\vec{U} = A\vec{N}$ into (5.4.33) and multiplying it by N_i, we obtain

$$\bar{M}_i^0(A\vec{N})N_i + i\omega B_{33}A = 0. \tag{5.4.35}$$

Here

$$B_{33} = \Delta a_{ijkl}p_l p_j N_i N_k = \alpha^{-2}\Delta a_{ijkl}N_i N_j N_k N_l. \tag{5.4.36}$$

Equation (5.4.35) can be suitably solved along any ray Ω^0 of the P wave constructed in the isotropic background medium. Using the expressions for $M_i(A\vec{N})N_i$ derived in Section

2.4.2, (5.4.35) can be expressed in the following form:

$$d \ln(A\sqrt{J\rho\alpha})/dT + \tfrac{1}{2} i\omega B_{33} = 0. \tag{5.4.37}$$

This gives the final solution for qP waves:

$$A(T) = \left[\frac{J(T_0)\rho(T_0)\alpha(T_0)}{J(T)\rho(T)\alpha(T)} \right]^{1/2} A(T_0)\mathcal{A}^{QP}(T, T_0), \tag{5.4.38}$$

where the quasi-isotropic correction factor $\mathcal{A}^{QP}$ is given by the relation

$$\mathcal{A}^{QP}(T, T_0) = \exp\left[-\tfrac{1}{2} i\omega \int_{T_0}^{T} B_{33}dT \right]. \tag{5.4.39}$$

Here the integral is taken along the ray Ω^0 of the P wave in the isotropic background medium, and T_0 denotes the initial travel time at an arbitrary point on Ω^0, at which $A(T_0)$ is known. Thus, the quasi-isotropic modification of qP waves consists in a multiplicative quasi-isotropic correction $\mathcal{A}^{QP}$, by which the amplitudes of P waves calculated in the background medium should be multiplied. If we compare (5.4.39) with (3.9.15) and take into account that $\vec{g} = \vec{N}$ for P waves in isotropic media, we can conclude that the factor $\mathcal{A}^{QP}(R, S)$ represents $\exp[i\omega\Delta T(R, S)]$, where $\Delta T(R, S)$ is the travel-time perturbation of P waves from isotropic to weakly anisotropic medium. Thus, the quasi-isotropic correction factor of qP waves $\mathcal{A}^{QP}$ takes into account the travel-time perturbation only.

2. qS WAVES

Inserting $\vec{U} = B\vec{e}^{(1)} + C\vec{e}^{(2)}$ into (5.4.33) and multiplying it by $e_i^{(K)}$, we obtain

$$\bar{M}_i^0\big(B\vec{e}^{(1)} + C\vec{e}^{(2)}\big)e_i^{(K)} + i\omega(B_{1K}B + B_{2K}C) = 0. \tag{5.4.40}$$

Here

$$B_{MN} = \Delta a_{ijkl}p_j p_l e_i^{(M)} e_k^{(N)} \tag{5.4.41}$$

are elements of the 2×2 weak-anisotropy matrix; see also Section 3.9.4. Note that $p_l = N_l/\beta$ for S waves. We now use equations for $\bar{M}_i^0(B\vec{e}^{(1)} + C\vec{e}^{(2)})e_i^{(K)}$ derived in Section 2.4.2 and introduce $B_0(T)$ and $C_0(T)$ by relations

$$B(T) = \left[\frac{J(T_0)\rho(T_0)\beta(T_0)}{J(T)\rho(T)\beta(T)} \right]^{1/2} B_0(T),$$

$$\tag{5.4.42}$$

$$C(T) = \left[\frac{J(T_0)\rho(T_0)\beta(T_0)}{J(T)\rho(T)\beta(T)} \right]^{1/2} C_0(T).$$

Here T_0 is the travel time corresponding to an arbitrary (initial) point on Ω^0. Then we obtain the system of two linear ordinary differential equations of the first order for B_0 and C_0, which can be solved along any ray Ω^0 of the S wave constructed in the isotropic background medium (common ray):

$$dB_0/dT + C_0 e_j^{(1)} de_j^{(2)}/dT + \tfrac{1}{2} i\omega(B_{11}B_0 + B_{12}C_0) = 0,$$

$$\tag{5.4.43}$$

$$dC_0/dT + B_0 e_j^{(2)} de_j^{(1)}/dT + \tfrac{1}{2} i\omega(B_{12}B_0 + B_{22}C_0) = 0.$$

As we can see, the *qS wave coupling system* (5.4.43) for B_0 and C_0 is coupled and frequency-dependent. The unit vectors $\vec{e}^{(1)}$ and $\vec{e}^{(2)}$ may be taken arbitrarily along the common ray

Ω^0; they only have to form a triplet of mutually perpendicular unit vectors with $\vec{N}$. Similar equations such as (5.4.43) with different choices of $\vec{e}^{(1)}$ and $\vec{e}^{(2)}$ are known from the literature. Kravtsov (1968) and Kravtsov and Orlov (1980) used the unit normal $\vec{n}$ and unit binormal $\vec{b}$ to Ω^0 as $\vec{e}^{(1)}$ and $\vec{e}^{(2)}$. Pšenčík (1998) and Zillmer, Kashtan, and Gajewski (1998) used the basis vectors $\vec{e}_1$ and $\vec{e}_2$ of the ray-centered coordinate system; see Section 4.1.1. In this choice, $\vec{e}^{(1)} d\vec{e}^{(2)}/dT = \vec{e}^{(2)} d\vec{e}^{(1)}/dT = 0$. The coupling equations derived by Coates and Chapman (1990b) using a generalized Born approximation are also similar to (5.4.43), with $\vec{e}^{(1)}$ and $\vec{e}^{(2)}$ representing the polarization vectors of qS1 and qS2 waves. In this choice, $B_{12} = 0$. For discussion of various choices, see Pšenčík (1998).

Note that the vectorial amplitude of the qS wave in a weakly anisotropic inhomogeneous medium is frequency-dependent. For this reason, we shall not use the term *vectorial ray amplitude* in this section but merely the term *qS vectorial amplitude*. Alternatively, it would be possible to speak of "zeroth-order quasi-isotropic approximation." The higher-order quasi-isotropic approximations are not investigated here. For the additional component of the first-order quasi-isotropic approximation, see Pšenčík (1998).

3. DECOMPOSITION OF qS VECTORIAL AMPLITUDES INTO RAY-CENTERED COMPONENTS

Here we shall use the choice of $\vec{e}^{(1)} = \vec{e}_1$ and $\vec{e}^{(2)} = \vec{e}_2$, where $\vec{e}_1$ and $\vec{e}_2$ are the basis vectors of the ray-centered coordinate system; see Section 4.1. As in (5.4.42), we introduce B_0^e and C_0^e by the relations:

$$B(T) = \left[\frac{J(T_0)\rho(T_0)\beta(T_0)}{J(T)\rho(T)\beta(T)} \right]^{1/2} B_0^e(T),$$

$$C(T) = \left[\frac{J(T_0)\rho(T_0)\beta(T_0)}{J(T)\rho(T)\beta(T)} \right]^{1/2} C_0^e(T). \tag{5.4.44}$$

As the second terms of equations (5.4.43) vanish, we obtain

$$\frac{d}{dT}\begin{pmatrix} B_0^e \\ C_0^e \end{pmatrix} = -\tfrac{1}{2} i\omega \mathbf{B}^e \begin{pmatrix} B_0^e \\ C_0^e \end{pmatrix}, \tag{5.4.45}$$

with

$$B_{IJ}^e = \Delta a_{ijkl} p_j p_l e_{Ii} e_{Jk}. \tag{5.4.46}$$

Here $\mathbf{B}^e$ is the 2×2 weak-anisotropy matrix (5.4.41), *expressed in terms of basis vectors* $\vec{e}_I$. The system (5.4.45) for B_0^e and C_0^e can be still simplified. We decompose matrix $\mathbf{B}^e$ as follows:

$$\mathbf{B}^e = \tfrac{1}{2}(\mathbf{B}^{ae} + \mathbf{B}^{se}), \qquad \mathbf{B}^{ae} = \begin{pmatrix} B_{11}^e + B_{22}^e & 0 \\ 0 & B_{11}^e + B_{22}^e \end{pmatrix},$$

$$\mathbf{B}^{se} = \begin{pmatrix} B_{11}^e - B_{22}^e & 2B_{12}^e \\ 2B_{12}^e & B_{22}^e - B_{11}^e \end{pmatrix}. \tag{5.4.47}$$

Here $\mathbf{B}^{ae}$ is the *average qS wave matrix*, and $\mathbf{B}^{se}$ is the *qS wave splitting matrix*. We further introduce

$$B_0^e = B^e \mathcal{A}^{QS}(T, T_0), \qquad C_0^e = C^e \mathcal{A}^{QS}(T, T_0), \tag{5.4.48}$$

where $\mathcal{A}^{QS}(T, T_0)$ is the quasi-isotropic qS wave correction factor, given by the relation

$$\mathcal{A}^{QS}(T, T_0) = \exp\left[-\tfrac{1}{4}i\omega \int_{T_0}^{T} \left(B_{11}^e + B_{22}^e \right) \mathrm{d}T \right].$$

(5.4.49)

The integral in (5.4.49) is taken along the ray Ω^0 of S wave in the background isotropic medium. The expression in the exponential function (5.4.49) represents the average qS wave travel-time perturbation ΔT^a (see (3.9.28)) and is analogous to $\mathcal{A}^{QP}(T, T_0)$ for qP waves, given by (5.4.39). Then we can write the *final equations* for the vectorial amplitude of the qS wave in weakly anisotropic media $\vec{U}(T)$ in the following form

$$\vec{U}(T) = \left[\frac{J(T_0)\rho(T_0)\beta(T_0)}{J(T)\rho(T)\beta(T)} \right]^{1/2} \mathcal{A}^{QS}(T, T_0)[B^e(T)\vec{e}_1(T) + C^e(T)\vec{e}_2(T)],$$

(5.4.50)

where $B^e(T)$ and $C^e(T)$ are solutions of the system of two coupled linear ordinary differential equations of the first order (the qS wave coupling system in ray-centered components):

$$\frac{\mathrm{d}}{\mathrm{d}T}\begin{pmatrix} B^e \\ C^e \end{pmatrix} = -\tfrac{1}{4}i\omega \mathbf{B}^s \begin{pmatrix} B^e \\ C^e \end{pmatrix}, \qquad \mathbf{B}^s = \mathbf{B}^{se},$$

(5.4.51)

with the initial conditions $B^e(T_0)$ and $C^e(T_0)$ at $T = T_0$ corresponding to the amplitude vector $\vec{U}(T_0) = B^e(T_0)\vec{e}_1(T_0) + C^e(T_0)\vec{e}_2(T_0)$. Consequently, the amplitudes B^e and C^e of qS waves are coupled and frequency-dependent.

The quasi-isotropic approach described here was used to compute synthetic seismograms in weakly anisotropic media by Pšenčík and Dellinger (2000). They used the anisotropic reflectivity modeling program by Mallick and Frazer (1990) to test the validity of the approach. The results show that the quasi-isotropic approach spans the gap between the isotropic and anisotropic ray methods. It can be used in isotropic regions (where it reduces to the isotropic ray method), in regions of weak anisotropy (where no ray method works properly), and even in regions of moderately strong anisotropy (in which the qS waves decouple and could be modeled using the anisotropic ray method).

4. DECOMPOSITION OF qS VECTORIAL AMPLITUDE INTO $\vec{g}^{(1)}$ AND $\vec{g}^{(2)}$ COMPONENTS

As in (3.9.17), we introduce vectors $\vec{e}^{(1)} = \vec{g}^{(1)}$ and $\vec{e}^{(2)} = \vec{g}^{(2)}$ by relations

$$g_k^{(1)} = a_J^{(1)} e_{Jk}, \qquad g_k^{(2)} = a_J^{(2)} e_{Jk},$$

(5.4.52)

where $\vec{e}_1$ and $\vec{e}_2$ are the basis vectors of the ray-centered coordinate system, and $\vec{a}^{(1)}$ and $\vec{a}^{(2)}$ are eigenvectors of the 2×2 weak anisotropy matrix $\mathbf{B}^e$; see (5.4.46). The summation is over $J = 1, 2$. The eigenvectors $\vec{a}^{(1)}$ and $\vec{a}^{(2)}$ are given by relations

$$a_1^{(1)} = a_2^{(2)} = 2^{-1/2}\left[1 + D^{-1}\left(B_{11}^e - B_{22}^e \right) \right]^{1/2},$$
$$a_2^{(1)} = -a_1^{(2)} = 2^{-1/2} \operatorname{sgn} B_{12}^e \left[1 - D^{-1}\left(B_{11}^e - B_{22}^e \right) \right]^{1/2},$$

(5.4.53)

with

$$D = \left[\left(B_{11}^e - B_{22}^e \right)^2 + 4\left(B_{12}^e \right)^2 \right]^{1/2};$$

(5.4.54)

see (3.9.22). We shall also use the transformation matrix

$$\mathbf{A} = \begin{pmatrix} a_1^{(1)} & -a_2^{(1)} \\ a_2^{(1)} & a_1^{(1)} \end{pmatrix}; \tag{5.4.55}$$

see (3.9.26).

We remind the reader that $\vec{g}^{(I)}$ are strictly perpendicular to the ray Ω^0 and that they are approximately equal to the projection of the polarization vectors of qS waves into the plane perpendicular to the ray Ω^0. We denote by $\mathbf{B}^g$ the 2×2 matrix with components

$$B_{IJ}^g = \Delta a_{ijkl} p_i p_l g_j^{(I)} g_k^{(J)}; \tag{5.4.56}$$

see also (3.9.25). Here $\mathbf{B}^g$ is the 2×2 weak-anisotropy matrix, expressed in terms of eigenvectors $\vec{g}^{(I)}$. Finally, we introduce B_0^g and C_0^g by relations:

$$B(T) = \left[\frac{J(T_0)\rho(T_0)\beta(T_0)}{J(T)\rho(T)\beta(T)} \right]^{1/2} B_0^g(T),$$

$$C(T) = \left[\frac{J(T_0)\rho(T_0)\beta(T_0)}{J(T)\rho(T)\beta(T)} \right]^{1/2} C_0^g(T), \tag{5.4.57}$$

as in (5.4.42). Then (5.4.43) reads

$$\frac{\mathrm{d}}{\mathrm{d}T} \begin{pmatrix} B_0^g \\ C_0^g \end{pmatrix} = \left[\begin{pmatrix} 0 & \gamma \\ -\gamma & 0 \end{pmatrix} - \tfrac{1}{2} i\omega \mathbf{B}^g \right] \begin{pmatrix} B_0^g \\ C_0^g \end{pmatrix}. \tag{5.4.58}$$

The *coupling function* $\gamma = \gamma(T)$ is given by the relation

$$\gamma(T) = \vec{g}^{(2)} \mathrm{d}\vec{g}^{(1)}/\mathrm{d}T = \vec{a}^{(2)} \mathrm{d}\vec{a}^{(1)}/\mathrm{d}T. \tag{5.4.59}$$

Here $\vec{a}^{(I)}$ are given by (5.4.53) and represent the eigenvectors of the 2×2 weak anisotropy matrix $\mathbf{B}^e$; see (5.4.46). As $\vec{g}^{(1)} \cdot \vec{g}^{(2)} = 0$, we also have $\gamma(T) = -\vec{g}^{(1)} \mathrm{d}\vec{g}^{(2)}/\mathrm{d}T$ and similarly $\gamma(T) = -\vec{a}^{(1)} \mathrm{d}\vec{a}^{(2)}/\mathrm{d}T$. Using the notation $a_1^{(1)} = \vec{g}^{(1)} \cdot \vec{e}_1 = \cos\varphi$, $a_2^{(1)} = \vec{g}^{(1)} \cdot \vec{e}_2 = \sin\varphi$, we also obtain

$$\gamma(T) = \mathrm{d}\varphi/\mathrm{d}T. \tag{5.4.60}$$

Thus, φ is the angle between $\vec{e}_1$ and $\vec{g}^{(1)}$ and the coupling function represents the velocity of the rotation of eigenvectors $\vec{g}^{(I)}$ about the ray Ω^0 as the wave progresses, in the frame specified by $\vec{e}_1$ and $\vec{e}_2$.

Matrix $\mathbf{B}^g$ in (5.4.58) is diagonal; see (3.9.27). Consequently, the term with $\mathbf{B}^g$ in (5.4.58) could be removed by a suitable substitution. Here we shall, however, proceed in a slightly different way. We again decompose $\mathbf{B}^g$ into two matrices, as in (5.4.47):

$$\mathbf{B}^g = \tfrac{1}{2}(\mathbf{B}^{ag} + \mathbf{B}^{sg}), \qquad \mathbf{B}^{ag} = \begin{pmatrix} B_{11}^g + B_{22}^g & 0 \\ 0 & B_{11}^g + B_{22}^g \end{pmatrix},$$

$$\mathbf{B}^{sg} = \begin{pmatrix} D & 0 \\ 0 & -D \end{pmatrix}. \tag{5.4.61}$$

Here $\mathbf{B}^{ag}$ represents average qS wave matrix, and $\mathbf{B}^{sg}$ represents the qS wave splitting matrix. They are related to $\mathbf{B}^{ae}$ and $\mathbf{B}^{se}$, given by (5.4.47), by relations $\mathbf{B}^{se} = \mathbf{A}\mathbf{B}^{sg}\mathbf{A}^T$ and $\mathbf{B}^{ae} = \mathbf{A}\mathbf{B}^{ag}\mathbf{A}^T$, where $\mathbf{A}$ is given by (5.4.55). Similarly, as in (5.4.48), we introduce B^g and C^g by relations

$$B_0^g(T) = B^g(T)\mathcal{A}^{QS}(T, T_0), \qquad C_0^g(T) = C^g(T)\mathcal{A}^{QS}(T, T_0), \tag{5.4.62}$$

where $\mathcal{A}^{QS}(T, T_0)$ is given by the relation

$$\mathcal{A}^{QS}(T, T_0) = \exp\left[-\tfrac{1}{4}i\omega \int_{T_0}^{T} \left(B_{11}^g + B_{22}^g\right)dT\right], \tag{5.4.63}$$

and the integral is taken along the ray Ω^0. Because the trace of matrix $\mathbf{B}^g$ is invariant with respect to the choice of basis vectors, $\mathcal{A}^{QS}(T, T_0)$ given by (5.4.63) is the same as that given by (5.4.49). Then we can write the *final expression* for the amplitude vector of the qS wave in a weakly anisotropic medium as follows:

$$\vec{U}(T) = \left[\frac{J(T_0)\rho(T_0)\beta(T_0)}{J(T)\rho(T)\beta(T)}\right]^{1/2}$$
$$\times \mathcal{A}^{QS}(T, T_0)\left[B^g(T)\vec{g}^{(1)}(T) + C^g(T)\vec{g}^{(2)}(T)\right], \tag{5.4.64}$$

where $B^g(T)$ and $C^g(T)$ are solutions of the system of two coupled linear ordinary differential equations of the first order (the qS wave coupling system in $\vec{g}^{(1)}$, $\vec{g}^{(2)}$ components):

$$\frac{d}{dT}\begin{pmatrix} B^g \\ C^g \end{pmatrix} = \mathbf{B}^d\begin{pmatrix} B^g \\ C^g \end{pmatrix}, \qquad \mathbf{B}^d = \begin{pmatrix} -\tfrac{1}{4}i\omega D & \gamma \\ -\gamma & \tfrac{1}{4}i\omega D \end{pmatrix}. \tag{5.4.65}$$

The initial conditions $B^g(T_0)$ and $C^g(T_0)$ at $T = T_0$ should correspond to the amplitude vector $\vec{U}(T_0) = B^g(T_0)\vec{g}^{(1)}(T_0) + C^g(T_0)\vec{g}^{(2)}(T_0)$. Thus, the amplitudes B^g and C^g of qS waves are also coupled and frequency-dependent.

We can also express the system matrix $\mathbf{B}^d$, given by (5.4.65), in an alternative form. Using (3.9.29), we obtain $D = 2d\Delta T^s/dT$, where ΔT^s is the time delay between the two split qS waves. Then we obtain

$$\mathbf{B}^d(T) = \begin{pmatrix} -\tfrac{1}{2}i\omega d\Delta T^s/dT & \gamma \\ -\gamma & \tfrac{1}{2}i\omega d\Delta T^s/dT \end{pmatrix}$$
$$= \frac{d}{dT}\begin{pmatrix} -\tfrac{1}{2}i\omega\Delta T^s(T) & \varphi(T) \\ -\varphi(T) & \tfrac{1}{2}i\omega\Delta T^s(T) \end{pmatrix}. \tag{5.4.66}$$

Here $\varphi(T)$ is the angle between $\vec{e}_1(T)$ and $\vec{g}^{(1)}(T)$.

Note that Equations (5.4.64) with (5.4.65) could be also obtained directly from (5.4.50) with (5.4.51), using the diagonalization relation $\mathbf{B}^{se} = \mathbf{A}\mathbf{B}^{sg}\mathbf{A}^T$ and the relation

$$\begin{pmatrix} B^g \\ C^g \end{pmatrix} = \mathbf{A}^T\begin{pmatrix} B^e \\ C^e \end{pmatrix}, \tag{5.4.67}$$

where the rotation matrix $\mathbf{A}$ is given by (5.4.55).

An important remark. As we can see from (5.4.66), the qS-wave coupling system (5.4.65) depends on the variations of ΔT^s and φ along the common ray Ω^0. The quantities ΔT^s and φ are calculated here using the quasi-isotropic approximation. The qS-wave coupling system (5.4.65) is, however, valid more generally; even if ΔT^s and φ are calculated along Ω^0 in a more sophisticated way (Coates and Chapman 1990b). Then the qS-wave coupling system (5.4.65) can yield more accurate results. This applies also to (5.4.86), with (5.4.80) and (5.4.83). For example, it is possible to consider the reference common ray Ω^0, corresponding to the average eigenvalue $G^{av} = \tfrac{1}{2}(G_1 + G_2)$ of the two quasi-shear waves in anisotropic media. Such a ray may be safely computed even in weakly anisotropic media and may be close to shear wave singularities. See Section 3.6.2.

5. QUASI-ISOTROPIC PROPAGATOR MATRICES

To express (5.4.50) and (5.4.64) in a more compact and flexible form, we shall use the propagator technique. The propagator technique is described in great detail in Section 4.3. Even though it is related there to 4×4 ray propagator matrices of the dynamic ray tracing system, we may use all equations and conclusions of Section 4.3 here; we merely replace the 2×2 submatrices of the 4×4 ray propagator matrix by scalars. Let us consider a ray Ω^0 in the background isotropic medium $\mathcal{M}^0$, and two points S and R situated on it. Assume that the ray Ω^0 is not in contact with any structural interface between S and R. The travel time is $T(S) = T_0$ at point S and $T(R) = T$ at R. Because the qS wave coupling systems (5.4.51) and (5.4.65) are linear, we can introduce the complex-valued frequency-dependent 2×2 propagator matrices $\mathbf{\Pi}^e(R, S)$ and $\mathbf{\Pi}^g(R, S)$ in such a way that

$$\begin{pmatrix} B^e(R) \\ C^e(R) \end{pmatrix} = \mathbf{\Pi}^e(R, S) \begin{pmatrix} B^e(S) \\ C^e(S) \end{pmatrix}, \qquad \begin{pmatrix} B^g(R) \\ C^g(R) \end{pmatrix} = \mathbf{\Pi}^g(R, S) \begin{pmatrix} B^g(S) \\ C^g(S) \end{pmatrix}.$$

$$(5.4.68)$$

The *quasi-isotropic propagator matrices* $\mathbf{\Pi}^e(R, S)$ and $\mathbf{\Pi}^g(R, S)$ are formed by two linearly independent solutions of systems of differential equations (5.4.51) and (5.4.65), with initial conditions $\mathbf{\Pi}^e(S, S) = \mathbf{I}$ and $\mathbf{\Pi}^g(S, S) = \mathbf{I}$, respectively. Because $\operatorname{tr} \mathbf{B}^s = 0$ and $\operatorname{tr} \mathbf{B}^d = 0$, the determinants of $\mathbf{\Pi}^e(R, S)$ and $\mathbf{\Pi}^g(R, S)$ equal unity for an arbitrarily situated point R on Ω^0 (Liouville's theorem; see Section 4.3.3). It is not difficult to see that both propagator matrices $\mathbf{\Pi}^e(R, S)$ and $\mathbf{\Pi}^g(R, S)$ are symplectic (see Section 4.3.2), that they satisfy the chain rule (see Section 4.3.4), and that their inverses can be computed by equations analogous to those given in Section 4.3.5. Particularly attractive for us is the chain rule. Assume that points $Q_1, Q_2, \ldots, Q_N$ are situated along Ω^0 between S and R. Then

$$\mathbf{\Pi}^e(R, S) = \mathbf{\Pi}^e(R, Q_N)\mathbf{\Pi}^e(Q_N, Q_{N-1}) \cdots \mathbf{\Pi}^e(Q_1, S). \qquad (5.4.69)$$

The same relation is valid also for propagator matrix $\mathbf{\Pi}^g(R, S)$. This chain relation is suitable for matching different solutions computed along different parts of the ray Ω^0. Note that the quasi-isotropic propagator matrices $\mathbf{\Pi}^e(R, S)$ and $\mathbf{\Pi}^g(R, S)$ are frequency-dependent. To simplify the notation, we shall not write the frequency ω among arguments of the propagator matrices.

Due to (5.4.67), propagator matrices $\mathbf{\Pi}^e(R, S)$ and $\mathbf{\Pi}^g(R, S)$ are mutually related:

$$\mathbf{\Pi}^e(R, S) = \mathbf{A}(R)\mathbf{\Pi}^g(R, S)\mathbf{A}^T(S). \qquad (5.4.70)$$

Here $\mathbf{A}$ is given by (5.4.55). Relation (5.4.67) can be also used if we wish to transform $(B^e, C^e)^T$ into $(B^g, C^g)^T$, or vice versa, at any point of the ray Ω^0. Again, this may be suitable for matching.

Now we shall write expressions for the qS amplitude matrices. We introduce

$$\mathbf{U}^{(q)}(R) = (B^e(R), C^e(R))^T, \qquad \mathbf{U}^{(g)}(R) = (B^g(R), C^g(R))^T, \qquad (5.4.71)$$

and similarly for $\mathbf{U}^{(q)}(S)$ and $\mathbf{U}^{(g)}(S)$. Then we can express relations (5.4.50) and (5.4.64) in a more suitable matrix form. We take into account that the phase shifts due to caustics are the same for both B^e and C^e (and for both B^g and C^g). See more details later. Then we obtain

$$\mathbf{U}^{(q)}(R) = \left[\frac{\rho(S)\beta(S)}{\rho(R)\beta(R)} \right]^{1/2} \frac{\mathcal{L}(S)}{\mathcal{L}(R)}$$
$$\times \exp[iT^c(R, S)]\mathcal{A}^{QS}(R, S)\mathbf{\Pi}^e(R, S)\mathbf{U}^{(q)}(S), \qquad (5.4.72)$$

and, similarly,

$$
\mathbf{U}^{(g)}(R) = \left[\frac{\rho(S)\beta(S)}{\rho(R)\beta(R)} \right]^{1/2} \frac{\mathcal{L}(S)}{\mathcal{L}(R)}
$$
$$
\times \exp[iT^c(R, S)]\mathcal{A}^{QS}(R, S)\mathbf{\Pi}^g(R, S)\mathbf{U}^{(g)}(S). \qquad (5.4.73)
$$

The symbols $\mathcal{L}(S)$, $\mathcal{L}(R)$, and $T^c(R, S)$ have the same meaning as in (5.2.18). Comparing (5.4.72) with (5.2.18), we can see that the expression for the qS amplitude matrix (5.4.72) in a weakly anisotropic medium is formally the same as the expression (5.2.18) for the amplitude matrix of the S wave propagating in an isotropic medium; only the initial amplitude matrix $\mathbf{U}^{(q)}(S)$ should be multiplied by the propagator matrix $\mathbf{\Pi}^e(R, S)$ and by $\mathcal{A}^{QS}(R, S)$. Consequently, (5.4.72) can be modified for a point source situated at S and for the elementary ray-theory Green functions in the same way as in Section 5.2. It can also be modified for many other special cases. See more details later.

Using (5.4.69), the propagator matrices can be chained into factors corresponding to shorter segments of the ray Ω^0. Along these segments, the propagator matrices sometimes may be computed analytically, either exactly or approximately, or in some alternative numerical way. We shall present now several such situations.

a. Vanishing perturbations. $\mathbf{B}^s = 0$, $\mathbf{B}^d = 0$, and we obtain

$$
\mathcal{A}^{QS}(R, S) = 1, \qquad \mathbf{\Pi}^e(R, S) = \mathbf{I}, \qquad \mathbf{\Pi}^g(R, S) = \mathbf{I}. \qquad (5.4.74)
$$

Equation (5.4.72) then reduces to (5.2.18) for an isotropic medium.

b. Isotropic background, isotropic perturbed medium. $B_{IJ} = 2\beta^{-1}\Delta\beta\delta_{IJ}$ so that $\mathbf{B}^s = \mathbf{0}$ and $\mathbf{B}^d = \mathbf{0}$. Consequently,

$$
\mathcal{A}^{QS}(R, S) = \exp\left[i\omega \int_S^R \Delta(1/\beta)\mathrm{d}s \right],
$$
$$
\mathbf{\Pi}^e(R, S) = \mathbf{I}, \qquad \mathbf{\Pi}^g(R, S) = \mathbf{I}. \qquad (5.4.75)
$$

Here the integral is taken along the ray Ω^0. Equation (5.4.72) again reduces to standard equation (5.2.18) for S waves in the isotropic case; only (5.2.18) should be multiplied by $\mathcal{A}^{QS}(R, S)$ given by (5.4.75). The function $\mathcal{A}^{QS}(R, S)$ fully expresses the influence of the isotropic first-order travel-time perturbation; see (3.9.9).

c. No coupling. If the eigenvectors $\vec{g}^{(I)}$ do not vary along Ω^0 with respect to $\vec{e}_1$ and $\vec{e}_2$, the coupling function $\gamma(T)$ vanishes. System (5.4.65) then decouples, and its solution is

$$
\mathbf{\Pi}^g(R, S) = \begin{pmatrix} \exp\left[-\tfrac{1}{4}i\omega \int_S^R D\mathrm{d}T\right] & 0 \\ 0 & \exp\left[\tfrac{1}{4}i\omega \int_S^R D\mathrm{d}T\right] \end{pmatrix}. \qquad (5.4.76)
$$

The integrals in (5.4.76) are taken along ray Ω^0 and represent twice the time delay between the two split qS waves; see (3.9.29).

As shown by (5.4.76), the propagator matrix $\mathbf{\Pi}^g(R, S)$ decouples. A similar decoupled equation analogous to (5.4.73) with (5.4.76) was also derived by Pšenčík (1998), using a different procedure. Pšenčík (1998) also proposed some qualitative criteria showing when (5.4.76) can be used as an approximation. This applies mostly to models close to homogeneous (weakly inhomogeneous background media).

The propagator matrix $\mathbf{\Pi}^e(R, S)$, corresponding to (5.4.76), is not decoupled. We obtain it from (5.4.70) and (5.4.76):

$$\mathbf{\Pi}^e(R, S) = \mathbf{I}\cos\left[\tfrac{1}{4}\omega\int_S^R DdT\right] - iD^{-1}\sin\left[\tfrac{1}{4}\omega\int_S^R DdT\right]\mathbf{B}^s.$$

(5.4.77)

Of course, Equations (5.4.76) and (5.4.77) can be used if the background isotropic and perturbed weakly anisotropic media are homogeneous. Then D is constant and $\int_{T_0}^T DdT = DT(R, S) = D(T(R) - T(S))$. Equation (5.4.77) then yields

$$\mathbf{\Pi}^e(R, S) = \mathbf{I}\cos\left[\tfrac{1}{4}\omega DT(R, S)\right] - \tfrac{1}{4}i\omega T(R, S)\,\mathrm{sinc}\left[\tfrac{1}{4}\omega DT(R, S)\right]\mathbf{B}^s.$$

(5.4.78)

Here $\mathrm{sinc}(x) = x^{-1}\sin(x)$.

d. Propagator matrices by the method of mean coefficients. The method of mean coefficients, described by Gilbert and Backus (1966), yields simple approximate expressions for the quasi-isotropic propagator matrices along short segments of the ray Ω^0, even for coupled equations. We shall consider the quasi-isotropic propagator matrix $\mathbf{\Pi}^g(T, T_0)$, controlled by (5.4.65): $d\mathbf{\Pi}^g/dT = \mathbf{B}^d\mathbf{\Pi}^g$. Let us consider points $T_0, T_1, \ldots, T_{l-1}, T_l, \ldots,$ $T_n = T$ along the ray and use the chain rule for $\mathbf{\Pi}^g(T, T_0)$, analogous to (5.4.69). We also denote $\Delta T_l = T_l - T_{l-1}$. For small ΔT_l, we can approximately express $\mathbf{\Pi}^g(T_l, T_{l-1})$ as follows: $\mathbf{\Pi}^g(T_l, T_{l-1}) = \exp[\mathbf{B}^d(\bar{T}_l)\Delta T_l]$, where $\bar{T}_l$ is some intermediate point of the interval $T_{l-1} < \bar{T}_l < T_l$. The method of mean coefficients is then represented by the following approximation for small ΔT_l:

$$\mathbf{\Pi}^g(T_l, T_{l-1}) \doteq \exp[\mathbf{B}^d(\bar{T}_l)\Delta T_l] \doteq \exp[\mathbf{S}(T_l, T_{l-1})],$$

(5.4.79)

where

$$\mathbf{S}(T_l, T_{l-1}) = \int_{T_{l-1}}^{T_l} \mathbf{B}^d(T)dT = \begin{pmatrix} -\tfrac{1}{4}i\omega\int_{T_{l-1}}^{T_l} DdT & \varphi_l - \varphi_{l-1} \\ -(\varphi_l - \varphi_{l-1}) & \tfrac{1}{4}i\omega\int_{T_{l-1}}^{T_l} DdT \end{pmatrix},$$

(5.4.80)

where $\varphi_l = \varphi(T_l)$ and $\varphi_{l-1} = \varphi(T_{l-1})$. Instead of $\int_{T_{l-1}}^{T_l} DdT$, we can also use $2[\Delta T^s(T_l) - \Delta T^s(T_{l-1})]$, where ΔT^s is the time delay between the two split qS waves; see (3.9.29). What remains is to calculate $\exp[\mathbf{S}(T_l, T_{l-1})]$. To do it, we shall use the Cayley-Hamilton theorem and Sylvester theorem; see Korn and Korn (1961, Sections 13.4–13.7). The characteristic equation of matrix $\mathbf{S}$ reads

$$\lambda^2 + Y = 0,$$

(5.4.81)

where Y is given by the relation

$$Y = \tfrac{1}{16}\omega^2\left(\int_{T_{l-1}}^{T_l} DdT\right)^2 + (\varphi_l - \varphi_{l-1})^2.$$

(5.4.82)

Alternatively, we can use $D = 2d\Delta T^s/dT$. Then the expression for Y reads

$$Y = \tfrac{1}{4}\omega^2(\Delta T^s(T_l) - \Delta T^s(T_{l-1}))^2 + (\varphi_l - \varphi_{l-1})^2.$$

(5.4.83)

According to the Cayley-Hamilton theorem, matrix $\mathbf{S}$ given by (5.4.80) satisfies its characteristic equation (5.4.81):

$$\mathbf{S}^2 + Y\mathbf{I} = 0. \tag{5.4.84}$$

This can also be simply verified directly. Then we can use the Sylvester theorem

$$\exp(\mathbf{S}) \doteq \exp(\lambda_1)\frac{\mathbf{S} - \lambda_2\mathbf{I}}{\lambda_1 - \lambda_2} + \exp(\lambda_2)\frac{\mathbf{S} - \lambda_1\mathbf{I}}{\lambda_2 - \lambda_1}. \tag{5.4.85}$$

Inserting $\lambda_{1,2} = \pm i Y^{1/2}$, we obtain the final equation for $\exp(\mathbf{S})$, and, consequently, also for $\mathbf{\Pi}^g(T_l, T_{l-1})$,

$$\mathbf{\Pi}^g(T_l, T_{l-1}) \doteq \mathbf{I} \cos Y^{1/2} + \mathbf{S} \operatorname{sinc} Y^{1/2}; \tag{5.4.86}$$

see (5.4.79).

Here are several notes to the final equation (5.4.86).

i. The expression (5.4.86) contains only one integral, $\int_{T_{l-1}}^{T_l} D dT = 2\Delta T^s$, where ΔT^s is the time delay between the two split qS waves.

ii. The values of φ (angle between $\vec{e}_1$ and $\vec{g}^{(1)}$) should be known only at end points, T_{l-1} and T_l. Consequently, the eigenvectors $\vec{g}^{(1)}$, $\vec{g}^{(2)}$ need not be computed along the ray between T_{l-1} and T_l.

iii. If there is no coupling, $\varphi_l = \varphi_{l-1}$. Then (5.4.86) yields (5.4.76).

iv. Using (5.4.70), we can also obtain $\mathbf{\Pi}^e(R, S)$ from (5.4.86).

e. **Solutions in terms of φ.** We shall use new amplitude functions B^φ and C^φ in (5.4.65), connected with B^g and C^g by relations:

$$\begin{aligned} B^g(T) &= B^\varphi(T)\exp\left[-\tfrac{1}{2}i\omega\Delta T^s(T)\right], \\ C^g(T) &= C^\varphi(T)\exp\left[\tfrac{1}{2}i\omega\Delta T^s(T)\right]. \end{aligned} \tag{5.4.87}$$

Here ΔT^s is the time delay between the two split qS waves. The two coupled linear ordinary differential equations of the first order for $B^\varphi(T)$ and $C^\varphi(T)$ are then as follows:

$$\frac{\mathrm{d}}{\mathrm{d}T}\begin{pmatrix} B^\varphi \\ C^\varphi \end{pmatrix} = \gamma\begin{pmatrix} 0 & \sigma(T) \\ -\sigma^*(T) & 0 \end{pmatrix}\begin{pmatrix} B^\varphi \\ C^\varphi \end{pmatrix}. \tag{5.4.88}$$

Here $\sigma(T)$ is given by the relation

$$\sigma(T) = \exp[i\omega\Delta T^s(T)], \tag{5.4.89}$$

and $\sigma^*(T)$ denotes the complex conjugate quantity. The 2×2 matrix on the RHS of (5.4.88) is unitary and antihermitean. It would again be possible to introduce the propagator matrix $\mathbf{\Pi}^\varphi$ for the coupled system (5.4.88). Consequently, the propagator matrices $\mathbf{\Pi}^g(R, S)$ and $\mathbf{\Pi}^e(R, S)$ can be also found by solving the system (5.4.88).

System of equations (5.4.88), in a slightly different form, was discussed in detail by Coates and Chapman (1990b). They also proposed to use φ as a variable along the ray instead of the variable T. Assume that the relation $\varphi = \varphi(T)$ is monotonic along the segment of ray Ω^0 between S and R. Using (5.4.60), we can express (5.4.88) in the following form:

$$\frac{\mathrm{d}}{\mathrm{d}\varphi}\begin{pmatrix} B^\varphi \\ C^\varphi \end{pmatrix} = \begin{pmatrix} 0 & \sigma(T) \\ -\sigma^*(T) & 0 \end{pmatrix}\begin{pmatrix} B^\varphi \\ C^\varphi \end{pmatrix}. \tag{5.4.90}$$

This system of two coupled linear ordinary differential equations of the first-order (5.4.90) is surprisingly simple, but it is still exact. The system matrix of (5.4.90) is unitary and antihermitean. As we can notice in (5.4.89), $|\sigma(T)| = 1$ for any T.

The systems (5.4.88) or (5.4.90) can be solved along the ray Ω^0 in various ways. It may again be suitable to divide the whole ray into short segments and use some approximate treatment along any short segment such as the method of mean coefficients. See, for example, Coates and Chapman (1990b) who also described a successful application of (5.4.90) in the vicinity of some shear wave singularities.

f. Propagator matrix through a caustic point. Let us consider a ray Ω^0 and two points S and R situated on Ω^0. Assume that a caustic point Q^c is situated between S and R. At the caustic point Q^c, the phase shift due to caustic is the same for both amplitude components B^e and C^e (or for B^g and C^g). Consequently, we can eliminate the phase shifts due to caustics from the propagator matrix and collect them in a common multiplicative factor. The results are analogous to the standard ray theory: the final expression for the amplitude matrix should be multiplied by $\exp[iT^c(R, S)]$, where $T^c(R, S)$ is the phase shift due to all caustics situated between S and R. Thus, we can compute the propagator matrix even through caustic points without any change. The relevant phase shift due to caustics $T^c(R, S)$ is taken into account separately; see the factor $\exp[iT^c(R, S)]$ in (5.4.72) and (5.4.73).

g. Propagator matrices across structural interfaces. Assume that ray Ω^0 is in contact with a structural interface between points S and R. As usual, we denote the point of incidence by Q, and the relevant R/T point by $\tilde{Q}$. Then the quasi-isotropic propagator matrices $\mathbf{\Pi}^e(R, S)$ and $\mathbf{\Pi}^g(R, S)$ can be chained as follows:

$$\mathbf{\Pi}^e(R, S) = \mathbf{\Pi}^e(R, \tilde{Q})\mathbf{\Pi}^e(\tilde{Q}, Q)\mathbf{\Pi}^e(Q, S),$$
$$\mathbf{\Pi}^g(R, S) = \mathbf{\Pi}^g(R, \tilde{Q})\mathbf{\Pi}^g(\tilde{Q}, Q)\mathbf{\Pi}^g(Q, S). \tag{5.4.91}$$

Thus, at a point of incidence, it is necessary to insert the *interface propagator* $\mathbf{\Pi}^e(\tilde{Q}, Q)$ (or $\mathbf{\Pi}^g(\tilde{Q}, Q)$). It is given by the relation

$$\mathbf{\Pi}^e(\tilde{Q}, Q) = \mathbf{R}^T(Q), \qquad \mathbf{\Pi}^g(\tilde{Q}, Q) = \mathbf{A}^T(\tilde{Q})\mathbf{R}^T(Q)\mathbf{A}(Q). \tag{5.4.92}$$

Here $\mathbf{A}$ is the 2×2 matrix given by (5.4.55), and $\mathbf{R}$ is the 2×2 matrix of S $\rightarrow$ S plane wave displacement reflection/transmission coefficients for isotropic structures; see also $\mathbf{R}_{S \rightarrow S}$ in (5.2.38). Finally, the superscript T denotes the transpose. As in the standard ray method, it is suitable to replace the 2×2 matrix of S $\rightarrow$ S R/T displacement coefficients $\mathbf{R}(Q)$ by the 2×2 matrix of S $\rightarrow$ S R/T normalized displacement coefficients $\mathcal{R}(Q)$:

$$\mathbf{\Pi}^e(\tilde{Q}, Q) = \mathcal{R}^T(Q), \qquad \mathbf{\Pi}^g(\tilde{Q}, Q) = \mathbf{A}^T(\tilde{Q})\mathcal{R}^T(Q)\mathbf{A}(Q); \tag{5.4.93}$$

see Section 5.2.4. The multiplicative factor obtained by this replacement is canceled with some other factors in the final expression for $\hat{\mathbf{U}}^{(q)}(R)$.

Note that the interface propagators $\mathbf{\Pi}^e(\tilde{Q}, Q)$ and $\mathbf{\Pi}^g(\tilde{Q}, Q)$ are not symplectic so that the chains (5.4.91) are not symplectic. Even in this case, however, the chain rule (5.4.91) can be safely used in (5.4.69). If we wish to preserve the symplecticity, it would be necessary to normalize $\mathbf{\Pi}^e(\tilde{Q}, Q)$ by $[\det \mathbf{\Pi}^e(\tilde{Q}, Q)]^{1/2}$ and $\mathbf{\Pi}^g(\tilde{Q}, Q)$ by $[\det \mathbf{\Pi}^g(\tilde{Q}, Q)]^{1/2}$ and to consider separately the products of all determinants.

h. Factorized anisotropic medium. In a factorized anisotropic medium, the perturbations Δa_{ijkl} are given by relation $\Delta a_{ijkl}(x_i) = A_{ijkl}\Delta f^2(x_i) + f^2(x_i)\Delta A_{ijkl}$; see (3.6.34). This relation immediately shows that the shear wave splitting matrix $\mathbf{B}^{se}$ in (5.4.51) and matrix $\mathbf{B}^d$ in (5.4.65) do not depend on structural perturbations $\Delta f(x_i)$ but only on anisotropy perturbations ΔA_{ijkl}. Consequently, the systems of equations (5.4.51), (5.4.65), and (5.4.88) and relevant propagator matrices $\mathbf{\Pi}^e(R, S)$ and $\mathbf{\Pi}^g(R, S)$ do not depend on structural perturbations $\Delta f(x_i)$ but only on position-independent anisotropy perturbations ΔA_{ijkl}. Thus, the coupling of S waves in a weakly anisotropic FAI medium is controlled only by position-independent anisotropy perturbations ΔA_{ijkl}, not by structural perturbations $\Delta f(x_i)$.

i. Strongly anisotropic medium. Formally, equations analogous to (5.4.73) can be used even if the model $\mathcal{M}$ is strongly anisotropic in some region outside shear wave singular directions. This, however, requires one to perform ray tracing and dynamic ray tracing for the strongly anisotropic medium in this region, using anisotropic ray tracing systems of Section 3.6 and anisotropic dynamic ray tracing systems of Section 4.14.2. Consequently, no perturbations are involved in this region. It is not difficult to combine isotropic and anisotropic ray tracing and dynamic ray tracing along different segments of the ray.

The rays of qS1 and qS2 in strongly anisotropic media are, of course, different. Thus, if we wish to perform ray tracing in a strongly anisotropic region, it is necessary to specify by a proper ray code which of the two qS waves we wish to compute. If we are interested also in the other qS wave, both rays should be computed independently. Only after both computations may these waves be combined to give the complete qS wave.

For ΔT^s exceeding some limit, such complete qS wave will give more accurate results than any approximation based on a common ray. This applies not only to strongly anisotropic media but also to weakly anisotropic media. The limit is, of course, frequency-dependent. For example, it may be chosen to be proportional to the prevailing period of the wave. Thus, the computation of qS waves in anisotropic media should be performed differently for ΔT^s not exceeding the limit and for ΔT^s exceeding the limit:

a.   In the region where ΔT^s does not exceed the limit, an approximation based on a common ray should be used (for example, the quasi-isotropic approximation).

b.   In the region where ΔT^s exceeds the limit, complete anisotropic qS wave computations should be performed. The problem how to choose the proper limit requires further investigation.

Let us consider a ray Ω, composed of segments situated in the background isotropic medium, and of segments situated in strongly anisotropic media. Consider one segment of Ω, situated in a strongly anisotropic medium, between points Q_{k-1} and Q_k on Ω. Assume that there is no contact with structural interfaces and with shear wave singularities along Ω between Q_{k-1} and Q_k. In other words, the eigenvalues G_1 and G_2 of the Christoffel matrix differ considerably along Ω between Q_{k-1} and Q_k. Then we can use (5.4.73), with

$$\mathcal{A}^{QS}(Q_k, Q_{k-1}) = 1, \qquad \mathbf{\Pi}^g(Q_k, Q_{k-1}) = \mathbf{I}. \tag{5.4.94}$$

To compute the qS1 wave, we must use $\mathbf{U}^g(Q_{k-1}) = (B^g(Q_{k-1}), 0)^T$, and to compute the qS2 wave, we need to use $\mathbf{U}^g(Q_{k-1}) = (0, C^g(Q_{k-1}))^T$. If the first and/or last segment of Ω is situated in a strongly anisotropic medium, $\beta(S)$ and/or $\beta(R)$ in (5.4.73) should be replaced by $\mathcal{U}(S)$ and/or $\mathcal{U}(R)$, where $\mathcal{U}$ is the relevant group velocity; see (5.4.4).

6. AMPLITUDE MATRICES IN 3-D LAYERED WEAKLY ANISOTROPIC MEDIA

Equations (5.4.72) and (5.4.73) give general expressions for the qS amplitude matrices in a smooth medium without interfaces. Using (5.4.91), these equations may be generalized even for qS wave amplitude matrices in layered media. Here we wish to give general expressions for 3×1 amplitude column matrices $\hat{\mathbf{U}}^{(q)}(R)$ and $\hat{\mathbf{U}}^{(g)}(R)$ of an arbitrary multiply-reflected, possibly converted, wave propagating in a weakly anisotropic layered structure.

We introduce 3×3 matrices $\hat{\mathbf{\Pi}}^e(R, S)$ and $\hat{\mathbf{\Pi}}^g(R, S)$ for a smooth segment of the ray Ω^0 between S and R (without structural interfaces) as follows:

$$\hat{\mathbf{\Pi}}^e(R, S) = \begin{pmatrix} \mathbf{\Pi}^e(R, S) & 0 \\ & 0 \\ 0 & 0 & 1 \end{pmatrix}, \qquad \hat{\mathbf{\Pi}}^g(R, S) = \begin{pmatrix} \mathbf{\Pi}^g(R, S) & 0 \\ & 0 \\ 0 & 0 & 1 \end{pmatrix}.$$

(5.4.95)

Due to (5.4.70), the mutual relation between the two matrices is

$$\hat{\mathbf{\Pi}}^e(R, S) = \hat{\mathbf{A}}(R)\hat{\mathbf{\Pi}}^g(R, S)\hat{\mathbf{A}}^T(S), \qquad \hat{\mathbf{A}} = \begin{pmatrix} \mathbf{A} & 0 \\ & 0 \\ 0 & 0 & 1 \end{pmatrix}. \quad (5.4.96)$$

Here the 2×2 matrix $\mathbf{A}$ is given by (5.4.55). For a layered medium, with N R/T points on Ω^0 between S and R, we obtain

$$\hat{\mathbf{\Pi}}^e(R, S) = \hat{\mathbf{\Pi}}^e(R, \tilde{Q}_N) \prod_{k=N}^{1} [\hat{\mathcal{R}}^T(Q_k)\hat{\mathbf{\Pi}}^e(Q_k, \tilde{Q}_{k-1})]. \quad (5.4.97)$$

Using (5.4.96), an analogous equation can also be obtained for $\hat{\mathbf{\Pi}}^g(R, S)$. Here Q_1, $Q_2, \ldots, Q_N$ are points of incidence, $\tilde{Q}_1, \tilde{Q}_2, \ldots, \tilde{Q}_N$ the relevant R/T points, and $\tilde{Q}_0 \equiv S$. $\hat{\mathcal{R}}(Q_k)$ is the 3×3 matrix of normalized displacement coefficients of reflection/transmission on structural interfaces between two isotropic media. It is different for reflection and for transmission. There are four types of the R/T matrix $\hat{\mathcal{R}}(Q_k)$: $\hat{\mathcal{R}}_{P \to P}$, $\hat{\mathcal{R}}_{P \to S}$, $\hat{\mathcal{R}}_{S \to P}$, and $\hat{\mathcal{R}}_{S \to S}$. The appropriate type of the R/T matrix must be consistent with the ray code. We remind the reader that certain components of individual R/T matrices vanish; see (5.2.38) for standard displacement R/T matrices for the isotropic case. For example, $\hat{\mathcal{R}}_{P \to P}$ has only one nonvanishing component $\mathcal{R}_{33}$; $\hat{\mathcal{R}}_{S \to S}$ has four nonvanishing components $\mathcal{R}_{11}$, $\mathcal{R}_{12}$, $\mathcal{R}_{21}$, and $\mathcal{R}_{22}$; and $\hat{\mathcal{R}}_{P \to S}$ and $\hat{\mathcal{R}}_{S \to P}$ have two nonvanishing components each. The matrices $\hat{\mathbf{\Pi}}^e$ on the RHS of (5.4.97) are given by (5.4.95). The complete matrix $\hat{\mathbf{\Pi}}^e(R, S)$ for a layered medium given by (5.4.97) is, however, more general than (5.4.95). Only for unconverted S waves can it again be expressed in the form of (5.4.95). For unconverted P waves, $\Pi_{33}^e(R, S) \neq 1$. Similarly, for converted R/T waves (PS, SP), some of the elements $\Pi_{13}^e(R, S)$, $\Pi_{23}^e(R, S)$, $\Pi_{31}^e(R, S)$, and $\Pi_{32}^e(R, S)$ may differ from zero. Points Q_k and $\tilde{Q}_k$ in (5.4.97) may also be situated on Ω^0 in a smooth medium, without any structural interface at Q_k. In this case, we use $\hat{\mathcal{R}}(Q_k) = \hat{\mathbf{I}}$. Such a choice may be useful if we wish to calculate the matrices $\hat{\mathbf{\Pi}}^e$ or $\hat{\mathbf{\Pi}}^g$ by different methods along two different segments of Ω^0 separated by point Q_k.

We shall also introduce function $\mathcal{A}^Q(R, S)$:

$$\mathcal{A}^Q(R, S) = \mathcal{A}^Q(R, \tilde{Q}_N)\mathcal{A}^Q(Q_N, \tilde{Q}_{N-1}) \cdots \mathcal{A}^Q(Q_1, S). \quad (5.4.98)$$

Here $\mathcal{A}^Q(Q_k, \tilde{Q}_{k-1})$ corresponds to $\mathcal{A}^{QP}(Q_k, \tilde{Q}_{k-1})$ given by (5.4.39) if the kth segment of the ray Ω^0 corresponds to a qP wave; it corresponds to $\mathcal{A}^{QS}(Q_k, \tilde{Q}_{k-1})$ given by (5.4.63) if the kth segment of the ray Ω^0 corresponds to a qS wave. We now introduce 3×1 amplitude matrices $\hat{\mathbf{U}}^{(q)}$ and $\hat{\mathbf{U}}^{(g)}$:

$$\hat{\mathbf{U}}^{(q)} = (B^e, C^e, A)^T, \qquad \hat{\mathbf{U}}^{(g)} = (B^g, C^g, A). \quad (5.4.99)$$

These matrices correspond to the vectorial decomposition of amplitudes $\vec{U} = B^e \vec{e}_1 + C^e \vec{e}_2 + A \vec{e}_3$ and $\vec{U} = B^g \vec{g}^{(1)} + C^g \vec{g}^{(2)} + A \vec{g}^{(3)}$. In weakly anisotropic media, $\vec{g}^{(I)}$ are given by (5.4.52), and $\vec{g}^{(3)} = \vec{N}$. In strongly anisotropic media, $\vec{g}^{(i)}$ represent eigenvectors of the Christoffel matrix for relevant anisotropic media. The final relation for the 3×1 amplitude matrix $\hat{\mathbf{U}}^{(q)}(R)$ in a layered weakly anisotropic medium is then given by the expression

$$\hat{\mathbf{U}}^{(q)}(R) = \left[\frac{V(S)\rho(S)}{V(R)\rho(R)}\right]^{1/2} \frac{\mathcal{L}(S)}{\mathcal{L}(R)}$$
$$\times \exp[iT^c(R, S)]\mathcal{A}^{\mathcal{Q}}(R, S)\hat{\mathbf{\Pi}}^e(R, S)\hat{\mathbf{U}}^{(q)}(S), \qquad (5.4.100)$$

which is similar to (5.2.45).

Here $\hat{\mathbf{\Pi}}^e(R, S)$ and $\mathcal{A}^{\mathcal{Q}}(R, S)$ are given by (5.4.97) and (5.4.98), and all other symbols have the same meaning as in (5.2.45). The velocities $V(S)$ and $V(R)$ correspond to the type of the wave at S and R (either α or β).

An alternative relation to (5.4.100) can be written for $\hat{\mathbf{U}}^{(g)}(R)$ in terms of $\hat{\mathbf{\Pi}}^g(R, S)$. Because we wish to use such an equation even if the source and receiver segments of Ω^0 are situated in strongly anisotropic media, we shall use group velocities $\mathcal{U}(S)$ and $\mathcal{U}(R)$ instead of $V(S)$ and $V(R)$:

$$\hat{\mathbf{U}}^{(g)}(R) = \left[\frac{\mathcal{U}(S)\rho(S)}{\mathcal{U}(R)\rho(R)}\right]^{1/2} \frac{\mathcal{L}(S)}{\mathcal{L}(R)}$$
$$\times \exp[iT^c(R, S)]\mathcal{A}^{\mathcal{Q}}(R, S)\hat{\mathbf{\Pi}}^g(R, S)\hat{\mathbf{U}}^{(g)}(S); \qquad (5.4.101)$$

see (5.4.4). For weakly anisotropic media, both equations (5.4.100) and (5.4.101) are alternative.

Let us now briefly discuss (5.4.100). As we can see, it differs from (5.2.45) only by the multiplicative factor $\mathcal{A}^{\mathcal{Q}}(R, S)$ and by the 3×3 matrix $\hat{\mathbf{\Pi}}^e(R, S)$, which replaces $\hat{\mathcal{R}}^C(Q)$. Consequently, (5.4.100) can be modified in many ways, in much the same way as (5.2.45) in Section 5.2. For example, we obtain point-source solutions from (5.4.100) if we replace $\hat{\mathbf{U}}^{(q)}(S)$ by the ray-centered radiation matrix $\hat{\mathcal{G}}^{(q)}(S; \gamma_1, \gamma_2)$, and $\mathcal{L}(S)/\mathcal{L}(R)$ by $1/\mathcal{L}(R, S)$. We shall present here only one very important relation following from (5.4.100), corresponding to the quasi-isotropic ray-theory elastodynamic Green function $G_{in}(R, S, \omega)$ for a weakly anisotropic layered medium:

$$G_{in}(R, S, \omega) = \frac{e_{ki}(R)e_{ln}(S)\mathcal{A}^{\mathcal{Q}}(R, S)}{4\pi[\rho(S)\rho(R)V(S)V(R)]^{1/2}\mathcal{L}(R, S)} \Pi^e_{kl}(R, S)$$
$$\times \exp[i\omega T(R, S) + iT^c(R, S)]. \qquad (5.4.102)$$

Here the summation over k and l should be specified properly; see Section 5.2.6. If the first element of the ray (at S) is P, we put $l = 3$, and if it is S, we put $l = L$ (with the summation over $L = 1, 2$). Similarly, if the last element of the ray (at R) is P, we put $k = 3$, and if it is S, we put $k = K$ (with the summation over $K = 1, 2$). The relation (5.4.102) can also be modified for waves generated at an initial surface or at an initial line, for receiver R situated at an structural interface and the like. The procedures are the same as in Section 5.2.

An alternative expression for $G_{in}(R, S, \omega)$ follows from (5.4.101). It reads

$$G_{in}(R, S, \omega) = \frac{g_i^{(k)}(R)g_n^{(l)}(S)\mathcal{A}^{\mathcal{Q}}(R, S)}{4\pi[\rho(S)\rho(R)\mathcal{C}(S)\mathcal{C}(R)]^{1/2}\mathcal{L}(R, S)} \Pi^g_{kl}(R, S)$$
$$\times \exp[i\omega T(R, S) + iT^G(R, S)]. \qquad (5.4.103)$$

Here $T^G(R, S)$ is the complete phase shift due to caustics in an anisotropic medium given by (5.4.25), $\mathcal{C}(S)$ and $\mathcal{C}(R)$ are phase velocities, and $\vec{g}^{(k)}$ are explained after (5.4.99).

Equations (5.4.100) and (5.4.101), with (5.4.97) and (5.4.98), and the relevant expressions (5.4.102) and (5.4.103) for the ray theory Green functions, can be used for any elementary multiply-reflected, possibly converted, wave propagating in a 3-D laterally varying layered structure. The model may be isotropic, perturbed isotropic, weakly anisotropic, and the like.

Equations (5.4.100) through (5.4.103) look surprisingly simple. We must, however, remember that the 3×3 matrices $\hat{\Pi}^e$ and $\hat{\Pi}^g$ are frequency-dependent and complex-valued along segments of the ray situated in an inhomogeneous weakly anisotropic medium. Thus, in the synthetic seismogram computations, it is necessary to compute the propagator matrices many times, for the whole range of frequencies under consideration (similarly as in the reflectivity method). Also function $\mathcal{A}^Q(R, S)$ is frequency-dependent, but its implementation is considerably simpler. It reduces to one numerical quadrature along the ray to evaluate the travel-time perturbation. This travel-time perturbation can be connected with travel time $T(R, S)$, so that no additional computations are required. An alternative is to work directly in the time domain; see Coates and Chapman (1990b).

5.4.7 R/T Coefficients and R/T Matrices

The determination of R/T coefficients of plane waves at a plane interface between two homogeneous anisotropic halfspaces was discussed in Section 2.3.3. In the seismic ray method, the R/T coefficients can be locally applied even to a nonplanar wave, incident at a curved interface separating two inhomogeneous anisotropic media. Here we shall first briefly recapitulate the main steps in the evaluation of the R/T coefficients and then express all relations in suitable matrix form.

We shall mostly use the same notations as in Section 2.3.3. We consider a plane interface Σ between anisotropic homogeneous halfspaces 1 ($\rho^{(1)}$, $c_{ijkl}^{(1)}$, $a_{ijkl}^{(1)}$) and 2 ($\rho^{(2)}$, $c_{ijkl}^{(2)}$, $a_{ijkl}^{(2)}$). We shall not use superscripts (1) or (2) denoting the selected halfspace in relations valid generally in both halfspaces. We denote the unit normal to Σ by $\vec{n}$ and orient it into any of the two halfspaces. Finally, we assume that the type (qS1, qS2, qP), the slowness vector, and the amplitude of the incident plane wave are known. We can then express the slowness vectors of all generated plane waves by relations $\vec{p} = \vec{a} + \sigma \vec{n}$. Here $\sigma \vec{n}$ represents the normal component of the slowness vector, and $\vec{a}$ represents the tangential component. The tangential components $\vec{a}$ of all generated plane waves are the same as the tangential component of the incident wave so that they are presumably known. Quantities σ, however, are different for different generated R/T waves. They are solutions of algebraic equations of the sixth-order:

$$\det[a_{ijkl}(a_j + \sigma n_j)(a_l + \sigma n_l) - \delta_{ik}] = 0. \tag{5.4.104}$$

Solving (5.4.104) for σ, we obtain six roots $\sigma^{(m)}$ ($m = 1, 2, \ldots, 6$), six slowness vectors $\vec{p}^{(m)} = \vec{a} + \sigma^{(m)}\vec{n}$, and six eigenvectors $\vec{g}^{(m)}$ of the Christoffel matrix $\Gamma_{ij} = a_{ijkl}p_j^{(m)}p_l^{(m)}$ (no summation over m):

$$a_{ijkl}(a_j + \sigma^{(m)}n_j)(a_l + \sigma^{(m)}n_l)g_k^{(m)} - g_i^{(m)} = 0; \tag{5.4.105}$$

see (2.2.32) with $G_m = 1$. The eigenvectors $\vec{g}^{(m)}$ represent polarization vectors of six individual waves. Finally, we obtain six group velocity vectors $\vec{\mathcal{U}}^{(m)}$ with Cartesian components given by the relation

$$\mathcal{U}_i^{(m)} = a_{ijkl}p_l^{(m)}g_j^{(m)}g_k^{(m)}; \tag{5.4.106}$$

see (2.2.65). Roots $\sigma^{(m)}$ may be real-valued (homogeneous plane waves) or appear in pairs of complex-conjugate quantities (inhomogeneous plane waves). We can divide the six roots $\sigma^{(m)}$ into two groups: $\sigma^{(1)+}$, $\sigma^{(2)+}$, $\sigma^{(3)+}$ and $\sigma^{(1)-}$, $\sigma^{(2)-}$, $\sigma^{(3)-}$. The first group includes the roots $\sigma^{(i)+}$, for which $\vec{\mathcal{U}}^{(i)+} \cdot \vec{n} > 0$ (waves propagating along $\vec{n}$), and the second group contains the roots $\sigma^{(i)-}$ for which $\vec{\mathcal{U}}^{(i)-} \cdot \vec{n} < 0$ (waves propagating against $\vec{n}$). For complex-valued roots, see the more detailed discussion in Section 2.3.3. In the same way, we also form relevant groups of eigenvectors $\vec{g}^{(i)+}$ and $\vec{g}^{(i)-}$ and of slowness vectors $\vec{p}^{(i)+}$ and $\vec{p}^{(i)-}$ ($i = 1, 2, 3$).

The system (2.3.50) of six linear algebraic equations of the first order for amplitudes of the generated R/T waves for the wave incident from the first halfspace may be suitably expressed in matrix form. It can be extended by six analogous equations for the wave incident from the second halfspace. We introduce four 3×3 matrices $\hat{\mathbf{H}}_1^+$, $\hat{\mathbf{H}}_1^-$, $\hat{\mathbf{H}}_2^+$, and $\hat{\mathbf{H}}_2^-$, formed by eigenvectors $\vec{g}^{(i)+}$ and $\vec{g}^{(i)-}$ in the first and second halfspaces. For example, the ith column of $\hat{\mathbf{H}}_1^+$ is formed by Cartesian components of the eigenvectors $\vec{g}^{(i)+}$ in the first halfspace. If we use the notation g_{ki}^+ for the ith Cartesian component of the eigenvector $\vec{g}^{(k)+}$, we obtain $(\hat{\mathbf{H}}^+)_{ik} = g_{ki}^+$. Within the individual matrices $\hat{\mathbf{H}}_1^+$, $\hat{\mathbf{H}}_1^-$, $\hat{\mathbf{H}}_2^+$, and $\hat{\mathbf{H}}_2^-$, the relevant three eigenvectors may be ordered in an arbitrary way. It is usual to arrange them in order of increasing phase velocity, that is, the first column for the slower qS wave, the second column for the faster qS wave, and the third column for the qP wave. Note that the 3×3 matrices $\hat{\mathbf{H}}_1^+$, $\hat{\mathbf{H}}_1^-$, $\hat{\mathbf{H}}_2^+$, and $\hat{\mathbf{H}}_2^-$ are not unitary because the individual eigenvectors forming these matrices correspond to different rays. We further introduce four 3×3 "traction" matrices $\hat{\mathbf{F}}_1^+$, $\hat{\mathbf{F}}_1^-$, $\hat{\mathbf{F}}_2^+$, and $\hat{\mathbf{F}}_2^-$, related to tractions $T_k = \tau_{jk}n_j$:

$$(\hat{\mathbf{F}}^+)_{ik} = -X_i^{(k)+}, \qquad X_i^{(k)+} = c_{ijnl}n_j g_n^{(k)+} p_l^{(k)+}; \tag{5.4.107}$$

see (2.1.3) and (2.3.51). Analogous equations define $\hat{\mathbf{F}}^-$. The minus sign in the first equation of (5.4.107) is introduced for formal reasons; it does not influence the boundary conditions. The individual columns in matrices $\hat{\mathbf{F}}$ must be arranged in the same order as in $\hat{\mathbf{H}}$.

We now add two analogous equations for the wave incident from the second halfspace to (2.3.50) and express the equations in matrix form. We obtain four matrix equations:

$$\begin{aligned}
\hat{\mathbf{H}}_1^- + \hat{\mathbf{H}}_1^+ \hat{\mathbf{R}}_{11}^T = \hat{\mathbf{H}}_2^- \hat{\mathbf{R}}_{12}^T, \qquad & \hat{\mathbf{H}}_2^+ + \hat{\mathbf{H}}_2^- \hat{\mathbf{R}}_{22}^T = \hat{\mathbf{H}}_1^+ \hat{\mathbf{R}}_{21}^T, \\
\hat{\mathbf{F}}_1^- + \hat{\mathbf{F}}_1^+ \hat{\mathbf{R}}_{11}^T = \hat{\mathbf{F}}_2^- \hat{\mathbf{R}}_{12}^T, \qquad & \hat{\mathbf{F}}_2^+ + \hat{\mathbf{F}}_2^- \hat{\mathbf{R}}_{22}^T = \hat{\mathbf{F}}_1^+ \hat{\mathbf{R}}_{21}^T.
\end{aligned} \tag{5.4.108}$$

These four matrix equations represent 36 scalar equations. The left-hand equations correspond to the wave incident from the first halfspace, and the right-hand equations correspond to the wave incident from the second halfspace. $\hat{\mathbf{R}}_{IJ}$ represent the 3×3 matrices of the R/T coefficients, with I specifying the incident wave halfspace and J specifying the generated wave halfspace. Consequently, $\hat{\mathbf{R}}_{11}$ and $\hat{\mathbf{R}}_{22}$ are reflection matrices, and $\hat{\mathbf{R}}_{12}$ and $\hat{\mathbf{R}}_{21}$ transmission matrices. Moreover, the individual elements of $\hat{\mathbf{R}}_{IJ}$, $(\hat{\mathbf{R}}_{IJ})_{kl}$, have the following meaning: k specifies the type of incident wave (qS1, qS2, qP), and l indicates the type of generated wave (qS1, qS2, qP). This numbering corresponds to the convention used in this book; see Sections 2.3 and 5.3. Due to this convention, it is necessary to use transposes of $\hat{\mathbf{R}}_{IJ}$ in (5.4.108). Note that a different convention for the indices in R/T matrices has been sometimes used in the seismological literature. Thus, it is necessary to be careful in comparing equations taken from different papers.

Equations (5.4.108) can be expressed in a more compact form using 6×6 matrices,

$$\mathbf{W}_1 \begin{pmatrix} \hat{\mathbf{R}}_{11}^T & \hat{\mathbf{R}}_{21}^T \\ \hat{\mathbf{I}} & \hat{\mathbf{0}} \end{pmatrix} = \mathbf{W}_2 \begin{pmatrix} \hat{\mathbf{0}} & \hat{\mathbf{I}} \\ \hat{\mathbf{R}}_{12}^T & \hat{\mathbf{R}}_{22}^T \end{pmatrix}, \tag{5.4.109}$$

where

$$\mathbf{W}_1 = \begin{pmatrix} \hat{\mathbf{H}}_1^+ & \hat{\mathbf{H}}_1^- \\ \hat{\mathbf{F}}_1^+ & \hat{\mathbf{F}}_1^- \end{pmatrix}, \qquad \mathbf{W}_2 = \begin{pmatrix} \hat{\mathbf{H}}_2^+ & \hat{\mathbf{H}}_2^- \\ \hat{\mathbf{F}}_2^+ & \hat{\mathbf{F}}_2^- \end{pmatrix}. \tag{5.4.110}$$

Taking the transpose of (5.4.109), we obtain, after some simple algebra,

$$\begin{pmatrix} -\hat{\mathbf{R}}_{22}\hat{\mathbf{R}}_{12}^{-1} & \hat{\mathbf{I}} \\ \hat{\mathbf{R}}_{12}^{-1} & \hat{\mathbf{0}} \end{pmatrix} \begin{pmatrix} \hat{\mathbf{R}}_{11} & \hat{\mathbf{I}} \\ \hat{\mathbf{R}}_{21} & \hat{\mathbf{0}} \end{pmatrix} = \mathbf{Q}, \qquad \mathbf{Q} = (\mathbf{W}_1^{-1}\mathbf{W}_2)^T. \tag{5.4.111}$$

This yields

$$\begin{pmatrix} \hat{\mathbf{R}}_{21} - \hat{\mathbf{R}}_{22}\hat{\mathbf{R}}_{12}^{-1}\hat{\mathbf{R}}_{11} & -\hat{\mathbf{R}}_{22}\hat{\mathbf{R}}_{12}^{-1} \\ \hat{\mathbf{R}}_{12}^{-1}\hat{\mathbf{R}}_{11} & \hat{\mathbf{R}}_{12}^{-1} \end{pmatrix} = \begin{pmatrix} \hat{\mathbf{Q}}_{11} & \hat{\mathbf{Q}}_{12} \\ \hat{\mathbf{Q}}_{21} & \hat{\mathbf{Q}}_{22} \end{pmatrix}.$$

Here $\hat{\mathbf{Q}}_{IJ}$ are 3×3 partitions of the 6×6 matrix $\mathbf{Q}$. This equation can be used to express $\hat{\mathbf{R}}_{IJ}$ in terms of $\hat{\mathbf{Q}}_{KL}$:

$$\begin{aligned} \hat{\mathbf{R}}_{11} &= \hat{\mathbf{Q}}_{22}^{-1}\hat{\mathbf{Q}}_{21}, & \hat{\mathbf{R}}_{21} &= \hat{\mathbf{Q}}_{11} - \hat{\mathbf{Q}}_{12}\hat{\mathbf{Q}}_{22}^{-1}\hat{\mathbf{Q}}_{21}, \\ \hat{\mathbf{R}}_{12} &= \hat{\mathbf{Q}}_{22}^{-1}, & \hat{\mathbf{R}}_{22} &= -\hat{\mathbf{Q}}_{12}\hat{\mathbf{Q}}_{22}^{-1}. \end{aligned} \tag{5.4.112}$$

It may be convenient to introduce the 6×6 matrix $\mathbf{R}$ of 36 R/T coefficients

$$\mathbf{R} = \begin{pmatrix} \hat{\mathbf{R}}_{11} & \hat{\mathbf{R}}_{12} \\ \hat{\mathbf{R}}_{21} & \hat{\mathbf{R}}_{22} \end{pmatrix} = \begin{pmatrix} \hat{\mathbf{Q}}_{22}^{-1}\hat{\mathbf{Q}}_{21} & \hat{\mathbf{Q}}_{22}^{-1} \\ \hat{\mathbf{Q}}_{11} - \hat{\mathbf{Q}}_{12}\hat{\mathbf{Q}}_{22}^{-1}\hat{\mathbf{Q}}_{21} & -\hat{\mathbf{Q}}_{12}\hat{\mathbf{Q}}_{22}^{-1} \end{pmatrix}. \tag{5.4.113}$$

This is the final result. Equations (5.4.112) or (5.4.113) can be used to compute any of the 36 R/T coefficients for a plane wave incident from any side at interface Σ. Note that we have kept to our convention: all 3×3 matrices have been marked with a circumflex above the letter, but the 6×6 matrices have not.

Equation (5.4.111) for $\mathbf{Q}$ indicates that it would be necessary to invert the 6×6 matrix $\mathbf{W}_1$ if we wished to compute the R/T coefficients. Fortunately, this is not necessary; some simple relation for $\mathbf{W}_1^{-1}$ can be derived. We introduce the two following 6×6 matrices:

$$\mathbf{I}_1 = \begin{pmatrix} \hat{\mathbf{0}} & \hat{\mathbf{I}} \\ \hat{\mathbf{I}} & \hat{\mathbf{0}} \end{pmatrix}, \qquad \mathbf{I}_2 = \begin{pmatrix} -\hat{\mathbf{I}} & \hat{\mathbf{0}} \\ \hat{\mathbf{0}} & \hat{\mathbf{I}} \end{pmatrix}, \tag{5.4.114}$$

and the 6×6 matrix $\mathbf{D}$ by the relation:

$$\mathbf{D} = \mathbf{W}^T \mathbf{I}_1 \mathbf{W} = \begin{pmatrix} \hat{\mathbf{H}}^{+T}\hat{\mathbf{F}}^+ + \hat{\mathbf{F}}^{+T}\hat{\mathbf{H}}^+ & \hat{\mathbf{H}}^{+T}\hat{\mathbf{F}}^- + \hat{\mathbf{F}}^{+T}\hat{\mathbf{H}}^- \\ \hat{\mathbf{H}}^{-T}\hat{\mathbf{F}}^+ + \hat{\mathbf{F}}^{-T}\hat{\mathbf{H}}^+ & \hat{\mathbf{H}}^{-T}\hat{\mathbf{F}}^- + \hat{\mathbf{F}}^{-T}\hat{\mathbf{H}}^- \end{pmatrix}. \tag{5.4.115}$$

It will be proved later in this section that $\mathbf{D}$ is a diagonal matrix:

$$\mathbf{D} = \mathrm{diag}\big(D^{(1)+}, D^{(2)+}, D^{(3)+}, D^{(1)-}, D^{(2)-}, D^{(3)-}\big). \tag{5.4.116}$$

Determining the diagonal elements $D^{(i)+}$ and $D^{(i)-}$ ($i = 1, 2, 3$) is not difficult. We realize that

$$(\hat{\mathbf{F}}^{+T}\hat{\mathbf{H}}^+)_{km} = -X_i^{(k)+}g_i^{(m)+} = -c_{ijnl}\,n_j g_n^{(k)+} g_i^{(m)+} p_l^{(k)+}$$

(no summation over k). We now use the known relation

$$c_{ijnl}\,g_n^{(k)+} g_i^{(m)+} p_l^{(k)+} = \rho\,\mathcal{U}_j^{(k)+} \qquad \text{for } m = k.$$

Analogous relations are obtained for $\hat{\mathbf{F}}^{-T}\hat{\mathbf{H}}^-$, and for $\hat{\mathbf{H}}^{+T}\hat{\mathbf{F}}^+ = (\hat{\mathbf{F}}^{+T}\hat{\mathbf{H}}^+)^T$, and for $\hat{\mathbf{H}}^{-T}\hat{\mathbf{F}}^- = (\hat{\mathbf{F}}^{-T}\hat{\mathbf{H}}^-)^T$. Together,

$$D^{(i)+} = -2\rho\vec{\mathcal{U}}^{(i)+} \cdot \vec{n}, \qquad D^{(i)-} = -2\rho\vec{\mathcal{U}}^{(i)-} \cdot \vec{n}. \tag{5.4.117}$$

Thus, $D^{(i)+}$ are negative, and $D^{(i)-}$ are positive. Using the diagonal matrix $\mathbf{D}_1$, corresponding to the first halfspace, we can simply determine $\mathbf{W}_1^{-1}$, see (5.4.115), and $\mathbf{Q}$, see (5.4.111):

$$\mathbf{W}_1^{-1} = \mathbf{D}_1^{-1}\mathbf{W}_1^T\mathbf{I}_1, \qquad \mathbf{Q} = \left(\mathbf{W}_1^{-1}\mathbf{W}_2\right)^T = \mathbf{W}_2^T\mathbf{I}_1\mathbf{W}_1\mathbf{D}_1^{-1}. \qquad (5.4.118)$$

Here $\mathbf{I}_1$ is given by (5.4.114). These are the final general relations for the inverse of $\mathbf{W}_1$ and for $\mathbf{Q}$.

Equation (5.4.113) for the R/T coefficients can also be used in isotropic media. The procedure then simplifies considerably. It is not necessary to solve (5.4.104) for σ because the solutions are analytical: $\sigma = \pm[1/V^2 - a_ia_i]^{1/2}$, with $V = \alpha$ for P waves and $V = \beta$ for S waves (double root). Moreover, the criteria to select upgoing and downgoing waves also simplify. The plus sign in the expression for σ indicates waves propagating along $\vec{n}$, and the minus sign indicates the waves propagating against $\vec{n}$. Finally, expressions for $\vec{g}^{(1)} = \vec{e}_1$, $\vec{g}^{(2)} = \vec{e}_2$, and $\vec{g}^{(3)} = \vec{e}_3$ can also be found analytically. Because the isotropic medium is a degenerate case of the anisotropic medium, the eigenvectors $\vec{e}_1$ and $\vec{e}_2$ of the S waves cannot be completely determined, but they form a right-handed triplet with $\vec{e}_3 = \beta\vec{p}$, where $\vec{p}$ is the slowness vector of the S wave. For more details, see Sections 2.3.2 and 5.3.

AN APPROACH BASED ON THE SOLUTION OF A 6×6 EIGENVALUE PROBLEM

The previous formulation is, in principle, based on the 3×3 matrices. Only later did we arrive at 6×6 matrices. An alternative approach is to use the 6×6 matrices from the very beginning, and to calculate $\sigma^{(m)}$, $m = 1, 2, \ldots, 6$, as the eigenvalues of some 6×6 matrix. This approach is closely connected with the 6×6 propagator technique developed for computing the wave fields in 1-D media. This technique has been broadly used in seismology, both for isotropic and anisotropic 1-D media. See Woodhouse (1974), Kennett, Kerry, and Woodhouse (1978), Kennett (1983), Fryer and Frazer (1984, 1987), Thomson, Clarke, and Garmany (1986), Frazer and Fryer (1989), Chapman (1994), and Thomson (1996a), where many other references can be found. The general 6×6 propagator technique can be used to determine R/T coefficients at any 1-D inhomogeneous layer, including a stack of homogeneous layers, and to study their properties and symmetries. The results may then be simply specified for the R/T coefficients at a single planar interface. Here our aim is more modest; we shall not discuss the R/T coefficients at transition layers but only at a single interface. For this reason, we shall not start with the 1-D propagator technique but rather discuss the problem of R/T coefficients at a single interface from the beginning.

Consider a plane wave propagating in a homogeneous elastic medium, in which the Cartesian components of the particle velocity $v_i = \dot{u}_i$ and the stress tensor τ_{ij} are given by relations

$$v_i = V_i \exp[-i\omega(t - p_kx_k)], \qquad \tau_{ij} = T_{ij} \exp[-i\omega(t - p_kx_k)].$$
$$(5.4.119)$$

Inserting these relations into (2.1.18), with source terms not considered, a system of 12 equations for V_i and T_{ij} is obtained:

$$T_{ij} + c_{ijkl}p_lV_k = 0, \qquad V_i + \rho^{-1}p_jT_{ij} = 0. \qquad (5.4.120)$$

If we eliminate T_{ij} from (5.4.120), the system (5.4.120) of 12 equations is reduced to

3 equations for V_i, that is $V_i - a_{ijkl}p_jp_lV_k = 0$, $i = 1, 2, 3$. These equations are fully equivalent to (2.2.35) for displacement amplitudes U_i. Here, however, we shall proceed in a different way and discuss system (5.4.120).

We specify some direction by unit vector $\vec{n}$ and express slowness vector $\vec{p}$ by the relation $\vec{p} = \vec{a} + \sigma\vec{n}$, where $\vec{a}$ is perpendicular to $\vec{n}$, and σ has the same meaning as in (5.4.104). Multiplying the first equation of (5.4.120) by n_j and putting $T_i = T_{ij}n_j$, we obtain a system of six equations for T_i and V_i:

$$T_i + c_{ijkl}n_ja_lV_k + c_{ijkl}n_jn_l\sigma V_k = 0,$$
$$V_i + \rho^{-1}\sigma T_i - \rho^{-1}c_{ijkl}a_ja_lV_k - \rho^{-1}c_{ijkl}a_j\sigma n_lV_k = 0.$$

Here T_i represents the ith Cartesian component of the vectorial amplitude of the traction acting on a surface element perpendicular to $\vec{n}$. These equations simplify if we use the matrix notation. We introduce four 3×3 matrices $\hat{\mathbf{C}}^{(1)}$, $\hat{\mathbf{C}}^{(2)}$, $\hat{\mathbf{C}}^{(3)}$, and $\hat{\mathbf{C}}^{(4)}$ with components $C_{ik}^{(1)}$, $C_{ik}^{(2)}$, $C_{ik}^{(3)}$, and $C_{ik}^{(4)}$ given by relations:

$$\begin{aligned} C_{ik}^{(1)} &= c_{ijkl}n_jn_l, & C_{ik}^{(2)} &= c_{ijkl}n_ja_l, \\ C_{ik}^{(3)} &= c_{ijkl}a_jn_l = C_{ki}^{(2)}, & C_{ik}^{(4)} &= c_{ijkl}a_ja_l. \end{aligned} \tag{5.4.121}$$

Then the system reads

$$\begin{aligned} \hat{\mathbf{T}} + \hat{\mathbf{C}}^{(2)}\hat{\mathbf{V}} + \hat{\mathbf{C}}^{(1)}\sigma\hat{\mathbf{V}} &= \hat{\mathbf{0}}, \\ \hat{\mathbf{V}} + \rho^{-1}\sigma\hat{\mathbf{T}} - \rho^{-1}\hat{\mathbf{C}}^{(4)}\hat{\mathbf{V}} - \rho^{-1}\hat{\mathbf{C}}^{(3)}\sigma\hat{\mathbf{V}} &= \hat{\mathbf{0}}, \end{aligned} \tag{5.4.122}$$

where $\hat{\mathbf{T}} = (T_1, T_2, T_3)^T$ and $\hat{\mathbf{V}} = (V_1, V_2, V_3)^T$. By expressing $\sigma\hat{\mathbf{V}}$ from the first equation and inserting it into the second equation, we arrive at the final form of the system:

$$\mathbf{A}\begin{pmatrix} \hat{\mathbf{V}} \\ \hat{\mathbf{T}} \end{pmatrix} = \begin{pmatrix} \hat{\mathbf{V}} \\ \hat{\mathbf{T}} \end{pmatrix}\sigma. \tag{5.4.123}$$

Here the 6×6 system matrix $\mathbf{A}$ is given by the relation

$$\mathbf{A} = \begin{pmatrix} \hat{\mathbf{A}}_{11} & \hat{\mathbf{A}}_{12} \\ \hat{\mathbf{A}}_{21} & \hat{\mathbf{A}}_{22} \end{pmatrix}, \tag{5.4.124}$$

with

$$\begin{aligned} \hat{\mathbf{A}}_{11} &= -\hat{\mathbf{C}}^{(1)-1}\hat{\mathbf{C}}^{(2)} = \hat{\mathbf{A}}_{22}^T, & \hat{\mathbf{A}}_{12} &= -\hat{\mathbf{C}}^{(1)-1}, \\ \hat{\mathbf{A}}_{21} &= -\rho\hat{\mathbf{I}} + \hat{\mathbf{C}}^{(4)} - \hat{\mathbf{C}}^{(3)}\hat{\mathbf{C}}^{(1)-1}\hat{\mathbf{C}}^{(2)}, & \hat{\mathbf{A}}_{22} &= -\hat{\mathbf{C}}^{(3)}\hat{\mathbf{C}}^{(1)-1}. \end{aligned} \tag{5.4.125}$$

System (5.4.123) is expressed in general Cartesian coordinates, with arbitrarily oriented unit normal $\vec{n}$. In the seismological literature, it is usual to use a local Cartesian coordinate system x_i, with the x_3-axis along $\vec{n}$. Then, $n_1 = n_2 = 0$, $n_3 = 1$, and $a_3 = 0$. We denote $a_I = p_I$ and introduce the 3×3 matrices $\hat{\mathbf{C}}_{jl}$ by relations $(\hat{\mathbf{C}}_{jl})_{ik} = c_{ijkl}$. Then

$$\hat{\mathbf{C}}^{(1)} = \hat{\mathbf{C}}_{33}, \quad \hat{\mathbf{C}}^{(2)} = p_L\hat{\mathbf{C}}_{3L}, \quad \hat{\mathbf{C}}^{(3)} = p_J\hat{\mathbf{C}}_{J3}, \quad \hat{\mathbf{C}}^{(4)} = \hat{\mathbf{C}}_{JL}p_Jp_L. \tag{5.4.126}$$

and

$$\begin{aligned} \hat{\mathbf{A}}_{11} &= -p_L\mathbf{C}_{33}^{-1}\hat{\mathbf{C}}_{3L} = \hat{\mathbf{A}}_{22}^T, & \hat{\mathbf{A}}_{12} &= -\hat{\mathbf{C}}_{33}^{-1}, \\ \hat{\mathbf{A}}_{21} &= -\rho\hat{\mathbf{I}} + p_Jp_L(\hat{\mathbf{C}}_{JL} - \hat{\mathbf{C}}_{J3}\hat{\mathbf{C}}_{33}^{-1}\hat{\mathbf{C}}_{3L}), & \hat{\mathbf{A}}_{22} &= -p_J\hat{\mathbf{C}}_{J3}\hat{\mathbf{C}}_{33}^{-1}. \end{aligned} \tag{5.4.127}$$

System (5.4.123) with (5.4.127) is well known from the seismological literature; see Chapman (1994) and Thomson (1996a), among others. Here, however, we prefer to work with (5.4.125) because it does not require c_{ijkl} to be transformed from the general to a local Cartesian coordinate system.

Note that the 6×6 system matrix $\mathbf{A}$ is not symmetrical but that the product $\mathbf{I}_1\mathbf{A}$ (where $\mathbf{I}_1$ is given by (5.4.114)) is.

As we can see from (5.4.123), σ represents an eigenvalue of the 6×6 matrix $\mathbf{A}$. We shall now prove that the same σ also represents a root of (5.4.104). Eliminating $\hat{\mathbf{T}}$ from (5.4.122), we obtain

$$\left[\hat{\mathbf{C}}^{(4)} + \sigma\left(\hat{\mathbf{C}}^{(2)} + \hat{\mathbf{C}}^{(3)}\right) + \sigma^2\hat{\mathbf{C}}^{(1)} - \rho\hat{\mathbf{I}}\right]\hat{\mathbf{V}} = \hat{\mathbf{0}}.$$

This system of six equations for $\hat{\mathbf{V}}$ has a nontrivial solution only if the determinant of the system vanishes:

$$\det\left[\hat{\mathbf{C}}^{(4)} + \sigma\left(\hat{\mathbf{C}}^{(2)} + \hat{\mathbf{C}}^{(3)}\right) + \sigma^2\hat{\mathbf{C}}^{(1)} - \rho\hat{\mathbf{I}}\right] = 0. \qquad (5.4.128)$$

Using the notation (5.4.121) in (5.4.104), we can see that (5.4.128) and (5.4.104) are exactly the same. Consequently, the six eigenvalues $\sigma^{(m)}$ ($m = 1, 2, \ldots, 6$) of matrix $\mathbf{A}$ can also be alternatively defined as six roots of (5.4.104).

We now denote $\hat{\mathbf{V}}$ and $\hat{\mathbf{T}}$ in (5.4.123), corresponding to an arbitrarily selected eigenvalue $\sigma^{(m)}$, by $\hat{\mathbf{V}}^{(m)}$ and $\hat{\mathbf{T}}^{(m)}$, and express (5.4.123) in the following form:

$$\mathbf{A}\mathbf{U}^{(m)} = \mathbf{U}^{(m)}\sigma^{(m)}, \qquad \mathbf{U}^{(m)} = \begin{pmatrix} \hat{\mathbf{V}}^{(m)} \\ \hat{\mathbf{T}}^{(m)} \end{pmatrix}, \qquad (5.4.129)$$

(no summation over m). $\mathbf{U}^{(m)}$ represents the eigenvector of matrix $\mathbf{A}$ corresponding to the eigenvalue $\sigma^{(m)}$. We shall now prove that $\mathbf{U}^{(m)}$ exists and can be expressed in terms of $g_i^{(m)}$ and $X_i^{(m)}$; see (5.4.107). Equation (5.4.129) yields

$$\hat{\mathbf{A}}_{11}\hat{\mathbf{V}}^{(m)} + \hat{\mathbf{A}}_{12}\hat{\mathbf{T}}^{(m)} = \hat{\mathbf{V}}^{(m)}\sigma^{(m)}, \qquad \hat{\mathbf{A}}_{21}\hat{\mathbf{V}}^{(m)} + \hat{\mathbf{A}}_{22}\hat{\mathbf{T}}^{(m)} = \hat{\mathbf{T}}^{(m)}\sigma^{(m)}.$$
$$(5.4.130)$$

Eliminating $\hat{\mathbf{T}}^{(m)}$ from (5.4.130) yields

$$\left(-\rho\hat{\mathbf{I}} + \hat{\mathbf{C}}^{(4)}\right)\hat{\mathbf{V}}^{(m)} + \left(\hat{\mathbf{C}}^{(3)} + \hat{\mathbf{C}}^{(2)}\right)\hat{\mathbf{V}}^{(m)}\sigma^{(m)} + \hat{\mathbf{C}}^{(1)}\hat{\mathbf{V}}^{(m)}\sigma^{(m)2} = \hat{\mathbf{0}}.$$
$$(5.4.131)$$

However, this is exactly the same equation as (5.4.105) for the eigenvector of the Christoffel matrix $\hat{\mathbf{g}}^{(m)} \equiv (g_1^{(m)}, g_2^{(m)}, g_3^{(m)})^T$. Now we determine $\hat{\mathbf{T}}^{(m)}$. The first equation of (5.4.130) yields

$$\hat{\mathbf{T}}^{(m)} = \left(-\hat{\mathbf{A}}_{12}^{-1}\hat{\mathbf{A}}_{11} + \hat{\mathbf{A}}_{12}^{-1}\sigma^{(m)}\right)\hat{\mathbf{V}}^{(m)} = \left(-\hat{\mathbf{C}}^{(2)} - \hat{\mathbf{C}}^{(1)}\sigma^{(m)}\right)\hat{\mathbf{V}}^{(m)}.$$
$$(5.4.132)$$

For $\hat{\mathbf{T}}^{(m)}$ given by (5.4.132), the second equation of (5.4.130) is automatically satisfied. Consequently, $\hat{\mathbf{T}}^{(m)}$ is related to $\hat{\mathbf{V}}^{(m)}$ as shown by (5.4.132). Inserting (5.4.121) into (5.4.132) yields the final expression for $T_i^{(m)}$:

$$T_i^{(m)} = -c_{ijkl}n_j(a_l + n_l\sigma^{(m)})g_k^{(m)} = -c_{ijkl}n_j p_l^{(m)}g_k^{(m)} = -X_i^{(m)}$$
$$(5.4.133)$$

(no summation over m); see (5.4.107). Thus, we have proved that the eigenvector $\mathbf{U}^{(m)}$ of the 6×6 matrix $\mathbf{A}$, corresponding to eigenvalue $\sigma^{(m)}$, exists and can be expressed in terms of $g_i^{(m)}$ and $X_i^{(m)}$ as follows:

$$\mathbf{U}^{(m)} = \left(g_1^{(m)}, g_2^{(m)}, g_3^{(m)}, -X_1^{(m)}, -X_2^{(m)}, -X_3^{(m)}\right)^T. \qquad (5.4.134)$$

This indicates that we can compute $\sigma^{(m)}$ and $U^{(m)}$ $(m = 1, 2, \ldots, 6)$, in two alternative ways.

a. Calculate them as eigenvalues and eigenvectors of the 6×6 matrix $\mathbf{A}$.

b. Calculate $\sigma^{(m)}$ $(m = 1, 2, \ldots, 6)$, as roots of (5.4.104) and construct the relevant slowness vectors $\vec{p}^{(m)} = \vec{a} + \sigma^{(m)}\vec{n}$. $\vec{g}^{(m)}$ is then the appropriate eigenvector of the 3×3 Christoffel matrix and represents the polarization vector of the selected wave. Finally, $X_i^{(m)} = c_{ijkl}n_j p_l^{(m)} g_k^{(m)}$ is the relevant traction component; see (5.4.107).

We now take into account the complete system of eigenvalues and eigenvectors $\sigma^{(m)}$, and $\mathbf{U}^{(m)}$ $(m = 1, 2, \ldots, 6)$ and express Equation (5.4.129) in the following general form:

$$\mathbf{AW} = \mathbf{W}\sigma, \qquad (5.4.135)$$

where 6×6 matrices σ and $\mathbf{W}$ are given by relations

$$\sigma = \begin{pmatrix} \hat{\sigma}^+ & \hat{\mathbf{0}} \\ \hat{\mathbf{0}} & \hat{\sigma}^- \end{pmatrix},$$

$$\mathbf{W} = \left(\mathbf{U}^{(1)}, \mathbf{U}^{(2)}, \mathbf{U}^{(3)}, \mathbf{U}^{(4)}, \mathbf{U}^{(5)}, \mathbf{U}^{(6)}\right) = \begin{pmatrix} \hat{\mathbf{H}}^+ & \hat{\mathbf{H}}^- \\ \hat{\mathbf{F}}^+ & \hat{\mathbf{F}}^- \end{pmatrix}. \qquad (5.4.136)$$

Here $\hat{\sigma}^+$ and $\hat{\sigma}^-$ are 3×3 diagonal matrices, with diagonal elements $\sigma^{(1)+}, \sigma^{(2)+}, \sigma^{(3)+}$ and $\sigma^{(1)-}, \sigma^{(2)-}, \sigma^{(3)-}$, representing the eigenvalues of $\mathbf{A}$. The columns of the 6×6 matrix $\mathbf{W}$ are represented by eigenvectors $\mathbf{U}^{(m)}$. Alternatively, we can express $\mathbf{W}$ in terms of 3×3 submatrices of polarization vectors $\hat{\mathbf{H}}^+$ and $\hat{\mathbf{H}}^-$, and of 3×3 traction submatrices $\hat{\mathbf{F}}^+$ and $\hat{\mathbf{F}}^-$, just as in (5.4.110). The eigenvalues $\sigma^{(m)}$ and relevant eigenvectors $\mathbf{U}^{(m)}$ in the 6×6 matrices σ and $\mathbf{W}$ should be arranged in the same order.

Because $\hat{\mathbf{V}}$ and $\hat{\mathbf{T}}$ are continuous across the interface, the foregoing equations can be used to solve the R/T problem. We consider interface Σ with unit normal $\vec{n}$. We again obtain the system of equations (5.4.109). Consequently, all the relations (5.4.110) through (5.4.118) are obtained exactly in the same way as before. For this reason, we shall not repeat the derivations.

Equations (5.4.135) are very convenient in studying certain symmetries of the R/T coefficients; see Chapman (1994) and Thomson (1996a). Such symmetries also apply to the R/T coefficients from transition layers, but here we shall apply them only to the R/T coefficients at a single interface. We shall first prove that the 3×3 matrix $\mathbf{D} = \mathbf{W}^T\mathbf{I}_1\mathbf{W}$ is diagonal. Thereafter, we shall discuss the reciprocity of the R/T coefficients.

First, we prove that $\mathbf{D} = \mathbf{W}^T\mathbf{I}_1\mathbf{W}$ is diagonal. Using (5.4.125), it is simple to see that $\mathbf{I}_1\mathbf{A}$ is symmetric, where $\mathbf{I}_1$ is given by (5.4.114). We multiply the transpose of (5.4.135) from the right by $\mathbf{I}_1$ and obtain $\mathbf{W}^T\mathbf{A}^T\mathbf{I}_1 = \sigma\mathbf{W}^T\mathbf{I}_1$. Because $\mathbf{I}_1\mathbf{A}$ is symmetric, we obtain $\mathbf{W}^T\mathbf{I}_1\mathbf{A} = \sigma\mathbf{W}^T\mathbf{I}_1$. Multiplying this by $\mathbf{W}$ from the right and taking into account (5.4.135), we finally obtain $\mathbf{W}^T\mathbf{I}_1\mathbf{W}\sigma = \sigma\mathbf{W}^T\mathbf{I}_1\mathbf{W}$. This shows that diagonal matrix σ commutes with $\mathbf{D} = \mathbf{W}^T\mathbf{I}_1\mathbf{W}$. Consequently, matrix $\mathbf{D} = \mathbf{W}^T\mathbf{I}_1\mathbf{W}$ is diagonal. This also proves that

$$\hat{\mathbf{H}}^{+T}\hat{\mathbf{F}}^- + \hat{\mathbf{F}}^{+T}\hat{\mathbf{H}}^- = \hat{\mathbf{0}}, \qquad \hat{\mathbf{H}}^{-T}\hat{\mathbf{F}}^+ + \hat{\mathbf{F}}^{-T}\hat{\mathbf{H}}^+ = \hat{\mathbf{0}}; \qquad (5.4.137)$$

see (5.4.115).

Note that the eigenvectors of the 6×6 matrix $\mathbf{A}$ are not orthogonal, but $\mathbf{W}^T\mathbf{I}_1\mathbf{W}$ is diagonal. This represents a generalization of the concept of orthogonality and may be called the $\mathbf{I}_1$-*orthogonality*. Consequently, the eigenvectors of the 6×6 matrix $\mathbf{A}$ are not orthogonal but are $\mathbf{I}_1$-orthogonal; see Frazer and Fryer (1989).

Now we shall study the reciprocity of the R/T coefficients for forward and backward propagation. As in Section 5.3.7, we shall mark all quantities corresponding to backward propagation with a bar above the quantity. In the backward propagation, we keep the Cartesian coordinate system and unit normal $\vec{n}$ the same as in the forward propagation. The only change is in the direction of the slowness vectors. This changes the signs of a_i in (5.4.121), so that $\hat{\bar{\mathbf{C}}}^{(1)} = \hat{\mathbf{C}}^{(1)}$, $\hat{\bar{\mathbf{C}}}^{(2)} = -\hat{\mathbf{C}}^{(2)}$, $\hat{\bar{\mathbf{C}}}^{(3)} = -\hat{\mathbf{C}}^{(3)}$, and $\hat{\bar{\mathbf{C}}}^{(4)} = \hat{\mathbf{C}}^{(4)}$. System matrix $\bar{\mathbf{A}}$ then reads

$$\bar{\mathbf{A}} = -\mathbf{I}_2 \mathbf{A} \mathbf{I}_2 = \begin{pmatrix} -\hat{\mathbf{A}}_{11} & \hat{\mathbf{A}}_{12} \\ \hat{\mathbf{A}}_{21} & -\hat{\mathbf{A}}_{22} \end{pmatrix}. \tag{5.4.138}$$

Here $\mathbf{I}_2$ is given by (5.4.114). Inserting this into (5.4.135) yields $\bar{\mathbf{A}} \mathbf{I}_2 \mathbf{W} = -\mathbf{I}_2 \mathbf{W} \sigma$, that is, $\bar{\mathbf{A}} \overline{\mathbf{W}} = \overline{\mathbf{W}} \bar{\sigma}$, where

$$\overline{\mathbf{W}} = \mathbf{I}_2 \mathbf{W}, \qquad \bar{\sigma} = -\sigma. \tag{5.4.139}$$

The relation $\bar{\sigma} = -\sigma$ in (5.4.139) expresses the central point symmetry of the slowness surfaces. Consequently, the first three columns in matrix $\overline{\mathbf{W}}$ correspond to the downgoing waves, and the next three columns correspond to the upgoing waves. If we take this into account, Equation (5.4.109) must be modified to read

$$\overline{\mathbf{W}}_1 \begin{pmatrix} \hat{\mathbf{I}} & \hat{\mathbf{0}} \\ \hat{\bar{\mathbf{R}}}_{11}^T & \hat{\bar{\mathbf{R}}}_{21}^T \end{pmatrix} = \overline{\mathbf{W}}_2 \begin{pmatrix} \hat{\bar{\mathbf{R}}}_{12}^T & \hat{\bar{\mathbf{R}}}_{22}^T \\ \hat{\mathbf{0}} & \hat{\mathbf{I}} \end{pmatrix}.$$

Taking the transpose of this equation and multiplying it by $\mathbf{I}_2 \mathbf{I}_1$ and by (5.4.109) from the right, we obtain

$$\begin{pmatrix} \hat{\mathbf{I}} & \hat{\bar{\mathbf{R}}}_{11} \\ \hat{\mathbf{0}} & \hat{\bar{\mathbf{R}}}_{21} \end{pmatrix} \overline{\mathbf{W}}_1^T \mathbf{I}_2 \mathbf{I}_1 \mathbf{W}_1 \begin{pmatrix} \hat{\mathbf{R}}_{11}^T & \hat{\mathbf{R}}_{21}^T \\ \hat{\mathbf{I}} & \hat{\mathbf{0}} \end{pmatrix} = \begin{pmatrix} \hat{\bar{\mathbf{R}}}_{12} & \hat{\mathbf{0}} \\ \hat{\bar{\mathbf{R}}}_{22} & \hat{\mathbf{I}} \end{pmatrix} \overline{\mathbf{W}}_2^T \mathbf{I}_2 \mathbf{I}_1 \mathbf{W}_2 \begin{pmatrix} \hat{\mathbf{0}} & \hat{\mathbf{I}} \\ \hat{\mathbf{R}}_{12}^T & \hat{\mathbf{R}}_{22}^T \end{pmatrix}. \tag{5.4.140}$$

Using (5.4.139) and (5.4.115), we obtain

$$\overline{\mathbf{W}}_1^T \mathbf{I}_2 \mathbf{I}_1 \mathbf{W}_1 = \mathbf{W}_1^T \mathbf{I}_1 \mathbf{W}_1 = \mathbf{D}_1.$$

Inserting this into (5.4.140) yields

$$\begin{pmatrix} \hat{\mathbf{I}} & \hat{\bar{\mathbf{R}}}_{11} \\ \hat{\mathbf{0}} & \hat{\bar{\mathbf{R}}}_{21} \end{pmatrix} \mathbf{D}_1 \begin{pmatrix} \hat{\mathbf{R}}_{11}^T & \hat{\mathbf{R}}_{21}^T \\ \hat{\mathbf{I}} & \hat{\mathbf{0}} \end{pmatrix} = \begin{pmatrix} \hat{\bar{\mathbf{R}}}_{12} & \hat{\mathbf{0}} \\ \hat{\bar{\mathbf{R}}}_{22} & \hat{\mathbf{I}} \end{pmatrix} \mathbf{D}_2 \begin{pmatrix} \hat{\mathbf{0}} & \hat{\mathbf{I}} \\ \hat{\mathbf{R}}_{12}^T & \hat{\mathbf{R}}_{22}^T \end{pmatrix}. \tag{5.4.141}$$

Equation (5.4.141) can be used to find various reciprocity relations between $\hat{\bar{\mathbf{R}}}_{IJ}$ and $\hat{\mathbf{R}}_{KL}$. It can be expressed in many alternative forms. In general, the reciprocity relations include the diagonal matrices $\mathbf{D}_1$ and $\mathbf{D}_2$, given by (5.4.116) with (5.4.117).

We shall not discuss the reciprocity relations for general $\mathbf{D}_1$ and $\mathbf{D}_2$ but shall focus on one specific, very important case. We shall simplify the expressions for $\mathbf{D}_1$ and $\mathbf{D}_2$ by suitably normalizing the eigenvectors. For simplicity, we shall assume that all eigenvalues $\sigma^{(m)}$ are real-valued. In our treatment, $\vec{g}^{(i)+}$ and $\vec{g}^{(i)-}$ are unit vectors. We can, however, normalize them in a different way. We introduce vectors $\vec{f}^{(i)+}$ and $\vec{f}^{(i)-}$ by relations:

$$\vec{f}^{(i)+} = \vec{g}^{(i)+} / \sqrt{2\rho |\vec{\mathcal{U}}^{(i)+} \cdot \vec{n}|}, \qquad \vec{f}^{(i)-} = \vec{g}^{(i)-} / \sqrt{2\rho |\vec{\mathcal{U}}^{(i)-} \cdot \vec{n}|}. \tag{5.4.142}$$

Analogous normalization is also performed in the expressions for $\hat{\mathbf{F}}^+$ and $\hat{\mathbf{F}}^-$ because they contain $\vec{g}^{(i)+}$ and $\vec{g}^{(i)-}$; see (5.4.107). We then obtain very simple expressions for matrices $\mathbf{D}_1$ and $\mathbf{D}_2$: $\mathbf{D}_1 = \mathbf{D}_2 = \mathbf{I}_2$, where $\mathbf{I}_2$ is given by (5.4.114). The R/T coefficients, of course, are also influenced by the normalization of the eigenvectors. We denote the R/T coefficients corresponding to the eigenvectors (5.4.142) by $\mathcal{R}_{ij}$ (instead of R_{ij}), $i, j = 1, 2, \ldots, 6$, and call them the *normalized displacement R/T coefficients*. It can be deduced from (5.4.142) that the relation between the standard displacement R/T coefficients R_{ij} and the normalized displacement R/T coefficients $\mathcal{R}_{ij}$ is

$$\mathcal{R}_{ij} = [(\rho|\vec{\mathcal{U}} \cdot \vec{n}|)_j / (\rho|\vec{\mathcal{U}} \cdot \vec{n}|)_i]^{1/2} R_{ij}. \tag{5.4.143}$$

Here i is the index corresponding to the incident wave, and j is the index corresponding to the generated R/T wave (no summation over i and j). Thus, we have again arrived at the normalized R/T coefficients introduced by (5.4.13).

The reciprocity relations for the normalized R/T coefficients $\mathcal{R}_{ij}$ follow immediately from (5.4.141), where we insert $\mathbf{D}_1 = \mathbf{D}_2 = \mathbf{I}_2$. After some simple algebra, we obtain the final result

$$\bar{\mathcal{R}}_{ij} = \mathcal{R}_{ji} \tag{5.4.144}$$

$(i, j = 1, 2, \ldots, 6)$. This important and very general reciprocity relation is due to Chapman (1994). It is valid for isotropic and anisotropic media and for reflected and transmitted waves. Moreover, it also remains valid for the R/T coefficients at a stack of homogeneous anisotropic layers and at a 1-D anisotropic transition layer. It should, however, be emphasized that the simple reciprocity relation (5.4.144) is valid only for the normalized R/T coefficients, not for the standard displacement R/T coefficients. This corresponds to the conclusions of Section 5.3.7 for isotropic media.

It may be useful to emphasize the differences between the general reciprocity relation (5.4.144) and the relation (5.3.31) derived for isotropic media. In deriving (5.4.144), only the direction of the slowness vector has been changed; the Cartesian system and the orientation of normal $\vec{n}$ are the same both in the forward and backward propagation. In deriving (5.3.31), however, we have used a different convention for the backward and forward propagation; see Section 4.4.9. Due to this convention, it was also necessary to transform orientation index ϵ; see Section 5.3.7. Otherwise, however, the reciprocity relations (5.4.144) and (5.3.31) are the same: the normalized R/T coefficient $\mathcal{R}_{ij}$ in the backward propagation equals the normalized R/T coefficient $\mathcal{R}_{ji}$ in the forward propagation.

5.4.8 Initial Ray-Theory Amplitudes at a Smooth Initial Surface. Elastic Kirchhoff Integrals

In this section, we shall closely follow the analogous treatment of initial ray-theory amplitudes of pressure waves at a smooth initial surface in a fluid medium, as discussed in detail in Section 5.1.11. For elastic waves, the derivation is practically the same, it is only formally more complex due to the vectorial character of the wavefield and the existence of three types of waves (qS1, qS2, qP). We shall consider general anisotropic inhomogeneous media. All the derived equations, however, will also be applicable to isotropic inhomogeneous media.

We consider a smooth initial surface Σ^0 in an elastic medium, which may correspond to a structural interface, a free surface, an auxiliary surface situated in a smooth medium, a wavefront, and the like. We assume that initial time T^0 is specified along Σ^0. We can then determine the initial slowness vectors and initial values of matrices $\mathbf{Q}^{(y)}$ and $\mathbf{P}^{(y)}$ of all waves generated along Σ^0. For isotropic media, the relevant equations were derived and discussed in detail in Section 4.5. For anisotropic media, the derivation would be analogous. Consequently, we can start ray tracing and dynamic ray tracing of all waves generated at any point of Σ^0. To determine the vectorial amplitudes along these rays, however, we must also know the initial ray-theory vectorial amplitudes at the initial points of rays along Σ^0. The determination of these initial amplitudes along Σ^0 is the main purpose of this section. In addition to this, we shall also discuss the elastic Kirchhoff integral. See also Haddon and Buchen (1981), Sinton and Frazer (1982), Frazer and Sen (1985), Zhu (1988), Tygel, Schleicher, and Hubral (1994), Ursin and Tygel (1997), Druzhinin (1998), Druzhinin et al. (1998), and Chapman (in press).

As the point of departure, we shall use the elastic Kirchhoff integral (2.6.4) and assume that the distribution of the displacement components u_i and of the traction components T_i along Σ^0 are given by the relations:

$$
\begin{aligned}
u_i(\vec{x}', \omega) &= U_i^0(\vec{x}') \exp[i\omega T^0(\vec{x}')], \\
T_i(\vec{x}', \omega) &= i\omega T_i^0(\vec{x}') \exp[i\omega T^0(\vec{x}')].
\end{aligned}
\tag{5.4.145}
$$

Here $\vec{x}'$ are the points along Σ^0, and $U_i^0(\vec{x}')$, $T_i^0(\vec{x}')$, and $T^0(\vec{x}')$ do not presumably depend on the frequency. Using (5.4.145) in (2.6.4), the elastic Kirchhoff integral reads:

$$
\begin{aligned}
u_n(\vec{x}, \omega) = \int_{\Sigma^0} &\left[i\omega T_i^0(\vec{x}') G_{in}(\vec{x}', \vec{x}, \omega) - U_i^0(\vec{x}') h_{in}(\vec{x}', \vec{x}, \omega) \right] \\
&\times \exp[i\omega T^0(\vec{x}')] d\Sigma^0(\vec{x}').
\end{aligned}
\tag{5.4.146}
$$

Integral (5.4.146) is still exact, subject to assumption (5.4.145). Unit normal $\vec{n}$ to Σ^0 is oriented outside the medium in which the receiver point $\vec{x}$ is situated.

Now we shall use the asymptotic expressions for the elementary ray-theory elastodynamic Green function $G_{in}(\vec{x}', \vec{x}, \omega)$ and for the corresponding "traction" Green function $h_{in}(\vec{x}', \vec{x}, \omega)$. In a layered medium, there would be a large number (perhaps infinite) of elementary waves and relevant elementary Green functions; see Section 5.4.5. It would be necessary to sum all these elementary contributions to construct the complete ray-theory Green function. Here we shall consider only one elementary wave specified by a proper ray code and the relevant elementary ray-theory Green function. We denote the ray of the selected elementary wave, connecting points $\vec{x}$ and $\vec{x}'$, by $\tilde{\Omega}(\vec{x}', \vec{x})$. As an approximation, only smooth-medium Green functions G_{in} and h_{in} will be used here. For the smooth-medium Green function $G_{in}(\vec{x}', \vec{x}, \omega)$, not influenced by Σ^0, we can use relation (5.4.24):

$$
G_{in}(\vec{x}', \vec{x}, \omega) = G_n(\vec{x}', \vec{x}) \tilde{g}_i(\vec{x}') \exp[i\omega T(\vec{x}', \vec{x})].
\tag{5.4.147}
$$

Here $T(\vec{x}', \vec{x})$ is the travel time along $\tilde{\Omega}(\vec{x}', \vec{x})$ from $\vec{x}$ to $\vec{x}'$, $\tilde{g}_i(\vec{x}')$ is the ith Cartesian component of the eigenvector $\vec{\tilde{g}}(\vec{x}')$ at $\vec{x}'$, corresponding to ray $\tilde{\Omega}(\vec{x}', \vec{x})$, and $G_n(\vec{x}', \vec{x})$ is given by the relation

$$
G_n(\vec{x}', \vec{x}) = \frac{\exp[i\omega T^G(\vec{x}', \vec{x})]}{4\pi [\rho(\vec{x})\rho(\vec{x}')\mathcal{C}(\vec{x})\mathcal{C}(\vec{x}')]^{1/2} \mathcal{L}(\vec{x}', \vec{x})} \mathcal{R}^C(\vec{x}', \vec{x}) g_n(\vec{x}).
\tag{5.4.148}
$$

All symbols in (5.4.148) have the same meaning as in (5.4.24) and correspond to ray $\tilde{\Omega}(\vec{x}', \vec{x})$. $\mathcal{R}^C(\vec{x}', \vec{x})$ denotes the product of all normalized R/T coefficients along $\tilde{\Omega}$ between $\vec{x}$ and $\vec{x}'$. Because the elementary wave under consideration may be converted at some intermediate interface between $\vec{x}$ and $\vec{x}'$, eigenvectors $g_n(\vec{x})$ and $\tilde{g}_i(\vec{x}')$ may correspond to different wave types (qS1, qS2, qP).

For $h_{in}(\vec{x}', \vec{x}, \omega)$, we obtain

$$h_{in}(\vec{x}', \vec{x}, \omega) = i\omega \tilde{X}_i(\vec{x}') G_n(\vec{x}', \vec{x}) \exp[i\omega T(\vec{x}', \vec{x})]. \tag{5.4.149}$$

Function $\tilde{X}_i$ transforms the displacement into traction at $\vec{x}'$. It is given by the relation analogous to (5.4.107):

$$\tilde{X}_i(\vec{x}') = \tilde{c}_{ijqs}(\vec{x}') n_j(\vec{x}') \tilde{g}_q(\vec{x}') \tilde{p}_s(\vec{x}'). \tag{5.4.150}$$

We use the tilde to emphasize the quantities corresponding to ray $\tilde{\Omega}(\vec{x}', \vec{x})$. Inserting (5.4.147) and (5.4.149) into (5.4.146) yields

$$u_n(\vec{x}, \omega) = -i\omega \int_{\Sigma^0} a(\vec{x}') G_n(\vec{x}', \vec{x}) \exp[i\omega(T^0(\vec{x}') + T(\vec{x}', \vec{x}))] d\Sigma^0(\vec{x}'), \tag{5.4.151}$$

where $a(\vec{x}')$ is the so-called weighting function and is given by the relation

$$a(\vec{x}') = U_i^0(\vec{x}') \tilde{X}_i(\vec{x}') - T_i^0(\vec{x}') \tilde{g}_i(\vec{x}'). \tag{5.4.152}$$

Equation (5.4.151) with (5.4.152) represents the final form of the elastic Kirchhoff integral for arbitrary initial conditions $U_i^0(\vec{x}')$ and $T_i^0(\vec{x}')$ along Σ^0; see (5.4.145).

1. THE KIRCHHOFF INTEGRAL FOR A WAVE INCIDENT ON Σ^0

We shall now specify the Kirchhoff integral (5.4.151) for the distribution of $U_i^0(\vec{x}')$ and $T_i^0(\vec{x}')$ corresponding to an arbitrary elementary wave incident at Σ^0. We shall assume that the wave is incident at Σ^0 from the first medium, described by medium parameters $c_{ijkl}^{(1)}$ and $\rho^{(1)}$, and denote the medium parameters in the second medium by $c_{ijkl}^{(2)}$ and $\rho^{(2)}$. All medium parameters may vary with the coordinates.

The smooth-medium incident wave at $\vec{x}'$, not influenced by surface Σ^0 (which may represent a structural interface, a free surface, etc.), is specified as follows:

$$\begin{aligned} u_i(\vec{x}') &= U_i^{inc}(\vec{x}') \exp[i\omega T^0(\vec{x}')], \\ T_i(\vec{x}') &= i\omega T_i^{inc}(\vec{x}') \exp[i\omega T^0(\vec{x}')]. \end{aligned} \tag{5.4.153}$$

Quantities U_i^{inc} and T_i^{inc} are given by relations

$$U_i^{inc}(\vec{x}') = g_i^{inc}(\vec{x}') U^{inc}(\vec{x}'), \qquad T_i^{inc}(\vec{x}') = X_i^{inc}(\vec{x}') U^{inc}(\vec{x}'), \tag{5.4.154}$$

where function

$$X_i^{inc}(\vec{x}') = c_{ijqs}^{(1)}(\vec{x}') n_j(\vec{x}') g_q^{inc}(\vec{x}') p_s^{inc}(\vec{x}'). \tag{5.4.155}$$

An arbitrary elementary ray-theory wave is considered. Eigenvector $\vec{g}^{inc}(\vec{x}')$ corresponds to the wave incident at $\vec{x}'$. In the following, we shall specify the type of wave incident at $\vec{x}'$ by the index l ($l = 1$ for the incident qS1 wave, $l = 2$ for the incident qS2 wave, and $l = 3$ for the incident qP wave at $\vec{x}'$).

The complete wavefield at the point of incidence $\vec{x}'$ on Σ^0 is composed of the smooth-medium incident wave and of three reflected waves (qS1, qS2, qP). On the opposite side of Σ^0, the complete wavefield is formed by the superposition of three transmitted waves. See

Section 5.2.7 for isotropic media. The final equations for $U_i^0(\vec{x}')$ and $T_i^0(\vec{x}')$ along Σ^0 are

$$U_i^0 = \left(g_i^{inc} + \sum_{m=1}^{3} g_i^{(m)} R_{lm}^r \right) U^{inc}, \qquad T_i^0 = \left(X_i^{inc} + \sum_{m=1}^{3} X_i^{(m)} R_{lm}^r \right) U^{inc},$$

(5.4.156)

where R_{lm}^r are displacement reflection coefficients. Alternative equations in terms of displacement transmission coefficients are

$$U_i^0 = \sum_{m=1}^{3} g_i^{(m)} R_{lm}^t U^{inc}, \qquad T_i^0 = \sum_{m=1}^{3} X_i^{(m)} R_{lm}^t U^{inc}.$$

(5.4.157)

In the reflection R_{lm}^r and transmission R_{lm}^t coefficients, the first index l ($l = 1, 2, 3$) corresponds to the incident wave, and the second index m ($m = 1, 2, 3$) corresponds to the generated wave. Summation over $m = 1,\ 2,\ 3$ is understood; each R/T coefficient (R_{lm}^r, R_{lm}^t) is multiplied by $g_i^{(m)}$ or $X_i^{(m)}$. The summation over $m = 1,\ 2,\ 3$ yields three terms; any one of them corresponds to one generated wave at Σ^0. Quantity $X_i^{(m)}$ is given by relation analogous to (5.4.150) and (5.4.155),

$$X_i^{(m)}(\vec{x}') = c_{ijqs}(\vec{x}') n_j(\vec{x}') g_q^{(m)}(\vec{x}') p_s^{(m)}(\vec{x}').$$

(5.4.158)

The eigenvector components $g_i^{(m)}$ and functions $X_i^{(m)}$ correspond to the first medium if they are connected with the reflection coefficients R_{lm}^r in (5.4.156). Similarly, they correspond to the second medium if they are connected with transmission coefficients R_{lm}^t; see (5.4.157). Analogously, we choose $c_{ijks}^{(1)}$ or $c_{ijks}^{(2)}$ instead of c_{ijks} in (5.4.158).

We shall now compute the weighting function $a(\vec{x}')$ of the Kirchhoff integral (5.4.151) for the wave incident at Σ^0. We factorize $a(\vec{x}')$ using the relation

$$a(\vec{x}') = \mathcal{A}(\vec{x}') U^{inc}(\vec{x}').$$

(5.4.159)

We shall also call $\mathcal{A}(\vec{x}')$ the weighting function. The Kirchhoff integral (5.4.151) then reads

$$u_n(\vec{x}, \omega) = -i\omega \int_{\Sigma^0} \mathcal{A}(\vec{x}') U^{inc}(\vec{x}') G_n(\vec{x}', \vec{x})$$
$$\times \exp[i\omega(T^0(\vec{x}') + T(\vec{x}', \vec{x}))] d\Sigma^0(\vec{x}').$$

(5.4.160)

Inserting (5.4.156) into (5.4.152) yields the expression for $\mathcal{A}(\vec{x}')$ in terms of reflection coefficients R_{lm}^r:

$$\mathcal{A}(\vec{x}') = g_i^{inc} \tilde{X}_i - X_i^{inc} \tilde{g}_i + \sum_{m=1}^{3} \left(g_i^{(m)} \tilde{X}_i - X_i^{(m)} \tilde{g}_i \right) R_{lm}^r.$$

(5.4.161)

Similarly, using (5.4.157) in (5.4.152) yields the expression for $\mathcal{A}(\vec{x}')$ in terms of transmission coefficients R_{lm}^t:

$$\mathcal{A}(\vec{x}') = \sum_{m=1}^{3} \left(g_i^{(m)} \tilde{X}_i - X_i^{(m)} \tilde{g}_i \right) R_{lm}^t.$$

(5.4.162)

All quantities in (5.4.161) and (5.4.162) are taken at $\vec{x}'$. Expressions (5.4.161) and (5.4.162) for $\mathcal{A}(\vec{x}')$ are fully equivalent. Any of them can be used for $\vec{x}$ situated in the first medium or in the second medium. Thus, the Kirchhoff integral for reflected waves ($\vec{x}$ in the first medium) can also be expressed in terms of transmission coefficients R_{lm}^t; see (5.4.162). Similarly, the Kirchhoff integral for transmitted waves ($\vec{x}$ in the second medium) can be expressed in terms of reflection coefficients; see (5.4.161).

We shall now express (5.4.161) and (5.4.162) in a more specific form, inserting there the expressions for $\tilde{X}_i$, X_i^{inc}, and $X_i^{(m)}$; see (5.4.150), (5.4.155), and (5.4.158). For $\vec{x}$ situated in the first medium, we use (5.4.161):

$$\mathcal{A}(\vec{x}') = c_{ijqs}^{(1)} \tilde{g}_q \left[g_i^{inc} \left(n_j \tilde{p}_s - n_s p_j^{inc} \right) + \sum_{m=1}^{3} g_i^{(m)} \left(n_j \tilde{p}_s - n_s p_j^{(m)} \right) R_{lm}^r \right].$$

(5.4.163)

Similarly, for $\vec{x}$ situated in the second medium, we use (5.4.162):

$$\mathcal{A}(\vec{x}') = c_{ijqs}^{(2)} \tilde{g}_q \sum_{m=1}^{3} g_i^{(m)} \left(n_j \tilde{p}_s - n_s p_j^{(m)} \right) R_{lm}^t.$$

(5.4.164)

Kirchhoff integral (5.4.160), with the weighting function $\mathcal{A}(\vec{x}')$ given by any of the relations (5.4.161) through (5.4.164), is very general and is valid for any incident wave. It is only necessary to specify $U^{inc}(\vec{x}')$ properly. The general relations for $U^{inc}(\vec{x}')$ can be found in Section 5.4.4. As a special, but very important case, we shall consider an incident wave, generated by a single-force point source, situated at point $\vec{x}_0$ in the first medium. We consider a single force at $\vec{x}_0$ oriented along the kth Cartesian axis, with unit amplitude. $U^{inc}(\vec{x}')$ can then be expressed similarly as in (5.4.148),

$$U^{inc}(\vec{x}') = G_k(\vec{x}', \vec{x}_0) = \frac{\exp[i\omega T^G(\vec{x}', \vec{x}_0)]}{4\pi [\rho(\vec{x}_0)\rho(\vec{x}')\mathcal{C}(\vec{x}_0)\mathcal{C}(\vec{x}')]^{1/2} \mathcal{L}(\vec{x}', \vec{x}_0)}$$
$$\times \mathcal{R}^C(\vec{x}', \vec{x}_0) g_k(\vec{x}_0).$$

(5.4.165)

The elastic Kirchhoff integral (5.4.160) then yields the *elastic Kirchhoff Green function* $G_{nk}^K(\vec{x}, \vec{x}_0, \omega)$:

$$G_{nk}^K(\vec{x}, \vec{x}_0, \omega) = -i\omega \int_{\Sigma^0} \mathcal{A}(\vec{x}') G_n(\vec{x}', \vec{x}) G_k(\vec{x}', \vec{x}_0)$$
$$\times \exp[i\omega(T^0(\vec{x}') + T(\vec{x}', \vec{x}))] d\Sigma^0(\vec{x}').$$

(5.4.166)

Note that functions $G_n(\vec{x}', \vec{x})$ and $G_k(\vec{x}', \vec{x}_0)$ do not include the components of the relevant eigenvectors at $\vec{x}'$; these are shifted to weighting function $\mathcal{A}(\vec{x}')$. In (5.4.161) and (5.4.162), the eigenvector of the wave incident at Σ^0 at $\vec{x}'$ (corresponding to $G_k(\vec{x}, \vec{x}_0)$) is denoted by $\vec{g}^{inc}(\vec{x}')$ and the eigenvector corresponding to $G_n(\vec{x}', \vec{x})$ by $\tilde{g}(\vec{x}')$.

In this case, the travel time $T^0(\vec{x}')$ in (5.4.166) corresponds to $T^0(\vec{x}', \vec{x}_0)$. We can modify $G_n(\vec{x}', \vec{x})$ and $G_k(\vec{x}', \vec{x}_0)$ as follows:

$$\bar{G}_n(\vec{x}', \vec{x}, \omega) = G_n(\vec{x}', \vec{x}) \exp[i\omega T(\vec{x}', \vec{x})],$$
$$\bar{G}_k(\vec{x}', \vec{x}_0, \omega) = G_k(\vec{x}', \vec{x}_0) \exp[i\omega T(\vec{x}', \vec{x}_0)].$$

(5.4.167)

The elastic Kirchhoff Green function (5.4.166) then formally simplifies to

$$G_{nk}^K(\vec{x}, \vec{x}_0, \omega) = -i\omega \int_{\Sigma^0} \mathcal{A}(\vec{x}') \bar{G}_n(\vec{x}', \vec{x}, \omega) \bar{G}_k(\vec{x}, \vec{x}_0, \omega) d\Sigma^0(\vec{x}').$$

(5.4.168)

This expression is very similar to the expression (5.1.75) for the pressure Kirchhoff Green function. Functions $\bar{G}(\vec{x}', \vec{x}, \omega)$ and $\bar{G}_k(\vec{x}, \vec{x}_0, \omega)$, however, do not represent elastodynamic ray-theory Green functions. They are related to these functions as follows:

$$G_{in}(\vec{x}', \vec{x}, \omega) = \tilde{g}_i(\vec{x}') \bar{G}_n(\vec{x}', \vec{x}, \omega),$$
$$G_{ik}(\vec{x}', \vec{x}_0, \omega) = g_i^{inc}(\vec{x}') \bar{G}_k(\vec{x}', \vec{x}_0, \omega).$$

(5.4.169)

2. INITIAL RAY THEORY AMPLITUDES AT Σ^0

As in the acoustic case in Section 5.1.11, we can apply the method of stationary phase to the Kirchhoff integral (5.4.151) to derive expressions for the initial-value amplitude $A(S)$ of an elementary wave generated at point S on the initial surface Σ^0. Such amplitudes can then be continued along the whole ray using (5.4.15). Up to three elementary waves can be generated at S on each side of initial surface Σ^0. We assume that the initial slowness vectors of these waves, $\vec{p}^{(m)}(S)$, $m = 1, 2, \ldots, 6$, have been determined. We can then also determine the appropriate eigenvectors $\vec{g}^{(m)}(S)$, the group velocity vectors $\vec{\mathcal{U}}^{(m)}(S)$, and the traction components $\vec{X}^{(m)}(S)$ with components $X_i^{(m)}(S) = c_{ijqs}(S)n_j(S)g_q^{(m)}(S)p_s^{(m)}(S)$ (no summation over m). We can also determine the initial values of the 2×2 matrices $\mathbf{Q}^{(y)}(S)$ and $\mathbf{P}^{(y)}(S)$ for dynamic ray tracing; see (4.14.70).

Any stationary point S of Kirchhoff integral (5.4.151) is defined so that the partial derivatives of $T^0(\vec{x}') + T(\vec{x}', x)$ along Σ^0 vanish at S. For a fixed receiver point R, the stationary points corresponding to the individual generated waves (with different $T(\vec{x}', \vec{x})$) have a different position on Σ^0. Here, however, we shall be interested only in the initial-value problem and determine $A(S)$ for six waves generated at S. We can again use the local principle. The derivation is practically the same as in Section 5.1.11, and we shall not repeat it here. For the initial ray-theory amplitude $A(S)$ of the nth wave, we obtain the following relation:

$$A(S) = a^{(n)}(S)/2\rho(S)(\vec{\mathcal{U}}^{(n)}(S) \cdot \vec{n}(S)). \tag{5.4.170}$$

Here $\rho(S)$ is the density at S corresponding to the nth wave, and $a^{(n)}(S)$ represents $a(S)$ given by (5.4.152), determined at the stationary point and specified for the nth wave. To find $a^{(n)}(S)$, we must determine $\tilde{X}_i(S)$ and $\tilde{g}_i(S)$, corresponding to the reciprocal direction, from the receiver R to S. We choose the eigenvectors in both directions so that they satisfy reciprocity relation (5.4.139). Then, for the nth wave, $\vec{\tilde{g}}(S) = -\vec{g}^{(n)}(S)$ and $\vec{\tilde{X}}(S) = \vec{X}^{(n)}(S)$. This yields

$$a^{(n)}(S) = U_i^0(S)X_i^{(n)}(S) + T_i^0(S)g_i^{(n)}(S). \tag{5.4.171}$$

Equation (5.4.170) with (5.4.171) represents the final expression for the initial ray-theory amplitude of any wave generated on Σ^0. It is valid both for isotropic and anisotropic media, assuming that the eigenvectors satisfy reciprocity condition (5.4.139).

Relations (5.4.170) with (5.4.171) can also be used if $U_i^0(S)$ and $T_i^0(S)$ in (5.4.171) correspond to a wave incident at Σ^0. In this case, however, (5.4.171) simplifies considerably. Using (5.4.159), we obtain

$$a^{(n)}(S) = \mathcal{A}^{(n)}(S)U^{inc}(S). \tag{5.4.172}$$

Here $\mathcal{A}^{(n)}(S)$ is determined from (5.4.161) or (5.4.162) at the stationary point, using $\tilde{X}_i = X_i^{(n)}(S)$ and $\tilde{g}_i = -g_i^{(n)}(S)$. This yields

$$\mathcal{A}^{(n)}(S) = g_i^{inc}X_i^{(n)} + X_i^{inc}g_i^{(n)} + \sum_{m=1}^{3}\left(g_i^{(m)}X_i^{(n)} + X_i^{(m)}g_i^{(n)}\right)R_{lm}^r,$$

$$\mathcal{A}^{(n)}(S) = \sum_{m=1}^{3}\left(g_i^{(m)}X_i^{(n)} + X_i^{(m)}g_i^{(n)}\right)R_{lm}^t. \tag{5.4.173}$$

Both expressions are alternative. Using the orthogonality relations (5.4.137), we obtain

$$g_i^{(m)}X_i^{(n)} + X_i^{(m)}g_i^{(n)} = 0 \qquad \text{for} \quad m \neq n,$$

$$g_i^{inc}X_i^{(n)} + X_i^{inc}g_i^{(n)} = 0,$$

and

$$g_i^{(m)} X_i^{(n)} + X_i^{(m)} g_i^{(n)} = 2\rho(\vec{\mathcal{U}}^{(n)} \cdot \vec{n}) \qquad \text{for } m = n. \tag{5.4.174}$$

Here $\vec{\mathcal{U}}^{(n)}$ is the group velocity vector corresponding to the nth wave at S. Using either of the two relations (5.4.161) and (5.4.162), we obtain

$$\mathcal{A}^{(n)}(S) = 2\rho(\vec{\mathcal{U}}^{(n)} \cdot \vec{n}) R_{ln} \tag{5.4.175}$$

(no summation over n). This relation is valid both for reflected waves ($R_{ln} = R_{ln}^r$) and transmitted waves ($R_{ln} = R_{ln}^t$). Inserting this into (5.4.170) and (5.4.172), we obtain very simple expressions for the initial ray-theory amplitude $A(S)$ of the selected generated wave:

$$\begin{aligned} A(S) &= R_{ln}^r(S) U^{inc}(S) && \text{for reflected waves,} \\ A(S) &= R_{ln}^t(S) U^{inc}(S) && \text{for transmitted waves.} \end{aligned} \tag{5.4.176}$$

Here $l = 1, 2, 3$ is the index of the incident wave, and $n = 1, 2, 3$ the index of the R/T wave. (5.4.176) is analogous to (5.1.85) for acoustic waves.

For completeness, we shall present the final initial ray-theory expressions for the elastic wavefield, generated at an initial surface Σ^0 situated in a laterally varying anisotropic elastic medium. Consider point S on Σ^0, and construct a ray Ω^0 of a selected wave from S. Also determine $\mathbf{Q}^{(y)}(S)$ and $\mathbf{P}^{(y)}(S)$, using (4.14.70), and perform dynamic ray tracing. We can then determine geometrical spreading $\mathcal{L} = |\det \mathbf{Q}^{(y)}|^{1/2}$ along the whole ray Ω^0. Equations (5.4.1), (5.4.2), and (5.4.15) yield

$$\begin{aligned} \vec{u}(R, \omega) &= A(R)\vec{g}(R) \exp[i\omega(T^0(S) + T(R, S))] \\ A(R) &= (\rho(S)\mathcal{U}(S)/\rho(R)\mathcal{U}(R))^{1/2}(\mathcal{L}(S)/\mathcal{L}(R))\mathcal{R}^C \exp[iT^c(R, S)]A(S). \end{aligned} \tag{5.4.177}$$

Here $A(S)$ is given by (5.4.170) or (5.4.176), and $\vec{g}(R)$ is the eigenvector at R. All other symbols have the same meaning as in (5.4.15).

3. SEVERAL COMMENTS ON THE DERIVED EQUATIONS

Let us consider an elementary wave, generated by a point source at S, incident at Σ^0, and a receiver situated at R. As in the acoustic case, it is possible to show that the method of stationary phase, applied to Kirchhoff integral (5.4.160), yields exactly the same results as the zeroth-order ray method. This remains valid even for the general inhomogeneous anisotropic layered model. In the derivation, it is again convenient to use certain properties of the Fresnel zone matrix $\mathbf{M}^F(Q; R, S)$; see Section 4.14.11 for the Fresnel zone matrix in anisotropic medium. Analogous ray-theory solutions can also be obtained from (5.4.177) with (5.4.176), using suitable expressions for $U^{inc}(S)$.

Consequently, the intermediate ray-theory solutions can be stored along arbitrary reference surfaces Σ^0 and used in further computations. The procedure is the same as in the acoustic case; see Section 5.1.11. Actually, it is sufficient to store the travel time $T^0(\gamma_1, \gamma_2)$ and the smooth-medium ray-centered amplitude $U^{inc}(\gamma_1, \gamma_2)$ of the incident wave along Σ^0. We also need to know the type of wave incident at S (specified by index $l = 1, 2,$ or 3), and the information from which side the incident wave approaches Σ^0. It is then possible to recover all the six generated elementary waves. Of course, the efficiency of computations may increase if some other quantities related to the incident wave are also stored along Σ^0 (slowness vector, matrices $\mathbf{Q}^{(y)}$ and $\mathbf{P}^{(y)}$, and so on). For more details, see Section 5.5 of Červený, Klimeš, and Pšenčík (1988b).

5.5 Weakly Dissipative Media

In nondissipative media, the amplitudes of seismic body waves decrease with increasing distance from the source due to geometrical spreading, reflection and transmission losses, and the like. Real media, however, are dissipative, and the amplitudes of seismic body waves also attenuate due to various *anelastic processes*, such as grain defects, grain-boundary processes, and thermoelastic effects. The investigation of such processes has been the subject of a broad research in material science, and has been reviewed in many books and papers. A collection of important papers devoted to absorption was published by Toksöz and Johnston (1981), which also contain many other references to this subject. See also Aki and Richards (1980).

Macroscopically, many anelastic processes, which play an important role in the absorption of seismic body waves, obey the linear stress-strain relations and may be well described within the framework of *linear viscoelasticity*. For a detailed theoretical exposition of linear viscoelasticity, see Hudson (1980a). In the frequency domain, the stress-strain relations in linear viscoelasticity can be formally expressed in the same way as in Hooke's law, but the real-valued, frequency-independent elastic moduli must be replaced by *viscoelastic moduli*, which are complex-valued and frequency-dependent. The imaginary parts of the viscoelastic moduli are then responsible for the attenuation of the amplitudes of seismic body waves. The imaginary parts of the viscoelastic moduli must be taken negative in our treatment due to the Fourier transform sign convention used here. The negative imaginary parts of viscoelastic moduli yield the *exponential decay* of amplitudes with increasing distance from the source.

We shall denote any viscoelastic modulus $M(\omega)$ and specify its real and imaginary parts as follows:

$$M(\omega) = M^R(\omega) + iM^I(\omega). \tag{5.5.1}$$

The viscoelastic modulus $M(\omega)$ may correspond to the bulk modulus k, modulus of torsion μ, Young modulus E, or even any anisotropic modulus c_{ijkl}. In inhomogeneous medium, of course, $M(\omega)$ depends on coordinates.

The concept of viscoelastic media (such as the Maxwell, Voigt, or general Boltzman medium) has found broad applications in geophysics. Viscoelastic media have been used to study both high-frequency seismic phenomena (such as the propagation of seismic body waves) and low-frequency tectonic phenomena (such as creep, convection, and stress relaxation). If we are interested only in certain particular problems, the general concepts may be simplified. Here we shall make *four simplifying assumptions*, related to the purpose of this section.

a. We shall discuss only high-frequency phenomena.
b. We shall assume that the frequency dependence of $M(\omega)$ is weak in the frequency range under consideration.
c. We shall consider only weakly dissipative media.
d. We shall consider only homogeneous waves for which the real part and the imaginary part of the slowness vector (that is, the propagation and attenuation vectors) are parallel.

In seismology, the most common measures of attenuation are the dimensionless *quality factor* Q and its inverse Q^{-1}, called the *loss factor*. A strict physical definition of quality factor Q as an intrinsic parameter of the medium can be found in Fung (1965), Aki and

Richards (1980), and Toksöz and Johnston (1981), among others. In the investigation of the dissipation of high-frequency seismic waves propagating in weakly dissipative media, quality factor Q can be simply related to the viscoelastic modulus $M(\omega)$ as follows:

$$Q(\omega) = -M^R(\omega)/M^I(\omega). \tag{5.5.2}$$

Equation (5.5.1) then becomes

$$M(\omega) = M^R(\omega)(1 - i/Q(\omega)). \tag{5.5.3}$$

The minus sign is due to a Fourier transform sign convention used here. For perfectly elastic media ($M^I(\omega) = 0$ and $M^R(\omega)$ is independent of frequency ω), loss factor Q^{-1} vanishes, and quality factor $Q(\omega)$ is infinite. Note that also the *logarithmic decrement of absorption* $\delta = \pi/Q$ has often been used instead of Q and Q^{-1}.

The quality factors $Q(\omega)$, corresponding to different viscoelastic moduli $M(\omega)$, are in general different. Thus, we have Q_k, Q_μ, Q_E, and so on. The expressions for viscoelastic moduli easily yield the expressions for the phase velocities $V(\omega)$ of plane waves, which are also complex-valued and frequency-dependent:

$$V^2(\omega) = (V^R(\omega))^2(1 - i/Q(\omega)),$$
$$V(\omega) = V^R(\omega)(1 - i/Q(\omega))^{1/2}. \tag{5.5.4}$$

For weakly dissipative media ($Q \gg 1$), we can use approximate relations:

$$V(\omega) = V^R(\omega)(1 - i/2Q(\omega)),$$
$$1/V(\omega) = (1/V^R(\omega))(1 + i/2Q(\omega)). \tag{5.5.5}$$

Here $V^R(\omega)$ and $Q(\omega)$ differ for different waves. The expressions for $V^R(\omega)$ and $Q(\omega)$ for P and S waves can easily be obtained from $M^R(\omega)$ and $Q(\omega)$ corresponding to the viscoelastic bulk modulus and modulus of torsion (Q_k, Q_μ, and relevant M^R). Let us briefly discuss the physical meaning of (5.5.4). Consider a homogeneous time-harmonic plane wave propagating in a viscoelastic homogeneous medium along the x-axis:

$$\begin{aligned} u_i(x, t) &= U_i \exp[-i\omega(t - x/V(\omega))] \\ &\doteq U_i \exp[-i\omega(t - x/V^R(\omega))] \exp[-\omega x/(2V^R(\omega)Q(\omega))]. \end{aligned} \tag{5.5.6}$$

The last factor can also be expressed as $\exp[-\alpha(\omega)x]$, where $\alpha(\omega)$ is known as the *coefficient of absorption*. It is related to $Q(\omega)$ as follows: $Q^{-1}(\omega) = 2\alpha(\omega)V^R(\omega)/\omega$. As we can see from (5.5.6), the amplitudes of the plane waves propagating in viscoelastic medium are attenuated for finite $Q(\omega)$. Moreover, wave (5.5.6) is *dispersive*; in other words, its real-valued velocity of propagation V^R depends on frequency.

Quantities $Q(\omega)$ and $V^R(\omega)$ are not independent, they are mutually related by so-called *dispersion relations*. The dispersion relations follow from *causality requirements*. The dispersion relations are well known from the theory of propagation of electromagnetic waves, where they are called the *Kramers-Krönig dispersion relations*. Even for Q independent of frequency, the dispersion relations require V^R to be frequency-dependent. Otherwise, the causality principle would not be satisfied. A detailed treatment of dispersion relations can be found in Aki and Richards (1980, pp. 173–5), Kennett (1983, Section 1.3.3), and Müller (1983, pp. 173–5), among others. Thus, the absorption of amplitudes is always intrinsically connected with the disperson of velocities. For this reason, we also speak of *causal absorption*.

In principle, it is possible to apply the high-frequency asymptotic concepts, such as the ray method, even to the solution of the viscoelastic equation of motion, which contains complex-valued frequency-dependent viscoelastic moduli. See Buchen (1974), Zhu and Chun (1994a), Hearn and Krebes (1990a,1990b), Caviglia and Morro (1992), and Thomson (1997a). If it is assumed that the frequency dependence of viscoelastic moduli is weak, ray solutions in the form of asymptotic series in inverse powers of ω can be sought. The method requires the computation of frequency-dependent rays in a complex phase space (complex rays) and frequency-dependent, complex-valued travel times. This is, of course, more time-consuming than real-valued ray tracing. In particular, two-point ray tracing in a complex space may be cumbersome, even if the source and the receiver are situated in real space. See the brief discussion in Section 5.6.8.

Here we shall discuss a considerably simpler approach, which is applicable to homogeneous waves in weakly dissipative media only. We shall work in the frequency domain; for the implementation of the results into synthetic seismograms in the time domain, see Section 6.3. The method is based on the evaluation of complex-valued, frequency-dependent travel times and the subsequent computation of certain *dissipation filters*, which include both the effects of attenuation of amplitudes and dispersion of velocities. The determination of dissipation filters from known complex-valued frequency-dependent travel times is simple. The complex-valued, frequency-dependent amplitudes of seismic body waves propagating in viscoelastic media are obtained by multiplying the relevant amplitudes of seismic body waves propagating in perfectly elastic media by the dissipation filter. The dissipation filter in a weakly dissipative medium can be computed very simply with the use of some additional quadratures along known real-valued rays, computed in the relevant nondissipative model.

There are several ways of deriving the dissipation filters. It is possible to start directly from the viscoelastic equation of motion or to compute the complex-valued frequency-dependent travel times by using perturbation methods. We shall derive the dissipation filter using a very simple technique based on the *perturbation approach*.

We shall modify (5.5.5) slightly by introducing a reference frequency ω^r,

$$\frac{1}{V(\omega)} = \frac{1}{V^R(\omega^r)} + \left[\frac{1}{V^R(\omega)} - \frac{1}{V^R(\omega^r)} + \frac{i}{2 V^R(\omega) Q(\omega)} \right]. \tag{5.5.7}$$

We assume that the unperturbed background medium is fully specified by $1/V^R(\omega^r)$, that is, by the real-valued slowness for reference frequency ω^r. The expression in brackets in (5.5.7) is considered to be the slowness perturbation. We can then use the well-known expressions of Section 3.9 to evaluate the perturbations of travel time T_d due to dissipation. The travel-time perturbations will be complex-valued and frequency-dependent. They can be computed by quadratures along unperturbed rays (calculated in the background medium). The dissipative filter then takes the simple form

$$D = \exp[i \omega T_d]. \tag{5.5.8}$$

The dissipative filter yields the attenuation of amplitudes along rays as the wave progresses, and the corresponding dispersion of velocities. However, it is not able to include some other effects of dissipation on amplitudes. For example, it does not introduce the changes of geometrical spreading due to the dissipative properties of the medium. This is simple to understand because geometrical spreading is a property of the ray field, and the proposed method uses only background rays. (For more details on geometrical spreading of seismic body waves in viscoelastic media, see Krebes and Hearn 1985.) Similarly, it does not introduce the reflection and transmission coefficients for viscoelastic media. See Section 5.5.4.

5.5.1 Noncausal Dissipation Filters

Velocity dispersion in weakly dissipative media is usually very small, and it is not necessary to take it into account in many applications. In addition, even Q may often be considered to be frequency-independent. We can then use a rough, noncausal approximation of (5.5.5),

$$\frac{1}{V} \doteq \frac{1}{V^R}\left(1 + \frac{i}{2Q}\right). \tag{5.5.9}$$

In this case, we speak of *noncausal absorption*.

Let us consider the real-valued ray Ω connecting points S and R, computed in the background medium described by real-valued slowness $1/V^R(x_i)$. We assume that the imaginary part in (5.5.9), $i(2V^R Q)^{-1}$, is a small perturbation of slowness $1/V^R$. Using (3.9.9), we obtain

$$T_d(R, S) = \frac{i}{2}\int_S^R \frac{ds}{V^R Q} = \frac{i}{2}\int_S^R \frac{dT}{Q}. \tag{5.5.10}$$

The integral is taken along ray Ω. Equation (5.5.10) yields dissipation filter $D(R, S)$ in the following form:

$$D(R, S) = \exp\left[-\tfrac{1}{2}\omega t^*(R, S)\right]; \tag{5.5.11}$$

see (5.5.8). Here quantity $t^*(R, S)$ is given by the integral

$$t^*(R, S) = \int_S^R \frac{ds}{V^R Q} = \int_S^R \frac{dT}{Q}. \tag{5.5.12}$$

The integral is again taken along ray Ω. The quantity "t-star" is also sometimes called the *global absorption factor*. It fully controls the dissipative decay of amplitudes of seismic body waves in weakly dissipative inhomogeneous elastic medium. The dimension of t^* is time. It is obvious that the noncausal dissipation filter (5.5.11) does not yield the dispersion of velocities.

For the noncausal dissipation operator in the time domain, corresponding to the dissipation filter (5.5.11), see Section 6.3.

5.5.2 Causal Dissipation Filters

We shall now discuss the complex-valued frequency-dependent slowness $1/V(\omega)$, given by (5.5.7). We assume that the relation is causal, i.e. that $V^R(\omega)$ is determined from $Q(\omega)$ using some dispersion relations that guarantee the causality of the results.

We shall now take the background medium corresponding to the real-valued, frequency-independent slowness $1/V^R(\omega^r)$; see (5.5.7). The expression in brackets in (5.5.7) then represents the perturbation of slowness $1/V^R(\omega^r)$.

Let us consider the real-valued ray Ω computed in the background medium and two points S and R situated on it. Using (3.9.9), we can compute the travel-time perturbation

$T_d(R, S)$ caused by the slowness perturbations:

$$T_d(R, S) = \int_S^R \left(\frac{1}{V^R(\omega)} - \frac{1}{V^R(\omega^r)} \right) ds + \frac{i}{2} \int_S^R \frac{ds}{V^R(\omega)Q(\omega)}. \quad (5.5.13)$$

This perturbation yields dissipation filter $D(R, S)$ in the following form:

$$D(R, S) = \exp\left[i\omega \int_S^R \left(\frac{1}{V^R(\omega)} - \frac{1}{V^R(\omega^r)} \right) ds \right]$$
$$\times \exp\left[-\frac{\omega}{2} \int_S^R \frac{ds}{V^R(\omega)Q(\omega)} \right]. \quad (5.5.14)$$

Here the integrals are taken along ray Ω. The first factor in the dissipation filter is responsible for dispersion; the second, for amplitude attenuation.

It remains to specify $V^R(\omega)$ corresponding to a given $Q(\omega)$ using a dispersion relation. We shall use two important dispersion relations: the Futterman (1962) dispersion relation and the Müller (1983) dispersion relation. For many other dispersion relations, see Szabo (1995) and a review by Toverud and Ursin (1998).

a. FUTTERMAN DISPERSION RELATION

This relation (Futterman 1962) is the classical and a very popular dispersion relation. It has a very interesting property in that $V^R(\omega)Q(\omega)$ is independent of ω so that $V^R(\omega)Q(\omega) = V^R(\omega^r)Q(\omega^r)$. It reads

$$\frac{1}{V^R(\omega)} = \frac{1}{V^R(\omega^r)} \left[1 - \frac{1}{\pi Q(\omega^r)} \ln \frac{\omega}{\omega^r} \right], \quad (5.5.15)$$

$$Q(\omega) = Q(\omega^r) \left[1 - \frac{1}{\pi Q(\omega^r)} \ln \frac{\omega}{\omega^r} \right]. \quad (5.5.16)$$

Here ω^r is the reference frequency. Equations (5.5.15) and (5.5.16) yield

$$1/V(\omega) = (1/V^R(\omega^r))\left[1 - \pi^{-1}Q^{-1}(\omega^r) \ln(\omega/\omega^r) + \tfrac{1}{2}iQ^{-1}(\omega^r) \right], \quad (5.5.17)$$

$$T_d(R, S) = t^*(R, S)\left(-\pi^{-1} \ln(\omega/\omega^r) + \tfrac{1}{2}i \right), \quad (5.5.18)$$

$$D(R, S) = \exp\left[-i\pi^{-1}\omega t^* \ln(\omega/\omega^r) - \tfrac{1}{2}\omega t^* \right], \quad (5.5.19)$$

where $t^* = t^*(R, S)$ is given by the relation

$$t^*(R, S) = \int_S^R \frac{ds}{V^R(\omega^r)Q(\omega^r)} = \int_S^R \frac{dT}{Q(\omega^r)}. \quad (5.5.20)$$

A disadvantage of the classical Futterman relation is that it does not satisfy the causality requirements exactly; some deviations may be observed at very low and very high frequencies.

b. MÜLLER DISPERSION RELATION

Müller (1983) studied the very important case of $Q(\omega)$ obeying a frequency power law,

$$Q(\omega) = Q(\omega^r)(\omega/\omega^r)^\gamma. \quad (5.5.21)$$

Here ω^r is a reference frequency, and γ is a constant, $0 \leq \gamma \leq 1$. Equation (5.5.21) includes, among others, the very important cases of $Q(\omega)$ independent of frequency ($\gamma = 0$) and

$Q(\omega)$ proportional to frequency ($\gamma = 1$). The frequency-dependent slowness for $Q \gg 1$ corresponding to (5.5.21) reads

$$\frac{1}{V^R(\omega)} = \frac{1}{V^R(\omega^r)}\left[1 - \frac{1}{2Q(\omega^r)}\left(1-\left(\frac{\omega^r}{\omega}\right)^{\gamma}\right)\cot\frac{\gamma\pi}{2}\right]. \tag{5.5.22}$$

The equation for the dissipation filter then reads

$$D(R, S) = \exp\left\{-\frac{\omega t^*}{2}\left[\left(\frac{\omega^r}{\omega}\right)^{\gamma} + i\cot\left(\gamma\frac{\pi}{2}\right)\left(1-\left(\frac{\omega^r}{\omega}\right)^{\gamma}\right)\right]\right\}. \tag{5.5.23}$$

Here $t^* = t^*(R, S)$ is again given by (5.5.20). See also Schmidt and Müller (1986).

Note that the power law (5.5.21) for $Q(\omega)$ has been recently reported both in laboratory experiments and in seismological applications.

The case of constant Q, independent of frequency, may be obtained from (5.5.23) by applying the limit $\gamma \to 0$. It is interesting to note that (5.5.22) and (5.5.23) yield for $\gamma \to 0$ the same expressions for $1/V^R(\omega)$ and $D(R, S)$ as the Futterman dispersion relations.

Exact dispersion relations for $\gamma = 0$ were derived by Kjartansson (1979). The Kjartansson exact dispersion relations have found important applications in seismology and seismic exploration.

5.5.3 Anisotropic Media

The procedures for finding the dissipation filters for homogeneous waves propagating in the anisotropic inhomogeneous weakly dissipative media are similar to those for isotropic media.

Let us first consider the noncausal, density-normalized, viscoelastic moduli, independent of frequency,

$$a_{ijkl} = a_{ijkl}^R + i\,a_{ijkl}^I. \tag{5.5.24}$$

We specify the background, unperturbed model by real-valued, frequency-independent, density-normalized moduli $a_{ijkl}^R(x_i)$ and compute the rays Ω in it. Then, $i\,a_{ijkl}^I$ represents the model perturbation. We use (3.9.15) and obtain the imaginary travel-time perturbation

$$T_d(R, S) = -\frac{i}{2}\int_S^R a_{ijkl}^I p_i\, p_l\, g_j^{(m)} g_k^{(m)}\,dT. \tag{5.5.25}$$

The integral is taken along ray Ω computed in the background medium. Quantities $\vec{p}$, $\vec{g}^{(m)}$ and dT also refer to the background medium.

The noncausal dissipation filter $D(R, S)$ can then be expressed as

$$D(R, S) = \exp[i\omega T_d(R, S)] = \exp[-\tfrac{1}{2}\omega t^*(R, S)], \tag{5.5.26}$$

where $t^*(R, S)$ represents the "t-star" quantity for the anisotropic medium,

$$t^*(R, S) = -\int_S^R a_{ijkl}^I p_i\, p_l\, g_j^{(m)} g_k^{(m)}\,dT. \tag{5.5.27}$$

As we can see, the noncausal dissipation filter $D(R, S)$ (5.5.26) is exactly the same for isotropic and anisotropic media (see (5.5.11)), only quantity $t^*(R, S)$ has a different meaning in both cases. Using (5.5.27), we can formally introduce the quality factor Q for the anisotropic weakly dissipative medium,

$$Q^{-1} = -a_{ijkl}^I\, p_i\, p_l\, g_j^{(m)} g_k^{(m)}. \tag{5.5.28}$$

Thus, the quality factor and dissipation filters in anisotropic media depend not only on position but also on the direction of propagation. This relation means that the waves propagating through an anisotropic dissipative medium in different directions are attenuated differently. This effect is known as *directional attenuation*. See Gajewski and Pšenčík (1992).

For causal dissipation, the quantities in (5.5.24) are frequency-dependent:

$$a_{ijkl}(\omega) = a_{ijkl}^R(\omega) + i\, a_{ijkl}^I(\omega). \tag{5.5.29}$$

As in isotropic medium, (5.5.29) can be modified to read

$$a_{ijkl}(\omega) = a_{ijkl}^R(\omega^r) + \left[a_{ijkl}^R(\omega) - a_{ijkl}^R(\omega^r) + i\, a_{ijkl}^I(\omega) \right]; \tag{5.5.30}$$

see (5.5.7). We specify the background medium by $a_{ijkl}^R(\omega^r)$ and compute the rays Ω and the relevant travel time $T(R, S)$ in it. The expression in brackets in (5.5.30) represents the model perturbation. The final expressions for the dissipation filters are obtained similarly as for isotropic media.

5.5.4 Waves Across Interfaces in Dissipative Media

In the process of reflection/transmission of inhomogeneous plane waves at a plane structural interface between two viscoelastic homogeneous halfspaces, three vectors play a basic role: the propagation vector $\vec{p}^R$ of the incident wave, the attenuation vector $\vec{p}^I$ of the incident wave, and the vector $\vec{n}$ normal to the interface. These three vectors are not, in general, coplanar. A consequence is that the system of six boundary equations cannot be decomposed into two subsystems (as in isotropic nondissipative media) but must be solved as a whole (as in anisotropic media). Solving the system of six linear equations three times, the complete set of nine reflection coefficients and nine transmission coefficients is obtained. The exception is only the situation in which $\vec{p}^R$ and $\vec{p}^I$ of the incident wave and $\vec{n}$ are coplanar. For example, this is the case of a homogeneous incident wave. Then the system can be decomposed.

It follows from boundary conditions that the tangential components of $\vec{p}^R$ and $\vec{p}^I$ for all generated plane waves must equal the tangential components of $\vec{p}^R$ and $\vec{p}^I$ of the incident wave. The consequent relations yield the Snell's law for dissipative media, which involves both the propagation and attenuation vector components of the incident wave and of the relevant R/T wave.

The computation of R/T coefficients at an interface between two dissipative media has been discussed broadly in the seismological literature. See, for example, Cooper and Reiss (1966), Cooper (1967), Silva (1976), Borcherdt (1977, 1982), Krebes and Hron (1980a, 1980b), Bourbiè and Gonzalez-Serrano (1983), Krebes (1983, 1984), Bourbiè (1984), Borcherdt, Glassmoyer, and Wennerberg (1986), Caviglia and Morro (1992), Samec and Blangy (1992), Carcione (1993), Nechtschein and Hron (1996, 1997), and many other references therein. For anisotropic dissipative media, see Carcione (1997) and Carcione, Helle, and Zhao (1998).

We shall briefly discuss the effects of weak dissipation on R/T coefficients for incident plane homogeneous waves. It is obvious that weak dissipation will affect the R/T coefficients only slightly in regions where the R/T coefficients are smooth. On the other hand, the effect may be strong in regions where the variations of R/T coefficients are abrupt, mainly in the vicinity of critical and Brewster angles. In general, weak absorption has a smoothing effect on the R/T coefficients; it removes the sharp edges and anomalies of the coefficients.

5.6 Ray Series Method. Acoustic Case

The complete equations for the amplitudes of high-frequency pressure waves propagating in general laterally varying fluid media were derived in Section 5.1. The expressions for amplitudes, however, are only approximate. We remind the reader that pressure wave $p(x_i, t)$ was described by a simple formula, $p(x_i, t) = P(x_i) \exp[-i\omega(t - T(x_i))]$ (see (2.4.4)), travel time $T(x_i)$ and amplitude $P(x_i)$ being frequency-independent. This trial solution cannot satisfy wave equation (2.4.3) exactly. Actually, inserting the foregoing trial solution into wave equation (2.4.3) yields (2.4.5), which consists of three terms. The first term has multiplier ω^2, the second has ω^1, and the third has ω^0. As we were seeking a high-frequency solution, we were interested particularly in the first two terms. The requirement that the two first terms must vanish yielded the eikonal and transport equation. The third term, $\nabla^2 P$, however, is in general nonvanishing and causes some errors in our solution, particularly for lower frequencies ω. Without changing the form of the foregoing trial solution, we cannot satisfy (2.4.3), including the third term, completely.

There are several ways to overcome this problem and to increase the accuracy of the solution. The classical and widely used method is to consider a solution in the form of ray series. The method was briefly outlined in Section 2.4.1; here we shall discuss it in considerably greater detail.

5.6.1 Scalar Ray Series. Amplitude Coefficients

We shall consider the wave equation for pressure $p(x_i, t)$ in a medium with smoothly varying velocity $c(x_i)$ and density $\rho(x_i)$,

$$\nabla \cdot \rho^{-1} \nabla p = (\rho c^2)^{-1} \ddot{p}; \qquad (5.6.1)$$

see (2.4.9). We shall seek the time-harmonic solution of this equation in the form of a *scalar ray series*:

$$p(x_i, t) = \exp[-i\omega(t - T(x_i))] \sum_{n=0}^{\infty} \frac{P^{(n)}(x_i)}{(-i\omega)^n}. \qquad (5.6.2)$$

Thus, the ray-series solution in the time-harmonic domain is represented by a series in inverse powers of frequency ω. We assume that travel time $T(x_i)$ and the *amplitude coefficients of the ray series* $P^{(n)}(x_i), n = 0, 1, 2, \ldots$, depend only on coordinates x_i, not on frequency.

A few words on the terminology. If we consider only the leading term of ray series (5.6.2), we usually speak of the *zeroth-order ray approximation*. The higher-order terms are then called the *higher-order ray approximations*. For example, the *first-order ray approximation* is specified by the relation $(-i\omega)^{-1} P^{(1)}(x_i) \exp[-i\omega(t - T(x_i))]$. This terminology, however, has not been accepted generally.

Ray series (5.6.2) is not a standard convergent infinite series, but presumably has the character of an *asymptotic series for $\omega \to \infty$*. Extensive literature is devoted to asymptotic series and various asymptotic approximations. For a very detailed treatment, see, for example, Bleistein (1984). Most of the books on mathematical physics and on wave propagation present at least a brief explanation of this subject. Here we shall not go into mathematical details; it will be sufficient to introduce the asymptotic series as follows: Equation (5.6.2) represents an asymptotic series for $\omega \to \infty$ if the following inequality is valid for arbitrary N,

$$\left| p(x_i, t) - \exp[-i\omega(t - T(x_i))] \sum_{n=0}^{N} \frac{P^{(n)}(x_i)}{(-i\omega)^n} \right| \leq \frac{a}{\omega^{N+1}}, \qquad (5.6.3)$$

where a is a constant independent of ω. It is obvious that (5.6.3) does not imply the convergence of ray series (5.6.2). It follows from the asymptotic character of ray series (5.6.2) that the accuracy of a finite number of terms of the asymptotic series (N fixed) may be arbitrarily increased by increasing ω. The situation is, however, quite different for ω fixed. For ω fixed, the required accuracy cannot, in general, be obtained by increasing the number of terms N. For $N \to \infty$, the asymptotic ray series is usually divergent. Thus, the ray series is seismologically meaningful only if a *finite number of terms is considered*. The infinite sign above the summation symbol in (5.6.2) has been used only for convenience; it means that N in (5.6.3) may be arbitrarily large.

The asymptotic character of ray series (5.6.2) has actually been proved for many particular cases. The general proof of the asymptotic character of (5.6.2), however, has not been given. In such cases, series (5.6.2) has a formal meaning only.

From the computational point of view, we are mainly interested in the behavior of asymptotic ray series (5.6.2) for fixed ω. Usually, the moduli of the individual terms of the ray series, $|P^{(n)}(x_i)/(-i\omega^n)|$, first decrease with increasing n, and for some $n = n_m$, they reach a minimum value. They then increase with increasing n. The best accuracy is usually obtained if we take the sum from $n = 0$ to $n = n_m$. Taking more terms than n_m does not improve the accuracy but *makes it worse*. Thus, for a fixed ω, asymptotic ray series (5.6.2) *always yields some error* that cannot be removed by increasing the number of terms in the series.

The ray theory based on trial solution (5.6.2) in the form of the asymptotic series for $\omega \to \infty$ is also often called the *asymptotic ray theory*, abbreviated ART. We then speak of the ART solution (5.6.2), of ART methods, and so on.

5.6.2 Recurrence System of Equations of the Ray Method

Inserting ray series (5.6.2) into the wave equation (5.6.1) yields

$$
e^{-i\omega(t-T)} \left\{ \sum_{k=-2}^{\infty} \frac{1}{(-i\omega)^k} N\left(P^{(k+2)}\right) \right.
$$

$$
\left. - \sum_{k=-1}^{\infty} \frac{1}{(-i\omega)^k} M\left(P^{(k+1)}\right) + \sum_{k=0}^{\infty} \frac{1}{(-i\omega)^k} L\left(P^{(k)}\right) \right\} = 0, \qquad (5.6.4)
$$

where symbols N, M, and L have the following meaning:

$$
\begin{aligned}
N\left(P^{(k)}\right) &= \rho^{-1} P^{(k)}[T_{,i}\, T_{,i} - 1/c^2], \\
M\left(P^{(k)}\right) &= \rho^{-1} T_{,i}\, P_{,i}^{(k)} + T_{,i}\left(P^{(k)}/\rho\right)_{,i} + \rho^{-1} T_{,ii}\, P^{(k)} \\
&= \rho^{-1/2}\left\{ T_{,ii}\left(P^{(k)}/\sqrt{\rho}\right) + 2T_{,i}\left(P^{(k)}/\sqrt{\rho}\right)_{,i} \right\}, \\
L\left(P^{(k)}\right) &= \left(P_{,i}^{(k)}/\rho\right)_{,i}.
\end{aligned}
\qquad (5.6.5)
$$

Equation (5.6.4) represents a power series in terms of $(1/i\omega)$. It may vanish only if the coefficients of all $(1/i\omega)^k$, $k = -2, -1, 0, 1, 2, \ldots$, vanish. Equation (5.6.4) then yields the infinite system of equations

$$
\begin{aligned}
N\left(P^{(0)}\right) &= 0, \\
N\left(P^{(1)}\right) - M\left(P^{(0)}\right) &= 0, \\
N\left(P^{(k)}\right) - M\left(P^{(k-1)}\right) + L\left(P^{(k-2)}\right) &= 0 \qquad \text{for } k \geq 2.
\end{aligned}
\qquad (5.6.6)
$$

This system of equations can be expressed in a more compact form if we formally introduce $P^{(-1)}$ and $P^{(-2)}$:

$$P^{(-1)} = P^{(-2)} = 0. \tag{5.6.7}$$

Hence,

$$N\big(P^{(k)}\big) - M\big(P^{(k-1)}\big) + L\big(P^{(k-2)}\big) = 0 \qquad \text{for } k = 0, \, 1, \, 2, \ldots. \tag{5.6.8}$$

This is the *basic recurrence system of equations* of the ray method for the acoustic case. The system can be used to determine successively $T(x_i)$, $P^{(0)}(x_i)$, $P^{(1)}(x_i)$, ..., if certain initial conditions are available.

For $k = 0$, (5.6.8) yields $N(P^{(0)}) = 0$. Assuming a nontrivial amplitude $P^{(0)}$, we immediately obtain the eikonal equation $T_{,i}\, T_{,i} - 1/c^2 = 0$; see (5.6.5). This remains valid even for $k > 0$, so that $N(P^{(k)}) = 0$ for any k, and (5.6.8) yields

$$M\big(P^{(k)}\big) - L\big(P^{(k-1)}\big) = 0 \qquad \text{for } k = 0, 1, \ldots. \tag{5.6.9}$$

Here again $P^{(-1)} = 0$.

5.6.3 Transport Equations of Higher Order and Their Solutions

Inserting (5.6.5) into (5.6.9), we obtain the equation

$$2T_{,i}\left(P^{(k)}\big/\sqrt{\rho}\right)_{,i} + \left(P^{(k)}\big/\sqrt{\rho}\right)T_{,ii} = \sqrt{\rho}\left(P_{,i}^{(k-1)}\big/\rho\right)_{,i}. \tag{5.6.10}$$

In vectorial form, (5.6.10) reads

$$2\nabla T \cdot \nabla\left(P^{(k)}\big/\sqrt{\rho}\right) + \left(P^{(k)}\big/\sqrt{\rho}\right)\nabla^2 T = \sqrt{\rho}\,\nabla \cdot \left(\rho^{-1}\nabla P^{(k-1)}\right). \tag{5.6.11}$$

As we can see from (2.4.11), the left-hand side of (5.6.11) for $P^{(k)}$ has exactly the same form as the transport equation (2.4.11) for $P^{(0)}$. Equation (5.6.11) for $P^{(k)}$, however, has a nonvanishing right-hand side, depending on $P^{(k-1)}$. For this reason, (5.6.11) is usually called the *transport equation of higher order*, or *higher order transport equation*. Of course, for $k = 0$ the higher order transport equation (5.6.10) reduces to the standard transport equation (2.4.11) for $P^{(0)}$, as $P^{(-1)} = 0$.

The transport equation (5.6.11) of higher order for any $k \geq 0$ can be simply transformed into an ordinary differential equation of the first order for $P^{(k)}$ if we solve it along the ray. We take into account that

$$2\nabla T \cdot \nabla \frac{P^{(k)}}{\sqrt{\rho}} = \frac{2}{c}\frac{d}{ds}\left(\frac{P^{(k)}}{\sqrt{\rho}}\right), \qquad \nabla^2 T = \frac{1}{J}\frac{d}{ds}\frac{J}{c}.$$

Here J denotes the ray Jacobian. The derivative d/ds is taken along the ray, ds being an elementary arclength along the ray. Using these two relations in (5.6.11), we obtain

$$\frac{d}{ds}\left(\frac{P^{(k)}}{\sqrt{\rho}}\right) + \frac{P^{(k)}}{\sqrt{\rho}}\frac{d}{ds}\ln\sqrt{\frac{J}{c}} = \frac{c\sqrt{\rho}}{2}\,\nabla \cdot \left(\frac{1}{\rho}\nabla P^{(k-1)}\right). \tag{5.6.12}$$

This is the final form of the ordinary differential equation of the first order for $P^{(k)}/\sqrt{\rho}$, representing the transport equation of the higher order. The equation can be integrated by well-known methods, see, for example, Kamke (1959). The solution can be written in various forms. Here we shall present two forms of the solution that may be useful in different applications.

First, we shall present a *continuation formula*. We assume that $P^{(k)}$ is known at one point of the ray, say at s_0. Then we can compute $P^{(k)}(s)$ along the whole ray,

$$P^{(k)}(s) = \sqrt{c(s)\rho(s)/J(s)} \left\{ P^{(k)}(s_0)\sqrt{J(s_0)/\rho(s_0)c(s_0)} \right.$$
$$\left. + \frac{1}{2} \int_{s_0}^{s} \sqrt{c(s')\rho(s')J(s')} \, \nabla \cdot \left(\rho^{-1}(s')\nabla P^{(k-1)}(s')\right) ds' \right\}. \quad (5.6.13)$$

This equation cannot be used for a *point source situated at the point s_0*. In this case, $J(s_0) \to 0$. Equation (5.6.13) can, however, be modified as in Section 5.1.2. If we express $J(s)$ and $J(s')$ in terms of $\det \mathbf{Q}_2(s, s_0)$ and $\det \mathbf{Q}_2(s', s_0)$, we obtain

$$P^{(k)}(s) = \sqrt{\frac{c(s)\rho(s)}{c(s_0)\rho(s_0)}} \frac{1}{\mathcal{L}(s, s_0)} \left\{ \mathcal{G}^{(k)}(s_0) \exp[\mathrm{i}T^c(s, s_0)] \right.$$
$$+ \frac{1}{2}\sqrt{c(s_0)\rho(s_0)} \int_{s_0}^{s} \sqrt{c(s')\rho(s')} \, \mathcal{L}(s', s_0) \exp[\mathrm{i}T^c(s', s_0)]$$
$$\left. \times \nabla \cdot \left(\rho^{-1}(s')\nabla P^{(k-1)}(s')\right) ds' \right\}. \quad (5.6.14)$$

Here $T^c(s, s_0)$ denotes the phase shift due to caustics on the ray between s_0 and s, and $\mathcal{G}^{(k)}(s_0)$ is the *radiation function of the kth order*, given by the relation

$$\mathcal{G}^{(k)}(s_0) = \lim_{s' \to s_0} \left\{ \mathcal{L}(s', s_0)P^{(k)}(s') \right\}. \quad (5.6.15)$$

The limit $s' \to s_0$ is taken along the ray.

Thus, to compute the kth order amplitude coefficient $P^{(k)}$ of the ray series of a pressure wave generated by a point source, it is necessary to know the radiation functions $\mathcal{G}^{(i)}$, $i = 0, 1, \ldots, k$. Such higher order radiation functions are, however, known only for some sources situated in a homogeneous medium. For a point source situated in an inhomogeneous medium, the problem of determination of $\mathcal{G}^{(i)}$, $i = 1$, is considerably more involved. With a few exceptions, the analytical expressions for $\mathcal{G}^{(i)}$ are not known. This fact considerably decreases the importance of higher order ray approximations in inhomogeneous media, both in numerical modeling of acoustic wavefields and in practical applications.

5.6.4 Reflection and Transmission

We shall now consider the problem of reflection and transmission of acoustic waves at a curved interface Σ between two inhomogeneous media. We denote the point of incidence by Q. (We do not need to introduce the points $\tilde{Q}$ of reflection/transmission in this section; we use Q for all generated waves.) We assume that the curvature of the interface Σ at Q is small; see (2.4.60). We shall consider the incident pressure wave in the form of a ray series:

$$p^{inc} = \exp[-\mathrm{i}\omega(t - T^{inc})] \sum_{n=0}^{\infty} P^{(n)inc}/(-\mathrm{i}\omega)^n, \quad (5.6.16)$$

where T^{inc} and $P^{(n)inc}$, $n = 0, 1, 2, \ldots$, are pressumably known. For reflected and transmitted waves, the trial solutions will again be written in the form of ray series:

$$p^r = \exp[-i\omega(t - T^r)] \sum_{n=0}^{\infty} P^{(n)r}/(-i\omega)^n, \tag{5.6.17}$$

$$p^t = \exp[-i\omega(t - T^t)] \sum_{n=0}^{\infty} P^{(n)t}/(-i\omega)^n. \tag{5.6.18}$$

Otherwise, we shall use the notation of Sections 2.3.1 and 2.4.5.

Inserting (5.6.16) through (5.6.18) into interface conditions, we can determine T^r, T^t, $P^{(n)r}$, and $P^{(n)t}$ ($n = 0, 1, 2, \ldots$) along Σ. The discussion of travel times T^r and T^t remains the same as in Section 2.3.1. We shall be interested here in the determination of amplitude coefficients of R/T waves, $P^{(n)r}$ and $P^{(n)t}$, at Q. Collecting the terms with the same power of frequency in the interface conditions at the point Q gives

$$P^{(n)r} - P^{(n)t} = -P^{(n)inc},$$

$$\rho_1^{-1} T,_i^r \, n_i \, P^{(n)r} - \rho_2^{-1} T,_i^t \, n_i \, P^{(n)t} = -\rho_1^{-1} T,_i^{inc} \, n_i \, P^{(n)inc} - \Delta^{(n-1)},$$

where $\Delta^{(n)}$ is given by the relation

$$\Delta^{(n)} = \rho_1^{-1} \left[P,_i^{(n)r} \, n_i + P,_i^{(n)inc} \, n_i \right] - \rho_2^{-1} \, P,_i^{(n)t} \, n_i. \tag{5.6.19}$$

Using the relation $T,_i^r \, n_i = -T,_i^{inc} \, n_i$, we obtain the final system of two equations for unknown $P^{(n)r}$ and $P^{(n)t}$, at the point Q,

$$P^{(n)r} - P^{(n)t} = -P^{(n)inc}$$

$$\rho_1^{-1} T,_i^{inc} \, n_i \, P^{(n)r} + \rho_2^{-1} T,_i^t \, n_i \, P^{(n)t} = \rho_1^{-1} T,_i^{inc} \, n_i \, P^{(n)inc} + \Delta^{(n-1)}.$$

$$\tag{5.6.20}$$

The solution of (5.6.20) for $P^{(n)r}$ and $P^{(n)t}$ at Q is

$$P^{(n)r} = R^r P^{(n)inc} + \frac{c_1 c_2 \rho_1 \rho_2 \Delta^{(n-1)}}{\rho_2 c_2 \cos i_1 + \rho_1 c_1 \cos i_2},$$

$$P^{(n)t} = R^t P^{(n)inc} + \frac{c_1 c_2 \rho_1 \rho_2 \Delta^{(n-1)}}{\rho_2 c_2 \cos i_1 + \rho_1 c_1 \cos i_2}. \tag{5.6.21}$$

Here R^r and R^t are standard pressure reflection/transmission coefficients given by (2.3.26); see also (5.1.21).

Let us emphasize again that Equations (5.6.21) yield the values of the higher order amplitude coefficients of reflected and transmitted waves, $P^{(n)r}$ and $P^{(n)t}$, only at the point Q situated directly on the interface Σ. They do not say anything about $P^{(n)r}$ and $P^{(n)t}$ in the vicinity of the interface Σ. To compute $P^{(n)r}$ and $P^{(n)t}$ in the vicinity of Σ, we must use the continuation formulae (5.6.13), in which the values of $P^{(n)r}$ and $P^{(n)t}$, calculated from (5.6.21), should be used as initial conditions.

We shall now briefly discuss some consequences of (5.6.21).

For $n = 0$, we obtain $\Delta^{(n-1)} = \Delta^{(-1)} = 0$ because $P^{(-1)r} = P^{(-1)t} = P^{(-1)inc} = 0$. Then (5.6.21) yields

$$P^{(0)r} = R^r P^{(0)inc}, \qquad P^{(0)t} = R^t P^{(0)inc}. \tag{5.6.22}$$

Here R^r and R^t represent the pressure reflection/transmission coefficients of plane waves at a plane interface between two homogeneous halfspaces. Thus, in the zeroth-order ray approximation, $P^{(0)r}/P^{(0)inc}$ and $P^{(0)t}/P^{(0)inc}$ at Q do not depend on the curvature of the interface Σ, the curvature of the wavefront of the incident wave, and the gradients of c and ρ at Q. They depend only on the local values of velocities and densities at Q on both sides of Σ and on the angle of incidence. See a detailed discussion in Section 2.4.5.

The situation, however, is changed drastically for $n \geq 1$ because $\Delta^{(n-1)} \neq 0$ in this case. The quantity $\Delta^{(n-1)}$ contains derivatives of $P^{(n-1)r}$ and $P^{(n-1)t}$ in the direction perpendicular to Σ. These derivatives depend on all the aforementioned factors, including curvature of the interface and of the wavefront of the incident wave and derivatives of velocity. Thus, the local (plane wave) approximation cannot be used to determine the higher amplitude coefficients of the ray series of reflected and transmitted waves, $P^{(n)r}$ and $P^{(n)t}$, for $n \geq 1$.

Equations (5.6.21) are very general; they remain valid even for *interfaces of higher order*, also called *weak interfaces*. For an interface of $(N + 1)$th order, we understand such a surface Σ across which the Nth derivatives of the velocity c and density ρ (or at least one of them) are discontinuous, whereas all lower derivatives $(N - 1, \ N - 2, \ldots)$ of these parameters are continuous. Thus, at the *interface of the first order*, the velocity c and/or the density ρ themselves are discontinuous. Similarly, the interface Σ at which the gradient of velocity c and/or density ρ are discontinuous, but the velocity and the density themselves are continuous, is called the *interface of the second order*. For an interface of higher order (weak interface), we understand any interface of the Nth order with $N \geq 2$.

At a weak interface, $c_1 = c_2$, $\rho_1 = \rho_2$, and $i_1 = i_2$ so that $R^r = 0$ and $R^t = 1$. Equations (5.6.21) then simplify considerably:

$$P^{(n)r} = \frac{c_1 \rho_1 \Delta^{(n-1)}}{2 \cos i_1}, \qquad P^{(n)t} = P^{(n)inc} + \frac{c_1 \rho_1 \Delta^{(n-1)}}{2 \cos i_1}. \tag{5.6.23}$$

This immediately yields $P^{(0)r} = 0$ and $P^{(0)t} = P^{(0)inc}$, as was expected.

Let us now briefly discuss the process of *reflection/transmission at an interface of Nth order*. In this case, $\Delta^{(n-1)} = 0$ for $n < N - 1$, but $\Delta^{(n-1)} \neq 0$ for $n = N - 1$. This yields,

$$P^{(n)r} = 0, \qquad P^{(n)t} = P^{(n)inc}, \qquad \text{for } n < N - 1.$$

Thus, the leading term of the reflected wave from an interface of Nth order corresponds to the $(N - 1)$th amplitude coefficient of the ray series, $P^{(N-1)r}$. Consequently, the ray series (5.6.17) for the reflected wave from the interface of the Nth order reads

$$p^r = \exp[-i\omega(t - T^r)] \sum_{n=N-1}^{\infty} P^{(n)r}/(-i\omega)^n. \tag{5.6.24}$$

In the terminology of the ray method, the reflected wave (5.6.24) described by the ray series with $P^{(0)r}$, $P^{(1)r}, \ldots, \ P^{(N-2)r}$ vanishing is called the *wave of $(N-1)$th order*. In general, the waves described by the ray series with the zeroth-order amplitude coefficient vanishing are called the *higher order waves*.

5.6.5 Alternative Forms of the Scalar Ray Series

In Section (5.6.1), we considered the ray series in frequency domain for time-harmonic waves. Seismic signals, however, are not time-harmonic but rather transient. It may be useful to consider alternatively other forms of the ray series, directly applicable for transient signals. We shall present here two such forms.

1. RAY SERIES FOR HIGH-FREQUENCY TRANSIENT SIGNALS

Let us consider a real-valued signal $x(t)$ with a Fourier spectrum $x(\omega)$. We remind the reader of our convention that the signal $x(t)$ and its Fourier spectrum $x(\omega)$ are denoted by the same letter. The signal $x(t)$ and its spectrum $x(\omega)$ are distinguished only by arguments t and ω; see Section 2.1.5. Under the *high-frequency signal* $x(t)$, we shall understand such signal, the Fourier spectrum $x(\omega)$ of which has the following property:

$$|x(\omega)| = 0 \qquad \text{for } 0 \le \omega \le \omega_0, \tag{5.6.25}$$

where ω_0 is high.

Because the ray series (5.6.2) is asymptotic, it has only a finite number of terms and can be integrated term by term. We multiply it by $x(\omega)$ and apply the integration $\pi^{-1} \int_0^\infty x(\omega) \ldots d\omega$ to all terms. Then we obtain the ray series (5.6.2) in the following form:

$$p(x_i, t) = \sum_{n=0}^{\infty} P^{(n)}(x_i)\, F^{(n)}(t - T(x_i)). \tag{5.6.26}$$

This represents the *ray series for high-frequency signals*. The complex-valued functions $F^{(n)}(\zeta)$ in (5.6.26) $n = 0, 1, 2, \ldots$, are defined by the relation

$$F^{(n)}(\zeta) = \pi^{-1} \int_0^\infty (-i\omega)^{-n} x(\omega)\, \exp[-i\omega\zeta] d\omega. \tag{5.6.27}$$

The functions $F^{(n)}(\zeta)$ satisfy the following *three properties*:

a. $F^{(n)}(\zeta)$ are *high-frequency signals*; that is, their Fourier spectra $(-i\omega)^{-n} x(\omega)$ satisfy (5.6.25).

b. $F^{(n)}(\zeta)$ are *analytical signals*; that is,

$$F^{(n)}(\zeta) = x^{(n)}(\zeta) + i\, g^{(n)}(\zeta), \tag{5.6.28}$$

where $x^{(n)}(\zeta)$ and $g^{(n)}(\zeta)$ are real-valued functions and form a Hilbert transform pair. See (A.2.4). They can be expressed in terms of $x(\omega)$ using (5.6.27):

$$x^{(n)}(\zeta) = \frac{1}{\pi} \text{Re} \int_0^\infty (-i\omega)^{-n} x(\omega)\, \exp[-i\omega\zeta] d\omega,$$

$$g^{(n)}(\zeta) = \frac{1}{\pi} \text{Im} \int_0^\infty (-i\omega)^{-n} x(\omega)\, \exp[-i\omega\zeta] d\omega. \tag{5.6.29}$$

c. $F^{(n)}(\zeta)$ satisfy the following relations:

$$dF^{(n)}(\zeta)/d\zeta = F^{(n-1)}(\zeta), \qquad n = 1, 2, \ldots, \tag{5.6.30}$$

or, alternatively,

$$F^{(n)}(\zeta) = \int_{-\infty}^{\zeta} F^{(n-1)}(\zeta')d\zeta'. \tag{5.6.31}$$

Relations (5.6.30) and (5.6.31) allow us to compute successively all $F^{(n)}(\zeta)$ ($n = 1, 2, \ldots$), as soon as $F^{(0)}(\zeta)$ is known.

2. RAY SERIES FOR DISCONTINUITIES OF THE WAVEFIELD

The ray series (5.6.26) for high-frequency transient signals can be generalized if we consider the distributions and Fourier transforms of distributions. Let us rewrite the definition (5.6.27) of the analytical signal $F^{(n)}(\zeta)$ in the following form:

$$F^{(n)}(\zeta) = \frac{1}{2\pi} \int_{-\infty}^{\infty} 2(-i\omega)^{-n} H(\omega) x(\omega) \exp[-i\omega\zeta] d\omega, \qquad (5.6.32)$$

where $H(\omega)$ is the Heaviside step function defined by the relation

$$H(\omega) = 0 \quad \text{for} \quad \omega < 0, \qquad H(\omega) = 1 \quad \text{for} \quad \omega > 0.$$
$$(5.6.33)$$

Then (5.6.32) yields an alternative relation for the analytical signal $F^{(n)}(\zeta)$:

$$F^{(n)}(\zeta) = F^{(0)}(\zeta) * h^{(n)}(\zeta). \qquad (5.6.34)$$

Here $h^{(n)}(\zeta)$ are defined by the following relations:

$$h^{(0)}(\zeta) = \delta(\zeta), \qquad h^{(1)}(\zeta) = H(\zeta), \dots,$$
$$h^{(n)}(\zeta) = \zeta^{n-1} H(\zeta)/(n-1)!. \qquad (5.6.35)$$

In (5.6.35), $\delta(\zeta)$ and $H(\zeta)$ have a standard meaning: $\delta(\zeta)$ is the Dirac delta function and $H(\zeta)$ the Heaviside step function.

Using (5.6.34), the ray series (5.6.26) can be written in a new form:

$$p(x_i, t) = F^{(0)}(t) * \sum_{n=0}^{\infty} P^{(n)}(x_i) h^{(n)}(t - T(x_j)). \qquad (5.6.36)$$

Alternatively, we can also write

$$p(x_i, t) = F^{(0)}(t - T(x_i)) * \sum_{n=0}^{\infty} P^{(n)}(x_i) h^{(n)}(t). \qquad (5.6.37)$$

The functions $h^{(n)}(t)$ for four n ($n = 0, 1, 2,$ and 3) are shown in Figure 5.12. As we can see, the most important is the leading (zeroth) term of the ray series (5.6.37) because $h^{(0)}(t)$ represents the highest order of discontinuity on the wavefront. The subsequent terms of the ray series (5.6.37) change more smoothly across the wavefront; therefore, they are not as distinct in the wavefield. In the convolution with $F^{(0)}(t - T(x_i))$, the functions $h^{(n)}(t)$ cause the nth order integration. Because integrations imply smoothing, the higher order terms are generally smoother than the lower order terms.

Any of the form of the ray series shown above, (5.6.2), (5.6.26), (5.6.36) and (5.6.37), includes exactly the same functions $T(x_j)$ and $P^{(n)}(x_j)$, $n = 0, 1, 2, \dots$. Thus, they can be used alternatively.

Whereas the ray series expansion in the frequency domain (5.6.2) has, in general, an asymptotic character for $\omega \to \infty$, the character of the ray series (5.6.36) in the time domain is different. It may be *convergent* in a vicinity of the wavefront (for small $t - T(x_j)$). For this reason, the ray series (5.6.36) is also often called the *near-wavefront expansion*. For great $t - T(x_j)$, the ray series (5.6.36) usually diverges and cannot be used. See Babich (1961b).

One important remark. The pressure $p(x_i, t)$ as determined from the previously shown ray series is a complex-valued function. Although we have written the ray series in complex-valued forms, only the real and imaginary parts of them (or, perhaps, a linear combination

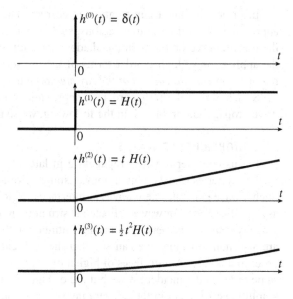

Figure 5.12. Functions $h^{(n)}(t)$, $n = 0$, 1, 2, 3, for the ray series in discontinuities. Here $h^{(0)}(t)$ represents the delta function, $h^{(1)}(t)$ represents the Heaviside function, and so on. The higher order functions $h^{(n)}(t)$ are smoother than the lower-order functions at $t = 0$.

of both) have a physical meaning. It may be useful to stress this fact by writing Equations (5.6.26), (5.6.36), and (5.6.37) as follows:

$$p(x_i, t) = \mathrm{Re} \sum_{n=0}^{\infty} P^{(n)}(x_j) \, F^{(n)}(t - T(x_j)), \qquad (5.6.38)$$

$$p(x_i, t) = \mathrm{Re} \left\{ F^{(0)}(t) * \sum_{n=0}^{\infty} P^{(n)}(x_j) \, h^{(n)}(t - T(x_j)) \right\}, \qquad (5.6.39)$$

$$p(x_i, t) = \mathrm{Re} \left\{ F^{(0)}(t - T(x_j)) * \sum_{n=0}^{\infty} P^{(n)}(x_j) \, h^{(n)}(t) \right\}. \qquad (5.6.40)$$

Here Re may also be replaced by Im or by some linear combination of Re and Im.

In the following, we shall again consider the ray series in their complex-valued forms and return to their real-valued forms (5.6.38) through (5.6.40) only in the final stage of computation, if necessary.

5.6.6 Applications of Higher Order Ray Approximations

In the numerical modeling of wavefields and in the solution of relevant inverse problems in inhomogeneous media, only the zeroth-order ray approximation has been traditionally used; the application of higher order ray approximations has been exceptional. The argument against the application of higher order ray approximations follows. In regular regions, where the validity conditions of the ray method are well satisfied, the zeroth-order ray approximation is usually sufficiently accurate, and the higher order ray approximations are small. Thus, it is not necessary to use them because they do not contribute considerably to the zeroth-order ray approximation. On the contrary, in singular regions, the zeroth-order ray approximation is highly inaccurate, but the higher order ray approximations do not increase the accuracy of the ray method; in most cases *they decrease it*. The higher order approximations are even more sensitive to the singularities than the zeroth-order ray approximation is.

In principle, this argumentation is correct. The inaccuracy of the zeroth-order ray approximation in the singular regions, such as the caustic region, the critical region, and the transition region between the shadow and illuminated zones, cannot be decreased by the application of higher order terms of the ray series. To increase the accuracy in such regions, some *modifications of the ray method* must be used; see Section 5.6.8.

Still, however, the higher order ray approximations may play an important role in certain wave propagation problems. In the following, we shall list several such applications.

1. HIGHER ORDER WAVES

Let us consider a wave propagating in laterally varying layered structure, described by the ray series (5.6.2), with a nonvanishing zeroth-order amplitude coefficient $P^{(0)}(x_i)$. Such a wave generates new waves at interfaces, which may again be described by relevant asymptotic series. For waves reflected at structural interfaces of the first order (at which the velocity c and/or the density ρ are discontinuous), the zeroth-order amplitude coefficients are also nonvanishing. We can say that the incident and reflected wave are of the same order in this case. At interfaces of higher order, however, the reflected waves are not of the same order as the incident wave but are of higher order. This means that the zeroth-order amplitude coefficient in the relevant ray series of the reflected wave vanishes and that the leading term of the ray series corresponds to some higher order amplitude coefficient, $n \geq 1$. For more details on waves reflected from higher order interfaces, refer to Section 5.6.4, particularly ray series (5.6.24).

Another typical example of a higher order wave is the *head wave*. Let us consider a wave generated by a point source, incident at a plane interface between two homogeneous halfspaces (with $c_2 > c_1$). Then, in addition to the standard reflected wave, the head wave is also generated at postcritical distances. The zeroth-order amplitude coefficient in the ray series of the head wave vanishes; only the first-order amplitude coefficient is nonvanishing. Thus, the head wave is the *first-order wave* in the terminology of the ray theory. For more details on head waves, see Section 5.6.7.

2. AMPLITUDES ALONG RAYS AT WHICH $P^{(0)}(x_i)$ VANISHES

We shall now consider the regular wave described by ray series (5.6.2). We assume that $P^{(0)}(x_i)$ is, in general, nonvanishing but that it vanishes along certain rays and is small in the close vicinity of these rays. This may happen, for example, if the *radiation function is zero* for a particular direction (nodal lines) or if the *reflection/transmission coefficient vanishes* for a particular angle of incidence (Brewster angle). The first-order ray approximation will then play an important role and will fill these gaps. The wavefield may then be described by the two terms of the ray series,

$$p(x_i, t) = \exp[-i\omega(t - T(x_i))][P^{(0)}(x_i) + (-i\omega)^{-1} P^{(1)}(x_i)], \qquad (5.6.41)$$

even in the close vicinity of such rays.

3. ESTIMATION OF THE ACCURACY OF THE ZEROTH-ORDER RAY APPROXIMATION

For a finite frequency ω, ray series (5.6.2) can be used in practical computations only if the moduli of the higher order approximations decrease with increasing n, at least for a few small n. We shall now devote our attention only to two terms of the ray series; see (5.6.41). Then, if we use the zeroth-order ray approximation in our computations, the first-order ray approximation roughly represents the error of computations. We denote the *relative error* caused by neglecting the second term of (5.6.41) ϵ. Hence,

$$\left| P^{(1)}(x_i) \big/ (-i\omega) P^{(0)}(x_i) \right| < \epsilon. \tag{5.6.42}$$

Calculating the expression $|P^{(1)}/P^{(0)}|$ may thus be very useful in accuracy considerations. The criterion of validity (5.6.42) is, of course, very rough because only two terms of the ray series are considered. Nevertheless, it may yield valuable estimates of accuracy of the zeroth-order ray computations. Brekhovskikh (1960) was probably the first one who used the outlined method to estimate the accuracy of the zeroth-order ray approximation in the problem of reflection of spherical waves at a plane interface. See also Popov and Camerlynck (1996).

5.6.7 Head Waves

The most important and famous example of seismic body waves of the higher order are head waves. The head waves have been used broadly in seismic exploration and in deep seismic sounding of the Earth's crust. They are also known as refracted waves or refraction arrivals, and the exploration method based on them as the refraction method. Many theoretical papers have been devoted to head waves generated by a spherical wave incident at a plane interface between two homogeneous halfspaces. Mostly wave methods, based on an exact integral representation of the spherical wave, have been used, and the head waves were obtained by an asymptotic high-frequency treatment of these integrals. Jeffreys (1926) was probably the first to derive approximate expressions for head waves in this way. The references are too numerous to be given here. For many classical references, see Brekhovskikh (1960), Červený and Ravindra (1971), and Drijkoningen, Chapman, and Thomson (1987). In the 1950s, the ray method was also successfully applied to derive and study acoustic head waves (Friedlander 1958). The application of the ray method to the investigation of head waves is explained in detail in Červený and Ravindra (1971), where many other references can also be found.

Let us consider a plane interface Σ of the first order between two homogeneous half-spaces, with propagation velocities c_1 and c_2. Assume a point source, generating a spherical acoustic wave, situated at point S in the first halfspace (with propagation velocity c_1), at distance h_S from interface Σ. We also assume that $c_1 < c_2$ and introduce the critical angle of incidence i_1^*, $\sin i_1^* = c_1/c_2$.

We shall first explain the generation of head waves using simple wavefront charts, see Figure 5.13. Assume that source S starts to generate a wave at time $t = 0$. For $t < h_S/c_1$, that is, before the wave impinges on the interface, there exists only this one wave. For $t = h_S/c_1$, the wavefront reaches interface Σ at point O (projection of S on Σ) and is tangent to Σ. As time increases further, reflected and transmitted waves are generated. We denote O^* the position of the *interface critical point*, corresponding to the point of incidence of the critical ray on Σ. The relevant interface critical time is given by the relation $T^* = h_S/(c_1 \cos i_1^*)$; see Figure 5.13. The wavefronts of the incident, reflected, and transmitted waves are mutually connected at point A on Σ if $h_1/c_1 < t < T^*$. The common point A is situated between O and O^* and moves along interface Σ with *apparent velocity* $c_A = c_1/\sin i_1(A)$, where $i_1(A)$ is the angle of incidence i_1 at point A.

Apparent velocity c_A decreases with increasing epicentral distance of point A (that is, with increasing $\overline{OA}$). At $A = O$, c_A equals ∞, and for $\overline{OA} \to \infty$, c_A approaches c_1. An important role in the generation of head waves is played by the interface critical point O^*. If $t < T^*$, point A is situated between O and O^*, and $c_A > c_2$. When $t = T^*$, $c_A = c_{O^*} =$

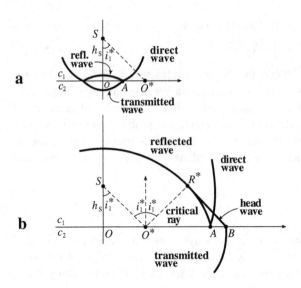

Figure 5.13. Explanation of the generation of a pressure head wave at a plane interface between two homogeneous fluid halfspaces. The incident wave is generated by a point source S situated in the lower-velocity halfspace. Head waves are generated only at postcritical distances. For more details, see text.

$c_1/\sin i_1^* = c_2$. Finally, if $t > T^*$, $c_A < c_2$. It is simple to see that if $t > T^*$, the transmitted wave propagating from point O^* along the interface in the second medium with velocity $c_2 > c_A$ will overtake the incident and reflected waves. At point A, only the wavefronts of the incident and reflected waves are connected, but the wavefront of the transmitted wave will reach a point B, advanced with respect to A. The wavefront of the transmitted wave is perpendicular to Σ at B, so that the ray of the transmitted wave is parallel to the interface between O^* and B.

The interface conditions must be satisfied between A and B. However, in the second halfspace, only the transmitted wave propagates along interface Σ because the reflected and incident waves are delayed. The transmitted wave itself cannot satisfy the interface conditions between A and B; an additional wave must also exist in the first halfspace. The generation of this wave can be explained simply by taking into account the Huyghens principle: the transmitted wave at interface Σ generates disturbances propagating into the first medium. This additional wave generated by the transmitted wave is called the *head wave*; see Figure 5.13(b). As velocities c_1 and c_2 are constant along interface Σ, the wavefront of the head wave is a straight line in Figure 5.13 (conical in 3-D). It passes through point B on interface Σ and is tangent to the wavefront of the reflected wave at critical point R^*, situated on the critical ray of the reflected wave. Because the wavefronts of the head wave are parallel straight lines, all the rays of head waves are parallel to the critical ray of the reflected wave.

Let us now briefly discuss the travel-time curves of the individual waves in our model. They may be determined using simple geometrical considerations. We assume that the source S and the receiver R are situated at distances h_S and h_R from the interface and denote the epicentral distance of R by r. The travel times of the direct and reflected waves, T^d and T^r, are

$$T^d(r) = [r^2 + (h_R - h_S)^2]^{1/2}/c_1, \qquad T^r(r) = [r^2 + (h_R + h_S)^2]^{1/2}/c_1.$$

$$(5.6.43)$$

Thus, both these travel-time curves are hyperbolical, but travel-time curve $T^d(r)$ is linear for $h_S = h_R$. The head wave exists only if $c_2 > c_1$, and if the epicentral distance of the

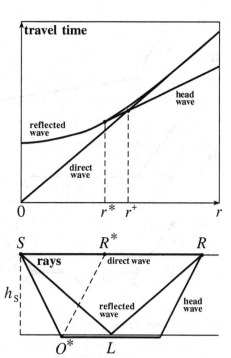

Figure 5.14. Travel times, rays, and amplitudes of pressure head waves generated by an incident spherical wave at a plane interface between two homogeneous halfspaces (schematically). Top figure: Travel times of direct, reflected, and head waves. The critical distance r^* and the crossover distance r^+ are shown. Head waves exist only at epicentral distances $r \geq r^*$. At $r = r^*$, the travel-time curve of the head wave is tangent to the travel-time curve of the reflected wave. Middle figure: Rays of direct, reflected, and head waves. The critical ray is shown by a dashed line. Bottom figure: Ray amplitudes of direct, reflected, and head waves. The dotted line shows schematically the amplitude of the interference reflected head wave in the critical region. This region is singular in the ray method.

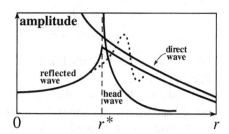

receiver is larger than critical distance r^*,

$$r^* = (h_S + h_R) \tan i_1^* = (h_R + h_S)n/\sqrt{1 - n^2}, \qquad (5.6.44)$$

where $n = c_1/c_2$. The travel-time curve of the head wave, $T^h(r)$, is then given by the equation

$$T^h(r) = T^r(r^*) + \frac{r - r^*}{c_2} = \frac{h_S + h_R}{c_1\sqrt{1 - n^2}} + \frac{r - r^*}{c_2}$$

$$= \frac{h_S + h_R}{c_1}\sqrt{1 - n^2} + \frac{r}{c_2}. \qquad (5.6.45)$$

Directly at the critical point $r = r^*$, the travel time is given by the relation

$$T^{h*} = T^h(r^*) = T^r(r^*) = (h_S + h_R)/c_1\sqrt{1 - n^2}. \qquad (5.6.46)$$

This travel time is also called the *critical travel time*. Figure 5.14 shows schematically the rays and travel times of direct, reflected, and head waves for $h_S = h_R$.

For completeness, we shall also present the relation for the *crossover distance* r^+, at which the travel-time curves of direct and head waves intersect, and for the relevant

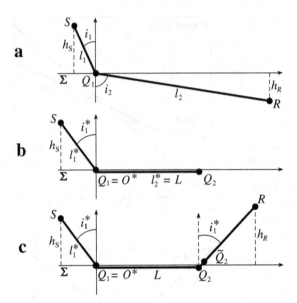

Figure 5.15. Definition of various symbols for the derivation of ray-theory amplitudes of head waves. For details, see text.

crossover travel time T^+:

$$r^+ = \frac{1}{\sqrt{1-n^2}}[n(h_S + h_R) + 2\sqrt{h_S h_R}],$$

$$T^+ = \frac{1}{c_1\sqrt{1-n^2}}[(h_S + h_R) + 2n\sqrt{h_S h_R}].$$

(5.6.47)

For $h_S = h_R$, (5.6.47) simplifies to $r^+ = 2(1+n)h_S/\sqrt{1-n^2}$, and $T^+ = 2h_S(1+n)/(c_1\sqrt{1-n^2})$.

We shall now show that the head wave is a first-order wave in the terminology of the ray method and derive expressions for its amplitude and waveform. To derive the expressions for the amplitudes of head waves, we first have to express the zeroth-order amplitude coefficient of transmitted wave $P^{(0)t}$ at the receiver point R situated in the second medium; see Figure 5.15(a). We use (5.1.17) and insert $\mathcal{G}(S; \gamma_1, \gamma_2) = c_1$ and $T^c(S, R) = 0$ because we are considering an omnidirectional unit directivity pattern of the source $\mathcal{F} = 1$ and no caustics exist in our situation. If we also use (5.1.22) for the normalized transmission coefficient $\mathcal{R}^t$ and (4.10.22) for $\mathcal{L}(R, S)$, we obtain

$$P^{(0)t}(R) = \frac{2c_1\rho_2c_2 P_1 P_2 \sqrt{p}}{(\rho_2 c_2 P_1 + \rho_1 c_1 P_2)[r(c_1 l_1 P_2^2 + c_2 l_2 P_1^2)]^{1/2}}.$$

(5.6.48)

Here l_1 and l_2 represent the lengths of the ray elements in the first and second medium, r is the epicentral distance between S and R, p is the ray parameter, $p = (\sin i_1)/c_1 = (\sin i_2)/c_2$, $P_1 = \cos i_1$, and $P_2 = \cos i_2$.

We shall now decrease the depth of receiver R and move it toward the interface; see Figure 5.15(b). We denote the receiver position directly at Σ by Q_2 (not R). We assume that point Q_2 is situated close to B in Figure 5.13(b), where only the transmitted and head waves exist. We also denote the point situated on the opposite side of Σ from Q_2 by $\tilde{Q}_2$. We assume that $r(Q_2) > h_S \tan i_1^*$, where i_1^* is the critical angle of incidence. Then Q_1 represents the interface critical angle O^*, $r(O^*) = h_S n/\sqrt{1-n^2}$, $l_1 = h_S/\sqrt{1-n^2}$, $l_2 = L = r(Q_2) - r(O^*)$. Thus, L is the distance of the point Q_2 from the interface critical point, that is, the length of the element of the ray of the transmitted wave parallel to Σ. Similarly, we obtain $\cos i_2 = P_2 = 0$; consequently, $P^{(0)t}(Q_2) = 0$ (see (5.6.48)). The zeroth-order amplitude coefficient of the transmitted wave then vanishes along the ray parallel to interface Σ beyond the interface critical point O^*. The higher order ray approximation, however, does not vanish.

There are several approaches to deriving the expressions for the amplitudes of head waves at $\tilde{Q}_2$. We shall use the simplest and most straightforward approach and derive them directly from the interface conditions. We shall not need to calculate the first-order ray approximation of the transmitted wave at Q_2 in this approach, but we shall obtain it as a by-product.

We denote the pressure head wave propagating in the first medium p^h and the pressure transmitted wave p^t. We shall consider only two terms of the ray series for both waves:

$$p^h = \exp[-i\omega(t - T^h)]\big(P^{(0)h} + P^{(1)h}/(-i\omega)\big),$$
$$p^t = \exp[-i\omega(t - T^t)]\big(P^{(0)t} + P^{(1)t}/(-i\omega)\big).$$

$$(5.6.49)$$

see (5.6.41). At points Q_2 and $\tilde{Q}_2$ of the interface Σ, only these two waves exist. The boundary conditions are $p^h(\tilde{Q}_2) = p^t(Q_2)$ and $(\rho_1^{-1}\partial p^h/\partial z)_{\tilde{Q}_2} = (\rho_2^{-1}\partial p^t/\partial z)_{Q_2}$. They yield, at Σ,

$$P^{(0)h} + (-i\omega)^{-1} P^{(1)h} = P^{(0)t} + (-i\omega)^{-1} P^{(1)t}, \qquad (5.6.50)$$

$$\frac{1}{\rho_1} i\omega \frac{\partial T^h}{\partial z} P^{(0)h} - \frac{1}{\rho_1} \frac{\partial T^h}{\partial z} P^{(1)h} + \frac{1}{\rho_1} \frac{\partial P^{(0)h}}{\partial z} - \frac{1}{i\omega\rho_1} \frac{\partial P^{(1)h}}{\partial z}$$
$$= \frac{1}{\rho_2} i\omega \frac{\partial T^t}{\partial z} P^{(0)t} - \frac{1}{\rho_2} \frac{\partial T^t}{\partial z} P^{(1)t} + \frac{1}{\rho_2} \frac{\partial P^{(0)t}}{\partial z} - \frac{1}{i\omega\rho_2} \frac{\partial P^{(1)t}}{\partial z}.$$

$$(5.6.51)$$

Boundary condition (5.6.50) yields $P^{(0)h}(\tilde{Q}_2) = P^{(0)t}(Q_2)$ and $P^{(1)h}(\tilde{Q}_2) = P^{(1)t}(Q_2)$. Because $P^{(0)t}(Q_2) = 0$ along Σ (see (5.6.48)), $P^{(0)h}(\tilde{Q}_2) = 0$. Using the continuation relations, we arrive at an important result $P^{(0)h} = 0$ in the whole first halfspace. Thus, the head wave is a wave of the higher order. This also yields $\partial P^{(0)h}/\partial z = 0$ along Σ. Finally, because the wavefront of the transmitted wave is perpendicular to Σ at Q_2, we have $\partial T^t/\partial z = 0$ at Q_2. Thus,

$$P^{(0)h}(\tilde{Q}_2) = \left(\frac{\partial P^{(0)h}}{\partial z}\right)_{\tilde{Q}_2} = 0, \qquad P^{(1)h}(\tilde{Q}_2) = P^{(1)t}(Q_2),$$
$$\left(\frac{\partial T^t}{\partial z}\right)_{Q_2} = 0.$$

$$(5.6.52)$$

Inserting (5.6.52) into (5.6.51) and neglecting the two terms with factors $1/i\omega$ yields

$$\rho_1^{-1}(\partial T^h/\partial z)_{\tilde{Q}_2} P^{(1)h}(\tilde{Q}_2) = -\rho_2^{-1}\big(\partial P^{(0)t}/\partial z\big)_{Q_2}.$$

Because $(\partial T^h/\partial z)_{\tilde{Q}_2} = P_1^*/c_1 = \sqrt{1 - n^2}/c_1$,

$$P^{(1)h}(\tilde{Q}_2) = -(\rho_1 c_1/\rho_2 P_1^*)\big(\partial P^{(0)t}/\partial z\big)_{Q_2}. \qquad (5.6.53)$$

It remains to calculate $\partial P^{(0)t}/\partial z$ at Q_2 from (5.6.48). The derivative $\partial P^{(0)t}/\partial z$ at Q can be computed using the simple formula

$$\big(\partial P^{(0)t}/\partial z\big)_{Q_2} = -L^{-1}\big(\partial P^{(0)t}/\partial P_2\big)_{P_2=0}.$$

Using this relation, we obtain

$$
\left(\frac{\partial P^{(0)t}}{\partial z}\right)_{Q_2} = -\frac{1}{L}\left[\frac{2c_1\rho_2 c_2 P_1\sqrt{p}}{\rho_2 c_2 P_1 + \rho_1 c_1 P_2}\frac{1}{\left[r\left(c_1 l_1 P_2^2 + c_2 l_2 P_1^2\right)\right]^{1/2}}\right]_{P_2=0}
$$

$$
= -\frac{2c_1}{c_2 P_1^* r^{1/2} L^{3/2}}. \tag{5.6.54}
$$

Here $L = r(Q_2) - r(O^*)$. Inserting (5.6.54) into (5.6.53) finally yields

$$
P^{(1)h}(\tilde{Q}_2) = \frac{2\rho_1 c_1 n}{\rho_2(1 - n^2)r^{1/2}L^{3/2}}. \tag{5.6.55}
$$

This is the final expression for $P^{(1)h}(\tilde{Q}_2)$, where $\tilde{Q}_2$ is situated on Σ. It is not difficult to extend the validity of (5.6.55) into the first medium using continuation formulae (5.6.13). See Figure 5.15(c). We remind the reader that all rays of head waves in the first halfspace are straight lines making the critical angle i_1^* with the normal to interface Σ. There is then no in-plane spreading. Transverse spreading, however, is different at $\tilde{Q}_2$ and R because the wavefront is conical in 3-D. Using (5.6.13), we obtain

$$
P^{(1)h}(R) = \frac{\mathcal{L}^{\perp}(\tilde{Q}_2, S)}{\mathcal{L}^{\perp}(R, S)}\, P^{(1)h}(\tilde{Q}_2) = \left(\frac{r(\tilde{Q}_2)}{r(R)}\right)^{1/2} P^{(1)h}(\tilde{Q}_2). \tag{5.6.56}
$$

Equations (5.6.55) and (5.6.56) yield

$$
P^{(1)h}(R) = \frac{2\rho_1 c_1 n}{\rho_2(1 - n^2)r^{1/2}L^{3/2}}, \qquad r = r(R), \qquad L = r(R) - r^*,
$$
$$
\tag{5.6.57}
$$

where r^* is the critical distance given by (5.6.44). Thus, L again denotes the length of the element of the ray of the transmitted wave parallel to Σ.

Equation (5.6.57) represents the final relation for the first-order amplitude coefficient of the head wave. For completeness, we shall also give the final relations for the head wave of the time-harmonic waves,

$$
p^h(R, t) = \frac{2i\rho_1 c_1 n}{\omega\rho_2(1 - n^2)r^{1/2}L^{3/2}}\exp[-i\omega(t - T^h(R, S))], \tag{5.6.58}
$$

and of transient signals,

$$
p^h(R, t) = \frac{2\rho_1 c_1 n}{\rho_2(1 - n^2)r^{1/2}L^{3/2}}\, F^{(1)}(t - T^h(R, S)). \tag{5.6.59}
$$

Here $T^h(R, S)$ is given by (5.6.45).

Hence, the Fourier spectrum of the head wave corresponds to the Fourier spectrum of the incident wave divided by frequency. For transient signals, the shape of the wavelet of the head wave corresponds to the integral of the wavelet of the incident wave.

As a by-product of our computations of head waves, we have also obtained the first-order amplitude coefficient $P^{(1)t}$ of the transmitted wave along interface Σ beyond the interface critical point. Because $P^{(1)h}(\tilde{Q}_2) = P^{(1)t}(Q_2)$, (5.6.55) also yields the relations for $P^{(1)t}(Q_2)$. In fact, it was simpler to calculate $P^{(1)t}(Q_2)$ in this way; the general continuation formula would have required more cumbersome computations.

We shall now discuss the ray amplitudes of acoustic head waves very briefly. Figure 5.14 shows the ray amplitude-distance curves of direct, reflected, and head waves schematically.

The largest amplitudes correspond to the direct waves. They decrease with epicentral distance as r^{-1}. For $c_1 < c_2$, the amplitudes of reflected waves display a more complicated behavior. At subcritical epicentral distances, they increase with epicentral distance and have a sharp maximum at the critical point. At postcritical distances, they decrease smoothly with epicentral distance, roughly as r^{-1}. The ray amplitudes of head waves are infinite at the critical point. Along the critical ray, the ray amplitudes cannot be calculated by the ray method. The critical region is singular both for head and reflected waves in the same way as the caustic region is for other types of waves. The amplitudes of head waves decrease with increasing epicentral distance as $r^{-1/2}(r - r^*)^{-3/2}$, that is, at large epicentral distances $r \gg r^*$ as r^{-2}. Thus, at large epicentral distances beyond the critical point, the amplitudes of head waves are, as a rule, considerably smaller than the amplitudes of direct and reflected waves.

In the critical region, it is necessary to take into account the fact that transient reflected and head waves mutually interfere. Thus, the detailed investigation of the wavefield in the critical region must take into account two facts.

a. The critical region is a singular region and the ray formulae for amplitudes cannot be used there.
b. The reflected and head waves must be considered jointly, not separately.

For a detailed investigation of the wavefield in the critical region, see, for example, Brekhovskikh (1960), Červený (1966a, 1966b), and Červený and Ravindra (1971). Many other references can be found there. The dotted line in Figure 5.14 shows also the amplitude-distance curve of the interference reflected-head wave in the critical region, calculated by more accurate methods. The most distinct feature of these accurate amplitude-distance curves is the *shift of the maximum* of the amplitude-distance curve *beyond the critical point*. The shift is frequency-dependent; it is smaller for higher frequencies and more pronounced for lower frequencies. Just at the critical point, the accurate amplitude-distance curve is smooth, with a continuous derivative.

Beyond the region of interference of reflected and head waves, the reflected waves and head waves are separated and propagate independently. The ray amplitudes of the head waves are considerably smaller than the amplitudes of the reflected waves there. This behavior, however, applies to a simple model consisting of a plane interface separating two homogeneous halfspaces. The amplitudes of head waves are extremely sensitive to structural deviations from this model, particularly to the curvature of the interface and to the positive velocity gradient in the bottom halfspace. In the case of a convex interface and/or in the case of a positive velocity gradient below the interface, the segment of the ray, parallel to the interface in the original model, may deviate from the interface. The "pure" head wave described in this section then changes to a so-called *slightly refracted wave* (diving wave). This wave is a zeroth-order wave in the terminology of the ray method, and its amplitude is considerably higher than the amplitude of pure head waves, even if the deviation of the ray from the interface is small. For details see Červený and Ravindra (1971, Chap. 6), Hill (1973), and Thomson (1990).

5.6.8 Modified Forms of the Ray Series

The four alternative forms of the ray series – (5.6.2), (5.6.26), (5.6.36), and (5.6.37) – may be further modified to study wave propagation in certain situations in which the standard forms of the ray series are not applicable. Several such modifications will be discussed in this section.

a. SPACE-TIME RAY SERIES

The zeroth-order approximation of the space-time ray method was briefly discussed in Section 2.4.6. In a similar way, it is possible to construct the space-time ray series. We shall present here the three most common forms of the space-time ray series for pressure waves in fluid media. The first is

$$p(x_i, t) = \exp[iq\theta(x_j, t)] \sum_{n=0}^{\infty} (iq)^{-n} P^{(n)}(x_j, t), \qquad (5.6.60)$$

where q is a large formal parameter. The second form reads

$$p(x_i, t) = \sum_{n=0}^{\infty} P^{(n)}(x_j, t) F^{(n)}(\theta(x_j, t)), \qquad (5.6.61)$$

where $F^{(n)}$, $n = 0, 1, \ldots$, are high-frequency analytical signals, satisfying the three conditions a, b, and c, discussed in Section 5.6.5. The third form is

$$p(x_i, t) = \exp\left[iq \sum_{n=0}^{\infty} q^{-n} \Phi^{(n)}(x_j, t) \right]. \qquad (5.6.62)$$

The application of the space-time ray series to the solution of the wave equation is analogous to the application of the space-ray series. The eikonal equation and the zeroth-order transport equation are obtained in the same way as in Section 2.4.6. Hamiltonian formalism can then be applied to the eikonal equation to compute space-time rays and the phase function $\theta(x_j, t)$ along these rays. Similarly, the amplitude coefficient $P^{(0)}(x_j, t)$ can be obtained by solving the zeroth-order transport equation along space-time rays. To determine $P^{(n)}(x_j, t)$ for $n \geq 1$, the higher order space-time transport equations should be derived.

For more details on various forms of the space-time ray series, on the space-time ray tracing and solution of space-time transport equations, see Babich, Buldyrev, and Molotkov (1985). The book also gives many other references. See also Section 2.4.6 herein.

b. RAY METHOD WITH A COMPLEX EIKONAL

We have assumed that eikonal $T(x_j)$ is a real-valued function. In certain problems of seismological interest, it may also be useful to consider the complex-valued eikonal $T(x_j)$, $T(x_j) = \operatorname{Re} T(x_j) + i \operatorname{Im} T(x_j)$. This is particularly suitable in anelastic media, where the incompressibility κ (or elastic parameters c_{ijkl} in anisotropic viscoelastic case) are complex-valued. Then we can seek the solution of the frequency-domain acoustic equation using the following ansatz:

$$p(x_j, \omega) = P(x_j, \omega) \exp[i\omega T(x_j, \omega)]. \qquad (5.6.63)$$

Here both $P(x_j, \omega)$ and $T(x_j, \omega)$ are in general complex-valued. Using (5.6.63), we again obtain the eikonal equation (3.1.1) or (3.1.2) and the ray tracing system (3.1.3) in the same form as before. The difference is that all quantities in (3.1.1) through (3.1.3) are complex-valued. The model should also be complex-valued, including interfaces. Consequently, the complex eikonal equations and complex rays should be treated in a 12-D phase space with complex p_i and x_i, $i = 1, 2, 3$. There is no problem in initial-value ray tracing of complex rays, but the boundary-value ray tracing (for example, two-point ray tracing) becomes considerably more complicated. The same approach can be applied even to isotropic and anisotropic viscoelastic media. See Budden (1961b), Suchy (1972), Hearn and Krebes (1990a, 1990b), Zhu and Chun (1994a), Chapman et al. (1998), and Kravtsov, Forbes, and Asatryan (1999), among others. For a detailed treatment of complex rays and their application in seismic wave propagation, see Thomson (1997a), which also gives a critical review of previous work and many references.

In fact, the ray-series solution (5.6.2) in the frequency domain may also be used for the complex-valued T, without any change, including the basic recurrence system of equations of the ray method (5.6.8). The definition of functions $F^{(n)}(\zeta)$ given by (5.6.27) also remains valid for complex ζ.

The ray method with a complex eikonal has found applications in various problems of seismological interest. It has been used to study the wavefield in shadow regions (such as in the caustic shadow) and close to acoustic axes in anisotropic media (shear wave singularities). The complex eikonal and complex rays have also been used in the theory of Gaussian beams; in the investigation of inhomogeneous waves, tunnel waves, surface waves, and leaking waves; and in the study of seismic body waves propagating in dissipative media. The ray method with the complex eikonal can also be combined with the space-time ray method, if the complex-valued phase function $\theta(x_j, t)$ in (5.6.60) is taken into account.

c. MORE COMPLEX ASYMPTOTIC EXPANSIONS

The ray series method based on the expansion (5.6.2) fails in the regions of singular behavior of the ray field, such as the caustic region, and the critical region, the shadow region, and various transition regions. Moreover, it cannot be used to describe certain types of waves, mainly various types of diffracted waves and the like. In all these cases, the trial solutions have a more complicated form than the ray series (5.6.2). They often include special functions, such as the Airy functions, Weber functions, and Bessel functions. Asymptotic series are not, in general, constructed in powers of ω^{-1} but rather in fractional powers of ω^{-1} (for example, in $\omega^{-1/3}$). Suitable forms of the trial solution are usually based on known analytical solutions of particular *canonical problems*, which include the singularity under consideration in pure form. It is often simpler to find the *local asymptotic expansions*, which are accurate in the close vicinity of the singularity but fail at larger distances from it. In practical applications, local asymptotic expansions must be combined with standard asymptotic expansions, which are accurate in nonsingular regions but lose their accuracy close to the singularity. Fortunately, the regions of validity of both expansions often overlap so that a combination of standard and local asymptotics may be used. A more complex problem is to find *uniform asymptotic expansions*, valid not only in the vicinity of the singularity but even at larger distances from it. Such uniform asymptotic expansions are usually analytically considerably more complex than the relevant local asymptotics.

The problems of local and uniform asymptotic expansions for waves propagating in laterally varying layered structures are not simple; extensive literature is devoted to them. See Section 5.9.2.

5.7 Ray-Series Method. Elastic Case

In this section, the vectorial ray series will be introduced and used to study the propagation of elastic waves in inhomogeneous elastic, isotropic, or anisotropic media. The basic principles of the vectorial ray-series approach remain the same as in the acoustic case, but the derivations and final equations are more cumbersome.

The ray-series solutions of the elastodynamic equation were first written and studied by Babich (1956) and by Karal and Keller (1959). See also Babich and Alekseyev (1958), Alekseyev and Gel'chinskiy (1959), Alekseyev, Babich, and Gel'chinskiy (1961),

Podyapol'skiy (1966a, 1966b), Červený and Ravindra (1971), Červený, Molotkov, and Pšenčík (1977), Goldin (1979, 1986), Červený and Hron (1980), Červený (1987b, 1989a), among others. For inhomogeneous anisotropic media, see Babich (1961a), Červený (1972), and Červený, Molotkov, and Pšenčík (1977).

5.7.1 Vectorial Ray Series. Vectorial Amplitude Coefficients

We consider a time-harmonic solution of the elastodynamic equation for an isotropic or anisotropic inhomogeneous medium (with $f_i = 0$) in the form of a *vectorial ray series*:

$$\vec{u}(x_j, t) = \exp[-i\omega(t - T(x_j))]\sum_{n=0}^{\infty}(-i\omega)^{-n}\vec{U}^{(n)}(x_j). \tag{5.7.1}$$

We shall refer to $\vec{U}^{(n)}(x_j)$, $n = 0, 1, 2, \ldots$, as the *vectorial amplitude coefficients of the ray series*. Similarly as $T(x_j)$, the vectorial amplitude coefficients $\vec{U}^{(n)}(x_j)$ depend only on coordinates x_j, not on time t and frequency ω.

As in the case of acoustic pressure waves, the leading term of (5.7.1),

$$\vec{u}(x_j, t) = \exp[-i\omega(t - T(x_j))]\vec{U}^{(0)}(x_j), \tag{5.7.2}$$

is usually called the *zeroth-order ray approximation*. The properties of the zeroth-order ray approximation have been studied in earlier chapters of this book, including Sections 5.1 through 5.4. In a similar way, the term

$$(-i\omega)^{-1}\exp[-i\omega(t - T(x_j))]\vec{U}^{(1)}(x_j)$$

is usually called the *first-order ray approximation*.

Vectorial ray series (5.7.1) is again assumed to have an asymptotic character for $\omega \to \infty$. In other words, it satisfies (5.6.3), where vectors $\vec{u}$ and $\vec{U}^{(n)}$ replace scalars p and $P^{(n)}$.

5.7.2 Recurrence System of Equations of the Ray Method

Inserting ray series (5.7.1) into the elastodynamic equation and equating to zero the coefficients at all powers of frequency yields the following infinite system of equations:

$$N_i(\vec{U}^{(0)}) = 0,$$
$$N_i(\vec{U}^{(1)}) - M_i(\vec{U}^{(0)}) = 0, \tag{5.7.3}$$
$$N_i(\vec{U}^{(k)}) - M_i(\vec{U}^{(k-1)}) + L_i(\vec{U}^{(k-2)}) = 0 \qquad \text{for } k \geq 2.$$

Here vectorial differential operators N_i, M_i, and L_i are given by (2.4.41) for an anisotropic medium and by (2.4.16) for an isotropic medium. System (5.7.3) can be expressed in a more compact form if we formally introduce $\vec{U}^{(-1)}$ and $\vec{U}^{(-2)}$ by the relations,

$$\vec{U}^{(-1)}(x_j) = \vec{U}^{(-2)}(x_j) = 0. \tag{5.7.4}$$

System (5.7.3) then becomes

$$N_i(\vec{U}^{(k)}) - M_i(\vec{U}^{(k-1)}) + L_i(\vec{U}^{(k-2)}) = 0 \qquad \text{for } k \geq 0. \tag{5.7.5}$$

This is the *basic recurrence system of equations of the ray method* for the elastic medium. System (5.7.5) is analogous to system (5.6.8) derived for the acoustic case. The most important difference is that (5.7.5) has vectorial character, whereas (5.6.8) has scalar character.

System (5.7.5) is valid both for anisotropic and isotropic media, only operators N_i, M_i, and L_i need to be properly specified.

The system has a recurrent character. It must first be solved for $k = 0$, then for $k = 1$, etc. The first equation of the system (for $k = 0$) is used to derive the decomposition of the wavefield into individual waves (P and S for isotropic; qP, qS1, and qS2 for anisotropic), the eikonal equations for these waves, and the polarization of the individual waves. The next equation ($k = 1$) can be used to find the relevant equations for $\vec{U}^{(0)}$. Similarly, the kth equation ($k = 2, 3, \ldots$) of the system can be used to find $\vec{U}^{(k-1)}$. The system needs to be solved successively. To determine $\vec{U}^{(k)}$ from the $(k + 1)$st equation of (5.7.5), the lower amplitude coefficients $\vec{U}^{(k-1)}$ and $\vec{U}^{(k-2)}$ must be known.

5.7.3 Decomposition of Vectorial Amplitude Coefficients

The amplitude coefficients of ray series $\vec{U}^{(n)}(x_i)$ have vectorial character in elastic media. To solve the basic recurrence system of equations of the ray method (5.7.5), it is useful to decompose the amplitude coefficients into components, the computation of which is relatively simpler. We shall first consider isotropic media and then anisotropic media.

1. ISOTROPIC MEDIA

As we know, in the ray theory approximation, two waves can propagate in smooth isotropic inhomogeneous media: the P and S waves. The eikonal equations for the travel times of these waves can be obtained from the first equation of (5.7.3) or (5.7.5), namely $N_i(\vec{U}^{(0)}) = 0$; see Section 2.4.2. For P waves, equation $N_i(\vec{U}^{(0)}) = 0$ yields eikonal equation $\nabla T \cdot \nabla T = 1/\alpha^2$, and for S waves, eikonal equation $\nabla T \cdot \nabla T = 1/\beta^2$, where α and β are the velocities of the P and S waves and are given by (2.4.23). In addition, equation $N_i(\vec{U}^{(0)}) = 0$ also allows us to determine the polarization of the zeroth-order amplitude coefficient of ray series $\vec{U}^{(0)}$. The P waves are polarized in the direction of the normal to the wavefront $\vec{N}$, $\vec{U}^{(0)} = A\vec{N}$, and the S waves are polarized in the plane tangent to the wavefront, $\vec{U}^{(0)} = B\vec{e}_1 + C\vec{e}_2$; see (2.4.26) and (2.4.28). Amplitudes A, B, and C can be computed using the transport equations derived in Section 2.4.

For higher-order amplitude coefficients $\vec{U}^{(n)}(x_j)$, however, the polarization is more complex. The amplitude coefficients $\vec{U}^{(n)}$ corresponding to P waves are not necessarily polarized along $\vec{N}$, and the coefficients $\vec{U}^{(n)}$ corresponding to S waves are not necessarily polarized in the plane tangent to the wavefront for $n \geq 1$. We shall consider the following general decomposition of $\vec{U}^{(n)}$ into ray-centered components,

$$\vec{U}^{(n)} = U_1^{(n)} \vec{e}_1 + U_2^{(n)} \vec{e}_2 + U_3^{(n)} \vec{e}_3, \tag{5.7.6}$$

where $\vec{e}_1$, $\vec{e}_2$, and $\vec{e}_3$ are the basis vectors of the ray-centered coordinate system. Thus, $\vec{e}_1$ and $\vec{e}_2$ are tangent to the wavefront, and $\vec{e}_3 \equiv \vec{N}$ is perpendicular to it.

It is common to use the following terminology:

a. *For P waves*, (5.7.6) will be expressed in the following form:

$$\vec{U}^{(n)} = U_3^{(n)} \vec{e}_3 + \vec{W}^{(n)}, \qquad \text{with } \vec{W}^{(n)} = U_1^{(n)} \vec{e}_1 + U_2^{(n)} \vec{e}_2. \tag{5.7.7}$$

Component $U_3^{(n)}\vec{e}_3$ has the same polarization as the zeroth-order amplitude coefficient $\vec{U}^{(0)} = A\vec{N}$, and is called the *principal component of $\vec{U}^{(n)}$*. The other component, $\vec{W}^{(n)}$, is tangent to the wavefront and is called the *additional component of $\vec{U}^{(n)}$*. See Figure 5.16.

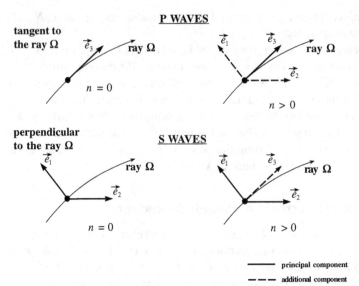

Figure 5.16. Principal and additional components of the amplitude coefficients $\vec{U}^{(n)}$ of the ray series in isotropic media. The additional components vanish for $n = 0$ (left). For P waves, the principal component is tangent to the ray, and the additional components are perpendicular to it. For S waves, the principal components are perpendicular to the ray, and the additional component is tangent to it.

b. *For* S *waves*, (5.7.6) will be expressed as

$$\vec{U}^{(n)} = U_1^{(n)}\,\vec{e}_1 + U_2^{(n)}\,\vec{e}_2 + \vec{W}^{(n)}, \qquad \text{with } \vec{W}^{(n)} = U_3^{(n)}\,\vec{e}_3. \qquad (5.7.8)$$

For S waves, $U_1^{(n)}\,\vec{e}_1 + U_2^{(n)}\,\vec{e}_2$ (tangent to the wavefront) is called the *principal component of* $\vec{U}^{(n)}$, and $\vec{W}^{(n)} = U_3^{(n)}\,\vec{e}_3$ the *additional component of* $\vec{U}^{(n)}$. See Figure 5.16.

Thus, the principal components of $\vec{U}^{(n)}$ have the same polarization as $\vec{U}^{(0)}$, but the additional components of $\vec{U}^{(n)}$ are perpendicular to $\vec{U}^{(0)}$.

Note that the principal components of $\vec{U}^{(n)}$ are also sometimes called the *main components of* $\vec{U}^{(n)}$ or the *normal components of* $\vec{U}^{(n)}$. Similarly, the additional components of $\vec{U}^{(n)}$ are often called the *anomalous components*.

One remark on the notation. In Section 5.2.2, we used superscript (q) to denote the ray-centered components of the displacement vector and vectorial amplitude function. Here we avoid superscripts (q) in expressions $U_1^{(n)}$, $U_2^{(n)}$, and $U_3^{(n)}$, even though they represent the ray-centered components of $\vec{U}^{(n)}$. The reason is that we wish to avoid notations that are too complicated. The author hopes that this will not cause any misunderstanding.

2. ANISOTROPIC MEDIA

Three seismic body waves can propagate in smooth anisotropic inhomogeneous media: one qP and two qS1 and qS2 waves. The eikonal equations for these waves can be derived from the first equation of (5.7.3), $N_i(\vec{U}^{(0)}) = 0$; see Section 2.4.3. For any of these three waves, the eikonal equations are given by the relation $G_m(x_i, p_i) = 1$, where G_m is an eigenvalue of the Christoffel matrix $\Gamma_{ik} = a_{ijkl}\,p_j p_l$. The zeroth-order amplitude coefficient of the selected mth wave $\vec{U}^{(0)}$ is linearly polarized along the appropriate mth eigenvector $\vec{g}^{(m)}$ of the Christoffel matrix, $\vec{U}^{(0)} = A\vec{g}^{(m)}$. For more details, refer to Section 2.4.3.

It is most usual to decompose the higher order coefficient $\vec{U}^{(n)}$ of the ray series in anisotropic media as

$$\vec{U}^{(n)} = U_1^{(n)} \vec{g}^{(1)} + U_2^{(n)} \vec{g}^{(2)} + U_3^{(n)} \vec{g}^{(3)}. \tag{5.7.9}$$

As in the isotropic case, we shall introduce the principal and additional components of the higher order amplitude coefficients $\vec{U}^{(n)}$ so that the principal component of $\vec{U}^{(n)}$ has the same polarization as $\vec{U}^{(0)}$, and the additional component is perpendicular to it. For example, let us consider the wave specified by $m = 1$. Then $\vec{U}^{(0)} = U_1^{(0)} \vec{g}^{(1)}$. Consequently, (5.7.9) will become

$$\vec{U}^{(n)} = U_1^{(n)} \vec{g}^{(1)} + \vec{W}^{(n)}, \qquad \text{with } \vec{W}^{(n)} = U_2^{(n)} \vec{g}^{(2)} + U_3^{(n)} \vec{g}^{(3)}. \tag{5.7.10}$$

The principal component of $\vec{U}^{(n)}$ is represented by $U_1^{(n)} \vec{g}^{(1)}$, and the additional component is represented by $\vec{W}^{(n)}$.

5.7.4 Higher Order Ray Approximations. Additional Components

In this section, we shall discuss the determination of the additional components $\vec{W}^{(n)}$ of the higher order amplitude coefficients $\vec{U}^{(n)}$ of the ray series. The procedure is practically the same for the anisotropic and isotropic case. For this reason, we shall first derive the basic equations for the additional components for anisotropic media and then specify these equations for isotropic media.

We shall consider one of the three seismic body waves propagating in an anisotropic media specified by equation $G_m(x_i, p_i) = 1$, with $m = 1$. Thus, $G_1 = 1$, but G_2 and G_3 in general differ from 1. The wave under consideration is polarized along unit vector $\vec{g}^{(1)}$. We can then use (5.7.10), where $\vec{W}^{(n)}$ is the additional component of $\vec{U}^{(n)}$. Inserting (5.7.10) into (5.7.5) yields

$$N_i\big(U_1^{(n)} \vec{g}^{(1)}\big) + N_i\big(\vec{W}^{(n)}\big) = M_i\big(\vec{U}^{(n-1)}\big) - L_i\big(\vec{U}^{(n-2)}\big). \tag{5.7.11}$$

We remind the reader that the vectorial operator $N_i(\vec{U}^{(n)})$ can be expressed as $N_i(\vec{U}^{(n)}) = \rho(\Gamma_{ik} U_k^{(n)} - U_i^{(n)})$. We shall also use relation $(\Gamma_{ik} - G_m \delta_{ik}) g_k^{(m)} = 0$ for $m = 1$, so that $g_i^{(1)} = G_1^{-1} \Gamma_{ik} g_k^{(1)} = \Gamma_{ik} g_k^{(1)}$. Similarly we obtain $\Gamma_{ik} g_k^{(2)} = G_2 g_i^{(2)}$ and $\Gamma_{ik} g_k^{(3)} = G_3 g_i^{(3)}$. Then

$$N_i\big(U_1^{(n)} \vec{g}^{(1)}\big) = \rho\big(\Gamma_{ik} U_1^{(n)} g_k^{(1)} - U_1^{(n)} g_i^{(n)}\big) = \rho U_1^{(n)} \big(\Gamma_{ik} g_k^{(1)} - g_i^{(1)}\big) = 0.$$

Similarly, for $N_i(\vec{W}^{(n)})$, we obtain

$$N_i\big(\vec{W}^{(n)}\big) = \rho\big(\Gamma_{ik} g_k^{(2)} - g_i^{(2)}\big) U_2^{(n)} + \rho\big(\Gamma_{ik} g_k^{(3)} - g_i^{(3)}\big) U_3^{(n)}$$
$$= \rho(G_2 - 1) U_2^{(n)} g_i^{(2)} + \rho(G_3 - 1) U_3^{(n)} g_i^{(3)}.$$

Then (5.7.11) yields

$$\rho(G_2 - 1) U_2^{(n)} g_i^{(2)} + \rho(G_3 - 1) U_3^{(n)} g_i^{(3)} = M_i\big(\vec{U}^{(n-1)}\big) - L_i\big(\vec{U}^{(n-2)}\big). \tag{5.7.12}$$

We assume that we know the RHS of (5.7.12) because it contains $\vec{U}^{(n-1)}$ and $\vec{U}^{(n-2)}$, and the system is recursive. Taking the scalar products of (5.7.12) with $\vec{g}^{(2)}$ and $\vec{g}^{(3)}$ yields

$$U_2^{(n)} = \frac{1}{\rho(G_2 - 1)}\big[\vec{M}(\vec{U}^{(n-1)}) - \vec{L}(\vec{U}^{(n-2)})\big] \cdot \vec{g}^{(2)},$$

$$U_3^{(n)} = \frac{1}{\rho(G_3 - 1)}\big[\vec{M}(\vec{U}^{(n-1)}) - \vec{L}(\vec{U}^{(n-2)})\big] \cdot \vec{g}^{(3)}. \tag{5.7.13}$$

These are the final expressions for the additional components of amplitude coefficient $\vec{U}^{(n)}$. Mutatis mutandis, (5.7.13) can also be used for $m = 2$ or $m = 3$.

For $n = 0$, the additional components $\vec{U}_2^{(0)}$ and $\vec{U}_3^{(0)}$ of $\vec{U}^{(0)}$ vanish, due to (5.7.4).

For $n = 1$, we use $\vec{U}^{(n-2)} = \vec{U}^{(-1)} = 0$ so that

$$U_2^{(1)} = \frac{1}{\rho(G_2 - 1)}\, \vec{M}(\vec{U}^{(0)}) \cdot \vec{g}^{(2)}, \qquad U_3^{(1)} = \frac{1}{\rho(G_3 - 1)}\, \vec{M}(\vec{U}^{(0)}) \cdot \vec{g}^{(3)}. \tag{5.7.14}$$

These equations can also be used for *isotropic media*; only the eigenvalues and eigenvectors must be properly specified. We remind the reader that $G_1 = G_2 = \beta^2 p_i p_i$, $G_3 = \alpha^2 p_i p_i$, $\vec{g}^{(1)} = \vec{e}_1$, $\vec{g}^{(2)} = \vec{e}_2$, and $\vec{g}^{(3)} = \vec{e}_3 \equiv \vec{N}$ in the isotropic case, where $\vec{N}$ is the normal to the wavefront, and $\vec{e}_1$ and $\vec{e}_2$ are tangent to the wavefront.

For P *waves*, $p_i p_i = 1/\alpha^2$ and $G_1 = G_2 = \beta^2 p_i p_i = \beta^2/\alpha^2$. Renumbering appropriately the eigenvalues and eigenvectors, Equation (5.7.13) yields

$$U_1^{(n)} = -\frac{\alpha^2}{\rho(\alpha^2 - \beta^2)}\big[\vec{M}(\vec{U}^{(n-1)}) - \vec{L}(\vec{U}^{(n-2)})\big] \cdot \vec{e}_1,$$

$$U_2^{(n)} = -\frac{\alpha^2}{\rho(\alpha^2 - \beta^2)}\big[\vec{M}(\vec{U}^{(n-1)}) - \vec{L}(\vec{U}^{(n-2)})\big] \cdot \vec{e}_2. \tag{5.7.15}$$

For S *waves*, $p_i p_i = 1/\beta^2$ and $G_3 = \alpha^2 p_i p_i = \alpha^2/\beta^2$. Equation (5.7.13) then yields

$$U_3^{(n)} = \frac{\beta^2}{\rho(\alpha^2 - \beta^2)}\big[\vec{M}(\vec{U}^{(n-1)}) - \vec{L}(\vec{U}^{(n-2)})\big] \cdot \vec{e}_3. \tag{5.7.16}$$

For $n = 0$, additional components $U_1^{(0)}$ and $U_2^{(0)}$ (for P waves) and $U_3^{(0)}$ (for S waves) again vanish. The expressions for the additional components of the *first-order ray approximation* yield

$$U_1^{(1)} = -\frac{\alpha^2}{\rho(\alpha^2 - \beta^2)}\, \vec{M}(\vec{U}^{(0)}) \cdot \vec{e}_1,$$

$$U_2^{(1)} = -\frac{\alpha^2}{\rho(\alpha^2 - \beta^2)}\, \vec{M}(\vec{U}^{(0)}) \cdot \vec{e}_2 \tag{5.7.17}$$

for P waves and

$$U_3^{(1)} = \frac{\beta^2}{\rho(\alpha^2 - \beta^2)}\, \vec{M}(\vec{U}^{(0)}) \cdot \vec{e}_3 \tag{5.7.18}$$

for S waves.

5.7.5 Higher Order Ray Approximations. Principal Components

The computation of the principal components of the higher order amplitude coefficients of the ray series is more involved than the evaluation of the additional components. In general, they can be computed by solving higher order transport equations, as in the acoustic case. In this section, we shall derive and discuss these higher order transport equations and their solutions.

We shall again start with anisotropic media. Multiplying (5.7.12) by $g_i^{(1)}$ and increasing index n by one, we obtain

$$\vec{M}(\vec{U}^{(n)}) \cdot \vec{g}^{(1)} = \vec{L}(\vec{U}^{(n-1)}) \cdot \vec{g}^{(1)}. \tag{5.7.19}$$

Using decomposition (5.7.10) and inserting it into (5.7.19) yields

$$\vec{M}(U_1^{(n)} \vec{g}^{(1)}) \cdot \vec{g}^{(1)} = [\vec{L}(\vec{U}^{(n-1)}) - \vec{M}(\vec{W}^{(n)})] \cdot \vec{g}^{(1)}. \tag{5.7.20}$$

Because $\vec{W}^{(n)}$ is the additional component, it can be determined from $\vec{U}^{(n-1)}$ and $\vec{U}^{(n-2)}$; see (5.7.13). Thus, for a given n, we can assume that the RHS of (5.7.20) is known because system (5.7.5) has a recurrent character. We shall use the following notation:

$$\zeta_m^{(n-1)} = \tfrac{1}{2}\rho^{-1/2}[L_i(\vec{U}^{(n-1)}) - M_i(\vec{W}^{(n)})]g_i^{(m)}, \qquad \zeta_m^{(-1)} = 0. \tag{5.7.21}$$

Then, (5.7.20) reads

$$M_i(U_1^{(n)}\vec{g}^{(1)})g_i^{(1)} = 2\sqrt{\rho}\,\zeta_1^{(n-1)}. \tag{5.7.22}$$

This equation will be used to derive the transport equation for $U_1^{(n)}$. Suitable expressions for $M_i(U_1^{(n)}\vec{g}^{(1)})g_i^{(1)}$ were derived in Section 2.4.3. For example,

$$M_i(U_1^{(n)}\vec{g}^{(1)})g_i^{(1)} = 2\rho\,\mathcal{U}_i U_{1,i}^{(n)} + U_1^{(n)}(\rho\mathcal{U}_i)_{,i}$$
$$= \sqrt{\rho}\,[2\mathcal{U}_i(\sqrt{\rho}U_1^{(n)})_{,i} + \sqrt{\rho}\,U_1^{(n)}\mathcal{U}_{i,i}]. \tag{5.7.23}$$

Inserting (5.7.23) into (5.7.22) yields

$$2\mathcal{U}_i(\sqrt{\rho}\,U_1^{(n)})_{,i} + \sqrt{\rho}\,U_1^{(n)}\mathcal{U}_{i,i} = 2\zeta_1^{(n-1)}. \tag{5.7.24}$$

This is the final form of the transport equation of higher order for $\sqrt{\rho}\,U_1^{(n)}$. In vectorial form it reads

$$2\vec{\mathcal{U}}\cdot\nabla(\sqrt{\rho}\,U_1^{(n)}) + \sqrt{\rho}\,U_1^{(n)}\nabla\cdot\vec{\mathcal{U}} = 2\zeta_1^{(n-1)}. \tag{5.7.25}$$

Transport equations (5.7.24) or (5.7.25) can be solved simply along the ray. We take into account (3.10.25) for $\nabla\cdot\vec{\mathcal{U}}$, $\nabla\cdot\vec{\mathcal{U}} = J^{-1}\mathrm{d}(\mathcal{U}J)/\mathrm{d}s$, and use relation $\vec{\mathcal{U}}\cdot\nabla(\sqrt{\rho}\,U_1^{(n)}) = \mathcal{U}\mathrm{d}(\sqrt{\rho}\,U_1^{(n)})/\mathrm{d}s$, and (5.7.25) yields

$$\frac{\mathrm{d}}{\mathrm{d}s}(\sqrt{\rho}\,U_1^{(n)}) + \frac{\sqrt{\rho}\,U_1^{(n)}}{2\mathcal{U}J}\frac{\mathrm{d}}{\mathrm{d}s}(\mathcal{U}J) = \frac{\zeta_1^{(n-1)}}{\mathcal{U}}. \tag{5.7.26}$$

An alternative form of (5.7.26) is

$$\mathrm{d}(\sqrt{\rho\mathcal{U}J}\,U_1^{(n)})/\mathrm{d}s = \sqrt{J/\mathcal{U}}\,\zeta_1^{(n-1)}. \tag{5.7.27}$$

This equation can be integrated very simply. Before we do so, we shall give another useful form of the transport equations. In anisotropic media, it is more common to use variable T along the ray instead of s, $\mathrm{d}T = \mathrm{d}s/\mathcal{U}$, and Jacobian $J^{(T)}$ instead of J, $J^{(T)} = \mathcal{U}J$. The transport equation of higher order (5.7.27) then reads

$$\mathrm{d}(\sqrt{\rho J^{(T)}}\,U_1^{(n)})/\mathrm{d}T = \sqrt{J^{(T)}}\,\zeta_1^{(n-1)}. \tag{5.7.28}$$

The solutions of (5.7.27) and (5.7.28) are easy. We shall first present the *continuation formulae*. Assume that $U_1^{(n)}$ is known at point s_0 of the ray and that we wish to determine $U_1^{(n)}$ at another arbitrary point s of the ray. Then the solution of (5.7.26) is

$$U_1^{(n)}(s) = \frac{1}{\sqrt{\rho(s)\mathcal{U}(s)J(s)}}$$
$$\times \left\{ \sqrt{\rho(s_0)\mathcal{U}(s_0)J(s_0)}\, U_1^{(n)}(s_0) + \int_{s_0}^{s} \sqrt{\frac{J(s')}{\mathcal{U}(s')}}\, \zeta_1^{(n-1)}(s')\mathrm{d}s' \right\}.$$
(5.7.29)

A similar solution is obtained from (5.7.28),

$$U_1^{(n)}(T) = \frac{1}{\sqrt{\rho(T)J^{(T)}(T)}}$$
$$\times \left\{ \sqrt{\rho(T_0)J^{(T)}(T_0)}\, U_1^{(n)}(T_0) + \int_{T_0}^{T} \sqrt{J^{(T)}(T')}\, \zeta_1^{(n-1)}(T')\mathrm{d}T' \right\}.$$
(5.7.30)

The integration parameters s' and T' in the integrals in (5.7.29) and (5.7.30) represent the arc length and the travel time along the ray.

The equations derived for the principal components of higher order amplitude coefficients of ray series (5.7.29) and (5.7.30) can be expressed in various alternative forms. They simplify considerably in certain important situations, particularly in homogeneous media. They can also be modified to consider a *point source* situated at $s = s_0$, as in the zeroth-order case. Let us consider, for a while, only (5.7.29). The first term in the brackets of (5.7.29) vanishes if $U_1^{(n)}(s_0)$ is finite because $J(s_0) = 0$ at the point-source. $U_1^{(n)}(s_0)$, however, may be infinite at the point source. As in the zeroth-order ray approximation, we can introduce the radiation function of higher order $\mathcal{G}_1^{(n)}(s_0; \gamma_1, \gamma_2)$,

$$\mathcal{G}_1^{(n)}(s_0; \gamma_1, \gamma_2) = \lim_{s' \to s_0} \left\{ \mathcal{L}(s', s_0)\, U_1^{(n)}(s') \right\}.$$
(5.7.31)

Equation (5.7.29) can then be modified to read

$$U_1^{(n)}(s) = \sqrt{\frac{\rho(s_0)\mathcal{U}(s_0)}{\rho(s)\mathcal{U}(s)}}\, \frac{1}{\mathcal{L}(s, s_0)} \left\{ \exp[\mathrm{i}T^c(s, s_0)]\, \mathcal{G}_1^{(n)}(s_0; \gamma_1, \gamma_2) \right.$$
$$\left. + \frac{1}{\sqrt{\rho(s_0)\mathcal{U}(s_0)}} \int_{s_0}^{s} \frac{\mathcal{L}(s', s_0)}{\sqrt{\mathcal{U}(s')}}\, \exp[\mathrm{i}T^c(s', s_0)]\zeta_1^{(n-1)}(s')\mathrm{d}s' \right\}.$$
(5.7.32)

Equation (5.7.30) would lead to a similar result.

For *isotropic media*, the equations derived above can again be used, but they must be properly modified. We need to replace $\vec{g}^{(1)}$, $\vec{g}^{(2)}$, and $\vec{g}^{(3)}$ by $\vec{e}_1$, $\vec{e}_2$, and $\vec{e}_3 \equiv \vec{N}$ and to use $\vec{\mathcal{U}} = \alpha \vec{N}$ for P waves and $\vec{\mathcal{U}} = \beta \vec{N}$ for S waves. Moreover, L_i and M_i in definition equation (5.7.21) are defined by (2.4.16). We also need to use proper numbering for P and S waves ($m = 3$ for P waves and $m = 1, 2$ for S waves). For the reader's convenience, we shall give the *isotropic version* of (5.7.29), separately for P and S waves.

For P *waves*, the principal component of $\vec{U}^{(n)}$ is $U_3^{(n)}$, and $\mathcal{U} = \alpha$. Hence,

$$
U_3^{(n)}(s) = \frac{1}{\sqrt{\rho(s)\alpha(s)J(s)}}
$$
$$
\times \left\{ \sqrt{\rho(s_0)\alpha(s_0)J(s_0)}\, U_3^{(n)}(s_0) + \int_{s_0}^{s} \sqrt{\frac{J(s')}{\alpha(s')}}\, \zeta_3^{(n-1)}(s')\mathrm{d}s' \right\}.
$$

(5.7.33)

For S *waves*, the principal components of $\vec{U}^{(n)}$ are $U_1^{(n)}$ and $U_2^{(n)}$, and $\mathcal{U} = \beta$. Hence

$$
U_{1,2}^{(n)}(s) = \frac{1}{\sqrt{\rho(s)\beta(s)J(s)}}
$$
$$
\times \left\{ \sqrt{\rho(s_0)\beta(s_0)J(s_0)}\, U_{1,2}^{(n)}(s_0) + \int_{s_0}^{s} \sqrt{\frac{J(s')}{\beta(s')}}\, \zeta_{1,2}^{(n-1)}(s')\mathrm{d}s' \right\}.
$$

(5.7.34)

For a point source situated at s_0, we can express the ray Jacobian $J(s)$ in terms of the relative geometrical spreading $\mathcal{L}(s, s_0)$. Then (5.7.32) yields

$$
U_i^{(n)}(s) = \left(\frac{V(s_0)\rho(s_0)}{V(s)\rho(s)} \right)^{1/2} \frac{\exp(iT^c(s, s_0))}{\mathcal{L}(s, s_0)} \mathcal{G}_i^{(n)}(s_0; \gamma_1, \gamma_2)
$$
$$
+ \frac{1}{\sqrt{\rho(s)V(s)}\mathcal{L}(s, s_0)} \int_{s_0}^{s} \frac{\mathcal{L}(s', s_0)}{\sqrt{V(s')}}
$$
$$
\times \exp(iT^c(s', s_0))\, \zeta_i^{(n-1)}(s')\mathrm{d}s'.
$$

(5.7.35)

Here $U_i^{(n)}(s)$ denote the principal components of the nth vectorial amplitude coefficients $\vec{U}^{(n)}$ of ray series (5.7.1). For P waves, $i = 3$, and $V = \alpha$. For S waves, $i = 1, 2$, and $V = \beta$. Moreover, $T^c(s, s_0)$ denotes the phase shift due to caustics and $\mathcal{G}_i^{(n)}(s_0, \gamma_1, \gamma_2)$ denotes the radiation pattern of the nth order, given as

$$
\mathcal{G}_i^{(n)}(s_0, \gamma_1, \gamma_2) = \lim_{s' \to s_0} \left\{ \mathcal{L}(s', s_0)\, U_i^{(n)}(s') \right\}.
$$

(5.7.36)

Limit $s' \to s_0$ is taken along the ray specified by ray parameters γ_1 and γ_2. For the zeroth-order approximation ($n = 0$), (5.7.35) yields (5.2.22), in slightly different notation.

Let us emphasize the basic difference between the computation of additional and principal components of the higher order amplitude coefficients $\vec{U}^{(k)}$, $k \geq 1$. The computation of additional components is *purely local*; they can be obtained from known $\vec{U}^{(k-1)}$ and $\vec{U}^{(k-2)}$ using the differential operators $\hat{M}$ and $\hat{L}$. Mostly, the spatial derivatives are calculated numerically, using the quantities known along several vicinity rays. No new initial data are required. The computation of the principal components of $\vec{U}^{(k)}(R)$ at a point R on ray Ω requires that we know $\vec{U}^{(k)}(S)$ at some point S on Ω and perform a numerical integration along Ω from S to R. Thus, new initial data must be known for each k. For a point source situated at S, the vectorial higher order radiation function $\mathcal{G}^{(k)}$ must be known at S. Similarly, as in the acoustic case, the higher order radiation functions are known only for some simple point sources situated in a homogeneous medium. For point sources situated in inhomogeneous media, suitable expressions for higher order radiation functions have not yet been derived. This represents the basic limitation of the practical applicability of Equation (5.7.32).

5.7.6 Reflection and Transmission

The problem of reflection and transmission of elastic waves at a curved interface between two inhomogeneous elastic media can be treated by the ray-series method in much the same way as the same problem was treated for acoustic waves in Section 5.6.4. Instead of two interface equations, however, we have six interface equations, expressing the continuity of displacement and traction across Σ. Representing the reflected and transmitted P and S waves by the relevant ray series, we need to determine *six principal components* of all these waves (two for P and four for S) successively for $n = 0, 1, 2, \ldots$. For each n, the system of six linear equations for six principal components is obtained by collecting the terms with the same power of frequency in the interface conditions.

For $n = 0$, the system has exactly the same form as (2.3.37) or (2.3.50). For $n \geq 1$, the left-hand side of the system is the same as in (2.3.37) or (2.3.50), but the right-hand side is considerably more complicated. As in the acoustic case, see (5.6.20), the right-hand side of the system for $n \geq 1$ contains the amplitude coefficients of the ray series of order $n - 1$. The right-hand side also contains the additional components of the amplitude coefficients of order n, but these may be expressed in terms of the amplitude coefficients of orders $n - 1$ and $n - 2$. The systems are solved successively, first for $n = 0$, then for $n = 1$, and so on. Thus, the right-hand sides of the system are assumed known for any n.

It would be simple to derive the system of six linear equations for any n, but the final equations are rather cumbersome. For this reason, we do not present them here. For isotropic media, the interested reader can find these systems in the book by Červený and Ravindra (1971). The discussion of the system for the elastic case remains practically the same as for the acoustic case. Again, the local (plane-wave) approximation can be used only in the zeroth-order approximation of the ray method, but *not for higher-order approximations*. As in the acoustic case, the waves reflected at an interface of the N-th order are of the $(N - 1)$-st order, in the terminology of the ray series method.

5.7.7 Alternative Forms of the Vectorial Ray Series

As in the scalar acoustic case, the vectorial ray series can be expressed in several alternative forms. For *high-frequency transient signals*, the vectorial ray series takes the form

$$u_k(x_j, t) = \sum_{n=0}^{\infty} U_k^{(n)}(x_j) \, F^{(n)}(t - T(x_j)). \tag{5.7.37}$$

The *ray series for discontinuities* reads

$$u_k(x_j, t) = F^{(0)}(t) * \sum_{n=0}^{\infty} U_k^{(n)}(x_j) \, h^{(n)}(t - T(x_j)) \tag{5.7.38}$$

or

$$u_k(x_j, t) = F^{(0)}(t - T(x_j)) * \sum_{n=0}^{\infty} U_k^{(n)}(x_j) \, h^{(n)}(t). \tag{5.7.39}$$

Here $F^{(n)}(\zeta)$ are high-frequency analytical signals, satisfying conditions (5.6.30) or (5.6.31). Similarly, $h^{(n)}(t)$ are given by relations (5.6.35). Functions $T(x_j)$ and $U_k^{(0)}(x_j)$, $U_k^{(1)}(x_j), \ldots$, are exactly the same as in the vectorial ray series for harmonic waves (5.7.1). Thus, ray series (5.7.1) and (5.7.37) through (5.7.39) can be used alternatively.

All the remarks and comments of Section 5.6.5, related to the scalar ray series, also apply to vectorial ray series. In practical applications, we again must use the real or imaginary parts of (5.7.37) through (5.7.39), or a linear combination of the real and imaginary part (depending on the initial and boundary conditions).

5.7.8 Exact Finite Vectorial Ray Series

Asymptotic vectorial ray series are, in general, only approximate. Only a finite number of terms of the asymptotic series has practical meaning. For a fixed ω, the asymptotic series always yields some unremovable error.

Nevertheless, there are some canonical situations in which the ray series is *finite and exact*. One of the very important examples is the elastodynamic Green function for a *homogeneous isotropic* medium. As we know, the elastodynamic Green function corresponds to a single-force point source. The analytical expressions for the elastodynamic Green function in a homogeneous isotropic medium are known well from the seismological literature (Aki and Richards 1980), and were also derived in Section 2.5.4. In the frequency domain, the exact analytical expression *consists of three terms*, multiplied successively by $(-i\omega)^0$, $(-i\omega)^{-1}$, and $(-i\omega)^{-2}$; see (2.5.56). It thus has the form of a finite asymptotic series in terms of $(-i\omega)^{-n}$ and is exact.

This example may be a good test of the ray-series equations derived in Sections 5.7.1 through 5.7.5. Indeed, it was proved by Vavryčuk and Yomogida (1995) that the ray-series equations yield the exact expressions in this case. Among others, the ray-series equations yield $\vec{U}^{(k)}(x_j) = 0$ for $k \geq 3$. The ray equations remain valid even in the static case. It is very comforting to know that the ray-series method yields exact expressions in certain important situations. Similarly, the exact expressions for the seismic wavefield generated by a center-of-dilatation (explosive) point source in a homogeneous isotropic medium, have two terms with factors $(-i\omega)^0$ and $(-i\omega)^{-1}$. In this case, the ray-series method again yields exact results.

The elastodynamic Green function may be expressed exactly by a finite ray series even in some special cases of an anisotropic homogeneous medium. See Vavryčuk and Yomogida (1996) for the SH wave Green function in a homogeneous transversely isotropic medium and Vavryčuk (1997) for the elastodynamic Green function in a homogeneous weakly transverse isotropic medium. The last reference offers an interesting discussion of this subject and many numerical examples.

5.7.9 Applications of Higher Order Ray Approximations. Two-Term Ray Method

In principle, everything discussed in Section 5.6.6 with regard to the applications of higher order ray approximations for the acoustic case also applies to the elastic vectorial case.

Higher order waves are again represented mainly by waves reflected from interfaces of higher order (see Section 5.7.6) and by head waves (see Section 5.7.10). The general situation, however, is considerably more complex in the elastic case than in the acoustic case because various types of converted waves may be generated at structural interfaces, in addition to unconverted waves.

The first-order ray approximation again plays an important role if the zeroth-order ray approximation *is vanishing along certain rays*. Particularly important is the case of the radiation function vanishing along a particular direction. Such rays exist for many important types of point sources, including the single-force point source. We remind the reader that the single-force point source corresponds to the elastodynamic Green function. As we know from Section 5.7.8, the ray series corresponding to the single-force point source is finite (three terms) and exact in homogeneous isotropic media. This is, however, not true for inhomogeneous layered media. As in a homogeneous medium, the equations derived in this section for the higher order ray approximation yield a nonvanishing wavefield even along nodal lines. This result also applies to other types of radiation functions, including the double-couple point source, which plays a very important role in seismology as the representative of natural earthquakes.

The first-order ray approximation may be conveniently applied to the case of reflected/transmitted waves along those rays along which the relevant R/T coefficients vanish. A very important case corresponds to the reflected/converted waves (P $\rightarrow$ S and S $\rightarrow$ P) for normal angle of incidence. In this case, the relevant reflection coefficient is zero so that the zeroth-order ray approximation vanishes for the normal angle of incidence. The first-order ray approximation, however, shows that the converted reflected waves even exist for normal angles of incidence. The application of the first-order ray approximation to the investigation of converted reflected waves for normal and near-normal angles of incidence is probably the most common application of higher order ray approximations in the seismological literature.

The first-order ray approximation can be also used to compute the dilatation ($\nabla \cdot \vec{u}$) for S waves and the rotation ($\nabla \times \vec{u}$) for P waves. In the zeroth-order ray approximation, these quantities vanish. They have been used in the discrimination of P and S waves and in other applications. See Kiselev and Rogoff (1998).

The first-order ray approximation has been also successfully used to study the properties of S waves generated by a point source in a homogeneous, transversely isotropic medium, propagating in directions close to that of a kiss singularity. Although the zeroth-order approximation is very inaccurate in this case, satisfactory results are obtained if also the first-order approximation is supplied. See Vavryčuk (1999).

The first-order ray approximation can be used to appreciate the *accuracy of the zeroth-order ray approximation*, as in the acoustic case.

In addition to the applications discussed previously, we must mention one very important vectorial application of the first-order ray approximation. It concerns the investigation of *polarization anomalies* of seismic body waves due to the inhomogeneities. The additional components of the first-order amplitude coefficients in inhomogeneous isotropic media show that the P wave is not polarized strictly perpendicular to the wavefront, and the S waves are not strictly polarized in the plane tangent to the wavefront. We shall now give the expressions for the seismic body waves propagating in isotropic inhomogeneous media, which contain the leading terms for all components of the displacement vector (both principal and additional):

$$u_k(x_j, t) = \exp[-i\omega(t - T(x_j))] \left(U_k^{(0)}(x_j) + (-i\omega)^{-1} W_k^{(1)}(x_j) \right). \quad (5.7.40)$$

Let us emphasize that $\vec{W}^{(1)}(x_j)$ is not the complete first-order amplitude coefficient $\vec{U}^{(1)}(x_j)$ but only its additional component. It is not necessary to include the principal component of the first-order amplitude coefficient $\vec{U}^{(1)}(x_j)$ because it does not represent the leading term but only a correction to the zeroth-order amplitude coefficient. The ray method based on (5.7.40) has also been known as the *two-term ray method* or the *two-component ray method* in the seismological literature. The two-term ray method is the lowest approximation of the ray-series method, which can be used to study the polarization anomalies of seismic body waves in inhomogeneous isotropic media.

It should be mentioned that the computation of the additional component $W_k^{(1)}(x_j)$ is fully local. $\vec{W}^{(1)}$ can be simply calculated from $\vec{U}^{(0)}$ by applying differential operator $\hat{M}$

to $\vec{U}^{(0)}$ (see (5.7.17) and (5.7.18)), without knowledge of the radiation function of the first order.

The two-term expression (5.7.40) may be of great interest in investigating anisotropic properties of the medium. Let us consider, for example, P waves. The deviation of the displacement vector of a P wave from the direction of the ray is commonly attributed to anisotropy, but this deviation may appear in a fully isotropic inhomogeneous media. It is caused by the additional component $W_k^{(1)}$ in (5.7.40).

Equation (5.7.40) can also be used for an anisotropic inhomogeneous medium. In this case, however, $\vec{W}^{(1)}$ is given by (5.7.10). The expression then represents the deviation of the displacement vector from the direction of eigenvector $\vec{g}^{(1)}$.

For more details and numerical computations, see Alekseyev and Mikhailenko (1982), Kiselev (1983), Daley and Hron (1987), Roslov and Yanovskaya (1988), Babich and Kiselev (1989), Goldin (1989), Kiselev and Roslov (1991), Hron and Zheng (1993), Goldin and Kurdyukova (1994), Popov and Camerlynck (1996), and Eisner and Pšenčík (1996). An up-to-date exposition of the two-term ray method and a detailed discussion of its possibilities can be found in Fradkin and Kiselev (1997).

5.7.10 Seismic Head Waves

As in the acoustic scalar case, head waves are the most typical example of higher order seismic body waves even in the vectorial elastic case. The generation of elastic head waves is practically the same as in the acoustic case. The only difference is that, in the elastic case, various types of head waves may propagate from the source to the receiver.

Theoretical investigation of seismic head waves started as early as in the 1930s, see Muskat (1933). The investigation by Muskat was based on an asymptotic treatment of exact integral expressions. The ray method was first applied to seismic head waves by Alekseyev and Gel'chinskiy (1961). For a detailed treatment of seismic head waves and for many other references, see Červený and Ravindra (1971).

We shall consider a plane interface Σ of the first order situated between two elastic homogeneous halfspaces, specified by medium parameters α_1, β_1, ρ_1 and α_2, β_2, ρ_2. We assume that a point source of spherical elastic waves with isotropic radiation patterns is situated at point S in the first halfspace, at distance h_S from Σ. The receiver is situated at point R, at distance h_R from Σ. Receiver R may be situated either in the first or in the second halfspace. The rays of R/T waves and head waves are planar in this case; they are fully situated in the plane of incidence. We can then consider the P-SV and SH case separately. We shall start with the P-SV *case*, but we shall call the SV waves simply S waves. Only later shall we briefly discuss the SH head waves.

We denote the velocity of the wave generated by the source by V_S (this may be either α_1 or β_1, depending on the type of wave generated), and the velocity of the wave arriving at the receiver V_R (this may be α_1, β_1, α_2, or β_2, depending on the position of the receiver and on the type of wave at R). Similarly, we denote the velocity of the head wave along the ray segment parallel to interface Σ by V^* (this may be either α_1, α_2, or β_2). A head wave specified by V_S, V_R, and V^* may be generated only if the two following conditions are satisfied:

$$V^* > V_S, \qquad V^* > V_R. \tag{5.7.41}$$

All possible types of seismic head waves generated at interface Σ for source S situated in

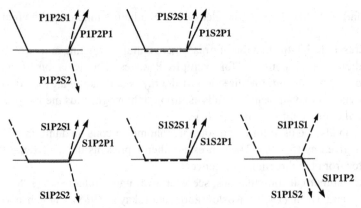

Figure 5.17. Possible types of seismic head waves generated at a plane structural interface between two homogeneous isotropic solid halfspaces.

the first (upper) halfspace are schematically shown in Figure 5.17. The first line corresponds to a P-wave source, and the second line corresponds to an S-wave source. The individual head waves are denoted by a simple, self-explanatory alphanumerical code in Figure 5.17. As we can see, 13 different types of head waves may be generated at an interface: 5 for the source of P waves and 8 for the source of S waves. For a specified model, the individual head waves are generated only if (5.7.41) is satisfied. Thus, all 13 head waves shown in Figure 5.17 cannot be generated at once.

The most favorable conditions for the generation of head waves exist if $\alpha_2 > \beta_2 > \alpha_1 > \beta_1$. In this case, 11 head waves will be generated: 5 for the P source and 6 for the S source. In fact, all head waves shown in Figure 5.17 are generated in this case, with the exception of head waves S1P1S2 and S1P1P2, shown in the last diagram in Figure 5.17.

The worst conditions for the generation of head waves exist if $\alpha_1 > \beta_1 > \alpha_2 > \beta_2$. In this case, only three head waves are generated; all of them for an S-wave source and none for a P-wave source. They correspond to the S1P1S1, S1P1S2, and S1P1P2 head waves, shown in the last diagram of Figure 5.17.

As we can see in Figure 5.17, the head waves may be generated both by reflected and transmitted waves. We can also see that the ray segment parallel to the interface may exist on either side of the interface. The only requirement is that (5.7.41) is satisfied.

Let us consider an arbitrarily selected head wave. We introduce the *R/T wave associated with the selected head wave*. There is always one R/T wave associated with the selected head wave. Both waves have one common ray, corresponding to the critical angle of incidence. Along this ray, the travel times of both waves are the same and the travel-time curves of both waves are mutually tangent. As an example, see Figure 5.14 for head wave P1P2P1 and associated reflected wave P1P1. The ray code of the R/T wave associated with the selected head wave is obtained simply. We remove the code corresponding to the middle element (parallel to the interface) from the ray code of the head wave. For example, the R/T waves associated with head waves P1P2S1, S1P2S2, S1P1S1, and S1P1P2 are the reflected wave P1S1, transmitted wave S1S2, reflected wave S1S1, and transmitted wave S1P2, respectively.

We shall now present, without a derivation, the equations for an *arbitrarily selected head wave*, corresponding to parameters h_S, h_R, V_S, V_R, and V^*, where V_S, V_R, and V^* satisfy (5.7.41). An important role for a given head wave is played by the relevant *critical* *angle of incidence* i_1^* and *critical distance* r^*, given by the relations

$$\sin i_1^* = \frac{V_S}{V^*}, \qquad r^* = \frac{h_S V_S}{\left(V^{*2} - V_S^2\right)^{1/2}} + \frac{h_R V_R}{\left(V^{*2} - V_R^2\right)^{1/2}}. \qquad (5.7.42)$$

The head wave exists only at *postcritical distances* $r > r^*$. The travel time T^h of the head wave is

$$T^h = \frac{r}{V^*} + \frac{h_S}{V_S} \left(1 - \left(\frac{V_S}{V^*}\right)^2\right)^{1/2} + \frac{h_R}{V_R} \left(1 - \left(\frac{V_R}{V^*}\right)^2\right)^{1/2}. \qquad (5.7.43)$$

At the critical point, travel time T^h is the same as the travel time of the associated R/T wave.

It is not difficult to derive the expression for the displacement vector $\vec{u}^{\,h}$ of the head wave. We assume that the radiation function of the source is omnidirectional and equals unity (for both P and S waves). The modification for other types of radiation functions should be straightforward. The displacement vector of the head wave at receiver point R, $\vec{u}^{\,h}(R, t)$

$$\vec{u}^{\,h}(R, t) = \frac{i \, V^* \Gamma \, \tan i_1^*}{\omega \, r^{1/2} \, L^{3/2}} \exp[-i\omega(t - T^h)] \, \vec{e}_R. \qquad (5.7.44)$$

Here T^h is given by (5.7.43), $L = r - r^*$ is the length of the ray segment parallel to Σ, $\tan i_1^* = V_S/(V^{*2} - V_S^2)^{1/2}$, Γ is the so-called *head-wave coefficient*, and $\vec{e}_R$ is the polarization vector of the head wave at receiver R. If the wave arrives at R as a P wave, $\vec{e}_R$ is the unit vector parallel to the ray of the head wave (perpendicular to its wavefront). If it arrives at R as an S wave, $\vec{e}_R$ is perpendicular to the ray (tangent to the wavefront). It is oriented with respect to the ray to the same side of the ray as the associated R/T wave.

Head-wave coefficient Γ may be computed easily in several ways from the expressions for the R/T coefficients given by (5.3.2) through (5.3.5). We use the following notation: $P = (1 - V^{*2} p^2)^{1/2}$, where $p = \sin i_1/V_S$ is the ray parameter. Note that P is one of the quantities P_1, P_3, P_4 defined by (5.3.5), depending on the type of head wave under consideration. Head-wave coefficient Γ is then given by

$$\Gamma = -(dR/dP)_{p=1/V^*}, \qquad (5.7.45)$$

where R is the R/T coefficient of the associated R/T wave. Alternatively, if the R/T coefficient of the associated R/T wave is expressed as

$$R = (R_1 + R_2 P)/(R_3 + R_4 P), \qquad (5.7.46)$$

the head-wave coefficient is given by the relation

$$\Gamma = \left[(R_4 R_1 - R_3 R_2)/R_3^2\right]_{p=1/V^*}; \qquad (5.7.47)$$

see (5.7.45). Analytical expressions for individual head-wave coefficients can be found in Červený and Ravindra (1971).

As an example, we shall apply the equations presented in this section to head wave P1P2P1 and compare them with the equivalent equations for the acoustic head waves derived in Section 5.6.7. We put $V_S = V_R = c_1$, $V^* = c_2$, $n = c_1/c_2$. Then (5.7.42) for r^* yields (5.6.44), and (5.7.43) for T^h yields (5.6.45). It is also obvious that (5.7.44) must yield (5.6.58), if we use $\beta_1 = \beta_2 = 0$ and eliminate $\vec{e}_R$ from (5.7.44). We remind the reader that the displacement P1P1 reflection coefficient yields the pressure reflection coefficient (5.1.21) for $\beta_1 = \beta_2 = 0$, where $P = (1 - V^{*2} p^2)^{1/2} = (1 - c_2^2 p^2)^{1/2}$ is represented by

P_2. If we express pressure reflection coefficient (5.1.21) in the form of (5.7.46), we obtain $R_1 = R_3 = \rho_2 c_2 P_1$ and $R_2 = -R_4 = -\rho_1 c_1$. Then (5.7.47) yields

$$\Gamma = 2\rho_1 c_1 / \rho_2 c_2 \sqrt{1 - n^2}. \tag{5.7.48}$$

This is the final expression for the *acoustic head-wave coefficient*. The same expression for Γ would be obtained from (5.7.45). Inserting the acoustic head-wave coefficient Γ given by (5.7.48) into (5.7.44) yields the amplitudes

$$\frac{iV^*\Gamma \tan i_1^*}{\omega r^{1/2} L^{3/2}} = \frac{2i\rho_1 c_1 n}{\omega \rho_2 (1 - n^2) r^{1/2} L^{3/2}}.$$

This fully coincides with the expression for the amplitudes of the acoustic head waves derived in Section 5.6.7; see (5.6.58).

Equation (5.7.44) also remains valid for SH *head waves*. In this case polarization vector $\vec{e}_R$ is represented by a unit vector perpendicular to the plane of incidence. There is only one type of SH head wave for a source situated in the first halfspace: S1S2S1. SH head wave S1S2S1 is generated only if $n = \beta_1/\beta_2 < 1$. The SH *head-wave coefficient* of this wave can be simply obtained from the SH reflection coefficient R_{22} given by (5.3.2), if we use (5.7.46) and (5.7.47). We obtain

$$\Gamma = 2\rho_2 \beta_2 / (\rho_1 \beta_1 \sqrt{1 - n^2}).$$

The final expression for the displacement vector of the SH head wave is then given by (5.7.44), where we insert this expression for Γ, and put $V^* = \beta_2$, $\tan i_1^* = n/\sqrt{1 - n^2}$.

We shall not discuss the derived equations and the properties of elastic head waves. In fact, most of the properties of head waves in the vectorial elastic case remain very similar to those of head waves in the scalar acoustic case; see Section 5.6.7. For more details, see Alekseyev and Gel'chinskiy (1961) and Červený and Ravindra (1971). A very detailed derivation of all equations for seismic head waves and a physical discussion of their properties can be found in the book by Červený and Ravindra (1971). The book also extends our treatment to an arbitrary structure of the overburden and discusses in detail the situation in the critical region. The book also devotes considerable attention to so-called *interference head waves*, generated at convex interfaces and/or in the case of velocity increasing with distance from the interface.

5.7.11 Modified Forms of the Vectorial Ray Series

We shall not go into details of the seismic space-time ray-series method and seismic complex eikonal method here; the principles of both methods are very similar to the relevant scalar acoustic methods; see Section 5.6.8. The interested reader is referred to Sections 2.4.6 and 5.6.8 for other suitable references and to Thomson (1997a). Similarly, extensive literature has been devoted to various local and uniform asymptotic expansions of the vectorial wavefields in singular regions. A more detailed treatment of these subjects, however, would extend the scope of this book inadmissibly.

5.8 Paraxial Displacement Vector. Paraxial Gaussian Beams

In this section, we shall construct approximate high-frequency solutions of the elastody-namic equations, valid not only along rays but also in the paraxial vicinity of these rays. We shall call them the paraxial ray approximations (for real-valued $\mathbf{Q}$, $\mathbf{P}$, and $\mathbf{M}$) and paraxial

Gaussian beams (for complex-valued $\mathbf{Q}$, $\mathbf{P}$, and $\mathbf{M}$). We shall also use these elementary solutions, connected with the individual rays, as cornerstones in the superposition integrals, representing more general solutions of the elastodynamic equation. The summation of paraxial ray approximations and the summation of paraxial Gaussian beams remove certain singularities of the standard ray method.

Let us choose an arbitrary ray Ω and call it the *central ray*. The ray may be reflected, transmitted, and converted at structural interfaces. We introduce the ray-centered coordinate system $q_1, q_2, q_3 \equiv s$ connected with ray Ω, and denote unit basis vectors along Ω by $\vec{e}_1$, $\vec{e}_2$, and $\vec{e}_3 \equiv \vec{t}$; see Section 4.1. Here s is the arclength along Ω, measured from some reference point on Ω, and $\vec{t} = \vec{t}(s)$ is the unit tangent vector to Ω. In addition to variable point s, we also consider one fixed point s_0 on Ω. Point s_0 may be chosen quite arbitrarily on Ω. We assume that dynamic ray tracing has been performed along Ω from s_0 to s and that ray propagator matrix $\mathbf{\Pi}(s, s_0)$ is known. We also introduce the submatrices $\mathbf{Q}_1(s, s_0)$, $\mathbf{Q}_2(s, s_0)$, $\mathbf{P}_1(s, s_0)$, and $\mathbf{P}_2(s, s_0)$ of ray propagator matrix $\mathbf{\Pi}(s, s_0)$; see (4.3.5).

Chapter 4 described the use of ray propagator matrix $\mathbf{\Pi}(s, s_0)$ to determine the travel-time field in the paraxial vicinity of Ω. The expansion (4.1.77) of *paraxial travel time* $T(q_1, q_2, s)$ is valid up to the quadratic terms in q_I. A similar expansion (4.6.29) for the *paraxial slowness vector components* $p_I^{(q)}(q_1, q_2, s)$ is valid up to the linear terms in q_I. In many applications, it would also be very useful to know the ray amplitudes in the paraxial vicinity of Ω. The situation is, however, more complex in this case. The amplitudes along the central ray are inversely proportional to $[\det \mathbf{Q}]^{1/2}$ so that it would be necessary to determine the expansion of quantity $[\det \mathbf{Q}]^{-1/2}$ in terms of q_I. This is, however, impossible in the framework of standard dynamic ray tracing. Dynamic ray tracing along Ω yields only $\mathbf{Q}(s) = [\mathbf{Q}(q_1, q_2, s)]_{q_1=q_2=0}$ along Ω, but not $\mathbf{Q}(q_1, q_2, s)$ in the vicinity of Ω. There are two possibilities of determining the distribution of $\mathbf{Q}$ in the vicinity of Ω.

- **a.** Computation of new rays in the vicinity of Ω. We can then compute $\mathbf{Q}'s$ by the dynamic ray tracing along the new rays, and perform interpolation between nearby rays. See Eisner and Pšenčík (1996).
- **b.** Application of higher order paraxial methods, outlined in Section 4.7.6. Such methods, however, have not yet been numerically investigated.

In layered media, there would be an additional difficulty connected with the amplitude variations, caused by the variations of the R/T coefficients from one ray to another. As a consequence of all these difficulties, paraxial amplitudes are usually considered constant in the paraxial vicinity of Ω and are considered equal to the amplitudes on the central ray. We shall call the final expressions for the paraxial displacement vector, constructed in this way, *paraxial ray approximations* and discuss them in Section 5.8.1.

Nevertheless, dynamic ray tracing yields certain important results regarding the direction of the vectorial amplitudes in the paraxial vicinity of Ω. We know that the slowness vector changes its direction in the vicinity of Ω due to the curvature of the wavefront. Because amplitude vector $\vec{U}^{(q)}$ is parallel to the slowness vector for P waves, and perpendicular to it for S waves, the direction of $\vec{U}^{(q)}$ also changes in the vicinity of Ω. Dynamic ray tracing can be used to determine the slowness vector in the vicinity of Ω, and it can thus also be used to determine the paraxial correction to the vectorial amplitudes of P waves there. The determination of paraxial corrections to the vectorial amplitudes of S waves is more involved; it requires two additional quadratures along the ray. These paraxial corrections are not discussed here. See Coates and Chapman (1990a) and Klimeš (1999b).

The equations for the time-harmonic paraxial displacement vector can also be used to find some more general solutions of the elastodynamic equation, closely concentrated around central ray Ω for high frequencies ω. Such solutions can be constructed in various ways. Here we shall briefly discuss one form of these solutions, known as *paraxial Gaussian beams*, or simply *Gaussian beams*. See Section 5.8.2.

Both the paraxial ray approximations and the paraxial Gaussian beams play an important role in various applications. The most important application resides in the construction of more general solutions of the elastodynamic equation by their summation. The integral superposition of paraxial ray approximations or of paraxial Gaussian beams yields only an approximate high-frequency solution of the elastodynamic equation but removes certain singularities of the standard ray method (caustic region, critical region, and the like). See Section 5.8.3 for a detailed explanation of superposition integrals, Section 5.8.4 for their discussion, Section 5.8.5 for the summation of paraxial ray approximations, Section 5.8.6 for the superposition integrals in 2-D models, and Section 5.8.7 for the alternative versions of the superposition integrals. Note that the summation of paraxial ray approximations in 3-D models (Section 5.8.5) and in 2-D models (Section 5.8.6) yields results close or equal to the Maslov-Chapman theory, and in 1-D models it yields the WKBJ integral. The paraxial ray approximations and the paraxial Gaussian beams offer also a very suitable tool to investigate the phase shift due to caustics and the KMAH index; see Section 5.8.8.

5.8.1 Paraxial Ray Approximation for the Displacement Vector

We introduce the time-harmonic *paraxial displacement vector*, connected with central ray Ω:

$$\vec{u}^{\text{par}}(q_1, q_2, s) = \vec{U}(s) \exp\left[-i\omega\left(t - T(s) - \tfrac{1}{2}\mathbf{q}^T \mathbf{M}(s)\mathbf{q}\right)\right]. \tag{5.8.1}$$

Here $\mathbf{q} = (q_1, q_2)^T$ and $\mathbf{M}(s) = \mathbf{P}(s)\mathbf{Q}^{-1}(s)$ is the 2×2 matrix of the second derivatives of the travel-time field with respect to ray-centered coordinates q_I; see (4.1.72). The amplitude vector $\vec{U}(s)$ is given by the relations derived in Section 5.2. Here we shall mostly use Equation (5.2.83), which expresses amplitude vector $\vec{U}(s)$ in terms of its Cartesian components $U_i^{(x)}(s)$. Using (5.2.83), we obtain

$$U_i^{(x)}(s) = U_i^{\Omega}(s)(\det \mathbf{Q}(s)/\det \mathbf{Q}(s_0))^{-1/2}. \tag{5.8.2}$$

We shall call vector $\vec{U}^{\Omega}(s)$ with Cartesian components $U_i^{\Omega}(s)$ the *spreading-free vectorial amplitude*. $U_i^{\Omega}(s)$ are given by the relation

$$U_i^{\Omega}(s) = [\rho(s_0)V(s_0)/\rho(s)V(s)]^{1/2} H_{ik}(s)\mathcal{R}_{kj}^C U_j^{(q)}(s_0). \tag{5.8.3}$$

Most quantities in (5.8.2) and (5.8.3) have the same meaning as in (5.2.83). Factor $H_{ik}(s)$ denotes $e_{ki}(s)$, the ith Cartesian component of the polarization vector $\vec{e}_k$ at s, corresponding to the elementary wave under consideration. If point s is situated on a structural interface, $H_{ik}(s)$ should be replaced by conversion coefficient $\mathcal{D}_{ik}(s)$. $U_j^{(q)}(s_0)$ represents the "initial" ray-centered amplitude of the elementary wave under consideration at s_0. It is related to $U_i^{\Omega}(s_0)$ as follows: $U_i^{\Omega}(s_0) = H_{ij}(s_0)U_j^{(q)}(s_0)$. The only difference between (5.8.2) with (5.8.3) and (5.2.83) is that we have used $(\det \mathbf{Q}(s)/\det \mathbf{Q}(s_0))^{-1/2}$ in (5.8.2) instead of $\mathcal{L}(s_0)/\mathcal{L}(s)$ in (5.2.83). Consequently, the phase shift due to caustics is tacitly included in $(\det \mathbf{Q}(s)/\det \mathbf{Q}(s_0))^{-1/2}$, and factor $\exp[i\omega T^c(s, s_0)]$ has been eliminated from (5.8.3).

Equations (5.8.2) with (5.8.3) are valid for any multiply-reflected, possibly converted, elementary wave propagating in a 3-D layered isotropic structure. We need to use the convention regarding indices j and k in (5.8.3) described in Section 5.2.9. If the first element of the ray at s_0 is P, we put $j = 3$ (no summation over j). If it is S, we put $j = J$ and perform the summation over $J = 1, 2$. Similarly, if the last element of the ray at s is P, we put $k = 3$ (no summation over k). If it is S, we put $k = K$, and perform the summation over $K = 1, 2$.

Now we shall express $[\det \mathbf{Q}(s)/\det \mathbf{Q}(s_0)]^{-1/2}$ and $\mathbf{M}(s)$ in terms of some initial quantities at s_0, specifying the behavior of the wavefront at that point. We use relations (4.6.2) and (4.6.6) to obtain

$$(\det \mathbf{Q}(s)/\det \mathbf{Q}(s_0))^{-1/2} = [\det(\mathbf{Q}_1(s, s_0) + \mathbf{Q}_2(s, s_0)\mathbf{M}(s_0))]^{-1/2},$$
$$\mathbf{M}(s) = [\mathbf{P}_1(s, s_0) + \mathbf{P}_2(s, s_0)\mathbf{M}(s_0)][\mathbf{Q}_1(s, s_0) + \mathbf{Q}_2(s, s_0)\mathbf{M}(s_0)]^{-1}.$$

$$(5.8.4)$$

Here $\mathbf{M}(s_0)$ is the 2×2 matrix of second derivatives of the travel-time field with respect to ray-centered coordinates q_I at s_0. It is related to the 2×2 matrix of the curvature of the wavefront $\mathbf{K}(s_0)$ by the relation $\mathbf{M}(s_0) = V^{-1}(s_0)\mathbf{K}(s_0)$. Further, $\mathbf{Q}_1(s, s_0)$, $\mathbf{Q}_2(s, s_0)$, $\mathbf{P}_1(s, s_0)$, and $\mathbf{P}_2(s, s_0)$ are 2×2 minors of the known 4×4 ray propagator matrix $\mathbf{\Pi}(s, s_0)$; see (4.3.5). It should be emphasized that $\det \mathbf{Q}(s)$ may vanish at some points between s_0 and s (caustic points). At these points, $U_i^{(x)}(s) = \infty$, but the spreading-free amplitude $U_i^{\Omega}(s)$ remains finite.

Thus, to construct any paraxial ray approximation (5.8.1) connected with the known central ray Ω, we must specify $\mathbf{M}(s_0)$, or $\mathbf{K}(s_0)$. Point s_0 on Ω may be chosen arbitrarily. Matrix $\mathbf{M}(s_0)$ must be symmetrical. A consequence of the symplectic properties of propagator matrix $\mathbf{\Pi}(s, s_0)$ is that $\mathbf{M}(s)$ is symmetrical along the whole central ray Ω. Consequently, $\mathbf{M}(s)$ has two real-valued eigenvalues, $M_1(s)$ and $M_2(s)$. We can relate eigenvalues $M_1(s_0)$ and $M_2(s_0)$ to the principal curvatures of the wavefront at s_0, $K_1(s_0)$, and $K_2(s_0)$ such that $K_I(s_0) = V(s_0)M_I(s_0)$.

As we can see, any value of $\mathbf{M}(s_0)$ specifies one paraxial ray approximation (5.8.1). Thus, the complete system of paraxial ray approximations (5.8.1), connected with central ray Ω, is three-parameteric. The relevant three parameters are $M_{11}(s_0)$, $M_{22}(s_0)$, and $M_{12}(s_0) = M_{21}(s_0)$.

The curvature of the wavefront of the paraxial ray approximation (5.8.1), connected with central ray Ω, equals $\mathbf{K}(s) = V(s)\mathbf{M}(s)$ at point s. Consequently, we have a three-parameteric system of paraxial wavefronts connected with the selected central ray Ω at point s. All these paraxial wavefronts are mutually tangent at s on Ω. Assume that central ray Ω belongs to an orthonomic system of rays, corresponding to some elementary wave. We can then also construct the actual wavefront, corresponding to the elementary wave under consideration at s on Ω. Denote its curvature at s on Ω by $\mathbf{K}^a(s)$. Then $\mathbf{K}^a(s)$ is one curvature of the three-parameteric system of paraxial curvatures $\mathbf{K}(s)$, and all other paraxial wavefronts are tangent to it at s on Ω. These paraxial wavefronts, tangent to the actual wavefront, are often called Snell's wavefronts. But there is no uniqueness in this terminology.

Analogous general equations for paraxial ray approximations can be determined even in anisotropic inhomogeneous layered media; see Section 5.4.4. We again consider a central ray Ω. Equation (5.8.1) then reads

$$\vec{u}^{\,\mathrm{par}}(y_1, y_2, s) = \vec{U}(s) \exp\left[-i\omega(t - T(s) - \tfrac{1}{2}\mathbf{y}^T \mathbf{M}^{(y)}(s)\mathbf{y})\right]. \qquad (5.8.5)$$

Here $\mathbf{y} = (y_1, y_2)^T$, and y_1, y_2 are wavefront orthonormal coordinates in the plane tangent to the wavefront at s, with $y_3 = 0$; see Sections 4.2.2 and 4.2.3. As in (5.8.2), the Cartesian component $U_i^{(x)}(s)$ of the paraxial amplitude vector $\vec{U}(s)$ is given by the expression

$$U_i^{(x)}(s) = U_i^{\Omega}(s)\left(\det \mathbf{Q}^{(y)}(s)/\det \mathbf{Q}^{(y)}(s_0)\right)^{-1/2} \tag{5.8.6}$$

(see (5.4.15)), where the spreading-free amplitude reads

$$U_i^{\Omega}(s) = [\rho(s_0)\mathcal{C}(s_0)/\rho(s)\mathcal{C}(s)]^{1/2} g_i(s) \mathcal{R}^C A(s_0). \tag{5.8.7}$$

Here $g_i(s)$ is the ith Cartesian component of polarization vector $\vec{g}(s)$ at s, and $A(s_0)$ represents the "initial" amplitude of the elementary wave under consideration at s_0. It is related to $U_i^{\Omega}(s_0)$ as follows: $U_i^{\Omega}(s_0) = g_i(s_0)A(s_0)$, where $\vec{g}(s_0)$ is the polarization vector of the elementary wave under consideration at initial point s_0. The spreading-free amplitude $U_i^{\Omega}(s)$ is the same for any paraxial ray approximation (5.8.5) with central ray Ω and is not influenced by the ray field in the vicinity of Ω. The continuation relations (5.8.4) for $\mathbf{M}^{(y)}(s)$ and $\det \mathbf{Q}^{(y)}(s)/\det \mathbf{Q}^{(y)}(s_0)$ are now as follows:

$$\left(\det \mathbf{Q}^{(y)}(s)/\det \mathbf{Q}^{(y)}(s_0)\right)^{-1/2} = \left[\det\left(\mathbf{Q}_1(s, s_0) + \mathbf{Q}_2(s, s_0)\mathbf{M}^{(y)}(s_0)\right)\right]^{-1/2},$$

$$\mathbf{M}^{(y)}(s) = \left[\mathbf{P}_1(s, s_0) + \mathbf{P}_2(s, s_0)\mathbf{M}^{(y)}(s_0)\right]\left[\mathbf{Q}_1(s, s_0) + \mathbf{Q}_2(s, s_0)\mathbf{M}^{(y)}(s_0)\right]^{-1}. \tag{5.8.8}$$

Here $\mathbf{Q}_1(s, s_0)$, $\mathbf{Q}_2(s, s_0)$, $\mathbf{P}_1(s, s_0)$, and $\mathbf{P}_2(s, s_0)$ are 2×2 minors of the known 4×4 anisotropic propagator matrix $\mathbf{\Pi}(s, s_0)$; see Section 4.14.3.

Thus, the complete system of paraxial ray approximations (5.8.5) in an anisotropic inhomogeneous medium, connected with the central ray Ω, is again three-parameteric. The three parameters are $M_{11}^{(y)}(s_0)$, $M_{22}^{(y)}(s_0)$, and $M_{12}^{(y)}(s_0) = M_{21}^{(y)}(s_0)$. Travel time $T(s)$ and spreading-free amplitude $U_i^{\Omega}(s)$ are the same in all these paraxial ray approximations.

More flexible equations for the paraxial displacement vector are obtained, if we express the travel time functions in (5.8.1) and (5.8.5) in global Cartesian coordinates. See (4.1.86) for isotropic media and (4.2.48) for anisotropic media.

5.8.2 Paraxial Gaussian Beams

We again consider central ray Ω situated in an isotropic elastic laterally varying layered structure and real-valued travel time $T(s)$ along central ray Ω. The paraxial high-frequency solution (5.8.1) of the elastodynamic equation can be generalized by allowing complex-valued solutions $\mathbf{Q}(s)$ and $\mathbf{P}(s)$ of the dynamic ray tracing system. Consequently, also matrix $\mathbf{M}(s) = \mathbf{P}(s)\mathbf{Q}^{-1}(s)$ and quantity $\det \mathbf{Q}(s)$ are complex-valued. We shall use the following notation:

$$\mathbf{M}(s) = \mathrm{Re}(\mathbf{M}(s)) + \mathrm{i}\,\mathrm{Im}(\mathbf{M}(s)). \tag{5.8.9}$$

Assuming that $\mathrm{Im}(\mathbf{M}(s))$ is positive definite, the paraxial high-frequency solution of the elastodynamic equation, alternative to (5.8.1), is closely concentrated about the central ray and *represents a beam*. The relevant displacement vector is given by the relation

$$\vec{u}^{\,\text{beam}}(q_1, q_2, s) = \vec{U}(s)\exp\left[-\mathrm{i}\omega(t - T(s) - \tfrac{1}{2}\mathbf{q}^T \mathbf{M}(s)\mathbf{q})\right]$$

$$= \vec{U}(s)\exp\left[-\mathrm{i}\omega(t - T(s) - \tfrac{1}{2}\mathbf{q}^T\,\mathrm{Re}(\mathbf{M}(s))\mathbf{q})\right]$$

$$\times \exp\left[-\tfrac{1}{2}\omega\mathbf{q}^T\,\mathrm{Im}(\mathbf{M}(s))\mathbf{q}\right]. \tag{5.8.10}$$

Here $\vec{U}(s)$ is expressed in terms of $U_i^{\Omega}(s)$ using (5.8.2) and (5.8.3). The spreading-free amplitude $U_i^{\Omega}(s)$ remains exactly the same as in (5.8.3); the matrices $\mathbf{M}(s)$, $\mathbf{Q}(s)$, and $\mathbf{Q}(s_0)$ in (5.8.2) and (5.8.10) are, however, complex-valued.

As in the case of paraxial displacement vector (5.8.1), it is sufficient to know $\mathbf{M}(s_0)$ to determine $(\det \mathbf{Q}(s)/\det \mathbf{Q}(s_0))^{-1/2}$ and $\mathbf{M}(s)$ along the whole ray Ω in (5.8.10). Relations (5.8.4) remain valid even for complex-valued $\mathbf{Q}(s)$ and $\mathbf{M}(s)$. Complex-valued matrices $\mathbf{M}(s_0)$ and $\mathbf{Q}(s_0)$, however, must satisfy three conditions.

a. $\mathbf{Q}(s_0)$ is regular; that is, $\det(\mathbf{Q}(s_0)) \neq 0$ and $\det(\mathbf{M}(s_0)) \neq \infty$.
b. $\mathbf{M}(s_0)$ is symmetrical.
c. $\mathrm{Im}(\mathbf{M}(s_0))$ is positive definite.

It can be proved that the following three important properties are then satisfied along the whole ray Ω, at any s.

i. $\mathbf{Q}(s)$ is regular everywhere; that is, $\det \mathbf{Q}(s) \neq 0$.
ii. $\mathbf{M}(s)$ is symmetrical.
iii. $\mathrm{Im}(\mathbf{M}(s))$ is positive definite.

The proof of these conclusions can be found in Červený and Pšenčík (1983b). The argument of square root $[\det \mathbf{Q}(s)/\det \mathbf{Q}(s_0)]^{-1/2} = [\det(\mathbf{Q}_1(s, s_0) + \mathbf{Q}_2(s, s_0)\mathbf{M}(s_0))]^{-1/2}$ in (5.8.4) is determined in the following way.

a. It equals zero for $s = s_0$.
b. It varies continuously along ray Ω.

The solutions (5.8.10) of the elastodynamic equation satisfying the foregoing three conditions are known as *paraxial Gaussian beams*, or simply *Gaussian beams*.

Thus, once $\mathbf{M}(s_0)$ satisfies the Gaussian beam conditions at any point s_0 of ray Ω, (5.8.10) represents the Gaussian beam along the whole ray Ω. *Once a Gaussian beam, always a Gaussian beam.* The Gaussian beam is regular everywhere, even at caustic points. Because the amplitudes are inversely proportional to $[\det \mathbf{Q}(s)]^{1/2} \neq 0$, they remain finite along the whole ray Ω, including the caustic points. This is the important difference between the paraxial ray approximation (5.8.1) and the paraxial Gaussian beam (5.8.10).

What is the meaning of matrices $\mathrm{Re}(\mathbf{M}(s))$ and $\mathrm{Im}(\mathbf{M}(s))$? Matrix $\mathrm{Re}(\mathbf{M}(s))$ describes the *geometrical properties of the phase front* of the Gaussian beam. Because $\mathrm{Re}(\mathbf{M}(s))$ is always symmetrical, its eigenvalues $M_1^R(s)$ and $M_2^R(s)$ are always real. Quantities $K_1(s) = V(s)M_1^R(s)$ and $K_2(s) = V(s)M_2^R(s)$ represent the principal curvatures of the phase front of the Gaussian beam.

Matrix $\mathrm{Im}(\mathbf{M}(s))$ controls the *amplitude profile of the Gaussian beam* in the cross section perpendicular to Ω at s. Because matrix $\mathrm{Im}(\mathbf{M}(s))$ is positive definite and symmetrical, it has two real-valued positive eigenvalues $M_1^I(s)$ and $M_2^I(s)$, and the amplitude profile is Gaussian in the paraxial vicinity of ray Ω. For this reason, the solutions (5.8.10), with $\mathrm{Im}\mathbf{M}(s) \neq \mathbf{0}$, are called *paraxial Gaussian beams*. Alternatively, they are also known as *solutions closely concentrated in the vicinity of ray* Ω. With the increasing square of distance from ray Ω, the amplitudes of the beam decrease exponentially. The exponential decrease is frequency-dependent; it is faster for higher frequencies. Quadratic curve $\frac{1}{2}\omega \mathbf{q}^T \mathrm{Im}(\mathbf{M}(s)) \mathbf{q} = 1$ in the plane perpendicular to Ω represents a *spot ellipse* for frequency ω. Along the spot ellipse, the amplitudes of the Gaussian beam equal $e^{-1} \vec{U}(s)$. Instead of $M_1^I(s)$ and $M_2^I(s)$, we can also introduce the quantities $L_1(s)$ and $L_2(s)$ given by relations

$$L_{1,2}(s) = \left(\pi M_{1,2}^I(s)\right)^{-1/2}. \tag{5.8.11}$$

Quantities $L_1(s)$ and $L_2(s)$ represent the half-axes of the spot ellipse for frequency $f = 1\,\mathrm{Hz}$ ($\omega = 2\pi$). We call them the *half-widths of the paraxial Gaussian beam*.

Thus, we can construct a six-parameteric system of paraxial Gaussian beams (5.8.10) connected with any one central ray Ω. Parameters $\mathrm{Re}\,M_{11}(s_0)$, $\mathrm{Re}\,M_{22}(s_0)$, and $\mathrm{Re}\,M_{12}(s_0)$ control the shape of the phase-front of the Gaussian beam at $s = s_0$. Parameters $\mathrm{Im}\,M_{11}(s_0)$, $\mathrm{Im}\,M_{22}(s_0)$, and $\mathrm{Im}\,M_{12}(s_0)$ control the width of the Gaussian beam at $s = s_0$. The real-valued travel time $T(s)$ and spreading-free amplitudes $U_i^\Omega(s)$ are the same in the whole system of Gaussian beams and also the same as in the three-parameteric system of paraxial approximations (5.8.1).

To guarantee the fulfilment of Gaussian beam conditions (a) through (c) at $s = s_0$, $L_1(s_0)$ and $L_2(s_0)$ must not vanish and must be finite. On the contrary, $M_1^R(s_0)$ and $M_2^R(s_0)$ must be finite, but both of them may vanish (plane phase-front at $s = s_0$, with a Gaussian windowing of amplitudes). For the limiting case of infinitely broad Gaussian beams ($L_1(s_0) \to \infty$ and $L_2(s_0) \to \infty$), the solution (5.8.10) does not represent the Gaussian beam but rather standard paraxial ray approximation (5.8.1).

Half-widths $L_1(s)$ and $L_2(s)$ vary along the central ray with s. We can determine these variations from (5.8.11) and (5.8.4). It is not difficult to calculate them analytically for a homogeneous medium without interfaces. In a homogeneous isotropic medium, $\mathbf{M}^{-1}(s) = \mathbf{M}^{-1}(s_0) + \mathbf{I}\sigma$, where $\sigma(s) = (s - s_0)V$. Equations (5.8.11) for $L_1(s)$ and $L_2(s)$ then show that these curves are hyperbolas, with minimum widths at some points $s = s_M$. With increasing $|s - s_M|$, the width of the Gaussian beam increases.

Consequently, the paraxial Gaussian beams may be narrow in some region of s but very broad in some other region. In the latter region, where the width of Gaussian beams is large, the validity conditions for the paraxial ray method are violated, and the accuracy of the paraxial expressions derived here is low.

The expressions for paraxial Gaussian beams concentrated at a ray Ω situated in an inhomogeneous anisotropic medium are analogous to (5.8.5):

$$\begin{aligned}
\vec{u}^{\,\mathrm{beam}}(y_1, y_2, s) &= \vec{U}(s) \exp\left[-\mathrm{i}\omega(t - T(s) - \tfrac{1}{2}\mathbf{y}^T \mathbf{M}^{(y)}(s)\mathbf{y})\right] \\
&= \vec{U}(s) \exp\left[-\mathrm{i}\omega(t - T(s) - \tfrac{1}{2}\mathbf{y}^T\,\mathrm{Re}(\mathbf{M}^{(y)}(s))\,\mathbf{y})\right] \\
&\quad \times \exp\left[-\tfrac{1}{2}\omega \mathbf{y}^T\,\mathrm{Im}(\mathbf{M}^{(y)}(s))\,\mathbf{y}\right].
\end{aligned} \tag{5.8.12}$$

Here the Cartesian components $U_i^{(x)}(s)$ of $\vec{U}(s)$ are given by (5.8.6) and (5.8.7). In (5.8.8), however, $\mathbf{M}^{(y)}(s_0)$ must be taken complex-valued and must satisfy conditions (a) through (c) given earlier. Spreading-free amplitude $U_i^\Omega(s)$ is exactly the same as in the paraxial ray approximation; see (5.8.7).

More flexible equations for the paraxial Gaussian beams are obtained, if we express the travel-time functions in (5.8.10) and (5.8.12) in global Cartesian coordinates. This is analogous to paraxial approximation; see Section 5.8.1.

As the travel time is complex-valued outside the central ray of the beam, the corresponding rays in the vicinity of the central ray can be interpreted as complex rays. In this way, the Gaussian beams may be understood as *bundles of complex rays* (Keller and Streifer 1971; Deschamps 1971; Felsen and Marcuvitz 1973; Felsen 1976b). Both the positions of the points along the ray and the ray-centered components of the slowness vector are complex-valued for complex rays. The interpretation of Gaussian beams as bundles of complex rays is closely connected with the idea of displacing a source into a complex coordinate space (Felsen 1976b, 1984; Wu 1985; Norris 1986).

Several alternative approaches can be used to derive the equations for high-frequency elastodynamic paraxial Gaussian beams. For a detailed derivation of 3-D paraxial Gaussian beams in a smooth isotropic elastic medium without interfaces, see Červený and Pšenčík (1983b). The elastodynamic HF Gaussian beams are derived there as asymptotic HF one-way solutions of the elastodynamic equation, concentrated close to the rays of P and S waves. In this case, the elastodynamic equation is reduced to a parabolic equation, which further leads to a matrix Riccati equation for complex-valued matrix $\mathbf{M}(s)$, and to the transport equation for the amplitude factor along the central ray. Finally, the matrix Riccati equation for $\mathbf{M}(s)$ yields the dynamic ray tracing system for $\mathbf{Q}(s)$ and $\mathbf{P}(s)$.

For more details on paraxial Gaussian beams see Babich (1968), Kirpichnikova (1971), Babich and Buldyrev (1972), Babich and Kirpichnikova (1974), Babich and Popov (1981), Popov (1982), Červený, Popov, and Pšenčík (1982), Červený and Pšenčík (1983a, 1983b, 1984a), and Hanyga (1986). By a simple extension of the foregoing approach, we obtain *Hermite-Gaussian beams*, which have a more complex amplitude profile in the plane perpendicular to the central ray; see Siegman (1973) and Klimeš (1983). They represent higher modes of Gaussian beams.

Another extension of the paraxial Gaussian beams yields *paraxial Gaussian wave packets*, moving along rays. See Arnaud (1971a, 1971b), Babich and Ulin (1981a, 1981b), Ralston (1983), Katchalov (1984), Klimeš (1984b), Norris, White, and Schrieffer (1987), and Klimeš (1989a). For anisotropic media, see Norris (1987). The most natural way to derive and study the Gaussian wave packets is to use the space-time ray method; see Section 2.4.6.

Let us add one important note regarding Gaussian beams. Whereas rays are only *mathematical trajectories*, Gaussian beams may represent *physical objects*. They may be generated by physical sources and investigated experimentally. Particularly important in practical applications are very narrow beams. The disadvantage of the elastodynamic Gaussian beams discussed here is their large width in certain regions. There is no physical mechanism in our treatment that would keep the Gaussian beam narrow along the whole ray. Such mechanisms, however, exist, but are mostly nonlinear. The most important nonlinear mechanism is based on the dependence of the propagation velocity on amplitudes, assuming the amplitudes are large. The velocities of propagation decrease with increasing amplitudes. Because the highest amplitudes are concentrated close to the center of the beam, a small low-velocity channel is formed along the central ray keeping the beam narrow even along long rays (waveguide effects).

5.8.3 Summation Methods

As we know from Section 2.5.1, the spherical wave in a homogeneous medium can be expressed as the superposition of plane waves using the Weyl integral, and as the superposition of cylindrical waves using the Sommerfeld integral. Both superposition integrals are exact. These integrals play an important role in 1-D isotropic media, where they can be used to compute numerically and/or asymptotically the wavefields generated by a point source and the relevant Green functions. Using the elastodynamic representation theorems, the

Green functions may further be applied to compute the wavefield in 1-D isotropic models even in more general situations.

In 2-D and 3-D laterally varying layered isotropic or anisotropic structures, such exact integral expressions for the wavefield generated by a point source are not generally available. It is, however, possible to construct useful expressions for the wavefield by integral superposition of asymptotic ray-based solutions. These expressions are not exact, but they often represent a uniform asymptotic solution of the problem under study, valid even in certain singular regions of the ray method.

The superposition integrals may be expressed either in the time domain or in the frequency domain. We shall present two general forms of the frequency-domain superposition integrals, which can be used in many applications. For time-domain versions of these integrals, see Section 6.2.6. The first of them is based on the summation of paraxial ray approximations. The time-domain versions of the resulting superposition integrals are then close or equal to the Maslov-Chapman integrals. In 1-D models, they yield the WKBJ integrals; see Section 5.8.6. The second form is based on the summation of paraxial Gaussian beams.

It should be emphasized that the individual contributions in the superposition integrals, corresponding to paraxial ray approximations or to paraxial Gaussian beams, represent approximate solutions of the elastodynamic equation. Consequently, the superposition integral also represents an approximate high-frequency solution of the elastodynamic equation.

Let us consider an elementary wave propagating in a laterally varying layered isotropic or anisotropic 3-D structure, and the relevant orthonomic system of rays $\Omega(\gamma_1, \gamma_2)$, parameterized by two ray parameters γ_1 and γ_2. On each ray, we specify one initial point S_γ, at which some initial conditions are specified. We assume that dynamic ray tracing has been performed along ray $\Omega(\gamma_1, \gamma_2)$ from initial point S_γ, and that the 4×4 ray propagator matrix $\mathbf{\Pi}(R_\gamma, S_\gamma)$ is known along $\Omega(\gamma_1, \gamma_2)$ at any point R_γ situated on $\Omega(\gamma_1, \gamma_2)$. The initial points S_γ of rays $\Omega(\gamma_1, \gamma_2)$ are distributed along a smooth initial surface Σ^0 or along a smooth initial line C^0 or coincide at a common point S^0 (central ray field with a point source at S^0). We also assume that the 2×2 matrices $\mathbf{Q}^a(S_\gamma)$, $\mathbf{P}^a(S_\gamma)$, and $\mathbf{M}^a(S_\gamma) = \mathbf{P}^a(S_\gamma)\mathbf{Q}^{a-1}(S_\gamma)$, corresponding to the actual ray field $\Omega(\gamma_1, \gamma_2)$, are known at S_γ. To emphasize the fact that these matrices correspond to the actual ray field, we use a in the superscript.

With each ray $\Omega(\gamma_1, \gamma_2)$, we can connect a three-parameteric system of paraxial ray approximations, specified by a 2×2 real-valued symmetric matrix $\mathbf{M}(S_\gamma)$ and/or a six-parameteric system of paraxial Gaussian beams, specified by a 2×2 complex-valued symmetric matrix $\mathbf{M}(S_\gamma)$ with a positive definite imaginary part; see Sections 5.8.1 and 5.8.2. The 2×2 matrices $\mathbf{Q}(S_\gamma)$, $\mathbf{P}(S_\gamma)$, and $\mathbf{M}(S_\gamma)$ correspond to any of these paraxial solutions. Let us emphasize that $\mathbf{Q}^a(S_\gamma)$, $\mathbf{P}^a(S_\gamma)$, and $\mathbf{M}^a(S_\gamma)$ correspond to the actual ray field and are fixed for the elementary wave under consideration. Matrices $\mathbf{Q}(S_\gamma)$, $\mathbf{P}(S_\gamma)$, and $\mathbf{M}(S_\gamma)$, however, should be specified in some other way. Their selection corresponds to the selection of the paraxial ray approximation or paraxial Gaussian beams we wish to use in the superposition integrals. In all superposition integrals we shall present, it is sufficient to specify $\mathbf{M}(S_\gamma)$, not $\mathbf{Q}(S_\gamma)$ and $\mathbf{P}(S_\gamma)$. We shall prove, however, that $\mathbf{M}(S_\gamma)$ must be specified in such a way that $\mathbf{M}(S_\gamma) \neq \mathbf{M}^a(S_\gamma)$.

1. SUPERPOSITION INTEGRALS

We wish to determine the wavefield of the elementary wave $\vec{u}(R, \omega)$ at a fixed receiver point R. We do not need to know the ray that passes through R. The wavefield at R is calculated by a weighted superposition of paraxial ray approximations or paraxial Gaussian beams connected with rays $\Omega(\gamma_1, \gamma_2)$ passing in the vicinity of R. Consequently, two-point ray tracing is not required in the whole procedure.

In the frequency domain, the superposition integral for the Cartesian component $u_i^{(x)}(R, \omega)$ of the displacement vector $\vec{u}(R, \omega)$ reads

$$u_i^{(x)}(R, \omega) = \iint_{\mathcal{D}} \Psi(\gamma_1, \gamma_2) U_i^{(x)}(R_\gamma) \exp[i\omega T(R, R_\gamma)] \mathrm{d}\gamma_1 \mathrm{d}\gamma_2. \qquad (5.8.13)$$

Factor $\exp[-i\omega t]$ is omitted in (5.8.13). The integral is over the rays specified by ray parameters γ_1 and γ_2; $\mathcal{D}$ denotes the region of ray parameters under consideration. We will not be too concerned with the size of the region $\mathcal{D}$; it may be different in different problems. However, we must remember that the boundaries of region $\mathcal{D}$ may generate spurious arrivals in the computation of synthetic seismograms; see Thomson and Chapman (1986) for a detailed treatment. Function $\Psi(\gamma_1, \gamma_2)$ is the weighting function and its various forms will be given later. Point R_γ is situated on the same ray $\Omega(\gamma_1, \gamma_2)$ as S_γ. It is convenient to choose R_γ on $\Omega(\gamma_1, \gamma_2)$ as close to the fixed point R as possible. The function $U_i^{(x)}(R_\gamma) \exp[i\omega T(R, R_\gamma)]$ in (5.8.13) represents the paraxial ray approximation (or the paraxial Gaussian beam) connected with ray $\Omega(\gamma_1, \gamma_2)$. In isotropic media, it is given by (5.8.1) through (5.8.4) for paraxial ray approximations and by (5.8.10) for paraxial Gaussian beams. In anisotropic media, we can use (5.8.5) through (5.8.8) for paraxial ray approximation and (5.8.12) for paraxial Gaussian beams. In all these expressions, point R_γ corresponds to s, and S_γ corresponds to s_0. The travel-time function $T(R, R_\gamma)$ represents the travel time at R, calculated by the paraxial ray methods from the travel time $T(R_\gamma)$ at R_γ, situated on a nearby ray $\Omega(\gamma_1, \gamma_2)$.

In isotropic media, we can express $T(R, R_\gamma)$ in ray-centered coordinates $q_I(R)$ of point R, connected with the (variable) ray $\Omega(\gamma_1, \gamma_2)$:

$$T(R, R_\gamma) = T(R_\gamma) + \tfrac{1}{2}\mathbf{q}^T(R)\mathbf{M}(R_\gamma)\mathbf{q}(R); \qquad (5.8.14)$$

see (5.8.1). In this case, point R_γ should be situated at the intersection of ray $\Omega(\gamma_1, \gamma_2)$ with the plane perpendicular to $\Omega(\gamma_1, \gamma_2)$, passing through R. In anisotropic media, we can use the orthonormal wavefront coordinates $y_I(R)$ of point R:

$$T(R, R_\gamma) = T(R_\gamma) + \tfrac{1}{2}\mathbf{y}^T(R)\mathbf{M}^{(y)}(R_\gamma)\mathbf{y}(R); \qquad (5.8.15)$$

see (5.8.5). In this case, point R_γ is situated at the intersection of ray $\Omega(\gamma_1, \gamma_2)$ with the plane tangent to the wavefront at R_γ, passing through R.

Let us emphasize again that the ray passing through R does not need to be known. Actually, such a ray need not exist, for example, if R is situated in the shadow region of the elementary wave under consideration. On the contrary, several such rays may exist in the case of multipathing. In both cases, superposition integral (5.8.13) is well defined.

In the actual computations, $T(R, R_\gamma)$ may be represented in various coordinate systems. As an example, see (4.6.24) in Cartesian coordinates. It is most customary to introduce a "target" surface Σ^R passing through R and consider points R_γ situated on this surface Σ^R. For more details refer to item 3 in this section.

Integral (5.8.13) expresses the wavefield at point R by a weighted superposition of the paraxial ray approximations or paraxial Gaussian beams, connected with the neighboring rays, passing close to R. Due to this, we can expect the integrals to smooth the singular behavior in the caustic region, critical regions, and the like. Moreover, we can expect the summation of synthetic seismograms not to be as sensitive to the minor details in the approximation of the structural model as the standard ray synthetic seismograms. This is, in fact, the reason why the superposition integrals are used.

The superposition integral (5.8.13) may be also expressed in a slightly different, alternative form. We use the standard zeroth-order ray-theory complex-valued vectorial amplitude $U_i^{ray}(R_\gamma)$ and take into account that

$$U_i^{(x)}(R_\gamma) = \Psi_0(\gamma_1, \gamma_2)U_i^{ray}(R_\gamma), \qquad (5.8.16)$$

where, for anisotropic media,

$$\Psi_0(\gamma_1, \gamma_2) = \left[\frac{\det \mathbf{Q}^{(y)a}(R_\gamma)}{\det \mathbf{Q}^{(y)a}(S_\gamma)} \right]^{1/2} \left[\frac{\det \mathbf{Q}^{(y)}(S_\gamma)}{\det \mathbf{Q}^{(y)}(R_\gamma)} \right]^{1/2}. \qquad (5.8.17)$$

For isotropic media, the expression for $\Psi_0(\gamma_1, \gamma_2)$ is analogous, only (y) is removed from the superscripts. The arguments of both square roots in (5.8.17) are determined in a standard way. The superposition integral (5.8.13) then reads

$$u_i^{(x)}(R, \omega) = \iint_{\mathcal{D}} \Phi(\gamma_1, \gamma_2)U_i^{ray}(R_\gamma) \exp[i\omega T(R, R_\gamma)]d\gamma_1, d\gamma_2. \qquad (5.8.18)$$

The relation between the weighting functions $\Psi(\gamma_1, \gamma_2)$ in (5.8.13) and $\Phi(\gamma_1, \gamma_2)$ in (5.8.18) is

$$\Phi(\gamma_1, \gamma_2) = \Psi(\gamma_1, \gamma_2)\Psi_0(\gamma_1, \gamma_2). \qquad (5.8.19)$$

In the following discussion, we shall mostly use the superposition integral (5.8.18) and determine directly the weighting function $\Phi(\gamma_1, \gamma_2)$. Strictly speaking, the superposition integral (5.8.18) does not represent the expansion into paraxial ray approximations or into paraxial Gaussian beams because the ray-theory amplitude factor $U_i^{ray}(R_\gamma)$ is different from $U_i^{(x)}(R_\gamma)$. The differences in amplitudes are, however, only formal (see (5.8.16)) and are compensated by the weighting functions. The most important exponential factor $\exp[i\omega T(R, R_\gamma)]$ in (5.8.18) is exactly the same as in (5.8.13). For this reason, we shall continue to speak about the summation of paraxial ray approximations and the summation of paraxial Gaussian beams, even if we consider the superposition integral (5.8.18).

2. DETERMINATION OF THE WEIGHTING FUNCTION $\Phi(\gamma_1, \gamma_2)$

We shall now present a very simple derivation of the weighting function $\Phi(\gamma_1, \gamma_2)$, based on the asymptotic high-frequency treatment of (5.8.18). In regular regions of the ray field, the asymptotic high-frequency treatment of superposition integral (5.8.18) should yield the same result as the standard ray method. Consequently, $\Phi(\gamma_1, \gamma_2)$ can be determined by matching both solutions.

In the derivation of the weighting function $\Phi(\gamma_1, \gamma_2)$, there is no large difference between anisotropic and isotropic media. For this reason, we shall consider here the more general case of anisotropic media, with $T(R, R_\gamma)$ given by (5.8.15). The final expressions for $\Phi(\gamma_1, \gamma_2)$ can then be easily specified even for isotropic media.

If we wish to determine weighting function $\Phi(\gamma_1, \gamma_2)$, we can assume point R to be situated in a region covered regularly by rays $\Omega(\gamma_1, \gamma_2)$. One of these rays passes through point R. We denote this ray by Ω_0, the relevant ray parameters by γ_{10} and γ_{20}, and the initial point of Ω_0 on Σ^0 by S.

Before we evaluate superposition integral (5.8.18) for high frequencies ω, we shall transform it to a more useful form, suitable for its asymptotic evaluation. Because the travel time $T(R_\gamma)$ in (5.8.15) is different for different rays $\Omega(\gamma_1, \gamma_2)$, we shall express it in terms of $T(R)$. Point R is presumably situated in the regular ray region so that we can use

$$T(R) = T(R_\gamma) + \tfrac{1}{2}\mathbf{y}^T(R)\mathbf{M}^{(y)a}(R_\gamma)\mathbf{y}(R),$$

where $\mathbf{M}^{(y)a}(R_\gamma)$ corresponds to the actual ray field. Inserting this into (5.8.15) yields

$$T(R, R_\gamma) = T(R) + \tfrac{1}{2}\mathbf{y}^T(R)\big[\mathbf{M}^{(y)}(R_\gamma) - \mathbf{M}^{(y)a}(R_\gamma)\big]\mathbf{y}(R).$$

The superposition integral (5.8.18) then reads

$$u_i^{(x)}(R, \omega) = \exp[i\omega T(R)] \iint_{\mathcal{D}} \Phi(\gamma_1, \gamma_2) U_i^{ray}(R_\gamma)$$
$$\times \exp\big[\tfrac{1}{2}i\omega\mathbf{y}^T \boldsymbol{\mathcal{M}}^{(y)}(R_\gamma)\mathbf{y}\big] d\gamma_1 d\gamma_2, \tag{5.8.20}$$

where the 2×2 matrix $\boldsymbol{\mathcal{M}}^{(y)}$ is given by the relation

$$\boldsymbol{\mathcal{M}}^{(y)}(R_\gamma) = \mathbf{M}^{(y)}(R_\gamma) - \mathbf{M}^{(y)a}(R_\gamma). \tag{5.8.21}$$

This integral, valid in regular ray regions, is very general. It may be used to calculate weighting function $\Phi(\gamma_1, \gamma_2)$ for both isotropic and anisotropic media, for the summation of paraxial ray approximations (real-valued $\mathbf{Q}^{(y)}$, $\mathbf{P}^{(y)}$, $\mathbf{M}^{(y)}$), and for the summation of paraxial Gaussian beams (complex-valued $\mathbf{Q}^{(y)}$, $\mathbf{P}^{(y)}$, $\mathbf{M}^{(y)}$).

For high frequencies ω, the main contribution of the superposition integral (5.8.20) comes from the vicinity of ray $\Omega_0(\gamma_{10}, \gamma_{20})$, passing through the receiver point R. Consequently, we can put approximately $U_i^{ray}(R_\gamma) \doteq U_i^{ray}(R)$ and $\Phi(\gamma_1, \gamma_2) \doteq \Phi(\gamma_{10}, \gamma_{20})$. As a target surface Σ^R, we shall use the plane y_1, y_2, tangent to the wavefront at R, with the origin at R. In this case, we can approximately use the following relation for $T(R, R_\gamma)$:

$$T(R, R_\gamma) = T(R) + \tfrac{1}{2}\mathbf{y}^T(R_\gamma)\boldsymbol{\mathcal{M}}^{(y)}(R)\mathbf{y}(R_\gamma).$$

The superposition integral (5.8.18) can then be approximately expressed in the following way:

$$u_i^{(x)}(R, \omega) \doteq \exp[i\omega T(R)]\, \Phi(\gamma_{10}, \gamma_{20})\, U_i^{ray}(R)$$
$$\times \iint_{\mathcal{D}} \exp\big[\tfrac{1}{2}i\omega\mathbf{y}^T(R_\gamma)\boldsymbol{\mathcal{M}}^{(y)}(R)\mathbf{y}(R_\gamma)\big] d\gamma_1 d\gamma_2. \tag{5.8.22}$$

To compute (5.8.22), we shall exploit the known integral

$$\iint_{-\infty}^{\infty} \exp\big[\tfrac{1}{2}i\omega\mathbf{y}^T\mathbf{W}\mathbf{y}\big] dy_1 dy_2 = (2\pi/\omega)[-\det \mathbf{W}]^{-1/2}. \tag{5.8.23}$$

Here y_1 and y_2 are Cartesian coordinates, $\mathbf{y} = (y_1, y_2)^T$, and $\mathbf{W}$ is a constant 2×2 matrix with $\det \mathbf{W} \neq 0$. Matrix $\mathbf{W}$ may also be complex-valued, with a positive-definite imaginary part. The argument of $[-\det \mathbf{W}]^{-1/2}$ in (5.8.23) is given by the following relations:

$$\mathrm{Re}[-\det \mathbf{W}]^{1/2} > 0 \qquad\qquad \text{for } \mathrm{Im}\, \mathbf{W} \neq \mathbf{0},$$
$$[-\det \mathbf{W}]^{1/2} = |\det \mathbf{W}|^{1/2} \exp\big[-i\tfrac{\pi}{4}\, \mathrm{Sgn}\, \mathbf{W}\big] \qquad \text{for } \mathrm{Im}\, \mathbf{W} = \mathbf{0}.$$
$$\tag{5.8.24}$$

As usual, $\mathrm{Sgn}\, \mathbf{W}$ denotes the signature of the real-valued matrix $\mathbf{W}$; it equals the number of its positive eigenvalues minus the number of its negative eigenvalues. Thus, it equals 2, 0, or -2.

Here are several comments to (5.8.23). For real-valued $\mathbf{W}$, the result (5.8.23) with (5.8.24) is well known from the method of stationary phase, see Bleistein (1984). For complex-valued $\mathbf{W}$, the computation of (5.8.23) is based on a simultaneous diagonalization of 2×2 matrices $\mathrm{Re}\,\mathbf{W}$ and $\mathrm{Im}\,\mathbf{W}$, with $\mathrm{Im}\,\mathbf{W}$ positive definite, and on a consequent transformation of the double integral (5.8.23) into a product of two relevant single integrals. Such a diagonalization is possible if $\mathrm{Im}\,\mathbf{W}$ is positive definite, but this is just the case of paraxial Gaussian beams. For more details on the simultaneous diagonalization of $\mathbf{W}$ and on the computation of (5.8.23), see Červený (1982b).

Transforming the integration variables γ_1 and γ_2 in (5.8.22) into integration variables y_1 and y_2 and using (5.8.23), we obtain

$$
u_i^{(x)}(R, \omega) \doteq (2\pi/\omega)\Phi(\gamma_{10}, \gamma_{20})\left[-\det \boldsymbol{\mathcal{M}}^{(y)}(R)\right]^{-1/2}
$$
$$
\times \left|\det \mathbf{Q}^{(y)a}(R)\right|^{-1} U_i^{ray}(R)\exp[i\omega T(R)]. \tag{5.8.25}
$$

Matching (5.8.25) with the standard zeroth-order ray-theory solution $u_i^{(x)}(R, \omega) = U_i^{ray}(R)$ $\times \exp[i\omega T(R)]$ yields the following expression for the weighting function $\Phi(\gamma_{10}, \gamma_{20})$:

$$
\Phi(\gamma_{10}, \gamma_{20}) = (\omega/2\pi)\left[-\det \boldsymbol{\mathcal{M}}^{(y)}(R)\right]^{1/2}\left|\det \mathbf{Q}^{(y)a}(R)\right|. \tag{5.8.26}
$$

Because point R may be situated on an arbitrary ray $\Omega(\gamma_1, \gamma_2)$, we can use the same expression for point R_γ, situated on the ray $\Omega(\gamma_1, \gamma_2)$,

$$
\Phi(\gamma_1, \gamma_2) = (\omega/2\pi)\left[-\det \boldsymbol{\mathcal{M}}^{(y)}(R_\gamma)\right]^{1/2}\left|\det \mathbf{Q}^{(y)a}(R_\gamma)\right|. \tag{5.8.27}
$$

This is the final expression for $\Phi(\gamma_1, \gamma_2)$. The argument of $[-\det \boldsymbol{\mathcal{M}}^{(y)}(R_\gamma)]^{1/2}$ is given by (5.8.24). Inserting (5.8.27) with (5.8.15) into (5.8.18) yields the final form of the superposition integral. See Section 5.8.4 for its more detailed discussion.

The integral (5.8.23) cannot be applied to the computation of the superposition integral (5.8.18) if $\det \boldsymbol{\mathcal{M}}^{(y)}(R) = 0$. In other words, the results are not valid for receiver points R, satisfying the relation

$$
\det \left[\mathbf{M}^{(y)}(R) - \mathbf{M}^{(y)a}(R)\right] = 0; \tag{5.8.28}
$$

see (5.8.21). Receiver points R, at which (5.8.28) is satisfied, are usually called the *pseudo-caustic points*. The pseudocaustic points play an important role only in the superposition of paraxial ray approximations. In the superposition of paraxial Gaussian beams, the pseudocaustic points cannot exist because $\mathrm{Im}\mathbf{M}^{(y)}(R)$ is always positive definite and $\mathrm{Im}\mathbf{M}^{(y)a}(R)$ is zero.

3. TRAVEL-TIME FUNCTION $T(R, R_\gamma)$

Travel-time function $T(R, R_\gamma)$ in the superposition integral (5.8.18) reads

$$
T(R, R_\gamma) = T(R_\gamma) + \tfrac{1}{2}\mathbf{y}^T(R, R_\gamma)\mathbf{M}^{(y)}(R_\gamma)\mathbf{y}(R, R_\gamma), \tag{5.8.29}
$$

with $\mathbf{y}(R, R_\gamma) = (y_1(R, R_\gamma),\ y_2(R, R_\gamma))^T$. It represents the travel time at R, calculated by the paraxial ray methods from the travel time $T(R_\gamma)$ at R_γ, situated on a nearby ray $\Omega(\gamma_1, \gamma_2)$. It is assumed here that the point R_γ is situated at the intersection of ray $\Omega(\gamma_1, \gamma_2)$ with the plane tangent to the wavefront at R_γ, passing through R. The origin of the wavefront orthonormal coordinates y_1 and y_2 is taken at R_γ.

For a given R, the computation of point R_γ, situated on the ray $\Omega(\gamma_1, \gamma_2)$, is not a simple task and may be rather cumbersome. For 2-D isotropic media, a suitable procedure to find

R_γ was proposed in Červený, Popov, and Pšenčík (1982). It is, however, more suitable to use the paraxial ray methods to transform (5.8.29) into a more suitable form.

Equation (5.8.29) can be simply transformed into general or local Cartesian coordinates x_i. Using (4.2.48), we obtain

$$T(R, R_\gamma) = T(R_\gamma) + (\hat{\mathbf{x}}(R) - \hat{\mathbf{x}}(R_\gamma))^T \hat{\mathbf{p}}^{(x)}(R_\gamma)$$
$$+ \tfrac{1}{2}(\hat{\mathbf{x}}(R) - \hat{\mathbf{x}}(R_\gamma))^T \hat{\mathbf{M}}^{(x)}(R_\gamma)(\hat{\mathbf{x}}(R) - \hat{\mathbf{x}}(R_\gamma)), \qquad (5.8.30)$$

where $\hat{\mathbf{M}}^{(x)}(R_\gamma)$ is expressed in terms of $\hat{\mathbf{M}}^{(y)}(R_\gamma)$ as

$$\hat{\mathbf{M}}^{(x)}(R_\gamma) = \hat{\mathbf{H}}(R_\gamma)\hat{\mathbf{M}}^{(y)}(R_\gamma)\hat{\mathbf{H}}^T(R_\gamma), \qquad (5.8.31)$$

and where $\hat{\mathbf{M}}^{(y)}(R_\gamma)$ is given by (4.2.44). $\hat{\mathbf{H}}(R_\gamma)$ is the 3×3 transformation matrix from general Cartesian coordinates x_i into wavefront orthonormal coordinates y_i at R_γ.

Equation (5.8.30) corresponds strictly to (5.8.29). The only difference is that Cartesian coordinates $x_i(R)$ and $x_i(R_\gamma)$ are used instead of wavefront orthonormal coordinates $y_I(R)$ and $y_I(R_\gamma)$. Note that $y_3(R) = y_3(R_\gamma) = 0$ because both points R and R_γ are presumably situated in the plane $y_3 = 0$. Thus, it would be again necessary to perform cumbersome computations of points R_γ situated at all rays $\Omega(\gamma_1, \gamma_2)$, for each receiver position R.

In the framework of the paraxial ray methods, however, Equation (5.8.30) with (5.8.31) is valid more generally. Actually, the position of point R_γ on the ray $\Omega(\gamma_1, \gamma_2)$ may be arbitrary; the only requirement is that the distance $|\vec{x}(R) - \vec{x}(R_\gamma)|$ be small and that the terms higher than quadratic may be neglected in the expansion (5.8.30) for $T(R, R_\gamma)$. Thus, point R_γ on the ray $\Omega(\gamma_1, \gamma_2)$ may be chosen arbitrarily, but close to R. This important conclusion allows us to consider smoothly curved, arbitrarily oriented target surface Σ^R.

Let us consider a smoothly curved target surface Σ^R, passing though the receiver point R, and assume that all termination points R_γ of rays $\Omega(\gamma_1, \gamma_2)$ are situated on this surface. Surface Σ^R may represent a formal surface in a smooth medium, a structural interface inside the model, or a free surface of the model. If a system of receivers is considered, it is suitable to choose the target surface Σ^R in such a way as to contain all receivers.

We shall now present one suitable equation for $T(R, R_\gamma)$, which will be useful in the following discussion. The travel-time expansion (5.8.30) may be expressed in local Cartesian coordinates z_i introduced in Section 4.4.1, which is devoted to the R/T problem at a curved interface. The origin of the local Cartesian coordinate system z_i is situated at R_γ on Σ^R, with the axes z_1 and z_2 situated in a plane tangent to the target surface Σ^R at R_γ. The position of receiver point R on the target surface Σ^R is then specified by $z_i(R)$, with $z_3(R)$ presumably small. The travel-time function $T(R, R_\gamma)$ is then approximately given by the relation (see 4.14.53),

$$T(R, R_\gamma) = T(R_\gamma) + \mathbf{z}^T(R)\mathbf{p}^{(z)}(R_\gamma) + \tfrac{1}{2}\mathbf{z}^T(R)\mathbf{F}(R_\gamma)\mathbf{z}(R)$$
$$= T(R_\gamma) + \hat{\mathbf{z}}^T(R)\hat{\mathbf{p}}^{(z)}(R_\gamma) + \tfrac{1}{2}\mathbf{z}^T(R)\bar{\mathbf{F}}(R_\gamma)\mathbf{z}(R). \qquad (5.8.32)$$

Here the 2×2 matrix $\mathbf{F}(R_\gamma)$ is given by the relation

$$\mathbf{F}(R_\gamma) = (\mathbf{G} - \mathbf{A}^{an})\mathbf{M}^{(y)}(\mathbf{G} - \mathbf{A}^{an})^T + \mathbf{E} - p_3^{(z)}\mathbf{D}; \qquad (5.8.33)$$

see (4.14.54). All quantities in (5.8.33) are taken at R_γ. The 2×2 matrices $\mathbf{G}(R_\gamma)$, $\mathbf{A}^{an}(R_\gamma)$, $\mathbf{M}^{(y)}(R_\gamma)$, $\mathbf{E}(R_\gamma)$, and $\mathbf{D}(R_\gamma)$ have the same meaning as in (4.14.54). Similarly, $\bar{\mathbf{F}}(R_\gamma)$ is given by (5.8.33), with the last term excluded,

$$\bar{\mathbf{F}}(R_\gamma) = (\mathbf{G} - \mathbf{A}^{an})\mathbf{M}^{(y)}(\mathbf{G} - \mathbf{A}^{an})^T + \mathbf{E}. \qquad (5.8.34)$$

The relation between the two equations (5.8.33) and (5.8.34) is obvious, if we take into account that $z_3 \doteq -\frac{1}{2} z_I z_J D_{IJ}(R_\gamma)$.

Using (5.8.33) or (5.8.34), we can also express $\mathcal{M}^{(y)}(R_\gamma)$ in a useful form:

$$
\begin{aligned}
\mathcal{M}^{(y)} &= (\mathbf{G} - \mathbf{A}^{an})^{-1}(\mathbf{F} - \mathbf{F}^a)(\mathbf{G} - \mathbf{A}^{an})^{-1T} \\
&= (\mathbf{G} - \mathbf{A}^{an})^{-1}(\bar{\mathbf{F}} - \bar{\mathbf{F}}^a)(\mathbf{G} - \mathbf{A}^{an})^{-1T};
\end{aligned}
\tag{5.8.35}
$$

see (5.8.21). The superscript a again specifies the actual ray field. Note also that $\det(\mathbf{G} - \mathbf{A}^{an}) = (\mathcal{U}/\mathcal{C})\cos\delta$, where δ is the angle between the ray and the normal to Σ^R at R_γ; see Section 4.14.8.

4. SPECIFICATION OF MATRIX M$^{(y)}$

Superposition integral (5.8.18) is influenced by the choice of the 2×2 matrix $\mathbf{M}^{(y)}$. As explained earlier, $\mathbf{M}^{(y)}$ may be specified at S_γ, at R_γ, or at any other point on ray $\Omega(\gamma_1, \gamma_2)$. Using the ray propagator matrix, $\mathbf{M}^{(y)}$ can be determined at any point of ray $\Omega(\gamma_1, \gamma_2)$ as soon as it is known at any other point of the ray. It is most common to specify it at R_γ or S_γ. If we specify it at R_γ, it is possible to use the superposition integral (5.8.18) with (5.8.30) and (5.8.27) directly. If it is specified at S_γ, we must supplement the preceding equations with the continuation relations,

$$
\begin{aligned}
\mathbf{M}^{(y)}(R_\gamma) &= \left[\mathbf{P}_1 + \mathbf{P}_2 \mathbf{M}^{(y)}(S_\gamma)\right]\left[\mathbf{Q}_1 + \mathbf{Q}_2 \mathbf{M}^{(y)}(S_\gamma)\right]^{-1}, \\
\mathbf{Q}^{(y)}(R_\gamma)\mathbf{Q}^{(y)-1}(S_\gamma) &= \mathbf{Q}_1 + \mathbf{Q}_2 \mathbf{M}^{(y)}(S_\gamma);
\end{aligned}
\tag{5.8.36}
$$

see (5.8.8). Here $\mathbf{P}_1 = \mathbf{P}_1(R_\gamma, S_\gamma)$, $\mathbf{Q}_1 = \mathbf{Q}_1(R_\gamma, S_\gamma)$, $\mathbf{P}_2 = \mathbf{P}_2(R_\gamma, S_\gamma)$, and $\mathbf{Q}_2 = \mathbf{Q}_2(R_\gamma, S_\gamma)$ are the 2×2 minors of the ray propagator matrix $\mathbf{\Pi}(R_\gamma, S_\gamma)$ for anisotropic media. Analogous relations, with the same $\mathbf{P}_1, \mathbf{Q}_1, \mathbf{P}_2$, and $\mathbf{Q}_2$, are valid even for $\mathbf{M}^{(y)a}$ and $\mathbf{Q}^{(y)a}$.

We shall first present two simple examples of summation integrals with $\mathbf{M}^{(y)}$ specified at S_γ. For the expansion of the wavefield at a smooth initial surface Σ^0 into locally plane waves, we use $\mathbf{M}^{(y)}(S_\gamma) = \mathbf{0}$. If we wish to expand the same wavefield into locally plane waves with Gaussian amplitude profiles, we use $\operatorname{Re}\mathbf{M}^{(y)}(S_\gamma) = \mathbf{0}$ and $\operatorname{Im}\mathbf{M}^{(y)}(S_\gamma)$ positive definite. The same choices can be used in the expansion of the wavefield generated by a point source into locally plane waves or into locally plane waves with Gaussian amplitude profiles.

It is however, very common to choose $\mathbf{M}^{(y)}$ at points R_γ, situated on the target surface Σ^R, $\mathbf{M}^{(y)}(R_\gamma)$. We can again consider locally plane waves at R_γ, with $\mathbf{M}^{(y)}(R_\gamma) = \mathbf{0}$, or locally plane waves with a Gaussian amplitude windowing, using $\operatorname{Re}\mathbf{M}^{(y)}(R_\gamma) = \mathbf{0}$ and $\operatorname{Im}\mathbf{M}^{(y)}(R_\gamma) \neq \mathbf{0}$. In this case, $\operatorname{Im}\mathbf{M}^{(y)}(R_\gamma)$ must be positive definite. Now we shall discuss in greater detail a choice of $\operatorname{Re}\mathbf{M}^{(y)}(R_\gamma)$, which removes the quadratic terms from the expansion (5.8.32) of $\operatorname{Re}T(R, R_\gamma)$. We consider (5.8.32) and choose $\operatorname{Re}\mathbf{M}^{(y)}(R_\gamma)$ in such a way that $\operatorname{Re}\bar{\mathbf{F}}(R_\gamma) = \mathbf{0}$. We take into account (5.8.34) and obtain

$$
\operatorname{Re}\mathbf{M}^{(y)}(R_\gamma) = -(\mathbf{G} - \mathbf{A}^{an})^{-1}\mathbf{E}(\mathbf{G} - \mathbf{A}^{an})^{-1T}.
\tag{5.8.37}
$$

All matrices in (5.8.37) are again taken at R_γ and have the same meaning as in (5.8.34). The choice (5.8.37) also removes approximately the quadratic terms from the expansion of $\operatorname{Re}T(R, R_\gamma)$ in Cartesian coordinates (5.8.30) so that (5.8.30) reads

$$
\begin{aligned}
T(R, R_\gamma) \doteq T(R_\gamma) &+ (\hat{\mathbf{x}}(R) - \hat{\mathbf{x}}(R_\gamma))^T \hat{\mathbf{p}}^{(x)}(R_\gamma) \\
&+ \mathrm{i}\tfrac{1}{2}\left(\hat{\mathbf{x}}(R) - \hat{\mathbf{x}}(R_\gamma)\right)^T \operatorname{Im}\hat{\mathbf{M}}^{(x)}(R_\gamma)(\hat{\mathbf{x}}(R) - \hat{\mathbf{x}}(R_\gamma)).
\end{aligned}
\tag{5.8.38}
$$

Thus, if the summation of paraxial ray approximation is considered, (5.8.38) yields a very simple relation

$$T(R, R_\gamma) = T(R_\gamma) + (\hat{\mathbf{x}}(R) - \hat{\mathbf{x}}(R_\gamma))^T \hat{\mathbf{p}}^{(x)}(R_\gamma). \tag{5.8.39}$$

In the summation of paraxial Gaussian beams, Im $\hat{\mathbf{M}}^{(x)}(R_\gamma) \neq \mathbf{0}$ in (5.8.38). We can choose a positive-definite 2×2 matrix Im $\mathbf{M}^{(y)}(R_\gamma)$ arbitrarily and use (5.8.31) to determine Im $\hat{\mathbf{M}}^{(x)}(R_\gamma)$. Matrix Im $\hat{\mathbf{M}}^{(x)}(R_\gamma)$ controls the width of Gaussian beams under consideration. For infinitely broad paraxial Gaussian beams, Im $\mathbf{M}^{(y)}(R_\gamma) = \mathbf{0}$, and the superposition of paraxial Gaussian beams becomes the superposition of paraxial ray approximations.

5.8.4 Superposition Integrals: Discussion

In this section, we shall present the final form of the superposition integral. It is valid both for the summation of paraxial ray approximations (Im $\mathbf{M}^{(y)}(R_\gamma) = \mathbf{0}$) and for the summation of paraxial Gaussian beams (Im $\mathbf{M}^{(y)}(R_\gamma)$ positive definite). It is assumed that all termination points R_γ of rays $\Omega(\gamma_1, \gamma_2)$ are situated along a target surface Σ^R, passing through the receiver(s) R. The target surface Σ^R may be smoothly curved. The superposition integral is then as follows:

$$u_i^{(x)}(R, \omega) = \frac{\omega}{2\pi} \iint_{\mathcal{D}} U_i^{ray}(R_\gamma) \left[-\det \mathcal{M}^{(y)}(R_\gamma) \right]^{1/2}$$
$$\times |\det \mathbf{Q}^{(y)a}(R_\gamma)| \exp[i\omega T(R, R_\gamma)] d\gamma_1 d\gamma_2; \tag{5.8.40}$$

see (5.8.18), (5.8.24), (5.8.27), and (5.8.30). Here

$$\mathcal{M}^{(y)}(R_\gamma) = \mathbf{M}^{(y)}(R_\gamma) - \mathbf{M}^{(y)a}(R_\gamma), \tag{5.8.41}$$

and the argument of $[-\det \mathcal{M}^{(y)}(R_\gamma)]^{1/2}$ is given by relations,

$$\operatorname{Re}\left[-\det \mathcal{M}^{(y)} \right]^{1/2} > 0 \qquad \text{for Im } \mathcal{M}^{(y)} \neq \mathbf{0},$$
$$\left[-\det \mathcal{M}^{(y)} \right]^{1/2} = \left| \det \mathcal{M}^{(y)} \right|^{1/2} \exp\left[-i\frac{\pi}{4} \operatorname{Sgn} \mathcal{M}^{(y)} \right] \tag{5.8.42}$$
$$\text{for Im } \mathcal{M}^{(y)} = \mathbf{0};$$

see (5.8.24). The travel-time function $T(R, R_\gamma)$, expressed in Cartesian coordinates, is

$$T(R, R_\gamma) = T(R_\gamma) + (\hat{\mathbf{x}}(R) - \hat{\mathbf{x}}(R_\gamma))^T \hat{\mathbf{p}}^{(x)}(R_\gamma)$$
$$+ \tfrac{1}{2}(\hat{\mathbf{x}}(R) - \hat{\mathbf{x}}(R_\gamma))^T \hat{\mathbf{M}}^{(x)}(R_\gamma)(\hat{\mathbf{x}}(R) - \hat{\mathbf{x}}(R_\gamma)). \tag{5.8.43}$$

The superposition integral (5.8.40), with (5.8.41) through (5.8.43), is valid for any inhomogeneous anisotropic elastic layered structure, which may be bounded or unbounded. It may be, however, applied even to isotropic elastic media and to fluid media. For isotropic media, we perform dynamic ray tracing in ray-centered coordinates and remove (y) from superscripts of all 2×2 matrices in (5.8.40) through (5.8.42). For pressure waves in fluid models, we only replace vectorial components $u_i^{(x)}(R, \omega)$ and $U_i^{ray}(R_\gamma)$ by scalars $p(R, \omega)$ and $P^{ray}(R_\gamma)$.

Actually, the superposition integral (5.8.40) with (5.8.41) through (5.8.43) can be applied to any other type of wavefield for which the ray-theory amplitudes $U_i^{ray}(R_\gamma)$ can be computed and the dynamic ray tracing can be performed along rays to compute $\mathbf{Q}^{(y)a}(R_\gamma)$ and $\mathbf{M}^{(y)a}(R_\gamma)$. For example, this applies to radiowaves.

The wavefield under consideration may correspond to any multiply-reflected zeroth-order ray-theory body wave. It may be also converted at structural interfaces. The wave may be generated at a smoothly curved initial surface, at a smooth initial line, or by a point source. The radiation function of a point source or of the line source may be arbitrary. If the radiation function corresponding to the ray-theory elastodynamic Green function is used in $U_i^{ray}(R_\gamma)$, the **"superposition" Green functions** are obtained. We can call them elastodynamic Green functions based on the superposition of paraxial ray approximations and on the superposition of paraxial Gaussian beams.

The receiver may be situated arbitrarily in the model, including structural interfaces and the Earth's surface. The expression for the ray amplitude $U_i^{ray}(R_\gamma)$ must contain the proper conversion coefficients in this case. For receivers situated along the Earth's surface, it is suitable to choose the Earth's surface as a target surface Σ^R. For receivers distributed along a borehole, any smooth surface containing the borehole may be chosen as the target surface Σ^R.

It is very interesting to see that the superposition integral (5.8.40) does not require the computation of the whole ray propagator matrices along rays $\Omega(\gamma_1, \gamma_2)$ from S_γ to R_γ, if $\mathbf{M}^{(y)}(R_\gamma)$ is specified at R_γ. In this case, it is sufficient to perform the dynamic ray tracing only once, for initial conditions $\mathbf{Q}^{(y)a}(S_\gamma)$ and $\mathbf{P}^{(y)a}(S_\gamma)$ at S_γ. For the specification of $\mathbf{Q}^{(y)a}(S_\gamma)$ and $\mathbf{P}^{(y)a}(S_\gamma)$ at a curved initial surface with an arbitrary distribution of the initial travel time, see (4.14.70). The computation of the whole ray propagator matrix $\mathbf{\Pi}(R_\gamma, S_\gamma)$ is needed only if we wish to specify $\mathbf{M}^{(y)}$ at S_γ (see (5.8.36)) or at any other point between S_γ and R_γ.

The integrand of the superposition integral (5.8.40) is finite everywhere, including the caustic point at R_γ. If R_γ is a caustic point, $|\det \mathbf{Q}^{(y)a}(R_\gamma)| = 0$ and $U_i^{ray}(R_\gamma)$ is infinite. The expression $U_i^{ray}(R_\gamma)[-\det \boldsymbol{\mathcal{M}}^{(y)}(R_\gamma)]^{1/2}|\det \mathbf{Q}^{(y)a}(R_\gamma)|$, however, remains finite for arbitrary $\mathbf{M}^{(y)}(R_\gamma)$.

In anisotropic media, the matrices $\mathbf{Q}^{(y)a}(R_\gamma)$ and $\mathbf{M}^{(y)a}(R_\gamma) = \mathbf{P}^{(y)a}(R_\gamma)\mathbf{Q}^{(y)a-1}(R_\gamma)$ in (5.8.40) and (5.8.41) can be directly computed by dynamic ray tracing in wavefront orthonormal coordinates; see (4.2.31). In isotropic media, the analogous 2×2 matrices $\mathbf{Q}^a(R_\gamma)$ and $\mathbf{M}^a(R_\gamma) = \mathbf{P}^a(R_\gamma)\mathbf{Q}^{a-1}(R_\gamma)$ can be computed by dynamic ray tracing in ray-centered coordinates; see (4.1.64). Instead of dynamic ray tracing in wavefront-orthonormal or ray-centered coordinates, we can also perform dynamic ray tracing in general Cartesian coordinates and determine $Q_{iJ}^{(x)} = (\partial x_i / \partial \gamma_J)_{T=\text{const.}}$ and $P_{iJ}^{(x)} = (\partial p_i^{(x)} / \partial \gamma_J)_{T=\text{const.}}$. From these 12 quantities, $Q_{IJ}^{(y)a}$, $P_{IJ}^{(y)a}$, and $M_{IJ}^{(y)a}$ can be determined and used in (5.8.40) and (5.8.41). Mutual relations between these quantities can be found in Section 4.14. See also more details in the next section.

Here are several comments to the summation of paraxial Gaussian beams and to the choice of the 2×2 symmetric matrix $\text{Im} \mathbf{M}^{(y)}$. The superposition integral (5.8.40) represents the summation of paraxial Gaussian beams if $\text{Im} \mathbf{M}^{(y)}$ is positive definite. Matrix $\text{Im} \mathbf{M}^{(y)}$ controls the amplitude profile of paraxial Gaussian beams used in the expansion, mainly their width. For $\text{Im} \mathbf{M}^{(y)} = 0$ (that is, for infinitely broad Gaussian beams), (5.8.40) represents the summation of paraxial ray approximations; see the next section. Even small values of $\text{Im} \mathbf{M}^{(y)}$, however, increase the stability of computations and suppress pseudocaustic points and spurious arrivals. The main disadvantage of the Gaussian beam summation solutions is that they depend on the free parameters (that is, on the widths of the Gaussian beams) in certain singular regions. With sufficiently broad Gaussian beams (small $\text{Im} \mathbf{M}^{(y)}(R_\gamma)$), they yield the wavefield in caustic and critical regions correctly, but they are not accurate for edge diffractions. The computation of edge diffractions requires the use of very narrow Gaussian beams close to the edge. Moreover, spurious saddle points can arise in strongly laterally inhomogeneous media and degrade the beam solutions (see White et al. 1987). The optimum choice of Im $\mathbf{M}^{(y)}$ that would minimize the error of the computation is not known. The problem of the choice of Im $\mathbf{M}^{(y)}$ in the superposition integrals has been broadly discussed in seismological literature, but it still requires further research. See, for example, Klimeš (1989b).

The expansion of a high-frequency wavefield into paraxial Gaussian beams was first proposed by Babich and Pankratova (1973). See also Katchalov and Popov (1981), Popov (1982), Červený, Popov, and Pšenčík (1982), Červený and Pšenčík (1983a, 1983b, 1984a), Červený (1983, 1985a, 1985c), Katchalov, Popov, and Pšenčík (1983), Červený and Klimeš (1984), Madariaga (1984), Konopásková and Červený (1984), Nowack and Aki (1984), Klimeš (1984a, 1984b, 1989b), Müller (1984), Fertig and Pšenčík (1985), Madariaga and Papadimitriou (1985), Norris (1986), White et al. (1987), Katchalov and Popov (1988), Weber (1988), and Wang and Waltham (1995). A review with extensive literature up to 1985 can be found in Červený (1985c). The application of paraxial Gaussian wave packets instead of paraxial Gaussian beams in the summation procedure requires additional integration in the superposition integrals; see Ralston (1983) and Klimeš (1984b, 1989a).

Alternatively to the summation of Gaussian beams, it is also possible to use the so-called *coherent-state transform*; see Klauder (1987) and Foster and Huang (1991). We then speak of the *coherent-state method* and of the *coherent-state approximation*. For a detailed explanation and many references and applications, see Thomson (in press).

5.8.5 Maslov-Chapman Integrals

We shall now discuss the superposition integral (5.8.40) for a special case, specified by the following requirements:

a. Summation of paraxial ray approximations is considered; that is, Im $\mathbf{M}^{(y)}(R_\gamma) = \mathbf{0}$. Then the second relation of (5.8.42) can be used to determine the argument of $[-\det \mathcal{M}^{(y)}(R_\gamma)]^{1/2}$ in the superposition integral (5.8.40).

b. The real part of $\mathbf{M}^{(y)}(R_\gamma)$ is specified at R_γ using (5.8.37). The second derivatives of the travel-time field along the target surface Σ^R then vanish, and $T(R, R_\gamma)$ is given by a simple linear expansion (5.8.39) (similarly as in transform methods).

c. The dynamic ray tracing is performed in Cartesian coordinates (see (4.2.4)) not in wavefront orthonormal coordinates (see (4.2.31)). In other words, the superposition integral (5.8.40) should be expressed in terms of 12 quantities $Q_{iJ}^{(x)} = (\partial x_i / \partial \gamma_J)_{T=\text{const.}}$ and $P_{iJ}^{(x)} = (\partial p_i / \partial \gamma_J)_{T=\text{const.}}$, not in terms of 8 quantities $Q_{IJ}^{(y)a} = (\partial y_I / \partial \gamma_J)_{T=\text{const.}}$, and $P_{IJ}^{(y)a} = (\partial p_I^{(y)} / \partial \gamma_J)_{T=\text{const.}}$.

The superposition integral satisfying the requirements (a) and (b) is

$$
\begin{aligned}
u_i^{(x)}(R, \omega) = \frac{\omega}{2\pi} \iint_{\mathcal{D}} & U_i^{ray}(R_\gamma) |\det \mathcal{N}(R_\gamma)|^{1/2} \\
& \times \exp\left[i\omega(T(R_\gamma) + (\hat{\mathbf{x}}(R) - \hat{\mathbf{x}}(R_\gamma))^T \hat{\mathbf{p}}^{(x)}(R_\gamma)) \right. \\
& \left. - i\tfrac{\pi}{4} \operatorname{Sgn} \mathcal{N}(R_\gamma) \right] d\gamma_1 d\gamma_2 .
\end{aligned}
\tag{5.8.44}
$$

Here the 2×2 matrix $\mathcal{N}(R_\gamma)$ is given by the relation

$$
\mathcal{N}(R_\gamma) = \mathbf{Q}^{(y)aT}(R_\gamma)\left[\mathbf{M}^{(y)}(R_\gamma) - \mathbf{M}^{(y)a}(R_\gamma) \right] \mathbf{Q}^{(y)a}(R_\gamma),
\tag{5.8.45}
$$

where $\mathbf{M}^{(y)}(R_\gamma)$ is specified by (5.8.37).

The superposition integral (5.8.44) represents the Maslov integral for anisotropic media, derived in a different way by Chapman (in press, Section 4.1). To obtain a full agreement, we only need to express $\mathcal{N}(R_\gamma)$ in terms of $Q_{iJ}^{(x)}$ and $P_{iJ}^{(x)}$, computed by dynamic ray tracing in Cartesian coordinates. See also the requirement (c).

If we choose $\mathbf{M}^{(y)}(R_\gamma) = \mathbf{0}$, (5.8.45) yields

$$\mathcal{N}(R_\gamma) = -\mathbf{Q}^{(y)aT}(R_\gamma)\mathbf{P}^{(y)a}(R_\gamma). \tag{5.8.46}$$

The components $\mathcal{N}_{IJ}(R_\gamma)$ of matrix $\mathcal{N}(R_\gamma)$ in (5.8.46) represent the second derivatives of the travel-time field with respect to γ_I and γ_J, taken along the wavefront. Using relations derived in Section 4.2.2, we obtain

$$\mathbf{Q}^{(y)aT}(R_\gamma)\mathbf{P}^{(y)a}(R_\gamma) = \hat{\mathbf{Q}}^{(x)T}(R_\gamma)\hat{\mathbf{P}}^{(x)}(R_\gamma), \tag{5.8.47}$$

where $\hat{\mathbf{Q}}^{(x)}$ and $\hat{\mathbf{P}}^{(x)}$ are 3×2 matrices with elements $Q_{iJ}^{(x)}$ and $P_{iJ}^{(x)}$, computed by dynamic ray tracing in Cartesian coordinates. They represent the derivatives of x_i and $p_i^{(x)}$ with respect to γ_J, taken along the wavefront.

Now we shall consider a target surface Σ^R inclined with respect to the wavefront at R_γ, which may be curved. We determine $(Q_{iJ}^{(x)})_{\Sigma^R}$ and $(P_{iJ}^{(x)})_{\Sigma^R}$, representing the derivatives with respect to γ_J, taken along the target surface Σ^R. The derivation is simple:

$$\begin{aligned}
(Q_{iJ}^{(x)})_{\Sigma^R} &= \left(\frac{\partial x_i}{\partial \gamma_J}\right)_{\Sigma^R} = \left(\frac{\partial x_i}{\partial \gamma_J}\right)_T + \left(\frac{\partial x_i}{\partial T}\right)_{ray}\left(\frac{\partial T}{\partial \gamma_J}\right)_{\Sigma^R} \\
&= Q_{iJ}^{(x)} + \mathcal{U}_i^{(x)}\left(\frac{\partial T}{\partial \gamma_J}\right)_{\Sigma^R}, \\
(P_{iJ}^{(x)})_{\Sigma^R} &= \left(\frac{\partial p_i^{(x)}}{\partial \gamma_J}\right)_{\Sigma^R} = \left(\frac{\partial p_i^{(x)}}{\partial \gamma_J}\right)_T + \left(\frac{\partial p_i^{(x)}}{\partial T}\right)_{ray}\left(\frac{\partial T}{\partial \gamma_J}\right)_{\Sigma^R} \\
&= P_{iJ}^{(x)} + \eta_i^{(x)}\left(\frac{\partial T}{\partial \gamma_J}\right)_{\Sigma^R}.
\end{aligned}$$

We now denote the normal unit vector to Σ^R at R_γ by $\vec{n}$. Multiplying the first equation by n_i, and taking into account that $n_i(\partial x_i/\partial \gamma_J)_{\Sigma^R} = 0$, we obtain

$$\left(\frac{\partial T}{\partial \gamma_J}\right)_{\Sigma^R} = -\frac{n_i}{n_j \mathcal{U}_j^{(x)}} Q_{iJ}^{(x)}.$$

This yields the final expressions for $(Q_{iJ}^{(x)})_{\Sigma^R}$ and $(P_{iJ}^{(x)})_{\Sigma^R}$:

$$(Q_{iJ}^{(x)})_{\Sigma^R} = \left(\delta_{ik} - \frac{\mathcal{U}_i^{(x)}n_k}{\mathcal{U}_j^{(x)}n_j}\right)Q_{kJ}^{(x)}, \qquad (P_{iJ}^{(x)})_{\Sigma^R} = P_{iJ}^{(x)} - \frac{\eta_i^{(x)}n_k}{\mathcal{U}_j^{(x)}n_j}Q_{kJ}^{(x)}. \tag{5.8.48}$$

We remind the reader that $Q_{kJ}^{(x)}$ and $P_{kJ}^{(x)}$ are computed by dynamic ray tracing in Cartesian coordinates. The final expression for matrix $\mathcal{N}(R_\gamma)$ is then given by the relation

$$\mathcal{N}(R_\gamma) = -\left(\hat{\mathbf{Q}}^{(x)}\right)_{\Sigma^R}^T\left(\hat{\mathbf{P}}^{(x)}\right)_{\Sigma^R}. \tag{5.8.49}$$

Here $(\hat{\mathbf{Q}}^{(x)})_{\Sigma^R}$ and $(\hat{\mathbf{P}}^{(x)})_{\Sigma^R}$ are 3×2 matrices with elements $(Q_{iJ}^{(x)})_{\Sigma^R}$ and $(P_{iJ}^{(x)})_{\Sigma^R}$, given by (5.8.48). The expression (5.8.49) corresponds to that given by Chapman (in press) for the weighting function in the Maslov integral in anisotropic media.

Note that the Maslov-Chapman integral (5.8.44) can be modified to allow complex-valued $\mathbf{M}^{(y)}(R_\gamma)$ with a positive definite $\mathbf{M}^{(y)}(R_\gamma)$. The expression (5.8.49) for the weighting function has then an additional term containing $\mathbf{M}^{(y)}(R_\gamma)$. Similarly, even the exponent has an additional imaginary quadratic term. Actually, the integral (5.8.44) then represents the summation of paraxial Gaussian beams. A detailed derivation and discussion of the relevant integrals can be found in Červený (2000). All quantities in the integral are expressed in global Cartesian coordinates, not in wavefront orthonormal coordinates.

The Maslov method was introduced to seismology and applied to the computation of synthetic seismograms (Maslov seismograms) by Chapman. The Maslov integrals are also known as *Maslov-Chapman integrals*, to acknowledge the basic contribution of C. H. Chapman to the application of the method to seismic wave propagation. The relevant expressions for the Maslov-Chapman integrals have mostly been derived in a more sophisticated way than outlined here, using pseudodifferential and Fourier integral operators in a mixed x_i-p_i-phase space. The Maslov-Chapman technique yields satisfactory results even in certain singular regions of the ray method (Airy caustics, critical regions).

For a more detailed derivation and discussion of the Maslov-Chapman integrals and for many applications in seismology, see Chapman and Drummond (1982), Klimeš (1984b), Ziolkowski and Deschamps (1984), Chapman (1985, in press), Thomson and Chapman (1985), Tromp and Dahlen (1993), Guest and Kendall (1993), Kendall and Thomson (1993), Brown (1994), Huang and West (1997), Huang, West, and Kendall (1998), and Huang, Kendall, Thomson, and West (1998). Klimeš (1984b) was the first who explained the Maslov-Chapman integrals as a limiting case of the Gaussian beam summation integrals, for $\operatorname{Im} \mathbf{M}^{(y)} \to \mathbf{0}$.

A note on the problem of pseudocaustic points. Integral (5.8.44) fails at pseudocaustic points at which $\det[\mathbf{M}^{(y)}(R_\gamma) - \mathbf{M}^{(y)a}(R_\gamma)] = 0$; see (5.8.28). We remind the reader that the standard ray-theory expressions fail at caustic points, where $\det \mathbf{Q}^{(y)a}(R_\gamma) = 0$. If the caustic and pseudocaustic points are well separated, it is possible to use the superposition integrals at caustic points (and close to them), and the standard ray-theory expressions at pseudocaustic points (and close to them). It is possible to obtain uniformly valid solutions by blending together the two asymptotic solutions with weighting functions (also called neutralizers); see Maslov (1965) and Chapman and Drummond (1982). Here we shall only consider the situation appropriate to our case. We introduce two smooth weighting functions $w_1(\gamma_1, \gamma_2)$ and $w_2(\gamma_1, \gamma_2)$ on target surface Σ^R, corresponding to the end points R_γ of the rays $\Omega(\gamma_1, \gamma_2)$ on Σ^R. We choose $w_1(\gamma_1, \gamma_2)$ and $w_2(\gamma_1, \gamma_2)$ so that $w_1(\gamma_1, \gamma_2) + w_2(\gamma_1, \gamma_2) = 1$ everywhere on Σ^R. We then assign the weighting functions to the two asymptotic solutions and take them to be unity where the relevant asymptotic solution is valid, to be zero where it is invalid, and to vary smoothly from 0 to 1 inbetween. This method has yielded very accurate results in many important applications. Nevertheless, the pseudocaustic points may appear very close to the caustic points in certain situations, and the phase partitioning becomes difficult. For a detailed discussion, refer to Kendall and Thomson (1993), where also other methods of treating the pseudocaustic problem are proposed.

5.8.6 Summation in 2-D Models

We shall now consider a 2-D model, in which the structural parameters do not depend on one Cartesian coordinate, say x_2, but may depend on the remaining coordinates x_1 and x_3. We shall further consider a one-parameteric system of planar rays $\Omega(\gamma_1)$, situated in plane $\Sigma^{\parallel}$, and assume that the ray plane $\Sigma^{\parallel}$ coincides with plane $x_2 = 0$. In isotropic media, such a system of planar rays $\Omega(\gamma_1)$ always exists; it is only necessary to consider the initial slowness vectors $\vec{p}_0$ in such a way that $p_{20} = 0$. In anisotropic media, however, the rays with $p_{20} = 0$ form a system of planar rays $\Omega(\gamma_1)$ situated in $\Sigma^{\parallel}$ only exceptionally, in case of some simpler types of anisotropy. In general, the rays in anisotropic media deviate from the plane $\Sigma^{\parallel}$ even if $p_{20} = 0$. See Section 3.6.5.

We further assume that all 2×2 matrices $\mathbf{Q}^{(y)}$, $\mathbf{P}^{(y)}$, $\mathbf{M}^{(y)}$, $\mathbf{Q}^{(y)a}$, $\mathbf{M}^{(y)a}$, and $\mathbf{P}^{(y)a}$ can be diagonalized along rays $\Omega(\gamma_1)$, so that $\det \mathbf{Q}^{(y)} = Q^{\parallel} Q^{\perp}$, $\det \mathbf{Q}^{(y)a} = Q^{\parallel a} Q^{\perp a}$, $\det \mathbf{P}^{(y)} = P^{\parallel} P^{\perp}$, and so on. For isotropic media, such a diagonalization is always possible; see Section 4.13. For anisotropic media, this is not the general case, but we shall consider here only such simple models in which this is possible. We can then introduce 2-D paraxial ray approximations and 2-D paraxial Gaussian beams by putting $Q^{\perp} = Q^{\perp a}$, $P^{\perp} = P^{\perp a}$, and $M^{\perp} = M^{\perp a}$. Thus, $Q^{\perp}$, $P^{\perp}$, and $M^{\perp}$ are real-valued even for paraxial Gaussian beams. Quantities $Q^{\parallel}$, $P^{\parallel}$, and $M^{\parallel}$, however, are real-valued for paraxial ray approximations, but complex-valued for paraxial Gaussian beams. The system of paraxial ray approximations connected with a selected ray $\Omega(\gamma_1)$ is then one-parameteric, with one free parameter $M^{\parallel}$. Similarly, the system of paraxial Gaussian beams connected with a selected ray $\Omega(\gamma_1)$ is two-parameteric, with two free parameters $\operatorname{Re} M^{\parallel}$ and $\operatorname{Im} M^{\parallel} > 0$. These parameters may be specified at any point of ray $\Omega(\gamma_1)$.

We shall now assume that the receiver point R is situated in plane $\Sigma^{\parallel}$. The superposition integral can be derived in the same way as in Section 5.8.3. Instead of (5.8.23), however, we shall exploit an analogous single integral

$$\int_{-\infty}^{\infty} \exp\left[\tfrac{1}{2}i\omega x^2 W\right] \mathrm{d}x = (2\pi/\omega)^{1/2}[-iW]^{-1/2}. \tag{5.8.50}$$

Here W is a real-valued or a complex-valued constant different from zero. If W is complex-valued, its imaginary part must be positive, $\operatorname{Im} W > 0$. The argument of $[-iW]^{1/2}$ is given by the following relations:

$$\begin{aligned}
\operatorname{Re}[-iW]^{1/2} &> 0 & &\text{for } \operatorname{Im} W > 0, \\
[-iW]^{1/2} &= |W|^{1/2} \exp\left[-i\tfrac{\pi}{4}\operatorname{sgn} W\right] & &\text{for } \operatorname{Im} W = 0.
\end{aligned} \tag{5.8.51}$$

The 2-D superposition integral, analogous to (5.8.40), then reads

$$\begin{aligned}
u_i^{(x)}(R, \omega) = (\omega/2\pi)^{1/2} \int_{\mathcal{D}} U_i^{ray}(R_\gamma)[-i\mathcal{M}^{\parallel}]^{1/2}|Q^{\parallel a}(R_\gamma)| \\
\times \exp[i\omega T(R, R_\gamma)]\mathrm{d}\gamma_1.
\end{aligned} \tag{5.8.52}$$

Here

$$\mathcal{M}^{\parallel}(R_\gamma) = M^{\parallel}(R_\gamma) - M^{\parallel a}(R_\gamma), \tag{5.8.53}$$

and the argument of $[-i\mathcal{M}^{\parallel}]^{1/2}$ is given by (5.8.51), for $W = \mathcal{M}^{\parallel}(R_\gamma)$. The travel-time function $T(R, R_\gamma)$, expressed in Cartesian coordinates, follows immediately from (5.8.43) for $x_2(R) - x_2(R_\gamma) = 0$, and from the appropriately simplified expression for $\hat{\mathbf{M}}^{(x)}(R_\gamma)$, given by (4.14.47) and (4.2.44). For isotropic media, see Section 4.13.5, particularly (4.13.58) and (4.13.39).

The discussion of the superposition integral (5.8.52) remains practically the same as the discussion of (5.8.40). The superposition integral (5.8.52) may be used both for the summation of paraxial ray approximations ($\operatorname{Im} M^{\parallel}(R_\gamma) = 0$) and for the summation of paraxial Gaussian beams ($\operatorname{Im} M^{\parallel}(R_\gamma) > 0$). The superposition of paraxial ray approximations fails at pseudocaustic points R, at which $M^{\parallel}(R) = M^{\parallel a}(R)$. The summation of paraxial Gaussian beams, however, remains regular even at pseudocaustic points because $\operatorname{Im} M^{\parallel}(R) > 0$ and $\operatorname{Im} M^{\parallel a}(R) = 0$ in this case.

The superposition integral (5.8.52) may be used both for a strictly 2-D case (a line source perpendicular to $\Sigma^{\parallel}$ in a 2-D model) and for a 2.5-D case (point source in the plane $\Sigma^{\parallel}$ in a 2-D model). The integral remains exactly the same in both cases; only the geometrical spreading in the expression for the ray amplitudes $U_i^{ray}(R_\gamma)$ are different. See Sections 4.13.4 and 5.2.13.

Commonly, the summation in (5.8.52) is performed along a target line C^R, at which the receiver R and the end points R_γ of rays $\Omega(\gamma_1)$ are situated. The specification of $M^{\parallel}$ is also fully analogous to the specification of matrix $\mathbf{M}^{(y)}$, discussed in Section 5.8.3.4. We only replace the 2×2 matrices in (5.8.37) by their upper diagonal elements.

Superposition integral (5.8.52) can be easily simplified for 1-D models in which the structural parameters depend on Cartesian coordinate x_3 only (say depth) and in which the ray plane $\Sigma^{\parallel}$ again coincides with plane $x_2 = 0$. Let us consider Im $M^{\parallel}(R_\gamma) = 0$, that is, the summation of paraxial ray approximations. The resulting superposition integrals are then usually called the WKBJ integrals. It is common to express these integrals directly in the time domain; see Section 6.2.6. The WKBJ integrals were first derived by Chapman (1976a, 1978), using the transform methods. Chapman also proposed that the resulting synthetic seismograms be called WKBJ seismograms. For more details, see also Dey-Sarkar and Chapman (1978), Chapman and Drummond (1982), Chapman and Orcutt (1985), Chapman (1985), and Garmany (1988). An efficient and general computer program to compute the WKBJ seismograms in vertically inhomogeneous isotropic layered structures is described by Chapman, Chu, and Lyness (1988). See Singh and Chapman (1988) for WKBJ method in anisotropic media and Wang and Dahlen (1994) for surface waves.

5.8.7 Alternative Versions of the Superposition Integral

The general superposition integral (5.8.40) can be expressed in many alternative forms. For example, $U_i^{ray}(R_\gamma)$ can be replaced by $U_i^{(x)}(R_\gamma)$ using (5.8.16), or by the spreading-free amplitude $U_i^{\Omega}(R_\gamma)$, using (5.8.6) and (5.8.7). Here we shall present one version of the superposition integral, which has been used in seismological literature.

Using (5.8.36), we express $\mathcal{M}^{(y)}(R_\gamma)$, given by (5.8.21), in the following form:

$$\mathcal{M}^{(y)}(R_\gamma) = \left[\mathbf{P}_1 + \mathbf{P}_2 \mathbf{M}^{(y)}(S_\gamma)\right]\left[\mathbf{Q}_1 + \mathbf{Q}_2 \mathbf{M}^{(y)}(S_\gamma)\right]^{-1}$$
$$-\left[\mathbf{P}_1 + \mathbf{P}_2 \mathbf{M}^{(y)a}(S_\gamma)\right]\left[\mathbf{Q}_1 + \mathbf{Q}_2 \mathbf{M}^{(y)a}(S_\gamma)\right]^{-1}.$$

Here $\mathbf{P}_1, \mathbf{P}_2, \mathbf{Q}_1$, and $\mathbf{Q}_2$ are the minors of the ray propagator matrix $\mathbf{\Pi}(R_\gamma, S_\gamma)$ for anisotropic media, as in (5.8.36). Because $\mathbf{M}^{(y)}(R_\gamma)$ is symmetric, $\mathbf{M}^{(y)}(R_\gamma) = \mathbf{M}^{(y)T}(R_\gamma)$, and the preceding equation yields

$$\mathcal{M}^{(y)}(R_\gamma) = \left(\mathbf{Q}_1 + \mathbf{Q}_2 \mathbf{M}^{(y)}(S_\gamma)\right)^{-1T}$$
$$\times \left[(\mathbf{P}_1 + \mathbf{P}_2 \mathbf{M}^{(y)}(S_\gamma))^T (\mathbf{Q}_1 + \mathbf{Q}_2 \mathbf{M}^{(y)a}(S_\gamma))\right.$$
$$\left. - (\mathbf{Q}_1 + \mathbf{Q}_2 \mathbf{M}^{(y)}(S_\gamma))^T (\mathbf{P}_1 + \mathbf{P}_2 \mathbf{M}^{(y)a}(S_\gamma))\right]$$
$$\times \left(\mathbf{Q}_1 + \mathbf{Q}_2 \mathbf{M}^{(y)a}(S_\gamma)\right)^{-1}.$$

Using the symplectic properties (4.3.16) of the ray propagator matrix, which are valid for both isotropic and anisotropic media, we can simplify the expression in the square brackets considerably; it reads $\mathbf{M}^{(y)}(S_\gamma) - \mathbf{M}^{(y)a}(S_\gamma)$. Using also the second equation in (5.8.36), we then obtain

$$
\mathcal{M}^{(y)}(R_\gamma) = \left(\mathbf{Q}^{(y)}(S_\gamma)\mathbf{Q}^{(y)-1}(R_\gamma)\right)^T \mathcal{M}^{(y)}(S_\gamma)\left(\mathbf{Q}^{(y)a}(S_\gamma)\mathbf{Q}^{(y)a-1}(R_\gamma)\right).
$$

$$(5.8.54)$$

Here $\mathcal{M}^{(y)}(S_\gamma)$ is again given by (5.8.21). Equation (5.8.54) represents the continuation relation for $\mathcal{M}^{(y)}$ along ray Ω. If we use (5.8.27), we obtain

$$
\Phi(\gamma_1, \gamma_2) = (\omega/2\pi)\left[-\det \mathcal{M}^{(y)}(S_\gamma)\right]^{1/2}\left|\det \mathbf{Q}^{(y)a}(S_\gamma)\right|\Psi_0(\gamma_1, \gamma_2),
$$

$$(5.8.55)$$

where $\Psi_0(\gamma_1, \gamma_2)$ is given by (5.8.17). If we use (5.8.19), Equation (5.8.55) yields a simple expression for the weighting function $\Psi(\gamma_1, \gamma_2)$:

$$
\Psi(\gamma_1, \gamma_2) = (\omega/2\pi)\left[-\det \mathcal{M}^{(y)}(S_\gamma)\right]^{1/2}\left|\det \mathbf{Q}^{(y)a}(S_\gamma)\right|.
$$

$$(5.8.56)$$

The superposition integral (5.8.40) then reads

$$
u_i^{(x)}(R, \omega) = (\omega/2\pi)\iint_{\mathcal{D}} U_i^{(x)}(R_\gamma)\left[-\det \mathcal{M}^{(y)}(S_\gamma)\right]^{1/2}\left|\det \mathbf{Q}^{(y)a}(S_\gamma)\right|
$$
$$
\times \exp[i\omega T(R, R_\gamma)]\,\mathrm{d}\gamma_1\,\mathrm{d}\gamma_2,
$$

$$(5.8.57)$$

where $\mathcal{M}^{(y)}(S_\gamma)$ is again given by (5.8.41) with (5.8.42), only it is taken at the initial point S_γ. The travel-time function $T(R, R_\gamma)$ is the same as in (5.8.43).

Recall that $U_i^{(x)}(R_\gamma)$ in the superposition integral (5.8.57) represents the amplitude of a paraxial ray approximation or of a paraxial Gaussian beam (see (5.8.2)), not the ray amplitude. Actually, (5.8.57) represents the original superposition integral (5.8.13).

The superposition integral (5.8.57) is more transparent than (5.8.40) if we wish to specify $\mathbf{M}^{(y)}$ at S_γ. The determination of $U_i^{(x)}(R_\gamma)$ is, however, slightly more involved than the determination of $U_i^{ray}(R_\gamma)$ and requires the computation of the whole ray propagator matrix along ray Ω. Another advantage of the superposition integral (5.8.40) is that the computer routines for the determination of $U_i^{ray}(R_\gamma)$ are broadly available in various program packages, but not the routines for $U_i^{(x)}$.

5.8.8 Phase Shift Due to Caustics. Derivation

The expressions for the paraxial ray approximations and paraxial Gaussian beams, derived in Sections 5.8.1 and 5.8.2, can also be used to derive simple expressions for contribution Δk to the KMAH index k, when the ray passes through a caustic point. The relevant phase shift at the caustic point is then $-(\pi/2)\Delta k$. Here we shall derive expressions for Δk at caustic points situated in general anisotropic inhomogeneous media. By simple specification of these expressions, we also obtain analogous (simpler) expressions for isotropic inhomogeneous media.

The importance of the KMAH index in computing high-frequency seismic wavefields along rays has been emphasized in several places in this book. Detailed recipes for computing it have also been presented. This applies mainly to Section 4.12 for isotropic inhomogeneous media, and to Section 4.14.13 for anisotropic inhomogeneous media. The derivation of these recipes, however, was not given in those sections because the necessary background for their derivation had not yet been developed there. We shall use the theory given in Sections 5.8.1 and 5.8.2 to present a simple derivation of convenient expressions for Δk. We shall follow mainly the approach proposed by Bakker (1998) because it is simple and transparent. For other relevant references, see Sections 4.12 and 4.14.13.

As we can see from (5.8.5) through (5.8.7), the singularities at caustic points are contained in the factor

$$[\det \mathbf{Q}^{(y)}]^{-1/2} \exp[\tfrac{1}{2}i\omega\mathbf{y}^T\mathbf{M}^{(y)}\mathbf{y}]$$
$$= [\det \mathbf{P}^{(y)}]^{-1/2}[\det \mathbf{M}^{(y)}]^{1/2} \exp[\tfrac{1}{2}i\omega\mathbf{y}^T\mathbf{M}^{(y)}\mathbf{y}],$$

because $\mathbf{M}^{(y)} = \mathbf{P}^{(y)}\mathbf{Q}^{(y)-1}$. The notation here is the same as in Section 5.8.1. We shall assume, for a while, that $\det \mathbf{P}^{(y)}(s_c) \neq 0$ at the caustic point $s = s_c$ on the ray. Consequently, we shall discuss the behavior of the amplitude factor

$$\mathcal{A}(s) = [\det \mathbf{M}^{(y)}(s)]^{1/2} \exp[\tfrac{1}{2}i\omega\mathbf{y}^T\mathbf{M}^{(y)}(s)\mathbf{y}], \tag{5.8.58}$$

in the vicinity $s \sim s_c$ of the caustic point. (We shall briefly return to the case of $\det \mathbf{P}^{(y)}(s_c) = 0$ later.) We also denote the two eigenvalues of the 2×2 symmetric matrix $\mathbf{M}^{(y)}(s)$ by $M_1^{(y)}(s)$ and $M_2^{(y)}(s)$, and the relevant normalized eigenvectors by $\mathbf{m}_1^{(y)}(s)$ and $\mathbf{m}_2^{(y)}(s)$, $\mathbf{m}_1^{(y)T}\mathbf{m}_1^{(y)} = 1$ and $\mathbf{m}_2^{(y)T}\mathbf{m}_2^{(y)} = 1$.

Amplitude factor $\mathcal{A}(s)$ remains formally the same both for the paraxial ray approximation and for the paraxial Gaussian beams. The only difference is that the 2×2 matrix $\mathbf{M}^{(y)}(s)$ is real-valued for the paraxial ray approximation, but complex-valued for paraxial Gaussian beams. There is no jump in $[\det \mathbf{M}^{(y)}(s)]^{1/2}$ at the caustic point $s = s_c$ for paraxial Gaussian beams as for the paraxial ray approximation; the factor $[\det \mathbf{M}^{(y)}(s)]^{1/2}$ varies quite smoothly and continuously for the Gaussian beams there.

Thus, in deriving the phase shift of the amplitude factor at a caustic point $s = s_c$, we can proceed in the following transparent way. We shall first study the amplitude factor $\mathcal{A}(s)$ in the vicinity of the caustic point $s = s_c$ for the paraxial ray approximation. We shall then extend the result to Gaussian beams considering small values of Im $\mathbf{M}^{(y)}(s)$, varying quite smoothly in the vicinity of the caustic point. Finally, we shall apply the limiting process for Im $\mathbf{M}^{(y)}(s) \to 0$. In this way, we shall obtain expressions for Δk at caustic points of any type, both in isotropic and anisotropic structures.

As we shall see later, we shall express Δk in terms of the 2×2 matrix $\mathbf{B}^{(y)}(s_c)$, representing one submatrix of the 4×4 system matrix of the dynamic ray tracing system; see (4.2.32). Alternatively, we can express it in terms of the 2×2 matrix $\mathbf{D}^S(s_c)$, representing the curvature matrix of the local slowness surface at $s = s_c$. In inhomogeneous anisotropic media, both matrices depend on position and the slowness vector components but not on the derivatives of material parameters. Consequently, $\mathbf{B}^{(y)}(s_c)$ and $\mathbf{D}^S(s_c)$ are locally related by (4.14.29), just as in a homogeneous medium. This equation also shows that matrices $\mathcal{U}^{-1}(s_c)\mathbf{B}^{(y)}(s_c)$ and $\mathbf{D}^S(s_c)$ are congruent. Because both matrices are symmetric, the eigenvalues are real-valued. Moreover, the Sylvester law of inertia shows that $\mathrm{Sgn}\,\mathbf{B}^{(y)}(s_c) = \mathrm{Sgn}\,\mathbf{D}^S(s_c)$. We shall not consider the exceptional cases of $\mathbf{D}^S(s_c)$ and $\mathbf{B}^{(y)}(s_c)$ being rank-deficient at the caustic point.

In the following discussion, we shall investigate the phase shifts at caustic points $s = s_c$ of the second order (rank $\mathbf{Q}^{(y)}(s_c) = 0$) and at caustic points $s = s_c$ of the first order (rank $\mathbf{Q}^{(y)}(s_c) = 1$) independently.

1. Caustic points of the second order (rank $\mathbf{Q}^{(y)}(s_c) = 0$). In this case, $\mathbf{Q}^{(y)}(s_c) = \mathbf{0}$, and also $\mathbf{M}^{(y)-1}(s_c) = \mathbf{Q}^{(y)}(s_c)\mathbf{P}^{(y)-1}(s_c) = \mathbf{0}$. The Riccati equation (4.14.50) for $\mathbf{M}^{(y)-1}$ at $s = s_c$ simplifies to $(d\mathbf{M}^{(y)-1}/dT)_{s=s_c} = \mathbf{B}^{(y)}(s_c)$. In the vicinity of caustic point $s = s_c$, this yields $\mathbf{M}^{(y)-1}(s) \doteq (s - s_c)\mathbf{B}^{(y)}(s_c)/\mathcal{U}(s_c)$. Consequently,

$$1/M_I^{(y)}(s) \doteq (s - s_c)\mathcal{U}^{-1}(s_c)B_I^{(y)}(s_c). \tag{5.8.59}$$

This yields the expression for the amplitude factor $\mathcal{A}(s)$,

$$\mathcal{A}(s) = \prod_{I=1}^{2}\left(M_I^{(y)}(s)\right)^{1/2} \exp\left[\tfrac{1}{2}i\omega M_I^{(y)}(s)\left|\mathbf{y}^T \mathbf{m}_I^{(y)}(s)\right|^2\right]; \tag{5.8.60}$$

see (5.8.58).

We shall now extend Equations (5.8.59) and (5.8.60), derived for paraxial ray approximations, to broad Gaussian beams. In this case, $M_I^{(y)}(s)$ has a small positive imaginary part along the whole ray; see Section 5.8.2. This can also be clearly verified in the exponential factor in (5.8.60). Consequently, $1/M_I^{(y)}(s)$ has a small negative imaginary part, and (5.8.59) should be modified to read:

$$1/M_I^{(y)}(s) \doteq (s - s_c)\mathcal{U}^{-1}(s_c)B_I^{(y)}(s_c) - i\Delta, \tag{5.8.61}$$

where Δ is a constant, $\Delta > 0$. We shall now discuss the cases of $B_I^{(y)}(s_c) > 0$ and $B_I^{(y)}(s_c) < 0$ independently.

a. For $B_I^{(y)}(s_c) > 0$, the real part of $1/M_I^{(y)}(s)$ is negative for $s < s_c$ and positive for $s > s_c$; see (5.8.61). The small imaginary part of $1/M_I^{(y)}(s)$ is negative, both for $s < s_c$ and $s > s_c$. Consequently, square root $(1/M_I^{(y)}(s))^{1/2}$ should be almost negative imaginary (with a small positive real part) for $s < s_c$ and almost positive real (with a small negative imaginary part) for $s > s_c$. In the limiting process for Im $\mathbf{M}^{(y)}$ vanishing, we obtain $(1/M_I^{(y)}(s))^{1/2}$ negative imaginary for $s < s_c$ and positive real for $s > s_c$. Consequently, $|1/M_I^{(y)}(s)|^{1/2}$ should be multiplied by $\exp(i\pi/2)$ at $s = s_c$ to account for the phase jump.

b. For $B_I^{(y)}(s_c) < 0$, the real parts of $1/M_I^{(y)}(s)$ have signs opposite to those in (a); see (5.8.59). In the same way as in (a) we obtain the final result that $|1/M_I^{(y)}(s)|^{1/2}$ should be multiplied by $\exp(-i\pi/2)$ at $s = s_c$ to account for the phase jump.

Amplitude factor $\mathcal{A}(s)$, however, has square roots $(M_I^{(y)}(s))^{1/2}$; not $(1/M_I^{(y)}(s))^{1/2}$; see (5.8.60). Consequently, square root $|M_I^{(y)}(s)|^{1/2}$ should be multiplied at $s = s_c$ by $\exp(-i\pi/2)$ for $B_I^{(y)}(s_c) > 0$, and by $\exp(+i\pi/2)$ for $B_I^{(y)}(s_c) < 0$, to account for the phase jump. Moreover, we must take into account both $(M_1^{(y)}(s))^{1/2}$ and $(M_2^{(y)}(s))^{1/2}$ in (5.8.60). Thus, the final expression for Δk at the caustic point of the second order is

$$\Delta k = \sum_{I=1}^{2}\operatorname{sgn}\left(B_I^{(y)}(s_c)\right) = \operatorname{Sgn}\left(\mathbf{B}^{(y)}(s_c)\right) = \operatorname{Sgn}\mathbf{D}^S(s_c). \tag{5.8.62}$$

All three expressions are alternative and give the same result. The phase shift factor at caustic point $s = s_c$ is then given by $\exp[-i(\pi/2)\Delta k]$. For the positive-definite slowness surface $\mathbf{D}^S(s_c)$, we obtain $\Delta k = 2$; for the negative definite, we obtain $\Delta k = -2$; and for eigenvalues of different signs, $\Delta k = 0$.

2. Caustic points of the first order (rank $\mathbf{Q}^{(y)}(s_c) = 1$). Because $\mathbf{P}^{(y)}(s_c)$ is assumed to be of full rank, rank $\mathbf{M}^{(y)-1}(s_c) = 1$. Only one of the two eigenvalues $M_1^{(y)}(s_c)$ and $M_2^{(y)}(s_c)$ is singular at s_c, say $M_1^{(y)}(s_c)$. The relevant eigenvector is denoted by $\mathbf{m}_1^{(y)}(s_c)$. Consequently, also $M_1^{(y)-1}(s_c) = 0$, $\mathbf{M}^{(y)-1}(s_c)\mathbf{m}_1^{(y)}(s_c) = \mathbf{0}$, $\mathbf{m}_1^{(y)T}(s_c)\mathbf{M}^{(y)-1}(s_c) = \mathbf{0}$. At

$s = s_c$, we further find that

$$\mathrm{d}\big(1/M_1^{(y)}(s)\big)\big/\mathrm{d}s = \mathrm{d}\big(\mathbf{m}_1^{(y)T}\mathbf{M}^{(y)-1}\mathbf{m}_1^{(y)}\big)\big/\mathrm{d}s = \mathbf{m}_1^{(y)T}\big(\mathrm{d}\mathbf{M}^{(y)-1}\big/\mathrm{d}s\big)\mathbf{m}_1^{(y)}.$$

At $s = s_c$, the matrix Riccati equation (4.14.50) then yields

$$\mathrm{d}\big(1/M_1^{(y)}(s)\big)\big/\mathrm{d}s = \mathcal{U}^{-1}\mathbf{m}_1^{(y)T}\mathbf{B}^{(y)}\mathbf{m}_1^{(y)},$$

and, finally, in the vicinity of caustic point $s = s_c$,

$$1\big/M_1^{(y)}(s) \doteq (s - s_c)\mathcal{U}^{-1}(s_c)\big(\mathbf{m}_1^{(y)T}(s_c)\mathbf{B}^{(y)}(s_c)\mathbf{m}_1^{(y)}(s_c)\big). \tag{5.8.63}$$

Equation (5.8.63) is analogous to (5.8.59). Consequently, we can proceed in the same way as for caustic points of the second order and obtain the final expression for Δk at the caustic point of the first order:

$$\Delta k = \mathrm{sgn}\big[\mathbf{m}_1^{(y)T}(s_c)\mathbf{B}^{(y)}(s_c)\mathbf{m}_1^{(y)}(s_c)\big] = \mathrm{sgn}\big[\mathbf{m}_1^{(y)T}(s_c)\mathbf{D}^S(s_c)\mathbf{m}_1^{(y)}(s_c)\big]. \tag{5.8.64}$$

Both expressions in (5.8.64) are alternative and give the same result. The phase shift at the caustic is given by the factor $\exp[-\mathrm{i}(\pi/2)\Delta k]$. For positive definite $\mathbf{D}^S$ (convex slowness surface), $\Delta k = 1$; for negative definite $\mathbf{D}^S$ (concave slowness surface), $\Delta k = -1$. For $\mathbf{D}^S(s_c)$ neither positive definite nor negative definite, $\Delta k = 1$ for certain directions of eigenvector $\mathbf{m}_1^{(y)}$ and $\Delta k = -1$ for its other directions, depending on (5.8.64).

Bakker (1998) proved that the same expressions for Δk as in (5.8.64) are obtained even if rank $\mathbf{P}^{(y)}(s_c) = 1$. In this case, the preceding approach cannot be used, but the result is again given by (5.8.64). The case of rank $\mathbf{P}^{(y)}(s_c) = 1$ plays an important role in 2-D computations. For more details, see Bakker (1998).

In isotropic media, the matrix of the curvature of the slowness surface is always positive definite. Thus, $\Delta k = 2$ at caustic points of the second order, and $\Delta k = 1$ at caustic points of the first order. In isotropic media, the (anomalous) phase shift $T^c = \frac{1}{2}\pi$, corresponding to $\Delta k = -1$, is not possible. The same applies to qP waves in anisotropic media. For a more detailed discussion, see Section 4.14.13.

5.9 Validity Conditions and Extensions of the Ray Method

The great virtue of the ray method resides in its universality, effectiveness, and conceptual clarity and its ability to investigate various waves of importance in seismology and seismic prospecting separately from other waves. Although its accuracy is limited, it is the only method that is able to give an approximate answer to many problems of HF seismic body wave propagation in structurally complex media.

The ray method also has another great advantage: it offers heuristic principles for many other asymptotic approximate methods of propagation, scattering, and diffraction of HF seismic body waves, which lead to more accurate results even in situations where the standard ray method fails. In all these methods, the rays form some coordinate frame for the evaluation of HF seismic wavefields, although they may lose their physical meaning of trajectories along which the energy of HF seismic body waves propagates.

In this section, we shall first discuss the validity conditions of the ray method. After this, we shall briefly discuss the computation of wavefields in various singular regions, and various modifications and extensions of the ray method.

5.9.1   Validity Conditions of the Ray Method

The ray method is only approximate. It is applicable only for high frequencies ω, or, alternatively, for small wavelengths λ. It would be very useful to know the validity conditions under which the ray method can be applied to compute seismic wavefields. These validity conditions usually are only of a qualitative, not quantitative character. We shall not try to discuss the validity conditions of the ray method in greater detail; we shall merely make some general remarks concerning this problem. The validity conditions we shall present are applicable to scalar wavefields. Analogical validity conditions, however, are applicable even to seismic vectorial wavefields. It is only necessary to supplement them by some additional validity conditions which deal with the separation of P and S waves. For a more detailed treatment of the validity conditions of the ray method, see Ben-Menahem and Beydoun (1985), Beydoun and Ben-Menahem (1985), Chapman (1985), Fradkin (1989), Popov and Camerlynck (1996), and particularly Kravtsov and Orlov (1980).

a. GENERAL VALIDITY CONDITIONS

We shall first describe some *general validity conditions* that are of a fully qualitative character. The three most important general validity conditions are usually formulated in the following way.

1. The wavelength λ of the wave under consideration must be considerably smaller than any characteristic quantity of length dimension l_j $(j = 1, 2, \ldots)$ in the problem under study,

$$\lambda \ll l_1, l_2, \ldots. \tag{5.9.1}$$

   For example, characteristic quantities l_j are radii of curvature of interfaces, some scale lengths of the inhomogeneity of the medium of type $V/|\nabla V|$, where V is the velocity, and similar scale lengths of the inhomogeneity of density. It follows from the study of certain canonical problems and from comparisons with exact solutions that the ray method can sometimes be applied even to situations in which some of the characteristic lengths of a problem are not much larger than the wavelength. Thus, the list of quantities l_j in condition (5.9.1) must be specified in detail for each problem under study.

2. The ray method fails in the vicinity of surfaces S along which the ray field of the wave under consideration is not regular. Let n be the distance from surface S. The validity condition then reads

$$\lambda \ll n. \tag{5.9.2}$$

   Examples of surfaces S are caustic surfaces and boundaries of shadow zones.

3. The ray method is not applicable when the length L of the ray trajectory of the wave under consideration between the source and the receiver is too large. The estimates based on the theorem of mean value lead to the condition

$$\lambda \ll l_0^2/L. \tag{5.9.3}$$

   Here l_0 has the same meaning as l_j in Condition 1.

Is it possible to remove or at least reduce the aforementioned restrictions? Because validity condition (5.9.1) expresses the high-frequency character of the wavefield, it is *not possible* to remove it in the framework of high-frequency asymptotics. It is only possible to specify the qualitative conditions in a more quantitative way, by comparing the ray solutions with the exact solutions for some canonical cases.

The problem of doing away with condition (5.9.3) has not been solved yet. Note that validity condition (5.9.3) practically coincides with the condition $L \ll l^2/\lambda$, which applies to waves in random media, and where l denotes certain measures of random inhomogeneities. Comparing these conditions, we can see a similar role of random and deterministic inhomogeneities in the validity conditions that restrict the length of the ray path of short waves. The specific criteria related to possible chaotic behavior of rays have not yet been developed.

To do away with condition (5.9.2), various modifications and extensions of the ray method have been proposed. These extensions give good results also in certain singular regions where the ray method fails. The problem of singularities, however, still remains one of the most serious problems in the application of the ray method. In laterally inhomogeneous media, various singular regions may overlap, and the application of local extensions becomes more complicated.

b. FRESNEL VOLUME VALIDITY CONDITIONS

Using the concept of Fresnel volumes, we can express the validity conditions in a more quantitative way. The validity conditions in relation to Fresnel volumes for scalar wavefields were investigated in great detail by Kravtsov and Orlov. We shall present a short review of their investigations, taken from the book by Kravtsov and Orlov (1980). They present *two validity conditions*.

1. The parameters of the medium and the parameters of the wave under consideration (amplitude and slowness vector) must not vary significantly over the cross section of the Fresnel volume.

   If we denote the maximum perpendicular dimension of the Fresnel volume r_F, we can write

   $$r_F \left| \frac{\nabla_\perp V}{V} \right| \ll 1, \qquad r_F \left| \frac{\nabla_\perp P}{P} \right| \ll 1, \qquad r_F \left| \frac{\nabla_\perp p_i}{p} \right| \ll 1, \qquad (5.9.4)$$

   and so on. Here $\nabla_\perp$ denotes the gradient perpendicular to the ray, P is the scalar amplitude, and p_i are the components of the slowness vector $\vec{p}$. The physical meaning of inequalities (5.9.4) is obvious. Criteria (5.9.4) are expressed in the form given by Kravtsov and Orlov. It would be possible to rewrite them in our notation, but we shall not do so here.

2. The Fresnel volumes of two rays belonging to one and the same congruence and reaching one and the same receiver point must not penetrate into one another significantly. The criterion may then be expressed in the following way:

   $$\delta V_F \ll V_F, \qquad (5.9.5)$$

   where V_F is the sum of Fresnel volumes corresponding to both rays arriving at the receiver point, and δV_F is the common part of these Fresnel volumes. The same validity condition may be used even for more rays arriving at the receiver point. For a more detailed explanation, see Kravtsov and Orlov (1980). Kravtsov and Orlov (1980) use the criteria (5.9.4) and (5.9.5) to compute and study the *caustic zones*, where the ray approximation fails.

According to Kravtsov and Orlov, criteria (5.9.4) and (5.9.5) are *universal and sufficient* in the scalar ray method. Kravtsov and Orlov investigated many special cases and found that validity conditions (5.9.4) and (5.9.5) can replace all other forms of validity conditions of the ray method proposed by other authors.

5.9.2 Singular Regions. Diffracted Waves

The ray field corresponding to an elementary wave specified by ray parameters γ_1 and γ_2 is called *regular* in spatial region S if it covers continuously and uniquely region S with rays, that is, if one and only one ray passes through any point of the region.

In smooth media without structural interfaces, the regularity of the ray field in S is connected mainly with the behavior of the ray Jacobian $J = \partial(x, y, z)/\partial(\gamma_1, \gamma_2, \gamma_3)$ at S. The *caustic surfaces*, along which J vanishes, are *singular*, and the vicinities of such surfaces represent *singular regions* of the ray method. The validity conditions of the ray method (5.9.2) are not satisfied along the singular surfaces.

In the case of media with structural interfaces, other singular surfaces, not connected with caustics, also play an important role. Let us name here the *critical surfaces* of reflected waves, separating the subcritical region from the postcritical region. In the postcritical region, the head wave exists in addition to the reflected wave. Validity condition is not satisfied along the critical surface. Other examples are the *boundary surfaces* separating the illuminated regions from the Fresnel shadows connected with some screening bodies or interfaces. Validity conditions are again not satisfied along the boundary surfaces so that the boundary surfaces are singular.

The three types of singular surfaces described here and the relevant singular regions will be briefly discussed in this section. Before we start the discussion, here are a few general remarks.

It is obvious that the ray-series solutions cannot be applied in singular regions. The zeroth-order approximation fails completely, and the higher order terms of the ray series do not improve the accuracy, but instead make it even worse. The problem of finding the high-frequency asymptotic in the singular regions of the ray method is not simple. Extensive literature has been devoted to uniform and local asymptotics valid in singular regions. *Uniform asymptotics* are valid everywhere, both in the singular region and outside it. In the regular region outside the singular region, it yields the ray solution. The uniform asymptotic expansion, however, is often rather complicated or is not known at all. It is then useful to look for simpler *local asymptotics*, valid in the singular region, but failing in the regular region outside it. Thus, the local asymptotic must be combined with standard ray formulae. The local asymptotic is used in the singular region and the ray equations outside it. Local asymptotics are very convenient if the regular region overlaps the region of validity of the local asymptotics: The matching of both solutions at the boundary of the singular region is then simple.

In general laterally varying layered structures, the algorithms based on the application of local asymptotics may sometimes be rather complicated. There are two main reasons for this.

i. There are *various types* of singular regions. Each singular region requires a different local asymptotic. This leads to cumbersome alorithms.

ii. The singular regions in 3-D laterally varying layered structures often *overlap*. In the region in which two singular regions overlap, the local asymptotics corresponding to the individual singular regions are usually no longer valid; more general asymptotics are required. Thus, the algorithms based on local asymptotics are useful in many simple situations, but, in principle, they cannot be complete. For more details, see items 1, 2, and 3 in the next part of this section. Under item 3, a very general approach based on the geometric theory of diffraction is also briefly described.

The wavefields in singular regions are also briefly discussed in other parts of this text. See Section 3.2.3 on anomalous rays in layered structures, Sections 5.6.8 and 5.7.11 for modified forms of the ray series, particularly for the ray method with a complex eikonal, and Section 5.8 for the uniform asymptotics based on the summation of paraxial Gaussian beams and on the summation of paraxial ray approximations.

1. CAUSTIC REGION

Point C of ray Ω at which the ray Jacobian vanishes, $J(C) = 0$, is called the *caustic point*. At the caustic point, the ray amplitude of the wave under consideration is infinite, and the wavefield is singular. The ray-series method cannot be used to calculate the ray amplitudes not only directly at the caustic point, but also in its vicinity, called the *caustic region*. Outside the caustic region, however, the zeroth-order ray expressions for amplitudes are again valid. These expressions at both sides of the caustic point may be connected using the KMAH index.

In space, caustic points form caustic surfaces. The shape of caustic surfaces depends on the medium and on the position of the source. Caustic surfaces represent the *envelopes of rays*. Caustic surfaces may have singularities like cusps and swallow-tails. The part of the caustic surface that is locally smooth in some region is usually called *a simple caustic* in that region.

From a physical point of view, caustic point C of ray Ω is situated at a point where ray Ω is tangent to the caustic surface. In the case of a simple caustic, the caustic surface separates an *illuminated region* from the *caustic shadow*. In the illuminated region, there are two rays at any point situated close to the caustic surface (the ray approaching and leaving the caustic). In the shadow region, there are no real-valued rays; only complex-valued rays may penetrate there. In the case of a more complicated caustic surface, there may be three or even more intersecting rays at any point situated on one side of the caustic surface, with a lower member of rays on the other side of the caustic surface.

Caustic surfaces are common even in smooth media without structural interfaces. In fact, they represent the only singular surfaces of the ray method in smooth isotropic media. In addition to these "smooth-medium caustics," the structural interfaces are usually responsible for additional caustic surfaces. As a well-known example, let us mention the caustic surfaces corresponding to the waves reflected from the interface of a syncline form.

The wavefield in the caustic region cannot be calculated by the zeroth-order ray approximation. Even the higher order terms of the ray series do not improve the accuracy; instead, they make it worse. The high-frequency wavefield in the caustic region may be calculated by asymptotic methods, but more sophisticated ansatz solutions must be used. In the case of a simple caustic, such solutions usually employ Airy functions (Ludwig 1966). We then speak of the Airy caustic. For more complex caustic surfaces, the solution may often be expressed in terms of two-parameteric Pearcey functions (Stamnes 1986).

The literature devoted to caustics is too extensive to be presented here. For a very detailed treatment and for other literature, the reader is referred to Babich and Buldyrev (1972), Stavroudis (1972), Felsen and Marcuvitz (1973), Babich and Kirpichnikova (1974), Kravtsov and Orlov (1980, 1993), Chapman (1985), Stamnes (1986), Hanyga (1995), and Hanyga and Helle (1995).

It should be emphasized again that the problems in the illuminated part of the caustic region consist only of the computation of amplitudes. The standard ray tracing and travel-time computation may be used there; neither theoretical nor numerical problems occur in the ray tracing procedures in the vicinity of caustics.

2. CRITICAL REGION

The critical region is the most important singular region of R/T waves at structural interfaces. It is situated in the close vicinity of the critical ray of R/T waves; see Sections 3.2.3, 5.3.4, 5.6.7, and 5.7.10. Critical rays are defined as follows. If any of the rays of the R/T waves corresponding to a selected incident wave is grazing the interface at the R/T point, the rays of all other remaining R/T waves corresponding to the same incident ray are called critical. The critical rays play a very important role in seismic measurements, particularly in wide-angle source-receiver configurations. On one side of the critical ray (subcritical region), only the reflected wave exists; on the other side of it (postcritical, also called overcritical, region) the head wave is also generated. The critical ray is common to the reflected and relevant head waves. For head waves, it also represents the boundary ray.

There are many types of critical rays and relevant critical regions, corresponding to various reflected and transmitted PP, PS, SP, and SS waves. The classification of all possible critical rays would, in fact, be the same as the classification of the relevant head waves; see Figure 5.17.

The zeroth-order ray amplitudes of R/T waves along the critical ray are finite, but their derivative with respect to the ray parameters is infinite there (from one ray to another). The amplitudes of the relevant head waves along the critical ray are infinite; see Figure 5.14. Thus, the validity conditions are not satisfied at the critical ray, and the ray is singular.

There is a large difference between the caustic singular region and critical singular region. The critical ray is singular as a whole; that is, any point situated on the critical ray is singular. On the contrary, the ray tangent to the caustic surface is not singular as a whole, but only at the caustic point where the ray is tangent to the caustic surface.

The amplitudes of reflected waves in the critical region cannot be calculated by the zeroth-order ray approximation. Even the higher order terms of the ray series do not improve the accuracy, rather they make it worse. The local asymptotics in the critical region employ Weber-Hermite functions. See Brekhovskikh (1960) for acoustic waves, and Červený (1966a, 1966b, 1967), Smirnova (1966), and Marks and Hron (1980) for elastic waves. A review with many other references can be found in Červený and Ravindra (1971). The most comprehensive and detailed treatment, applicable even to curved interfaces between inhomogeneous media, was given by Thomson (1990).

The behavior of travel times, rays, and ray amplitudes in the critical region of pressure reflected and head waves is displayed in Figure 5.14. We emphasize that the computation of the rays and travel times of reflected waves in the critical region is quite easy; there is only a problem with the computation of amplitudes. Figure 5.14 also shows schematically the typical form of the amplitude-distance curves of interference pressure reflected-head waves in the critical region, calculated by local asymptotic methods.

3. FRESNEL SHADOWS

Fresnel shadows are the regions in space where no rays of the elementary wave under consideration penetrate. The Fresnel shadow is, as a rule, connected with a discontinuous wavefront of the wave. In the illuminated region, where the wavefront exists, the ray field is regular. The surface that separates the illuminated region from the shadows region is called the *boundary surface*. It is formed by *boundary rays*, corresponding to the limiting rays in the illuminated region.

The boundary surface is singular. From one side of the boundary surface, the ray theoretical amplitudes vanish (shadow). From the other side of the boundary surface, the regular zeroth-order approximation of the ray method is nonvanishing (illuminated region).

Thus, the validity conditions of the ray method are not satisfied along the boundary surface, and the boundary surface is singular. The zeroth-order approximation of the ray method cannot be used even in the vicinity of the boundary surface, which is usually called the *transition region* (between the illuminated region and shadow). The higher order terms of the ray series do not improve the situation; a new ansatz solution must be used.

Shadow zones are usually connected with some bodies in the medium, which screen the regular rays. The two main types of shadow zones, corresponding to the different types of screening bodies, are as follows.

a. Shadow zones connected with *edges and vertices* in structural interfaces (including the Earth's surface).

b. Shadow zones connected with *smooth interfaces* (Thomson 1989).

There are two diffraction effects connected with either of these two situations.

i. The boundary surface between the shadow and the illuminated region is singular, and a new ansatz solution must be used. It usually employs Fresnel functions.

ii. New diffracted waves are generated. The diffracted waves generated at edges and vertexes are called the *edge waves* and the *vertex waves* (or *tip waves*). The diffracted waves generated by rays tangent to smooth interfaces are called *smooth interface diffracted waves*. Both of these waves not only penetrate into shadow zones but also affect the illuminated region close to the boundary surface.

The asymptotic high-frequency calculation of edge and smooth interface diffracted rays is based on the computation of *diffracted rays*; see Section 3.2.3. The diffracted rays play a basic role in the extension of the ray method, known as the *geometric theory of diffraction*, proposed by Keller (1958, 1962, 1963); see also James (1976). Note that diffracted rays can be calculated by standard ray tracing and that only the initial conditions must be properly chosen; see Section 3.2.3. The computation of amplitudes of diffracted waves requires the evaluation of so-called *diffraction coefficients*, which are in general direction- and frequency-dependent. Typical examples are the diffraction coefficients of edge and vertex waves; see Klem-Musatov (1980, 1994).

Extensive literature is devoted to the computation of amplitudes of edge and smooth interface diffractions. We shall mention only a few selected references: Friedrichs and Keller (1955), Keller, Lewis, and Seckler (1956), Lewis, Bleistein, and Ludwig (1967), Trorey (1970, 1977), Babich and Buldyrev (1972), Červený, Molotkov, and Pšenčík (1977), Felsen and Marcuvitz (1973), James (1976), Babich and Kirpichnikova (1974), Klem-Musatov (1980, 1994), Aizenberg and Klem-Musatov (1980), Achenbach, Gautesen, and McMaken (1982), Bleistein (1984), Felsen (1984), Klem-Musatov and Aizenberg (1984, 1985), Chapman (1985), Thomson (1989), Bakker (1990), Sun (1994), Hanyga (1995, 1996a), and Luneva (1996). These publications also contain many other references. Very good reviews can be found in James (1976), Thomson (1989), Klem-Musatov (1994), and Hanyga (1995).

5.9.3 Inhomogeneous Waves

The standard ray method considers only elementary seismic body waves with travel times real-valued along the whole ray. As soon as the travel times and relevant slowness vectors of some elementary wave are complex-valued, the wave cannot be treated by the standard ray method; the ray method with a complex eikonal should be used. This applies even to elementary waves for which the travel times and slowness vectors are complex-valued only along some segment(s) of the ray and are real-valued along other segments. We shall call here such waves *inhomogeneous*, generalizing in this way the term *inhomogeneous plane waves*, which was introduced in Section 2.2.10. We emphasize that the inhomogeneous wave in this terminology is not a plane wave. To call the wave inhomogeneous, it is sufficient if the travel time is complex-valued only along certain part of the ray, not necessarily passing through the receiver. In frequency domain, the amplitudes of inhomogeneous waves *decay exponentially* with increasing frequency. In nondissipative isotropic media, the exponential decay is along the wavefront. Note that the inhomogeneous waves, or at least certain types of them, are also often called *evanescent waves*; see Felsen (1976a).

Various methods have been used to study numerically the inhomogeneous waves. We shall name several of them here.

- **a.** Traditionally, exact integral expressions have been used to study certain simple types of inhomogeneous waves. Such exact solutions are, however, available practically only for 1-D models such as the models composed of homogeneous layers separated by plane-parallel interfaces. In laterally varying structures with curved interfaces, such solutions cannot be used, or can be used only locally.
- **b.** The ray method with a complex eikonal can be used even in laterally varying structures with curved interfaces. See Sections 5.6.8.B and 5.7.11 and Thomson (1997a).
- **c.** The space-time ray method with a complex-valued phase function is also applicable to general 2-D and 3-D structures. See Section 2.4.6.
- **d.** Perturbation ray theory may be very useful in such situations where the effects leading to inhomogeneous waves are weak (for example, weakly dissipative media). The imaginary part of the travel time, yielding the exponential decay of amplitudes, can be then computed by quadratures along real ray in the nonperturbed reference model. See Sections 5.5 and 6.3.
- **e.** Various hybrid methods can be used if the "inhomogeneous" complex-valued segment of the ray is very short in comparison with the prevailing wavelength. See Section 5.9.4 for hybrid ray-matrix method.
- **f.** More complex asymptotic approximations can be used to describe certain inhomogeneous waves. See Section 5.6.8C.

Let us now present several examples of inhomogeneous waves.

1. Waves in dissipative media. They can be treated by complex rays or by the space-time ray method; see Sections 2.4.6 and 5.6.8. See also Caviglia and Morro (1992). In weakly dissipative media, perturbation methods can also be applied. Then the standard ray method can be used in the reference nondissipative medium, and the dissipative effects (amplitude decay, dispersion) are obtained by simple quadratures along the real ray. Consequently, the approach can be easily incorporated into computer codes based on the standard ray method. See Section 5.5.

2. Postcritically transmitted waves. For a wave incident postcritically on an interface, the transmitted wave has a complex-valued travel time and the relevant ray is complex. See Figure 3.6(c). The PP inhomogeneous transmitted wave often has been called "the direct wave root." Not only transmitted waves, but also SP reflected waves are inhomogeneous for angles of incidence greater than $\arcsin(\beta/\alpha)$. All these inhomogeneous transmitted and reflected waves are well known from the solution of a classical seismological problem of reflection and transmission of a spherical wave at a plane interface between two homogeneous elastic halfspaces. Their implementation into routine ray-theory computer programs for the computation of seismic wavefields in 2-D and 3-D laterally varying layered structures, however, is not straightforward. Usually, they are not considered in these programs at all. Application of complex rays would help in this case.

3. Tunneling of waves through a high-velocity thin layer. When the wave is incident postcritically on the layer, it penetrates through the thin layer along a complex ray. The relevant travel time and slowness vector are complex-valued inside the layer. Leaving the thin high-velocity layer, the complex ray may again change into a real ray. Consequently, the waves transmitted through the layer may again have a real-valued travel-time and slowness vector and represent a standard elementary wave. The only difference is that its amplitude contains a factor exponentially decaying with increasing frequency. The tunneling of waves was investigated by the reflectivity method (Fuchs and Schulz 1976), by the generalized ray method (Drijkoningen and Chapman 1988; Drijkoningen 1991a), and by the hybrid ray-matrix method (Červený and Aranha 1992). Note that the hybrid ray-matrix method can be easily incorporated into computer codes based on the standard ray method if the thickness of the layer is considerably less than the prevailing wavelength of the wavefield under consideration. See Červený (1989b).

4. Pseudospherical waves. S* waves. Spherical waves generated by a point source can be expanded into plane waves using the Weyl integral; see Section 2.5.1. This expansion also contains inhomogeneous plane waves. If the point source is situated close to a structural interface in a higher velocity halfspace, the inhomogeneous wave propagating from the point source may generate a regular transmitted wave in the lower velocity halfspace. The wave has a roughly spherical wavefront, with its center at the projection of the source to the interface. For this reason, it is often called the pseudospherical wave. The pseudospherical waves were probably first obtained by the asymptotic treatment of exact integrals by Ott (1942) and Brekhovskikh (1960) for fluid media and by Červený (1957) for elastic media. They were also identified by laboratory modeling using the schlieren method (see Červený, Kozák, and Pšenčík 1971; Červený and Kozák 1972). The amplitudes of pseudospherical waves decrease exponentially with increasing d/λ, where d is the distance of the point source from the interface. For a detailed asymptotic treatment, see also Daley and Hron (1983b).

For an explosive source, generating P waves, situated close to the Earth's surface, pseudospherical S waves are generated at the Earth's surface (in addition to standard PS converted waves). They are known as S* waves and were investigated by finite-differences by Hron and Mikhailenko (1981); see also Daley and Hron (1983a). The S* waves also play an important role in S-wave directivity patterns of explosive sources situated close to the Earth's surface; see Jílek and Červený (1996).

The hybrid ray-matrix method can be incorporated into computer codes based on the standard ray method to compute even pseudospherical waves if d/λ is very small.

Even some other inhomogeneous waves may be generated at a structural interface if the point source is situated close to it. Let us mention here the leaking, head-wave-type waves and Stoneley waves. See Gilbert and Laster (1962), Tsvankin, Kalinin, and Pivovarov (1983), Tsvankin and Kalinin (1984), and Kiselev and Tsvankin (1989), and other literature given therein.

Many other types of inhomogeneous waves (in our terminology) may propagate in realistic complex structures. They are briefly discussed in other places in this book. Let us mention the inhomogeneous waves penetrating into caustic shadows (see Section 5.9.2 and

Thomson 1997a), the surface waves and inhomogeneous waves connected with waveguides (see Section 5.9.4), and so on.

5.9.4 Waves Propagating in a Preferred Direction

In the zeroth-order approximation of the ray method, individual elementary waves in 3-D laterally varying media propagate along rays. The high-frequency part of the energy flux vector of these waves is oriented along appropriate rays. Because the ray approximation does not consider the back-scattering at inhomogeneities, we also speak of one-way propagation along rays.

To study the high-frequency one-way propagation along a certain preferred direction (not necessarily the ray), various approaches have been used. In 2-D, the preferred direction may represent an axis of the waveguide, the surface of the Earth, a structural interface, and the like. If the preferred direction differs from the ray, the resulting wave under investigation is more complex than the elementary ray-theory wave. Among others, the amplitude of this wave is frequency-dependent. Sometimes, it may be possible to construct it by a superposition of many elementary waves or by many multiple arrivals of the same elementary wave. The resulting interference wave, however, has properties that may be quite different from the properties of individual elementary waves; only the whole interference complex has a good physical meaning. Often, the wavefield can be hardly computed by the summation of individual ray contributions, and it is preferable to use some extensions of the ray method.

One of the most common approaches to study the high-frequency waves propagating along a certain preferred direction is based on the *parabolic wave equation method* (Leontovich and Fock 1946; Fock 1965). The method of the parabolic wave equation, also known as the parabolic approximation method, has been applied to many wave propagation problems such as waves propagating in waveguides, whispering gallery waves, and beam propagation. A detailed historical survey of various applications of the parabolic wave equation can be found in Tappert (1977). For elastic waves, see McCoy (1977) and Hudson (1980b, 1981). The parabolic wave equation was used even in seismic exploration; see Landers and Claerbout (1972), Claerbout (1976, 1985), and Sutton (1984). The method was also applied to the derivation of paraxial Gaussian beams in 3-D structures (see Section 5.8.2 and Červený and Pšenčík 1983b). In most applications, the narrow-angle propagation was assumed.

The most general approach to the investigation of one-way wave propagation in a preferred direction, which does not assume a narrow-angle propagation, is based on the *factorization of the wave equation*. The factorization yields two factors, one of them corresponding to the forward propagation and the second to the backward propagation. By this factorization, a differential equation for one-way propagation is obtained. The differential equation is of the first order in the preferred direction and of a higher order in the transverse direction. For the Helmholtz equation, see Fishman and McCoy (1984, 1985), and for the elastodynamic equation for inhomogeneous anisotropic medium, see Thomson (1999). Thomson also gives a high-frequency alternative to the one-way elastodynamic equation. The Thomson one-way equation is very general; it includes the qP wave, two qS waves, and the coupling between the two qS waves (for example, in weakly anisotropic media). It remains valid at S-wave singularities (for example, at conical points), at caustics, and at other singularities.

The Thomson one-way elastodynamic equation can be solved in two ways. First, it can be solved analytically, by using so-called *path integrals* (Schulman 1981; Weigel 1986; Schlottmann 1999). The path integrals may be further reduced by the application of stationary-phase method to the ray-theory, Maslov-Chapman, and Kirchhoff representations. Second, it can be solved numerically, by "forward-stepping" FD algorithms. Thomson (1999) also discusses several narrow-angle approximations, which would considerably increase the efficiency of FD computations. They represent anisotropic analogues of the well-known $15°$ and $45°$ one-way differential equations for acoustic waves (Claerbout 1985). The wide-angle form of the equation is also close to the equations of the phase-screen method (Wu 1994; Wild and Hudson 1998).

We shall now present several important examples of waves of interference character.

Typical examples of waves of interference character are the *surface waves*, like Rayleigh and Love waves. Certain ray approaches may be useful even in the investigation and processing of seismic surface waves propagating in smoothly laterally varying Earth, particularly in surface wave tomography. This applies mainly to *surface-wave ray tracing*. See Section 3.12 for more details.

Waves propagating in *low-velocity smooth waveguides* are very important in ocean acoustics and the ionosphere. As a rule, waves propagating in such waveguides carry a large amount of energy to great distances from the source. If the source is situated in the waveguide, many rays are trapped in the waveguide, and the ray field in the waveguide forms many caustics. At large distances from the source, the caustic regions inside the waveguide are extensive and mutually overlap. The ray field in the waveguide may have a chaotic character. The ray method is not then suitable for evaluating the wavefield inside the waveguide, and various extensions have to be used. In many cases, it is more suitable to use the *normal mode method*. For more details, see Budden (1961a, 1961b), Babich and Buldyrev (1972), Miklowitz (1978), Kravtsov and Orlov (1980), and Abdullaev (1993), and other references given therein.

The waveguide may also be formed by a *homogeneous low-velocity layer bounded by plane interfaces of the first order* or by the surface of the Earth. Caustics are then not formed, but other difficulties appear. At large distances from the source, the number of multiply-reflected waves is very large, particularly due to PS and SP conversions in elastic media. Moreover, the critical regions corresponding to individual multiply-reflected waves may be very extensive. Finally, the zeroth-order contributions of the ray method of the individual waves may cancel one another due to destructive interference, and the non-ray and higher order ray contributions become important. Thus, at large distances from the source, the normal mode method again becomes more efficient.

In the preceding two examples, we have considered a low-velocity smooth waveguide and a homogeneous low-velocity layer bounded by interfaces. In the real Earth, waveguides may be considerably more complex. Instead of one homogeneous low-velocity layer, there may be a stack of thin layers, and/or a combination of smooth velocity variations with structural interfaces of the first order. The interfaces may be curved. The difficulties in treating such structures by the ray method are even more serious than in the preceding two simplified examples.

Other waves of interference character are *interference head waves* (see the end of Section 5.6.7 and more details in Červený and Ravindra 1971), and whispering gallery waves (see Babich and Buldyrev 1972; Babich and Kirpichnikova 1974; Popov and Pšenčík 1976a, 1976b; and Thomson 1990).

Several *hybrid methods* have also been proposed to investigate seismic waves of inter-ference character.

The hybrid *ray-finite difference method* was proposed by Lecomte (1996). The ray method is used in smooth parts of the model, and the finite-difference method is applied locally to a structurally very complex (but small) part of the model. See also Piserchia et al. (1998).

The *hybrid ray-mode method* combines the modal and ray approaches in computing the wavefield in a waveguide. Several modal and several ray contributions are used to compute the complete wavefield. See Felsen (1981), Kapoor and Felsen (1995), Zhao and Dahlen (1996), and Lu (1996), and other references given therein.

The *hybrid ray-matrix method*, also called *hybrid ray-reflectivity method*, was proposed to compute high-frequency seismic body waves propagating in media containing thin layers. The ray method is used in smooth parts of the model, and the matrix method is used at points of interaction of the ray with a thin layer. The matrix method is used at the points of interaction to compute the frequency-dependent R/T coefficients at thin layers. For 1-D media, the method was proposed by Ratnikova (1973); see also Daley and Hron (1982). For laterally varying structures, see Červený (1989b). The method was also used to investigate the tunneling of seismic body waves through thin high-velocity layers by Červený and Aranha (1992) and to study the effect of a thin surface low-velocity layer on the wavefield recorded at the Earth's surface by Červený and Andrade (1992).

5.9.5 Generalized Ray Theory

The wavefield generated by a point source situated in a one-dimensional medium consisting of homogeneous layers separated by plane-parallel interfaces can be expressed *exactly* in terms of closed-form integrals. Analytical expressions for the integrands of these integrals can be found for the model containing one structural interface (or a free surface). For a higher number of structural interfaces, matrix methods have been usually used to find the integrand. The integrals are usually of an oscillatory character and represent the complete wavefield, not separated into individual elementary waves. The relevant integrals can be ex-pressed in many alternative forms. This is a classical problem of theoretical seismology and wave physics. A broad literature, including many textbooks, gives details of the derivation of such integrals. See Lamb (1904), Sommerfeld (1909), Love (1944), Brekhovskikh (1960), Ewing, Jardetzky, and Press (1957), Červený and Ravindra (1971), Officer (1974), Pilant (1979), Aki and Richards (1980), Ben-Menahem and Singh (1981), and Kennett (1983).

Various methods have been proposed to treat these integrals numerically. At present, the most popular and efficient method consists of their direct numerical computation in frequency domain and is known as the *reflectivity method*. It was proposed by Fuchs (1968a, 1968b) and further developed by Fuchs and Müller (1971), Kennett (1983), and Müller (1985). For a detailed tutorial treatment, see Aki and Richards (1980, Chap. 9) and Müller (1985). For anisotropic layered media, see Booth and Crampin (1983), Fryer and Frazer (1984, 1987), Mallick and Frazer (1987, 1990, 1991), and Kelly, Baltensperger, and McMechan (1997).

The integrals representing the complete wavefield can also be expanded into contri-butions corresponding to rays of individual multiply-reflected/transmitted waves. Each of these contributions is characterized by an arrival time in such a way that it vanishes for times less than its arrival time. There is always a finite number of contributions (if any) with arrival times less than arbitrarily chosen time. Consequently, the summation of individual contributions yields an exact result if the contributions are treated exactly and if a sufficient number of contributions is used. For this reason, we speak about *generalized ray theory* or *exact ray theory*. It should be emphasized that these ray contributions have a more general meaning than the zeroth-order approximation of the ray method. Each of them also may include the head waves related to the multiple reflection under consideration, leaking mode contributions, and Stoneley waves. For this reason, we also speak about an expansion into *generalized ray contributions*, or simply into *generalized rays*.

The generalized ray contributions may be computed in many different ways, either in time or in frequency domain. The time-domain computations play a particularly important role. In time domain, the integrals corresponding to generalized ray contributions may be transformed to simpler nonoscillatory integrals along a suitably chosen contour in a complex plane. Basic ideas of computations were proposed by Cagniard (1939, 1962) and extended by de Hoop (1960). For this reason, the method is also known as the *Cagniard-deHoop method* (or Cagniard-deHoop technique), and the relevant integration contours are known as *Cagniard- deHoop contours*. In Russia, a similar method has been proposed by Smirnov and Sobolev; see a short description in Smirnov (1964). The method was further developed by Zvolinskiy (1957, 1958, 1965).

In this book, we are interested in the computation of seismic wavefields in 3-D laterally varying isotropic and anisotropic layered structures with curved interfaces. The generalized ray method cannot be applied to such computations; it can be used only for 1-D models consisting of plane-parallel layers. For this reason, we shall not discuss here the Cagniard-deHoop technique, we shall give only several references for further reading. See Bortfeld (1967), Helmberger (1968), Chapman (1974), Müller (1968, 1970), and a tutorial explanation in Aki and Richards (1980, Chap. 6). For anisotropic stratified media, see van der Hijden (1988). The method may be used even for vertically inhomogeneous media, simulated by a system of thin homogeneous layers (similarly as the reflectivity method). For the Epstein profile, see Drijkoningen (1991b). The method was used to study various nonray effects and behavior of seismic body waves in singular regions. As an example, let us name here the studies devoted to tunneling of waves, see Drijkoningen and Chapman (1988) and Drijkoningen (1991a). The method was also modified for radially symmetric media (see Gilbert and Helmberger 1972), and for models with nonparallel interfaces (see Hong and Helmberger 1977; Kühnicke 1996). For a comparison of synthetic seismograms computed by generalized ray theory and reflectivity method, see Burdick and Orcutt (1979). The generalized ray method was applied to many important seismological investigations by D. Helmberger and his group.

The exact integral expressions corresponding to generalized rays can also be treated *in the frequency domain*. Such integrals were broadly used in the investigation of the reflection and transmission of spherical waves at a plane interface. The main idea of computation consists of a suitable deformation of integration contours in a complex plane, and of approximate, or numerical, evaluation of the integrals along these new contours. Integration contours close to the steepest descent trajectory have been mostly used because they suppress the oscillations of the integrand considerably. For more details, see Ewing, Jardetzky, and Press (1957), Červený and Ravindra (1971), and Aki and Richards (1980, Chap. 6). The method has also been effectively used to find suitable local asymptotic approximations in singular regions, where the standard zeroth-order approximation of the ray method fails. Particular attention was devoted to the critical region of reflected and head waves. See Brekhovskikh (1960) for acoustic waves and Červený (1957, 1966a, 1966b, 1967) for elastic waves. Similar local asymptotic approximations in the critical region were also found for multiply-reflected waves and were applied successfully in the computer codes for synthetic seismograms for vertically inhomogeneous structures; see Červený (1979b). The integrals were also used to study various nongeometrical effects such as pseudospherical waves and leaking modes (Ott 1942; Červený 1957; Tsvankin and Kalinin 1984) and to compare the exact and asymptotic computations (Kiselev and Tsvankin 1989). For an excellent review of methods of the computation of body wave synthetic seismograms in laterally homogeneous media, see Chapman and Orcutt (1985).

CHAPTER SIX

Ray Synthetic Seismograms

U nder *ray synthetic seismograms*, we understand time-domain, high-frequency asymptotic solutions of the elastodynamic equation by the ray method. Ray synthetic seismograms are represented by a superposition of elementary ray synthetic seismograms, corresponding to the individual elementary body waves arriving at the receiver within a specified time window. In a layered 3-D structure, there may be many seismic body waves that travel from the source to the receiver along different ray trajectories. They correspond to various reflected, refracted, multiply-reflected, converted, and other seismic body phases.

In this chapter, we shall consider only standard, *zeroth-order elementary waves* in the construction of ray synthetic seismograms. Waves of higher order (such as head waves), diffracted, inhomogeneous, and other waves that cannot be described by standard ray approaches will not be included in the superposition. Moreover, we shall not consider the higher order terms in the ray series for individual elementary waves; all elementary waves will be represented only by their zeroth-order ray approximation. Thus, it would be more precise to speak of *zeroth-order ray synthetic seismograms*. We shall, however, avoid the term *zeroth-order* to make the terminology simpler. We believe this will not cause any misunderstanding. Of course, the ray synthetic seismograms studied here may also be extended to include certain types of waves not included among zeroth-order ray approximation waves. This would, however, require special treatment. See a brief exposition in Section 6.2.5.

If the individual waves forming the ray synthetic seismograms are separated and do not interfere mutually in certain regions, it is possible to study them separately. For example, their amplitudes may be fully determined by the equations of Sections 5.1, 5.2, and 5.4. In certain regions, however, elementary waves interfere, particularly if the signals are longer. In these regions, the expressions for the amplitudes of the individual waves cannot fully describe the wavefield, and various interference effects must be considered. Ray synthetic seismograms provide the proper description of such interference effects. This is the main advantage of ray synthetic seismograms in comparison with elementary ray synthetic seismograms.

Although the computations of ray synthetic seismograms are only approximate and the ray method may fail partially or completely in certain situations, ray concepts and ray synthetic seismograms have been found very useful in many important applications of great interest in seismology and in seismic exploration.

We shall discuss the construction of elementary ray synthetic seismograms in Section 6.1 and the construction of ray synthetic seismograms in Section 6.2. We shall exploit certain properties of the Hilbert transform and of analytical signals to propose efficient algorithms for the computation of synthetic seismograms. The necessary equations related to the Hilbert transform and analytical signals can be found in Appendix A. The construction of ray synthetic seismograms in the frequency domain and in the time domain is discussed. In the frequency domain, a *fast frequency response* (FFR) algorithm is described in detail; this algorithm increases the efficiency of the frequency domain computations considerably. For completeness, time-domain versions of certain integral solutions, derived in Chapter 5, are discussed in Section 6.2.6. They represent the time domain versions of Kirchhoff integrals and of various superposition integrals.

In Sections 6.1 and 6.2, only nondissipative media are considered. Section 6.3 generalizes these results for dissipative media. We consider only weakly dissipative media in which it is possible to obtain the amplitudes of seismic body waves by a simple extension of the equations for the amplitudes of waves propagating in nondissipative media along real ray trajectories; see Section 5.5. The computation of synthetic seismograms of seismic body waves propagating in strongly dissipative media would require the computation of new, complex rays. Both noncausal and causal absorption models are considered in Section 6.3.

There is no difference between the construction of ray synthetic seismograms for scalar (acoustic) and vectorial (elastic) wavefields. For this reason, we shall consider only the vectorial (elastic) wavefield, represented by displacement vector $\vec{u}(x_j, t)$. The scalar acoustic wavefield is not discussed explicitly; the relevant equations for the scalar wavefields are exactly the same as for vectorial wavefields. We have only to replace the vectorial expressions $(\vec{u}, \vec{U})$ by the relevant scalar expressions (p, P).

The only section that is fully devoted to vectorial wavefields is Section 6.4, which discusses the polarization of vectorial seismic wavefields. The main attention in Section 6.4 is devoted to the construction of *ray synthetic particle ground-motion diagrams*. After the ray synthetic seismograms for the Cartesian components of the displacement vector are known, it is simple to construct the particle-motion diagrams in any Cartesian coordinate plane. It is shown that the ray synthetic particle ground motion of compressional waves is linear but that it may be elliptic, quasi-elliptic, or even more complex for shear waves. The relevant algorithms are briefly described and discussed in Section 6.4.

For more details on the construction of ray synthetic seismograms in complex laterally varying *isotropic structures* see a review paper by Červený (1985b). The paper also gives many examples of computation and references for further reading and applications. For *anisotropic structures* see Gajewski and Pšenčík (1987b, 1990, 1992), Thomson, Kendall, and Guest (1992), Guest and Kendall (1993), and Alkhalifah (1995).

6.1 Elementary Ray Synthetic Seismograms

In this section, we shall consider one arbitrarily selected "elementary" wave, propagating in a 3-D laterally varying layered structure. The wave may be of any type, including multiply-reflected and converted waves. We shall call the expression for any component of the displacement vector of this wave in the time domain the *elementary ray synthetic seismogram*. A superposition of various elementary waves arriving along various ray trajectories from the source at the receiver will be considered in Section 6.2.

The elementary ray synthetic seismogram can be computed in two ways: in the frequency domain and in the time domain. We shall describe both approaches in some detail.

6.1.1 Displacement Vector of an Elementary Wave

We can express any component of displacement vector $u_n(R, t)$ at point R, by

$$u_n(R, t) = \text{Re}\{U_n(R)F(t - T(R))\}, \tag{6.1.1}$$

where $U_n(R)$ is the nth component of the vectorial ray theory complex-valued amplitude function, $T(R)$ is the scalar-valued travel time, $F(\zeta)$ is the high-frequency analytical signal, and $\text{Re}\{\ldots\}$ stands for the real part of $\{\ldots\}$. Components u_n and U_n may be given in any coordinate system.

We shall call (6.1.1) the elementary ray synthetic seismogram or simply the elementary synthetic seismogram.

We know how to calculate both $U_n(R)$ and $T(R)$. We must, however, discuss two other problems:

a. How to select the relevant analytical signal $F(\zeta)$.
b. Which method would be efficient in computing the elementary synthetic seismogram given by (6.1.1).

6.1.2 Conservation of the Analytical Signal Along the Ray

Before we discuss these two problems in greater detail, we shall draw two important conclusions from Equation (6.1.1). As we can see from (6.1.1), *analytical signal $F(t - T(R))$ is conserved along the whole ray* (even across structural interfaces). The actual form of the signal may change along the ray due to phase changes of $U_n(R)$, caused by caustics and postcritical incidence at structural interfaces and at the Earth's surface, but the analytical signal remains the same. This basic law is, of course, valid only in nondissipative media and in the zeroth-order approximation of the ray method.

This law also implies that the shape of the envelope of the displacement vector is conserved along the whole ray in nondissipative media; see (A.3.24). If we introduce a *normalized envelope* (the maximum of the envelope is reduced to unity), the normalized envelope of the displacement vector remains the same along the whole ray. Thus, the normalized envelope will be the same at points S and R.

6.1.3 Analytical Signal of the Elementary Wave. Source Time Function

The analytical signal is given by the relation

$$F(t - T) = x(t - T) + \text{i}g(t - T), \tag{6.1.2}$$

where $x(\zeta)$ is some real-valued function and $g(\zeta)$ is its Hilbert transform; see (A.3.1). It is sufficient to know one of the functions x, g, and F to determine the two other. Because analytical signal $F(\zeta)$ is preserved along the ray, it is sufficient to know $x(\zeta)$ (or $g(\zeta)$) at one reference point of the ray, and we can determine the analytical signal $F(\zeta)$ along the whole ray.

In all ray computations, the analytical signals (or $x(\zeta)$ or $g(\zeta)$) corresponding to the individual elementary waves must be specified at one point of the ray. Usually, they are specified at the initial point S of the ray. They may be considered the same for all elementary waves, but this is not necessary. For example, the analytical signals of P waves generated by a point source may be different from the analytical signals of S waves generated by the source.

Let us consider, for a while, a point source in a homogeneous medium, situated at S. The ray-centered components of displacement vector $u_n(R, t)$ are then obtained from (6.1.1)

$$u_n(R, T) = \text{Re}\left\{\frac{F(t - l/V)}{\mathcal{L}(R, S)}\mathcal{G}_n(S)\right\}, \tag{6.1.3}$$

where $\mathcal{L}(R, S) = Vl(R, S)$ is the relative geometrical spreading in a homogeneous medium, $l(R, S)$ is the distance of R from S, and $\mathcal{G}_n(S)$ is the nth ray-centered component of the radiation function. As an example, see (5.4.23) for the radiation function corresponding to a single-force point source in an anisotropic medium. Note that $\mathcal{L}(R, S)$ is always real. Often $\mathcal{G}_n(S)$ is also real-valued. Then, (6.1.3) yields

$$u_n(R, T) = \frac{\mathcal{G}_n(S)}{\mathcal{L}(R, S)}x(t - l/V). \tag{6.1.4}$$

For this reason, the real-valued time function $x(\zeta)$ is usually called the *source time function*. If $\mathcal{G}_n(S)$ is not real-valued, signal $x(\zeta)$ is *phase-shifted*. For simplicity, we shall call $x(\zeta)$ the source-time function even in this case, considering that the source-time function may be phase shifted directly from the source.

Several signals have been broadly used in the numerical modeling of seismic wavefields by ray methods to represent the source-time functions of a point source. The most popular are the Ricker signal (Hosken 1988; Dietrich and Bouchon 1985), Gabor signal (Gabor 1946; Červený 1976, 1985b; Červený, Molotkov, and Pšenčík 1977, pp. 47–50), Küpper signal (Küpper 1958; Müller 1970), Berlage signal (Aldridge 1990), and Rayleigh signal (Hubral and Tygel 1989). They are defined by the following equations.

  a. **The Ricker signal:**

$$x(t) = \left(1 - 2\beta^2(t - t_i)^2\right)\exp\left[-\beta^2(t - t_i)^2\right], \tag{6.1.5}$$

    with two free parameters, β and t_i.

  b. **The Gabor signal:**

$$x(t) = \exp\left[-(2\pi f_M(t - t_i)/\gamma)^2\right]\cos[2\pi f_M(t - t_i) + \nu], \tag{6.1.6}$$

    with four free parameters, f_M, γ, ν, and t_i. Note that this signal is also known under different names such as the Gaussian envelope signal and the Puzyrev signal.

  c. **The Küpper signal:**

$$\begin{aligned} x(t) &= 0 &&\text{for } t \leq t_i \text{ and } t \geq t_i + T, \\ &= \sin\frac{N\pi(t - t_i)}{T} - \frac{N}{N+2}\sin\frac{(N+2)\pi(t - t_i)}{T} \\ &&&\text{for } t_i < t < t_i + T, \end{aligned} \tag{6.1.7}$$

    with three free parameters, T, N and t_i. At present, the signal is mostly known as the Müller signal.

  d. **The Berlage signal:**

$$\begin{aligned} x(t) &= 0 &&\text{for } t < t_i, \\ &= (t - t_i)^N\exp[-\beta(t - t_i)]\sin[2\pi f_M(t - t_i)] &&\text{for } t > t_i, \end{aligned} \tag{6.1.8}$$

    with four free parameters f_M, β, N, and t_i.

  e. **The Rayleigh signal:**

$$x(t) = \frac{1}{\pi}\frac{\epsilon}{(t - t_i)^2 + \epsilon^2}, \tag{6.1.9}$$

    with two free parameters, t_i and $\epsilon > 0$.

Several important signals $x(t)$, together with their Hilbert transforms $g(t)$, can be seen in Figure 6.1.

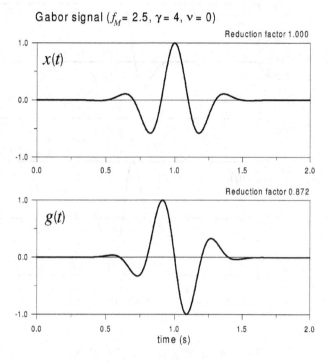

Figure 6.1. Examples of signals $x(t)$ of seismic body waves, often used in numerical modeling of seismic wavefields, and their Hilbert transforms $g(t)$. (a) Gabor signals for $\gamma = 4$ and $\gamma = 6$ (this page). (b) Ricker signal and Berlage signal. (c) Müller signal and the box car function.

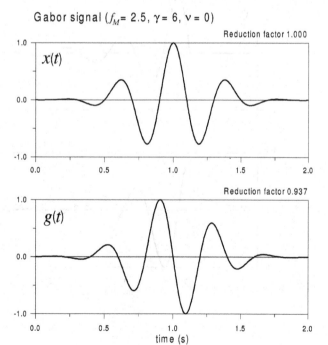

Ricker signal ($\beta = 8$)

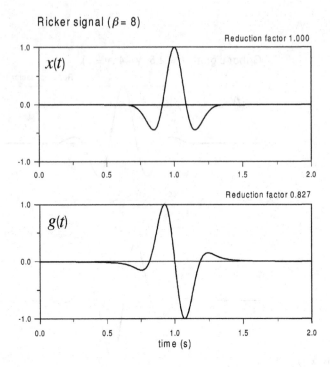

Figure 6.1(b)

Berlage signal ($f_M = 2.5$, $\beta = 3$, $N = 0$)

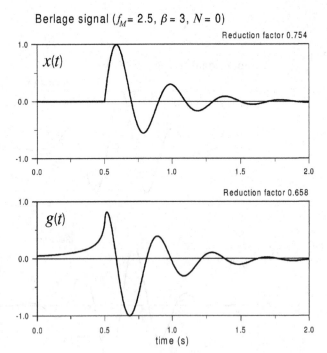

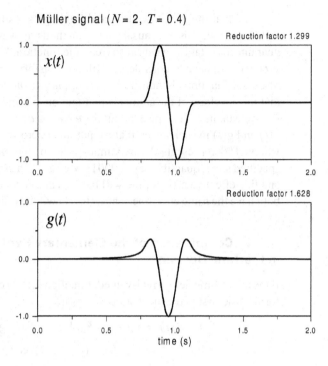

Figure 6.1(c)

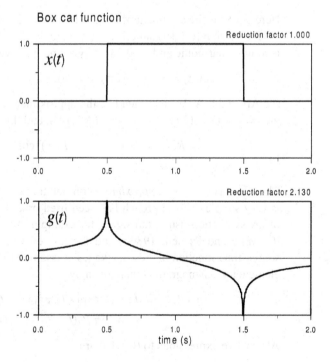

Note that the Ricker, Gabor, and Rayleigh signals are noncausal but that the Berlage and Küpper signals are causal. In ray methods, however, the causality of the results is not guaranteed by taking a causal source time function. In addition to the source time function $x(t)$, we also have to consider its Hilbert transform $g(t)$. In general, the Hilbert transform of causal functions is noncausal. Thus, *the ray method is, in principle, noncausal*. Note that the envelope of the source-time function is always noncausal.

We can, however, speak about the *effective causality* of the ray method. Let us modify $x(t)$ and $g(t)$ in such a way that we put them zero for all times t at which $|F(t)| < A^{max}\delta$, where A^{max} represents the maximum value of the envelope of the signal, and δ is some specified small quantity (say, 0.0001). We can then simply define the effective arrival time, and this effective arrival time will be the same for the signal and its Hilbert transform, as both have the same envelope. Thus, (6.1.1) will be effectively causal at any point of the ray.

6.1.4 Computation of the Elementary Synthetic Seismogram in the Time Domain

If the travel-time field is real-valued, Equation (6.1.1) can be expressed in several alternative forms. The first is the most straightforward

$$u_n(R, t) = x(t - T(R)) \operatorname{Re} U_n(R) - g(t - T(R)) \operatorname{Im} U_n(R)$$
$$= |U_n(R)|\{x(t - T(R)) \cos \chi_n(R) - g(t - T(R)) \sin \chi_n(R)\}.$$
$$(6.1.10)$$

Here $\chi_n(R)$ is the argument of $U_n(R)$, $U_n(R) = |U_n(R)| \exp(i\chi_n(R))$.

Expression (6.1.10) simplifies considerably if we consider a source-time function that is roughly harmonic and modulated by a smooth envelope $a(t)$,

$$x(t) = a(t) \cos(2\pi f_M t + \nu); \qquad (6.1.11)$$

see Appendix A. We can then use the approximate expression for the Hilbert transform, $g(t) \doteq -a(t) \sin(2\pi f_M t + \nu)$ (see (A.2.14)), and this leads to

$$u_n(R, t) \doteq |U_n(R)|a(t - T(R)) \cos[2\pi f_M(t - T(R)) + \nu - \chi_n(R)].$$
$$(6.1.12)$$

Thus, in this case, the *approximate* computation of the elementary synthetic seismogram is very simple. This approach has been used broadly for a long time in the computation of ray synthetic seismograms; see, for example, the SEIS83 computer program packages (Červený and Pšenčík 1984b) and the revised version of SEIS83, called SEIS88. The source-time function in these packages is assumed to be the Gabor signal; see (6.1.6). The elementary seismogram is then given by

$$u_n(R, t) = |U_n(R)| \exp\left[-(2\pi f_M(t - t_i - T(R))/\gamma)^2\right]$$
$$\times \cos[2\pi f_M(t - t_i - T(R)) + \nu - \chi_n(R)]. \qquad (6.1.13)$$

Alternative expressions to (6.1.10) are

$$u_n(R, t) = \operatorname{Re}\{F(t) * U_n(R)\delta(t - T(R))\}, \qquad (6.1.14)$$

or

$$u_n(R, t) = \operatorname{Re}\{x(t) * U_n(R)\delta^{(A)}(t - T(R))\}, \qquad (6.1.15)$$

see (6.1.1) and (A.3.14). The star (∗) denotes the convolution.

Equation (6.1.15) can also be expressed in the following forms:

$$u_n(R, t) = x(t) * \operatorname{Re}\{U_n(R)\delta^{(A)}(t - T(R))\}$$

$$= x(t) * \left\{\delta(t - T(R)) \operatorname{Re} U_n(R) + \frac{1}{\pi(t - T(R))} \operatorname{Im} U_n(R)\right\}.$$

$$(6.1.16)$$

For elementary seismograms, these relations are obvious. We shall, however, see in Section 6.2 that they will lead to different algorithms in computing complete ray synthetic seismograms.

6.1.5 Elementary Synthetic Seismograms for Complex-Valued Travel Times

For complex-valued T, we replace $\delta^{(A)}(t - T)$ in (6.1.15) by $-\mathrm{i}/[\pi(t - T)]$; see (A.3.21). Then (6.1.15) yields

$$u_n(R, t) = \operatorname{Re}\{x(t) * U_n(R)\delta^{(A)}(t - T(R))\} = \operatorname{Re}\left\{x(t) * \frac{-\mathrm{i}U_n(R)}{\pi(t - T(R))}\right\}$$

$$= x(t) * \operatorname{Im}\left\{\frac{U_n(R)}{\pi(t - T(R))}\right\} = \operatorname{Re}\left\{F(t) * \frac{1}{\pi}U_n(R) \operatorname{Im} \frac{1}{t - T(R)}\right\}.$$

$$(6.1.17)$$

These relations are important particularly in the computation of synthetic seismograms of waves propagating in weakly dissipative media.

6.1.6 Computation of Elementary Synthetic Seismograms in the Frequency Domain

In certain situations, it is useful to compute the elementary ray synthetic seismograms in the frequency domain. We shall first compute the *elementary frequency response*. By elementary frequency response, we understand the Fourier spectrum of the relevant component of the displacement vector at R, for the source-time function corresponding to the Dirac delta function, $x(t) = \delta(t)$. The elementary frequency response is given by the relation

$$u_n(R, f) = U_n(R) \exp[2\mathrm{i}\pi f T(R)].$$ $$(6.1.18)$$

To obtain the elementary synthetic seismogram from the elementary frequency response (6.1.18), we must multiply (6.1.18) by the Fourier spectrum $x(f)$ of the source-time function $x(t)$ and use the inverse Fourier transform.

As we can see from (6.1.18), the frequency response for one elementary wave has a very simple meaning. The concept of the frequency response will be more important in the computation of complete ray synthetic seismograms, by summation of elementary frequency responses. The frequency response is independent of the actual source-time function, and the same frequency response can be used to evaluate ray synthetic seismograms for various source-time functions.

6.1.7 Fast Frequency Response Algorithm

If we compute the synthetic seismograms in the frequency domain using the Fourier transform, we must determine (6.1.18) for a series of frequencies f_k ($k = 1, 2, \ldots, K$), with $f_{i+1} - f_i = \Delta f = \text{const}$. The most time-consuming step in the computation of the frequency response is the evaluation of the trigonometric functions $\cos(2\pi f_k T)$ and $\sin(2\pi f_k T)$. A considerably simpler procedure, which practically eliminates the computation of these trigonometric functions, can be found. For any selected frequency $f_k = f_1 + (k-1)\Delta f$,

$$
\begin{aligned}
u_n(R, f_k) &= U_n(R) \exp[2i\pi f_k T(R)] \\
&= U_n(R) \exp[2i\pi (f_1 + (k-1)\Delta f) T(R)] \\
&= U_n(R) \exp[2i\pi f_1 T(R)][\exp(2i\pi \Delta f T(R))]^{k-1}.
\end{aligned}
$$

If

$$
A = \exp[2i\pi \Delta f T(R)], \tag{6.1.19}
$$

then

$$
u_n(R, f_k) = u_n(R, f_{k-1})A = u_n(R, f_1)A^{k-1}. \tag{6.1.20}
$$

Thus, the fast frequency response algorithm, which eliminates the calculation of the trigonometric functions, follows. Contribution $U_n(R) \exp(2i\pi f T(R))$ is directly calculated only for the first frequency $f = f_1$. In addition, we also compute A given by (6.1.19). For every other frequency ($f_2, f_3, \ldots$), only *one complex-valued multiplication* is needed to obtain the frequency response (see (6.1.20)); no trigonometric function needs to be computed.

The FFR algorithm increases the efficiency of the computation of synthetic seismograms in the frequency domain tremendously. Instead of the computation of two trigonometric functions and several algebraic operations, only one complex-valued multiplication by a constant quantity is needed for each frequency.

The FFR algorithm can also be used for complex-valued $T(R)$, but $T(R)$ must be *independent of frequency*. Thus, the FFR algorithm can be used for noncausal dissipation. For causal dissipation, $T(R)$ depends on frequency, and the FFR algorithm cannot be used. See Section 6.3.

6.2 Ray Synthetic Seismograms

In Section 6.1, we considered only one elementary wave specified by an appropriate alphanumeric code, and its ray Ω passing through R. In a layered medium, however, there may be many body waves that travel from the point source at S or from the initial surface Σ to point R along various trajectories Ω. They correspond to various reflected, refracted, converted, and other seismic body phases. Moreover, even one elementary wave may travel from S (or from Σ) to R along different ray trajectories (so-called multiple rays).

To distinguish between the displacement vector $u_n(R, t)$ corresponding to one selected elementary wave, and the displacement vector of the complete wavefield, obtained by the superposition of various elementary waves, we shall use a bar above the letter for the displacement vector of the complete wavefield, $\bar{u}_n(R, t)$. We shall again consider only one (nth) component of the displacement vector, but we shall speak of the displacement vector, for simplicity.

Ray synthetic seismograms have found applications in the interpretation of observed seismograms, mainly in structural studies of the Earth's crust and the uppermost mantle.

They have usually been used to improve successively the structural model, comparing the observed and synthetic seismograms. The ray synthetic seismograms cannot compete in accuracy with some other synthetic seismograms generating the complete wavefield, like finite-difference synthetic seismograms. However, they have one great advantage when compared with them. Any signal in ray synthetic seismograms can be simply assigned to a structure in a very limited part of the model, corresponding to the vicinity of the relevant ray. Consequently, we know which part of the model should be corrected to obtain a better fit. The finite-difference synthetic seismograms do not offer such a possibility because they yield the complete wavefield. (Exception are the finite-difference movies that may be used to correlate individual signals on successive time levels.) The ray synthetic seismograms also offer some advantages with respect to the independent treatment of travel times and amplitudes of individual elementary waves. They include, in one picture, three pieces of information related to individual signals: travel times, amplitudes, and waveforms. Moreover, they also give a proper description of various interference effects in regions, where individual elementary waves interfere and cannot be treated separately.

In the 1960s and 1970s, the ray synthetic seismograms were evaluated mostly for *1-D crustal models* composed of parallel thick homogeneous layers; see, for example, Hron and Kanasewich (1971). Several modifications of the ray method, such as the Weber-Hermite modification for the critical regions, were included to increase the accuracy of ray computations; see Červený and Ravindra (1971). The computer programs were also modified for vertically inhomogeneous layered structures, simulating them by thin homogeneous layers (similar to the reflectivity method) and computing only primary unconverted reflections from individual interfaces. See the detailed expositions of these methods, with many examples, in the book by Červený, Molotkov, and Pšenčík (1977), in Červený (1979b), and Hron et al. (1986). Very fast and efficient algorithms and computer programs for 1-D ray synthetic seismograms based on polynomial rays (see Section 3.7) are described in Červený and Janský (1985) for vertically inhomogeneous layered medium and in Zedník, Janský, and Červený (1993) for radially symmetric layered structure.

In 1-D models, the standard ray method is not as accurate as the WKBJ, reflectivity and generalized ray methods. Consequently, ray synthetic seismograms have found most natural applications in *laterally varying layered models*, where the three mentioned methods cannot be used. The first ray synthetic seismograms for 2-D crustal structures, mostly composed of homogeneous isotropic nondissipative layers separated by curved interfaces, were computed in 1970s. The key problem in the algorithms consists in the two-point ray tracing. As soon as the problem of two-point ray tracing was reliably solved for general 2-D laterally varying layered structures, it was possible to produce more-or-less universal and routine computer packages for 2-D ray synthetic seismograms. See, for example, computer package SEIS83, distributed by World Data Center (A) for Solid Earth Geophysics, Boulder (Červený and Pšenčík 1984b). Alternative algorithms have been proposed by many seismologists. 2-D ray synthetic seismograms have been used in deep seismic sounding of the Earth's crust, and in seismic exploration, in strong motion seismogram computations, in vertical seismic profiling, in cross-well computations, among others. The algorithms, details of computations, and many numerical examples of 2-D ray synthetic seismograms can be found in Červený and Pšenčík (1977), Červený, Molotkov, and Pšenčík (1977), Hron, Daley, and Marks (1977), Chapman (1978), Hong and Helmberger (1978), May and Hron (1978), Červený (1979a), Marks and Hron (1980), McMechan and Mooney (1980), Cassel (1982), Langston and Lee (1983), Lee and

Langston (1983a, 1983b), Bernard and Madariaga (1984), Červený and Pšenčík (1984b), Cormier and Mellen (1984), Cormier and Spudich (1984), Spence, Whittal, and Clowes (1984), Spudich and Frazer (1984), Madariaga and Bernard (1985), Fertig and Pšenčík (1985), Moczo, Bard, and Pšenčík (1987), Brokešová (1996), and Iversen, Vinje, and Gelius (1996). For a more detailed review with many numerical examples and references, see Červený et al. (1980), and particularly Červený (1985b). At present, the computation of ray synthetic seismograms in 2-D laterally varying layered isotropic structures is a well-understood and routine problem, and relevant computer packages are available at most seismological institutions. The recent references would be too numerous to be given here.

In 3-D laterally varying layered structures, ray synthetic seismograms can be computed practically in the same way as in 2-D; see Červený and Klimeš (1984), Azbel et al. (1984), Musil (1989), and Chen (1998), among others. There are, however, two serious additional problems. The first problem is connected with the construction of sufficiently general 3-D models of laterally varying layered and blocked media, which would satisfy the validity conditions of the ray method in the range of prevailing frequencies under consideration. The second, very critical problem consists of fast, efficient, and safe algorithms for two-point ray tracing, yielding all multiple rays. The two-point ray tracing in 3-D structures is considerably more complicated than in 2-D structures. Suitable program packages for 3-D ray synthetic seismograms exist but are usually not available in the public domain. One of the popular 3-D packages, designed by L. Klimeš, is called the Complete Ray Tracing package and is based on algorithms described in detail by Červený, Klimeš, and Pšenčík (1988b), supplemented by the two-point ray tracing algorithm by Bulant (1996, 1999).

In anisotropic inhomogeneous layered structures, the main principles of ray synthetic seismograms remain the same as in isotropic inhomogeneous media. Because the ray fields in anisotropic inhomogeneous layered structures are intrinsically 3-D, the 2-D ray synthetic seismograms do not play as important a role in practical applications there as they do in isotropic media. For this reason, 3-D models usually have been considered. The two problems of 3-D computations discussed here (model, two-point ray tracing) play a similarly important role in anisotropic media as in isotropic media. An additional problem is connected with shear wave singularities and with weak anisotropy, where the standard anisotropic ray method fails. See Gajewski and Pšenčík (1987a, 1987b, 1990, 1992) and their program package ANRAY, as well as Guest and Kendall (1993) and Alkhalifah (1995).

We have discussed here only the ray synthetic seismograms computed by standard ray methods. The synthetic seismograms for laterally varying layered structures, based on asymptotic high-frequency integral solutions of the elastodynamic equation are discussed in Section 6.2.6. Equations of Section 6.2.6 may be applied to the Kirchhoff integrals and to various other superposition integrals. For weakly dissipative media, see Section 6.3.

6.2.1 Ray Expansions

The displacement vector component of the complete wavefield at R, $\bar{u}_n(R, t)$,

$$\bar{u}_n(R, t) = \sum_{(\Omega)} u_n(R, t), \qquad (6.2.1)$$

where the summation runs over all (or some selected) rays from S (or from the initial surface Σ), arriving at R within a specified time window. Contributions $u_n(R, t)$ correspond to the individual rays and are given by the equations of Section 6.1.

Formula (6.2.1) is usually called the *ray expansion formula*. Considerable attention in the seismological literature has been devoted to the automatic generation of numerical codes of multiply reflected waves in horizontally layered media with homogeneous layers; see, for example, Hron (1971, 1972) and Červený, Molotkov, and Pšenčík (1977). In such media, the elementary waves may be effectively grouped into families of kinematic and dynamic analogues. In this case, only a *finite number* of elementary waves arrives at the receiver within a specified time window. (The number of elementary waves, however, may be tremendously high.) In inhomogeneous layered structures, an infinite number of elementary waves may arrive at the receiver, even if a time window of finite length is considered. Thus, a ray synthetic seismogram (6.2.1) *cannot be generally complete*, but only *partial ray expansion* is possible. Moreover, in laterally varying layered structures, the elementary waves must be treated individually; they cannot be grouped into families of kinematic and dynamic analogues. The ray method is the most effective in situations in which only a small number of elementary waves needs to be computed. A medium composed of thick layers separated by smooth interfaces may serve as a good example, particularly if the epicentral distances of the receivers are not too large.

6.2.2 Computation of Ray Synthetic Seismograms in the Time Domain

Several methods can be used if the synthetic seismograms are to be computed in the time domain.

The first method is based on the *summation of elementary synthetic seismograms*. It evaluates the elementary synthetic seismograms for a given source-time function one after another and performs summation (6.2.1). This method is conceptually very simple, but it is numerically efficient only if the number of elementary synthetic seismograms is small. Moreover, it requires a very effective algorithm to compute the individual elementary seismograms. It has been used traditionally mainly in connection with the approximate computation of the ray synthetic seismograms using relation (6.1.12), particularly for the Gabor source-time function; see (6.1.13). For example, the method is used in the 2-D program packages SEIS83 and SEIS88; see Červený and Pšenčík (1984b). For a large number of elementary waves, the method is not efficient.

A more efficient method is based on *impulse synthetic seismograms*, see Equation (6.1.14). We first compute the complex-valued time series

$$\bar{u}_n^I(R, t) = \sum_{(\Omega)} U_n(R)\delta(t - T(R)) \tag{6.2.2}$$

and obtain

$$\bar{u}_n(R, t) = \mathrm{Re}\left\{F(t) * \bar{u}_n^I(R, t)\right\} = \mathrm{Re}\left\{F(t) * \sum_{(\Omega)} U_n(R)\delta(t - T(R))\right\}. \tag{6.2.3}$$

Therefore, we need to determine two *real-valued impulse synthetic seismograms* for $\mathrm{Re}\sum U_n(R)\delta(t - T(R))$ and $\mathrm{Im}\sum U_n(R)\delta(t - T(R))$; see (6.2.2). Convolving the impulse synthetic seismograms with the relevant analytical signal and taking its real part, we obtain the final result.

Alternatively, we can use equation $F(t) * \delta(\zeta) = x(t) * \delta^{(A)}(\zeta)$, see (A.3.14), to obtain

$$\bar{u}_n(R, t) = \text{Re}\left\{ x(t) * \sum_{(\Omega)} U_n(R)\delta^{(A)}(t - T(R)) \right\}$$

$$= x(t) * \text{Re}\left\{ \sum_{(\Omega)} U_n(R)\delta^{(A)}(t - T(R)) \right\}. \qquad (6.2.4)$$

The sum in (6.2.4) represents a *complex-valued impulse synthetic seismogram* (for the source-time function given by the Dirac delta function). The application of (6.2.3), how-ever, will usually be more efficient because it uses only the real-valued δ functions in the summation and does not require the analytical delta signal $\delta^{(A)}(t - T)$ to be calculated.

6.2.3 Computation of Ray Synthetic Seismograms for Complex-Valued Travel Times

If the model under consideration is dissipative, we obtain complex-valued travel times T. The equations of the preceding section must then be modified. In this case, the simplest approach to derive the equations for the synthetic seismograms is to insert (A.3.21) into (6.2.4). We then obtain

$$\bar{u}_n(R, t) = \text{Re}\left\{ x(t) * \sum_{(\Omega)} U_n(R)\left(\frac{-\text{i}}{\pi(t - T(R))} \right) \right\}$$

$$= \text{Im}\left\{ x(t) * \frac{1}{\pi}\sum_{(\Omega)} \frac{U_n(R)}{t - T(R)} \right\} = x(t) * \text{Im}\frac{1}{\pi}\sum_{(\Omega)} \frac{U_n(R)}{t - T(R)}, \qquad (6.2.5)$$

where $T(R)$ is complex-valued. Alternatively, we can also use $x(t) * (-\text{i}/\pi(t - T)) = F(t) * \text{Im}(1/\pi(t - T))$ to arrive at

$$\bar{u}_n(R, t) = \text{Re}\left\{ F(t) * \frac{1}{\pi}\sum_{(\Omega)} U_n(R)\text{Im}\frac{1}{t - T(R)} \right\}. \qquad (6.2.6)$$

The last equation seems to be most efficient for complex-valued travel times.

6.2.4 Computation of Ray Synthetic Seismograms in the Frequency Domain

In the frequency domain, it is convenient to compute first *frequency response* $Y_n(R, f)$:

$$Y_n(R, f) = \sum_{(\Omega)} U_n(R) \exp[2\text{i}\pi f T(R)]. \qquad (6.2.7)$$

If we multiply frequency response $Y_n(R, f)$ by the Fourier spectrum of the source-time function, we obtain the actual spectrum of the synthetic seismogram. Using the inverse Fourier transform, we obtain the ray synthetic seismogram. If travel time $T(R)$ is complex-valued, we can again use (6.2.7), without any modification.

The frequency domain approach is efficient particularly if we use the FFR algo-rithm. If the FFR algorithm cannot be used, the frequency domain approach becomes time-consuming, particularly if we are treating a large number of elementary waves. Nevertheless, the method allows various frequency-dependent effects to be included, which can hardly be included in the time-domain approach.

6.2.5 Modified Frequency-Response Expansions

The frequency-response ray expansion (6.2.7) can be modified to include a considerably broader class of applications by allowing $U_n(R)$ to be frequency-dependent:

$$Y_n(R, f) = \sum_{(\Omega)} U_n(R, f) \exp[2i\pi f T(R)]. \tag{6.2.8}$$

The expressions $U_n(R, f)$ may include even possible frequency-dependent effects corresponding to the travel time $T(R)$. As examples, we can name here the quasi-isotropic correction factor (5.4.39) of qP waves propagating in weakly anisotropic media or dissipation filters $D(R, S)$ introduced in Section 5.5. The expansion (6.2.8) simplifies if $U_n(R, f)$ can be factorized, $U_n(R, f) = U_n(R)X_n(f)$ (no summation over n), where $X_n(f)$ are some frequency filters, different for different elementary waves. Then

$$Y_n(R, f) = \sum_{(\Omega)} U_n(R)X_n(f) \exp[2i\pi f T(R)]. \tag{6.2.9}$$

Finally, if the filter is the same for all elementary waves, (6.2.9) yields

$$Y_n(R, f) = X_n(f) \sum_{(\Omega)} U_n(R) \exp[2i\pi f T(R)]. \tag{6.2.10}$$

The modified expansions (6.2.8) through (6.2.10) can be used in various applications, where (6.2.7) fails. We shall discuss here briefly several of them.

    a. Dissipative effects. Because these effects play a very important role in practical applications, Section 6.3 will be devoted to them.

    b. Singular regions. Diffracted waves. In caustic, critical, and transition regions, various modifications of the ray method can be used. The amplitudes of elementary waves under consideration are then usually frequency-dependent. Similarly, amplitudes of diffracted waves, pseudospherical waves, inhomogeneous waves, and the like are also frequency-dependent. Expansion (6.2.8), however, can be used even in these cases.

    c. Higher order approximations. Head waves. In this case, it is suitable to write the higher order approximations and higher order waves as independent elementary waves. Then expansion (6.2.9) can be used. The frequency filter $X_n(f)$ for individual contributions equals either 1 for the zeroth-order approximation, $(-2i\pi f)^{-1}$ for the first-order approximation, $(-2i\pi f)^{-2}$ for the second-order approximation, and so on. The contributions with $X_n(f) = (-2i\pi f)^{-k}$, for k fixed, can then be grouped, and (6.2.10) can be used for these groups. See Eisner and Pšenčík (1996).

    d. Hybrid combination of the ray method with other methods. Here the amplitudes of individual waves are usually frequency-dependent, and (6.2.8) should be used. As an example, let us name here the hybrid combination of the ray method with the matrix method, in which the R/T coefficients and conversion coefficients are frequency-dependent. Such hybrid combination has been successfully used for models with thin transition layers and with thin subsurface layering. See Červený (1989b).

e. **Point sources with different source-time functions.** If sources with different source-time functions are considered independently, standard ray expansions (6.2.7) can be used. Modifications, however, are required if such sources should be used in one computation, together. As an example, let us consider single-force point sources and moment-tensor point sources; see Section 5.2.3. A similar example are single-force point sources and explosive point sources; see Section 5.2.3. See also Douglas, Young, and Hudson (1974).

f. **Point sources with frequency-dependent radiation functions.** In this case, (6.2.8) or (6.2.9) should be used. As an example, let us present the radiation functions of point sources situated close to structural interface and/or to the Earth's surface; see Jílek and Červený (1996).

g. **Point and line sources.** If point and line sources are considered independently, standard ray expansions (6.2.7) can be used in both cases. The modification is required if we wish to consider both sources together, in one computation, or to compare ray synthetic seismograms with those computed by finite-difference method for a 2-D model with a line source. See more details in Sections 2.6.3, 5.1.12, and 5.2.15.

h. **Frequency filters of recording instruments.** The frequency-domain characteristics of recording instruments can be introduced in ray-theoretical computation using (6.2.10). The filter $X_n(f)$ can also be used to transform computed seismograms (in displacements) to velocigrams ($X_n = -2i\pi f$) or to accelerograms ($X_n = -4\pi^2 f^2$). Moreover, frequency filters corresponding to the local structure close to the recording equipment can also be introduced in this way.

The preceding list of applications of modified frequency-response expansions is far from being complete. It would be possible to present here many other examples. Of course, many of the foregoing frequency filters may be suitably replaced by a convolution in the time domain or by direct computations in the time domain.

6.2.6 Time-Domain Versions of Integral Solutions

Certain more general solutions of the elastodynamic equation, presented in this book, have been constructed by integral superposition of asymptotic ray-based solutions. In the frequency domain, they can be mostly expressed in the following integral form:

$$u_n(R, f) = \iint_{\mathcal{D}} A_n(R, \gamma_I) \exp[i2\pi f \theta(R, \gamma_I)] d\gamma_I. \tag{6.2.11}$$

Here $f = \omega/2\pi$ is the frequency, R denotes the position of the receiver, γ_1 and γ_2 are ray parameters, $\mathcal{D}$ is some region in the ray parameter domain, $A_n(R, \gamma_I)$ are components of a vectorial amplitude function, and $\theta(R, \gamma_I)$ is a scalar phase function. Each of $A_n(R, \gamma_I)$ and $\theta(R, \gamma_I)$ may be either real-valued or complex-valued.

The frequency-domain solution can be transformed into the time-domain solution using the Fourier integral:

$$u_n(R, t) = 2 \operatorname{Re} \int_0^\infty X(f) u_n(R, f) \exp[-i2\pi f t] df$$

$$= 2 \operatorname{Re} \iint_{\mathcal{D}} A_n(R, \gamma_I)$$

$$\times \left(\int_0^\infty X(f) \exp[-i2\pi f(t - \theta(R, \gamma_I))] df \right) d\gamma_I;$$

see (A.1.3). $X(f)$ is the complex Fourier spectrum of the signal $x(t)$ under consideration, for example, of the source-time function. Using (A.3.15), we obtain

$$u_n(R, t) = \text{Re} \iint_\mathcal{D} A_n(R, \gamma_I) x^{(A)}(t - \theta(R, \gamma_I)) d\gamma_I. \tag{6.2.12}$$

This is the time-domain analogue of the frequency-domain solution (6.2.11), and $x^{(A)}(t - \theta)$ is the analytical signal corresponding to $x(t - \theta)$; see (A.3.2).

The time-domain solution (6.2.12) can be expressed in several alternative forms. Using relation $x^{(A)}(t - \theta) = x(t) * \delta^{(A)}(t - \theta)$, (6.2.12) yields

$$u_n(R, t) = x(t) * \text{Re} \iint_\mathcal{D} A_n(R, \gamma_I) \delta^{(A)}(t - \theta(R, \gamma_I)) d\gamma_I. \tag{6.2.13}$$

$\delta^{(A)}(t - \theta)$ denotes the analytical delta function; see (A.3.6). Finally, we can use (A.3.21) in (6.2.13) and obtain

$$u_n(R, t) = x(t) * \pi^{-1} \text{Im} \iint_\mathcal{D} \frac{A_n(R, \gamma_I)}{t - \theta(R, \gamma_I)} d\gamma_I. \tag{6.2.14}$$

The time-domain integrals (6.2.12) through (6.2.14) can be used both for A_n and θ complex-valued.

The frequency-domain superposition integral (6.2.11) may contain some simple frequency-dependent multiplicative factor, for example factor $-i\omega = -i2\pi f$. We do not consider such factors here because they can be connected with the Fourier spectrum $X(f)$. For example, factor $-i\omega$ yields the Fourier spectrum $-i2\pi f X(f) = \mathcal{F}(\dot{x})$, where $\dot{x} = dx/dt$. Thus, if (6.2.11) is multiplied by $(-i\omega)$, $x^{(A)}(t - \theta)$ in (6.2.12) should be replaced by $\dot{x}^{(A)}(t - \theta)$, and $x(t)$ in (6.2.13) and (6.2.14) should be replaced by $\dot{x}(t)$.

With properly specified functions $A_n(R, \gamma_I)$ and $\theta(R, \gamma_I)$, (6.2.11) represents various integral solutions of the elastodynamic equation derived in this book. It represents the solutions based on the summation of paraxial ray approximations and on the summation of paraxial Gaussian beams; see Section 5.8. Consequently, it also represents various forms of the Maslov-Chapman integrals, derived in different ways. In this case, integration is performed in the ray parameter domain. It also represents Kirchhoff integrals; see Sections 5.1.11 and 5.4.8. In this case, integration is performed along surface Σ^0, which may again be parameterized by ray parameters. In the summation of paraxial Gaussian beams, phase function $\theta(R, \gamma_I)$ is complex-valued; in all other cases, it is real-valued. Expressions (6.2.11) through (6.2.14) remain valid for a one-parameteric system of rays, with ray parameter γ_1. In this case, the double integral $\iint_\mathcal{D}$ must, of course, be replaced by a single integral. The foregoing integrals then also represent the superposition integrals in the 2-D and 1-D models; see Section 5.8.6.

For complex-valued phase function $\theta(R, \gamma_I)$ (for example, in the summation of paraxial Gaussian beams), the most efficient way is to perform the computation of synthetic seismograms in the frequency domain. The computation is very fast and effective because the fast frequency response algorithm can be used. For this reason, we shall consider only *real-valued phase function* $\theta(R, \gamma_I)$ in the following. We can then use the relation $x^{(A)}(t - \theta) = x^{(A)}(t) * \delta(t - \theta)$ in (6.2.12) and obtain

$$u_n(R, t) = \text{Re}\left\{ x^{(A)}(t) * \iint_\mathcal{D} A_n(R, \gamma_I) \delta(t - \theta(R, \gamma_I)) d\gamma_I \right\}. \tag{6.2.15}$$

Thus, the integral over $\mathcal{D}$ reduces to a line integral in $\mathcal{D}$ along the *equal-phase line* $t = \theta(R, \gamma_I)$, also called the *isochrone*. The relevant line integral along the isochrone in $\mathcal{D}$ can be modified in two ways.

1. IMPULSE INTEGRALS

We introduce a variable γ along the isochrone $t = \theta(R, \gamma_I)$ in $\mathcal{D}$ and substitute γ_1, $\gamma_2 \to \theta$, γ in (6.2.15). We can introduce γ so that $d\gamma = [d\gamma_1^2 + d\gamma_2^2]^{1/2}$ along the isochrone in $\mathcal{D}$. Using also the relation $(\partial\theta/\partial\gamma_I)(\partial\gamma_I/\partial\gamma) = 0$, we obtain

$$d\theta d\gamma = \det\begin{pmatrix} \partial\theta/\partial\gamma_1 & \partial\theta/\partial\gamma_2 \\ \partial\gamma/\partial\gamma_1 & \partial\gamma/\partial\gamma_2 \end{pmatrix} d\gamma_1 d\gamma_2$$

$$= \left(\left(\frac{\partial\theta}{\partial\gamma_1}\right)^2 + \left(\frac{\partial\theta}{\partial\gamma_2}\right)^2 \right)^{1/2} d\gamma_1 d\gamma_2.$$

Then (6.2.15) yields

$$u_n(R, t) = \mathrm{Re}\left\{ x^{(A)}(t) * \int_{t=\theta(R,\gamma_I)} A_n(R, \gamma_I) \left[\left(\frac{\partial\theta}{\partial\gamma_1}\right)^2 + \left(\frac{\partial\theta}{\partial\gamma_2}\right)^2 \right]^{-1/2} d\gamma \right\}. \tag{6.2.16}$$

The integral is taken along the isochrone $t = \theta(R, \gamma_I)$ in $\mathcal{D}$. Consequently, γ_I depend on time t, and also the integrand of (6.2.16) is a function of t.

Now we shall consider one-parameteric single integrals, with one ray parameter γ_1 only. The example is the WKBJ integral; see Section 5.8.6. In this case, the integral (6.2.16) along the isochrone in $\mathcal{D}$ is replaced by the sum over the individual points at which $t = \theta(R, \gamma_1)$. Also, $[(\partial\theta/\partial\gamma_1)^2 + (\partial\theta/\partial\gamma_2)^2]^{1/2}$ should be replaced by $|\partial\theta/\partial\gamma_1|$. Equation (6.2.16) then yields

$$u_n(R, t) = \mathrm{Re}\left\{ x^{(A)}(t) * \Sigma_{t=\theta(R,\gamma_1)} A_n(R, \gamma_1)/|\partial\theta(R, \gamma_1)/\partial\gamma_1| \right\}. \tag{6.2.17}$$

The equation analogous to (6.2.17) for vertically inhomogeneous media was first derived by Chapman (1976a, 1976b) and by Wiggins (1976). The Wiggins' (1976) derivation was based on physical argumentation; he speaks of the "disk ray theory." For a detailed derivation and discussion, see Chapman (1978).

2. BAND-LIMITED INTEGRALS

We shall now derive an important approximate expression for $u_n(R, t)$, which includes some smoothing. We use the box car window $B(t) = \frac{1}{2}(H(t + 1) - H(t - 1))$, where $H(t)$ is the Heaviside function (see (A.1.15)), and introduce a new window:

$$\frac{B(t/\Delta t)}{\Delta t} = \frac{1}{2\Delta t}\left[H\left(\frac{t + \Delta t}{\Delta t}\right) - H\left(\frac{t - \Delta t}{\Delta t}\right) \right]. \tag{6.2.18}$$

Δt denotes the digitalization interval of the discrete time series. Instead of $u_n(R, t)$, we shall compute $u_n^a(R, t) = u_n(R, t) * B(t/\Delta t)/\Delta t$. It can be proved that $u_n^a(R, t)$ is the average value of $u_n(R, t)$ over time interval $2\Delta t$:

$$u_n^a(R, t) = u_n(R, t) * \frac{B(t/\Delta t)}{\Delta t} = \frac{1}{2\Delta t}\int_{t-\Delta t}^{t+\Delta t} u_n(R, \tau)d\tau. \tag{6.2.19}$$

If we take into account that $\delta(t - \theta) * B(t/\Delta t)/\Delta t = B((t - \theta)/\Delta t)/\Delta t$, we can express (6.2.15) in the following form:

$$u_n^a(R, t) = \frac{1}{2\Delta t} \operatorname{Re}\left\{ x^{(A)}(t) * \iint_{t\pm\Delta t=\theta} A_n(R, \gamma_I)d\gamma_I \right\}. \tag{6.2.20}$$

The integration is not performed along the whole of $\mathcal{D}$, but only along the strip $t - \Delta t \leq \theta(R, \gamma_I) \leq t + \Delta t$ in $\mathcal{D}$.

Indeed, integral (6.2.20) is surprisingly simple. For a given time t, only the vectorial amplitude function $A_n(R, \gamma_I)$ is integrated over the strip $t \pm \Delta t = \theta$, situated along isochrone $t = \theta$. Suitable numerical algorithms were proposed by Spencer, Chapman, and Kragh (1997). They are based on the triangulation of the strip, with the ray-theory results known at the apexes of the triangular elements, and on the linear interpolation of ray-theory quantities inside the individual triangles.

For one-parametric single integrals (for example, for the WKBJ integrals, see Section 5.8.7), Equation (6.2.20) yields

$$u_n^a(R, t) = \frac{1}{2\Delta t} \operatorname{Re}\left\{ x^{(A)}(t) * \int_{t\pm\Delta t=\theta} A_n(R, \gamma_1)d\gamma_1 \right\}. \tag{6.2.21}$$

The integral is taken over all γ_1-intervals, satisfying for a given R and t the conditions $t - \Delta t \leq \theta(R, \gamma_1) \leq t + \Delta t$.

If $A_n(R, \gamma_1)$ varies slowly and smoothly within individual γ_1-intervals, (6.2.21) yields approximately

$$u_n^a(R, t) = \frac{1}{2\Delta t} \operatorname{Re}\left\{ x^{(A)}(t) * \sum A_n(R, \bar{\gamma}_1)\Delta\gamma_1 \right\}. \tag{6.2.22}$$

Here the summation is again over all γ_1-intervals. Within each interval, $A_n(R, \gamma_1)$ is taken constant, that is, $A_n(R, \gamma_1) = A_n(R, \bar{\gamma}_1)$, where $\bar{\gamma}_1$ is determined by solving the equation $t = \theta(R, \bar{\gamma}_1)$. Further, $\Delta\gamma_1$ is the width of the γ_1-interval (measured along a straight line of constant t). In other words, interval $\Delta\gamma_1$ is defined by $t \pm \Delta t = \theta(R, \gamma_1)$. Equation (6.2.21) was first derived by Chapman (1978), where its detailed discussion can also be found. The relevant algorithms and computer programs are described in Chapman, Chu, and Lyness (1988). Chapman's algorithm to compute the WKBJ seismograms is now quite standard in seismology and seismic exploration.

A great advantage of the band-limited integrals is that they are not as sensitive to the approximation of the medium and to minor details of the model as the ray method itself. For a more detailed discussion, see Chapman (1978, Section 3.4).

A final note. Until now, we have discussed only single and double integrals. Analogous methods can, however, also be used for volume integrals, with the integrands expressed is the same form as in (6.2.11). The most important example is the Born scattering integral (2.6.18), with the ray-theory expressions for the displacement and Green function used in the integrand. A suitable numerical procedure to evaluate the time-domain version of the Born scattering integral is described in Spencer, Chapman, and Kragh (1997).

6.3 Ray Synthetic Seismograms in Weakly Dissipative Media

As in Section 5.5, we shall consider only homogeneous waves in *weakly dissipative media*. The method is based on the application of certain dissipation frequency filters to the amplitudes of the individual elementary waves. The dissipation filters can be computed by quadratures along real-valued rays of these elementary waves, calculated in nondissipative models. Thus, the method does not require the computation of complex-valued rays, which would be necessary for strongly dissipative media. For more details, see Krebes and Hron (1980a, 1980b, 1981), Krebes and Hearn (1985), Hearn and Krebes (1990a, 1990b), and Thomson (1997a), among others.

6.3.1 Dissipation Filters

It was shown in Section 5.5 that the effect of weak absorption on the amplitudes of seismic waves in the ray method may be expressed in terms of the dissipation frequency filter $D(R, S)$, given by (5.5.8),

$$D(R, S) = \exp[i\omega T_d(R, S)]. \tag{6.3.1}$$

Here $T_d(R, S)$ is a complex-valued frequency-dependent travel-time perturbation. It represents the difference between the travel time in the *viscoelastic (weakly dissipative) model* under consideration and the travel time in a specified frequency-independent elastic *background model*. Both travel times are calculated along a ray Ω^0 from S to R, *determined in the background model*. If $T_d(R, S)$ is complex-valued, but frequency-independent, we speak of *noncausal* absorption. Under *causal absorption*, $T_d(R, S)$ must be complex-valued and frequency-dependent. This yields the *material dispersion*, which is intrinsically connected with dissipation. Using the Kramers–Krönig dispersion relations, it can be proved that dissipation filter (6.3.1) *must be frequency-dependent* so that the existence of absorption always implies the dispersion of velocities. The dispersion, however, may be small if we do not consider large distances between S and R. Thus, the dispersion may be neglected in certain practical cases and only noncausal absorption can be considered.

In this section we shall describe the computation of synthetic seismograms in weakly dissipative media. We shall distinguish two cases.

a. Noncausal absorption, that is, $T_d(S, R)$ independent of frequency (but complex-valued).

b. Causal absorption, that is, $T_d(S, R)$ both complex-valued and frequency-dependent.

In the last part of the section, we shall describe one particularly simple case: wave propagation in a constant-Q model. If quality factor Q is constant in the whole model under consideration, the dissipation filter is extremely simple and may be applied to the wavefield as a whole, not divided into individual elementary waves.

6.3.2 Noncausal Absorption

For each elementary wave, quantity $T_d(R, S)$ can be expressed in terms of the *global absorption factor t^** (also called *t*-star) using the relation

$$T_d(R, S) = \tfrac{1}{2} i t^*(R, S). \tag{6.3.2}$$

Here $t^*(R, S)$ is given by (5.5.12) and can be evaluated simply along a known ray of the elementary wave under consideration. Because $t^*(R, S)$ is real-valued, $T_d(R, S)$ is imaginary-valued. Note that quantities $t^*(R, S)$ differ for different rays connecting S and R. The final expression for the elementary seismogram corresponding to a selected elementary wave is obtained by considering the complex-valued frequency-independent travel time $T(R, S) + \tfrac{1}{2} i t^*(R, S)$ instead of the standard real-valued travel time $T(R, S)$.

The final expressions for the displacement vector can be obtained in three ways:

a. **In the frequency domain.** In this case, the algorithm of fast frequency response can be used; see Section 6.1.7. It is used independently for each elementary wave, and the fact that $t^*(R, S)$ are different for different rays is of no consequence. The computation of synthetic seismograms for waves propagating in a noncausal model is practically as fast as in a perfectly elastic model. The only thing we have to do is to compute $t^*(R, S)$ along all the rays under consideration connecting S and R. This is, however, fast and easy, once the rays are known.

b. **In the time domain.** We can use any of equations (6.1.17), but again we use them separately for each elementary wave. Particularly simple is the first expression in (6.1.17). Putting $U_n(R) = U_n^R + iU_n^I$ and using (A.3.19), we obtain

$$u_n(R, t) = \mathrm{Re}\left\{x(t) * U_n(R)\delta^{(A)}(t - T(R))\right\}$$

$$= \frac{1}{\pi}U_n^R \int_{-\infty}^{\infty} \frac{x(u)T^I\,du}{(t - T^R - u)^2 + (T^I)^2}$$

$$+ \frac{1}{\pi}U_n^I \int_{-\infty}^{\infty} \frac{x(u)(t - T^R - u)du}{(t - T^R - u)^2 + (T^I)^2}. \tag{6.3.3}$$

Integrals in (6.3.3), divided by π, are also known as *Poisson integrals*. They have simple limits for $T^I \to 0$. The first has the limit $x(t - T^R)$, and the second has $-g(t - T^R)$, where g is the Hilbert transform of x. Thus, for $T^I \to 0$, (6.3.3) yields (6.1.10). In fact, Poisson integrals represent the analytical continuation of x and g from the real axis into a complex plane. See also the similar treatment for inhomogeneous waves in Section 2.2.9, particularly Equations (2.2.83) and (2.2.84).

c. **Approximate expressions.** Regarding monochromatic high-frequency signals with a broad envelope $a(t)$, the expressions for the signals propagating in a weakly dissipative media can also be derived approximately. They will not be given here because they can be obtained simply from the more general approximate expressions for causal signals presented in the next section; see (6.3.6).

6.3.3 Causal Absorption

In the case of causal absorption, $T_d(R, S)$ in (6.3.1) is frequency-dependent. See, for example, Futterman's dissipation filter (5.5.19) or Müller's dissipation filter (5.5.23). These dissipation filters are, as a rule, again functions of $t^*(R, S)$, but the exponent of $D(R, S)$ is not a linear function of frequency, as in the case of noncausal models. The simple approaches of Section 6.3.2 cannot be used to construct synthetic seismograms, but more time-consuming approaches must be used.

a. **In the frequency domain**, the fast frequency response cannot be used. We must calculate dissipation frequency filter $D(R, S)$ separately for all waves, frequencies, and receiver positions. Thus, the procedure becomes rather cumbersome. It may simplify considerably only if $T_d(R, S)$ is taken to be independent of frequency in the range of frequencies being considered. The FFR algorithm can then be used.

b. **In the time domain**, it would be necessary to find the time domain version of the dissipation filter. The results may then be obtained by convolution. The time version of the dissipation filter, $d(t; R, S)$, can be obtained by the inverse Fourier transform from dissipation filter $D(R, S)$:

$$d(t; R, S) = \tfrac{1}{2}\pi^{-1} \int_{-\infty}^{\infty} D(R, S)\exp[-i\omega t]d\omega. \tag{6.3.4}$$

This dissipation time function $d(t; R, S)$ can be computed easily for any dissipation filter $D(R, S)$. The graphical representations of $d(t; R, S)$ for various $D(R, S)$ can be found in many seismological publications.

c. **Approximate expressions.** If a high-frequency harmonic carrier with a broad smooth envelope $a(t)$ is involved, the expression for the signal propagating in a weakly dissipative media can be derived approximately. These expressions were derived by Červený and Frangié (1980, 1982) and used in computer packages to calculate synthetic seismograms in inhomogeneous weakly dissipative media. A very general form of the source-time function was considered:

$$x(t) = a(2\pi f_M(t - t_i)/\gamma)\cos(2\pi f_M(t - t_i) + v). \tag{6.3.5}$$

Here $a(\zeta)$ is assumed to be the broad and smooth envelope of the signal. Here we shall consider only a special case of (6.3.5), in which envelope $a(\zeta)$ is Gaussian, $a(\zeta) = \exp(-\zeta^2)$. Signal (6.3.5) then corresponds to the Gabor signal (6.1.6). The parameter controlling the width of the envelope is dimensionless quantity γ. For small γ, the envelope is narrow, and for large γ the envelope is broad. In applications, values of γ close to 4 are very common. See the examples in Figure 6.1. Parameter f_M is the frequency of the harmonic carrier, which is close to the prevailing frequency of the signal for sufficiently broad envelopes ($\gamma > 3$). For simplicity, we shall call f_M the prevailing frequency of the Gabor source-time function. Phase shift v vanishes for cosine-like signals and equals $-\pi/2$ for sine-like signals. Quantity v may, however, be arbitrary.

Now assume that the wave is generated by a point source situated at S and recorded at receiver R. Consider an arbitrary elementary wave propagating in a general 3-D layered weakly dissipative structure from S to R along ray Ω. For simplicity, we shall consider Futterman's dissipative model, but the same approach can also be applied to other dissipative models. We put $\omega^r = 2\pi f_M$, $V = V(f_M)$, and $Q = Q(f_M)$. Thus, the background model corresponds to frequency f_M. The travel time from S to R is $T(R, S) = \int_S^R V^{-1}ds$, and the global absorption factor is $t^*(R, S) = \int_S^R (VQ)^{-1}ds$. Both integrals are taken along ray Ω^0 in the background medium. We further denote the nth ray-centered component of the amplitude in the background medium by $U_n(R)$ and introduce the phase of the amplitude, $\chi_n(R)$ by $U_n(R) = |U_n(R)|\exp[i\chi_n(R)]$.

Then the nth ray-centered component of the displacement vector of the elementary wave under consideration corresponding to the Gabor source-time function, propagating in an inhomogeneous weakly dissipative layered medium from S to R is given by a simple approximate relation:

$$u_n(R, t) \doteq |U_n(R)|\exp\left[-(2\pi f_M(t - T^a)/\gamma)^2 - (\pi f_M t^*/\gamma)^2 - \pi f^* t^*\right]$$
$$\times \cos[2\pi f^*(t - T^a - \pi^{-1}t^*) + v - \chi_n(R)]. \tag{6.3.6}$$

Here

$$T^a = T(R, S) - \frac{t^*(R, S)}{\pi}\left(1 + \ln\frac{f^*}{f_M}\right) + t_i,$$
$$f^* = f_M(1 - 2\pi f_M t^*/\gamma). \tag{6.3.7}$$

For small $t^*(R, S)$ (say, $t^*(R, S) < 1$), (6.3.6) provides a good description of all the effects of causal absorption on the Gabor signal with a broad envelope (say, $\gamma > 3$) propagating in weakly dissipative media. The decrease of amplitudes with increasing t^* is mainly due to the term $-\pi f^* t^*$ in the exponent of (6.3.6). The envelope, however, varies in shape as the wave progresses. The prevailing frequency f^* of the signal decreases as the wave progresses; see the second equation of (6.3.7). The decrease of f^* with t^* depends on γ; it is larger for small γ, higher f_M, and higher t^*. The carrier $\cos[\ldots]$ propagates with a lower velocity than the envelope; see the additional term $-\pi^{-1} t^*$ in the expression for $\cos[\ldots]$. Numerical investigation of expression (6.3.6) and the investigation of its accuracy can be found in Červený and Frangié (1982). They show that signal (6.3.6) satisfies the condition of causality very well.

6.3.4 Constant–Q Model

The expressions for dissipation filter $D(R, S)$ simplify considerably if quality factor Q is constant in the whole model. Then $t^*(R, S) = Q^{-1} \int_S^R dT = Q^{-1} T(R, S)$. Thus, for constant Q, $t^*(S, R)$ is proportional to the travel time from S to R, and the dissipation filter reads

$$D(R, S) = \exp[-\omega T(R, S)/Q]. \tag{6.3.8}$$

For a given travel time $T(R, S)$, the dissipation filter does not depend on the type of wave, on the ray trajectory between S and R, or even on the position of point R.

For the source-time function of a short duration, the complete wavefield is composed of short signals, and their arrival times $T(R, S)$ may be approximately replaced by running time. Such filters can be effectively used even in finite-difference computations, where they can be directly applied to the successive time levels. See Zahradník (1982) and Zahradník, Jech, and Moczo (1990a, 1990b) for more details and various modifications of the filter. Some modifications can be applied even to the final complete wavefield. If the complete wavefield is computed for a perfectly elastic (nondissipative) medium, the modified filter may be applied to it a posteriori.

6.4 Ray Synthetic Particle Ground Motions

The study of the polarization of vectorial wavefields is a classical problem of physics, particularly of optics (Born and Wolf 1959; Grant and West 1965; Kravtsov and Orlov 1980). In seismology, polarization studies have a long tradition mainly in the investigations of surface waves. It has been known for a long time that the Rayleigh waves propagating along the surface of a homogeneous halfspace are elliptically polarized in the saggital plane and that the ratio of the vertical to the horizontal half-axis is close to 1.5 (Bullen and Bolt 1985; Pilant 1979). In seismic body wave studies, the main attention has been devoted to the effects of the Earth's surface on the polarization of these waves (Gutenberg 1952; Malinovskaya 1958; Nuttli 1959, 1961; Nuttli and Whitmore 1962; Mendiguren 1969; Herrman 1976; Evans 1984; Booth and Crampin 1985), particularly to the nonlinear polarization of S waves due to a postcritical incidence on the Earth's surface. Structural effects on the polarization of seismic body waves have been studied only exceptionally (Cormier 1984; Liu, Crampin, and Yardley 1990). The nonlinear polarization of S waves, however, may be caused not only by postcritical incidence of the S wave at the Earth's surface, but also by postcritical incidence of the S wave at an inner structural interface.

The interest in the polarization of seismic body waves has recently increased considerably, mainly in seismic exploration. There are, at least, three reasons for this increased interest. First, seismic prospecting has been tradionally based on P waves. Because P waves are, as a rule, linearly polarized, it was not necessary to consider more complex, nonlinear polarization. Recently, however, S waves have also been used frequently in various seismic methods (Dohr 1985; Puzyrev, Trigubov, and Brodov 1985; Danbom and Domenico 1986). These studies require a better knowledge of the nonlinear polarization of seismic S waves. Second, polarization studies require three-component (or, at least, two-component) recording equipment. In seismic prospecting for oil, three-component measurements were, until recently, more or less exceptional. Now, three-component measurements are being used more frequently (Gal'perin 1984). Third, polarization studies play an important role studying seismic anisotropy. The increased interest in seismic anisotropy has also stimulated interest in polarization measurements, and consequently, in the particle motions of seismic body waves in general structures (Crampin 1981, 1986).

In the investigation of particle motions of seismic body waves in complex structures, we shall consider only the zeroth-order approximation of the ray method. It is well known that the zeroth-order ray approximation may fail in certain singular regions. Of course, in these regions it will be necessary to support the ray method by a more sophisticated investigation. Various waves that cannot be described by the zeroth-order approximation of the ray method, such as head, diffracted, and inhomogeneous waves, are not considered either. In certain applications, it would be useful to apply the two-term ray method; see Section 5.7.9.

Seismic body waves have a transient character. In some other branches of physics (eletromagnetic waves), the wavefields are mostly time-harmonic. In the case of time-harmonic wavefields, the vectorial wavefield is polarized mostly linearly or elliptically. Usually, the waveform of seismic signals is roughly quasi-harmonic; that is, it is represented by a harmonic carrier with a smooth envelope (Gabor signal, Berlage signal). In general, however, seismic body wave signals are neither harmonic nor quasi-harmonic. It would be useful to study the polarization of general signals, not just of harmonic or quasi-harmonic ones.

In this section, we shall discuss the ray-theoretical polarization diagrams of elementary seismic body waves. The approach is based on the complex-valued representation of seismic signals (analytical signals). It is no more complicated than the approach used for harmonic waves; nevertheless, it yields considerably more general results, specifically two basic types of polarization of seismic body waves: linear and quasi-elliptical. The *quasi-elliptical polarization* is more complex than the standard elliptical polarization. The quasi-ellipses have the form of *elliptical spirals*. The elliptical spiral is affected considerably by the *envelope* of the signal under consideration. Thus, the envelope of the signal plays a very important role in the polarization studies of transient seismic body waves.

6.4.1 Polarization Plane

Consider the displacement vector $\vec{u}(R, t)$ at point R. The *polarization diagram* is then the locus of the end points of the displacement vector $\vec{u}(R, t)$, constructed at R for time t varying. We shall use the complex-valued representation (6.1.1) of the displacement vector of the seismic body wave under consideration. In vectorial form, (6.1.1) reads

$$\vec{u}(R, t) = \mathrm{Re}\{\vec{U}(R)F(t - T(R))\}. \tag{6.4.1}$$

Here $F(\zeta)$ is a high-frequency analytical signal, $F(\zeta) = x(\zeta) + ig(\zeta)$, where $x(\zeta)$ and $g(\zeta)$ are real-valued functions forming a Hilbert transform pair; see (A.3.1). We shall express the analytical signal in the following form:

$$F(\zeta) = a(\zeta)\exp(-i\phi(\zeta)), \tag{6.4.2}$$

where $a(\zeta)$ is the *envelope* of both $x(\zeta)$ and $g(\zeta)$, and $\phi(\zeta)$ is the so-called *phasogram*,

$$a(\zeta) = [x^2(\zeta) + g^2(\zeta)]^{1/2}, \qquad \phi(\zeta) = -\arctan(g(\zeta)/f(\zeta));$$

see (A.3.23) and (A.3.24). We shall consider only *normalized analytical signals* $F(\zeta)$ for which the maximum of envelope function $a(\zeta)$ is 1. We also speak of the *normalized envelope*. For harmonic waves with frequency f, $a(\zeta) = 1$ and $\phi(\zeta) = 2\pi f\zeta + \nu$, where ν is a real-valued constant.

Now

$$\vec{U}(R) = \vec{U}^R(R) + i\vec{U}^I(R), \tag{6.4.3}$$

where $\vec{U}^R(R)$ and $\vec{U}^I(R)$ are two real-valued vectors. Inserting (6.4.2) and (6.4.3) into (6.4.1) yields

$$\vec{u}(R, t) = a(t - T(R))\{\vec{U}^R(R)\cos\phi(t - T(R)) + \vec{U}^I(R)\sin\phi(t - T(R))\}. \tag{6.4.4}$$

This vectorial equation describes the ray synthetic particle ground motion, corresponding to seismic body wave (6.4.1) in a 3-D space at point R. Evidently, the particle ground motion is confined to a plane specified by real-valued vectors $\vec{U}^R(R)$ and $\vec{U}^I(R)$.

Equation (6.4.4) immediately shows that the particle motion trajectory is linear only in the following three cases:

$$\vec{U}^R(R) = 0 \quad \text{or} \quad \vec{U}^I(R) = 0 \quad \text{or} \quad \vec{U}^R(R) = c\,\vec{U}^I(R), \tag{6.4.5}$$

where c is a real-valued constant. In all the other cases, the polarization will be nonlinear.

Thus, with the exception of cases (6.4.5), the particle motion is nonlinear and confined to plane Σ_p specified by $\vec{U}^R$ and $\vec{U}^I$. We shall call Σ_p *the plane of polarization* or *polarization plane*. The plane of polarization is fully defined by point R and by unit normal $\vec{N}_p$ to the plane Σ_p:

$$\vec{N}_p = (\vec{U}^R \times \vec{U}^I)/|\vec{U}^R \times \vec{U}^I|. \tag{6.4.6}$$

6.4.2 Polarization Equations

In the numerical modeling of seismic wavefields and in practical measurements, we do not treat the displacement vector $\vec{U}$ directly but rather treat its components in a suitably chosen local or global Cartesian coordinate system (for example, vertical, radial, and transverse). Consequently, we usually study the projection of the particle motion into the relevant coordinate planes (vertical-radial, vertical-transverse, transverse-radial). For simplicity, we shall speak of *polarization equations in the relevant planes*.

In this section, we shall derive the polarization equations in an arbitrary plane passing through point R. The plane may fully coincide with polarization plane Σ_p, but it may also be arbitrarily inclined with respect to Σ_p, or even perpendicular to it. The polarization equations we shall derive will also be applicable to points R situated on an interface or on the Earth's surface.

Let us consider a Cartesian coordinate system x_1, x_2, x_3 with three unit basis vectors $\vec{i}_1$, $\vec{i}_2$, and $\vec{i}_3$ and with its origin at R. Hence,

$$\vec{U} = U_1\vec{i}_1 + U_2\vec{i}_2 + U_3\vec{i}_3, \qquad U_k = U_k^R + iU_k^I. \tag{6.4.7}$$

Components U_1, U_2, and U_3 are, in general, complex-valued. In view of (6.4.3),

$$\vec{U}^R = U_1^R\vec{i}_1 + U_2^R\vec{i}_2 + U_3^R\vec{i}_3, \qquad \vec{U}^I = U_1^I\vec{i}_1 + U_2^I\vec{i}_2 + U_3^I\vec{i}_3. \tag{6.4.8}$$

We are interested in the particle ground motion in an arbitrary plane passing through R. Without loss of generality, we choose coordinate plane x_1-x_2, specified by unit vectors $\vec{i}_1$ and $\vec{i}_2$. Because the Cartesian coordinate system x_i may be chosen arbitrarily at R, this choice represents an arbitrary plane passing through R. To avoid unnecessary subscripts and superscripts, we shall denote

$$U_1 = B \exp[i\beta], \qquad U_2 = C \exp[i\gamma]. \tag{6.4.9}$$

Here the amplitudes of the individual components B and C are real-valued, which also applies to phase-shifts β and γ. The components U_1 and U_2 are said to be *phase-shifted*, if the following relation is valid:

$$\beta - \gamma \neq k\pi, \qquad k = 0, \pm 1, \pm 2, \ldots. \tag{6.4.10}$$

The *polarization equations* can now be expressed in the following simple form:

$$u_1 = aB \cos(\phi - \beta), \qquad u_2 = aC \cos(\phi - \gamma); \tag{6.4.11}$$

see (6.4.1), (6.4.2) and (6.4.9). Remember that a and ϕ are functions of the argument $\zeta = t - T(R)$ and that B, C, β, and γ are functions of position R only.

We shall now discuss the shape of the particle motion diagram described by polarization equations (6.4.11) and try to find some of its general characteristics. It is not difficult to eliminate ϕ from (6.4.11):

$$\mathbf{u}^T\mathbf{H}\mathbf{u} = a^2 B^2 C^2 \sin^2(\gamma - \beta), \qquad \mathbf{u} = (u_1, u_2)^T, \tag{6.4.12}$$

where the *polarization matrix* $\mathbf{H}$ reads

$$\mathbf{H} = \begin{pmatrix} C^2 & -BC \cos(\gamma - \beta) \\ -BC \cos(\gamma - \beta) & B^2 \end{pmatrix}. \tag{6.4.13}$$

Equation (6.4.12) does not represent a polarization equation and cannot be used to compute the particle motion trajectory. It is, however, an equation very convenient in discussing certain general properties of the particle ground motion diagram.

The polarization matrix $\mathbf{H}$ is real-valued and symmetric. Its eigenvalues H_1 and H_2 are as follows:

$$H_{1,2} = \tfrac{1}{2}(B^2 + C^2) \mp \tfrac{1}{2}\Delta,$$
$$\Delta = \left[(B^2 + C^2)^2 - 4B^2C^2 \sin^2(\gamma - \beta)\right]^{1/2}. \tag{6.4.14}$$

For $BC \sin(\gamma - \beta) \to 0$, we obtain an approximate relation for H_1:

$$H_1 \doteq \frac{B^2C^2}{B^2 + C^2} \sin^2(\gamma - \beta)\left(1 + \frac{B^2C^2}{(B^2 + C^2)^2} \sin^2(\gamma - \beta)\right). \tag{6.4.15}$$

We shall now introduce ϕ_1 and ϕ_2 to specify the direction of eigenvectors $\vec{h}^{(1)}$ and $\vec{h}^{(2)}$ of **H**. Angle ϕ_1 is measured from unit vector $\vec{i}_1$ to $\vec{h}^{(1)}$, and the angle ϕ_2 is measured from $\vec{i}_1$ to $\vec{h}^{(2)}$. They are given by the relations

$$\tan \phi_1 = \frac{\Delta + C^2 - B^2}{2BC \cos(\gamma - \beta)}, \qquad \phi_2 = \phi_1 + \pi/2. \qquad (6.4.16)$$

We shall take ϕ_1 within the limits $-\pi/2 \le \phi_1 \le \pi/2$, and, consequently, $0 \le \phi_2 \le \pi$.

If $BC \cos(\gamma - \beta) \to 0$, Equation (6.4.16) is not suitable for computations. Expanding Δ in terms of $BC \cos(\gamma - \beta)$ yields

$$\tan \phi_1 \doteq \frac{(C^2 - B^2)^2(1 + \text{sgn}(C - B)) + 2B^2C^2 \cos^2(\gamma - \beta) \, \text{sgn}(C - B)}{2BC(C^2 - B^2) \cos(\gamma - \beta)}.$$

$$(6.4.17)$$

For $B > C$, this yields

$$\tan \phi_1 = BC \cos(\gamma - \beta)/|B^2 - C^2|.$$

Consequently, if $BC \cos(\gamma - \beta) = 0, \phi_1 = 0$. This has, of course, been expected. If $\gamma - \beta = \pm(1 + 2k)\pi/2$ and $B = C$, angle ϕ_1 can be taken arbitrarily (circular polarization).

Let us now introduce new axes x_1' and x_2', related to eigenvectors $\vec{h}^{(1)}$ and $\vec{h}^{(2)}$, respectively. In this new coordinate system, Equation (6.4.12) can be expressed in diagonalized form,

$$u_1'^2/a_1^2 + u_2'^2/a_2^2 = a^2, \qquad (6.4.18)$$

where

$$a_1 = BC|\sin(\gamma - \beta)|/\sqrt{H_1}, \qquad a_2 = BC|\sin(\gamma - \beta)|/\sqrt{H_2}. \qquad (6.4.19)$$

For $BC \sin(\gamma - \beta) \to 0$, the expression (6.4.19) for a_1 is not suitable because $H_1 \to 0$, too. We can, however, insert (6.4.15) for H_1. Directly for $BC \sin(\gamma - \beta) = 0$, the quasi-elliptic polarization degenerates to linear polarization,

$$a_1 = (B^2 + C^2)^{1/2}, \qquad a_2 = 0.$$

For *time-harmonic waves*, the normalized envelope $a(t - T(R))$ in (6.4.18) is independent of time, $a(t - T(R)) = 1$. Equation (6.4.18) then represents an ellipse with the half-axes a_1 and a_2 given by (6.4.19). It is simple to see that a_1 corresponds to the greater half-axis and that a_2 corresponds to the smaller half-axis of the ellipse. The directions of the axes of the ellipse are determined by eigenvectors $\vec{h}^{(1)}$ and $\vec{h}^{(2)}$. Thus, for harmonic waves, we can call ellipse (6.4.18) with $a = 1$ the *polarization ellipse*. For $a_1 \ne 0$ and $a_2 \ne 0$, we speak of the *elliptical polarization* of time-harmonic waves.

In the case of transient signals corresponding to seismic body waves, the particle motion trajectory is more complicated. For time $t = t_m$ corresponding to the maximum of the envelope, $a(t_m - T(R)) = 1$; the particle motion diagram has a tangent point with the time harmonic polarization ellipse. For all other times, $a(t - T(R)) \le 1$. Thus, the complete polarization diagram of a transient seismic body wave is *bounded by the time-harmonic polarization ellipse*. For this reason, we shall, in this case, call ellipse (6.4.18) with $a = 1$ the *boundary polarization ellipse*. Inside the boundary ellipse, the polarization diagram has spiral character. In general, the particle motion trajectory of a transient seismic body wave has the shape of an *elliptic spiral*. We speak of *quasi-elliptical polarization* and of the *polarization quasi-ellipse*.

The reader is reminded that a_1 and a_2 are the half-axes of the boundary polarization ellipse. Similarly, eigenvectors $\vec{h}^{(1)}$ and $\vec{h}^{(2)}$ represent the directions of the axes of the boundary ellipse. We shall refer to a_1, a_2, $\vec{h}^{(1)}$, and $\vec{h}^{(2)}$ as the *principal parameters of the polarization quasi-ellipse*.

The typical shapes of polarization quasi-ellipses for several commonly used seismic signals are shown in Figure 6.2. The first two examples correspond to the Gabor signal (6.1.6) with $f_M = 4.5\,\text{Hz}$, $\nu = 0$, $t_i = 1\,\text{s}$, and two different widths, $\gamma = 4$ and $\gamma = 8$. The third example corresponds to the Ricker signal (6.1.5), with $\beta = 16\,\text{s}^{-1}$, $t_i = 1\,\text{s}$. The last example corresponds to the Berlage signal (6.1.8) with $f_M = 4.5\,\text{Hz}$, $\beta = 8\,\text{s}^{-1}$, $N = 0$, and $t_i = 1\,\text{s}$. To compute and plot the polarization diagrams in Figure 6.2, Equations (6.4.11) are used, with u_1 corresponding to the horizontal axis and u_2 corresponding to the vertical axis. Parameters B, C, β, and γ in all diagrams are chosen as follows: $B = 0.3$, $C = 1$, $\beta = -\frac{1}{2}\pi$, and $\gamma = 0$. In this case, Equations (6.4.11) read: $u_1(\zeta) = -0.3a(\zeta)\sin\phi(\zeta) = 0.3g(\zeta)$, $u_2(\zeta) = a(\zeta)\cos\phi(\zeta) = x(\zeta)$, with $\zeta = t - T(R)$; see (6.4.2).

The principal parameters of the polarization quasi-ellipse, a_1, a_2, $\vec{h}^{(1)}$, and $\vec{h}^{(2)}$, as introduced here, correspond to the boundary polarization ellipse. For signals with a broad envelope $a(\zeta)$, they roughly represent the actual size of the quasi-ellipse. For signals with narrower envelopes $a(\zeta)$, however, the effect of the envelope on the quasi-ellipse is greater, and the quantities a_1 and a_2 do not necessarily describe the size of the relevant quasi-ellipse properly.

Equations (6.4.19) for the half-axes a_1 and a_2 of the polarization quasi-ellipse, and Equation (6.4.17) for the direction of the principal axis of the quasi-ellipse (corresponding to the greater half-axis) are valid quite universally, even for the degenerate case of linear polarization.

The *eccentricity* of the boundary polarization ellipse $\epsilon = \sqrt{a_1^2 - a_2^2}/a_1 = \sqrt{1 - H_1/H_2}$. Let us discuss two limiting situations. For linear polarization ($B = 0$ or $C = 0$ or $\beta = \gamma$), $\epsilon = 1$. If $B = C \neq 0$ and $\gamma - \beta = \pm(2k + 1)\pi/2$, $\epsilon = 0$ (that is, circular polarization).

The particle motion can be clockwise or counterclockwise. We shall determine the conditions when it is clockwise and when it is counterclockwise. We know that ϕ is a linear function of t for harmonic waves, $\phi = 2\pi f_m(t - T(R))$. It is given by the same relation for quasi-harmonic waves. For more complex signals, ϕ is not necessarily a linear function of time, but it usually increases with time. We shall now consider only such signals with ϕ increasing with time. We introduce a polar angle ν by the relation,

$$\tan\nu = u_2/u_1 = C\cos(\phi - \gamma)/B\cos(\phi - \beta).$$

For $d\nu/d\phi$, we obtain

$$\frac{d\nu}{d\phi} = \frac{C\cos^2\nu\sin(\gamma - \beta)}{B\cos^2(\phi - \beta)}. \tag{6.4.20}$$

Thus, we can conclude:

    a. If $\sin(\gamma - \beta)$ is positive, $d\nu/d\phi$ is also positive, and the particle motion is counterclockwise. We also speak of a *retrograde movement*.

    b. If $\sin(\gamma - \beta)$ is negative, $d\nu/d\phi$ is also negative and the particle motion is clockwise. We speak of a *prograde movement*.

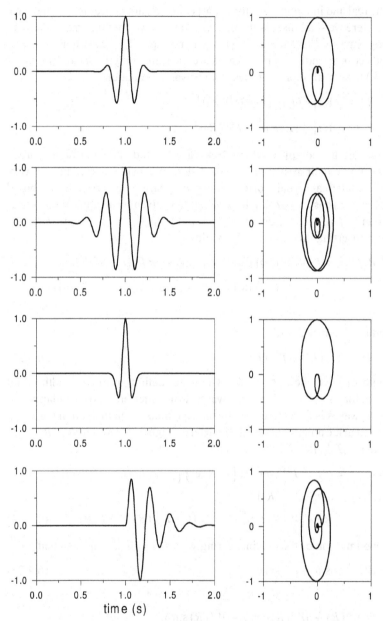

Figure 6.2. Examples of ray-theoretical polarization diagrams (particle ground motion trajectories) of an elementary wave for several simple signals. The polarization diagram at a point R is the locus of the end points of the displacement vector $\vec{u}(R, t)$, constructed at R for time t varying. It is either linear or has the shape of an elliptic spiral. In the latter case, we speak of a quasi-elliptic polarization and of the polarization quasi-ellipse. Each of the 2-D diagrams shows a projection of the particle ground motion into a plane. For details, see text.

6.4.3 Polarization of Interfering Signals

In the preceeding sections, we have shown that the polarization of a seismic body wave is quasi-elliptical if the real and imaginary vectorial parts of $\vec{U}(R)$ are nonvanishing and have different directions. Here we shall discuss the polarization of a wavefield corresponding to two interfering body waves $\vec{u}^{(1)}(R, t)$ and $\vec{u}^{(2)}(R, t)$. For simplicity, we shall assume that the analytical signals corresponding to both waves are the same, only $\vec{U}(R)$ and $T(R)$ are different. As in (6.4.1), we shall represent these two waves by equations

$$\vec{u}^{(1)}(R, t) = \mathrm{Re}\{\vec{V}(R)F(t - T^{(1)}(R))\},$$
$$\vec{u}^{(2)}(R, t) = \mathrm{Re}\{\vec{W}(R)F(t - T^{(2)}(R))\}.$$

(6.4.21)

Here $\vec{V}(R)$ and $\vec{W}(R)$ are the complex-valued vectorial amplitude factors of both waves, $T^{(1)}(R)$ and $T^{(2)}(R)$ denote their arrival times. We shall also consider a specific, but very common form of the analytical signal, corresponding to a harmonic carrier with a broad envelope $a(\zeta)$, $x(\zeta) = a(\zeta)\cos 2\pi f_M \zeta$. In this case, $g(\zeta)$ is given by an approximate relation $g(\zeta) \doteq -a(\zeta)\sin 2\pi f_M \zeta$, and we obtain $F(\zeta) \doteq a(\zeta)\exp(-2\mathrm{i}\pi f_M \zeta)$. Here f_M is the prevailing frequency. Inserting this into (6.4.21) yields

$$\vec{u}^{(1)}(R, t) = a(t - T^{(1)}(R))\,\mathrm{Re}\{\vec{V}(R)\exp[-2\mathrm{i}\pi f_M(t - T^{(1)}(R))]\},$$
$$\vec{u}^{(2)}(R, t) = a(t - T^{(2)}(R))\,\mathrm{Re}\{\vec{W}(R)\exp[-2\mathrm{i}\pi f_M(t - T^{(2)}(R))]\}.$$

(6.4.22)

We shall now introduce quantity $\Delta T(R)$

$$\Delta T(R) = T^{(2)}(R) - T^{(1)}(R)$$

(6.4.23)

and denote the width of envelope $a(\zeta)$ by h. (The exact definition of the width is not important here.) Then, for $|\Delta T(R)| > h$, the two waves do not interfere, and the polarization of both noninterfering waves is fully described by the equations given in Section 6.4.1. We are, however, interested in the polarization of the interfering signals for which $|\Delta T(R)| < h$. If $\Delta T(R)$ is small, $a(t - T^{(1)}(R)) \doteq a(t - T^{(2)}(R))$. Then in view of (6.4.22),

$$\vec{u}^{(1)}(R, t) = a(t - T^{(1)}(R))\,\mathrm{Re}\{\vec{V}(R)\exp[-2\mathrm{i}\pi f_M(t - T^{(1)}(R))]\},$$
$$\vec{u}^{(2)}(R, t) \doteq a(t - T^{(1)}(R))$$
$$\times\,\mathrm{Re}\{\vec{W}(R)\exp[-2\mathrm{i}\pi f_M(t - T^{(1)}(R) - \Delta T(R))]\}. \quad (6.4.24)$$

For the particle ground motion of the two interfering waves, $\vec{u} = \vec{u}^{(1)} + \vec{u}^{(2)}$ so that

$$\vec{u} = a\{\vec{U}^R \cos\phi + \vec{U}^I \sin\phi\},$$

(6.4.25)

where

$$\vec{U}^R = \vec{V}^R(R) + \vec{W}^R(R)\cos\delta - \vec{W}^I(R)\sin\delta,$$
$$\vec{U}^I = \vec{V}^I(R) + \vec{W}^I(R)\cos\delta + \vec{W}^R(R)\sin\delta.$$

(6.4.26)

We have also used the notation,

$$\phi = 2\pi f_M(t - T^{(1)}(R)), \qquad \delta = 2\pi f_M \Delta T(R), \qquad a = a(t - T^{(1)}(R)).$$

(6.4.27)

The motion is confined to polarization plane Σ_p specified by vectors $\vec{U}^R$ and $\vec{U}^I$ given by (6.4.26).

If we compare (6.4.25) with (6.4.4), we can see that all the formulae of the preceding section remain valid, only vectors $\vec{U}^R$ and $\vec{U}^I$ have a different meaning.

It is obvious from (6.4.26) that the superposition of two interfering signals will be, in general, quasi-elliptically polarized. It is polarized linearly only if one of the conditions (6.4.5) is satisfied, where $\vec{U}^R$ and $\vec{U}^I$ are given by (6.4.26). This is, however, not very common.

Relations (6.4.25) through (6.4.27) can be simply generalized for N interfering signals. The main conclusions remain valid even in this case.

In the derivation of (6.4.25) through (6.4.27), we have made several simplifying assumptions. Two assumptions are most important.

a. The analytical signals $F(\zeta)$ of both waves are the same.
b. $a(t - T^{(1)}(R)) \doteq a(t - T^{(2)}(R))$.

If these two assumptions are not satisfied, the particle ground motion may be even more complex than the quasi-elliptical.

6.4.4 Polarization of Noninterfering P Waves

The equations of Sections 6.4.1 through 6.4.3 are quite general and may be used to study the ray synthetic polarization of noninterfering P and S waves propagating in 3-D laterally varying structures.

For P waves, the treatment is simple. Let us first consider a *smooth medium*. In this case, $\vec{U}(R) = A(R)\vec{N}(R)$, where $\vec{N}$ is the unit vector normal to the wavefront and A is a scalar, possibly a complex-valued function. This implies that $\vec{U}^R(R) = c\,\vec{U}^I(R)$, where $c = \mathrm{Re}(A)/\mathrm{Im}(A)$ (for $\mathrm{Im}(A) \neq 0$). In view of (6.4.5), the noninterfering P wave in a smooth medium is always linearly polarized along the normal $\vec{N}$ to the wavefront.

Let us now discuss the particle ground motion at point R situated on a *structural interface*, where R is the point of incidence of a P wave. At point R situated at a structural interface, we need to take into account conversion matrices $\mathcal{D}_{ik}(R)$; see (5.2.67). The displacement vector at R will always be confined to the plane of incidence in the case of an incident P wave. We thus can only discuss the situation in the plane of incidence. We can consider only two elements $\mathcal{D}_{13}$, and $\mathcal{D}_{33}$ of conversion matrix $\mathcal{D}_{ik}(R)$; see (5.2.116). If the angle of incidence is subcritical, all these elements are real-valued, but for postcritical angles of incidence they are complex-valued.

This implies that the particle ground motion at point R situated on a structural interface *is linear* for a subcritically incident P wave at R. It is confined to the plane of incidence, but its direction deviates from the direction of the incident P wave at R.

For postcritical angles of incidence, however, the particle ground motion at R is, in general, *quasi-elliptical in the plane of incidence*. The reason is that $\mathcal{D}_{13}(R)$ and $\mathcal{D}_{33}(R)$ are not only complex-valued but are also phase-shifted.

As we know, the critical angle of incidence i_1^* is given by relation $\sin i_1^* = \alpha_1/\alpha_2$, where α_1 is the P wave velocity of the incident wave at R and α_2 is the P wave velocity at R, but on the other side of the interface. Thus, postcritical angles of incidence may exist only if $\alpha_2 > \alpha_1$.

At the Earth's surface, the incidence of P waves is always subcritical so that the polarization at point R situated at the Earth's surface is *always linear* assuming that a P wave is incident at R.

In the *vicinity of a structural interface or of the Earth's surface*, the incident P wave always interferes with the reflected PP and PS waves on one side of the interface, and the transmitted PP and PS waves interfere on the other side of the interface. Thus, the ray synthetic particle ground motion is, in general, quasi-elliptical in the vicinity of the interface; see Section 6.4.3. The exception is the situation corresponding to normal incidence.

The conclusions of this section are fully based on the zeroth-order approximation of the ray method. If *higher-order terms of the ray series* are also taken into account, the conclusions are different, the particle ground motion of a noninterfering P wave may be quasi-elliptical even in a smooth medium. This can be clearly demonstrated on the simplest case of the ray series method: the two-term ray method; see (5.7.40). The two-term ray method considers the additional component of the first-order approximation $\vec{W}^{(1)}$, in addition to the zeroth-order approximation $\vec{U}^{(0)}$, that is, $\vec{U}^{(0)} + (-i\omega)^{-1}\vec{W}^{(1)}$. Here $\vec{U}^{(0)}$ and $\vec{W}^{(1)}$ are mutually perpendicular. Even for the real-valued $\vec{U}^{(0)}$ and $\vec{W}^{(1)}$, there is a phase shift of 90° between the two components due to factor $(-i\omega)^{-1}$. Thus, in the two-term ray method, the noninterfering P wave is, in general, quasi-elliptical even in a smooth medium. The polarization plane contains $\vec{N}$, the normal to the wavefront. Of course, the shape of the boundary polarization ellipse depends considerably on the mutual relation between $\vec{U}^{(0)}$ and $(-i\omega)^{-1}\vec{W}^{(1)}$. In most cases, term $(-i\omega)^{-1}\vec{W}^{(1)}$ is small with respect to $\vec{U}^{(0)}$. The eccentricity of the boundary polarization ellipse is then close to unity, and the ellipse is very thin, close to linear polarization. The quasi-elliptic polarization of P waves may be more pronounced only if $\vec{U}^{(0)}$ is small, for example, close to nodal lines.

6.4.5 Polarization of Noninterfering S Waves in a Smooth Medium

In a smooth medium, the complex-valued vectorial amplitude function $\vec{U}(R)$ of a noninterfering S wave is given by the relation

$$\vec{U}(R) = U_1^{(q)}(R)\vec{e}_1(R) + U_2^{(q)}(R)\vec{e}_2(R); \tag{6.4.28}$$

see (5.2.4) and (5.2.7). Here $\vec{e}_1(R)$ and $\vec{e}_2(R)$ are the basis vectors of the ray-centered coordinate system, situated in the plane tangent to the wavefront at R. Note that the three unit vectors $\vec{e}_1$, $\vec{e}_2$, and $\vec{e}_3 \equiv \vec{N}$ form a mutually orthogonal, right-handed triplet of unit vectors. Scalar functions $U_1^{(q)}(R)$ and $U_2^{(q)}(R)$ represent the ray-centered components of vectorial amplitude function $\vec{U}(R)$ and are, in general, complex-valued.

Thus, the polarization plane Σ_p of a noninterfering S wave propagating in a smooth medium is represented by a plane tangent to the wavefront at R. We shall now investigate the particle ground motion of S waves directly in plane Σ_p. In fact, we can use the results of Section 6.4.2 directly, if we introduce the Cartesian coordinate system at R so that $\vec{i}_1 \equiv \vec{e}_1$, $\vec{i}_2 \equiv \vec{e}_2$, and $\vec{i}_3 \equiv \vec{N}$. Polarization plane Σ_p then coincides with the Cartesian coordinate plane x_1-x_2, where the x_1-axis is taken along $\vec{e}_1$ and the x_2-axis is taken along $\vec{e}_2$. It is obvious that, in this case, the x_1 and x_2 coordinates coincide with ray-centered coordinates q_1 and q_2.

If we use the notation $U_1^{(q)} = B\exp[i\beta]$, $U_2^{(q)} = C\exp[i\gamma]$, where B, C, β, and γ are real-valued (see (6.4.9)), we can conclude that the S wave in a smooth medium is *linearly polarized only if*

$$BC\sin(\beta - \gamma) = 0. \tag{6.4.29}$$

Thus, if (6.4.29) is satisfied, the particle ground motion of a noninterfering S wave in a smooth medium is *linear* in plane Σ_p, tangent to the wavefront at R.

If (6.4.29) is not satisfied, the particle ground motion of a noninterfering S wave in a smooth medium is *quasi-elliptical* in plane Σ_p tangent to the wavefront at R. The half-axes of the boundary polarization ellipse, a_1 and a_2, are given by (6.4.19) and are oriented along the eigenvectors $\vec{h}^{(1)}$ and $\vec{h}^{(2)}$ of polarization matrix **H**; see (6.4.13).

Thus, the particle ground motion of S waves in a smooth medium is quasi-elliptical only if both B and C are nonvanishing ($B \neq 0$, $C \neq 0$) and if $U_1^{(q)}$ and $U_2^{(q)}$ are phase-shifted ($\beta - \gamma \neq k\pi$).

Now we shall investigate the *continuation of the particle ground motion diagrams* of noninterfering S waves along the ray in a smooth medium. We shall first mention three facts.

a. Polarization plane Σ_p is perpendicular to the ray at any point of the ray.

b. The normalized envelope of the wave propagating in a nondissipative medium does not change as the wave progresses along the ray; see Section 6.1.2.

c. The ray-centered components $U_1^{(q)}$ and $U_2^{(q)}$ of an S wave propagating in a smooth medium vary along the ray in the same way.

Here (c) follows from the fact that the transport equations for $U_1^{(q)}$ and $U_2^{(q)}$ are the same; see (2.4.37) and (2.4.38). Thus, we can formulate the *conservation law of the normalized polarization quasi-ellipse*: The principal parameters of the normalized polarization quasi-ellipse of an S wave propagating in a nondissipative isotropic smooth medium do not change along the ray. By the principal parameters of the normalized polarization quasi-ellipse we understand the half-axes a_1 and a_2 of the normalized polarization boundary ellipse and the eigenvectors $\vec{h}^{(1)}$ and $\vec{h}^{(2)}$, specifying the direction of both axes of the boundary ellipse, *with respect to unit vectors $\vec{e}_1$ and $\vec{e}_2$.*

If the smooth part of the ray under consideration does not contain any caustic point, the conservation law may even be formulated more generally. The *shape* of the normalized polarization quasi-ellipse of an S wave propagating in a nondissipative isotropic smooth medium and its orientation with respect to unit vectors $\vec{e}_1$ and $\vec{e}_2$ do not change along the smooth part of the ray, which does not contain caustic points.

Note that the principal parameters of the normalized polarization quasi-ellipse do not change as the S wave passes through a caustic point, but the shape of the normalized polarization quasi-ellipse does change. It is rotated by $\frac{1}{2}\pi$ (or by π for focus) inside the boundary ellipse, changing its shape appropriately.

The conservation law remains valid even for linearly polarized S waves. Thus, if a linearly polarized S wave has the direction of $\vec{e}_1$ at any reference point of the ray, it has the direction of $\vec{e}_1$ along the whole smooth part of the ray. Similarly, if the displacement vector of an S wave makes an angle θ with $\vec{e}_1$ at any reference point of the ray, it makes the same angle along the whole smooth part of the ray.

Thus, the basis vectors $\vec{e}_1$ and $\vec{e}_2$ of the ray-centered coordinate system play a very important role in the investigation of the variations of the polarization of S waves along the ray in a smooth medium: the polarization of S waves *remains fixed* with respect to the unit vectors $\vec{e}_1$, $\vec{e}_2$. For this reason, $\vec{e}_1$ and $\vec{e}_2$ are also called *polarization vectors.*

It should be emphasized that the polarization vectors of S waves, $\vec{e}_1$ and $\vec{e}_2$, rotate along the ray with respect to the unit normal $\vec{n}$ and unit binormal $\vec{b}$ as the wave progresses, if a 3-D nonplanar ray with a nonzero torsion is considered. In other words, the polarization quasi-ellipse of the S wave rotates with respect to $\vec{n}$ and $\vec{b}$. The relevant angle between the polarization vector and $\vec{n}$ or $\vec{b}$ can also be determined by the classical Rytov law; see (4.1.14).

Let us now discuss the *projection of the particle ground motion of an S wave* propagating in a smooth medium into various planes passing through R. We shall only study the situation in which the polarization of S waves in polarization plane Σ_p is quasi-elliptical. (The case of linear polarization in Σ_p is elementary.) In general, the projection into any plane passing through R is again quasi-elliptical. There is, however, one important exception. The projection of the quasi-ellipse into a plane perpendicular to plane Σ_p *is linear*.

The foregoing conclusion has a very important consequence in 1-D and 2-D studies. Assume that ray Ω of an S wave is fully situated in plane $\Sigma^{\|}$. Plane $\Sigma^{\|}$ is often vertical. For simplicity, we shall also treat it as a vertical plane. It is then usual to decompose the S wave into two components: SV and SH. The SH component is perpendicular to $\Sigma^{\|}$, and the SV component is confined to $\Sigma^{\|}$ (but perpendicular to Ω). The S wave in a smooth medium is quasi-elliptically polarized if the following three conditions are satisfied:

1. The amplitude of the SH component is nonvanishing.
2. The amplitude of the SV component is nonvanishing.
3. The amplitudes of the SH and SV components are phase-shifted.

Polarization plane Σ_p is perpendicular to Ω, so that it is also perpendicular to the plane of ray $\Sigma^{\|}$. The consequence is that the ray theoretical polarization of a noninterfering S wave propagating in a smooth medium along a planar ray Ω situated in plane $\Sigma^{\|}$ is *always linear in plane $\Sigma^{\|}$*. Thus, in some way we can say that the quasi-elliptical polarization of S waves in a smooth medium is a typical 3-D phenomenon. If we investigate the S wavefield in plane $\Sigma^{\|}$ only, the polarization of the S waves is linear. It can be quasi-elliptical only in planes not coinciding with $\Sigma^{\|}$.

Finally, we need to emphasize that all the conclusions of this section are valid only in the *zeroth-order approximation* of the ray method. As for P waves, the two-term ray method also yields quasi-elliptical polarization of S waves even if the zeroth-order approximation of the ray method yields linear polarization of the S waves. Polarization plane Σ_p, however, is not perpendicular to the ray. The two-term ray method yields quasi-elliptical polarization in a smooth medium even in the 2-D case in plane $\Sigma^{\|}$ of the ray. In most cases, however, the eccentricity of the polarization ellipse is close to unity, and the polarization ellipse is very thin and close to linear polarization.

6.4.6 Polarization of S Waves at Structural Interfaces

If the receiver is situated on a structural interface, we must consider conversion matrices $\mathcal{D}_{ik}(R)$; see (5.2.68). We shall first discuss the polarization in the *plane of incidence*. We again decompose the S wave into SH and SV components. In the plane of incidence, we need to consider only two elements $\mathcal{D}_{11}$ and $\mathcal{D}_{31}$ of conversion matrix $\mathcal{D}_{ik}(R)$. For subcritical angles of incidence, these elements are real-valued, but for postcritical angles of incidence, they are complex-valued, with $\mathcal{D}_{11}$ and $\mathcal{D}_{31}$ mutually phase-shifted. Thus, we can draw a conclusion related to an S wave incident at a structural interface at point R. The particle ground motion in the plane of incidence is linear if the S wave is incident subcritically; it is quasi-elliptical, if the S wave is incident postcritically.

For incident S waves, quasi-elliptical polarization at structural interfaces is more frequent than for incident P waves. The main reason is that the velocity of P waves α is always higher than the velocity of S waves β. Whereas angles of incidence can be critical only if $\alpha_2 > \alpha_1$ for incident P waves, they may be critical for any interface for incident S waves. One critical angle exists *always* and is given by the relation $i_1^* = \arcsin(\beta_1/\alpha_1)$. Of course, there may be one or two additional critical angles of incidence for the incident S wave, such as $\arcsin(\beta_1/\alpha_2)$ and $\arcsin(\beta_1/\beta_2)$, but they do not exist always. Thus, the critical angle of incidence $i_1^* = \arcsin(\beta_1/\alpha_1)$ is quite universal and is important even for a receiver situated at the Earth's surface.

We have so far only considered polarization in the plane of incidence. In general, however, the quasi-elliptic polarization due to the phase shift of the two components (horizontal and vertical) in the plane of incidence is combined with the quasi-elliptic polarization due to the phase shift of the SV and SH components. The final polarization plane Σ_p is neither perpendicular to ray Ω, nor coincident with plane $\Sigma^{\|}$; it is situated more generally. The particle ground motion in Σ_p is, however, again quasi-elliptic.

Close to structural interfaces, several waves always interfere, so that the polarization is, as a rule, quasi-elliptical. The exception are the normally incident S waves.

6.4.7 Polarization of S Waves at the Earth's Surface

We shall consider an S wave incident at point R situated on the Earth's surface. We again decompose the S wave at R into two components; one component (SH) is perpendicular to the plane of incidence, and the other (SV) is confined to the plane of incidence and is perpendicular to the ray.

As we know from Section 6.4.6, there is always one critical angle of incidence $i_1^* = \arcsin(\beta_1/\alpha_1)$, if an S wave is incident at the Earth's surface. Under standard conditions, the critical angle of incidence equals $25°-35°$. We call the range of angles of incidence $i_1 < i_1^*$ the *shear wave window*; see Crampin (1989). For angles of incidence within the shear wave window ($i_1 < i_1^*$), the polarization in the plane of incidence at point R is linear, but outside the shear wave window ($i_1 > i_1^*$) it is quasi-elliptical. This conclusion also applies if the SH component of the S wave vanishes at R.

For a nonvanishing SH component, particularly if the SH and SV components are phase-shifted, the polarization plane Σ_p again has a more general position; it does not coincide with the plane of incidence and is not perpendicular to the ray.

6.4.8 Causes of Quasi-Elliptical Polarization of Seismic Body Waves in Isotropic Structures

As explained in Sections 6.4.1 through 6.4.7, the polarization of seismic body waves may be nonlinear for several reasons. Let us first consider an elementary, noninterfering wave. The wave is quasi-elliptically polarized if vectorial amplitude factor $\vec{U}$ is complex-valued, $\vec{U} = \vec{U}^R + i\vec{U}^I$, with both $\vec{U}^R \neq 0$ and $\vec{U}^I \neq 0$, and if the directions of $\vec{U}^R$ and $\vec{U}^I$ are different. This can also be formulated otherwise: the wave is linearly polarized if $\vec{U}^R = 0$ or $\vec{U}^I = 0$ or if $\vec{U}^R = c\vec{U}^I$, where c is a real-valued nonvanishing constant.

Alternatively, we can also express $\vec{U}$ in a component form; see (6.4.9). We then can say that the wave is quasi-elliptically polarized in plane x_1-x_2 if both the components U_1 and U_2 are nonvanishing and if they are phase-shifted ($\beta - \gamma \neq k\pi$).

Evidently, the complex-valued character of $\vec{U}$ is not sufficient to generate a quasi-elliptical polarization of the wave. It is also required that the directions of $\vec{U}^R$ and $\vec{U}^I$ differ, or, alternatively, that the Cartesian complex-valued components of $\vec{U}$ are phase-shifted. We shall present two examples. Let us assume that a P wave with a real-valued $\vec{U} = \vec{U}^R$ passes through a caustic point. Beyond the caustic, the vectorial amplitude factor is imaginary-valued, $\vec{U} = -i\vec{U}^R$. The wave, however, remains linearly polarized. The example for the postcritically reflected PP wave is similar. Even though the reflection coefficient is complex-valued, the wave is linearly polarized.

In this section, we shall discuss all the possible causes of quasi-elliptical polarizations of seismic body waves. All explanations are based fully on ray theory considerations and assume the validity of the zeroth approximations of the ray method. Our conclusions cannot be applied to singular regions of the ray method and to waves that cannot be described by the ray method. The polarization may then be more complicated.

There are four main causes of quasi-elliptic polarization of seismic body waves in isotropic structures. They are as follows:

1. Complex-valued directivity patterns of the source.
2. Postcritical incidence of an S wave at the Earth's surface.
3. Postcritical incidence of an S wave at an inner structural interface.
4. Interference of two or more signals.

We shall now briefly discuss these causes.

1. COMPLEX-VALUED DIRECTIVITY PATTERNS OF THE SOURCE

We shall consider a point source of seismic body waves. Within the framework of the ray method, the amplitudes of seismic body waves generated by a point source are proportional to the directivity patterns of the source. In the ray method, the directivity patterns are usually introduced in such a way that they represent an angular distribution of the vectorial complex-valued amplitude factor along a sphere whose center is at the source. We denote it $\vec{U}_0$. Vector $\vec{U}_0$ has, in general, three components: a normal (radial) component, corresponding to P waves, and two tangential mutually perpendicular components, corresponding to S waves. Let us call them S1 and S2.

It is obvious that the P wave generated by a point source will always be linearly polarized, even if the relevant directivity patterns are complex-valued. The reason is that, in this case, $\vec{U}_0^R$ and $\vec{U}_0^I$ have the same direction.

The S wave generated by the point source, however, may be quasi-elliptically polarized. The conditions for the quasi-elliptical polarization of the generated S wave are: (a) both S1 and S2 components are nonvanishing, and (b) they are mutually phase-shifted. Polarization plane Σ_p is then a plane perpendicular to the ray of the generated S wave.

For directivity patterns of simple types of point sources such as a single force, the S1 and S2 components are usually real-valued. The generated S wave will then be linearly polarized. This situation, however, may be different if more complex directivity patterns are considered. It may also be different if a source is situated on (or close to) a structural interface, or the surface of the Earth. For example, the directivity patterns of a single-force source situated on the Earth's surface are complex-valued for radiation angles i larger than critical angle $i^* = \arcsin(\beta/\alpha)$. This implies that the S waves generated by an inclined or horizontal single force situated on the Earth's surface should be elliptically polarized at receivers situated in certain regions.

We have considered only a point source. For sources of finite extent, the generation conditions for S waves will be more favorable for the quasi-elliptic, or even more complex polarization.

2. POSTCRITICAL INCIDENCE OF AN S WAVE AT THE EARTH'S SURFACE

There are two polarization effects related to the postcritical incidence of an S wave at the Earth's surface.

The first effect is the quasi-elliptical polarization at point R situated on the Earth's surface. This effect was explained in detail in Section 6.4.7. The quasi-elliptical polarization in the plane of incidence does not depend on the SH component of the S wave and exists even if the SH component vanishes.

The second effect is related to the polarization of generated reflected SS waves. Assume that both SV and SH components of the incident S wave are nonvanishing and that they are not phase-shifted. The polarization of the incident S wave is then linear, perpendicular to the ray Ω of the incident S wave. The polarization of the reflected S wave remains linear if the angle of incidence is subcritical. This is obvious because both SV and SH reflection coefficients are in this case real-valued. If, however, the angle of incidence is postcritical, the polarization of the reflected SS wave will be quasi-elliptical in plane Σ_p, perpendicular to the ray of the reflected SS wave. The reason is that the SH reflection coefficient at the surface of the Earth is always real-valued and equals -1, whereas the SV reflection coefficient is complex-valued for postcritical incidence. Consequently, the SV and SH components for postcritically reflected S waves are phase-shifted, and this will cause the quasi-elliptical polarization.

Thus, the postcritical incidence of an S wave at the Earth's surface is one of the most important causes of quasi-elliptical polarization of generated reflected S waves.

3. POSTCRITICAL REFLECTIONS OF S WAVES AT STRUCTURAL INTERFACES

The two polarization effects related to the postcritical incidence of an S wave at a structural interface remain practically the same as described for the Earth's surface. The first effect is related to the quasi-elliptical polarization at point R on the interface, in the plane of incidence. For details refer to Section 6.4.6. The second effect is related to the quasi-elliptical polarization of generated reflected and transmitted SS waves.

Thus, the postcritical incidence of an S wave at a structural interface is one of the most important causes of quasi-elliptical polarization of generated reflected and transmitted S waves.

4. INTERFERING SIGNALS

The interference of two or more signals is very common, but it usually has a local character. There is, however, one very important exception: interference of seismic body waves close to structural interfaces and close to the surface of the Earth. In these regions, several waves usually interfere: either incident, reflected P and reflected S, or transmitted P and transmitted S.

It is not difficult to estimate the extent of the region of interference. Assume, for example, that we have a linearly polarized incident P wave and reflected P waves. The thickness of the layer close to the interface in which the incident and reflected waves interfere is

$$H \doteq \tfrac{1}{2}\alpha h \cos i,$$

where h is the width of the envelope of the signal under consideration, i is the angle of incidence, and α is the P-wave velocity. In this region, the two waves will interfere, and the polarization will be quasi-elliptical (even though the incident wave is compressional).

There are some exceptions. *For normal incidence*, the polarization is linear because the displacement vectors of the incident wave and of both monotypic reflected and transmitted waves have the same direction. Note that, in this case, converted waves are not generated. Only under oblique incidence is it quasi-elliptical.

At the interface itself, the polarization is linear if the incident wave is subcritical because $\Delta T(R) = 0$ (see (6.4.23)). The polarization is then quasi-elliptical only at some distance from the interface. If the *incident wave is postcritical*, however, the polarization is quasi-elliptical, even though $\Delta T(R) = 0$. This applies to any incident wave, including the incident P wave. This is *the only case* in which the P *wave is quasi-elliptically polarized*: the P wave must be postcritically incident at the structural interface and the receiver must be situated itself on the interface. The reason is that in this case the P wave is, in fact, composed of three waves with complex-valued amplitudes.

This type of quasi-elliptical polarization may play an important role in VSP studies. Quasi-elliptical polarization is a diagnostic information that indicates the position of the interface.

6.4.9 Quasi-Elliptical Polarization of Seismic Body Waves in Layered Structures

In this section, we shall discuss the polarization of an arbitrary multiply-reflected seismic body wave propagating in an isotropic laterally varying layered structure. The wave may be unconverted P or S along the whole ray or converted (several segments P, several S). We shall consider only *receivers situated inside a locally smooth structure*, not on a structural interface or on the Earth's surface. The effects related to the position of the receiver on a structural interface or on the Earth's surface are well known from Sections 6.4.6 and 6.4.7. We shall also consider only the zeroth-order ray theoretical polarization, but not the effects caused by the higher order terms.

If the wave arrives at receiver R as a P wave, the situation is simple. The wave is always linearly polarized in the direction perpendicular to the wavefront. This applies even if certain segments of the ray correspond to S waves. Even in the case of converted waves, it is fully sufficient for the wave to arrive as P at the receiver to have linear polarization at R.

If the wave arrives as an S wave at the receiver, the situation is more complex: it may be polarized linearly or quasi-elliptically. In the discussion of this case, we shall consider independently planar and nonplanar 3-D rays.

Let us first consider a *planar ray*. Unit vector $\vec{e}_2$ can then be conveniently chosen perpendicular to the plane at the initial point of the ray. It then remains perpendicular to the plane *along the whole ray* even across interfaces. As usual, we denote one relevant component of the S wave U_{SH} and the other U_{SV}. The SV and SH components are then fully separated along the whole ray and do not affect each other. The wave is then quasi-elliptically polarized at the receiver only if it is an unconverted S wave *along the whole ray* between the source and receiver, and only if both components U_{SV} and U_{SH} are nonvanishing at the source. If the wave is linearly polarized at the source ($\arg U_{SV} - \arg U_{SH} = k\pi$), then there must be at least one postcritical reflection/transmission point along the ray. If it is elliptically polarized directly from the source, all reflection/transmission points may be subcritical. Let us emphasize that one P element is sufficient to eliminate fully the quasi-elliptic polarization from the remaining part of the ray. The polarization plane is perpendicular to the ray. This implies that the quasi-elliptic polarization will be observed only in vertical-transverse and transverse-radial planes. The polarization due to postcritical reflection/transmission at structural interfaces will always be *linear in the vertical-radial plane*. (We again emphasize that receiver R is considered to be situated in a smooth medium and not on a structural interface or on the surface of the Earth.)

In 3-D models with 3-D, *nonplanar rays*, such strict constraints are not required because the vector $\vec{e}_2$ perpendicular to the plane of incidence at one R/T point is not necessarily perpendicular to the plane of incidence at the next R/T point. There may be two reasons for this.

a. The rotation of $\vec{e}_2$ along the ray due to the torsion of the ray.
b. Different planes of incidence.

We shall speak of the *coupling of S1 and S2 components*. The coupling effect implies that an S wave with only one nonvanishing component U_1 or U_2 at one point of incidence may have both nonvanishing components at the next point of incidence. Even P elements may change to S elements with nonvanishing S1 and S2 components: At one point of incidence, the incident P wave generates an S1 wave, and the S1 wave may generate nonvanishing S1 and S2 at the next point of incidence due to the coupling effect.

Let us draw some conclusions related to 3-D layered models. Assume first that the wave is *unconverted S along the ray*. Then, if the wave is quasi-elliptically polarized at the source, it will be quasi-elliptically polarized along the whole ray, including the receiver (with the exception, perhaps, of the Brewster angles). If a linearly polarized S wave is generated by the source, the requirement is that the incident wave has nonzero components U_1 and U_2 at least at one R/T point and that the angle of incidence is postcritical at that point. Nonzero components U_1 and U_2 may be generated directly by the source or by coupling effects at some R/T points.

We now consider a converted wave, which has at least one P element. The P element may even be the first. (This means that a P source is being considered.) The wave will be quasi-elliptically polarized at the receiver if the following two requirements are fulfilled:

a. The chain of several (at least two) last elements of the ray corresponds to an S wave.
b. At least at one R/T point within this chain, the incident S wave has nonzero components U_1 and U_2, and the angle of incidence is postcritical.

6.4.10 Polarization of Seismic Body Waves in Anisotropic Media

Three seismic body waves (qS1, qS2, and qP) can propagate in an anisotropic inhomogeneous medium without interfaces. If these waves are well separated and propagate independently, *they are polarized linearly* along eigenvectors $\vec{g}^{(1)}$, $\vec{g}^{(2)}$, and $\vec{g}^{(3)}$ of the Christoffel matrix Γ_{ik}. Because these eigenvectors can be uniquely determined at any point on the ray, the direction of the linear polarization of the three waves can also be uniquely determined, even for qS waves. Thus, the qS waves in anisotropic media are not quasi-elliptically polarized as they may be in an isotropic media, but they are polarized linearly. This is the great difference between the polarization of shear waves in isotropic and anisotropic media.

As we have shown in Section 2.2.5, the eigenvectors $\vec{g}^{(1)}$, $\vec{g}^{(2)}$, and $\vec{g}^{(3)}$ corresponding to three plane waves qS1, qS2, and qP, propagating with common normal in a homogeneous anisotropic medium, are mutually perpendicular. Thus, the qS1 and qS2 waves are polarized in a plane perpendicular to $\vec{g}^{(3)}$ and are also mutually perpendicular. In an inhomogeneous anisotropic medium, however, the situation may be different. The three waves qS1, qS2, and qP propagate along their own rays. The direction of the slowness vectors of the three waves are, in general, different at the receiver. This implies that eigenvectors $\vec{g}^{(1)}$, $\vec{g}^{(2)}$, $\vec{g}^{(3)}$ are not necessarily mutually perpendicular at the receiver. In practical applications, however, they are usually nearly perpendicular. The particle ground-motion analysis exploits this fact to discriminate the two qS waves and to study shear-wave splitting.

In certain situations, however, the polarization of qS waves may be considerably more complex. We shall briefly discuss two such situations here.

1. **Shear-wave singularities.** For certain directions of the slowness vector, the phase velocities of the qS1 and qS2 waves coincide; see Section 2.2.8. Eigenvectors $\vec{g}^{(1)}$ and $\vec{g}^{(2)}$ cannot then be uniquely determined, and the polarization of the qS1 and qS2 waves is not well defined. The polarization of qS waves along and close to these directions is in general anomalous and may be very complicated, particularly in the vicinity of a point singularity; see Crampin (1981) and Rümpker and Thomson (1994). The standard ray method cannot be used to investigate these polarization anomalies.

2. **Weakly anisotropic media.** In inhomogeneous, but weakly anisotropic media (quasi-isotropic), the phase velocities of qS1 and qS2 waves are globally close to each other. In such cases, the two qS waves cannot be treated independently, and the standard ray method, based on independent qS1 and qS2 waves, cannot be applied. The polarization may be very complex in this case. For more details on qS waves in weakly anisotropic media, see Section 5.4.6.

APPENDIX A

Fourier Transform, Hilbert Transform, and Analytical Signals

I n this appendix, we shall present, without derivation, certain properties of Fourier and Hilbert transforms and of analytical signals. We shall not go into mathematical details and complexity of the treatment; we shall concentrate only on the aspects related to the subject of this book. For a detailed treatment, we recommend Papoulis (1962), Bracewell (1965), and many other books devoted to the Fourier transform.

A.1 Fourier Transform

We shall consider function $x(t)$ of real-valued time t and denote its *Fourier spectrum $X(f)$*, where f is the frequency:

$$
\begin{aligned}
X(f) &= \mathcal{F}(x(t)) = \int_{-\infty}^{\infty} x(t) \exp[2i\pi f t] \mathrm{d}t, \\
x(t) &= \mathcal{F}^{-1}(X(f)) = \int_{-\infty}^{\infty} X(f) \exp[-2i\pi f t] \mathrm{d}f.
\end{aligned}
\tag{A.1.1}
$$

The symbol $\mathcal{F}(x(t))$ is used to denote the *Fourier transform* of $x(t)$ and $\mathcal{F}^{-1}(X(f))$ to denote the *inverse Fourier transform* of $X(f)$. $x(t)$ and $X(f)$ are said to form a Fourier transform pair.

We shall mostly consider functions $x(t)$ to be finite signals of finite duration. In this case, integrals (A.1.1) always exist in their standard sense. In certain cases, however, we shall apply the Fourier transform to more complex functions $x(t)$, particularly to distributions such as the Dirac delta function $\delta(t)$. For a more detailed discussion of the applicability of (A.1.1) in such cases, refer to the preceding references.

In the text of this book, we have denoted the Fourier spectrum of function $x(t)$ by $x(f)$ (not by $X(f)$) and distinguished time function $x(t)$ from its spectrum $x(f)$ only by arguments t and f; see Section 2.1.5. This convention was very useful in expressing the acoustic and elastodynamic wave equation in a similar form in the time domain and in the frequency domain. The convention also decreased the number of necessary symbols considerably. In this appendix, however, we shall distinguish strictly between lowercase letters for time function $x(t)$ and uppercase letters for their Fourier spectrum $X(f)$. The exception is the symbol $F(t)$ for the analytical signal.

Often, circular frequency $\omega = 2\pi f$ is used instead of frequency f. Fourier transform (A.1.1) then reads

$$S(\omega) = \int_{-\infty}^{\infty} x(t) \exp[i\omega t]dt, \qquad x(t) = \frac{1}{2\pi} \int_{-\infty}^{\infty} S(\omega) \exp[-i\omega t]d\omega.$$

$$(A.1.2)$$

Thus, the relation between Fourier spectra $X(f)$ and $S(\omega)$ is $S(2\pi f) = X(f)$ (or $S(\omega) = X(\omega/2\pi)$). Unless otherwise stated, by the Fourier spectrum of the function $x(t)$ we shall understand function $X(f)$.

Fourier transforms (A.1.1) and (A.1.2) could also be written with opposite signs in their exponents. For real-valued $x(t)$, this would yield complex-conjugate expressions for spectra $X(f)$ and $S(\omega)$. In this book, we have used systematically the *sign convention* corresponding to (A.1.1) and (A.1.2). This convention is traditional in the seismic ray method.

For *real-valued signals* $x(t)$ of real-valued variable t, the Fourier transform pair may be expressed in a slightly simpler version:

$$X(f) = \int_{-\infty}^{\infty} x(t) \exp[2i\pi f t]dt,$$

$$x(t) = 2\,\mathrm{Re} \int_{0}^{\infty} X(f) \exp[-2i\pi f t]df.$$

$$(A.1.3)$$

Thus, the integration in the inverse Fourier transform runs over nonnegative frequencies f only.

For the reader's convenience, we shall give several useful relations of Fourier transform pairs; however, we give only those relations used in book so that the list is far from complete. We assume $X_1(f) = \mathcal{F}(x_1(t))$, $X_2(f) = \mathcal{F}(x_2(t))$, and α_1, α_2, α, and t_0 are arbitrary real-valued constants.

$$\mathcal{F}(\alpha_1 x_1(t) + \alpha_2 x_2(t)) = \alpha_1 X_1(f) + \alpha_2 X_2(f), \tag{A.1.4}$$

$$\mathcal{F}(x(\alpha t)) = |\alpha|^{-1} X(f/\alpha), \tag{A.1.5}$$

$$\mathcal{F}(x(t - t_0)) = X(f) \exp(2i\pi f t_0), \tag{A.1.6}$$

$$\mathcal{F}(x_1(t) * x_2(t)) = X_1(f) X_2(f), \tag{A.1.7}$$

$$\mathcal{F}(x(-t)) = X(-f). \tag{A.1.8}$$

Here and in the following equations, the star $(*)$ denotes the convolution. In addition, we shall also give several useful expressions for the direct and inverse Fourier transform:

$$\mathcal{F}(\exp[-\pi t^2]) = \exp[-\pi f^2], \qquad \mathcal{F}^{-1}(\exp[-\pi f^2]) = \exp[-\pi t^2],$$

$$(A.1.9)$$

$$\mathcal{F}(\delta(t)) = 1, \qquad\qquad\qquad \mathcal{F}^{-1}(\delta(f)) = 1, \tag{A.1.10}$$

$$\mathcal{F}(\mathrm{sgn}\, t) = i/\pi f, \qquad\qquad \mathcal{F}^{-1}(\mathrm{sgn}\, f) = -i/\pi t, \tag{A.1.11}$$

$$\mathcal{F}(H(t)) = \tfrac{1}{2}(\delta(f) + i/\pi f), \qquad \mathcal{F}^{-1}(H(f)) = \tfrac{1}{2}(\delta(t) - i/\pi t), \tag{A.1.12}$$

$$\mathcal{F}(\sqrt{2}\, H(t) t^{-1/2}) = |f|^{-1/2} \exp\left[i\frac{\pi}{4}\,\mathrm{sgn}\, f\right] \tag{A.1.13}$$

$$\mathcal{F}\big(H(t - a)(t^2 - a^2)^{-1/2}\big) = \tfrac{1}{2} i\pi\, H_0^{(1)}(2\pi f a). \tag{A.1.14}$$

Here $H(t)$ is the Heaviside function

$$H(t) = 0 \quad \text{for } t < 0, \qquad H(t) = 1 \quad \text{for } t > 0,$$
$$H(t) = \tfrac{1}{2} \quad \text{for } t = 0.$$

(A.1.15)

Function sgn t can be expressed using the Heaviside function and relation sgn $t = 2(H(t) - \frac{1}{2})$. In (A.1.14), $H_0^{(1)}$ is the Hankel function of the first kind and zeroth-order; see Abramowitz and Stegun (1970).

A.2 Hilbert Transform

We again consider a real-valued function $x(t)$ of real-valued variable t. The *Hilbert transform* $g(t) = \mathcal{H}(x(t))$ of $x(t)$ is then defined by the following integral:

$$g(t) = \mathcal{H}(x(t)) = \frac{1}{\pi} \, \mathcal{P.V.} \int_{-\infty}^{\infty} \frac{x(\zeta)}{\zeta - t} d\zeta.$$

(A.2.1)

Function $x(t)$ can then be expressed in terms of $g(t)$ using the *inverse Hilbert transform*:

$$x(t) = \mathcal{H}^{-1}(g(t)) = -\frac{1}{\pi} \, \mathcal{P.V.} \int_{-\infty}^{\infty} \frac{g(\zeta)}{\zeta - t} d\zeta.$$

(A.2.2)

Here $\mathcal{P.V.}$ denotes the Cauchy principal value of the integral. Alternative expressions for (A.2.1) and (A.2.2) are

$$g(t) = -\frac{1}{\pi t} * x(t), \qquad x(t) = \frac{1}{\pi t} * g(t),$$

(A.2.3)

and,

$$g(t) = 2 \, \text{Im} \int_0^{\infty} X(f) \exp[-2i\pi f t] df,$$
$$x(t) = 2 \, \text{Re} \int_0^{\infty} X(f) \exp[-2i\pi f t] df.$$

(A.2.4)

Here $X(f) = \mathcal{F}(x(t))$.

These are a few useful Hilbert transform pairs:

$$\mathcal{H}(\delta(t)) = -1/\pi t, \qquad\qquad \mathcal{H}(1/\pi t) = \delta(t), \qquad \text{(A.2.5)}$$
$$\mathcal{H}(\cos(t)) = -\sin t, \qquad\qquad \mathcal{H}(\sin t) = \cos(t), \qquad \text{(A.2.6)}$$
$$\mathcal{H}\left(\frac{1}{\pi} \frac{\epsilon}{t^2 + \epsilon^2}\right) = -\frac{t}{\pi(t^2 + \epsilon^2)}, \qquad \mathcal{H}\left(\frac{1}{\pi} \frac{t}{t^2 + \epsilon^2}\right) = \frac{1}{\pi} \frac{\epsilon}{t^2 + \epsilon^2},$$

(A.2.7)

$$\mathcal{H}(H(t)t^{-1/2}) = H(-t)(-t)^{-1/2}.$$

(A.2.8)

where ϵ is a real-valued constant. Thus, for Rayleigh signal (6.1.9), the Hilbert transform is known exactly; see (A.2.7).

Several useful properties of the Hilbert transform pair follow:

$$\int_{-\infty}^{\infty} (x(t))^2 dt = \int_{-\infty}^{\infty} (g(t))^2 dt,$$

(A.2.9)

$$\int_{-\infty}^{\infty} x(t)g(t)dt = 0,$$

(A.2.10)

$$dg(t)/dt = \mathcal{H}(dx(t)/dt),$$

(A.2.11)

$$g(at + b) = \mathcal{H}(x(at + b)), \tag{A.2.12}$$

$$\mathcal{H}(x(t) * y(t)) = \mathcal{H}(x(t)) * y(t) = x(t) * \mathcal{H}(y(t)). \tag{A.2.13}$$

The approximate relations follow:

$$\mathcal{H}(a(t)\cos(2\pi f_M t + v)) \doteq -a(t)\sin(2\pi f_M t + v), \tag{A.2.14}$$

$$\mathcal{H}\big(\exp\big(-(2\pi f_M t/\gamma)^2\big)\cos(2\pi f_M t + v)\big)$$
$$\doteq -\exp\big(-(2\pi f_M t/\gamma)^2\big)\sin(2\pi f_M t + v). \tag{A.2.15}$$

Here $a(t)$ is a broad smooth envelope of the harmonic carrier with frequency f_M. Equation (A.2.15) shows the Hilbert transform of Gabor signal (6.1.6) as a special case of (A.2.14). For (A.2.15) to be sufficiently accurate, we need to choose γ roughly larger than 3. For the derivation of (A.2.14) and (A.2.15) and for many numerical examples see Červený (1976) and Červený, Molotkov, and Pšenčík (1977).

The Hilbert transform has found important applications not only in the time domain but also in the frequency domain. For causal functions $x(t)$ (satisfying the relation $x(t) = 0$ for $t < 0$), the real and imaginary parts of Fourier spectrum $X(f)$ form a Hilbert transform pair.

A.3 Analytical Signals

The analytical signal may appear in some form in any high-frequency asymptotic solution of seismic wave propagation problems. It is important in the construction of synthetic seismograms using the ray method and its extensions. The concept of the analytical signal is closely connected with the Hilbert transform.

There is nothing surprising in the concept of the analytical signal. We know that it is often considerably simpler to solve some problems involving the time-dependent real-valued harmonic function $\cos \omega t$ if we use the complex-valued exponential function $\exp(-i\omega t)$ instead of $\cos \omega t$ and return to the real-valued functions only in the final solution of the problem. The same applies to function $\sin \omega t$. A similar approach is useful even in the ray method and in its extensions. Instead of the real-valued source-time function $x(t)$, we use the relevant complex-valued analytical signal $x^{(A)}(t)$ and return to the real-valued solutions only in the final equations.

We shall now introduce the complex-valued analytical signal $F(t)$ corresponding to the real-valued signal $x(t)$:

$$F(t) = x(t) + ig(t) = x(t) + i\,\mathcal{H}(x(t)). \tag{A.3.1}$$

In addition to the general notation $F(t)$ for the complex-valued analytical signal corresponding to any real-valued function, we shall also use a more specific notation, $x^{(A)}(t)$, for the analytical signal corresponding to the real-valued function $x(t)$,

$$x^{(A)}(t) = x(t) + ig(t) = x(t) + i\mathcal{H}(x(t)). \tag{A.3.2}$$

Alternatively, it would be the minus sign $(-)$ in (A.3.1) and (A.3.2). The sign convention we use in (A.3.1) and (A.3.2) corresponds to the sign convention in Fourier transform (A.1.1). We have used the plus sign in the expressions for analytical signals systematically throughout the whole book.

From (A.2.4) and (A.3.1), we obtain an alternative expression for the analytical signal using the Fourier integral

$$x^{(A)}(t) = 2 \int_0^\infty X(f)\exp[-2i\pi f t]\mathrm{d}f. \tag{A.3.3}$$

Thus, the Fourier spectrum of the analytical signal vanishes for negative frequencies. It is given by the relation

$$\mathcal{F}\big(x^{(A)}(t)\big) = 2H(f)X(f). \tag{A.3.4}$$

The last alternative definition of analytical signal $x^{(A)}(t)$ follows from (A.2.3) and (A.3.1)

$$x^{(A)}(t) = (\delta(t) - \mathrm{i}/\pi t) * x(t) = \delta^{(A)}(t) * x(t). \tag{A.3.5}$$

Function

$$\delta^{(A)}(t) = \delta(t) - \mathrm{i}/\pi t \tag{A.3.6}$$

is called the *analytical delta function* and plays a very important role in the seismic ray method. As we can check in (A.2.5), analytical delta function $\delta^{(A)}(t)$ is the analytical signal corresponding to delta function $\delta(t)$.

Three other important *exact expressions* for the analytic signals correspond to the trigonometric functions, to Rayleigh function (6.1.9), and to the inverse square root function:

$$(\cos \omega t)^{(A)} = \exp[-\mathrm{i}\,\omega t], \tag{A.3.7}$$

$$\left(\frac{1}{\pi}\frac{\epsilon}{t^2 + \epsilon^2}\right)^{(A)} = \frac{1}{\pi}\frac{\epsilon - \mathrm{i}t}{(t^2 + \epsilon^2)} = -\frac{\mathrm{i}}{\pi}\frac{1}{t - \mathrm{i}\epsilon}, \tag{A.3.8}$$

$$(H(t)t^{-1/2})^{(A)} = H(t)t^{-1/2} + \mathrm{i}H(-t)(-t)^{-1/2}. \tag{A.3.9}$$

Now we shall write two *approximate expressions* for the analytical signals, corresponding to a harmonic carrier with a broad envelope $a(t)$:

$$[a(t)\cos(2\pi f_M t + v)]^{(A)} \doteq a(t)\exp[-\mathrm{i}(2\pi f_M t + v)], \tag{A.3.10}$$

$$\big[\exp\big(-(2\pi f_M t/\gamma)^2\big)\cos(2\pi f_M t + v)\big]^{(A)}$$
$$\doteq \exp\big[-(2\pi f_M t/\gamma)^2 - \mathrm{i}(2\pi f_M t + v)\big]. \tag{A.3.11}$$

Thus, the analytical signal (A.3.11) corresponding to Gabor signal (6.1.6) can be approximately calculated analytically, see (A.3.11). The accuracy is good if γ is larger roughly than 3.

Three very useful relations regarding analytical signals $F(t)$, corresponding to any function $x(t)$, follow:

$$\int_{-\infty}^{\infty} F^2(t)\mathrm{d}t = \int_{-\infty}^{\infty} F^{*2}(t)\mathrm{d}t = 0, \tag{A.3.12}$$

$$\int_{-\infty}^{\infty} F(t)F^*(t)\mathrm{d}t = \int_{-\infty}^{\infty} (x^2(t) + g^2(t))\mathrm{d}t$$
$$= 2\int_{-\infty}^{\infty} x^2(t)\mathrm{d}t = 2\int_{-\infty}^{\infty} g^2(t)\mathrm{d}t, \tag{A.3.13}$$

$$[x(t) * y(t)]^{(A)} = [x(t)]^{(A)} * y(t) = x(t) * [y(t)]^{(A)}. \tag{A.3.14}$$

Using definition equation (A.3.3), we can also generate *analytical signals for complex-valued variable* t. The analytical signals for complex-valued variable t are very useful in problems that consider complex-valued travel times (dissipative media, inhomogeneous waves, and Gaussian beams). Consider a real-valued function $x(t)$, its complex Fourier spectrum $X(f)$, and derive the analytical signal $x^{(A)}(t - T)$ corresponding to $x(t)$, where T is complex-valued, $T = T^R + iT^I$. Using (A.3.3), we obtain

$$x^{(A)}(t - T) = 2 \int_0^\infty X(f) \exp[-2i\pi f(t - T)] \mathrm{d}f. \tag{A.3.15}$$

This can be expressed in a convolutory form

$$x^{(A)}(t - T) = x(t) * \mathcal{F}^{-1}(2H(f) \exp(2i\pi fT)). \tag{A.3.16}$$

The inverse Fourier transform in (A.3.16) can be simply computed, assuming $T^I \geq 0$,

$$\mathcal{F}^{-1}(2H(f) \exp(2i\pi fT)) = 2 \int_0^\infty \exp[-2i\pi f(t - T)] \mathrm{d}f = \frac{-i}{\pi(t - T)}. \tag{A.3.17}$$

Thus, finally

$$x^{(A)}(t - T) = x(t) * (-i/\pi(t - T)). \tag{A.3.18}$$

Alternatively,

$$x^{(A)}(t - T) = \frac{1}{\pi} \int_{-\infty}^\infty \frac{T^I x(u) \mathrm{d}u}{(T^I)^2 + (t - T^R - u)^2}$$
$$- \frac{i}{\pi} \int_{-\infty}^\infty \frac{(t - T^R - u)x(u) \mathrm{d}u}{(T^I)^2 + (t - T^R - u)^2}. \tag{A.3.19}$$

If $t - T$ is real-valued, (A.3.5) becomes

$$x^{(A)}(t - T) = x(t) * \delta^{(A)}(t - T), \tag{A.3.20}$$

where $\delta^{(A)}(\zeta)$ is the analytical delta function; see (A.3.6). Thus, all expressions derived in the time domain for the real-valued travel time T in terms of $\delta^{(A)}(t - T)$ can be applied even for complex-valued T, if we put

$$\delta^{(A)}(t - T) \to -i/\pi(t - T). \tag{A.3.21}$$

Finally, we shall show that (A.3.21) has a very close relation to the properties of Rayleigh signal (6.1.9). If we put $T = T^R + iT^I$, (A.3.21) can be expressed in the following form:

$$\delta^{(A)}(t - T) = \frac{T^I}{\pi[(t - T^R)^2 + (T^I)^2]} - \frac{i}{\pi} \frac{t - T^R}{[(t - T^R)^2 + (T^I)^2]}. \tag{A.3.22}$$

The analytical signal has many important applications. Among others, it can be used to evaluate the *envelope of signal* $x(t)$ and its *instantaneous frequency*. We use

$$x^{(A)}(t) = |x^{(A)}(t)| \exp[-i\Phi(t)], \tag{A.3.23}$$

where

$$|x^{(A)}(t)| = [x^2(t) + g^2(t)]^{1/2}, \qquad \Phi(t) = -\arctan(g(t)/x(t)). \tag{A.3.24}$$

Thus, function $|x^{(A)}(t)|$ represents the envelope of signal $x(t)$ (and also of its Hilbert transform $g(t)$). The instantaneous frequency $f_I(t_0)$ of signal $x(t)$ at time t_0 then reads

$$f_I(t_0) \doteq (2\pi)^{-1} (\mathrm{d}\Phi(t)/\mathrm{d}t)_{t=t_0}. \tag{A.3.25}$$

References

Abdullaev, S. S. (1993). *Chaos and dynamics of rays in waveguide media*. Amsterdam: Gordon and Breach.

Abgrall, R., and Benamou, J.-D. (1999). Big ray-tracing and eikonal solver on unstructured grids: Application to the computation of a multivalued traveltime field in the Marmousi model. *Geophysics* **64**, 230–9.

Abramowitz, M., and Stegun, I. A. (1970). *Handbook of mathematical functions*. New York: Dover.

Achenbach, J. D. (1975). *Wave propagation in elastic solids*. Amsterdam: Elsevier.

Achenbach, J. D., Gautesen, A. K., and McMaken, H. (1982). *Ray methods for waves in elastic solids*. London: Pitman.

Aizenberg, A. M., and Klem-Musatov, K. D. (1980). Calculation of wave fields by the method of superposition of edge waves. *Geology and Geophysics* **21**, 92–108.

Aki, K., Christoffersen, A., and Husebye, E. S. (1977). Determination of three-dimensional structure of the lithosphere. *J. Geophys. Res.* **82**, 277–296.

Aki, K., and Richards, P. (1980). *Quantitative seismology. Theory and methods*. San Francisco: Freeman.

Al-Chalabi, M. (1997a). Instantaneous slowness versus depth functions. *Geophysics* **62**, 270–3.

Al-Chalabi, M. (1997b). Time-depth relationships for multilayer depth conversion. *Geophys. Prospecting* **45**, 715–20.

Aldridge, D. F. (1990). The Berlage wavelet. *Geophysics* **55**, 1508–11.

Alekseyev, A. S., Babich, V. M., and Gel'chinskiy, B. Ya. (1961). Ray method for the computation of the intensity of wave fronts. In *Problems of the dynamic theory of propagation of seismic waves* (in Russian), vol. 5, ed. G. I. Petrashen, pp. 3–24. Leningrad: Leningrad Univ. Press.

Alekseyev, A. S., and Gel'chinskiy, B. Ya. (1958). Determination of the intensity of head waves in the theory of elasticity by the ray method (in Russian). *Dokl. Akad. Nauk SSSR* **118**, 661–4.

Alekseyev, A. S., and Gel'chinskiy, B. Ya. (1959). On the ray method of computation of wave fields for inhomogeneous media with curved interfaces. In *Problems of the dynamic theory of propagation of seismic waves* (in Russian), vol. 3, ed. G. I. Petrashen, pp. 107–60. Leningrad: Leningrad Univ. Press.

Alekseyev, A. S., and Gel'chinskiy, B. Ya. (1961). The ray method of computation of the intensity of head waves. In *Problems of the dynamic theory of propagation of seismic waves* (in Russian), vol. 5, ed. G. I. Petrashen, pp. 54–72. Leningrad: Leningrad Univ. Press.

Alekseyev, A. S., and Mikhailenko, B. G. (1982). "Nongeometrical phenomena" in the theory of propagation of seismic waves (in Russian). *Dokl. Akad. Nauk SSSR* **267** (5), 1079–83.

Alkhalifah, T. (1995). Efficient synthetic-seismogram generation in transversely isotropic, inhomogeneous media. *Geophysics* **60**, 1139–50.

Arnaud, J. A. (1971a). Mode coupling in first-order optics. *J. Opt. Soc. Am.* **61**, 751–8.

Arnaud, J. A. (1971b). Modes of propagation of optical beams in helical gas lenses. *Proc. IEEE* **59**, 1378–9.

Arnold, V. I. (1967). Characteristic classes entering in quantization conditions. *Funct. Anal. Appl.* **1**, 1–13.

Arnold, V. I. (1974). *Mathematical methods of classical mechanics* (in Russian). Moscow: Nauka. (Translated to English by Springer, Berlin, 1978.)

Asakawa, E., and Kawanaka, T. (1993). Seismic ray tracing using linear traveltime interpolation. *Geophys. Prospecting* **41**, 99–112.

Asatryan, A. A., and Kravtsov, Yu. A. (1988). Fresnel zone of hyperbolic type from the physical point of view. *Wave Motion* **10**, 45–57.

Audebert, F., Nichols, D., Rekdal, T., Biondi, B., Lumley, D. E., and Urdaneta, H. (1997). Imaging complex geologic structure with single-arrival Kirchhoff prestack depth migration. *Geophysics* **62**, 1533–43.

Auld, B. A. (1973) *Acoustic fields and waves in solids*. New York: Wiley.

Azbel, I. Ya., Dmitrieva, L. A., Gobarenko, V. S., and Yanovskaya, T. B. (1984). Numerical modelling of wavefields in three-dimensional inhomogeneous media. *Geophys. J. R. astr. Soc.* **79**, 199–206.

Babich, V. M. (1956). Ray method of the computation of the intensity of wave fronts (in Russian). *Dokl. Akad. Nauk SSSR* **110**, 355–7.

Babich, V. M. (1961a). Ray method of the computation of the intensity of wave fronts in elastic inhomogeneous anisotropic medium. In *Problems of the dynamic theory of propagation of seismic waves* (in Russian), vol. 5, ed. G. I. Petrashen, pp. 36–46. Leningrad: Leningrad Univ. Press. (Translation to English: *Geophys. J. Int.* **118**, 379–83, 1994.)

Babich, V. M. (1961b). On the convergence of series in the ray method of calculation of the intensity of wave fronts. In *Problems of the dynamic theory of propagation of seismic waves* (in Russian), vol. 5, ed. G. I. Petrashen, pp. 25–35. Leningrad: Leningrad Univ. Press.

Babich, V. M. (1968). Eigenfunctions, concentrated in the vicinity of a closed geodesics. In *Mathematical problems of the theory of propagation of waves* (in Russian), vol. 9, ed. V. M. Babich, pp. 15–63. Leningrad: Nauka.

Babich, V. M. (1979). On the space-time ray method in the theory of elastic waves (in Russian). *Izv. Akad. Nauk SSSR, Fizika Zemli*, No. 2, 3–13.

Babich, V. M., and Alekseyev, A. S. (1958). On the ray method of the computation of the intensity of wave fronts (in Russian). *Izv. Akad. Nauk SSSR, Geophys. Series*, No. 1, 17–31.

Babich, V. M., and Buldyrev, V. S. (1972). *Asymptotic methods in problems of diffraction of short waves* (in Russian). Moscow: Nauka. (Translated to English by Springer, Berlin, 1991, under the title *Short-wavelength diffraction theory*.)

Babich, V. M., Buldyrev, V. S., and Molotkov, I. A. (1985). *Space-time ray method. Linear and nonlinear waves* (in Russian). Leningrad: Leningrad Univ. Press.

Babich, V. M., Chikhachev, B. A., and Yanovskaya, T. B. (1976). Surface waves in vertically inhomogeneous elastic halfspace with a weak horizontal inhomogeneity (in Russian). *Izv. Akad. Nauk SSSR, Fizika Zemli*, No. 4, 24–31.

Babich, V. M., and Kirpichnikova, N. Y. (1974). *The boundary-layer method in diffraction problems* (in Russian). Leningrad: Leningrad Univ. Press. (English translation by Springer, 1979).

Babich, V. M., and Kiselev, A. P. (1989). Non-geometrical waves – Are there any? An asymptotic description of some "non-geometrical" phenomena in seismic wave propagation. *Geophys. J. Int.* **99**, 415–20.

Babich, V. M., and Pankratova, T. F. (1973). On discontinuities of the Green function of mixed problems for wave equation with variable coefficients. In *Problems of mathematical physics* (in Russian), vol. 6, ed. V. M. Babich, pp. 9–27. Leningrad: Leningrad Univ. Press.

Babich, V. M., and Popov, M. M. (1981). Propagation of concentrated acoustical beams in three-dimensional inhomogeneous media (in Russian). *Akust. Zh.* **27**, 828–35.

Babich, V. M., and Ulin, V. V. (1981a). Complex-valued ray solutions and eigenfunctions, concentrated in the vicinity of a closed geodesic. In *Mathematical problems of the theory of propagation of waves* (in Russian), vol. 11, ed. V. M. Babich, pp. 6–13. Leningrad: Nauka.

Babich, V. M., and Ulin, V. V. (1981b). Complex space-time ray method and "quasiphotons." In *Mathematical problems of the theory of propagation of waves* (in Russian), vol. 12, ed. V. M. Babich, pp. 5–12. Leningrad: Nauka.

Babuška, V., and Cara, M. (1991). *Seismic anisotropy in the Earth*. Dordrecht: Kluwer.

Backus, G. E. (1962). Long-wave elastic anisotropy produced by horizontal layering. *J. Geophys. Res.* **67**, 4427–40.

Backus, G. E. (1965). Possible forms of seismic anisotropy of the uppermost mantle under oceans. *J. Geophys. Res.* **70**, 3429–39.

Bakker, P. M. (1990). Theory of edge diffraction in terms of dynamic ray tracing. *Geophys. J. Int.* **102**, 177–89.

Bakker, P. M. (1996). Theory of anisotropic dynamic ray tracing in ray-centred coordinates. *PAGEOPH* **148**, 583–9.

Bakker, P. M. (1998). Phase shift at caustics along rays in anisotropic media. *Geophys. J. Int.* **134**, 515–18.

Belonosova, A. V., Tadzhimukhamedova, S. S., and Alekseyev, A. S. (1967). Computation of travel-time curves and geometrical spreading in inhomogeneous media. In *Certain methods and algorithms in the interpretation of geophysical data* (in Russian), ed. A. S. Alekseyev, pp. 124–36. Moscow: Nauka.

Benamou, J. D. (1996). Big ray tracing: Multivalued traveltime computation using viscosity solutions of the eikonal equation. *J. Comput. Phys.* **128**, 463–74.

Ben-Hador, R., and Buchen, P. (1999). Love and Rayleigh waves in non-uniform media. *Geophys. J. Int.* **137**, 521–34.

Ben-Menahem, A. (1990). SH waves from point sources in anisotropic inhomogeneous media. *Geophysics* **55**, 488–91.

Ben-Menahem, A., and Beydoun, W. B. (1985). Range of validity of seismic ray and beam methods in general inhomogeneous media. I. General theory. *Geophys. J. R. astr. Soc.* **82**, 207–34.

Ben-Menahem, A., and Gibson, R. L. (1990). Scattering of elastic waves by localized anisotropic inclusions. *J. Acoust. Soc. Am.* **87**, 2300–9.

Ben-Menahem, A., Gibson, R. L., Jr., and Sena, A. G. (1991). Green's tensor and radiation patterns of point sources in general anisotropic inhomogeneous elastic media. *Geophys. J. Int.* **107**, 297–308.

Ben-Menahem, A., and Sena, A. G. (1990). Seismic source theory in stratified anisotropic media. *J. Geophys. Res.* **95**, 15395–427.

Ben-Menahem, A., and Singh, S. J. (1981). *Seismic waves and sources.* Heidelberg: Springer.

Berkhaut, A. J. (1987). *Applied seismic wave theory.* Amsterdam: Elsevier.

Bernard, P., and Madariaga, R. (1984). A new asymptotic method for the modeling of near-field accelerograms. *Bull. Seismol. Soc. Am.* **74**, 539–57.

Beydoun, W. B., and Ben-Menahem, A. (1985). Range of validity of seismic ray and beam methods in general inhomogeneous media. II. A canonical problem. *Geophys. J. R. astr. Soc.* **82**, 235–62.

Beydoun, W. B., and Keho, T. H. (1987). The paraxial ray method. *Geophysics* **52**, 1639–53.

Beydoun, W. B., and Mendes, M. (1989). Elastic ray-Born l_2-migration/inversion. *J. Geophys. Res.* **97**, 151–60.

Beydoun, W. B., and Tarantola, A. (1988). First Born and Rytov approximations: Modeling and inversion conditions in a canonical example. *J. Acoust. Soc. Am.* **83**, 1045–55.

Beylkin, G., and Burridge, R. (1990). Linearized inverse scattering problems in acoustics and elasticity. *Wave Motion* **12**, 15–52.

Biondi, B. (1992). Solving the frequency dependent eikonal equation. 62nd Annual SEG Meeting Expanded Abstracts, New Orleans, 1315–19.

Blangy, J. P. (1994). AVO in transversely isotropic media – An overview. *Geophysics* **59**, 775–81.

Bleistein, N. (1984). *Mathematical methods for wave phenomena.* New York: Academic Press.

Bleistein, N., Cohen, J. K., and Hagin, F. G. (1987). Two and one-half dimensional Born inversion with an arbitrary reference. *Geophysics* **52**, 26–36.

Booth, D. C., and Crampin, S. (1983). The anisotropic reflectivity technique: Theory. *Geophys. J. R. astr. Soc.* **72**, 755–66.

Booth, D. C., and Crampin, S. (1985). Shear wave polarizations on a curved wavefront at an isotropic free-surface. *Geophys. J. R. astr. Soc.* **83**, 31–45.

Borcherdt, R. D. (1977). Reflection and refraction of Type-II S waves in elastic and anelastic media. *Bull. Seismol. Soc. Am.* **67**, 43–67.

Borcherdt, R. D. (1982). Reflection-refraction of general P- and type-I S-waves in elastic and anelastic solids. *Geophys. J. R. astr. Soc.* **70**, 621–38.

Borcherdt, R. D., Glassmoyer, G., and Wennerberg, L. (1986). Influence of welded boundaries in anelastic media on energy flow, and characteristics of P, S-I and S-II waves: Observational evidence for inhomogeneous body waves in low-loss solids. *J. Geophys. Res.* **91**, **B11**, 11503–18.

Borejko, P. (1996). Reflection and transmission coefficients for three-dimensional plane waves in elastic media. *Wave Motion* **24**, 371–93.

Born, M., and Wolf, E. (1959). *Principles of optics*. New York: Pergamon Press.

Bortfeld, R. (1961). Approximations to the reflection and transmission coefficients of plane longitudinal and transverse waves. *Geophys. Prospecting* **9**, 485–502.

Bortfeld, R. (1967). Elastic waves in layered media. *Geophys. Prospecting* **15**, 644–50.

Bortfeld, R. (1989). Geometrical ray theory: Rays and traveltimes in seismic systems (second-order approximations of the traveltimes). *Geophysics* **54**, 342–9.

Bortfeld, R., and Kemper, M. (1991). Geometrical ray theory: Line foci and point foci in the anterior surface of seismic systems (second-order approximation of the traveltimes). *Geophysics* **56**, 806–11.

Bourbiè, T. (1984). Effects of attenuation on reflections: Experimental test. *J. Geophys. Res.* **89**, 6197–202.

Bourbiè, T., and Gonzalez-Serrano, A. (1983). Synthetic seismograms in attenuating media. *Geophysics* **48**, 1575–87.

Bracewell, R. (1965). *The Fourier transform and its application*. New York: McGraw Hill.

Brekhovskikh, L. M. (1960). *Waves in layered media*. New York: Academic Press.

Brekhovskikh, L. M., and Godin, O. A. (1989). *Acoustics of layered media* (in Russian). Moscow: Nauka. (Translated to English by Springer, Berlin, in two volumes: I, 1990; II, 1992.)

Bretherton, F. P. (1968). Propagation in slowly varying waveguides. *Proc. R. Soc. London* **A302**, 555–76.

Brokešová, J. (1994). 2 1/2-D approach in ray-based seismic wavefield modelling. In *Seismic waves in complex 3-D structures, Report 1*, pp. 45–50. Prague: Faculty of Mathematics and Physics, Charles Univ.

Brokešová, J. (1996). Construction of ray synthetic seismograms using interpolation of travel times and ray amplitudes. *PAGEOPH* **148**, 503–38.

Brown, M. G. (1994). A Maslov-Chapman wavefield representation for wide-angle one-way propagation. *Geophys. J. Int.* **116**, 513–26.

Brown, M. G., and Tappert, F. D. (1987). Catastrophe theory, caustics and travel time diagrams in seismology. *Geophys. J. R. astr. Soc.* **88**, 217–29.

Brown, M. G., Tappert, F. D., and Goni, G. J. (1991). An investigation of sound ray dynamics in a range dependent model of the ocean volume using an area-preserving mapping. *Wave Motion* **14**, 93–9.

Brown, M. G., Tappert, F. D., Goni, G. J., and Smith, K. B. (1991). Chaos in underwater acoustics. In *Ocean Variability and Acoustic Propagation*, eds. J. Potter and A. Warn-Varnas, pp. 139–60. Dordrecht: Kluwer.

Brühl, M., Vermeer, G. J. O., and Kiehn, M. (1996). Fresnel zones for broadband data. *Geophysics* **61**, 600–4.

Buchen, P. W. (1974). Application of the ray series method to linear viscoelastic wave propagation. *PAGEOPH* **112**, 1011–30.

Buchwald, V. T. (1959). Elastic waves in anisotropic media. *Proc. R. Soc. London A* **253**, 563–80.

Budden, K. G. (1961a). *The wave-guide mode theory of wave propagation*. London: Logos.

Budden, K. G. (1961b). *Radio waves in the ionosphere*. Cambridge: Cambridge Univ. Press.

Bulant, P. (1996). Two-point ray tracing in 3-D. *PAGEOPH* **148**, 421–47.

Bulant, P. (1999). Two-point ray-tracing and controlled initial-value ray-tracing in 3-D heterogeneous block structures. *J. Seismic Exploration* **8**, 57–75.

Bulant, P., and Klimeš, L. (1999). Interpolation of ray theory traveltimes within ray cells. *Geophys. J. Int.* **139**, 273–82.

Bulant, P., Klimeš, L., and Pšenčík, I. (1999). Comparison of ray methods with the exact solution in the 1-D anisotropic "twisted crystal" model. In *Seismic waves in complex 3-D structures, Report 8*, pp. 119–26. Prague: Department of Geophysics, Charles University.

Bullen, K. E. (1965). *An introduction to the theory of seismology*. Cambridge: Cambridge Univ. Press.

Bullen, K. E., and Bolt, B. A. (1985). *An introduction to the theory of seismology*. Cambridge: Cambridge Univ. Press.

Burdick, L. J., and Orcutt, J. A. (1979). A comparison of the generalized ray and reflectivity methods of waveform synthesis. *Geophys. J. R. astr. Soc.* **58**, 261–78.

Burridge, R. (1967). The singularity on the plane lids of the wave surface of elastic media with cubic symmetry. *Q. J. Mech. Appl. Math.* **20**, 41–56.

Burridge, R. (1976). *Some mathematical topics in seismology.* New York: New York Univ., Courant Inst. of Math. Sci.

Burridge, R., and Knopoff, L. (1964). Body force equivalents for seismic dislocations. *Bull. Seismol. Soc. Am.* **54**, 1875–88.

Burridge, R., and Weinberg, H. (1977). Horizontal rays and vertical modes. In *Wave propagation and underwater acoustics. Lecture Notes in Physics*, vol. 70, eds. J. B. Keller and J. S. Papadakis, pp. 86–152. Berlin: Springer Verlag.

Buske, S. (1996). Finite-difference solution of the transport equation: First results. *PAGEOPH* **148**, 565–81.

Cagniard, L. (1939). *Réflexion et réfraction des ondes séismiques progressives.* Paris: Gauthier-Villars.

Cagniard, L. (1962). *Reflexion and refraction of progressive seismic waves.* New York: McGraw Hill.

Cao, S., and Greenhalgh, S. (1993). Calculation of the seismic first-break time field and its ray path distribution using a minimum traveltime tree algorithm. *Geophys. J. Int.* **114**, 593–600.

Cao, S., and Greenhalgh, S. (1994). Finite-difference solution of the eikonal equation using an efficient, first-arrival, wavefront tracking scheme. *Geophysics* **59**, 632–43.

Carcione, J. M. (1993). Seismic modeling in viscoelastic media. *Geophysics* **58**, 110–20.

Carcione, J. M. (1997). Reflection and transmission of qP-qS plane waves at a plane boundary between viscoelastic transversely isotropic media. *Geophys. J. Int.* **129**, 669–80.

Carcione, J. M., Helle, H. B., and Zhao, T. (1998). Effects of attenuation and anisotropy on reflection amplitude versus offset. *Geophysics* **63**, 1652–8.

Carter, J. A., and Frazer, L. N. (1984). Accommodating lateral velocity changes in Kirchhoff migration by means of Fermat's principle. *Geophysics* **49**, 46–53.

Cassel, B. R. (1982). A method for calculating synthetic seismograms in laterally varying media. *Geophys. J. R., astr. Soc.* **69**, 339–54.

Caviglia, G., and Morro, A. (1992). *Inhomogeneous waves in solids and fluids.* Singapore: World Scientific.

Červený, V. (1957). The reflection of spherical elastic waves at a plane boundary. In *Geofyzikální sborník*, vol. 4, ed. A. Zátopek, pp. 343–66. Praha: NČSAV.

Červený, V. (1966a). The dynamic properties of reflected and head waves around the critical point. In *Geofyzikální sborník*, vol. 13, ed. A. Zátopek, pp. 135–245. Praha: NČSAV.

Červený, V. (1966b). On dynamic properties of reflected and head waves in the n-layered Earth's crust. *Geophys. J. R. astr. Soc.* **11**, 139–47.

Červený, V. (1967). The amplitude-distance curves of waves reflected on a plane interface for different frequency ranges. *Geophys. J. R. astr. Soc.* **13**, 187–96.

Červený, V. (1972). Seismic rays and ray intensities in inhomogeneous anisotropic media. *Geophys. J. R. astr. Soc.* **29**, 1–13.

Červený, V. (1976). Approximate expressions for the Hilbert transform of a certain class of functions and their applications in the ray theory of seismic waves. *Studia Geoph. et Geod.* **20**, 125–32.

Červený, V. (1979a). Ray theoretical seismograms for laterally inhomogeneous structures. *J. Geophys.* **46**, 335–42.

Červený, V. (1979b). Accuracy of ray theoretical seismograms. *J. Geophys.* **46**, 135–49.

Červený, V. (1980). A new approximation of the velocity-depth distribution and its application to the computation of seismic wave fields. *Studia Geoph. et Geod.* **24**, 17–27.

Červený, V. (1982a). Direct and inverse kinematic problems for inhomogeneous anisotropic media – Linearization approach. *Contrib. Geophys. Inst. Slov. Acad. Sci.* **13**, 127–33.

Červený, V. (1982b). Expansion of a plane wave into Gaussian beams. *Studia Geoph. et Geod.* **26**, 120–131.

Červený, V. (1983). Synthetic body wave seismograms for laterally varying layered structures by the Gaussian beam method. *Geophys. J. R. astr. Soc.* **73**, 389–426.

Červený, V. (1985a). The application of ray tracing to the numerical modeling of seismic wavefields in complex structures. In *Handbook of geophysical exploration, Section I: Seismic exploration*, eds. K. Helbig and S. Treitel, vol. 15, *Seismic shear waves, Part A: Theory*, ed. G. Dohr, pp. 1–124. London: Geophysical Press.

Červený, V. (1985b). Ray synthetic seismograms for complex two-dimensional and three-dimensional structures. *J. Geophys.* **58**, 2–26.

Červený, V. (1985c). Gaussian beam synthetic seismograms. *J. Geophys.* **58**, 44–72.

Červený, V. (1987a). Ray tracing algorithms in three-dimensional laterally varying layered structures. In *Seismic tomography with applications in global seismology and exploration geophysics*, ed. G. Nolet, pp. 99–133. Dordrecht: Reidel.

Červený, V. (1987b). *Ray methods for three-dimensional seismic modelling*. Lecture Notes. Continuing Education Course, pp. 1–870. Trondheim: Univ. Trondheim, NTH and Mobil Exploration Norway, Inc.

Červený, V. (1989a). Seismic ray theory. In *The encyclopedia of solid earth geophysics*, ed. D. E. James, pp. 1098–108. New York: Van Nostrand Reinhold.

Červený, V. (1989b). Synthetic body wave seismograms for laterally varying media containing thin transition layers. *Geophys. J. Int.* **99**, 331–49.

Červený, V. (1989c). Ray tracing in factorized anisotropic inhomogeneous media. *Geophys. J. Int.* **99**, 91–100.

Červený, V. (1990). *Seismic waves in anisotropic media*. Lecture Notes. Cursos Advançados em Exploração Geofísica. Rio de Janeiro: Petrobrás.

Červený, V. (2000). Summation of paraxial Gaussian beams and of paraxial ray approximations in inhomogeneous anisotropic layered structures. In *Seismic waves in complex 3-D structures, Report 10*, pp. 121–59. Prague: Faculty of Mathematics and Physics, Charles Univ.

Červený, V., and de Andrade, F. C. M. (1992). Influence of a near-surface structure on seismic wave fields recorded at the Earth's surface. *J. Seismic Exploration* **1**, 107–16.

Červený, V., and Aranha, P. R. A. (1992). Tunneling of seismic body waves through thin high-velocity layers in complex structures. *Studia Geoph. et Geod.* **36**, 115–38.

Červený, V., and Coppoli, A. D. M. (1992). Ray-Born synthetic seismograms for complex structures containing scatterers. *J. Seismic Exploration* **1**, 191–206.

Červený, V., and Firbas, P. (1984). Numerical modelling and inversion of travel times of seismic body waves in inhomogeneous anisotropic media. *Geophys. J. R. astr. Soc.* **76**, 41–51.

Červený, V., and Frangié, A. B. (1980). Elementary seismograms of seismic body waves in dissipative media. *Studia Geoph. et Geod.* **24**, 365–72.

Červený, V., and Frangié, A. B. (1982). Effects of causal absorption on seismic body waves. *Studia Geoph. et Geod.* **26**, 238–53.

Červený, V., Fuchs, K., Müller, G., and Zahradník, J. (1980). Theoretical seismograms for inhomogeneous media. In *Problems of the dynamic theory of propagation of seismic waves* (in Russian), vol. 20, ed. G. I. Petrashen, pp. 84–109. Leningrad: Nauka.

Červený, V., and Hron, F. (1980). The ray series method and dynamic ray tracing system for three-dimensional inhomogeneous media. *Bull. Seismol. Soc. Am.* **70**, 47–77.

Červený, V., and Janský, J. (1983). Ray amplitudes of seismic body waves in inhomogeneous radially symmetric media. *Studia Geoph. et Geod.* **27**, 9–18.

Červený, V., and Janský, J. (1985). Fast computation of ray synthetic seismograms in vertically inhomogeneous media. *Studia Geoph. et Geod.* **29**, 49–67.

Červený, V., and Jech, J. (1982). Linearized solutions of kinematic problems of seismic body waves in inhomogeneous slightly anisotropic media. *J. Geophys.* **51**, 96–104.

Červený, V., and Klimeš, L. (1984). Synthetic body wave seismograms for three-dimensional laterally varying media. *Geophys. J. R. astr. Soc.* **79**, 119–33.

Červený, V., Klimeš, L., and Pšenčík, I. (1984). Paraxial ray approximation in the computation of seismic wavefields in inhomogeneous media. *Geophys. J. R. astr. Soc.* **79**, 89–104.

Červený, V., Klimeš, L., and Pšenčík, I. (1988a). Applications of the dynamic ray tracing. *Phys. Earth Planet. Int.* **51**, 25–35.

Červený, V., Klimeš, L., and Pšenčík, I. (1988b). Complete seismic-ray tracing in three-dimensional structures. In *Seismological algorithms*, ed. D. J. Doornbos, pp. 89–168. New York: Academic Press.

Červený, V., and Kozák, J. (1972). Experimental evidence and investigation of pseudospherical waves. *Z. Geophys.* **38**, 617–26.

Červený, V., Kozák, J., and Pšenčík, I. (1971). Refraction of elastic waves into a medium of lower velocity–Pseudospherical waves. *PAGEOPH* **92**, 115–32.

Červený, V., Langer, J., and Pšenčík, I. (1974). Computation of geometrical spreading of seismic body waves in laterally inhomogeneous media with curved interfaces. *Geophys. J. R. astr. Soc.* **38**, 9–19.

Červený, V., Molotkov, I. A., and Pšenčík, I. (1977). *Ray method in seismology.* Praha: Universita Karlova.

Červený, V., Pleinerová, J., Klimeš, L., and Pšenčík, I. (1987). High-frequency radiation from earthquake sources for laterally varying layered structures. *Geophys. J. R. astr. Soc.* **88**, 43–79.

Červený, V., Popov, M. M., and Pšenčík, I. (1982). Computation of wave fields in inhomogeneous media – Gaussian beam approach. *Geophys. J. R. astr. Soc.* **70**, 109–28.

Červený, V., and Pretlová, V. (1977). Computation of ray amplitudes of seismic body waves in vertically inhomogeneous media. *Studia Geoph. et Geod.* **21**, 248–55.

Červený, V., and Pšenčík, I. (1972). Rays and travel-time curves in inhomogeneous anisotropic media. *Z. Geophys.* **38**, 565–77.

Červený, V., and Pšenčík, I. (1977). Ray theoretical seismograms for laterally varying layered structures. In *Publ. Inst. Pol. Acad. Sci.*, pp. 173–85. Warszawa-Lodz: PWN.

Červený, V., and Pšenčík, I. (1979). Ray amplitudes of seismic body waves in laterally inhomogeneous media. *Geophys. J. R. astr. Soc.* **57**, 97–106.

Červený, V., and Pšenčík, I. (1983a). Gaussian beams in two-dimensional elastic inhomogeneous media. *Geophys. J. R. astr. Soc.* **72**, 417–33.

Červený, V., and Pšenčík, I. (1983b). Gaussian beams and paraxial ray approximation in three-dimensional elastic inhomogeneous media. *J. Geophys.* **53**, 1–15.

Červený, V., and Pšenčík, I. (1984a). Gaussian beams in elastic 2-D laterally varying layered structures. *Geophys. J. R. astr. Soc.* **78**, 65–91.

Červený, V., and Pšenčík, I. (1984b). SEIS83 – Numerical modelling of seismic wave fields in 2-D laterally varying layered structures by the ray method. In *Documentation of earthquake algorithms, Report SE-35*, ed. E. R. Engdahl, pp. 36–40. Boulder: World Data Center A for Solid Earth Geophysics.

Červený, V., and Ravindra, R. (1971). *Theory of seismic head waves.* Toronto: Univ. of Toronto Press.

Červený, V., and Simões-Filho, I. A. (1991). The traveltime perturbations for seismic body waves in factorized anisotropic inhomogeneous media. *Geophys. J. Int.* **107**, 219–29.

Červený, V., and Soares, J. E. P. (1992). Fresnel volume ray tracing. *Geophysics* **57**, 902–15.

Chander, R. (1975). On tracing seismic rays with specified end points. *J. Geophys.* **41**, 173–7.

Chapman, C. H. (1971). On the computation of seismic ray traveltimes and amplitudes. *Bull. Seismol. Soc. Am.* **61**, 1267–74.

Chapman, C. H. (1974). Generalized ray theory in vertically inhomogeneous media. *Geophys. J. R. astr. Soc.* **36**, 673–704.

Chapman, C. H. (1976a). Exact and approximate generalized ray theory in vertically inhomogeneous media. *Geophys. J. R. astr. Soc.* **46**, 201–33.

Chapman, C. H. (1976b). A first-motion alternative to geometrical ray theory. *Geophys. Res. Letters* **3**, 153–6.

Chapman, C. H. (1978). A new method for computing synthetic seismograms. *Geophys. J. R. astr. Soc.* **54**, 481–518.

Chapman, C. H. (1985). Ray theory and its extensions: WKBJ and Maslov seismograms. *J. Geophys.* **58**, 27–43.

Chapman, C. H. (1994). Reflection/transmission coefficient reciprocities in anisotropic media. *Geophys. J. Int.* **116**, 498–501.

Chapman, C. H. (in press). Seismic ray theory. In *Handbook of earthquake and engineering seismology*, eds. W. H. K. Lee, H. Kanamori, and P. C. Jennings. New York: Academic Press. (Preprint.)

Chapman, C. H., Chu, J.-Y., and Lyness, D. G. (1988). The WKBJ seismogram algorithm. In *Seismological algorithms*, ed. D. J. Doornbos, pp. 47–74. New York: Academic Press.

Chapman, C. H., and Coates, R. T. (1994). Generalized Born scattering in anisotropic media. *Wave Motion* **19**, 309–41.

Chapman, C. H., and Drummond, R. (1982). Body-wave seismograms in inhomogeneous media using Maslov asymptotic theory. *Bull. Seismol. Soc. Am.* **72**, S277–S317.

Chapman, C. H., and Orcutt, J. A. (1985). The computation of body wave synthetic seismograms in laterally homogeneous media. *Revs. Geophys.* **23**, 105–63.

Chapman, C. H., and Pratt, R. G. (1992). Traveltime tomography in anisotropic media: I. Theory. *Geophys. J. Int.* **109**, 1–19.

Chapman, C. H., and Shearer, P. M. (1989). Ray tracing in azimuthally anisotropic media: II. Quasi-shear wave coupling. *Geophys. J.* **96**, 65–83.

Chapman, S. J., Lawry, J. M. H., Ockendon, J. R., and Tew, R. (1998). *On the theory of complex rays,* vol. III. New York: Plenum.

Chen, H.-W. (1998). Three-dimensional geometrical ray theory and modelling of transmitted seismic energy of data from the Nevada Test Site. *Geophys. J. Int.* **133**, 363–78.

Chen, K. C., and Ludwig, D. (1973). Calculation of wave amplitudes by ray tracing. *J. Acoust. Soc. Am.* **54**, 431–6.

Cheng, N., and House, L. (1996). Minimum traveltime calculation in 3-D graph theory. *Geophysics* **61**, 1895–8.

Chernov, L. A. (1960). *Wave propagation in random media.* New York: McGraw Hill.

Claerbout, J. F. (1976). *Fundamentals of geophysical data processing.* New York: McGraw Hill.

Claerbout, J. F. (1985). *Imaging the Earth's interior.* Oxford: Blackwell.

Clarke, R. A., and Jannaud, L. R. (1996). Raytracing in the overthrust model. 66th Annual SEG Meeting Expanded Abstracts, Denver, 687–90.

Cline, A. K. (1981). FITPACK – Software package for curve and surface fitting employing splines under tension. Austin: Dept. of Comp. Sci., Univ. of Texas.

Coates, R. T., and Chapman, C. H. (1990a). Ray perturbation theory and the Born approximation. *Geophys. J. Int.* **100**, 379–92.

Coates, R. T., and Chapman, C. H. (1990b). Quasi-shear wave coupling in weakly anisotropic 3-D media. *Geophys. J. Int.* **103**, 301–20.

Coates, R. T., and Chapman, C. H. (1991). Generalized Born scattering of elastic waves in 3-D media. *Geophys. J. Int.* **107**, 231–63.

Coates, R. T., and Charrette, E. E. (1993). A comparison of single scattering and finite difference synthetic seismograms in realizations of 2-D elastic random media. *Geophys. J. Int.* **113**, 463–82.

Coddington, E. A., and Levinson, N. (1955). *Theory of ordinary differential equations.* New York: McGraw Hill.

Cohen, J. K., Hagin, F. G., and Bleistein, N. (1986). Three-dimensional Born inversion with an arbitrary reference. *Geophysics* **51**, 1552–8.

Comer, R. P. (1984). Rapid seismic ray tracing in a spherically symmetric Earth via interpolation of rays. *Bull. Seismol. Soc. Am.* **74**, 479–92.

Cooper, H. F., Jr. (1967). Reflection and transmission of oblique plane waves at a plane interface between viscoelastic media. *J. Acoust. Soc. Am.* **42**, 1064–9.

Cooper, H. F., Jr., and Reiss, E. L. (1966). Reflection of plane viscoelastic waves from plane boundaries. *J. Acoust. Soc. Am.* **39**, 1133–8.

Cormier, V. (1984). The polarization of S waves in a heterogeneous isotropic Earth model. *J. Geophys.* **56**, 20–3.

Cormier, V., and Mellen, M. H. (1984). Application of asymptotic ray theory to vertical seismic profiling. In *Vertical seismic profiling: Advanced concepts,* eds. M. N. Toksöz and R. R. Stewart, pp. 28–44. London: Geophys. Press.

Cormier, V., and Spudich, P. (1984). Amplification of ground motion and waveform complexity in fault zones; examples from the San Andreas and Calaveras Faults. *Geophys. J. R. astr. Soc.* **79**, 135–52.

Coultrip, R. L. (1993). High-accuracy wavefront tracing traveltime calculation. *Geophysics* **58**, 284–92.

Courant, R., and Hilbert, D. (1966). *Methods of mathematical physics.* New York: Wiley.

Crampin, S. (1981). A review of wave motion in anisotropic and cracked elastic media. *Wave Motion* **3**, 343–91.

Crampin, S. (1986). Crack porosity and alignments from shear wave VSPs. In *Shear wave exploration,* eds. S. H. Danbom and S. N. Domenico, pp. 227–51. Tulsa: Society of Exploration Geophysicists.

Crampin, S. (1989). Suggestions for a consistent terminology for seismic anisotropy. *Geophys. Prospecting* **37**, 753–70.

Crampin, S., and Kirkwood, S. C. (1981). Velocity variations in systems of anisotropic symmetry. *J. Geophys.* **49**, 35–42.

Crampin, S., and Yedlin, M. (1981). Shear-wave singularities of wave propagation in anisotropic media. *J. Geophys.* **49**, 43–6.

Dahlen, F. A., and Tromp, J. (1998). *Theoretical global seismology*. Princeton: Princeton Univ. Press.

Daley, P. F., and Hron, F. (1977). Reflection and transmission coefficients for transversely isotropic media. *Bull. Seismol. Soc. Am.* **67**, 661–75.

Daley, P. F., and Hron, F. (1979). Reflection and transmission coefficients for seismic waves in ellipsoidally anisotropic media. *Geophysics* **44**, 27–38.

Daley, P. F., and Hron, F. (1982). Ray-reflectivity method for SH-waves in stacks of thin and thick layers. *Geophys. J. R. astr. Soc.* **69**, 527–35.

Daley, P. F., and Hron, F. (1983a). High-frequency approximation to the non-geometrical S^* arrivals. *Bull. Seismol. Soc. Am.* **73**, 109–23.

Daley, P. F., and Hron, F. (1983b). Nongeometric arrivals due to highly concentrated sources adjacent to plane interfaces. *Bull. Seismol. Soc. Am.* **73**, 1655–71.

Daley, P. F., and Hron, F. (1987). Reflection of an incident spherical P wave on a free surface (near-vertical incidence). *Bull. Seismol. Soc. Am.* **77**, 1057–70.

Daley, P. F., Marfurt, K. J., and McCarron, E. B. (1999). Finite-element ray tracing through structurally deformed transversely isotropic formations. *Geophysics* **64**, 954–62.

Danbom, S. H., and Domenico, S. N., eds. (1986). *Shear wave exploration*. Tulsa: Society of Exploration Geophysicists.

Davis, J. L. (1988). *Wave propagation in solids and fluids*. New York: Springer.

DeSanto, J. A. (1992). *Scalar wave theory. Green's functions and applications*. Berlin, Heidelberg: Springer.

Deschamps, G. A. (1971). Gaussian beam as a bundle of complex rays. *Electr. Lett.* **7**, 684–5.

Deschamps, G. A. (1972). Ray techniques in electromagnetics. *Proc. IEEE* **60**, 1022–35.

Dey-Sarkar, S. K., and Chapman, C. H. (1978). A simple method for the computation of body-wave seismograms. *Bull. Seismol. Soc. Am.* **68**, 1577–93.

Dietrich, M., and Bouchon, M. (1985). Measurements of attenuation from vertical seismic profiles by iterative modeling. *Geophysics* **50**, 931–49.

Dijkstra, E. W. (1959). A note on two problems in connection with graphs. *Numerische Mathematik* **1**, 269–71.

Docherty, P. (1985). *A fast ray tracing routine for laterally inhomogeneous media*. In: *Report CWP-018*. Golden: Centre for Wave Phenomena, Colorado School of Mines.

Docherty, P. (1991). A brief comparison of some Kirchhoff integral formulas for migration and inversion. *Geophysics* **56**, 1164–9.

Dohr, G., ed. (1985). *Seismic shear waves*. London: Geophysical Press.

Douglas, A., Young, J. B., and Hudson, J. A. (1974). Complex P-wave seismograms from simple earthquake sources. *Geophys. J. R. astr. Soc.* **37**, 141–50.

Drijkoningen, G. G. (1991a). Tunnelling and the generalized ray method in piecewise homogeneous media. *Geophys. Prospecting* **39**, 757–81.

Drijkoningen, G. G. (1991b). Generalized ray theory for an Epstein profile. *Geophys. J. Int.* **104**, 469–77.

Drijkoningen, G. G., and Chapman, C. H. (1988). Tunneling rays using the Cagniard-de Hoop method. *Bull. Seismol. Soc. Am.* **78**, 898–907.

Drijkoningen, G. G., Chapman, C. H., and Thompson, C. J. (1987). On head-wave amplitudes. *J. Math. Phys.* **28** (8), 1729–31.

Druzhinin, A. B. (1996). Nonlinear ray perturbation theory with its applications to ray tracing and inversion in anisotropic media. *PAGEOPH* **148**, 637–83.

Druzhinin, A. B. (1998). Explicit Kirchhoff-Helmholtz integrals and their accurate numerical implementation. *Wave Motion* **28**, 119–48.

Druzhinin, A. B., Pedersen, H., Campillo, M., and Kim, W. (1998). Elastic Kirchhoff-Helmholtz synthetic seismograms. *PAGEOPH* **151**, 17–45.

Duff, G. F. D. (1960). The Cauchy problem for elastic waves in an anisotropic medium. *Phil. Trans. R. Soc. London A* **252**, 249–73.

Eaton, D. W. S. (1993). Finite difference traveltime calculation for anisotropic media. *Geophys. J. Int.* **114**, 273–80.

Eaton, D. W. S., Stewart, R. R., and Harrison, M. P. (1991). The Fresnel zone for converted P-SV waves. *Geophysics* **56**, 360–4.

Eisner, L., and Pšenčík, I. (1996). Computation of additional components of the first-order ray approximation in isotropic media. *PAGEOPH* **148**, 227–53.

Ettrich, N., and Gajewski, D. (1996). Wave front construction in smooth media for prestack depth migration. *PAGEOPH* **148**, 481–502.

Evans, R. (1984). Effects of the free surface on shear wavetrains. *Geophys. J. R. astr. Soc.* **76**, 165–72.

Every, A. G., and Kim, K. Y. (1994). Time domain dynamic response functions of elastically anisotropic solids. *J. Acoust. Soc. Am.* **95**, 2505–16.

Ewing, W. M., Jardetzky, W. S., and Press, F. (1957). *Elastic waves in layered media.* New York: McGraw Hill.

Faria, E. L., and Stoffa, P. L. (1994). Traveltime computation in transversely isotropic media. *Geophysics* **59**, 272–81.

Farra, V. (1989). Ray perturbation theory for heterogeneous hexagonal anisotropic medium. *Geophys. J. Int.* **99**, 723–37.

Farra, V. (1990). Amplitude computation in heterogeneous media by ray perturbation theory: A finite element approach. *Geophys. J. Int.* **103**, 341–54.

Farra, V. (1992). Bending method revisited: a Hamiltonian approach. *Geophys. J. Int.* **109**, 138–50.

Farra, V. (1993). Ray tracing in complex media. *J. Appl. Geophysics* **30**, 55–73.

Farra, V. (1999). Computation of second-order traveltime perturbation by Hamiltonian ray theory. *Geophys. J. Int.* **136**, 205–17.

Farra, V., and Le Bégat, S. (1995). Sensitivity of qP-wave traveltimes and polarization vectors to heterogeneity, anisotropy and interfaces. *Geophys. J. Int.* **121**, 371–84.

Farra, V., and Madariaga, R. (1987). Seismic waveform modeling in heterogeneous media by ray perturbation theory. *J. Geophys. Res.* **92** (B3), 2697–712.

Farra, V., Virieux, J., and Madariaga, R. (1989). Ray perturbation theory for interfaces. *Geophys. J. Int.* **99**, 377–90.

Fedorov, F. I. (1968). *Theory of elastic waves in crystals.* New York: Plenum.

Felsen, L. B. (1976a). Evanescent waves. *J. Opt. Soc. Am.* **66**, 751–60.

Felsen, L. B. (1976b). Complex-source-point solutions of the field equations and their relation to the propagation and scattering of Gaussian beams. *Ist. Naz. Alta Matem., Symp. Math.* **18**, 39–56.

Felsen, L. B. (1981). Hybrid ray-mode fields in inhomogeneous waveguides and ducts. *J. Acoust. Soc. Am.* **69**, 352–61.

Felsen, L. B. (1984). Geometrical theory of diffraction, evanescent waves, complex rays and Gaussian beams. *Geophys. J. R. astr. Soc.* **79**, 77–88.

Felsen, L. B., and Marcuvitz, N. (1973). *Radiation and scattering of waves.* Englewood Cliffs, NJ: Prentice Hall.

Fertig, J., and Pšenčík, I. (1985). Numerical modeling of P and S waves in exploration seismology. In *Handbook of geophysical exploration, Section I: Seismic exploration*, eds. K. Helbig and S. Treitel, vol. 15, *Seismic shear waves, Part A: Theory*, ed. G. Dohr, pp. 226–82. London: Geophysical Press.

Firbas, P. (1984). Travel time curves for complex inhomogeneous slightly anisotropic media. *Studia Geoph. et Geod.* **28**, 393–406.

Fischer, R., and Lees, J. M. (1993). Shortest path ray tracing with sparse graphs. *Geophysics* **58**, 987–96.

Fishman, L., and McCoy, J. J. (1984). Derivation and application of extended parabolic wave theories. I. The factorized Helmholtz equation. *J. Math. Phys.* **25**, 285–96.

Fishman, L., and McCoy, J. J. (1985). A new class of propagation models based on a factorization of the Helmholtz equation. *Geophys. J. R. astr. Soc.* **80**, 439–61.

Flatté, S. M., Dashen, R., Munk, W. H., Watson, K. M., and Zachariasen, F., eds. (1979). *Sound transmission through a fluctuating ocean.* Cambridge: Cambridge Univ. Press.

Fock, V. A. (1965). *Electromagnetic diffraction and propagation problems.* New York: Pergamon Press.

Fokkema, J. T., and van den Berg, P. M. (1993). *Seismic applications of acoustic reciprocity*. Amsterdam: Elsevier.

Foster, D. J., and Huang, J.-I. (1991). Global asymptotic solutions of the wave equation. *Geophys. J. Int.* **105**, 163–71.

Fradkin, L. Yu. (1989). Limits of validity of geometrical optics in weakly irregular media. *J. Opt. Soc. Am.* **A6**, 1315–19.

Fradkin, L. Yu., and Kiselev, A. P. (1997). The two-component representation of time-harmonic elastic body waves in the high- and intermediate-frequency regimes. *J. Acoust. Soc. Am.* **101**, 52–65.

Frazer, L. N., and Fryer, G. J. (1989). Useful properties of the system matrix for a homogeneous anisotropic visco-elastic solid. *Geophys. J.* **97**, 173–7.

Frazer, L. N., and Sen, M. K. (1985). Kirchhoff-Helmholtz reflection seismograms in a laterally inhomogeneous multi-layered elastic medium. Part I: Theory. *Geophys. J. R. astr. Soc.* **80**, 121–47.

Friedlander, F. G. (1958). *Sound pulses*. London: Cambridge Univ. Press.

Friedrichs, K. O., and Keller, J. B. (1955). Geometrical acoustics, II: Diffraction, reflection and refraction of a weak spherical or cylindrical shock at a plane interface. *J. Appl. Phys.* **26**, 961–6.

Fryer, G. J., and Frazer, L. N. (1984). Seismic waves in stratified anisotropic media. *Geophys. J. R. astr. Soc.* **78**, 691–710.

Fryer, G. J., and Frazer, L. N. (1987). Seismic waves in stratified anisotropic media. II. Elastodynamic eigensolutions for some anisotropic systems. *Geophys. J. R. astr. Soc.* **91**, 73–101.

Fuchs, K. (1968a). The reflection of spherical waves from transition zones with arbitrary depth-dependent elastic moduli and density. *J. Phys. Earth* **16**, 27–41.

Fuchs, K. (1968b). Das Reflexions- und Transmissionvermögen eines geschichteten Mediums mit beliebiger Tiefen-Verteilung der elastischen Moduln und der Dichte für schrägen Einfall ebener Wellen. *Z. Geophys.* **34**, 389–413.

Fuchs, K., and Müller, G. (1971). Computation of synthetic seismograms with the reflectivity method and comparison with observations. *Geophys. J. R. astr. Soc.* **23**, 417–33.

Fuchs, K., and Schulz, K. (1976). Tunneling of low-frequency waves through the subcrustal lithosphere. *J. Geophys.* **42**, 175–90.

Fung, Y. C. (1965). *Foundations of solid mechanics*. Englewood Cliffs, NJ: Prentice Hall.

Futterman, W. I. (1962). Dispersive body waves. *J. Geophys. Res.* **67**, 5279–91.

Gabor, D. (1946). Theory of communication. *J. Inst. Elec. Eng.* **93**, 429–57.

Gajewski, D. (1993). Radiation from point sources in general anisotropic media. *Geophys. J. Int.* **113**, 299–317.

Gajewski, D., and Pšenčík, I. (1987a). Computation of high-frequency seismic wavefields in 3-D laterally inhomogeneous anisotropic media. *Geophys. J. R. astr. Soc.* **91**, 383–411.

Gajewski, D., and Pšenčík, I. (1987b). Ray synthetic seismograms for a 3-D anisotropic lithospheric structure. *Phys. Earth Planet. Int.* **51**, 1–23.

Gajewski, D., and Pšenčík, I. (1990). Vertical seismic profile synthetics by dynamic ray tracing in laterally varying layered anisotropic structures. *J. Geophys. Res.* **95**, 11301–15.

Gajewski, D., and Pšenčík, I. (1992). Vector wavefields for weakly attenuating anisotropic media by the ray method. *Geophysics* **57**, 27–38.

Gal'perin, E. I. (1984). *The polarization method of seismic exploration*. Dordrecht: Reidel.

Garmany, J. (1988). Seismograms in stratified anisotropic media. Part I: WKBJ theory. *Geophys. J.* **92**, 365–77. Part II: Uniformly asymptotic approximations. *Geophys. J.* **92**, 379–89.

Gassman, F. (1964). Introduction to seismic travel time method in anisotropic media. *PAGEOPH* **58**, 63–112.

Gebrande, H. (1976). A seismic-ray tracing method for two-dimensional inhomogeneous media. In *Explosion seismology in Central Europe. Data and results*, eds. P. Giese, C. Prodehl, and A. Stein, pp. 162–7. Berlin: Springer.

Gel'chinskiy, B. Ya. (1961). An expression for the geometrical spreading. In *Problems of the dynamic theory of propagation of seismic waves* (in Russian), vol. 5, ed. G. I. Petrashen, pp. 47–53. Leningrad: Leningrad Univ. Press.

Gelchinsky, B. Ya. (1985). The formulae for calculation of the Fresnel zones or volumes. *J. Geophys.* **57**, 33–41.

Geoltrain, S., and Brac, J. (1993). Can we image complex structures with first-arrival traveltime? *Geophysics* **58**, 564–75.

Gibson, R. L., and Ben-Menahem, A. (1991). Elastic wave scattering by anisotropic obstacles: Application to fractured volumes. *J. Geophys. Res.* **96**, 19905–24.

Gibson, R. L., Sena, A. G., and Toksöz, M. N. (1991). Paraxial ray tracing in 3-D inhomogeneous anisotropic media. *Geophys. Prospecting* **39**, 473–504.

Gilbert, F., and Backus, G. E. (1966). Propagator matrices in elastic wave and vibration problems. *Geophysics* **31**, 326–32.

Gilbert, F., and Helmberger, D. V. (1972). Generalized ray theory for a layered sphere. *Geophys. J. R. astr. Soc.* **27**, 57–80.

Gilbert, F., and Laster, S. J. (1962). Excitation and propagation of pulses on an interface. *Bull. Seismol. Soc. Am.* **52**, 299–319.

Gilmore, R. (1981). *Catastrophe Theory For Scientists And Engineers*. New York: Wiley-Interscience.

Gjevik, B. (1973). A variational method for Love waves in nonhorizontally layered structures. *Bull. Seismol. Soc. Am.* **63**, 1013–23.

Gjevik, B. (1974). Ray tracing for seismic surface waves. *Geophys. J. R. astr. Soc.* **39**, 29–39.

Gjøystdal, H., Reinhardsen, J. E., and Ursin, B. (1984). Traveltime and wavefront curvature calculations in three-dimensional inhomogeneous layered media with curved interfaces. *Geophysics* **49**, 1466–94.

Goldin, S. V. (1979). *Interpretation of seismic reflection data* (in Russian). Moscow: Nedra.

Goldin, S. V. (1986). *Seismic travel time inversion*. Tulsa: Society of Exploration Geophysicists.

Goldin, S. V. (1989). Physical analysis of the additional components of seismic waves in the first approximation of ray series. *Geology and Geophysics* **30**, 128–32.

Goldin, S. V. (1991). Decomposition of geometrical approximation of a reflected wave. *Geology and Geophysics* **32**, 128–37.

Goldin, S. V., and Kurdyukova, T. V. (1994). On computation of additional components of seismic body waves. *Geology and Geophysics* **35**, 56–67.

Goldin, S. V., and Piankov, V. N. (1992). On geometrical spreading and the number of singularities on a ray. *Geology and Geophysics* **33**, 90–105.

Goldstein, H. (1980). *Classical mechanics*. Reading, MA: Addison-Wesley.

Graebner, M. (1992). Plane-wave reflection and transmission coefficients for a transversely isotropic solid. *Geophysics* **57**, 1512–19.

Grant, F. S., and West, G. F. (1965). *Interpretation theory in applied geophysics*. New York: McGraw Hill.

Gray, S. H., and May, W. P. (1994). Kirchhoff migration using eikonal equation traveltimes. *Geophysics* **59**, 810–17.

Grechka, V. Y., and McMechan, G. A. (1996). 3-D two-point ray tracing for heterogeneous, weakly transversely isotropic media. *Geophysics* **61**, 1883–94.

Grechka, V. Y., and Obolentseva, I. R. (1993). Geometrical structure of shear wave surfaces near singularity directions in anisotropic media. *Geophys. J. Int.* **115**, 609–16.

Grechka, V. Y., Tsvankin, I., and Cohen, J. K. (1999). Generalized Dix equation and analytic treatment of normal-moveout velocity for anisotropic media. *Geophys. Prospecting* **47**, 117–48.

Green, A. G. (1976). Ray paths and relative intensities in one- and two-dimensional velocity models. *Bull. Seismol. Soc. Am.* **66**, 1581–607.

Gregersen, S. (1974). Surface waves in isotropic, laterally inhomogeneous media. *PAGEOPH* **114**, 821–32.

Grinfeld, M. A. (1980). A new system of equations for calculation of geometrical spreading. In *Computational seismology* (in Russian), vol. 13, eds. V. I. Keilis-Borok and A. L. Levshin, pp. 127–33. Moscow: Nauka.

Gritsenko, S. A. (1984). The derivatives of time fields. *Geology and Geophysics* **25**, 141–67.

Gudmundsson, O. (1996). On the effect of diffraction on traveltime measurements. *Geophys. J. Int.* **124**, 304–14.

Guest, W. S., and Kendall, J-M. (1993). Modelling seismic waveforms in anisotropic inhomogeneous media using ray and Maslov asymptotic theory: Application to exploration seismology. *Canad. J. Expl. Geophys.* **29**, 78–92.

Guest, W. S., Thomson, C. J., and Kendall, J-M. (1992). Displacements and rays for splitting S-waves: Perturbation theory and the failure of the ray method. *Geophys. J. Int.* **108**, 372–8.

Guiziou, J. L. (1989). Enhanced ray tracing techniques for 3-D applications. 59th Annual SEG Meeting Expanded Abstracts, Dallas, 1102–4.

Guiziou, J. L., and Haas, A. (1988). Three-dimensional traveltime inversion in anisotropic media. 58th Annual SEG Meeting Expanded Abstracts, Anaheim, 1089–91.

Guiziou, J. L., Mallet, J. L., and Madariaga, R. (1996). 3-D seismic reflection tomography on top of the GOCAD depth modeler. *Geophysics* **61**, 1499–510.

Gutenberg, B. (1952). SV and SH. *Trans. Am. Geophys. Un.* **33**, 573–84.

Haddon, R. A. W., and Buchen, P. W. (1981). Use of Kirchhoff's formula for body wave calculations in the Earth. *Geophys. J. R. astr. Soc.* **67**, 587–98.

Haines, A. J. (1983). A phase-front method: I. Narrow-frequency-band SH-waves. *Geophys. J. R. astr. Soc.* **72**, 783–808.

Haines, A. J. (1984a). A phase-front method: II. Broad-frequency-band SH-waves. *Geophys. J. R. astr. Soc.* **77**, 43–64.

Haines, A. J. (1984b). A phase-front method: III. Acoustic waves, P- and S-waves. *Geophys. J. R. astr. Soc.* **77**, 65–104.

Hanyga, A. (1982a). Dynamic ray tracing in an anisotropic medium. *Tectonophysics* **90**, 243–51.

Hanyga, A. (1982b). The kinematic inverse problem for weakly laterally inhomogeneous anisotropic media. *Tectonophysics* **90**, 253–62.

Hanyga, A. (1984). Point source in anisotropic elastic medium. *Gerlands Beitr. Geophys.* **93**, 463–79.

Hanyga, A. (1986). Gaussian beams in anisotropic media. *Geophys. J. R. astr. Soc.* **85**, 473–503.

Hanyga, A. (1988). Numerical methods of tracing rays and wavefronts. In *Seismological algorithms*, ed. D. J. Doornbos, pp. 169–233. New York: Academic Press.

Hanyga, A. (1995). Asymptotic edge-and-vertex diffraction theory. *Geophys. J. Int.* **123**, 277–90.

Hanyga, A. (1996a). Diffraction by plane sectors and polygons. *PAGEOPH* **148**, 137–53.

Hanyga, A. (1996b). Point-to-curve ray tracing. *PAGEOPH* **148**, 387–420.

Hanyga, A., and Helle, H. B. (1995). Synthetic seismograms from generalized ray tracing. *Geophys. Prospecting* **43**, 51–75.

Hanyga, A., Lenartowicz, E., and Pajchel, J. (1984). *Seismic wave propagation in the Earth*. Amsterdam: Elsevier.

Hanyga, A., and Pajchel, J. (1995). Point-to-curve ray tracing in complicated geological models. *Geophys. Prospecting* **43**, 859–72.

Hearn, D. J., and Krebes, E. S. (1990a). On computing ray-synthetic seismograms for anelastic media using complex rays. *Geophysics* **55**, 422–32.

Hearn, D. J., and Krebes, E. S. (1990b). Complex rays applied to wave propagation in a viscoelastic medium. *PAGEOPH* **132**, 401–15.

Helbig, K. (1994). *Foundations of anisotropy for exploration seismics*. Oxford: Pergamon.

Helmberger, D. V. (1968). The crust-mantle transition in the Bering Sea. *Bull. Seismol. Soc. Am.* **58**, 179–214.

Henneke, E. G. (1972). Reflection-refraction of a stress wave at a plane boundary between anisotropic media. *J. Acoust. Soc. Am.* **51**, 210–17.

Herrman, R. B. (1976). Some more complexity of S-wave particle motion. *Bull. Seismol. Soc. Am.* **66**, 625–9.

Hill, D. P. (1973). Critically refracted waves in a spherically symmetric radially heterogeneous Earth model. *Geophys. J. R. astr. Soc.* **34**, 149–79.

Hole, J. A., and Zelt, B. C. (1995). 3-D finite-difference reflection traveltimes. *Geophys. J. Int.* **121**, 427–34.

Hong, T.-L., and Helmberger, D. V. (1977). Generalized ray theory for dipping structure. *Bull. Seismol. Soc. Am.* **67**, 995–1008.

Hong, T.-L., and Helmberger, D. V. (1978). Glorified optics and wave propagation in nonplanar structure. *Bull. Seismol. Soc. Am.* **68**, 1313–30.

de Hoop, A. T. (1960). A modification of Cagniard's method for solving seismic pulse problems. *Appl. Sci. Res. B* **8**, 349–56.

Hörmander, L. (1971). Fourier integral operators. *Acta Math.* **127**, 79–183.

Hosken, J. W. J. (1988). Ricker wavelets in their various guises. *First Break* **6**, 24–33.

Hrabě, J. (1994). Paraxial ray theory in general coordinates. *Studia Geoph. et Geod.* **38**, 157–67.

Hron, F. (1971). Criteria for selection of phases in synthetic seismograms for layered media. *Bull. Seismol. Soc. Am.* **61**, 765–79.

Hron, F. (1972). Numerical methods of ray generation in multilayered media. In *Methods in computational physics*, vol. 12, ed. B. A. Bolt, pp. 1–34. New York: Academic Press.

Hron, F., Daley, P. F., and Marks, L. W. (1977). Numerical modelling of seismic body waves in oil exploration and crustal seismology. In *Computing methods in geophysical mechanics*, ed. R. P. Shaw, pp. 21–42. New York: American Soc. of Mechanical Engineers.

Hron, F., and Kanasewich, E. R. (1971). Synthetic seismograms for deep seismic sounding studies using asymptotic ray theory. *Bull. Seismol. Soc. Am.* **61**, 1169–1200.

Hron, F., May, B. T., Covey, J. D., and Daley, P. F. (1986). Synthetic seismic sections for acoustic, elastic, anisotropic, and vertically inhomogeneous layered media. *Geophysics* **51**, 710–35.

Hron, F., and Mikhailenko, B. G. (1981). Numerical modeling of nongeometrical effects by the Alekseyev-Mikhailenko method. *Bull. Seismol. Soc. Am.* **71**, 1011–29.

Hron, F., and Zheng, B. S. (1993). On the longitudinal component of the particle motion carried by the shear PS wave reflected from the free surface at normal incidence. *Bull. Seismol. Soc. Am.* **83**, 1610–16.

Huang, X., Kendall, J.-M., Thomson, C. J., and West, G. F. (1998). A comparison of the Maslov integral seismogram and the finite-difference method. *Geophys. J. Int.* **132**, 584–94.

Huang, X., and West, G. F. (1997). Effects of weighting functions on Maslov uniform seismograms: A robust weighting method. *Bull. Seismol. Soc. Am.* **87**, 164–73.

Huang, X., West, G. F., and Kendall, J.-M. (1998). A Maslov-Kirchhoff seismogram method. *Geophys. J. Int.* **132**, 595–602.

Hubral, P. (1979). A wave-front curvature approach to computing of ray amplitudes in inhomogeneous media with curved interfaces. *Studia Geoph. et Geod.* **23**, 131–7.

Hubral, P. (1980). Wave front curvatures in 3-D laterally inhomogeneous media with curved interfaces. *Geophysics* **45**, 905–13.

Hubral, P. (1983). Computing true amplitude reflections in a laterally inhomogeneous earth. *Geophysics* **48**, 1051–62.

Hubral, P., and Krey, Th. (1980). *Interval velocities from seismic reflection time measurements*. SEG Monograph Series No. 3, Tulsa: Society of Exploration Geophysicists.

Hubral, P., Schleicher, J., and Tygel, M. (1992). Three-dimensional paraxial ray properties: Part I. Basic relations. *J. Seismic Exploration* **1**, 265–79. Part II. Applications. *J. Seismic Exploration* **1**, 347–62.

Hubral, P., Schleicher, J., Tygel, M., and Hanitzsch, C. (1993). Determination of Fresnel zones from traveltime measurements. *Geophysics* **58**, 703–12.

Hubral, P., and Tygel, M. (1989). Analysis of the Rayleigh pulse. *Geophysics* **54**, 654–8.

Hubral, P., Tygel, M., and Schleicher, J. (1995). Geometrical-spreading and ray-caustic decomposition of elementary seismic waves. *Geophysics* **60**, 1195–202.

Hudson, J. A. (1980a). *The excitation and propagation of elastic waves*. Cambridge, GB : Cambridge Univ. Press.

Hudson, J. A. (1980b). A parabolic approximation for elastic waves. *Wave Motion* **2**, 207–14.

Hudson, J. A. (1981). A parabolic approximation for surface waves. *Geophys. J. R. astr. Soc.* **67**, 755–70.

Hudson, J. A., and Heritage, J. R. (1981). Use of the Born approximation in seismic scattering problems. *Geophys. J. R. astr. Soc.* **66**, 221–40.

Iversen, E., Vinje, V., and Gelius, L.-J. (1996). Raytracing in a smooth velocity model. *J. Seismic Exploration* **5**, 129–40.

Jacob, K. H. (1970). Three-dimensional seismic ray tracing in a laterally heterogeneous spherical Earth. *J. Geophys. Res.* **75**, 6675–89.

James, G. L. (1976). *Geometrical theory of diffraction for electromagnetic waves*. Stevenage, GB: Peregrinus.

Janský, J., and Červený, V. (1981). Computation of ray integrals and ray amplitudes in radially symmetric media. *Studia Geoph. et Geod.* **25**, 288–92.

Jech, J. (1983). Computation of rays in an inhomogeneous transverselly isotropic medium with a non-vertical axis of symmetry. *Studia Geoph. et Geod.* **27**, 114–21.

Jech, J., and Pšenčík, I. (1989). First-order perturbation method for anisotropic media. *Geophys. J. Int.* **99**, 369–76.

Jech, J., and Pšenčík, I. (1992). Kinematic inversion for qP- and qS-waves in inhomogeneous hexagonally symmetric structures. *Geophys. J. Int.* **108**, 604–12. Erratum (1992), *Geophys. J. Int.* **110**, 397.

Jeffreys, H. (1926). On compressional waves in two superposed layers. *Proc. Cambridge Phil. Soc.* **23**, 472–81.

Jeffreys, H. (1970). *The Earth, its origin, history and physical constitution*. 5th ed. Cambridge, GB: Cambridge Univ. Press.

Jeffreys, H., and Jeffreys, B. S. (1966). *Methods of mathematical physics*. Cambridge, GB: Cambridge Univ. Press.

Jiang, Z-Y., Pitts, T. A., and Greenleaf, J. F. (1997). Analytic investigation of chaos in a class of parabolic systems. *J. Acoust. Soc. Am.* **101**, 1971–80.

Jílek, P., and Červený, V. (1996). Radiation patterns of point sources situated close to structural interfaces and to the Earth's surface. *PAGEOPH* **148**, 175–225.

Jobert, N. (1976). Propagation of surface waves on an ellipsoidal Earth. *PAGEOPH* **114**, 797–804.

Jobert, N., and Jobert, G. (1983). An application of ray theory to the propagation of waves along a laterally heterogeneous spherical surface. *Geophys. Res. Letters* **10**, 1148–51.

Jobert, N., and Jobert, G. (1987). Ray tracing for surface waves. In *Seismic tomography with applications in global seismology and exploration geophysics*, ed. G. Nolet, pp. 275–300. Dordrecht: Reidel.

Julian, B. R., and Gubbins, D. (1977). Three-dimensional seismic ray tracing. *J. Geophys.* **43**, 95–113.

Kamke, E. (1959). *Differentialgleichungen. Lösungsmethoden und Lösungen, vol. 1, Gewöhnliche Differentialgleichungen*. Leipzig: Akademische Verlagsgessellschaft Geest und Portig.

Kampfmann, W. (1988). A study of diffraction-like events on DEKORP2-S by Kirchhoff theory. *J. Geophys.* **62**, 163–74.

Kapoor, T. K., and Felsen, L. B. (1995). Hybrid ray-mode analysis of acoustic scattering from a finite, fluid-loaded plate. *Wave Motion* **22**, 109–31.

Karal, F. C., and Keller, J. B. (1959). Elastic wave propagation in homogeneous and inhomogeneous media. *J. Acoust. Soc. Am.* **31**, 694–705.

Katchalov, A. P. (1984). Coordinate system for description of "quasi-photons". In *Mathematical problems of the theory of propagation of waves* (in Russian), vol. 14, ed. V. M. Babich, pp. 73–6. Leningrad: Nauka.

Katchalov, A. P., and Popov, M. M. (1981). The application of the Gaussian beam summation method to the computation of high-frequency wave fields (in Russian). *Dokl. Akad. Nauk SSSR* **258 (5)**, 1097–1100.

Katchalov, A. P., and Popov, M. M. (1988). Gaussian beam method and theoretical seismograms. *Geophys. J.* **93**, 465–75.

Katchalov, A. P., Popov, M. M., and Pšenčík, I. (1983). On the applicability of the method of summation of Gaussian beams to problems with edge points on interfaces. In *Mathematical problems of the theory of wave propagation* (in Russian), vol. 13, ed. V. M. Babich, pp. 65–71. Leningrad: Nauka.

Kaufman, H. (1953). Velocity functions in seismic prospecting. *Geophysics* **18**, 289–97.

Kawasaki, I., and Tanimoto, T. (1981). Radiation patterns of body waves due to the seismic dislocation occuring in an anisotropic source medium. *Bull. Seismol. Soc. Am.* **71**, 37–50.

Kazi-Aoual, M. N., Bonnet, G., and Jouanna, P. (1988). Response of an infinite elastic transversely isotropic medium to a point force. An analytical solution in Hankel space. *Geophys. J.* **93**, 587–90.

Keers, H., Dahlen, F. A., and Nolet, G. (1997). Chaotic ray behaviour in regional seismology. *Geophys. J. Int.* **131**, 361–80.

Keho, T. H., and Beydoun, W. B. (1988). Paraxial ray Kirchhoff migration. *Geophysics* **53**, 1540–6.

Keith, C. M., and Crampin, S. (1977). Seismic body waves in anisotropic media: Reflection and refraction at a plane interface. *Geophys. J. R. astr. Soc.* **49**, 181–208.

Keller, H. B., and Perozzi, P. J. (1983). Fast seismic ray tracing. *SIAM J. Appl. Math.* **43**, 981–92.

Keller, J. B. (1958). A geometrical theory of diffraction. In *Calculus of variations and its applications*, ed. L. M. Graves, pp. 27–52. New York: McGraw Hill.

Keller, J. B. (1962). Geometrical theory of diffraction. *J. Opt. Soc. Am.* **52**, 116–30.

Keller, J. B. (1963). Geometrical methods and asymptotic expansions in wave propagation. *J. Geophys. Res.* **68**, 1182–3.

Keller, J. B., Lewis, R. M., and Seckler, B. D. (1956). Asymptotic solution of some diffraction problems. *Commun. Pure Appl. Math.* **9**, 207–65.

Keller, J. B., and Streifer, W. (1971). Complex rays with an application to Gaussian beams. *J. Opt. Soc. Am.* **61**, 40–3.

Kelly, S., Baltensperger, P., and McMechan, G. A. (1997). P-to-S conversion for a thin anisotropic zone produced by vertical fracturing. *Geophys. Prospecting* **45**, 551–70.

Kendall, J-M., Guest, W. S., and Thomson, C. J. (1992). Ray-theory Green's function reciprocity and ray-centered coordinates in anisotropic media. *Geophys. J. Int.* **108**, 364–71.

Kendall, J-M., and Thomson, C. J. (1989). A comment on the form of the geometrical spreading equations, with some examples of seismic ray tracing in inhomogeneous, anisotropic media. *Geophys. J. Int.* **99**, 401–13.

Kendall, J-M., and Thomson, C. J. (1993). Maslov ray summation, pseudo-caustics, Lagrangian equivalence and transient seismic waveforms. *Geophys. J. Int.* **113**, 186–214.

Kennett, B. L. N. (1983). *Seismic wave propagation in stratified media*. Cambridge, GB: Cambridge Univ. Press.

Kennett, B. L. N. (1995). Approximation for surface waves propagating in laterally varying media. *Geophys. J. Int.* **122**, 470–8.

Kennett, B. L. N., Kerry, N. J., and Woodhouse, J. H. (1978). Symmetries in the reflection and transmission of elastic waves. *Geophys. J. R. astr. Soc.* **52**, 215–29.

Kim, K. Y., Wrolstad, K. H., and Aminzadeh, F. (1993). Effects of transverse isotropy on P-wave AVO for gas sands. *Geophysics* **58**, 883–8.

Kirpichnikova, N. J. (1971). Construction of solutions concentrated close to rays for the equations of elasticity theory in an inhomogeneous isotropic space. In *Mathematical problems of theory of diffraction and propagation of waves* (in Russian), vol. 1, ed. V. M. Babich, pp. 103–13. Leningrad: Nauka. (Translation to English by AMS, 1974.)

Kirpichnikova, N. J., and Popov, M. M. (1983). Reflection of space-time ray amplitudes from a moving boundary. In *Mathematical problems of the theory of propagation of waves* (in Russian), vol. 13, ed. V. M. Babich, pp. 72–88. Leningrad: Nauka.

Kiselev, A. P. (1983). Additional components of elastic waves (in Russian). *Izv. Akad. Nauk SSSR, Fizika Zemli*, No. 8, 51–6.

Kiselev, A. P. (1994). Body waves in weakly anisotropic medium: I. Plane waves. *Geophys. J. Int.* **118**, 393–400.

Kiselev, A. P., and Rogoff, Z. M. (1998). Dilatation of S waves in smoothly inhomogeneous isotropic elastic media. *J. Acoust. Soc. Am.* **104**, 2592–5.

Kiselev, A. P., and Roslov, Yu. V. (1991). Use of additional components for numerical modelling of polarization anomalies of elastic body waves (in Russian). *Sov. Geol. Geophys.* **32**, 105–14.

Kiselev, A. P., and Tsvankin, I. D. (1989). A method of comparison of exact and asymptotic wave field computations. *Geophys. J.* **96**, 253–8.

Kjartansson, E. (1979). Constant Q – Wave propagation and attenuation. *J. Geophys. Res.* **84**, 4737–48.

Klauder, J. R. (1987). Global uniform asymptotic wave-equation solutions for large wavenumbers. *Ann. Phys.* **180**, 108–51.

Klem-Musatov, K. D. (1980). *The theory of edge waves and its applications in seismology* (in Russian). Novosibirsk: Nauka.

Klem-Musatov, K. D. (1994). *Theory of seismic diffractions*. Tulsa: Society of Explor. Geophysicists.

Klem-Musatov, K. D., and Aizenberg, A. M. (1984). The ray method of the theory of edge waves. *Geophys. J. R. astr. Soc.* **79**, 35–50.

Klem-Musatov, K. D., and Aizenberg, A. M. (1985). Seismic modelling by methods of the theory of edge waves. *J. Geophys.* **57**, 90–105.

Klimeš, L. (1983). Hermite-Gaussian beams in inhomogeneous elastic media. *Stud. Geoph. et Geod.* **27**, 354–65.

Klimeš, L. (1984a). Expansion of a high-frequency time-harmonic wavefield given on an initial surface into Gaussian beams. *Geophys. J. R. astr. Soc.* **79**, 105–18.

Klimeš, L. (1984b). The relation between Gaussian beams and Maslov asymptotic theory. *Stud. Geoph. et Geod.* **28**, 237–47.

Klimeš, L. (1989a). Gaussian packets in the computation of seismic wavefields. *Geophys. J. Int.* **99**, 421–33.

Klimeš, L. (1989b). Optimization of the shape of Gaussian beams of a fixed length. *Stud. Geoph. et Geod.* **33**, 146–63.

Klimeš, L. (1994). Transformations for dynamic ray tracing in anisotropic media. *Wave Motion* **20**, 261–72.

Klimeš, L. (1996). Grid travel-time tracing: Second-order method for the first arrivals in smooth media. *PAGEOPH* **148**, 539–63.

Klimeš, L. (1997a). Phase shift of the Green function due to caustics in anisotropic media. 67th Annual SEG Meeting Expanded Abstracts, Dallas, 1834–7.

Klimeš, L. (1997b). Calculation of the third and higher travel-time derivatives in isotropic and anisotropic media. In *Seismic waves in complex 3-D structures, Report 6*, pp. 157–66. Prague: Faculty of Mathematics and Physics, Charles Univ.

Klimeš, L. (1997c). Phase shift of the Green function due to caustics in anisotropic media. In *Seismic waves in complex 3-D structures, Report 6*, pp. 167–73. Prague: Faculty of Mathematics and Physics, Charles Univ.

Klimeš, L. (1999a). Lyapunov exponents for 2-D ray tracing without interfaces. In *Seismic waves in complex 3-D structures, Report 8*, pp. 83–96. Prague: Faculty of Mathematics and Physics, Charles Univ.

Klimeš, L. (1999b). Perturbation of the polarization vectors in the isotropic ray theory. In *Seismic waves in complex 3-D structures, Report 8*, pp. 97–102. Prague: Faculty of Mathematics and Physics, Charles Univ.

Klimeš, L. (2000). Calculation of geometrical spreading from gridded slowness vectors in 2-D. In *Seismic waves in complex 3-D structures, Report 10*, pp. 115–20. Prague: Faculty of Mathematics and Physics, Charles Univ.

Klimeš, L., and Kvasnička, M. (1994). 3-D network ray tracing. *Geophys. J. Int.* **116**, 726–38.

Kline, M., and Kay, I. W. (1965). *Electromagnetic theory and geometrical optics*. New York: Interscience.

Knapp, R. W. (1991). Fresnel zones in the light of broadband data. *Geophysics* **56**, 354–9.

Knott, C. G. (1899). Reflection and refraction of seismic waves with seismological applications. *Phil. Mag.* **48**, 64–97.

Koch, M. (1985). A numerical study on the determination of the 3-D structure of the lithosphere by linear and non-linear inversion of teleseismic travel times. *Geophys. J. R. astr. Soc.* **80**, 73–93.

Koefoed, O. (1962). Reflection and transmission coefficients for plane longitudinal incident waves. *Geophys. Prospecting* **10**, 304–51.

Kogan, S. Ya. (1975). *Seismic energy and methods of its determination* (in Russian). Moscow: Nauka.

Koketsu, K., and Sekine, S. (1998). Pseudo-bending method for three-dimensional seismic ray tracing in a spherical earth with discontinuities. *Geophys. J. Int.* **132**, 339–46.

Konopásková, J., and Červený, V. (1984). Numerical modelling of time-harmonic seismic wave fields in simple structures by the Gaussian beam method. Part I. *Stud. Geoph. et Geod.* **28**, 19–35. Part II. *Stud. Geoph. et Geod.* **28**, 113–28.

Korn, G. A., and Korn, T. M. (1961). *Mathematical handbook for scientists and engineers*. New York: McGraw Hill.

Körnig, M. (1995). Cell ray tracing for smooth, isotropic media: A new concept based on generalized analytic solution. *Geophys. J. Int.* **123**, 391–408.

Kravtsov, Yu. A. (1968). "Quasiisotropic" approximation to geometrical optics (in Russian). *Dokl. Akad. Nauk SSSR* **183**, 74–7.

Kravtsov, Yu. A. (1988). Rays and caustics as physical objects. In *Progress in optics*, vol. 26, ed. E. Wolf, pp. 227–348. Amsterdam: North Holland.

Kravtsov, Yu. A., Forbes, G. W., and Asatryan, A. A. (1999). Theory and applications of complex rays. In *Progress in optics*, vol. 39, ed. E. Wolf. Amsterdam: Elsevier.

Kravtsov, Yu. A., Naida, O. N., and Fuki, A. A. (1996). Waves in weakly anisotropic 3-D inhomogeneous media: Quasiisotropic approximation of geometrical optics (in Russian). *Usp. Fiz. Nauk* **166**, 141–67.

Kravtsov, Yu. A., and Orlov, Yu. I. (1980). *Geometrical optics of inhomogeneous media* (in Russian). Moscow: Nauka. (Translation to English by Springer, Berlin, 1990.)

Kravtsov, Yu. A., and Orlov, Yu. I. (1993). Caustics, catastrophes and wavefields. Berlin: Springer.

Krebes, E. S. (1983). The viscoelastic reflection/transmission problem: Two special cases. *Bull. Seismol. Soc. Am.* **73**, 1673–83.

Krebes, E. S. (1984). On the reflection and transmission of viscoelastic waves – Some numerical results. *Geophysics* **49**, 1374–80.

Krebes, E. S., and Hearn, D. J. (1985). On the geometrical spreading of viscoelastic waves. *Bull. Seismol. Soc. Am.* **75**, 391–6.

Krebes, E. S., and Hron, F. (1980a). Ray-synthetic seismograms for SH waves in anelastic media. *Bull. Seismol. Soc. Am.* **70**, 29–46.

Krebes, E. S., and Hron, F. (1980b). Synthetic seismograms for SH waves in a layered anelastic medium by asymptotic ray theory. *Bull. Seismol. Soc. Am.* **70**, 2005–20.

Krebes, E. S., and Hron, F. (1981). Comparison of synthetic seismograms for anelastic media by asymptotic ray theory and the Thompson-Haskell method. *Bull Seismol. Soc. Am.* **71**, 1463–8.

Kühnicke, E. (1996). Three-dimensional waves in layered media with unparallel and curved interfaces: A theoretical approach. *J. Acoust. Soc. Am.* **100**, 709–16.

Kuo, J. T., and Dai, T.-F. (1984). Kirchhoff elastic wave migration for the case of noncoincident source and receiver. *Geophysics* **49**, 1223–38.

Küpper, F. J. (1958). Theoretische Untersuchungen über die Mehrfachaufstellung von Geophonen. *Geophys. Prospecting* **6**, 194–256.

Kvasnička, M., and Červený, V. (1994). Fresnel volumes and Fresnel zones in complex laterally varying structures. *J. Seismic Exploration* **3**, 215–30.

Kvasnička, M., and Červený, V. (1996). Analytical expressions for Fresnel volumes and interface Fresnel zones of seismic body waves. Part 1: Direct and unconverted reflected waves. *Studia Geoph. et Geod.* **40**, 136–55. Part 2: Transmitted and converted waves. Head waves. *Studia Geoph. et Geod.* **40**, 381–97.

Lafond, C. F., and Levander, A. R. (1990). Fast and accurate dynamic raytracing in heterogeneous media. *Bull. Seismol. Soc. Am.* **80**, 1284–96.

Lamb, M. (1904). On the propagation of tremors over the surface of an elastic solid. *Phil. Trans. R. Soc. London A* **203**, 1–42.

Lambaré, G., Lucio, P. S., and Hanyga, A. (1996). Two-dimensional multivalued traveltime and amplitude maps by uniform sampling of a ray field. *Geophys. J. Int.* **125**, 584–98.

Landau, L. D., and Lifschitz, E. M. (1965). *Theory of elasticity* (in Russian). Moscow: Nauka.

Landau, L. D., and Lifschitz, E. M. (1974). *Quantum mechanics – Nonrelativistic theory* (in Russian). Moscow: Nauka.

Landers, T., and Claerbout, J. F. (1972). Numerical calculation of elastic waves in laterally inhomogeneous media. *J. Geophys. Res.* **77**, 1476–82.

Langan, R. T., Lerche, I., and Cutler, R. T. (1985). Tracing of rays through heterogeneous media: An accurate and efficient procedure. *Geophysics* **50**, 1456–65.

Langston, C. A., and Lee, J.-J. (1983). Effects of structure geometry on strong ground motions: The Duwamish River Valley, Seattle, Washington. *Bull. Seismol. Soc. Am.* **73**, 1851–63.

Lecomte, I. (1993). Finite difference calculation of first traveltimes in anisotropic media. *Geophys. J. Int.* **113**, 318–42.

Lecomte, I. (1996). Hybrid modeling with ray tracing and finite difference. 66th Annual SEG Meeting Expanded Abstracts, Denver, 699–702.

Lee, J.-J., and Langston, C. A. (1983a). Three-dimensional ray tracing and the method of principal curvature for geometrical spreading. *Bull. Seismol. Soc. Am.* **73**, 765–80.

Lee, J.-J., and Langston, C. A. (1983b). Wave propagation in a three-dimensional circular basin. *Bull. Seismol. Soc. Am.* **73**, 1637–53.

Lee, W. H. K., and Stewart, S. W. (1981). *Principles and applications of microearthquake networks.* New York: Academic Press.

Leontovich, M. A., and Fock, V. A. (1946). Solution of the problem of propagation of electromagnetic waves along the Earth's surface using the parabolic equation method (in Russian). *ZETF* **16**, 557–73.

Levshin, A. L., Yanovskaya, T. B., Lander, A. V., Bukchin, B. G., Barmin, M. P., Ratnikova, L. I., and Its, Ye. N. (1987). *Surface seismic waves in horizontally inhomogeneous Earth* (in Russian). Nauka: Moscow. (Translated to English by Kluwer, Dordrecht, 1989.)

Lewis, R. M. (1965). Asymptotic theory of wave-propagation. *Archive for Rational Mechanics and Analysis* **20**, 191–250.

Lewis, R. M., Bleistein, N., and Ludwig, D. (1967). Uniform asymptotic theory of creeping waves. *Commun. Pure Appl. Math.* **20**, 295–328.

Li, X.-G., and Ulrych, T. J. (1993a). LTI formulations and application to curved wave fronts. *J. Seismic Exploration* **2**, 239–46.

Li, X.-G., and Ulrych, T. J. (1993b). Traveltime computation in discrete heterogeneous layered media. *J. Seismic Exploration* **2**, 305–18.

Lighthill, M. J. (1960). Studies on magneto-hydrodynamic waves and other anisotropic wave motions. *Phil. Trans. R. Soc. London A* **252**, 397–430.

Lindsey, J. P. (1989). The Fresnel zone and its interpretative significance. *The Leading Edge* **8**, 33–9.

Liu, E., Crampin, S., and Yardley, G. (1990). Anomalous behaviour of polarization of reflected shear-waves in an isotropic structure. Abstract. *Geophys. J. Int.* **101**, 291.

Liu, E., and Tromp, J. (1996). Uniformly valid body-wave ray theory. *Geophys. J. Int.* **127**, 461–91.

Loewenthal, D., and Hu, L.-Z. (1991). Two methods for computing the imaging condition for common-shot prestack migration. *Geophysics* **56**, 378–81.

Love, A. E. H. (1944). *A treatise on the mathematical theory of elasticity*. New York: Dover.

Lu, I.-T. (1996). A hybrid ray-mode (wavefront resonance) approach for analyzing acoustic radiation and scattering by submerged structures. *J. Acoust. Soc. Am.* **99**, 114–32.

Lucio, P. S., Lambaré, G., and Hanyga, A. (1996). 3D multivalued travel time and amplitude maps. *PAGEOPH* **148**, 449–79.

Ludwig, D. (1966). Uniform asymptotic expansion at a caustic. *Commun. Pure Appl. Math.* **29**, 215–50.

Luneburg, R. K. (1964). *Mathematical theory of optics*. Berkeley: Univ. of California Press.

Luneva, M. (1996). Application of the edge wave superposition method. *PAGEOPH* **148**, 113–36.

Madariaga, R. (1984). Gaussian beam synthetic seismograms in a vertically varying medium. *Geophys. J. R. astr. Soc.* **79**, 589–612.

Madariaga, R., and Bernard, P. (1985). Ray theoretical strong motion synthesis. *J. Geophys.* **58**, 73–81.

Madariaga, R., and Papadimitriou, P. (1985). Gaussian beam modelling of upper mantle phases. *Ann. Geophys.* **3**, 799–812.

Malinovskaya, L. N. (1958). The dynamic features of totally reflected transverse waves (in Russian). *Izv. Akad. Nauk SSSR, Geophys. Series*, No. 2, 184–95.

Mallick, S., and Frazer, L. N. (1987). Practical aspects of reflectivity modeling. *Geophysics* **52**, 1355–64.

Mallick, S., and Frazer, L. N. (1990). Computation of synthetic seismograms for stratified azimuthally anisotropic media. *J. Geophys. Res.* **95**, B6, 8513–26.

Mallick, S., and Frazer, L. N. (1991). Reflection/transmission coefficients and azimuthal anisotropy in marine seismic studies. *Geophys. J. Int.* **105**, 241–52.

Mandal, B. (1992). Forward modeling for tomography: Triangular grid-based Huygens' principle method. *J. Seismic Exploration* **1**, 239–50.

Mao, W., and Stuart, G. W. (1997). Rapid multi-wave-type ray tracing in complex 2-D and 3-D isotropic media. *Geophysics* **62**, 298–308.

Marks, L. W., and Hron, F. (1980). Calculation of synthetic seismograms in laterally inhomogeneous media. *Geophysics* **45**, 509–10.

Martin, B. E., and Thomson, C. J. (1997). Modelling surface waves in anisotropic structures. II. Examples. *Phys. Earth Planet. Int.* **103**, 253–79.

Maslov, V. P. (1965). *Theory of perturbations and asymptotic methods* (in Russian). Moscow: Moscow State Univ. Press.

Matsuoka, T., and Ezaka, T. (1992). Ray tracing using reciprocity. *Geophysics* **57**, 326–33.

Matyska, C. (1999). Note on chaos in ray propagation. In *Seismic waves in complex 3-D structures, Report 8*, pp. 71–82. Prague: Faculty of Mathematics and Physics, Charles Univ.

May, B. T., and Hron, F. (1978). Synthetic seismic sections of typical petroleum traps. *Geophysics* **43**, 1119–47.

Mazur, M. A., and Gilbert, K. E. (1997). Direct optimization methods, ray propagation and chaos. I. Continuous media. *J. Acoust. Soc. Am.* **101**, 174–83. II. Propagation with discrete transitions. *J. Acoust. Soc. Am.* **101**, 184–92.

McCamy, K., Meyer, R. P., and Smith, T. J. (1962). Generally applicable solutions of Zoeppritz' amplitude equations. *Bull. Seismol. Soc. Am.* **52**, 923–55.

McCoy, J. J. (1977). A parabolic theory of stress wave propagation through inhomogeneous linearly elastic solids. *J. App. Mech.* **44**, 462–82.

McMechan, G. A., and Mooney, W. D. (1980). Asymptotic ray theory and synthetic seismograms for laterally varying structures: Theory and application to the Imperial Valley, California. *Bull. Seismol. Soc. Am.* **70**, 2021–35.

Mendiguren, J. A. (1969). Study of focal mechanism of deep earthquakes in Argentina using non-linear particle motion of S waves. *Bull. Seismol. Soc. Am.* **59**, 1449–73.

Mensch, T., and Farra, V. (1999). Computation of qP-wave rays, traveltimes and slowness vectors in orthorhombic media. *Geophys. J. Int.* **138**, 244–56.

Miklowitz, J. (1978). *The theory of elastic waves and waveguides*. Amsterdam: North-Holland.

Mochizuki, E. (1987). Ray tracing of body waves in an anisotropic medium. *Geophys. J. R. astr. Soc.* **90**, 627–34.

Mochizuki, E. (1989). Ray tracing on an ellipsoid. *Bull. Seismol. Soc. Am.* **79**, 917–20.

Moczo, P., Bard, P.-Y., and Pšenčík, I. (1987). Seismic response of two-dimensional absorbing structures by the ray method. *J. Geophys.* **62**, 38–49.

Montagner, J. P. (1996). Surface waves on a global scale – Influence of anisotropy and anelasticity. In *Seismic modelling of Earth structure*, eds. E. Boschi, G. Ekström, and A. Morelli, pp. 81–148. Bologna: Editrice Compositori.

Morse, P. M., and Feshbach, H. (1953). *Methods of theoretical physics*. New York: McGraw Hill.

Moser, T. J. (1991). Shortest path calculation of seismic rays. *Geophysics* **56**, 59–67.

Moser, T. J. (1992). The shortest path method for seismic ray tracing in complicated media. PhD. Thesis, Inst. of Geophysics, Utrecht University.

Moser, T. J., Nolet, G., and Snieder, R. (1992). Ray bendig revisited. *Bull. Seismol. Soc. Am.* **82**, 259–88.

Müller, G. (1968). Theoretical seismograms for some types of point-sources in layered media. Part I. Theory. *Z. Geophys.* **34**, 15–35. Part II. Numerical calculations. *Z. Geophys.* **34**, 147–62.

Müller, G. (1970). Exact ray theory and its application to the reflection of elastic waves from vertically inhomogeneous media. *Geophys. J. R. astr. Soc.* **21**, 261–84.

Müller, G. (1983). Rheological properties and velocity dispersion of a medium with power-law dependence of Q on frequency. *J. Geophys.* **54**, 20–9.

Müller, G. (1984). Efficient calculation of Gaussian-beam seismograms for two-dimensional inhomogeneous media. *Geophys. J. R. astr. Soc.* **79**, 153–66.

Müller, G. (1985). The reflectivity method: A tutorial. *J. Geophys.* **58**, 153–74.

Müller, G., Roth, M., and Korn, M., (1992). Seismic-wave traveltimes in random media. *Geophys. J. Int.* **110**, 29–41.

Mura, T. (1982). *Micromechanics of defects in solids*. London: Martinus Nijhoff Publishers.

Musgrave, M. J. P. (1970). *Crystal acoustics*. San Francisco: Holden Day.

Musil, M. (1989). Computation of synthetic seismograms in 2-D and 3-D media using the Gaussian beam method. *Studia Geoph. et Geod.* **33**, 213–29.

Muskat, M. (1933). The theory of refraction shooting. *Physics* **4**, 14–28.

Muskat, M., and Meres, M. W. (1940). Reflection and transmission coefficients for plane waves in elastic media. *Geophysics* **5**, 115–48.

Nafe, J. E. (1957). Reflection and transmission coefficients at a solid-solid interface of high velocity contrast. *Bull. Seismol. Soc. Am.* **47**, 205–19.

Najmi, A.-H. (1996). Closed form solutions for geometrical spreading in inhomogeneous media. *Geophysics* **61**, 1189–97.

Nakanishi, I., and Yamaguchi, K. (1986). A numerical experiment on nonlinear image reconstruction from first-arrival times for two-dimensional island arc structure. *J. Phys. Earth* **34**, 195–201.

Nechtschein, S., and Hron, F. (1996). Reflection and transmission coefficients between two anelastic media using asymptotic ray theory. *Canad. J. Expl. Geophys.* **32**, 31–40.

Nechtschein, S., and Hron, F. (1997). Effects of anelasticity on reflection and transmission coefficients. *Geophys. Prospecting* **45**, 775–93.

Neele, F., and Snieder, R. (1992). Topography of the 400 km discontinuity from observations of long period P400P phases. *Geophys. J. Int.* **109**, 670–82.

Nichols, D. E. (1996). Maximum energy traveltimes calculated in the seismic frequency band. *Geophysics* **61**, 253–63.

Nolet, G. (1987). Seismic wave propagation and seismic tomography. In *Seismic tomography with applications in global seismology and exploration geophysics*, ed. G. Nolet, pp. 1–23. Dordrecht: Reidel.

Nolet, G., and Moser, T. J. (1993). Teleseismic delay times in a 3-D Earth and a new look at the S discrepancy. *Geophys. J. Int.* **114**, 185–95.

Norris, A. N. (1986). Complex point-source representation of real point sources and the Gaussian beam summation method. *J. Opt. Soc. Am.* **A3**, 2005–10.

Norris, A. N. (1987). A theory of pulse propagation in anisotropic elastic solids. *Wave Motion* **9**, 509–32.

Norris, A. N., White, B. S., and Schrieffer, J. R. (1987). Gaussian wave packets in inhomogeneous media with curved interfaces. *Proc. R. Soc. London* **A412**, 93–123.

Nowack, R., and Aki, K. (1984). The 2-D Gaussian beam synthetic method: Testing and applications. *J. Geophys. Res.* **89**, 1466–94.

Nowack, R. L., and Lutter, W. J. (1988). Linearized rays, amplitude and inversion. *PAGEOPH* **128**, 401–22.

Nowack, R. L., and Pšenčík, I. (1991). Perturbation from isotropic to anisotropic heterogeneous media in the ray approximation. *Geophys. J. Int.* **106**, 1–10.

Nuttli, O. (1959). The particle motion of the S wave. *Bull. Seismol. Soc. Am.* **49**, 49–56.

Nuttli, O. (1961). The effect of the Earth's surface on the S wave particle motion. *Bull. Seismol. Soc. Am.* **51**, 237–46.

Nuttli, O., and Whitmore, J. D. (1962). On the determination of the polarization angle of the S wave. *Bull. Seismol. Soc. Am.* **52**, 95–107.

Obolentseva, I. R., and Grechka, V. Y. (1988). Two-point algorithms for calculating rays in stratified homogeneous anisotropic media. *Soviet Geology and Geophysics* **29**, 97–104.

Officer, Ch. B. (1974). *Introduction to theoretical geophysics*. New York: Springer.

Ojo, S. B., and Mereu, R. F. (1986). The effect of random velocity functions on the travel times and amplitudes of seismic body waves. *Geophys. J. R. astr. Soc.* **84**, 607–18.

Ott, H. (1942). Reflexion und Brechung von Kugelwellen: Effekte 2. Ordnung. *Ann. Phys.* **41**, 443–66.

Palmer, D. R., Brown, M. G., Tappert, F. D., and Bezdek, H. F. (1988). Classical chaos in non-separable wave propagation problems. *Geophys. Res. Letters* **15**, 569–72.

Papoulis, A. (1962). *The Fourier Integral and Its Applications*. New York: McGraw Hill.

Passier, M. L., and Snieder, R. K. (1995). On the presence of intermediate-scale heterogeneity in the upper mantle. *Geophys. J. Int.* **123**, 817–37.

Payton, R. G. (1983). *Elastic wave propagation in transversely isotropic media*. The Hague: Martinus Nijhoff Publishers.

Pereyra, V. (1988). Numerical methods for inverse problems in three-dimensional geophysical modeling. *Appl. Num. Math.* **4**, 97–139.

Pereyra, V. (1992). Two-point ray tracing in general 3D media. *Geophys. Prospecting* **40**, 267–87.

Pereyra, V. (1996). Modeling, ray tracing and block nonlinear travel time inversion in 3D. *PAGEOPH* **148**, 345–86.

Pereyra, V., Lee, W. H. K., and Keller, H. B. (1980). Solving two-point seismic-ray tracing problems in a heterogeneous medium. Part 1. A general adaptive finite difference method. *Bull. Seismol. Soc. Am.* **70**, 79–99.

Petrashen, G. I. (1980). *Wave propagation in anisotropic elastic media* (in Russian). Leningrad: Nauka.

Petrashen, G. I., and Kashtan, B. M. (1984). Theory of body-wave propagation in inhomogeneous anisotropic media. *Geophys. J. R. astr. Soc.* **76**, 29–39.

Pilant, W. L. (1979). *Elastic waves in the Earth*. Amsterdam: Elsevier.

Pilipenko, V. N. (1979). Numerical method of time fields for the construction of seismic interfaces. In *Inverse kinematic problems of explosion seismology* (in Russian), ed. S. M. Zverev, pp. 124–81. Moscow: Nauka.

Pilipenko, V. N. (1983). Numerical example of time-field method in up-to-date seismic prospecting (in Russian). *Izv. Akad. Nauk SSSR, Fizika Zemli*, No. 1, 36–42.

Piserchia, P.-F., Virieux, J., Rodrigues, D., Gaffet, S., and Talandier, J. (1998). Hybrid numerical modelling of T-wave propagation: Application to the Midplate experiment. *Geophys. J. Int.* **133**, 789–800.

Podvin, P., and Lecomte, I. (1991). Finite difference computation of traveltimes in very contrasted velocity models: a massively parallel approach and its associated tools. *Geophys. J. Int.* **105**, 271–84.

Podyapol'skiy, G. S. (1959). On a certain formula connecting the coefficients of head waves with reflection and refraction coefficients. *Izv. Akad. Nauk SSSR, Geophys. Series*, No. 11, 1108–13.

Podyapol'skiy, G. S. (1966a). Physics of elastic waves. In *Seismic Exploration* (in Russian), eds. I. I. Gurvich and V. P. Nomokonov, pp. 28–96. Moscow: Nedra.

Podyapol'skiy, G. S. (1966b). A ray series expansion for reflected and transmitted waves (in Russian). *Izv. Akad. Nauk SSSR, Fizika Zemli*, No. 6, 347–63.

Popov, M. M. (1977). On a method of computation of geometrical spreading in inhomogeneous medium containing interfaces (in Russian). *Dokl. Akad. Nauk SSSR* **237**, 1059–62.

Popov, M. M. (1982). A new method of computation of wave fields using Gaussian beams. *Wave Motion* **4**, 85–95.

Popov, M. M., and Camerlynck, C. (1996). Second term of the ray series and validity of the ray theory. *J. Geophys. Res.-Solid Earth* **101**, 817–26.

Popov, M. M., and Pšenčík, I. (1976a). Whispering gallery waves in the vicinity of the bend of the interface (in Russian). *Dokl. Akad. Nauk SSSR* **230**, 822–5.

Popov, M. M., and Pšenčík, I. (1976b). Numerical solution of the whispering gallery waves problem in the neighbourhood of a boundary-flex point. In *Mathematical problems of the theory of propagation of waves* (in Russian), vol. 8, ed. V. M. Babich, pp. 207–19. Leningrad: Leningrad Univ. Press.

Popov, M. M., and Pšenčík, I. (1978a). Ray amplitudes in inhomogeneous media with curved interfaces. In *Geofyzikální sborník*, vol. 24, ed. A. Zátopek, pp. 111–29. Praha: Academia.

Popov, M. M., and Pšenčík, I. (1978b). Computation of ray amplitudes in inhomogeneous media with curved interfaces. *Studia Geoph. et Geod.* **22**, 248–58.

Popov, M. M., and Tyurikov, L. G. (1981). On two approaches to the evaluation of geometrical spreading in inhomogeneous isotropic media. In *Problems of the dynamic theory of propagation of seismic waves* (in Russian), vol. 20, ed. G. I. Petrashen, pp. 61–8. Leningrad: Nauka.

Pratt, R. G., and Chapman, C. H. (1992). Traveltime tomography in anisotropic media. II. Application. *Geophys. J. Int.* **109**, 20–37.

Pretlová, V. (1976). Bicubic spline smoothing of two-dimensional geophysical data. *Studia Geoph. et Geod.* **20**, 168–77.

Prothero, W. A., Taylor, W. J., and Eickemeyer, J. A. (1988). A fast, two-point, three-dimensional ray tracing algorithm using a simple step search method. *Bull. Seismol. Soc. Am.* **78**, 1190–8.

Pšenčík, I. (1972). Kinematics of refracted and reflected waves in inhomogeneous media with non-planar interfaces. *Studia Geoph. et Geod.* **16**, 126–52.

Pšenčík, I. (1979). Ray amplitudes of compressional, shear and converted seismic body waves in 3D laterally inhomogeneous media with curved interfaces. *J. Geophys.* **45**, 381–90.

Pšenčík, I. (1998). Green's functions for inhomogeneous weakly anisotropic media. *Geophys. J. Int.* **135**, 279–88.

Pšenčík, I., and Dellinger, J. (2000). Quasi-shear waves in inhomogeneous weakly anisotropic media by the quasi-isotropic approach: A model study. In *Seismic waves in complex 3-D structures, Report 10*, pp. 203–25. Prague: Faculty of Mathematics and Physics, Charles Univ.

Pšenčík, I., and Teles, T. N. (1996). Point source radiation in inhomogeneous anisotropic structures. *PAGEOPH* **148**, 591–623.

Pulliam, J., and Snieder, R. (1996). Fast, efficient calculation of rays and travel times with ray perturbation theory. *J. Acoust. Soc. Am.* **99**, 383–91.

Pulliam, J., and Snieder, R. (1998). Ray perturbation theory, dynamic ray tracing and the determination of Fresnel zones. *Geophys. J. Int.* **135**, 463–9.

Puzyrev, N. N. (1959). *Interpretation of the data of seismic exploration by the method of reflected waves* (in Russian). Moscow: Gostoptechizdat.

Puzyrev, N. N., Trigubov, A. V., and Brodov, L. Yu. (1985). *Seismic prospecting by the method of shear and converted waves* (in Russian). Moscow: Nedra.

Qin, F., Luo, Y., Olsen, K. B., Cai, W., and Schuster, G. T. (1992). Finite difference solution of the eikonal equation along expanding wavefronts. *Geophysics* **57**, 478–87.

Qin, F., and Schuster, G. T. (1993). First-arrival traveltime calculation for anisotropic media. *Geophysics* **58**, 1349–58.

Ralston, J. (1983). Gaussian beams and the propagation of singularities. In *Studies in partial differential equations. MAA studies in mathematics*, vol. 23, ed. W. Littman, pp. 206–48. Washington: The Mathematical Association of America.

Ratnikova, L. I. (1973). *Methods of computation of seismic waves in thin-layered media* (in Russian). Moscow: Nauka.

Reinsch, C. H. (1967). Smoothing by spline functions. *Numerische Mathematik* **10**, 177–83.

Reshef, M. (1991). Prestack depth imaging of three-dimensional shot gathers. *Geophysics* **56**, 1158–63.

Reshef, M., and Kosloff, D. (1986). Migration of common-shot gathers. *Geophysics* **51**, 324–31.

Riahi, M. A., and Juhlin, C. (1994). 3-D interpretation of reflected arrival times by finite-difference techniques. *Geophysics* **59**, 844–9.

Riznichenko, Yu. V. (1946). Geometrical seismics of layered media. In *Works of Inst. Theor. Geophysics* (in Russian), Vol. II, ed. O. Yu. Schmidt, pp. 1–114. Moscow: Izd. AN SSSR.

Riznichenko, Yu. V. (1985). *Seismic prospecting of layered media* (in Russian). Moscow: Nedra.

Rockwell, D. W. (1967). A general wavefront method. In *Seismic refraction prospecting*, ed. A. W. Musgrave, pp. 363–405. Tulsa: Society of Exploration Geophysicists.

Rokhlin, S. I., Bolland, T. K., and Adler, L. (1976). Reflection and refraction of elastic waves on a plane interface between two generally anisotropic media. *J. Acoust. Soc. Am.* **79**, 906–18.

Romanov, V. G. (1972). *Some inverse problems for hyperbolic equations* (in Russian). Novosibirsk: Nauka.

Romanov, V. G. (1978). *Inverse problems for differential equations* (in Russian). Novosibirsk: Novosibirsk State Univ.

Roslov, Yu. V., and Yanovskaya, T. B. (1988). Estimation of the contribution of the first approximation in wave field reflected from the free surface. In *Problems of the dynamic theory of propagation of seismic waves*, vol. 27, ed. G. I. Petrashen, pp. 117–33. Leningrad: Nauka.

Roth, M., Müller, G., and Snieder, R. (1993). Velocity shift in random media. *Geophys. J. Int.* **115**, 552–63.

Rueger, A. (1997). P-wave reflection coefficients for transversely isotropic models with vertical and horizontal axes of symmetry. *Geophysics* **62**, 713–22.

Rümpker, G., and Thomson, C. J. (1994). Seismic-waveform effects of conical points in gradually varying anisotropic media. *Geophys. J. Int.* **118**, 759–80.

Rytov, S. M., Kravtsov, Yu. A., and Tatarskii, V. I. (1987). *Principles of statistical radiophysics*. Berlin: Springer.

Ryzhik, L., Papanicolaou, G., and Keller, J. B. (1996). Transport equations for elastic and other waves in random media. *Wave Motion* **24**, 327–70.

Saito, H. (1989). Travel times and ray paths of first arrival seismic waves: Computation method based on Huygens' principle. 59th Annual SEG Meeting Expanded Abstracts, Dallas, 244–7.

Sambridge, M. S., and Kennett, B. L. N. (1990). Boundary value ray tracing in a heterogeneous medium: a simple and versatile algorithm. *Geophys. J. Int.* **101**, 157–68.

Sambridge, M. S., and Snieder, R. (1993). Applicability of ray perturbation theory to mantle tomography. *Geophys. Res. Letters* **20**, 73–6.

Samec, P., and Blangy, J. P. (1992). Viscoelastic attenuation, anisotropy and AVO. *Geophysics* **57**, 441–50.

Samuelides, Y. (1998). Velocity shift using the Rytov approximation. *J. Acoust. Soc. Am.* **104**, 2596–603.

Savarenskiy, E. F., and Kirnos, D. P. (1955). *Elements of seismology and seismometry* (in Russian). Moscow: GITTL.

Sayers, C. M. (1994). P-wave propagation in weakly anisotropic media. *Geophys. J. Int.* **116**, 799–805.

Scales, J. A., and Van Vleck, E. S. (1997). Lyapunov exponents and localization in randomly layered media. *J. Comput. Phys.* **133**, 27–42.

Schleicher, J., Hubral, P., Tygel, M., and Jaya, M. S. (1997). Minimum apertures and Fresnel zones in migration and demigration. *Geophysics* **62**, 183–94.

Schleicher, J., Tygel, M., and Hubral, P. (1993). 3-D true-amplitude finite-offset migration. *Geophysics* **58**, 1112–26.

Schlottmann, R. B. (1999). A path integral formulation of acoustic wave propagation. *Geophys. J. Int.* **137**, 353–63.

Schmidt, T., and Müller, G. (1986). Seismic signal velocity in absorbing media. *J. Geophys.* **60**, 199–203.

Schneider, W. A., Jr. (1995). Robust and efficient upwind finite-difference traveltime calculations in three dimensions. *Geophysics* **60**, 1108–17.

Schneider, W. A., Jr., Ranzinger, K. A., Balch, A. H., and Kruse, C. (1992). A dynamic programming approach to first arrival traveltime computation in media with arbitrarily distributed velocities. *Geophysics* **57**, 39–50.

Schoenberg, M., and Protázio, J. (1992). "Zoeppritz" rationalized and generalized to anisotropy. *J. Seismic Exploration* **1**, 125–44.

Schulman, L. S. (1981). *Techniques and applications of path integration*. New York: Wiley.

Scott, P., and Helmberger, D. (1983). Applications of the Kirchhoff-Helmholtz integral to problems in seismology. *Geophys. J. R. astr. Soc.* **72**, 237–54.

Sethian, J. A., and Popovici, A. M. (1999). 3-D traveltime computation using the fast marching method. *Geophysics* **64**, 516–23.

Shah, P. M. (1973a). Ray tracing in three dimensions. *Geophysics* **38**, 600–4.

Shah, P. M. (1973b). Use of wavefront curvature to relate seismic data with subsurface parameters. *Geophysics* **38**, 812–25.

Sharafutdinov, V. A. (1979). On the geometrical spreading. In *Mathematical methods of the interpretation of geophysical observations* (in Russian), ed. A. S. Alekseyev, pp. 161–74. Novosibirsk: Acad. Sci. USSR, Computing Center.

Sharafutdinov, V. A. (1994). Quasi-isotropic approximation in dynamic elasticity and some problems of geotomography. *Russian Geology and Geophysics* **35**, 58–71.

Shearer, P. M., and Chapman, C. H. (1989). Ray tracing in azimuthally anisotropic media. I. Results for models of aligned cracks in the upper crust. *Geophys. J.* **96**, 51–64.

Sheriff, R. E. (1989). *Geophysical methods*. Englewood Cliffs, NJ: Prentice Hall.

Sheriff, R. E., and Geldard, L. P. (1982). *Exploration seismology*. Cambridge, GB: Cambridge Univ. Press.

Siegman, A. E. (1973). Hermite-Gaussian functions of complex argument as optical beam eigenfunctions. *J. Opt. Soc. Am.* **63**, 1093–4.

Šilený, J. (1981). Anomalous behaviour of seismic waves at interfaces between anisotropic media. *Studia Geoph. et Geod.* **25**, 152–9.

Silva, W. (1976). Body waves in layered anelastic solid. *Bull. Seismol. Soc. Am.* **66**, 1539–54.

Singh, S. C., and Chapman, C. H. (1988). WKBJ seismogram theory in anisotropic media. *J. Acoust. Soc. Am.* **84**, 732–41.

Sinton, J. B., and Frazer, L. N. (1982). A Kirchhoff method for the computation of finite frequency body wave synthetic seismograms in laterally varying media. *Geophys. J. R. astr. Soc.* **71**, 37–55.

Smirnov, V. I. (1953). *Course of higher mathematics* (in Russian). vol. 4. Moscow: GITTL.

Smirnov, V. I. (1964). *A course of higher mathematics*, vol. 4, trans. D. E. Brown and I. N. Sneddon. Oxford: Pergamon Press.

Smirnova, N. S. (1966). On the character of the field of reflected waves near the critical point. In *Problems of the dynamic theory of propagation of seismic waves* (in Russian), vol. 8, ed. G. I. Petrashen, pp. 16–22. Moscow-Leningrad: Nauka.

Smith, K. B., Brown, M. G., and Tappert, F. D. (1992). Ray chaos in underwater acoustics. *J. Acoust. Soc. Am.* **91**, 1939–49.

Snieder, R. (1996). Surface wave inversions on a regional scale. In *Seismic modelling of Earth structure*, eds. E. Boschi, G. Ekström, and A. Morelli, pp. 149–81. Bologna: Editrice Compositori.

Snieder, R., and Aldridge, D. F. (1995). Perturbation theory for travel times. *J. Acoust. Soc. Am.* **98**, 1565–9.

Snieder, R., and Chapman, C. (1998). The reciprocity properties of geometrical spreading. *Geophys. J. Int.* **132**, 89–95.

Snieder, R., and Lomax, A. (1996). Wavefield smoothing and the effect of rough velocity perturbations on arrival times and amplitudes. *Geophys. J. Int.* **125**, 796–812.

Snieder, R., and Sambridge, M. (1992). Ray perturbation theory for traveltimes and ray paths in 3-D heterogeneous media. *Geophys. J. Int.* **109**, 294–322.

Snieder, R., and Sambridge, M. (1993). The ambiguity in ray perturbation theory. *J. Geophys. Res.* **98**, 22021–34.

Snieder, R., and Spencer, C. (1993). A unified approach to ray bending, ray perturbation and paraxial ray theories. *Geophys. J. Int.* **115**, 456–70.

Sommerfeld, A. (1909). Über die Ausbreitung des Wellen in der Drahtlosen Telegraphie. *Ann. Phys.* **28**, 665–736.

Sorrells, G. G., Crowley, J. B., and Veith, K. F. (1971). Methods for computing ray paths in complex geological structures. *Bull. Seismol. Soc. Am.* **61**, 27–53.

Spence, G. D., Whittall, K. P., and Clowes, R. M. (1984). Practical synthetic seismograms for laterally varying media calculated by asymptotic ray theory. *Bull. Seismol. Soc. Am.* **74**, 1209–23.

Spencer, C. P., Chapman, C. H., and Kragh, J. E. (1997). A fast, accurate integration method for Kirchhoff, Born and Maslov synthetic seismogram generation. 67th Annual SEG Meeting Expanded Abstracts, Dallas, 1838–41.

Spudich, P., and Frazer, L. N. (1984). Use of ray theory to calculate high-frequency radiation from earthquake sources having spatially variable rupture velocity and stress drop. *Bull. Seismol. Soc. Am.* **74**, 2061–82.

Stamnes, J. J. (1986). *Waves in focal regions*. Bristol and Boston: Adam Hilger.

Stavroudis, O. N. (1972). *The optics of rays, wavefronts and caustics*. New York: Academic Press.

Suchy, K. (1972). Ray tracing in an anisotropic absorbing medium. *J. Plasma Physics* **8**, 53–65.

Sun, J. (1994). Geometrical ray theory: Edge-diffracted rays and their traveltimes (second-order approximation of the traveltimes). *Geophysics* **59**, 148–55.

Sun, J. (1996). The relationship between the first Fresnel zone and the normalized geometrical spreading factor. *Geophys. Prospecting* **44**, 351–74.

Sutton, G. R. (1984). The effect of velocity variations on the beam width of seismic wave. *Geophysics* **49**, 1649–52.

Synge, J. L. (1954). *Geometrical optics. An introduction to Hamilton's method*. London: Cambridge Univ. Press.

Synge, J. L., and Schild, A. (1952). *Tensor calculus*. Toronto: Univ. Toronto Press.

Szabo, T. L. (1995). Causal theories and data for acoustic attenuation obeying a frequency power law. *J. Acoust. Soc. Am.* **97**, 14–24.

Tanimoto, T. (1987). Surface-wave ray tracing equations and Fermat's principle in an anisotropic earth. *Geophys. J. R. astr. Soc.* **88**, 231–40.

Tappert, F. D. (1977). The parabolic approximation method. In *Wave propagation and underwater acoustics. Lecture notes in Physics*, vol. 70, eds. J. B. Keller and J. S. Papadakis, pp. 224–87. Berlin: Springer-Verlag.

Tappert, F. D., and Tang, X. (1996). Ray chaos and eigenrays. *J. Acoust. Soc. Am.* **99**, 185–95.

Thom, R. (1972). *Stabilite structurelle et morphogénese*. Reading, MA: W. A. Benjamin.

Thomsen, L. (1986). Weak elastic anisotropy. *Geophysics* **51**, 1954–66.

Thomson, C. J. (1989). Corrections for grazing rays in 2-D seismic modelling. *Geophys. J.* **96**, 415–46.

Thomson, C. J. (1990). Corrections for critical rays in 2-D seismic modelling. *Geophys. J. Int.* **103**, 171–210.

Thomson, C. J. (1996a). *Notes on Rmatrix, a program to find the seismic plane-wave response of a stack of anisotropic layers*. Kingston: Queen's University, Dept. of Geol. Sci.

Thomson, C. J. (1996b). *Notes on waves in layered media to accompany program Rmatrix*. Kingston: Queen's University, Dept. of Geol. Sci.

Thomson, C. J. (1997a). Complex rays and wave packets for decaying signals in inhomogeneous, anisotropic and anelastic media. *Studia Geoph. et Geod.* **41**, 345–81.

Thomson, C. J. (1997b). Modelling surface waves in anisotropic structures I. Theory. *Phys. Earth Planet. Int.* **103**, 195–206.

Thomson, C. J. (1999). The 'gap' between seismic ray theory and 'full' wavefield extrapolation. *Geophys. J. Int.* **137**, 364–80.

Thomson, C. J. (in press). Seismic coherent states and ray geometrical spreading. *Geophys. J. Int.*

Thomson, C. J., and Chapman, C. H. (1985). An introduction to Maslov's asymptotic method. *Geophys. J. R. astr. Soc.* **83**, 143–68.

Thomson, C. J., and Chapman, C. H. (1986). End-point contributions to synthetic seismograms. *Geophys. J. R. astr. Soc.* **87**, 285–94.

Thomson, C. J., Clarke, T., and Garmany, J. (1986). Observations on seismic wave equation and reflection coefficient symmetries in stratified media. *Geophys. J. R. astr. Soc.* **86**, 675–86.

Thomson, C. J., and Gubbins, D. (1982). Three-dimensional lithospheric modelling at Norsar: Linearity of the method and amplitude variations from the anomalies. *Geophys. J. R. astr. Soc.* **71**, 1–36.

Thomson, C. J., Kendall, J-M., and Guest, W. S. (1992). Geometrical theory of shear-wave splitting: Corrections to ray theory for interference in isotropic/anisotropic transitions. *Geophys. J. Int.* **108**, 339–63.

Thore, P. D., and Juliard, C. (1999). Fresnel zone effect on seismic velocity resolution. *Geophysics* **64**, 593–603.

Thornburgh, H. R. (1930). Wavefront diagrams in seismic interpretation. *Bull. Am. Assoc. Petrol. Geol.* **14**, 185–200.

Thurber, C. H. (1986). Analysis methods for kinematic data from local earthquakes. *Revs Geophys.* **24**, 793–805.

Thurber, C. H., and Ellsworth, W. L. (1980). Rapid solution of ray tracing problems in heterogeneous media. *Bull. Seismol. Soc. Am.* **70**, 1137–48.

Toksöz, M. N., and Johnston, D. H., eds. (1981). *Seismic wave attenuation*. Geophysics Reprint Series, No. 2, Tulsa, OK: Society of Exploration Geophysicists.

Tooley, R. D., Spencer, T. W., and Sagoci, H. F. (1965). Reflection and transmission of plane compressional waves. *Geophysics* **30**, 552–70.

Toverud, T., and Ursin, B. (1998). Comparison of different seismic attenuation laws. *Extended Abstract Book, Modern Exploration and Improved Oil and Gas Recovery Methods, Sept. 1998*, Kraków, 130–2.

Tromp, J., and Dahlen, F. A. (1993). Maslov theory for surface wave propagation on a laterally heterogeneous earth. *Geophys. J. Int.* **115**, 512–28.

Trorey, A. W. (1970). A simple theory for seismic diffractions. *Geophysics* **35**, 762–84.

Trorey, A. W. (1977). Diffractions for arbitrary source-receiver locations. *Geophysics* **42**, 1177–82.

Tsvankin, I. (1995). Body-wave radiation patterns and AVO in transversely isotropic media. *Geophysics* **60**, 1409–25.

Tsvankin, I., and Chesnokov, E. (1990). Synthesis of body wave seismograms from point sources in anisotropic media. *J. Geophys. Res.* **95**, 11317–31.

Tsvankin, I., and Kalinin, A. V. (1984). Non-geometrical effects in the generation of converted seismic waves (in Russian). *Izv. Akad. Nauk SSSR, Fizika Zemli*, No. 2, 34–40.

Tsvankin, I., Kalinin, A. V., and Pivovarov, B. L. (1983). Refraction of a spherical wave for a source adjacent to an interface (in Russian). *Izv. Akad. Nauk SSSR, Fizika Zemli*, No. 10, 32–45.

Tverdokhlebov, A., and Rose, J. (1988). On Green's functions for elastic waves in anisotropic media. *J. Acoust. Soc. Am.* **83**, 118–21.

Tygel, M., and Hubral, P. (1987). *Transient waves in layered media*. Amsterdam: Elsevier.

Tygel, M., Schleicher, J., and Hubral, P. (1992). Geometrical spreading corrections of offset reflections in a laterally inhomogeneous earth. *Geophysics* **57**, 1054–63.

Tygel, M., Schleicher, J., and Hubral, P. (1994). Kirchhoff-Helmholtz theory in modelling and migration. *J. Seismic Exploration* **3**, 203–14.

Um, J., and Thurber, C. (1987). A fast algorithm for two-point seismic ray tracing. *Bull. Seismol. Soc. Am.* **77**, 972–86.

Ursin, B. (1982a). Quadratic wavefront and traveltime approximations in inhomogeneous layered media with curved interfaces. *Geophysics* **47**, 1012–21.

Ursin, B. (1982b). A new derivation of the wavefront curvature transformation at an interface between two inhomogeneous media. *Geophys. Prospecting* **30**, 569–79.

Ursin, B. (1983). Review of elastic and electromagnetic wave propagation in horizontally layered media. *Geophysics* **48**, 1063–81.

Ursin, B. (1990). Offset-dependent geometrical spreading in a layered medium. *Geophysics* **55**, 492–6.

Ursin, B., and Tygel, M. (1997). Reciprocal volume and surface scattering integrals for anisotropic elastic media. *Wave Motion* **26**, 31–42.

van der Hijden, J. H. M. T. (1988). *Propagation of transient elastic waves in stratified anisotropic media*. Amsterdam: North-Holland Elsevier.

van Trier, J., and Symes, W. W. (1991). Upwind finite-difference calculation of traveltimes. *Geophysics* **56**, 812–21.

Vasco, Don W., Peterson, J. E., Jr., and Majer, E. L. (1995). Beyond ray tomography: Wavepaths and Fresnel volumes. *Geophysics* **60**, 1790–804.

Vasil'yev, Y. I. (1959). Certain consequences of an analysis of reflection and refraction coefficients of elastic waves (in Russian). *Trudy Inst. Fiziki Zemli, Akad. Nauk SSSR* **6**, 52–80.

Vavryčuk, V. (1997). Elastodynamic and elastostatic Green tensors for homogeneous weak transversely isotropic media. *Geophys. J. Int.* **130**, 786–800.

Vavryčuk, V. (1999). Properties of S waves near a kiss singularity: A comparison of exact and ray solutions. *Geophys. J. Int.* **138**, 581–9.

Vavryčuk, V. (2000). Ray tracing for anisotropic media in singular directions of S waves. In *Seismic waves in complex 3-D structures, Report 10*, pp. 161–89. Prague: Faculty of Mathematics and Physics, Charles Univ.

Vavryčuk, V., and Yomogida, K. (1995). Multipolar elastic fields in homogeneous isotropic media by higher-order ray approximations. *Geophys. J. Int.* **121**, 925–32.

Vavryčuk, V., and Yomogida, K. (1996). SH-wave Green tensor for homogeneous transversely isotropic media by higher-order approximations in asymptotic ray theory. *Wave Motion* **23**, 83–93.

Vesnaver, A. L. (1996a). Irregular grids in seismic tomography and minimum-time ray tracing. *Geophys. J. Int.* **126**, 147–65.

Vesnaver, A. L. (1996b). Ray tracing based on Fermat's principle in irregular grids. *Geophys. Prospecting* **44**, 741–60.

Vidale, J. E. (1988). Finite-difference calculation of travel times. *Bull. Seismol. Soc. Am.* **78**, 2062–76.

Vidale, J. E. (1989). Finite-difference calculation of traveltimes in 3-D. *59th Annual SEG Meeting Expanded Abstract*, Dallas, 1096–8.

Vidale, J. E. (1990). Finite-difference calculation of traveltimes in three dimensions. *Geophysics* **55**, 521–6.

Vidale, J. E., and Houston, H. (1990). Rapid calculation of seismic amplitudes. *Geophysics* **55**, 1504–7.

Vinje, V., Iversen, E., Åstebøl, K., and Gjøystdal, H. (1996). Estimation of multivalued arrivals in 3D models using wavefront construction. Part I. *Geophys. Prospecting* **44**, 819–42; Part II. Tracing and interpolation. *Geophys. Prospecting* **44**, 843–58.

Vinje, V., Iversen, E., and Gjøystdal, H. (1992). Traveltime and amplitude estimation using wavefront construction. *Extended Abstracts of Papers, EAEG 54th Meeting and Technical Exhibition*, Paris, 504–5.

Vinje, V., Iversen, E., and Gjøystdal, H. (1993). Traveltime and amplitude estimation using wavefront construction. *Geophysics* **58**, 1157–66.

Vinje, V., Iversen, E., Gjøystdal, H., and Åstebøl, K. (1993). Estimation of multivalued arrivals in 3D models using Wavefront Construction. *Extended Abstracts of Papers, EAEG 55th Meeting and Technical Exhibition*, Stavanger.

Virieux, J. (1989). Perturbed ray tracing on a heterogeneous sphere. *Geophys. Res. Letters* **16**, 405–8.

Virieux, J. (1991). Fast and accurate ray tracing by Hamiltonian perturbation. *J. Geophys. Res.* **96**, 579–94.

Virieux, J. (1996). Seismic ray tracing. In *Seismic modelling of Earth structure*, eds. E. Boschi, G. Ekström, and A. Morelli, pp. 223–304. Bologna: Editrice Compositori.

Virieux, J., and Ekström, G. (1991). Ray tracing on a heterogeneous sphere by Lie series. *Geophys. J. Int.* **104**, 11–27.

Virieux, J., and Farra, V., (1991). Ray tracing in 3-D complex isotropic media: An analysis of the problem. *Geophysics* **56**, 2057–69.

Virieux, J., Farra, V., and Madariaga, R. (1988). Ray tracing for earthquake location in laterally hetero-geneous media. *J. Geophys. Res.* **93**, 6585–99.

Vlaar, N. J. (1968). Ray theory for an anisotropic inhomogeneous elastic medium. *Bull. Seismol. Soc. Am.* **58**, 2053–72.

Waltham, D. A. (1988). Two-point ray tracing using Fermat's principle. *Geophys. J.* **93**, 575–82.

Wang, C.-Y., and Achenbach, J. D. (1993). A new method to obtain 3-D Green's function for anisotropic solids. *Wave Motion* **18**, 273–89.

Wang, C.-Y., and Achenbach, J. D. (1994). Elastodynamic fundamental solutions for anisotropic solids. *Geophys. J. Int.* **118**, 384–92.

Wang, C.-Y., and Achenbach, J. D. (1995). Three-dimensional time-harmonic elastodynamic Green's functions for anisotropic solids. *Proc. R. Soc. London A* **449**, 441–58.

Wang, X., and Waltham, D. (1995). The stable-beam seismic modelling method. *Geophys. Prospecting* **43**, 939–61.

Wang, Y., and Houseman, G. A. (1995). Tomographic inversion of reflection seismic amplitude data for velocity variation. *Geophys. J. Int.* **123**, 355–72.

Wang, Z., and Dahlen, F. A. (1994). JWKB surface wave seismograms on a laterally heterogeneous Earth. *Geophys. J. Int.* **119**, 381–401.

Wang, Z., and Dahlen, F. A. (1995). Validity of surface-wave ray theory on a laterally heterogeneous earth. *Geophys. J. Int.* **123**, 757–73.

Weber, M. (1988). Computation of body-wave seismograms in absorbing 2-D media using the Gaussian beam method: Comparison with exact methods. *Geophys. J.* **92**, 9–24.

Weigel, F. W. (1986). *Introduction to path integral method in physics and polymer science*. Singapore: World Scientific.

Wesson, R. L. (1970). A time integration method for computation of the intensities of seismic rays. *Bull. Seismol. Soc. Am.* **60**, 307–16.

Wesson, R. L. (1971). Travel-time inversion for laterally inhomogeneous crustal velocity models. *Bull. Seismol. Soc. Am.* **61**, 729–46.

Weyl, H. (1919). Ausbreitung elektromagnetischer Wellen über einem ebenen Leiter. *Ann. Phys.* **60**, 481–500.

White, B. S., Norris, A., Bayliss, A., and Burridge, R. (1987). Some remarks on the Gaussian beam summation method. *Geophys. J. R. astr. Soc.* **89**, 579–636.

White, J. E. (1983). *Underground sound. Application of seismic waves*. Amsterdam: Elsevier.

Whitham, G. B. (1965). A general approach to linear and nonlinear dispersive waves using a Lagrangian. *J. Fluid Mech.* **22**, 273–83.

Whitham, G. B. (1974). *Linear and nonlinear waves*. New York: Wiley-Interscience.

Whittal, K. P., and Clowes, R. M. (1979). A simple, efficient method for the calculation of travel-times and ray paths in laterally inhomogeneous media. *Canad. J. Expl. Geophys.* **15**, 21–9.

Wielandt, E. (1987). On the validity of the ray approximation for interpreting delay times. In *Seismic tomography with applications in global seismology and exploration geophysics*, ed. G. Nolet, pp. 85–98. Dordrecht: Reidel.

Wiggins, J. W. (1984). Kirchhoff integral extrapolation and migration of nonplanar data. *Geophysics* **49**, 1239–48.

Wiggins, R. A. (1976). Body wave amplitude calculations II. *Geophys. J. R. astr. Soc.* **46**, 1–10.

Wild, A. J., and Hudson, J. A. (1998). A geometrical approach to the elastic complex screen. *J. Geophys. Res.* **103**, 707–25.

Will, M. (1976). Calculation of traveltimes and ray paths for lateral inhomogeneous media. In *Explosion seismology in Central Europe. Data and results*, eds. P. Giese, C. Prodehl, and A. Stein, pp. 168–77. Berlin: Springer Verlag.

Witte, O., Roth, M., and Müller, G. (1996). Ray tracing in random media. *Geophys. J. Int.* **124**, 159–69.

Woodhouse, J. H. (1974). Surface waves in a laterally varying layered structure. *Geophys. J. R. astr. Soc.* **37**, 461–90.

Woodhouse, J. H. (1996). Long period seismology and the Earth's free oscillations. In *Seismic modelling of Earth structure*, eds. E. Boschi, G. Ekström, and A. Morelli, pp. 31–80. Bologna: Editrice Compositori.

Woodhouse, J. H., and Wong, Y. K. (1986). Amplitude, phase and path anomalies of mantle waves. *Geophys. J. R. astr. Soc.* **87**, 753–73.

Wright, J. (1987). The effects of transverse isotropy on reflection amplitude versus offset. *Geophysics* **52**, 564–7.

Wu, R.-S. (1985). Gaussian beams, complex rays and analytic extension of Green's function in smoothly inhomogeneous media. *Geophys. J. R. astr. Soc.* **83**, 93–110.

Wu, R.-S. (1989a). Seismic wave scattering. In *The encyclopedia of solid earth geophysics*, ed. D. E. James, pp. 1166–87. New York: Van Nostrand Reinhold.

Wu, R.-S. (1989b). The perturbation method in elastic wave scattering. *PAGEOPH* **131**, 605–37.

Wu, R.-S. (1994). Wide-angle elastic wave one-way propagation in heterogeneous media and an elastic wave complex-screen method. *J. Geophys. Res.* **90**, 751–66.

Wu, R.-S., and Aki, K. (1985). Elastic wave scattering by a random medium and the small-scale inhomogeneities in the lithosphere. *J. Geophys. Res.* **90**, 10261–73.

Yacoub, N. K., Scott, J. H., and McKeown, F. A. (1970). Computer ray tracing through complex geological models for ground motion studies. *Geophysics* **35**, 586–602.

Yan, J., and Yen, K. K. (1995). A derivation of three-dimensional ray equations in ellipsoidal coordinates. *J. Acoust. Soc. Am.* **97**, 1538–44.

Yanovskaya, T. B. (1966). Algorithm for the calculation of reflection and refraction coefficients and head wave coefficients. In *Computational seismology* (in Russian), vol. 1, ed. V. I. Keilis-Borok, pp. 107–11.

Yeatts, F. R. (1984). Elastic radiation from a point source in an anisotropic medium. *Phys. Rev. B* **29**, 1674–84.

Yeliseyevnin, V. A. (1964). Calculation of rays propagating in an inhomogeneous medium (in Russian). *Akust. Zh.* **10**, 284–8.

Yomogida, K. (1985). Gaussian beams for surface waves in laterally slowly-varying media. *Geophys. J. R. astr. Soc.* **82**, 511–33.

Yomogida, K. (1988). Surface waves in weakly heterogeneous media. In *Mathematical geophysics*, eds. N. J. Vlaar, G. Nolet, M. J. R. Wortel, and S. A. P. L. Cloething, pp. 53–75. Dordrecht: Reidel.

Yomogida, K. (1992). Fresnel zone inversion for lateral heterogeneities in the Earth. *PAGEOPH* **138**, 391–406.

Yomogida, K., and Aki, K. (1985). Waveform synthesis of surface waves in a laterally heterogeneous Earth by the Gaussian beam method. *J. Geophys. Res.* **89**, 7797–819.

Zahradník, J. (1982). Seismic response analysis of two-dimensional absorbing structures. *Studia Geoph. et Geod.* **26**, 24–41.

Zahradník, J., Jech, J., and Moczo, P. (1990a). Approximate absorption corrections for complete SH seismograms. *Studia Geoph. et Geod.* **34**, 185–96.

Zahradník, J., Jech, J., and Moczo, P. (1990b). Absorption correction for computations of a seismic ground response. *Bull. Seismol. Soc. Am.* **80**, 1382–7.

Zedník, J., Janský, J., and Červený, V. (1993). Synthetic seismograms in radially inhomogeneous media for ISOP applications. *Computers & Geosciences* **19**, 183–7.

Zhao, L., and Dahlen, F. A. (1996). Mode-sum to ray-sum transformation in a spherical and an aspherical earth. *Geophys. J. Int.* **126**, 389–412.

Zhu, H. (1992). A method to evaluate three-dimensional time-harmonic elastodynamic Green's functions in transversely isotropic media. *J. Appl. Mech.* **59**, S96–S101.

Zhu, T. (1988). A ray-Kirchhoff method for body wave calculations in inhomogeneous media: Theory. *Geophys. J.* **92**, 181–93.

Zhu, T., and Chun, K.-Y. (1994a). Complex rays in elastic and anelastic media. *Geophys. J. Int.* **119**, 269–76.

Zhu, T., and Chun, K.-Y. (1994b). Understanding finite-frequency wave phenomena: Phase-ray formulation and inhomogeneity scattering. *Geophys. J. Int.* **119**, 78–90.

Zillmer, M., Kashtan, B. M., and Gajewski, D. (1998). Quasi-isotropic approximation of ray theory for anisotropic media. *Geophys. J. Int.* **132**, 643–53.

Ziolkowski, R. W., and Deschamps, G. A. (1980). *The Maslov method and the asymptotic Fourier transform: Caustic analysis*. Electromagnetic Laboratory Scientific Rep. No. 80-9. Urbana-Champaign, IL: Univ. Illinois.

Ziolkowski, R. W., and Deschamps, G. A. (1984). Asymptotic evaluation of high frequency fields near a caustic: An introduction to Maslov' s method. *Radio Science* **19**, 1001–25.

Zöppritz, K. (1919). Über Erdbebenwellen, VIIb. *Göttingen Nachrichten* **1**, 66–84.

Zvolinskiy, N. V. (1957). The reflected and head wave arising at a plane interface between two elastic media (in Russian). I. *Izv. Akad. Nauk SSSR, Geophys. Ser.*, No. 10, 1201–18.

Zvolinskiy, N. V. (1958). The reflected and head wave arising at a plane interface between two elastic media (in Russian). II, III. *Izv. Akad. Nauk SSSR, Geophys. Ser.*, No. 1, 3–16 and No. 2, 165–74.

Zvolinskiy, N. V. (1965). Wave problems in the theory of elasticity of the continuum (in Russian). *Izv. Akad. Nauk SSSR, Ser. Mekhanika* **1**, 109–23.

Index

Certain subentries appear in the text many times. In such cases, only the pages where the subentry is explained (or defined) are given.

The name entry referring to the name of the author of the book is not included.

Abdullaev, S. S., 371, 617
Abgrall, R., 187
Abramowitz, M., 97, 663
absorption
    causal, 543, 545–7, 622, 641–2
    effects on amplitudes, 542–8
    effects on synthetic seismograms, 639–43
    noncausal, 545, 547, 622, 640–1
Achenbach, J. D., 9, 54, 84, 92, 613
acoustic wave equation, 14–15
Adler, L., 51
Aizenberg, A. M., 613
Aki, K., 8, 9, 12, 13, 16, 38, 81, 90, 91, 92, 93,
        118, 135, 142, 160, 174, 189, 229, 230, 372,
        455, 478, 491, 542, 543, 577, 599, 618, 619
Al-Chalabi, M., 170
Aldridge, D. F., 189, 624
Alekseyev, A. S., 54, 213, 234, 392, 567, 579,
        582
Alkhalifah, T., 622, 632
Aminzadeh, F., 51
amplitude coefficients of the ray series
    additional components of, 569–72
    continuation along a ray, 552, 574
    decomposition of, 569–70
    principal components of, 569–75
    scalar, 549–50
    vectorial, 568–77
analytical delta function, 20, 637, 665
analytical signal, 20, 623, 645, 664–5
    approximate expressions for, 664
    conservation along a ray, 623
    normalized, 645
    properties of, 664–5
    sign convention of, 665

Andrade, F. C. M., de, 618
angle of incidence, 42, 424
    Brewster, 426, 495, 558
    critical, 121, 425–8, 491–4
    grazing, 426, 427
    minimum critical, 492
    normal, 425–7, 488–91
    postcritical, 121, 425–8, 494
    subcritical, 121, 425–8, 494
anisotropy systems, 10–12
    hexagonal, 11, 12
    orthorhombic, 10, 12
    triclinic, 10
Aranha, P. R. A., 615, 618
Arnaud, J. A., 589
Arnold, V. I., 214
Asakawa, E., 185
Asatryan, A. A., 377, 567
asymptotic ray theory (ART), 1, 550
asymptotics
    local, 567, 610, 611, 612, 619
    uniform, 567, 610
Audebert, F., 179
Auld, B. A., 8, 10, 12, 17, 38
Azbel, I. Ya., 632

Babich, V. M., 1, 6, 54, 64, 70, 72, 73, 103, 111,
        148, 230, 339, 556, 566, 567, 568, 579, 589,
        599, 611, 613, 617
Babuška, V., 11
Backus, G. E., 17, 194, 279, 282, 283, 288, 521
backward propagation along a ray, 308–9, 502–3,
        534–5
Bakker, P. M., 401, 412, 415, 416, 605, 607, 613
Baltensperger, P., 618

Bard, P.-Y., 632
Belonosova, A. V., 234
Benamou, J. D., 187
bending method, 223–8
    based on fitting ray equations, 225
    based on minimizing the travel time, 226–7
    based on paraxial ray approximation, 225–6
    based on structural perturbations, 227–8
    Hamiltonian approach to, 227–9
    Lagrangian approach to, 228
    pseudo-bending method, 226
Ben-Hador, R., 73, 230
Ben-Menahem, A., 8, 84, 95, 372, 455, 508, 511,
    608, 618
Berkhaut, A. J., 15
Bernard, P., 438, 632
Beydoun, W. B., 92, 95, 234, 372, 608
Beylkin, G., 95
Biondi, B., 56
Blangy, J. P., 51, 548
Bleistein, N., 15, 74, 92, 95, 96, 97, 100, 103,
    394, 440, 441, 443, 549, 594, 613
Bolland, T. K., 51
Bolt, B. A., 6, 8, 12, 160, 174, 175, 177, 178, 643
Bonnet, G., 84
Booth, D. C., 618, 643
Borcherdt, R. P., 548
Borejko, P., 478
Born, M., 17, 643
Born approximation, 8, 56, 89, 93–96, 639
Bortfeld, R., 305, 355, 478, 619
Bouchon, M., 624
boundary value ray tracing, 217–28, 348–56,
    396–7, 414
    bending method, *see* bending method
    paraxial method, 348–56
    point-to-curve ray tracing, 220
    shooting method, 220–3
    two-point ray tracing, 219–20, 396–7
Bourbiè, T., 548
Brac, J., 182
Bracewell, R., 661
Brekhovskikh, L. M., 14, 38, 229, 230, 559, 565,
    612, 615, 618, 619
Bretherton, F. P., 230
Brewster angle, 426, 495, 558
Brillouin, L., 163
Brodov, L. Yu., 644
Brokešová, J. (*see also* Pleinerová, J.), 128,
    394, 632
Brown, M. G., 214, 371, 601
Brühl, M., 372
Buchen, P. W., 73, 92, 230, 536, 544
Buchwald, V. T., 85, 87

Budden, K. G., 566, 617
Bulant, P., 220, 222, 223, 512, 632
Buldyrev, V. S., 6, 70, 73, 103, 111, 339, 566,
    589, 611, 613, 617
Bullen, K. E., 6, 8, 12, 160, 174, 175, 177, 178,
    643
Burdick, L. J., 619
Burridge, R., 17, 84, 92, 95, 107, 229, 230
Buske, S., 188

Cagniard, L., 619
Cagniard-deHoop method, 619
Camerlynck, C., 559, 579, 608
canonical coordinates, 104
canonical equations, 104
canonical vector, 104
Cao, S., 185, 187
Cara, M., 11
Carcione, J. M., 548
Carter, J. A., 92
Cassel, B. R., 135, 631
caustic points along a ray, 380–1, 415–16
    of the first order, 213, 380–1, 415–16, 606–7
    of the second order, 213, 380–1, 415–16,
        605–6
caustic region, 611
caustics, 122–3, 213–14, 611
    classification of, 213–14
caustic shadow, 122, 611
caustic surface, 122, 214, 608, 610, 611
Caviglia, G., 36, 544, 548, 614
Chander, R., 220
chaotic behavior of rays, 123, 370–2
Chapman, C. H., 6, 14, 51, 92, 95, 96, 104, 135,
    148, 156, 157, 163, 173, 189, 194, 207, 214,
    357, 415, 438, 501, 509, 512, 515, 518, 522,
    523, 527, 530, 532, 533, 535, 536, 559, 583,
    591, 600, 601, 603, 608, 611, 613, 615, 619,
    620, 631, 638, 639
Chapman, S. J., 566
Charrette, E. E., 96
Chen, H.-W., 632
Chen, K. C., 356
Cheng, N., 185
Chernov, L. A., 372
Chesnokov, E., 84
Chikhachev, B. A., 230
Christoffel equation, 22
Christoffel matrix, 22–4, 85
    for anisotropic media, 22
    eigenvalues of, 23–4, 26
    eigenvalues of qS waves, average, 152
    eigenvectors of, 23, 27
    for isotropic media, 26

properties of, 22–4
space-time, 71
Christoffel symbols, 147
Christoffersen, A., 135
Chu, J.-Y., 603, 639
Chun, K.-Y., 56, 544, 566
Claerbout, J. F., 616, 617
Clarke, R. A., 220
Clarke, T., 530
Cline, A. K., 172
Clowes, R. M., 135, 632
Coates, R. T., 14, 95, 96, 415, 512, 515, 518,
    522, 523, 527, 583
Coddington, E. A., 279, 282
coefficients
    of absorption, 543
    amplitude, see amplitude coefficients of the
        ray series
    of conversion, see conversion coefficients
    of diffraction, 613
    of head waves, 581–2
    of reflection/transmission, see
        reflection/transmission coefficients
Cohen, J. K., 95, 394, 401
coherent state transform, 599
Comer, R. P., 143
congruency of rays, 199
conical singularity of qS waves, 33
conservation
    of analytical signal along the ray, 623
    of energy flux in the ray tube, 216–17
    of normalized envelope of the signal along
        the ray, 623
    of polarization quasi-ellipse along the ray, 653
conversion coefficients
    at a free surface, 462, 501–3
    of pressure waves, 432–3
    of P-SV and SH waves, 501–3
    at a structural interface, 460–2, 501–3
Cooper, H. F., Jr., 548
coordinates
    canonical, 104
    curvilinear nonorthogonal, 145–7
    curvilinear orthogonal, 137–8
    cylindrical, 212
    Gaussian, 229, 311
    local Cartesian at an interface, 290–2
    local ray-centered Cartesian, 246–7
    nonorthogonal ray centered, 274
    phase-space, 104
    ray, 201–2
    ray-centered, see ray-centered coordinate
        system
    spherical, 141, 183, 212

wavefront orthonormal, see wavefront
    orthonormal coordinate system
Coppoli, A. D. M., 95
Cormier, V., 244, 632, 643
Coultrip, R. L., 187
coupling
    of qS waves, 149, 512–27, 616, 659
    of transport equations of S waves, 61, 244
coupling ray theory, 512
Courant, R., 103, 111
Crampin, S., 10, 11, 33, 34, 51, 618, 643, 644,
    655, 660
critical angle, 121–2, 425–8, 491–4
Crowley, J. B., 136
curvature
    of the initial line, 321
    of the ray, 123–4
curvature matrix
    of group velocity surface, 406–8
    of initial surface, 315
    of interface, 291–2
    of slowness surface, 88, 406–8
    of wavefront, 326–8, 391, 406–8, 410
Cutler, R. T., 135

Dahlen, F. A., 6, 73, 142, 230, 234, 372, 601,
    603, 618
Dai, T.-F., 92
Daley, P. F., 51, 187, 579, 615, 618, 631
Danbom, S. H., 644
Davis, J. L., 8
Debye procedure, 513
degenerate case of qS waves, 194
delay time, 163, 176, 184
Dellinger, J., 516
DeSanto, J. A., 76, 97
Deschamps, G. A., 213, 214, 588, 601
Dey-Sarkar, S. K., 603
Dietrich, M., 624
Dijkstra, E. W., 185
directional attenuation, 548
directivity patterns of a point source, 422–3,
    453, 507
    relation to the radiation matrix, 422, 453, 507
dispersion relations, 71, 543–8
    causal, 546–7
    Futterman, 546
    Kramers-Krönig, 543, 640
    Müller, 546–7
    noncausal, 545
    in the space-time ray method, 71
    of surface waves, 228–31
dissipation filters, 544, 639–40
Docherty, P., 92, 220, 227

Dohr, G., 644
Domenico, S. N., 644
Douglas, A., 636
Drijkoningen, G. G., 559, 615, 619
Drummond, R., 214, 601, 603
Druzhinin, A. B., 512, 536
Duff, G. F. D., 84
dynamic ray tracing
    analytical, 341–6
    for backward propagation, 284, 308–9
    across a curved interface, 300–1, 333–5,
        411–12
    Hamiltonian approach to, 259–78
    in layered structures, 289–303, 333–5
    along a planar ray, 384–9
    solutions in terms of ray propagator matrix,
        284
    surface-to-surface, 289, 304–5
    2-dimensional, 393
    2.5-dimensional, 394
dynamic ray tracing system, 234–7
    for anisotropic media, 260–78, 401–4
    in Cartesian coordinates, 261–4, 331–3, 401–3
    constraint relations of, 262, 266, 271, 275
    in curvilinear coordinates, 270–8
    fundamental matrix of, 279
    inhomogeneous, 288–9
    for isotropic media, 253–7
    noneikonal solution of, 264, 266, 272
    in nonorthogonal ray-centered coordinates,
        274–7
    propagator matrix of, see ray propagator
        matrix
    in ray-centered coordinates, 253–7
    ray-tangent solution of, 263, 266, 272
    reduced, 273
    symmetry relations of, 262, 272, 276
    system matrix of, 254, 279, 281
    transformations of, 273–4
    uniqueness theorem for the solution of, 279
    in wavefront orthonormal coordinates, 264–7,
        403–4

Earth flattening transformation, 145
Earth surface flattening transformation, 146
Eaton, D. W. S., 187, 375
Eickemeyer, J. A., 220, 226
eikonal equation
    in anisotropic media, 63, 149
    in a Hamiltonian form, 103–6
    in nonorthogonal curvilinear coordinates, 146
    in orthogonal curvilinear coordinates, 138–9
    for pressure waves, 55
    for P waves, 59

space-time, 71
    for S waves, 59
    for surface waves, 230
Eisner, L., 579, 583, 635
Ekström, G., 230
elastic moduli, 10
    density-normalized, 21
elastic tensor, 10
elastodynamic equation, 9
    for anisotropic media, 13–14
    for isotropic media, 14
ellipsoidal anisotropy, 156
Ellsworth, W. L., 220
energy, 16–18
    acoustic, 18, 28, 64–5
    elastic, 17, 28–30, 66, 71–2
    kinetic, 17, 28–30, 66, 71–2
    strain, 16, 28–30, 66, 71–2
    time-averaged, 17, 72
    time-integrated, 18, 28–30, 66
energy equation, 17, 72
energy flux, 17, 28–30, 66, 71–2
    velocity of, 18, 29–30
Ettrich, N., 188
Euler's equation, 111
    in a parameteric form, 111
Euler's theorem for homogeneous functions, 22,
    23, 406
Evans, R., 643
Every, A. G., 84
Ewing, W. M., 618, 619
exact ray theory, 619
existence conditions of plane waves
    in anisotropic media, 24
    in fluid media, 19
    for P and S waves in isotropic media, 26–7
extension of the ray method, 3, 4, 5, 418–19,
    607, 609, 611–13
    in the caustic region, 611
    coupling ray theory, 512
    in the critical region, 612
    in the Fresnel shadow region, 612–13
    Gaussian beam summation, 3, 5, 419, 584,
        590–605, 637
    Gaussian wave packet summation, 3, 599
    geometric theory of diffraction, 613
    Kirchhoff integrals, 436–44, 535–41
    Maslov-Chapman method, 3, 5, 419, 584, 590,
        599–601, 637
    for propagation in a preferred direction,
        516–17
    quasi-isotropic ray theory, 512–27
    ray method with the complex eikonal, 566–7
    space-time ray method, 4, 70–3, 566

summation of paraxial ray approximations, 5, 419, 584, 590–605, 637
superposition integrals, 590–4, 597–9
WKBJ method in 1-D models, 590, 603, 638–9
Ezaka, T., 187

factorization
  of anisotropic inhomogeneous medium, 157–9
  of elastodynamic equation, 616
  of geometrical spreading, 306, 364–5, 414
  of wave equation, 616
Faria, E. L., 187
Farra, V., 104, 136, 148, 156, 189, 190, 198, 199, 220, 226, 227, 237, 334, 335, 337, 356, 401
fast frequency response algorithm, 622, 630
Fedorov, F. I., 10, 51
Felsen, L. B., 6, 73, 160, 588, 589, 611, 613, 614, 618
Fermat's functional, 110–12
Fermat's principle, 1–2, 110–12, 219
Fertig, J., 599, 632
Feshbach, H., 15, 103
Firbas, P., 148, 192, 194, 199
Fischer, R., 185
Fishman, L., 616
Flatté, S. M., 126
Fock, V. A., 616
focus point, 213
Fokkema, J. T., 14
Forbes, G. W., 566
forward star, 185
Foster, D. J., 599
Fourier transform, 16
  properties of, 661–3
  sign convention of, 16, 662
Fradkin, L. Yu., 579, 608
Frangié, A. B., 642, 643
Frazer, L. N., 92, 438, 516, 530, 534, 536, 618, 632
Frenet's formulae, 123, 321
frequency response, 629, 635–6
Fresnel shadow, 612–13
Fresnel volume, 114–17, 372–80
  analytical computations of, 373–5
  of first arriving waves, 377–8
  in a homogeneous medium, 115–16
  by network ray tracing, 377–80
  paraxial, 375
Fresnel volume ray tracing, 373, 375–80
Fresnel zone, 115, 372–80
  analytic expressions for, 373–5
  elliptic, 376
  hyperbolic, 376–7
  in-plane half-axis of, 373–4

interface, 115, 372, 373–4, 377, 379
  off-ray shift of, 374, 379
  transverse half-axis of, 373–4
Fresnel zone matrix, 306–8, 346–8, 376–7, 414, 443
Friedlander, F. G., 559
Friedrichs, K. O., 613
Fryer, G. J., 530, 534, 618
Fuchs, K., 76, 615, 618
Fuki, A. A., 512
Fung, Y. C., 9, 542
Futterman, W. I., 546

Gabor, D., 624
Gajewski, D., 51, 64, 148, 188, 401, 508, 512, 515, 548, 622, 632
Gal'perin, E. I., 644
Garmany, J., 415, 530, 603
Gassman, F., 160
Gaussian beams, 419, 582–4, 586–9
  existence conditions of, 587
  half-width of, 588
  higher modes of, 589
  as physical objects, 589
  spot ellipse of, 587
  summation of, 589–604
Gaussian curvature
  of group velocity surface, 34, 408
  of slowness surface, 34, 87–8, 408
  of wavefront, 328–9, 408
Gaussian wave packets, 589, 599
Gautesen, A. K., 54, 613
Gebrande, H., 135
Gel'chinskiy, B. Ya., 54, 117, 213, 356, 376, 392, 567, 579, 582
Geldard, L. P., 372
Gelius, L.-J., 632
Geoltrain, S., 182
geometrical spreading (see also relative geometrical spreading), 356–8
  continuation along a ray, 358, 392–3
  definition of, 206, 357, 409
  determination from travel time data, 365–9, 397–400
  factorization of, 306, 364–5, 414
  in-plane, 392–4
  relation to ray Jacobian, 206
  relation to wavefront curvature, 363–4
  relative, 359–63, 393–4, 409
  transverse, 392–4
Gibson, R. L., 95, 401, 512
Gilbert, F., 279, 282, 283, 288, 521, 615, 619
Gilbert, K. E., 371
Gilmore, R., 214

Gjevik, B., 230
Gjøystdal, H., 213
Glassmoyer, G., 548
global absorption factor, 418, 545, 640
Godin, O. A., 14, 38, 229, 230
Goldin, S. V., 213, 357, 364, 368, 369, 381, 383, 568, 579
Goldstein, H., 104, 111, 112
Goni, G. J., 371
Gonzalez-Serrano, A., 548
Graebner, M., 51
Grant, F. S., 643
Gray, S. H., 92, 182
Grechka, V. Y., 33, 220, 401
Green, A. G., 356
Green function (*see also* ray-theory Green function)
    in anisotropic homogeneous media, 84–8
    elastic Kirchhoff, 539–40
    in fluid homogeneous media, 77–9
    in isotropic homogeneous media, 80–4
    pressure Kirchhoff, 439
    superposition, 598
    2-D, 96–8
Greenhalgh, S., 185, 187
Greenleaf, J. F., 371
Gregersen, S., 230
Grinfeld, M. A., 357
Gritsenko, S. A., 368
group velocity surface, 31–4
    curvature matrix of, 408
    cuspoidal edges in, 34
    Gaussian curvature of, 34, 408
    multivaluedness of, 34
    relation to slowness surface, 34–5
group velocity vector
    in anisotropic media, 29, 35, 63, 66, 152, 159
    definition of, 29, 63
    in isotropic media, 30, 35, 66
    paraxial, 410
    surface wave, 231
Gubbins, D., 142, 220, 225
Gudmundsson, O., 372
Guest, W. S., 84, 276, 401, 512, 601, 622, 632
Guiziou, J. L., 220, 227
Gutenberg, B., 643

Haas, A., 227
Haddon, R. A. W., 92, 536
Hagin, F. G., 95, 394
Haines, A. J., 179
Hamilton canonical equations, 104
Hamilton function, 104
    reduced, 107–9

Hamilton-Jacobi equation, 108, 230
Hanyga, A., 6, 84, 156, 160, 188, 189, 193, 220, 227, 276, 401, 419, 508, 589, 611, 613
Harrison, M. P., 375
Hearn, D. J., 544, 545, 566, 640
Helbig, K., 10, 22, 31, 33
Helle, H. B., 548, 611
Helmberger, D. V., 92, 619, 631
Helmholtz equation, 16
Henneke, E. G., 51
Heritage, J. R., 95
Hermite-Gaussian beams, 589
Herrman, R. B., 643
Hilbert, D., 103, 111
Hilbert transform, 20, 622, 623, 628, 663–4
    approximate expressions for, 664
    properties of, 663–4
Hill, D. P., 565
history function of the ray, 120
Hole, J. A., 187
Hong, T.-L., 619, 631
Hooke's law, 10, 12
    generalized, 10
Hoop, A. T., de, 619
Hörmander, L., 214
Hosken, J. W. J., 624
House, L., 185
Houseman, G. A., 220
Houston, H., 188, 366
Hrabě, J., 145
Hron, F., 51, 54, 135, 213, 235, 246, 255, 357, 548, 568, 579, 612, 615, 618, 631, 633, 640
Hu, L.-Z., 179
Huang, J.-I., 599
Huang, X., 438, 601
Hubral, P., 6, 76, 92, 136, 213, 235, 305, 307, 308, 355, 357, 367, 369, 375, 376, 383, 536, 624
Hudson, J. A., 8, 90, 92, 95, 98, 542, 616, 617, 636
Husebye, E. S., 135
Huygens principle, 178
hybrid methods, 615, 618, 635
    ray-finite difference, 618
    ray-matrix, 615, 618
    ray-mode, 618
hypereikonal, 56

index of ray trajectory, *see* KMAH index
inhomogeneous plane waves, 36–7, 76
    attenuation vector of, 36
    planes of constant amplitudes of, 36
    planes of constant phases of, 36
    propagation vector of, 36

inhomogeneous waves, 121–2, 613–15
initial conditions for dynamic ray tracing, 263, 310–22
  at an initial line, 318–22
  at an initial surface, 313–17
  noneikonal, 263–4, 276
  normalized plane wavefront, 280, 404–5
  normalized point source, 280, 404–5
  at a point source, 317–18
  ray-tangent, 264, 276
  standard paraxial, 263, 266, 276
initial conditions for ray tracing, 117–18
  in anisotropic media, 153–4
  at an initial line, 318–19
  at an initial surface, 311–13
  at a point source, 317
  for surface waves, 231
initial surface
  choice of ray parameters at, 200–1, 311, 315
  curvature matrix of, 315
  with edges and vertexes, 322
  Gaussian coordinates along, 311
  initial conditions for amplitudes at, 436–44, 474–5, 535–41
  initial conditions for dynamic ray tracing at, 313–16
  initial conditions for ray tracing at, 311–13
  initial time field along, 311
  local Cartesian coordinates at, 315
initial-value ray tracing, 101–2, 218–19
  controlled, 223
instantaneous frequency, 666
interface
  curvature matrix of, 291–2
  with edges and vertexes, 122
  of the first order, 554
  geometry of, 289–92
  of the higher order, 122, 554
  point source situated at, 433–5, 462–5
  receiver situated at, 431–2, 460–2
  unit vector normal to, 290
interface propagator matrix, 301–2, 334–6, 412–13
Iversen, E., 632

Jacob, K. A., 143
Jacobian of transformation, 202–3
James, G. L., 613
Jannaud, L. R., 220
Janský, J., 172, 178, 363, 631
Jardetzky, W. S., 618, 619
Jech, J., 148, 156, 189, 194, 196, 197, 643
Jeffreys, B. S., 12, 15, 111, 203
Jeffreys, H., 12, 15, 111, 160, 163, 203

Jiang, Z.-Y., 371
Jílek, P., 462, 615, 636
Jobert, G., 145, 230, 233
Jobert, N., 145, 230, 233
Johnston, D. H., 542, 543
Jouanna, P., 84
Juhlin, C., 187
Julian, B. R., 142, 220, 225
Juliard, C., 372

Kalinin, A. V., 615, 620
Kamke, E., 279, 282, 551
Kampfmann, W., 92
Kanasewich, E. R., 631
Kapoor, T. K., 618
Karal, F. C., 1, 54, 567
Kashtan, B. M., 148, 512, 515
Katchalov, A. P., 589, 599
Kaufman, H., 170
Kawanaka, T., 185
Kawasaki, I., 507
Kay, I. W., 6, 103, 104, 112, 160
Kazi-Aoual, M. N., 84
Keers, H., 372
Keho, T. H., 92, 234
Keith, C. M., 51
Keller, H. B., 220, 227
Keller, J. B., 1, 54, 214, 220, 225, 372, 567, 588, 613
Kelly, S., 618
Kemper, M., 305
Kendall, J.-M., 84, 276, 357, 401, 438, 512, 601, 622, 632
Kennett, B. L. N., 13, 14, 16, 90, 92, 93, 220, 230, 455, 479, 530, 543, 618
Kerry, N. J., 530
Kiehn, M., 372
Kim, K. Y., 51, 84
Kirchhoff integral, 419, 622, 637
  acoustic, 92, 435–44
  elastic, 92, 535–41
Kirkwood, S. C., 10
Kirnos, D. P., 160, 174
Kirpichnikova, N. J., 72, 589, 611, 613, 617
Kiselev, A. P., 512, 578, 579, 615, 620
kiss singularity of qS waves, 34
Kjartansson, E., 547
Klauder, J. R., 599
Klem-Musatov, K. D., 613
Klimeš, L., 119, 125, 145, 185, 187, 222, 223, 234, 276, 339, 357, 367, 371, 372, 381, 382, 400, 401, 415, 416, 418, 442, 455, 511, 512, 541, 583, 589, 599, 601, 632
Kline, M., 6, 103, 104, 112, 160

KMAH index, 214, 217, 380–4, 611
  in anisotropic media, 415–16, 604–7
  decomposition of, 383–4
  determination by dynamic ray tracing, 382–3
  reciprocity of, 381, 416
Knapp, R. W., 372
Knopoff, L., 92
Knott, G. G., 477
Koch, M., 135
Koefoed, O., 478
Kogan, S. Ya., 17
Koketsu, K., 220, 226
Konopásková, J., 599
Korn, G. A., 140, 314, 521
Korn, M., 372
Korn, T. M., 140, 314, 521
Körnig, M., 101, 134, 135, 136
Kosloff, D., 186
Kozák, J., 615
Kragh, J. E., 438, 639
Kramers, H. A., 163
Kravtsov, Yu. A., 6, 73, 95, 115, 116, 160, 167,
    169, 182, 194, 214, 372, 373, 377, 415, 512,
    515, 566, 608, 609, 611, 617, 643
Krebes, E. S., 544, 545, 548, 566, 640
Krey, Th., 6, 136, 213, 235
Kühnicke, E., 619
Kuo, J. T., 92
Küpper, F. J., 624
Kurdyukova, T. V., 579
Kvasnička, M., 117, 185, 187, 373, 374, 375,
    377–8

Lafond, C. F., 343
Lagrangian, 110
Lamb, M., 618
Lambaré, G., 188
Lamé's elastic moduli, 12
Landau, L. D., 9, 194, 196
Landers, T., 616
Langan, R. T., 135
Langer, J., 234, 356
Langston, C. A., 136, 213, 631, 632
Laster, S. J., 615
Le Bégat, S., 148, 189, 190, 198, 199, 227, 237,
    334, 335, 356, 401
Lecomte, I., 187, 618
Lee, J.-J., 136, 213, 220, 225, 631
Lee, W. H. K., 220, 225
Lees, J. M., 185
Legendre transformation, 112
Lenartowicz, E., 6, 160
Leontovich, M. A., 616
Lerche, I., 135

Levander, A. R., 343
Levinson, N., 279, 282
Levshin, A. L., 230
Lewis, R. M., 415, 613
Li, X.-G., 187
Lifschitz, E. M., 9, 194, 196
Lighthill, M. J., 84
Lindsey, J. P., 372
linear elastodynamics, 8–15
line singularity of qS waves, 34
line source
  choice of ray parameters at, 201, 318–19
  curvature of, 321
  in a homogeneous medium, 96–8
  initial conditions for amplitudes at, 444–9,
    475–7
  initial conditions for dynamic ray tracing at,
    318–22
  initial conditions for ray tracing at, 318–19
  radiation function for a, 447, 476
  ray-theory solutions for a, 446–7, 476–7
  torsion of, 321
  2-D Green function, 96–8, 449, 477
Lippman-Schwinger equation, 95
Liu, E., 643
Liu, X.-F., 143, 144, 220, 234
Loewenthal, D., 179
logarithmic decrement of absorption, 543
Lomax, A., 189
loss factor, 542
Love, A. E. H., 9, 618
Lu, I.-T., 618
Lucio, P. S., 188
Ludwig, D., 356, 611, 613
Luneburg, R. K., 244
Luneva, M., 613
Lutter, W. J., 189
Lyapunov exponent, 123, 371
Lyness, D. G., 603, 639

Madariaga, R., 104, 136, 189, 199, 220, 227,
    438, 599, 632
Majer, E. L., 372
Malinovskaya, L. N., 643
Mallet, J. L., 220
Mallick, S., 516, 618
Mandal, B., 185
Mao, W., 220
Marcuvitz, N., 6, 73, 160, 588, 611, 613
Marfurt, K. J., 187
Marks, L. W., 135, 612, 631
Martin, B. E., 229, 230, 231
Maslov, V. P., 214, 601
Matsuoka, T., 187

Matyska, C., 371
May, B. T., 182, 631
May, W. P., 92
Mazur, M. A., 371
McCamy, K., 478
McCarron, E. B., 187
McCoy, J. J., 616
McKeown, F. A., 136
McMaken, H., 54, 613
McMechan, G. A., 220, 618, 631
medium
    anisotropic, 9–12
    background, 93
    dissipative, 622, 639–43
    factorized anisotropic inhomogeneous (FAI),
        157–9
    fluid, 12–13
    isotropic, 12
    perturbed, 93
    radially symmetric, 160, 174–8
    transversely isotropic, 11
    vertically inhomogeneous, 160–73
    viscoelastic, 13, 37, 542
    weakly anisotropic, 11, 512–27
    weakly dissipative, 418, 512–27, 542–8, 622,
        639–43
Mellen, M. H., 632
Mendes, M., 95
Mendiguren, J. A., 643
Mensch, T., 148, 156
Mercator transformation, 144
Meres, M. W., 478
Mereu, R. F., 372
method
    of horizontal rays and vertical modes, 229
    perturbation, 93–5, 189–99
    phase matching, 68, 289, 296
    phase-screen, 617
    reflectivity, 76, 618
    of stationary phase, 86–7, 438, 440–1, 444,
        445, 594
    of two-scale expansion, 229
metric tensor, 112–13, 145–6, 231, 240
Meyer, R. P., 478
Mikhailenko, B. G., 579, 615
Miklowitz, J., 617
Mochizuki, E., 148, 233
Moczo, P., 632, 643
Molotkov, I. A., 6, 54, 61, 64, 70, 73, 103, 107,
    119, 148, 156, 339, 356, 363, 425, 462, 566,
    568, 613, 624, 631, 633, 664
Montagner, J. P., 230
Mooney, W. D., 631
Morro, A., 36, 544, 548, 614

Morse, P. M., 15, 103
Moser, T. J., 185, 220, 224, 226
Müller, G., 76, 135, 174, 372, 543, 546, 547,
    599, 618, 619, 624
Mura, T., 8
Musgrave, M. J. P., 10
Musil, M., 632
Muskat, M., 478, 579

Nafe, J. E., 478
Naida, O. N., 512
Najmi, A.-H., 357
Nakanishi, I., 185
near-wavefront expansion, 556
Nechtschein, S., 548
Neele, F., 377
Nichols, D. E., 179
Nolet, G., 185, 189, 220, 224, 226, 372
Norris, A. N., 589, 599
Nowack, R., 189, 194, 599
Nuttli, O., 643

Obolentseva, I. R., 33, 220
Officer, Ch. B., 618
Ojo, S. B., 372
Orcutt, J. A., 95, 603, 619, 620
orientation index, 480, 483
Orlov, Yu. I., 6, 73, 115, 116, 160, 167, 169, 182,
    194, 214, 372, 373, 415, 512, 515, 608, 609,
    611, 617, 643
orthonomic system of rays, 102, 199, 268–71,
    277–8
Ott, H., 615, 620

Pajchel, J., 6, 160, 220
Palmer, D. R., 371
Pankratova, T. F., 599
Papadimitriou, P., 599
Papanicolaou, G., 372
Papoulis, A., 661
parabolic wave approximation, 616
parallel transport of a vector along a curve,
    147, 242
parameters of the ray, see ray parameters
paraxial amplitudes, 583
paraxial ray approximation, 2, 3, 419, 582–6
paraxial rays, 235, 250–3
    close to a plane ray, 395–6
paraxial ray tracing system, 234–5
    for anisotropic media, 262–4
    in Cartesian coordinates, 262–4
    constraint equation for, 262
    in curvilinear coordinates, 271
    initial conditions for, 253, 263

paraxial ray tracing system (*cont.*)
  for isotropic media, 250–3, 262–4
  in ray-centered coordinates, 250–3
  reduced, 273
  in wavefront orthonormal coordinates, 267
paraxial slowness vector, 298–9, 330, 410
paraxial travel times, 235, 269–70, 278, 328–30, 394, 410
  accuracy of, 330
  in Cartesian coordinates, 258–9, 269
  close to a planar ray, 394
  in curvilinear coordinates, 278
  hyperbolic, 329
  in local ray-centered Cartesian coordinates, 257–8
  in nonorthogonal ray-centered coordinates, 278
  parabolic, 329
  in ray-centered coordinates, 256–9
  along a surface crossing the ray, 295–6
  in wavefront orthonormal coordinates, 269
particle acceleration, 9
particle ground motion, *see* polarization
particle velocity, 9
Passier, M. L., 220
path integral, 617
Payton, R. G., 51
Pereyra, V., 220, 224, 225
Perozzi, P. J., 220, 227
perturbation methods
  for displacement in anisotropic media, 93–5
  first-order, 93
  for pressure in fluid media, 94–5
  for rays, 336–8
  for travel times, *see* travel-time perturbations
Peterson, J. E., 372
Petrashen, G. I., 17, 51, 148
phase shift due to caustics, 213–14, 217, 380–5, 420, 452, 506–7, 604–7
  in anisotropic media, 415–16, 506–7, 511, 604–7
  anomalous, 416, 607
  complete, 511
  in isotropic media, 380–5, 452
  reciprocity of, 416, 511
  relation to KMAH index, 217, 380
phase space, 104
phase velocity surface, 30–4
phase velocity vector, 20, 30, 35, 152
Piankov, V. N., 383
Pilant, W. L., 8, 12, 38, 92, 160, 174, 618, 643
Pilipenko, V. N., 186
Piserchia, P.-F., 618

Pitts, T. A., 371
Pivovarov, B. L., 615
plane of incidence, 42, 290
plane waves, 19–37
  acoustic, 19–22
  in anisotropic media, 25–6
  elastic, 21–30
  energy considerations for, 28–30
  existence conditions of, 19, 24, 26
  in fluid media, 19–22
  group velocity of, 29–35
  inhomogeneous, 19, 36–7
  in isotropic media, 26–7
  phase velocity of, 19–20, 25, 29–35
  across a planar structural interface, 37–53
  transient, 20–24, 52–3
Pleinerová, J. (*see also* Brokešová, J.), 455
Podvin, P., 187
Podyapol'skiy, G. S., 478, 568
point source
  choice of ray parameters at, 199–200, 317
  directivity patterns of, 422, 453, 507
  in a homogeneous medium, 8, 73–88
  initial conditions for amplitudes at, 421–2, 452–5, 506–7
  initial conditions for dynamic ray tracing at, 317–18
  initial conditions for ray tracing at, 317
  at an interface, 433–5, 462–5
  moment-tensor, 455
  omnidirectional, 76, 80, 453
  radiation function of, 421–2, 430, 452–5, 506–7, 624
  single-force, 453–4
point source solutions
  in homogeneous anisotropic media, 85–6
  in homogeneous fluid media, 74–6
  in homogeneous isotropic media, 79–80
  ray-theory, 421–2, 452–5, 506–7
polarization, 643–60
  in anisotropic media, 25, 35–6, 659–60
  at the Earth's surface, 643, 651–2, 655
  elliptic, 4, 647
  of interfering signals, 650–1
  in isotropic media, 27, 35–6, 643–59
  in layered structures, 658
  linear, 4
  of noninterfering P waves, 27, 651–2
  of noninterfering S waves, 27, 644, 652–4
  quasi-elliptical, 4, 27, 644, 648, 651, 653
  at a structural interface, 651, 654–5
polarization equations, 645–8
polarization matrix, 646
polarization plane, 644–5

polarization vectors, 25, 243–5
 of S waves, 61, 244, 653
Popov, M. M., 72, 235, 242, 246, 254, 356, 357,
  559, 579, 589, 595, 599, 608, 617
Popovici, A. M., 179
Pratt, R. G., 156, 189, 194
Press, F., 618, 619
Pretlová, V., 170, 172
Protázio, J., 51, 478
Prothero, W. A., 220, 226
Pšenčík, I., 6, 51, 54, 61, 64, 107, 119, 125, 145,
  148, 156, 189, 194, 196, 222, 234, 235, 242,
  246, 251, 254, 356, 357, 363, 367, 382, 401,
  418, 425, 442, 455, 462, 508, 512, 515, 516,
  520, 541, 548, 568, 579, 583, 587, 589, 595,
  599, 613, 615, 616, 617, 622, 624, 628, 631,
  632, 633, 635, 664
pseudocaustic points, 594, 601
Pulliam, J., 225, 228, 372, 376
Puzyrev, N. N., 160, 170, 624, 644

Q factor, 542–3
Qin, F., 187
quasi-degenerate case of qS waves, 194
quasi-elliptical polarization, causes of, 655–8
quasi-elliptical polarization spirals, 648
quasi-isotropic approximation, 512
quasi-isotropic case of qS waves, 194
quasi-isotropic ray theory, 512–27
 average qS wave matrix of, 515, 517
 common ray of qS waves of, 514, 518
 correction factor of, 514, 516
 coupling function of, 517
 across an interface, 523
 in layered media, 524–6
 propagator matrix of, 519–24
 qS wave coupling system of, 514, 516, 518,
  522
 qS wave splitting matrix of, 515, 517

radiation function, 421–2, 506–8, 624, 636
 generalized, 435, 463–5
 of higher order, 552, 574
 of a line source, 444, 447, 476
 omnidirectional, 421
 of a point source, 421–2, 430, 506–8, 624
 relation to the directivity patterns, 422, 453
radiation matrix, 452–5
 of a center of dilatation source, 455
 of an explosive source, 455
 of a moment-tensor source, 455
 omnidirectional, 453
 relation to directivity patterns, 453
 of a single-force source, 453

for a source at an interface, 463–5
 transformations of, 453–4
radiation plane, 312, 319
Ralston, J., 589, 599
Ratnikova, L. I., 618
Ravindra, R., 6, 54, 61, 107, 212, 213, 216, 356,
  363, 392, 462, 478, 559, 565, 568, 576, 579,
  581, 582, 612, 617, 618, 619, 631
ray
 in anisotropic media, 149–56
 anomalous, 121–3
 boundary, 122, 612
 as a characteristic of eikonal equation, 100,
  103–9
 complex-valued, 5, 622
 critical, 122, 612
 curvature of, 123–4
 in curvilinear coordinates, 137–40, 145–7
 diffracted, 613
 edge, 122
 as an energy flux trajectory, 114
 as an extremal of Fermat's functional, 109–12
 frequency-dependent, 56
 generalized, 619
 as a geodesic in Riemannian space, 112–13
 grazing, 122
 multiple, 218
 network, 185
 in a parabolic layer, 169
 paraxial, 235, 250–3
 physical, 114–17
 polynomial, 132–3, 170–3, 177–8
 relation to wavefronts, 109
 space-time, 72
 successful, 120, 222
 surface-wave, 228–33
 tangent to interface, 122
 torsion of, 123–4
 turning points of, 109, 162
 two-point, *see* two point ray tracing
ray amplitudes
 in anisotropic media, 504–11
 continuation along a ray, 420, 451–2, 506
 at an initial line, 444–9, 474–7
 at an initial surface, 435–44, 474–5, 535–41
 across an interface, 422, 455–7, 508–9
 in layered structures, 428, 457–8, 465–6,
  509
 paraxial, 583
 along a planar ray, 430, 471–5
 of P-SV waves in 2-D, 472–4
 of P waves, 449, 466–70
 scalar, 419–36
 of SH waves in 2-D, 472

ray amplitudes (*cont.*)
  of S waves, 449, 470–1
  vectorial, 449–77, 505–11
  in weakly dissipative media, 418, 542–8
ray approximation
  first-order, 549, 568
  higher order, 54, 549, 577–9
  paraxial, 584–6
  zeroth-order, 54, 418, 549, 568
ray cell, 223
ray-centered coordinate system, 237–43
  basis vectors of, 238–9, 245–6
  definition of, 237–8
  dynamic ray tracing in, 253–7
  local Cartesian, 246–7
  orthogonality of, 240–1
  ray propagator matrix in, 280–5
  ray tracing in, 250–3
  region of validity of, 240
  relation to ray coordinates, 242–3
  relation to wavefront orthonormal coordinates, 265
  scale factors of, 240
ray chaos, 123, 370–2, 617
ray code of an elementary wave, 118–19, 218
ray coordinates, 201–2
ray estimator, 224
ray expansion, 632–3
  partial, 633
ray field, 102, 199–202
  density of, 203, 206
  regular, 202
  singular, 202
ray Jacobian, 205–13
  analytical computation of, 210–12
  in anisotropic media, 408–9
  dynamic ray tracing computations of, 209, 408–9
  FD computations of, 208–9
  for homogeneous media, 210
  across an interface, 209
  in orthogonal curvilinear coordinates, 209–10
  in radially symmetric media, 212
  relation to cross-sectional area of the ray tube, 208–9
  relation to geometrical spreading, 206
  relation to relative geometrical spreading, 359–60
  relation to $\nabla^2 T$, 207–8
  relation to $\nabla \cdot \mathcal{U}$, 206–7
  in vertically inhomogeneous media, 212
  in terms of wavefront curvature, 212–13
  in 1-D isotropic media, 211–12
Rayleigh function, 483

ray method
  with a complex eikonal, 4, 566–7
  extensions of, *see* extensions of the ray method
  generalized, 618–21
  space-time, 70–3, 566
  two-term, 577–9, 652
  validity conditions of, 607–9
ray parameter domain, 222
  homogeneous subdomain of, 222
  triangularization of, 222
ray parameter p in 1-D media, 161, 174, 211–12, 424, 481
ray parameters
  at an initial surface, 200–1, 311, 315
  for a line source, 201, 318–19
  for a point source, 199–200, 317
ray propagator matrix, 280
  in anisotropic media, 285–7, 404–8, 412–13
  for backward propagation, 284, 309
  in Cartesian coordinates, 285–6
  chain rule for, 283, 303
  in curvilinear coordinates, 286
  determination from travel-time data, 369–70
  eigenvalues of, 282
  in-plane, 387–9
  across an interface, 300–1
  inverse of, 283–4, 303
  in isotropic media, 279–84
  in layered media, 302–3, 333–6, 412–13
  Liouville's theorem for, 282, 303
  in nonorthogonal ray-centered coordinates, 287–8
  physical explanation of, 285, 286, 405
  along a planar ray, 387–9
  in ray-centered coordinates, 279–84
  reduced, 344–6
  surface-to-surface, 304–5, 413
  symplectic properties of, 281–2, 303
  transformation of, 286–7
  transverse, 387–9
  in wavefront orthonormal coordinates, 288
  4×4, 404–8, 412–13
  6×6, 285–7, 333–6
ray series, 1, 53–4, 56, 549–82
  with a complex eikonal, 566–7
  exact, 577
  for field discontinuities, 556–7, 575
  finite, 577
  higher order terms of, 2, 8, 54, 549, 577–9, 621
  for high-frequency signals, 555, 575
  across an interface, 552–4, 576
  leading term of, 3, 54, 549, 568
  modified forms of, 565–7, 582

recurrence system of equations for, 550–1, 568–9
scalar, 549–67
space-time, 70, 566
vectorial, 568–82
for time-harmonic waves, 549–52, 568
zeroth-order term of, 3, 8, 54, 549, 568
ray signature, 118
ray synthetic seismograms, 621–43
approximate, 628, 642
complete, 630–7
for complex-valued travel times, 629, 634
elementary, 621, 622–30
in frequency domain, 629–30, 634–5, 641
impulse, 633–4
modified forms of, 635–6
in time domain, 628–9, 633–4, 641
in weakly dissipative media, 622, 639–43
ray-theory Green function
in anisotropic media, 510–11
in fluid media, 430–1, 435–6
in isotropic media, 458–60, 466
of P-SV waves along a planar ray, 473
quasi-isotropic, 526–7
reciprocity of, 431, 436, 460, 511
of SH waves along a planar ray, 472
of unconverted P waves, 466–7
of unconverted S waves, 470–1
2-D, pressure, 449
2-D, P-SV, 477
2-D, SH, 477
ray tracing, 124–9
analytical, 101, 129–31, 163–9, 176–8
in anisotropic media, 149–53
approximate, 137
big, 187
boundary-value, 2, 101, 102, 217–28
cell, 135–6
controlled initial value, 223
in factorized anisotropic inhomogeneous media, 157–9
initial-value, 101–2, 218–19
across an interface, 119
in layered and block structures, 118–20, 154
network, 185
numerical, 101, 124–5
in radially symmetric models, 102, 174–8
semianalytical, 101, 136
in simpler types of anisotropic media, 127–9, 154–6
in simpler types of isotropic media, 127–9
surface-wave, 228–33, 617
two-point, 101, 219

in vertically inhomogeneous models, 102, 160–73
2-dimensional, 129
2.5-dimensional, 128
ray tracing system, 100, 104
in anisotropic media, 149–53
in curvilinear nonorthogonal coordinates, 146–7
in curvilinear orthogonal coordinates, 139–40
in a Hamiltonian form, 103–8, 112
in a Lagrangian form, 112
in modified spherical coordinates, 142–4
paraxial, 234, 250–3, 262, 267, 273
relation to Snell's law, 113–14
in spherical coordinates, 141–5
surface-wave, 230–3, 617
2-dimensional, 129
2.5-dimensional, 128
ray tube, 203–6
cross-sectional area of, 205
decomposition into ray cells, 223
elementary, 203–6
homogeneous, 223
section by a wavefront, 204
ray velocity vector, 114, 159
receiver continuation strategy, 224
reciprocity
of KMAH index, 381, 416
of normalized R/T coefficients, 424, 499–501, 534–5
of phase shift due to caustics, 416, 511
of ray-theory Green function, 412, 460, 511
of relative geometrical spreading, 409
reflection/transmission coefficients
in anisotropic media, 49, 527–35
for backward propagation, 502–3, 534–5
complete, 428, 458, 510
discussion of, 423–8, 485–95
displacement, 45–7, 49, 477–505
in dissipative media, 548
energy, 478
in fluid media, 41–2, 423–8
from a free-surface, 482, 495
in isotropic media, 45–7, 477–505
for normal incidence, 425, 427, 488–91
normalized displacement, 456–7, 478, 483–5, 509
normalized pressure, 423
particle velocity, 42
potential, 478
P-SV and SH, 46–7, 480–501
reflection/transmission matrices, 45–7, 49, 456, 458, 495–9, 527–35
Reinhardsen, J. E., 213

Reinsch, C. H., 172
Reiss, E. L., 548
relative geometrical spreading, 359–63, 409, 415
  definition of, 359
  by dynamic ray tracing, 359–60
  in homogeneous media, 360
  in radially symmetric media, 362
  reciprocity of, 359, 409
  relation to ray Jacobian, 359–60
  by surface-to-surface dynamic ray tracing,
    360
  in vertically inhomogeneous media, 360–2
representation theorems, 8, 89–95, 445
Reshef, M., 186
Riahi, M. A., 187
Riccati equation, 235, 255, 323, 327, 410
Richards, P., 8, 9, 12, 13, 16, 38, 81, 90, 91, 92,
    93, 118, 142, 160, 174, 189, 229, 455, 478,
    491, 542, 543, 577, 618, 619
Riznichenko, Yu. V., 178
Rockwell, D. W., 178
Rogoff, Z. M., 578
Rokhlin, S. I., 51
Romanov, V. G., 189, 193
Rose, J., 84
Roslov, Yu. V., 579
Roth, M., 372
Rueger, A., 51
Rümpker, G., 33, 34, 660
Rytov, S. M., 95, 372
Rytov angle, 242, 245
Rytov approximation, 95
Ryzhik, L., 372

Sagoci, H. F., 478
Saito, H., 185
Sambridge, M. S., 189, 220, 225
Samec, P., 548
Samuelides, Y., 95, 372
Savarenskiy, E. F., 160, 174
Sayers, C. M., 513
scale factors, 137, 240
Scales, J. A., 372
scattering
  Born, 93–5, 639
  generalized Born, 95–6
  single-scattering approximations, 94
scattering integrals, 14, 94–5
Schild, A., 112, 145, 147
Schleicher, J., 92, 305, 307, 308, 355, 357, 369,
    372, 383, 536
Schlottmann, R. B., 617
Schmidt, T., 547
Schneider, W. A., Jr., 187, 226

Schoenberg, M., 51, 478
Schrieffer, J. R., 589
Schulman, L. S., 617
Schulz, K., 615
Scott, J. H., 136
Scott, P., 92
Seckler, B. D., 613
seismic systems, theory of, 305
Sekine, S., 220, 226
Sen, M. K., 92, 536
Sena, A. G., 84, 401, 508, 512
Sethian, J. A., 179
shadow zone, 608, 612–13
Shah, P. M., 136, 213
Sharafutdinov, V. A., 356, 512
Shearer, P. M., 148, 157, 512
shear wave singularities, 33–4, 36
shear wave splitting, 35
shear wave window, 503, 655
Sheriff, R. E., 372
shooting method, 220–3
  controlled, 223
  standard, 220–2
Siegman, A. E., 589
signal
  analytical, see analytical signal
  Berlage, 624, 644
  causal, 628
  effectively causal, 628
  envelope of, 623, 666
  Gabor, 624, 628, 633, 642, 643
  high-frequency, 555
  Küpper, 624
  noncausal, 628
  Rayleigh, 624
  Ricker, 624
Šílený, J., 51
Silva, W., 548
Simões-Filho, I. A., 157, 197
Singh, S. C., 603
Singh, S. J., 8, 455, 618
singular directions of qS waves, 33, 36
singular lines, 34
singular points, 33
singular region, 418, 609, 610–13, 635
Sinton, J. B., 92, 536
slowness surface, 30–4
  curvature matrix of, 88, 407–8
  Gaussian curvature of, 34, 87–8, 408
  relation to the group velocity surface, 34–5
slowness surface of qS waves, 30–4
  kiss singularity of, 34
  line singularity of, 34
  point (conical) singularity of, 33

slowness vector, 20, 58
  paraxial, 298–9, 330, 410
Smirnov, V. I., 103, 111, 207, 619
Smirnov and Sobolev method, 619
Smirnova, N. S., 612
Smirnov's lemma, 207
Smith, K. B., 371
Smith, T. J., 478
Snell's law, 2, 42, 44, 69, 161, 244, 411,
    424, 548
  generalized, 175, 211
Snieder, R., 189, 220, 224, 225, 226, 228, 230,
    357, 372, 376, 377
Soares, J. E. P., 117, 373, 375, 376
Sommerfeld, A., 618
Sorrells, G. G., 136
source continuation strategy, 224
source term, 9, 14, 93–4, 445, 475
source-time function, 623–4, 629, 633, 636
Spence, G. D., 632
Spencer, C. P., 438, 639
Spencer, T. W., 189, 225, 228, 478
Spudich, P., 438, 632
Stamnes, J. J., 611
Stavroudis, O. N., 611
Stegun, I. A., 97, 663
Stewart, R. R., 375
Stewart, S. W., 220, 225
Stoffa, P. L., 187
Streifer, W., 588
stress-strain relations, 9–13
Stuart, G. W., 220
Suchy, K., 148, 566
summation
  of paraxial Gaussian beams, 590–604
  of paraxial ray approximations, 590–604
Sun, J., 377, 613
surface
  anterior, 289, 304, 413, 415
  boundary, 608, 610, 612
  caustic, 608, 610, 611
  critical, 610
  initial, 310
  posterior, 289, 304, 413, 415
  singular, 608, 610
  target, 591
surface-to-surface dynamic ray tracing, 289,
    304–5, 388
Sutton, G. R., 616
Symes, W. W., 187
Synge, J. L., 6, 112, 145, 147
synthetic seismograms (see also ray synthetic
    seismograms)
  Gaussian beam summation, 584, 590–604

Gaussian packet summation, 599
generalized ray, 619
Kirchhoff, 637
Maslov-Chapman, 601
reflectivity, 618, 619
summation of paraxial ray approximations,
    590–604
WKBJ, 603, 638–9
Szabo, T. L., 546

Tadzimukhamedova, S. S., 234
Tang, X., 371
Tanimoto, T., 230, 507
Tappert, F. D., 214, 371, 616
Tarantola, A., 95
Tatarskii, V. I., 95, 372
Taylor, W. J., 220, 226
Teles, T. N., 508, 512
Thom, R., 214
Thomsen, L., 11
Thomson, C. J., 33, 34, 51, 70, 84, 104, 207,
    225, 229, 230, 231, 276, 357, 401, 512,
    530, 532, 533, 544, 559, 565, 567, 582,
    591, 599, 601, 612, 613, 614, 616, 617,
    622, 640, 660
Thore, P. D., 372
Thornburgh, H. R., 178
Thurber, C. H., 137, 220, 226
time delay between qS waves, 197
time-depth relationships, 170
Toksöz, M. N., 401, 542, 543
Tooley, R. D., 478
torsion
  of an initial line, 321
  of a ray, 123–4
Toverud, T., 546
transformation across an interface
  of dynamic ray tracing in Cartesian
    coordinates, 334–5
  of dynamic ray tracing in curvilinear
    coordinates, 335–6
  of dynamic ray tracing in ray-centered
    coordinates, 299–300
  of ray tracing, 118–20
  of slowness vector, 40–1, 43–4, 50–1, 69
transport equation
  in anisotropic media, 63–4, 216
  coupled, 61, 216, 244
  frequency dependent, 56
  higher order, 551
  for a pressure wave, 55, 214–15
  for a P wave, 60, 215
  space-time, 72
  for an S wave, 61–2, 215–16

travel time
  critical, 561
  crossover, 562
  elementary, 100
  first-arrival, 100, 179–82
  linearized, 193, 194
  ray-theory, 100, 179–82
travel-time approximation
  curved wavefront, 323
  hyperbolic, 329, 370
  local plane wave, 322
  parabolic, 329
  quadratic, 323
travel-time computations
  along rays, 126
  direct, 178–88
  by fast marching method, 179
  by finite-difference method, 186–7
  by Huygens construction, 178
  by incomplete separation of variables, 184
  by method of expanding cube, 186
  by method of expanding halfspace, 186
  by method of expanding square, 186
  by method of time fields, 178
  by method of wavefronts, 178
  by separation of variables, 182–4
  by shortest path method, 185
  by wavefront construction method,
    187–8
travel-time derivatives
  first-order, 322
  higher-order, 339–41
  mixed second-order, 354, 399–400
  second-order, 255, 268–70, 277, 323, 390–1,
    409–10, 415
travel-time perturbations, 189–99
  in anisotropic media, 193–7, 336–7
  average qS, 196–7
  of the first order, 190–2
  due to interfaces, 198–9
  in isotropic media, 192, 336–7
  qS time delay, 197
  of the second order, 337–9
  in weakly anisotropic media, 194–7
Trigubov, A. V., 644
Tromp, J., 6, 73, 142, 143, 144, 220, 230,
    234, 601
Trorey, A. W., 613
t-star (t^*), 545, 547, 640
Tsvankin, I. D., 84, 401, 508, 615, 620
tunneling of waves, 618, 619
Tverdokhlebov, A., 84
two-point eikonal
  in anisotropic media, 414

  in isotropic media, 350, 354
  close to a planar ray, 396–7
two-point ray tracing
  by bending methods, 223–6
  in Cartesian coordinates, 350–4
  by paraxial methods, 225–6, 348–56
  by perturbation methods, 227–8
  in ray-centered coordinates, 349–50
  by shooting methods, 220–3
  in surface-to-surface ray tracing, 355
Tygel, M., 76, 92, 95, 305, 307, 308, 355, 357,
    369, 383, 536, 624
Tyurikov, L. G., 357

Ulin, V. V., 73, 589
Ulrych, T. J., 187
Um, J., 220, 226
Ursin, B., 92, 95, 213, 279, 357, 536, 546

validity conditions of the ray method, 607–9
  general, 608–9
  in terms of Fresnel volumes, 609
van den Berg, P. M., 14
van der Hijden, J. H. M. T., 619
van Trier, J., 187
Van Vleck, E. S., 372
Vasco, Don W., 372
Vasil'yev, Y. I., 478
Vavryčuk, V., 84, 152, 408, 577, 578
Veith, K. F., 136
Vermeer, G. J. O., 372
Vesnaver, A. L., 227
Vidale, J. E., 186, 187, 188, 366
Vinje, V., 187, 632
Virieux, J., 6, 136, 189, 199, 220, 227, 230
Vlaar, N. J., 156

Waltham, D. A., 220, 599
Wang, C.-Y., 84
Wang, X., 599
Wang, Y., 220
Wang, Z., 230, 603
wave
  compressional, 26, 59
  converted, 43
  diffracted, 610–13, 621, 635
  diffracted at edges and vertexes, 613
  diffracted at smooth objects, 613
  in a dissipative medium, 542–8, 614, 639–44
  diving, 565
  edge, 122, 613
  elementary, 119
  evanescent, 614
  head, 122, 559–65, 579–82, 612, 619, 621, 635

higher order, 122, 554, 558, 621, 635
inhomogeneous, 36–7, 613–15, 621
interference head, 582, 617
longitudinal, 27, 59
Love, 228
P, 26, 59
plane, *see* plane waves
postcritically transmitted, 614–15
in a preferred direction, 616–18
pseudospherical, 615, 620
qP, 25, 62, 505
qS, 25, 62, 505
Rayleigh, 228, 643
S, 28, 59–60
S*, 615
shear, 28, 59
slightly refracted, 565
spherical, 74–84
Stoneley, 615, 619
surface, 228, 616, 617, 643
tip, 122, 613
transverse, 28, 59
tunnel, 615, 618
unconverted, 43
vertex, 122, 613
whispering gallery, 616
wavefront, 19, 55
    curvature matrix of, 326–8, 391, 406–8, 410
    Gaussian curvature of, 328, 408
    mean curvature of, 328
    paraxial, 256
    quadratic, 327
    radii of curvature of, 327
    relation to rays, 109
wavefront construction method, 187–8
wavefront orthonormal coordinate system, 260, 264–5
    basis vectors of, 264–5
    definition of, 264–5
    dynamic ray tracing in, 264–7
    ray propagator matrix in, 288
    relation to ray-centered coordinates, 265
wave impedance, 20, 426
weakly anisotropic media, 512–27
    quasi-isotropic ray theory in, 512–27
    travel-time perturbation method in, 194–7
    weak anisotropy matrix, 195
Weber, M., 135, 599
Weigel, F. W., 617
Weinberg, H., 229, 230

Weingarten equations, 314
Wennerberg, L., 548
Wentzel, K., 163
Wesson, R. L., 220, 356
West, G. F., 438, 601, 643
Weyl, H., 76
Weyl integral, 74, 76, 589, 615
White, B. S., 589, 599
White, J. E., 462
Whitham, G. B., 73
Whitham's variational principle, 73
Whitmore, J. D., 643
Whittal, K. P., 135, 632
Wielandt, E., 372
Wiggins, R. A., 92, 638
Wild, A. J., 617
Will, M., 135
Witte, O., 372
WKB solution, 70
Wolf, E., 17, 643
Wong, Y. K., 230
Woodhouse, J. H., 230, 530
Wright, J., 51
Wrolstad, K. H., 51
Wu, R.-S., 95, 372, 589, 617

Yacoub, N. K., 136
Yamaguchi, K., 185
Yan, J., 140
Yanovskaya, T. B., 230, 478, 579
Yardley, G., 643
Yeatts, F. R., 84
Yedlin, M., 33, 34
Yeliseyevnin, V. A., 107
Yen, K. K., 140
Yomogida, K., 84, 230, 372, 408, 577
Young, J. B., 636

Zahradník, J., 643
Zedník, J., 178, 631
Zelt, B. C., 187
Zhao, L., 618
Zhao, T., 548
Zheng, B. S., 579
Zhu, H., 84
Zhu, T., 56, 92, 536, 544, 566
Zillmer, M., 512, 515
Ziolkowski, R. W., 214, 601
Zöppritz, K., 477
Zvolinskiy, N. V., 619